THE ARROW OF TIME: **The theme for the second**

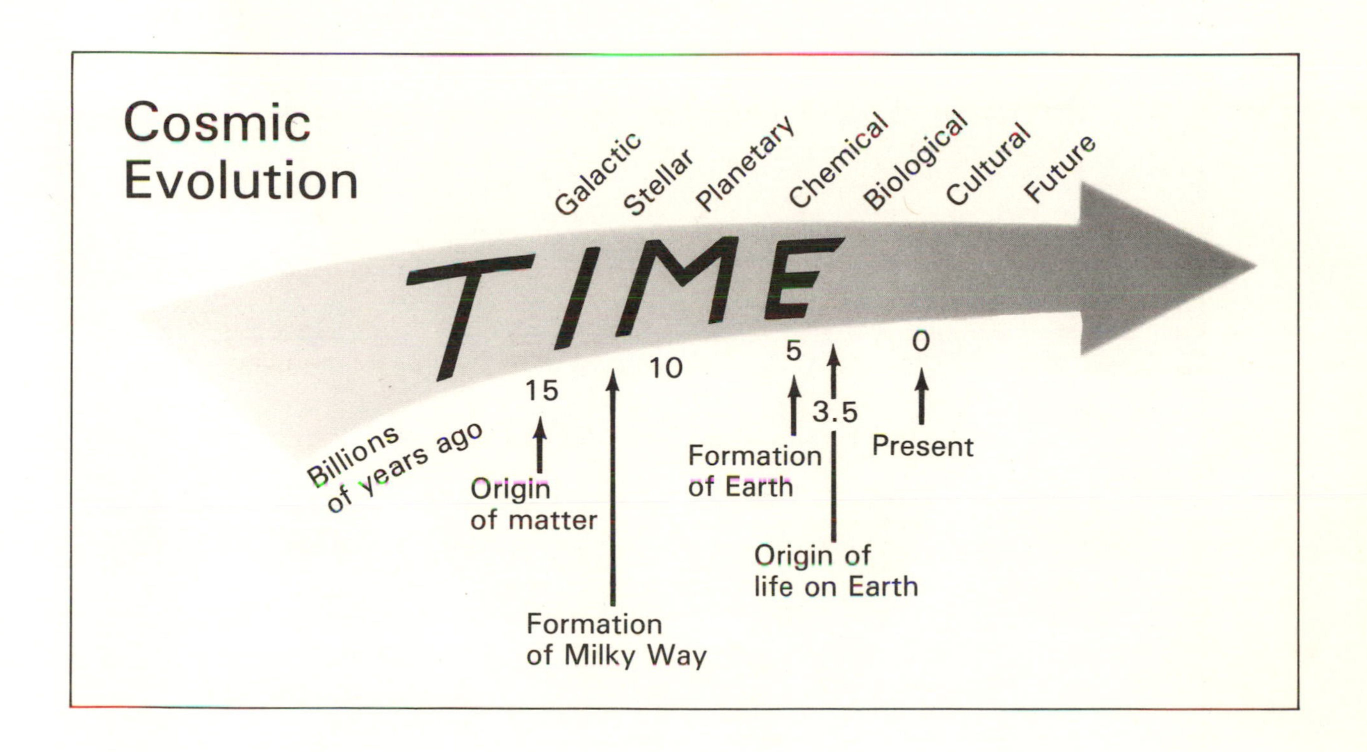

UNIVERSE
An Evolutionary Approach to Astronomy

ERIC CHAISSON
Senior Scientist,
Space Telescope Science Institute

Prentice Hall, Englewood Cliffs, New Jersey 07632

Library of Congress Cataloging-in-Publication Data

Chaisson, Eric.
Universe: an evolutionary approach to astronomy.
Bibliography: p.
Includes index.
1. Astronomy. 2. Cosmology. 3. Space and time.
4. Matter. 5. Life—Origin. I. Title.
QB43.2.C46 1988 520 88-4192
ISBN 0-13-938391-3

To Those Scientists Who Value the Need to Teach

Editorial/production supervision: Maria McColligan
Interior design: Jules Perlmutter
Cover art: Jon Lomberg
Cover design: Ben Santora
Manufacturing buyer: Paula Massenaro

Printed in the United States of America

10 9 8 7 6 5 4 3 2 1

ISBN 0-13-938391-3

Prentice-Hall International (UK) Limited, *London*
Prentice-Hall of Australia Pty. Limited, *Sydney*
Prentice-Hall Canada Inc., *Toronto*
Prentice-Hall Hispanoamericana, S.A., *Mexico*
Prentice-Hall of India Private Limited, *New Delhi*
Prentice-Hall of Japan, Inc., *Tokyo*
Simon & Schuster Asia Pte. Ltd., *Singapore*
Editora Prentice-Hall do Brasil, Ltda., *Rio de Janeiro*

Other books by the author:

COSMIC DAWN: The Origins of Matter and Life
LA RELATIVITÀ
THE INVISIBLE UNIVERSE: Probing the Frontiers of Astrophysics (coauthored with George B. Field)
THE LIFE ERA: Cosmic Selection and Conscious Evolution
RELATIVELY SPEAKING: Relativity Theory, Black Holes, and the Fate of the Universe

BRIEF CONTENTS

CONTENTS

PREFACE

This book can best be described as a study of space and time, of matter and life. Herein we explore our Universe, our planet, ourselves. We strive to summarize where science stands today in its search to answer some of the most basic questions.

Who are we? Where did we come from? How did everything around us on planet Earth arise? What are the origins of the stars and planets? How do we, as living creatures, relate to the rest of the Universe? Where are we, as intelligent beings, probably headed in the future? In short, what are our origin and destiny? What are the origins and destinies of the Earth, the Sun, the Universe?

This text is written for nonscientists. I have endeavored to distill valid, modern science to a level comprehensible to college students having little science background. Accuracy has not been sacrificed, and a feeling for the frontiers of science has been included.

Students should realize that answers to the questions posed above (among many others) are not yet crystal clear. Even among technical peers, scientists are often unable to provide precise and detailed solutions for the broad and profound questions. Only within the past couple of decades have we invented the technological tools needed to transfer these questions from the realm of philosophy to that of science. Accordingly, researchers are finding the frontier of science to resemble a thinning haze rather than a sharp boundary. The reason is that science is now moving at a fast clip, acquiring new knowledge at a phenomenal rate, and requiring new interdisciplinary efforts to put it straight. Furthermore, much of it involves human beings, each with different ideas and often personal biases. Consequently, we might say that a pencil sketch of the answers to some of the basic questions is now at hand, but that many details need to be discovered.

The main theme of our text is cosmic evolution—the study of the gradual changes in the assembly and composition of energy, matter, and life throughout the Universe; it is these changes that have produced galaxies, stars, planets, and life forms. Indeed, the phenomenon of change seems to be the hallmark in the development of all things.

Using hardly any mathematics, I describe the prevailing scientific view that the atoms in our bodies relate to the Universe in general. In the process, we first study the broadest view of the biggest picture, after which we analyze some of the most basic questions of all—neither the most relevant nor the most important questions, perhaps, for twentieth-century society, but the most fundamental ones.

We should not be surprised that our text is based largely on the subject of astronomy. After all, erect, inventive creatures have resided on Earth for less than one-tenth of 1 percent of the age of the Universe; Earth itself has existed for less than one-third of cosmic history. To be sure, we do discuss aspects of many other subjects, including physics, chemistry, geology, biology, and anthropology. But mostly in the second half of our text, after thoroughly studying the simpler forms of matter, such as galaxies, stars, and planets, do we meet the more complex forms of matter, such as life and intelligence.

This textbook forms the essence of a course that I have taught mostly at Harvard University since the mid-1970s. Over the years, I have class-tested the materials presented here in various ways and at a few other institutions.

Experience has shown that the subject can be usefully arranged into three parts. Part I contains a few introductory chapters. Originally, these basic issues were not part of the course, but were added in more recent years at the repeated request of students. Much of this introductory material need only be skimmed by students who have had an adequate course in high school science. For others, these chapters should provide some familiarity with the basic ideas and techniques needed to understand cosmic evolution.

With the preliminaries covered, the bulk of the textbook follows two designs. Part II is a **space format,** in which we essentially inventory all the known types of matter in the Universe. I have used a progressive ("in → out") approach, first describing Earth and then moving out gradually to the limits of the observable Universe. Part II builds on Part I and is mostly astronomy.

Similarly, Part III builds on Part II, but ranges over many scientific disciplines. This part follows a **time format,** wherein we use an evolutionary theme to study various highlights in the history of the Universe. Also adopting a progressive pattern, we proceed from the creation of the Universe to the present time in which we live. Part III is more than an inventory, however. In it, I try to show how all the contents of the Universe, including ourselves, interrelate in an almost storybook fashion according to the grand scenario of cosmic evolution. I aim to bring together the essential features of astrophysics and biochemistry, for these two subjects more than any others are making an enormous impact on the way that we view ourselves and our place in the Universe.

The text is suitable for classroom use for either a one-term or two-term course. Originally, all the material was covered in a single semester, but little time remained for outside readings; one 90-minute class meeting was essentially devoted to each chapter. In more recent years, I have used the material in a two-term sequence, as the design of the book lends itself nicely to such a full-year course; Parts I and II are studied in the first term and form an exciting course in modern astronomy, whereas Part III is studied in the second term by those students wishing to pursue the subject further, especially the renaissance relationship of astronomy to many other scientific endeavors. During the less-accelerated, full-year version, roughly one chapter is covered per week and ample time remains to sample many additional readings, including some of those listed at the end of each chapter. (Readings judged more challenging are marked with an asterisk.)

A final note on text suitability: I have endeavored to tailor the various chapters to be as self-consistent as possible; few chapters lean heavily on others and skipping around the book is entirely feasible. Uninhibited instructors and students should be able to use the text as a menu, picking and choosing their favorite topics. Although I prefer not to belabor the obvious, a few brief examples will suffice. Those instructors and students wedded to more conventional astronomy courses that discuss few, if any, aspects of life can profitably use the book simply by omitting the last few chapters. Those preferring to cover the topic of stellar evolution immediately after a discussion of stars can do so by going directly to Chapter 20 (and Chapters 19 and 21, as well) after Chapter 11. Similarly, a desire to study planetary science in a single gulp can be accommodated by following up Chapters 5, 6, 7, and 8 with Chapters 23 and 24. And those who feel pressed for time can skip some chapters; for example, a briefer course could, without too much penalty, omit Chapters 4, 5, 6, 12, 18, 21, 22, and so on. Numerous other combinations abound. Clearly, the usefulness of a large and diverse text is limited only by the reader's imagination.

Pedagogy has been stressed throughout. An overview provides the philosophy of approach and a guiding motivation, urging us to recognize living systems as an integral part of the astronomical perspective. A set of learning goals begins each chapter, while a summary of the highlights and a list of key terms follow each chapter. Those key terms appear in **boldface** type as they are introduced in the text, and their (more formal) definitions are collected in a Glossary at the end of the book. Every figure (including the colored plates) is referred to and discussed in the body of the text. Ten questions follow each chapter, those more challenging marked with an asterisk. The metric system is used throughout the text, although some English units are occasionally noted in parentheses when they aid learning.

I have tried to minimize the technical jargon of the many varied scientific specialties—jargon that often does no more than "keep the beginners out." For example, the use of logarithms is avoided as being unnecessarily tricky for nonscientists. The stellar magnitude scale is virtually omitted since many nonscientists and scientists alike find it confusing. And the official names of the various geological epochs are virtually ignored, for knowledge of them requires excessive memorization that often detracts from the central issues discussed. Furthermore, the text uses a minimum of mathematics, certainly none beyond high school algebra. Occasionally, when a math concept is introduced, I have written out the equation in English words; mathematical symbols are essentially banned from this book. Numerical values are everywhere rounded off, for the central concepts that I wish to stress can often be addressed adequately with order-of-magnitude estimates.

I have also ignored, completely and without apology, discussions of astrology, flying saucers, creationism, telepathy, magic, and the like. With this paragraph I dismiss from the book these and other pseudoscientific topics as being totally inconsistent with the modern scientific method, which combines a mixture of proven facts and *testable* ideas.

Finally, I should also acknowledge my deemphasis of the history of science in this text. This omission is due partly to space limitations, for room is needed to fully explore the renaissance nature of the evolutionary scenario. But this deemphasis is also partly due to my personal prejudice that our modern efforts to teach *introductory* science have become too historical, philosophical, and even political. Fascinating though these disciplines may be in their own realm of history or literature, I suggest that the relevance to science of the history and philosophy of science has been exaggerated. What is paramount in science is the totality of findings and generalizations available today. The operational content of science at any one moment best illustrates the method of scientific discovery, for it is the active body of knowledge that has served the tests of time, criticism, and debate.

The informal style of writing, the occasional and intentional repetitiveness, and the balance of heavier scientific concepts with lighter cultural interludes are pedagogical aids that all result from more than a decade of teaching specifically this subject to nonscientific groups having a curiosity about nature. Indeed, the anonymous legions of some three thousand Harvard, Radcliffe, Haverford, Bryn Mawr, and Wellesley college students have contributed much to the clarity and spirit of this book.

Many people have helped crystallize my thoughts on the presentation of this material. Foremost among them is my wife, Lola, who has generously devoted time and effort to help me discover a meaningful approach to science for nonscientists. During the long writing effort, she has been my best editor. Of greater importance, perhaps, she has been the driving force behind many of the illustrations that adorn the text. Those illustrations have been carefully designed to supplement and liven the descriptive words of our text.

Three secretaries, Donia Carey, Patricia McVity, and Lillian Dietrich, have done yeoman service while typing the many drafts of a large and weighty manuscript. Over the years, numerous colleagues have helped minimize errors at various parts of the script. I appreciate the corrections and criticisms offered by the following reviewers who read the entire script: Barbara A. Gage, Prince George's Community College; Alan P. Marscher, Boston University; Michael Lieber, University of Arkansas; Matthew Malkan, UCLA; and John J. Cowan, University of Oklahoma. And not least, a handful of "spiritual advisors," mostly at Harvard, offered encouragement in an educational environment not especially tolerant of teachers or synthesizers, and especially not both: Bart Bok, George Field, Leo Goldberg, Edward Lilley, Edward Purcell, George Wald, and E. O. Wilson.

Finally, I am indebted to the Atlantic Monthly Press for permission to draw heavily, in Part III, on one of my previous works, *Cosmic Dawn: The Origins of Matter and Life*, and to W. W. Norton, Inc., for permission to use aspects of another of my trade books, *Relatively Speaking*.

Eric J. Chaisson

OVERVIEW: *An Interdisciplinary Approach*

More than 10,000 years ago, the ancestors of our civilization began seriously thinking about very basic issues. Men and women contemplated themselves and probed their environment. They wondered who they were and where they came from. They longed for an understanding of the starry points of light in the nighttime sky, of the surrounding plants and animals, of the air, rivers, and mountains. They thought about their origin and their destiny. They thought, and thought some more.

All these basic issues were treated as secondary, for the primary concern seemed to be well in hand: Earth was assumed to be the stable hub of the Universe. After all, the Sun, Moon, and stars all appear to revolve around our planet. It was natural to conclude, not knowing otherwise, that home and self were special. This centrality led to a feeling of security or at least contentment—a belief that the origin, operation, and destiny of the Universe were governed by something more than natural, something supernatural.

The idea of Earth's centrality and the reliance on supernatural beings were shattered only a few hundred years ago. During the fifteenth- and sixteenth-century Renaissance, humans began to inquire more critically about themselves and the Universe. They realized that thinking about nature was no longer sufficient. Looking at it was also necessary. Experiments became a central part of the process of inquiry. To be effective, theories had to be tested experimentally, either to refine them if experiment favored them, or to reject them if it did not. A whole new method of investigation was thus embraced—one that combined thinking and doing, that is, theory and experiment. This combination is the most powerful technique ever conceived for the advancement of factual information. Modern science had arrived.

Today, all physical and biological scientists throughout the world use this **scientific method**. In a three-step process, they usually gather some data or ideas, then form a theory, and finally test that theory. This is a rational, methodical approach used to investigate all natural phenomena. Employed properly over a period of time, the scientific method enables us to arrive at conclusions that are mostly free of the personal bias and human values of any one scientist. It's designed to yield an objective view of the many varied aspects of our Universe.

Humans now query along the same lines as did the ancients. We ask the same basic questions: Who are we? Where did we come from? What are our origin and our destiny? But now our attempts to answer them are aided by the experimental tools of modern technology. We have astronomical telescopes to improve our vision of the macroscopic Universe of planets, stars, and even larger assemblages called galaxies. We have biological microscopes to aid our view of the microscopic world of cells, atoms, and even smaller entities called particles. We have human-made satellites to gather data unavailable from our vantage point on Earth. And we have sophisticated computers to keep pace with the tremendous increase of new data, theories, and experimental tests.

We live in an age of technology. And even though technology threatens to doom us, that same technology is now providing us with an unprecedentedly rich view of ourselves and our Universe.

The learning goals for this overview are:

- to understand the scientific method
- to appreciate the size and scale of things in space, and the great durations of time
- to recognize the central ideas comprising the grand scenario of cosmic evolution

FIGURE 0.1 Earth is a planet, a mostly solid and molten object, though it has some liquid in its oceans and gas in its atmosphere. (Discernible here are the African continent and parts of the Middle East.)

FIGURE 0.2 The Sun is a star, an extremely hot ball of gas, much bigger than Earth; it is held together by its own gravity.

FIGURE 0.3 A galaxy is a collection of hundreds of billions of stars, each separated by vast regions of nearly empty space. This galaxy, called The Great Spiral in Andromeda (which we shall refer to many times in our text as simply the Andromeda Galaxy), measures about a billion billion kilometers across. [See also Plate 2(a).] Our Sun is a rather undistinguished star near the edge of *another* such galaxy, called the Milky Way.

Prologue

Of all the science accomplished since Renaissance times, one result stands out most boldly: Our planet is neither central nor special. Application of the scientific method has demonstrated that we do not inhabit a unique place in the Universe. Research, especially within the past few decades, strongly suggests that we live on what seems to be an ordinary rocky **planet** called Earth, one planet orbiting an average **star** called the Sun, one **Solar System** in the suburbs of a much larger group of stars called the Milky Way **Galaxy**, one galaxy among countless billions of others spread throughout the observable abyss called the **Universe**.

Universe—the title of this text—can be defined as the vast tracts of empty space and enormous stretches of time populated sparsely by stars and galaxies glowing in the dark. More succinctly, the Universe is the totality of all space, matter, and energy. Consult Figures 0.1 through 0.3, and place some of these objects into perspective by studying Figure 0.4.

Now, in this last quarter of the twentieth century, experimental science is helping us unravel the details of the big picture. We are beginning to appreciate how all these objects—from atoms to galaxies, from cells to people—are interrelated. We are attempting to decipher the scenario of **cosmic evolution**. Broadly sketched in Figure 0.5, cosmic evolution describes a grand synthesis of a long series of changes in the assembly and composition of energy, matter, and life—changes that have occurred throughout almost incomprehensible space and nearly incomprehensible time. It is these changes that have produced our Galaxy, our Sun, our planet, and ourselves.

Modern science now has a reasonably good understanding of how countless billions of stars were born and have died to create the matter composing our world. It also suggests how life arose as a natural result of the evolution of matter. Accordingly, we can identify a clear thread linking the evolution of simple atoms into galaxies and stars, the evolution of stars into heavy elements, the evolution of those elements into the molecular building blocks of life, of those molecules into life, of life into intelligence, and of intelligent life into cultured and technological civilization.

To answer the basic questions, Who are we? and Where did we come from?, we must penetrate far back into the past—before the average 60 years of our human lifetimes, before the start of modern science centuries ago, before the onset of language and civilization tens of thousands of years ago, before our ancestral human-apes who emerged from the forests several million years ago, even before complex life began to flourish on our planet some billion years ago.

A thousand (1000), a million (1,000,000), a billion (1,000,000,000), and even a trillion (1,000,000,000,000) can be spoken of easily in words. (Note that we use "billion" in the American sense to mean a "thousand million.") But you should strive to understand the

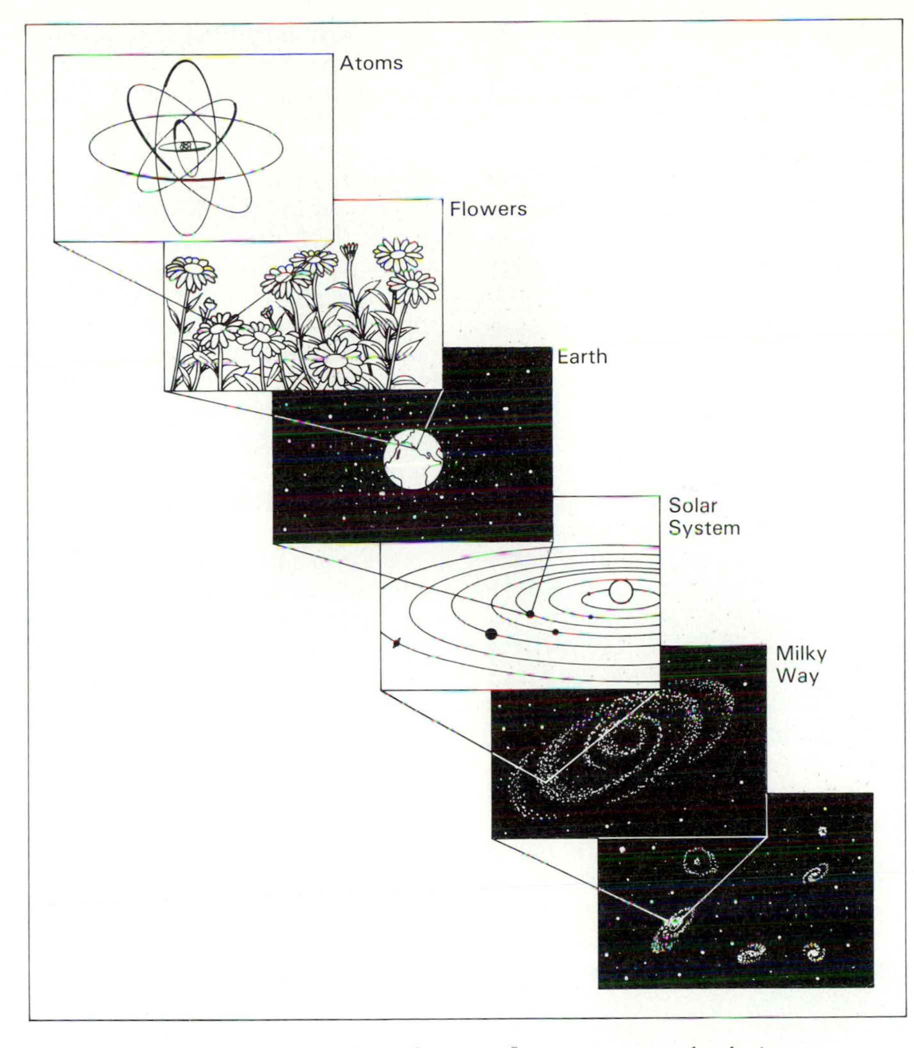

FIGURE 0.4 An artist's conception of atoms, flowers, stars, and galaxies.

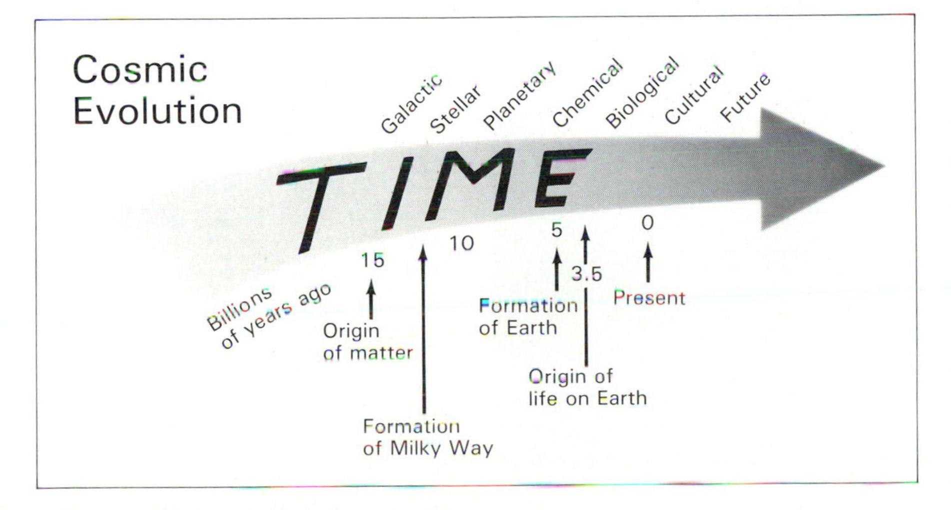

FIGURE 0.5 Some highlights of cosmic history are noted along this arrow of time, from the beginning of the Universe to the present. Cosmic evolution is the study of the myriad changes in the assembly and composition of energy, matter and life, as well as the reasons for those changes.

magnitude of these truly huge numbers, as well as the vast differences between them. For example, 1000 seems easy enough to understand; at the rate of one number per second, you could count to 1000 in about 15 minutes. However, if you wanted to reach a million, you would need 2 weeks of counting at the rate of one number per second, 16 hours per day (allowing 8 hours per day for sleep). Furthermore, to count from 1 to a billion at the same rate of one number per second and 16 hours per day would take an entire lifetime. Imagine, an entire lifetime is needed *just to count* to a billion!

Yet we will consider throughout our text time intervals spanning millions and billions of, not only seconds, but also years. We will discuss objects housing not only trillions of atoms but also trillions of stars. Hence you must become accustomed to gargantuan numbers of things, enormous intervals of space, and extremely long durations of time. Recognize especially that a million is much larger than a thousand, and a billion much, much larger still.

To appreciate cosmic evolution, then, we must broaden our horizons, expand our minds, and visualize what it was like long, long ago. We must go way back into the past: go back, for example, to 5 billion years ago, when there was no life on planet Earth. Why? Because there was no Earth, no Sun, no Solar System. These objects were only beginning to form out of a giant, swirling gas cloud near one edge of a vast complex of older stars that had already existed in one form or another for eons of time before that.

To trace the specific steps that led to our existence, modern science combines a wide variety of subjects—astronomy, physics, chemistry, geology, biology, and anthropology, among others. The ultimate objective is to unravel the most basic problems of all: the origin of matter, the origin of life, and how matter and life change with time. If we can understand the scenario of cosmic evolution, perhaps we can determine precisely who we are, specifically how life arose on planet Earth, and, incredibly enough, how living organisms have evolved to the point of invading land, generating language, creating culture, devising science, exploring space, and even studying themselves.

Universe

Our Milky Way system is a genuine galaxy—an enormous collection of about 100 billion stars held loosely together by gravity. This is more stars than people who have ever lived on Earth. Beyond this Milky Way Galaxy lie billions of other galaxies—each a gargantuan assemblage of matter having roughly 100 billion stars. Light radiation travels outward in all directions from these galaxies, carrying the message of their existence. A minute portion of that light is intercepted at Earth, to be captured occasionally in photographs like those of Figure 0.6 and Plate 2(a).

Light does not travel infinitely fast; it moves at a finite speed—the velocity of light. Consequently, even light needs time to travel through the almost unimaginably vast regions that separate objects in the Universe. The farther an object is from Earth, the longer it takes for its light to reach us. By studying that radiation, we can learn what conditions were like long ago, when distant objects emitted their light. Our perception of the Universe is therefore delayed. We see the Universe as it was, not as it is.

Looking out from Earth, then, is equivalent to looking back into time. Telescopes are time machines, and astronomers are historians. Much like archeologists who dig for information buried in decayed bones and ancient artifacts, astronomers sift through "old" radiation just arriving at planet Earth. Yet astronomers study more than the origins of men and women. By looking out far enough into space—with the best telescopes—astronomers can potentially address the origins of matter itself.

Light is only one kind of radiation. Radio, infrared, ultraviolet, x-ray, and gamma-ray waves all comprise invisible radiation. Regardless of the type, radiation is energy. It is also information—a most basic kind of information. Only by means of this one-way information flow can we infer anything about objects far beyond our planet. With the exception of the Moon missions and planetary probes of the Space Age, as well as meteorites that occasionally strike the ground, radiation is our only connection with the Universe beyond Earth. Whether visible starlight is sampled by our eyes, or other kinds of radiation by complex equipment, this radiative information is the sole way to fathom the depths of space.

FIGURE 0.6 Very distant galaxies can be observed through the world's largest telescopes. Mind-boggling at first sight, many of the objects in this photograph are vast galaxies unto themselves. Each such galaxy resembles the Andromeda Galaxy or our own Milky Way Galaxy, and each contains about a hundred billion stars held together in a swarm by the pull of gravity.

Radiation from distant galaxies suggests that our Universe began in a cataclysmic explosion—a "bang"—approximately 15 billion years ago. Unimaginably hot at first, the fireball of this cosmic bomb gradually cooled as it spread out. We don't know for sure if the Universe will continue to expand forever, or if it will someday stop expanding and eventually contract back to a single point much like that from which it arose. What we do know for sure is that the Universe is not static; it's changing with time; it's evolving.

This initial explosion seems to be the ultimate origin of all things. In the first few moments of the fireball, conditions were apparently right for the construction of the subatomic elementary particles from which all matter is made. However, energy was severe at the time, completely overwhelming matter, and breaking it apart as soon as it tried to assemble into anything substantive. Intense radiation prohibited even the simplest elementary particles from combining into the type of matter that we now call atoms.

Sometime later, the energy weakened and the matter cooled enough to allow some of the elementary particles to unite, thus forming the simplest and most abundant element, hydrogen. At about the same time, the next heaviest element, helium, materialized when hydrogen collided with one another. But once the Universe had expanded and cooled a little more, conditions were no longer suitable for the formation of elements heavier than helium. The elements composing this page you are now reading, the air we breathe, and the coins in our pockets were not created in the aftermath of the initial explosion. There simply wasn't enough time; events at the start of the Universe happened very rapidly.

Galaxies

Although astronomers do not yet understand precisely how the galaxies formed, all the data we currently have suggest that the galaxies originated during the first few billion years after the bang. We surmise this partly because observations have never clearly demonstrated evidence for galaxies forming at the present time. Evidently, the great turbulence, the hot gas, and the intense radiation of the early fireball—physical conditions favorable to the formation of galaxies—have changed substantially now, more than 10 billion years after the bang.

Yet galaxies exist. We see them in great abundance virtually everywhere in the Universe. Somehow they got there, and this text will explore some of the plausible ways how. But our inability to explain precisely how is the biggest missing link in the entire scenario of cosmic evolution. How galaxies change or evolve over the course of time is another major problem that astronomers have not yet solved.

Stars

Although current conditions seem inappropriate for the formation of galaxies, within those galaxies the conditions are ripe for the formation of stars. At numerous places within our own Milky Way Galaxy, giant parcels of gas are known to be assembling into stars. We now have direct observational evidence that huge quantities of matter are needed for gravity to hold together a pocket of stellar gas. How much matter? No less than a thousand billion billion billion billion billion atoms. That's obviously a large number, in fact much larger than all the grains of sand on all the beaches of the world or all the atoms in the entire Earth. It's larger than anything with which we're familiar because there's simply nothing on Earth comparable to a star.

The lifetime of a star, once formed, depends on its mass. Our Sun is a star of medium mass and has been emitting energy for almost 5 billion years. Based on our knowledge of stellar evolution, the Sun should endure for roughly another 5 billion years, continuing to provide that constant source of heat and light needed to maintain future generations of life on Earth.

Toward the end of its duration as a star, our Sun will become unstable and expand in one last gasp, extending its "life" perhaps several million years. Some of the interior planets, including Mercury, probably Venus, and perhaps even Earth, will be engulfed in the Sun itself as it evolves toward this swollen phase close to "death." Thus we can be sure that our planet is eventually doomed. It cannot last forever, and will finally contract back to become a dark, dead clinker in space. Just like humans and other forms of life, stars and planets are born, mature, and die. Be assured, though, our Sun will not lose its cool in this way for about another 5 billion years—time enough, if we can survive as a civilization, to undertake galactic engineering projects literally out of this world.

During the final death rattles of a star, elements heavier than hydrogen and helium are created in the star's insides. The "heavies" are virtually cooked within the hearts of stars. In small or medium stars such as the Sun, these heavy elements remain trapped in the stellar interior. But stars that are more massive than our Sun do not perish so simply. Bigger stars die catastrophically like that shown in Figure 0.7 and Plate 8(a), heaving their newly created elements into the surrounding regions of space. In this way, galaxies are regularly enriched by exploding stars that eject heavy elements needed in the formation of later-generation stars, planets, and other interesting things, such as life and intelligence.

FIGURE 0.7 This photograph shows the remnants of a massive star that exploded long ago, spilling its newly created elements into space. Stellar evolution includes the study of the origin of nature's elements, a process that mostly occurs in the hearts of stars. This subject combines the study of astronomy and nuclear physics. [See also Plate 8(a).]

Planets

Given the orderly arrangement of objects in our Solar System as it's known today, the planets did not likely originate by a collision between stars or by some other such rare accident. Instead, the birth of planets is thought to be a natural and frequent by-product of star formation. Although astronomers do not yet agree on the specific details, they do concur on the following broad outline. A huge ball of gas and dust must have become flattened by rotation, after which it fragmented into boulder-sized chunks orbiting at various distances from the centrally forming Sun. Far from the young Sun, where the temperature was low, chemistry favored the production of ices and gases that later accumulated into giant gassy balls like Jupiter; closer to the young Sun, conditions were too warm to form ices but just right to form rocks which then collided, stuck, and became the small rocky planets like Earth. It all happened some 5 billion years ago.

What information do we have concerning the earliest stages of Earth? Not much, unfortunately. Geological hints about the first billion years or so have been eroded away steadily by the weather as well as by violent surface activity in the form of erupting volcanos and rocky meteorites colliding with our planet. The early Earth was surely barren, with shallow, lifeless seas washing upon grassless, treeless continents. The atmosphere probably contained chemicals poisonous to most present-day life forms; oxygen and nitrogen were hardly as abundant as they are now. And the Sun's piercing radiation blazed strongly to the ground; the protective atmospheric layers had not yet fully formed. Although no life as we now know it could have existed on the primitive Earth, these harsh conditions apparently permitted the chemical steps needed to originate life.

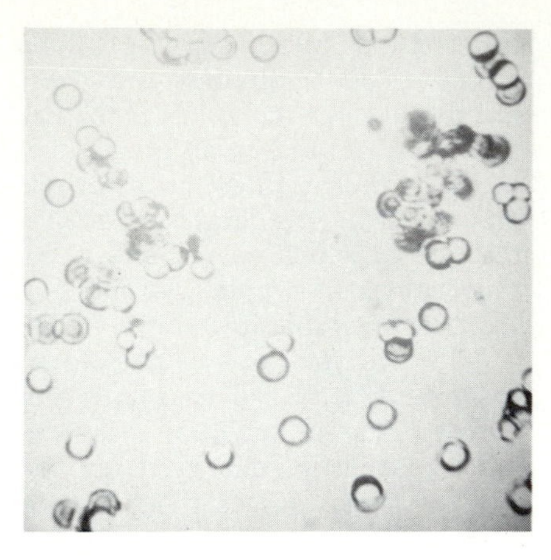

FIGURE 0.8 When heated, simple chemicals that must have existed on primitive Earth tend to cluster into carbon-rich droplets. Resembling simple cells in many ways, these droplets can be photographed as in this case with a high-powered microscope. Chemical evolution is the study of the production of complex chemicals from simple chemicals, and of the physical and chemical conditions that gave rise to the origin of life. This subject involves some astronomy and biology, but mostly physics and chemistry.

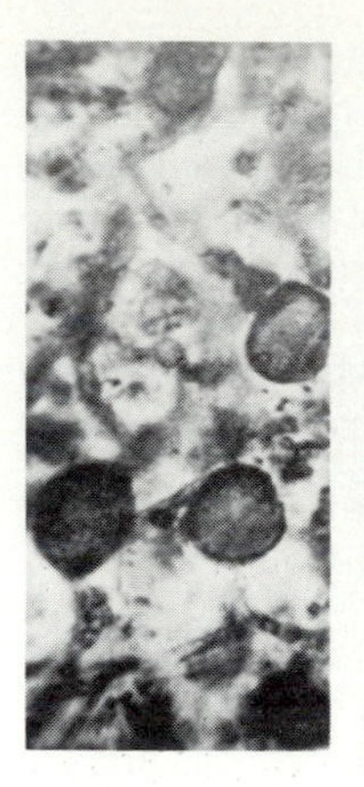

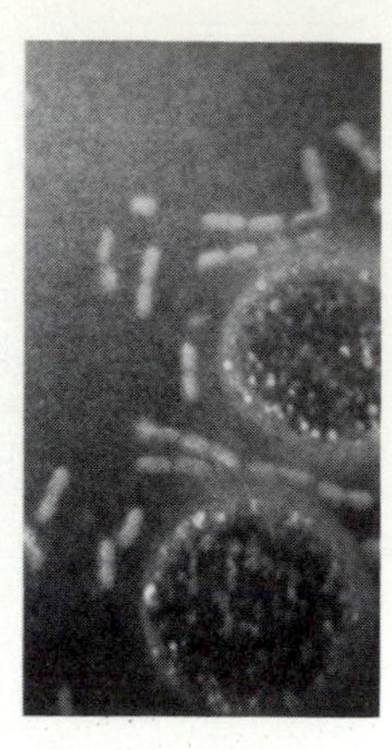

FIGURE 0.9 Fossils are the dead remains of organisms that once lived. Among the oldest fossils, those photographed at left through a microscope are dated to be nearly 3.5 billion years old. They show striking resemblances to the carbon-rich droplets (Figure 0.8) produced in laboratory simulations of the primitive Earth, and to modern blue-green algae photographed (at right) near a backyard stream.

Life

Experiments conducted in laboratories during the past few decades have demonstrated that energy causes simple chemicals to change into much more complex chemicals. A simple mixture of ammonia, methane, carbon dioxide, and water, "cooked" for about a week, produces some of the building blocks of life. The result is not life itself but the basic chemical ingredients needed to form all life from simple cells to complex humans.

More recent experiments have further probed the origin of life. We now know that additional heat causes these building blocks of life to group into oily compounds that resemble simple biological cells. These carbon-rich compounds are fascinating to study under a microscope, as shown in Figure 0.8. Curiously, they seem to represent matter within the fuzzy domain between living and nonliving. The carbon-rich droplets also bear a certain resemblance to the fossilized remains of some of Earth's oldest living organisms found in the sedimentary rocks of our planet. These fossils, radioactively dated to be about 3.5 billion years old, appear to have structures similar to those of modern blue-green algae, such as those photographed in Figure 0.9.

Thus it seems that scientists now have reliable information *suggesting* that life is hardly more than a combination of chemicals of matter. Furthermore, we can theorize that the origin of life is merely a natural (though not necessarily inevitable) result of the evolution of that matter. These are the *working ideas* adopted in our text. If correct, the general picture of chemical evolution is in place. Only the details remain to be unraveled.

Information about advanced life forms is even more complete. The record of the fossils shows how life became widespread and diversified. We no longer have any reasonable doubt that biological evolution has occurred and is continuing to occur. Scientists now recognize that accidental changes in the genes enable some organisms to attain the best available niche within a gradually changing environment. The result is that some species perish and become extinct; but many others thrive, multiply, and change some more, as suggested in Figure 0.10.

Studies of the fossils of dead organisms, as well as of the behavior of living organisms, help us gain perspective regarding the major events of biological evolution. Briefly, the facts show the following: Simple, one-celled life began on our planet at least 3.5 billion years ago, but didn't advance to more complex forms until roughly 2 billion years ago. Single cells combined into multicellular organisms less than a billion years ago, after which rapidly flourished a wide variety of increasingly complex organisms—plants, flowers, birds, reptiles, mammals. Surprisingly, humans have existed for only the past few million years, a very short time

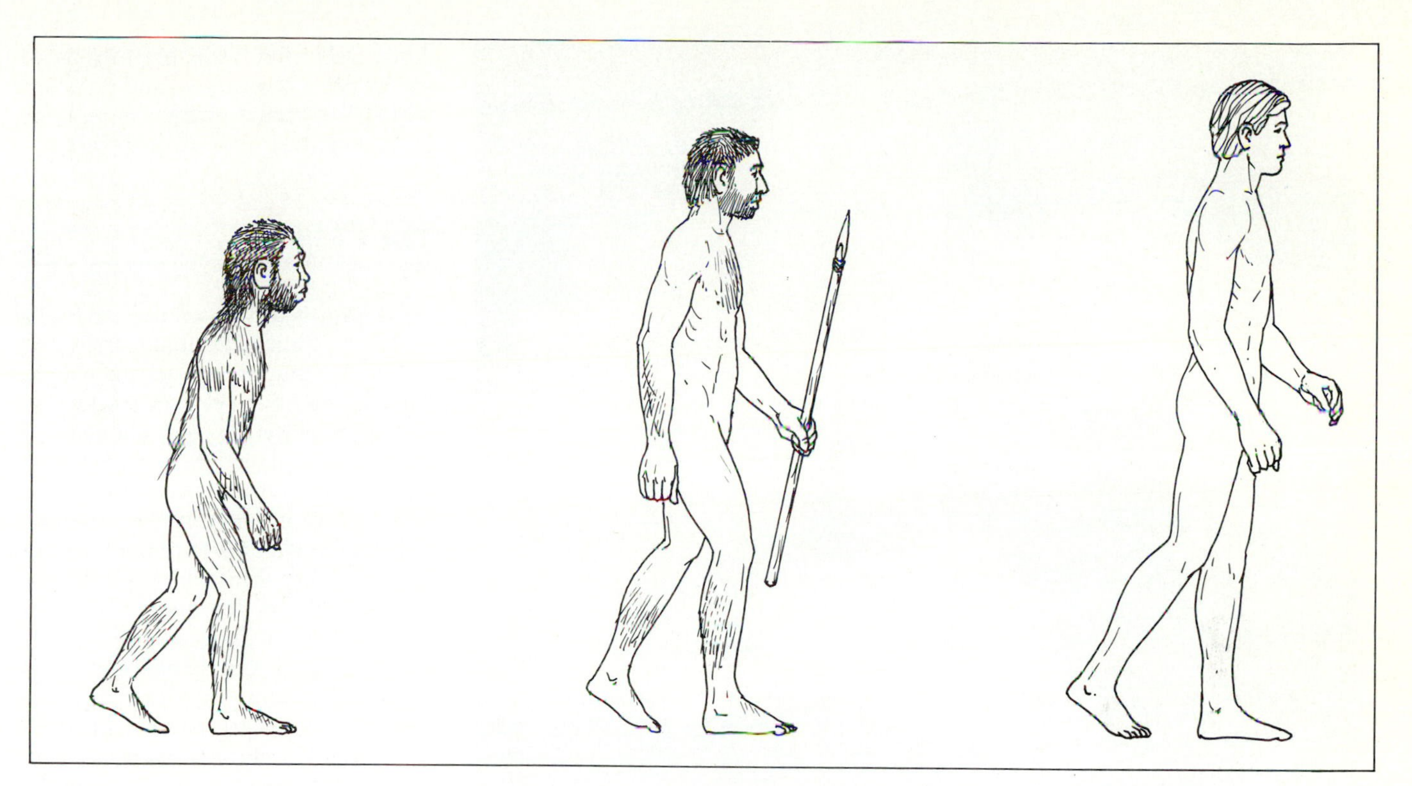

FIGURE 0.10 Accidental changes in the basic structure of some DNA chemicals (whose functional units are the so-called genes) enable species to change and life to diversify. Fossils clearly show substantial changes over long durations of time, as exemplified by this relatively recent line of ascent toward humans. Biological evolution is the study of the developmental changes, from generation to generation, experienced by life forms throughout the history of life on Earth. This subject involves a combination of biology and chemistry, and a little geology.

compared to the entire scenario of cosmic evolution.

Some people have trouble comprehending millions and billions of years. These are big numbers, and after a while they lose their meaning. To better understand Earth's history, consider an analogy. Imagine the entire lifetime of Earth to be 50 years rather than 5 billion years. This time scale is then comparable to a human life span, making various highlights of Earth's history more understandable.

With this analogy in mind, we can say that there is no record whatever for the first 10 years of Earth's existence. Rocks hardened and life arose quickly thereafter. In particular, life originated at least 35 years ago, when Earth was only about 15 years old in our analogy. Our planet's middle age is largely a mystery, although we can be sure that life continued to evolve, and that mountain chains and oceanic trenches steadily built up and eroded down. Not until about 5 years ago, in our 50-year analogy, did abundant life flourish throughout Earth's oceans. Life came ashore about 4 years ago, plants and animals mastered the land about 2 years ago, and dinosaurs reached their peak about 1 year ago, only to die suddenly about 8 months ago. Human-like apes changed into ape-like humans only last week, and the latest major ice ages occurred only a few days ago. *Homo sapiens*—our species—did not emerge in this 50-year analogy of Earth's history until about 4 hours ago. In fact, the invention of agriculture is only about 1 hour old, and the Renaissance only 3 minutes old!

Apparently, it takes time—lots and lots of time—to construct life, intelligence, and civilization.

Intelligence

Cosmic evolution has guided matter from a state of simplicity to a state of complexity. Advancing over the course of time, evolutionary change has fashioned intelligent life on at least one planet—Earth—and possibly on others as well. Since earliest times, evolution has consistently favored those organisms able to gather and understand information efficiently. This is apparently true whether the organism was a primitive microbe swimming in the early primordial sea, or a semicultured ape-man roaming the ancient forests of our planet.

The fossils document the increase in brain size of our ancestors. Some 25 million years ago, there were the small brains of our long-ago ancestors who came down out of the trees, the so-called *Aegyptopithecus*

FIGURE 0.11 The controlled use of fire by humans, as well as their inventions of tools, language, and agriculture, were among the most important cultural and social advances that helped mold technological intelligent life. Cultural evolution is the study of the changes in the ways, ideas, and behavior of society, especially among higher forms of life on Earth. This subject involves a wide variety of disciplines, including anthropology, geology, biology, and sociology.

ape-like creatures. Then there were the moderately sized brains of the best candidate for the "missing link," the so-called *Australopithecus* ape-men of several million years ago. Finally, there are the larger brains of our streetwise friends, namely *Homo sapiens sapiens,* or the wise guys of modern times.

Competition for survival apparently favored monkey-like creatures having an ability, for example, to judge accurately the distance to a banana, as well as to grasp branches and reach out to get that food. Those who successfully secured the food were able to survive and reproduce; those who could not, either fell off the tree and died or stayed in the tree and starved. Only those creatures having traits adaptable to changing environments were favored for survival. Our more recent ancestors were also granted advantageous traits while inventing fire, tools, language, and agriculture, as well as developing the qualities of foresight, memory, and curiosity. Many of these evolutionary accomplishments, one of which is sketched in Figure 0.11, had an effect on the brain: It got bigger.

The maze of cells within our skull—the brain—is the most complex clump of matter known anywhere in the Universe. As best we can tell, the human brain is an extraordinary example of the marvelous extent to which matter in the Universe has evolved. The brain is a living machine enabling us to acquire information, to store it as memory, and to pass it on to succeeding generations. It is, in fact, the human brain that allows us to maintain civilization, to appreciate the arts, to invent technology, and to unlock secrets of the Universe. Above all, our brain permits us to study the cosmic contents from which we arose—which is, by the way, a central goal of this book.

The Challenge

So the most advanced information-gathering animal—that's us—has come to dominate life on our planet. The process has been an evolutionary one, wherein the concept of change has played a key role. But now, another basic change is under way—a most significant change. After billions of years of nature's evolution, the dominant species on planet Earth is beginning to tinker with evolution. Whereas previously, environments and genes had governed evolution, humans are suddenly learning to control each of these factors. We are now tampering with matter, all the while causing the environment to change. And we stand on the threshold of altering life, potentially changing the genetic makeup of human beings.

For sure, elemental evolution continues unabated in the hearts of stars everywhere. Chemical evolution still occurs in remote places such as galactic clouds and perhaps other planets. Biological evolution persists for all Earth species. And cultural evolution no doubt continues in many corners of our world. But for modern men and women, the essence of evolution itself seems to be evolving.

What is the cause of this dramatic change—or, we should say, this dramatic change in the way things change? The answer, for the most part, is technology. We've become very intelligent—a lot smarter than any other known species in the Universe. But our technological youth and inexperience are causing problems unlike any faced by previous societies on Earth. Intelligence aside, just how *wise* are we?

Doomsayers regard technological intelligence as the end of nature's evolution on our planet. They argue that many of today's global problems threaten to end Earth civilization and perhaps all of Earth life itself, and that eventually one such problem will do us in. Optimists, on the other hand, do not view technological intelligence as the peak of material development. Instead, they view technology as a natural stepping stone—an intermediate stage within the grand cosmic evolutionary scenario—enabling us to attain heights of consciousness heretofore unimagined.

Which path will be taken? We don't know, of course. What we do know is that matter has come full cycle. Life now contemplates life. It contemplates matter. It seeks answers to our origin and our destiny. It quests for new knowledge.

Cosmic evolution has brought us forth, and now, having done so, it enables us to study that very same scenario of cosmic evolution. In a very real sense, we have gained such intellectual prowess as to be able to reflect back upon the material contents that gave life to us.

Provided that we realize our position in the cosmic scheme of things, that we embrace wisdom as well as intelligence, and above all, that we remain curious, perhaps we can better appreciate and analyze the various options as our planet, our life form, and our civilization move forward toward an uncertain future.

In this text we attempt to develop an appreciation for our rich universal heritage. We seek to decipher the nature and behavior of matter on the grandest scale of all. We try, literally, to put life into perspective—a cosmic perspective. And we explain where science now stands in our understanding of the truly big picture—the Universe-view known as cosmic evolution.

SUMMARY

Change is the hallmark in the development of all things. From the start of the Universe roughly 15 billion years ago, changes in the composition and assembly of energy, matter, and life have helped create, in turn, galaxies, stars, planets, and intelligence.

As technological beings, we now strive to discover the nature of these changes, as well as the reasons for them. Our efforts comprise the study of cosmic evolution—the central subject of our book.

KEY TERMS

cosmic evolution
galaxy
planet
scientific method
Solar System
star
Universe

FOR FURTHER READING

AUDOUZE, J., AND ISRAEL, G. (eds.), *The Cambridge Atlas of Astronomy*. Cambridge: Cambridge University Press, 1985.

CHAISSON, E., *Cosmic Dawn*. Boston: Atlantic–Little, Brown, 1981.

CHAISSON, E., *The Life Era*. New York: Atlantic Monthly Press, 1987.

FIELD, G., PONNAMPERUMA, C., AND VERSCHUUR, G., *Cosmic Evolution*. Boston: Houghton Mifflin, 1978.

PREISS, B. (ed.), *The Universe*. New York: Bantam, 1987.

REEVES, H., *Atoms of Silence*. Boston: MIT Press, 1984.

SAGAN, C., *Cosmos*. New York: Random House, 1980.

SEIELSTAD, G., *Cosmic Ecology*. Berkeley: University of California Press, 1983.

WEISSKOPF, V., *Knowledge and Wonder*. Boston: MIT Press, 1985.

1 RADIATIVE INFORMATION

The Overview displayed a variety of cosmic objects in our spectacular Universe. These included our Sun emitting bright and colorful light, explosive stars ejecting matter and energy, and powerful galaxies flickering in the depths of space. Through a telescope on a dark, moonless night, every object seems a superb example of cosmic architecture—a real jewel of the night.

But astronomical objects are more than works of art, more than things of beauty. Planets, stars, and galaxies are of vital significance if we are to realize our place in the big picture. Each is a source of information about the material aspects of our Universe. They light up the far away and the long ago.

In Part II we describe each of these and other kinds of objects, and in Part III we discuss their origins and evolution. But first, in a few introductory chapters of Part I, we study some basic concepts that underlie much of modern astronomy. Knowledge of these basic ideas is essential to appreciate the grand design of the Universe.

The learning goals for this chapter are:

- to distinguish astronomical observations from laboratory experiments
- to realize that radiation sometimes behaves as a wave, while at other times as a particle
- to understand how radiation moves, and thus transfers information, through the nearly void of outer space
- to know the different kinds of radiation, of which light is only one kind
- to understand how the strength or intensity of radiation can help to determine the temperature of an object
- to appreciate how motion can change our perception of the basic properties of radiation

Information

ACQUIRING INFORMATION: How do astronomers know anything about stars or galaxies far away from Earth? How do they obtain detailed, numerical information about any of the objects noted in the Overview? The answer is that we use the laws of physics to interpret the light emitted by these objects. In much the same way that scientists can discover many properties of an earthly piece of matter that emits light, astronomers acquire information by studying the light emitted by matter far beyond planet Earth.

But there's a basic difference in the methods used to study **terrestrial** matter on Earth, and those used to probe **extraterrestrial** matter in outer space. In a laboratory on Earth, we can manipulate terrestrial objects in order to determine their properties. An object's size, shape, temperature, weight, and many other properties can be discovered by working with either the matter being studied or the equipment used to study it.

For example, the scientists in Figure 1.1 are using complex laboratory equipment to study the properties of mineral ore. During their research, they could choose a variety of rock samples having different colors, sizes, and textures. They could change the position of the ore within the experimental apparatus. They

FIGURE 1.1 Laboratory scientists can manipulate the terrestrial matter being studied, as well as the experimental equipment used to study it.

could heat the ore, or cool it, and even subject it to a variety of forces that make it glow. In these ways, they can learn much about the mineral ore by carefully studying its emitted light. Thus the terrestrial environment can be changed to help us study any piece of matter on Earth.

No such manipulation is possible when studying matter far from Earth. We cannot change the extraterrestrial environment. We cannot probe distant objects directly; they are simply too far away. Instead, astronomers must work with the light that just happens to be intercepted by Earth. As starlight moves from one point to another in the Universe, if Earth gets in the way, we see the light; if Earth doesn't, we don't see it.

The paragraph above was once true in all cases. In recent years, however, technological advances have provided a few exceptions, enabling us to manipulate a few samples of nearby extraterrestrial matter: meteorites occasionally discovered in Earth's crust, Moon rocks returned to Earth during the American *Apollo* and Soviet *Lunakhod* programs, and Martian soil studied by the American *Viking* robot spacecraft. Yet it's likely to be many centuries before our descendants can make on-site studies of extraterrestrial matter far beyond our planetary system. Given the vastness of space, our descendants might never become smart enough to do so.

INTERPRETING INFORMATION: Figure 1.2 and Plate 2(a) show one of our neighboring spiral galaxies, called Andromeda in this text. Despite the adjective "neighboring," the Andromeda galaxy is roughly 2 million light-years from Earth. This is a great distance once we realize that a **light-year** is the *distance* traveled by light in a full year. After all, light moves at the fastest velocity known—the velocity of light. *A light-year, then, is a distance.* It equals about 10 trillion kilometers (or 6 trillion miles).

Although a light-year is a truly huge distance, be sure to realize that Andromeda is still much more distant than a single light-year. In fact, as noted above, Andromeda is 2 million such light-years away. Consequently, it's impractical to send a space probe to Andromeda for a close-up study. This object is just too far away. Even if a spacecraft could miraculously travel at the speed of light, 2 million years would be needed to reach its destination. Thus nearly all that we know about Andromeda, and other galaxies like it, results from studies of the light they naturally emit.

The fact that the velocity of light is the ultimate speed limit in the Universe yields some interesting consequences. Since the Andromeda galaxy is about 2 million light-years away, its light must take about 2 million years to travel from it to us. Nothing can exceed the velocity of light, so Andromeda's light could not get here any faster. Thus it takes time—often lots of time—for light to travel through the vast realms of space. While gazing at the nighttime sky, the light we see from Andromeda must have left that galaxy 2 million years ago—roughly the time that our human ancestors first emerged on planet Earth. That light has been racing through space ever since.

Even more fascinating, the light now reaching us from some of the Universe's really distant objects—like the truly remote galaxies clustered in Figure 1.3—left those objects before any life even existed on the lands of planet Earth. Their light has traveled through space for nearly a billion years before being captured in the photograph shown there.

FIGURE 1.2 The pancake-shaped Andromeda galaxy is about 2 million light-years away, and contains about a hundred billion stars. [See also Plate 2(a).]

FIGURE 1.3 This group of distant galaxies, each more than ½ billion light-years from Earth, is called the Hercules Cluster. Like many cosmic photographs in this book, it is a rendering of ancient times—in this case a fair fraction of a billion years into the past.

Light from distant objects, then, is a tool to study the past—a useful probe enabling us to learn more about earlier times, and occasionally abut the origins of things. In fact, light now reaching Earth from the most distant known objects must originally have been emitted from those objects in the earliest epochs of the Universe. By studying their light, we may someday discover the key to an understanding of the origin of the Universe itself.

So, as we study cosmic objects in this and succeeding chapters, we must always remember that the light now seen left those objects eons ago. *Looking out into space is equivalent to looking back into time.* We can never observe the Universe as it is—only as it was.

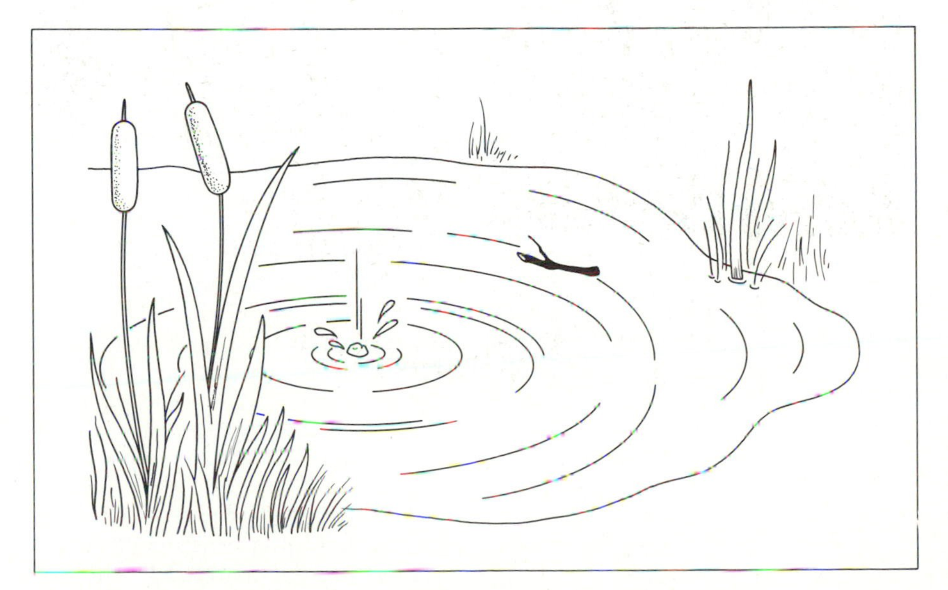

FIGURE 1.4 Waves can be used to transfer information from one place to another. In the case of this pond of water, information moves from the place of impact of a pebble to the place at which a twig is floating.

Wave Nature of Radiation

WAVE MOTION: **Light** is a special kind of information. To be precise, light is the kind of information to which our human eyes are sensitive. As light enters our eye, the cornea and lens focus the information onto the retina, whereupon small chemical reactions send electrical impulses to the brain to record the sensation of light.

A more general word for light is **radiation**. In this chapter we strive to understand radiation in some detail. Just what is radiation? How does it travel? Especially relevant for us in our text, how can it travel through the emptiness of outer space? In short, how does radiation tell us virtually all that we know about everything in the Universe beyond Earth?

Experiments have demonstrated that radiation is **energy**—the sort of energy (or ability to do work) that normally travels between any two places. In doing so, radiation usually takes the form of a **wave**. By means of wave motion, information can move from one place to another. Let's consider some familiar examples of wave motion, especially the concept of information exchange.

Imagine a twig floating in a calm pond of water. A pebble, thrown at some distance from the twig, will generate waves from the point of impact, as shown in Figure 1.4. From common experience, everyone knows that the twig will bob up and down when the wave reaches it. In this way, information can be transferred from the place of the pebble's impact to the location of the twig.

Sound waves are another example of information exchange. Normally, sound travels from the mouth of one person to the ear of another, transferring information as the particles of air collide with one another. Information of this type is transferred at a velocity that depends on the medium through which the waves pass. For example, waves travel more slowly across a bowl of molasses than across a bowl of water. The material medium—molasses or water—determines the velocity at which a wave, or therefore information, can be exchanged between any two places. Similarly, thin, cold air on high mountaintops causes weaker, more garbled sound to arrive at a slower rate than normal. High altitudes are just not suitable for crisp conversations.

In a total vacuum, without any air particles at all, no sound could possibly exist. That's why astronauts, apart from obvious breathing problems, cannot talk face to face while outside the artificial environment of their spacecraft or their spacesuits. Outer space is very tenuous, and hence very quiet. Can you imagine an astronaut of the future watching the completely silent collision of two orbiting satellites? The twisting and crunching of metal would make no audible sound in the near-perfect vacuum of space. Under zero-gravity conditions, such collisions would provide surrealistic scenes indeed!

WAVE PROPERTIES: Figure 1.5 shows how waves are classified. We classify them not only by the velocity with which they move but also by the length of their cycle. **Wavelength** is defined as the length of an *individual* wave cycle. This can be the distance between two adjacent crests, two adjacent troughs, or any two similar points along adjacent wave cycles.

If a wave moves quickly (with a large velocity), then the number of crests or cycles passing any given point per unit time, defined as the **frequency**, will be reasonably large. Conversely, a sluggish wave moves slowly, with only a few crests passing per unit time; we say that such a wave has a low frequency.

Simple experiments, even ones that we could perform with a metal Slinky or a rope tied to a doorknob,

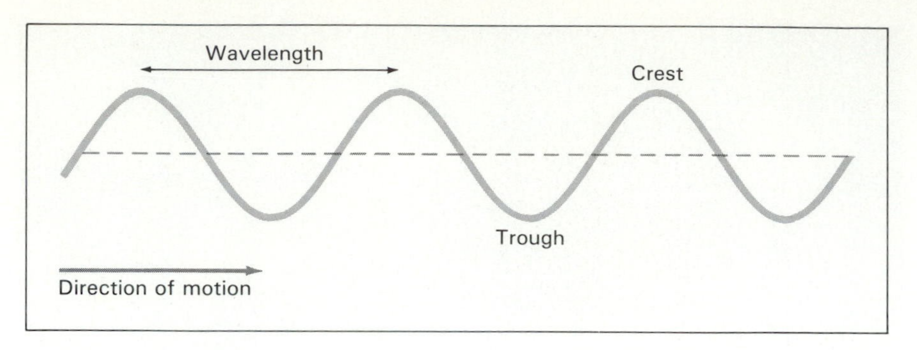

FIGURE 1.5 A typical wave.

demonstrate that the wavelength and wave frequency are *inversely* related. Together, their product equals the wave velocity:

$$\text{wavelength} \times \text{frequency} = \text{wave velocity}.$$

Consequently, a wave has a high frequency when its wavelength is small and a low frequency when its wavelength is large. This is the meaning of the inverse relationship between wavelength and frequency: For a given wave velocity, if the wave crests are close together, more of them pass by a given point each second; when the crests are far apart, few of them pass by per unit time.

Not surprisingly, wavelength has units of length, often expressed in centimeters. Units of velocity amount to length divided by time, for example, centimeters/second. And frequency has units of inverse time, or cycles/second, often termed hertz, in honor of a nineteenth-century German scientist.

Force Fields

Waves of radiation differ fundamentally from water waves, sound waves, or any other waves traveling through a material medium. The difference is simply that radiation needs no material medium. When light radiation travels from Andromeda galaxy or any other cosmic object, it moves through the virtual vacuum of empty space.

Sound waves, for example, cannot do this; if we removed all the air from a room, oral conversation would be impossible. But communication via flashlight or radio intercom would be entirely feasible. Waves of radiation can travel through nothing; a medium is unnecessary. In fact, only by means of radiation can astronauts communicate outside their spacecraft.

The ability of light or any other kind of radiation to travel through empty space was formerly a great mystery of the natural world. The notion of light moving through nothing seemed to violate common sense and challenged the best scientific minds of a century ago. The solution forms one of the cornerstones of modern physics. Let's discuss it in some detail.

A CHARGED PARTICLE: To understand the true nature of radiation, consider a charged particle, or any piece of matter having a net electric charge. Extending outward in all directions from this charge is an **electric force field**. Such a force field is a property of space itself–in this case, the influence of one electric charge on another charge. It's much like a field of heat surrounding a fire—a field that we can measure with a thermometer. (More advanced texts distinguish between "forces" and "fields;" in this elementary book, we make no such distinction and in fact use these words interchangeably when it aids learning.)

Experiments show that the strength of the electric force field decreases with increasing distance from the charge. That is, the field weakens at large distances. In much the same way, temperature decreases with distance from a fire; even grav-

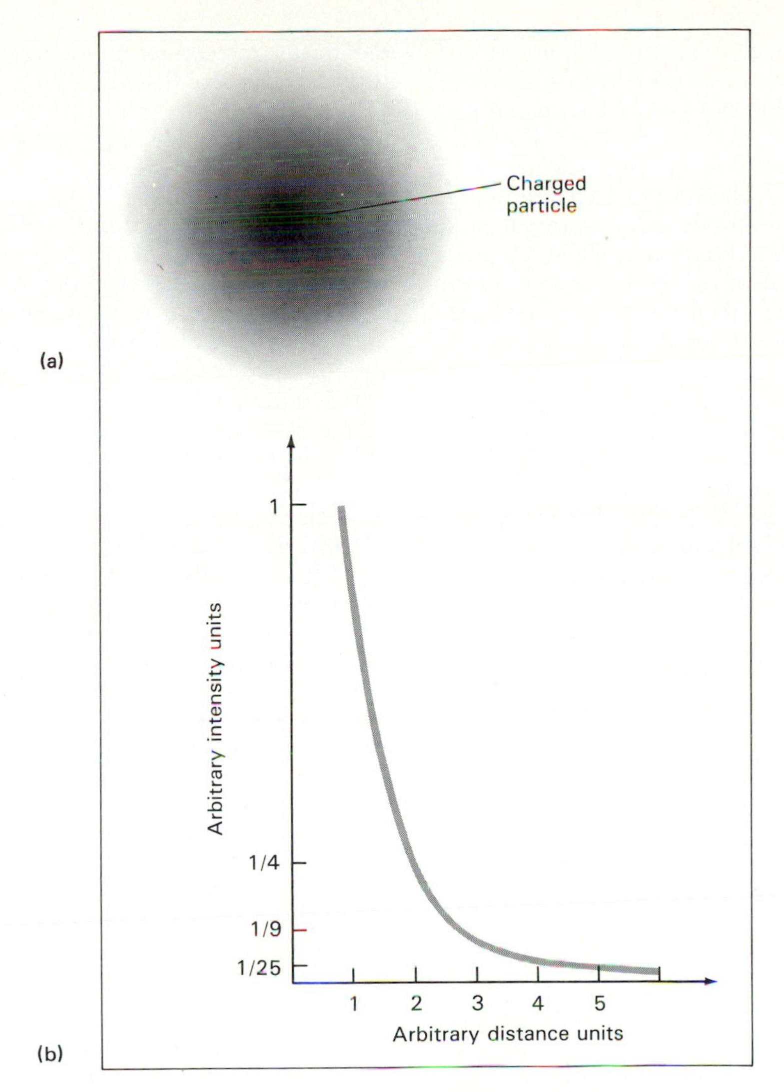

FIGURE 1.6 Inverse-square force fields rapidly weaken with distance from the source of that field. Here, in (a), the force field is stronger or more intense near the charged particle where the shading is heavier. The graph in (b) illustrates just how fast such forces decrease with distance.

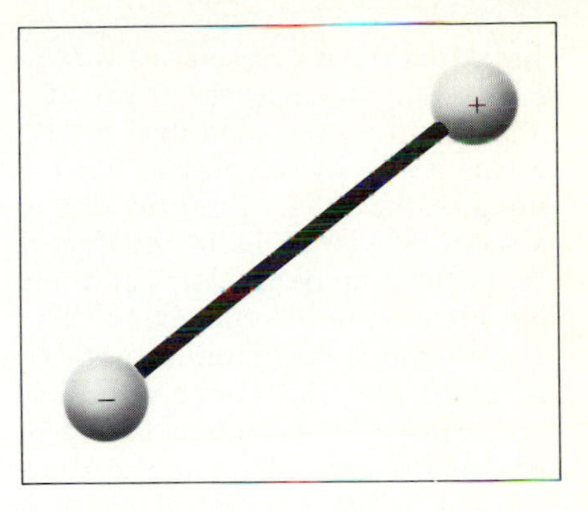

FIGURE 1.7 Two equal but opposite electric charges, restrained by a rubber tube.

ity weakens with distance from any object.

Force fields that decrease with distance from their source are encountered throughout all of science. Most of them, including electric and gravity forces, decrease as the *square* of the distance. They are said to obey the **inverse-square law**. Figure 1.6 shows the change in strength of such a force field. Inverse-square force fields decrease rapidly with distance from their source, becoming, for example, 9 or 25 times weaker at distances three or five times greater, respectively. Despite this rapid decrease, however, the strength of such a force field never quite reaches zero. Accordingly, the electric and gravitational force fields of objects having some charge and mass can never be extinguished, even in the empty space of a virtual vacuum.

TWO CHARGED PARTICLES: Consider now two equal but opposite charges. Call one ⊕ and the other ⊖. If left alone, they would mutually attract, crashing together. Opposite charges attract, while like charges repel.

Imagine, however, that the two charges are restrained, for example at the ends of a rubber tube like the dumbbell of Figure 1.7. At rest, each charge calmly resides with its own electric force field decreasing with distance. Furthermore, each charge is bathed in the electric force field of the opposite charge. Now suppose that one charge, say ⊖, begins oscillating up and down a little, perhaps because it becomes heated or collides with some other object. The change of one charge's position causes its associated force field in turn to change ever so slightly. Since ⊕ resides in ⊖'s electric force field, ⊕ will experience the force field change and will respond to that change by mimicking the vibration of ⊖. The opposite phenomenon can also occur: ⊕ wiggles; its electric force field changes; ⊖ knows it and responds to it since ⊖ sits in ⊕'s force field.

The important point here is that an oscillating electric charge induces a change in its own electric force field. This change can then be detected by other electric charges. In this way, information can be transferred from one place to another. By what means? By the change of the force field. How fast is this information transmitted? That's another problem.

Of prime importance is the question: How quickly does ⊖ feel the change in the electric force field when ⊕ begins to oscillate? Is the information transmitted at some measurable velocity, or is it an instantaneous flash? Well, both theory and experiment demand that this information move in the form of a wave at a very specific velocity. This is the **velocity of light**. We can imagine the wave of information as a series of vibrations in the electric force field, moving along at about 30 billion centimeters/second. This equals the more familiar 186,000 miles/second, and can be written in scientific notation as 3×10^{10} centimeters/second. We can also write it as 3×10^5 kilometers/second, since there are 10^5 centimeters in a kilometer. (If you are unfamiliar with scientific notation, consult Interlude 1-1.)

The velocity of light does not precisely equal 300,000 kilometers/second. The exact value, $299{,}793 \pm 1$ kilometers/second in a vacuum (and slightly less in material substances such as air or water), has, like most quantities in this text, been rounded off for convenience. Don't clutter up your mind with exact numbers; they can be looked up in reference books. It's more important to understand the concepts, one of which is that light travels fast. In the time needed to snap a finger—about a tenth of a second—light radiation can travel once around the entire Earth. If the currently known laws of physics are correct, the velocity of light is the fastest speed possible.

Basically the same events occur for real cosmic objects. When some of their charged contents move around, their electric force fields are changed, and we can detect that change. The resulting electric force field vibrations travel outward in a wave-like motion. These waves carry information; they are waves of radiation. And no material medium is involved. Small charged particles, either in our eyes or in experimental equipment, eventually respond to such force field changes by vibrating in tune with the received radiation. It's much like the example above, when the ⊕ particle eventually vibrated with the same frequency at which the ⊖ particle vibrated initially.

Figure 1.8 shows a familiar example of information being transferred by radiation. There, a television transmitter forces electric charges to oscillate up and down a metal rod near the tower's top, thereby radiating waves of information into the air. This radiation can then be detected by roof-top antennas having metal booms in which electric charges respond by vibrating in tune with the transmitted wave frequency. Radio waves of different wavelengths (corresponding to different AM or FM stations or TV channels) transfer information from transmitter to receiver by changing their electric force fields in different ways. Between any two places on Earth, this information can be transferred in less than a second.

Remember, though, wave motions from distant cosmic objects take longer to reach Earth. In fact, we can correctly say that the Andromeda galaxy's light, viewable in tonight's sky, results from changes among charged particles that occurred within that galaxy about 2 million years ago. Although light radiation travels at the fastest velocity known, it still takes time to get here. Accordingly, we see a delayed image of Andromeda. We can know nothing about this galaxy at the present epoch, namely, the last quarter of the twentieth-century Earth time. For all we know, Andromeda might no longer even exist! Only our descendants 2 million years into the future can know for sure if it really exists now.

ELECTROMAGNETISM: We need to discuss one more feature about force fields. A **magnetic force field** accompanies every *changing* electric force field. A stationary electric force field, like that near a motionless electric charge, has no associated magnetic force field. But whenever the strength of an electric force field

INTERLUDE 1-1 ***Scientific Notation***

Astronomy is a subject that spans knowledge of both very small (microscopic) things and very large (macroscopic) things. Atomic nuclei have sizes of about 0.0000000000001 centimeter, while galaxies measure some 100,000,000,000,000,000,000,000 centimeters. To avoid writing so many zeros, scientists use a shorthand notation, where the number of zeros is denoted by a superscript power of 10. This superscript is negative if the number is smaller than unity (1) and the zeros lie to the right of the decimal point; the superscript is positive if the number exceeds unity and the zeros lie to the left of the decimal.

For example, we can shorten the number above describing the typical size of atomic nuclei to 10^{-13} centimeter, and that describing galaxies to 10^{23} centimeters. In this way we avoid carrying around as excess baggage all those zeros. Be assured, it's nothing at all to write a zero, but scientists like to compact things to avoid errors.

Some other examples of scientific notation are:

- distance to a typical galaxy = 1,000,000,000 light-years = 1×10^9 light-years
- size of a common virus = 0.00001 centimeter = 1×10^{-5} centimeter
- mass of hydrogen atom = 0.00000000000000000000000167 gram = 1.67×10^{-24} gram. (A gram is nearly 500 times smaller than an American pound.)

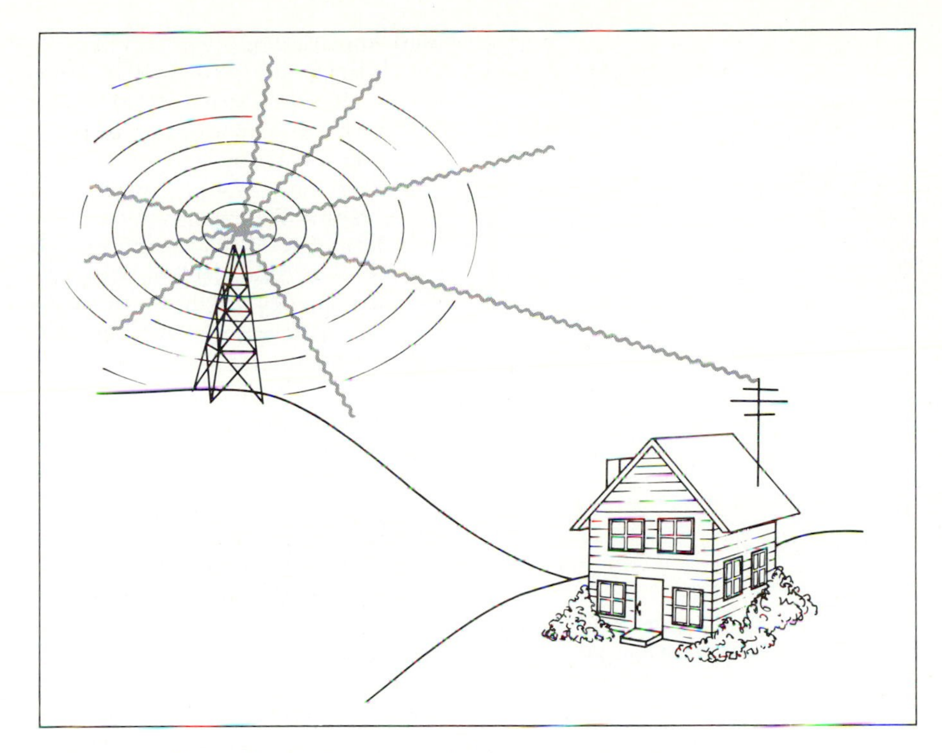

FIGURE 1.8 Charged particles in an ordinary household television antenna vibrate in response to the charged particle movements in the transmitter tower that sends out waves of radiation.

changes (upon movement of some charge), a magnetic force field is created. It obeys the same laws as the electric force field; in particular, its strength decreases as the distance squared. The principal difference is that a magnetic force field governs the influence of one magnet on another magnet.

Here, then, is the way that radiative waves emanate from any terrestrial or extraterrestrial object. As suggested in Figure 1.9, both the electric and magnetic force field vibrations are tied together inseparably, each moving at the velocity of light. The waves of electricity and magnetism do not really exist as independent quantities; instead, they are different aspects of a single physical phenomenon—**electromagnetism**. For this reason, radiative waves, or electric waves, or magnetic waves are collectively termed "electromagnetic waves." Thus it is this electromagnetic radiation that transfers information (i.e., energy) from one place to another.

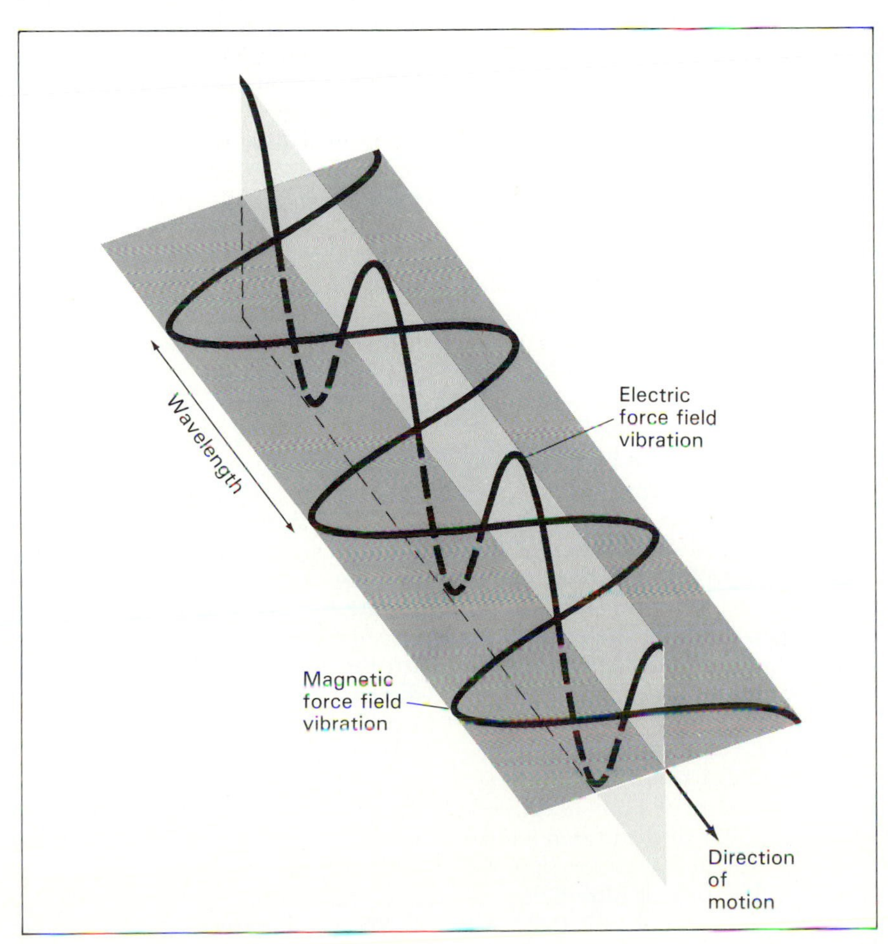

Components of Light

White light is made of various colors—red, orange, yellow, green, blue, and violet. We can identify each of these colors, as shown in Figure 1.10, by passing light through a triangular block of glass known as a prism. Each color is one component of white light radiation, and each is perceived differently by our human eyes because of the colors' different wavelengths. (Much the same thing happens when sunlight passes through a cloud of raindrops, each of which acts as a minute prism to form a rainbow.)

Experiments prove that red light has a wavelength of about 7×10^{-5} centimeter, nearly twice as long as violet light whose wavelength

FIGURE 1.9 Electric and magnetic force fields vibrate perpendicular to each other. Together they form electromagnetic radiation that always moves at the velocity of light.

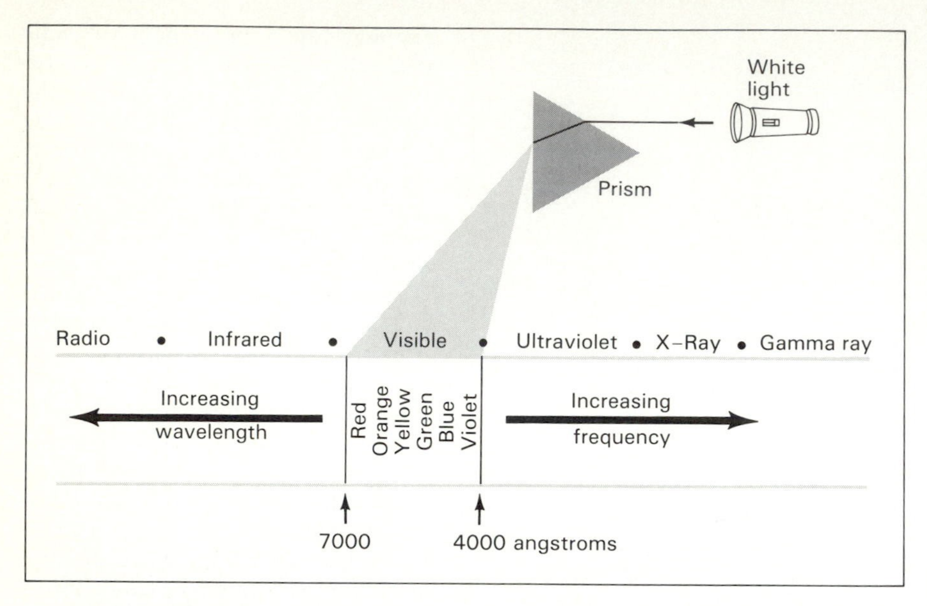

FIGURE 1.10 While passing through a prism, white light splits into its component colors, spanning red to violet in the visible part of the electromagnetic spectrum. [See also Plate 1(a).]

measures about 4×10^{-5} centimeter. Wavelengths have also been measured for the other colors, thus specifying the whole **visible spectrum**, as shown in Figure 1.10. Human eyes are insensitive to radiation of wavelength shorter than 4×10^{-5} centimeter or longer than 7×10^{-5} centimeter. Radiation outside this range is invisible.

For convenience, we often use another unit—called the angstrom—when describing the wavelength of light. It's a unit of length, named after a nineteenth-century Swedish physicist. By definition, there are 10^8 angstroms in 1 centimeter. Accordingly, the visible spectrum spans the wavelength range from 4000 to 7000 angstroms. The kind of radiation to which our eyes are most sensitive has a wavelength near the middle of this spectrum, at about 5500 angstroms.

We can use the equation written earlier in this chapter to derive a wave frequency for visible radiation. We simply divide the velocity of light by the wavelength. The result, nearly 10^{15} hertz, is an extremely fast wave vibration, impossible to observe without advanced laboratory equipment.

Note in Figure 1.10 that the frequency of light radiation increases toward the right, while the inversely related wavelength increases toward the left. We keep this standard throughout the book.

Electromagnetic Spectrum

Light is only one kind of radiation. It's the kind with which we are most familiar, for it helps us to see things. In short, light is that radiation to which human eyes are sensitive.

In addition, there are also other kinds of radiation—**invisible radiation**—to which our eyes are completely insensitive. These are radio, infrared, and ultraviolet waves, as well as x-rays and gamma-rays. Despite these different names, including "rays," "radiation," and "waves" (as well as "photons" to be explained in the next section), all these terms refer to the same phenomenon of electromagnetic radiation; the various names are historical accidents.

Figure 1.11 plots the entire spectrum of all radiation. To the long-wavelength side of light lie the regimes of radio and infrared radiation. At smaller wavelengths we have the domains of ultraviolet, x-ray, and gamma-ray radiation. All these spectral regions, including the visible spectrum, collectively comprise the **electromagnetic spectrum**.

A large amount of useful information is tucked into Figure 1.11. It's worth studying carefully. Note how the wave frequency (in hertz) is plotted increasing toward the right, whereas the wavelength (in centimeters) increases toward the left. These wave properties behave in opposite ways because, as noted earlier, they are inversely related.

Note also that the wavelength and frequency scales are not linear; that is, they do not increase by equal intervals of 10. Instead, the horizontal scale plots *factors* of 10. We are often forced to use this type of plot in science in order to condense a very large range of some quantity. Had we used a linear scale for the same wavelength range shown in Figure 1.11, we would have had to publish a figure that folded out many light-years long! Throughout the text we'll often find it useful to compact a wide range of some quantity onto a single easy-to-view plot by using a scale that varies by equal factors of 10 instead of equal *intervals* of 10.

Figure 1.11 shows the great range of wavelengths for the many kinds of electromagnetic radiation. Wavelengths extend from the typical size of mountains for radio radiation to sizes comparable to the atomic nucleus for gamma-ray radiation. Other familiar objects are also drawn adjacent to the wavelength scale to help you visualize the actual size of radiative waves in different parts of the spectrum.

Since the spectrum is drawn to scale in Figure 1.11, we can quickly note the very small size of the visible domain. Our eyes are simply not sensitive to any kind of radiation outside the range 4000 to 7000 angstroms. Thus we begin to appreciate the fact that our eyes perceive only a minute portion of the many kinds of radiation. If objects in the Universe emit invisible radiation—and many of them do—there must be a wealth of knowledge to be gained by studying all kinds of electromagnetic radiation.

Because Figure 1.11 is so basic to

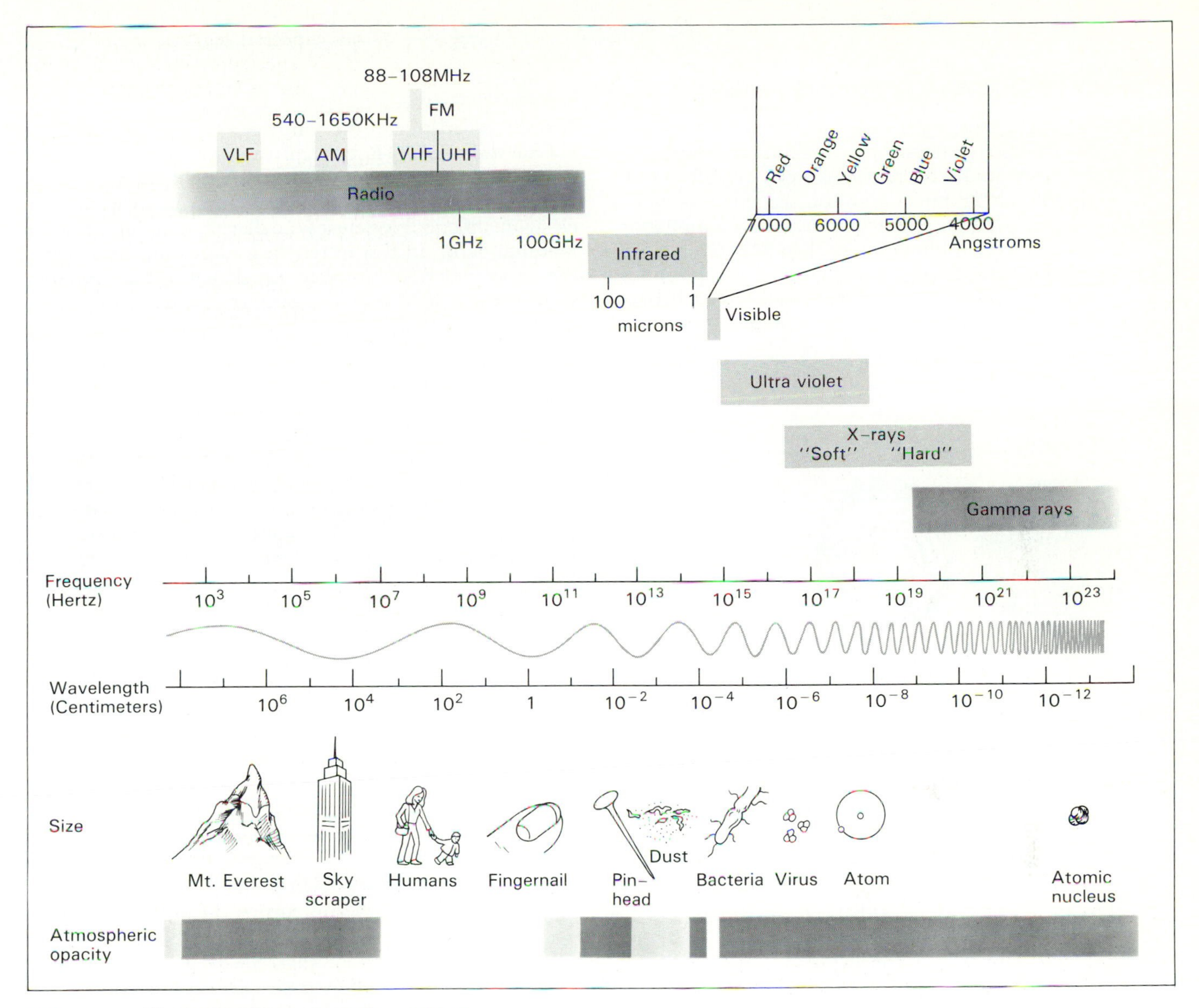

FIGURE 1.11 The entire electromagnetic spectrum.

much of the science discussed herein, we have reproduced it on the inside back cover. Feel free to refer often to this chart of the electromagnetic spectrum; it's surely one of the most important figures in the book.

ATMOSPHERIC BLOCKAGE: Why is the visible spectrum so small compared to the other parts of the entire electromagnetic spectrum? Why do our eyes respond to only a minute portion of the many different kinds of radiation known to exist?

To answer these questions, we need to consider the **opacity** of Earth's atmosphere—that is, the extent to which radiation is attenuated or blocked while passing through parcels of air; opacity is the opposite of transparency. At the bottom of Figure 1.11, the atmospheric opacity is plotted along the wavelength and frequency scales. The extent of shading is proportional to the opacity. For example, the atmosphere is completely opaque to radiation having frequencies where the shading is greatest; no radiation can get in or out. Where there is no shading at all, the atmosphere is completely transparent, meaning either that extraterrestrial radiation from space can reach Earth's surface or that terrestrial radiation from transmissions made by humans can pass unhindered into space.

What causes opacity to vary along the spectrum? In other words, how can visible radiation penetrate our atmosphere, but not ultraviolet radiation? Similarly, how can relatively long-wavelength radio waves (those of about 1 to 1000 centimeters wavelength) pass unhindered through the atmosphere, whereas short-wavelength radio waves (those of about 0.1 to 1 centimeter wavelength) cannot? The answers are found by studying the gaseous contents of Earth's atmosphere.

Certain gases are known to absorb radiation at some frequencies more

INTERLUDE 1-2 *More on Size and Scaling*

Put aside our discussion of the electromagnetic spectrum for a moment. Figure 1.11 includes a general scale of objects found on Earth. We can usefully extend this scale to include much larger objects encountered in space. For example, the sketch below is a continuation of the size scale for objects larger than Mt. Everest. These extend from Earth-sized objects about 10 billion or 10^{10} centimeters in diameter to galaxy superclusters which, at 10^{26} centimeters, are the largest known assemblies of matter in the Universe.

Our scale here is not meant to imply that some kinds of radiation have wavelengths comparable to these cosmic systems. Scientists are currently unaware of electromagnetic frequencies smaller than about a few hundred hertz. In this interlude we have merely taken the opportunity to complete our sketch of the size and scale—from 10^{-14} to 10^{26} centimeters—of all things known in the Universe.

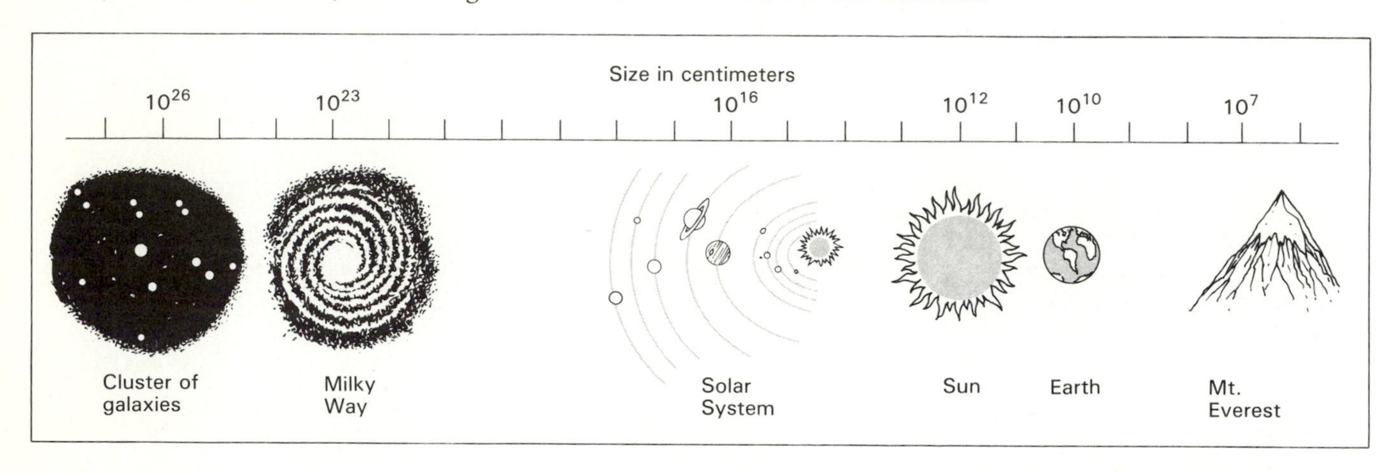

efficiently than others. For example, water vapor (H_2O) and oxygen (O_2), major constitutents of our atmosphere, absorb radio radiation having wavelengths less than about a centimeter. Similarly, water vapor and carbon dioxide (CO_2), important trace gases of Earth's atmosphere, are strong absorbers of infrared radiation; they make it hard to transmit or receive infrared radiation through our atmosphere. A third example is the blockage of visible light by atmospheric clouds; lots of water vapor often produces large, dense clouds (and with them rainy weather), thus preventing the direct passage of light radiation from celestial objects to our ground-based telescopes.

Ultraviolet, x-ray, and gamma-ray radiation is also completely blocked by Earth's atmosphere, although here the culprit is the **ozone layer**. High in Earth's atmosphere, oxygen gas, normally made of two bound oxygen atoms (O_2), breaks apart into their individual atoms (O) when hit with ultraviolet radiation emitted by the Sun. Almost immediately the free oxygen atoms recombine with the remaining O_2 to form a thin layer of ozone (O_3) which itself absorbs ultraviolet radiation. Thus all kinds of high-frequency radiation trying to pass through our atmosphere are intercepted by O_2 or O_3 gas (as well as by nitrogen gas, N_2).

This layer of ozone, by the way, is one of several insulating spheres that serve to protect life on Earth from the harsh realities of outer space. Not long ago, scientists judged space to be hostile to advanced life forms because of what is missing out there—breathable air and a warm environment. Now, most scientists regard outer space harsh because of what is present out there—fierce radiation and energetic particles, both of which are injurious to health. The ozone layer is one of our planet's umbrellas; without it, advanced life (at least on the surface of Earth) would be at best unlikely and probably impossible.

EYESIGHT: This is a good point to discuss further the limited usefulness of the human eye. Notice in Figure 1.11 that the visible spectrum is aligned almost exactly with a narrow domain in which the atmospheric opacity is minimized. In fact, this atmospheric "window," through which radiation can pass unhindered, spans 10^{-4} to 10^{-5} centimeter. What sort of radiation gets through? Well, the dominant source of radiation near Earth is the Sun. And the wavelength at which solar radiation is maximized equals just about 5000 angstroms. Most of the Sun's infrared radiation and all of its ultraviolet radiation is blocked by our atmosphere. But the Sun's visible radiation pours right in. It is no coincidence that the human eye has its greatest sensitivity in this window; our eyes have evolved over millions of years so as to use efficiently the most intense radiation reaching Earth's surface. This is the radiation that we call visible (sun)light. In this way, humans and most other life forms can perceive radiation that illuminates surface objects most clearly, thereby allowing us to identify objects and to get around on the surface of our planet.

If our eyes, for some reason, responded only to infrared radiation, they would be much less useful to us. Why? Because the Sun's infrared radiation cannot penetrate our atmosphere very well, and thus could at best only dimly illuminate objects on Earth's surface. Even worse, if our eyes were sensitive only to ultraviolet radiation or to x-rays, Earth's surface would be full of darkness.

Why aren't human eyes sensitive to radio radiation? After all, Figure 1.11 shows that Earth's atmosphere is largely transparent to most radio radiation. Why can't we perceive radio waves, or at least use them to judge the position of objects around us? The answer is that the Sun's radio radiation is much weaker than its visible radiation. Consequently, our eyes have evolved so as to take advantage of the *maximum* surface radiation provided by our nearest cosmic source. Had there been a source of radio radiation more intense than our Sun's visible radiation, life forms might have learned to perceive objects and to go about our business on the surface of Earth by listening to the scatterings of radio waves, instead of by looking at the reflections of visible light.

Toward the end of the book, when we study the prospects for extraterrestrial life, we'll need to remember that other planets might have atmospheric gases different from Earth's. (Other planets might also orbit stars much hotter or cooler than our Sun.) The nearby planet Venus is a good example. There, with thick clouds covering the planet at all times, the most intense solar radiation (namely, visible light) cannot penetrate well to the surface. Accordingly, should there be any organisms on Venus, they might need to depend partly on solar infrared or radio radiation that can filter through its atmosphere. Consequently, their "eyes" would have probably evolved to become sensitive to infrared or radio radiation. Since evolution accumulates small changes over the course of millions of years, humans should not expect to be able to adapt our eyesight to perceive infrared or radio radiation when they someday land on Venus. (Actually, this Venusian problem might not be as severe as described here; Venus' surface is illuminated roughly as well as Earth on the most heavily overcast day—which still provides enough light for human eyes to see things, albeit dimly.)

IONOSPHERIC REFLECTION: Absorption of radiation by molecular gases is not the only cause of Earth's partly opaque atmosphere. Some gases absorb so much radiation that they establish a thin, electrically conducting layer high in Earth's atmosphere. From this layer, called the **ionosphere**, long-wavelength radio waves are reflected. In fact, the ionosphere reflects long radio waves (1000 centimeters or greater wavelengths) as well as a mirror reflects light waves. In this way, extraterrestrial waves are kept out, and terrestrial waves are kept in.

Earth's ionosphere is used, by the way, to aid radio broadcasting. As shown in Figure 1.12, long-wavelength AM signals are transmitted up to the ionosphere, where they are

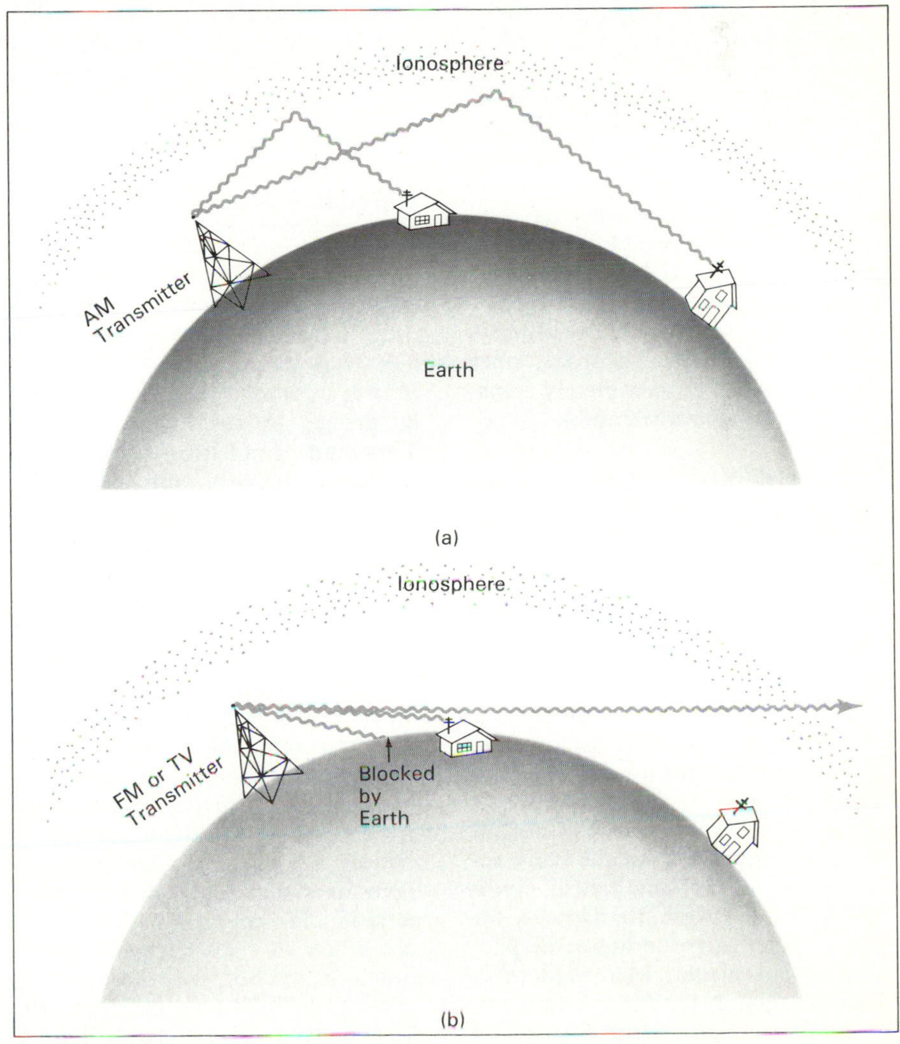

FIGURE 1.12 AM radio stations have greater ranges because their signals are bounced from the ionosphere (a). The ionosphere also prohibits cosmic radiation of wavelengths longer than about 1000 centimeters from reaching the surface of our planet. By contrast, FM and TV signals pass readily through the ionosphere, thus these signals must be sent directly from transmitter to receiver (b).

reflected back to ground receiving stations. In this way, transmissions can be sent over greater distances and thus reach wider audiences than would otherwise be possible along a direct "line of sight." And since the height of the ionosphere is often greater at night due to the absence of solar radiation pushing on it, AM reception from distant stations is often better once the Sun sets. For example, someone in Boston could easily tune in a Baltimore AM station (especially at night when the ionosphere is higher), but they would find it impossible to receive FM signals from Baltimore. (Short-wave "ham" transmissions can be heard around the world using such ionospheric reflections. The name "short-wave," however, is a misnomer; these man-made signals are really of long wavelength.) Shorter-wave FM signals, as well as TV signals, must be transmitted directly from station to station along the line of sight, thus limiting their range to less than 100 kilometers. Should FM or TV signals be transmitted up toward the ionosphere, Figures 1.11 and 1.12 show that they would pass right through, out into space, since our atmosphere is completely transparent at those wavelengths.

ATMOSPHERIC POLLUTION: Polluting our air with gases and other foreign matter not normally present in Earth's atmosphere could be harmful to life on our planet. In the first place, chemicals capable of absorbing or reflecting sunlight would cause surface objects to be illuminated less well, thereby degrading our perception of the world around us. But there's also a more severe problem.

Gases released as wastes into our atmosphere because of the daily activities of our technological civilization could begin to damage the ozone layer surrounding and protecting our planet. Man-made pollutants (such as the Freon in our refrigerators, and the so-called chlorofluorocarbons, which propel chemicals from some spray cans) thus become a socially relevant problem, since a depleted ozone layer would allow harmful ultraviolet, x-ray, and gamma-ray radiation to reach Earth's surface.

Interestingly enough, energy itself could possibly pollute our air and ultimately cause sunlight to be dimmed on Earth's surface. Increasing amounts of energy, produced by any means to support the daily needs of an ever-increasing population, could begin to melt the polar ice caps, thereby releasing more water vapor into the atmosphere. The result, a real possibility a few centuries into the future, would be heavier cloud cover, perhaps even a stifling, murky, totally enshrouded Earth. Thus, when we think small, the "energy crisis" points toward not enough energy; but by thinking big (over long time scales), we see that such a crisis might result from our civilization someday producing *too much* energy.

Particle Nature of Radiation

Some kinds of radiation are harmful, especially to living organisms, while other kinds seem completely harmless. For example, radiation surrounds us at all times, as can readily be proved by turning on a radio. This kind of radiation is not harmful, nor is the light radiation that illuminates our homes and offices. Nor is the infrared radiation that heats us. But if we were similarly immersed in x-ray radiation, we'd be sorry people. As a rule of thumb, infrared radiation will heat us, ultraviolet radiation will burn us, x-ray radiation will sterilize us, and gamma-ray radiation will kill us.

To understand the various effects of radiation on matter and life, we must study further the nature of radiation. For radiation not only behaves as a wave, but also acts as a particle. To appreciate this, recall that when we see a television signal fading in and out, it is often the result of interference between the wave coming directly from the TV station and another wave reflected from, say, a passing plane or a nearby building; such interference phenomena are characteristic of waves. On the other hand, when a Geiger counter exposed to a source of x-rays emits a "click," we know that the x-ray has registered as a particle.

We can gain a better feeling for the particle nature of radiation by studying the results of a classical experiment performed early in this century at the dawn of modern physics—one whose explanation won a Nobel Prize for a man named Einstein in 1921. The essence of this experiment is depicted in Figure 1.13, where light is shown shining on a metal surface. When high-frequency blue light hits the metal, a nearby detector picks up bursts of minute particles (called electrons in Chapter 2). The metal not merely reflects the incident radiation, but apparently absorbs some of it, which liberates the particles from the metal like one billiard ball hitting another. If we now decrease the frequency of the light, namely change its color from blue to, say, red, the detector still records bursts of particles, but now their energy is less. (This is true since the particles' velocity, hence their energy, is measured to be smaller.) Further experimentation proves that the vehemence with which the particles are torn from the metal depends only on the color of the light and not at all on its intensity. If the light source is removed to a considerable distance and dimmed to a faint glow, the ejected electrons are fewer in number but their velocity is undiminished. The important point is this: Even though the incident radiation shines on the metal in a steady, continuous way, both the absorbed radiation and the liberated particles have the form of discrete, quantized packets. The name given these bullets or particles of radiation is **photons**. They are essentially microscopic bundles of energy.

The phenomenon described above is termed the **photoelectric effect** and occurs whenever radiation interacts with matter. (You can see it at work, for example, on an activated TV screen; a received radio wave caused particles to be freed from a metal inside the TV and are then beamed toward a fluorescent screen which glows as a visual picture.) We'll see in the next chapter

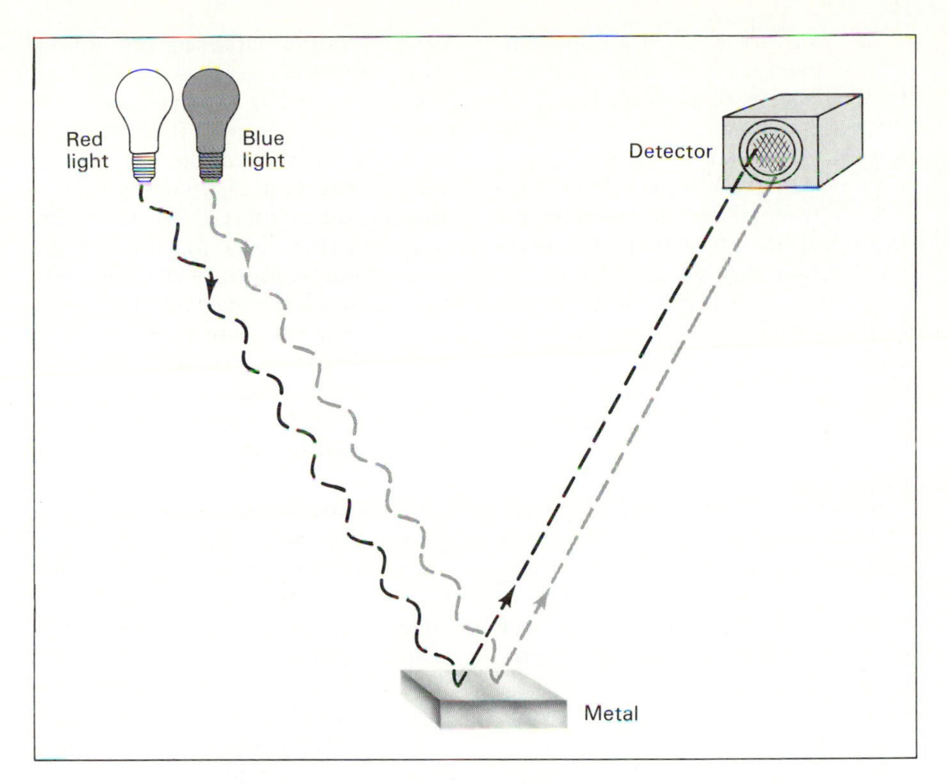

FIGURE 1.13 Schematic diagram of the equipment used to study the Photoelectric Effect. When energy (upper left) is directed toward metal, minute particles are ejected from the metal and can be recorded by a detector (at upper right.) Only the concept of discrete particles (photons) of radiation can explain the results of this experiment; a smooth wave of radiation cannot.

that minute quantities of matter (called atoms) within the metal are responsible for the sudden absorption of radiation and emission of particles. For now, we need only realize that in addition to a smooth, continuous stream, radiation can also act in a jerky, discontinuous way. The environmental conditions ultimately determine which way—wave or particle—the radiation behaves. But we have a general rule of thumb: In the macroscopic realm of large bulk objects, radiation is often more usefully described as a wave, whereas in the microscopic domain of atoms, radiation is best characterized as a series of particles.

Note how the photoelectric effect also demonstrates that the energy contained within a photon is proportional to the frequency of the radiation. In the experiment just described, the higher-frequency radiation succeeded in liberating particles of greater energy from the metal. Consequently, we say that

$$\text{photon energy} \propto \text{frequency of radiation}$$

where the symbol "∝" is a proportionality sign. Low-frequency radiation means small photon energy, and conversely. Stated somewhat more colloquially, "red" photons carry roughly half the energy of "blue" photons.

This relationship explains why, for example, x-ray radiation is more harmful than radio radiation; high-frequency x-rays carry more energy in every photon. So it's partly the *frequency* of radiation, in addition to the strength of that radiation, that harms living things. For these and other reasons, that part of the electromagnetic spectrum to the high-frequency (i.e., short-wavelength) side of the visible spectrum is often called the "high-energy" domain. Scientists studying high-frequency radiation from space are known as high-energy astronomers. Those studying lower-frequency radiation are often called radio or infrared astronomers.

Many people find confusing the idea that radiation sometimes behaves as a wave while at other times as a series of particles. To be truthful, modern physicists don't yet understand *why* nature prefers such a wave-particle duality. Nonetheless, we have experimental evidence for each of these concepts of radiation. We are stuck with the notion that some experiments are best explained when radiation is considered a wave, while others are understood only if radiation is considered a particle. In frustration, some scientists facetiously call radiation a "wavicle."

Intensity

All objects emit radiation, regardless of their size, shape, or chemical makeup. Burning wood, hot stoves, warm asphalt, electric bulbs, even human bodies, along with virtually all other objects around us, emit some type of radiation. They radiate mainly because heat causes their microscopic charged particles to wiggle around. And remember, whenever charges change their positions, electromagnetic radiation is emitted.

Some objects emit more radiation than others. A 100-watt bulb, for instance, radiates four times more energy than a 25-watt bulb. Similarly, radiation is greater near a hearty fire than near a smoldering log. **Intensity** is the term used to specify the amount or strength of radiation. Like frequency or wavelength, intensity is a major property of radiation.

SPREAD OF RADIATION: The spread of the intensity of radiation provides much insight about an emitting object. From it, we can gain direct information about an object's temperature, and often indirect information about its nature. Generally, the intensity of radiation is maximized at some frequency, while falling off to lesser values on either side of the peak. (This sort of spread mimics that of course grades; most students

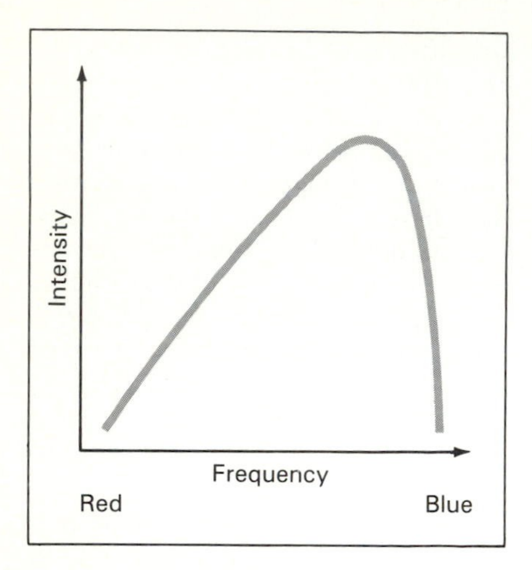

FIGURE 1.14 The Planck curve represents the spread of the intensity of radiation arising from any heated object.

receive average grades, whereas fewer get high and low grades.)

Figure 1.14 shows the standard spread of radiation emitted by any heated object. We call it the "curve of a perfect thermal emitter" or simply the **Planck curve**, named after a German physicist who first quantified these effects at the start of the twentieth century. Note that the curve is not shaped like a symmetrical bell that normally declines evenly on both sides of the peak; instead, the Planck curve decreases slowly at low frequencies and rapidly at high frequencies. This peculiar shape is characteristic of the emission of radiation from all heated objects.

The frequency at which the intensity peaks depends on the temperature of an emitting object. In fact, the entire Planck curve shifts toward higher frequencies as an object's temperature increases. Even so, the characteristic shape of the curve remains the same. This shifting of radiation's intensity with temperature agrees with the notion familiar to us all: Glowing objects, such as fire or stars, emit visible radiation, whereas cooler objects, such as warm rocks or smoldering wood, emit invisible radiation. The latter objects, warm to the touch but not glowing hot to the eye, actually emit their radiation most intensely as low-frequency infrared and radio waves.

We can further illustrate the shift of the Planck curve with temperature by means of a simple example. Suppose that we place a piece of metal into a hot furnace having a window that allows us to look inside. The metal first becomes warm, though it doesn't change its appearance. Upon further heating, it begins to glow dull red, then orange, brilliant yellow, and finally white. How do we explain this? At first, when the metal was warm, it emitted invisible radio and infrared radiation, much like a warm pot on the kitchen stove. But as the metal became even hotter and glowed visibly, the peak of the metal's emitted radiation shifted toward higher frequencies, producing much radiation in the visible domain. The Planck curve for an object of several thousand kelvin, regardless of its chemical composition, peaks in the visible spectrum, and as it becomes even hotter, its color gradually changes from red to orange to yellow. (For a discussion of temperature units, consult Interlude 1-3.)

Objects eventually become white hot (provided that they are not destroyed at high temperatures) because Planck curves peaked in the green or blue part of the spectrum still have a low-frequency "tail." This ensures that all colors are emitted, and together they combine to produce the neutral color white. This is just the reverse of the dispersion of white light into its component colors noted earlier; white light can be produced once again by passing the entire red-to-violet color spectrum of Figure 1.10 through a second, oppositely oriented prism.

Actually, no known natural *terrestrial* objects can become hot enough to emit high-energy radiation. The few inventions of our civilization that do, like x-ray machines, are carefully licensed because such energetic radiation can damage living tissue. On the other hand, many extraterrestrial objects do emit copious quantities of ultraviolet and x-ray radiation. Fortunately, these objects are far from Earth and, furthermore, our planet's protective atmosphere shields us from them.

APPLICATIONS: The beauty of the discussion above is that we can turn it around. That is, we can use Planck curves to determine the temperatures of objects without needing to make on-site measurements with, for example, a thermometer. In fact, this is the main method used by astronomers to measure the temperature of the Sun's surface. Observations of the radiation from our Sun at many different frequencies yield a Planck curve shaped like the one shown in Figure 1.14. And as we might expect, the Sun's curve peaks in the visible part of the electromagnetic spectrum. As shown in Figure 1.15(c) on page 26, the Sun also emits much infrared and a little ultraviolet radiation. The overall curve suggests that the surface of the Sun is approximately 6000 kelvin. This is the temperature of the Sun's *surface* only, not of its interior.

Many stars, including all those seen on a clear night, obviously emit visible radiation. Most of these stars have surface temperatures much like the Sun's. However, in addition to normal stars, many cosmic objects have surfaces much cooler or hotter than our Sun's. These objects emit radiation that is peaked somewhere in the invisible spectrum. For example, much cooler regions, like those of very young stars, measure about 600 kelvin and emit mostly infrared radiation. On the other hand, some older and much hotter objects often have surface temperatures of 60,000 kelvin and hence emit mostly ultraviolet radiation.

Figure 1.15 compares the Planck curves for a variety of objects: a typical galactic cloud, a young star's atmosphere, our Sun's surface, and a decrepit (though brilliant) old star. Be sure to realize that the temperatures of these cosmic objects pertain only to the part of the object emitting the observed radiation. In the case of stars, for example, such Planck curves suggest that *surface* temperatures generally range from about 2000 to 50,000 kelvin.

These surface temperatures, however, fall far short of the extremely hot, millions-of-kelvin temperatures near the cores of normal stars.

INTERLUDE 1-3 *The Kelvin Temperature Scale*

Temperature is a measure of the heat of an object. Like the archaic English system used to measure length (in feet) and mass (in pounds), the familiar Fahrenheit temperature scale is of dubious value. The "degree Fahrenheit" is a peculiarity of American society. Fortunately, everyone in the world will soon be using the more reasonable Celsius (or centigrade) scale of temperature measurement. The benchmarks of the Celsius scale are easy to remember: water freezes at 0 degrees Celsius (more sensible than 32°F), and it boils at 100 degrees Celsius (also more sensible than 212°F). The accompanying figure might help in making these comparisons.

Temperature can go below the freezing point of water, reaching as low as −273 degrees Celsius. Although we know of no matter anywhere in the Universe to be this cold, all atomic and molecular motion is expected to virtually stop at this lowest possible temperature.

For convenience, scientists have adopted an even more reasonable temperature scale. Known as the **Kelvin scale,** in honor of a nineteenth-century British physicist, it differs from the Celsius scale by some 273 degrees. As used in this book,

$$\text{kelvin} = \text{degrees Celsius} + 273.$$

Consequently:

- motion ceases at 0 kelvin,
- water freezes at 273 kelvin,
- water boils at 373 kelvin.

To put things into perspective, we can compare the temperatures of, for example, a typical galactic cloud (60 kelvin, and thus much colder than the freezing point of water), a young star's atmosphere (600 kelvin, or a little hotter than the boiling point of water), the Sun's surface (6000 kelvin, or well above water's boiling point), and a hot old star (60,000 kelvin, and very hot indeed, at least compared to anything familiar to us on Earth). Incidentally, these temperatures are characteristic, respectively, of the four Planck curves drawn in Figure 1.15.

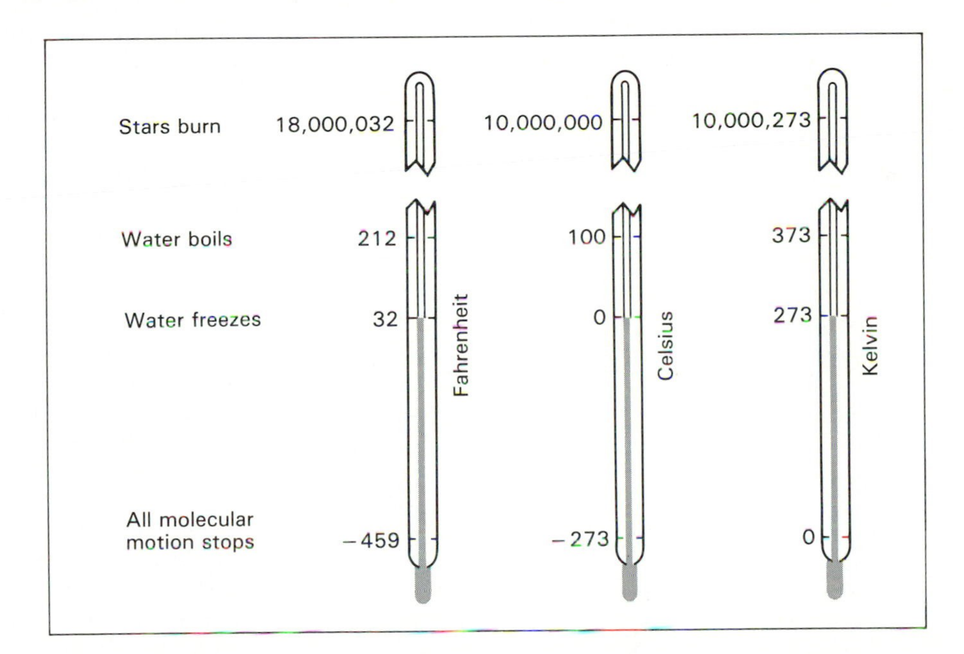

Fortunately for us, radiation from the extremely hot interior of any star doesn't reach Earth directly. Otherwise, we'd be flooded with absolutely lethal x-ray and gamma-ray radiation. Instead, the observed stellar radiation arises from relatively cooler surface layers.

So, by measuring the frequency spread of emitted radiation, we can infer the surface temperature of *any* object—terrestrial or extraterrestrial. Again, this method is not a direct measurement, but rather an inference. Provided that nothing interferes with the emitted radiation, the Planck curve yields a reasonably accurate temperature for whatever matter emits the radiation. In effect, the Planck curve acts as a thermometer to help us estimate the temperatures of many remote objects in the Universe.

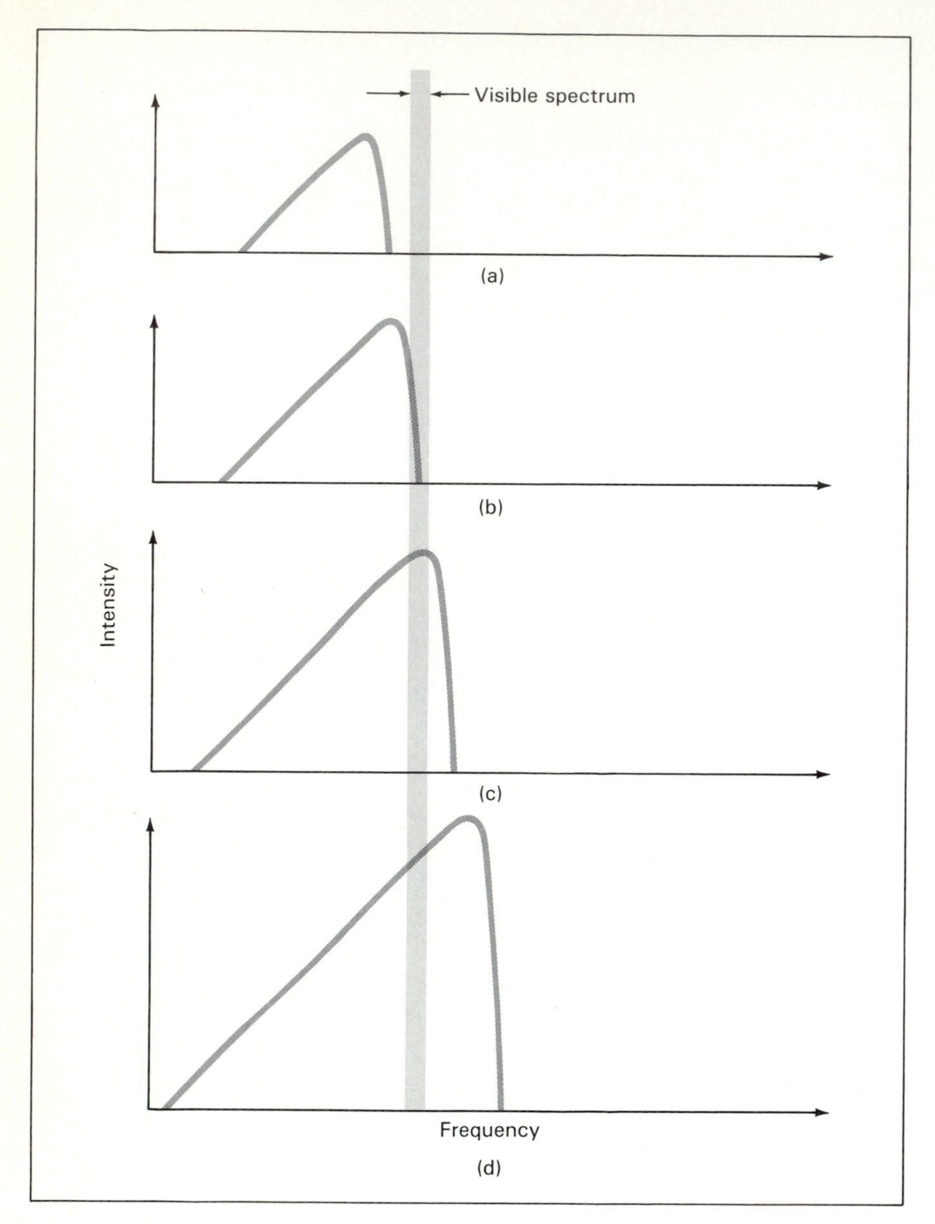

FIGURE 1.15 Comparison of Planck curves for four cosmic objects having different temperatures: (a) a cool, invisible galactic cloud at 60 kelvin, (b) a young star's dim atmosphere at 600 kelvin, (c) the bright Sun's surface at 6000 kelvin, and (d) an old yet luminous star at 60,000 kelvin.

Polarization

Most natural sources of radiation like the Sun or a candle flame, or even a man-made electric light bulb, emit electromagnetic waves that vibrate in many planes—not only the plane of this page, but also perpendicular to it, and at all other angles as well. Such a group of randomly oriented waves is said to be "unpolarized." By this, we mean that the waves have no preferred planes of vibration.

Some sources of radiation emit waves having only certain planes of vibration. Such radiation is said to be "polarized." **Polarization** is a subtle, though important property of radiation whereby a wave's plane of oscillation is oriented in a special way. We can explain it best by considering a familiar example.

Polarized radiation is obtained by shining light through a Polaroid filter. These filters are made of long-chain molecules stacked parallel to one another like a series of long vertical strips. Acting like slats of a venetian blind, the molecules allow passage of only those wave vibrations having one orientation—namely, those parallel to the long molecules.

Figure 1.16(a) shows the difference between polarized light and unpolarized light. Ordinary unpolarized light from a flashlight can become polarized by passing through a filter. Not all the unpolarized light manages to get through, however. Only those light waves having vibrations parallel to the filter's long molecules make it; the other waves are blocked. The intensity of such polarized light is smaller than the unpolarized light initially shining on the Polaroid filter.

If a second Polaroid filter is oriented with its long molecules also parallel to those of the other filter, the resultant light is not decreased further. But the intensity of light diminishes continuously as one filter is rotated relative to the other. In fact, the resultant light is completely blocked whenever the two filters are perpendicular, as illustrated in Figure 1.16(b). In this case, no light makes it through the two filters, regardless of the initial extent of polarization or the initial intensity of the radiation.

In nature, radiation that is reflected or scattered often becomes polarized to some extent. In particular, sunlight reflected from Earth's surface is usually horizontally polarized. In this way, Polaroid sunglasses, with their long-chain molecules oriented vertically, usefully decrease the glare on a bright sunny day.

Radiation captured from space is also sometimes polarized. Generally, polarization can be caused either intrinsically by a peculiar emission mechanism of some cosmic object. Or polarization can be caused by some intervening matter through which radiation passes on its way toward Earth. In later chapters we'll come to appreciate how studies of polarization give ad-

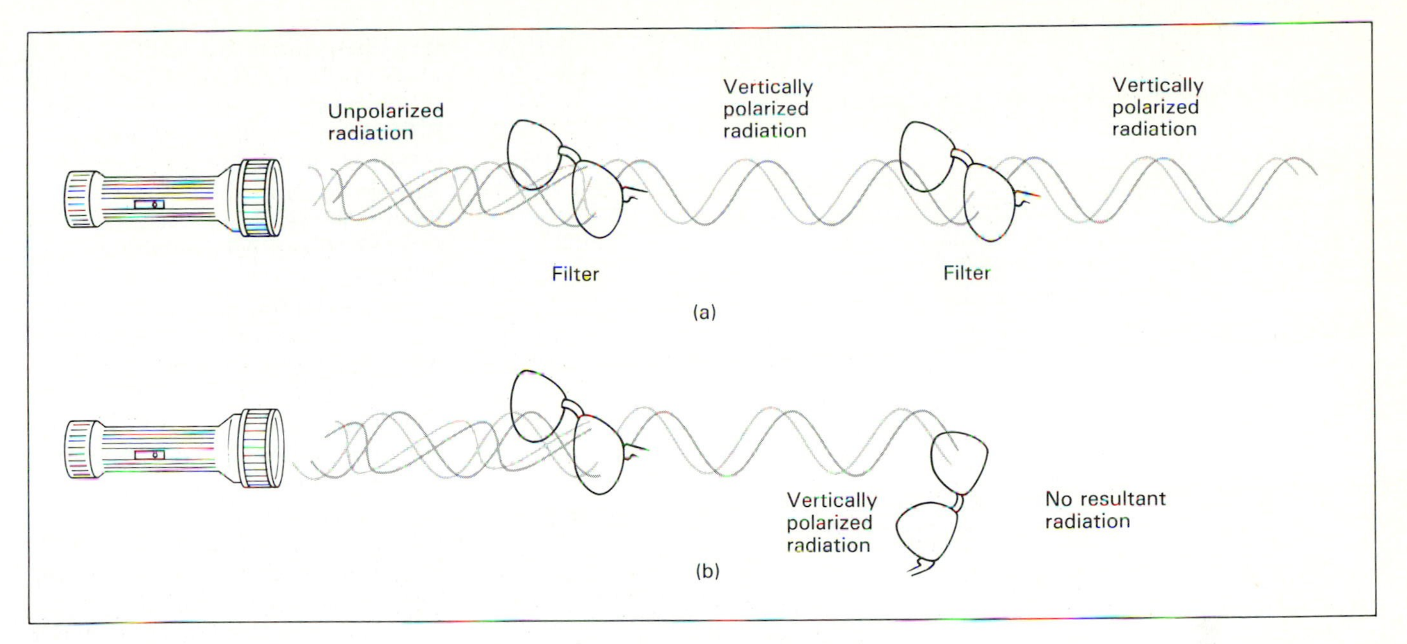

FIGURE 1.16 When ordinary unpolarized radiation passes through appropriate filters, it can change into polarized radiation. In (a), only waves having vertical vibrations get through two sets of similarly oriented polarizing lenses. By contrast, in (b), two oppositely oriented polarizing lenses allow no light to pass.

ditional clues about the nature of cosmic objects or of intervening galactic matter.

Doppler Effect

Radiation can change in many ways. Wavelength, frequency, intensity, and polarization can each change as environmental circumstances change. However, the speed of radiation never changes; it always equals the velocity of light.

In addition, our *perception* of radiation can sometimes change. This is a phenomenon more subtle than anything yet discussed in this chapter. In this case the radiation itself does not change; only our perception of it changes. Let's use a hypothetical example to illustrate what we mean.

A DESCRIPTION: Suppose that a rocket ship of the future is launched from Earth with enough fuel to allow it to accelerate constantly. Moving faster and faster, the rocket would eventually reach a very high velocity. On-board travelers, looking out from the front of the spacecraft, would note that the light emitted from the star system toward which they traveled seemed to be getting bluer. At first glance, they might think that the star's surface temperature was increasing, emitting more radiation in the blue part of the spectrum. But this is not the reason for the change in color. As the travelers would soon realize, *all* stars seem to become bluer than normal when approached at high speed. The greater the velocity of approach, the greater the color change. And whenever the spacecraft slowed down and came to rest, the stars would seemingly emit normally once again, just as they appeared through Earth-based telescopes prior to the voyage.

Furthermore, if the travelers looked back at several stars as the spacecraft surged forward at high speed, they would notice the opposite effect: The stars behind them seem redder than normal. Again, they would easily note that the greater the velocity of recession, the greater the color change.

Consequently, since all the stars in the spacecraft's forward direction seem bluer than normal, and all those behind redder than normal, the travelers would be forced to conclude that the stars changed their colors because of the spacecraft's motion, and not at all because of a real change in any of the physical properties of the stars.

Indeed, *motion* is the key here. Motion toward or away from a source of radiation can change the way we perceive that radiation. Such motion along any line of sight is known as **radial motion**; it should be distinguished from motion perpendicular to the line of sight. Only radial motions induce changes in waves.

We stress that radial motion does not really change the intrinsic properties of radiation—just our (or our equipment's) perception of it. Specifically, the wavelength (or frequency) is affected, since blue light differs from red light only by its shorter wavelength (or higher frequency). Radial motion doesn't change our perception of any other property of a radiative wave; in particular, radiation still travels at the velocity of light.

Change due to motion is not restricted to radiation or to fast space voyages. It's familiar to anyone who has, for example, waited at a railroad crossing for a train to pass. Recall that we hear the pitch of the engine's horn change from high shrill to low roar as the train approaches and then recedes. Since high-pitched sound (treble) has shorter wavelength than low-pitched sound (bass), this well-known sound change is similar to the light change experienced by the space travelers. In fact, if the train had a white light atop its engine, we could, in principle, witness the color change of this light, from blue to red, as the train passed by. In practice, however, the light's color change would be impossible to perceive, for trains travel much too slowly for appreciable color changes.

This motion-induced change in the wavelength (or frequency) of an observed wave—regardless if it's a radiative light wave or an acoustical sound wave—is known as the **Doppler effect** in honor of the nineteenth-century Austrian physicist who first explained it. The reason for it is not hard to understand, and since it plays such a crucial role in much of astronomy, we need to study it further.

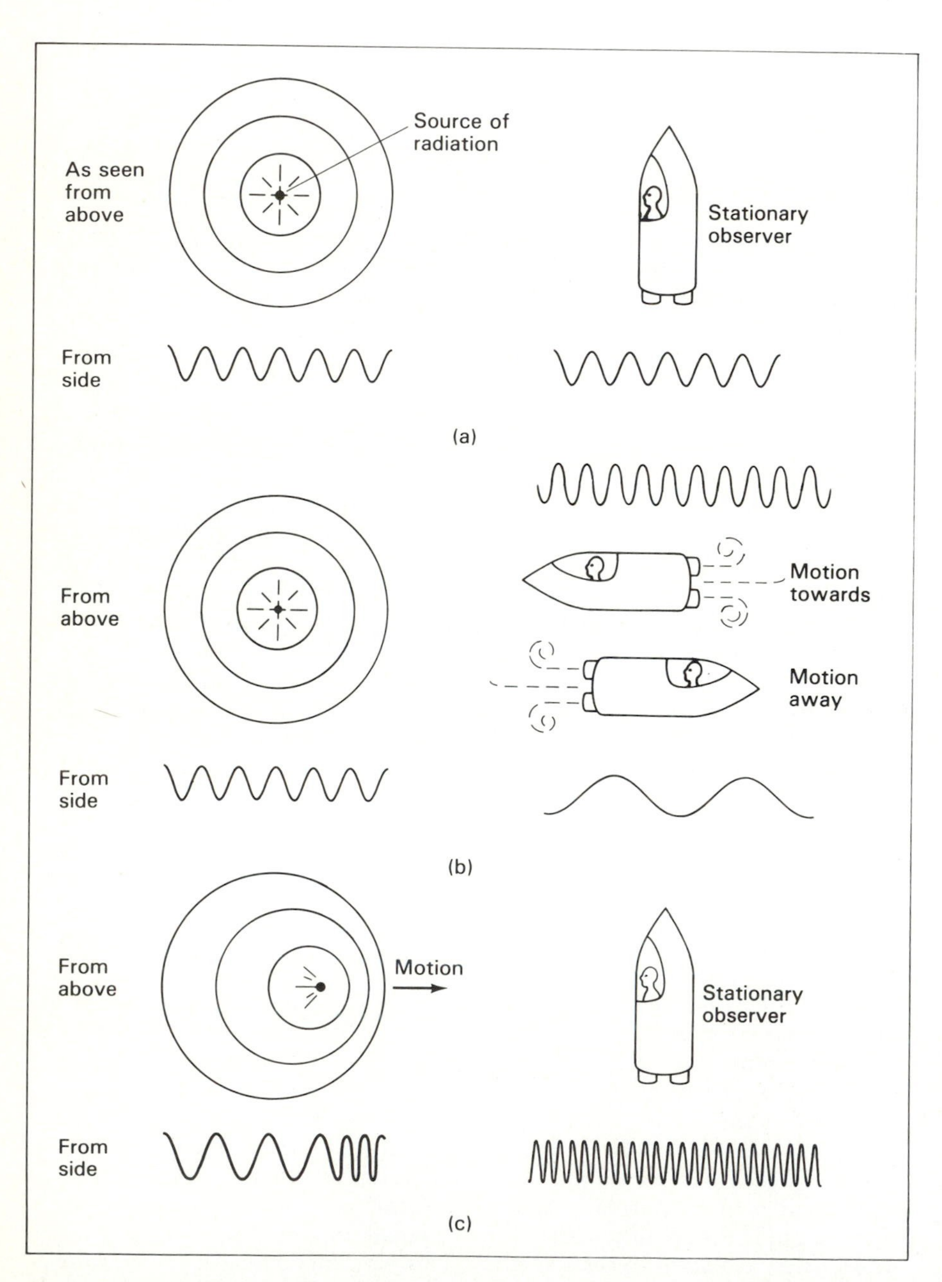

FIGURE 1.17 (a) Wave motion from a fixed source toward a fixed observer. (b) Wave motion from a fixed source toward a moving observer. (c) Wave motion from a moving source toward a fixed observer.

AN EXPLANATION: Imagine a wave motion from the place where it's generated toward those of us in an unmoving rocket ship, like that sketched in Figure 1.17(a). By counting the number of crests passing us per unit time, we could measure the frequency or the wavelength.

Suppose, however, that the ship is moving toward the source of the wave, as in Figure 1.17(b). This case is analogous to the space travelers approaching a distant star system. The number of wave crests now counted per unit time would be larger than for a fixed observer, because, after all, we are moving toward the oncoming waves. The *observed* wave will seem to have a higher frequency, or equivalently a shorter wavelength. We say that the wave has become "blue shifted," since blue radiation is of higher frequency than red radiation. (This convention holds even for invisible radiation for which red and blue color have no meaning; any shift toward shorter wavelengths is still called a blue shift, and conversely.) The greater the observer's velocity toward the source of the waves, the greater the increase in frequency and the greater the extent of blue shift.

The same sort of change would also be noted if the source of the waves were moving toward a fixed observer, as in Figure 1.17(c). In fact, **blue shift** (or shortening of the wavelength) occurs whenever the source and the observer have *any* net motion of *approach* between them.

The opposite phenomenon is true for recessional velocities. If the source of the waves moves away from us, or if we move away from a fixed source [as also shown in Figure 1.17(b)], or if there is *any net recessional motion* involved, we would count fewer wave crests per unit time. The observed wave is then said to have a **red shift** (or to be lengthened in wavelength).

In summary, we conclude that the extent of Doppler shift is directly proportional to the net radial velocity. We emphasize the word "net": If the net velocity is zero (either for a source and an observer both at rest, or for both moving in the same direction with the same velocity), no Doppler shift would be observed. Nor is the Doppler effect considered here applicable to motion perpendicular to the line of sight. Finally, be sure to note that Doppler effect does not depend on distance in any way. It depends strictly on the net radial velocity.

Of what value is the Doppler effect? In principle, we could use the extent of blue shift or red shift to measure the motion of some cosmic object along the line of sight. In practice, however, it's hard to measure Doppler shifts of radiation of the type studied in this chapter. Planck curves are widely spread over many frequencies, and small shifts of this curve are extremely hard to measure with any accuracy. Still, the situation is not hopeless. Precise measurements of Doppler effect *can* be made using more narrowly defined radiation, a phenomenon that we'll study in Chapter 2.

SUMMARY

Astronomical observations differ basically from laboratory experiments. In astronomy, we must rely only on the radiation that just happens to travel near Earth. Only by means of radiation do we know anything about most of the Universe beyond our planet.

Light is one such kind of radiation—the most familiar kind that we perceive with our eyes. There are other kinds as well, including radio, infrared, ultraviolet, x-ray, and gamma-ray radiation. Sometimes radiation behaves as a wave; at other times, as a particle. Either way, radiation always travels at the speed of light.

Radiation has four main properties: wavelength, frequency, intensity, and polarization. These differ for different kinds of radiation. By measuring each of these properties, we can determine a great deal about cosmic objects. In later chapters we'll use everything discussed here to study the various kinds of radiation, matter, and life known to exist in the Universe.

KEY TERMS

blue shift
Doppler effect
electric force field
electromagnetic spectrum
electromagnetism
energy
extraterrestrial
frequency
intensity
inverse-square law
invisible radiation
ionosphere
Kelvin scale
light
light-year
magnetic force field
opacity
ozone layer
photoelectric effect
photon
Planck curve
polarization
radial motion
radiation
red shift
temperature
terrestrial
velocity of light
visible spectrum
wave
wavelength

QUESTIONS

1. What is light?
2. Why don't cosmic x-rays reach Earth's surface?
3. In what regions of the electromagnetic spectrum is the atmosphere transparent enough to make observations from the ground?
4. Pick One: Cosmic x-rays (a) occasionally reach Earth's surface where telescopes can detect them, (b) must *always* be observed with instruments above Earth's atmosphere, (c) are sometimes emitted by the clouds in our atmosphere, (d) often escape into space from our Sun's interior, (e) are used by dentists to study astronomically sized molars.
5. A group of amateur astronomers discovers a new star. At their disposal are color filters, a telescope, a compass, and a simple gadget to measure light intensity. Have the astronomers got enough equipment to determine the surface temperature of the star? If so, explain how they would do it; if not, what else do they need, and how would they do it?
6. Why are x-rays generally harmful to life forms, whereas radio waves are generally harmless?
7. It is well known that radiation often takes the form of a wave. Cite and discuss briefly the main piece of evidence suggesting that radiation can also act as a particle.

*8. Calculate the wavelength of the radiative information transmitted by television's Channel 5. (The frequency of all Channel 5's in the United States is approximately 78×10^6 hertz.)

9. Do you think that it would be more valuable to be able to view the entire Universe as it is *now* than to be able to "look back" into the past as is actually possible? Why or why not?
10. If you have two pairs of Polaroid sunglasses (or one pair and don't mind taking them apart), you can verify the variable diminution of light through the filter by altering the orientation of one lens relative to another. Try it.

FOR FURTHER READING

*FAZIO, G., "Infrared Astronomy," in *Frontiers of Astrophysics,* E. Avrett (ed.). Cambridge, Mass.: Harvard University Press, 1976.

FIELD, G., AND CHAISSON, E., *The Invisible Universe.* Boston: Birkhäuser Boston, 1985.

FRIEDMAN, H., *The Amazing Universe.* Washington, D.C.: National Geographic Society, 1975.

LEDERMAN, L., "The Value of Fundamental Science." *Scientific American,* November 1984.

SPILLER, E., AND FEDER, R., "The Optics of Long-Wavelength X-Rays." *Scientific American,* November 1978.

2
THE INNER WORKINGS OF ATOMS

In Chapter 1 we introduced the basic nature of radiation. We discussed how bulk matter, on macroscopic scales, usually radiates a smooth, undulating wave. But we also noted that on microscopic scales, matter seems to emit and absorb radiation in a discontinuous way. The implication is that matter is basically made of individual building blocks. These building blocks are called **atoms**.

In this chapter we continue our study of radiation. Here we examine the microscopic world of atoms to better understand the basic properties of matter. We discuss especially the methods of **spectroscopy**, an observational technique enabling scientists to infer the nature of matter by the way it emits and absorbs radiation. Using this technique, we have no need to examine individual atoms directly. This is fortunate, for the direct study of atoms is hardly possible (owing to their small sizes).

Spectroscopic studies are done with the aid of an apparatus known as a spectroscope or spectrometer. This device can be as simple as the prism studied earlier to disperse light. Or it can be a more complex machine capable of dissecting invisible radiation. Basically, the modern research spectroscope comprises a telescope (to capture the spectral features), a dispersing device (to spread them out), and a detector (to record them), although much additional gear is needed for proper operation. We'll study that experimental operation in some detail in Chapter 3. For now, suffice it to note that spectroscopy has become an indispensable tool of modern astronomy, enabling researchers to analyze the fine details of the radiation emitted by any cosmic object. In this way we can infer the physical and chemical properties of truly remote objects without having to make direct measurements of those objects—another task that is currently impossible (owing to their great distances).

Simple spectroscopes of the prism variety have been used for more than 100 years. However, a spectroscope alone is not enough to unravel the fine details of radiation. Without a modern theory of atomic physics, spectroscopic measurements could be neither analyzed nor understood. Hence astronomers at the start of the twentieth century knew hardly more about the physics of stars than did the Renaissance pioneers Galileo and Newton. Indeed, early researchers held little hope that their observational data could ever be used to determine the makeup of stars, let alone how stars shine.

With much new insight and many novel experiments, however, scientists began to understand the atom's inner workings about 50 years ago. Coupling spectroscopy as a technique with atomic physics as a theory, we now know how each type of atom emits or absorbs in a very special way. Modern spectroscopy has thus become a powerful observational technique, not only revealing the chemical composition of stars and knowledge of how they shine, but also yielding a wealth of information about the birth, evolution, and death of myriad objects throughout the Universe.

The learning goals for this chapter are:

- to appreciate the size and scale of matter at the microscopic level
- to understand some of the basic concepts of modern atomic theory
- to realize that many different kinds of atoms exist in nature
- to understand how electron changes within atoms produce

emission and absorption features in the spectra of those atoms

- to appreciate the wealth of information obtained by analyzing spectral features
- to recognize that atoms can chemically bind together into larger clusters called molecules

Scale of Things

Throughout our text, we study large intervals of space and time. You might find it useful, during any discussion, to keep in mind the spatial and temporal domains being considered. Here, in the case of atoms, we need to appreciate the scale of things at the microscopic level.

Imagine for a moment something familiar, like an ordinary coin. Dimensions of about a centimeter (little less than half an inch) are easily seen. By magnifying the coin, we could see smaller parts of it. Figure 2.1 illustrates such a series of magnifications.

Suppose, for example, that we first magnified the coin by a factor of 1000. [This is about the best magnification achievable with ordinary ("light") microscopes.] Crystals, the basic substructure of all solids, would be seen at the scale of about one-thousandth of a centimeter, or 10^{-3} centimeter. A further magnification of 1000 would enable us to recognize clusters of atoms, that is, molecules, at the scale of 10^{-6} centimeter. Nowadays, we can resolve features at this level by means of high-powered ("electron") microscopes.

To reach down to the size of individual atoms, however, we would need to magnify by an additional factor of 100. Atoms themselves have typical sizes of 10^{-8} centimeter, or 1 angstrom. They thus require a magnification of some 100 million times more than normally viewable at, say, the scale of an ordinary coin. Microscopes have not yet been invented to see individual atoms clearly.

The basic composition of matter does not end with atoms. These structures are in turn made of even smaller building blocks, called **elementary particles**. To see them, however, would require atoms themselves to be magnified another 10^5 times; elementary particles have sizes on the order of 10^{-13} centimeter. So atoms cannot be seen directly, nor can elementary particles. Nonetheless, by smashing the particles together inside large laboratory machines, called accelerators, physicists can study the debris to *infer* the nature of matter at levels of about 10^{-13} centimeter. In effect, accelerators, like the one shown in Figure 2.2, are extremely powerful microscopes.

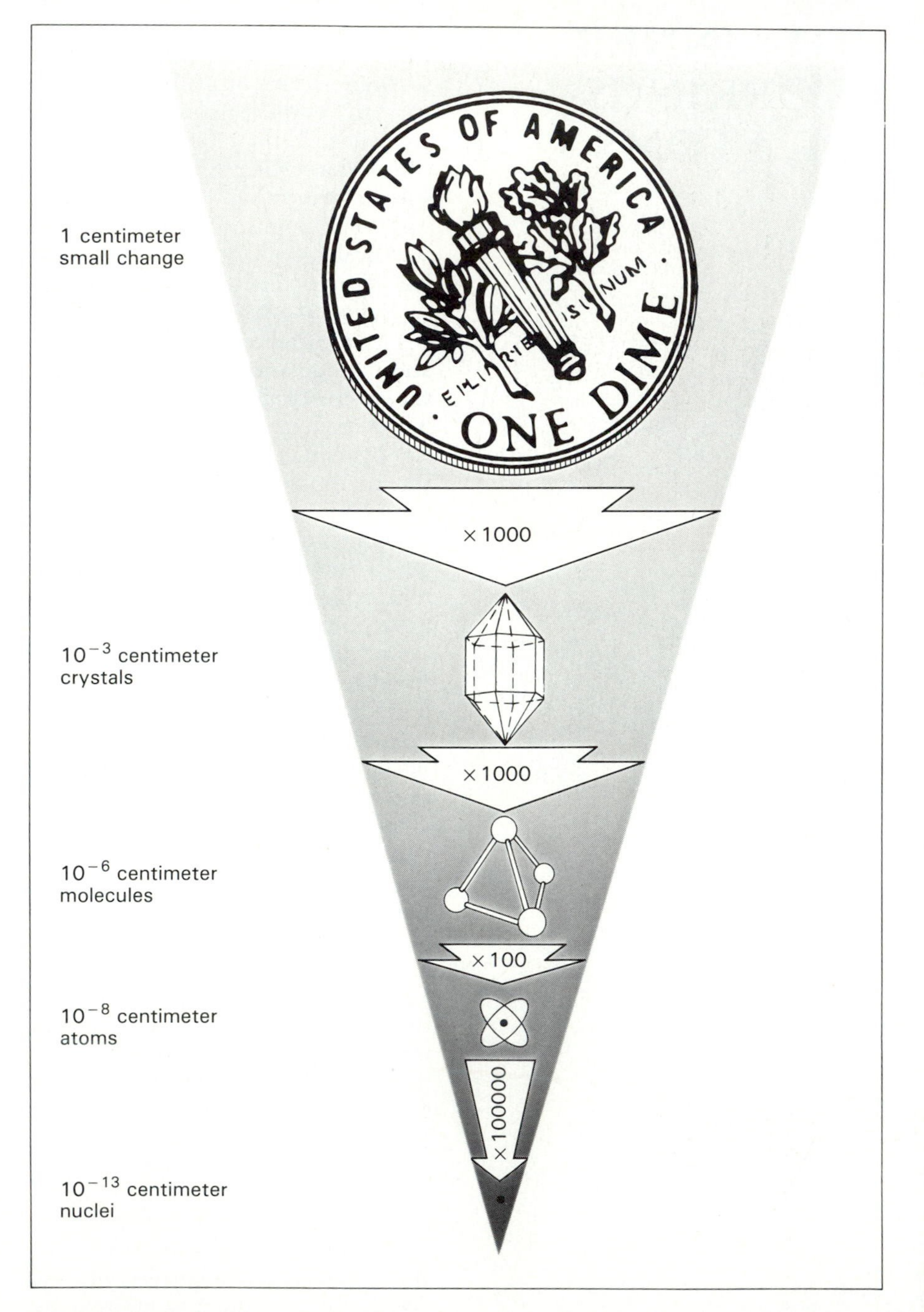

FIGURE 2.1 The basic structure of matter becomes clearer when we use greater magnification.

FIGURE 2.2 A laboratory accelerator, like this one in Geneva, is a ring-shaped device that boosts elementary particles to great speeds. Though the particles cannot be seen directly, collisions among them yield debris that can be used to infer some properties of the particles. The process can be likened to smashing together two watches, and then studying the debris of springs, cogs, and gears (or, nowadays, silicon chips, integrated circuits, and liquid crystal displays) in order to determine how watches work.

Atomic Models

SIMPLEST OF ALL: To appreciate the structure and behavior of atoms, let's start with the simplest of all—hydrogen. Figure 2.3 diagrams the hydrogen atom. As shown, the relationship of its two elementary particles—the central **proton** and its orbiting **electron**—is analogous to a grain of salt and a spherical shell the size of a football field. This gives us a feeling for the relative size and scale of things at the subatomic level.

In the classical model of the hydrogen atom, then, a negatively charged electron orbits a positively charged proton. This centrally located proton forms the **nucleus** of the hydrogen atom. The charges of the electron and proton are equal in magnitude, but opposite in kind, thus making the whole hydrogen atom neutral.

Actually, scientists are unsure *why* the proton and electron charges are identically equal, although if they were not, atoms probably wouldn't exist. Nor would we exist to worry about it. Indeed, much of atomic physics is hard, and sometimes even impossible, to appreciate deeply. We simply do not yet have a full understanding of *why* many fundamental things are as they are. Our description of atoms in this chapter may not even represent physical reality. Rather, atoms are probably only mathematical models used to explain observations of terrestrial or extraterrestrial matter.

The equal and opposite charges of the proton nucleus and the orbiting electron exert an electromagnetic force field that binds them together within the hydrogen atom. These two elementary particles are separated by about $\frac{1}{2}$ angstrom when the hydrogen atom is in its normal state. As shown in Figure 2.4, we can visualize a normal hydrogen atom to be about 1 angstrom across.

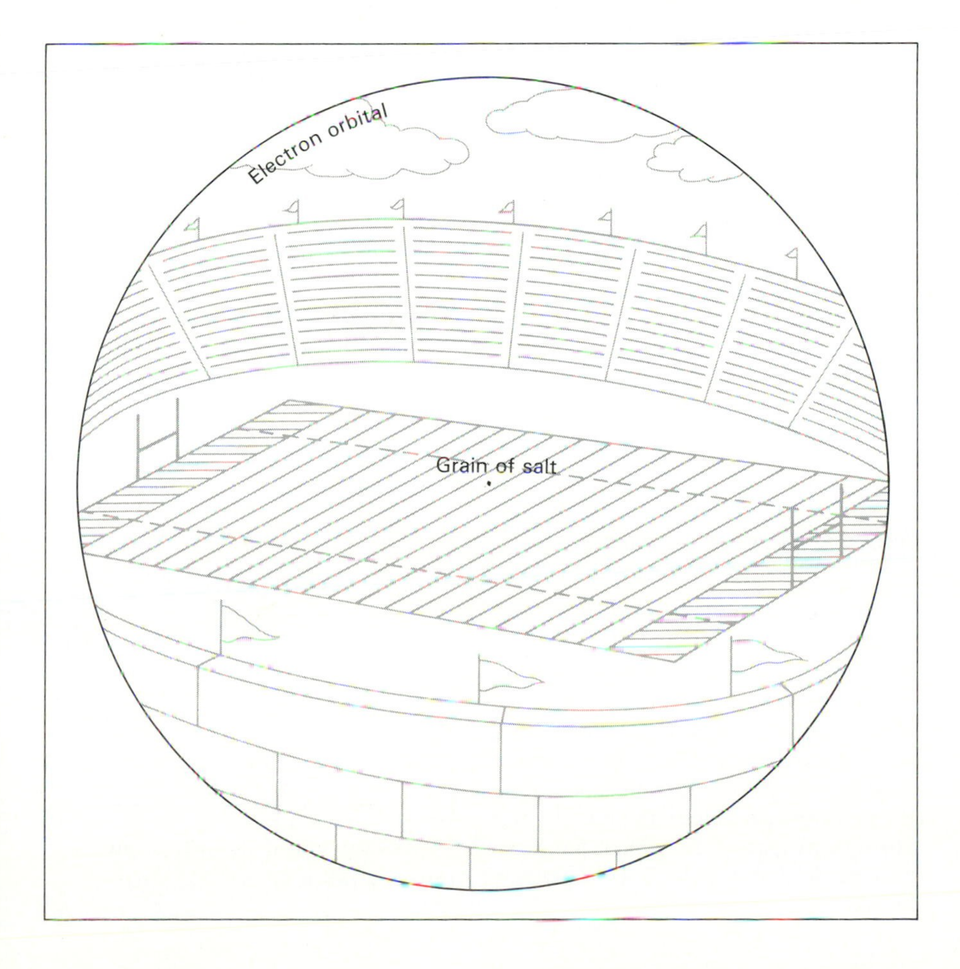

FIGURE 2.3 If the size of a hydrogen atom were expanded to the scale of a football field, the nucleus on this same scale would be only the size of a grain of salt.

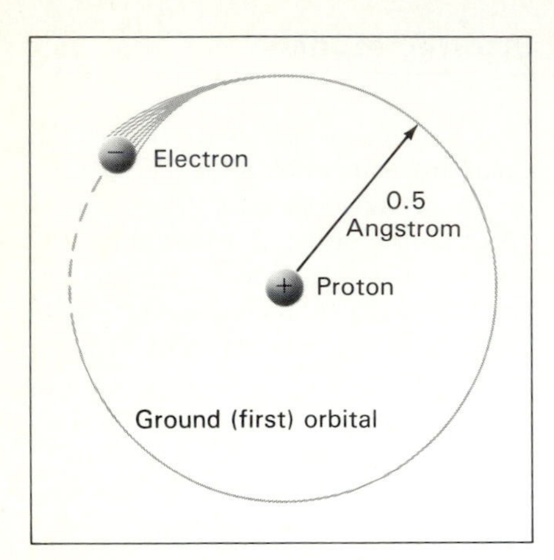

FIGURE 2.4 Schematic diagram of a hydrogen atom in its normal, ground state.

Despite their equivalent charge, the electron and proton are not at all alike in mass. A proton's mass is nearly 2000 times greater than an electron's. Taking these mass values into account, we could then visualize a better analogy than the one in Figure 2.3. We could envision the hydrogen atom to consist of a negatively charged mosquito orbiting about a positively charged baseball at a distance of some 4 kilometers. Regardless of the analogy used, the important point is that most of an atom is empty space—virtually all of an atom's mass is concentrated within its nucleus.

SOME RESTRICTIONS: This classical model of the atom, with electrons orbiting about protons, brings to mind the usual model of our Solar System with planets orbiting about the central Sun. However, like the analogies above, this one is also superficial.

Certain restrictions govern microscopic matter, while not affecting macroscopic matter. These restrictions are empirical. By that we mean that they result mainly from the inferred behavior of atoms in laboratory experiments, not necessarily from any deep theoretical understanding of atoms. They are:

1. Electrons can orbit nuclei at only specific distances, called **orbitals**.
2. The number of electrons permitted in each orbital is limited.

These restrictions contrast with intuition. Why does nature compel electrons to reside at only special distances from their parent nuclei? After all, man-made satellites can be launched into *any* Earth orbit and at *any* distance from our planet. Furthermore, with accurate rocketry techniques now known, we could boost hundreds, even thousands, of satellites into the same orbital flight path. Why are there no restrictions at the macroscopic level? We do not have a good answer to this question, apart from any orbital spacing being minute in the macroscopic realm. All we can say is that at the microscopic level, the overriding need for theory and experiment to agree forces us to adopt the restrictions above.

We should note one further point about atomic modeling: A central assumption in our text, as well as throughout all of science, is that atoms are identical everywhere. They are constructed—or at least can be modeled—in the same way regardless of whether they comprise Earth, another planet, some star, or a faraway galaxy. We regard the foregoing empirical rules governing the behavior of atoms to apply to all matter in every nook and cranny of the Universe. We simply have no other basis on which to proceed than to assume that atoms everywhere in the Universe obey the same rules as those studied terrestrially. Scientists regard this assumption as a good one, since we have no reason to suppose that it's wrong.

More Complex Atoms

Hydrogen is not the only kind of atom. While all hydrogen atoms have basically the same structure, there are many other kinds of atoms with different structures. Different kinds of atoms are called **elements**. And it is the number of protons that determines the element which an atom represents. Let's generalize our atomic model to include some larger atoms.

HELIUM AND CARBON: Consider helium, an atom having a slightly greater complexity than hydrogen. The central nucleus of helium is made of two protons, about which two electrons orbit, ensuring a zero net charge for this atom. In addition, the helium nucleus contains two **neutrons**, another kind of elementary particle having a mass slightly larger than that of a proton, but having no electric charge at all. As shown in Figure 2.5, both helium electrons reside simultaneously in the first orbital, roughly an angstrom from the nucleus. This is the usual structure of any helium atom in its normal state.

Not all the electrons of even more complex atoms, such as carbon, nitrogen, or oxygen, can reside in the

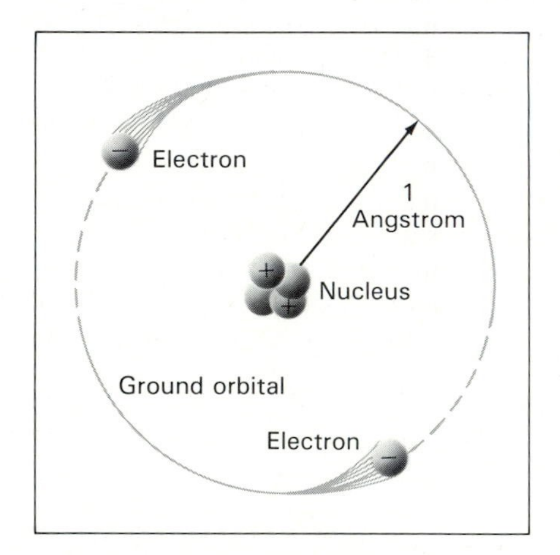

FIGURE 2.5 Schematic diagram of a helium atom in its normal, ground state.

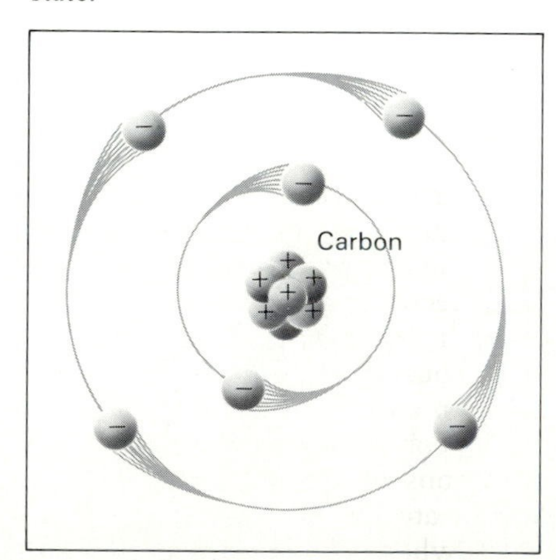

FIGURE 2.6 Schematic diagram of a carbon atom in its normal, ground state.

INTERLUDE 2-1 ***Fundamental Matter***

Exploring the basic composition of matter is not new. At least as far back as ancient Greece, attempts were made to unravel the fundamental makeup of all things. Although clearly great thinkers, the Greek philosophers were deeply in error; they presumed that *thinking* about nature was better than *looking* at it. Still, their ideas prevailed for more than 2000 years, culminating in the witchcraft and magic that befuddled the efforts of medieval astrologers and dark-age alchemists.

Only with the rise of logical, deductive reasoning during Renaissance times did the notion of "experimental philosophy" become fashionable. At last, a proper balance between thinking and looking was achieved. The principle is simple: Thoughts (theoretical work) are to be taken seriously only if confirmed by tests (experimental work). Modern science thereby emerged, and with it the scientific method—the most powerful technique ever conceived for the advancement of factual information.

Using this method, early twentieth-century science proved that atoms are not the most basic component of matter. All atoms of every different kind—that is, all elements—are made of negatively charged electrons whirling around positively charged nuclei. Each neutral atom has equal numbers of electrons and protons, as well as a similar number of neutrons. The protons and neutrons contain virtually all the mass of any atom, and together they comprise the atom's nucleus.

For much of this century, electrons, protons, and neutrons, along with photons of radiation, were considered the most basic particles of nature. However, during the past few decades, physicists have discovered a bewildering array of additional elementary particles. These newer particles are not really any more "elementary" or basic than the better-known protons, neutrons, and electrons. Rather, each elementary particle seems to play its own role in the subatomic realm.

Well over 100 elementary particles are currently known. Some behave like lightweight electrons, while others resemble heavyweight protons. Still others possess bizarre properties not yet understood. Many particles exist for only fleeting moments during fierce atomic collisions induced in laboratory accelerators. These particles literally materialize from the energy of the collision process. Usually, after a microsecond or so, they change back into energy, but not before leaving behind momentary traces on the accelerators' detectors.

The history of efforts to decipher the building blocks of nature is full of false claims. Each time researchers thought they had discovered a truly basic entity of matter, they have been proved wrong. With molecules now known to be made of atoms, and atoms in turn made of elementary particles, another question comes to mind: Are the elementary particles made of even more fundamental subparticles that have some identity or existence of their own? Current theory and some data suggest "yes."

Popular consensus has it that protons, neutrons, and many other elementary particles are made of subparticles called **quarks**. (The name derives from a meaningless word coined by novelist James Joyce in his book *Finnegan's Wake*.) Quarks are minute particles having only a fraction of the electric charge of a proton or electron.

Experimental evidence to date suggests the existence of at least six different kinds of quarks. And we have no good reason to prohibit nature from having more of them. There's only one problem: The very proliferation of quarks threatens to topple the central idea that they are the most fundamental constituents of matter. Perhaps the quarks are made of something even smaller. . . .

first orbital. For example, as shown in Figure 2.6, four of the six electrons in a normal carbon atom reside in a second orbital farther from the first orbital that houses the other two electrons. The six electrons in any carbon atom thus circle about a nucleus containing six protons and six neutrons. Carbon atoms, like all other atoms, have a net charge of zero when in their normal states.

Modern atomic theory demands that the second orbital can house a maximum of eight electrons. Elements heavier than neon, which has 10 electrons, must then begin to populate a third orbital.

The fact that the second orbital can handle only eight electrons grants carbon great strength when bonded to other atoms. By sharing the four electrons in carbon's second orbital with four other electrons of some other atoms, carbon can effectively gain its full complement of eight electrons in its outer orbital. For example, Figure 2.7 shows how the carbon atom can share the electrons of four hydrogen atoms. Electron sharing of this kind provides a strong chemical bond among atoms, especially when the outer orbital is filled with the maximum permitted number of electrons.

Electron sharing among groups of atoms will become an important factor when we study the origin of life and the prospects for extraterrestrial life later in Chapters 25 and 29. We shall then come to recognize that of all the known kinds of atoms, carbon is best suited for bonding on all

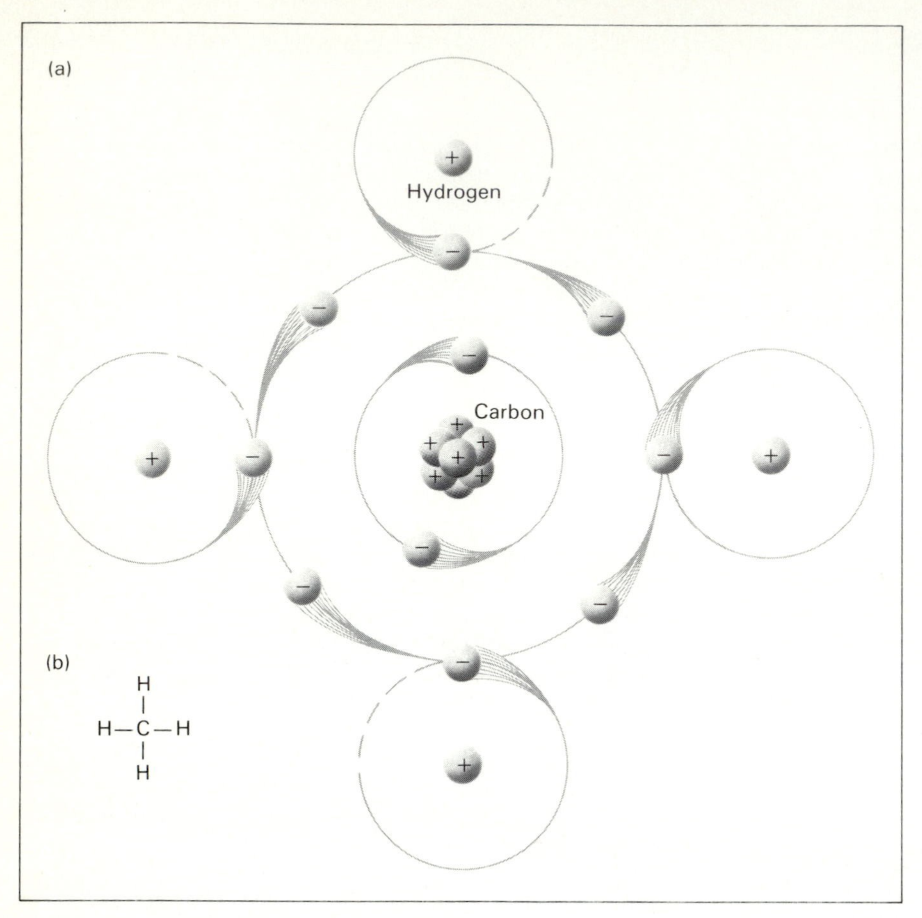

FIGURE 2.7 The single electron in each of four hydrogen atoms and the four outer electrons in a single carbon atom can form a very strong chemical bond. The drawing in (a) is the classical model of this substance, CH_4, called methane; frame (b) is a shorthand schematic diagram of methane.

four sides. Accordingly, carbon—and perhaps only carbon—has the ability to form very long chains of atoms needed for the diversity of life as we know it.

MUCH GREATER COMPLEXITY: Our atomic models can be extended to include the most complex atoms. For example, the well-known element uranium has 92 electrons in various orbitals at different distances from its nucleus. The nucleus itself contains 92 protons and 146 neutrons. Figure 2.8 diagrams the electron placements, while Figure 2.9 shows the three-dimensional complexity of such a model.

All the different kinds of known atoms can be grouped into the **periodic table of the elements**, shown in Figure 2.10. Though uranium is the heaviest element found naturally on Earth, elements even more complex have been made artificially in laboratory accelerators during the past few decades. The heaviest kind of atom confirmed to date is element 104, kirchatovium, named after the Soviet chemist who discovered it early in the 1970s. Like the other elements heavier than uranium, kirchatovium exists only for fleeting moments within the bowels of laboratory accelerators; the heaviest elements decay rapidly into some other less complex element. Even heavier elements will almost surely be discovered in the future, although perhaps in only very small amounts and for only minute periods of time.

Summarizing, two useful properties of atoms should be remembered at this point. First, in their normal states, atoms are always neutral; the total charge of their orbiting electrons perfectly balances the total charge of their proton nuclei. Second, the number of protons in the nucleus determines the number of electrons in any atom.

Furthermore, as we proceed, we'll gradually recognize that the *physical properties* of an atom, that is, its inherent behavior as a basic entity of matter, depends on the *number* of electrons. On the other hand, the *chemical properties* of an atom, that is, its ability to combine with other atoms, depends more on the *arrangement* of those electrons about its parent nucleus.

Electron Changes

POTENTIAL ENERGY: In our previous discussions we've referred to the *normal* states of atoms. This is the usual state, also called the **ground state**. It's a state of minimum energy, when the electrons are as close as possible to their parent nuclei. Several examples were given in Figures 2.4, 2.5, 2.6, and 2.8.

The ground state is analogous to the rather uninteresting case of, for example, a pencil lying on a table. The pencil is relatively inactive; it's stable. But if we pick up the pencil and hold it above the tabletop, it becomes more unstable *relative to the tabletop*. Why? Because it has gained some "potential energy" relative to the tabletop. It has even more potential energy relative to the floor below.

By potential energy we mean the amount of energy that potentially exists should we permit the pencil to fall. **Potential energy** is the energy possessed by an object because of its *position*. If the pencil did fall, the potential energy would be converted into **kinetic energy**, which is the energy possessed because of an object's *motion*.

All material systems, whether they be small atoms or larger chunks of matter, have some states that are more stable than others. A pencil has some potential energy relative to a tabletop; similarly, an electron has

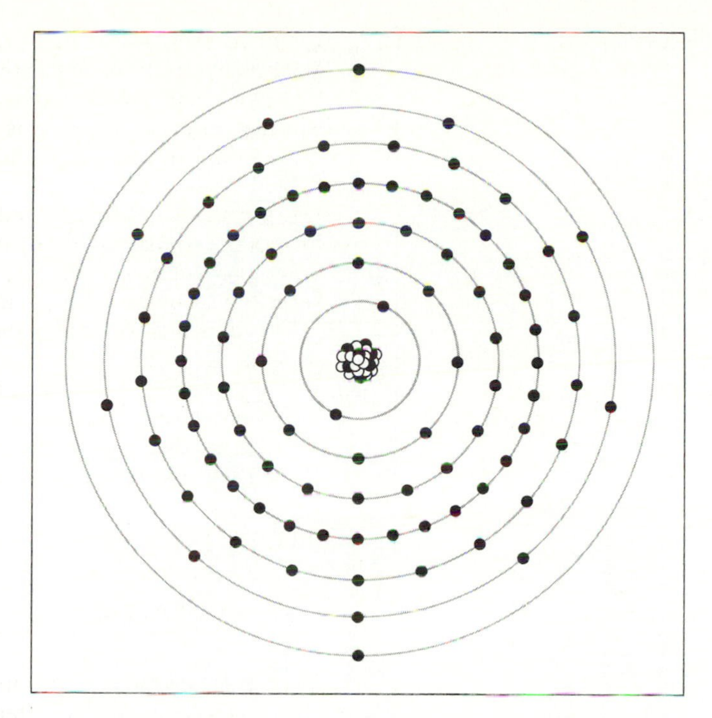

FIGURE 2.8 Schematic diagram of a uranium atom in its normal, ground state.

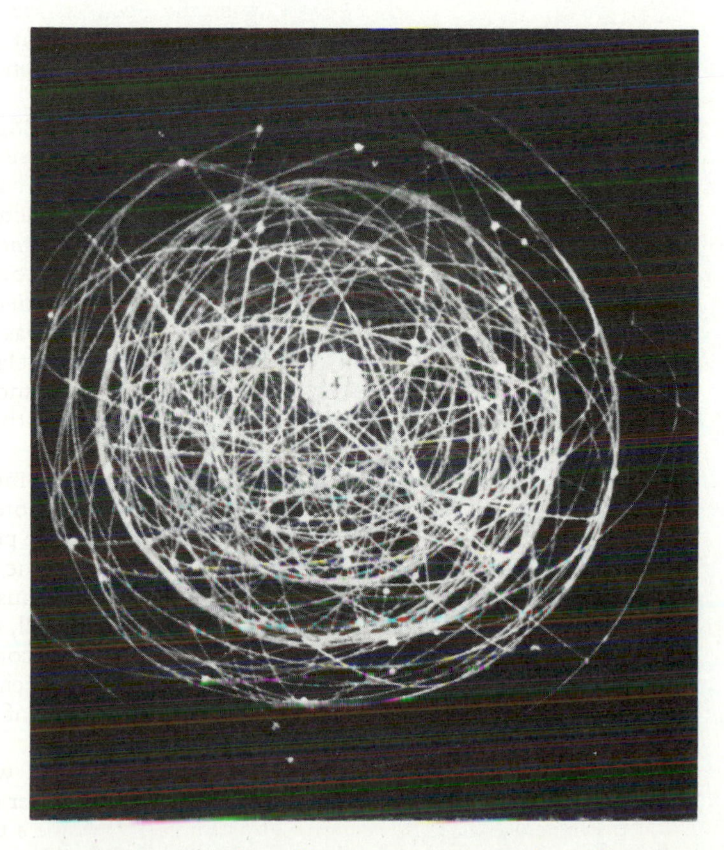

FIGURE 2.9 This three-dimensional model of the various electron orbitals within the uranium atom clearly demonstrates the complexity of this heavy element.

some potential energy relative to its parent nucleus. Each orbital within an atom has a certain amount of energy associated with it. The greater the separation of an electron from its parent nucleus, the greater the potential energy of that electron.

EXCITATION: Atoms do not always remain in their normal states; they can sometimes become "excited." An **excited state** is a somewhat abnormal condition and occurs when electrons temporarily reside at greater-than-normal distances from their parent nuclei. Accordingly, such excited-state atoms have greater-than-normal amounts of energy.

Atoms heavier than hydrogen, such as carbon or oxygen, can have many electrons in higher-than-normal orbitals; these atoms are still called excited-state atoms. However, when one or more electrons are completely removed from an atom, in effect separated far enough to be no longer bound to its parent nucleus, the atom is called something else—an ionized atom, or **ion** for short. Ions therefore carry an electrical charge; unlike atoms, they are not neutral.

Completely ionized matter is alternatively termed a **plasma**, a condition also known as the "fourth state of matter." The other three states are solid, liquid, and gas.

Although most matter on Earth is in the ground state, or perhaps in a slightly excited state, we'll see later that most matter in the Universe is ionized. Matter comprising Earth and the other planets is rather unexcited compared to the much hotter, more active atoms and ions in the rest of the Universe.

How can atoms become excited? Generally, they do so by one of two ways: They can become *radiatively excited* by absorbing energy from a source of radiation, or they can become *collisionally excited* by colliding with a nearby atom or elementary particle.

An example of radiative excitation is the heat gained by our hands while holding them above a home radiator; absorbing some infrared radia-

Atomic number
Atomic name
Atomic mass

1 Hydrogen 1.008																	2 Helium 4.002
3 Lithium 6.94	4 Beryllium 9.01											5 Boron 10.81	6 Carbon 12.01	7 Nitrogen 14.01	8 Oxygen 15.99	9 Fluorine 18.99	10 Neon 20.18
11 Sodium 22.98	12 Magne-sium 24.31											13 Aluminum 26.98	14 Silicon 28.08	15 Phos-phorus 30.97	16 Sulfur 32.06	17 Chlorine 35.45	18 Argon 39.94
19 Potassium 39.10	20 Calcium 40.08	21 Scandium 44.95	22 Titanium 47.90	23 Vanadium 50.94	24 Chromium 51.99	25 Manga-nese 54.93	26 Iron 55.84	27 Cobalt 58.93	28 Nickel 58.71	29 Copper 63.54	30 Zinc 65.37	31 Gallium 69.72	32 Germa-nium 72.59	33 Arsenic 74.92	34 Selenium 78.96	35 Bromine 79.90	36 Krypton 83.80
37 Rubidium 85.47	38 Strontium 87.62	39 Yttrium 88.90	40 Zirconium 91.22	41 Niobium 92.90	42 Molyb-denum 95.94	43 Techne-tium (99)	44 Ruthe-nium 101.07	45 Rhodium 102.90	46 Palladium 106.4	47 Silver 107.87	48 Cadmium 112.40	49 Indium 114.82	50 Tin 118.69	51 Antimony 121.75	52 Tellurium 127.60	53 Iodine 126.90	54 Xenon 131.30
55 Cesium 132.90	56 Barium 137.34	57 Lantha-num 138.91	72 Hafnium 178.49	73 Tantalum 180.94	74 Tungsten 183.85	75 Rhenium 186.2	76 Osmium 190.2	77 Iridium 192.2	78 Platinum 195.09	79 Gold 196.96	80 Mercury 200.59	81 Thallium 204.37	82 Lead 207.19	83 Bismuth 208.98	84 Polonium (210)	85 Astatine (210)	86 Radon (222)
87 Francium (223)	88 Radium (226)	89 Actinium (227)	104 Kircho-tovum (257)														

58 Cerium 140.12	59 Praseo-dymium 140.90	60 Neo-dymium 144.24	61 Pro-methium (147)	62 Samarium 150.35	63 Europium 151.96	64 Gado-linium 157.25	65 Terbium 158.92	66 Dyspro-sium 162.50	67 Holmium 164.93	68 Erbium 167.26	69 Thulium 168.93	70 Ytterbium 173.04	71 Lutetium 174.97
90 Thorium 232.03	91 Protac-tinium (231)	92 Uranium 238.03	93 Neptu-nium (237)	94 Plutonium (242)	95 Americium (243)	96 Curium (247)	97 Berkelium (247)	98 Califor-nium (249)	99 Einstein-ium (254)	100 Fermium (253)	101 Mendel-evium (256)	102 Nobelium (254)	103 Lawren-cium (257)

FIGURE 2.10 The Periodic Table of the Elements is a compact way to classify the physical and chemical properties of all the known kinds of atoms. The atomic number (at the top of each box) equals the number of protons or electrons, whereas the atomic mass (at the bottom) includes neutrons in the atom's nucleus as well.

INTERLUDE 2-2 ***Basic Forces of Nature***

As best we can tell, the behavior of all matter in the Universe—from elementary particles to clusters of galaxies—is ruled by just a few cosmic forces. Forces cause changes, and they are *fundamental* to *everything* in the Universe. In a sense, the search to understand the nature of the Universe is synonymous with the search to understand the nature of these forces. The cosmic forces are the key to unlocking secrets of the Universe.

The **gravitational force** is perhaps the best known force. Gravity binds galaxies, stars, and planets; in fact, it holds humans on Earth. Like the other forces, its strength decreases with distance from any object. However, that's only half of the law of gravity. Its strength is also proportional to mass. This means that gravity is terribly weak near, for example, a puny atom, but enormously powerful near a huge galaxy. Nothing can cancel the attractive pull of gravity; there is no such thing as antigravity that repels objects. Even the peculiar stuff known as antimatter has gravity, not antigravity. Consequently, the gravitational forces of all objects—even our own bodies—extend to the limits of the Universe.

The **electromagnetic force** is another of nature's basic agents. Its strength also decreases with distance and, like gravity, obeys the inverse-square law. This force acts as the cement for all ordinary materials. Any particle having a net charge, like an atom's electron and proton, has an electromagnetic force. All things around us—tables, chairs, books, mountains—are held together by the electromagnetic force. Because this force also binds the atoms within all life forms, some biologists call it the "life force." Unlike gravity, however, electromagnetic forces can repel (between like charges) as well as attract (between opposite charges). Such forces can then sometimes cancel one another. Similar numbers of positive and negative charges tend to neutralize the electromagnetic force, thus diminishing its influence. For example, although human bodies are made of vast numbers of charged particles, they comprise almost equal mixtures of positive and negative charges; our bodies therefore exert hardly any electromagnetic force.

A third cosmic force is simply termed the **weak force**. It's much weaker than either of the two forces above, and its influence is more subtle. We don't study it much in this text, except to note that the weak force helps to change one kind of elementary particle into another kind. Among other things, the weak force governs the emission of radiation from radioactive atoms. Many scientists now accept that the weak force is not really a separate force at all; rather, it's probably another form of the electromagnetic force acting under peculiar circumstances. As such, we now often speak of the "electroweak force," an issue to which we shall return in Chapter 17.

A stronger force than any of the above is the **nuclear force**. It glues atomic nuclei together and, in effect, serves as the source of energy in the Sun and stars. Like the weak force, and unlike the inverse-square law forces of gravity and electromagnetism, the nuclear force operates only at very close range; it's useless when matter is separated by more than a millionth of a millionth (10^{-12}) of a centimeter. But within this range, as is the case for all atomic nuclei, particles are bound with enormous strength—stronger, in fact, than any other force known—at least until recently.

Research done in the 1970s suggests that two forces can perhaps overwhelm the nuclear force. One is a special form of gravity, namely, that which occurs when large, massive objects are squeezed into very small places; such super-dense, super-attractive regions probably lurk in the hearts of black holes, a most bizarre phenomenon that we'll study in Chapter 21. Nature might also have another stronger-than-nuclear force, one that is sometimes called a "quark force." This force is the glue that binds the subatomic elementary particles. What's more basic than elementary particles? The quark is, as discussed in Interlude 2-1. The quark force binds quarks on a scale less than 10^{-13} centimeter. Some researchers currently suspect that the quark force may be merely another form of the nuclear force in disguise. Ultimately, all the basic forces of nature may be shown to be different manifestations of a single force, as we shall discuss further in Chapter 17.

tion emitted by the warm metal, the electrons in our hands are boosted to higher orbitals.

An example of collisional excitation is the mutual collision of atoms while rubbing our hands together; friction generates heat, which in turn causes the atoms to move around at high speed, collide often, absorb energy, and consequently boost their electrons to higher orbitals.

Since electrons are restricted to reside only at specific, fixed distances from their central nuclei, we conclude that atoms can absorb only specific, quantized amounts of energy. Accordingly, atoms can become excited by robbing certain amounts of energy either from a nearby source of radiation or from a colliding particle. The amount of energy absorbed must correspond precisely to the difference in potential energy of the two orbitals. Here's an example of what we mean.

Figure 2.11 illustrates our model of the hydrogen atom once again.

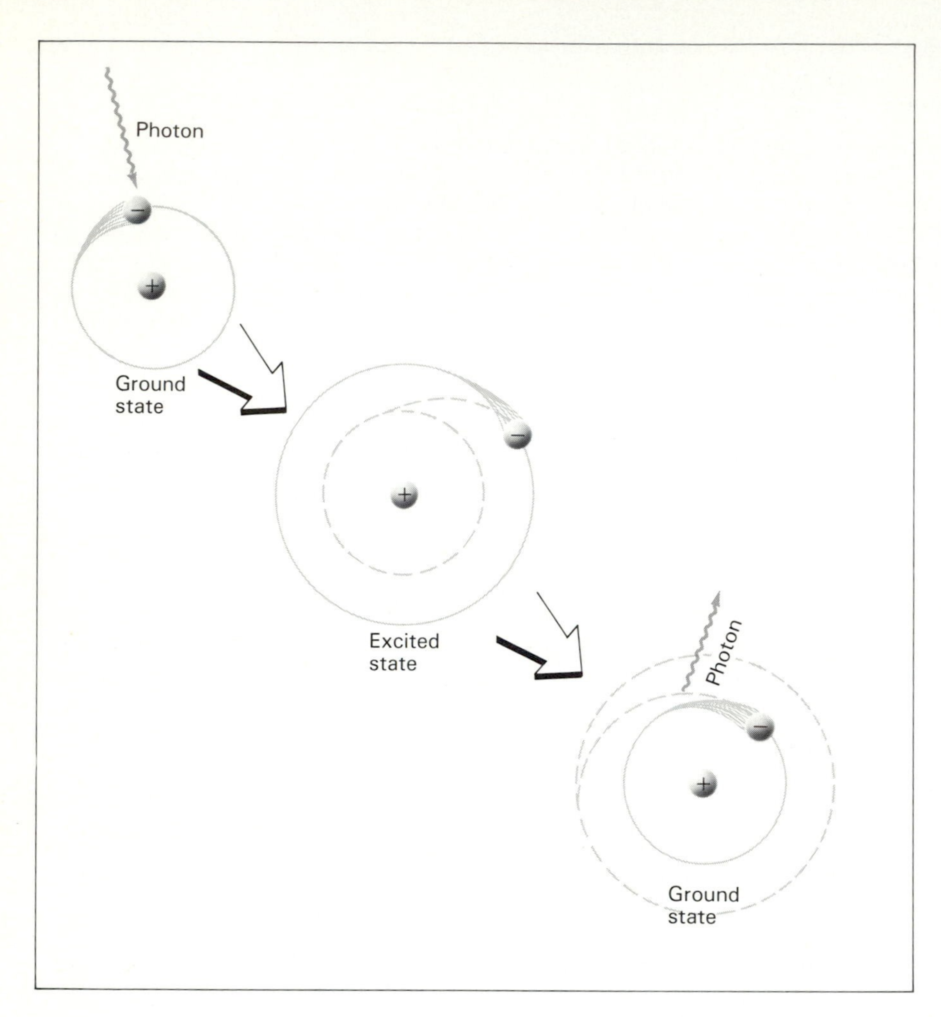

FIGURE 2.11 Schematic diagram of a photon being absorbed by a hydrogen atom (top), causing the momentary excitation of that atom (center). Eventually, the atom returns to its ground state, accompanied by the emission of a photon (bottom).

The frames at the top and center show that after absorbing some radiation, the electron transfers to a higher orbital, farther removed from its parent nucleus. The atom in the center frame has more energy than normal; it has become excited. But the electron has no desire to stay "out of place" in this higher orbital. Instead, it returns to its normal ground-state orbital, as shown in the frame at the bottom. And it does so rapidly—in about 0.01 microsecond, or 10^{-8} second.

Our example is not yet finished. When the electron returns to its lower orbital, as shown in the bottom frame of Figure 2.11, the atom must rid itself of the excess energy. It does so by emitting a photon, though not just any photon. The emitted photon must have an energy precisely equal to the difference in potential energy of the two orbitals involved.

Thus electrons are free to move among the different orbitals within atoms. The outward movement of any electron is always accompanied by an absorption of energy from the surrounding environment; in this case, the atom is said to be in a high-energy state. Alternatively, the downward movement of any electron is invariably accompanied by an emission of energy to the surrounding environment; the result is a lower-energy state.

Note one other feature of Figure 2.11. The finally emitted photon is not necessarily ejected in the same direction from which the initially absorbed photon came. This will be an important point as we now study the details of spectroscopy.

Spectra

CHARACTERISTIC FINGERPRINT: Be sure to recognize that the positions and energies of the electron orbitals are identical for all atoms of the same kind. They differ, however, for atoms of different kinds. All carbon atoms throughout the Universe have the same electron configuration. Oxygen or nitrogen atoms, on the other hand, have completely different electron configurations from carbon and from each other. Put yet another way, each element, or kind of atom, has its own particular set of electron orbitals and its own characteristic number of protons within the nucleus.

The movement of electrons within atoms also depends on the strength of the electromagnetic force field exerted by the nucleus. These two factors—orbital arrangement and electromagnetic force field—combine to make electron transitions among the various orbitals unique for each element. Consequently, as electrons return to lower orbitals, the emitted photons carry energies characteristic of only one kind of atom. In effect, individual atoms emit flashes of radiation having energies characteristic of that element—*and only that element.*

Furthermore, since a photon's energy is directly proportional to the frequency, we can equivalently say that atoms emit flashes of radiation at very specific frequencies (or wavelengths). These precise frequencies are characteristic of certain elements—so characteristic that they can be used to identify that element. In this way, each element, or kind of atom, has its own personality or fingerprint, much like the telltale, characteristic fingerprints of human beings. And similar to the extensive FBI fingerprint files in Washington, scientists have accumulated huge catalogues of characteristic frequencies at which all the different kinds of atoms emit radiation.

This is the powerful essence of spectroscopy. By focusing on specific frequencies, this technique enables us to study one kind of atom to the exclusion of all others, even though all the different kinds might be intermixed within a gas.

SPECTRUM OF HYDROGEN: Let's examine the details of emitted radiation as seen through a prism. Consider once again a hydrogen atom in its ground state. If left alone, the atom will neither emit nor absorb energy. But if, for example, we excite the atom (by heating a gas, for instance), the electron can reach a higher orbital. The atom will then almost immediately return to its normal state, in the process radiating a photon, as depicted in Figure 2.12(a).

Numerically, this emitted photon has a wavelength of 1216 Angstroms. Since the visible spectrum spans 7000 angstroms for red light to 4000 angstroms for violet light, this flash of radiation cannot be seen visually; as shown in Figure 2.12(b), the flash occurs in the invisible ultraviolet part of the spectrum.

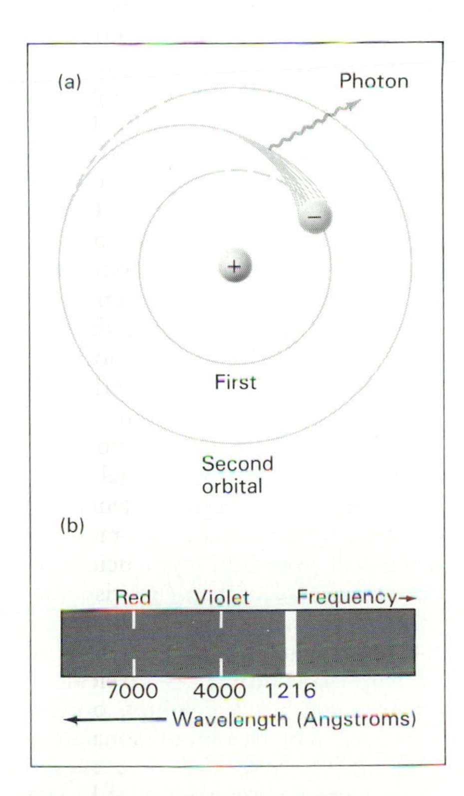

FIGURE 2.12 Orbital arrangement for a slightly excited hydrogen atom, in the process of returning to its ground state (a); the resulting spectrum of the emitted radiation (b).

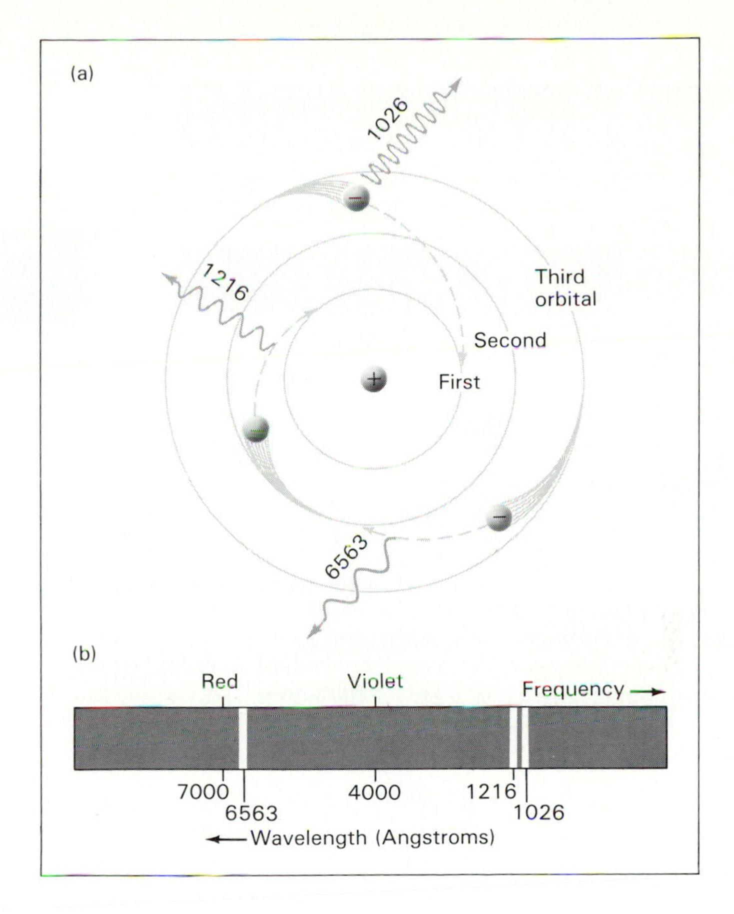

FIGURE 2.13 Orbital arrangement for a moderately excited hydrogen atom (a); resulting spectrum of the emitted radiation (b).

Now, imagine exciting a hydrogen atom enough to boost the electron from the ground orbital to an even higher orbital. Depicted schematically in Figure 2.13(a), this might be a third orbital far removed from the nucleus. As usual, the electron will return very rapidly to the ground orbital, but this time it can do so in two possible ways. Either it can proceed directly from the third orbital to the first (ground) orbital, in the process emitting a photon at an ultraviolet wavelength of 1026 angstroms. Or it can cascade down one orbital at a time, emitting two photons, one with an energy difference between the third and second orbitals, the other with an energy difference between the second and first orbitals.

As illustrated in Figure 2.13, the second step of the cascade process simply emits a 1216-Angstrom ultraviolet photon, as noted previously in Figure 2.12. But the first step of the cascade process—the electron change between the third and second orbitals—produces a photon of wavelength 6563 angstroms. That's in the visible spectrum, and the photon can be seen as red light. An individual atom—if one could be isolated—would emit a momentary bright red flash.

Because these emissions occur at very specific frequencies or wavelengths, we call them **spectral lines**. We say "lines" because the radiation is aligned with a narrow slit before passing through the prism, as dis-

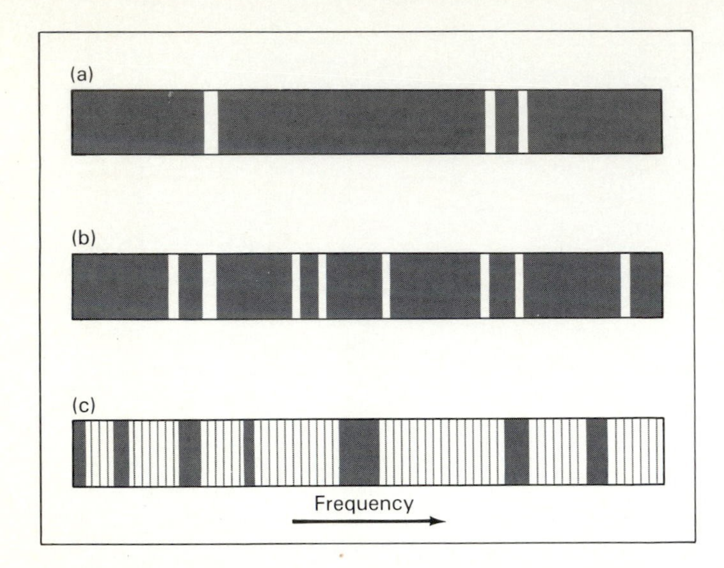

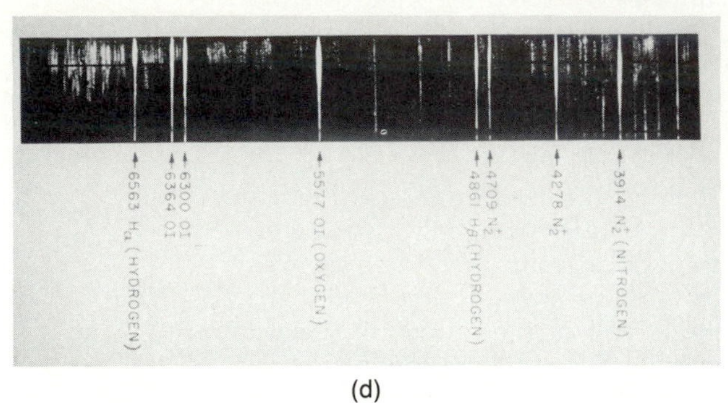

(d)

FIGURE 2.14 (a) The spectrum of a moderately excited hydrogen atom [which is reproduced from Figure 2.13(b)]. (b) The spectrum of either an extremely excited small atom such as hydrogen, or a moderately excited complex atom heavier than hydrogen. (c) The nearly complete color spectrum of an excited group of mixed gases. Frames (a), (b), and (c) are artist's renditions of spectra; for realism, frame (d) shows an actual spectrum of a hot gas with several prominent emission lines identified.

cussed in Chapter 3. Groups of such lines, like those depicted in Figure 2.14 and Plate 1, resemble the bar (price) codes printed on many of today's market products; and like spectral features that characterize only specific elements, each bar code designates a specific price. Look for the codes the next time you go shopping.

Additional heating (or absorption of any type of energy) could boost the electron to even higher orbitals within a hydrogen atom. Upon cascading, the atom would emit many photons, each with a different energy and hence a different wavelength. The resulting spectrum, as illustrated in Figure 2.14(b), would show many spectral lines.

Eventually, intense heating produces so many emitted photons that the visible spectrum becomes nearly filled with spectral lines. Under some conditions (especially in solids and dense gases), very highly excited atoms would emit so many lines as to almost touch one another; accordingly, the visible spectrum can sometimes become virtually continuous with colors stretching from red to violet, as shown in Figure 2.14(c).

Actually, a single hydrogen atom, or even a large group of hydrogen atoms, could not produce the many lines of Figure 2.14(c). With only one electron per hydrogen atom, even intensely heated hydrogen emits no more than several intense lines. Under special circumstances, however, excited *heavier* atoms can yield a rich spectrum of many lines. Because of their large number of electrons and strong electromagnetic force fields, heavy atoms have a much larger number of possible electron changes among their orbitals. [Consult Figure 2.14(d) and Plate 1(c) for some realistic examples.]

The easiest way to produce a rich spectrum of many lines is to heat a *mixture* of different kinds of atoms, that is, a whole group of different elements. This partly explains the complete color spectrum often emitted by complex objects such as our Sun. (Other, more subtle effects are also involved, such as the radiation emitted when free electrons glide past free ions in the Sun's plasma.)

An analogy might help here. If a single spectral line can be imagined to resemble a single musical tone, then a nearly complete spectrum [like that of Figure 2.14(c)] is akin to the noise rendered when many musical tones are struck simultaneously. It's perhaps a little like the difference between listening to the beautiful tones of a renowned pianist and the jumble of many tones produced by several preschool children banging on the keyboard together. In fact, the analogy can be made more appropriate to the present discussion: A bag full of different tuning forks, when shaken (i.e., energized), will produce a nearly complete spectrum of audible tones as the forks collide; a container full of many different atoms, when shaken (or otherwise energized), will produce a nearly complete spectrum of radiation as the atoms collide.

Heavy atoms, in fact atoms heavier than the two-body (electron and proton) hydrogen atom, cannot be modeled precisely. We just don't know how to solve the mathematics describing three or more interacting objects—whether they be atoms, billiard balls, or stars. No one is smart enough to know how to do it, not even with the aid of computers. We must rely instead on spectroscopy experiments conducted in laboratories around the world. There, the wavelengths of spectral features emitted by complex atoms can be identified and measured. The resulting wavelengths can then be compared against the observed spectrum of some extraterrestrial object in order to identify its spectral features. In this way we can determine not only the chemical composition of a cosmic source, but many of its physical properties as well.

BEYOND THE VISIBLE REGIME: Be sure to realize that spectral-line features occur throughout the whole electromagnetic spectrum. Usually, electron changes among the lowest orbitals of the lighter elements produce visible and ultraviolet spectral lines; changes among these same orbitals in heavier, more complex elements produce x-ray spectral lines. These spectral lines have been observed in modern laboratories with the aid of complicated equipment. Some of the lines are also observed in stars and other cosmic objects.

Electrons are not restricted to low-lying orbitals. Upon absorbing sufficient energy, they can range far from an atom's nucleus while remaining bound to it. Electrons moving from, say, orbital 20 to 19 would produce an emission spectral line in the infrared region of the spectrum. Infrared equipment has only recently become sophisticated enough to observe these lines in cosmic sources.

This doesn't mean that the electrons extend to great distances by our usual standards. We're still talking about the microscopic realm. An electron, for example, in the fortieth orbital of a hydrogen atom would be about 1000 angstroms from its nucleus. That's a relatively large atom.

The largest atoms of all are those where the electrons populate the very high-lying orbitals. Electron changes among these oribtals produce spectral lines within the radio part of the electromagnetic spectrum. The orbitals involved can be numbered in the hundreds, and often reach tens of thousands of angstroms from the central nucleus. The cool, dense properties of Earth make it impossible to detect these radio features in our laboratories, but they *can* be routinely observed from some hot, tenuous cosmic objects.

ABSORPTION SPECTRA: Sometimes, dark spectral lines can be seen superposed on a spectrum of bright emission. Figure 2.15 and Plate 1(a) and (b) are examples of such a spectrum in the visible domain. Dark spectral features of this type are produced by the *absorption* of light radiation at specific wavelengths. Let's examine the nature of such absorption spectra in greater detail.

Imagine a hot source of radiation, such as a light bulb. We noted earlier that some light bulbs (such as neon signs) contain gaseous mixtures of different elements. Each element is excited by colliding with charged particles oscillating in the bulb's filament because of the electricity that powers the bulb. As a result, the bulb's trillions of atoms constantly absorb energy, become excited, return to their ground states, and all the while emit spectral-line radiation. Since the many different kinds of excited atoms within the light bulb emit myriad spectral lines at all times, the emission features nearly merge to yield a complete color spectrum, red to violet. Emission is also present well outside, in the invisible domains—one can clearly feel a bulb's infrared heat, for example. Figure 2.16(a) shows this case, where a slit is used to align the light prior to its passage through a

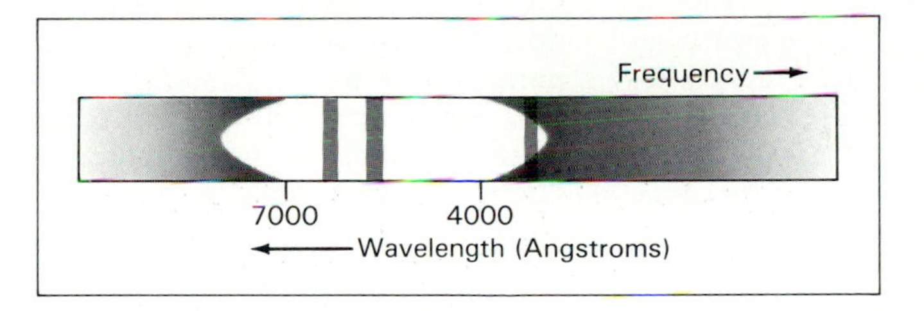

FIGURE 2.15 Dark absorption lines superposed on an otherwise continuous bright emission spectrum. (See also Plate 1.)

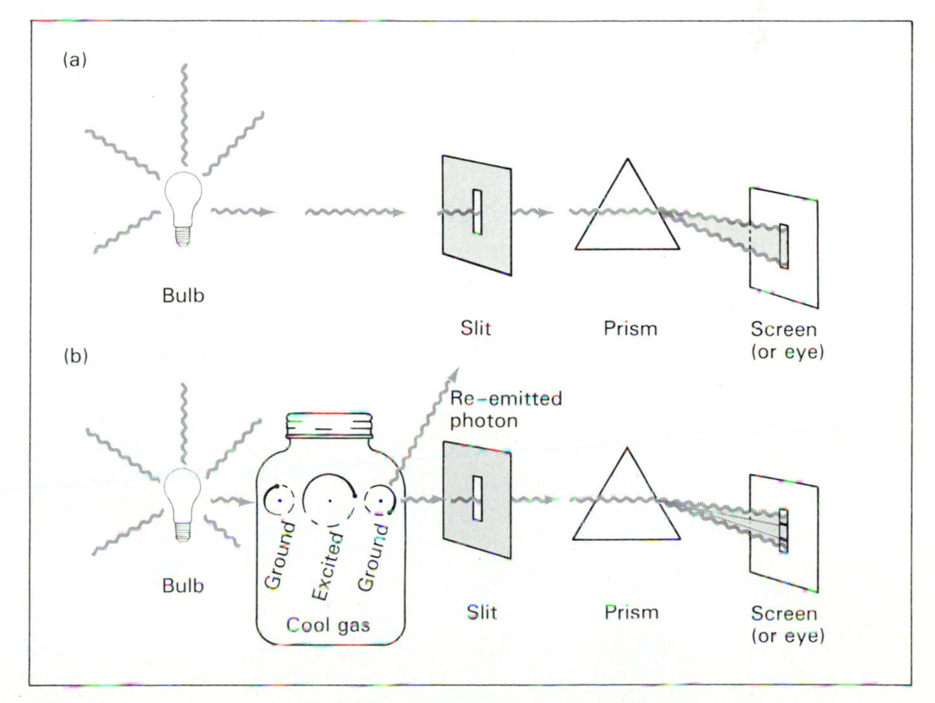

FIGURE 2.16 (a) A hot light bulb, containing many different elements, yields a colorful emission spectrum. (b) When some cool gas is placed between the bulb and the detector, the resulting emission spectrum is then peppered with dark absorption lines. (See also Plate 1.)

conventional prism. (Recognize also that the inside of a bulb's glass is coated with a material that scatters the filament's intense radiation, making the light appear to emanate from the entire bulb.)

Now, imagine a container of cool gas between the light bulb and your eye, as shown in Figure 2.16(b). Assume, furthermore, that billions upon billions of atoms comprise this cool gas. Virtually all these atoms exist in their ground states until the bulb's light passes through. When this happens, radiation once again interacts with matter according to the usual rules of atomic physics that we've been discussing.

Photons having precisely the correct energy to cause an electron to change within any of the atoms inside the container will be rapidly absorbed. For example, if the container is full of pure hydrogen, many of the bulb's photons having energy equaling the precise energy difference between the first and second orbitals (see Figure 2.12) will be absorbed, boosting hydrogen's only electron to the second orbital. (All those photons having energy not equaling *precisely* any of the potential energy differences among orbitals are not absorbed and hence pass unhindered through the cool gas.) In this way, external radiation can excite the gas of an intervening region. In the process, some photons are removed from the normal flow of radiation from the light bulb.

As we noted earlier, atoms rapidly return to their normal ground states. Accordingly, they emit a photon—in fact, a photon having precisely the energy absorbed in the first place. So we might reason that although some photons from the bulb are absorbed by the intervening cool gas, their reemission would cancel the effect of the absorption, and we would never observe the act of absorption. But this is wrong. It's wrong because although the photons not absorbed by the intervening cool gas follow a clear path from the bulb to the detector, the photons reemitted by the cool gas do not follow that path at all. The reemitted photons leave the cool gas at the velocity of light, but they can do so in *any* direction. No one direction is preferred. Consequently, only very few of the total number of reemitted photons would pass within the small viewing angle of the detector. So we would still detect the absorption of the bulb's radiation by the cool intervening gas. In other words, many of the photons normally viewable in the absence of the intervening cool gas now become absorbed and reemitted at unviewable angles. These photons are scattered out of our viewing line of sight. They're effectively lost.

The detector then records an emission spectrum except at those precise wavelengths for which some photons have been subtracted from our line of sight. Figure 2.17(a) shows a typical dark-line spectrum, often called an absorption spectrum. The dark lines are characteristic of the *intervening gas*. The bright emission surrounding the dark lines results from the hot bulb.

Be sure to recognize that absorption spectral lines occur at the precise wavelengths at which the same gas emits lines if heated. For example, Figure 2.17(b) shows two lines characteristic of the element sodium, one at 5890 angstroms and the other at 5896 angstroms. When hot enough or energized in some way, sodium atoms continuously emit two photons on their own accord. Their emission can be clearly seen as two bright lines in the yellow part of the spectrum.

We could generate sodium's characteristic yellow light by throwing some salt into a lit fireplace (since salt is the chemical sodium chloride). Or, better yet, take note of the new, very bright yellow street lamps that brighten many American cities. These "sodium vapor lamps" glow yellow from the emission of vast quantities of photons having the foregoing wavelengths.

Should some *cool* sodium, on the other hand, be placed between a source of radiation and our eye, then two sharp, dark *absorption* lines will be noted in the color spectrum at precisely the same wavelengths as the two lines normally emitted by hot sodium.

THE THREE CASES AGAIN: The concept of spectral-line radiation can be tricky for people unfamiliar with the subtleties of atomic physics. Once

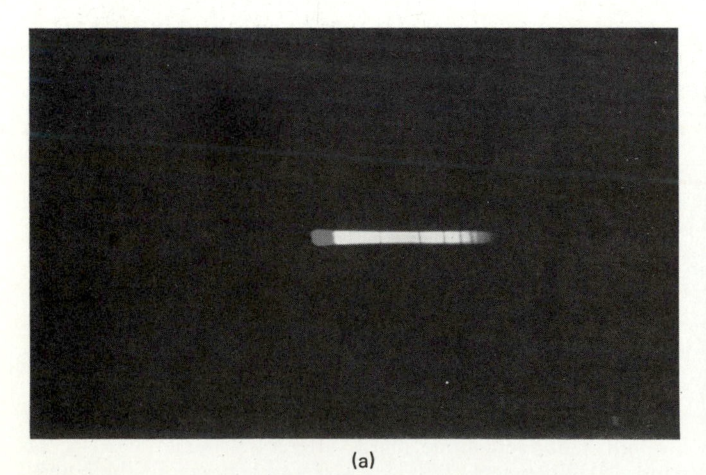

(a)

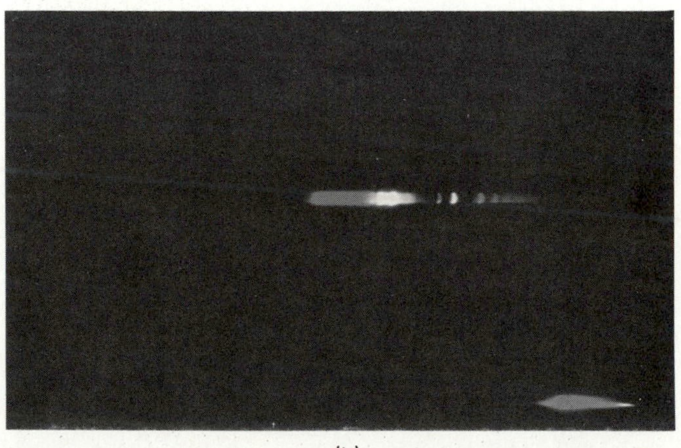

(b)

FIGURE 2.17 (a) A typical spectrum resulting from the case shown in Figure 2.16(b), showing dark absorption features amidst a bright, continuous spectrum. (b) Characteristic emission lines of a sodium atom. The pair of bright lines in the center appear in the yellow part of the spectrum. These are real spectra, not just artistic renditions.

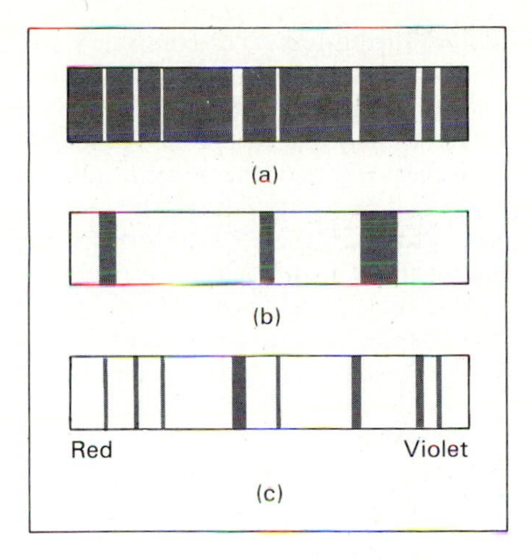

FIGURE 2.18 (a) A bright-line emission spectrum. (b) A nearly continuous emission spectrum. (c) A dark-line absorption spectrum.

more, let's summarize the basic cases that produce spectral lines. Figure 2.18 shows the three types of visible spectra often encountered under different environmental conditions. The three cases are said to obey Kirchhoff's laws, a set of spectroscopic rules developed by a nineteenth-century German physicist.

Figure 2.18(a) shows a bright-line spectrum. This results from the normal emission of radiation from a moderately heated gas containing a huge quantity of atoms of one or a few elements.

A bright-line spectrum can become complex in either of two ways: by intensely heating a gas having only a few elements or by moderately heating a gas having a mixture of many elements. First, a very hot gas has many atoms with electrons in high orbitals, thereby producing a complex emission spectrum from the huge number of possible electron changes. Second, a spectrum can become complex whenever the gas is made of different elements; after all, each element emits its own characteristic set of emission lines. If a gas containing a large number of different elements is heated sufficiently, the multitude of emission lines can become increasingly complex, in some cases nearly blending together to produce a color spectrum like that shown in Figure 2.18(b).

A third type of spectrum that can result from the interaction of radiation with matter is one that yields dark absorption lines. Such a spectrum occurs when radiation emitted from some hot object passes through a cooler gas. Figure 2.18(c) shows an example of an absorption spectrum characteristic of the intervening gas, whose dark lines are superposed on the bright spectrum of the hot background object. Such spectra can become rather complex when many different elements are present in the cool gas. Usually, though, the dark lines could never become numerous enough to blend together and entirely obliterate the bright spectrum. Even if the intervening gas were heated, the absorption lines themselves would become less dark, until eventually the lines disappeared when the temperature of the intervening gas equaled that of the background source. Continued heating of the intervening gas to still higher temperatures would eventually yield an *emission-line* spectrum characteristic of the intervening gas.

Figure 2.14(d) illustrates an actual spectruum observed toward a genuine cosmic object, as does Plate 1(b) and (c).

Spectral-Line Analysis

Thus far, we've used rather simple case studies to be sure to grasp the basic concepts of spectroscopy. As we move forward in our study of the cosmos, we'll see that the same general phenomena occur even when a nearby star or a distant galaxy takes the place of the bulb of the previous examples, a galactic cloud or stellar atmosphere plays the role of the intervening cool gas, and an advanced piece of equipment attached to a telescope becomes the prism and detector.

Let's now consider some realistic situations. These will be important if we are to appreciate the types of spectra often encountered throughout our text.

REAL SPECTRA: Stars are very hot, especially deep down inside. So many photons of great energy are produced in stellar cores that electrons within a star's atoms easily become excited. In fact, all stellar matter is ionized, with most electrons stripped from their parent nuclei. The interior of a star is a true plasma, containing only ionized gas and no neutral atoms.

Matter near the cooler surface of a star is different. There, some atoms remain intact, allowing many electrons to change their orbitals. Electrons in a wide variety of elements cascade down through the orbitals, emitting photons at many wavelengths. Averaged over the entire star, these spectral emissions are so numerous that a star produces a nearly complete color spectrum from red to violet in the visible domain, with much spectral emission in the infrared and ultraviolet domains as well. *Individual* emission lines cannot be distinguished; the spectroscopic details are lost because of the severe crowding of the emission lines. (Again, as noted in the preceding section, other effects—such as electrons gliding past ions—are also partly responsible for a star's continuous spectrum.)

Thus, when we observe a star through a prism attached to a telescope, the result is a nearly continuous emission spectrum. But that's not all. Superposed on this color spectrum are many dark lines. These absorption lines are produced in the star's atmosphere—cooler gas that absorbs some of the photons emitted from the star's surface.

Figure 2.19 exemplifies the spectrum emitted by the nearest star, namely the Sun. As shown, literally thousands of dark absorption lines litter the Sun's visible spectrum; nearly 800 lines are produced by variously excited atoms of only one element, namely iron.

Atoms of a single element like iron can yield so many lines for two reasons. First, iron has a large number of electrons, 26, each of which can move among the orbitals of the many iron atoms in the Sun's gas. Second, much of the iron is ionized, with *some* of the 26 electrons stripped away. Hence most of the hundreds

FIGURE 2.19 The visible spectrum of the Sun shows thousands of dark absorption lines superposed on a bright emission spectrum. Here, the scale extends from long wavelengths (red color) at the upper left to short wavelengths (blue color) at the lower right.

of observed lines of iron result from absorption of photons within this one element, which is in different stages of ionization and excitation at various places in the Sun's atmosphere. When we observe the entire Sun, all the iron ions absorb simultaneously to yield the rich spectrum of lines. Besides iron, many other elements, also in different stages of excitation and ionization, absorb photons at visible wavelengths.

You should now appreciate how typical spectra contain much information about the physical and chemical state of any gas. To extract that information requires a good knowledge of physics and many hard calculations. In a nutshell, we can unravel spectra, despite their complexity, mainly because each atom or ion has spectral features characteristic of a particular kind of atom or ion, and *of only that kind of atom or ion*.

INTENSITY: Astronomers classify spectral-line radiation in much the same way as we did earlier for continuous radiation in Chapter 1. A spectral line's wavelength (or frequency) can be measured directly from the spectrum, provided that observations are made with accurate telescopes. Polarization is seldom encountered in spectral-line work, although occasionally, astronomers derive some useful information from polarized lines. The final property characterizing radiation—intensity—is quite useful in spectral-line research; it merits further discussion.

Intensity of a spectral line depends partly on the *abundance of the atoms* giving rise to the line. That's because intensity is proportional to the number of photons emitted or absorbed. And the number of photons in turn relates to the number of electrons in a particular orbital of some element. That's not all, however. Line intensity also depends on the *temperature of the atoms*—that is, the temperature of the entire gas of which the atoms are members.

Over the years, astronomers have developed mathematical formulas that relate the number of emitted or absorbed photons to the temperature of the atoms, as well as to the population of electrons in each orbital. Once an object's spectrum is measured, we try to interpret the spectrum by matching the observed intensities of the spectral lines with those theoretically predicted by the formulas. We often begin by guessing the gas temperature. We then calculate the intensity of a particular spectral line of some element. All the while, we continually adjust our theoretical estimate of the number of electrons in whichever orbital produces the line in question.

Using certain temperature and abundance guesses, astronomers are often able to explain the intensities of all the observed lines. If so, we've successfully determined both the gas temperature and the atomic abundance. If we cannot explain the observed spectral-line intensities, we then guess the temperature again, adjust the electron populations in the many orbitals, and solve the mathematical formulas again. We keep guessing until the mathematics and the observations agree. This is a central goal of modern science: the agreement of theory and experiment. The final result yields both the elemental abundances and the temperature of the gas producing the spectral lines.

Fortunately, there's not as much drudgery as implied here, for large computers can now rapidly guess and reguess until the correct physical and chemical conditions are found. (Not that computers are smarter than we are; in fact, they are clearly only as bright as the humans programming them. But computers are capable of performing computations with astounding speed—much faster than humans. Perhaps we should regard them as incredibly fast morons.)

COSMIC ABUNDANCE: Spectroscopy is very important if we are to determine the composition of cosmic objects. It's the main technique used to measure elemental abundances throughout the Universe. Figure 2.20 plots the **cosmic abundances** or relative numbers of all but the heaviest elements, averaged for many stars near Earth. As shown, the light elements are much more abundant than the heavy ones. In fact, the lightest element, hydrogen, comprises about 90 percent of all the atoms in the Universe. The two light elements, hydrogen and helium, together comprise more than 99 percent of universal matter. All the other elements of the periodic table combine to make up no more than 1 percent of the cosmic atoms.

The relative abundances shown in Figure 2.20 are generally valid for all stars, even though spectra often differ from one star to another. However, the differences in spectra are

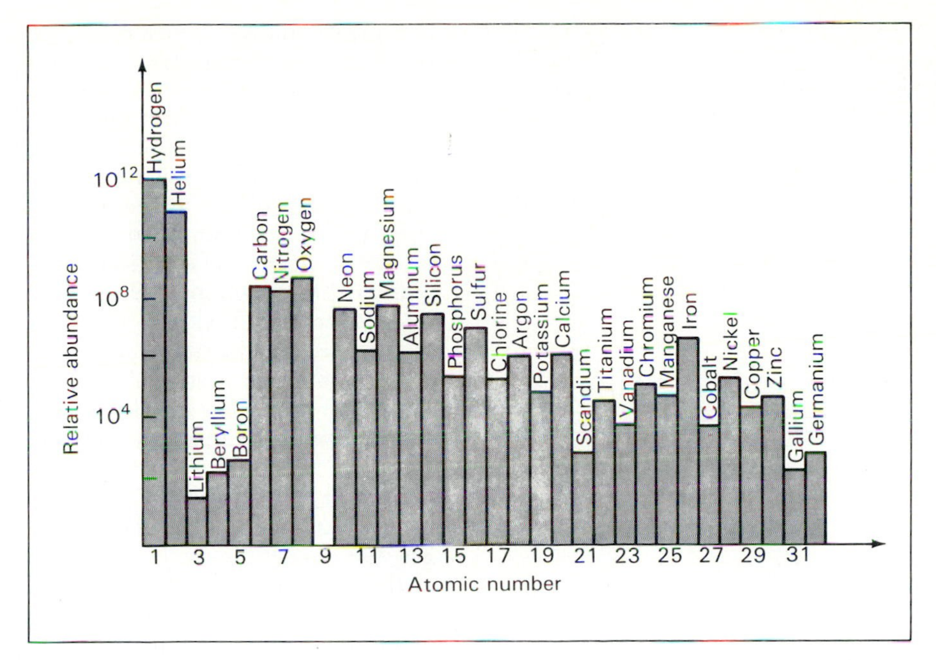

FIGURE 2.20 A plot of the cosmic abundances of some of the elements, averaged for many stars near Earth (including the Sun). Note that the abundance of hydrogen is more than a thousand times greater than that of most other elements, except helium.

usually not due to differences in chemical composition as much as to differences in stellar temperature. For example, we've already noted that the 6563-angstrom absorption line of hydrogen results from electrons changing from the second to the third orbital. Since the Sun's lower atmosphere is rather cool, relatively few atoms have electrons in the second orbital; most are in the ground (first) orbital. Hence, in the Sun, the 6563-angstrom line is weak (i.e., of low intensity), even though a mathematical analysis of the entire solar spectrum proves that hydrogen is by far the most abundant element.

Since most of hydrogen's electrons in the Sun's atmosphere reside in the first orbital, photons emitted from the surface will be absorbed in great numbers while passing through the atmosphere. This causes an intense ultraviolet absorption line at 1216 angstroms, as suggested by Figure 2.12. Recent observations of our Sun indeed confirm this to be the case.

On the other hand, stars with warmer atmospheres have more hydrogen atoms with electrons in the second orbital. Most atoms are still in the ground state to be sure, but a fair fraction are in an excited state. Stellar photons, absorbed under such conditions, thereby produce a 6563-angstrom line that is stronger (more intense) than observed in our Sun.

So we emphasize that the hydrogen abundance is roughly the same in most stars. Small changes in a star's atmospheric temperature can produce substantial changes in its stellar spectrum. As a rule of thumb, gas temperatures affect observed spectra more than elemental abundances.

Remember, Figure 2.20 is a plot of the *average* elemental abundances. It's not valid for any one object in particular, just for all objects in general. It's a true indicator of the elemental abundances *in stars*. But some cosmic objects have elemental proportions much different from this average. For example, the abundances of the elements familiar to us on Earth do not really mimic those plotted in Figure 2.20. That's because many planets do not at all represent most matter in the Universe. Planets are often small, cold, and rocky, in contrast to the larger, hotter balls of plasma called stars. We'll have more to say about the relative abundances of the elements as we proceed throughout our text, particularly when considering the origin of the elements in Chapter 22.

LINE PROPERTIES: Much useful information is buried within spectral-line data. In addition to the elemental abundances and gas temperatures mentioned above, spectral lines can tell us a great deal more about the physical state of the gas. The *shapes* of spectral lines are especially useful. Let's consider a typical example.

Imagine for a moment an absorption spectral line like that drawn in Figure 2.21(a). The line seems dark, though careful study shows its darkness to be greatest at the center, while tapering off toward each side. For example, Figure 2.21(b) is a plot of relative darkness across the line, sometimes called a line profile. As shown, the intensity (or brightness) is least at the line center (i.e., where the line is darkest), just as we expected by looking at Figure 2.21(a). The center of the line marks the wavelength (or frequency) at which most photons are absorbed by some intervening group of atoms. Fewer photons are absorbed at wavelengths slightly higher and lower than line center. The usual result is the bell-shaped profile shown. Spectral lines like this one are normally classified by their *intensity* (i.e., height or depth at line center), their *width* (i.e., breadth), and their *centroid* (i.e., central wavelength).

Astronomers easily trace the shape of a line of this type by scanning a thin beam of light (often a laser) across a photographic plate containing a spectrum. They then record the amount of light passing through the plate. Where the absorption line peaks, hardly any light gets through; near the edges of the line, much light penetrates the plate. As an example, Figure 2.22(a) displays a more realistic case of a complex spectrum being scanned by a narrow light beam. The resulting detailed tracing, shown in Figure 2.22(b), can then be analyzed mathematically with greater accuracy than is possible by working only with the

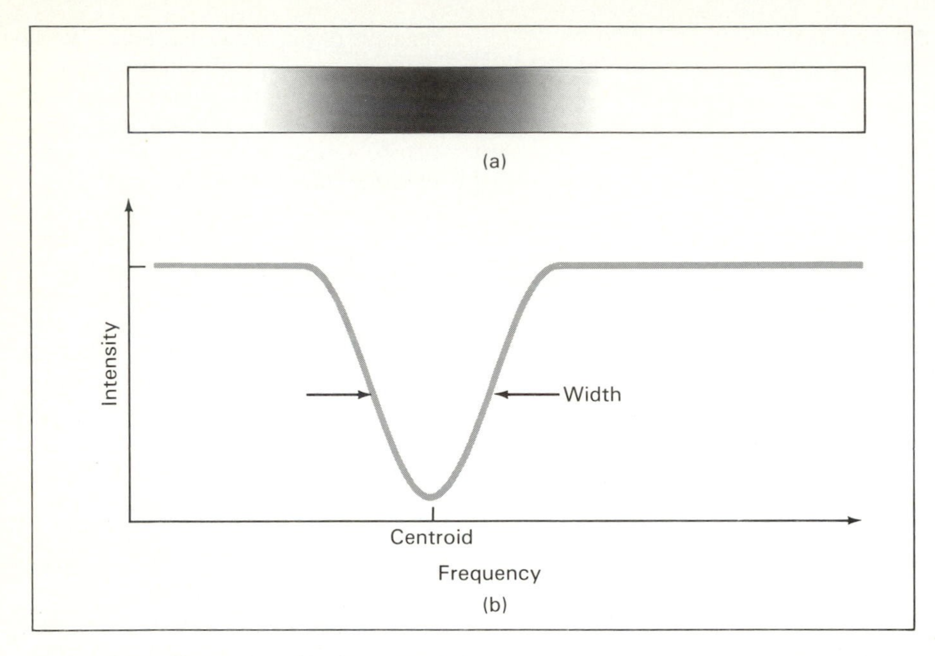

FIGURE 2.21 By tracing the changing darkness across a typical absorption spectral line (a), we obtain a convenient graph of its intensity plotted against frequency or wavelength (b).

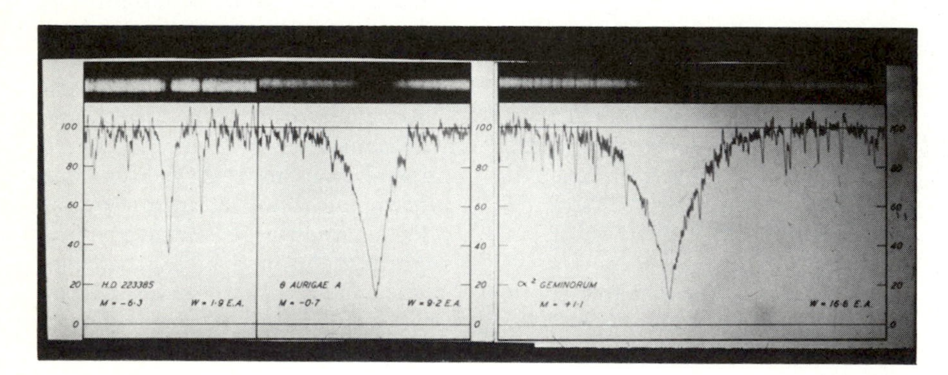

FIGURE 2.22 A laboratory apparatus known as a densitometer is used to make tracings of spectra. Here, a set of complex spectra (top) have been traced by passing a narrow light beam across a photographic plate, thus revealing (bottom) the intricate details of the spectral features.

fuzzy shades of darkness on a photograph. Plate 1(c) shows yet another example of a spectral-line tracing.

LINE BROADENING: We might be surprised that spectral lines have any shape or structure at all, especially since we earlier stressed that photons are emitted and absorbed at precise wavelengths. In other words, why aren't spectral lines extremely narrow, occurring at specific wavelengths, and at only those wavelengths? Well, it *is* true that photons emitted or absorbed by means of electron changes have precise energies, but the environment in which the emission or absorption occurs often changes our perception of a photon's energy.

Several physical mechanisms can broaden spectral lines. The most important of these is the Doppler effect. Recall from Chapter 1 that if a radiative source and an observer have some net recessional velocity, the perceived radiation is shifted from its normal wavelength toward the red end of the spectrum. Similarly, radiation is blue shifted if there is any net approach velocity between a radiative source and an observer.

To understand how the Doppler effect can broaden a spectral line, imagine a hot gas cloud. Each atom within the cloud has some motion, although this is random, chaotic motion. Figure 2.23(a) schematically illustrates this case. Individual atoms move to and fro, back and forth, just as the contents of a pot of water jostle around when heated. That's because heat is energy, and this energy affects atoms comprising the gas. The heat energy is changed into kinetic energy, causing the atoms to move chaotically. The hotter the gas, the swifter the random motions of the atoms.

Now, if a photon is emitted by an atom *in motion,* the frequency of the detected photon is changed by the Doppler effect. In other words, should an atom be, for example, moving away from our eye (or from a detector) while in the process of emitting a photon, that photon is red shifted. The photon is not recorded at the precise wavelength predicted by atomic physics, but rather at a slightly longer wavelength. According to our previous discussion of Doppler effect, the extent of this red shift is proportional to the recessional velocity.

In a cloud of gas, some atoms have radial motions toward us, some away from us. Still others have transverse or lateral motions that are unaffected by the Doppler effect. Throughout the whole cloud, completely random motions occur in every possible way. But the important point is that many atoms emit or absorb photons at slightly different wavelengths than would normally be the case if all the atoms were motionless.

Studies show that most atoms in a typical cloud have very small radial velocities; even atoms with large transverse velocities can still have small radial motions. As a result, most atoms emit or absorb radiation that is Doppler shifted very little, while only a minority of atoms cause larger shifts. Consequently, the center of any spectral line is more pronounced than either of its wings, the

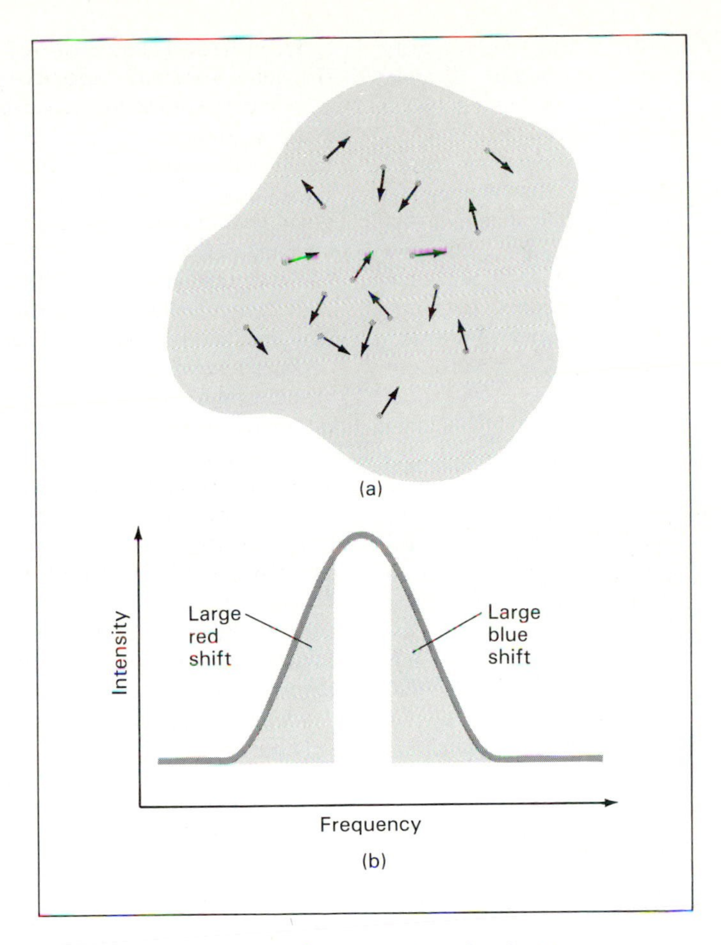

FIGURE 2.23 Atoms moving around randomly (a) emit broadened spectral lines, like the one sketched in (b).

result being a bell-shaped spectral feature like that depicted in Figure 2.23(b).

The neat thing about this Doppler interpretation is that the analysis can be reversed. That is, by measuring a line's width, we can estimate the temperature of a gas. The wider the line, the greater the spread of Doppler motions and therefore the hotter the gas.

So, in principle, careful study of the width of just one spectral line can yield the temperature of a group of atoms. However, in practice, analyses of spectral lines are not as easy as we've described here. Why not? Because heat is not the only environmental factor that broadens spectral lines. Under certain circumstances, several other physical factors can influence the observed width of spectral lines.

Another line-broadening mechanism is gas turbulence. By turbulence, we mean large-scale bulk motions of clumps of matter—clumps larger than a collection of numerous atoms but smaller than an entire cloud. For example, imagine a cluster of atoms being pushed around by some external force—not heated, just pushed around mechanically. Should the atoms have any radial motions while in the process of emitting photons, those photons will be Doppler shifted. Often, bulk motions of this type cause an *entire* spectral line to be shifted. But sometimes, clumps of atoms move around within a gas, each producing a small amount of Doppler broadening—broadening caused by turbulence, and having nothing to do with heat.

Rotation is a third influence on spectral-line width. For example, should a star or a gas cloud be spinning, opposite edges of the object move in opposite directions. Should the object be oriented so that the spin causes some radial motion along our line of sight, then, once again, the Doppler effect can be a factor. Photons emitted from opposite sides of a rotating object are therefore red and blue shifted a little. This causes broadening of the spectral lines radiated by the whole object. In this way, rotation is really a Doppler broadening mechanism, but here the culprit is neither heat nor turbulence.

The fourth and fifth broadening mechanisms differ basically from those above, for they do not depend directly or indirectly on Doppler effect. One such mechanism results from collisions among atoms. If electrons are moving between orbitals while in the act of colliding with other atoms, the energy of the emitted or absorbed photons can change slightly, thus "blurring" the resulting spectral lines. This mechanism is of greatest importance when the density of atoms is large, for that is when collisions are most frequent.

Magnetism is yet another cause of spectral-line broadening. The electrons and nuclei within atoms behave as tiny, spinning magnets. Accordingly, the basic emission and absorption rules of atomic physics change ever so slightly whenever atoms are immersed in a magnetic force field. Generally, the greater the amount of magnetism, the more pronounced the spectral-line broadening.

Deciphering the extent to which each of these five broadening mechanisms affects real spectral lines is often a difficult task. If some lines in a spectrum were broadened by only one mechanism, and other lines by another mechanism, the problem would be easily solvable. Astronomers could systematically analyze several lines and straightforwardly estimate the temperature, turbulence, rotation, density, and magnetism of the observed object. The problem, however, is that many of these broadening mechanisms occur simultaneously. Any observed spectral line is likely to be affected by a combination of several of the mechanisms described above. Sometimes, all five mechanisms broaden a given line. The challenge astronomers face, then, is to unravel the extent to which each mechanism contributes to spectral-line profiles, and thereby derive meaningful information about the gas producing the lines. In many cases, the various contributions can indeed be unraveled, although in other cases a clear solution is virtually impossible.

Molecules

We cannot end a chapter on atoms without saying something about molecules. A **molecule** is a tightly bound group of atoms. Sometimes called **chemical compounds**, molecules are indeed made of 2 or more chemical elements. We've already noted one such molecule in Figure 2.7.

Molecules are held together by the electromagnetic force fields of their atoms. To understand this, recall the notion of chemical bonds introduced earlier in this chapter. These bonds result from electrons that are shared among different atoms. Some molecules are simple and contain only two atoms, thereby forming *diatomic* molecules such as molecular oxygen (O_2), molecular hydrogen (H_2), or cyanogen (CN). Others are more complex, with several atoms bound together as in the *polyatomic* molecules water (H_2O), ammonia (NH_3), or methane (CH_4). Still others are extremely complex, combining thousands of atoms that often form the *macromolecules* prominent in living things. Figure 2.24 shows the simplified structures of a few molecules.

Like atoms, molecules can emit or absorb photons, thereby producing emission and absorption spectral lines. The photon energies are governed by the rules of molecular physics and, much like atomic physics, those rules often tend to clash with common sense. Molecular states are restricted much like atomic states—not a surprising statement since molecules are nothing more than clusters of atoms. However, since molecules are more complex than individual atoms, the rules of molecular physics are even more complex. Indeed, scientists currently find it impossible to completely understand the inner workings of large molecules.

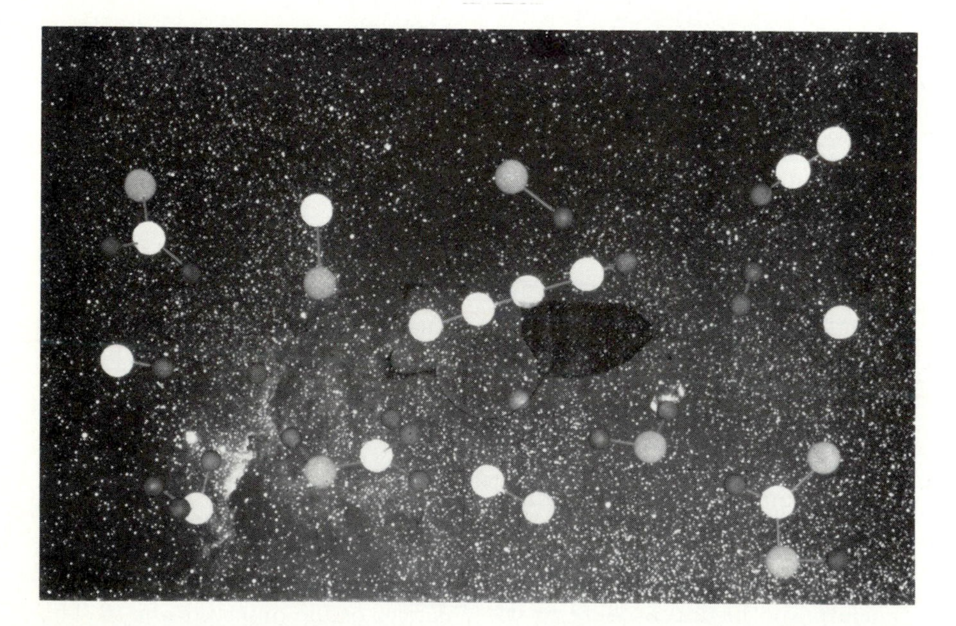

FIGURE 2.24 A fanciful illustration of several molecules. The filled circles represent atoms comprising the molecules. Those displayed include water (H_2O), carbon monoxide (CO), and formaldehyde (H_2CO).

Even so, we can empirically study the behavior of molecules in the laboratory. During the past few decades, experimental work has shown the precise wavelengths (or frequencies) at which millions of molecules emit and absorb radiation. This laboratory research was motivated by the chemical need to develop industrial compounds and by the biological need to solve medical problems.

These studies prove that molecules, like atoms, produce emission or absorption spectral lines because of electron changes among the orbitals of the clustered atoms. In addition, molecules also yield such lines because of the two other kinds of changes not possible in atoms. Molecules can rotate, and they can vibrate. Figure 2.25 schematically illustrates these basic molecular motions.

Molecules rotate and vibrate in specific ways; only certain spins and vibrations are allowed. Nature fixes these properties, just as it does electron arrangements within atoms. But when a molecule *changes* its rotational state, or similarly its vibrational state, a photon is emitted or absorbed. Consequently, any gas having lots of elements and chemicals is sure to have many molecules undergoing changes in electron, rotation, or vibration states. The result is spectral lines characteristic of different kinds of molecules. These are molecular fingerprints, analogous to their atomic counterparts, enabling researchers to study one kind of molecule to the exclusion of all others. Finally, like atoms, molecular spectral lines are often broadened by some or all of the five mechanisms noted in the preceding section.

As a rule of thumb, we can say that electron changes within molecules produce visible and ultraviolet spectral-line features; changes in molecular vibration produce infrared spectral features; and changes in molecular rotation produce spectral lines in the radio part of the electromagnetic spectrum.

Proceeding through our text, we'll come to recognize that molecules comprise the bulk of nonstellar matter that is not ionized in the Uni-

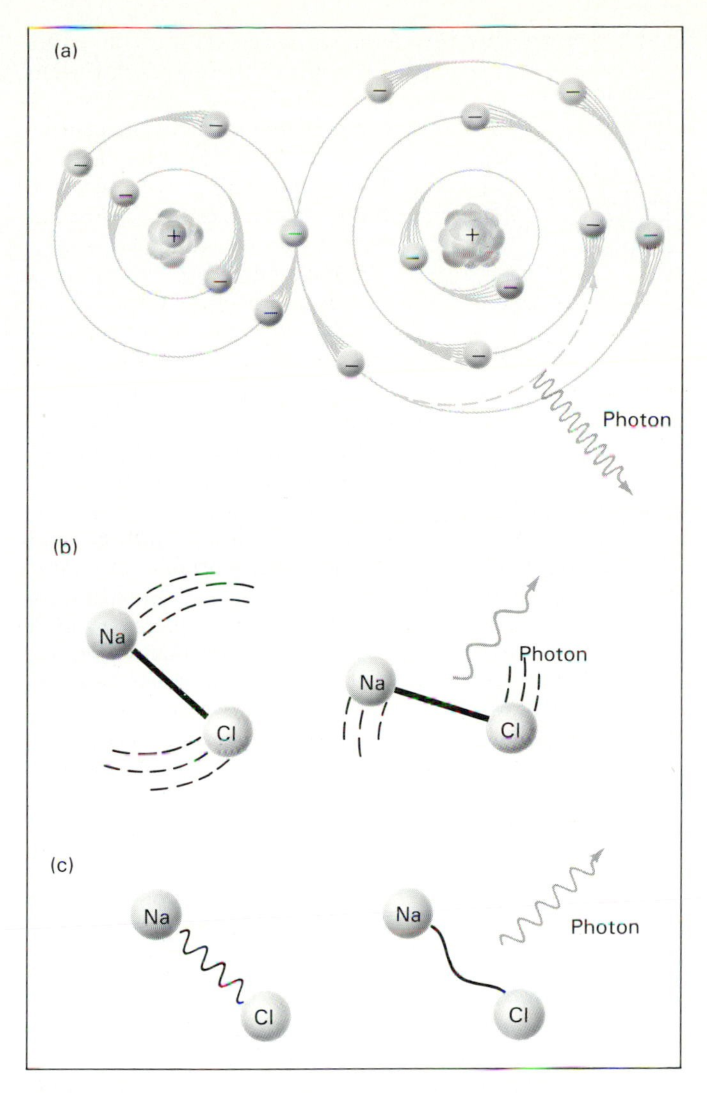

FIGURE 2.25 Molecules can change in three ways while emitting or absorbing electromagnetic radiation. Sketched here is the molecule sodium chloride (NaCl, or common table salt) experiencing (a) a change in electron arrangement, (b) a change in rotation, and (c) a change in vibration.

verse. Molecular spectroscopy, therefore, is an important diagnostic tool in our study of the origin and evolution of galactic clouds, planets, life, and intelligence—all of which we'll discuss in later chapters.

SUMMARY

Spectroscopy is a powerful technique enabling us to infer the nature of matter on a submicroscopic level. What we have found is that all matter is made of atoms, which are in turn made of elementary particles. Every kind of atom, that is, every element, has a unique structure; no two elements are alike. Even molecules, which are clusters of atoms, are built in certain, specific ways. It is these microscopic structures that give atoms and molecules their physical and chemical properties.

When atoms and molecules change their normal states, they can emit or absorb photons of radiation. This results in emission and absorption features that appear as narrow lines in spectra. Every kind of atom and molecule has its own characteristic spectral "fingerprint," enabling us to study their changes either in terrestrial laboratories or in cosmic objects.

In short, spectroscopy allows us to study the basic nature of matter by means of its ceaseless *changes*. Here, as elsewhere in the text, the concept of change is central to our arguments.

KEY TERMS

atom
chemical compound
cosmic abundances
electron
element
elementary particle
electromagnetic force
excited state
gravitational force
ground state
ion
kinetic energy
molecule
neutron
nuclear force
nucleus
orbital
periodic table of the elements
plasma
potential energy
proton
quark
spectral line
spectroscopy
weak force

QUESTIONS

1. Name and briefly explain at least three of the basic forces operating in our terrestrial world.
2. In a couple of paragraphs, describe how astronomers can extract information about a cosmic object from a study of its spectrum.
3. List three different kinds of physical properties of an astronomical object that can be detected via spectroscopy.
4. A thin, cool cloud passes between you and a hot star. Describe the apparent change in the optical spectrum.
5. Explain the relation between the submicroscopic structure of a hydrogen atom and the formation of its spectral lines. Draw a diagram if necessary to clarify your answer.

*6. Explain how both heat and rotation of an object can broaden a spectral line emitted by that object.

7. Speculate about the properties of matter on a scale smaller than an atomic nucleus. Do you think that physicists will ever discover a truly fundamental elementary particle, or will the search for nature's building blocks just keep reaching deeper and deeper into the submicroscopic world?
8. Name the two most important factors that combine to make electron movements within atoms, and therefore spectral lines, unique for each kind of atom.
9. Explain how a single element such as iron can yield more than 700 individual spectral lines within the Sun's visible spectrum alone. How do astronomers go about identifying which of the thousands of absorption lines present are attributable to iron?

*10. How might the heat of a molecular gas alter the electron, rotation, and vibration states of the molecules? If the heating is sufficient, what might be the appearance of the resultant spectrum?

FOR FURTHER READING

ALLER, L., *Atoms, Stars, and Nebulae*. Cambridge, Mass.: Harvard University Press, 1971.

HAMILTON, J., AND MARUHN, J., "Exotic Atomic Nuclei." *Scientific American,* July 1986.

JACKSON, J., TIGNER, M., AND WOJCICKI, K., "The Superconducting Supercollider." *Scientific American,* March 1986.

LEVY, D., "The Spectroscopy of Supercooled Gases." *Scientific American,* February 1984.

PHILLIPS, W., AND METCALF, H., "Cooling and Trapping Atoms." *Scientific American,* March 1987.

QUIGG, C., "Elementary Particles and Forces." *Scientific American,* April 1985.

3 THE EQUIPMENT OF SCIENCE

Knowledge of the cosmos usually advances in three phases: collection and detection of radiation from space, storage of the resulting data, and analysis of that data. The first phase is normally accomplished with a telescope of some sort, the second might employ photographic film or magnetic tape, while the third requires us to apply the laws of physics to create a mental image, or model, that explains the acquired data. Of course, theoretical work plays an important role, and often suggests the acquisition of crucial data, but more often than not for an empirical science like astronomy, observations of cosmic phenomena precede their theoretical prediction.

In this third introductory chapter, we study some of the equipment used by astronomers to extract information from the radiation emitted by cosmic objects. Once the radiation is gathered, focused to a small place, and recorded—as sketched in Figure 3.1—researchers can analyze the cosmic information, attempting to unravel secrets of the Universe.

The learning goals for this chapter are:

- to understand the advantages and disadvantages of the many different types of telescopes
- to appreciate the need for large telescopes
- to recognize that radiation collected in each part of the electromagnetic spectrum contributes to our understanding of cosmic objects
- to appreciate the various kinds of tools needed to capture and analyze invisible radiation
- to study further how spectroscopy works for visible and invisible radiation

Optical Instruments

In this section we study some of the apparatus used to sample visible radiation emitted by extraterrestrial objects. The main device is called an **optical telescope**.

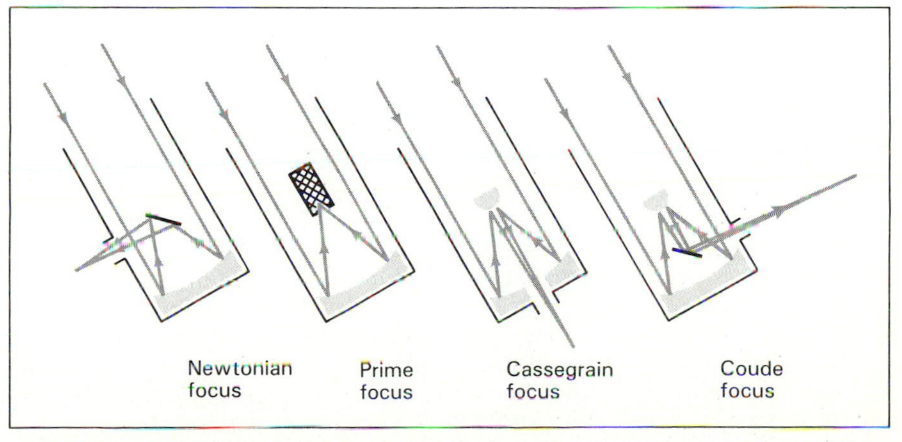

FIGURE 3.1 Schematic diagrams of the essentials of any apparatus used to collect, focus, and record cosmic information. Shown here are four different telescope designs. Each uses a large mirror at the bottom of the telescope to capture radiation, which is then directed along different paths for analysis.

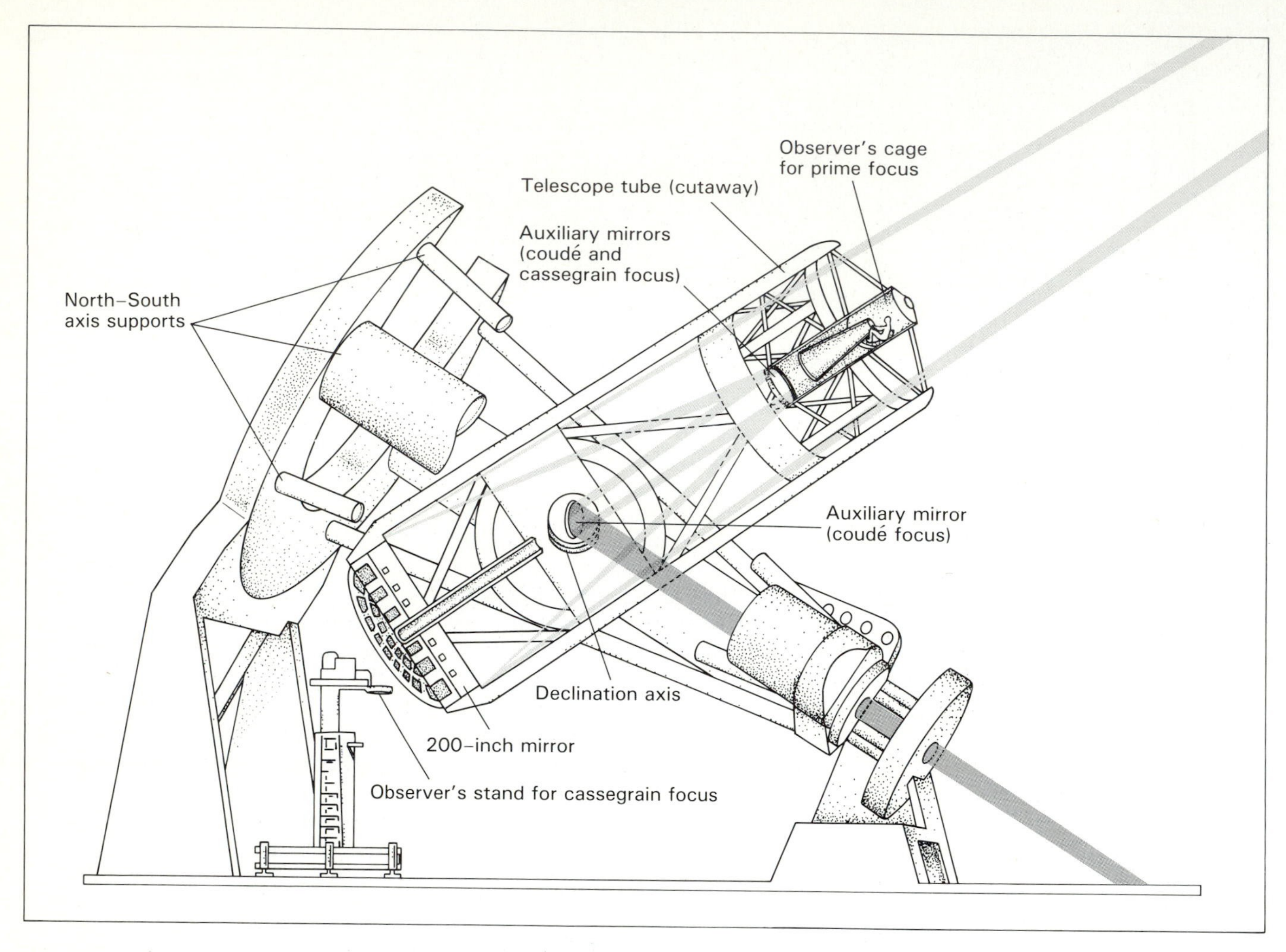

FIGURE 3.2 An artist's illustration of the 5-meter-diameter optical telescope on Palomar Mountain in California.

THE BASIC MACHINE: Figure 3.2 is an artist's conception of the 5-meter (200-inch)-diameter optical telescope on California's Palomar Mountain. Note the main mechanical features of the telescope: A large horseshoe-shaped yoke rotates around the "polar" mount or principal axis. This simple design permits any telescope to compensate for Earth's rotation and to follow with great precision the apparent motions of astronomical objects. Accurately pointing telescopes, some weighing many tons, are among the great marvels of modern engineering.

As shown in Figure 3.2, a minute fraction of a star's radiation enters through the slit in the observatory dome, passes down through the main telescope tube, reflects from the curved surface of the 5-meter mirror, and is finally focused to a small point near the top of the tube. This is the **prime focus**, where astronomers can collect and analyze light from distant cosmic objects.

Alternatively, astronomers may choose to work on a rear platform where they can use equipment too heavy to hoist to the prime focus of such delicately balanced telescopes. In this case, also shown in Figures 3.1 and 3.2, light proceeds from a star toward the 5-meter "primary" mirror, reflects up to the prime focus, where a smaller "secondary" mirror reflects it back down to the platform through a small hole at the center of the 5-meter mirror. Figure 3.3 shows a researcher adjusting photographic equipment at this so-called "Cassegrain focus."

Another, more complex observational method requires starlight to reflect from several mirrors. The radiation first reflects normally from the primary mirror toward the focus, after which a secondary mirror reflects it back down the tube. A third (much smaller) mirror then deflects the light into an environmentally controlled laboratory. Known as the Coudé Room (after the French word for "bent"), and partly illustrated in Figures 3.1 and 3.2, this laboratory is separate from the telescope itself, enabling astronomers to use very heavy and finely tuned equipment that couldn't possibly be lifted to either the prime focus or the elevated platform at the rear of the telescope.

Telescopes are often made more sophisticated by inserting, in front of the detector, devices such as filters

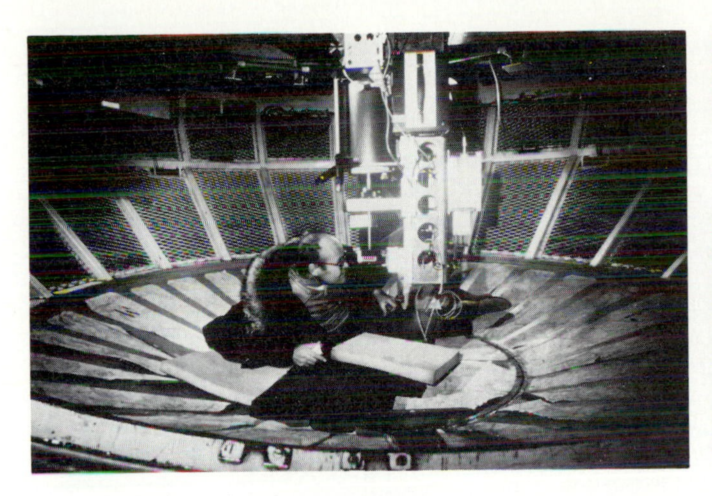

FIGURE 3.3 Some astronomers work at the Cassegrain focus of a telescope (i.e., underneath the main mirror).

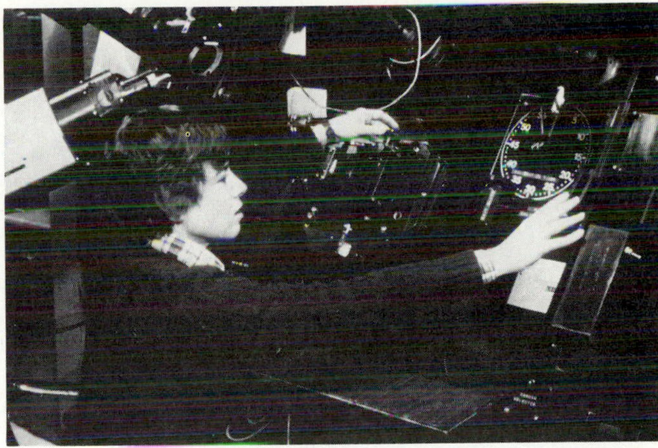

FIGURE 3.4 An astronomer adjusts the large spectrometer in the Coudé Room of the 4-meter telescope at the Kitt Peak National Observatory, near Tucson, Arizona.

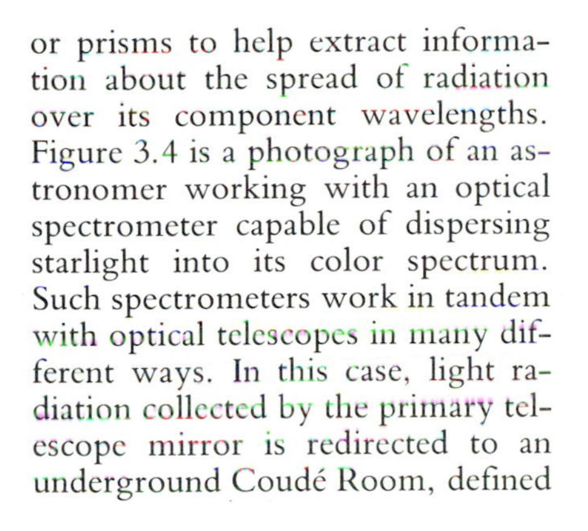

or prisms to help extract information about the spread of radiation over its component wavelengths. Figure 3.4 is a photograph of an astronomer working with an optical spectrometer capable of dispersing starlight into its color spectrum. Such spectrometers work in tandem with optical telescopes in many different ways. In this case, light radiation collected by the primary telescope mirror is redirected to an underground Coudé Room, defined by a narrow slit, passed through a prism, and projected onto a screen. There the spectrum can be studied in real time (as it happens), or stored on a photographic plate or in a computer for post-time analysis (at a later time).

TYPES OF TELESCOPES: Most of the world's large telescopes are **reflectors** like the one atop Palomar Mountain. However, another, basically different type of telescope is known as a **refractor**. Figure 3.5 compares the basic designs of these two kinds of telescope systems.

The 5-meter Palomar machine noted earlier is called a reflector telescope simply because the light-gathering device is a mirror *from* which light is reflected to a focus to be sampled by the eye, a camera, or nowadays a computer. On the other hand, a refractor telescope uses a lens as its light-gathering device, *through* which radiation must bend on its way to the detector.

The need to pass light through a lens is a major disadvantage of refracting telescopes. Large lenses cannot be made homogeneous enough to allow uniform passage of light throughout the entire lens. Various parts of the lens focus red and blue light differently, with the whole lens acting as a complex array of prisms. In short, large lenses cannot be focused well and often produce astronomical headaches. To be sure, nearly uniform glass can be manufactured routinely for small lenses such as eyeglasses, but it's virtually impossible to achieve the necessary quality for large lenses. Figure 3.6 shows the world's largest refractor: Located at the Yerkes Observatory in Wisconsin, it has a lens diameter of only 1 meter (40 inches).

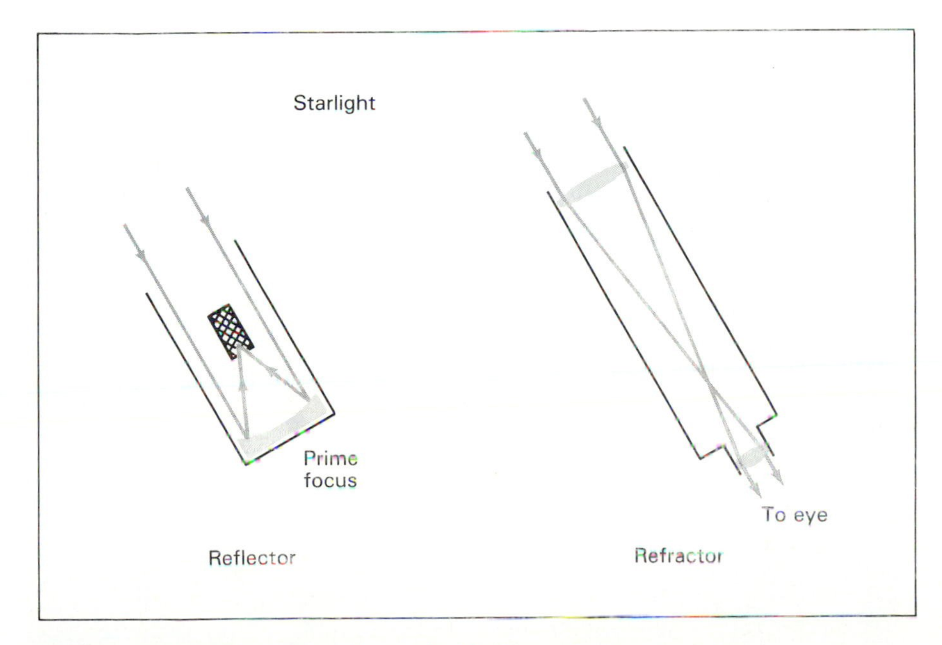

FIGURE 3.5 Schematic comparison of reflecting and refracting telescope systems. Both types are used to gather and focus cosmic radiation—to be either observed by human eyes, or recorded on photographs or in computers.

FIGURE 3.6 Photograph of the Yerkes Observatory's 1-meter-diameter refractor telescope.

Although refractors are hard to build, large reflectors are not exactly easy. Mirrors are usually made from large blocks of quartz, glass, or some other type of polishable material capable of withstanding large temperature changes without much expansion or contraction. The construction process begins by pouring the molten material into a large cast, and then cooling it slowly over the course of several years so as to avoid cracking while it transforms from liquid to solid. A couple more years are then normally needed to grind and polish the surface into the required curve. Finally, the surface is coated with a thin film of aluminum. Remember, reflecting mirrors do not pass light but simply redirect its path, focusing an image via multiple reflections.

SOME MODERN TELESCOPES: The largest reflector currently operational is the 6-meter-diameter telescope atop the Caucasus Mountains in the Soviet Union. Similar to the Palomar telescope, this Soviet telescope is shown in Figures 3.7 and 3.8. Like some of our cathedrals, it took more than a decade to build.

Figure 3.9 is a spectacular photograph of the U.S. national observatory for optical astronomy in the northern hemisphere. Located high on Kitt Peak near Tucson, Arizona, the site was chosen because of its many dry, clear nights.

Figure 3.10 is an aerial view of the major U.S. observatory capable of studying the skies of the southern hemisphere; it's located at Cerro Tololo, in the Chilean Andes. Numerous domes house optical telescopes of different sizes, each with varied supporting equipment, making this observatory the most versatile cosmic laboratory south of Earth's equator. Economic difficulties in the United States and political difficulties in Chile have postponed further construction of several major telescopes at this spectacularly clear site.

A radically different type of telescope was recently built on Mt. Hopkins, in Arizona, a few miles from Kitt Peak. Figure 3.11 shows the Multiple Mirror Telescope, called the MMT by the astronomers who built it. The primary optical device of this reflector is not a single huge and heavy mirror, but rather six smaller ones. These mirrors, each 1.8 meters (72 inches) in diameter, work in tandem to approximate the capabilities of a single mirror nearly 5 meters in diameter. The smaller mirrors are much easier and cheaper to construct, although the complex paths the incoming light waves must follow to reach a common focus make the telescope tricky and time consuming to align. Even so, now that we know that such a novel apparatus really works, optical telescopes made of many small mirrors arrayed like segments of a much larger mirror would seem to be the way of the future.

FIGURE 3.7 A distant view of the world's largest (6-meter-diameter) reflector telescope in the Caucasus, U.S.S.R. (Notice the people standing atop the dome.)

FIGURE 3.8 A view of the large Russian reflecting mirror prior to installation to the mid-1970s.

FIGURE 3.9 Several telescope domes adorning Kitt Peak National Observatory can be seen by the reflected light of an atmospheric storm.

FIGURE 3.10 An aerial photograph of a major southern-hemisphere observatory. Located high in the Andes Mountains, the Cerro Tololo Inter-American Observatory is operated by the same consortium of U.S. universities that oversees the Kitt Peak National Observatory.

FIGURE 3.11 Photograph of a large optical telescope of radical design—the Multiple Mirror Telescope atop Mt. Hopkins, Arizona. Each of its six mirrors can be clearly seen.

Telescope Size

COLLECTING AREA: Large telescopes are usually more advantageous than small ones. We say this for two reasons: First, a larger telescope has a greater **collecting area**, which is the total area of a telescope capable of capturing radiation. The larger the telescope (either reflecting mirror or refracting lens), the stronger the signals observed, and the easier we can measure and study an object's radiative properties. After all, optical telescopes are merely "light buckets" designed to collect numbers of photons per second. Accordingly, large telescopes can gather more radiation and bring it to a point (the focus), thereby increasing the strength of the signal to be studied. For this reason, weak cosmic sources can be observed only with larger telescopes. Astronomers who take time exposures with small telescopes, by building up a weak star's image on a photograph, find the process not only time consuming but also impractical.

The following example demonstrates this first advantage of large telescopes. The observed strength of any astronomical object is directly proportional to the *square* of the telescope diameter. Thus a 5-meter mirror can measure a cosmic source, to the same accuracy, 25 times quicker than a 1-meter mirror. That's because a mirror diameter five times larger has a telescope collecting area some 25 times larger. Expressed in another way, a 1-hour time exposure with a 1-meter mirror is roughly equivalent to a 2.5-minute time exposure with a 5-meter mirror. For this reason alone, the weakest sources in the Universe are studied exclusively by astronomers

INTERLUDE 3-1 ***Angular Measure***

The size and scale of things are often specified by measuring lengths and angles. The concept of *length measurement* is fairly intuitive; we've already noted the dimensions of some objects found in the Universe, ranging from elementary particles with sizes of 10^{-13} centimeter to galaxy clusters with sizes of 10^{26} centimeters.

The concept of *angular measurement* may seem less intuitive when first encountered, but it too can become a commonsense notion by remembering a few simple facts diagrammed below. One of these facts is obvious, namely, that a full circle contains 360 **arc degrees**. By slicing any circle into 360 equal pie-shaped pieces, we obtain 360 one-degree increments.

Each 1-degree increment can be further subdivided into fractions of an arc degree. We call these smaller fractions **arc minutes**; there are 60 arc minutes in 1 arc degree. For example, both the Sun and the Moon subtend an angular size of 30 arc minutes on the sky. Your thumb, held at arm's length, does just about the same, covering a rather small slice of the entire 180 arc degrees from horizon to horizon.

Finally, an arc minute can be further sliced into 60 equal **arc seconds**. Put another way, an arc minute is $\frac{1}{60}$ of an arc degree, and an arc second is $\frac{1}{60} \times \frac{1}{60}$, or $\frac{1}{3600}$ of an arc degree. So an arc second is an extremely small unit of angular measure. How small? Well, it's the angle projected by a centimeter-sized object at a distance of about 2 kilometers. Expressed in more familiar terms, an arc second is the angle subtended by an American dime when viewed from a distance of nearly 3 miles away. That's pretty small!

Don't be confused by the units used to measure angles. Arc minutes and arc seconds have nothing to do with the concept of time, nor do arc degrees have anything to do with temperature. Arc degrees, arc minutes, and arc seconds are simply convenient ways to help measure the size, shape, and position of objects in the Universe.

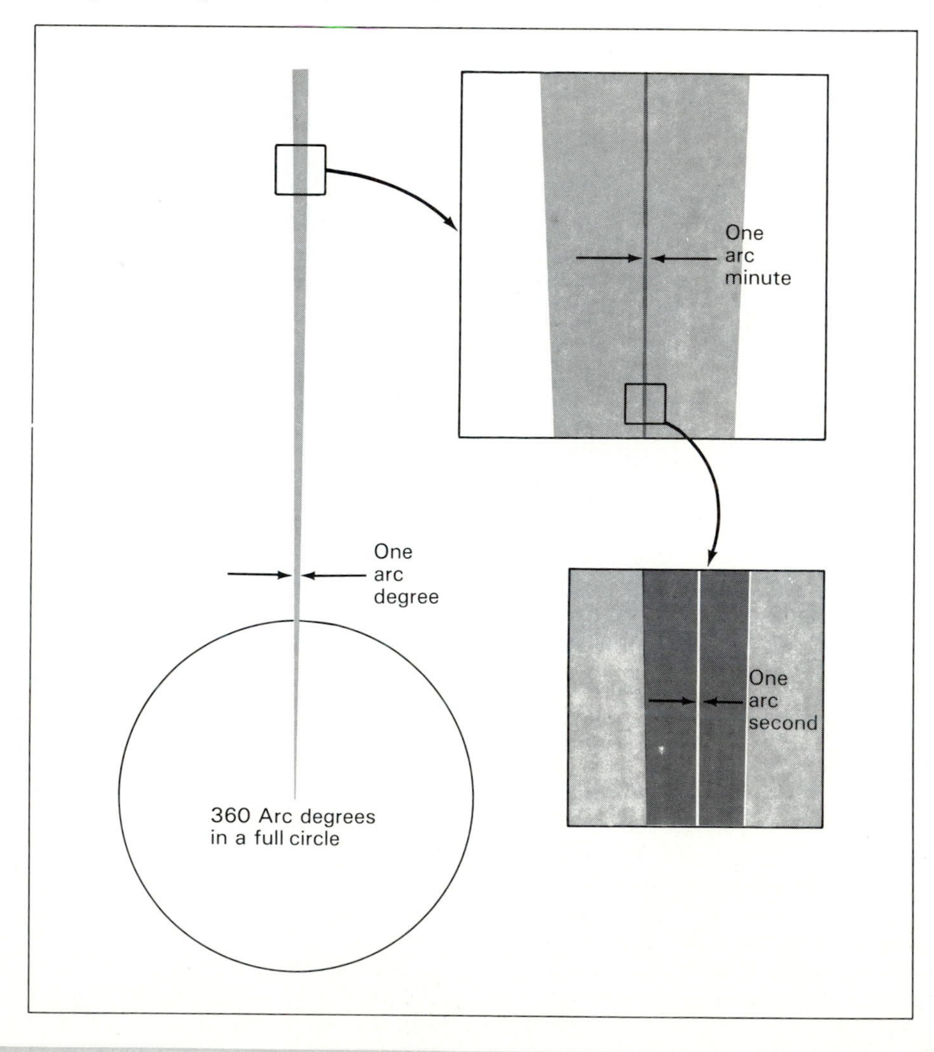

having access to the world's largest telescopes.

ANGULAR RESOLUTION: The second advantage of large telescopes is resolution. By **angular resolution** we mean the ability to distinguish between two adjacent objects on the sky (see Interlude 3-1). The finer the angular resolution, the better we can make such a distinction; furthermore, the better we can see the details of any one object.

Anyone who has used a biological microscope knows that larger instruments generally provide better resolution. More detail can be noted while viewing through a larger, more powerful microscope. The same holds true for astronomical telescopes; while studying astronomical objects, just like biological specimens, we strive to make the angular resolution as fine as possible.

Figure 3.12(a)–(c) illustrates how large telescopes can resolve adjacent sources of radiation that are unresolvable with small telescopes. Here, two sources are observed with a fixed wavelength (i.e., visible light) but with variable telescope size. The human eye, with a resolution of 1 arc minute, would see this double source as essentially a single object. However, aided by a telescope, we can do much better. Similarly, Figures 3.12(d)–(g) show how the Andromeda Galaxy appears with progressively higher resolution.

Angular resolution can be made finer, not only by increasing the telescope size, but also by decreasing the radiation's wavelength. How can that be done? Well, in principle, we could observe the short-wavelength radiation from cosmic objects. But in practice, this presents a problem: As noted in Chapter 1, the shortest-wavelength emissions—for example, ultraviolet waves or x-rays—do not penetrate Earth's atmosphere. Accordingly, until advanced high-energy observatories are deployed in orbit, we can best achieve high resolution only by using large ground-based telescopes.

Even large optical telescopes have their limitations. For example, although the 5-meter optical telescope on Palomar Mountain should be able to attain an angular resolution nearly as fine as 0.01 arc second, in practice it cannot do better than about 1 arc second. Aside from some special techniques now being developed to examine some bright stars, no ground-based optical telescope can resolve astronomical objects much finer than an arc second. Why not? Because the turbulence (or random motions) of Earth's air tends to blur the paths of light waves passing through our atmosphere from a cosmic source to a telescope. This irregular atmospheric turbulence causes the familiar "twinkling" of stars in the night sky. It's a little like shining light through a layer of turbulent (wavy) water in order to illuminate a submerged rock; the image can be severely distorted. Once the turbulence decreases and the water stills, our view of the rock becomes clearer. But the basic point here is that air turbulence is ever-present; our constantly changing weather ensures it. Thus stars continue twinkling, blurring their images, and making their apparent sizes (few arc seconds) much larger than their real images (which are really a great deal smaller than an arc second).

Optical telescopes in orbit about the Earth or emplaced on the Moon can overcome this limitation. Without atmospheric blurring, extremely fine resolution can be achieved, subject only to the engineering restrictions of building large structures in space. For example, the *Hubble Space Telescope*, named after one of America's most notable astronomers and scheduled to be launched soon into

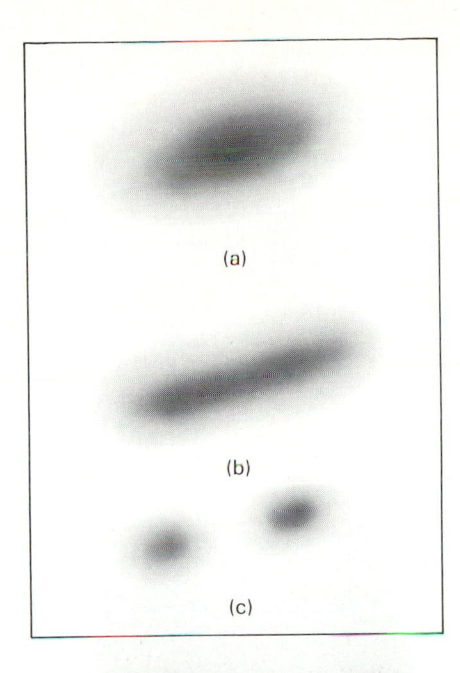

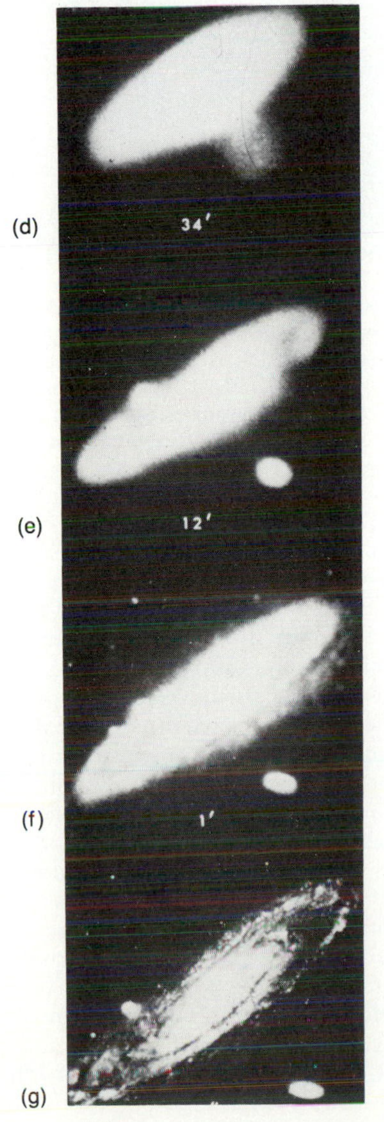

FIGURE 3.12 Two comparably bright light sources become progressively clearer when viewed by (a) a small telescope, (b) a moderate-sized telescope, and (c) a large telescope. The angular resolution is best with the largest telescope, making the two sources distinguishable. Similarly, detail can be more readily discerned on the Andromeda Galaxy as the angular resolution is improved some 2000 times, from (d) about ½ arc degree, (e) 12 arc minutes, (f) 1 arc minute, and (g) 1 arc second.

Earth orbit by NASA's *Space Shuttle*, should be able to resolve objects with an accuracy as fine as 0.03 arc second, even though its 2.4-meter mirror is not as large as the largest telescopes on the ground. Thus this orbiting observatory should—nearly overnight—give us a view of much of the Universe some 30 times sharper than ever before available.

Radio Astronomy

In addition to visible radiation that normally penetrates Earth's atmosphere (provided that it's not cloudy), radio radiation also reaches the ground. In fact, the radio window in the electromagnetic spectrum is much wider than the optical window, as we noted in Chapter 1. Recognizing that the atmosphere is no hindrance to long-wavelength radiation, radio astronomers have built many **radio telescopes** capable of detecting cosmic radio waves. Since all these radio devices have been constructed only within the past few decades, radio astronomy is a younger subject than optical astronomy.

ESSENTIALS OF RADIO TELESCOPES: Figure 3.13(a) shows a typical radio telescope, this particular one being the large 43-meter (140-foot)-diameter device located at the National Radio Astronomy Observatory. Figure 3.13(b) schematically diagrams its basic reception gear.

Although much larger than any reflecting optical telescope, radio telescopes are built in basically the same way. They have a large polar axis about which rotates a horseshoe-shaped mount, which in turn supports the large metal curved "dish." In its simplest observing mode, the collecting area captures cosmic radio waves and reflects them from the dish to the focus, where a receiver detects the signals.

Whereas optical telescopes resemble human eyes, radio telescopes are more like human ears. Eyes (or photographs) mix the effects of all visible frequencies and record instantaneously the notion of color. Ears, however, perceive different frequencies separately (although we can hold many musical tones in our consciousness). Similarly, radio telescopes normally receive radiation within only a narrow band of frequencies. To detect radiation at another radio frequency, we must retune the equipment much as we tune our TV sets to different channels.

(a)

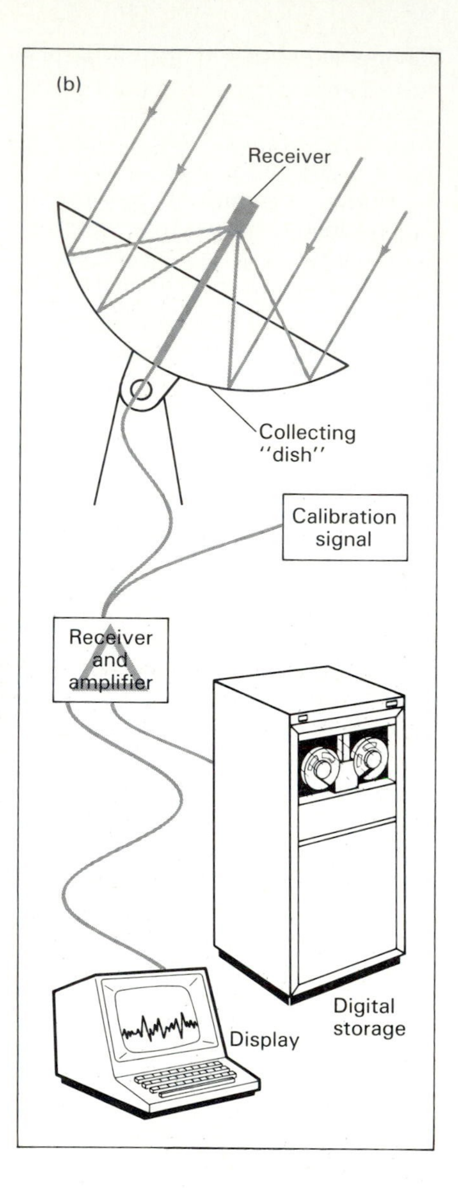

FIGURE 3.13 (a) Photograph of the 43-meter-diameter radio telescope at the National Radio Astronomy Observatory in Green Bank, West Virginia. (b) Schematic diagram of the basic radio receiver and data processing devices used at all radio telescopes.

Not surprisingly, astronomers often prefer large radio telescopes, since the great collecting area ensures both strong source strength and the finest possible angular resolution. (The arguments of the preceding section hold true for radio as well as optical telescopes.) But while the observed strength of some radio sources can sometimes be quite large, the angular resolution is often poor compared to that for optical astronomy. To be sure, *the* major disadvantage of all radio telescopes is their relatively poor resolving power, despite the enormous size of many radio dishes. The problem is that the typical wavelength of radio waves is roughly six orders of magnitude (i.e., about a million times) larger than those of light, and these longer wavelengths impose a corresponding crudeness in angular resolution.

The best angular resolution of a single radio telescope is about 10 arc seconds, an order of magnitude coarser than the 1 arc second possible with the largest optical mirrors. Resolution of about 1 arc minute is sometimes achieved with the 43-meter radio telescope of Figure 3.13, although it's designed to detect radio waves with an angular resolution of about 6 arc minutes, a mere tenth of an arc degree. Even larger radio telescopes, such as the 100-meter-diameter German dish shown in Figure 3.14, a device about the size of a football field, can achieve no finer resolution than 30 arc seconds. Despite this obvious limitation, the study of cosmic radio radiation provides us with a wealth of useful information about the Universe—especially about the "invisible" Universe.

You may wonder why radio telescopes can be built so much larger than their optical counterparts. After all, the primary collecting areas of most radio telescopes literally dwarf even the largest optical mirror. The reason for this concerns the accuracy to which the metal dishes and glass mirrors can be machined. The reflecting surface need not be as smooth for long-wavelength radio waves as for short-wavelength light waves. Provided that surface irregularities (e.g., dents, bumps, and the like) are much smaller than the wavelength of the waves to be detected, the surface will reflect them efficiently. Consequently, since the wavelength of visible radiation is small (approximately 10^{-4} centimeter), very smooth mirrors are needed to reflect the waves properly. Radio waves are much longer, however, and can often be reflected well enough even from rough surfaces.

For example, when light shines on a curved surface having irregularities of about a millimeter, the light scatters and does not focus well; the image of the light source is severely distorted. You could test this statement by looking at your own blurred face in a piece of rough, unpolished metal. But radio waves of centimeter or longer wavelength are not scattered at all by slightly rough surfaces; instead, they are reflected properly to an accurate focus. Very long radio waves of, say, 1-meter wavelength can reflect perfectly well from surfaces having irregularities even as large as your fist.

This point can be further understood by considering the following tennis-court analogy. If a clay court is finely groomed, its surface is smooth. Tennis balls, bouncing off the clay (i.e., reflecting), will volley true to form from one player to another. But suppose that the court surface is littered with irregularities such as small pebbles or sneaker gashes about a centimeter in size. When tennis balls bounce from such an irregular surface, they slightly deflect from their intended paths. On the other hand, if we used a larger ball, say a basketball, on the same irregular clay surface—in analogy with a longer-wavelength radio wave—the ball would travel true to form despite the surface flaws. Provided that the ball is much larger than the typical size of the irregularities, it cannot be deflected appreciably. Thus surfaces too rough for small balls (or short-wavelength visible radiation) may be smooth enough for larger balls (or longer-wavelength radio radiation).

Figures 3.15 and 3.16 show an even larger radio telescope, in fact the world's largest. Approximately 300 meters (1000 feet) in diameter, the surface of the Arecibo telescope spans nearly 100 acres. Provided that the openings between adjacent strands of wire are much smaller than the long-wavelength radio waves to be detected, even a rough chicken-wire surface is perfectly ad-

FIGURE 3.14 Photograph of the 100-meter-diameter radio telescope at the Max Planck Institute near Bonn, West Germany.

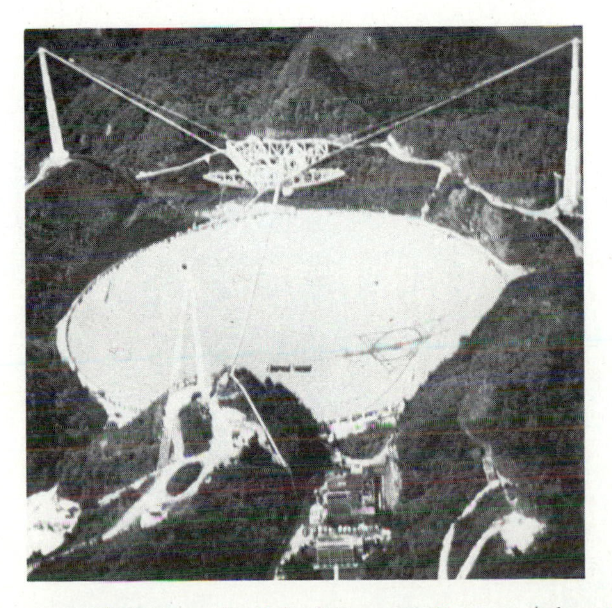

FIGURE 3.15 An aerial photograph of the 300-meter-diameter dish at the National Astronomy and Ionospheric Center near Arecibo, Puerto Rico. The receivers that detect the focused radiation are suspended nearly 300 meters (about 80 stories) above the center of the dish.

FIGURE 3.16 Maintenance crews at Arecibo get around the surface to repair the telescope by wearing large foot pads resembling snow shoes.

equate for proper reflection. Indeed, the Arecibo dish was originally constructed of chicken wire. Despite the fact that light radiation can pass right through such a telescope collector, the chicken-wire surface reflects long-wavelength radio waves just as well as a finely polished mirror reflects short-wavelength visible light.

The entire Arecibo dish was resurfaced several years ago with thin metal plates. As a result, the telescope can now be used to detect shorter-wavelength radio radiation. Even so, useful observations are still restricted to radiation of wavelength greater than 10 centimeters, making its angular resolution no better than that of smaller radio telescopes, despite Arecibo's enormous size. The huge size of the dish, in fact, creates one distinct disadvantage: The Arecibo telescope can't be pointed very well to follow cosmic objects across the sky. The dish is literally strung among several limestone hills, restricting its observations to those objects passing nearly overhead.

Arecibo, then, is an example of a roughly surfaced telescope capable of detecting long-wavelength radio radiation. At the other extreme, Figure 3.17 shows an example of a radio telescope having an extremely smooth surface. The 36-meter-diameter Haystack dish in northeastern Massachusetts maintains a parabolic curve to an accuracy of about a millimeter all the way across its solid metal surface (see Interlude 3-2). Constructed of polished aluminum, you could see a rough reflection of your face on the surface (consult Figure 3.18). We might say that

FIGURE 3.17 Cutaway diagram of the Haystack radio observatory, showing the large, precision-surfaced dish inside the fully enclosed radome.

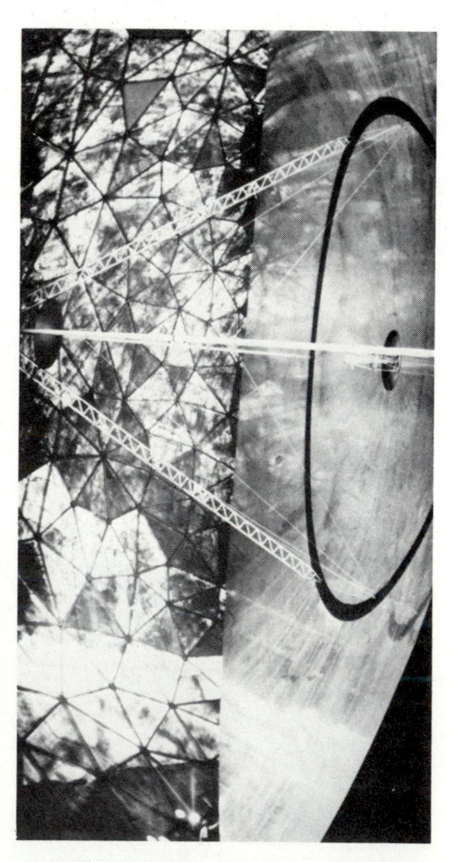

FIGURE 3.18 Photograph of the Haystack dish, taken from inside the radome. For scale, note the engineer at the bottom. Also note the dull shine on the telescope surface, indicating its smooth construction.

INTERLUDE 3-2 ***The "Eye" of a Radio Telescope***

We can make an interesting comparison between large radio telescopes and human eyes. In the following table we use the properties of the Haystack telescope, shown in Figures 3.17 and 3.18.

As we readily noted in this one-to-one correspondence, the human eye is more efficient than any radio telescope. No wonder; the eye has been around for a long time. Evolution has refined it into an exquisite piece of matter.

Nonetheless, the eye has a major limitation—it can't see in the dark. And in a Universe where much of the matter is invisible, that can be a problem. Fortunately, this limitation has been virtually eliminated by our technological civilization, whose recent cultural inventions, including radio telescopes, enable us to perceive and study the invisible regions of our multivaried Universe.

PROPERTY	HAYSTACK	HUMAN EYE
Diameter	3660 centimeters, fixed	0.2 centimeter, adjustable
Operational wavelength	2 centimeters	5.5×10^{-5} centimeter
Reception interval	10^8 hertz	4000–7000 angstroms $\cong 10^{14}$ hertz
Angular resolution	1 arc minute	1 arc minute
Focus	Reflector	Refractor
Detector	Array of 1000 radio receivers	Array of 100 million light-sensitive devices
Protection	Radome	Eyelid
Field of view	7000 square degrees scanned in 135 hours	7000 square degrees scanned in 0.1 second (21,000 square degrees with peripheral vision)
Cost	Several million dollars	Free of charge at birth
Manufacture	Several years, and considerable aggravation	9 months, and no problem at all

Haystack is a poor optical mirror but a superb radio telescope. Accordingly, it can be used to reflect and accurately focus radiation having short radio wavelengths, even as small as a fraction of a centimeter.

Recognize that Figure 3.17 is not an actual photograph of the Haystack telescope. It's really a cutaway diagram designed to show the telescope inside its protective shell or radome; to many aircraft pilots, it seems like a gigantic golf ball. Such a radome is needed to protect the accurate surface from the harsh wind and weather of New England. The radome acts much like the protective dome of an optical telescope. Even so, there is no slit that opens in the Haystack dome. Astronomers detect the incoming cosmic radio signals that pass virtually unimpeded through the fiberglass material of which the radome is made.

THE VALUE OF RADIO ASTRONOMY: Despite the inherent disadvantage of relatively poor angular resolution, radio astronomy enjoys many advantages. For example, we can observe 24 hours a day; darkness is not needed to receive radio signals, as we can easily demonstrate by turning on a home radio any time during the day or night. Also, radio observations can often be made through cloudy skies; the longest-wavelength radio waves can even be detected during rain or snow storms. Poor weather causes few problems, since the wavelength of most radio waves is much larger than the typical size of atmospheric raindrops or snowflakes. Adverse weather of this sort, however, is disastrous to optical astronomy; that's because light's wavelength is smaller than rain or snow, or even minute water droplets in a fluffy cloud.

To illustrate this important point further, consider another example. Suppose that an atmospheric cloud separates an aircraft and an air-traffic controller, as depicted in Figure 3.19. The pilot and the controller wouldn't be able to see one another or to transmit light messages. But radio waves could easily penetrate the cloud; the aircraft and the tower could still communicate by wireless telephone, walkie-talkie, or some other radio device. In this way, aircraft use radio beacons to help them land safely during poor weather.

The very same lesson is learned by driving an automobile in a thick fog at night. We may instinctively put on the high beams to see farther, but we quickly realize that additional light doesn't help at all. The scattering of waves from the headlights prohibits us from seeing much of the road ahead. That's because the water droplets comprising the fog are comparable in size to the wavelength of light. The light therefore scatters in every direction, often decreasing its usefulness as a means of providing information.

These examples point up one of the main advantages of radio astronomy: its ability to probe the cosmic darkness. Quite frankly, there's nothing to see in a dark region de-

FIGURE 3.19 This diagram illustrates how long-wavelength radio radiation can penetrate a cloud that otherwise blocks short-wavelength visible radiation.

void of any light. But such regions usually emit radio radiation which can make its way through dust clouds in space, just as radio signals on Earth can penetrate cloudy or foggy weather, thereby enabling radio astronomers to construct images of regions completely hidden from view.

Consider a typical case. Figure 3.20(a) shows an optical photograph of the Orion Nebula taken with a 4-meter telescope on Kitt Peak Mountain. Figure 3.20(b) shows a contour map of the radio strength of this same Orion region; such a map depicts the type of image that radio astronomers observe toward most cosmic objects. The vertical and horizontal axes of both the optical and radio images define a plane in the celestial sky, much like latitude and longitude specify localities on Earth. By aiming a radio telescope at the nebula, radio astronomers effectively *listen* to the strength of its radio emission. Scanning the nebula back and forth permits us to measure an array of many radio signals. A map is then drawn as a series of contour lines connecting locations of equal radio strength. These radio contours are similar to pressure contours drawn by meteorologists on weather maps, or height contours drawn by geographers on mountain maps.

The inner contours usually represent relatively strong radio signals, the outside ones weak signals. We can appreciate this by looking at Figure 3.20 from another point of view. For example, Figure 3.21 shows the same radio contours of Figure 3.20(b) viewed from a different perspective. The large peak at the nebula's center represents the strongest radio signal; the small, horizontal ripples around the nebula resemble noisy static on a home radio receiver. The result is a radio image of the Orion source that appears like an island in a sea of surrounding static. The intense radio signal contains information; the weak radio static does not.

The maps of radio strength shown in Figures 3.20(b) and 3.21 are similar to the spread of visible brightness across the nebula; the light is clearly brightest near the center of the optical image of Figure 3.20(a), while declining toward the nebular edge. But there are also some subtle differences between the radio and optical images of the Orion Nebula. Figure 3.22 shows a direct superposition of Figure 3.20(b) onto Figure 3.20(a). The two maps differ mainly toward the upper left, where light seems to be absent, despite the existence of radio waves. How could radio waves be detected from locations not showing any light emis-

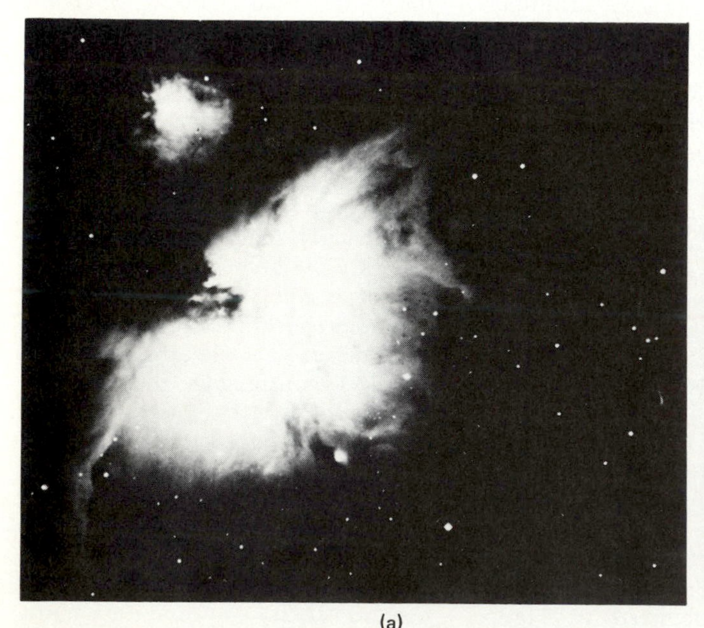

(a)

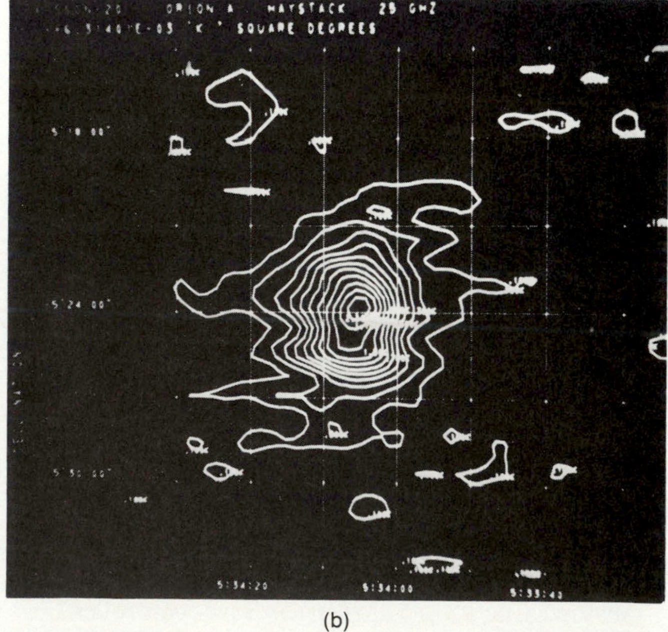

(b)

FIGURE 3.20 (a) Optical photograph of the Orion Nebula. (b) Radio contour map of the Orion Nebula, drawn to the same scale as in (a).

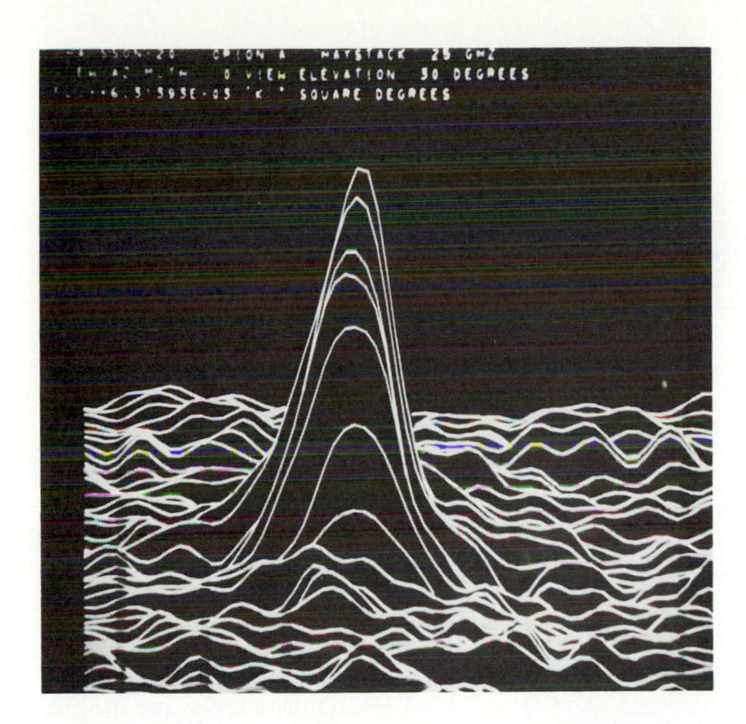

FIGURE 3.21 The same radio map as shown in Figure 3.20(b), plotted from a different point of view. A computer was used to make this map, since humans are restricted to only one perspective—the view from our home planet Earth.

FIGURE 3.22 Superposition of the radio contours of Figure 3.20(b) onto the optical photograph of Figure 3.20(a).

sion? Well, in this case, the nebular region is known to be especially dusty in its top-left quadrant. The dust obscures the short-wavelength visible radiation but not the long-wavelength radio radiation. (We shall study the nature of this dust in Chapter 12.)

Radio maps yield the true spread of a cosmic source. Optical images can be, and often are, distorted by intervening dust somewhere along our line of sight. In fact, many important objects and regions of the Universe cannot be seen at all; the very center of our Milky Way Galaxy is a prime example of a totally invisible region. Our only knowledge of such regions results almost entirely from analyses of their longer-wavelength radio and infrared emissions.

Finally, we should stress one more point: Radio telescopes normally do *not* transmit radio waves toward astronomical objects, any more than optical telescopes observe the stars with searchlight beams. Most people have the mistaken impression that radio telescopes transmit signals toward the stars. They don't. Radio astronomers just listen to the incoming radiation in much the same way that optical astronomers just look. (We'll study some minor exceptions to this in Chapters 4 and 7.)

Spectroscopy

Spectral-line radiation can be observed in the invisible as well as visible parts of the electromagnetic spectrum. Since we've been discussing radio astronomy, we'll illustrate this powerful technique for the case of radio waves. (This technique also pertains to other types of invisible radiation.)

Recall, in the case of the optical spectrometer discussed earlier, that the main device was a telescope-mounted prism that spread light and its spectral lines onto a photograph. Radio spectrometers are similar and measure the wavelength (or frequency) components of radio radiation, in analogy with the various colors of visible radiation. The result is a series of spectral-line features of radio emission or absorption much like the narrow bright or dark lines of the visible spectrum. Sophisticated radio receivers attached to a telescope can detect and measure these radio lines, thus providing much information about the nature of the emitting and absorbing matter.

Figure 3.23(a) shows schematically how a radio spectrometer works. In much the same way that we can hear commercial stations by tuning across the dial of an ordinary home radio receiver, radio astronomers can tune spectrometers to the specific wavelengths or frequencies at which individual atoms or molecules emit or absorb radiation in space. Figure 3.23(b) depicts the formation of spectral emission lines caused by electrons cascading among atomic orbitals, or molecules rotating and vibrating, as discussed in Chapter 2. Radio spectral lines are characterized by the same strength, width, and centroid properties, just like visible spectral lines.

In reality, spectral lines are not as "clean" and well defined as those shown in Figure 3.23(b). Figure 3.24 shows a better example of an actual radio emission feature; this one was observed toward Orion with a radio spectrometer attached to the Haystack telescope. Note the "noise," spread across the spectrum as small

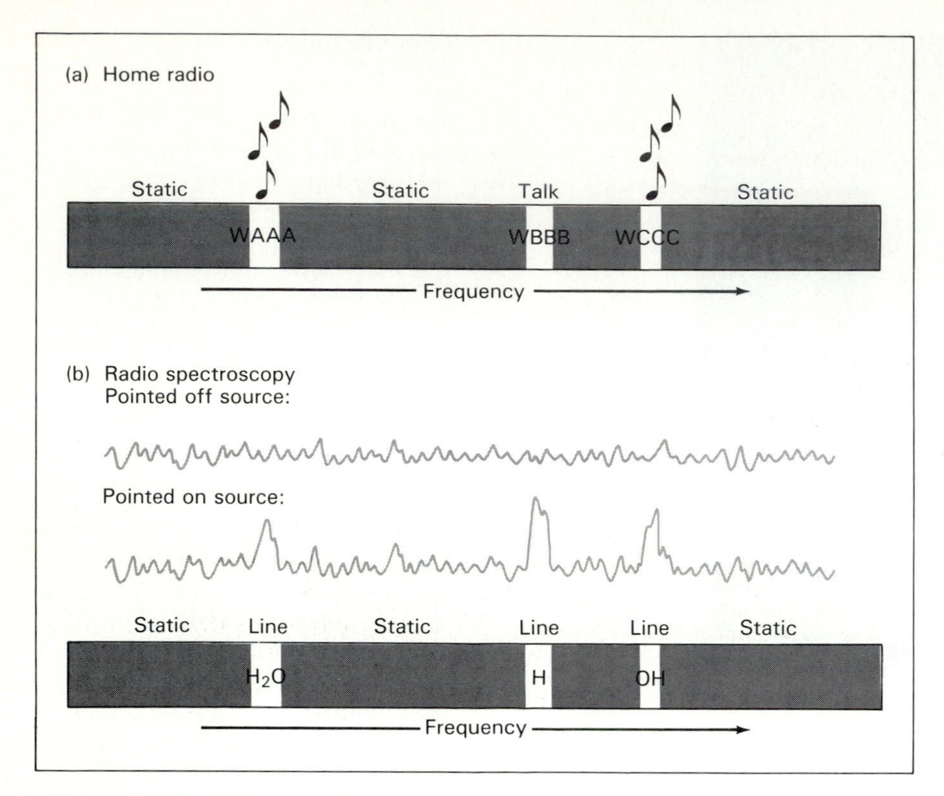

FIGURE 3.23 A home radio receiver that detects music, news, and static (a) is a simplified version of a more complex radio-astronomical spectrometer that detects cosmic spectral lines and static (b).

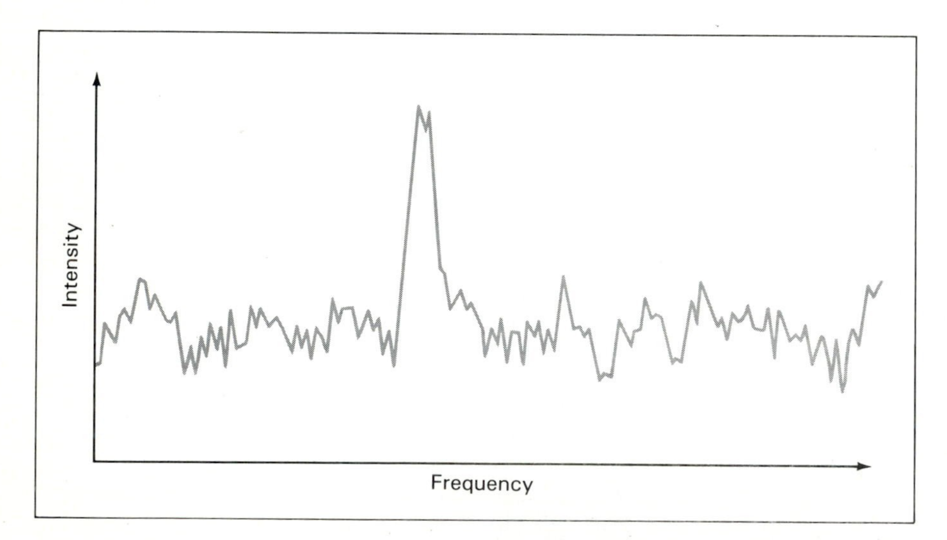

FIGURE 3.24 This radio spectral line results from the combined emissions from many hydrogen atoms in the Orion Nebula. The line is centered at 8309383000 Hertz (or 3.6104 centimeter wavelength), precisely that expected for electrons cascading from the 93rd to the 92nd orbital within hydrogen atoms.

wiggles; noise is anything that corrupts the integrity of a message, such as static on an AM radio or distortion of the picture on a TV screen. In the case of a spectral line, the noise results partly from collisions among the warm atoms and molecules inside radio receivers; it's also caused by extraneous gas along the line of sight, even in our own atmosphere. Nonetheless, the basic bell-shaped spectral feature clearly extends above the noise, as expected for an abundant gas having numerous atoms, each with a different velocity that tends to broaden the line because of the Doppler effect.

Radio spectroscopy is a relatively new science. Unlike the visible spectra of stars, which often contain literally thousands of emission and absorption lines, the radio spectra of cosmic objects displayed only a few lines until recently. Electronic advances of the 1970s, however, dramatically improved receiver technology and spectrometer sensitivity; radio and other "invisible astronomies" have greatly benefited from Space Age inventions. Currently, nearly 200 spectral lines have been observed in the radio domain. We'll discuss in Chapters 12 and 13 how radio spectroscopy of our Galaxy is now providing whole new insights literally undreamed of a decade ago.

Interferometry

Radio astronomy has become a valuable tool in our modern efforts to understand the invisible parts of the Universe. Its main disadvantage, as mentioned earlier, is lack of good angular resolution. Figure 3.25 further illustrates this primary drawback. There we show two radio maps of the *same* astronomical object, each map having been made from radiation detected by the same antenna. The different appearances of the maps result from different angular resolutions used to observe the object's radiation. In accord with our previous discussion, we can perceive more source detail with the finer resolution [Figure 3.25(b)]. Still, the resolution of a single radio telescope remains poor compared to the few arc seconds achievable with even small optical telescopes.

Not to be outdone, radio astronomers have invented ways to overcome the problem of poor resolution. By using special experimental techniques, the angular resolution of some radio maps can be greatly improved. Remarkably, this technique permits us to effectively "listen" to

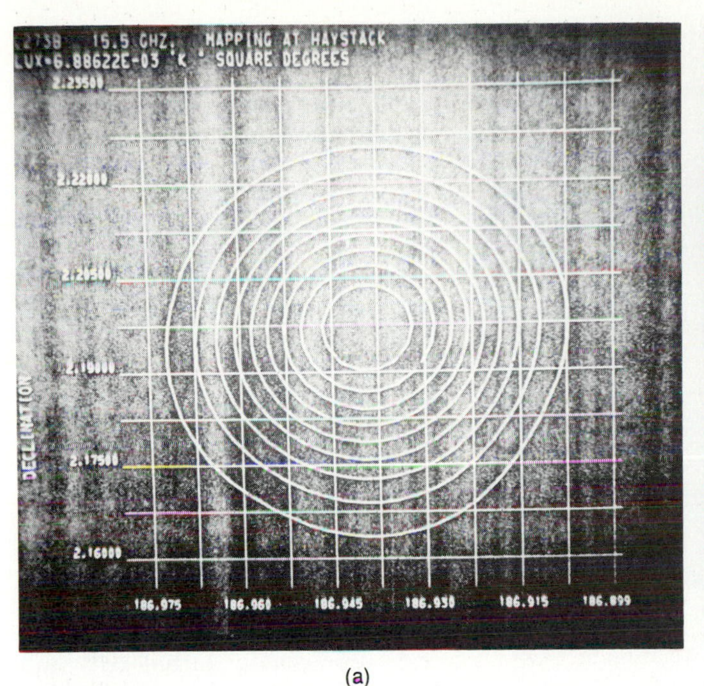

(a)

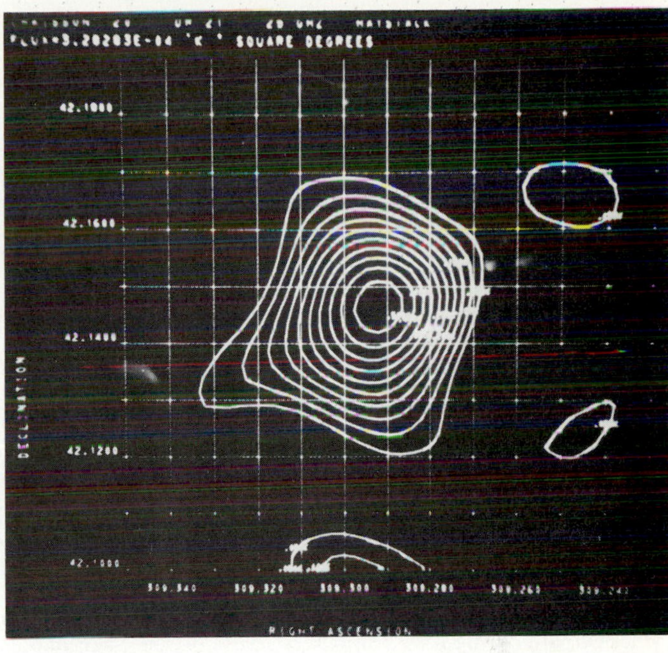

(b)

FIGURE 3.25 Comparison of two radio maps of the same cosmic object observed with the same radio telescope but at two different wavelengths. The angular resolution in (a) is 2 arc minutes, while that in (b) is 1 arc minute.

astronomical objects with much greater precision than can be seen with our telescope-aided eyes. Here's how we do it.

When two or more radio telescopes are used in tandem to observe the *same* object at the *same* wavelength and at the *same* time, the combined apparatus is called an **interferometer**. Figure 3.26 shows such an instrument—several separated radio telescopes that work together as a team. Via electronic cables or radio links, the waves received by each antenna are sent to a central computer that analyzes how the waves "interfere" when added together (hence the name of the device). If the detected waves are in step when added, they combine positively to form a strong radio signal; but if they are not in step, they destructively interfere and cancel each other. As the antennas track their target, a pattern of peaks and troughs emerges, which astronomers can use to translate into an image of the observed object.

FIGURE 3.26 This radio interferometer in Cambridge, England, is made of several dishes that are separated by 5 kilometers.

The great advantage of an interferometer is that it becomes a crude substitute for a huge antenna; the effective telescope diameter of an interferometer equals the distance between its outermost dishes. In other words, two small dishes can act as opposite ends of an imaginary but huge single radio telescope. This dramatically improves the angular resolution. For example, resolution of a few arc seconds can be achieved at typical radio wavelengths (say, 10 centimeters) either by using a single radio telescope 5 kilometers in diameter (which is quite impossible to build) or by using two or more much smaller dishes *separated* by 5 kilometers (like the one shown in Figure 3.26).

Large interferometers made of many dishes, like the instrument shown in Figure 3.27, can now rou-

FIGURE 3.27 This large interferometer is made of 27 dishes, spread along a Y-shaped pattern about 30 kilometers across on the Plain of San Augustin in New Mexico. The most sensitive radio device in the world, it is called the Very-Large-Array, or VLA for short.

tinely attain radio resolution comparable to that of optical photographs. Figure 3.28 compares such a radio map of a nearby galaxy with an ordinary photograph of that same galaxy. The radio clarity is superb and has little of the fuzziness or loss of detail noted, for example, in the radio maps of Figure 3.22 or 3.25.

The larger the distance separating the telescopes—known as the baseline—the better the resolution attainable. Scientists have now built radio interferometers over truly great distances, first across the North American continent, and then even between continents. A typical very-long-baseline-interferometry experiment (known by the acronym VLBI) might use radio telescopes, for example, in North America, Europe, Australia, and the Soviet Union (see Figure 3.29) to achieve angular resolution on the order of 0.001 arc second. This resolution is about 1000 times better than most results of conventional optical astronomy.

The technique of radio interferometry works largely because Earth's atmosphere is more transparent for radio radiation than for optical radiation. Given the nature of atmospheric turbulence (air inhomogeneities comparable in size to the wavelength of light), stars and other cosmic objects will always twinkle when viewed *optically* through our atmosphere. On the other hand, radio radiation has wavelengths much larger than typical atmospheric turbulence, and hence is unaffected by it. Angular resolution at radio wavelengths is thus limited only by the apparatus, not by our atmosphere. Radio interferometers, then, could conceivably be constructed using the largest possible baseline on Earth, namely the diameter of the entire planet—something that the American–Soviet VLBI linkup has just about done already! In fact, in a recent test, radio astronomers successfully used an antenna in orbit, together with several

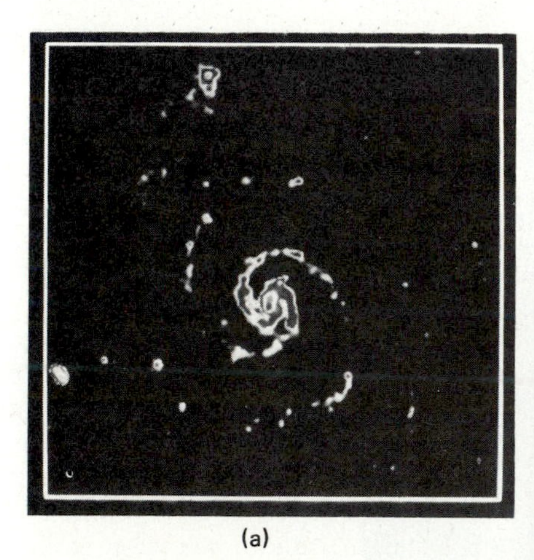

(a)

(b)

FIGURE 3.28 VLA radio "photograph" (or radiograph) of a spiral galaxy, observed *at radio frequencies* with an angular resolution of a few arc seconds (a), shows nearly as much detail as an actual (light) photograph of that same galaxy (b) made with the 4-meter Kitt Peak optical telescope.

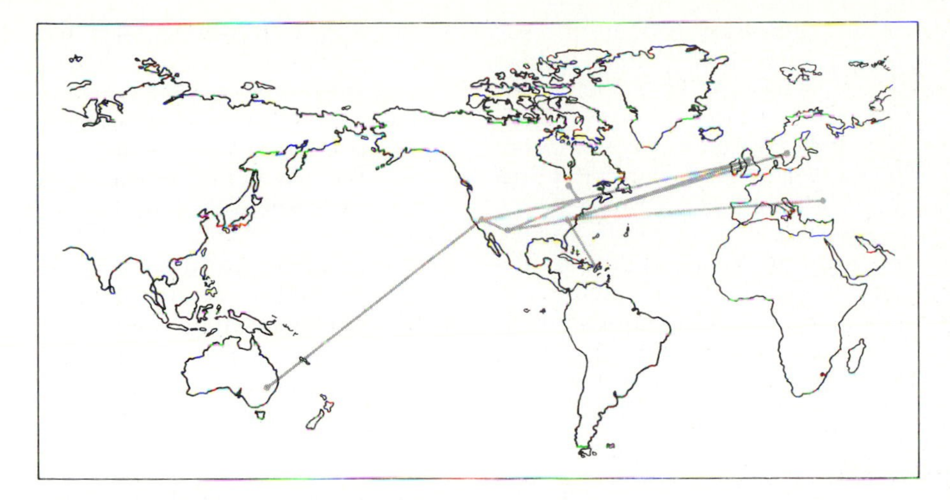

FIGURE 3.29 World-wide network of some of the radio telescopes occasionally used to form part of an enormous radio interferometer extending over intercontinental distances.

antennas on the ground, to construct a longer baseline and thus achieve yet better resolution. It seems that even Earth's diameter is no limit.

As might be imagined, interferometry also has its drawbacks. There are advantages and disadvantages associated with all experimental techniques used in science. For example, interferometers can observe spectral lines only when radiation is rather intense, their usefulness is restricted to certain parts of the radio domain, and they are less sensitive than single dishes when observing smoothly distributed, low-density gas. These problems are further complicated by the astropolitics often needed to arrange such international experiments. Accordingly, single radio telescopes are still highly valued pieces of instrumentation. They contribute daily to our storehouse of cosmic knowledge.

FIGURE 3.30 Photograph of the highest-altitude (ground-based) observatory at Mauna Kea, Hawaii. The domes house the Canada-France-Hawaii 3.6-meter telescope (left), the 2.2-meter telescope of the University of Hawaii (center), and Britain's 3.8-meter infrared facility (right). The foreground device is a 3-meter infrared telescope owned and operated by NASA.

Other Astronomies

INFRARED ASTRONOMY: Infrared techniques comprise another branch of the many modern observational tools used to aid our understanding of the Universe. Fortunately, there are a few "windows" in the high-frequency part of the infrared spectrum (see Figure 1.11) for which the atmospheric opacity is low enough to permit ground-based observations. In fact, some of the most useful infrared observing is done from the ground with conventional telescopes, although the radiation is somewhat affected by our atmosphere.

Figure 3.30 is a photograph of the world's highest ground-based observatory, perched more than 4 kilometers (about 14,000 feet) above sea level atop an extinct volcano at Mauna Kea, Hawaii. Despite the logistic problems of such a site, this observatory is used heavily throughout the year, demonstrating that astronomers are willing to go to great lengths (or heights) to get above as much of the atmosphere as possible; the thin air at high altitudes guarantees less atmospheric absorption and hence a clearer view than from sea level. Mauna Kea is one of the finest locations for ground-based infrared astronomy, but the air is so thin that astronomers occasionally wear oxygen masks while performing their tasks.

Better infrared observations can be made from above Earth's atmosphere. However, our civilization has only recently learned to hoist equipment to truly great heights, so the experimental techniques of infrared astronomy are not yet quite as advanced as those of optical and radio astronomy. Still, balloon, rocket, and satellite technology is rapidly making infrared research a powerful new tool with which to study the Universe.

Generally, infrared telescopes pretty much resemble optical telescopes, but they are sensitive to radiation of wavelength a little longer than light. As might be expected, the size of infrared gear that must be carried above the atmosphere is necessarily a good deal smaller than massive ground-based optical telescopes; it's just impractical to launch huge pieces of equipment into orbit.

Figure 3.31 shows a typical infrared telescope, this one being part of an unmanned balloon payload usually lofted to an altitude of about 30 kilometers (or 100,000 feet). The 1-meter-diameter mirror, clearly seen at the base of the payload, is one of the largest telescopes yet to get above *most* of Earth's atmosphere.

At infrared wavelengths, such a mirror provides an angular resolution of approximately 1 arc minute. Hence contour diagrams showing the spread of the invisible infrared radiation across cosmic sources often resemble those made with large radio telescopes. Figure 3.32 shows a typical infrared map, this one of the Orion Nebula made with an orbiting satellite. The contours (which are filled in here) are smooth, much like the radio observations of Figure 3.22; at nearly 1-arc-minute angular resolution, the fine details of Orion cannot be perceived. Nonetheless, useful information about this object and others like it can be extracted from such observations.

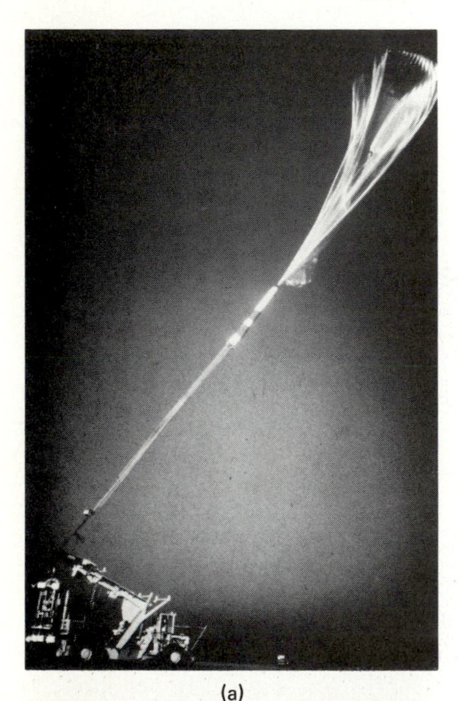

(a)

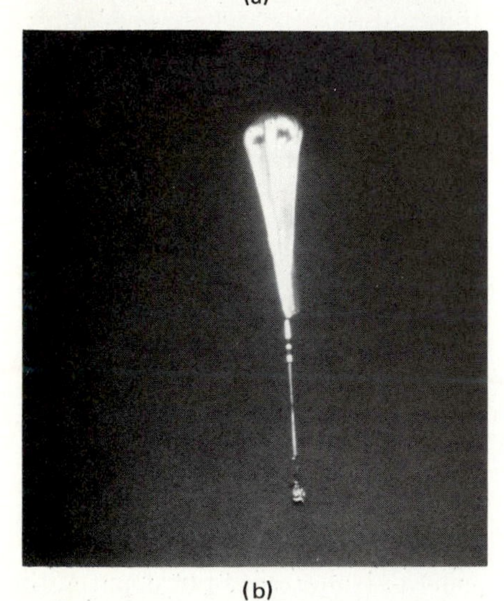

(b)

FIGURE 3.31 This balloon (a) is used to hoist a payload containing a mirror high above most of Earth's atmosphere. The small package below the balloon shown rising into the atmosphere (b) comprises an infrared telescope.

FIGURE 3.32 This infrared image of the Orion Nebula and its surrounding environment was made by the *Infrared Astronomy Satellite.* Here, the brighter regions denote greater strength of infrared radiation. (The scale of this figure is considerably larger than that of Figure 3.22; the dark spike at the top is an instrumental flaw.)

A clear advantage of infrared radiation is its longer-than-optical wavelength. The longer wavelength often allows better "seeing," enabling us to perceive objects partially hidden to optical view. In this sense, infrared radiation can provide information about invisible cosmic objects, although not quite as well as can even longer-wavelength radio radiation. Figure 3.33 exemplifies this added insight gained by studying infrared images recorded on special photographic film. Shown there is a dusty and hazy region in California, hardly viewable optically, but better perceived by detecting infrared radiation.

The most advanced facility to function in this part of the spectrum is the *Infrared Astronomy Satellite*, called *IRAS* for short. Launched into Earth orbit in 1983 but now inoperative, this British–Dutch–U.S. satellite houses a 0.6-meter mirror that focuses the incoming infrared radiation with an angular resolution as fine as 30 arc seconds, and occasionally finer. During its 10-month lifetime, *IRAS* contributed especially well to our knowledge of clouds of galactic matter that seem destined to become stars (and possibly planets)—regions composed of warm gas that can be neither seen

FIGURE 3.33 An infrared "photograph" taken near San Jose, California (right) is here compared with an optical photo of the same area taken at the same time (left).

with optical telescopes nor adequately studied with radio telescopes. Throughout the text we shall encounter many recent findings made with this satellite, be they observations of comets, stars, and galaxies, or just dust and rocky debris scattered across relatively uncharted realms of the Universe. Provided that such objects have some heat, they are sure to "glow" in the infrared. [Plates 2(b) and 4(b) are examples that we shall study later.]

ULTRAVIOLET ASTRONOMY: To the short-wavelength side of the visible spectrum lies the ultraviolet domain, only relatively recently explored, owing to advances in space astronomy. Since Earth's atmosphere is totally opaque shortward of about 4000 angstroms, no ultraviolet observing can be done from the ground, not even from the highest mountaintop. Rockets, balloons, or satellites are a necessary part of any **ultraviolet telescope**, which is a device sensitive to radiation of wavelength a little shorter than light.

One of the most successful ultraviolet space missions is the *International Ultraviolet Explorer*, called *IUE* for short. Depicted in Figure 3.34, this satellite was inserted into Earth orbit in the late-1970s and is still functioning as designed. A truly collaborative effort, several hundred astronomers from the United States and Europe have used *IUE* to explore a variety of phenomena in planets, stars, and galaxies, especially the surprisingly hot gases in the atmospheres of the latter two types of objects. Spectroscopic sensitivity is particularly good between 1000 and 3000 angstroms, where many superheated gases emit characteristic spectral features. At several places in the text, we shall learn how this relatively "new window" on the Universe has taught us much about the activity and even violence that seems to pervade the cosmos. (For more information regarding this remarkable ultraviolet-sensing satellite, consult Interlude 12-1.)

HIGH-ENERGY ASTRONOMY: Not to be outdone by radio, infrared, optical, and ultraviolet astronomy, scientists have also used modern equipment to develop the subject of high-energy astronomy. How can we detect radiation of very short wavelengths? First, high-energy radiation must be captured high above Earth's atmosphere. Second, we need to use equipment that is basically different from that used to capture the low-energy radiation studied to this point in this chapter.

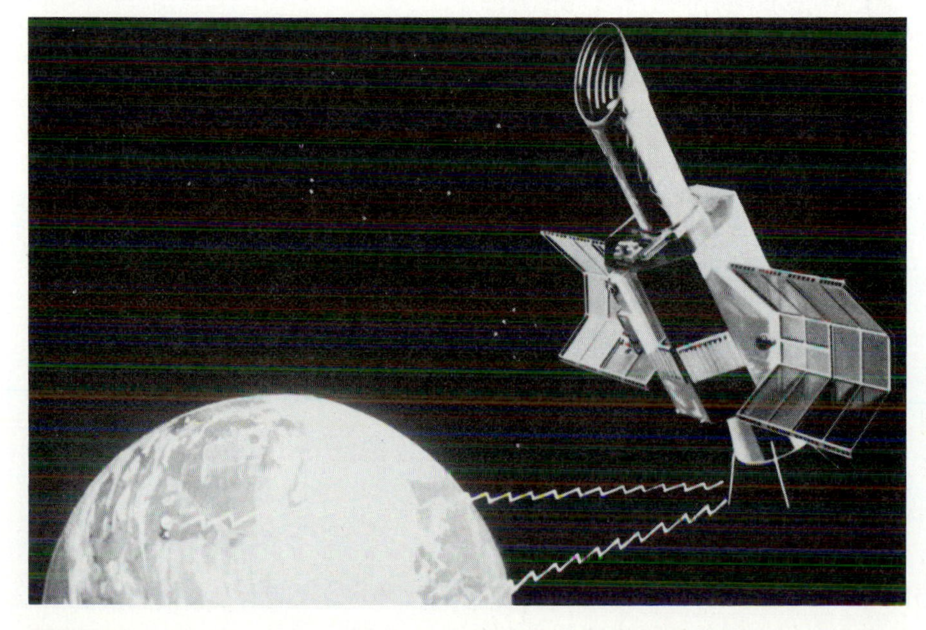

FIGURE 3.34 Artist's conception of the *International Ultraviolet Explorer* spacecraft in Earth orbit some 36,000 kilometers high above our atmosphere. The large panels on each side of the craft are solar collectors that gather energy from the Sun to power the satellite. The 0.5-meter-diameter telescope itself is housed in the tube that helps focus the short-wavelength radiation onto a variety of detectors at the core of the satellite. Resolution is usually a few arc seconds.

High-energy radiation—x-ray and gamma-ray radiation—cannot be easily reflected by any kind of surface; cosmic x-rays normally go straight through lenses or mirrors, just as they do in medical x-rays of our bodies. Normally, curved dishes cannot be made smooth enough to reflect radiation of such extremely small wavelength, often less than 100 angstroms. Methods akin to photography do not work well. Instead, individual x-ray and gamma-ray photons are counted by electronic devices called **high-energy telescopes**, flown in satellites or hoisted for shorter flights above the atmosphere by rockets or balloons. The information detected is then converted into radio signals by on-board equipment, transmitted through the atmosphere and received by conventional radio telescopes on the ground, and finally transferred into electronic signals that can be viewed and analyzed on a video screen.

Remember, the phrases "high-*frequency*" domain and "high-*energy*" domain of the electromagnetic spectrum are synonymous. As we discussed in Chapter 1, a photon's energy increases in direct proportion to the frequency of the radiation. However, the number of photons in the Universe seems to be inversely related to frequency; while billions of visible (starlight) photons arrive at Earth each second, hours or even days are often needed for a single gamma-ray photon to be recorded. Needless to say, the highest-energy photons are treated with a lot more respect than are their run-of-the-mill low-energy counterparts.

Figure 3.35 shows an instrument package designed to detect x-ray radiation. Now in Earth orbit, but inactive owing to lack of battery power, this high-energy telescope is on board the *Small Astronomy Satellite #1* (*SAS-1*). The honeycomb

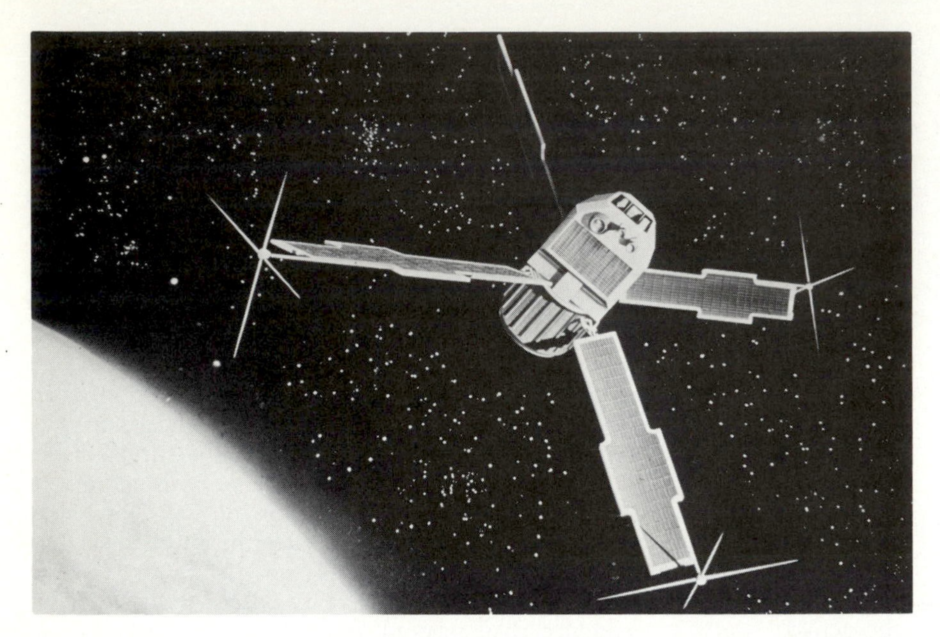

FIGURE 3.35 An artist's conception of the *Small Astronomy Satellite #1* placed into near circular orbit by NASA in the early 1970s in order to study cosmic objects that emit high-energy radiation. Since the rocket launch occurred from a sea-going platform off the coast of Kenya, which was then celebrating its independence, astronomers nicknamed this first x-ray satellite Uhuru—which means "freedom" in Swahili.

structure on the front (looking like a car radiator) serves to reject all x-ray photons except those coming nearly straight at the detector. This is how the telescope points toward a field of view which, for *SAS-1*, equals a few arc degrees across. However, because the position of the satellite could be accurately controlled by radio commands sent by engineers on the ground, cosmic sources of x-ray emission were scanned and studied to an accuracy of a few arc minutes. With a series of messages sent daily to the satellite, this orbiting telescope was able to scan the entire sky in about 2 months.

Continuous detection and counting of photons yields the positions and strengths of x-ray objects. Observations also check for time variations of the x-ray strength. And by employing electronic devices capable of filtering out most cosmic photons, satellites like this one can provide information about photons within narrow parts of the x-ray spectrum—crude spectroscopy of sorts. All this x-ray information is then radioed to the ground, where it's analyzed and examined for clues about the nature of the high-energy cosmic sources.

Toward the end of the 1970s, a new generation of x-ray and gamma-ray telescopes was launched into Earth orbit. Called the *High-Energy Astronomy Observatories*, or *HEAO* for short, these spacecraft have made startling advances concerning the nature of high-energy phenomena throughout the Universe. Having greater accuracy and sensitivity than all earlier high-energy satellites, these advanced spacecraft have nearly done for x-ray astronomy what the largest optical and radio telescopes normally do for longer-wavelength radiation.

Figure 3.36 is a photograph of the *HEAO-2* spacecraft, renamed the *Einstein Observatory* in honor of the birth centenary of the great scientist in 1979, the year the satellite began scientific work. Although the collecting diameter of *Einstein* is only 0.6 meter, the short wavelength of

(a)

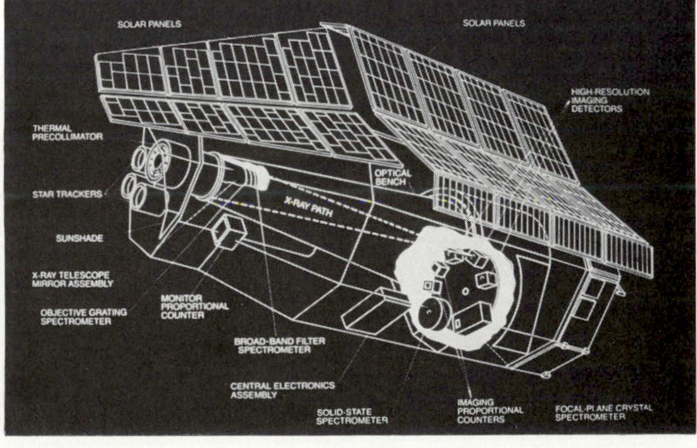

(b)

FIGURE 3.36 The x-ray telescope (a) onboard the *Einstein Observatory* is the largest ever built. The basic features of the telescope's construction are sketched in (b). Note how a revolving turret can position a variety of detectors and spectrometers to capture cosmic x-rays that graze along the sides of the variably curved and highly polished metal, gently reaching a focus where they are detected.

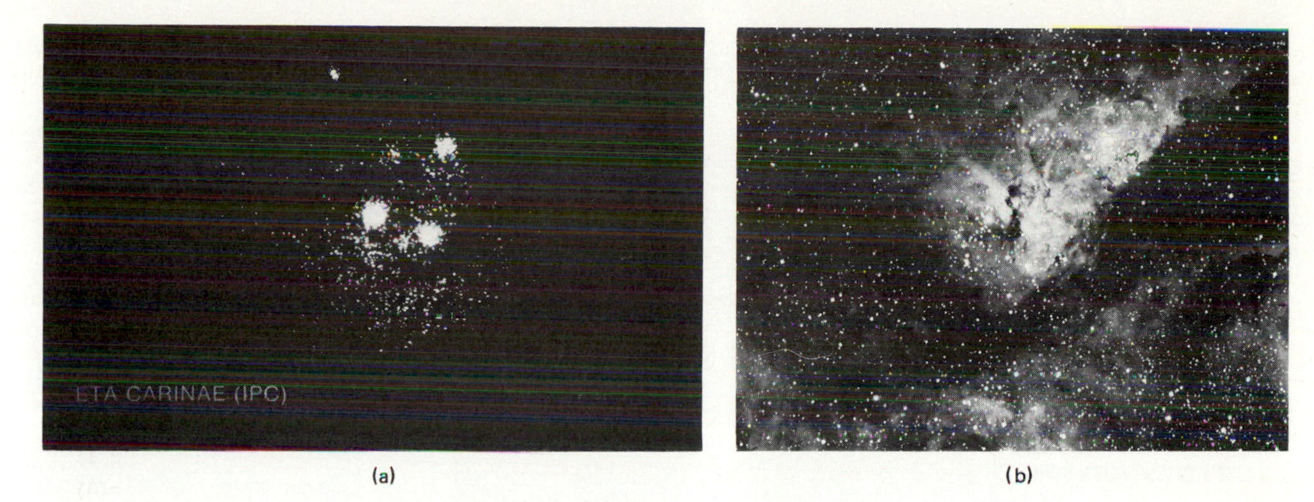

FIGURE 3.37 This x-ray image (a) of the Eta Carina Nebula [shown visually in (b)] highlights the nebula's hottest stars. By contrast, little of the nebula's gas detectable at optical, radio, or infrared wavelengths is hot enough to emit x-rays.

the x-rays makes its angular resolution a fine 3 arc seconds. Accordingly, this spacecraft can observe images of whole fields of view, much like that of ordinary photographs. Figure 3.37 is such an x-ray image, this one showing some of the many hot stars in and around the Eta Carina Nebula, a region much like the Orion Nebula discussed earlier in this chapter. In this "x-ray," the hottest stars stand out most clearly, for the Planck curve of their emission peaks well up in the high-energy domain. [Plates 6(a) and 8(c) are other examples to be studied later.]

One of the three *HEAO* spacecraft now in Earth orbit was designed to do some primitive x-ray spectroscopy. However, this instrument lacks sensitivity, and can collect photons of only strong emission lines from the brightest sources. At any rate, all but one of these high-energy satellites (*IUE*) either have run out of the gas needed to orient themselves or have scanned the skies to the limits of their sensitivities.

As we proceed through the text, we'll discuss more fully the wealth of new, and sometimes bizarre information that these and other high-energy satellites have provided. And we shall come to know some additional, prominent pieces of scientific apparatus.

SUMMARY

In this chapter we have introduced some of the basic techniques and equipment used by astronomers to study the Universe. Besides the familiar optical telescopes that collect light from cosmic objects, many other tools are needed to capture invisible radiation emitted by a variety of celestial sources. Sometimes, these "invisible astronomies" are crucial in our study of objects that are totally obscured from view. Radio astronomy was somewhat emphasized in this chapter because it is the oldest of the nonvisible subjects.

All the equipment together comprises an arsenal of devices available to modern astronomers. The electromagnetic spectrum is so large that it's unrealistic to expect to be able to use any one piece of equipment to detect radiation of all wavelengths. Different tools address different objectives. And some do the job better than others. In the end, they all supplement one another, helping to accumulate an ever-growing store of basic astronomical knowledge about our richly endowed Universe.

KEY TERMS

angular resolution
arc degree
arc minute
arc second
collecting area
high-energy telescope
infrared telescope
interferometer
optical telescope
prime focus
radio telescope
reflector
refractor
ultraviolet telescope

QUESTIONS

1. Cite two technical reasons that astronomers are constantly campaigning for larger telescopes.
2. Which type of telescope—radio or optical—is more advantageous for observing a cold (about 50 kelvin) interstellar cloud, and why?

3. Compare and contrast the capture of radio waves and x-rays by modern telescopes. Feel free to illustrate your answer, if helpful.
4. Draw a labeled sketch of an x-ray telescope, showing in particular how high-energy photons are captured.
5. Our eyes perceive visible radiation with an angular resolution of 1 arc minute. What might it be like if our eyes detected only infrared radiation with 1-arc-*degree* angular resolution? Would we be able to make our way around on Earth's surface? to read? to sculpture? to create a technology?
6. Why is it that all the world's largest optical telescopes use the reflector design?
7. What is the main reason that our technology allows us to construct radio telescopes much larger than their optical counterparts?
8. Compare and contrast the relative advantages and disadvantages of the Arecibo and Haystack radio telescopes.

*9. Verify that the spectral line shown in Figure 3.24 at a frequency of 8,309,383 kilohertz has an equivalent wavelength of 3.6104 centimeters.

10. Consult your library for information about the *Hubble Space Telescope*, the permanent observatory placed into Earth orbit by NASA. What are its main advantages? Does it have any disadvantages compared to ground-based telescopes?

FOR FURTHER READING

BAHCALL, J., AND SPITZER, L., "The Space Telescope." *Scientific American*, July 1982.

BINNING, G., AND ROHRER, H., "The Scanning Tunneling Microscope." *Scientific American*, August 1985.

*BURBIDGE, G., AND HEWITT, A. (eds.), *Telescopes for the 1980s*. Palo Alto, Calif.: Annual Reviews, 1981.

GIACCONI, R., "The Einstein X-Ray Observatory." *Scientific American*, February 1980.

KRISTIAN, J., AND BLANKE, M., "Charge-Coupled Devices in Astronomy." *Scientific American*, October 1982.

PHILLIPS, T., AND RUTLEDGE, D., "Superconducting Tunnel Detectors in Radio Astronomy." *Scientific American*, May 1986.

TUCKER, W., *The Star Splitters*, NASA SP-466. Washington, D.C.: U.S. Government Printing Office, 1984.

TUCKER, W., AND TUCKER, K., *The Cosmic Inquirers*. Cambridge, Mass.: Harvard University Press, 1986.

4 THE SIZE AND SCALE OF THINGS

Living in the Space Age, we often find it hard to imagine that Earth is anything but spherical. A quick glance at views of Earth taken from space, like the one shown in Figure 4.1, leaves no doubt in our minds that our planet is round. Yet there was a time, not too long ago, when our ancestors maintained that the Earth is flat. Indeed, until the beginning of the fifteenth-century Renaissance, most people believed Earth to be flat. There were some exceptions, however. These were people schooled in **geometry**, the study of the size, shape, and scale of things.

Geometers of old, at least as far back as ancient Greece, and probably including the Sumerians and Babylonians before that, realized that Earth, as well as many other cosmic objects, was spherical. Paintings on cave walls imply that even some people of the Stone Age may have had some idea of the true shape of our planet. We can't be sure that they recognized Earth to be a near-perfect sphere. But they surely had a strong interest in the phenomena of the skies.

FIGURE 4.1 A photograph of Earth, taken from an *Apollo* spacecraft, some 40,000 kilometers out in space.

The learning goals for this chapter are:

- to realize that some ancient Greeks knew the size and shape of Earth, despite the popular belief until Renaissance times that our planet is flat
- to understand the technique of triangulation, whereby we can indirectly estimate the distance to an object too far away or too inconvenient to measure directly
- to understand how astronomers measure the distances of the nearby stars
- to know the experimental method used to determine the distance from Earth to the Sun
- to appreciate the simple geometrical principles that form the foundation of the distance scale for the entire Universe

Early Notions

Aristotle, a philosopher of the fourth century before Christ, proposed one of the earliest and clearest geometrical arguments about our planet. He noticed that during a **lunar eclipse**, when Earth is positioned between the Sun and the Moon, our planet casts a *curved* shadow onto the surface of the Moon. Figure 4.2 shows a series of photographs taken during a recent lunar eclipse; clearly Earth's shadow, projected onto the Moon's surface, is curved. This is essentially what Aristotle must have seen and recorded 24 centuries ago.

On the basis of his observation,

FIGURE 4.2 In this series of photographs taken of the Moon during a lunar eclipse, Earth's shadow, shown projected on the Moon, is curved. Not only is it curved, but that curvature seems to be part of a circle, or sphere.

Aristotle theorized that all lunar eclipses would show that Earth's shadow is always curved, regardless of the orientation of our planet. Furthermore, since he argued that this observed shadow is always an arc of the same circle, he was able to conclude that Earth must be round.

In drawing this conclusion, Aristotle made a basic assumption. He presumed that the geometrical relationships of terrestrial objects with which he worked in his everyday experience—balls, sticks, planes, triangles—also held true for much larger objects with which he was not familiar. That is, he assumed that the terrestrial geometry he knew, namely **Euclidean geometry**, could be extrapolated into the astronomical domain. This is the same type of simple, straightforward geometry taught to all of us in high school.

For example, Aristotle knew well the shape of the shadow cast by a round stone. Regardless of the orientation of a round stone, its shadow is always curved. Guided by his assumption that the relationship between the shape of an object and the shape of its shadow is independent of the size of an object, he correctly reasoned that Earth is curved and not flat. (Aristotle was not the first person to argue that Earth is round, but he was apparently the first to offer a proof of it using the lunar-eclipse method.)

Aristotle further argued that if Earth were really round, then distant stars in the nighttime sky should appear at different positions to observers at different latitudes. He quoted the experience of travelers as confirmation of his theory, especially the then well-known fact that a bright star named Canopus is visible in Egypt but not farther north.

The reasoning procedure used by Aristotle forms the basis of all scientific inquiry today. He first made an observation. He then formulated a theory. And finally, future testing proved the validity of the theory. *Observation, theory, testing*—these are the cornerstones of the scientific method, a technique to which we will constantly refer, at least implicitly, throughout all aspects of the text.

Although pioneered by the ancient Greeks, the scientific method was not widely used until Renaissance times. Since then it has formed the basis of a standard, methodical, and logical procedure used by all modern scientists as they strive to investigate the nature of matter and life.

Earth Dimensions

Not until about 200 B.C. did another Greek by the name of Eratosthenes use the ideas of Aristotle in order to measure the physical size of the Earth. It's useful to consider how he did this.

Eratosthenes knew that while some observers noted the Sun directly overhead at a certain location on a certain day of the year, other people at other locations observed the Sun at some angle displaced from direct overhead. In particular, he knew that on the first day of summer, the Sun passed *directly* overhead at Syene, Egypt; at midday, the sunlight cast no shadows whatsoever, as shown in Figure 4.3. He then measured the angular displacement of the Sun from the overhead to be 7.2 arc degrees at Alexandria, a city 5000 stadia to the north. (The stadium is a Greek unit of length, equal to about 0.16 kilometer.)

Now, if Euclidean geometry is valid, rays of light from a very distant object such as the Sun can be assumed to travel parallel to one another. Consequently, as shown in Figure 4.3, the angle measured at Alexandria between the Sun's rays and the overhead is also proportional to the small portion of Earth's total circumference between Syene and Alexandria.

Since there are 360 arc degrees in a full circle, we easily note that 7.2 arc degrees is $\frac{1}{50}$ of a full circle. The entire Earth's circumference can then be estimated by multiplying the distance between the two cities by a factor of 50. Consulting Figure 4.3 again, we can also express this reasoning in the form of an analogy:

7.2 arc degrees is to 360 arc degrees as 5000 stadia is to ?

Since 360 equals 50 times 7.2, the

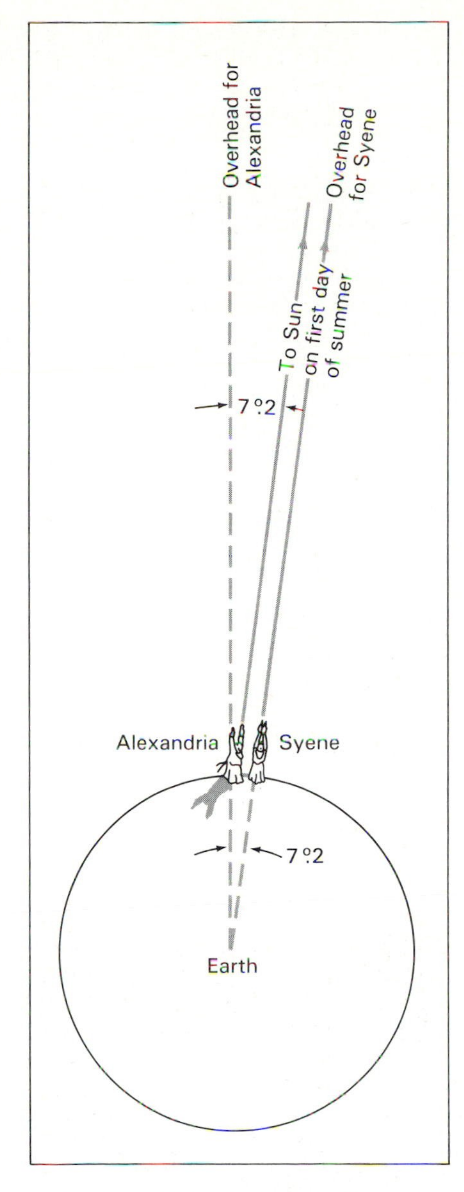

FIGURE 4.3 When the Sun is directly overhead at Syene, it subtends an angle of 7.2 arc degrees relative to the vertical at Alexandria. Simple geometry then shows that the "angular separation" between the two cities is 7.2 arc degrees.

correct answer is 50 × 5000 stadia or 250,000 stadia. Furthermore, since a stadium equals 0.16 kilometer, we can understand how Eratosthenes was able to estimate the full circumference of Earth to be, in modern units, about 40,000 kilometers. The correct value, now measured accurately by orbiting spacecraft, is 40,030 kilometers.

Eratosthenes' reasoning was a remarkable achievement, especially when considering it was done more than 20 centuries ago. His arguments enabled him to estimate the circumference of planet Earth to within 1 percent accuracy. Even more remarkable, he accomplished it using only simple Euclidean geometry—the geometry of terrestrial familiarity.

But Eratosthenes did not stop there. He went one step further because he knew that the ratio of the circumference of a circle to its diameter is π, a constant number equal to 3.1416. Solving for Earth's radius,

$$\text{radius} = \frac{\text{circumference}}{2\pi},$$

he estimated Earth's radius to be 6350 kilometers, also accurate to within 1 percent of the best modern value of 6371 kilometers.

Be sure to realize that the accomplishments of Aristotle, Eratosthenes, and other ancient Greek geometers are based solely on the principles of Euclidean geometry, a mathematical science developed by Euclid more than 2000 years ago. It's the geometry of "flat space"—a commonsense geometry that demands that a flat plane has a length and a width. It's also a geometry that specifies that any triangle has three angles whose sum equals 180 arc degrees, that any circle has a circumference equal to a product of π and its diameter, and that the shortest distance between any two points is a straight line. (Consult Interlude 4-1.)

Triangulation

TERRESTRIAL APPLICATIONS: The method of Eratosthenes relies on the same principles used in modern-day **triangulation**. Today's engineers, especially surveyors, use these age-old ideas to measure indirectly the distance to some faraway object.

For example, imagine trying to measure the distance to a tree on the other side of a river. The most direct method is to lay a tape across the river, but that's not the simplest way. A smart surveyor would make the measurement by visualizing an *imaginary* triangle. He could do this while sighting the tree on the far side of the river from two positions on the near side. Since the simplest possible triangle is a right triangle, where one of the angles is exactly 90 arc degrees, it's often advantageous to set up one observing position directly opposite the object, as at point *A* of Figure 4.4. The surveyor then paces off toward another observing position at point *B*, noting the distance covered between points *A* and *B*. This distance between points *A* and *B* is called the **baseline** of the imaginary triangle. Finally, the surveyor sights toward the tree whose distance is to be measured and notes the angle subtended at point *B*. No further observations are required. The rest of the problem involves calculations. Knowing the value of one side and one angle of the right triangle, the surveyor can then geometrically construct the remaining sides of the triangle, and thereby establish the distance to the tree on the opposite side of the river.

To accomplish such an experiment in reality, surveyors must have some familiarity with trigonometry, the mathematics of geometrical angles. But we need not bother with trigonometry here, or even to be

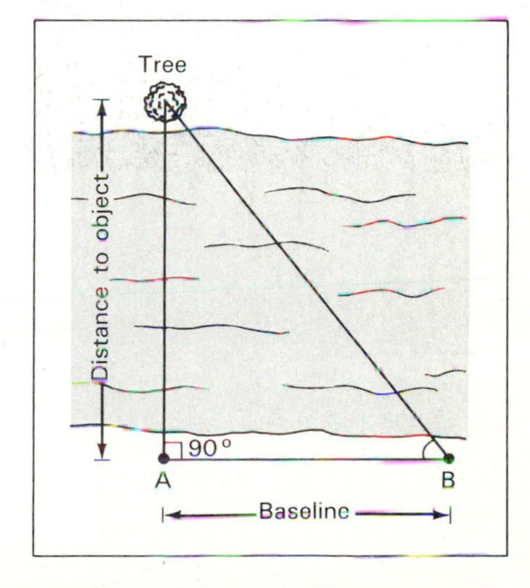

FIGURE 4.4 Surveyors often use simple geometry and trigonometry to estimate the distance to an object on the opposite side of a river.

INTERLUDE 4-1 ***Curved Space***

Euclidean geometry is the geometry of flat space—the kind now taught in high schools everywhere. It's the geometry of terrestrial familiarity, the geometry of everyday experience. Houses are usually built with straight (flat) floors. Writing tablets and blackboards are also flat. We understand how to work best with straight objects. That's because the straight line is the shortest distance between any two points; it's the path that light takes in its travels. This shortest path is called a geodesic.

In constructing houses or any other straight-walled buildings on the surface of Earth, the other basic axioms of Euclid's geometry would also apply: Parallel lines never meet even when extended to infinity; the angles of any triangle always sum to 180 arc degrees; the circumference of a circle equals π times its diameter. If these rules were not obeyed, the four walls and roof would never meet to form a house!

In reality, the geometry of Earth's surface is not really flat. It's curved. It's obviously curved; we live on the surface of a sphere. And on the *surface* of such a sphere, Euclidean geometry breaks down. It's wrong. Instead, the rules of a "positively curved" sphere are governed by the laws of Riemann geometry, named after a nineteenth-century German mathematician. For example, there are no parallel lines (or curves) on a sphere's surface; any lines drawn on the surface and around the full circumference will eventually intersect. (Lines of latitude are not parallel, as they do not cut through the sphere's center.) The sum of a triangle's angles, when drawn on the surface of a sphere, exceeds 180 arc degrees. And the circumference of a circle is less than π times its diameter.

So the curved surface of a sphere is governed by the spherical geometry of Riemann. It differs very much from the flat-space geometry of Euclid. These two geometries are approximately the same only at a small part of the sphere—at any locality where a flat plane can be laid tangent to a curved surface. Otherwise, the two geometries differ considerably. In other words, at any one location, say a small house sitting on the surface of a large sphere, Euclidean geometry is valid. There, simple Euclidean geometry is a good approximation to complex Riemann geometry—but only at individual, localized positions.

When working with larger parts of Earth, on the other hand, Euclidean geometry must be abandoned. World navigators are fully aware of this. Aircraft do not fly along what you might regard as a straight-line path from one point to another; instead they follow a "great circle" which, on the curved surface of a sphere, *is* the shortest distance between two points. For example, a flight from Boston to London does not proceed across the Atlantic Ocean as you might expect from looking at a flap map. Such a flight goes far to the north, over Newfoundland, near Iceland, almost to the

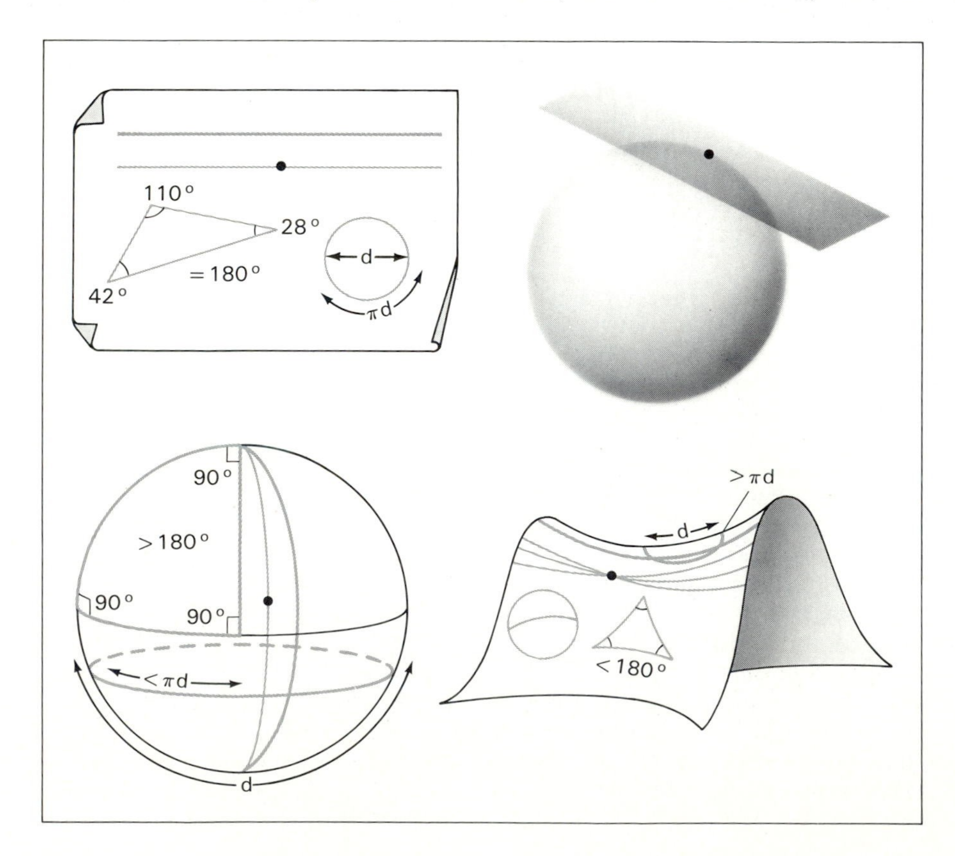

Arctic Circle, finally coming in over Scotland for a landing at London. This is the shortest path between Boston and London on the curved surface of the Earth; this great circle is the geodesic. It is, in fact, the path that light radiation would follow in curved space.

The positively curved space of Riemann is not the only kind of possible departure from flat space. Another type of geometry is the "negatively curved" space first studied by Lobechevsky, a nineteenth-century Russian mathematician. In this geometry, there are an infinite number of parallel lines, the sum of a triangle's angles is less than 180 arc degrees, and the circumference of a circle is greater than π times its diameter. Instead of the surface of a flat plane or a curved sphere, this type of space can be described by the *surface* of a curved saddle. It's admittedly a hard geometry to visualize.

The important point to realize here is that humans are prejudiced toward the simpler Euclidean geometry because it works well in small, localized areas with which we are familiar. The local realm of almost any object on Earth's surface is small compared to Earth's overall size. When venturing outside that local realm, however, the geometry of terrestrial familiarity falls apart, and we must resort to more sophisticated mathematics. What is fine for an architect, a carpenter, or even a chemist working in a laboratory, is not fine for an aircraft pilot, a sea captain, or an astronomer observing the Universe.

Most of the local realm of the Universe (including the Solar System, the neighboring stars, and even our Milky Way Galaxy) are correctly described by Euclidean geometry. To be sure, that's what we used in this chapter. But as we push out to greater realms of the Universe, Euclidean geometry must be abandoned. The description of the size and shape of our local realm discussed in this chapter will remain valid, but only because it's a minute niche within the grander Universe. As we shall see in Chapter 16, more advanced, non-Euclidean, curved-space geometries will be needed to describe and model the full expanse of the entire Universe.

able to make such a calculation. In fact, the whole problem could be solved on a grid, as shown in Figure 4.5, without knowing any trigonometry at all. Pacing off the baseline, letting it equal so many intervals on a piece of graph paper, and setting the angle at point *B*, we can draw the complete triangle. The number of graphed intervals from point *A* to the tree then approximates the desired distance.

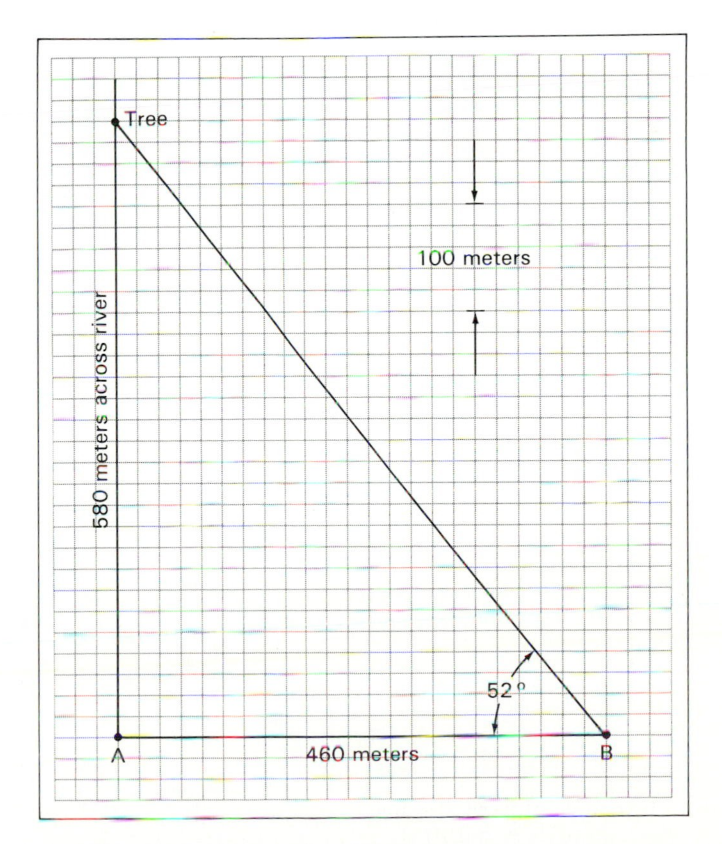

FIGURE 4.5 Knowledge of trigonometry is not even needed to estimate distances indirectly. Scaled estimates on a piece of graph paper often suffice.

One further point. A right triangle is unnecessary to estimate distance in this way. Generally, we could use any triangle, although an additional measurement is needed when one of the angles of the triangle does not equal 90 arc degrees. In other words, all the sides and all the angles of a triangle can be indirectly determined simply by geometry, provided that we know either two sides and the included angle, or two angles and the included side.

The upshot of the foregoing discussion is this: Nothing more complex than simple geometry is needed to infer the distance, size, and shape of some object too far away or too inconvenient to measure directly.

We stress again that this type of modeling assumes that the geometry of *real* terrestrial objects, rivers and trees, for example, mimics the geometry of much smaller *scale* models, like those drawn on paper in Figures 4.4 and 4.5. That is, we assume that the principles of geometry are independent of an object's size. This is the assumption used centuries ago by the Greek geometers. And it's the assumption that we'll continue to make now as we extend the imaginary triangle to cover larger spaces, and thus measure larger distances to extraterrestrial objects.

EXTRATERRESTRIAL APPLICATIONS: Triangles with larger baselines are needed if we are to measure greater distances. Figure 4.6 shows a triangle having a fixed baseline between two observing positions at points *A* and *B*. Note how the triangle becomes narrower as an object's distance becomes progressively greater. Narrow triangles cause problems, since the angles at points *A* and *B* are hard to measure accurately. The measurements can be made easier by "fattening" the triangle—in other words, by lengthening the baseline.

Consider an imaginary triangle extending from Earth to a nearby object in space. (It might be the Moon or a neighboring planet.) Figure 4.7(a) illustrates this case where Earth's diameter, measured from point *A* to point *B*, is the baseline. Two observers could, in principle, sight the object from opposite sides of the Earth and thus measure the triangle's angles at point *A* and point *B*. But in practice, these angles cannot be easily measured with any de-

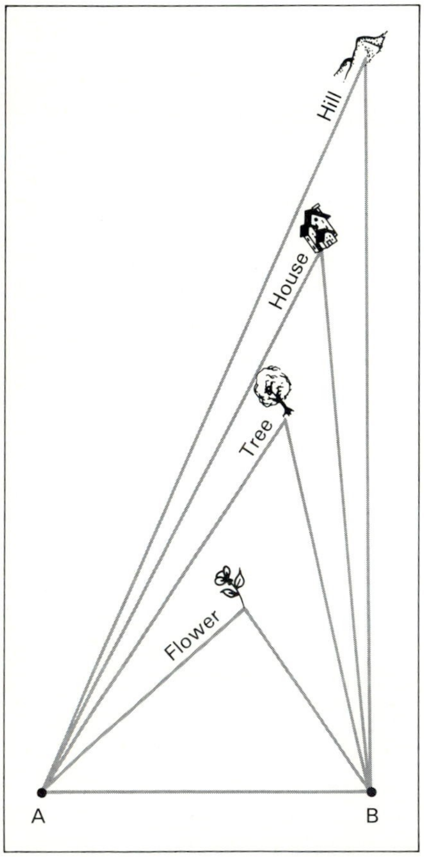

FIGURE 4.6 A triangle of fixed baseline (distance between points A and B) is more narrow for far-away objects. As shown here, the imaginary triangle is a lot thinner when estimating the distance to a remote hill compared to a nearby flower.

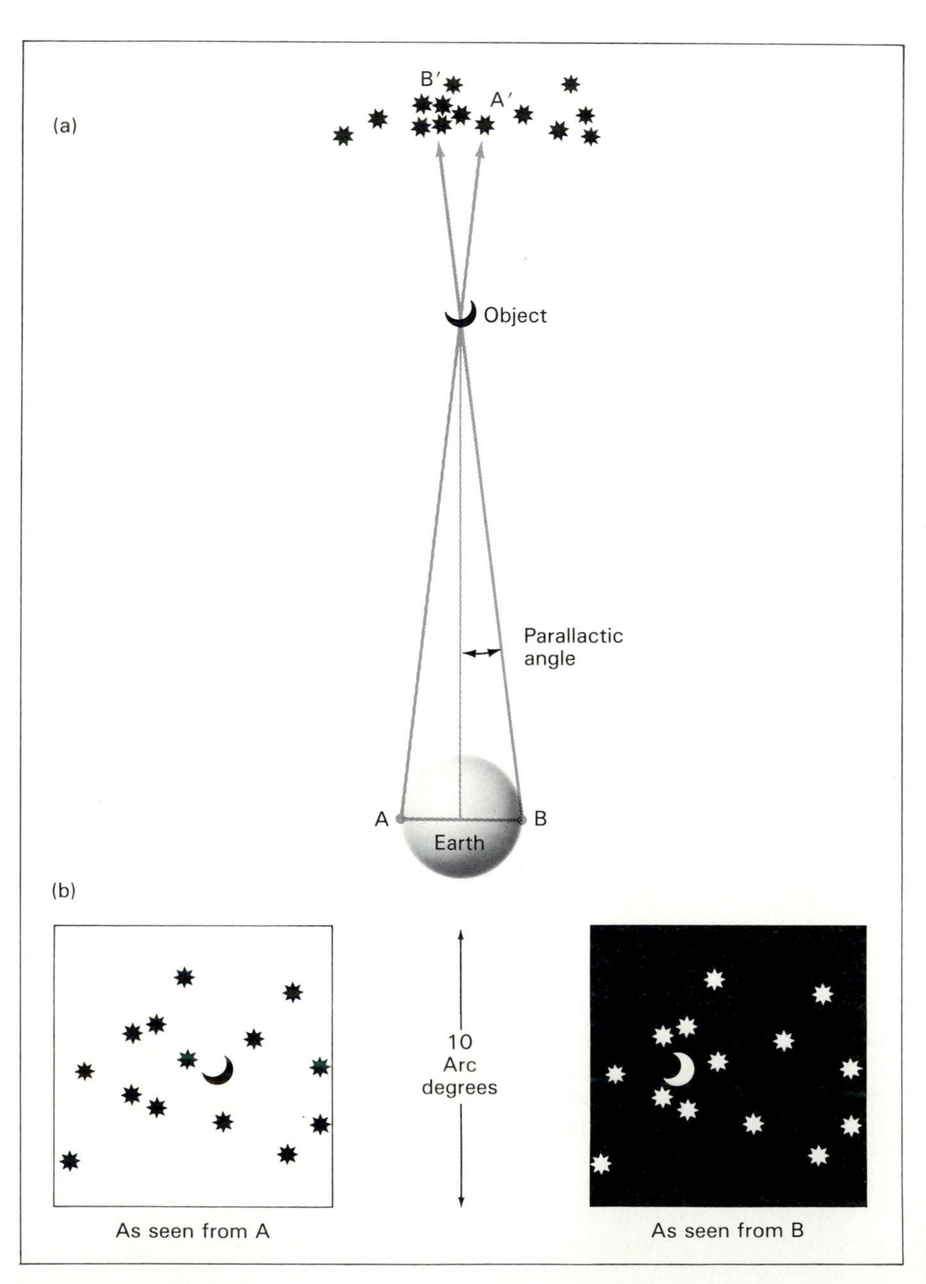

FIGURE 4.7 (a) This imaginary triangle extends from Earth to a nearby object in space (such as a planet). The group of stars at the top represents a background field of very distant stars. (b) Hypothetical photographs of a star field showing the nearby object's apparent displacement or shift relative to more distant, undisplaced stars.

gree of accuracy. The imaginary triangle is extremely long and narrow even for the nearest cosmic objects.

It's actually easier to measure the third angle of the imaginary triangle, namely the very small one near the object. This may seem paradoxical since we need to do so from our point of view at Earth. Yet a special observational technique enables astronomers to measure this narrow angle accurately. Here's how it's done.

Observers on each side of Earth sight toward the object, taking note of its position *relative to some distant stars* seen on the plane of the sky. The observer at point A really sees the object of interest projected against a field of very distant stars. Call it point A', as shown in Figure 4.7a. Similarly, the object of interest appears projected at point B' to an observer at point B on Earth. If each observer takes a photograph of the appropriate region of the sky, the object should appear at a slightly different place in each photograph. In other words, the object's photographic image should be slightly displaced or shifted *relative* to the field of very distant background stars which otherwise appear undisplaced because of their much greater distance.

Figure 4.7(b) shows a hypothetical illustration of an object's apparent displacement or shift on a pair of photographs, one of which was taken from point A, the other from point B. This shift equals the very small angle of the imaginary triangle. For historical reasons, one-*half* of this angular difference is called the **parallactic angle**. It can also be noted in Figure 4.7(a).

Parallax is defined as the *apparent* shift of an object as seen from two different points of observation. The closer an object is to the observer, the fatter is the imaginary triangle and the larger the parallax.

To understand the concept of parallax, hold a pencil vertically in front of your nose, as sketched in Figure 4.8. Concentrate on some remote view, say a distant wall. Now, close one eye; then open it while closing the other. By blinking in this way, you should be able to see a large shift

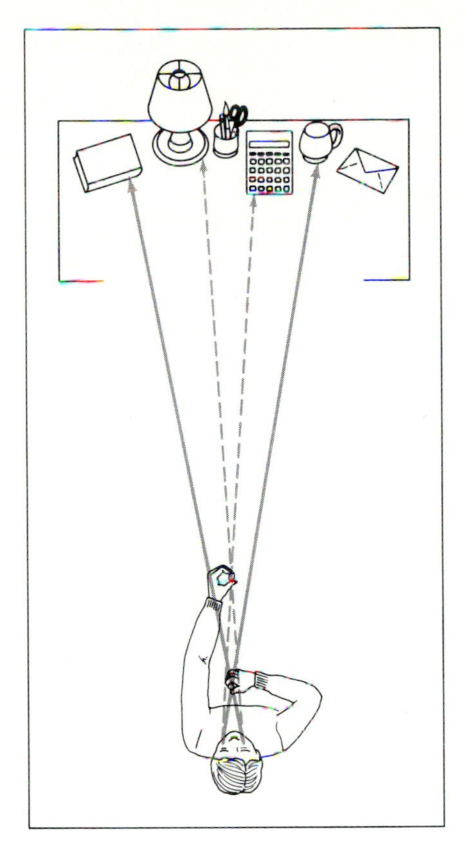

FIGURE 4.8 Parallax is inversely proportional to an object's distance; an object near your nose subtends a much larger parallactic angle than one at arm's length.

of the apparent position of the pencil projected onto the distant wall. In this example, one eye corresponds to point A, the other eye to point B, the distance between eyeballs the baseline, the pencil a nearby object, and the distant wall a remote field of stars. If you now hold the pencil at arm's length, corresponding to the case of a more distant object, but one still not as far away as the wall, the apparent shift of the pencil will be less upon blinking. Furthermore, if you were to paste the pencil to the wall, corresponding to the case where the object of interest is as far away as the background star field, blinking would show no apparent shift of the pencil at all.

Clearly, then, the amount of parallax is inversely proportional to an object's distance. Small parallax implies large distance, and conversely. Knowing the amount of parallax and the length of the baseline, we can derive that distance via triangulation.

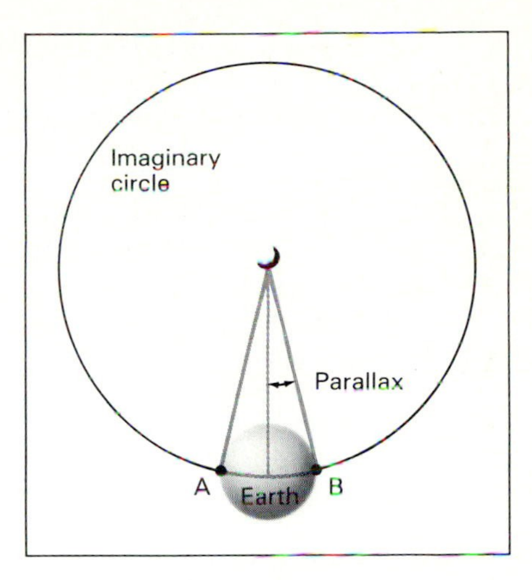

FIGURE 4.9 An imaginary circle drawn around an object of interest, and passing through the Earth, has a radius equal to the distance from Earth to the object.

STELLAR PARALLAX: Consider now some simple geometrical arguments, to make this technique a little more useful. Imagine, as shown in Figure 4.9, that the baseline (which in this case equals Earth's diameter) is only a small part of an enormous circle, at whose center lies the object of interest. If the object's distance is much greater than the baseline (as is always the case for astronomical applications), the length of the circle's arc AB approximately equals the distance from point A to point B. Note that this is a good approximation only because the triangle is so thin.

Now we can construct a geometrical analogy similar to that considered earlier for Figure 4.3. Looking at Figure 4.9, we note that the parallactic angle is to one-half of the baseline as 360 arc degrees is to the entire circumference of the imaginary circle. Quantitatively, we can express this as:

$$\frac{\text{parallactic angle}}{0.5 \text{ baseline}} = \frac{360 \text{ arc degrees}}{\text{circumference}}.$$

The neat thing about this relationship is that it involves the object's distance since, after all, the circumference of the imaginary circle equals π times the circle's diameter (or equivalently twice π times the radius of the imaginary circle). The object's distance equals the imaginary circle's radius, as shown in Figure 4.9.

Hence we can manipulate the foregoing equation for the object's distance using elementary algebra:

$$\text{object's distance} = \frac{28.5 \times \text{baseline}}{\text{parallactic angle (arc degrees)}}.$$

To use this equation, the parallactic angle must be expressed in arc degrees. However, in reality, this angle is never as large as an arc degree, so astronomers often recast the preceding equation in terms of the much smaller unit of angular measure, namely the arc second. Since there are 60 arc minutes in an arc degree, and 60 arc seconds in an arc minute (consult Interlude 3-1), the relationship above then becomes

$$\text{distance} = \frac{103{,}000 \times \text{baseline}}{\text{parallactic angle (arc seconds)}}.$$

With Earth's diameter (that's the baseline) known, a measure of the parallactic angle made from opposite sides of the Earth yields the object's distance. Indeed, the distance to the Moon, our nearest astronomical neighbor, has been determined in this rather simple way. The simplicity ends quickly, however. More distant objects have smaller parallaxes than nearby ones, so that eventually the technique no longer works when the parallactic angle becomes too small to measure. Unfortunately, the parallactic angle for even the *nearest* star is too small to measure accurately using the technique outlined in Figure 4.7.

The basic problem here is that the imaginary triangle is just too narrow to measure the distance to objects beyond our Solar System. When the baseline equals Earth's diameter, we have an extreme case of the illustration shown in Figure 4.6. For even nearby stars, each of the two angles of the imaginary triangle (those at points *A* and *B*) would measure more than 89 arc degrees, 59 arc minutes, 59.9999 arc seconds, while the third angle would measure no more than 0.0001 arc second. That's too small an angle to measure, even as a displacement on a photograph. After all, 1 arc degree contains 3600 arc seconds, making an arc second quite small in its own right; 0.0001 arc second is 10,000 times smaller yet!

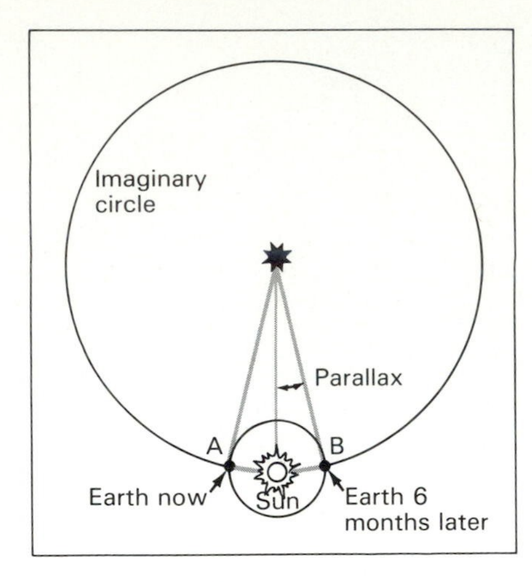

FIGURE 4.10 The same geometrical diagram as Figure 4.9, with the exception that now the baseline has increased from the diameter of the Earth itself to the diameter of the Earth's orbit about the Sun.

Notice, though, that the imaginary triangle *can* be "fattened." A larger baseline is available. What baseline exceeds Earth's diameter? The size of Earth's orbit about the Sun. In other words, we can recast the geometrical diagram drawn in Figure 4.9, with the exception that now the baseline from point *A* to point *B* becomes the diameter of Earth's orbit. As shown in Figure 4.10, we can still draw an imaginary circle centered on the distant object in question, and once again the imaginary circle's radius equals the object's distance. In this way, the triangle is widened simply because the baseline in Figure 4.10 is much larger than in Figure 4.9. After all, the diameter of Earth's orbit, equal to twice the distance from Earth to the Sun, greatly exceeds the diameter of Earth itself.

The previous mathematical equation can easily be modified to account for the increased baseline. Defining the average Earth–Sun distance to be the **astronomical unit**, we then have

$$\text{distance} = \frac{206{,}000 \times \text{astronomical unit}}{\text{parallactic angle (arc seconds)}}.$$

And if we make one further convenient definition, namely that there are approximately 206,000 astronomical units in a new distance unit called the **parsec** ("*par*allax in arc-*sec*onds"), we have

$$\text{distance (parsec)} = \frac{1}{\text{parallactic angle (arc seconds)}}.$$

This is a very simple relation between the distance to a star and its measured parallax. It agrees completely with our earlier "pencil experiment" that demonstrated the inverse nature of distance and parallax.

This specific equation holds valid only when the baseline equals twice the Earth–Sun distance. Hence it requires measurement of an object's apparent position relative to a more distant starfield at 6-month intervals, since the Earth orbits the Sun every 12 months. When the baseline equals twice the astronomical unit, the measured parallax is often given the special name **stellar parallax**, to distinguish it from parallax measurements made with other baselines.

The preceding equation is used as an operational definition of the parsec. In other words, a parsec is the distance of an object whose stellar parallax equals 1 arc second. It's a contrived distance unit—contrived to make the relation between distance and parallax simple. At any rate, the parsec equals approximately 206,000 times the mean distance between the Sun and the Earth.

Even with this larger baseline, stellar parallaxes are hard to measure at 6-month intervals. No such stellar parallax can be noted with the naked eye. Indeed, Aristotle and other geometers of ancient Greece used the apparent lack of stellar parallax to argue that Earth is stationary. Of course, they were eventually proved wrong—not because a sixteenth-century Polish churchman named Copernicus suggested that a heliocentric (Sun-centered) model of the Solar System more simply accounts for the observed planetary motions, but because a nineteenth-century German astronomer named Bessel succeeded in measuring the parallax of a nearby star.

Observation of stellar parallax thus proves that Earth moves about the Sun. The lack of easily observed stellar parallax proves that even the nearby stars have great distances.

As an example, Figure 4.11 is a near-superposition of two photographs of a certain region of the sky, each taken 6 months apart. Just as expected for two plates not precisely superposed, each star seems to have a double image. But the star in the middle clearly has a different shift than most of the others. In addition to the horizontal displacement of all the stars, this star is noticeably displaced vertically as well. The small horizontal displacement exhibited by all the stars is caused by the imperfect alignment of the two photographs; this displacement would disappear if the plates were slid together and perfectly superposed. But the vertical displacement would not disappear in the case of the central star, for this star has a measurable parallax.

FIGURE 4.11 Two real photographs of a typical star field, taken six months apart and nearly superposed, show the apparent displacement or shift of some nearby stars. If none of the stars showed any parallax, then all their double images, laid side by side, would be identical. But the star near the center of the frame, for example, shows a displacement ever so different, betraying its parallax. Can you spot any others?

The closest star to us (excluding the Sun) is called Proxima Centauri, a small member of a triple-star system known as the Alpha Centauri complex. Accordingly, this star displays the largest stellar parallax. Proxima Centauri's measured stellar parallax, 0.76 arc second, then implies (via the preceding equation) that this star is about 1.3 parsecs away, or equivalently about 270,000 astronomical units. That's the *nearest* star to Earth—a few hundred thousand times the distance from Earth to the Sun!

Small values of stellar parallax provide our first indication for the immensity of space beyond Earth. It won't be the last. In many places this text will reinforce the notion that our Universe is nearly incomprehensibly huge.

Vast distances can sometimes be grasped only by considering analogies. Here's one that might be helpful: Imagine Earth as a grain of sand orbiting a golfball-sized Sun at a distance of about 1 meter. The nearest star, also a golfball-sized object, is then more than 100 *kilo*meters distant. Except for a few other planets ranging in size from grains of sand to small marbles, nothing of consequence exists in the hundred kilometers separating the two "basketball" objects. Such is the void of space.

(There is some possibility that closer stellar neighbors roam the space between us and the Alpha Centauri system. Very dim objects, as well as extremely old stars that have run out of fuel, could well have escaped detection thus far. It seems that only exploration outside our Solar System by instrumented spacecraft will be capable of revealing such objects. However, only one such vehicle, *Pioneer 10,* has left our Solar System to date; its sister-craft, *Pioneer 11,* is also on its way out of the Solar System in nearly the opposite direction.)

The next nearest neighbor to the Sun beyond the Alpha Centauri system is called Barnard's Star. It displays a stellar parallax of 0.55 arc second, and is accordingly at a distance of 1.8 parsecs. Increasing numbers of stars are found at greater distances, although only about 30 known stars lie within 4 parsecs, nearly a million astronomical units from Earth. Figure 4.12 is a map of these nearby stars.

On a typical photograph, 1 arc second is roughly the width of 10 human hairs stacked side by side. Astronomers have special equipment resembling microscopes that can now measure stellar parallaxes as small as 0.03 arc second, but seldom smaller. This means that stars must be within 30 parsecs of Earth to have a well measured stellar parallax.

Only about 700 stars are within 30 parsecs distance. And most of these are not the naked-eye objects seen on a clear dark night. The majority of relatively "nearby" stars are dimmer than our Sun and are thus invisible to the naked eye. These stars can be seen and their parallaxes measured only with the aid of a telescope. Most of the bright stars in the night sky are really much brighter than our Sun; this accounts for their vis-

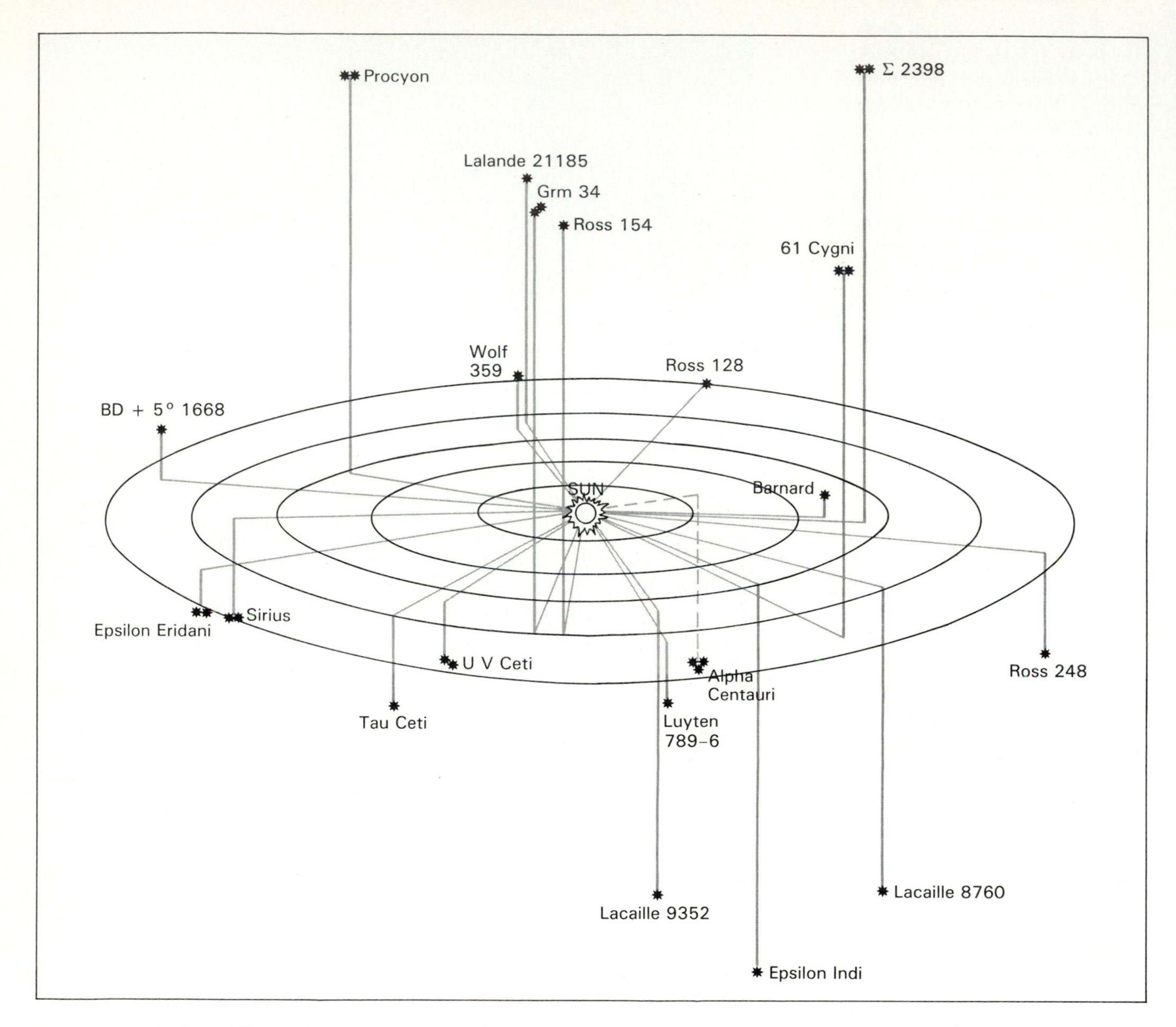

FIGURE 4.12 A plot of the 30 closest stars, projected in such a way as to reveal their three-dimensional relationships. Note that many are members of multiple-star systems. All lie within 4 parsecs (or about 13 light-years) of Earth.

ibility despite their great distance. Indeed, the vast majority of stars comprising our Galaxy are far more distant than 30 parsecs.

Solar System Dimensions

Up until now, we've discussed stellar distances in terms of the parsec, an unusual unit of distance. Just how large is a parsec? We already said that it equals some 206,000 astronomical units. And since everyone knows that the Earth–Sun distance is far greater than anything familiar on Earth, we immediately gain the feeling that the parsec distance unit must be enormous. We'll shortly realize that virtually *all* distances in the Universe are vast by everyday standards. So we must become more precise, and a little more quantitative. Let's consider for a moment just how large the average distance is from Earth to the Sun.

We might propose the technique of triangulation to measure the distance to the Sun. However, it doesn't work well. Attempting to measure the Sun's parallax using Earth's diameter as a baseline, we would find it nearly impossible to distinguish any apparent displacement of the Sun relative to the field of distant stars; the Sun is too bright, too big, too fuzzy. To measure the Sun's distance, we need some indirect method.

As noted previously, we can measure the Moon's parallax using Earth's diameter as a baseline. Accordingly, the lunar distance can be expressed in some familiar unit, for example, kilometers, or in whatever unit we choose for Earth's diameter. But the distance to the Moon is not related to the astronomical unit. After all, the Moon follows right along with Earth's orbital motion around the Sun. Thus we have no easy way to determine how many Earth–Moon distance units comprise one Earth–Sun distance unit. Another method is needed.

Parallax can also be measured for some of the neighboring planets, especially when we use the larger baseline of Earth's orbit. Of course, individual planetary motions, including that of Earth, must be taken into account. But even here, the distance to each planet can be expressed only in terms of the astronomical unit. Why? Because the baseline used is twice the astronomical unit, the absolute value of which we've not yet determined. In this way we can construct a perfectly good model of our Solar System's geometry, but the *absolute* (i.e., real) distance to each planet remains unspecified. The result would be analogous to an exquisitely detailed road map showing only the *relative* positions of cities and towns—a beautiful guide but one without the all-important scale marker showing, for example, the number of Solar System kilometers in each centimeter of map.

Similarly, parallax measurements can provide information about the relative positions of the planets and nearby stars, but all these distances are expressed in terms of the astronomical unit, which is, after all, the baseline used to measure these parallaxes in the first place. Somehow, we must find the absolute value of the astronomical unit.

MEASURING THE ASTRONOMICAL UNIT: The modern method to derive the absolute scale of the Solar System uses radar. The word **radar** is an acronym for *ra*dio *d*etection *a*nd *r*anging. In this technique, radio radiation is transmitted toward an object, after which a return echo is received indicating the object's direction and range, that is, its distance.

Unfortunately, the Sun cannot easily be used as a direct target for this experiment. The Sun has no solid surface from which radio waves can bounce to yield a clear echo. (Echos do return but from variable heights in the solar atmosphere.) Instead, the target often used in this experiment is the planet Venus. This planet is closest to Earth, and therefore returns a detectable echo, although the strength of the received echo is much weaker than the transmitted signal. The actual experiment is done by transmitting a pulse of radio radiation from a large parabolic antenna, such as the Haystack telescope shown in Figures 3.17 and 3.18 or the Arecibo telescope shown in Figure 3.15.

(As noted in Chapter 3, most radio telescopes are equipped only to listen to incoming cosmic radio waves. A few of these large dishes, however, can also transmit intense radio signals, especially for the purpose of communicating with orbiting satellites. As we'll discuss further in Chapter 7, radar astronomy has yielded much useful information about the physical properties of such relatively nearby objects as the planets Mercury, Venus, and Mars. The returned echo, however, is too weak for the radar technique to work well toward the remote planets, let alone the vastly more distant stars.)

Figure 4.13 is an idealized diagram of the Sun–Earth–Venus orbital geometry. We've drawn the planetary orbits as circles here, but in reality they are slight ellipses. This is a subtle difference and can be corrected using detailed knowledge of orbital motions. The important point to realize is that transmissions of radio waves at various times of the year toward Venus can yield a distance between Venus and Earth at any point in their orbits about the central Sun. Timing the electromagnetic signal's round trip, and knowing the value of the light velocity with which it moves, we can find the distance to Venus with great accuracy.

This technique resembles the calculation needed to derive the distance between Boston and Philadelphia if our car's speed and travel time are known. Twice the distance to Venus (back and forth) equals the velocity of light multiplied by the time elapsed between transmission of the signal and reception of the echo. The round-trip travel time can be measured with high precision, in fact, good enough to determine the dimensions of Venus' orbit to an accuracy of 1 kilometer.

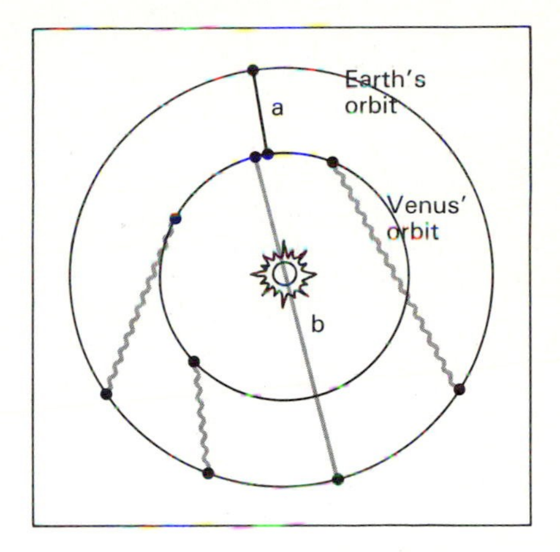

FIGURE 4.13 Simplified geometry of the orbits of Earth and Venus as they move around the Sun. The wavy lines represent typical paths along which radar signals are transmitted toward Venus and received back at Earth.

Let's not lose track of the objective here. We're not really concerned about the details of Venus' orbit as much as the absolute value of the astronomical unit. Ignoring the fact that planetary orbits are slightly elliptical, as opposed to perfect circles, we see from Figure 4.13 that twice the astronomical unit equals the sum of two terms: the distance from Earth to Venus when Venus is farthest from Earth, labeled b, plus the Earth–Venus distance when Venus is closest to Earth, labeled a. In other words, the astronomical unit equals $(a + b)/2$, which, because of precise radar ranging toward Venus, is now known to equal an average value of 149,603,500 kilometers. In this text, we round off this number to a value of 1.5×10^8 kilometers.

We can express this Earth–Sun distance in many ways. Here's another one. Since the velocity of light is 3×10^5 kilometers/second, a simple manipulation proves that the Sun is about 8 light-*minutes* from Earth. That's a distance—8 light-minutes.

SIZE AND SCALE: Having now the value of the astronomical unit, we can reexpress the previously mentioned stellar parallaxes in terms of some familiar unit such as kilometers, or even light-years. The entire scale of the Solar System can then be calibrated, and the absolute distances among the nine planets and their parent Sun determined to rather good precision.

Distances to the nearby stars, expressed earlier in terms of the parsec unit, can also now be cast into more familiar units. For example, 1 parsec equals the product of 206,000 astronomical units and 1.5×10^8 kilometers per astronomical unit, or thus about 3×10^{13} kilometers. This is such a large distance that astronomers prefer to express it in terms of the light-year unit. Some easy manipulation leads us to the conclusion that *1 parsec equals approximately 3.3 light-years*. Accordingly, we can express the distance to our nearest neighbor, Proxima Centauri, as 1.3 parsecs or 4.3 light-years. Think of that for a moment. The *nearest* star is some 4 light-*years* from Earth. Likewise, Barnard's Star is 1.8 parsecs or about 6 light-years away. Contrast these distances with the 8 light-minutes separating the Earth and the Sun.

You should now be getting a feeling for the vastness of space. Alas, we've only touched the tip of the iceberg; we're still very much confined to the local realm. Remember, the stellar parallax technique works well only for objects closer than about 30 parsecs or 100 light-years. Most objects in the Universe are far, far beyond.

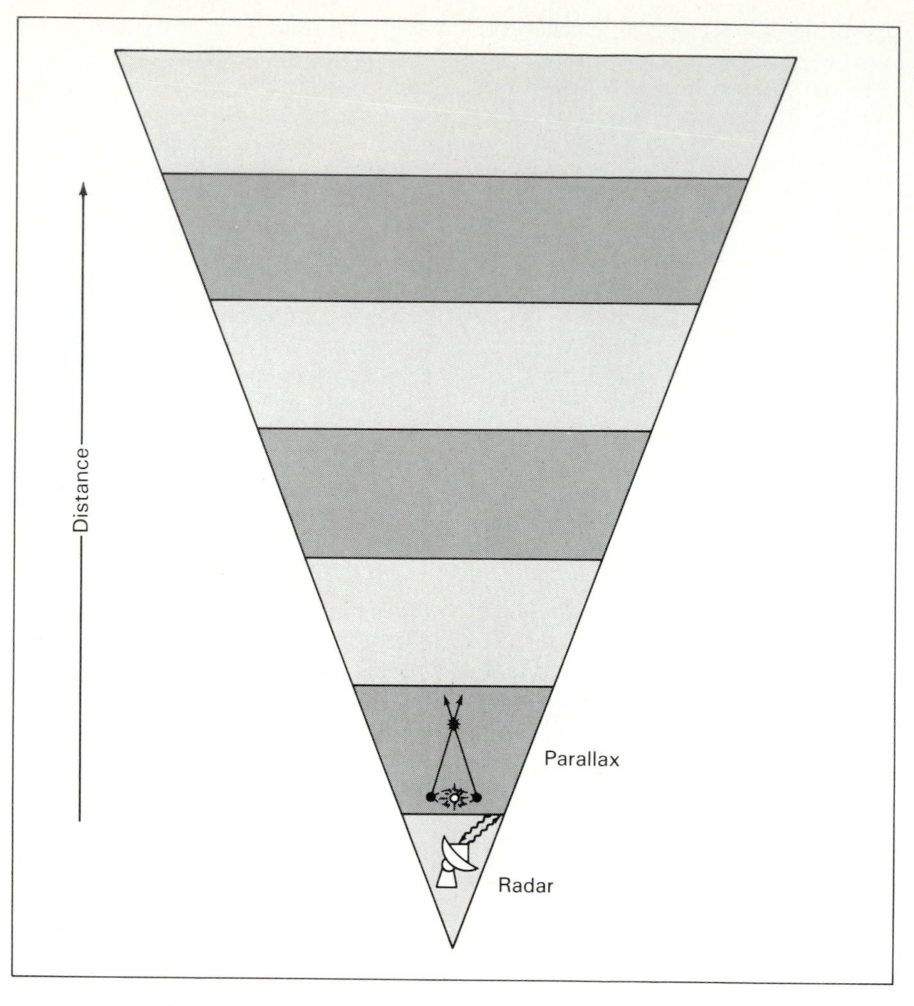

FIGURE 4.14 The distance scale: As distance increases, so does the interval of space.

Foundations of the Distance Scale

So the technique of stellar parallax is of limited usefulness. It provides the size and scale of matter within only about 100 light-years from Earth. You are thus no doubt wondering why an entire chapter has been devoted to it. Most students feel that parallax is a pain. But it's not a waste of time; it's important.

As we probe matter at progressively greater realms of the Universe in Part II, we'll study new methods used by astronomers to chart truly remote objects. With each new distance technique, we shall encounter larger volumes of space, much like the inverted pyramid of Figure 4.14. A succession of many techniques will take us from Earth to the limits of the observable Universe. But each new technique is calibrated on the basis of the previous one; objects whose distances have already been measured by a previous method are used to calibrate in turn each newer technique, on up the pyramid.

The entire method of determining size and scale in the Universe depends on a few distances measured by the method of stellar parallax. This parallax technique is in turn calibrated by radar ranging toward the nearby planets. Indeed, the very foundation of the distance scale depends on radar transmissions and a little geometry.

The elegance of it all is that no advanced mathematics are needed either to understand the local structure of the Universe or to appreciate the critically important scientific basis of the larger scale of things to come in Part II.

SUMMARY

We use more mathematics in this chapter than in any other chapter. This math, however, has not been complex, based only on simple geometry and some common sense. The whole idea has been to share with you two things: first, to provide some early demonstrations of the power of the scientific method, and second, to show how very simple arguments can be used to determine the size, shape, and scale of our local niche in the Universe.

Although perhaps hard to appreciate so early in the text, the reasoning used here is an integral part of the story of cosmic evolution. Be sure to understand the details, but be sure also to grasp the larger importance of basing our story on careful logic, real data, and decisive tests.

KEY TERMS

astronomical unit
baseline
Euclidean geometry
geometry
lunar eclipse
parallactic angle
parallax
parsec
radar
stellar parallax
triangulation

QUESTIONS

1. Pick one: In 1 second, light leaving Manhattan will reach approximately as far as (a) the Moon, (b) Venus, (c) Proxima Centauri, (d) London, (e) Brooklyn.
2. Explain how Aristotle argued that Earth must be stationary in space. Where did he go wrong?
3. Carefully explain the main piece of observational evidence which proves that Earth moves through space.
4. Consider two pictures of the Moon and some stars near it on the sky. One picture is taken from Boston, the other, at the same time, from San Francisco. Explain how these pictures could be used to find the distance to the Moon.
5. Pick one: History records that Earth's radius was first measured by (a) Aristotle, (b) the Sumerians, (c) Eratosthenes, (d) NASA, (e) anonymous Stone Age peoples.
6. Why have astronomers been able to measure the stellar parallaxes for only the nearest few hundred stars?
7. Explain, with the aid of a diagram if helpful, the modern method of measuring the distance from Earth to the Sun.
8. There are numerous everyday experiences where you actually use the scientific method in the process of going about your daily business. Can you name a few?
9. Using a spherical world globe and a piece of string, find the shortest path from Washington to Moscow. Does it extend into the Arctic Circle?

*10. What are the distances of objects whose parallax measurements amount to 0.4 arc second and 0.04 arc second? Express these distances in astronomical units.

5 THE EARTH: *Our Home in Space*

Looking at the nighttime sky, we can see an almost bewildering array of stars. Each star shines at a particular moment in its life cycle. Some are old, some are young; some large, some small; others are dim, still others bright; many have different colors. Despite this complexity, we are fortunate to be able to observe and study such a wide variety of stars. In fact, it's only *because* of the large number of different stars that we can piece together the evolutionary cycle of a typical star. In this way, we seek to understand the birth, maturity, and death of stars everywhere.

Planets differ greatly from typical stars. Nor do we have the luxury of being able to observe many of them. The rock we call Earth is cooler, smaller, and denser than most stars. That much we can say for sure. Actually, we know a great deal more about some aspects of Earth, but a complete understanding of our planet still eludes us.

Earth *as it exists today* is reasonably well explored. Probing our planet, we have the obvious advantage that we live on it. We can study the soil, bang the rocks, sample the water and perform many other direct experiments with earthly matter. To be sure, Earth is now examined and monitored in nearly every conceivable way—with aircraft in the atmosphere, satellites in orbit, gauges on the land, submarines within the ocean, and drilling gear below the rocky crust.

Despite this onslaught of research, our studies of Earth are limited for a very simple reason: We know of only a few such planets. Unlike the legions of stars, we are aware of only a single planetary system. Modern space probes are beginning to show that each of the known planets has some variety, as we'll see in Chapters 7 and 8, but the relevance of comparing Earth to the other planets is currently unclear.

In short, there is only one Earth. We can study it now. Yet to fathom what Earth might have been like long ago is not easy. To imagine what it might be like in the future is nearly impossible. So despite the fact that we inhabit Earth, we know less about the evolutionary cycle of our own planet than about a typical star. But more on this aspect of our planet in Chapters 23 and 24.

In this chapter we study Earth in some detail largely because it's our home. Equally important, Earth is an integral part of the cosmos, and that's a fact we shouldn't forget. From the matter of our planet sprang life, intelligence, culture, and the technology that we now use to explore the cosmos. We ourselves are made of "earthstuff" as much as are rocks, trees, or even the air. And Earth still nurtures us, providing a remarkably comfortable abode for life. Indeed, if we are to appreciate the Universe, we must come to know our planet. Our study of astronomy begins at home.

The learning goals for this chapter are:

- to know some of the average physical and chemical properties of our planet
- to understand that the atmosphere hugging Earth's surface helps to heat us as well as protect us from the harsh realm of outer space
- to recognize the benefit of being nearly surrounded by magnetism
- to appreciate some of the ex-

perimental techniques used to infer the structure of Earth's interior

• to realize that Earth's surface activity—quakes, volcanoes, and so on—are mere remnants of a hotter, more violent era

Earth in Bulk

Our planet can be divided into six main regions. As shown in Figure 5.1, a zone of trapped particles called the **magnetosphere** lies high above the **atmosphere** of air. At the surface, we have the **lithosphere**, comprising the solid continents and seafloor, as well as the **hydrosphere**, containing the liquid oceans, lakes, and rivers. In the interior, a large **mantle** surrounds a small **core**. Earth, then, is made of familiar material substances—solids, liquids, and gases.

DIMENSIONS: Although the circumference of our planet has been known reasonably well since the time of the ancient Greeks, space satellites are now used to derive an accurate value for the solid size of planet Earth. It measures approximately 40,000 kilometers all the way around. Knowing that the circumference equals twice π times the radius, we find the rounded-off radius to be 6500 kilometers. The volume, which equals $\frac{4}{3} \times \pi \times \text{radius}^3$, then follows directly; for Earth it amounts to about 10^{27} cubic centimeters.

Throughout most of the text we use the cubic centimeter as a unit of volume. Sketched in Figure 5.2, a cubic centimeter is simply a cube having a centimeter length on all sides. [A centimeter is a hundred thousand (10^5) times smaller than a kilometer, a hunderd (10^2) times smaller than a meter, or a little less than half an inch.] You can think of a cubic centimeter as roughly the volume of a squirrel's acorn or an old-fashioned sewing thimble. Thus, Earth's total volume would fill the equivalent of approximately a billion, billion, billion acorns! This calculation applies only to the solid and liquid parts of Earth, and does not include the gaseous atmosphere and magnetosphere.

MASS: We can find the mass of Earth by observing its gravitational influence on some other object. The law of gravity, to be studied in Chapter 7, explains the motions of any large objects in the vicinity of other large objects. This statement is true whether the objects are baseballs, rockets, or planets. By studying the dynamical behavior of objects near our planet, we can compute Earth's mass—in effect, weigh the Earth.

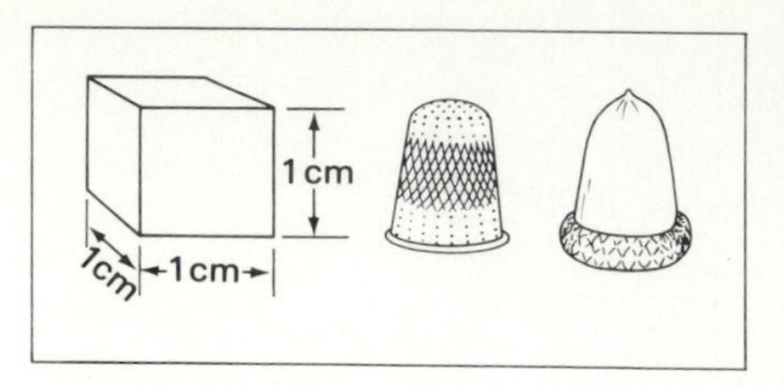

FIGURE 5.2 A cubic centimeter is a unit of volume having dimensions of 1 centimeter on each side. Thus, a cubic centimeter approximately equals the volume contained within a sewing thimble or a small acorn.

Mass can be broadly defined as the total amount of matter contained within an object. In the case of Earth, the result is approximately 6×10^{27} grams. A gram is a convenient unit of mass; on Earth, about 450 grams equals 1 pound. A 150-pound person, for instance, has a mass of about 68,000 grams. Virtually all of Earth's mass is contained within the surface and interior, the atmosphere and magnetosphere contributing hardly anything at all.

DENSITY: One of the most useful properties of any object is density. **Density** is computed by dividing the mass of an object by its volume; as such, it's a measure of "compactness" of the matter within an object. Dividing Earth's mass by its volume, we obtain 6×10^{27} grams/10^{27} cubic centimeters, or thus an average density of approximately 6 grams/cubic centimeter. Again, this density pertains to the mostly rocky planet, excluding the atmosphere and magnetosphere. *On average*, then, there are some 6 grams of Earth matter in every cubic centimeter of Earth volume.

Earth's average density can be compared to a value of 1 gram/cubic centimeter for water, proving that although much of Earth's surface is made of water, denser matter must lie deep within the interior. By contrast, ordinary rock beneath us on the continents has densities varying from 2 to 4 grams/cubic centimeter.

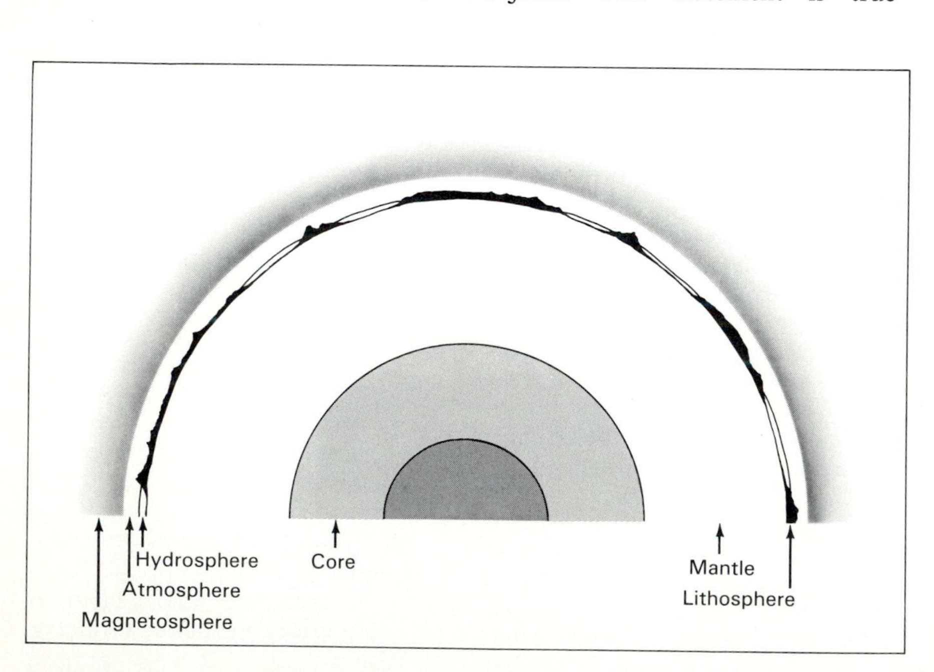

FIGURE 5.1 Six main regions of planet Earth.

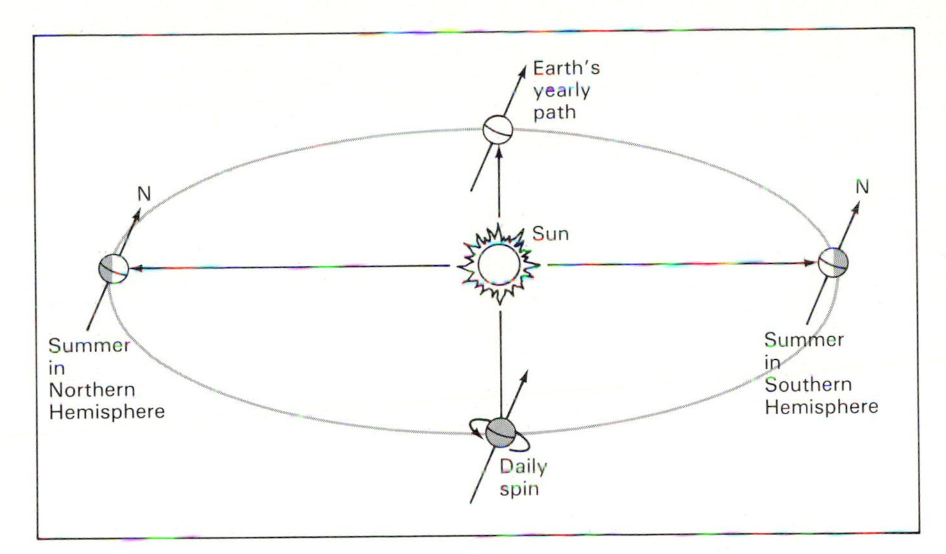

FIGURE 5.3 Earth's rotation around a north-south axis produces the day/night cycle. Revolution of the *tilted* Earth around the central Sun produces the four seasons. Exaggerated here for clarity, the change in the Earth-Sun distance due to the ellipticity of our planet's orbit is too slight to affect our annual climate. Earth, in fact, is *farthest* from the Sun during the northern hemisphere's summer. It's the tilt of the Earth that causes the seasons.

Hence we should expect that much of Earth's interior is made of very dense matter, apparently even more compact than the densest continental rocks on the surface.

SHAPE: Earth is not a perfect sphere. It bulges slightly at the equator; the equatorial diameter is about 40 kilometers larger than the polar diameter. This specific number is not as important to remember as is the fact that Earth definitely departs from sphericity. Even so, the bulge is not pronounced for an object of its size; 40 kilometers is not very large compared to the full Earth's diameter of nearly 13,000 kilometers. In fact, relative to its overall dimensions, Earth is smoother and more spherical than a billiard ball.

To understand this slight "flattening," examine Figure 5.3. As illustrated there, our planet spins around a north-south axis while simultaneously orbiting the Sun. Earth's orbital motion, called **revolution**, completes a full cycle around the Sun once every year; together with the tilt of Earth's axis (see Figure 5.3), this orbital motion produces the various seasons on our planet. Earth's spinning motion on its axis, called **rotation**, completes a cycle once every Earth day; it produces the continual procession of daylight and nighttime.

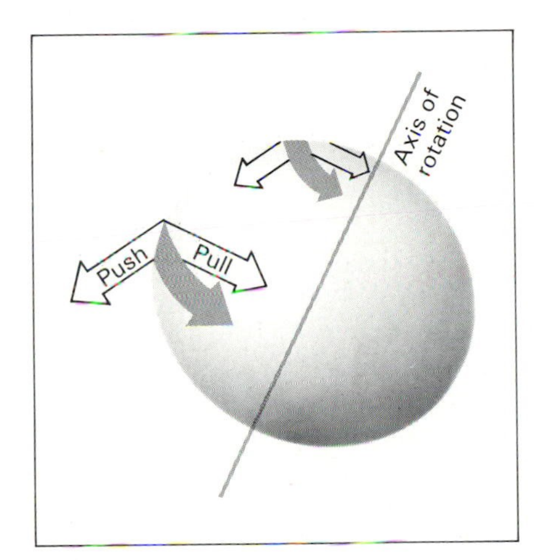

FIGURE 5.4 All spinning objects tend to develop a bulge since rotation causes matter to push outward against the inward-pulling gravity. The amount of the bulge depends on the strength of the matter and the rate of rotation.

Figure 5.4 shows how a spinning object develops a bulge around its midsection. This bulge can be large either for objects whose matter is loosely bound, or for objects that spin fast. In the case of some objects made only of gas or loosely packed matter, large spin rates can produce a pronounced bulge; even faster rates can completely tear matter from the object. The rings of the planet Saturn, for example (see Figure 8.8), may have originated as matter was left behind when this rapidly rotating ball of gas first formed; and the relatively fast spin of the Saturnian system (about 10 hours) guarantees that the matter in the rings does not fall into the main body of the planet itself. The present spin rate of Earth (24 hours) is large enough to cause some departure from sphericity, but not nearly fast enough to dislodge solid or liquid matter from its surface.

SLOWING SPIN: Earth's rotation on its axis is gradually slowing down. We now have direct evidence that the day is lengthening, although at an excruciatingly slow pace. How slowly? By about 0.002 second every century. That's clearly a small time increase compared to the scale of a human lifetime. But over very long durations—say over millions of years—this steady slowing of Earth's rotation adds up. At this rate of slowing, a conventional clock would gain several hours of time in about 500 million years.

How do we know for sure that Earth's spin is decreasing? Every time Earth spins on its axis, a growth mark is deposited on a certain type of coral plant in the reefs off the Bahamas. These growth marks mimic those of annual tree rings, although in the case of coral the marks are made daily. Coral growing at the present time shows 365 marks, but ancient coral shows many more growth deposits. In fact, fossilized reef dated to be several hundred million years old contains coral with nearly 400 deposits per yearly growth. At that time, there must have been 35 more days in each Earth year, implying that Earth must have spun more rapidly in the past. Thus nearly 500 million years ago, the day was apparently only 22 hours long and the year contained 400 days.

What causes the day to lengthen steadily? The answer is the tides. Both the Moon and the Sun exert strong gravitational forces on our planet. Although we shall study the subject of gravity in detail in Chapter 7, we can use common sense to address the issue of tides right now.

The strength of gravity depends on the distance separating any two objects. Different parts of Earth feel slightly different pulls of the Moon's gravity, depending on their actual distance from the Moon. For example, the gravitational attraction is greater for the side of Earth facing the Moon than for the opposite side, some 13,000 kilometers farther away. This variation in gravitational pull on all parts of Earth can produce some distortion in our planet's shape. Similarly, the Sun's gravitational pull on different parts of Earth can also cause tides, although the greater distance of the Sun ensures that this Sun-induced distortion is only half that caused by the Moon.

The Moon and the Sun have their greatest influence on the loose liquid parts of Earth. Three-quarters of our planet's surface is covered with water, where the average depth of this hydrosphere is several kilometers. Only 2 percent of the water is contained within lakes, rivers, clouds, and glaciers; the remainder makes up the freely flowing oceans. The height of a typical tide on the open ocean amounts to about a meter, but if this tide is funneled into a narrow opening such as the mouth of a river, the height of the tide can become much greater. The Bay of Fundy on the Maine–Canada border is one of the best examples; there high tide can reach nearly 20 meters (which is approximately 60 feet or the height of a six-story building!).

As noted in Figure 5.5, the side of Earth opposite the Moon also experiences a tidal bulge. The different gravitational pulls—greatest on that part of Earth closest to the Moon, weaker on Earth's interior, and weakest of all on Earth's opposite side—accounts for the fact that the average tides on both sides of our planet are of approximately equal height. On the near side, the ocean water is pulled slightly toward the Moon, whereas on the opposite side, the ocean water is literally left behind the underlying Earth. Thus high tide occurs every 12 hours, not every 24 hours.

The main culprit in Earth's slowing spin is the friction created by the continual sloshing of Earth's oceans. This is especially true near shores and in shallow seas. Just as friction tends to slow the steady movement of any object—a ball rolling along the ground, a plane traveling through the air—tidal friction acts to decrease our planet's rotation.

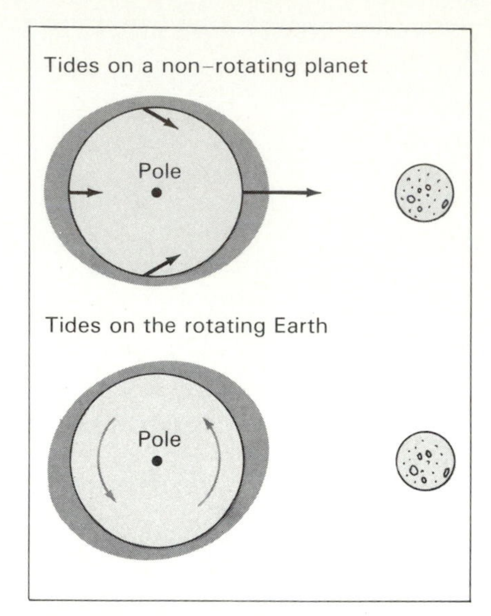

FIGURE 5.5 This exaggerated illustration shows how the Moon induces tides on both the near and far sides of Earth. The lengths of the straight arrows indicate the relative strengths of the Moon's gravitational pull on various parts of Earth. Friction results as Earth rotates beneath the tides.

Next, let's examine some of Earth's major regions in greater detail.

Atmosphere

As noted in Chapter 1, Earth's atmosphere protects us from most of the harsh radiation emitted by the Sun and other cosmic objects—particularly the high-frequency and often harmful ultraviolet and x-ray radiation. The atmosphere also guards us from most rocky debris (called "meteoroids") falling in from space. Furthermore, it warms us. In short, the atmosphere acts as a protective blanket, making the surface of our planet a rather comfortable place to live.

HEATING: Some of the Sun's radiation manages to penetrate Earth's atmosphere, eventually reaching the ground. The great majority of this penetrating energy is solar visible radiation—ordinary sunlight. (Most radio waves also reach Earth's surface, although by comparison the hot Sun emits comparatively little of this kind of radiation.) All the solar radiation not absorbed or reflected from the upper atmosphere shines directly on Earth's surface, including the soil, oceans, trees, houses, automobiles, people, and so on. The result is that our planet's surface and most objects on it heat up considerably during the daylight hours; anything exposed to sunshine becomes hotter than normal. But the surface can't possibly absorb this solar energy indefinitely. Otherwise, surface objects would soon become hot enough to melt (or perhaps explode)—and all life forms would be cooked.

Fortunately, the surface reradiates much of its absorbed energy. In doing so, the radiation obeys the usual Plank curve studied earlier in Chapter 1. Since Earth's surface and most objects on it have temperatures of nearly 300 kelvin (room temperature), Figure 1.15(b) shows that most of Earth's absorbed energy must be reemitted as infrared radiation. But as noted in Figure 1.11, infrared radiation penetrates our atmosphere only partially. A gas, made mostly of molecular nitrogen (N_2; 78 percent), molecular oxygen (O_2; 21 percent), and trace amounts of other gases, partially blocks infrared radiation. [Smog, a mixture of water (H_2O), carbon dioxide (CO_2), and nitrogen dioxide (NO_2), is an especially effective absorber of infrared waves.] Consequently, only some of the reemitted radiation escapes back into space; the remainder is trapped within our atmosphere.

The upshot is this: Solar visible radiation, which easily gets through the atmosphere from space, is effectively converted into infrared radiation, a portion of which cannot get out. In this way, Earth's atmosphere becomes thermally balanced—heated by incoming solar radiation, cooled by outgoing Earth radiation.

The result is a temperature comfortable for life as we know it.

This partial trapping of solar radiation is called the **greenhouse effect**. In a florist's greenhouse, where sunlight passes relatively unhindered through glass panes, much of the infrared (heat) radiation reemitted by the plants within cannot get out. The glass blocks circulation and some of the infrared waves trying to escape, thus preventing the warm air on the inside from mixing with the cold air on the outside. Consequently, the warm enclosure can be used to grow agricultural and floral products even on cold wintery days. The same sort of greenhouse effect warms and stagnates the air within enclosed automobiles on summer days (especially if the interior upholstery is darkly colored, causing much of the sunlight to be trapped when the windows are closed).

On Earth, the natural greenhouse effect is relatively mild. Most of the radiation reemitted by the surface manages eventually to escape the atmosphere. The escape of energy to space is slowed, however, thus helping to keep Earth's surface temperature from dropping greatly at night, despite the absence of sunlight. All things considered, Earth's greenhouse effect makes our planet some 25 kelvin hotter than would normally be the case without it. Hence it is the greenhouse effect that allows our oceans to remain in the liquid state, a condition perhaps critical for the origin, maintenance, and development of life.

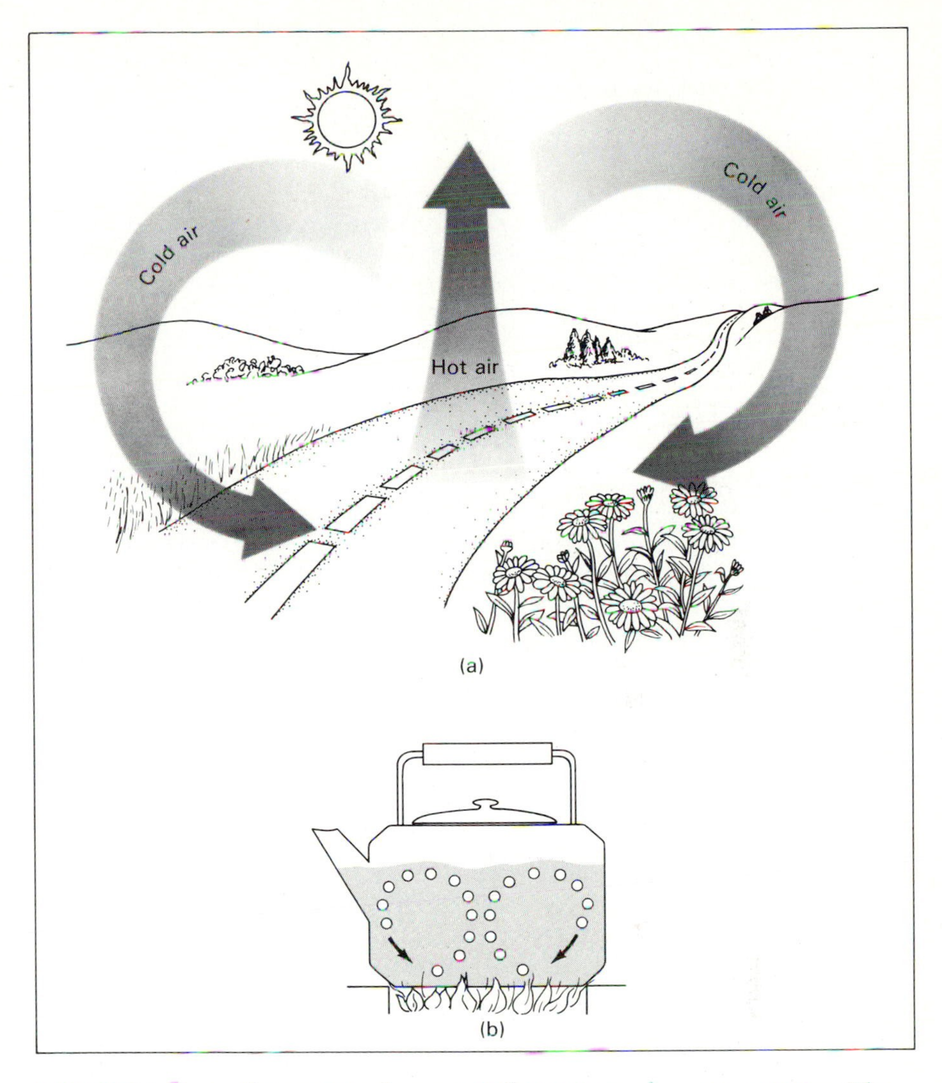

FIGURE 5.6 Convection occurs whenever cool matter overlays warm matter. The resulting circulation currents are familiar to us either (a) in Earth's atmosphere because of the winds caused by the solar heated ground, or (b) in a pot of warm water because of the upwelling water motions caused by the heat of a stove.

CONVECTION: Earth's air is heated not only by absorbing solar radiation. Another important heating mechanism is termed **convection**. Sketched in Figure 5.6, convection occurs when hot matter rises, thus physically transporting heat from a lower to a higher level.

In Figure 5.6(a), part of Earth's surface is schematically shown being heated by the Sun. The air immediately above the warmed surface heats, expands a little, and becomes less dense. As a result, the hot air becomes buoyant and rises to higher altitudes. (The fact that hot air rises can be proved by visiting the attic of a house in midsummer; it's stifling. Hot air rising from a burning building is another case in point.)

At the higher altitudes, the opposite effect occurs: The air gradually cools, grows more dense, and sinks back to the ground. Cool air at the surface rushes in to replace the hot buoyant air. In this way, a circulation pattern is established. Such convection cells of rising and falling air not only contribute to atmospheric heating, but also cause surface winds.

Atmospheric convection can also create clear-air turbulence—the bumpiness experienced by many of us while on aircraft flights. Ascending and descending parcels of air, especially below fluffy clouds, often cause a choppy ride.

The same convection process occurs in a pot of water warmed on a stove. Depicted in Figure 5.6(b), water heated at the bottom of a pot becomes buoyant, rises, cools, and eventually falls back to the bottom, where it's heated again. Steady circulation patterns with rising and falling currents are established and maintained, provided that the source of heat (the stove in the case of the water; the Sun in the case of the atmosphere) remains intact.

Convection is an important physical process. It plays a critical role for many objects in the Universe. We'll return to it again and again, when-

INTERLUDE 5-1 ***Why Is the Sky Blue?***

Is the sky blue because it reflects the color of the ocean deep, or is the ocean blue given the color of the surrounding sky? The latter is surely the case and the reason has to do with the way that light is scattered by air molecules and minute dust particles. (By "scattering" we mean the process whereby radiation is absorbed and then reradiated by some material medium.)

When waves of sunlight pass through our atmosphere, they are affected by microscopic particles in the air we breathe. But because the average wavelength of light (thousands of angstroms) is much larger than the typical size of a molecule (a few angstroms), the scattering process is very sensitive to wavelength. This phenomenon was first investigated by the British physicist Lord Rayleigh about a century ago and thus bears his name—Rayleigh scattering.

Rayleigh found that blue light scatters a good deal more than red light; this is so because blue light's wavelength is closer is size to that of the air molecules. (Technically, since the amount of scattering is inversely proportional to the fourth power of the wavelength, blue light is scattered $[7000/4000]^4$ or some 10 times more efficiently than red light.) Consequently, with the Sun at a reasonably high elevation, the blue component of incoming sunlight will scatter more than any other color, thus filling the sky with high-frequency bluishness. (This also explains why the sky is black in outer space, where there is no atmosphere to scatter the sunlight.)

At dawn or dusk, however, with the Sun near the horizon, sunlight must pass through much more atmosphere before reaching our eyes. Accordingly, the blue component of the Sun's light is scattered so heavily as to be nearly extinguished from view. Instead of a deep blue sky overhead, we tend to see the remaining colors comprising light; this is especially true along the line of sight to the Sun since the other colors scatter less well. Accordingly, the Sun itself appears orangish—a mixture of its normal yellow color and a reddishness caused by the subtraction of blue. At the end of a particularly dusty day, when human activity during the daytime hours has raised excess particles into the air, short-wavelength Rayleigh scattering can be so heavy as to give the Sun a brilliant red appearance. Reddening is often especially evident when looking at the westerly "sinking" summer Sun over the ocean, where sea water molecules have been evaporated into the air, or during the weeks and months after an active volcano has released huge quantities of gas and dust particles into the air—as is the case for North American inhabitants when the Mexican volcano, El Chicon, erupts every so often.

ever we have reason to suspect that cool matter overlays warm matter.

WHY AIR STICKS AROUND: Figure 5.7 shows a cross section of Earth's atmosphere. Compared to the overall dimensions of our planet, the extent of our atmosphere is not large. Half of it lies within 5 kilometers of the surface, and all but 1 percent exists below 30 kilometers. Both temperature and density decrease with altitude in these low-atmospheric zones where all the weather occurs. Climbing even a modest mountain demonstrates this cooling and thinning of the air.

In the thin regions near the middle of the atmosphere lies the ozone layer. There, at an altitude of 30–50 kilometers (roughly 100,000–150,000 feet), the air temperature slightly increases as the incoming solar ultraviolet radiation is absorbed by atmospheric oxygen (O_2) to form ozone (O_3), as explained in Chapter 1. The remaining 1 percent of the atmosphere has been traced up to approximately 100 kilometers, where there begins the electrically conducting ionosphere discussed in Chapter 1.

Some planets have thicker atmospheres than that of Earth. Others have thinner ones. This leads us to wonder why planets have atmospheres at all. Why does a layer of air hover just above the surface of Earth? After all, experience shows that most gas naturally expands to fill all the volume available. Perfume in a room, tear gas in a riot, or steam from a teakettle, all rapidly disperse until we can hardly sense them. Why, then, doesn't our atmosphere similarly disperse by just floating away into outer space? The answer is that gravity holds it down.

The gravitational force field of Earth exerts a pull on every atom and molecule in our atmosphere, thereby preventing them from escaping. On the other hand, gravity can't be the only influence on the air. Other agents must compete with Earth's gravity to keep the atmosphere buoyant; otherwise, gravity alone would force all the air to the ground. Obviously, this is not the case. The atmosphere just hangs in there!

Heat is the other influence that competes with gravity. Solar heating of our planet gives the atmospheric molecules some random motions, to and fro, back and forth, much like the gas within a hot-air balloon or the molecules in a pot of boiling water. The constant swirling movements of the heated molecules prohibit gravity from collapsing our atmosphere completely.

It's worth detailing the effect of these two competing influences—gravity and heat—for we'll eventually come to recognize that an atmosphere is a terribly important requirement for life. We can best

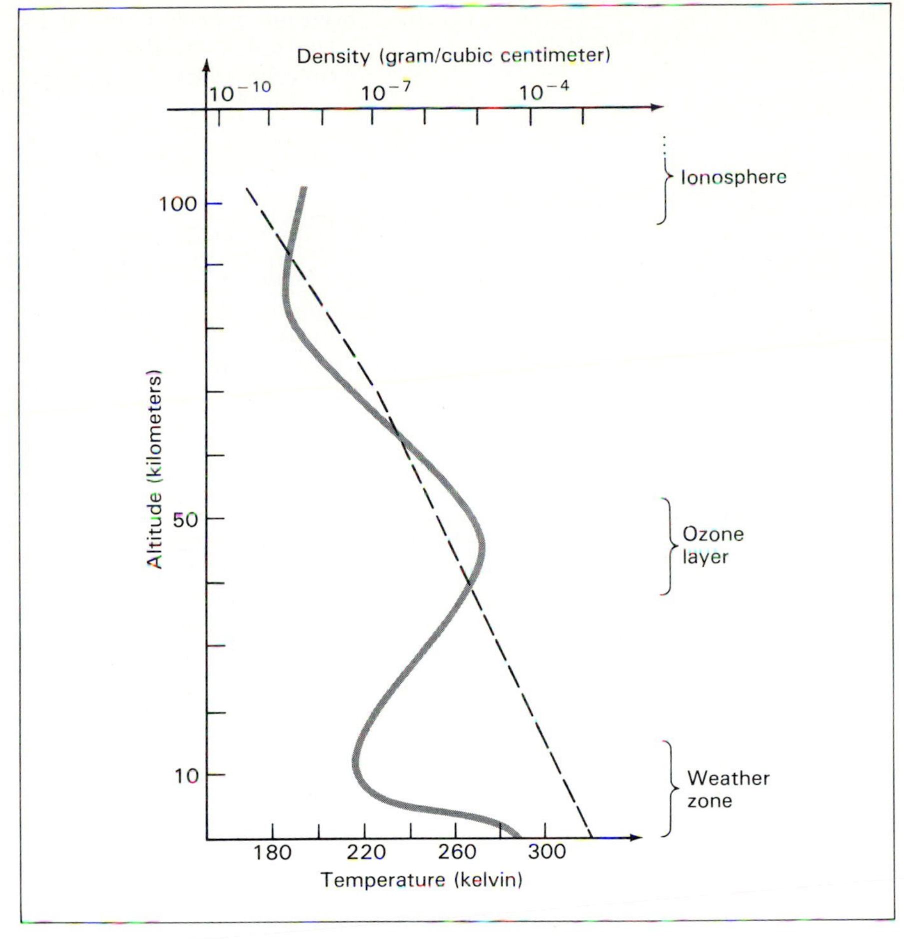

FIGURE 5.7 Schematic diagram of Earth's atmosphere, showing the changes of temperature (solid line) and density (dashed line) from the surface to the bottom of the ionosphere.

appreciate the atmosphere of any cosmic object by comparing two kinds of motion of the individual gas molecules.

The **escape velocity** is the speed needed for a small object to escape completely the gravitational pull of a large object. It depends on the mass and size of the larger or parent object (often a planet) in this way:

$$\text{escape velocity} \propto \sqrt{\frac{\text{mass of parent object}}{\text{radius of parent object}}}.$$

In the case of Earth, the escape velocity is about 11 kilometers/second. This is the minimum velocity needed for any object—molecules, baseballs, or rockets—to escape from Earth's surface. By comparison, slightly larger velocities are attained by many civilian rockets designed to propel robot space probes toward the other planets. On the other hand, military intercontinental ballistic missiles are designed to achieve less than escape velocity, for unless they return to Earth's surface, they could not fulfill their missions. A speeding bullet attains a maximum velocity of only about 1 kilometer/second—some 10 times less than escape velocity—hence the reason that bullets don't leave the Earth when guns are fired.

Note that the relationship above predicts that the escape velocity depends only on the physical properties of the parent object from which the escape is made. Accordingly, parent objects of relatively small mass generally have small escape velocities. For example, the Moon, being much less massive than Earth, has an escape velocity of only about 2.5 kilometers/second. Very massive objects usually have a large gravitational pull, making it harder to escape from their surfaces.

To fully appreciate planetary atmospheres, this escape velocity must be contrasted with the **molecular velocity**, which is the average speed of the gas particles. This molecular motion depends on the temperature and the mass of the individual molecules:

$$\text{molecular velocity} \propto \sqrt{\frac{\text{temperature of molecule}}{\text{mass of molecule}}}$$

As indicated by this relationship, the hotter the gas, the larger the average velocity of the molecules. This is reasonable since molecules within a container of gas move most rapidly when heat is applied, as cited by our earlier example of a pot of warmed water. The mass of the individual molecules also affects their motions, as shown by this relationship. The more massive molecules are predicted to have smaller molecular velocities. For example, molecules of molasses, being more massive than molecules of water, tend to move around sluggishly, even when heated.

Some atmospheric molecules either gain or lose energy (and therefore velocity) by bumping into one another or by colliding with other objects near the ground. Even so, whole groups of molecules in a gas can be characterized by an average molecular velocity, despite the individual molecules of the gas having different velocities. Figure 5.8 depicts the status of a large number of molecules in a typical gas. Plotted there is the number of molecules versus the velocity of those molecules. In this way, we can easily estimate the fraction of molecules having any velocity. As can be seen, the curve peaks near the middle and then falls off on both sides. The falloff is not the same on each side, however; as shown, a long "tail" dominates the high-velocity side. The shape of the curve means that most molecules have velocities close to an average value, while fewer molecules have velocities much larger or smaller

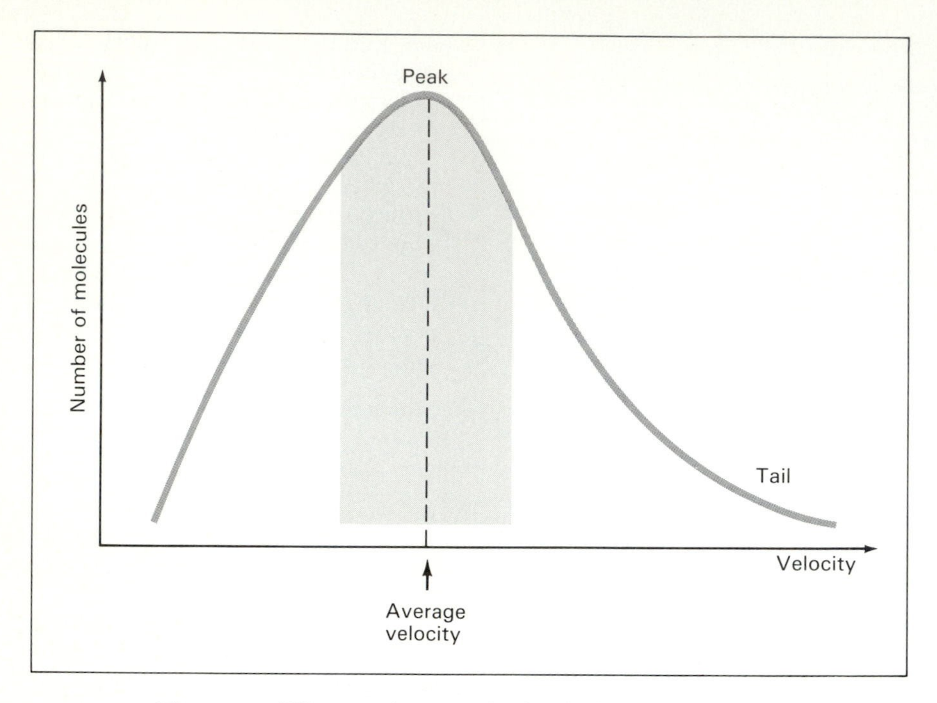

FIGURE 5.8 The many different velocities of individual molecules within a gas can be represented by a curve like this one. The majority of molecules (shaded area) have velocities close to the most probable value (dashed line drawn at the peak of the curve). A small number of molecules (those out in the tail) have velocities much larger than this average.

than this average. Of particular importance, a substantial number of molecules are represented by the tail of the curve and thus have abnormally large velocities.

As we can note from the relationship above, the peak of the curve—that is, the most probable molecular velocity—depends on the temperature of the molecules. For the case of Earth's atmosphere, where the temperature near the surface is nearly 300 kelvin, the curve peaks at about 0.5 kilometer/second. Compared to the velocity of 11 kilometers/second needed to escape Earth, this average molecular velocity is small—far smaller than that needed for escape. Even molecules having individual velocities that are 20 times larger (those represented by the tail of the curve of Figure 5.8) do not have enough velocity to escape. Thus Earth is able to retain its atmosphere. On the whole, the gravity of our planet simply has more influence than the heat of our atmosphere.

We can extend this type of analysis to any cosmic object. For example, consider the case where the mass of the planet and of the atmospheric molecules are similar to Earth's, but for which the atmospheric temperature is higher than Earth's. The curve of molecular velocities would have the same shape as in Figure 5.8, but it would peak at a larger velocity. The increased heat guarantees that the average molecular velocity is greater than on Earth, whereas the escape velocity for this hypothetical planet is similar to Earth's. Some molecules, especially those represented by the tail of the curve, would then have motions comparable to the escape velocity. Under these conditions we should not expect such a planet to have an appreciable atmosphere. As we'll see in Chapter 7, this argument almost certainly explains the lack of any kind of air on the extremely hot planet Mercury.

Another prominent object without an atmosphere is our Moon. There the culprit is not heat as much as a lack of enough gravity. If the Moon originally had an atmosphere—and many astronomers think it did—we might expect it to have been solar heated, much like Earth's air today. With an *average* molecular velocity of 0.5 kilometer/second on the Moon, we then expect that several of the *high-velocity* molecules would have attained values larger than the 2.5 kilometers/second needed to escape. Over long periods of time, the Moon would thus have gradually lost its atmosphere.

Throughout the text, we shall often compare escape velocities and gas particle velocities while studying varied objects in the Universe. Of special concern, toward the end of Part III, we'll note that the state of an atmosphere is an important factor when evaluating the prospects for life on alien worlds. The temperature and the mass of planets are key physical quantities. Together, they determine the likelihood of any planet having an appreciable atmosphere—to warm life, to protect it, to enable it to thrive.

Magnetosphere

Another part of Earth also helps protect us from the harsh realities of outer space. Discovered by some of the early man-made satellites launched in the late 1950s, this region lies far above the atmosphere. Known as the magnetosphere, Figure 5.9 shows how this invisible region envelopes our planet except near the north and south poles.

Instrumented spacecraft have found that the magnetosphere is dominated by two doughnut-shaped zones of radiation, one about 3000 and the other 20,000 kilometers above Earth's surface. Called the **Van Allen belts**, these zones are named for the American physicist whose instruments on board one of the first artificial satellites detected them. We call them "belts" because these two concentric zones are most pronounced near the Earth's equator or midsection, and because they completely surround the planet, as shown in the figure.

Both the atmosphere and the magnetosphere consist of gas. However, there their resemblance ends.

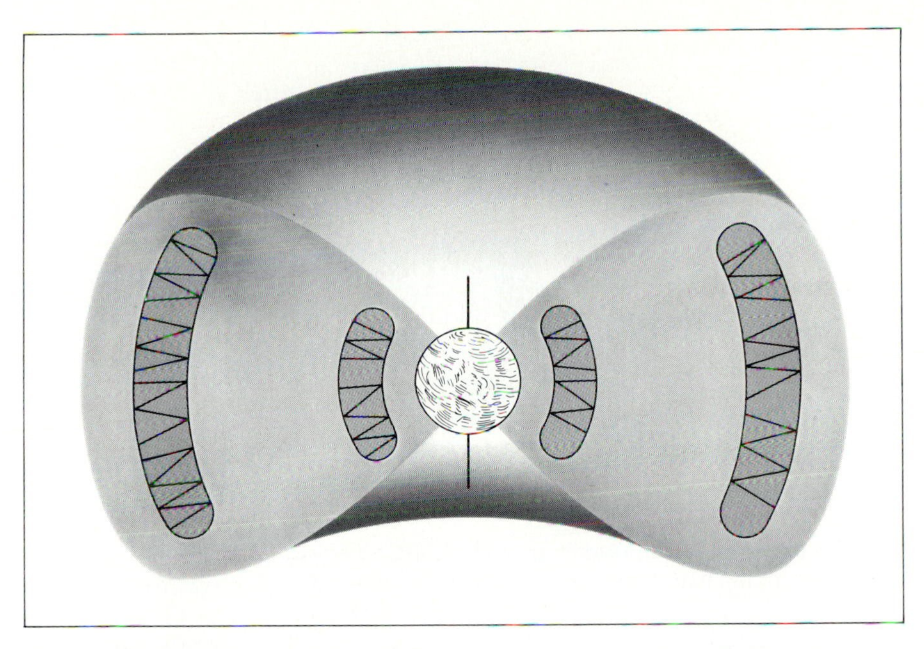

FIGURE 5.9 High above Earth's atmosphere, the magnetosphere (lightly shaded area) contains at least two doughnut-shaped regions (heavily shaded areas) of magnetically trapped charged particles. These are the Van Allen Belts.

While the low-lying atmosphere contains mostly neutral molecules, the high-lying magnetosphere is made exclusively of charged elementary particles, namely plasma, as discussed in Chapter 2.

The atmosphere and magnetosphere also differ for another reason. Whereas humans and other life forms need the atmosphere for warmth and protection, we could never survive in the Van Allen belts. Much of the magnetosphere is full of intense and harmful radiation, chiefly because the individual charged particles have very large velocities.

In the preceding section, we noted that the velocities of the molecules in our atmosphere are at most a few kilometers/second. They collide with our bodies all the time, sometimes heating us. But they do so sluggishly. Still, on a hot summer day, the air molecules can transfer surprising amounts of energy, thus heating our bodies enough to make us uncomfortable. Fortunately, biological mechanisms take over automatically, forcing our bodies to cool by sweating; in this way, our bodies are thermostatically controlled for maximum comfort.

None of this could occur in the Van Allen belts. There the individual charged particles have much greater velocities than the molecules in our atmosphere. Colliding violently with an unprotected human body, the charged particles would deposit lots of energy where they make contact, causing severe heat damage to living organisms. Without sufficient shielding on the *Apollo* spacecraft, for example, the astronauts might not have survived the passage through the magnetosphere on their journey to the Moon.

What is the origin of the magnetosphere and the belts within? Recognize first of all that our entire planet has a magnetic force field. Like the dynamos that run our industrial machines, the spin of electrically conducting metals deep inside Earth induces magnetism. Sketched in Figure 5.10, the resulting force field mimics that of a bar magnet. Having a north and a south pole, the magnetic force field surrounds our planet three-dimensionally.

Traveling through space, neutral particles and electromagnetic radiation are unaffected by Earth's magnetism. But electrically *charged* particles are strongly influenced. Again, it's a case of electricity and magnetism going hand in hand; one affects the other, and conversely. In fact, charged particles incident on Earth—mostly protons and electrons from the Sun and other, more distant cosmic objects—can become trapped by Earth's magnetism. In this way, the magnetic force field sketched in Figure 5.10 exerts elec-

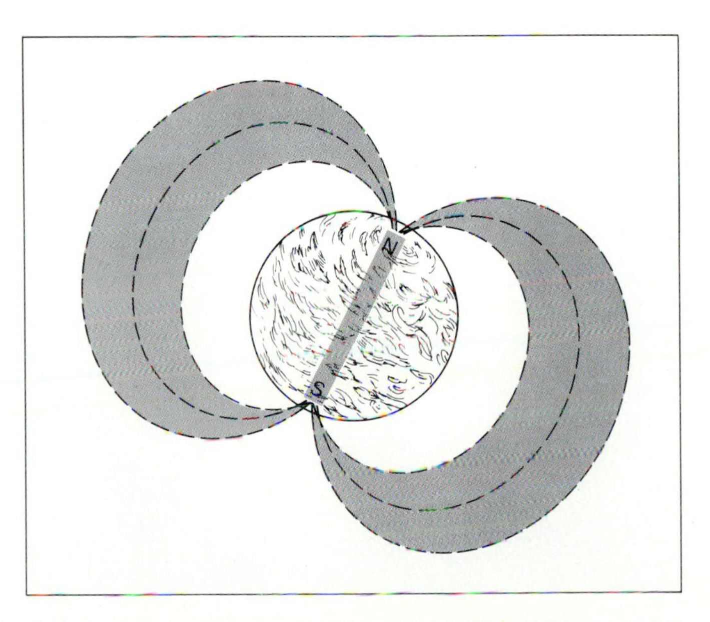

FIGURE 5.10 Earth's magnetic force field (dashed lines) can be imagined as though a bar magnet were inside our planet.

tromagnetic control over the particles, herding them into the shape of the Van Allen belts shown in Figure 5.9.

Once trapped, the charged particles move up and down in the belts, oscillating repeatedly from north to south and back to north. Gradually, they lose some of their energy, slow down, and leak out of the magnetosphere. Some return to space, while others plunge into our atmosphere. Having lost much of their original energy, most of the particles are no longer harmful to us. Even so, their collisions with the atmosphere rip apart some molecules, creating spectacular light shows. Called **aurora** and shown in Figure 5.11, these colorful displays result when the atmospheric molecules, excited upon collision with the charged particles, relax back to their ground states. Since the charged particles often escape from the magnetosphere near the north and south poles, and because energetic particles from the Sun can enter there most easily, auroras are most brilliant at high latitudes, especially within the Arctic and Antarctic Circles. On occasion, though, particularly after a storm on the Sun, the Van Allen belts become "overloaded" with more particles than normal, forcing some to escape prematurely, thus producing auroras viewable at lower, more populated latitudes.

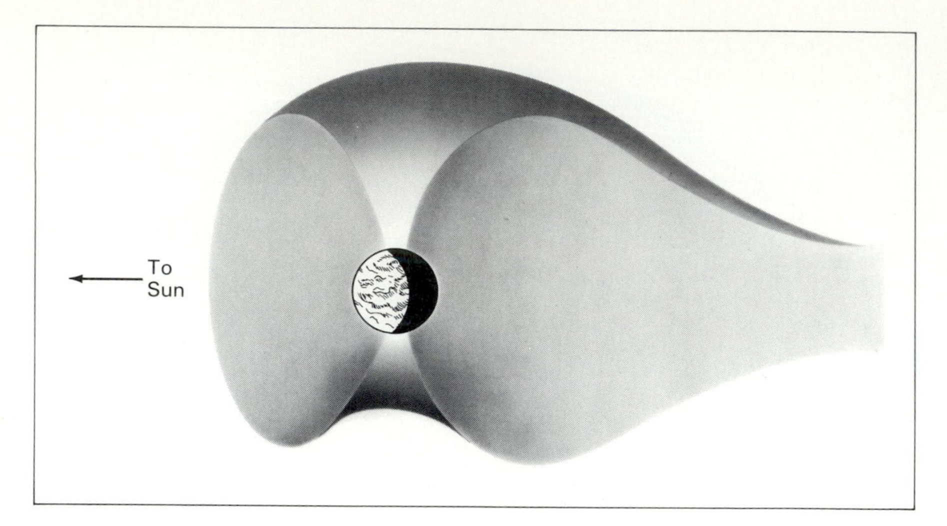

FIGURE 5.12 Earth's real magnetosphere is greatly distorted, with a long tail extending from the nighttime side of Earth well into space.

Figure 5.9 is really an idealized diagram of our magnetosphere. Actually, this region is not as symmetric as the one shown in that figure. More advanced satellites of the 1960s and 1970s managed to map the true shape of the magnetosphere. Shown in Figure 5.12, the entire region of trapped particles is rather distorted, forming a teardrop-shaped cavity. On the sunlit (or daytime) side of Earth, the magnetosphere is compressed by the steady flow of radiation and matter from the Sun. On the side opposite the Sun, the belts are extended, with a long tail often reaching beyond the distance to the Moon. The activity of the Sun and the "wind" of particles blowing from it, as we'll see in Chapter 9, greatly influences the irregularly changing shape of Earth's magnetosphere.

Of greatest importance, realize that Earth's magnetic force field electromagnetically controls the potentially destructive charged particles that venture near our planet. Without the magnetosphere, Earth's atmosphere and surface would be bombarded with these harmful particles, effectively damaging many forms of life on our planet. In fact, had the magnetosphere not existed in the first place, life might not have ever arisen at all on planet Earth.

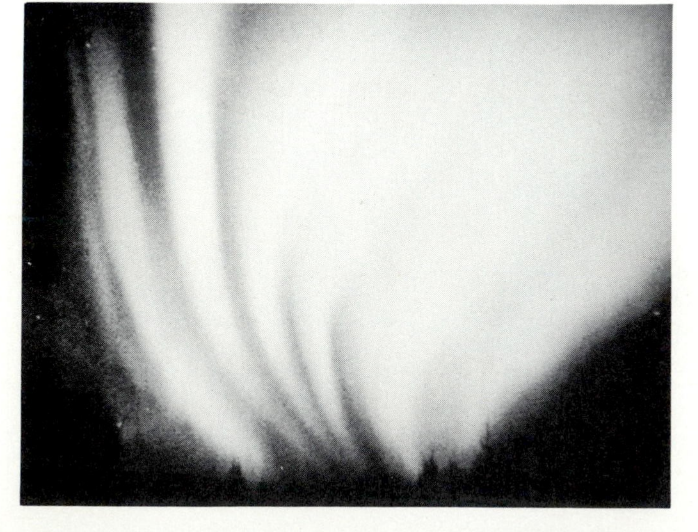

FIGURE 5.11 Colorful aurora result from the emission of light radiation after the collision of magnetospheric particles with atmospheric molecules. The aurora rapidly flash across the sky like huge wind-blown curtains glowing in the dark.

Interior

Although we reside on Earth, we cannot easily probe our planet's interior. Drilling gear can penetrate rock only so far before breaking; no substance yet fashioned by our civilization—even diamond, the hardest known material—can withstand the pressure below a depth of about 8 kilometers. That's rather shallow compared to Earth's 6500-kilometer radius. Fortunately, geologists have invented techniques that indirectly probe the deep recesses of our planet.

PLANETARY VIBRATIONS: Just as medical doctors use electrocardiograms to diagnose patients' heart conditions without actually operating, geologists can monitor the natural vibrations of Earth to study its interior without drilling a hole.

A sudden dislocation of rocky material near Earth's surface—an **earthquake**—causes the entire planet to vibrate a little. It literally rings like a bell. These are not random vibrations, but systematic waves that move outward from the site of the quake. They are given the special name **seismic waves**. And like all waves, they carry information. This information can be detected and analyzed using sensitive equipment—called a seismograph—capable of monitoring Earth tremors.

Decades of earthquake research have demonstrated two types of seismic waves. First, there are "shear waves" that normally travel through Earth's interior at about 3 to 4 kilometers/second. Seismic shear waves cannot travel through liquid; if they try to do so, they are quickly absorbed. Second, there are "pressure waves" that alternately expand and contract the material medium through which they move. Seismic pressure waves usually travel at velocities ranging from 5 to 6 kilometers/second, and can travel through liquids and solids. The exact velocity with which each type of wave travels depends on the density of the matter through which they pass. Consequently, if we can measure the time taken for the waves to move from the site of an earthquake to one or more monitoring stations on Earth's surface, we can infer the density of matter in some parts of Earth's interior.

Figure 5.13 schematically illustrates typical paths followed by shear and pressure waves from the site of an earthquake. Seismographs located around the world measure the times of arrival as well as the strengths of the seismic waves. Both

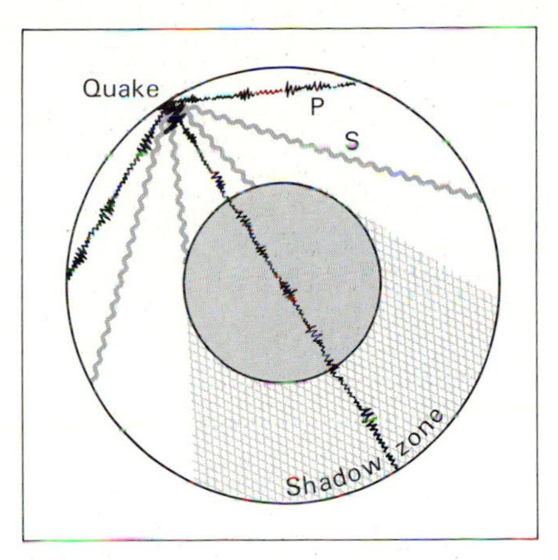

FIGURE 5.13 Earthquakes generate pressure (P) waves detectable at all seismographic stations throughout the world. Shear (S) waves are also produced, but they are not detected by stations "shadowed" by the liquid core of the Earth.

observations contain useful information—not so much about the earthquake itself, but rather, about Earth's interior through which the waves pass.

A particularly important result emerged after numerous quakes were monitored several decades ago: Seismic stations on the side of the world opposite a quake never detect shear waves; some shear waves seem to be blocked by material within Earth's interior. By contrast, pressure waves never fail to arrive at all seismographic stations after an earthquake. Most geologists favor the interpretation that the shear waves are not really blocked as much as they are absorbed—absorbed by liquid at the center of the Earth. The fact that every earthquake exhibits a "shadow zone" (see Figure 5.13), where shear waves are never detected, is the best evidence that the core of our planet is hot enough to be in the liquid state.

MODELING: Since quakes often occur at widespread places across the globe, geologists have accumulated a large amount of data about shadow zones and seismic-wave properties. They have used these data, along with direct knowledge of surface rocks, to build models of Earth's interior. Thus these models are not strictly theoretical—no one is just sitting back in a chair thinking about what the insides of Earth might be like. Modern geologists combine keen theoretical insight with the latest data acquired by monitoring stations around the world. Using computers, they then try to find the numerical solution most consistent with all the data.

Figure 5.14 presents the results of a recent model accepted by most scientists. As shown by the graphs, the two most useful physical quantities—density and temperature—both vary with depth. Specifically, from Earth's surface to its very center, the density increases from about 3 to a little more than 10 grams/cubic centimeter. The well-known properties of rock at the surface of our planet then suggest that an inner zone must be rich in nickel and iron; under the heavy pressure of the overlying layers, these metals can be compressed to the abnormally high densities predicted by the model. This inner zone, or core, measures about 3500 kilometers in radius, as sketched in Figure 5.14.

Most theoretical models predict that the core must be nearly pure nickel or iron. However, without direct observations, we cannot be absolutely sure. At any rate, all geologists agree that the bulk of the core must be in the liquid state. The existence of the shadow zone demands that.

Furthermore, some researchers claim that the pressure near the very center is great enough to raise the melting point of all matter, forcing the iron there back into the solid state. Although conceivable, there is currently no direct evidence to support this conjecture.

The core is surrounded by a thick mantle and topped with a thin crust. The mantle extends nearly 3000 kilometers below the surface, comprising the bulk (80 percent) of our planet's volume. The rigid crust, on the other hand, has an average thickness of only 10 kilometers. And since, as mentioned earlier, geologists have been unable to drill deeper than about 8 kilometers, we recognize a startling fact: No experiment has yet succeeded in piercing Earth's

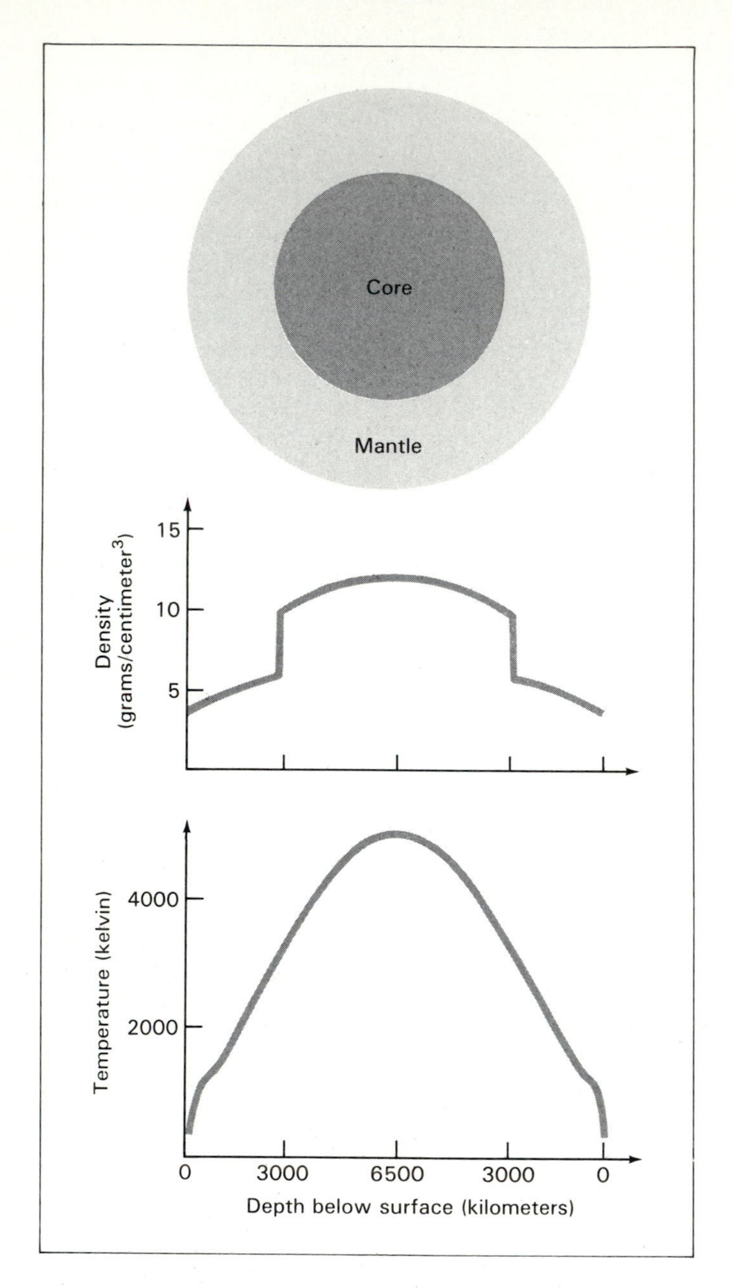

FIGURE 5.14 Computer models of Earth's interior imply that the density and temperature vary considerably through the mantle and core. Note the sharp density discontinuity between Earth's core and mantle.

crust to recover a sample of the mantle.

Fortunately, we are not entirely ignorant of the mantle's properties. In a **volcano**, hot lava upwells from somewhere below the crust, thus giving us some inkling of what it's like down there. The density and temperature of the newly emerged lava agree well with the predictions of the model sketched in Figure 5.14.

Models of Earth's interior suggest that much of the mantle has a density generally midway between those of the core and crust—namely, about 5 grams/cubic centimeter. Since this is the density of the usual iron–magnesium–silicate mixtures found in many surface rocks, we conclude that much of the mantle must be made of dense rock. Geologists often call such iron–magnesium–silicate mixtures by another name—basalt. You may have seen some dark gray basaltic rocks scattered across Earth's surface, especially near volcanoes. This type of rock is essentially cooled and solidified lava that upwelled from Earth's interior.

Basalt contrasts with the less dense (3 grams/cubic centimeter) granite type of rock comprising most of the crust. Granite is richer than basalt in the light elements silicon and aluminum, thus explaining why the surface continents never "sink" into the interior. Instead, granite's low-density composition tends to "float" atop the denser matter of the mantle and core below.

So Earth is not a homogeneous ball of rock. Its density changes from low values at the surface to high values at the core. Such variation in density and composition is termed **differentiation**. Why isn't our planet just one big, rocky ball of uniform density? The answer is that Earth must have been entirely molten at some time in the past. As a result, the higher-density matter sank toward the core, while the lower-density matter was displaced toward the surface.

The effect of this ancient heating can still be seen in the computer modeling of Earth's interior as it exists today. Figure 5.14 shows a sharp increase in temperature from the 300-kelvin values at the surface to an extremely hot 5000 kelvin at the center. In fact, the heat is so great within Earth's core to guarantee the melting of even strongly bound elements such as iron.

Recognize that Earth's interior heat is not measured directly. Instead, it is inferred from numerical experiments performed on computers. Although these calculations predict very high temperatures deep down below our feet, the models seem reasonable since the seismic-wave observations strongly imply a molten core under great pressure. Actually, we shouldn't be too surprised, for miners have always known that the deeper they dig, the hotter it gets. And active volcanoes surely imply a great inferno well beneath our feet. What is surprising is the magnitude of the heat in the core. Earth's central temperature nearly equals that of the Sun's surface!

CAUSE OF DIFFERENTIATION: Only the core of Earth is molten now. The temperature of today's mantle is not hot enough to melt its rocky matter. Yet for Earth to have become differentiated, as suggested by the models, the *entire* planet must have been molten at some ancient time. Only a fluid can rearrange itself in this way, with the density increasing toward the core.

How could Earth have liquefied completely? What process was responsible for melting the entire planet? To understand the answers to these questions, we must visualize the past. Shortly after Earth formed, it must have been bombarded with all sorts of debris left over from the formative stages of our Solar System of nine planets and one Sun. The Moon still shows the scars of this early bombardment. Collisions of boulders typically the size of automobiles with the primitive surface of Earth would surely have released large amounts of energy. Since collisions produce friction, some heat would have been generated. This early bombardment was probably severe enough to have melted the entire *surface*.

For a long while, researchers thought that Earth's interior became heated in this way. But modern studies of heat transfer in rocks prove that surface bombardment is an unlikely cause of Earth's interior melting. Even if the bombardment had been extremely heavy, the energy deposited at the surface by collisions with foreign debris could not have melted the planet to a depth of more than a few tens of kilometers. Rock simply doesn't conduct heat well; when we pick up an exposed rock on a hot sunny day we find that the bottom of it is still cold.

The only feasible process that could have heated the entire Earth is **radioactivity**, which is the release of energy by certain kinds of rare, heavy elements, such as uranium, thorium, and plutonium. They emit the energy as their complex, heavy nuclei decay into simpler, lighter nuclei. Figure 5.15 illustrates the breakdown of several well-known radioactive elements. As shown, some heavy uranium or thorium nuclei change into lighter lead nuclei, while some plutonium decays into bismuth. This change happens spontaneously, without any external influence.

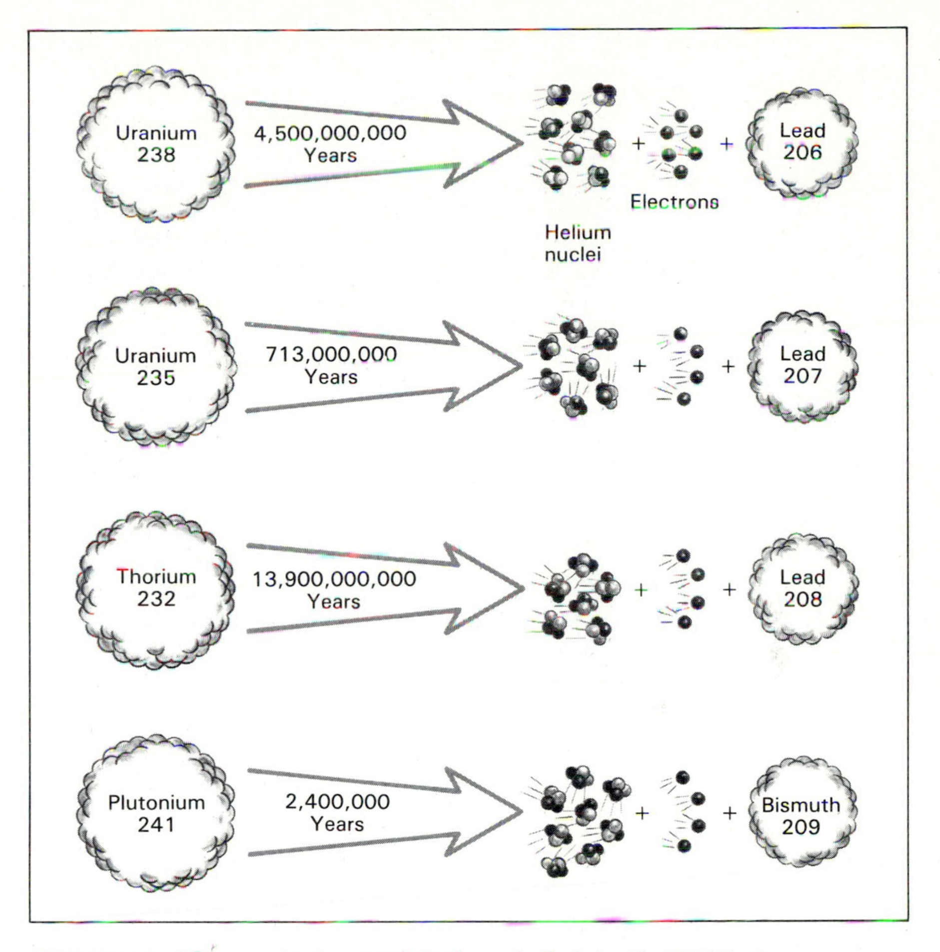

FIGURE 5.15 After certain characteristic times (called the "half-life", shown numerically in the arrows), *half* of all heavy, unstable radioactive elements change into lighter more stable elements (consult Interlude 5-2). Also emitted in the process are elementary particles that collide with, and thus heat, the surrounding rock.

Radioactive decay occurs whenever nuclei are inherently unstable. The nuclei achieve greater stability by disintegrating into lighter nuclei. But they do not change immediately. Each type of nucleus needs a specific amount of time to decay. The "half-life" is the time for which half of the original parent nuclei disintegrate into daughter nuclei. For example, if a billion radioactive nuclei were embedded in a rock, a half-billion nuclei would remain after one half-life, a quarter-billion after a second half-life, and so on.

For some radioactive elements, the half-life is short, even as brief as a few seconds. Other nuclei have decay times of a million years or more. Several elements noted in Figure 5.15 have half-lives of billions of years. Every kind of radioactive element has its own characteristic half-life, and most of them are now well known from studies in terrestrial laboratories during the past few decades.

Even so, how could minute atoms heat an object as large as Earth? The answer is that radioactive disintegrations produce subatomic particles that zip away with very fast velocities at the moment of transformation. Should these elementary particles collide with something else, their kinetic energy of motion is converted into heat. And this heat can then be absorbed by surrounding matter.

As you might imagine, the amount of heat generated by each radioactive decay is rather small. A rock the size of a golf ball contains enough radioactive elements to heat a thimble of water only 1 kelvin after several hundred years. But we need to stress two important points here: First, Earth has a lot of matter, vastly more than a single golf ball; second, much time is available, not just a few centuries. Provided that enough radioactive elements were originally spread throughout the primitive Earth like raisins in a cake, the entire planet—crust to core—probably melted over the course of about a billion years. That's surely a long time by human standards, but not such a long time in the cosmic scheme of things. Evidence that Earth began solidifying roughly a billion years after its original formation (see Interlude 5-2) further supports this heating scenario early in our planet's history.

EARTH'S GEOLOGICAL HISTORY: The arguments above imply that our entire planet was molten during most of its first billion years of existence. It differentiated during that time, as much of the denser matter sank toward the core while the lighter matter rose to the surface.

Radioactive heating did not stop after the first billion years or so. It continued even after Earth's surface cooled and solidified. However, radioactivity occurs in only one direction, always producing lighter elements from heavier ones. Once decayed, the heavy and rare radioactive elements are not easily replenished.

The early source of heat then decreased with time, steadily cooling our planet during the past several billion years. In so doing, it cooled from the outside in, much like an ice cube or a hot potato. In this way, the surface became crust-like, and the differentiated interior became semihardened into the positions now implied by seismic studies.

Today, radioactive heating continues throughout the Earth. However, there's probably not enough of it to melt any part of our planet. The inferno now at the core is mainly the trapped remnant of a much hotter Earth that existed eons ago. By contrast, the continental matter has completely cooled and solidified, making the surface a reasonably comfortable place from which technological intelligent life can now unravel the rich history of our planet.

INTERLUDE 5-2 ***Radioactive Dating***

The decay of unstable radioactive nuclei into more stable nuclei is an important phenomenon of nature. Not only does it heat and thus differentiate planets such as Earth, but it also provides a useful tool to date any rocks we can get our hands on.

How can geologists determine the age of a rock? The first step is to measure the amount of stable nuclei of a given kind, for example, lead. This is then compared to the amount of remaining unstable "parent" nuclei (e.g., uranium or thorium) from which the stable "daughter" nuclei descended. The third step involves knowing the rate (or "half-life") at which the disintegration occurs. Then, by making some assumptions about the original abundances of parent and daughter nuclei, the age follows directly.

For example, if half of the parent nuclei of some element has decayed, the age of the rock equals the half-life of the radioactive nucleus studied. Similarly, if only a quarter of the parent nuclei remains, the rock's age is twice the half-life of that nucleus.

In this way, the most ancient rocks on Earth are dated to be nearly 4 billion years old (3.8 billion years old, to be more precise). These rare specimens comprise much of Greenland and Labrador. However, Earth itself must be older than this, for we know that all of it must have been molten early in its history. Otherwise, the planet could not have differentiated. Since the radioactive dating technique gives us the time elapsed since the rocks solidified, this 4-billion-year value must represent only a portion of the true age of our planet.

A variety of indirect arguments (to be discussed at various places in the text) suggests that 4.5 billion years ago is a special date. For example, all meteorites (undifferentiated rocky debris that once traveled through space before accidentally colliding with Earth) are radioactively dated to be 4.5 billion years old. Furthermore, the oldest Moon rocks, which also apparently never melted, have an age close to 4.5 billion years. Clearly, something important happened in the vicinity of Earth, and perhaps throughout the entire Solar System, nearly 5 billion years ago. That something, astronomers reason, was the formation of the Sun and planets.

SUMMARY

Planet Earth is made of solids, liquids, and gases. The interior is metal, surrounded by rocky matter, much of it partly molten. The exterior is gaseous, most of it in the atmosphere that engulfs us, yet some of it in the magnetosphere far above us. Both of these "spheres" act as protective umbrellas for life on Earth.

Although hard to recognize, change is an important feature on Earth. Earthquakes and volcanoes are examples of obvious change. But subtle change occurs as well. The spin of Earth is slowing, the core of the Earth is cooling, mountain chains and ocean trenches are constantly being shaped. And as we'll note in Part III, Earth's environment and life itself slowly change over the course of time.

Earth is our home. In some ways, it seems special; in other ways, ordinary. Whichever, Earth is the cosmic platform from which we strive to decipher the nature of our diverse Universe.

KEY TERMS

atmosphere
aurora
convection
core
density
differentiation
earthquake
escape velocity
greenhouse effect
hydrosphere
lithosphere
magnetosphere
mantle
mass
molecular velocity
radioactivity
revolution
rotation
seismic wave
Van Allen belts
volcano

QUESTIONS

1. Explain in some detail the operation of the greenhouse effect on Earth, and speculate about its implications for the future of humankind.
2. Pick one: Earth is (a) perfectly spherical, (b) highly differentiated, (c) uniformly spinning, (d) convectively cooled, (e) photoelectrically heated.
3. Pick one: Seasons on Earth are caused by the (a) variable orbital motion of Earth, (b) precession of Earth on its axis, (c) variable tilt of Earth's axis relative to the Sun, (d) changing distance of Earth from the Sun, (e) variable emission from the Sun.
4. Explain how both Earth's atmosphere and magnetosphere protect life from the harsh realities of outer space. Be specific for each region.
5. What mechanism(s) causes aurora, and why is this phenomenon most intense following a storm on the Sun?
6. How do we know that Earth is differentiated, and in particular that the core of our planet is made of molten metal?
7. Draw a diagram illustrating the basic features of Earth's interior, and briefly explain the method used by geologists to conclude that Earth is differentiated.
8. Explain why shear seismic waves are not observed by seismic instruments in Antarctica following an earthquake in Alaska.
9. What is meant by density? Estimate and compare the densities of at least a half dozen objects near you. Does a dense object necessarily have to be a massive object?
10. Explain why the interior of our planet is much hotter and denser than the surface.

FOR FURTHER READING

ANDERSON, D., AND DZIEWONSKI, P., "Seismic Tomography." *Scientific American*, October 1984.

CARTER, W., AND ROBERTSON, D., "Studying the Earth by Very-Long-Baseline Interferometry." *Scientific American*, November 1986.

HONES, E., "The Earth's Magnetotail." *Scientific American*, March 1986.

JUDSON, S., KAUFFMAN, M., AND LEET, L., *Physical Geology*. Englewood Cliffs, N.J.: Prentice-Hall, 1987.

PRESS, F., AND SIEVER, R., *Earth*. San Francisco: W. H. Freeman, 1974.

SIEVER, R., "The Dynamic Earth." *Scientific American*, September 1983.

WEINER, J., *Planet Earth*. New York: Bantam Books, 1986.

6
THE MOON: *Age of Apollo*

The Moon is the only natural satellite of planet Earth. Despite its nearness, the Moon is a very different kind of object. It has no air, no sound, no water. Weather is nonexistent, with clouds, rainfall, and blue sky all absent. Boulders and pulverized dust litter the landscape. Virtually everything seems frozen in time.

For the past several billion years rocks have impacted, and during the past two decades Earth men and their machines have landed. But aside from these, nothing has interrupted this magnificent desolation. The Moon shows strikingly little activity. There are no obvious movements of the crust, probably no volcanoes at the present time. Apparently, the Moon is a dead object in space. It's dead geologically, and it's dead biologically.

FIGURE 6.1 Since the Moon does not emit its own visible radiation, we can see it only by the reflected light of the Sun.

The first people to aim telescopes at the Moon, among them the seventeenth-century Italian scientist Galileo, noted large dark areas resembling Earth's oceans. (Some of these are clear even to the naked eye.) They also saw light-colored areas resembling the continents. Both types of regions are clear in Figure 6.1, a photograph of the full Moon, made possible by sunlight reflected toward Earth. Such light and dark surface features are also evident to the naked eye; together they create the face of the familiar "Man-in-the-Moon."

Today we recognize that the dark areas are not oceans at all, although they are still called **maria**, a Latin word meaning seas. The lighter areas are now known to be elevated above the maria and are accordingly called the lunar **highlands**, "luna" being the Latin word for Moon. However, they are neither chemically nor structurally similar to Earth's continents or mountains.

Telescopic observations further resolve each of these surface features into numerous bowl-shaped depressions, or **craters**, most of which apparently formed eons ago in the Moon's history.

To an astronaut standing on the Moon's surface, Earth appears almost stationary in the sky. The Moon's rotation (spin) period about its axis nearly equals its revolution (orbital) period about the Earth, ensuring that the Moon keeps roughly the same side facing Earth all the time. Since these periods roughly equal a month, the Sun rises and sets at 2-week intervals. Aside from this, there are virtually no other changes.

At local noontime on the Moon, rock temperatures rise as high as 400 kelvin, higher than water's boiling point. But at night or in the shade, the temperature plummets to as cold as 100 kelvin, well below water's freezing point. The Moon has virtually no thermal convection or transfer of heat. Warmth does not spread at its surface, for lunar soil is made of pulverized dust that doesn't distribute heat well. Nor does heat move through lunar air, for the Moon has no air. Without an artificial Earth habitat such as a spacecraft or spacesuit, the Moon is completely alien to the needs of earthlings.

If the Moon is so hostile and desolate, why bother to explore and study it? It seems like a terribly uninteresting place. Not so. The Moon might contain some important clues about the origin of our Solar System. The fact that it hasn't changed much means that the Moon must be a truly primitive object. In contrast to Earth and the other nearby planets, the bulk of the lunar surface is ancient and uneroded. As such, its pristine condition suggests that the Moon is a vital key in our continuing quest to unlock the secrets of our Solar System.

The learning goals for this chapter are:

- to begin to compare and contrast the properties of cosmic objects beyond Earth, in this case the Moon
- to appreciate the magnificent desolation of our nearest neighbor
- to recognize that the Moon has changed very little during the past several billion years
- to realize that the Moon's surface (as we now see it) was formed and molded by dynamic events early in its history
- to understand how we can piece together an outline of the Moon's history, based largely on studies of rocks returned by the *Apollo* astronauts

Moon in Bulk

As done earlier for Earth, let's begin our study of the Moon by describing its bulk properties.

DISTANCE: We can determine the distance to the Moon rather easily. As noted in Chapter 4, parallax methods can be used successfully in the case of the Moon, provided that Earth's diameter is the baseline. A good result can be obtained in this way, although its experimental error is somewhat larger than those of more modern techniques.

A more accurate lunar distance can be measured using the radar technique. As mentioned in Chapter 4, precise timing of the outbound transmission and return echo yields a measure of the distance traveled by radio signals moving at the velocity of light. Since the Moon is much closer than any of the planets, the returned echo is quite strong and is easily received with a radio telescope after about a 2.5-second wait. And since the velocity of the transmitted radio signal is 300,000 kilometers/second, and the Earth–Moon distance equals one-half the distance traveled in a round trip, the Moon's distance must be about 300,000 kilometers/second multiplied by 1.25 seconds, or about 375,000 kilometers.

The transmit-echo round trip, which can be measured to within microsecond accuracy, actually takes a little more than 2.5 seconds; the real Earth–Moon distance is close to 384,000 kilometers. Be sure to recognize that this is an *average* distance. At any moment, the actual distance is somewhat more or less, depending on the Moon's position in its slightly elliptical orbit about Earth. Repeated radar bombardment of the Moon has enabled astronomers to model the Moon's orbit to an accuracy of a few meters. Such precision was needed to program unmanned spacecraft to soft-land successfully at specific locations on the lunar surface. (See also Interlude 6-1.)

SIZE: Observations prove the Moon has an angular diameter of 0.5 arc degree. You could verify this by holding up a compass to the sky. Knowing the Moon's distance and a few principles of basic geometry, we can easily calculate the real lunar

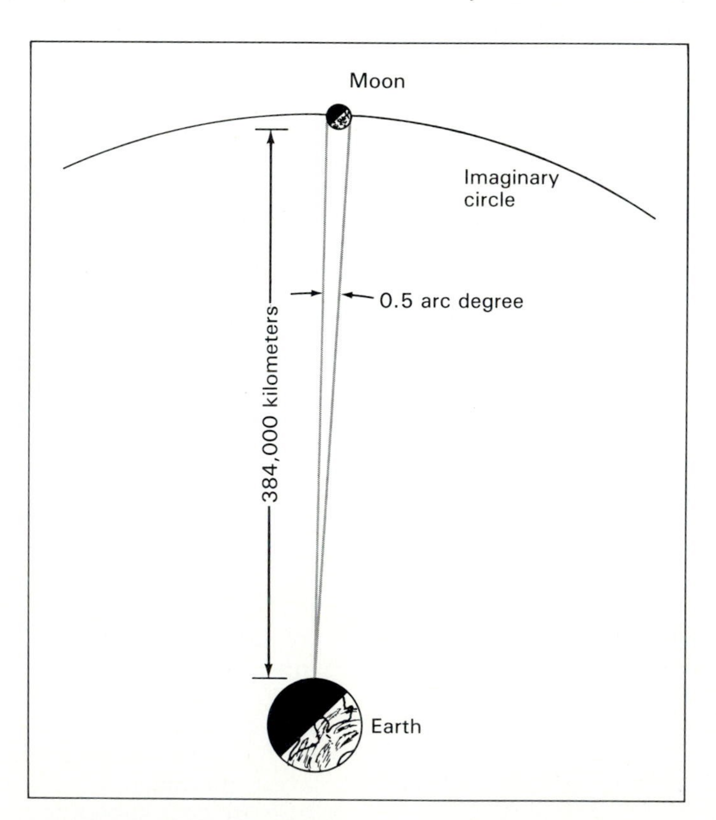

FIGURE 6.2 This simple geometrical construction of the Earth-Moon system, including an imaginary circle having a radius equal to the Earth-Moon distance, permits a calculation of the Moon's size.

INTERLUDE 6-1 ***Lunar Laser Ranging***

Several manned *Apollo* (U.S.) and unmanned *Luna* (U.S.S.R.) missions left equipment on the Moon. One of the most interesting devices left by the Americans is an array of mirrors designed to intercept light pulses launched from Earth and to reflect them back toward Earth. Each small mirror resembles those used on highway posts and signs to reflect automobile headlight beams. The whole array of reflectors is not much larger than this book.

This equipment can be used to measure the Moon's distance by a method similar to the radar technique described in Chapter 4 and noted elsewhere in this chapter. It differs from the radar technique, however, in two important ways. First, the lunar-laser-ranging method does not use radio signals; instead, it uses a *laser* which transmits highly concentrated *light* pulses. (The word "laser" is an acronym for *l*ight *a*mplification by *s*timulated *e*mission of *r*adiation.) Second, the reflectors are positioned at well-known locations on the Moon, allowing researchers to determine precisely from where the light echoes arise. (Radar echoes bounce from much larger areas on the Moon's surface and are thus a little less accurate. On the other hand, ordinary sunlight reflected from the Moon does not interfere with the radio signal, enabling the radar technique to be more versatile.)

This experiment is not easy. Aiming the laser at the reflector on the Moon is comparable to hitting an American dime with a rifle at a distance of about a kilometer. And like the radar technique, the return echo of light is a lot weaker than the transmitted light; about one photon is detected upon return from among a billion billion (10^{18}) photons transmitted in every successful burst of light sent toward the target.

This optical technique can currently determine the distance to the Moon, and hence model its orbit, to an accuracy of about 6 centimeters. Further refinements in equipment during the next few years should enable researchers to decrease the error to within a centimeter or two.

Interestingly enough, this increased accuracy should permit the lunar-laser-ranging technique to measure any movements in Earth's crust. Using the lunar reflectors as reference points in space, observers at strategically placed laser stations on Earth will soon be able to determine their relative positions so precisely that the rate and direction of crustal motions should be measurable within a few years. This information will almost certainly help us better understand the forces that produce earthquakes and volcanoes on our planet.

size. Figure 6.2 schematically diagrams this simple geometrical argument. As shown, 0.5 arc degree is related to the lunar diameter in much the same way as a full 360 arc degrees relate to the entire circumference of an imaginary circle cutting through the Moon. It then follows directly that the Moon's radius is about 1750 kilometers. We thus conclude that the Moon is about one-fourth the radius of Earth.

It is interesting that despite the much larger size of the Sun, both the Moon and the Sun subtend an angle of nearly 30 arc minutes when viewed from Earth. In other words, the *apparent* size of the full Moon just about equals the *apparent* size of the glowing Sun. This match ensures perfect blockage of the Sun by the Moon during a total solar eclipse. It's a convenient accident of nature, and one that provides us with a most spectacular event.

Be sure to keep in mind the notion of distance when viewing an object's *appearance* projected onto the sky. Although often hard to measure, distance is an important factor when measuring the absolute (true) physical size of any object. It's crucial to distinguish absolute size from apparent size; absolutely large objects can appear small when far away, whereas absolutely small objects can appear large when nearby.

MASS: We can determine the mass of the Moon either by studying the Moon's response to the pull of Earth's gravity, or by observing the motion of man-made spacecraft in the vicinity of the Moon. Each of these methods requires a knowledge of the law of gravity, which we'll study in the next chapter. For now, recognize that if we know both the mass of an orbiting spacecraft and its distance from the Moon, we can find the Moon's mass. The result is 7.5×10^{25} grams, a mass approximately 80 times less than Earth's.

It's a good idea to place things into perspective while proceeding through the text. At this point, for example, note that Earth's mass, radius, and volume are approximately 80, 4, and 60 times those of the Moon, respectively. Accordingly, Earth is a much larger astronomical object than the Moon. Both are dwarfed, however, by the Sun, as we'll see several chapters hence. Always compare astronomical objects by using some convenient common denominator: Mass, size, and volume are good indicators.

DENSITY: Having the Moon's size enables us to find its volume and, with its mass known, we can estimate the average lunar density. The result, 3 grams/cubic centimeter, contrasts with the average Earth value of nearly 6 grams/cubic centimeter.

Recall that terrestrial studies imply that Earth is differentiated, with the density increasing considerably from the surface to the core. However, the density variation throughout the lunar interior is probably less than for Earth; the Moon seemingly never heated enough to allow complete differentiation. We don't yet know for sure, but the current consensus is that the entire Moon is made largely of basalt, a cooled lava type of rock having a density of about 3 grams/cubic centimeter. At any rate, basalt is the main component of the 400,000 grams (nearly ½ ton) of lunar rocks returned to Earth during the *Apollo* (U.S.) and *Luna* (Soviet) programs of the 1970s.

SHAPE: As for Earth, the Moon is not perfectly spherical. The Moon's equatorial diameter exceeds its polar diameter, as expected for a massive rotating object. But as in most areas of astronomy, there's a puzzle here as well.

The lunar bulge, as measured by spacecraft orbiting the Moon, is too large to be explained by the current rate of lunar spin. The equatorial diameter is nearly 4 kilometers larger than the polar diameter. Given the Moon's rather sluggish rotation—just about once around its axis per month—we theoretically expect its bulge to be a mere 0.1 kilometer.

The easiest explanation of the extra bulge suggests that the Moon rotated more rapidly long ago, much as we noted for Earth in Chapter 5. Part of he bulge might also be caused by the tidal forces of Earth pulling on the Moon. Indeed, most of the Moon's bulge points directly toward Earth.

The surprisingly large lunar bulge might also explain why only one side of the Moon faces Earth. Over the course of billions of years, gravity has gradually pulled more on that part of the Moon closest to Earth. It is natural, then, for the Moon's bulge to be preferentially pointed toward Earth, virtually locked into that position by Earth's strong gravitational force field.

LUNAR AIR: What about the Moon's atmosphere? That's easy; there is none! All of it apparently escaped. Recall from Chapter 5 that massive objects have a greater chance to retain their atmospheres, since large velocities are needed for atoms and molecules to escape. An escape velocity of about 2.5 kilometers/second for the Moon contrasts with the 11 kilometers/second derived earlier for Earth. The low lunar escape velocity results mainly from the relative lack of lunar mass. Although the Moon is only 4 times smaller than Earth, it's 80 times less massive. Thus the Moon has a lot less pulling power.

Over a period of time—how long, no one knows—the Moon's atmosphere drifted away. And once a planet or any other massive object loses its air, it's hard to regenerate it. (In Chapter 24, however, we'll argue that Earth did just that; the atmosphere currently surrounding Earth is probably a secondary atmosphere—one replenished by volcanic gases. Earth's primary atmosphere was probably lost in an earlier epoch when our planet was very much hotter than it is today.)

Impacting Meteorites

DEBRIS: Despite the lack of activity on the Moon, the lunar surface is not entirely changeless. Even in the absence of wind and water, there is some evidence of slight **erosion**, which is the wearing away of surface matter. Figure 6.3 shows some examples of this surface wear and tear, especially the well-rounded mountaintops. Note also that the smaller craters seem partially filled with either lunar or foreign material, whereas the larger craters have rims that are clearly smoothed to some extent.

The primary source of erosion is the **interplanetary matter** that collides with the lunar surface. This is mainly debris strewn throughout the Solar System, much of it rocky objects just wandering around until they accidentally collide with some planet or moon. Such collisions often dislodge large amounts of lunar material, scarring, cratering, and generally reshaping the landscape in a chaotic manner. Even the microscopic particles escaping the Sun (mostly high-velocity protons having masses of about 10^{-24} gram, as discussed in Chapter 9) chip away at the lunar surface, eroding crater shapes, smoothing their edges, and even helping to fill in some of the smaller ones.

In the present-day Solar System, interplanetary matter is generally of two types: large, meter-sized, but scarce **meteoroids**, and small, millimeter-sized, and more abundant

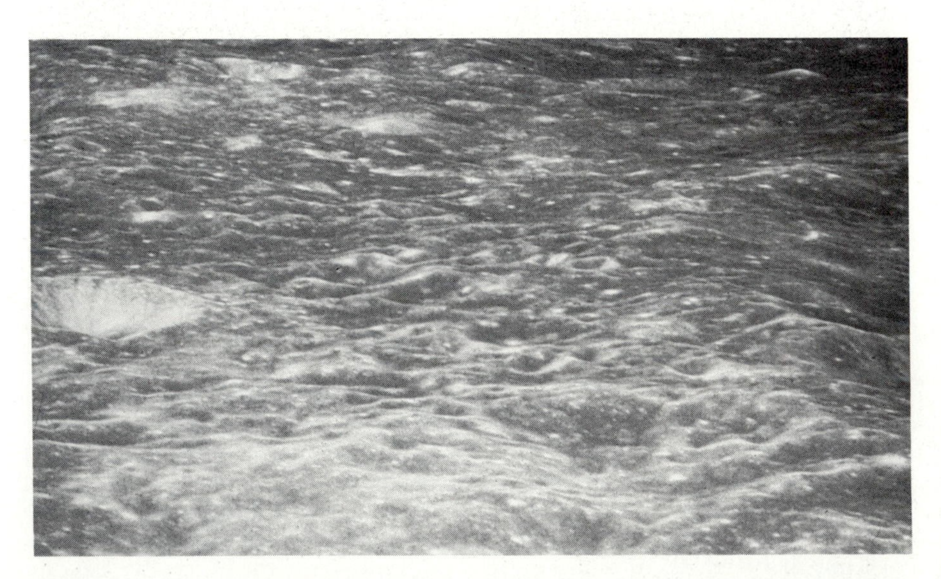

FIGURE 6.3 Despite the complete lack of wind and water on the airless Moon, the surface has still eroded a little under the constant "rain" of impacting meteorites and especially micrometeorites.

micrometeoroids. The larger objects generally span the range from 10 to 1000 grams (about 0.02 to 2 pounds) and are called **meteorites** when and if they impact with the surface of a planet or one of its moons. A few percent of these can be even larger, reaching about 10^9 grams (or about 1000 tons). The smaller objects, called micrometeoroids while in space, and **micrometeorites** upon impact, generally vary in mass from 10^{-12} to 10 grams. Both types are made of rocky and rocky-metallic matter.

The word "meteor" stems from the Greek root meaning "things high in the air." In its modern connotation, **meteor** refers to a heated, glowing object streaking through Earth's atmosphere but not yet having hit the surface. Centuries ago, this word referred to an entire repertoire of strange atmospheric effects, such as rainbows, lightning, snowfall, and so on—hence the science of meteorology, which is, by the way, the study of atmospheric weather and not at all of meteors. In any case, since the Moon lacks an atmosphere, no meteors are possible there—just large and small meteoroids prior to impact, and the remains of large and small meteorites after impact.

Even the miniature micrometeoroids can cause surprising amounts of damage. A single micrometeoroid, traveling through space with a typical velocity of 30 kilometers/second (i.e., tens of thousands of miles/hour), has an explosive power, pound for pound, about 100 times that of an equivalent mass of TNT. When such a micrometeoroid or a larger meteoroid strikes the surface of a planet or a moon, its kinetic (motion) energy is converted into mechanical (disruptive) energy, thermal (heat) energy, and acoustical (sound) energy. The inevitable result is a deep crater surrounded by a raised rim of upthrusted and ejected matter.

Impact with the surface causes sudden and tremendous pressures to build up, heating the normally brittle rock and deforming the ground like heated plastic. The instant of impact probably mimics the opening of the petals of a flower, with previously flat layers of rock pushed up and out as a result of the meteoritic explosion. The net effect is to pulverize the lunar surface, ejecting an entire array of debris from fine dust to large boulders.

One of the best examples of a large meteorite crater is the Arizona Crater *on Earth*. Shown in Figure 6.4(a), this crater somewhat resembles the one illustrated in Figure 6.4(b)—the aftermath of an explosion of a manmade nuclear bomb. Another large crater, this one on the Moon, is pictured in Figure 6.5.

The extent and depth of crater sizes can tell us something about meteorites as well as about the Moon. Massive meteorites produce larger craters. Impact velocity is also important. Atmosphere is a third factor affecting crater sizes on an impacted object, at least for those objects having an atmosphere.

In the case of Earth, the majority of meteoroids—and all the micrometeoroids—are vaporized during passage through our atmosphere. They just burn up. Only if the initial mass (before hitting the atmosphere) of the meteoroid exceeds about 10 kilograms will it survive passage through Earth's atmosphere without totally vaporizing. The result is a crater when some fraction of it strikes the ground.

The Moon, without an atmosphere, has no such protection against interplanetary matter. Large meteoroids just zoom in, collide with the surface, and produce huge lunar craters. In addition, a constant "rain" of micrometeoroids chips away at the structure of those cra-

(a)

(b)

FIGURE 6.4 (a) The Arizona Crater in North America's Southwest is 1.2 kilometers in diameter, 0.2 kilometer deep, and is thought to have been made by an interplanetary meteorite about 25,000 years ago. (b) The crater resulting from the blast of a thermonuclear warhead at a Nevada test site suggests many similarities between the way meteorites and nuclear bombs transfer their energy to the surrounding matter. (The manmade crater measures roughly 0.5 kilometer across and 0.1 kilometer deep.)

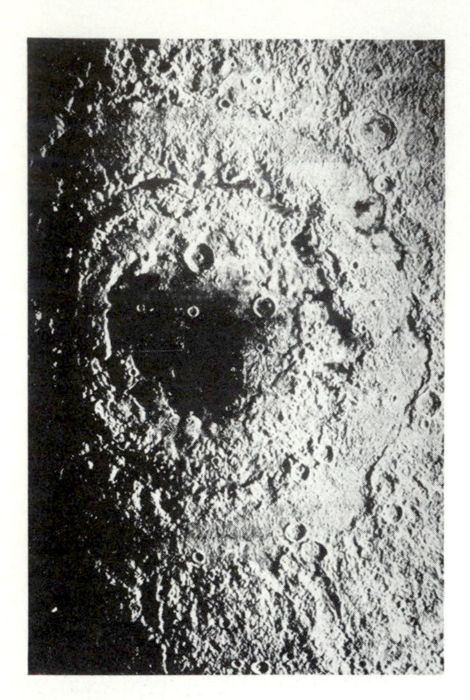

FIGURE 6.5 This large lunar crater (called Mare Orientale) shows its interior basin, now filled with solidified lava. The meteorite that produced this crater upthrust much surrounding matter which can be seen as concentric rings of cliffs. The outermost ring is nearly 1000 kilometers in diameter and is comparable to the size of New England.

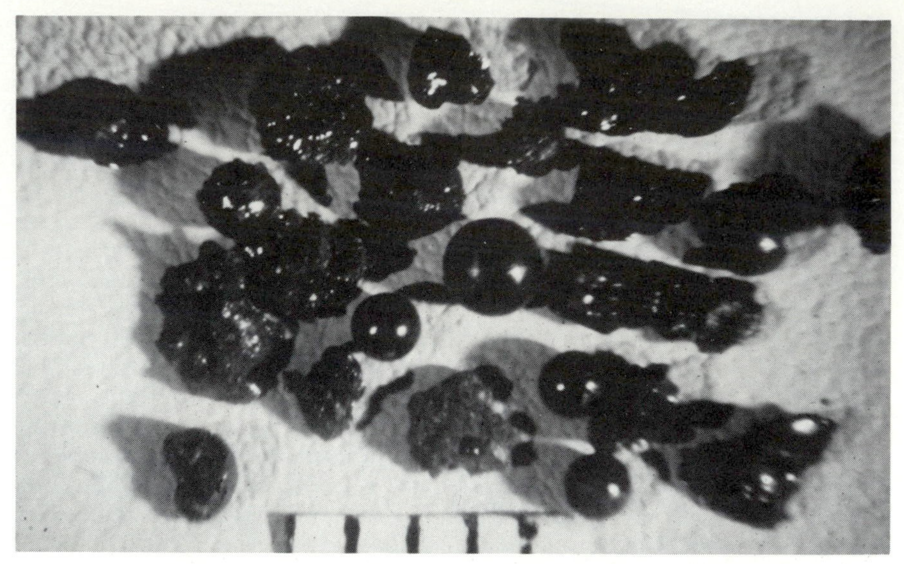

FIGURE 6.6 Craters of all sizes litter the lunar landscape. Some shown here embedded in glassy beads retrieved by American astronauts measure only 0.01 millimeter across. (The scale at the bottom is that of millimeters.)

ters. All of them together help to erode the lunar surface.

LUNAR DUST: The extent of erosion caused either by the micrometeorite sandblasting or by the larger meteorite collisions can be made a little more quantitative. Here's how we do it. Lunar photographs taken from spacecraft show very few craters having a depth between 1 and 20 meters. Craters shallower than about a meter are constantly being created across the lunar surface. This is especially true of the fresh, miniature craters seen in microscope photographs like Figure 6.6. But the lack of moderately sized craters less than about 20 meters in depth is significant. It allows us to estimate the rate of lunar erosion, since craters less than 20 meters have presumably been filled in by lunar and interplanetary matter. Radioactive dating of samples returned from the *Apollo* missions suggests that the large lunar craters were formed early in the Moon's history—about 4 billion years ago. So we can judge that the lunar erosion rate is about 20/4 or about 5 meters per billion years. This is an *average* rate, the depth of debris being thinner in the lunar highlands and thicker in the mare depressions. Apparently, most particles gradually slide downhill to the lower-lying terrain.

This average erosion rate is about 10,000 times less than on Earth. The big difference results from an abundance of prime erosion sources on our planet—especially running water and atmospheric wind. In fact, the Arizona Crater of Figure 6.4(a), only 25,000 years old and already decaying, will probably disappear in a mere million years. That's a long time compared to human life spans, but quite a short time geologically.

So meteoroid collisions with the Moon are mainly responsible for the pulverized ejecta, called **regolith** (meaning "fine rocky layer"), spread across the lunar landscape. A close look at this surface material is interesting. Figure 6.7 shows an astronaut's bootprint in the extremely fine lunar dust. Such microscopic dust has a typical particle size of about 0.01 millimeter and accounts for most of the lunar surface debris. Unlike coarse beach sand, the regolith has some definite cohesiveness, though it lacks strength; it's much like baby powder or ready-mix dry mortar.

FIGURE 6.7 A photograph of an *Apollo* astronaut's bootprint in the lunar dust, compacted to a depth of a few centimeters.

Although the regolith is often called lunar "soil," you should recognize that Moon dust differs totally from soils formed on Earth by wind, water, and life. Soil of the lunar type can form only on the surface of an airless body. Owing to the very small rate of lunar erosion, even those shallow bootprints will remain intact for millions of years, virtually frozen into the lunar landscape.

A final point is worth noting about the dusty regolith. It and other surface debris are not made exclusively of old meteorites. In fact, very little of the lunar dust is pulverized meteoritic matter. Since each meteorite dislodges several hundred times more lunar matter than its own mass, the regolith is really a mixture containing only a small fraction of meteoritic matter. Studies of the *Apollo* samples show that meteoritic matter comprises less than 1 percent of all the lunar soil. Hence, of the 20 meters of lunar debris, no more than several centimeters result from the direct accumulation of meteoritic matter.

Surface Composition

We've mentioned earlier that the Moon's surface is made mostly of basalt, which is a type of rock formed by the cooling of molten lava. However, the rocky surface, including the dusty regolith, is not made of precisely the same basaltic matter everywhere. Depending on the place, the density and chemical composition vary. As a rule of thumb, Moon rocks seem to be made of the same chemical elements as their earthly counterparts, although the proportions differ.

Generally, two kinds of rock exist on the Moon. The *Apollo* program demonstrated a clear contrast between the less-dense highlands (2.9 grams/cubic centimeter) and the more-dense maria (3.3 grams/cubic centimeter). The highlands are made largely of rocks rich in aluminum, giving it its lighter color and less density; whereas the maria's basaltic matter contains lots of iron, giving it its darker color and greater density.

Actually, long before the astronauts retrieved direct samples, researchers suspected that the lunar composition was mostly basalt of some type. This early conjecture was based on the nature of Earth's composition, which at a depth of a few hundred kilometers has a density of about 3 grams/cubic centimeter (see Figure 5.14). And as noted in Chapter 5, the terrestrial matter of Earth's mantle is largely basalt.

In contrast to Earth's soil, lunar regolith contains no organic matter like that ordinarily produced by biological organisms. No life whatsoever exists on the Moon. Nor were any fossils found, suggestive of extinct life. In fact, lunar rocks are so barren of life that the astronauts were not even quarantined after returning from the last few *Apollo* landings.

The prospects for the future development of life on the Moon are also bleak. Without an atmosphere, there can be no water—at least none at the surface. Any water on the Moon's surface would evaporate rapidly, there being neither an atmospheric shield to absorb solar ultraviolet radiation nor atmospheric pressure capable of "holding a lid on" any vaporized water molecules. During a single lunar day—approximately 2 weeks of Earth time—any water conceivably present *on the surface* would evaporate. This doesn't mean that subsurface ice is impossible. Radar studies in fact tend to support this possibility; the returned echos show evidence of colder material a few meters below the surface.

Despite this optimism for subsurface water or ice, all the lunar samples returned by the American and Soviet Moon programs were absolutely bone dry. Lunar basalt doesn't even contain minerals with water molecules locked within their crystal structure. These findings contrast with terrestrial rocks, which almost always have a percent or two of embedded water. Still, we can argue that these retrieved lunar samples represent only the surface material and hence do not necessarily conflict with the radar suggestions of ice well below the surface.

Why worry so much about water? Well, without water, it's hard to imagine the existence of life as we know it. Even if all the basic ingredients for life were placed on the surface of the Moon, we could hardly conceive of life arising. Without a freely flowing fluid—and water is the most abundant liquid known—the molecular ingredients could not possibly interact chemically. So we can be sure that the Moon is at a cosmic evolutionary dead end. No advanced forms of matter (such as life or intelligence) will ever arise from the lunar soil. It's too dry, too barren, too desolate.

Lunar Interior

A surprising result of NASA's *Apollo* program was the discovery of several mass concentrations, called **mascons** for short, just below the lunar surface. This was an unanticipated finding, one that resulted from a study of the motions of spacecraft in the vicinity of the Moon.

As spacecraft orbit the Moon, they sometimes speed up or slow down unexpectedly. Careful analysis shows that these accelerations happen while passing over each of the five large, lava-filled basins on the Moon's near side. (For unknown reasons, the Moon has no large maria on its far side.) Apparently, these mare basins are either impact sites of huge meteorite fragments buried beneath the surface, or regions where molten lava forced its way up near the surface, degassed, and thus became denser than the surrounding lunar rock. The word "degas" (or "outgas") refers to the sudden loss of trapped gas as warm rocks rise near the surface; the gas escapes into space, causing the solid rock to more-or-less contract, thus making it more compact and dense.

Figure 6.8 shows these two possible origins for the mascons. Whichever is the correct solution, the concentration of dense matter just below the lunar surface suggests that at least the outer layers of the Moon's interior have been rigid for some time. It is this rigidity—or lack of deformable basalt—that has presumably prohibited the higher-density mascons from sinking toward the lunar center. This contrasts with

Earth, for which such mascons would easily sink through the warm "buttery" terrestrial basalt. The Moon is just too cold.

The general lack of lunar heat means that the Moon's interior is largely undifferentiated, as noted earlier. The average lunar density, about 3 grams/cubic centimeter, mimics that of ordinary rock, virtually eliminating any chance of the Moon having a large, massive, and very dense nickel-iron core like that within planet Earth. Nor is there any evidence for a lunar magnetic field at the present time, again suggesting the absence of a molten core.

Modeling of all the available data implies that the Moon is rather homogeneous, containing mostly primitive material that never melted. As depicted schematically in Figure 6.9, these models suggest a large, uniform core extending to about half of the lunar radius. This pristine region of ancient matter is then surrounded by an 800-kilometer mantle, topped by a thin 60-kilometer crust. Near the Moon's center, the *current* temperature is probably no more than about 1500 kelvin, too cool to melt rock.

We cannot be sure that the lunar interior was *never* molten at any time in its history. The mantle and crust were, to be sure, at least *partially* molten during some past epoch. Otherwise, the thin crust wouldn't have differentiated into the lower-density, brighter highlands and the higher-density, darker maria. But, as noted in Interlude 6-2, the evidence for a past or present molten core is sketchy at best.

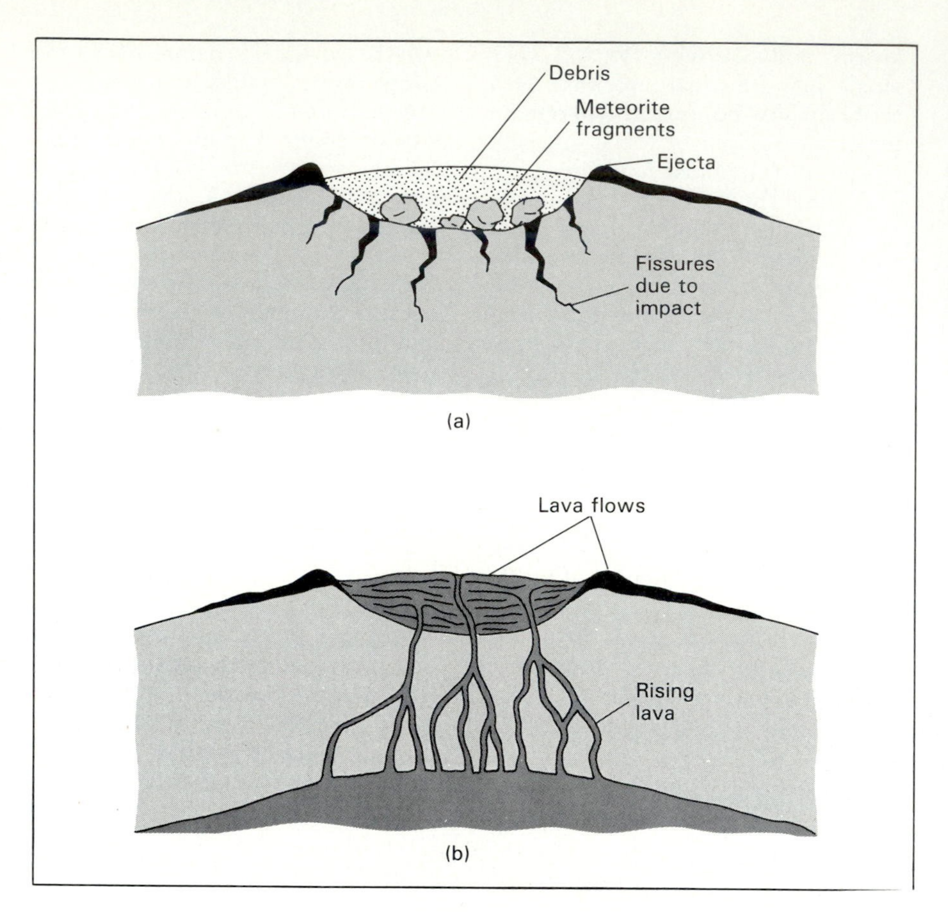

FIGURE 6.8 Schematic diagrams of possible mascon origins by (a) meteoritic impact or (b) degassed lava. Lunar experts currently favor (b).

Volcanism

Most of the lunar craters are meteoritic in origin. But some of them are not. For example, Figure 6.10 shows an intriguing alignment of several craters in a **crater-chain** pattern so straight that it could not have been

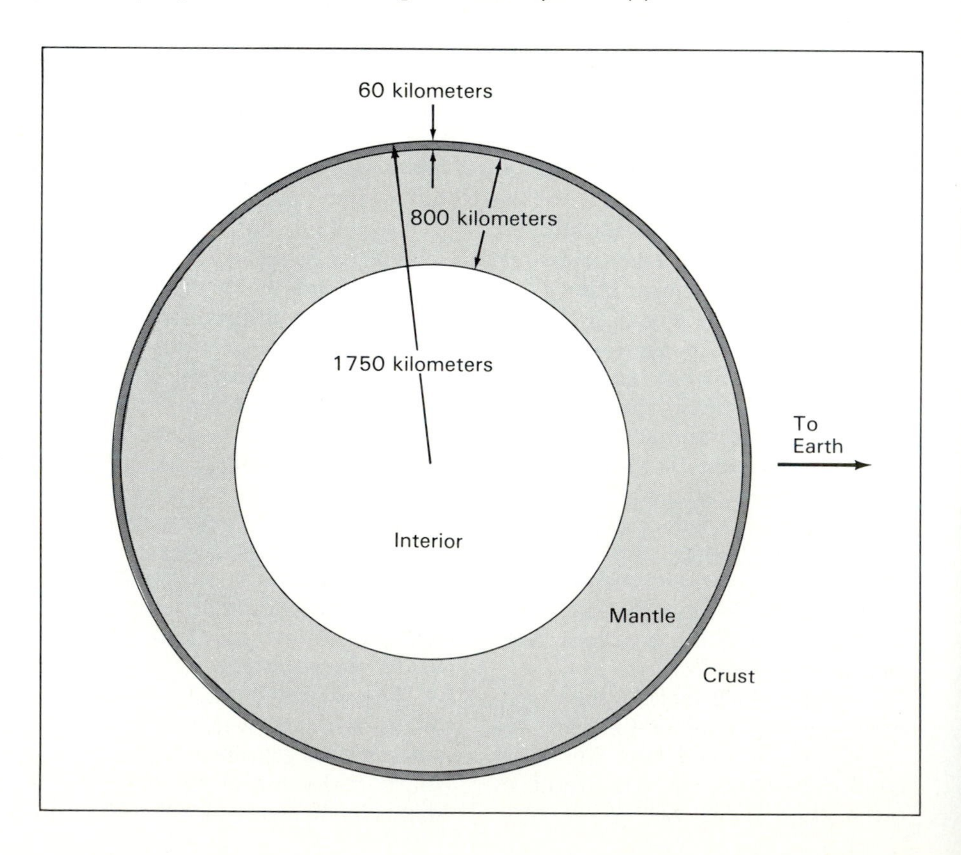

FIGURE 6.9 Cross-sectional diagram of the Moon in bulk.

INTERLUDE 6-2 *Moon Quakes*

In addition to reflecting mirrors and many other instruments left on the Moon, the American astronauts deployed several seismographs capable of measuring lunar vibrations. The figure below shows one of these drum-shaped instruments, powered by panels of solar cells that capture sunlight and convert it into electricity.

Since their deployment in the early 1970s, the seismographs have been rather inactive. Some slight vibrations—moonquakes—have been recorded and radioed back to Earth, but these indicate only very weak activity. Even if you stood directly above one of these quakes, the vibrations would hardly be felt. The total energy released in a year of moonquaking is nearly a billion times less than in a year of earthquaking. Expressed another way, the average moonquake releases about as much energy as a firecracker.

This hardly perceptible seismic action merely confirms that the Moon is just about dead. There are no large quakes, volcanoes, or lava flows at the present time. On the other hand, the fact that *some* seismic waves are detected once in a while implies that the Moon is still settling a little. We shouldn't be surprised, since the Moon's interior is hotter than its surface. Furthermore, Earth exerts a large tidal strain on the Moon. With time, the Moon will continue to cool, adjusting itself further, although at a much slower rate than Earth.

Despite the near lack of seismic activity, weak lunar vibrations can still be used to infer useful information about the Moon's interior. Signals transmitted back to Earth indicate:

1. The quakes occur at surprisingly great depths—roughly 800 kilometers below the surface, compared to about 100-kilometer depths for earthquakes. This implies that the Moon is securely capped with an 800-kilometer-thick shell of strong and rigid rock.
2. The seismic waves change speed while moving through several, apparently different layers of rock about 60 kilometers beneath the surface. This suggests that the rocky lunar crust is chemically differentiated somewhat, not a terribly surprising result in view of the early surface melting that must have occurred via meteoritic bombardment and perhaps radioactive heating.
3. The Moon *might* have a small, partially molten core. This tentative suggestion, not at all confirmed yet, results from the previously discussed (Chapter 5) fact that seismic shear waves cannot move through liquid. Indeed, *Apollo* seismographs operating on the near side do not detect shear waves when small meteoroids or discarded rocket stages hit the far side of the Moon. Additional seismographs will be needed, especially on the far side, before this suggestion becomes widely accepted by space scientists.

possibly produced by meteoroids colliding with the surface. The chance that interplanetary rocks could impact with such straight-line regularity is vanishingly small. Instead, the crater chain marks the location of a subsurface fault—a place where cracking or shearing of the surface allowed molten matter to upwell from below.

We can easily locate other examples of lunar volcanism. Figure 6.11 shows what seem to be lava flows. Figure 6.12 shows inescapable evidence for a volcanic **rille**, a ditch where molten lava once flowed. All these examples make a strong case for surface volcanism at *some* time in the Moon's history.

The best evidence for lunar volcanism was discovered by the *Apollo 15* astronauts along the inside walls of one of the rilles. Figure 6.13 shows this surface feature, called Hadley Rille. Deep down inside the rille, as shown in Figure 6.14, we can see very definite layering of lunar deposits. The most reasonable explanation is that a series of volcanoes must have occasionally flooded the area with lava at various times in the past.

As noted in Chapter 5, the basic

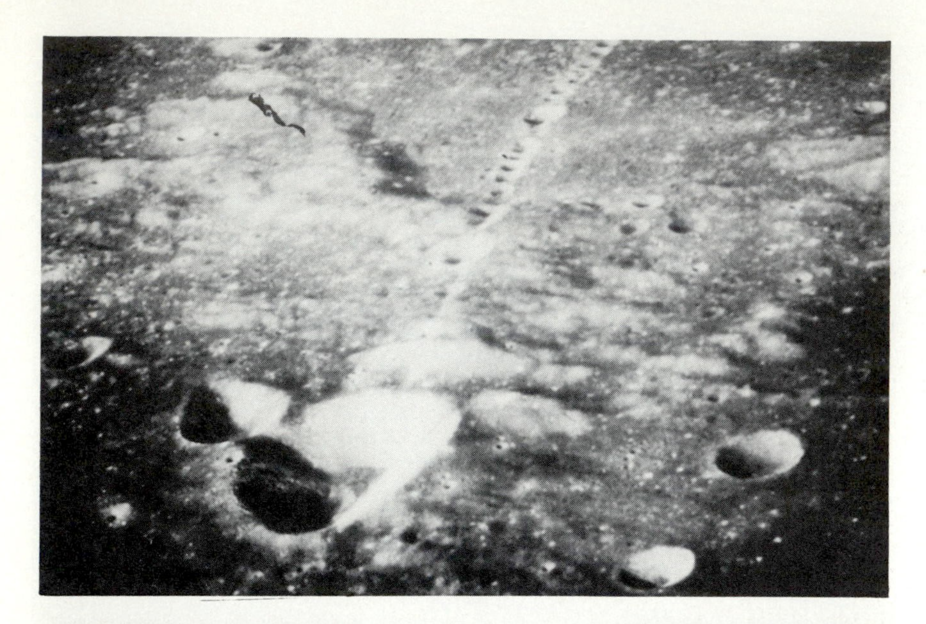

FIGURE 6.10 This "chain" of well-ordered craters was photographed by an *Apollo 14* astronaut.

FIGURE 6.11 The lunar lava beds seen here form the bulk of the maria known as Oceanis Procellarum. Rising above the lava plain are unusual lava domes called the Marius Hills.

FIGURE 6.12 A volcanic rille, photographed from an *Apollo* spacecraft orbiting the Moon, can be clearly seen here winding its way through some of the maria. Called Hadley Rille, this system of valleys runs along the base of the Apennine Mountains.

FIGURE 6.13 A photograph of Hadley Rille in the distance, taken by one of the *Apollo* astronauts while on the Moon's surface. The width of the rille is about 1.5 kilometers (or about a mile) and its depth more than 300 meters (or about 1000 feet).

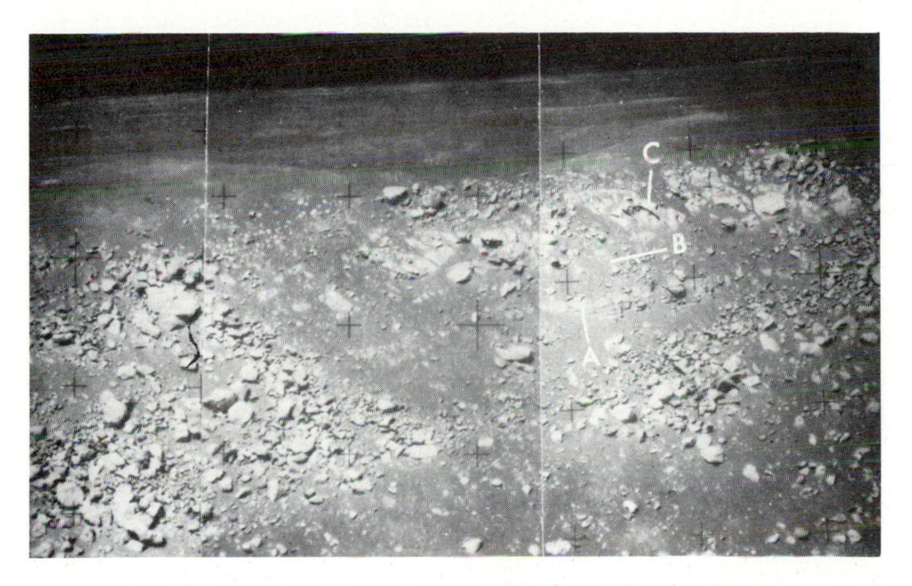

FIGURE 6.14 This mosaic of three photographs of a small part of an interior wall of Hadley Rille shows evidence for subsurface horizontal layering. Each distinct layer (labeled A, B, and C) is a few meters deep and presumably represents successive lava flows that helped form an extensive lava plain called Mare Imbrium.

causes of volcanism are only partially understood. Certainly, in the case of Earth, we know that volcanoes result from hot interior matter upwelling through fissures, cracks, and other geological faults. But many researchers argue that the lack of a differentiated lunar interior rules out large-scale heating of the Moon's insides at any time in its history. If so, what is the origin of the cold and hardened lunar lava seen on the surface?

We are forced to conclude that much of the lunar volcanism—especially that in the maria—must have originated in the upper mantle which was indeed molten at some time in the past. That past era must have been very long ago, since several of the lunar lava flows have been radioactively dated to be 3 to 4 billion years old. And remember, the radioactivity clock doesn't begin ticking until the rock solidifies.

Whatever volcanic activity existed on the Moon must have occurred long ago. The low-density lunar highlands are dated to be *at least* 4 billion years old (and some are as old as 4.5 billion years), whereas the high-density maria are in all cases found to be a little younger. Nowhere on the Moon are rocks known to be younger than 3 billion years. Apparently, the molten basalt solidified into rock at least 3 billion years ago, and the Moon has been dormant ever since.

Such chronological dating and chemical analysis of the lunar rocks at various locations on the Moon are among the most important findings of the *Apollo* program.

Lunar History

Given all the data, can we construct a reasonably consistent history of the Moon? The answer seems to be "yes." Many specifics are still debated, but a general consensus is at hand. Refer to Figure 6.15 while studying the following details.

The Moon apparently originated as a *cold* object about 4.5 billion years ago, the approximate date of the oldest rocks discovered in the lunar highlands. The ages of these ancient rocks match those of the oldest meteorites that have fallen to Earth, suggesting that most of the objects in the Solar System formed nearly 5 billion years ago.

During the earliest phases of the Moon's existence—roughly the first 0.5 billion years or so—meteoritic bombardment must have been frequent enough to heat, and perhaps melt, much of the *surface* of the Moon. The early Solar System was surely populated with lots of interplanetary matter, much of it as boulder-sized fragments capable of colliding with planets and their moons. But the intense heat derived from such collisions could not have transferred very deeply into the lunar interior; rock simply does not conduct heat well.

FIGURE 6.15 Paintings of the Moon (a) about 4 billion years ago, after much of the meteoritic bombardment had subsided and the surface had somewhat solidified; (b) about 3 billion years ago after molten lava had made its way up through surface fissures to fill the low-lying impact basins and thus create the smooth maria; and (c) at present, with much of the originally smooth maria now heavily pitted with craters formed at various times within the past 3 billion years.

So while some *heating* spread throughout much of the Moon, any *melting* was probably confined to a depth of a few hundred kilometers. This situation resembles the surface melting we suspect occurred on Earth from meteoritic impacts during the first billion years or so. But unlike Earth, the Moon is much less massive and does not contain enough radioactive elements to heat it much further. Radioactivity must have heated the Moon a *little,* although not sufficiently to transform it from a warm object to a liquid one.

The younger age of the maria—3.5 billion years old—apparently indicates the time when the meteoritic and volcanic activity subsided and the surface solidified. In other words, sometime during the first billion years or so after the Moon's formation, surface activity must have reached its peak because of a combination of intense meteoritic impacts and limited radioactive heating. All the surface layers heated, some melted, and some even experienced limited volcanism. The mare basins, in particular, are the sites of the last extensive lava flows on the Moon some 3.5 billion years ago. Their smoothness, compared to the older, more rugged highlands, disguise their great age.

Being rather small and having no atmosphere, the Moon rapidly lost its heat to space. Once the Moon cooled, the volcanic activity ended and the surface solidified. With the exception of 20 meters of erosion from eons of meteoritic bombardment, the lunar landscape has remained more or less structurally frozen for the past 3.5 billion years. The Moon is now dead, and it's been dead for a long time.

SUMMARY

The Moon is our nearest cosmic object. Smaller and much less massive than Earth, the Moon never heated extensively and thus never fully differentiated. After some initial surface melting several billion years ago, the Moon experienced some volcanism. Since then, the Moon has been essentially changeless, except for occasional meteoroids that fall to its surface, thus producing most of the craters that scar our cosmic neighbor.

KEY TERMS

crater
crater chain
erosion
highlands
interplanetary matter
maria
mascon
meteor
meteorite
meteoroid
micrometeorite
micrometeoroid
regolith
rille

QUESTIONS

1. Name and discuss at least eight differences between the Moon and the Earth.
2. Why do we see only one side of the Moon from Earth?
3. What, in your opinion, is the most important reason for studying the Moon? Justify your answer.
4. Why doesn't the Moon have an atmosphere? Did it ever have one? Explain.
5. Why doesn't the Moon have a large, dense, nickel–iron core like Earth's?
6. Compare and contrast the compositions and ages of the lunar maria and the lunar highlands. Explain their differences.
7. Name and discuss the two most important findings of the U.S. *Apollo* program.
8. How do we know that volcanism occurred on the Moon at some time in its history?
9. If the Earth and the Moon originated at the same time, explain the present geological inactivity of the Moon.
10. If the Earth and the Moon have always traveled together in space, give two reasons why the lunar surface is more cratered than that of the Earth.

FOR FURTHER READING

French, B., *The Moon Book*. New York: Penguin Books, 1977.

Moore, P., *New Guide to the Moon*. New York: W. W. Norton, 1976.

Taylor, S., *Lunar Science*. New York: Pergamon Press, 1975.

7
THE INNER SOLAR SYSTEM: *The Terrestrial Planets*

Expanding our inventory of matter, we now move farther from planet Earth. In the process, we encounter an array of many new objects: rocky planets of varying sizes and descriptions, some with atmospheres and moons, some not; gaseous planets much larger than Earth, each having several moons no two of which seem similar; chunks of rock and ice wandering among the planets of our Solar System.

In less than a single generation, namely since the mid-1960s, we have learned more about the planetary environment than in all the centuries that went before. Instruments aboard unmanned robots have taken close-up photographs and in some cases made on-site measurements, helping us to appreciate the planets as more than just points of light in the evening sky. Indeed, the most recent era of Solar System exploration has revealed the planets and many of their moons to be worlds unto themselves. They are alien worlds to be sure, with unearthly conditions and histories, but at the same time, we have come to know them as forbidding yet beautiful.

In this first of two chapters on our planetary system, we examine mostly the interior planets close to the Sun. From a brief study of the motions of the planets in their orbits, we shall become familiar with some basic physical laws long known by scientists, including the concept of gravity. We shall then study the physics, chemistry, and geology of the individual interior planets, in the process gaining a richer perspective on our own home in space. All the while, without directly addressing evolutionary issues (for that is more properly the theme of Part III), we shall use the powerful and newly emerging subject known as comparative planetology to compare and contrast the bulk properties of these diverse worlds.

The learning goals for this chapter are:

- to appreciate how the observed motions of the planets led to our modern view of the Solar System centered on the Sun
- to recognize the role gravity plays in guiding planets around the Sun, moons around the planets, and man-made spacecraft now journeying throughout the Solar System
- to gain a brief though clear understanding of the Solar System in bulk
- to compare and contrast the various planets far beyond our earthly world
- to appreciate the vast store of new planetary knowledge accumulated during the brief quarter-century of the Space Age

Modeling

OLDER VIEWS: The Greeks long ago, and perhaps civilizations before them, tried to build a model of the Solar System. In doing so, they used only what Aristotle had taught was the perfect curve, namely the circle. However, they quickly ran into trouble, for the actual orbits of the planets are not quite that simple. Planets tend to change in brightness—and hence distance from Earth—during their orbits. And as shown partly in Figure 7.1, some planets loop back and forth while traveling across the sky during the

FIGURE 7.1 The movements of several planets over the course of many months are reproduced here on the inside dome of a planetarium.

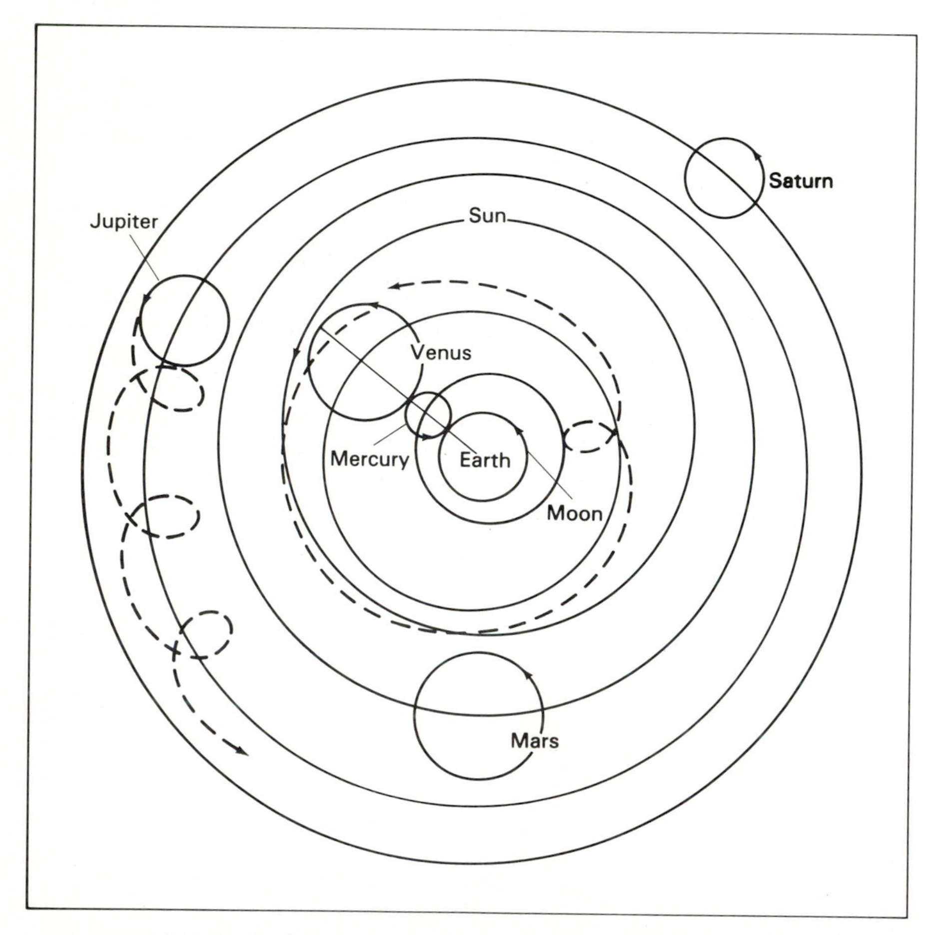

FIGURE 7.2 This sketch shows the basic features of a geocentric model of the Solar System that enjoyed widespread popularity prior to the Renaissance. Each planet followed a small circular orbit about an imaginary point which itself traveled in a large circular orbit about the Earth; to avoid confusing the diagram completely, the resultant paths of only two planets, Venus and Jupiter, have been drawn as dashed curves.

year. They often seem to wander against the backdrop of distant stars, not at all following the regular procession of stars across the sky. In fact, the word "planet" derives from the Greek word *planetes*, meaning "wanderer."

Despite the difficulties plaguing models of the Solar System based strictly on circles, all popular models were so constructed until about 500 years ago. Most of these models were **geocentric**, that is, centered on the Earth. They required a complex series of large and small circles, with the planets theorized to have circular paths that spun around the edges of larger circles, which in turn rolled like wheels across the sky. Still, these models were able to predict the approximate positions of the planets, at least to the accuracy of the observations at the time.

A second century (A.D.) Greek geometer named Ptolemy constructed a geocentric model that explained surprisingly well the paths of the then-known five planets (as well as the Sun and Moon). Figure 7.2 is an example of this complex geocentric view. Unfortunately, it was based on a series of no less than 80 circles, with some planets restricted to move in small circles superposed at the edges of larger circles, which in turn were superposed at the edges of even larger circles. To account for the paths of all the currently known planets and one Sun would presumably require a vast array of circles within circles within circles.

THE MODERN VIEW: Much of this complexity was eliminated in Renaissance times. Looking (observation) and thinking (theory) were merged to achieve a simpler view than that deduced by the ancients. The sixteenth-century Polish cleric Copernicus recognized that a **heliocentric** model—one centered on the Sun—improved the harmony and organization of the tangled geocentric models previously imagined by the Greeks and Romans of antiquity.

Despite much supporting observational data, the simpler model of Copernicus was not easy to accept even as recently as 300 years ago.

Heliocentricity rubbed the grain of much previous thinking. And it violated the religious teachings of the time. Above all, it relegated Earth to a noncentral and undistinguished place within the Solar System and the Universe. Earth became just one of several planets.

We now recognize that Renaissance researchers were correct, but none of them could prove at the time that our system is centered on the Sun, or even that Earth moves. Proof of the latter was unambiguously obtained in the mid-nineteenth century when the first parallax observations were made, as discussed in Chapter 4. To be sure, measurement of trigonometric parallax denoted a final vindication of the Copernican world-view, the correctness of which no one doubted any longer. And, heliocentricity of the Solar System has been gradually verified over the years by ever-increasing experimental tests, culminating with the expeditions of our unmanned space probes of the 1960s, 1970s, and 1980s.

Be sure to note that a prime motivation for the heliocentric model was simplicity (although even Copernicus, still influenced by Greek thinking, clung to the idea of circles to model the planets' motion, thereby retaining unnecessary complexity). Heliocentricity provides a more ordered (and more accurate) explanation of the observed facts than can any geocentric model. Even today, scientists are guided by simplicity, symmetry, and beauty in modeling all aspects of the Universe.

The development and eventual acceptance of the heliocentric model was an important milestone in our thinking as human beings. Understanding the framework of our planetary system freed us from an Earth-centered view of the Universe, and eventually enabled us to realize that Earth orbits only one of myriad similar stars in the Milky Way Galaxy.

Orbital Motion

KEPLER'S LAWS: Table 7-1 lists some useful numbers describing the orbits of each of the nine known planets. These properties for the innermost six planets were known to Renaissance workers, and it is on the basis of such data that they were able to construct the currently accepted heliocentric model of the Solar System.

The middle column tabulates each planet's average distance from the Sun. The right-hand column gives the orbital period, that is, the time needed for a complete orbit around the Sun. Note that each tabulated quantity is expressed in terms of Earth values. For example, since the orbital period of our planet can be written as 1.0 Earth year, the right column shows that Venus' orbital period is about 0.6 of an Earth year, or roughly 220 days. This relatively brief "year" may be compared to that of Pluto, which equals nearly 2.5 Earth centuries. Similarly, the middle column shows that the Saturn–Sun distance averages 9.5 astronomical units. The main point to be grasped from this table is that the farther a planet is from the Sun, the greater is its orbital period.

By examining these and other data for the six interior planets (acquired by the naked eyes of the eccentric, yet wealthy Danish observer, Brahe, from his private island), the seventeenth-century German astronomer Kepler was able to formulate three laws of planetary motion—called, not surprisingly, **Kepler's laws**. The *first law of Kepler* is easy: *The orbital paths of the planets are elliptical (not circular) with the Sun at one focus.* By taking into account the relative speeds and positions of the planets about the Sun, this law can explain the variable brightnesses and the peculiar to-and-fro motions observed for some of the planets. Gone were the circles within circles that rolled across the sky; Kepler's adjustment of Copernicus's ideas caused the model of the Solar System to be greatly simplified.

Figure 7.3 illustrates the shape of an **ellipse**, which is merely a distorted or elongated circle. We could draw almost any ellipse by varying the length of a string attached by tacks at each of the two foci. None of the planets' elliptical orbits are as pronounced as that shown in the figure. Most, in fact, are quite close to a circle; with the exception of two planetary orbits (those of Mercury and Pluto), our eyes would have trouble distinguishing a real planetary orbit from a true circle.

The *second law of Kepler* is sketched schematically in Figure 7.4: *An imaginary line connecting the Sun to any planet sweeps out equal areas of the ellipse in equal intervals of time.* In other words, a planet, while orbiting the Sun, traces the orbital intervals labeled A, B, and C in equal times. Consequently, when a planet is close to the Sun, as in sector C, it moves much faster than when farther away, as in sector A. This law is not restricted to planets; it applies to any orbiting object. Spy satellites, for example, move very rapidly as they swoop close to Earth's surface—not because they are propelled with powerful on-board rockets but rather, because that's the normal be-

TABLE 7-1
Some Solar System Dimensions

PLANET	AVERAGE DISTANCE FROM SUN (ASTRONOMICAL UNITS)	ORBITAL PERIOD (EARTH YEARS)
Mercury	0.4	0.2
Venus	0.7	0.6
Earth	1.0	1.0
Mars	1.5	1.9
Jupiter	5.2	11.9
Saturn	9.5	29.5
Uranus	19.2	84.1
Neptune	30.1	164.8
Pluto	39.4	248.4

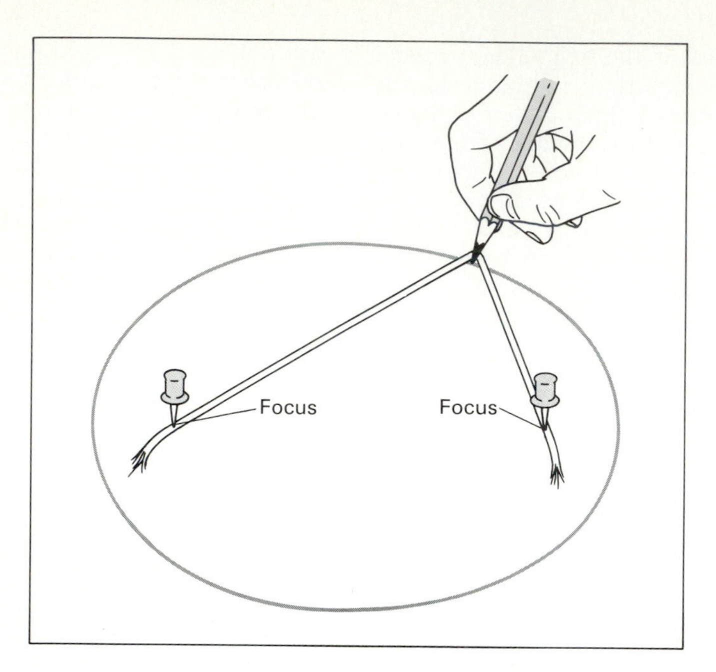

FIGURE 7.3 Any ellipse can be constructed with a string, a pencil, and two tacks.

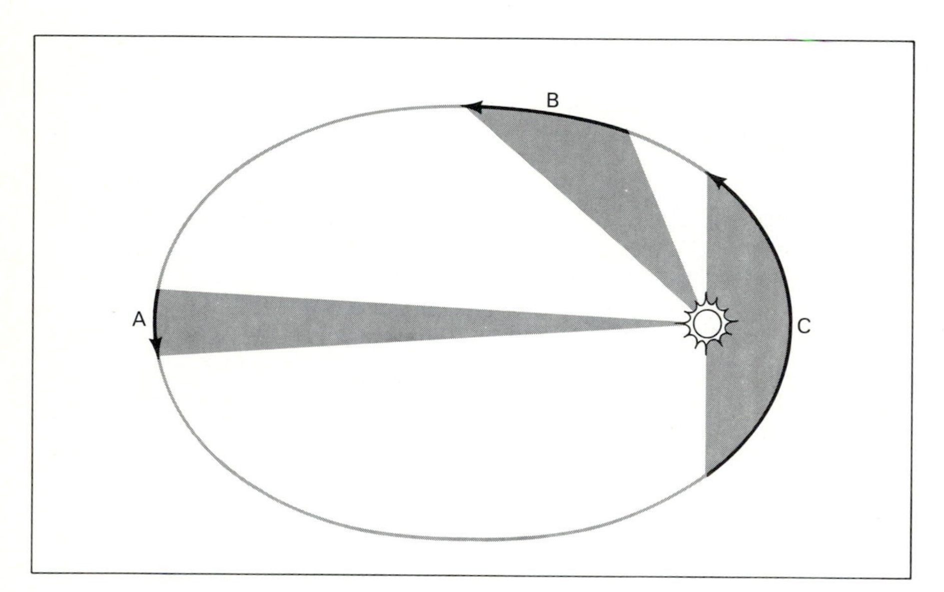

FIGURE 7.4 A schematic diagram illustrating Kepler's second law: Equal areas are swept out in equal intervals of time. Thus, an object would travel the length of each of the three arrows in the same amount of time.

havior of any object in a highly elliptical orbit.

A *third law of Kepler* can also be formulated strictly on the basis of the observational data of Table 7-1. Upon careful analysis, you will find, as Kepler did, that

$$\text{orbital size}^3 \propto \text{orbital period}^2.$$

That is, *the cube of any planet's average distance from the Sun is proportional to the square of its orbital period about the Sun.* In other words, a planet's period of revolution about the Sun must increase more rapidly than the size of its orbit. This accords well with those numbers given above, namely, that although Pluto is "only" 40 astronomical units from the Sun its orbital period is a good deal more than 40 years; in fact, it is amost 250 years. (Quantitatively expressed for the case of Pluto, $39.4^3 = 248.4^2$.)

NEWTON'S LAWS: The three laws of Kepler are empirically founded. In other words, they are based solely on analyses of data, without much recourse to theory or mathematical underpinnings. Indeed, Kepler did not have much understanding of the physical mechanisms underlying his laws. Nor did Copernicus appreciate the fundamental reasons for his heliocentric model of the Solar System.

A basic understanding of planetary motions involves orbital stability and energy. What prevents the planets from flying off into space, or alternatively, from falling into the Sun? What causes the planets to revolve, apparently endlessly? To be sure, the motions of the planets obey Kepler's three laws, but only by considering something more fundamental—Newton's laws—can those motions be truly understood.

One of the basic laws of the seventeenth-century British mathematician Newton is this: An object at rest or steadily moving in a straight line remains that way until some force field acts on it. Figure 7.5 illustrates the essence of this law. As shown, a moving object will, in principle, move forever in a straight line unless a *momentary* force field changes its direction of motion. Note that we are neglecting here the familiar notion of friction, an effect that slows, for example, balls rolling along the ground or pencils on a tabletop. At any rate, friction is not an issue for the planets; there is no appreciable friction in outer space.

In addition to momentary force fields, force fields can also act *continuously*. These are governed by another of Newton's laws of motion: Any object having mass always exerts an inward, attractive force field. Gravity is a good example of such a force field. For example, Earth's gravity pulls continuously on a baseball thrown upward. Figure 7.6 illustrates how the ball's trajectory changes continuously. (The baseball, having some mass of its own, also exerts a gravitational pull on the Earth. But the more massive Earth

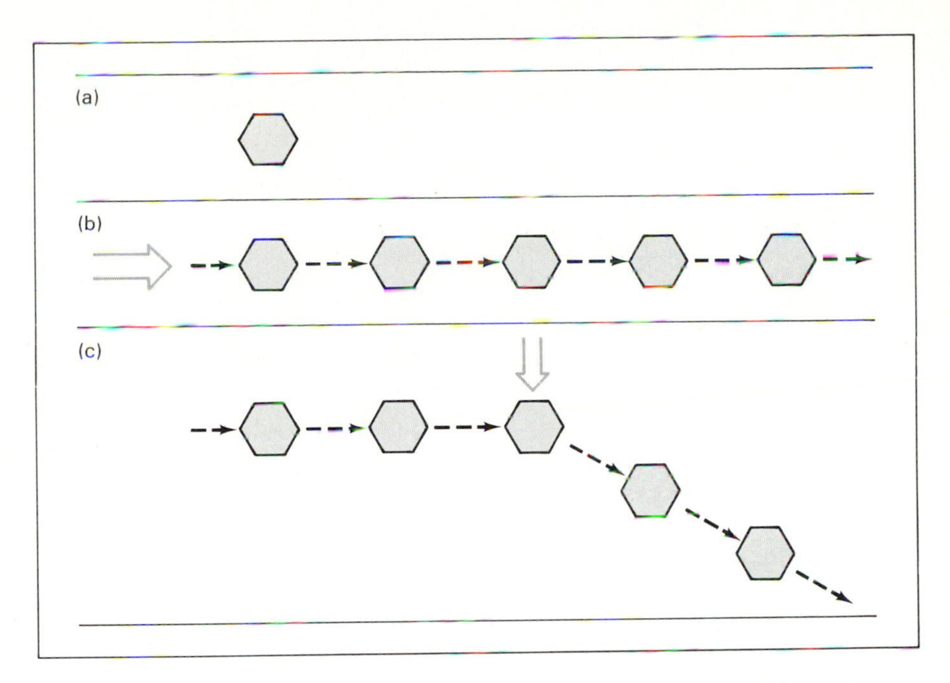

FIGURE 7.5 An object at rest will remain at rest (a) until some force field momentarily acts on it (b). It will then remain in that state of uniform motion until another force field acts on it. The arrow in (c) shows a second force field acting at a different direction than the first, causing the object to change direction.

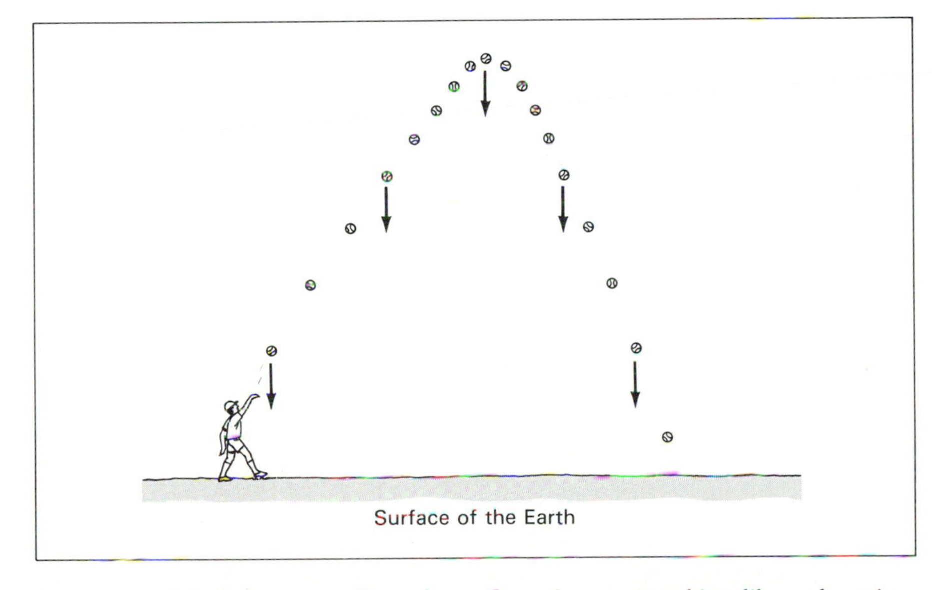

FIGURE 7.6 A ball thrown up from the surface of a massive object like a planet is pulled continuously by the gravitational force field of that planet (and conversely).

has a much greater effect on the baseball than conversely.)

Likewise, consider the behavior of a baseball batted from the surface of the Moon. Because the pull of gravity is about six times weaker on the Moon than on Earth, a baseball's trajectory changes more slowly near the Moon. A typical home run in a terrestrial ballpark would travel nearly a kilometer on the Moon! That's because the Moon, being less massive than Earth, has less gravitational influence on the baseball.

So the magnitude of the gravitational force field depends on the mass of the attracting bodies. Theoretical insight, as well as detailed experiments, prove that this force field is directly proportional to the product of the two masses.

A second aspect of this same type of gravitational force field can be derived by studying the motions of the planets about the Sun. These studies show that at all locations equidistant from the Sun's center, the gravitational force field has the same strength. Furthermore, that force field is always directed toward the Sun. As we might expect, the force field's strength decreases with distance. Figure 7.7 shows that the strength of the force field is four times greater at half the distance from the Sun; at a third the distance, the force field is nine times greater. Additional observations prove that the strength of the force field is inversely proportional to the square of the distance from the Sun. Once again, we have an example of a force field governed by the inverse-square law studied in Chapter 1.

The foregoing arguments can be combined to form a law of gravity that dictates the way that all material objects attract each other. As a proportionality, the law is:

$$\text{gravity force field} \propto \frac{\text{mass of object \#1} \times \text{mass of object \#2}}{\text{distance}^2}$$

This relationship is a compact way of stating that the gravitational pull between two objects is directly proportional to the product of their masses and inversely proportional to the square of the distance separating them.

It is precisely this mutual gravitational attraction of the Sun and the planets that produces elliptical orbits. Since the Sun is much more massive than any of the planets, it inevitably dominates the interaction; we might say the Sun "controls" the planets, not the other way around. As the Sun pulls, it tries to capture the planets straightaway. But this gravitational pull is counteracted somewhat by each planet's forward momentum, which would, in the absence of gravity, cause them to escape into deep space. Figure 7.8

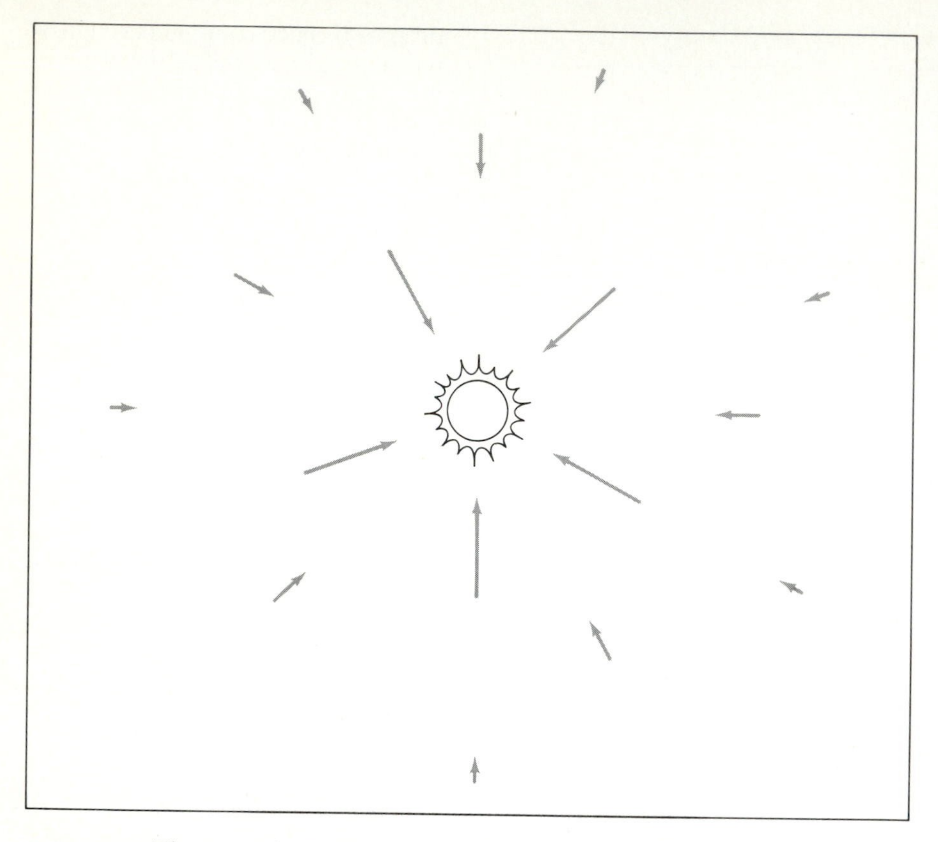

FIGURE 7.7 The strength of the gravitational force field (length of the arrows) decreases with the square of the distance from the Sun.

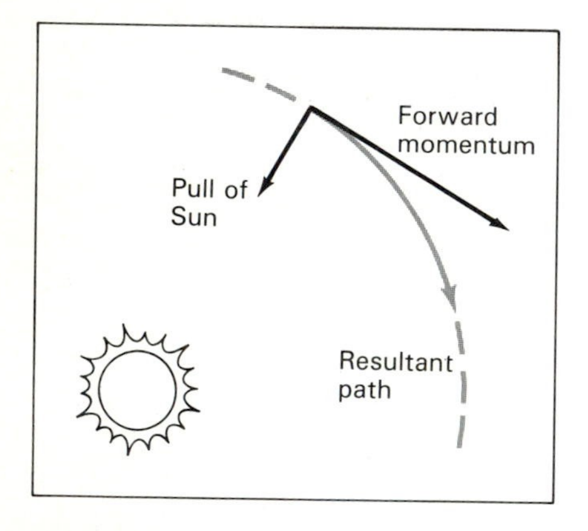

FIGURE 7.8 The Sun's inward pull of gravity on a planet is counteracted continuously by the forward momentum of that planet. These two effects combine, causing the planet to move smoothly along an intermediate path. This repeated tug-of-war between the Sun's gravity and the planet's forward momentum results in a stable orbit.

depicts this gravity/momentum competition.

The situation here resembles the case of a rock at the end of a string whirled above your head. When you suddenly release it, the rock flies away with a great deal of forward momentum. In this analogy, the Sun's gravitational force field is your hand and the string, the planet is the rock at the end of that string, and the release of the rock mimics the sudden elimination of the Sun's gravity. In reality, though, the Sun's gravity will continue to pull as long as the Sun remains intact as a massive object.

At this very moment, Earth is under the influence of these two effects, one being its forward momentum and the other the Sun's gravity. The net result is the orbital stability that all of us share—an Earth that feels "rock solid," despite its continuous and rapid motion through space. (In fact, Earth orbits the Sun with a velocity of about 30 kilometers/second, or some 70,000 miles/hour.)

Both of these influences on planets orbiting the Sun apply equally well to natural moons and artificial satellites orbiting any planet. All of Earth's human-made satellites move along paths governed by a combination of the inward pull of Earth's gravity and the forward momentum gained during the rocket launch. If the rocket initially imparts enough momentum to the satellite, it successfully counteracts gravity and goes into orbit. Those not given enough momentum at launch [such as intercontinental ballistic missiles (ICBMs)] fail to achieve orbit, and (unfortunately) fall back to Earth. (Technically, ICBMs actually do orbit Earth's attracting center, but their orbits intersect Earth's surface.)

Thus Newton's laws explain the paths of objects moving at any point in space near a gravitating body. These laws provide a firm physical foundation for Copernicus's heliocentric model of the Solar System and for Kepler's observational laws of planetary motions. Newtonian gravitation not only governs the planets, moons, and satellites in their elliptical orbits, but also, as discussed in Interlude 7-1, guides the more complex motions taken by spacecraft traveling throughout our Solar System.

Our Planetary System in Brief

As currently explored, our Solar System contains 1 star, 9 planets, 54 moons (at last count), 4 asteroids larger than 300 kilometers in diameter, more than 3000 smaller but well-observed asteroids, myriad comets a few kilometers in diameter, and countless meteoroids less than a meter across. The near-void in and among all these objects is termed **interplanetary space**. We'll study the star (our Sun) in Chapters 9 and 10, and the asteroids, comets, and meteoroids toward the end of Chapter 8. The rest of this chapter concentrates on the four interior planets: Mercury, Venus, Earth, and Mars. In the next chapter we take up the remaining, outer planets—in order of increasing distance from the Sun, Jupiter, Saturn, Uranus, Neptune, and

INTERLUDE 7-1 ***Interplanetary Navigation***

"Celestial mechanics" is the study of the motions of gravitationally interacting objects. These days, this subject combines Newton's laws and powerful computers in order to understand the intricate movements of astronomical objects. It enables astronomers to calculate the orbits of the planets with high precision, and even aided the discovery of the outermost planets, Neptune and Pluto, solely by studying the distortion of Uranus' orbit.

Celestial mechanics has become an essential tool as scientists and engineers navigate manned and unmanned spacecraft throughout the Solar System. Robot probes can now be sent on stunningly accurate trajectories, often expressed in the trade with slang phrases such as "sinking a corner shot on a billion-kilometer pool table." In fact, near-flawless rocket launches, aided by occasional midcourse changes in flight paths, now enable interplanetary navigators to steer spacecraft remotely into an imaginary aperture just a few kilometers wide and a billion kilometers away.

Sophisticated knowledge of celestial mechanics can also aid navigation of a single space probe toward several planets. For example, in 1974, the *Mariner 10* spacecraft was guided into that part of Venus' gravitational force field which swung it around to precisely the right path for an additional trek toward Mercury. Venus itself propelled the probe in a new direction, a course alteration that required zero fuel. Such a "slingshot" maneuver has also been used several times in subsequent years as, for example, the *Voyager 2* spacecraft (launched in late 1977) closely bypassed Jupiter (in 1979), Saturn (in 1981), Uranus (in 1986), and is now on its way toward Neptune and the outermost reaches of the Solar System. The gravity force fields of the giant planets whipped the craft around at each visitation, in turn enabling flight controllers to get a rather large amount of "mileage" out of *Voyager.* Instruments on board then radioed back information about each of these planets, as discussed in this and the next chapter.

Gravity-assisted flights like these may well be the way of the future. No additional rocket fuel (or money) is needed for a spacecraft to be redirected in this way from one planet to another. The only difficulty is knowing enough celestial mechanics to decide when and where to aim the space probe into the strong gravitational force field of the appropriate planet. A small mistake in calculation can lead to an enormous positional error in space—especially when compounded over the large distances between planets. For example, a 0.1-arc-second mistake at launch from Earth can produce nearly a 1000-kilometer error in the position of a spacecraft by the time it reaches Saturn.

With better knowledge of planetary masses, accurate radio tracking of space vehicles, and a fortuitous positioning of the planets, we might someday be able to send a single robot probe of the future on a grand tour to visit each of the planets of our Solar System. We would then probably enter into a whole new era of planetary exploration, acquiring hints and clues about our cosmic neighbors for a small investment of money.

Pluto. [Very seldom, as is currently (1979 to 1999) the case, Pluto ventures inside Neptune's orbit.]

The four innermost planets are often called the **terrestrial planets** because of their physical and chemical similarities to rocky Earth. The larger, outer planets—Jupiter through Neptune—are often labeled the **Jovian Planets** because of their resemblance to gassy Jupiter. Pluto doesn't fit well into either category; it might once have been a moon of Neptune (or even a large comet), and not originally a planet at all.

Figure 7.9 is an artist's rendition of the planetary system as future generations of space voyagers might perceive it from a distant vantage point. All the planets orbit the Sun in nearly the same plane as Earth

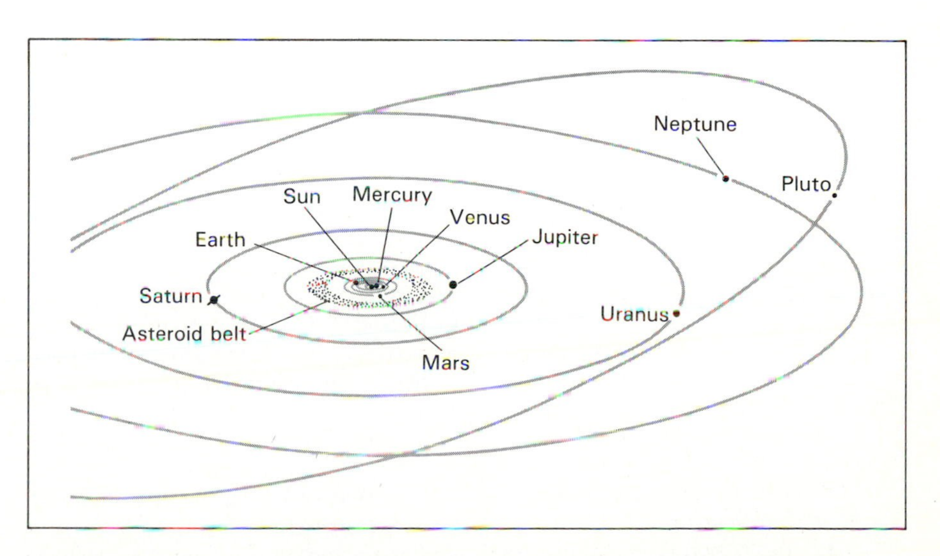

FIGURE 7.9 Schematic diagram of the planetary orbits drawn to scale. With the Earth-Sun distance equal to 1 astronomical unit, the entire Solar System extends end to end for nearly 80 astronomical units. That may sound large, but it's only about a thousandth of a light-year.

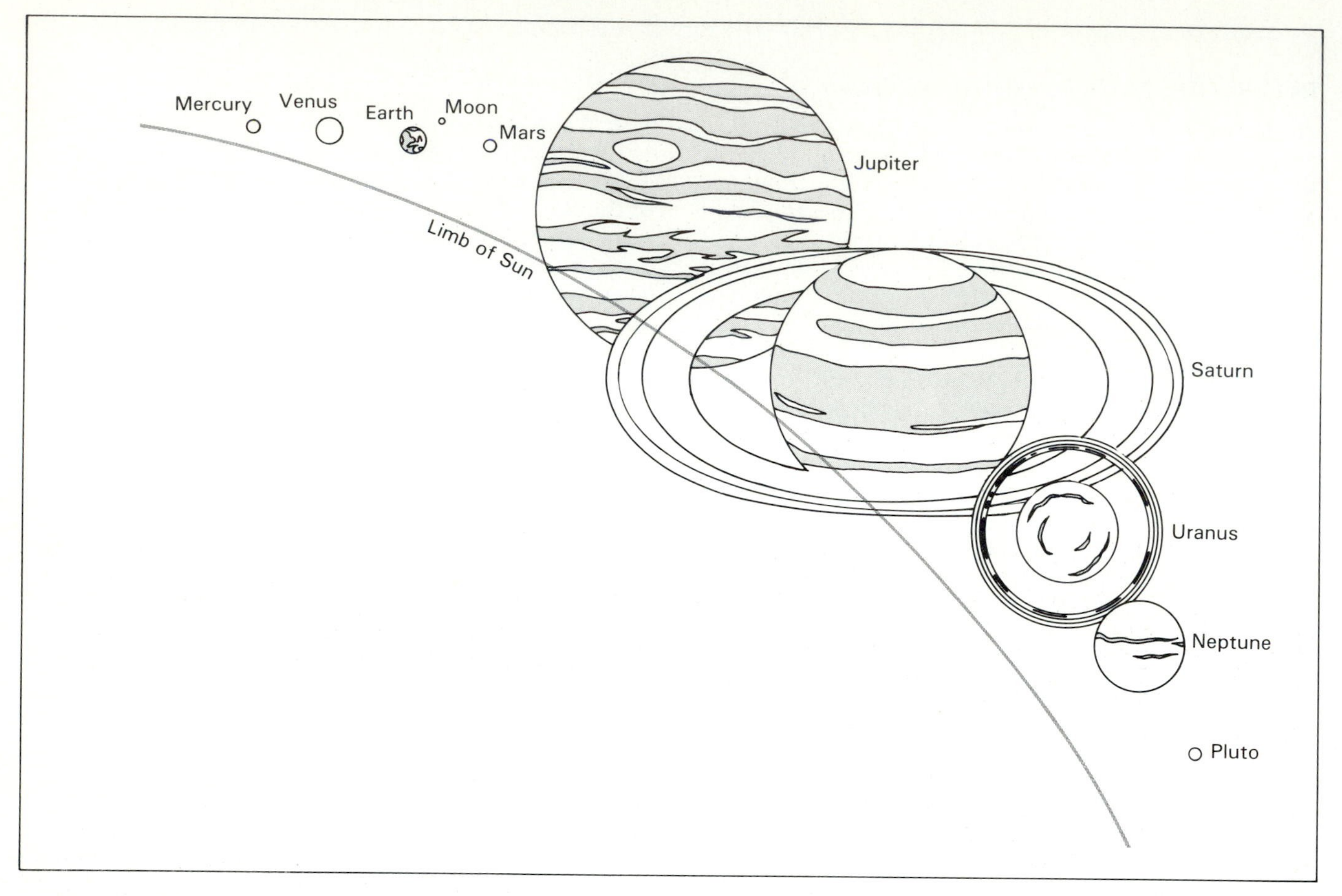

FIGURE 7.10 Schematic diagram, drawn to scale, of the relative sizes of the planets and our Sun.

(called the "ecliptic plane"), and they all orbit in the same direction (counterclockwise from terrestrial north).

Despite this organization, the planets themselves comprise a varied lot. Figure 7.10 shows a schematic of the relative sizes of the planets compared to the Sun. The Sun dominates, with Jupiter an inferior second; Earth is hardly noticeable. Our star has about 1000 times Jupiter's mass, and about 700 times the mass of the entire Solar System, including Jupiter. Thus the Sun contains more than 99.9 percent of all the matter in the Solar System. Everything else, especially the small Terrestrial Planets, constitute a collection of nearly insignificant matter.

In the next several sections we consider some of the main features of each of the interior planets. A few of their properties are listed numerically in Table 7-2 for ease in comparison.

The Terrestrial Planets

MERCURY: Mercury is the second smallest planet; it's hardly larger than Earth's Moon. Because of its small size and proximity to the Sun, Mercury is extraordinarily difficult to observe from Earth. Even with a telescope, it can be seen merely as a slightly pinkish disk only at dawn or dusk.

Figure 7.11 is one of the few photographs of Mercury taken from Earth that shows some evidence for surface markings. Although somewhat speculative in the (pre-1970s) days before the arrival of robot spacecraft, these faint darkish markings are now regarded by astronomers to be much like those seen when gazing casually at Earth's Moon. That's an indication of Mercury's great distance; the largest telescopes can resolve features on the surface of Mercury no better than we can perceive features on our Moon with our naked eyes.

That these dark regions are lava-covered areas akin to the dark maria of Earth's Moon was proved during the 1974 flyby of the U.S. *Mariner 10* spacecraft. Figure 7.12 shows a picture radioed back to Earth from a distance of 200,000 kilometers from Mercury, while Figure 7.13 shows a higher resolution photograph of the planet from a distance of 20,000 kilometers. There is no sign of clouds, rivers, dust storms, or other aspects of weather. The surface is largely cratered, very much like our Moon. Ancient lava flows are seen in and around numerous

TABLE 7-2
Some Planetary Properties[a]

PLANET	AVERAGE DISTANCE FROM SUN (ASTRONOMICAL UNITS)	MASS (EARTH MASSES)	KNOWN SATELLITES	SIZE (EARTH RADII)	ROTATION (EARTH DAYS)	DENSITY (EARTH DENSITIES)
Mercury	0.4	0.05	0	0.4	59	0.9
Venus	0.7	0.8	0	0.9	243	1.0
Earth	1.0	1.0	1	1.0	1.0	1.0
Mars	1.5	0.1	2	0.5	1.03	0.7
Jupiter	5.2	318.0	17	11.1	0.4	0.2
Saturn	9.5	95.2	20	9.5	0.4	0.1
Uranus	19.2	14.6	15	3.7	0.7	0.3
Neptune	30.1	17.3	2	3.5	0.6	0.4
Pluto	39.4	0.01	1	0.25	6.4	0.2

[a] To find the physical values in this table, each of the entries above should be multiplied by the following equivalent Earth properties: distance = 1.5×10^8 kilometers; mass = 6×10^{27} grams; radius = 6400 kilometers; rotation = 24 hours; density = 5.5 grams/cubic centimeter.

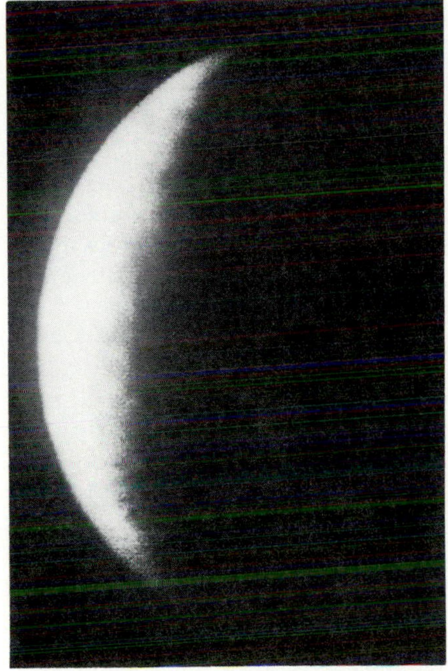

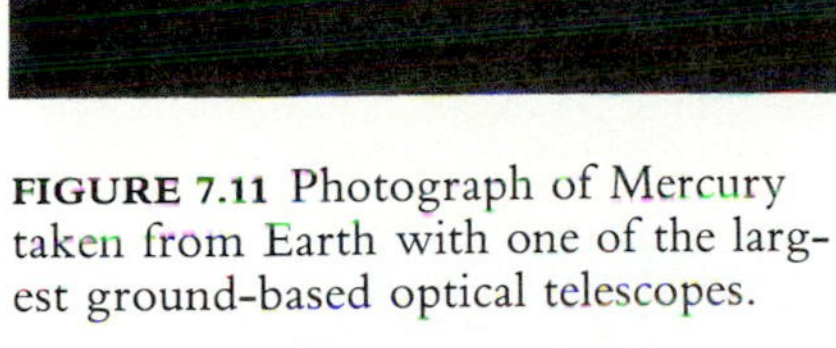

FIGURE 7.11 Photograph of Mercury taken from Earth with one of the largest ground-based optical telescopes.

FIGURE 7.12 Mercury is imaged here as a mosaic of photographs taken by the *Mariner 10* spacecraft in the mid-1970s. At the time, the spacecraft was some 200,000 kilometers away from the planet.

FIGURE 7.13 Another photograph of Mercury by the *Mariner 10* spacecraft, this time 20,000 kilometers away from the planet.

craters. Evidence abounds for both meteoritic bombardment and planetary volcanism; both these kinds of geological upheaval are old. Much of the planet is covered with a layer of fine pulverized dust. However, Mercury shows no evidence for crustal motions; its faults (or "escarpments") probably formed when the crust cooled and shrank long ago, like wrinkles on the skin of an old apple.

All in all, most of the discussion in Chapter 6 about the *surface* of Earth's Moon apparently applies equally well to Mercury. This planet has a rather flat, dusty terrain, perhaps with pebbles and even an occasional boulder strewn about. Here and there, as on the Moon, the broad Mercurian plains are interrupted by a mountain ridge, a canyon, or more likely, an impact crater. Figure 7.14 shows what was probably the last great event in the geologic history of

FIGURE 7.14 Mercury's most prominent geological feature—the Caloris Basin—measures about 1300 kilometers across and is ringed by concentric mountain ranges that reach more than 3 kilometers high in places.

Mercury—an immense bull's-eye crater called the Caloris Basin, formed eons ago by the impact of a large asteroid. Like the Moon, it seems that Mercury has been a geologically dead world for roughly the past 4 billion years. The two objects cannot be entirely alike, though, since Mercury's average density (5 grams/cubic centimeter) differs substantially from that of the Moon (3 grams/cubic centimeter).

Mercury's rotation period is slow, although not slow enough to match its 88-day orbital period. (Whenever "day" is used in our text, we mean an Earth day unless otherwise specified.) If these two rates—rotation and revolution—were the same, the same side of Mercury would always face the Sun, much as our Moon perpetually keeps one side facing Earth. To the embarrassment of optical observers who were convinced since the 1880s that this was in fact the case, radar astronomers in 1965 discovered that Mercury's rotation period is 59 days. Accordingly, the Sun stays "up" in the black Mercury sky for almost 3 (Earth) months at a time, after which follows nearly 3 months of darkness. Thus, at any one time, half of Mercury is subject to severe solar heating, with temperatures routinely reaching 400 to 600 kelvin, depending on the nature of the surface soil. (Remember, water freezes at 273 kelvin and boils at 373 kelvin, while lead melts at roughly 600 kelvin.) And with virtually no atmosphere, this heat cannot conduct well to the night side of the planet, where the predawn equatorial temperature measures about 100 kelvin.

Why not much atmosphere? Mercury's small mass and high temperature virtually guarantee that any gas should be able to escape. Observationally, no appreciable atmosphere has ever been detected either during Mercury's transits across the face of the Sun or during the *Mariner 10* rendezvous. Although *Mariner 10* found a trace of what was at first thought to be an atmosphere, this gas is now known to be trapped helium emitted by the Sun and lost to Mercury a few weeks after capture. In addition to meteoroids, ultraviolet radiation must constantly rain down on the planet.

Mercury's magnetic force field, discovered by *Mariner 10*, is about 100 times weaker than Earth's. Nonetheless, Mercury's magnetism is sufficient to deflect matter particles (mostly electrons and protons) emanating from the Sun. And this planet's weak magnetic force field as well as its large average density together imply that Mercury is chemically differentiated. Current models suggest that most of its interior is dominated by a giant core of heavy iron-rich matter. Less dense, lunar-like basalt probably lies atop this to a depth of about 500 kilometers. Mercury must therefore have heated *and melted* at some time in its distant past, allowing its heavy matter to sink toward the center, all the while compressing its global contents and becoming more dense than, for example, our Moon, which never melted entirely.

Despite its huge metal core, Mercury does not have a strong magnetic force field probably because it rotates too slowly; the difference in spin rates of Mercury's core and mantle must not be very large, thus lessening any dynamo action that is widely regarded as the origin of Earth's magnetic force field. Actually, Mercury's current weak magnetism might be a mere remnant of an extinct dynamo no longer active. To be sure, given Mercury's rather small size, its metallic core might well be "frozen" solid. Models are simply inconclusive on this issue. And no spacecraft are scheduled to revisit Mercury in the foreseeable future.

Mercury, then, is undoubtedly a sizzling planet on the solar side and a frigid one on the opposite side. It's airless, waterless, and heavily scarred. It's clearly unfit for humans. And as we'll note in more detail toward the end of this book, such a hostile environment is not very suited to the origin and evolution of any form of life. Most likely, Mercury could not harbor anything resembling life as we know it.

VENUS: Because Venus most nearly matches Earth in size, mass, and density, and because its orbit is closest to us, this second planet is often called Earth's sister planet. Unlike Mercury, Venus has a dense atmosphere, rising approximately 70 kilometers above the surface (compared to 8 kilometers on Earth). Much sunlight is *reflected* from the top of its atmosphere, making Venus seems the third brightest object in the sky (after the Sun and our Moon). It can best be observed from Earth when low on our horizon, a few hours either after the setting Sun or before the rising Sun, depending on where Venus is in its orbit. For this reason, Venus is often called the "evening star" or the "morning star," although it appears more than 10 times brighter than the brightest real star, Sirius. You can even see Venus in the daytime if you know exactly where to look.

The atmosphere of Venus is nearly opaque to visible radiation, making its surface completely invisible from the outside. Figure 7.15 shows one of the best modern photographs of Venus taken with a large telescope on Earth. The planet resembles a white-yellow disk, and shows rare hints of cloud circulation patterns.

These atmospheric patterns are much more prominent when examined with equipment capable of detecting only ultraviolet radiation; some of Venus' atmospheric constituents absorb this high-frequency radiation, thereby increasing the

FIGURE 7.15 This photograph, taken from Earth, shows Venus with its creamy yellow mask of clouds.

FIGURE 7.16 Venus is pictured here by the *Pioneer* spacecraft cameras some 200,000 kilometers away from the planet. This image was made by capturing solar ultraviolet radiation reflected from the planet's clouds, which are composed mostly of sulfuric acid droplets much like the corrosive acid in a car's battery. (In fact, the only kind of precipitation on Venus is entirely "acid rain".) In this way, Venus' circulation patterns can be best studied; they might some day help us better understand Earth's weather.

cloud contrast. Figure 7.16 is a photograph—really an ultraviolet image—taken in 1979 by America's *Pioneer* spacecraft when at a distance of 200,000 kilometers from Venus. The large, fast-moving cloud patterns resemble Earth's high-altitude jet stream more than the great cyclonic whirls characteristic of Earth's low-altitude clouds. In fact, the upper deck of Venus rotates around the planet in a rapid 4 hours, which is much, much faster than the underlying surface.

Prior to the mid-1960s, the rotation period of Venus was a mystery. Attempts to determine its rotation by watching motions of the cloud markings as seen from Earth were frustrated by the rapidly changing nature of the clouds themselves. Some astronomers argued for a 25-day period, while others favored a 24-hour period. Controversies raged until, to the surprise of optical astronomers, radar observers announced that the Doppler broadening of their returned echoes implied a sluggish 243-day rotation period. Furthermore, Venus' spin was found to be "retrograde"—namely, opposite that of Earth and most other Solar System objects—a fact astronomers do not understand.

So if you could stand on the surface of Venus and could see the Sun—neither of which is possible, by the way—the Sun would rise in the west, and then set in the east nearly 2 Earth months later. In fact, since Venus' orbital period is only 225 Earth days, a Venusian "day" is more than half as long as a Venusian "year." Furthermore, the peculiar filtering effects of whatever light does penetrate the clouds would make Venus' air and ground peach-colored. What's more, the thick atmosphere would cause the landscape to appear strangely distorted, as if viewed through a fishbowl. Strange place, to say the least.

Although the clouds are extremely thick and the surface totally shrouded, we are not entirely ignorant of Venus' surface. Radar astronomers have bombarded the planet with radio signals launched from Earth and from the *Pioneer* (U.S.) and *Venera* (U.S.S.R.) spacecraft now in Venusian orbit. The long wavelength of the radio radiation enables the signals to penetrate to the surface, as we discussed in Chapter 3 (see Figure 3.19). The echoes can then be analyzed and a map made of the planet's surface. As Figure 7.17 illustrates, such early maps suffered from poor resolution, but more recent ones have provided some tantalizing hints of the global features of Venus. The results suggest that its surface is mostly smooth, perhaps resembling rolling plains with modest highlands and lowlands. Only two or three continental-sized features adorn the landscape, and these contain mountain peaks comparable to those on Earth. Such geological features imply surface activity—if not currently, then at least at some time in the past. And since most meteoroids would probably burn up in Venus' thick atmosphere before reaching its surface, volcanism seems a more reasonable explanation. To be sure, the *Pioneer* spacecraft measured some chemical anomalies in Venus' atmosphere indirectly suggestive of volcanic outgassings as well as radio waves apparently caused by lightning discharges normally found in the plumes of erupting volcanoes.

Figure 7.18 depicts one of Venus'

(a)

(b)

FIGURE 7.17 (a) This radar image of the *surface* of Venus was made by transmitting manmade signals and detecting the returned echos with a large radio telescope on Earth. (b) The resolution of the radar image of Venus has improved dramatically in recent years by using a transmitter and receiver onboard the *Pioneer* spacecraft now in orbit about Venus. The two continent-sized landmasses are named Ishtar Terra (upper left) and Aphrodite (lower right).

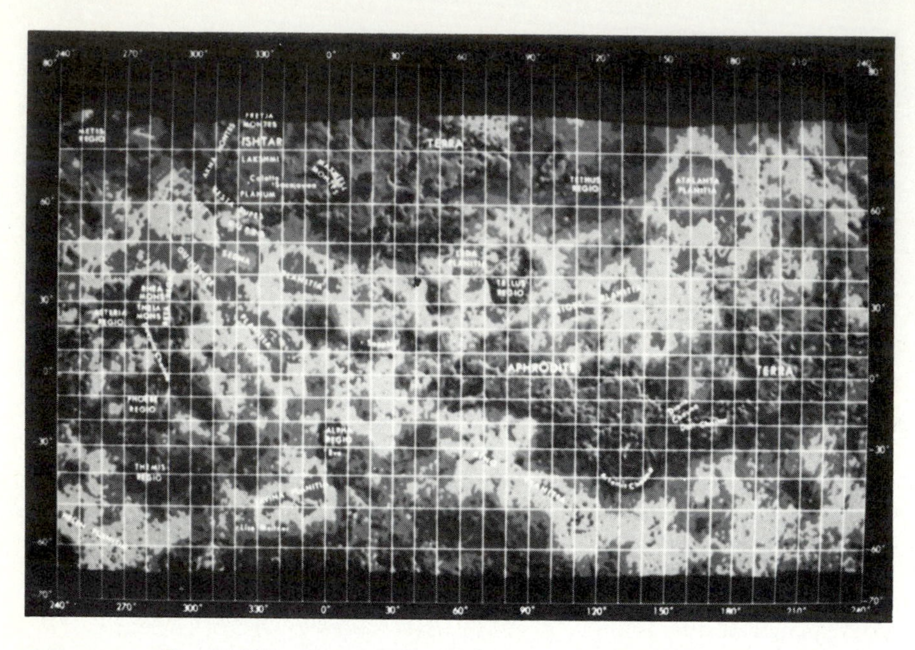

FIGURE 7.18 Venus' Ishtar Terra region is shown at the upper left in this radar image. Its size is comparable to that of Australia.

continental-sized regions in the northern high latitudes. An extensive uplifted plateau called Ishtar Terra, this landmass houses a great volcano and the highest mountains on Venus, Maxwell Montes, which reach an altitude higher than our own Mount Everest above sea level. The other continental-sized formation, called Aphrodite Terra, is comparable in size to Africa and might well have some active volcanoes. These volcanic-looking features seem to be localities of the observed lightning on Venus, again *suggestive* of active volcanism on the planet today.

In all, some 20 spacecraft have visited Venus during the last two decades, far more than have spied any other planet. Of note, the early American *Mariner 2* and *Mariner 5* missions passed to within 35,000 kilometers in 1962 and 1967, the *Mariner 10* craft grazed Venus at a distance of 6000 kilometers in 1974, and the Soviet *Venera 4* through *Venera 12* probes parachuted into its atmosphere between 1967 and 1978. In the most recent efforts, two American *Pioneer* crafts went into low orbit some 150 kilometers above Venus' surface in 1978, one mother ship dispatching several probes to the surface. And several more Soviet probes of the *Venera* series of spacecraft arrived at Venus in the 1980s. All these spacecraft radioed back data indicating the surface temperature to be a sweltering 750 kelvin. That's enough heat to melt lead, or roughly twice as hot as a kitchen oven. The *Venera* and *Pioneer* probes that landed (or tried to land) were quickly either cooked by this inferno or crushed by an atmospheric pressure nearly 100 times greater than on Earth. This Venusian pressure equals that at an (Earthly) underwater depth of about 1 kilometer (or 3000 feet); unprotected humans can hardly dive a tenth of this depth.

Under such high pressures and temperatures, liquid water cannot possibly exist on Venus' surface; nor could much water vapor exist in its atmosphere. Furthermore, the temperature is nearly as high at the poles as at the equator, and not much different on the "night side" opposite the Sun; the atmosphere is so thick that it spreads energy efficiently, making it impossible to escape the blazing heat. Consequently, life of any sort is a remote prospect on Venus, unless it's somehow suspended at high altitudes where the temperatures and pressures are much less than on the ground.

Given the (100-million-kilometer) distance of Venus from the Sun, the planet was not expected to be such a pressure cooker. Since it's closer to the Sun than Earth, astronomers originally thought that Venus' surface temperature would be a little warmer than on Earth. Surely, it was reasoned, Venus should be no hotter than the 500 kelvin values that characterize the hot side of the innermost planet Mercury (and perhaps might even be less than on Earth, owing to the fact that Venus reflects so much sunlight). However, numerous observations—from Earth and via probe—demonstrate that Venus' surface is much hotter than that of any other planet. How can this be? The culprit, it turns out, is the greenhouse effect.

Recall from our discussion in Chapter 5 that greenhouses serve to trap heat from the Sun. The sunlight first penetrates the glass and is then absorbed by soil in the greenhouse, whereupon a little of it is used to nourish plants. Most of the absorbed radiation is reemitted, but not as visible radiation. It's reemitted as infrared radiation, according to the Planck curve for any warm object. Since infrared radiation cannot easily penetrate the glass panes of the greenhouse, especially because the glass prevents the heated air from escaping, some of the radiation becomes trapped. And since radiation is energy, the interior of the greenhouse becomes warmer than the outside air.

The same sort of effect naturally occurs on Venus, where the atmospheric constituents—about 96 percent carbon dioxide (CO_2), along with trace amounts of nitrogen, (N_2), water vapor (H_2O), sulfuric acid (H_2SO_4) and other chemicals—allow some solar radiation to get in but prohibit some of the planetary radiation from getting out. The thick blanket of CO_2 is the main culprit; as in Earth's atmosphere, this molecule is a voracious absorber of infrared radiation. In fact, CO_2 absorbs 99 percent of all the infrared

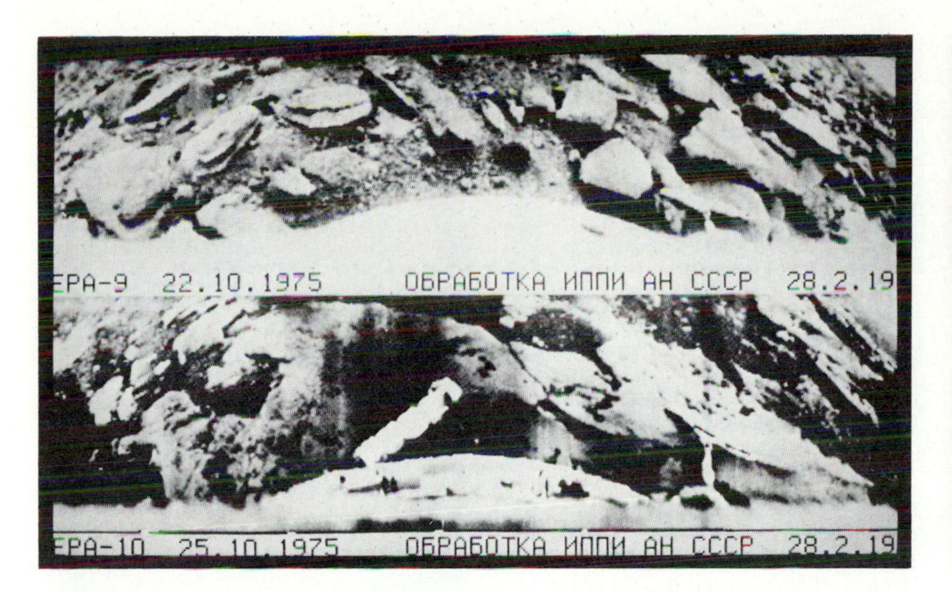

FIGURE 7.19 The first direct view of the surface of Venus, radioed back to Earth via the Russian *Venera 9* spacecraft which made a soft-landing in 1975. The amount of sunlight penetrating Venus' cloud cover apparently resembles that on a heavily overcast day on Earth.

FIGURE 7.20 Photograph of Earth from 200,000 kilometers out in space—often called NASA's greatest contribution to science.

radiation released from the surface of Venus. Hence this planet has heated greatly while gravity prevents the gases from escaping by holding the atmosphere under high pressure. (Note that the planet, like any greenhouse, does not trap heat indefinitely, for an equilibrium temperature is eventually reached; otherwise, both the atmosphere and the greenhouse would explode.)

That Venus' surface is dry and dusty was established directly by the *Venera 9* and *Venera 10* Soviet spacecraft that landed successfully on the surface in 1975. Figure 7.19 shows one of the first photos of the surface of Venus, radioed back to Earth just prior to the spacecraft's demise. Each craft lasted about an hour before overheating, their electronic circuitry literally melting in this planetary oven. Typical rocks in the photo measure about 50 centimeters across by 20 centimeters high—a little like flagstones on Earth. Having sharp edges and a slablike character, the Venusian rocks show hardly any evidence for erosion. Apparently, they are young rocks, implying that Venus is geologically active.

Not much theoretical modeling of Venus' interior has been accomplished to date. An average density similar to Earth's implies that Venus is probably made of similar mixtures of basaltic rocks. But Venus has no detectable magnetic force field, almost surely because of its extremely slow rotation. As further data are gathered, especially via instrumented spacecraft, astronomers will be especially eager to compare the interior of Venus to that of Earth, since the two planets have nearly equal masses and radii but very different environmental conditions.

All in all, Venus is rocky, hot, and miserable—a truly forbidding planet. If there exists a "hell" anywhere in the Universe, Venus is as good a candidate as any!

EARTH: Our planet has already been studied in considerable detail in Chapter 5. Still, we can usefully describe its "big picture" in parallel with those of the other planets of this chapter.

As Figure 7.20 shows, Earth is prominently characterized by variable white cloud patterns superposed on a generally bluish background. These bluish areas are the oceans of liquid water, comprising almost three-quarters of our planet's surface. Only a remarkable combination of temperature and pressure, unlike that on any other planet in our Solar System, permits large amounts of *liquid* water to exist. Tannish and greenish areas can often be seen on Earth, indicative of rocky continents that undergo exceedingly slow drifts on its surface.

Earth's atmosphere is mainly composed of nitrogen, oxygen, and water vapor, which together with meteorological phenomena near the surface, form the clouds that dominate our planet as seen from afar. Liquid and solid water frequently precipitate onto the surface and cause substantial erosion of the silica-rich soils and rocks. The interior is active, causing continued mountain building and volcanic activity, even at the present time. Clearly, our planet is geologically alive.

The most unique feature of Earth is, without doubt, its life. Earth is also biologically alive. Plant and animal life is widespread, both on the land and in the sea, although evidence of this life is not easy to detect from space. Men and women, in particular, have appeared within only the last 0.1 percent of Earth's history. They are well adapted to the planet, although slight changes in the geological, chemical, thermal, or political environment could render the planet uninhabitable.

MARS: The red planet, named Mars by the ancient Romans for their god of war, is for some people the most intriguing of all celestial objects. Many of Mars' physical properties have for a long time been known to approximate those of Earth. Its rotation period—24.7 hours—nearly equals Earth's, making its day much like ours. Its spin axis is tilted relative to the Sun, producing seasons (although each season is twice as long as on Earth since the Martian orbital period is nearly 2 years). Foremost in our minds, Mars is the planet most often thought potentially to house life beyond Earth.

Viewed from Earth, the most prominent Martian features are its bright polar caps. Figure 7.21 compares a photograph of Mars taken through one of Earth's best optical telescopes with that taken by a U.S. spacecraft en route to the planet. The caps are now known to be mostly frozen carbon dioxide (i.e., dry ice), not water ice as at Earth's north and south poles. The caps grow or diminish according to the seasons, almost disappearing at the time of Martian summer.

Generally, surface temperatures on Mars average about 50 kelvin cooler than on Earth. Predawn equatorial temperatures often drop to a frigid 200 kelvin, far below the freezing point of water and close to the lowest temperatures ever recorded on the surface of Earth. But in summer, midday equatorial temperatures can briefly reach a comfortable 300 kelvin, close to the usual setting for home thermostats on Earth.

Mars' reddish surface features vary from time to time, although their variability probably has little to do with the melting of the polar ice caps. The dark markings seen in Figure 7.21, once claimed to be part of a network of "canals" dug by Martians for irrigation purposes (see Interlude 7-2), are now recognized as highly cratered and eroded areas around which surface dust occasionally blows. Repeated covering and uncovering of these landmarks gives the impression (from a distance) of surface variability, but it's only the thin dust cover that varies. A powdery material, the surface dust is borne aloft by strong winds that often reach hurricane proportions (i.e., hundreds of kilometers/hour). In fact, when the American *Mariner 9* spacecraft went into orbit around Mars in 1971, a planet-wide dust storm obscured the entire landscape. Fortunately, the storm subsided, enabling the craft to radio home detailed information about the surface.

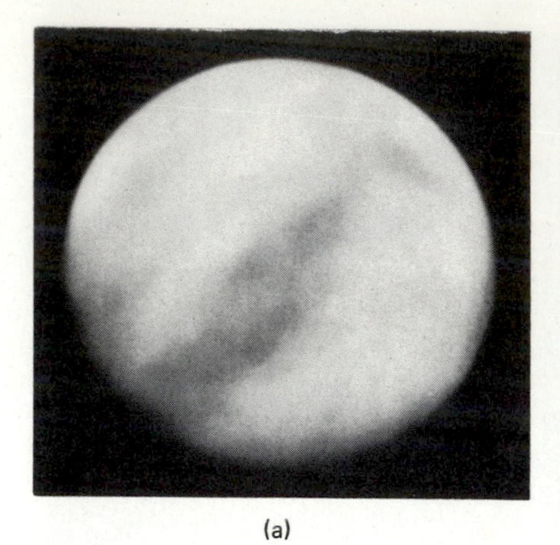

(a)

(b)

FIGURE 7.21 (a) A typical photograph of Mars taken from Earth. One of its polar caps can be seen at upper left. (b) A view of Mars taken from a *Viking* spacecraft on its approach pattern.

Storms are apparently common on Mars. They are not, however, accompanied by rain or snow. Instead, strong winds at the surface sweep up the dry dust, carry it aloft, and eventually spread it somewhere else on the planet. At its greatest fury, a Martian storm can inundate the atmosphere with dust, making the worst storm we could imagine on Earth's Sahara Desert seem inconsequential by comparison.

Such fierce dust storms are possible only because the atmosphere is so thin. The density and pressure on the surface of Mars are much less than the rarefied air atop Earth's Mount Everest, enabling small particles to remain suspended for long periods. Specifically, Mars' surface pressure is about 1 percent that of Earth's, as was first measured by the *Mariner 4* spacecraft in 1965 and later confirmed by the American *Viking 1* and *Viking 2* spacecraft that landed on Mars in 1976. The atmosphere is composed mostly of carbon dioxide gas (95 percent) along with trace amounts of nitrogen, argon, oxygen, and water vapor. Even if Mars' atmosphere were completely conducive to our respiratory system, namely, made of an oxygen–nitrogen mixture, it would still be too thin for earthlings to breathe normally. There is just not enough gas to breathe. And at times, it's too dusty.

The *Mariner* series of U.S. spacecraft demonstrated that the Martian surface and its two moons are pitted with impact craters formed by meteoroids falling in from space. Some of these craters can be seen in Figures 7.22 and 7.23. As with our Moon, the smaller craters are often filled with surface matter, confirming that Mars is a dusty and desert-like planet. However, these craters are not the most interesting features on Mars. There are huge volcanoes, deep canyons, vast dune fields, among other signs of a geological wonderland, although no direct evidence for crustal activity as on Earth. Figures 7.23 to 7.26 show some of these interesting surface features.

The widespread existence of extinct volcanoes and eroded fissures strongly suggests geological activity

INTERLUDE 7-2 ***Canals on Mars?***

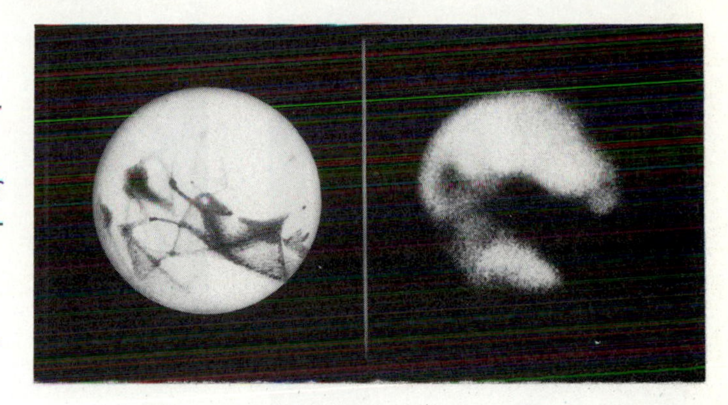

The year 1877 was an important one in the human study of the planet Mars. Then the red planet came unusually close to Earth, affording astronomers an especially good view. Of particular note was the discovery, by U.S. Naval Observatory astronomer Asaph Hall, of the two moons circulating Mars. But of greatest excitement, the Italian astronomer Schiaparelli reported observing a network of linear markings which he termed "canali," which in Italian means "grooves" or "channels." Observations of these features became sensationalized in the world's press (especially in the United States) and some astronomers began drawing elaborate maps of Mars, showing oases and lakes where the canals met in desert areas.

Percival Lowell, a successful Boston businessman (and brother of the poet Amy Lowell and Harvard president Abbott Lawrence Lowell) became fascinated by these reports. He abandoned his business, purchased a clear sight at Flagstaff, Arizona, where he built a major observatory, and devoted the rest of his life to achieving a better understanding of the Martian "canals." In doing so, he heavily championed the idea that Mars was drying out and that an intelligent society had constructed the canals to transport water from the wet poles to the arid equatorial deserts.

The figures below show how surface features (which were probably genuinely observed by astronomers at the turn of the century) might be imagined to be connected. The figure at right is a photograph, that at left is an interpretive sketch of a similar observation. The human eye, under great physiological stress, tends to interpolate linearly, in the process often bridging dimly observed yet distinctly separated features.

Alas, the Martian valleys and channels photographed by robot spacecraft during the 1970s are far too small to be the "canali" that Schiaparelli, Lowell, and others thought they saw on Mars. The entire episode represents a classical case in the history of science—a case whereby well-intentioned observers, perhaps obsessed with the notion of life on other worlds, let their personal opinions and prejudices seriously affect their interpretations of reasonable data.

The chronicle of the Martian canals is also a case study of how the scientific method relies heavily on the acquisition of data to sort out the sense from the nonsense, the fact from the fiction. Rather than believing the existence of Martian canals as claimed, other scientists demanded further observations to test Lowell's hypothesis. The result is that the canals were totally disproved by the *Mariner* and *Viking* exploratory missions to the red planet nearly a century after all the hoopla began. Although it often takes time, the scientific method does in fact make progress toward reality.

on Mars—at least in its past. There is ample evidence for extensive lava plains—much more extensive than on Earth or the Moon—presumably caused by eruptions involving great volumes of lava. These are strewn with blocks of volcanic rock as well as boulders blasted out of impact areas by infalling meteoroids (there being hardly any atmosphere to stop them). Figure 7.26 is an overhead view of the most prominent volcano and Figure 7.27 shows a typically littered landscape.

Photographic evidence for dried-up river beds (Figure 7.23) and interconnected canyons further strengthen the idea that Mars once experienced surface activity, including running water. Yet no activity seems present now, except for some occasional landsliding along the edge of the Mariner Valley. Furthermore, no spacecraft, including the Viking probes that landed on the Martian surface in 1976, has found any evidence for *liquid* water. What little water does exist is apparently in the form of ice, either locked up in the polar caps or spread under parts of the surface as permafrost. (Not more than 0.1 percent of the atmosphere is water vapor.)

What could have caused the valleys, canyons, and tributaries now seen on Mars? The larger valleys (Figure 7.24) were probably formed by faulting and quaking of the Martian crust. But the smaller depressions and channels are probably remnants of an earlier Martian epoch—an ancient time when the atmosphere was thicker, the surface warmer, and liquid water widespread. Flash floods might have carved some of the channels and teardrop-shaped islands (resembling miniature versions seen in the wet sand of our beaches at low tide). Even rainfall may have been common in Mars' past.

Mars' size has undoubtedly played a central role in the planet's history. Being smaller than Earth, its pull of

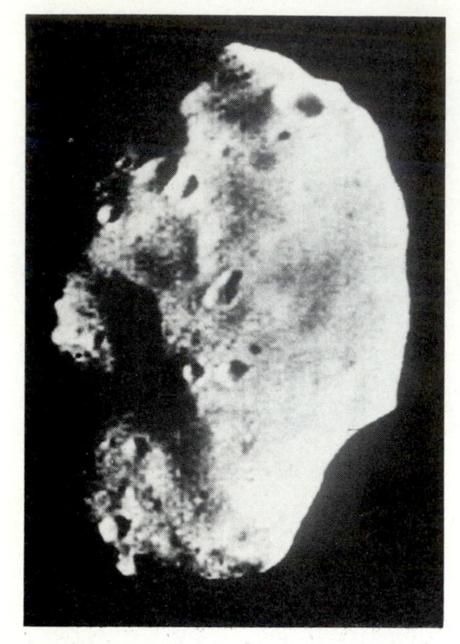

FIGURE 7.22 Mars is accompanied in its trek around the Sun by two tiny moons. Deimos (Greek for "terror") and Phobos ("fear") average 10 and 20 kilometers in diameter. Both are shaped like potatoes, with lots of cratering. This figure is a *Mariner-9* photograph of the irregularly shaped Phobos, not much larger than Manhattan Island.

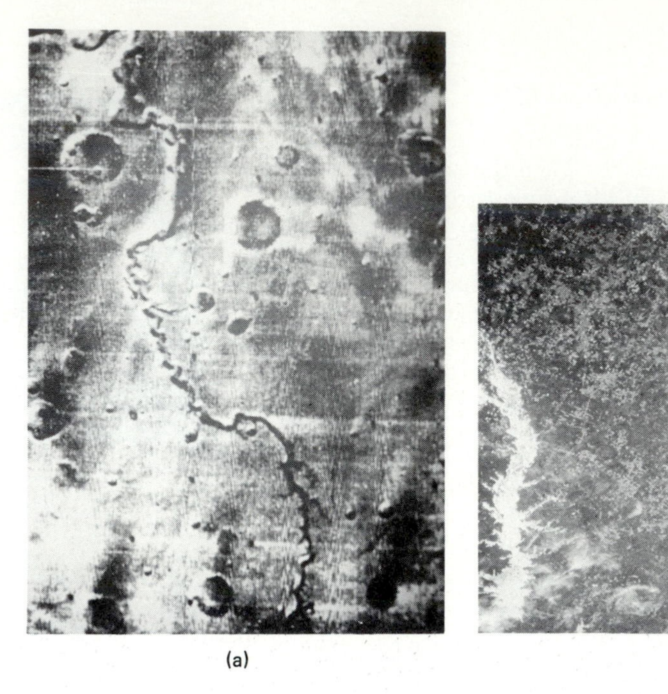

(a) (b)

FIGURE 7.23 This sinuous valley on Mars (a) measures about 400 kilometers long and 5 kilometers wide. Here, we compare it to a photograph of the Red River (b) running from Shreveport, Louisiana to the Mississippi River. The two differ mainly in that there is currently no liquid water in this, or any other, Martian valley.

(a)

(b)

FIGURE 7.24 This truly vast Martian canyon (a), called the Mariner Valley, measures at least 4000 kilometers long, 120 kilometers wide, and 5 kilometers deep in some places. We compare it here to Earth's Grand Canyon (b), only 20 kilometers wide and 2 kilometers deep.

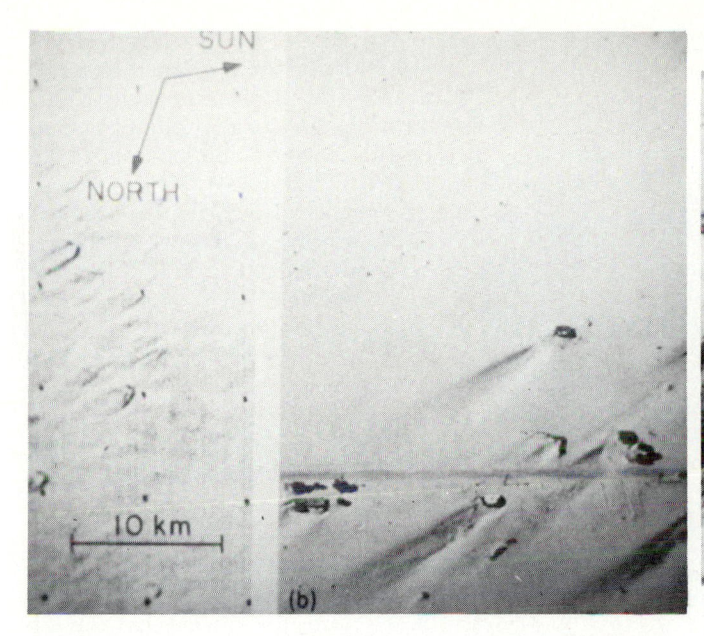

FIGURE 7.25 Martian sand dunes (a) compared to Earth's Sahara Desert (b). Small dune fields, with regularly shaped parallel ridges, occur almost everywhere on Mars, especially inside craters.

FIGURE 7.26 Olympus Mons, the largest volcano known either on Mars or anywhere else. Nearly three times taller than Mount Everest on Earth, this Martian mountain measures about 500 kilometers at the base and extends 25 kilometers at the peak. It seems currently inactive and has probably been extinct for at least several hundred million years. (By comparison, the largest volcano on Earth, Hawaii's Moana Loa, measures a mere 120 kilometers across and peaks about 9 kilometers above the ocean floor.)

gravity is less than on our planet. This allows surface features such as Olympus Mons to rise to immense heights. Its smaller gravity also means that Mars must have had trouble retaining an atmosphere. Once its "air" thinned by escaping into space, the surface cooled for lack of greenhouse heating, allowing the liquid water either to evaporate and thus also escape, or to freeze and thus become locked up as permafrost below the surface or in the polar regions. The polar caps are estimated to be at least several hundred meters thick, containing enough ordinary ice to evenly cover the entire planet with about ½ meter of water if it were released all at once (although most of the caps are made of "dry ice," which is frozen CO_2).

The small size of Mars also implies that the radioactive heating of its interior was less than for Earth. The evidence noted above for ancient surface activity, especially volcanism, suggests that at least parts of Mars' interior must have melted and possibly differentiated at some time in the past. But the lack of current activity, the absence of a magnetic force field, the relatively low density (4 grams/cubic centimeter), and a normal abundance of iron at the surface all suggest that Mars never melted as extensively as did Earth. The planet is probably made of rather uniform basalt throughout most of its interior.

From the time when Mars lost its atmosphere and cooled, thus becoming inactive and devoid of liquids, the surface has remained essentially unchanged. Apparently, Mars has looked much like that in the *Viking 1* views reproduced as Figure 7.27 and Plate 3(b) for quite some time. How long? The lack of fresh lava and the extent of meteoritic cratering of old lava near Martian volcanoes suggest that Mars "died" about 0.5 billion years ago. It's as though the entire planet entered an ice age from which it has yet to recover.

The present lack of liquid water on Mars dims the chances for life there now. Indeed, the *Viking 1* and *Viking 2* spacecraft were unable to find clear evidence for even microbial life on the surface. Even so, as we'll discuss in Chapter 29, the *Viking* robots did detect peculiar and unknown substances that mimic astonishingly well the basic properties of living organisms. Since geological activity on Mars apparently ended rather recently, perhaps—just perhaps—any biological activity on the planet ended then as well.

Despite *Viking*'s lack of clear evidence for life on today's Mars, it is still possible that this most intriguing planet harbors fossilized evidence for a life that flourished during some previous epoch. This is only conjecture, however, and remains for future exploration to prove or disprove.

Comparative Planetology

Not to be overlooked is the fact that by studying the other planets, we are gaining a deeper understanding of many natural phenomena occurring

FIGURE 7.27 Panoramic view from the perspective of the *Viking 1* spacecraft now parked on the surface of Mars. The fine-grained soil and the rock-strewn terrain stretching toward the horizon is reddish in color. Containing substantial amounts of iron ore, the surface of Mars is literally rusting away. The sky is a pale pink, the result of airborne dust. [See also Plate 3(b).]

within the Earth, in its atmosphere, and in its oceans. In the newly emerging subject of "comparative planetology," the Solar System is considered a giant laboratory where many of the natural events native to Earth are also recognized to occur in other planetary settings. These events often transpire under diverse physical conditions and usually at different evolutionary stages. A few examples will suffice until we reach the appropriate place—Chapters 23 and 24—to describe more fully the varied evolutionary processes at work.

Consider the atmospheres of Mars and Venus. Direct measurements of meteorological conditions on Mars' surface (by the *Viking* landers) have led to an increased understanding of the seasonal Martian variations that have long been observed from Earth. The details of Martian dust storms and the overall circulation of its lower atmosphere turn out to have considerable similarities to such processes on Earth. Also, curious (Y-shaped) markings seen in the ultraviolet images of Venus' upper cloud deck seem to be driven by a global system of atmospheric waves possibly akin to Earth's jet stream. However, continuous buffeting by the solar wind makes it hard to unravel the specific mechanisms responsible. Whatever the process, it may eventually help us better understand the upper-level weather patterns that seemingly dominate the spread of high- and low-pressure zones on our own planet.

Of some future significance, perhaps, is the fact that the upper atmosphere of Mars displays an ozone chemistry closely analogous to what would be expected on a highly polluted Earth. Venus, too, might be a model of a future-polluted Earth: Although CO_2 currently comprises only 0.03 percent of our planet's atmosphere, even this minute amount manages to produce a small greenhouse effect on Earth, increasing the temperature from near the freezing point of water to more comfortable temperate values. We can now begin to appreciate the predicament our descendants may face someday should our civilization continue to burn fossil fuels and thus dirty Earth's air with increasing quantities of man-made CO_2.

Consider also the history and interior heating of several terrestrial-like bodies. In the preceding section we found that Mars' general properties are intermediate between those of Earth and the Moon. This is largely because of size. Mars has approximately twice the diameter of the Moon and one-half the diameter of Earth; its mass is roughly 10 times the Moon's mass and one-tenth the Earth's mass. We can then begin to understand why Earth has a thick atmosphere, Mars has a thin atmosphere, and the Moon has no atmosphere at all. Similarly, Earth is still geologically alive, Mars is apparently inactive now but shows signs of past activity, and the Moon has been geologically dead for much of its history. Since we've already reasoned that the Moon's activity was probably confined to the first billion years after its formation, we can surmise that Mars, being a larger object, was active for the first several billion years of its history.

Our newly acquired data about the makeup of the Terrestrial Planets' atmospheres, surfaces, and interiors (including those of planet Earth) mark the beginning of our efforts to construct comprehensive evolutionary models at least for the inner parts of the Solar System. Although only in its early stages of development, the subject of comparative planetology, to which we shall return in Part III, might well become as powerful a framework for the evolution of matter as is Darwinian biological evolution for the evolution of life—another subject that we shall address in Part III. In any case, we can anticipate that a further broadening of our experience with an increasingly diverse range of planetary conditions will lead to a keener understanding of the rock that is our home in space.

SUMMARY

The past few decades of spacecraft exploration have seen extraordinary advances in our understanding of our cosmic neighborhood. During this time, humans have begun to recognize other planets as genuine worlds. Equally important, we have gained a better appreciation for our own world as a planet.

Electronic eyes and an arsenal of instruments aboard our robot spacecraft have presented us with an embarrassment of riches. And what we have found, foremost among many discoveries, is that diversity reigns, even among the Terrestrial Planets. From planet to planet, sizes differ as do colors, atmospheres, surfaces, interiors, spin rates, and climates, among a host of other properties. Yet it is this diversity that unites the world of humans—our planet Earth—to the other planets and even to the stars beyond.

The U.S. and Soviet programs of Solar System investigation by "smart" spacecraft comprise perhaps the grandest technological achievement of our generation. The age of exploration represented by robots named *Mariner, Viking,* and *Venera* is at least equal to, and perhaps greater than, the era of Columbus, Magellan, and da Gama. Although funding constraints have recently slowed the pace of planetary missions, there is no denying the fact that humankind has begun its long climb out of its cosmic cradle.

KEY TERMS

ellipse
geocentricity
gravity
heliocentricity
interplanetary space
Jovian Planets
Kepler's laws
Terrestrial Planets

QUESTIONS

1. Pick one: Venus' suspected volcanoes have best been studied to date by (a) optical photos taken on the surface by a Soviet *Venera* craft, (b) radar observations from the U.S. *Pioneer* craft in Venusian orbit, (c) radar observations using radio dishes on Earth's surface, (d) infrared signals that emerge from the volcanoes' mouths, (e) radio studies made by the U.S. *Mariner* series of spacecraft.
2. Define escape velocity, and explain briefly how it depends on the mass of an object.
3. Given that the Moon is colder than Mars, why does Mars have an atmosphere, whereas the Moon does not?
4. The velocity needed to escape from Mars is virtually the same as that from Mercury. Why, then, does Mars have an appreciable atmosphere, whereas Mercury does not?
5. Pick one: At birth, the greatest gravitational pull on a baby is caused by (a) the planet Mars, (b) all the planets combined, (c) the nearest star, (d) the accumulation of the nearby constellations, (e) the obstetrician in the delivery room.
6. Is the Martian sky reddish or bluish, and why?
7. Explain to what extent the greenhouse effect operates on Earth, Mars, and Venus.
8. Mars and our own Moon have been geologically dead for several billion years and probably less than a billion years, respectively, while Earth (and possibly Venus) is still active geologically. Explain this variable feature of comparative planetology.
9. In a few paragraphs, describe what environmental changes you might experience during a 2-week visit to each of the Terrestrial Planets.
10. Explain why orbiting satellites do not fall into the Earth, or for that matter Earth does not fall into the Sun.

FOR FURTHER READING

ARVIDSON, R., BINDER, A., AND JONES, K., "The Surface of Mars." *Scientific American*, March 1978.

*HARTMANN, W., *Moons and Planets*. Belmont, Calif.: Wadsworth, 1983.

HOBERLE, R., "The Climate of Mars." *Scientific American*, May 1986.

JONES, B., *The Solar System*. New York: Pergamon Press, 1984.

PETTENGILL, G., CAMPBELL, D., AND MASURSKY, H., "The Surface of Venus." *Scientific American*, August 1980.

PRINN, R., "The Volcanoes and Clouds of Venus." *Scientific American*, March 1985.

SKINNER, B. (ed.), *The Solar System and Its Strange Objects*. Los Altos, Calif.: William Kaufmann, 1981.

8
THE OUTER SOLAR SYSTEM: *The Jovian Environment*

During the last decade, robotic cameras have taken us into the deep outer realms of the Solar System. As U.S. spacecraft skirted past many of the dimly lit giant planets far from the Sun, they revealed these worlds in stunning clarity and detail that could only have been dreamed of by three centuries of Earth-bound astronomers. Indeed, we saw many worlds up close for the first time. With a diverse array of moons and a multitude of rings circling each of the Jovian Planets, we have confirmed our earlier suspicion that the distant regions of our planetary system resemble little of our home or of the other Terrestrial Planets in our cosmic neighborhood. Those faraway planets that have captured our imaginations for centuries—especially perhaps Saturn—are even more spectacular up close. They are also a good deal more baffling, for pioneering exploration often yields both a rich lode of answers and a host of new questions.

In this chapter we continue taking an inventory of the planets. Here, in the far reaches of the Solar System, we encounter objects very much different from the more mundane Terrestrial Planets. Huge gas balls, highly peculiar moons, ring-like structures, and a wide variety of physical and chemical properties that are simply not understood seem to be par for the course far from the Sun. To be sure, the remote worlds of the outer Solar System are utterly fascinating in their own right, if only *because* they are so alien from what we have come to regard as terrestrial familiarity.

And although according to classical definitions in the textbooks (including this one) there are only nine planets, we should keep in mind that there are several thousand other celestial bodies currently known to be moving in well-determined orbits around the Sun. Admittedly, all these asteroids and comets (which we study toward the end of this chapter) are small, yet each is a world of its own. Furthermore, based on statistical inferences, astronomers estimate a total population of more than a billion such objects still to be discovered! Rocky and icy "debris" they may seem, but these are the objects, more so than the planets, that hark back to the formative stages of the Solar System. Many of these are pristine, uneroded bodies that may well one day tell us most about our local origins—a subject to which we shall return in Chapter 23.

The learning objectives for this chapter are:

- to gain an understanding of the planetary residents of the dark and frigid regions within the outer Solar System
- to understand how greatly the Jovian Planets differ from the Terrestrial Planets
- to appreciate the highly varied physical and chemical properties of the myriad moons orbiting the giant planets
- to realize that meteorites are stray asteroids
- to know the physical properties of the comets
- to recognize that the nonplanetary stray bodies probably contain hints and clues as to the origin of our Solar System

The Jovian Planets

JUPITER: Named after the most powerful god of the Roman pantheon, Jupiter is by far the largest planet; in fact, it has more mass than all the other planets combined. It even houses more than three times the mass of the next largest planet, Saturn, although only a slightly larger diameter; more dramatically stated, more than 1000 Earths would be needed to equal the volume of Jupiter. Jupiter is such a large planet that many celestial mechanicians—those researchers concerned with the motions of interacting cosmic objects—regard our Solar System as having primarily two objects—the Sun and Jupiter. To be sure, in this day and age of sophisticated spacecraft navigation, the gravitational influence of all objects must be considered. But in the broadest sense, our Solar System is a two-object system plus a lot of debris. As massive as Jupiter is, though, it's still some 1000 times less massive than the Sun. This makes studies of Jupiter all the more important, for here we have an object intermediate in size between the Sun and the Terrestrial Planets.

Jupiter is visually dominated by two features, one a series of ever-changing atmospheric bands arranged parallel to the equator and the other an oval atmospheric blob called the Great Red Spot. The cloud bands, seen clearly in Figure 8.1 and Plate 3(a), display different colors—pale yellows, light blues, deep browns, drab tans, and vivid reds among others—and must be caused by Jupiter's various chemical compounds. However, the chemistry creating these striking colors is not well understood.

Some atmospheric gases are known from spectroscopic studies of reflected sunlight, although none of these gases alone are thought to cause the observed hues. The most abundant gas is molecular hydrogen (H_2), followed by helium (He); together they comprise 99 percent of Jupiter's atmosphere. Trace amounts of ammonia (NH_3), methane (CH_4), and water vapor (H_2O), as well as a few other heavy chemicals have also been observed. The gravitational pull of the larger Jovian Planets is sufficient to retain hydrogen as well as several of the other light chemicals, but most of them probably escaped from the atmospheres of the less massive Terrestrial Planets.

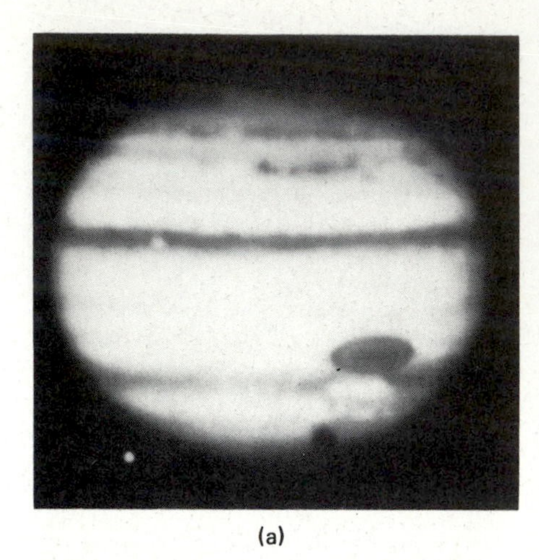

(a)

(b)

FIGURE 8.1 (a) Photograph of Jupiter made with an Earth-based telescope. One of its moons, called Ganymede, can be seen at lower left, as can the moon's shadow on Jupiter itself at bottom center. (b) Jupiter's belts, bands, and spots can be more clearly discerned in this view during *Voyager's* approach in 1979.

The Red Spot, shown in Figure 8.2 in a close-up photograph taken as the *Voyager 1* spacecraft glided past in 1979, presents an even greater mystery than the nature of the cloud bands surrounding it. First reported by the British astronomer Hooke in the mid-seventeenth century, we can be sure of its existence in one form

FIGURE 8.2 This *Voyager-1* photograph of Jupiter's Red Spot (upper left) was taken from a distance of a few million kilometers. Note the complex turbulence patterns to the right of both the Red Spot and the smaller white oval vortex south of it.

or another for more than 300 years. And it might well be much older than that. Modern observations show the spot to be characterized by swirling circulations much like a whirlpool or a terrestrial hurricane. Indeed, the Red Spot might be nothing more than an atmospheric storm—but a persistent and vast storm at that. The size of the spot is known to vary a bit, although it usually averages about twice the diameter of Earth. The origin of its color is unknown, as is its source of energy—even if it is "only" an atmospheric storm.

Actually, storms may be common on Jupiter. The *Voyager* mission discovered many smaller light-and-dark-colored spots that are also apparently circulation cells. More startling, spacecraft photos of the dark side of the planet revealed both a display of auroral activity and bright flashes closely resembling lightning.

Unlike Earth's atmosphere, different parts of Jupiter's atmosphere move independently; visual observations and Doppler-shifted spectral lines prove that the equatorial zones rotate faster (9^h50^m period) than the higher latitudes (9^h56^m period). That's fast for such a large object; it causes a pronounced bulge at the equator. Probably neither of these rotation rates represent the true rotation of whatever lies beneath the clouds. Nor can the Red Spot itself, drifting around with a variable rotation period that averages about 9^h56^m, be related to any fixed surface features.

The nature of the underlying surface of Jupiter, should there be one, is controversial. Surely, deep down in the clouds lurks total darkness, strong gravity, and extremely high pressures and temperatures. Unfortunately, we have little conclusive knowledge about the structure of this largest planet below the upper cloud deck. The average density of the entire planet is low enough—1.3 grams/cubic centimeter—to rule out much of a rocky composition like that of the Terrestrial Planets. In fact, Jupiter's density is so small as to practically guarantee a mixture of almost exclusively the lighest elements (although some researchers suspect that its core has some rockiness to it).

Jupiter differs from the Terrestrial Planets in many other ways. Foremost among them is the heat source buried deep inside this enormous planet. Unlike the smaller planets but similar to the Sun (and probably Saturn and Uranus as well), Jupiter emits radiation of its own accord. In other words, in addition to the *reflected* sunlight by which we see Jupiter as a bright object in the sky, the planet itself also *emits* a substantial amount of radiation, mostly of the radio and infrared type. Measurements of this emission describe a classical Planck curve for 125 kelvin. Although this directly indicates that the topmost atmospheric layers are cold, this value is still higher than that theoretically expected for a planet so far from the Sun. Put simply, Jupiter emits about twice as much energy as it receives from the Sun. The only possible explanation is that Jupiter must have its own internal energy source. Popular conjecture suggests that this energy results from collisions and thus frictional heating among the particle fragments that gathered long ago to form the planet. The original heat was accordingly trapped beneath Jupiter's heavy atmospheric blanket, but it has been leaking out slowly ever since. The whole planet might still be contracting, or settling, a little even today, thus converting its gravitational potential energy into heat which is radiated away.

Surely, then, the temperature of Jupiter must be higher deep down within the cloud cover. After all, the greenhouse effect operates on Jupiter as well as on Earth and Venus. The atmospheric density must also increase with depth, although it's unlikely that a solid surface exists anywhere inside. Instead, Jupiter's gas just grows more and more dense, largely because of the pressure of the overriding layers. There's probably no chemically distinct crust or core, but rather a thick gaseous or possibly liquid interior having a central density of about 4 grams/cubic centimeter and a pressure approximately 100 million times that on Earth's surface. The central temperature is roughly as high as 20,000 kelvin.

Some theoretical models of Jupiter's interior predict that under such odd physical conditions, hydrogen may have a peculiar metallic form, creating a hot, molten, metal "surface" in the form of an Earth-sized ball deep below the clouds. The explanation is straightforward: As the pressure increases, the atoms and molecules must be pressed so close together that their electrons are no longer associated with individual atoms, but are free to wander throughout the material as they do in a metal. Above the metallic hydrogen, there probably exists a slushy mixture of liquid hydrogen and ice crystals. These theories are debatable, however, as no one on Earth has ever seen or worked with metallic hydrogen.

On the other hand, these theoretical models have some observational support. In particular, if Jupiter is made of cold gas atop hot gas, we can confidently expect to find some evidence for convection. Recall that convection is a natural phenomenon whereby heated matter physically upwells in a pot of boiling water, in Earth's atmosphere and interior, or in the Sun's interior, as we'll note in the next chapter. Sure enough, astronomers have gathered experimental evidence for convection on Jupiter as well. Figure 8.3 is a photograph of numerous convection cells near one of Jupiter's poles where its atmosphere is thin.

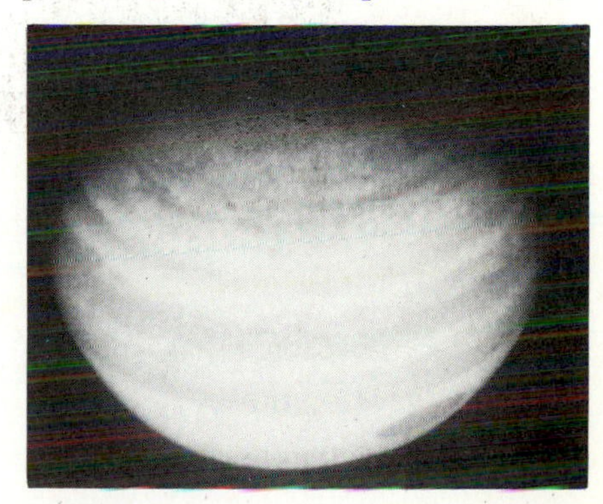

FIGURE 8.3 The *Pioneer 11* spacecraft took this photograph of convection cells while gliding past Jupiter in 1974. The probe was about 100,000 kilometers from the cloud tops at the time.

The circular, reddish patches bubble up like hot-air pockets in a pot of boiling oatmeal, essentially enabling the planet to "breathe" or vent its heat. Convection of this sort is probably the source of the day-to-day variations of Jupiter's clouds that can even be observed with a small telescope. In fact, much of the planet seems to be in a state of convection, with pockets of gases continuously bubbling up and sinking down.

The observed convection is important because it proves that temperatures do indeed increase deep within Jupiter's clouds. As such, there must be some atmospheric layers at which life could conceivably exist, provided that life could somehow anchor itself at an intermediate altitude to avoid death either by freezing at higher altitudes or by cooking at lower ones. We shall return to this issue in Chapter 29.

Discovery of Jupiter's magnetic force field further supports the theoretical models suggesting a hot, dense, and molten interior. For decades, ground-based radio telescopes have monitored radiation leaking from Jupiter's magnetosphere. But it wasn't until the *Pioneer* spacecraft reconnoitered the planet in the mid-1970s that the true extent of the magnetosphere was realized. Jupiter, as it turns out, is surrounded by a vast sea of energetic charged particles (mostly electrons and protons) similar to Earth's Van Allen Belts. Direct spacecraft measurements show Jupiter's magnetosphere to be a million times more voluminous than Earth's magnetosphere, and to have a long tail extending away from the Sun at least as far as Saturn's orbit. As sketched in Figure 8.4, this is a minimum extension of 4 astronomical units. Jupiter's electromagnetic belts would be even more impressive if it were not for several of the large moons that have orbits within the magnetosphere, and which sweep up many of the high-energy particles trapped there. Still, the radiation associated with Jupiter's belts is several thousand times more intense than Earth's, creating a tremendous hazard for manned and unmanned space vehicles. The upshot here is that all the ground-based and space-borne observations can be explained provided that Jupiter itself has a magnetic force field some 20 times larger than Earth's. This, in turn, strongly implies the presence of an electrically conducting liquid interior made of metallic hydrogen, thus bolstering the theoretical models described above.

An important point not to be overlooked is that Jupiter nearly became a star. The composition and structure of this giant planet—and possibly all the Jovian Planets; certainly Saturn—are star-like. But none of them is big enough to "ignite." If Jupiter were about 50 times more massive, its central temperature would just about equal that needed to begin nuclear reactions, converting it into a small dim star. Thus our Solar System could conceivably have formed as a double-star system, a situation that would have produced severe temperature fluctuations on all the planets, making Earth life probably impossible. You might say that we owe a debt of gratitude to the Sun for lighting up, and to Jupiter for not!

As if it doesn't have enough peculiarities already, Jupiter is endowed with still more. Having at least 17 moons, the entire Jovian system resembles a miniature Solar System. Its four large moons, each about the size of Earth's Moon, are often called the Galilean satellites, having been discovered in the early seventeenth century by Galileo himself; they are also known by their proper names, Io, Europa, Ganymede, and Callisto, after the mythical associates of the Roman god Jupiter. Several of the smaller moons have wildly elliptical orbits unlike the near-circular orbits of the Galilean satellites; a few even orbit in a direction opposite the larger moons,

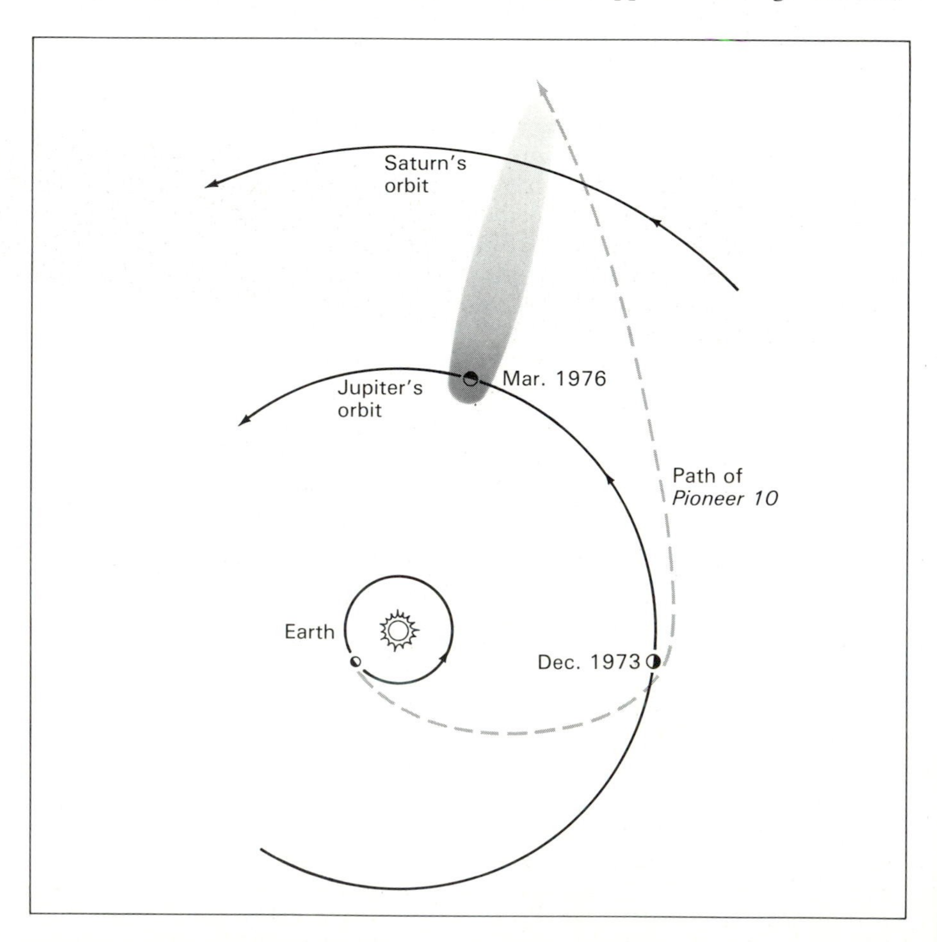

FIGURE 8.4 The *Pioneer 10* spacecraft was unable to detect any solar particles while moving behind Jupiter. Accordingly, as sketched here, Jupiter's magnetosphere apparently extends well beyond the orbit of Saturn.

leading to the suspicion that some of the smaller moons might have been captured by Jupiter's strong gravitational force field long after the planets and the largest moons originally formed.

Since the *Voyager 1* spacecraft passed very close to the Galilean moons, it was able to send back to Earth some remarkably detailed photographs, a few of which are shown in Figures 8.5 and 8.6 and also Plate 3(a). Ganymede and Callisto both seem to be mixtures of rock and ice; their densities are only about 2 grams/cubic centimeter, suggesting that they harbor substantial amounts of ice throughout, not just icy or snowy surfaces. Like Earth's Moon, Ganymede has a variety of light and dark mottlings, including craters and maria; as shown in the figure, this moon also has a system of grooves and ridges that probably result from crustal motions and faulting much like during Earth's quakes. This is reasonable since Ganymede is the largest of all moons (exceeding not only Earth's Moon but also the smallest planets, Mercury and Pluto), and its original radioactivity probably helped to heat and differentiate its partly rocky body, after which it cooled and cracked its crust. Callisto is similar, although with more craters and fewer fault lines. Its most obvious feature is a huge series of concentric ridges surrounding each of two large basins; the ridges resemble the ripples made as a stone impacts with water, but in Callisto's case probably resulted from a cataclysmic impact with a meteorite that partly melted the upthrusted ice, after which the matter quickly resolidified before the ripples had a chance to subside. Today, both these rigid crusts are frigid ice and show no obvious signs of geological activity.

FIGURE 8.5 The *Voyager 2* spacecraft photographed each of the four Galilean moons of Jupiter. Shown here to scale, they are, clockwise from upper left, Io, Europa, Callisto, and Ganymede. [See also Plate 3(a).]

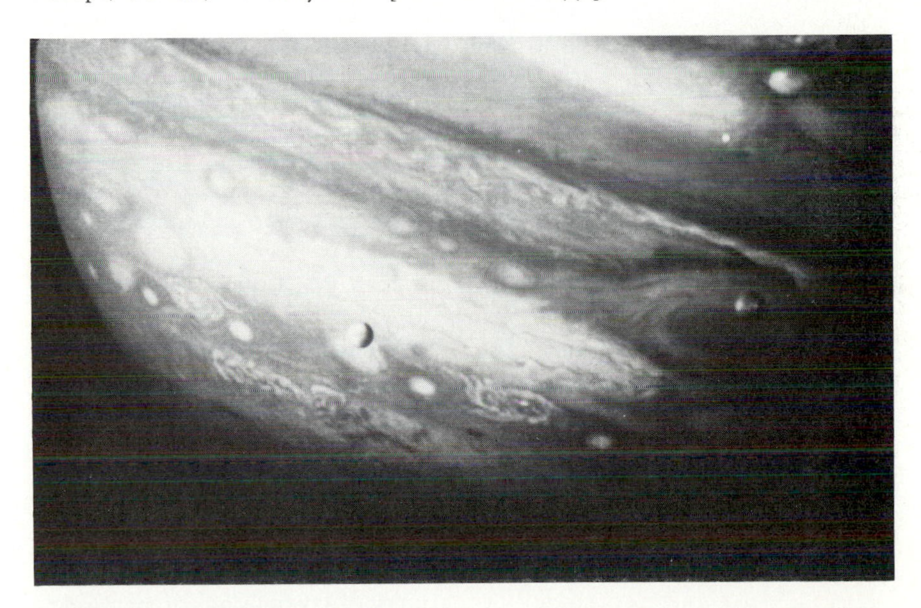

FIGURE 8.6 *Voyager 1* took this photo of Jupiter with Io on the right and Europa on the left. Note the scale of objects here: Io and Europa are each comparable to the size of our Moon, while the Red Spot is roughly twice as big as Earth.

Europa is very different from Ganymede and Callisto, nearly devoid of craters and therefore suggesting geologic youth. Surprisingly, its surface displays a vast network of lines crisscrossing otherwise bright, clear fields of water ice. Some of these linear "bands" or fractures appear to extend halfway around the satellite, and resemble the pressure ridges that develop in ice floes on Earth's polar oceans. Some researchers in fact have theorized that Europa is covered completely by an ocean of liquid water whose top is frozen, owing to the low temperatures that prevail at this moon's great distance from the Sun. If the markings truly are fault lines of ice, then this moon is probably still quite active—a notion that is not unreasonable given the tidal stress of Jupiter on Europa and the high density of Europa (implying lots of rock), both of which can lead to heating, cooling, and cracking.

Io, the densest of the Galilean moons, is definitely active—in fact, by all accounts the most geologically active object known in the Solar System. Shown in Figure 8.5, its surface is a collage of mottled reds, yellows, and blackish browns—resembling a pizza in the minds of many startled *Voyager* scientists. As the spacecraft glided past Io, an outstanding discovery was made: Io has several active volcanoes! There can be no mistake about it, for the volcanoes were photographed during eruption. Illustrated in Figure 8.7, one volcano is shown ejecting matter to the great altitude of 200 kilometers. "Squirting" may be a more apt verb, as the gases are thrown up at 2000 kilometers/second, not like the sluggish ooze emanating from Earth's insides. The orange color immediately surrounding the volcano implies that sulfur compounds dominate much of the ejected material. In stark contrast to the other Galilean moons, Io's surface is neither cratered nor streaked. Its surface is exceptionally smooth, apparently the result of molten matter constantly filling in the "dents and cracks." Accordingly, we can conclude that this remarkable moon must have the youngest surface of any known object. Of further significance, Io also has a thin, temporary atmosphere, presumably the result of gases ejected by volcanic activity.

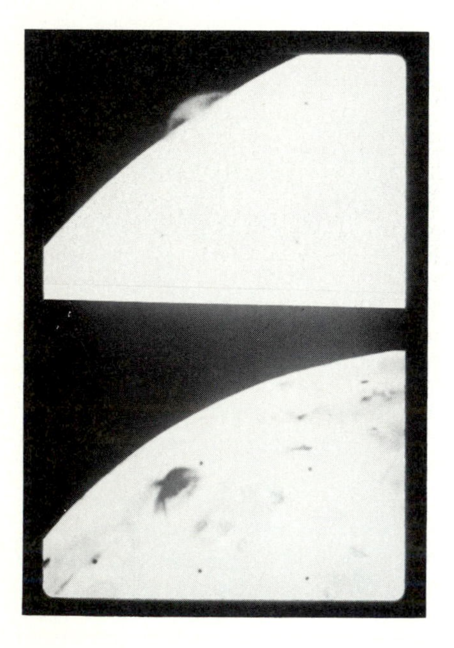

FIGURE 8.7 One of Io's volcanoes was caught in the act while the *Voyager* spacecraft bypassed this absolutely fascinating moon (which has a size and density close to those of Earth's Moon). Surface features here are resolved to within a few kilometers. At the top, the volcano's umbrella-like profile can be clearly seen against the darkness of space. At the bottom, several dark jets can be discerned as *Voyager* prepared to "overfly" another such volcano.

What is the cause of such astounding activity on a distant object the size of our Moon? Originally, it was thought that Jupiter's magnetosphere might be the culprit, as Io orbits in the midst of the most intense region of charged particles girdling Jupiter. (And it is true that the magnetosphere does tend to rip off much of Io's atmosphere; however, those atmospheric gases usually do not escape the Jovian system itself, which now explains why astronomers have observed for years that Io's orbit is enshrouded by sulfur gas.) But the real source of Io's energy is almost certainly gravity. Relentless gravitational tides exerted by huge Jupiter on relatively minute Io (one of its closest moons) would tend to distort the moon one way when closest to Jupiter and another way when farthest away in its orbit. Just as the repeated back-and-forth bending of a wire can produce heat via friction, Io is thought to be constantly energized because of the ever-changing distortion of this rather hapless moon. Hapless or not, the result is one of the most fascinating objects in our Solar System, indeed in the known Universe.

A final spectacular finding of the 1979 *Voyager* mission was the discovery of several rings of matter circling Jupiter's equator. Extremely thin, these rings extend roughly 50,000 kilometers above the cloud tops of this giant planet. The small, dark particles comprising the ring might well have been chipped off by meteorite impacts on two small moons also discovered very close to the ring by *Voyager*. The nature of Jupiter's rings can be best understood by studying those of the most famous ringed planet—Saturn.

SATURN: To many people, Saturn is the most beautiful and enchanting of all astronomical objects. Indeed, the rings of Saturn make this planet a breath-taking sight when viewed through even a small telescope. Figure 8.8 shows Saturn with its thin rings and its Jupiter-like cloud bands. Three or four obvious rings dominate thousands of delicate ringlets; although each major ring is only (incredibly) a few tens of meters thick, they have diameters ranging from 150,000 to 350,000 kilometers. (For comparison, the diameter of the Saturn gas ball itself is about 120,000 kilometers; and the Earth–Moon distance is almost 400,000 kilometers.) The rings are so thin that when viewed edge-on, they disappear altogether. Like automobile headlights penetrating an open-weave curtain, stars can occasionally be seen through the rings. Figure 8.9 shows their complex structure when viewed up close (see also Interlude 8-1).

The rings do not form a solid plate of matter. Rather, both Doppler shifts of spectral lines of sunlight reflected from them as well as direct *Voyager* imaging shows the rings to consist of large aggregates of billions of particles, each moving according to Kepler's laws in an independent orbit of period close to 10 hours. The typical particle in a ring resembles a large hailstone or a tennis-ball-sized chunk, but their sizes apparently range from micron infinitesimals (10^{-6} meter) to boulders the size of a house. Spectroscopic observations show the particles to be made of water, snow, ice, or ice-coated matter. The rings represent either debris left from moons that broke up, or, more likely, formative material that never coalesced into one or more moons.

Apart from its spectacular rings, Saturn itself is much like Jupiter but smaller. (Even so, its volume still approximates that of 1000 Earths.) Shown in Figure 8.10, yellowish and tan cloud belts parallel the equator, but display less atmospheric structure than on Jupiter. Ammonia crystals high in Saturn's atmosphere give the planet its dull, butterscotch hue.

(a)

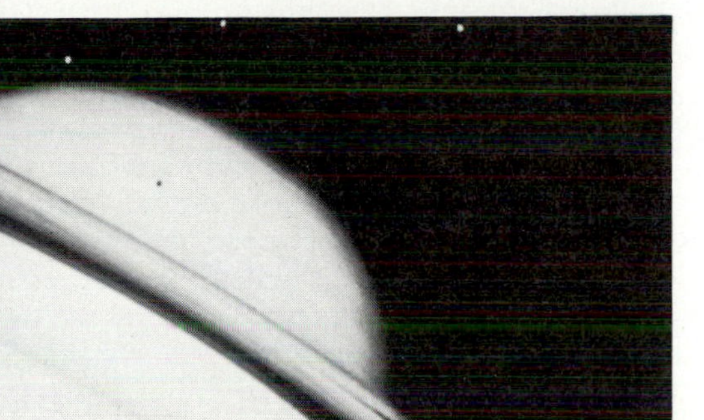

(b)

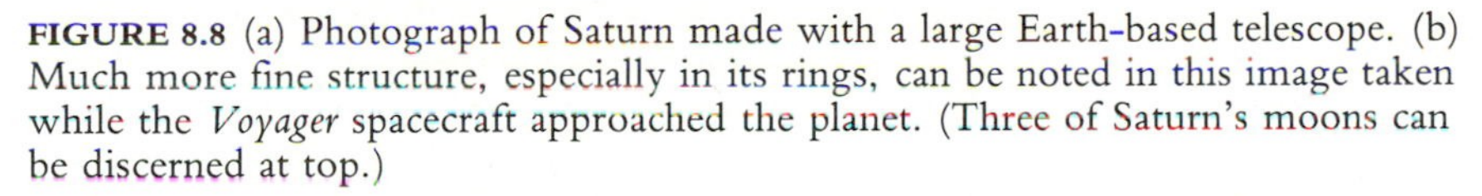

FIGURE 8.8 (a) Photograph of Saturn made with a large Earth-based telescope. (b) Much more fine structure, especially in its rings, can be noted in this image taken while the *Voyager* spacecraft approached the planet. (Three of Saturn's moons can be discerned at top.)

FIGURE 8.9 The *Voyager 2* cameras took this closeup of the ring structure just before plunging through the tenuous outer rings of Saturn. Earth is superposed, to proper scale, for a size comparison.

FIGURE 8.10 Saturn, its rings, and one of its associated moons, Titan (farthest away), are shown in this photograph taken by the *Pioneer 11* spacecraft as it sped past in 1979.

No large "spots" adorn Saturn, although its gaseous composition is pretty much the same as for Jupiter; hydrogen greatly dominates since this most abundant element never escaped, owing to either planet's large mass and great distance from the Sun. Saturn's interior properties also mimic those of Jupiter, including perhaps the bizarre metallic hydrogen liquid, but its core regions probably have less extreme temperature, density, and pressure than Jupiter. And, like Jupiter, Saturn rotates once around in nearly 11 hours, spawns an extensive magnetosphere enveloping the rings and most of its moons, and radiates nearly twice as much heat as it receives from the Sun. Apparently, Saturn is still shrinking as part of the mechanism that formed it billions of years ago, in the process maintaining a fast spin rate while liberating heat.

Saturn's most peculiar feature is its extreme lightweight matter. Its average density, 0.7 gram/cubic centimeter, is even less than water's 1 gram/cubic centimeter. Accordingly, if you could get an *average* piece of Saturn into your bathtub, it would float like an icecube! Certainly, Saturn is not made of the higher-density rocky matter found on the Terrestrial Planets. Even the ices and frozen chemicals characterizing the other Jovian Planets must not be

INTERLUDE 8-1 ***Braids, Spirals, and Gaps***

The *Voyager* mission to Saturn has changed forever our view of this spectacular region in our cosmic backyard. In particular, it has revealed the rings to be exceedingly more complex than heretofore imagined. Gone is the apparent simplicity of the Saturnian disk observed first by Galileo in 1610 and thereafter by legions of Earth-based observers right up until the 1980–81 encounter with the *Voyager* spacecraft. What a difference a close-up camera can make!

Voyager's much increased resolution proved that the entire width of the rings displays an intricate structure in the form of tens of thousands of concentric ringlets. Some 20 or so gaps embed the rings as true voids, probably swept clean by small moons not yet seen. Most remarkable, this fine structure is not fixed but varies with time and position. Although the process is only partly understood, the mutual gravitational attraction of the myriad ring particles (as well as gravitational tidal effects of the inner moons) enable waves of matter to form and move within the plane of the rings like ripples on the surface of a pond. The wave crests typically wrap around and across the rings, often in the form of spiral patterns resembling grooves in a huge celestial phonograph record.

Like a flight of birds bobbing and weaving in their formation, the ring's particles—from snowflakes to boulders—randomly weave about among themselves, owing to an extraordinarily complex gravity force field. The rings are constantly shifting as particles collide, stick, grow, break apart, and grow again in a cyclical fashion. The resulting fine structure of the rings, always changing, is a nightmare to model mathematically.

Even more fascinating (and challenging to understand), some of the strands of particles in the tenuous outermost ring appear to twist around each other like a braid. The cause of this totally unexpected finding is thought to be the gravitational interaction of two small moons that orbit on each side of the ring. Although each moon is a mere few tens of kilometers across, these moons are apparently able to shepherd the ring particles into a thin ribbon whose strands can fashion some very bizarre structure.

All this dynamic activity in Saturn's rings implies to some researchers that those rings are quite young—perhaps no more than 50 million years, which is 100 times younger than the age of the Solar System. There's just too much going on there, they argue, to have remained stable for billions of years. If true, then the ring debris we see today might well have been caused by a small Saturnian moon that was hit by a big comet. Astronomers normally prefer not to invoke such catastrophic events to explain observed phenomena, but the more we examine our cosmic neighborhood the more we realize that catastrophe has probably played an important role in much of the Universe around us.

present here in large quantities. Instead, to have such a low density, Saturn must be a gas bag of predominantly hydrogen and helium, without any solid surface beneath.

Saturn has the most extensive, and apparently the most complex, system of natural satellites. At least 20 moons have been spotted to date. Their reflected light suggests that all of them are covered with snow and ice; many of them are probably made entirely of water ice. Even so, they comprise a curious and varied lot. For example, Hyperion, a moon that orbits in the outer part of the Saturnian system (well beyond the rings), is known to have a "chaotic" rotation—throughout its eccentric trek around Saturn, its spin speeds up and slows down apparently at random. Enceladus, merely 500 kilometers in diameter and orbiting just outside the extremely tenuous outermost ring, is so bright and shiny that its surface must be coated with crystals of ice so fine as to act like the glassy beads on a movie projection screen; not coincidentally, the nearby ring seems to be made of just such fine particles and furthermore must be unstable, though whether the ring's particles are falling onto Enceladus or Enceladus is erupting to produce the ring is unknown. What's more, this moon does bear evidence of large-scale volcanic activity, but the "lava" observed by *Voyager* almost surely is frozen water once liquefied by recent internal upheavals. But why so much geological activity if this moon is so small? No one knows. One more case will suffice: Iapetus is a two-faced moon, as shown in Figure 8.11. Only one side of its icy, cratered surface—the side that always faces forward in its orbit around Saturn—displays a huge, nearly circular black spot. Such black deposits seen elsewhere in the Solar System are thought to be organic (i.e., carbon-containing) in nature, but how such material could adorn only one side of the object is just about anyone's guess.

The most intriguing of all of Saturn's moons, indeed more fascinating than any other moon in the whole Solar System, is Titan, shown in Figure 8.12. At 5000 kilometers across, Titan is not the largest of all known moons (Jupiter's Ganymede is just a bit bigger), but it certainly has by far the thickest and richest atmosphere of any moon in our planetary system. In fact, spectroscopic observations have conclusively established that Titan's atmosphere is

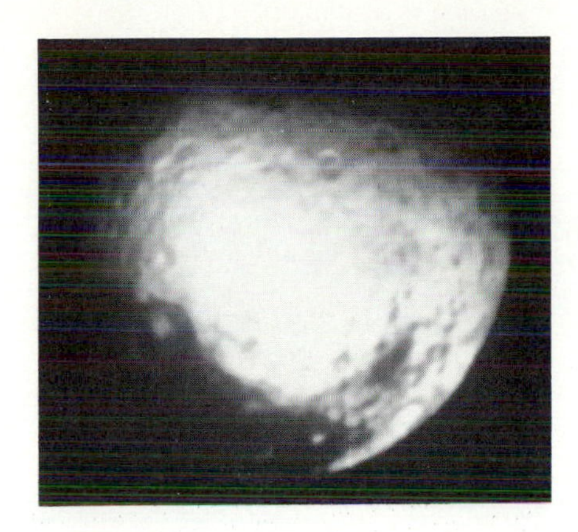

FIGURE 8.11 Another of Saturn's moons, photographed up close by *Voyager*. This one, called Iapetus, is one of the most peculiar worlds known. The contrast is clearly evident here between its light (icy) surface at top and center, and the black hemisphere at bottom.

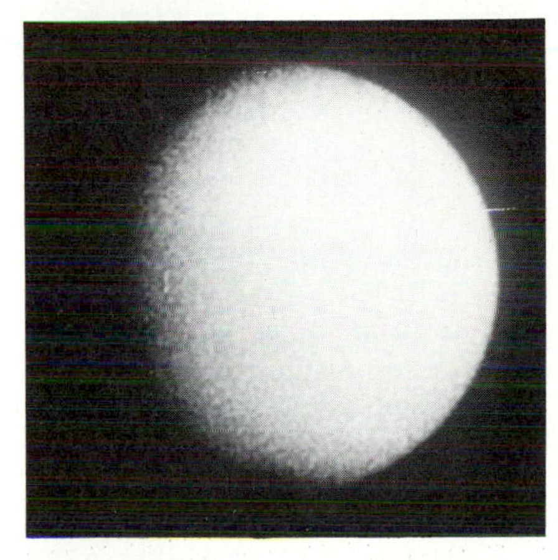

FIGURE 8.12 Titan, larger than the planet Mercury and roughly half the size of Earth, was photographed here from only 4000 kilometers away as the *Voyager* spacecraft glided by in 1980. All that can be seen here is Titan's upper cloud deck, adding to its mystery.

surprisingly similar to that of Earth. Having a blanket of partly nitrogen, much like the air on Earth yet some 10 times thicker, this moon must surely experience some greenhouse heating. But is it hot enough to support life as we know it? *Voyager* settled this key issue by measuring a surface temperature close to 100 kelvin; much too cold by virtually any living standard, this temperature (despite a small greenhouse boost) is in keeping with the fact that Titan is just too far away from the Sun. All things considered, Titan's atmosphere seems to be acting like a gigantic chemical factory. Also rich in methane and other complex carbon compounds, this moon's gases are thought to be undergoing, by the energy of sunlight, a complex series of chemical reactions that ultimately result in smog. Accordingly, its atmosphere is thick with haze, its unseen surface covered with organic sediment settled down from the clouds. Speculation runs the gamut from oceans of liquid hydrocarbons, especially ethane (gasoline), to valleys laden with petrochemical sludge. Second only to Mars as the most desirable place to concentrate future missions of spacecraft exploration, Titan could well present an opportunity to study the kind of chemistry thought to have occurred billions of years ago on Earth—the kind of chemical, prebiotic changes that ultimately led to life on our planet.

URANUS: The three outer planets were unknown to the ancients. The first of these, Uranus, discovered by the British amateur astronomer Herschel in 1781, resembles Jupiter and Saturn. Named after Urana, the Greek muse of astronomy, this planet is only half the size of Saturn, but still four times that of Earth and more than 10 times the mass of our planet. Even through a large optical telescope on Earth, Uranus appears hardly more than a tiny pale greenish disk like that shown in Figure 8.13. With the recent flyby of the *Voyager 2* spacecraft in 1986, our detailed knowledge of Uranus has increased dramatically. Figure 8.14 is a closeup visible image, a remarkable technological achievement given that Uranus, at about 20 astronomical units away, is roughly half the distance to Pluto.

Not surprisingly, early observers reported dusky atmospheric bands parallel to the equator; some even claimed to have spotted bright spots reminiscent of the large one on Jupiter. More recently, rings similar to those around Saturn and Jupiter were discovered, although only indirectly, while they momentarily blocked the faint light of a passing background star. Each of Uranus' 10 main rings is only a few kilometers wide—hundreds of times smaller than Saturn's. Although their widths subtend angles less than 0.001 arc second as seen from Earth,

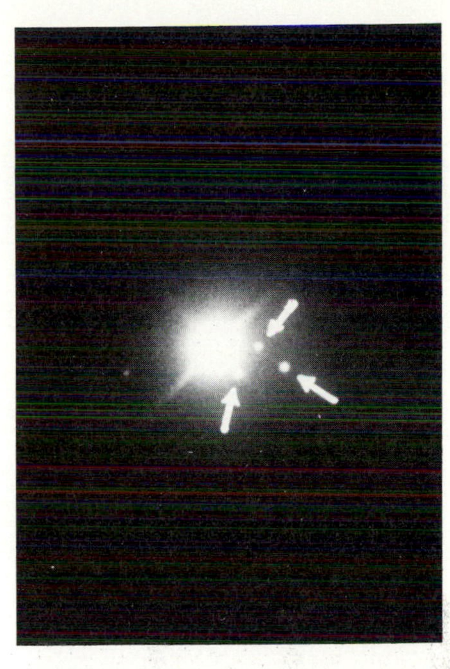

FIGURE 8.13 Details are barely visible on photographs of Uranus made with large Earth-based telescopes. (Arrows denote three of its moons.)

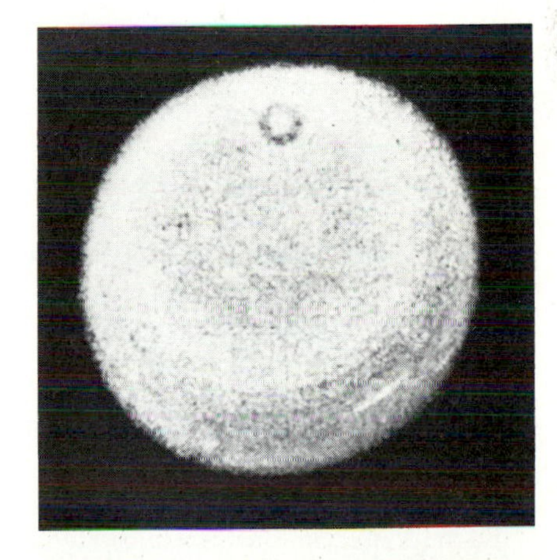

FIGURE 8.14 A closeup view of Uranus, sent back to Earth by the *Voyager 2* spacecraft while whizzing past this giant planet at 10 times the speed of a rifle bullet. The image is that of Uranus' upper atmosphere, not of its surface, and was taken when *Voyager* was about 12 million kilometers away.

Voyager managed to photograph these rings as well as another extremely thin, broad ring just inside the main ring system. In addition, this spacecraft (launched in 1977 and still going strong) photographed hundreds of new ringlets—partial rings or "ring arcs" of fine dust particles (and not chunks like at Saturn), implying that some of the rings are quite young or at least constantly evolving.

Spectra of sunlight reflected from Uranus' dense atmosphere also mimic those of Jupiter and Saturn, in that they reveal the most abundant chemical to be molecular hydrogen (H_2), with lesser amounts of helium (He), methane (CH_4) and ammonia (NH_3). The planet can't be made totally of such lightweight matter, however, for the average density of Uranus—2.8 grams/cubic centimeter—is more than twice that of Jupiter or Saturn. We can accordingly theorize that Uranus has a central core of heavier, perhaps Earth-like composition surrounded by a uniform mantle of various slushy hydrogen mixtures such as methane, ammonia, and water vapor. But we don't know for sure.

Uranus' rotation period seems to be, like the other Jovian planets, surprisingly rapid. Analyses of the Doppler shifts of spectral lines, as observed from Earth, imply that Uranus rotates once every 10 to 20 hours. The length of the Uranian "day" was determined much better when a radio-astronomy instrument aboard *Voyager* was able to track radio signals associated with Uranus' aurora: The answer is 17 hours.

Each planet in our Solar System seems to have some outstanding peculiarity, and Uranus is no exception. Its particular anomaly concerns its axis of rotation. Unlike all the other planets that have their spin axes nearly perpendicular to the plane of the Solar System, Uranus' rotational axis lies almost within that plane. We might say that Uranus lies on its side—or is tipped over! As a result, the "north" (spin) pole of Uranus, at some time in its orbit, points toward the Sun; half a Uranian "year" later, its "south" pole faces the Sun. What's more, *Voyager* discovered that Uranus' magnetic field is tilted some 60 arc degrees from the planet's axis of rotation; this is as though on Earth the North (magnetic) Pole were somewhere in the Caribbean, suggesting that the magnetic field on Uranus might be undergoing a reversal. (For more on magnetic field reversals, you can consult Chapter 24.)

The five well-known moons of Uranus (as well as 10 smaller ones discovered by *Voyager*) share its odd rotation feature. Two of them are shown in Figure 8.15, illustrating how they revolve in orbits parallel to the planet's equator, hence nearly perpendicular to the plane of our Solar System. Strangest of the icy Uranian moons is Miranda, shown in Figure 8.16. Displaying ridges, valleys, and large oval-shaped faults among other tortuous geological features, this baffling object has seemingly been catastrophically disrupted several times (from within or without). The new (*Voyager*-detected) mini-moons and probably other as-yet undiscovered moons might all be part of the previously mentioned ring system surrounding Uranus—a vast array, perhaps, of small moons, boulders, ringlets, and other odds and ends orbiting this faraway and alien world.

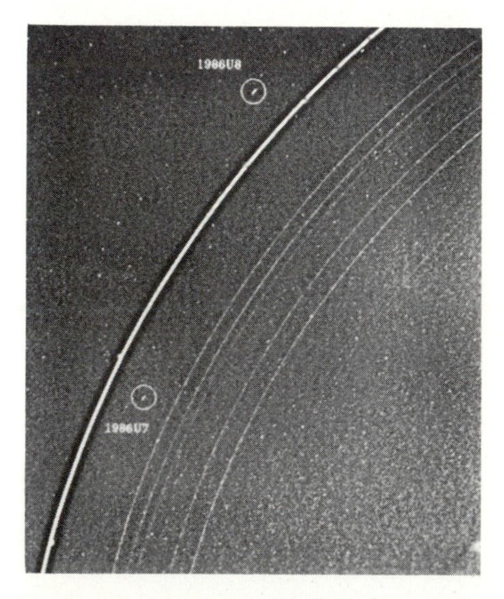

FIGURE 8.15 These two moons, newly discovered by the *Voyager* spacecraft in 1986, tend to "shepherd" one of Uranus' rings, thus keeping it from diffusing away.

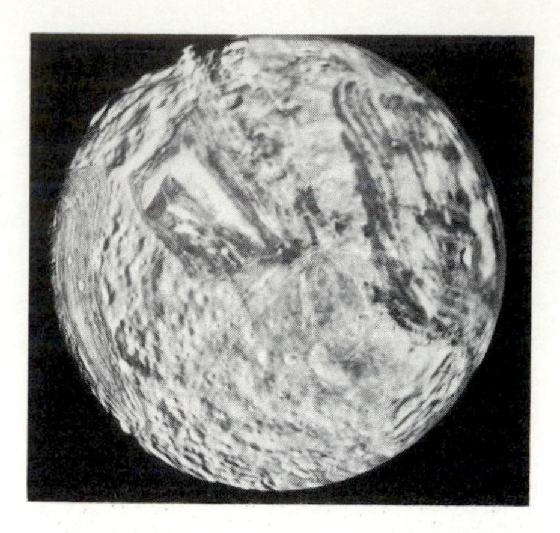

FIGURE 8.16 Miranda, an asteroid-sized moon of Uranus as photographed by *Voyager,* has a (presumably) volcanic feature that resembles a giant race track. Resolution is about 10 kilometers.

NEPTUNE: Soon after Uranus was discovered, astronomers sought to determine its orbit but, as hard as they tried, they could not predict its precise path on the sky. Theorists then argued that another planet must exist near Uranus—a planet whose gravitational attraction causes Uranus to depart slightly from its expected orbit. However, not until 1846, after mathematicians Adams of England and LeVerrier of France had predicted the orbit of the unexpected planet, did Galle of Germany manage to find the farthest of the giant planets—Neptune, some 30 astronomical units from the Sun.

Figure 8.17 shows a long exposure of Neptune and one of its moons, Triton, which is actually larger than the planet Mercury. Neptune is so distant that any features are virtually impossible to discern. Only with the best observing conditions can a few markings be seen, and not surprisingly these are suggestive of multicolored cloud bands; light bluish hues seem to dominate. (Hope is on the way, as *Voyager 2* is due at Neptune in late 1989.)

Neptune is the smallest of the Jovian planets, being slightly smaller though more massive and hence more dense than its neighbor Uranus. Neptune's average density, 3.6 grams/cubic centimeter, implies that its interior is much like Uranus, with apparently a substantially larger rocky core. At least superfi-

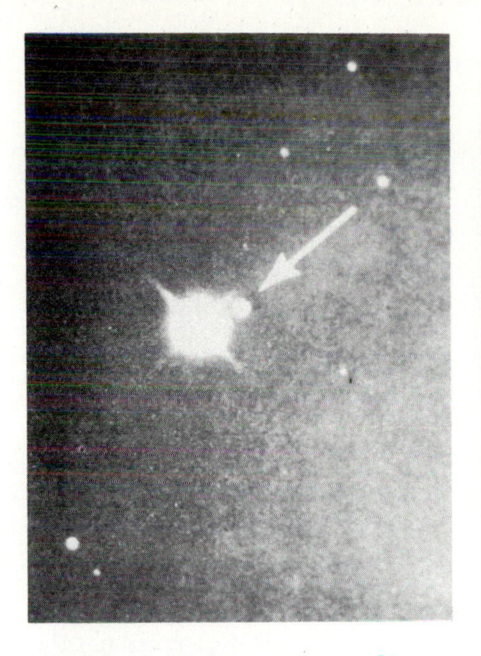

FIGURE 8.17 Neptune and one of its moons, Triton (arrow), photographed with a large Earth-based telescope.

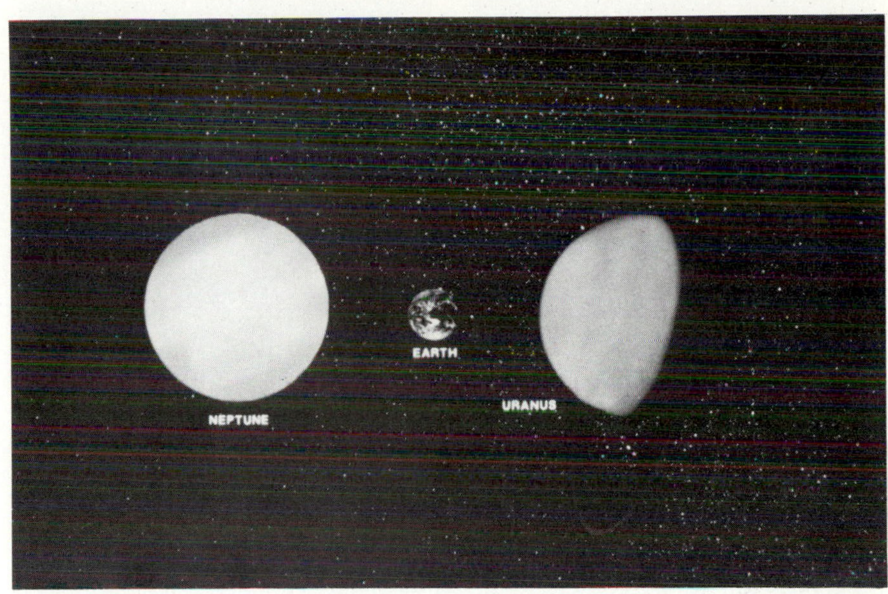

FIGURE 8.18 Neptune, Earth, and Uranus, illustrated to scale.

cially, as depicted in Figure 8.18, Neptune resembles Uranus in many ways.

Spectroscopic studies reveal an atmosphere around Neptune made mainly of hydrogen, helium, and trace amounts of ammonia and especially methane, the principal chemicals of all the Jovian planets. However, the abundances of gaseous ammonia and methane vary among the Jovian planets. For example, Jupiter has much more gaseous ammonia than methane. Proceeding outward, the more distant planets have decreasing amounts of ammonia but relatively stronger amounts of methane. This abundance variation results from the progressively greater freezing (at about 70 kelvin) of ammonia gas into ammonia ice crystals in the colder atmospheres of the more distant planets; it provides an important key to our understanding of the origin of the Solar System, a subject to which we shall return in Chapter 23. (The slight excess of methane in Neptune probably also accounts for its bluish color.)

These same spectral-line features of ammonia and methane can be used—indeed must be used in the absence of reliably observed surface markings—to reveal a rotation period between 15 and 25 hours. This is an indirect measurement, inferred from analyses of the Doppler broadening of spectral lines. Hence, it's subject to some uncertainty.

Finally, here is Neptune's contribution to Solar System peculiarities: Both of its known moons revolve in highly eccentric orbits. In fact, they orbit in opposite directions. The larger one, Triton, is said to have a retrograde orbital revolution, traveling around Neptune itself in a direction opposite to the spin of its parent planet; in this respect it is unique among the large moons of our Solar System.

Pluto

Actually, the mass and orbit of Neptune did not fully explain the irregularities of Uranus' orbit; furthermore, Neptune's orbit also displays its own set of irregularities, implying that some other object nudges and pulls Neptune slightly off its normal path around the Sun. Another planet, Pluto, was therefore sought for decades and finally found by the American astronomer Tombaugh in 1930. (However, this most distant planet is too small to fully explain the observed irregularities, prompting speculation that yet another planet—"Planet X"—*might* reside beyond Pluto.)

Pluto, named for the Roman god of the dead who presided over eternal darkness, is so far away that little is known of its physical nature. At nearly 40 astronomical units from the Sun, it is often hard to distinguish this planet from the background stars. Shown in the two photographs of Figure 8.19, Pluto is actually fainter than many stars. Studies of its reflected light variations suggest a rotation period of nearly a week, but measurements of its mass and diameter are still uncertain. The best current values imply an object about a fourth the diameter of Earth, having only about 1 percent of Earth's mass, and thus a density of about 1 gram/cubic centimeter.

Although Pluto is almost certainly made of mostly ordinary (water) ice, spectroscopy shows that one of its main constituents is *frozen* methane (CH_4)—the only planet on which methane exists in the solid state, implying that the surface temperature on Pluto is at least as cold as 50 kel-

FIGURE 8.19 These two photographs, taken one night apart, show motion of the planet Pluto (arrow) projected against a field of much more distant stars.

vin. Given its great distance from the Sun, it's probably a lot colder even than that. Clearly, Pluto is in the deep freeze—but as such, it may be the only planet to have survived in its pristine state, preserving a memory of the conditions when our Solar System formed eons ago.

Some researchers suspect that Pluto is not an original planet at all, but simply a comet. Pluto does have an oddball orbit, more elliptical and more inclined relative to the plane of the Solar System than any of the other planets. Furthermore, Pluto occasionally cuts inside the orbit of Neptune, as is currently the case. But it never gets close enough to the Sun to develop a tail—if it is a comet.

Other astronomers contend that Pluto might be an escaped moon of Neptune. Theoretical studies of the motions of Pluto, Neptune, and Neptune's large moon Triton support the possibility that Pluto and Triton could have had a close encounter sometime in the past. Such a violent gravitational interaction of two sizable moons of Neptune might then explain Triton's unusual retrograde orbit about Neptune, as well as Pluto's ejection into a new and peculiar orbit around the Sun. Figure 8.20 schematically illustrates what the encounter could have looked like. But the 1978 discovery of a moon around Pluto makes such a theory less likely; the gravitational forces needed to rip Pluto from the Neptunian system almost surely would have been sufficient to disrupt any object orbiting about Pluto.

There are no known planets beyond Pluto.

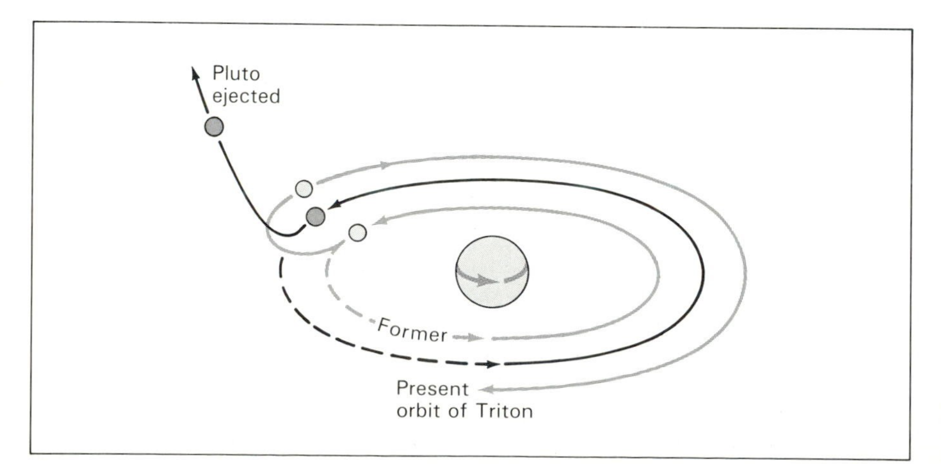

FIGURE 8.20 This schematic diagram shows how a near collision between two of Neptune's moons could have sent Pluto into its own orbit as the ninth planet and Triton into retrograde revolution. This idea is only hypothetical and in fact seems less viable now that Pluto is known to have its own moon.

Asteroids

Planets and their moons are not the only objects in the Solar System. Literally billions of less massive objects are estimated to roam interplanetary space. Some are called asteroids, others comets, still others meteoroids. Knowledge of them is important if we are to inventory all the different types of matter in the neighboring Universe, and in particular if we are to address the Solar System's origin in Chapter 23.

Sometimes called "minor planets" or "planetoids," **asteroids** are relatively small, rocky objects that revolve around the Sun. Their name literally means "starlike bodies," but they are surely not stars. Asteroids are even too small to be classified as planets. The largest one is a mere 0.0001 times Earth's mass; called Ceres, it extends about 1000 kilometers across. Only a few are larger than 300 kilometers, and most of the others are far smaller. Together, the asteroids amount to probably no more than 0.001 Earth mass, and hence do not contribute much to the total mass of the Solar System.

Asteroids differ from the larger planets and the smaller micrometeoroids first by their size, which ranges from a tenth of a kilometer to a few hundred kilometers, and second by their orbits, which generally lie between those of Mars and Jupiter.

ORBITAL PROPERTIES: The first asteroids were discovered early in the nineteenth century by European astronomers searching for additional planets. By the start of the twentieth century, several hundred asteroids were catalogued with well-determined orbits. The current list numbers more than 3000, the great majority confined to the **asteroid belt**, located at a distance ranging from 2.1 to 3.3 astronomical units from the Sun—roughly halfway between Mars at 1.5 astronomical units and Jupiter at 5.2 astronomical units. Almost assuredly, there exist innumerable, perhaps hundreds of thousands of asteroids whose orbits have not yet been determined.

Such a compact concentration of asteroids in a well-defined ring or belt implies that they are either the fragments of a planet broken up long ago, or primal rocks that just never managed to accumulate into a genuine planet. Researchers don't know for sure, but the latter view is favored, if only because no one can imagine how an entire planet could be ruptured and scattered against its own gravity. Instead, the strong gravity force field of Jupiter probably continuously disturb the motions of the chunks of primitive matter, nudging, pulling, and thereby prohibiting them from aggregating.

Orbits of the asteroids are much more elliptical than the almost circular planetary orbits, but not as elliptical as those of most comets. Parts of the asteroid belt are also highly inclined, by as much as 30 arc degrees, to the plane of the Solar System. In contrast, none of the planets (save Pluto) deviates from this plane by more than 3 arc degrees, as we noted earlier in Figure 7.9. Despite these differences, all except one of the known asteroids revolve about the Sun in harmony with the planets.

A few stray asteroids have very elliptical orbits, presumably caused by the gravitational influences of nearby Mars and especially Jupiter. These planets can disturb the normal orbits of certain asteroids and fling them into the inner Solar System. Those whose paths cross the orbit of Earth are termed Apollo Asteroids. Although we are currently aware of only about three dozen such objects, they are among the most famous asteroids because of their occasional close encounters with Earth. For example, the Apollo Asteroid named Icarus periodically comes to within 0.2 astronomical unit of the Sun; on its way by Earth in 1968, it missed our planet by "only" 6 million kilometers—a close call by cosmic standards.

The potential for collision with Earth is real. Some astronomers argue that Earth must be struck roughly every million years by three Apollo Asteroids, two of which probably end up in the ocean. Several dozen large land basins and eroded craters on our planet are suspected to be sites of ancient asteroid collisions. (The Arizona Crater of Figure 6.4 is the premier example.) For a head-on collision with Earth, the orbital conditions must of course be just right (or wrong!), but calculations imply that most Apollo Asteroids will in fact eventually collide with Earth. Fortunately, most known Apollo Asteroids are relatively small—about a kilometer in diameter (although one of 10 kilometers in diameter has been identified). Still, a visit of a kilometer-sized asteroid to Earth would be truly impressive—perhaps even catastrophic by human standards. Even such a small asteroid packs enough kinetic energy to devastate an area some 100 kilometers in diameter. And a fatal shock wave would doubtlessly inundate a much larger area than that. Should asteroids bang our planet hard enough, they might even be responsible for the extinction of life forms (consult Interlude 26-2).

PHYSICAL PROPERTIES: With few exceptions, asteroids are too small to be resolved by Earth-based telescopes. We must rely on indirect methods to find their sizes, shapes, and compositions. Consequently, only a few of their well-known physical and chemical properties are known.

One well-established fact is that all asteroids vary the amount of sunlight they reflect. This observation is normally attributed to rotation or tumbling of an oblong object, an explanation that was verified directly in the case of the asteroid Eros, which made a close approach to Earth in 1975. In addition, astronomers found spectroscopic evidence for iron and silicate material. Infrared observations of Eros implied a highly porous, rocky surface, covered with some "dust." And radar observations showed its surface to be very much rougher than any planetary or moon surface. If Eros is a typical asteroid, then they must all be pocked with surface holes and embedded chunks—not a terribly surprising description in view of the fact that asteroids normally reside in what is surely the most congested part of our Solar System. We shall return a few pages hence to discuss the sizes, shapes, and compositions of asteroids in the section on meteoroids.

Comets

Another type of resident of the Solar System is the **comet**. Renowned for their long wispy tails, comets derive their name from the Greek *kome*, or the Latin *cometes*, meaning "hair." They are usually discovered as faint, fuzzy patches of light on the sky, while several astronomical units away from the Sun in whose general direction they are racing. Far from the Sun, they are visible only by reflected sunlight. But as comets near the Sun, they can emit radiation of their own. Traveling in a highly elliptical orbit with the Sun at one focus, a comet brightens and develops a tail as its matter becomes heated and excited. As depicted in Figure 8.21, cometary matter boils away, thus creating the familiar tail.

As a comet departs from the Sun's vicinity, its brightness and tail diminish until it becomes once again a faint point of light receding into the distance. Comets that survive such a close and violent encounter with the Sun—some do break up entirely—continue their outward journey, reaching the extreme outer edges of the Solar System. Their highly elliptical orbits take many comets far beyond Pluto, perhaps

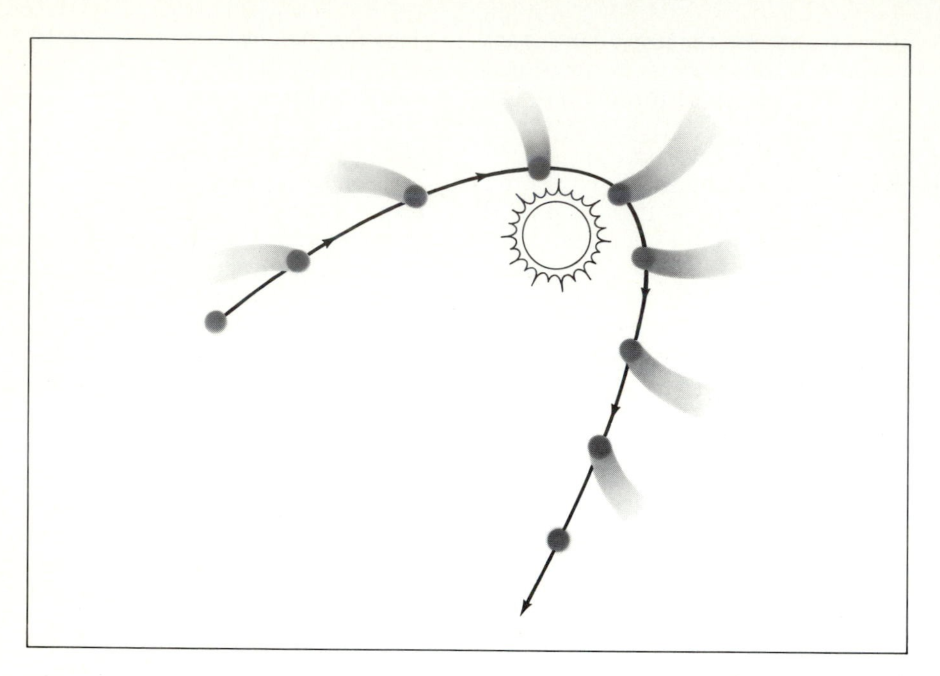

FIGURE 8.21 Schematic diagram of part of the orbit of a typical comet. Note how its tail always points away from the Sun.

even as far as 50,000 astronomical units, where, in accord with Kepler's second law, they spend most of their time as invisible objects. Most take hundreds of thousands, sometimes even millions of years for a round trip (although a few comets return for another encounter within a human life span). Accordingly, some astronomers reason that there must exist a huge "cloud" of comets far from the Sun, named the Oort Cloud after a Dutch astronomer who first wrote of the possibility of such a vast and distant reservoir of inactive, frozen comets. Despite such great distances, (50,000 astronomical units still equal less than a light-year, though), all comets remain as part of our Solar System, in contrast to even more remote galactic objects not bound to the Sun.

Perhaps the most famous comet is named for the seventeenth-century British astronomer Halley. It was the first such object to be recognized as periodic—in this case, 76 years—and hence was proved to be in orbit about the Sun. Historical records show that Halley's Comet has been observed at every passage since 240 B.C. Reportedly sometimes rivaling a total solar eclipse for nature's most spectacular show, the tail of Halley's Comet can extend almost a full astronomical unit, reaching clear across much of the sky. As shown by the sketch of its orbit in Figure 8.22, its next scheduled visit is in 2061.

TYPES OF COMET TAILS: Comets generally come in two types, each distinguished mostly by the geometry and material composition of their tails. Type I tails are approximately straight, often made of glowing, linear streamers like those of Figure 8.23(a). Their spectra show emission lines of numerous *ionized* molecules, such as carbon monoxide (CO^+), diatomic nitrogen (N_2^+), carbon dioxide (CO_2^+), hydroxyl (OH^+), methalydyne (CH^+), and several others. It's easy to remember that type I tails are composed mainly of ionized molecules by associating the "I" of the type with the first letter of the word "ionized."

Type II tails are usually broad, diffuse and gently curved [see Figure 8.23(b)]. Their tails are rich in microscopic dust particles that reflect sunlight. Thus it is the dust that causes this type of comet to be visually conspicuous, despite the tail's lack of emitted radiation.

Some comets' tails are mixtures of types I and II. Comet Kohoutek, which appeared in 1975, is a typical example. This comet caused much fanfare, for astronomers predicted it to be very bright in the twilight sky. To everyone's great dismay, however, Kohoutek was hardly visible to the naked eye despite its abundance of ionized matter and emitted radiation. The reason for its disappointing appearance was an unusually large amount of dust that caused the light from the tail of the comet to be scattered and thus dimmed.

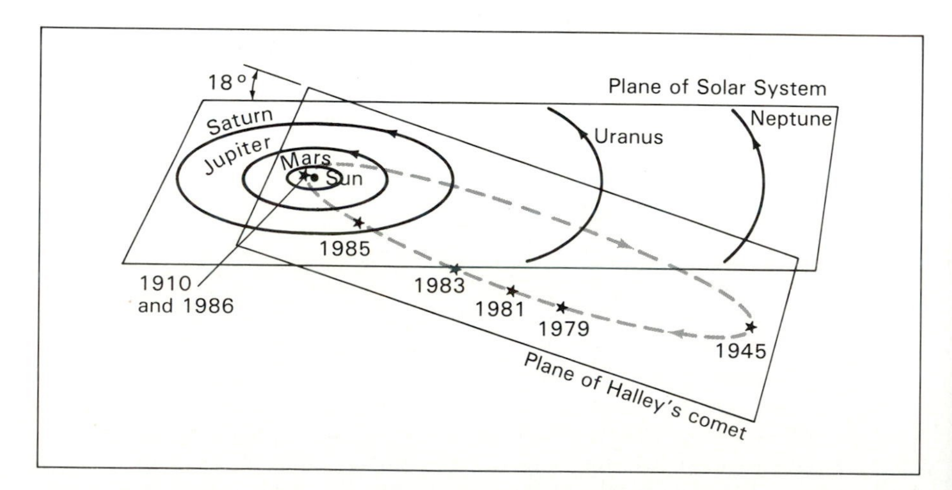

FIGURE 8.22 The orbital path of Halley's Comet (dashed) is abnormal in the sense that its ellipse is small and its period short. At sometime in the past, it must have encountered a Jovian planet which threw it into a tighter orbit that extends not to the Oort Cloud but merely a little beyond Neptune. Well observed in the eighteenth century, this comet provided the first practical application of Newton's law of gravity.

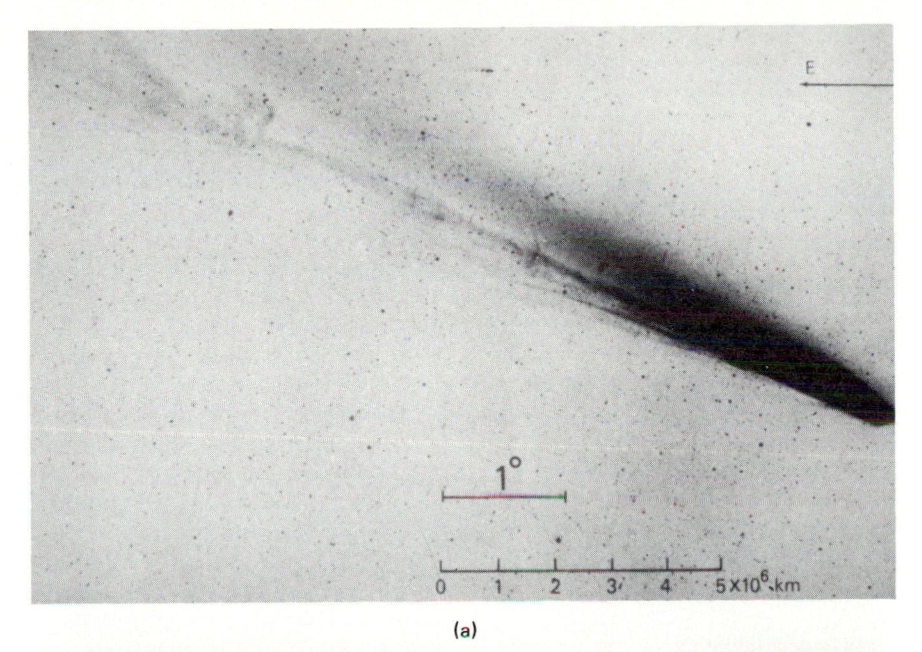

(a)

(b)

FIGURE 8.23 (a) Negative photograph of a comet with a type I tail, this one called Comet Kohoutek. The "tail" is not a sudden streak in time across the sky, as in the case of meteors or fireworks; instead, the tail travels sedately along with the head of the comet. (b) Photograph of a comet having (mostly) a type II tail, showing both its gentle curvature and inherent fuzziness. This is Comet West.

So comets are made of gas and dust evaporated from the main body (or "nucleus") of the object. (The correct word is not "evaporate" but "sublimate" since, in space, solid ices change directly into a gas without passing through the liquid phase.) The tails are in all cases directed *away* from the Sun by a "wind" of invisible matter and radiation escaping the Sun. Consequently, as depicted in Figure 8.21, the tail leads during that part of a cometary orbit which is outbound from the Sun. Every tiny particle in space—including those in comet tails—follows an orbit determined by gravity and radiation. As comets near the Sun, the outward push of solar radiation often overwhelms the inward gravitational pull on a dust particle, expelling it, and giving rise to the slightly curved tail of the type II variety. In other words, while obeying Kepler's second law, dust particles most distant from the Sun (far out in the tail) revolve about the Sun more slowly and thus lag behind, as can be seen in Figure 8.23(b). The interplay of many effects—the strength of gravity as well as the intensity of solar radiation—makes every comet tail appear different.

Even with large telescopes, the main bodies of all comets seem no more than minute points of light, proving that the actual sizes of comets (apart from their tails) are very small, surely much smaller than a planet and probably approximately a few kilometers in diameter. Even so, these small bodies can produce a diffuse halo (or "head" or "coma") of dust and evaporated gas, often extending 100,000 kilometers in diameter—physically much larger than planets. It is this halo that is the source of most of a comet's light. Beyond this, an invisible hydrogen halo engulfing the comet head stretches across several million kilometers, and is often distorted by the solar wind. Comet tails are generally much larger still, sometimes spanning distances comparable to an astronomical unit. From Earth, only the head and tail of a comet are visible to the naked eye.

In 1986, when comet Halley last rounded the Sun, a small armada of spacecraft launched by the Soviet Union, Japan, and a consortium of western European countries examined the comet up close. One of the Russian craft, *VeGa 2*, traveled through the comet's halo, coming to within some 8000 kilometers of the nucleus. ("VeGa" is a Russian acronym for "Venus-Galley," so named because the craft bypassed Venus earlier, dropping probes into the Venusian atmosphere before continuing on toward Halley's Comet.) Using positional knowledge of the comet gained from the Soviet craft encounters, the European *Giotto* craft (named after the Italian artist who painted Halley's image not long after its appearance in the year A.D. 1301) was navigated to within a mere 600 kilometers of the nucleus—a daring trajectory that in fact damaged the craft's camera, but not before sending home much data. (At 70 kilometers/second, the speed of the craft relative to the comet, a colliding dust particle becomes a devastating bullet.) Figure 8.24 shows a couple of high-tech images of the comet's nucleus radioed back to Earth by these spacecraft.

The results were somewhat surprising. Halley's irregular, potato-shaped nucleus is apparently larger than previously estimated; spacecraft measurements gave 15 kilometers long by as much as 10 kilometers wide. Also, the nucleus appeared "jet black"—as dark as finely ground charcoal or carbon black. This solid nucleus was seen to be enveloped by a cloud of dust, which causes light to be scattered throughout the head of the comet (much as a dirty windshield can blindingly scatter light from the headlights of an oncoming auto). Partly because of this scattering and partly because of direct obscuration by the dust, none of the spacecraft were able to discern surface detail on the comet.

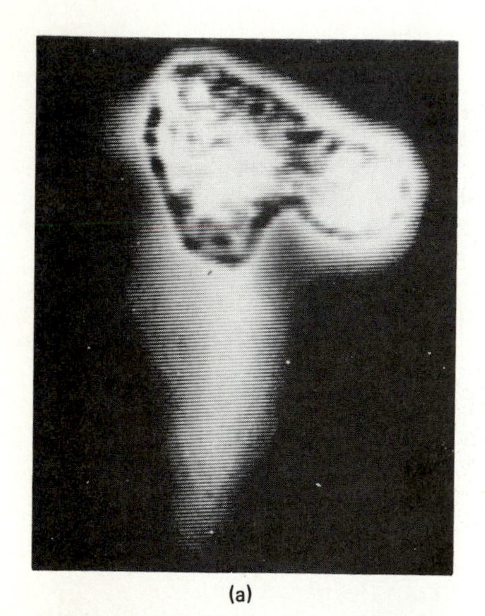

(a)

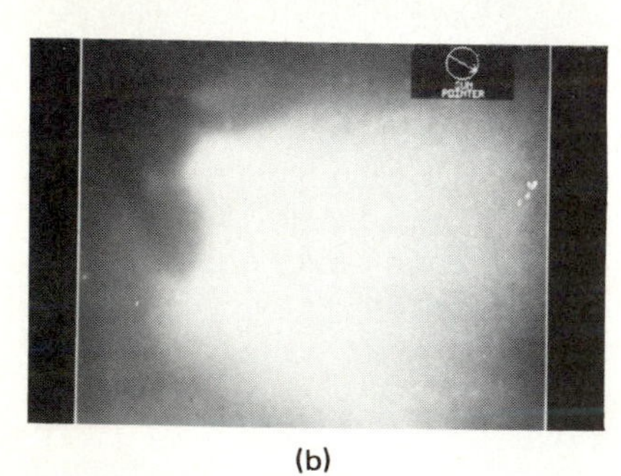

(b)

FIGURE 8.24 (a) *VeGa 2's* camera here imaged comet Halley's nucleus, seemingly showing a jet (or jets) of evaporated matter spewing downward. (b) The *Giotto* spacecraft better resolved the comet, showing its nucleus (upper left) to be very dark, although heavy dust in the area obscured any surface features. (Resolution here is about 50 meters—half the size of a football field.)

PHYSICAL PROPERTIES: The mass of a comet, most of which resides in its main body, can sometimes be estimated either by watching how its orbit is disturbed by other Solar System objects or by measuring the rate of loss of dust particles. Both methods are crude but yield masses ranging from 10^{15} to 10^{19} grams, comparable to masses of small asteroids. Complicating such a calculation, the spacecraft visiting comet Halley showed direct evidence for several jets of matter emanating from the nucleus. Instead of evaporating uniformly from the whole surface, gas and dust apparently vent from small areas on the sunlit side of the comet's nucleus. These jets might be largely responsible for the inferred spin of the comet, consistent with a 53-hour rotation period. Figure 8.25 is a composite of several findings concerning Halley's nucleus.

Actually, a comet's mass is variable, since some of it is lost each time it rounds the Sun. For some comets that travel within an astronomical unit of the Sun, this evaporation rate can reach as high as 10^{30} molecules/second, or equivalently 10^7 grams/second. That's a loss of nearly 10 million grams (or 10 tons) *every second.* However, this rapid evaporation does not last long since, in accord with Kepler's second law, comets speed up in their orbits when near the Sun.

A few comets break up while near the Sun. Returning in later years, they can be seen as pairs or swarms of smaller comets. Some comets have even disappeared entirely after

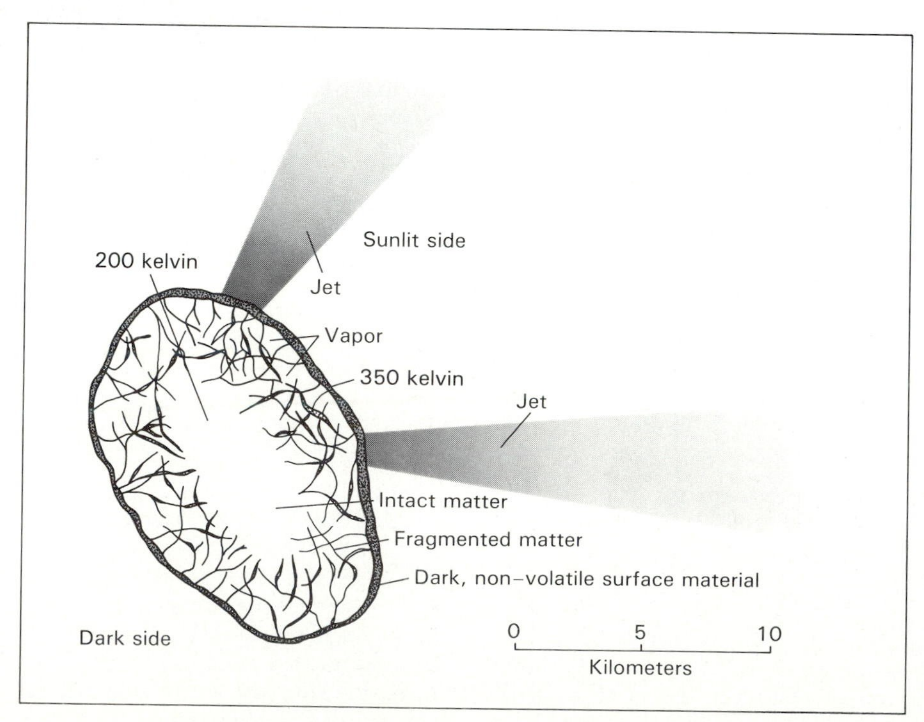

FIGURE 8.25 A drawing of Halley's nucleus, depicting its size, shape, jets, and other physical and chemical properties.

an especially close approach to the Sun; at least a couple of them are known to have fallen into the Sun. In the act of breaking up, these Sun-grazing comets often emit prominent spectral lines of simple atoms as well as more complex metals such as iron, nickel, and sodium that give additional clues about comet composition.

Seeking the physical makeup of a cometary body itself, astronomers are guided by the observation that comets have dust that reflects light, as well as gas that emits spectral lines of the abundant elements, hydrogen, nitrogen, carbon, and oxygen. Even as the atoms, molecules, and dust particles boil off, thus creating the halo and tail, the main body itself remains a cold mixture of gas and dust. Hidden by their gaseous halos, comet bodies are hardly more than balls of loosely packed ice having a density of about 0.1 gram/cubic centimeter and a temperature of only a few tens of kelvin. Experts now consider cometary bodies to be largely made of silicate and metallic dust particles trapped within a mixture of methane (CH_4), ammonia (NH_3), and ordinary water ice (H_2O). Not surprisingly, comets are often referred to as "dirty snowballs."

Puzzlingly, though, infrared heat sensors aboard the *VeGa* craft imply that the comet nucleus could be as hot as 400 kelvin—too hot for bare ice—although the icy core might be surrounded by a mantle of warmer material. Or, the heat measurements might refer to the ejected dust just outside the nucleus. Still, roughly 80 percent of the molecules in Halley's tail were found to be water (most of the remaining 20 percent being carbon dioxide), so there is little doubt that comets are mostly ice. That Halley is classified as a moderately dusty comet was also confirmed, since about a tenth of the mass of matter in its tail was found to be dust.

Comets, then, result from a wearing away or melting of cold, icy objects probably left over from the earliest epochs of the Solar System. When a comet approaches to within a few astronomical units of the Sun, its surface becomes too warm for the ices to remain stable. The ices then change directly from the solid to the gas phase, creating a hot gaseous halo surrounding the still-frozen central body. Some of these gases absorb additional amounts of solar radiation, become ionized, and get trapped in the constant stream of particles and radiation emitted by the Sun. The gases move outward along the comet's tail along with the dust particles loosened by the melting of the ices. In this way, comets repeatedly lose matter when in the vicinity of the Sun, explaining why some of them eventually disappear.

Finally, be sure to realize that comets are more than great light shows. Among the planets and Sun that originally formed in our Solar System, some bundles of matter were doubtlessly left over—debris that did not cluster into a major planet or the central Sun. Comets seem to be such debris and, as such, they may be the only true representatives of the ancient building blocks of our Solar System.

Meteoroids

Meteors, or "shooting stars," are familiar to anyone who has watched the night sky for more than a few minutes. We should be able to see a few of them every hour on a clear night. Meteors are essentially swift objects that frictionally heat and excite the air molecules while plowing through Earth's atmosphere. The molecules in the atmosphere then relax back to their ground states by emitting light. We see the effect of this as a sudden, bright streak. (To avoid confusion, recall our Chapter 6 definitions: Before plunging into a planetary atmosphere, meteors are termed meteoroids, whereas after having crashed to the ground, they are called meteorites.)

We have already discussed in Chapter 6 some of the consequences of meteoroid collisions with the surface of astronomical objects, particularly the Moon. Here we note that some of the meteoroids are broken-up comets, whereas others seem to be stray asteroids.

COMETARY FRAGMENTS: Large numbers of meteors can often be seen burning their way through Earth's atmosphere, sometimes at the rate of several meteors per minute. Meteor showers, such as that shown in Figure 8.26, are composed of enormous numbers of meteors and are usually named for the constellation that lies in the direction from which they come. For exam-

FIGURE 8.26 This photograph shows several streaks in Earth's atmosphere caused by micrometeoroids of the Leonid meteor shower.

ple, the Perseid shower is seen to emanate from the constellation Perseus. It can last for several days but is maximized every year on the morning of 12 August, when upward of 50 meteors per hour can be observed.

The velocity and direction of a meteor's flight can be used to compute its interplanetary trajectory. This is how meteoroid swarms have come to be identified with well-known comet orbits. For example, the Perseid shower shares the same orbit as Comet 1862III, the third comet discovered in the year 1862 (some say this is comet Swift-Tuttle). Thus the meteoroids travel around the Sun in elliptical orbits just like comets, asteroids, and planets. All these objests obey Kepler's laws, as they all move under the gravitational guidance of the Sun.

Table 8-1 lists some prominent meteor showers, the dates they are visible from Earth, and the comet from which they are thought to originate. See also Interlude 8-2 for a discussion of miscellaneous planetary debris.

TABLE 8-1
Some Prominent Meteor Showers

MORNING OF MAXIMUM ACTIVITY	SHOWER NAME	ROUGH HOURLY COUNT	PARENT COMET
21 April	Lyrid	10	1861 I
12 August	Perseid	50	1862 III
10 October	Draconids	Up to 500	Giacobini-Zinner
21 October	Orionid	30	Halley
7 November	Taurid	10	Encke
16 November	Leonid	Up to 100,000	1866 I (Tempel)
12 December	Geminid	50	?

On each pass of a comet near the Sun, some cometary fragments dislodge from the main body. The fragments initially travel in a tight-knit swarm of dust or pebble-sized objects while moving in the same orbit as the parent comet. But over the course of time, the swarm gradually disperses along the same orbit so that eventually the micrometeoroids become more or less smoothly spread all the way around the parent comet's orbit. Whenever Earth intersects the cometary orbit of a young, relatively undispersed cluster of meteoroids, a spectacular meteor shower

INTERLUDE 8-2 ***Miscellaneous Interplanetary Debris***

Planets, moons, asteroids, comets, and meteoroids are all well-known, if not yet well-understood inhabitants of our Solar System. As if these were not enough, additional objects apparently roam the planetary system as well. An example is a 1977 discovery of an object several hundred kilometers in diameter and orbiting between Saturn and Uranus. It seems to be neither a comet (too big in size and too sharp of an image photographically) nor an asteroid (too far outside the conventional asteroid belt). This new type of object has been named Chiron after one of the centaurs of Greek mythology. Its orbital period is measured to be about 50 years, ranging from 8 to 20 astronomical units from the Sun. The sketch shows its interplanetary trajectory as currently understood.

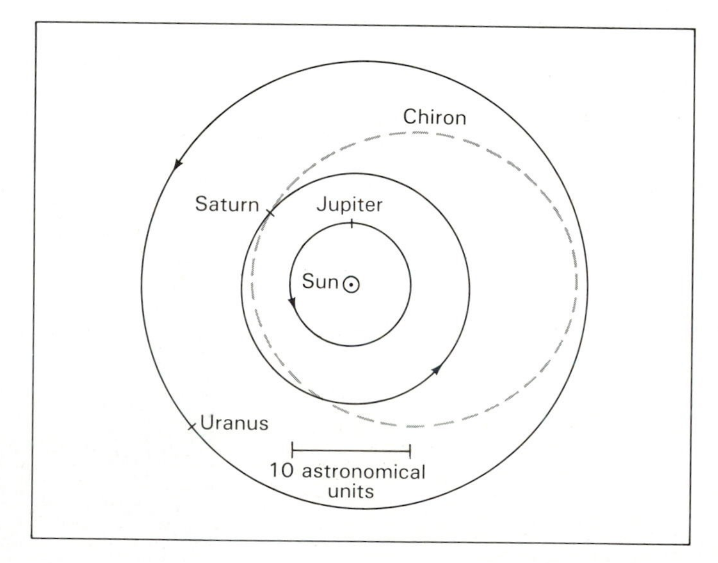

Just what kind of an object is Chiron? Could it be a maverick asteroid that has strayed far outside the usual asteroid belt, or is it perhaps an ancient comet whose volatile elements have all escaped after so many previous trips around the Sun? All things considered, it doesn't fit either category well. Maybe Chiron is an escaped moon of Saturn. Or perhaps it is a member of a newly discovered outer belt of asteroids. Astronomers might be able to tell when it comes closest to Earth (8 astronomical units) in 1996.

Regardless of Chiron's true nature, it seems hard to avoid the impression that the Solar System is full of debris—the scattered remains of a more violent, but at the same time more formative, era in the local history of our galactic environment. To understand the whole puzzle will require us to fit all the pieces together. But at this point, we are apparently still in the process of gathering those pieces.

FIGURE 8.27 This cluster of micrometeoroids was collected by a reconnaissance (*U-2*) aircraft at a high altitude of about 20 kilometers (nearly 70,000 feet). The size of this photograph is 10^{-2} centimeter, but, if reconstituted, these fluffy structures would make an object about 10^{-3} centimeter.

can result. Since Earth's motion takes it across a given comet's orbit only once a year, this explains the regular appearance of certain meteor showers at the same time each year.

Astronomers use the deceleration, or slowing down, of a meteor caused by atmospheric friction to estimate the velocity and density of meteoroids. Most are very tenuous, fragile objects varying in density from 0.01 to about 1 gram/cubic centimeter. With an average velocity of about 25 kilometers/second, a typically small (10^{-4}-centimeter) micrometeoroid has no chance of reaching Earth's surface. In fact, as far as we know, no particle in *any* meteor shower has even been large enough or dense enough to survive the atmospheric encounter and reach the ground. However, some have been collected during rocket flights. Figure 8.27 shows a typical low-density, irregularly shaped micrometeoroid collected at an altitude of about 20 kilometers (or 70,000 feet). Having a frothy, nonmetallic structure, they often resemble bits of burned newspaper or charred toast.

STRAY ASTEROIDS: Larger meteors, *not* associated with swarms of cometary debris, also occasionally speed through our atmosphere. Regarded as stray asteroids, these objects have produced most of the cratering on the surfaces of the Moon, Mercury, Mars, and some of the moons of the Jovian Planets. These large meteoroids enter Earth's atmosphere with a typical velocity of nearly 20 kilometers/second, producing energetic shock waves or "sonic booms" as well as a bright sky streak and a dusty trail of discarded debris. Faster incident velocities mean greater frictional heating, hotter meteor surfaces, and quicker evaporation times; a few large meteors pass through the atmosphere with such a great velocity (about 75 kilometers/second) that they either fragment or disperse entirely at high altitudes.

As noted in Chapter 6, large meteors (each at least a million grams and a meter across) impact with the surface by converting their kinetic energy (motion) into mechanical (damage), thermal (heat), and acoustical (sound) energy. This combination is inevitably explosive, and produces a crater such as the kilometer-wide Arizona Crater of Figure 6.4. From the size of this crater, we can estimate that the meteoroid responsible must have had a mass of about 50 billion grams (i.e., 50,000 tons). Only 25 million grams of iron meteorite fragments have been found at the crash site, the remainder having been apparently destroyed by the explosion or buried in the landscape.

Currently, Earth is scarred with nearly 100 craters larger than 0.1 kilometer in diameter. Most of these are so heavily eroded by weather and distorted by crustal activity that they can be identified only by means of satellite photography, as shown in Figure 8.28. One of the most recent

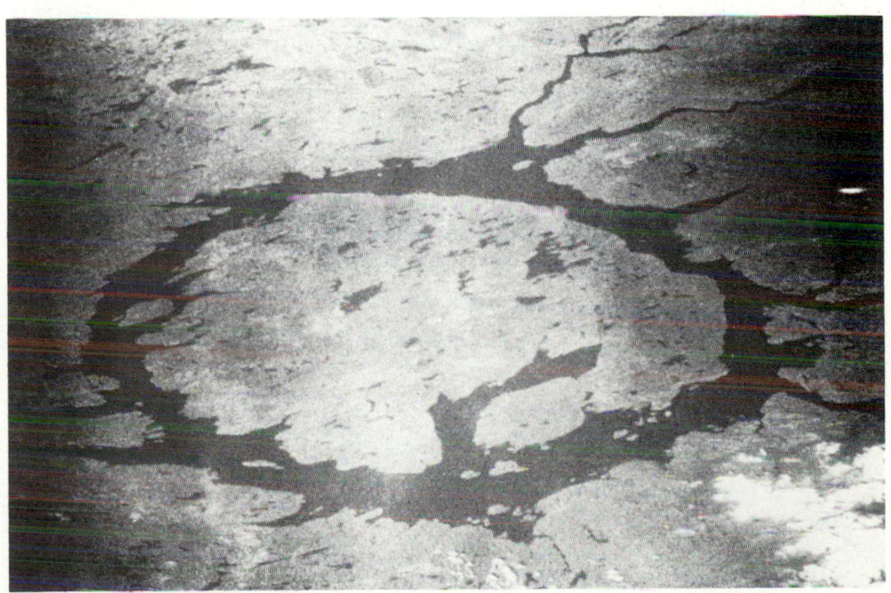

FIGURE 8.28 This photograph, taken from orbit by the U.S. *Skylab* space station, clearly shows the ancient impact basin that forms Quebec's Manicouagan Reservoir. A large meteorite must have landed there long ago. The most ancient such terrestrial crater is nearly 2 billion years old.

impacts occurred in central Siberia on 30 June 1908. However, the presence of only a shallow depression as well as a complete lack of fragments implies that this Siberian event was a collision with an icy comet. (The comet's slushy nature produced no deep crater, and its icy fragments melted away; or perhaps the comet exploded several kilometers above the ground, naturally leaving a blasted depression at ground level but no well-formed crater.) At any rate, such major collisions between Earth and large meteoroids are thought to be rare events now, 4.5 billion years after the origin of the Solar System. Given the number of large objects and the vastness of space, another major collision will probably not occur for hundreds of thousands of years.

Orbits of large meteorites that survive their plunge through Earth's atmosphere can be reconstructed using techniques similar to those we discussed earlier for meteor showers. There's only one problem: Accurate observations of large meteors dashing through Earth's atmosphere are quite rare. In fact, astronomers have been able to determine the orbits for only a few recovered meteorites. Figure 8.29 diagrams the reconstructed orbits of 3 of them: the Pribram fall (Czechoslovakia, 7 April 1959), the Lost City fall (Oklahoma, 3 January 1970), and the Innisfree fall (Alberta, 5 February 1977). As shown, the most distant part of each of their orbits intersects the asteroid belt. This reconstructed orbital information for these three fallen meteorites is the strongest evidence that large meteorites were probably once part of the asteroid belt before being redirected into an Earth-crossing orbit that led to collisions with our planet.

All in all, several thousand meteorites have been recovered over the years. Some are accidental "finds," whereas others are discovered because of an observed "fall." Antarctica has recently become an especially good place to seek relatively pristine meteorites since the subfreezing temperatures and oft-renewed snow cover protect them from erosion and contamination that normally alter their properties. Every so often, strong winds remove the snow cover, enabling researchers to extract the meteorite fragments from the deep freeze.

Actually, some meteorites found in Antarctica may well have come from the Moon and even Mars. At least one meteorite is identical to those returned by the *Apollo* astronauts from the lunar highlands. It was probably chipped off the Moon by a giant meteor that struck the Moon with such force that surface material was hurled out into space; it then gradually gravitated toward Earth and finally landed as a meteorite in the Antarctic ice flows. Another class of meteorites, very few in number, are distinctly younger than all the rest. Only about 1.5 billion years old (compared to the usual 4.5 billion years), this age coincides with the time when the outer crust of Mars was solidifying from molten rock. Accordingly, some researchers have speculated that a small fraction of all meteorites might well have come from Mars.

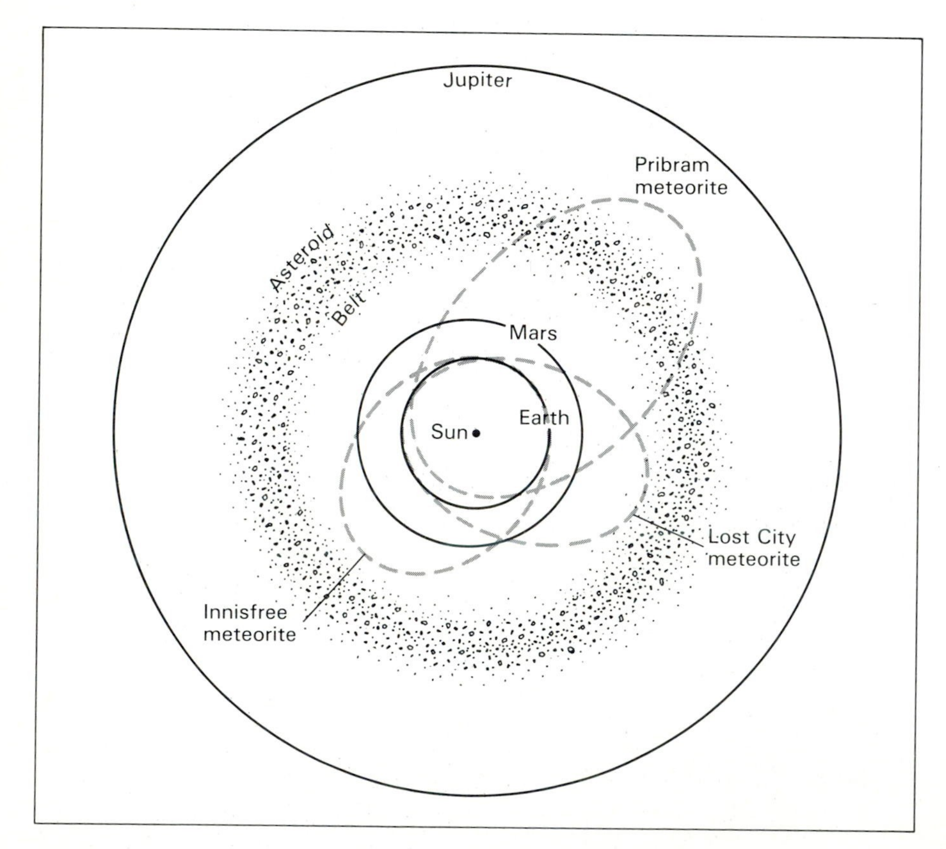

FIGURE 8.29 Reconstructed orbits of three large meteorites that collided with Earth during the past few decades.

COMPOSITION: One other feature distinguishes the small micrometeoroids that burn in Earth's atmosphere from the large meteoroids that manage to reach the ground. That feature is composition. The density of all meteoritic fireballs too small to reach the ground averages 0.5 gram/cubic centimeter, but in no case exceeds 1 gram/cubic centimeter. Such low densities typify those of comets which are made of loosely packed ice and dust. They are much less dense than the approximately 5 grams/cubic centimeter values derived for those meteorites that reach Earth's surface, and which are apparently typical of the more dense asteroids.

Meteorites like that shown in Figure 8.30 have been subjected to close scrutiny in terrestrial laboratories. Prior to the Space Age, they were the only type of extraterrestrial matter to which we had access. Their basic composition is much like the

FIGURE 8.30 This blackened rock is the "Lost City meteorite" which was detected in 1970 by a network of cameras as it plunged through Earth's atmosphere. It was recovered in Oklahoma 6 days later. (The tape measures inches.)

stoney inner planets and the Moon, and include a wide variety of familiar compounds and minerals. Some meteorites, called "irons," contain blobs of pure iron, suggesting heating and differentiation at some previous time in their history. Others are called "carbonaceous chondrites" since they are black or dark gray stones housing small globules rich in carbon. By and large, though, when meteorites are heated in the laboratory, and thus broken down into their various elements, their atomic abundances usually agree with the standard cosmic abundances noted in Figure 2.20. The only exception is an underabundance of some of the light elements such as hydrogen and oxygen that presumably boiled away into space when the meteoroids were molten at some past epoch.

Finally, almost all meteorites are old. Direct radioactive dating proves most of them to be 4.5 billion years old—roughly the age of the oldest lunar rocks. Consequently, meteorites, along with some lunar rocks, comets, and perhaps the planet Pluto, provide important clues about the original state of matter in our parent gas cloud from which our Solar System formed several billion years ago.

Overall View

Completing these two chapters on our planetary environment, you might be struck by the wide variety of planets, moons, asteroids, comets, and meteoroids in the Solar System. Our cosmic neighbors display a vast range of physical and chemical properties. The planets, in particular, sport a long list of interesting features and peculiarities. In all, our astronomical neighborhood may seem more like a great junkyard than a smoothly running planetary system. Can we really make any sense out of the entire collection of Solar System matter? Astronomers think the answer is yes.

Every time a new discovery is made, we learn a little more about the properties and role of each type of matter in our planetary system. The robot space probes of the past decade have made especially startling advances—in some cases forcing astronomers to completely revise our perceptions of some celestial objects. It's becoming clear that each of the planets and moons, as well as the random debris, contributes to our emerging appreciation of the entire Solar System. What we have found is that almost every planet or moon is now in a different stage of development: The Jovian Planets resemble galactic fragments frozen in time, not massive enough to become stars, and yet too massive to develop into huge rocks. They have preserved to varying extents the pristine properties of the primitive Solar System. The asteroidal, cometary, and meteoritic objects have probably changed least of all, and thus harbor the best clues about the formative stages of our Solar System nearly 5 billion years ago. Midway in size between the big planets and the small debris are the Terrestrial Planets that have evolved a great deal, generating hard surfaces, moderate atmospheres, and in at least one case, life.

SUMMARY

We should not regard the Solar System as a random array of tattered matter. Every object tells us something. Each time a new space probe reconnoiters a planet—and some probes are on their way right now—we learn more about that planet. In particular, we learn how that planet fits into the general scheme of Solar System science.

A study of the planets—*all* the planets taken together—should eventually allow us to decipher the origin, evolution, and destiny of our little niche within this vast Universe. In Part III we shall return to the issue of comparative planetology and thereby learn more about the planets' role in the general evolutionary pattern of all things. For now, though, we must continue to take inventory of the varied material objects elsewhere in the Universe. We need to gain a complete appreciation for the Universe's contents before synthesizing our grand scenario of cosmic evolution.

KEY TERMS

asteroid

asteroid belt

comet

QUESTIONS

1. Compare and contrast the chemical compositions, densities, and masses of Saturn and Earth.
2. Discuss volcanism on the Earth, Moon, Mars, and Io.
3. Compare and contrast four qualitative differences between the Terrestrial Planets and the Jovian Planets (not including distance from the Sun).
4. The planet Mercury and Saturn's moon, Titan, have comparable masses, yet the latter has an atmosphere whereas the former does not. State clearly and explicitly the reason(s) for this.
5. Name a peculiar characteristic of each of the nine planets.
6. Why is Mars thought to harbor the best chance for extraterrestrial life within our Solar System? How about Titan, one of the interesting moons of Saturn?
7. Can you imagine living on Earth if Jupiter had also become a small star? Explain what it would be like, assuming that life could survive.
8. Why are comet tails always directed away from the Sun?
9. Of what value are meteorite analyses? What might we learn if we could intercept a comet or asteroid and return some of it to Earth?
10. Name five properties of asteroids and note in each case how they differ from comets.

FOR FURTHER READING

BEATTY, J., O'LEARY, B., AND CHAIKIN, A., *The New Solar System.* Cambridge: Cambridge University Press, 1982.

BRANDT, J., AND NIEDNER, M., "The Structure of Comet Tails." *Scientific American*, June 1986.

CHAPMAN, R., AND BRANDT, J., *The Comet Book*, Cambridge: Jones and Bartlett, 1984.

GORE, R., "The Planets." *National Geographic*, January 1985.

INGERSOLL, A., "Uranus." *Scientific American*, January 1987.

MILLER, R., AND HARTMANN, W., *The Grand Tour.* New York: Workman Publishing Co., 1981.

OWEN, T., "Titan." *Scientific American*, February 1982.

PREISS, B. (ed.), *The Planets.* New York: Bantam Books, 1985.

9 THE SUN: *An Average Star*

Living in the Solar System, we have the chance to study, at rather close range, perhaps the most common type of cosmic object—a star. Our Sun, of course, is a star, and an average star at that. But the unique feature about the Sun is that it's so close—some 300,000 times closer than our next nearest stellar neighbor, the Alpha-Centauri star system. Another way to appreciate the relative closeness of the Sun is to note that while Alpha-Centauri is 4.3 light-years distant, the Sun is only 8 light-*minutes* away. Consequently, astronomers know far more about the detailed properties of our Sun than about any of the other distant points of light in the Universe. In fact, a good deal of our total astrophysical knowledge is directly or indirectly based on modern studies of the Sun.

Recall our previous notion that the Sun, and probably the planets as well, originated from a dense pocket of galactic gas. Such pockets apparently come and go repeatedly because of the normal random motions of the atoms and molecules in space. Provided that the collective gravitational attraction exerted on each gas particle by all the other particles is great enough, the pocket is said to have become gravitationally bound. Almost immediately, such a gaseous blob begins contracting, falling inward on itself under the relentless pull of gravity. During infall, the atoms and molecules collide with one another ever more frequently and more violently, in the process heating the pocket of gas by means of friction. (The process is much the same when we use friction to warm our hands by rubbing them together.) Eventually, in a few million years, the core of the gaseous blob becomes hot enough to emit infrared radiation. This dense pocket, often called a protostar, continues to contract via gravity for tens of millions of years until the cloud has heated sufficiently to ignite nuclear burning. The blob by that time has become a genuine star, after which it stops contracting and settles into a long period of equilibrium between the inward pull of gravity and the outward pressure of the hot gas.

The Sun and most other stars burn steadily in this way, provided that the gravity–pressure equilibrium is maintained. We estimate that this period of equilibrium will last about 10 billion years for a star like our Sun. And since we date the origin of the Earth and Solar System to be nearly 5 billion years old, we can surmise that our Sun is a middle-aged star, just about halfway between birth and death.

In our Milky Way Galaxy, we have evidence for young stars as well as old stars. To be sure, we see stars of all ages. On the one hand, we have direct observational evidence that stars are forming in numerous galactic locations even at the present time. The scenario of star formation described above is apparently an ongoing process. On the other hand, many stars are known to be much older than 5 billion years. They must have formed long before the Sun. In fact, the theory of stellar evolution, to be studied in Part III, suggests that myriad stars came and went before the Sun even originated.

Indeed, stars come and go, much like people. Guided by gravity, stars are born by means of a natural process, endure for a certain amount of time prescribed largely by their mass, and then die. Most stars fade away gently, but the more massive ones tend to perish catastrophically, exploding as supernovae. During the act of death, the more massive stars expel most of the heavy elements created within their stellar cores, thereby populating space with an abundance of heavy elements from which new stars can subsequently form.

Since solar spectroscopy shows that the Sun *already* contains some heavy elements (even elements as heavy as iron), we conclude that our Sun cannot be a **first-generation star**, which is one having no heavies. The Sun could not have formed exclusively from the primordial hydrogen and helium gases present in the earliest epochs of the Universe. Instead the Sun must be at least a **second-generation star**, which is one having some heavies; the Sun might even be an *N*th-generation star, where *N* could well be a number larger than 2. Nor could the heavy elements observed in the Sun have been produced in the early evolutionary phases of the Sun itself. They, along with the heavy elements comprising Earth and the other planets, must have been formed in the fiery cores of long-gone stars that blew up eons ago.

Despite these intriguing facts about our Sun, we emphasize throughout this book that our star seems to be in no way special or unique—at least when compared to other individual stars known throughout the Universe. Of course, you may want to object by stating that our Sun is indeed special because it's the sole source of light and heat for the maintenance of life on Earth. But this is an anthropocentric opinion based on narrow-minded notions of our own biological survival—not a broad opinion based on modern astronomical knowledge.

Furthermore, the fact that entire civilizations of our ancestors revered the Sun means little to the modern solar astronomer trying to fathom the nature of this nearest star. Sun worship may perhaps be of cultural curiosity, but surely it's of no influence whatever on the disposition of the Sun as a star. The point is that the average physical and chemical properties of our Sun seem to mimic those of all other stars now shining, regardless of when and where they formed. Accordingly, we sometimes say that our Sun is "magnificently mediocre."

The learning goals for this chapter are:

- to appreciate the average properties of the nearest star, our Sun
- to understand how energy travels from the solar core, through its interior, and out into space
- to recognize the general differences between the quiet Sun and the active Sun
- to know how astronomers model the entire Sun's temperature and density, from the core through the atmosphere
- to realize that many of the planets are embedded in the Sun's outer atmosphere of radiation and high-energy particles
- to understand some of the similarities and differences among the various types of solar activity—sunspots, flares, prominences, and the like

Sun in Bulk

In Chapter 5 we sketched the six main zones comprising planet Earth. Similarly, here we delineate several broad regions of the Sun.

MAKEUP: First, recognize that the Sun has a surface of sorts—not a solid surface, but that part of the sunny gas ball seen with our eyes or through a (filtered) telescope. This "surface" is called the **photosphere**. Just above the photosphere is a lower atmosphere, called the **chromosphere**. And far beyond that, engulfing many—perhaps all—of the planets in an astronomical umbrella, is a higher atmosphere called the **corona**. Figure 9.1 shows the approximate dimensions of each of these bulk regions, although the exact size of the chromosphere and corona are hard to determine with accuracy; the chromosphere gradu-

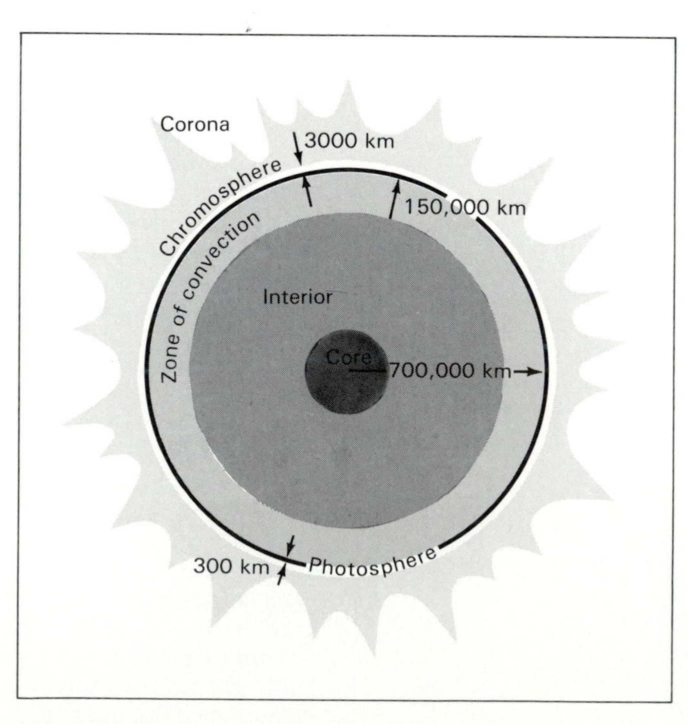

FIGURE 9.1 Six bulk regions of the Sun, not drawn to scale, although labeled with proper physical dimensions.

ally melds with the corona and the corona gradually peters out somewhere in the outer reaches of the Solar System.

Figure 9.1 also shows three more regions, all inside the photosphere. One lies just below the surface; it's called the **convection zone**, and we'll study it more in the next section. Another region lies at the heart of the Sun; known as the **solar core**, this central region is the site of powerful nuclear reactions that we'll study extensively in Chapter 10. In between exists a large region of gas simply called the **solar interior**.

DISTANCE: From our vantage point at Earth, we can determine a number of useful things about the Sun. We already know that the Earth–Sun distance—the astronomical unit—is about 150 million kilometers. Recall, astronomers have derived this distance indirectly by studying the response of radar signals sent toward some of the nearby planets, particularly Venus. In this way, and with knowledge of Newtonian gravity, we can construct a model of the Solar System having correct distances among all of the planets and the Sun. Note, however, that the Earth–Sun distance is not precisely constant; since Earth's orbit is slightly elliptical, the Earth–Sun distance changes a little throughout the year. (This distance variation, however, does not cause the different seasons on Earth. Instead, as noted in Chapter 5, the tilt of Earth's axis is responsible.)

ENERGETICS: We can now ask how much solar energy is radiated *isotropically* (i.e., in all directions), recognizing that Earth intercepts only a minute portion of the Sun's total energy. In other words, what is the total energy budget of the Sun? We can answer this question by holding a small light-sensitive device, for instance a solar cell of centimeter dimensions, perpendicular to the Sun's rays. In this way, the Sun's radiation strikes the flat surface of this square-centimeter detector, making it easy to measure the amount of solar energy received per square centimeter per second. This amount of solar energy received per unit area per unit time is known as the **solar constant**.

When we express the energy in terms of the commonly used unit called an *erg*, and correct the result for absorption by Earth's atmosphere, we find that the solar constant equals about 1.5×10^6 ergs/square centimeter/second. Actually, 1 erg is a very small amount of energy; if an insect were to crawl along with a velocity of about 1 centimeter/second, it would have a kinetic energy of about 1 erg. Or, if a fly were to collide with and splat against a wall, it would release an energy of about 1 erg in the process. Thus a single erg is a rather small unit of energy. But with regard to the solar constant, we're talking about more than a million ergs. And, remember, this 1.5 million ergs is the amount of solar energy incident only on our little square-centimeter detector, and every second at that.

Let's now inquire about the total amount of energy radiated in all directions from the Sun. Figure 9.2 shows how this is measured. First imagine, surrounding the entire Sun, a large three-dimensional sphere whose edge intercepts the Earth and whose radius equals the astronomical unit. By calculating the surface area of this imaginary sphere, we can easily find the total number of square centimeters on this sphere at the distance of Earth. Using the well-known relation that the surface area of a sphere equals 4π times its radius squared, we find the surface area of our imaginary sphere to be approximately 2.7×10^{27} square centimeters. In other words, there are more than a billion billion billion 1-square-centimeter areas on the entire imaginary sphere.

Multiplying the value of the solar energy incident on one of these 1-square-centimeter areas (i.e., the value of the solar constant) by the total surface area of the entire imaginary sphere, we can find the total isotropic rate of solar energy received at the distance of Earth. That number turns out to be 4×10^{33} ergs/second. We call this total *rate* of electromagnetic energy leaving the Sun's surface the **luminosity** of the Sun. As the units imply, luminosity is a measure of the total energy radiated each second of time.

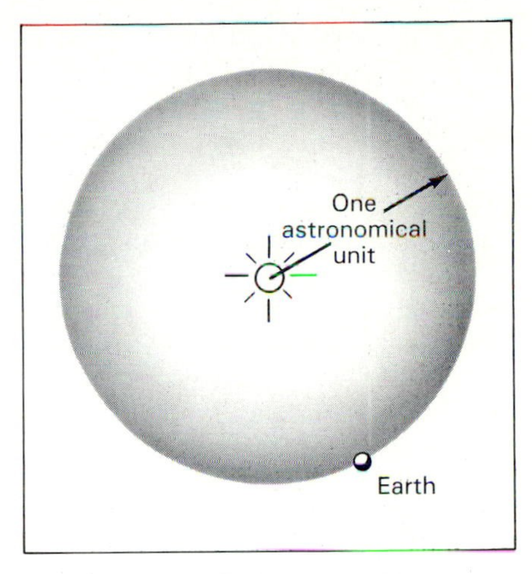

FIGURE 9.2 An imaginary sphere can be drawn around the Sun so that the sphere's edge passes through Earth's mid-section. The radius of this imaginary sphere then equals the astronomical unit.

The solar luminosity is a useful number to remember, for it's one of the truly basic properties used to characterize stars. As we study other stars, we'll be able to make comparisons among them by simply measuring their luminosities detected at Earth.

Thus the Sun, not surprisingly, is a powerful source of energy. Its luminosity can be alternatively expressed as a measure of power equal to about 4×10^{26} watts. (Power is the *rate* of energy expenditure; and there are some 10^7 ergs/second in 1 watt.) Note that this solar power is equivalent to more than a trillion trillion 100-watt light bulbs shining simultaneously. It also equals well over a trillion American dollars worth of energy radiated at current Earth-based rates—each second! Indeed, there are many impressive ways to alternatively express the total rate of solar energy output. The bottom line is that it's enormous.

Keep in mind that Earth intercepts an extremely small fraction of the Sun's total energy—no more than a billionth of the total energy emitted. After all, Earth subtends a very small angle (18 arc seconds) as seen from the Sun. Nonetheless, this rather small fraction of the total solar energy budget is sufficient to provide for us a reasonably warm and comfortable abode. Should our civilization learn how to harness the Sun's radiation, we clearly have a lot to gain. No chemical pollutants plague solar energy, it's constantly renewable, and we don't have to waste money ransacking our own planet!

SIZE: Another solar property easily found is the physical size of the ball of gas associated with the Sun proper. Even a casual observer could note that the solar disk subtends a visual angular extent of about 0.5 arc degree. Knowing the absolute value of the astronomical unit, we can calculate the physical diameter of the Sun. Simple geometry shows it to be about 1.4×10^6 kilometers. That's about 100 times the size of Earth. Since the volume of a sphere scales as the cube of its radius, we can then quickly figure that the solar volume is about 100^3 or thus a million times Earth's volume. Think about that for a moment. The Sun is a truly vast astronomical object—at least by terrestrial standards.

MASS: The mass of the Sun can be determined by studying the gravitational behavior of another object, for instance a planet, near the Sun. The easiest planet to use is Earth. By analyzing Earth's orbital motion around the Sun, we can gain an appreciation for the magnitude of the Sun's gravitational force field. Combined with the known mass of the Earth and the Earth–Sun distance, an estimate of the solar mass follows directly from the essentials of Newtonian gravity, where, recall, the gravitational force field varies directly as the product of the two masses (the Earth and Sun) and inversely as the square of the distance separating them (the astronomical unit). In this way, the solar mass is inferred to be about 2×10^{33} grams. By comparison, Earth is about 300,000 times less massive.

DENSITY: So the Sun is truly enormous. But its enormity includes not only large mass but also large size. Expressed another way, the Sun is massive alright, but not all that massive considering its huge size. In fact, the total amount of solar matter packed within its volume, namely, the average solar density, is not at all atypical of other terrestrial or astronomical objects studied thus far. The solar density, found by dividing the solar mass by the solar volume, averages 1.5 grams/cubic centimeter. This number simply means, as discussed before, that about 1.5 grams of *average* solar material would fill a typical sewing thimble. It certainly wouldn't weigh very much by terrestrial standards; on Earth, about 450 grams equal a pound, a common unit of weight in the United States. So a cubic centimeter of average solar stuff is quite lightweight, certainly lighter than a thimbleful of lead (11 grams/cubic centimeter) or even of average Earth material (5.5). It's heavier, though, than a thimbleful of water (1.0).

TEMPERATURE: A final solar property that we can easily determine is temperature, that is, temperature of the solar surface, not temperature of the core. The method used has already been discussed in Chapter 1 when we studied the total intensity of solar radiation as a function of wavelength (or frequency). Remember, the spread of radiation mimicked the shape of a theoretical Planck curve for a 6000-kelvin emitter. This is indeed the average temperature—sometimes called the effective temperature—of the Sun's photosphere.

Interestingly enough, it is because of this rather hot surface that humans can perceive the Sun with our eyes. If the Sun were cooler at the surface, the intensity of its emitted radiation would peak in the infrared part of the electromagnetic spectrum, and we might not be able to see it. And, of course, terrestrial objects would appear a lot dimmer. More fundamentally, given the physical and chemical makeup of Earth's atmosphere, intelligent beings probably wouldn't exist to experience 24 hours of nighttime anyway. It would seem that advanced species might need, for their very existence, some expertise with high-frequency radiation whose angular resolution is good enough to develop a sight-oriented intelligence. At any rate, we can be sure that earthlings, as currently "constructed," need sunlight—not suninfrared or sunultraviolet, but sun*light*.

Interior

Although the Sun has been observed to vibrate like a complex carillon of bells, the interpretation of those observations has proved difficult. Accordingly, there is no well-established solar equivalent to the terrestrial seismic waves that enable geologists to derive models of Earth's interior. Instead, solar physicists must rely on theoretical calculations to infer the density and temperature within the interior of the Sun. Let's examine how this is done.

MODELING: First, the bulk solar properties determined via Earth-based observations, as discussed above, are normally inserted as boundary conditions into a numerical computer program designed to infer the physical parameters of the solar interior. These boundary conditions must be matched by the computer model, for theory and observation to agree—a central objective of any modern science. Along with the boundary conditions, we need to provide the computer with values for the relative abundances of the various elements comprising solar material, as well as estimates for the rates at which nuclear reactions occur in the core. The elemental abundances can be derived from photospheric spectral-line observations, to be discussed in the next section; the appropriate nuclear reactions have been extensively studied in terrestrial accelerators, as we'll note in Chapter 10. These computer models then predict the bulk physical properties of, for example, av-

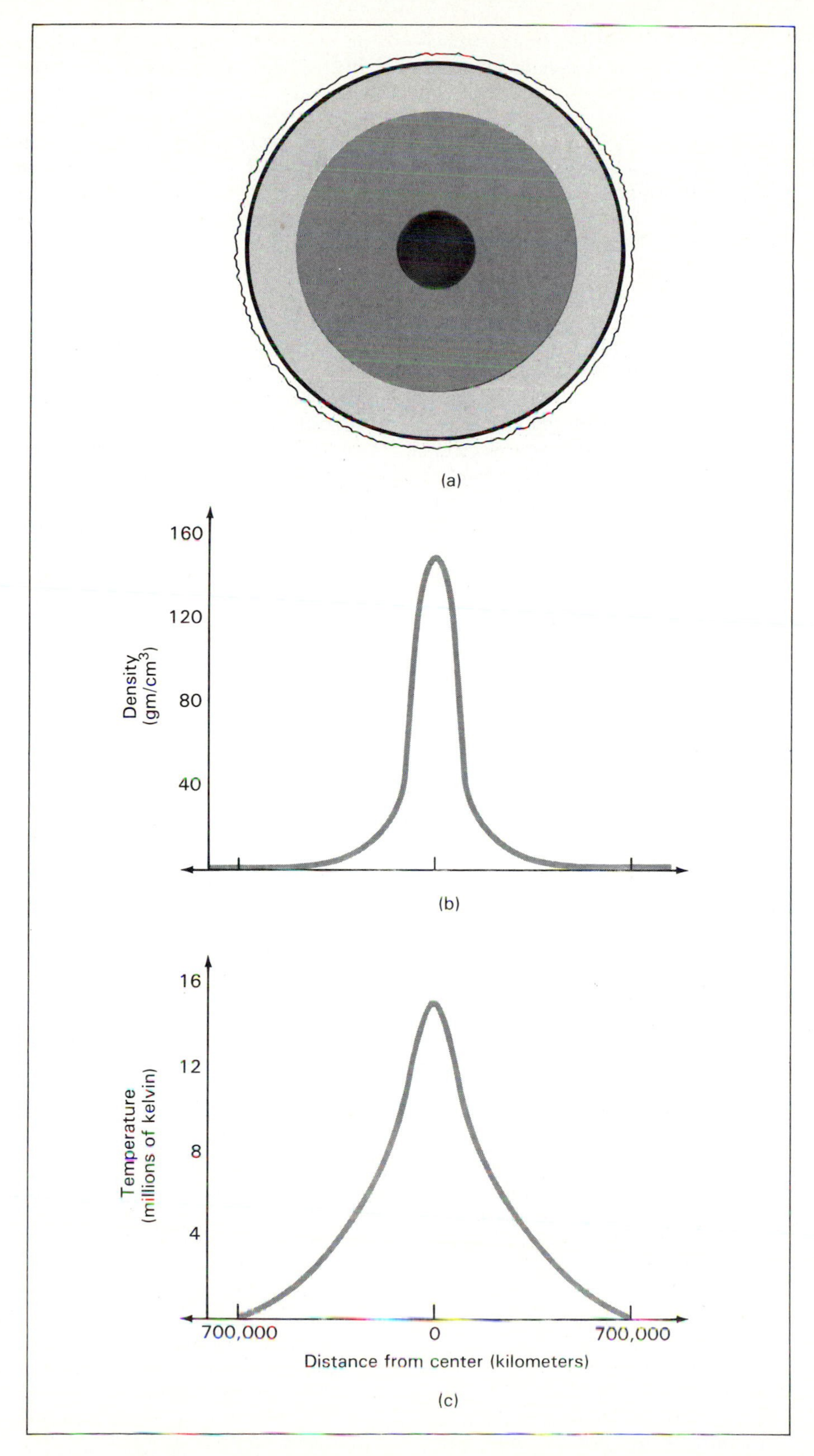

FIGURE 9.3 Theoretically modelled profiles of density (b) and temperature (c) for the interior of the Sun, seen for perspective in (a). All three figures pertain to a cross-sectional cut *through the core* of the Sun.

erage solar density and effective surface temperature. Those models which predict bulk solar properties that are inconsistent with the observations are discarded as physically unrealistic.

By number-crunching all this information on a large computer—the observed boundary conditions, the solar elemental abundances, the nuclear reaction rates, and the basic physics describing a gas—theoreticians have arrived at a general consensus concerning the run of density and temperature throughout the solar interior. Figure 9.3 shows each of these important thermodynamic properties plotted as a function of distance from the core of the Sun. Note how the density drops rather precipitously at first, and then decreases more slowly near the solar photosphere some 700,000 kilometers from the center. This variation of density is large, ranging from a core value of about 150 grams/cubic centimeter, which is 20 times the density of iron, to a midway (350,000 kilometers) value of about 1 gram/cubic centimeter, which is the density of water, to an extremely small photospheric value of 10^{-7} gram/cubic centimeter, which is about 10,000 times less dense than air on Earth's surface. But remember, the *average* density of that entire core-to-photosphere ball of gas we call the Sun is 1.5 grams/cubic centimeter, a little larger than the density of water. This peculiar variation in density means that 90 percent of the Sun's mass is compacted within the inner half of its radius.

The solar density continues to decrease out beyond the photosphere, reaching values as low as 10^{-20} gram/cubic centimeter far into the corona. Gas of such low density is just about as thin as some of the best vacuums that physicists can make in terrestrial laboratories. [However, galactic gas from which the Sun originated is even thinner (i.e., less dense) than the solar corona.]

The temperature profiled in Figure 9.3(c) also decreases with solar radius, although not as quickly as the density. Computer models predict a temperature of about 15 million kelvin at the core, consistent with the minimum 10 million kelvin needed to initiate the nuclear reactions known to power all stars. The modeled solar temperature nonetheless decreases rapidly enough to equal about 6000 kelvin at the photosphere, in agreement with the observed value derived from the Planck curve of solar intensity.

CONVECTION: Be sure to recognize that the computer modeling of the interior solar temperature and density includes aspects of the convection zone lying just below the photosphere. In the Sun, the physical process of convection results simply from the direction of the temperature change within the solar interior: hotter gas underlies cooler gas.

Recall that we've encountered the notion of convection several times previously. First, we've noted its action in the lower atmosphere of Earth; sunlight incident on a piece of unshaded ground heats the air immediately above, causing the warm air to rise while cool air rushes in from the sides. The rising air then loses some of its warmth to the upper atmosphere, whereupon it sinks, allowing more hot air to rise. In this way, a convection cycle is set up, causing the wind to blow at or near Earth's surface. Second, we've noted the process of convection in a pan of boiling water, whereupon the water heated at the bottom of the pan becomes less dense and rises; the upwelled water then loses some heat to the air above, after which it cools and subsequently drifts to the pan's bottom where it's heated once again, reinforcing a general circulation pattern. Third, we've discussed convection as a candidate mechanism that probably powers the slow motions of the terrestrial continents; hot molten rock, when given the chance, will rise through fissures in Earth's mantle, literally pushing aside terrestrial plates as part of a very large but slow circulation pattern. And fourth, we've seen direct evidence for convection near the polar regions of Jupiter's atmosphere. In each case, convection results from cooler material overriding warmer material, the result being a rather well-defined circulation pattern that strives to even out the temperature.

Convection works no differently in the case of the Sun. But since the process of solar convection is so critically important for the transport of energy to the Sun's surface, we need to understand it in some detail.

The very hot solar interior ensures violent and frequent collisions among the gas particles. Particles move every which way with extremely large kinetic energies, bumping into one another unceasingly. The result of this continued jostling is that hardly any electrons remain bound to their parent atoms within the interior. All but the innermost electrons of the heaviest atoms are stripped from their parent nuclei; all the light atoms are completely stripped of their electrons. Such a gas containing so many free electrons and nuclei comprises a near-perfect plasma. (At the very core, the extremely high temperatures guarantee just that—an absolutely perfect plasma.)

With so few atoms intact, radiation produced via nuclear reactions at the core readily travels outward, because there are relatively few atoms deep in the interior capable of absorbing photons in the normal way by boosting electrons from ground atomic states to excited atomic states. Consequently, the deep solar interior is rather transparent to photon absorption (although the photons suffer many scatterings); such a region is known as a **radiation zone**.

The situation changes as the solar photons continue to travel through the interior toward the photosphere. Near the solar periphery, just below the photosphere, the gas temperature is cool enough to allow many of the atoms to remain intact—excited and partially ionized yes, but partially intact with bound electrons. After all, the cooler temperatures in the outer parts of the solar interior ensure less frequent and less violent collisions among the atoms, allowing some of them to retain electrons about their nuclei. The heavier atoms, in fact, tend to keep most of their electrons. In general, the closer the interior gas is to the photosphere, the greater the number of atoms having a partial complement of electrons.

As photons move outward from the core and through the radiation zone, they eventually encounter the relatively cooler gas underlying the photosphere which can absorb them. This absorption of solar photons is so complete that *none* of the originally produced photons reach the surface. This is true, despite copious numbers of photons synthesized in nuclear reactions at the core. The outer layers of any sphere contain most of the sphere's volume. And in the case of the Sun, these layers contain enough intact atoms to absorb all the photons released by the nuclear events at the core. At some distance from the center, then, the transport of energy by photons comes to an end. The radiation zone terminates.

Solar energy, however, has to get out somehow. If it didn't, the Sun would have blown up long ago. Indeed, we see sunlight, proving that some solar energy escapes. That energy manages to reach the surface via convection. Hot solar gas, beginning at a depth of nearly 200,000 kilometers below the photosphere, *physically* moves upward, while the cooler gas above it sinks to the sides, creating a regular pattern of convection cells. All through the upper interior, energy is transported in this way to the surface by physical movements of the solar gas. Figure 9.4 is a schematic diagram of numerous solar convection cells where columns of hot gas rise, cool off, and descend. Thus this region transports energy in a fundamentally different way than in the radiation zone; for that reason, we call it the convection zone.

In reality, the zone of convection is likely to be more complex than described here. There is probably an entire hierarchy of convection cells in different tiers at different depths. The deepest tier, at about 200,000 kilometers below the surface, is thought to contain massive cells some tens of thousands of kilometers in diameter. Heat is then successively carried upward through a

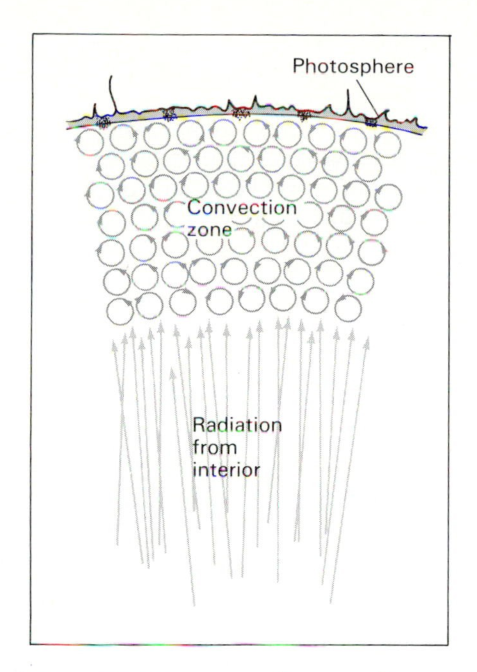

FIGURE 9.4 Schematic diagram of a simplified version of the physical transport of solar energy in the Sun's convection zone. In this way, we can visualize its upper interior as a boiling, seething sea of gas. Each convection loop is about 1000 kilometers across.

set of tiers having cells of progressively smaller size stacked one upon another until, at a depth of about 1000 kilometers, individual cells extend about 1000 kilometers in diameter. Actually, the top of this uppermost tier of convection constitutes the visible surface, where the cell sizes can be directly observed. Further information about the zone of convection is not observed but only inferred from the computer modeling of the solar interior.

Surface

At some distance from the core, the solar gas becomes too thin to sustain further upwelling via convection. Theory suggests that this distance roughly coincides with the photospheric surface that is observed visually.

Convection, then, does *not* proceed into the solar atmosphere. There's simply not enough gas to do so; the density is too low. Another mechanism must take over the transport of solar energy once the hot gas reaches the photosphere. After all, we feel the Sun's energy every day on Earth. It must escape somehow. It does so by the simplest mechanism: radiation. Since the photosphere is hot—6000 kelvin from direct observations—it emits thermal radiation like any other hot object.

So, out beyond the convection zone, another radiation zone is established. It's a zone through which photons can readily travel without much absorption. Here then are the flight paths of typical solar photons: Created in the Sun's core, photons suffer some scattering while bypassing free electrons and nuclei populating the inner radiative zone, reaching the convection zone in a few years. There, the photons are absorbed and reemitted so many times that nearly a million years are needed for them to reach the photosphere. Once free in the outer radiation zone, they travel outward without hesitation, reaching Earth in about 8 minutes.

Be sure to realize that sunlight starts its journey *from the photosphere*. None of the solar photons reaching Earth actually arise in the deep interior where the nuclear reactions generate energy in the first place. Lucky for us, those original photons cannot escape; at the extremely hot temperature of the solar core, those photons are of the harmful x-ray and gamma-ray variety. Thus we owe our existence in part to the convection zone, which effectively transforms copious amounts of lethal high-energy radiation into the more comfortable, lower-energy type of radiation. In short, the solar convection zone acts like an astronomical filter or transformer.

Figure 9.5 is a high-resolution photograph of the solar surface taken with instruments aboard NASA's *Skylab* space station, while orbiting above much of Earth's atmosphere in the mid-1970s. The visible surface is seen to be highly mottled or granulated with bright and dark gas. Each bright **granule** measures about 1000 kilometers across and has a lifetime between 5 and 10 minutes; each forms the topmost part of a convection cell. Together, several million granules comprise the top layer of the convection zone, which we alternatively call the photosphere. Incredibly enough, Earth, by comparison, is not much larger than the typical size of several such convection cells.

You should strive to visualize these hot, granulated, gaseous cells

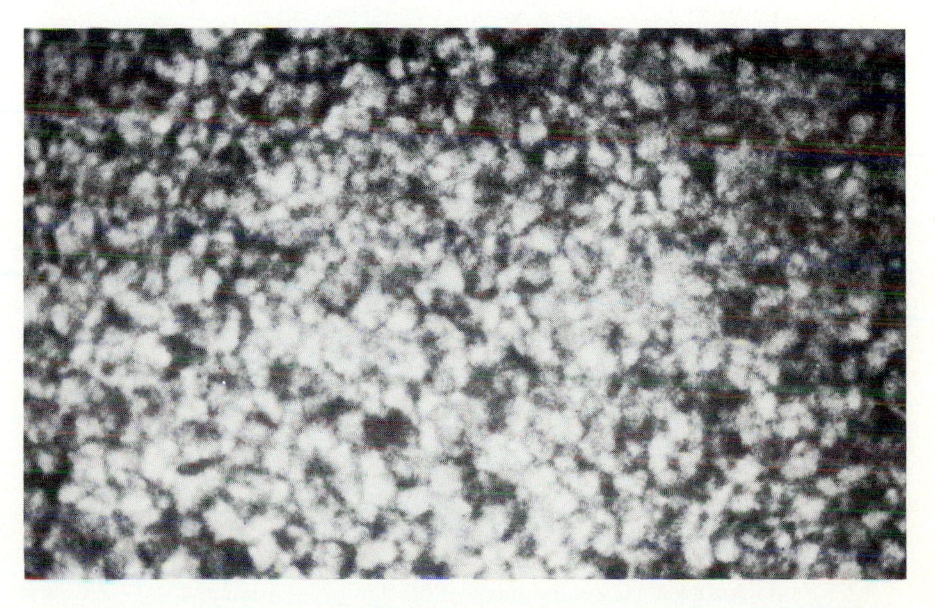

FIGURE 9.5 *Skylab* photograph of the granulated solar photosphere. Typical solar granules are comparable in size to Earth's continents.

"boiling" up from within, each coming and going on a time scale of minutes—sluggishly analogous to warm, rising bubbles in a pot of boiling molasses. Indeed, spectroscopic observations of the gas within and around these bright granules show direct evidence for the upward motion of gas, proving that convection really does occur at or below the photosphere. Characteristic spectral lines detected from the bright granules appear bluer than normal, indicating Doppler-shifted matter coming toward us with a velocity of about 1 kilometer/second. Conversely, spectroscopes focused only on the darker portions of the granulated photosphere show the same spectral lines to be red shifted, indicating matter moving away. Hence the solar convection zone is not based on theoretical inference; rather, its existence has been spectroscopically proved.

No appreciable density variations have ever been measured across the granulated solar surface; instead, the different brightnesses (i.e., light intensities) of the granules result strictly from temperature variations. The upwelling hotter gas obeys the usual Planck emission and thus radiates at somewhat shorter wavelengths, making it more visible than the downwelling cooler gas that emits less visible radiation. The bright and dark gases appear to contrast considerably in their juxtaposed positions, but in reality their temperature difference is only about 1000 kelvin.

Finally, note that despite the steady decrease in density and temperature from the interior to the atmosphere, the Sun does display a reasonably sharp limb. Figure 9.6 shows the entire Sun's disk photographed through a telescope filter, much like that seen on a hazy day. Its edge is not at all fuzzy, as we might expect for a gas that gradually thins out with distance from the core. Instead, the solar disk is sharp because the overwhelming majority of visible photons arise in the extremely shallow photosphere. Slightly below the photosphere, the gas is still convective and not radiative; slightly above, it's too thin to emit appreciably. Recent estimates for the depth of the photosphere suggest a value of no more than 500 kilometers. Although this is large by terrestrial standards, it's quite small compared to the full dimensions of the Sun. The photosphere is less than one-tenth of 1 percent of the Sun's radius. Physiologically, then, we perceive the Sun to have a rather sharp edge, despite the fact that it's a ball of loose gas—a little like a fluffy cumulous cloud made of mere drops of water yet which displays a well-defined "edge."

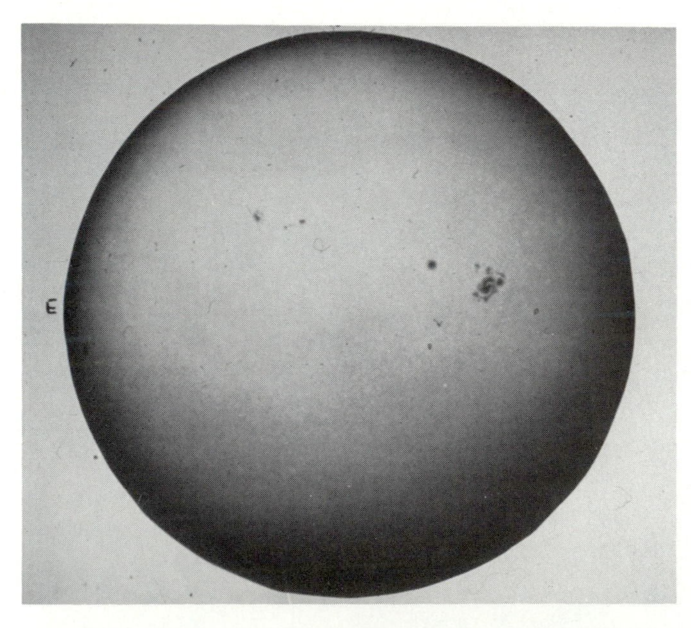

FIGURE 9.6 This photograph of the Sun taken through a filter shows a sharp solar limb despite the fact that our star, like all stars, is made of a gradually thinning gas.

Atmosphere

CHROMOSPHERIC REGIONS: Figure 9.7 is a detailed spectrum of the Sun, obtained for the full range of visual wavelengths, 4000 to 7000 angstroms. We can glean a good deal of information about the solar atmosphere from an analysis of these *absorption* lines that arise in the upper photosphere and lower chromosphere. Recall from Chapter 2 that photons radiating from the surface can be absorbed by the cooler gas in the lower solar atmosphere. The result is an absorption spectrum characteristic of the density and temperature of the region containing the absorbing atoms and ions.

Actually, thousands of spectral lines have been observed and cataloged in a typical solar spectrum, although there are not nearly this many elements in the Sun. Some of the heavier, multielectron elements can absorb a variety of photons having different energies. In fact, hundreds of lines are attributed to the one element iron in various stages of excitation and ionization, as discussed in Chapter 2. In all, there are some 67 elements currently identified in the Sun. Undoubtedly, many more elements exist in the Sun, perhaps all 92 that normally occur on Earth. But the expected abundances of the heaviest ones often fall below our current limits of instrumental sensitivity, so we shouldn't find it surprising that they haven't yet been observed. Solar elemental abundances, as might be expected, mimic pretty closely those of the cosmic abundances listed in Figure 2.19—hydrogen is by far the most abundant, followed by helium and so on.

Figure 9.8 is a series of doctored images showing the spreads of spe-

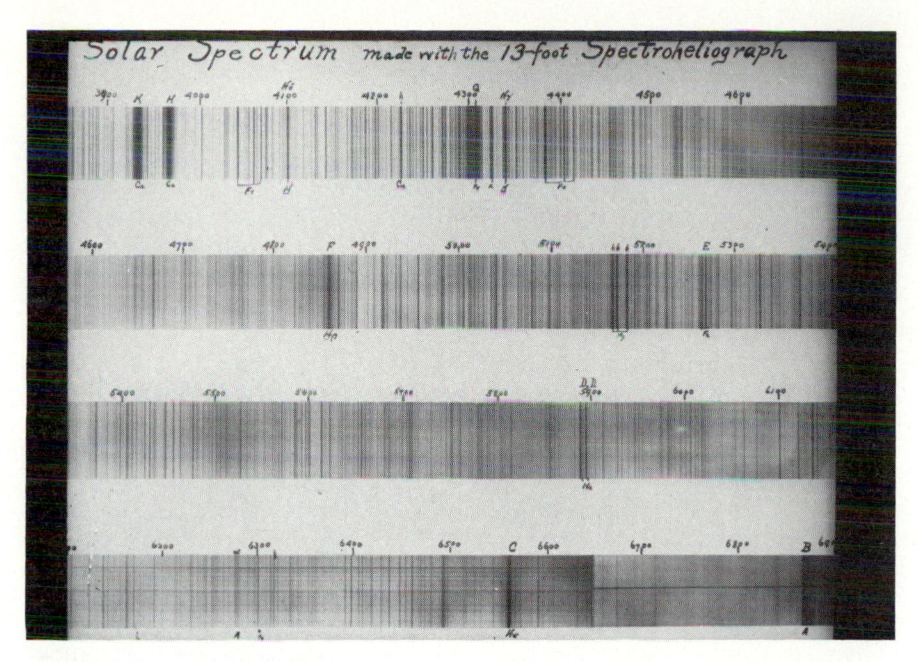

FIGURE 9.7 A detailed spectrum of our Sun in the visible domain shows thousands of spectral lines indicative of some 67 different elements in various stages of excitation and ionization in the lower solar atmosphere. These absorption features are sometimes called Fraunhoffer lines after the nineteenth-century German physicist who first identified several hundred of them.

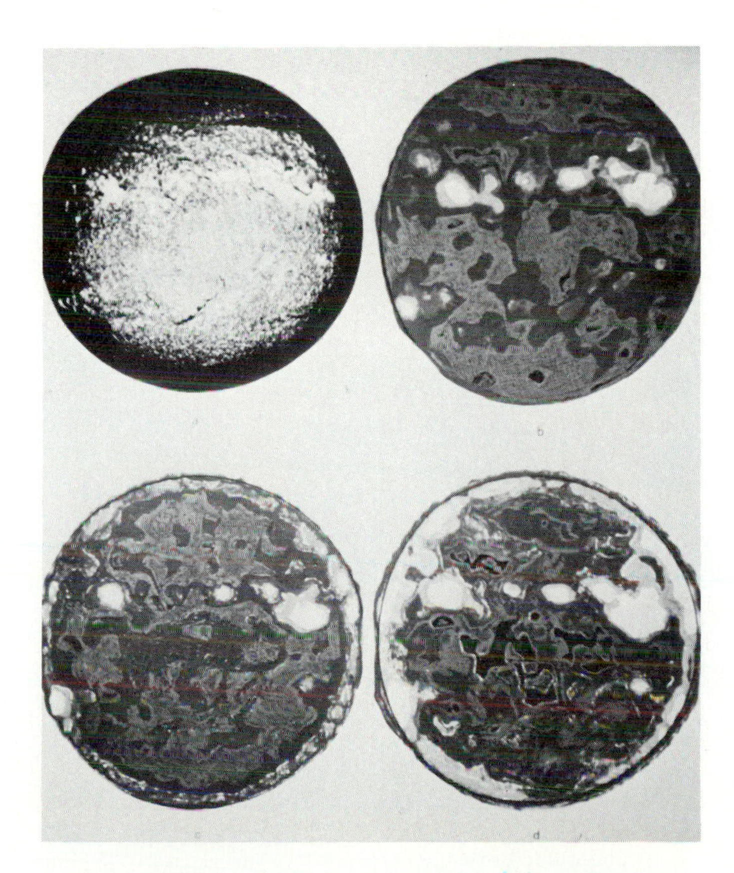

FIGURE 9.8 Maps of elemental abundances help to determine changes in temperature and density across the surface of the Sun: (a) hydrogen, (b) calcium, (c) iron, and (d) carbon.

cific elements across the Sun's surface. By filtering out all the radiation except that within a small spectral interval containing one of the lines in Figure 9.7, astronomers can study one particular element to the exclusion of all others. The maps of Figure 9.8 show just that. Narrow spectral intervals of four different lines were observed across the entire Sun. The different shadings of each map are then used by researchers to help compare how the different elements are distributed. Actually, the intensity of the shading is directly proportional to the intensity (darkness) of the individual absorption lines. However, you should remember that these solar maps only partly indicate (small) variations in the elemental abundances; gas temperature plays a more important role in the variations among line intensities in stars.

The powerful tool of spectroscopy permits us to derive several physical and chemical properties. For example, we can obtain temperature, density, and elemental abundances from spectral lines observed at one solar locality, or from the extended maps such as those of Figure 9.8. These physical conditions characterize only the lower chromospheric region where the photons are absorbed; they are not directly related to any other solar region. Certainly, the temperature and density each drastically differ both above and below the chromosphere. In contrast, however, elemental abundances are not thought to differ much anywhere in the Sun. Although astronomers cannot normally detect distinct spectral lines from the photosphere and interior (too hot) or from the corona (too tenuous), theory implies that the same proportions of atoms ought to be present throughout all solar zones, albeit in considerably different stages of excitation and ionization. And observations of convection, as noted earlier, strongly imply that the gases in the solar surface and interior regions are well mixed. So we'll assume throughout this book that the elemental composition of the chromosphere mimics reasonably well the abundances throughout the Sun, with the exception of the core where nuclear reactions are known to be processing large quantities of hydrogen into helium.

Now the fact that the solar spectral lines are of the absorption type tells us something important about the gas temperature in the lower solar atmosphere. As we might expect, the temperature continues to decrease into the upper photosphere and lower chromosphere. Detailed spectral-line observations imply that the temperature of the gas that absorbs photospheric photons measures just a little less than the photosphere's 6000 kelvin. The actual value is close to about 5000 kelvin. Analysis of the lines also proves that the density decreases with increasing altitude in the chromosphere.

We might expect these two thermodynamic quantities to continue to decrease with distance from the Sun. Indeed, the density does, but surprisingly the temperature does not. The rise in temperature in the outer atmospheric layers merits further inquiry.

CORONAL REGIONS: In the past couple of decades, there has been a surge of interest in solar regions far beyond the photosphere. This interest has been stimulated largely by the Space Age discovery that Earth is engulfed by the solar corona. In fact, the solar coronal regions have surprising influence on our planet, especially on the magnetosphere and atmosphere, as described in Chapter 5.

Neither the inner solar atmosphere—the chromosphere—nor the outer solar atmosphere—the corona—can be observed visually under normal conditions. The photospheric regions of the Sun are, by comparison, too bright, dominating the visible radiation arising in the much less dense solar atmosphere. This relative dimness of the corona results from a general lack of photons since, after all, large numbers of photons cannot be expected to be emitted (or absorbed) in a very tenuous gas containing extremely small numbers of atoms per unit volume.

On the other hand, should the intense light arising from the more dense photosphere be blocked, we could clearly observe the emission of radiation from the Sun's atmosphere. This is precisely what happens during a total **solar eclipse**, when the Moon completely blocks the Sun's disk from view. Figure 9.9 shows the plumes of hot gas extending well out into the solar corona.

During the brief moments of an eclipse, when light from the solar photosphere is fully blocked, something interesting happens. The pattern of spectral lines changes dramatically. Intensities of the usual spectral lines alter, suggesting changes in elemental abundances and/or gas temperature. Most lines shift from absorption to emission since the usual background hot source is now blocked. Furthermore, entirely new spectral lines, not seen in the photospheric absorption spectrum, suddenly appear. These new coronal (and in some cases chromospheric) lines were first observed during eclipses in the 1920s, and for many years afterwards, researchers attributed them to a nonterrestrial element called "coronium."

We now recognize that the new lines do not indicate any new kind of atom; coronium does not really exist. Rather, these new spectral lines are caused by electrons transitioning within highly ionized atoms. By "highly ionized atoms" we mean atoms from which *several* electrons have been removed, making them much more ionized than the usual ions residing in the photosphere. For example, lines have now been identified from iron ions having as many as 13 of their normal 26 electrons missing. That's a highly ionized piece of matter indeed!

All things considered, the most reasonable cause of the extensive electron stripping is an increase in temperature. Higher temperatures inevitably produce greater numbers of collisions, which in turn can remove many electrons from their parent atoms. Thus the peculiar spectra observed during solar eclipses compel us to conclude that the gas temperature of the upper chromosphere must exceed that of the photosphere. Furthermore, the temperature of the solar coronal regions must be even higher than in the chromosphere. Otherwise, the

FIGURE 9.9 This photograph of a total solar eclipse shows clearly the emission of radiation from the corona. If you ever have a chance to experience a solar eclipse, do so, for nature will reward you with one of its greatest light shows! The experience is awesome.

observed spectra cannot be understood.

Figure 9.10 plots the variation of gas temperature with altitude from the photosphere, outward into the corona. The temperature profile drawn results from many observations at different distances from the limb of the Sun. As shown, the temperature decreases to a minimum of nearly 4000 kelvin some 500 kilometers above the photosphere, after which it climbs steadily. At about 3000 kilometers above the photosphere, the gas temperature rises rapidly, reaching more than 1 million kelvin some 10,000 kilometers beyond the photosphere. Thereafter, it remains roughly constant well out from the Sun. This temperature profile is what really distinguishes the photosphere from the chromosphere and from the corona; the chromosphere or lower solar atmosphere extends from the top of the photosphere to approximately 3000 kilometers, beyond which lies the corona or upper solar atmosphere.

The cause of this rapid temperature rise has been a fairly difficult problem to crack. After all, the temperature profile runs contrary to intuition. Normally, heat flows from hot to cold, not the reverse. For example, moving away from a fire, we expect the heat to diminish. And for terrestrial fires, it does. But the Sun is apparently different. There must be another source of heating for the corona.

Briefly, coronal heating is thought to be caused by small-scale disturbances on the Sun's surface. These disturbances occur in many ways, but all of them tend to "upset" the Sun. That is, in addition to the relatively inactive Sun studied above, our Sun is also surprisingly active at times. This activity has little bearing on the Sun's bulk makeup. But solar activity can affect the corona, and that in turn can affect us at planet Earth. We need to learn more about solar activity; it is an area of practical astrophysics that affects our daily lives.

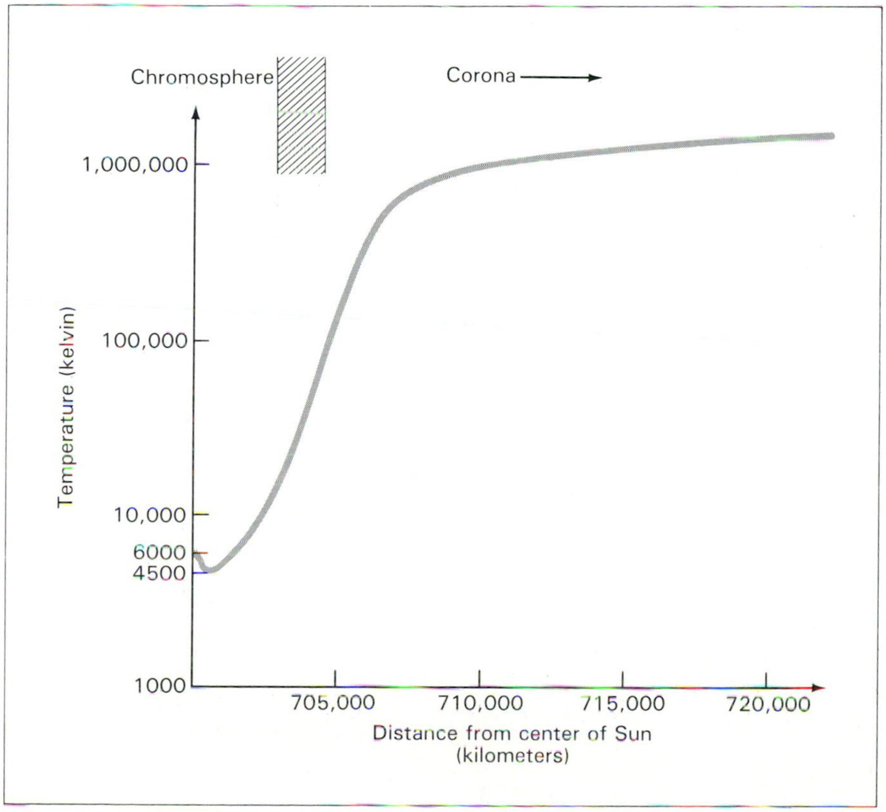

Activity

Most of the Sun's luminosity results from the rather steady emission of solar photons emitted continuously from the photosphere. This regularly emitted radiation arises from what we call the **quiet sun**, which is the predictable condition of the Sun that blazes forth on a daily basis. It contrasts with the sporadic, unpredictable radiation of the **active sun**, which is a more irregular condition prone to violent eruptions.

The active component of solar radiation contributes little to the Sun's total luminosity, and presumably has little effect on the evolution of the Sun as a star. Indeed, these transitory disturbances would be virtually undetectable if the Sun were at a distance of the next nearest star. Solar activity is a bit like weather on planet Earth; here we recognize the phenomena—clouds, winds, thunderstorms, tornadoes—yet we are unable to predict well when and where such phenomena will occur. But unlike meteorological activity on Earth, activity on the Sun can affect the more fragile, nonsolar regions (such as tiny planets) well beyond the surface of the Sun.

CORONAL HEATING: Every few minutes, small solar storms overwhelm the gravitational force field at the photosphere and literally expel particles of matter into the surrounding atmosphere. Figure 9.11 shows direct photographic evidence for such gas jets known as **spicules**. Like posts in a picket fence, these long (several thousand kilometer), thin spikes of flaming hot matter and radiation escape from the Sun's surface at a typical velocity of about 100 kilometers/second. The particulate matter—mostly protons and electrons—carry large amounts of kinetic energy; colliding with the surrounding gas in the solar atmosphere, the particles serve to

FIGURE 9.10 The change of gas temperature in the lower solar atmosphere is dramatic.

FIGURE 9.11 In this photograph of a field of solar spicules, these short-lived (minutes) narrow jets of gas can be clearly seen spouting up through the solar chromosphere.

heat the gas well beyond the Sun itself. Radiation is also expelled, but it's the high-energy particles that really heat the corona.

The expelled particles deposit some of their energy in the corona by creating pressure waves while speeding through the coronal gas. The process is a little like the never-ending crashing of ocean waves on Earth's shores, a well-known way to transport energy. Such ocean waves carry energy—in this case mechanical energy—and continually transfer their energy onto the shore. They do this even though the coastal water may be warmer than that offshore. In this way, ocean surf can transport energy from a cold body (the ocean) to a hot body (the shore). Similarly, we can think of waves of matter in the lower solar atmosphere as analogous to ocean surf; when the waves "break," lots of energy is suddenly dumped into the solar atmosphere, thus heating the corona. The spicules seem to be the solar analog of terrestrial surf.

SOLAR WIND: Electromagnetic radiation *and* fast-moving particles escape the Sun all the time. The radiation moves away from the photosphere at the velocity of light, forming sunlight and other types of solar radiation; it takes 8 minutes to reach Earth. The particles, by contrast, travel more slowly, although at the still considerable (supersonic) velocity of about 500 kilometers/second (which is about a million miles/hour); these protons and electrons reach Earth in a few days.

The constant stream of escaping solar particles is termed the **solar wind**. Although predicted for many decades just by looking at comets (whose tails always point away from the Sun, even on their outbound journeys), the solar wind was not directly detected until the U.S. *Mariner 2* spacecraft encountered it on its way to Venus in the 1960s. The wind extends far into space, bathing the planets of our Solar System in an extended solar corona or "heliosphere," and reaching to at least the orbit of Pluto. This much we know since *Pioneer 11*, which recently escaped our planetary system, is still radioing back information about the solar wind.

Be sure to realize that the solar wind is an extremely thin medium. Even though it blows constantly, radiating away roughly a trillion grams (a million tons) of solar matter each second, we estimate that less than 0.1 percent of the Sun has been lost since it formed billions of years ago. So the Sun is evaporating, yes, but it's losing only a negligible part of its huge bulk. That's why we say that solar activity hardly affects the Sun itself—only those more fragile abodes like ours nearby.

Actually, the solar wind is another example of an astronomical object unable to retain its atmosphere. Given the temperature of the corona, the escape velocity of a particle from the Sun is comparable to the average molecular motion of those particles. Despite the great solar mass, thereby making the escape velocity large, the coronal temperature is also very large. The net result is that the Sun sheds its atmosphere. But unlike planets or moons, where a lost atmosphere is rarely reestablished, the atmospheres of stars like our Sun are regularly replenished. They have to be; if it were not, the solar corona would disappear in about a day. Accordingly, we have the simple scenario: Convection cells constantly "splash" matter and energy up through the spicules, which in turn fuel the super hot corona that pushes solar matter off into space at supersonic velocities. The replenishment process occurs continuously.

What sort of radiation is emitted by a gas of 1 million kelvin? The answer is simple—thermal emission

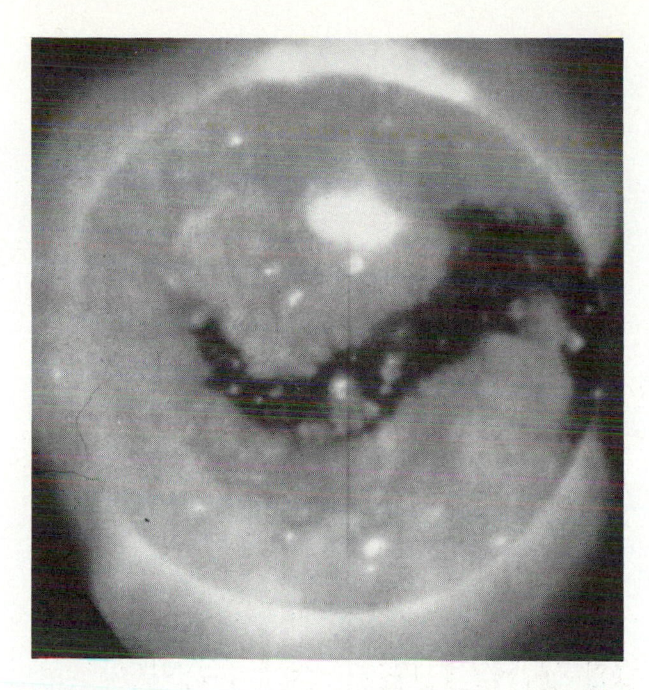

FIGURE 9.12 An image of the x-ray emission from the Sun observed by the *Skylab* space station. Note the coronal hole to the right where the x-ray observations outline in dramatic detail the abnormally thin regions through which the high-speed solar wind streams forth. [See also Plate 6(a).]

like that from any hot object. The Planck curves of Figure 1.15 predict that radiation from a million-kelvin gas must peak primarily in the x-ray portion of the electromagnetic spectrum. Unlike the 6000-kelvin photosphere that emits most strongly in the visual part of the spectrum, the hotter coronal gas radiates at higher frequencies. For this reason, x-ray instruments, hoisted above Earth's atmosphere, have in recent years become one of the principal tools used to study the solar corona.

Figure 9.12 and Plate 6(a) show a photograph—really an x-ray image—of the Sun. Images of such high-frequency radiation highlight the hottest solar regions, particularly the corona, and enable us to study the invisible solar plasma. The full corona extends well beyond even those regions shown, but the density of coronal particles emitting the radiation diminishes rapidly with distance from the Sun; the intensity of x-ray radiation farther out is too dim to be detectable in this image. In fact, the coronal density near Earth is roughly 10^{-20} gram/cubic centimeter, more than a dozen orders of magnitude thinner than the gas density at the photosphere.

In the mid-1970s, instruments aboard NASA's *Skylab* space station revealed that the solar wind escapes mostly through solar "windows" called **coronal holes**. The dark area to the right side of Figure 9.12 illustrates one of these newly discovered phenomena. Not really holes, they only seem to lack matter. We now know that coronal holes are vast regions of the Sun's atmosphere where the density is about 10 times lower than the already tenuous, normal corona. Like apertures in a fisherman's net, the holes are theorized to be underabundant in matter largely because the coronal plasma there is almost entirely unconfined; the gas within coronal holes streams forth with especially high speeds and is an important source of the solar wind. Constant monitorings in recent years prove that large coronal holes are not a regular feature of solar activity, although neither are they rare. Small holes seem to erupt every few hours.

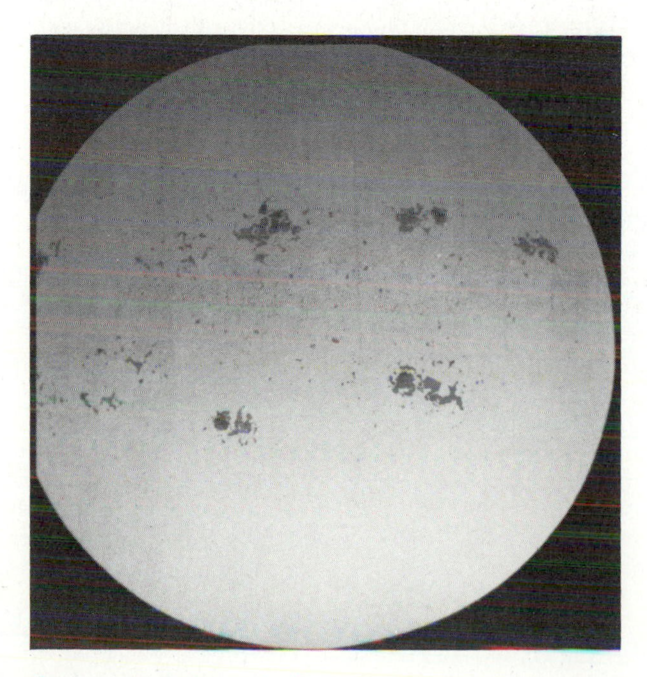

FIGURE 9.13 Photograph of the entire Sun showing clear evidence of sunspot activity.

SUNSPOTS: So, the solar wind blows *constantly*. But, in addition to this uniform, steady ejection of solar material, the Sun also ejects matter explosively on occasion. It's worth taking a closer look at some of the most active solar regions, especially those photospheric sites where storms rage with much violence.

Figure 9.13 is an optical photograph of the entire Sun. Shown there are some obvious dark blemishes. First studied in detail at the time of the Renaissance, these "spots" provided one of the first clues that the Sun, and therefore the Universe itself, is not a perfect creation, but a place of turmoil and constant change. Such small dark areas are indeed disturbed regions of an otherwise well-behaved photosphere. Called **sunspots**, they typically measure about 10,000 kilometers across, or about the size of Earth. Furthermore, as shown, they often occur in clusters. At any one time, the Sun may have hundreds of sunspots, or it may have none at all.

Rest assured, sunspots are not evil gods as thought by the ancient East Asians who first recorded them about 20 centuries ago. They are not planets inside the orbit of Mercury. Nor are they solar mountains protruding through the solar atmosphere. Not even are they thunder clouds floating in the solar atmosphere, a proposal made by Galileo when he first noted them telescopically early in the seventeenth century. Sunspots are simply *cooler* regions of the photospheric gas.

Careful studies of these spots provide clear evidence for a dark umbra toward the center, surrounded by a grayish penumbra. The close-up of a pair of sunspots in Figure 9.14 shows how each of these darkened areas are easily distinguished from one another, as well as from the brighter undisturbed photosphere nearby. This apparent gradation in spot darkness is really a gradual change in photospheric temperature. Direct observations show the umbra to be about 4500 kelvin while the penumbra measures about 5500 kelvin. The spots are, therefore, composed of hot gases, but only seem dark because of their superposition on an otherwise uniformly brighter (i.e., hotter, 6000-kelvin) background. If we could magically remove a sunspot from the Sun (or block out the rest of the Sun around it), the spot would glow brightly, just as any 4000- to 6000-kelvin emitter. Contrasted with the even brighter background, however, the spots *appear* to lose their emissive qualities and remiss into darkness, like an underexposed photograph.

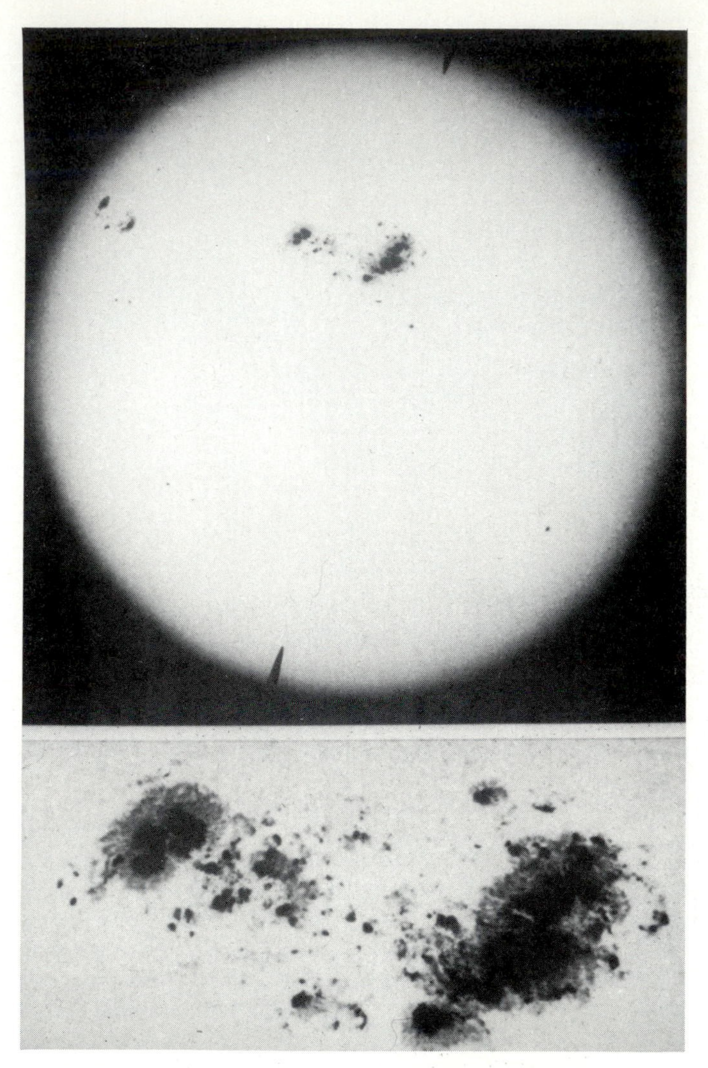

FIGURE 9.14 Close-up photograph of a pair of sunspots (bottom) that can also be seen in their broader context (above).

Sunspots are not steady; we often observe them to wiggle around. Most change their size and shape. They all come and go. Figure 9.15 shows a time sequence where several spots are seen to vary, sometimes growing, sometimes dissipating. Individual spots may last anywhere from 1 to 100 days, with large groups of spots typically lasting 50 days.

The spots also move across the face of the Sun. We take this movement to be part of the general rotation of the Sun on its axis. These same sunspot movements, however, indicate that the Sun does not rotate as a solid body. Instead, it spins like a gas—faster at the equator, slower at the poles. Spot movements imply that the photosphere rotates once in about 27 days at the equator, but only once in 31 days at the poles.

Not only do sunspots come and go with time, but they also change in a fairly regular fashion. Centuries of observations have established a clear **sunspot cycle**. Figure 9.16 shows a plot of the average annual number of sunspots as a function of time. The periodicity can easily be noted; the average number of spots maximizes every so often. We can use these data to deduce a sunspot

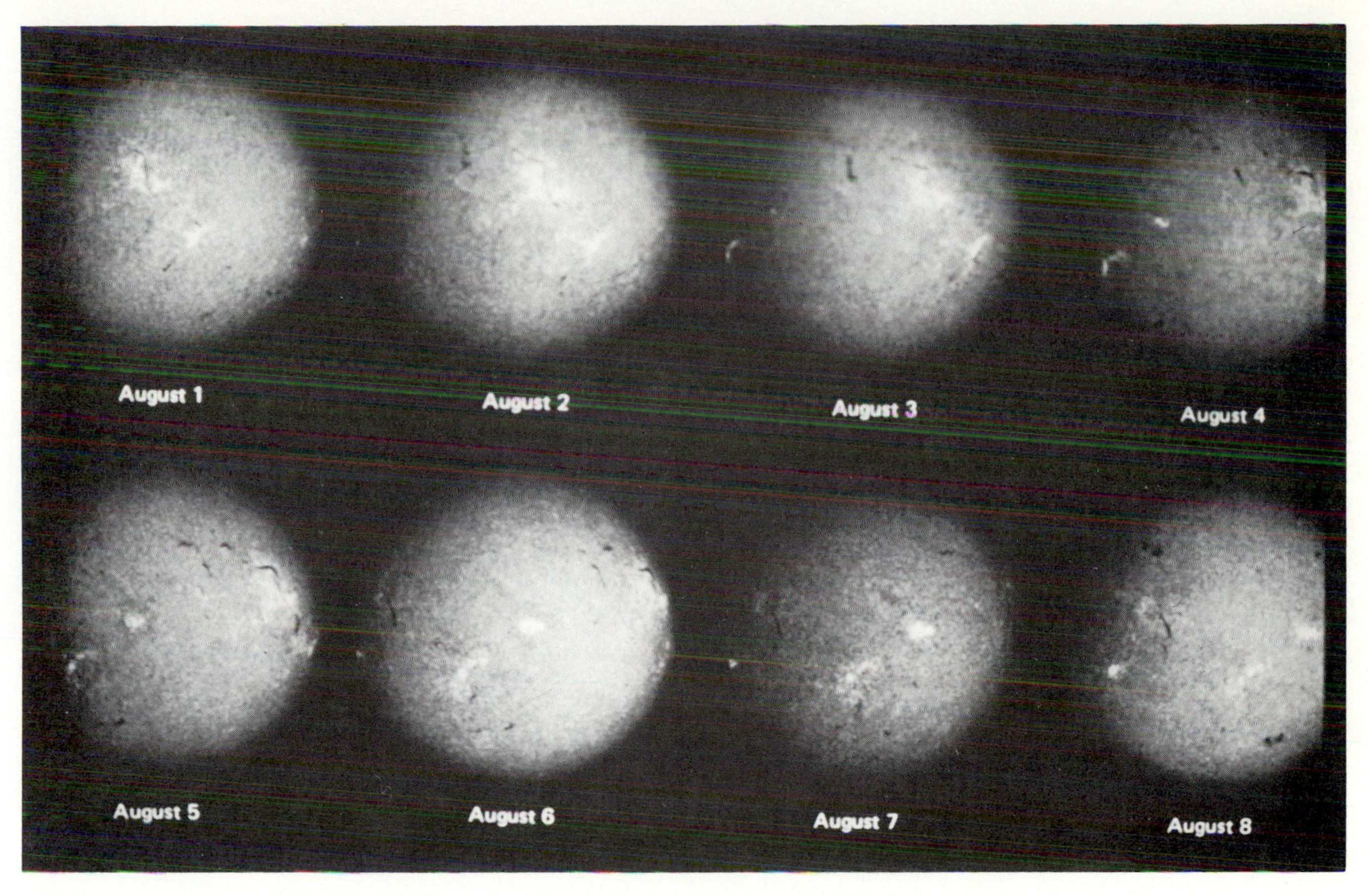

FIGURE 9.15 Photographic time sequence of the evolution of some sunspots. About a day of elapsed time separates each frame.

period of more or less 11 years—"more or less" since any individual cycle varies from 7 to 15 years.

A further peculiarity of sunspots is the observed fact that the latitude of most spots also seems to vary. And this variation also mimics the solar cycle. This point is demonstrated in Figure 9.17, which shows a plot of sunspot latitude as a function of time. At the start of each 11-year cycle (the time of **solar mini-**

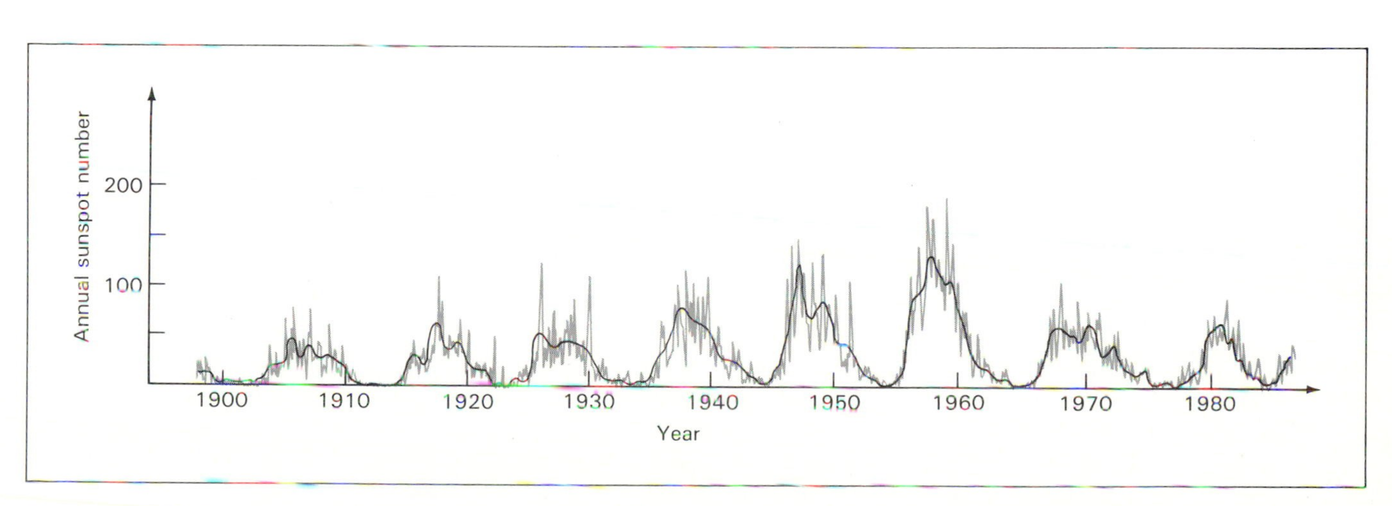

FIGURE 9.16 The sunspot cycle can be determined, as has been done here, by counting the annual number of sunspots throughout this century.

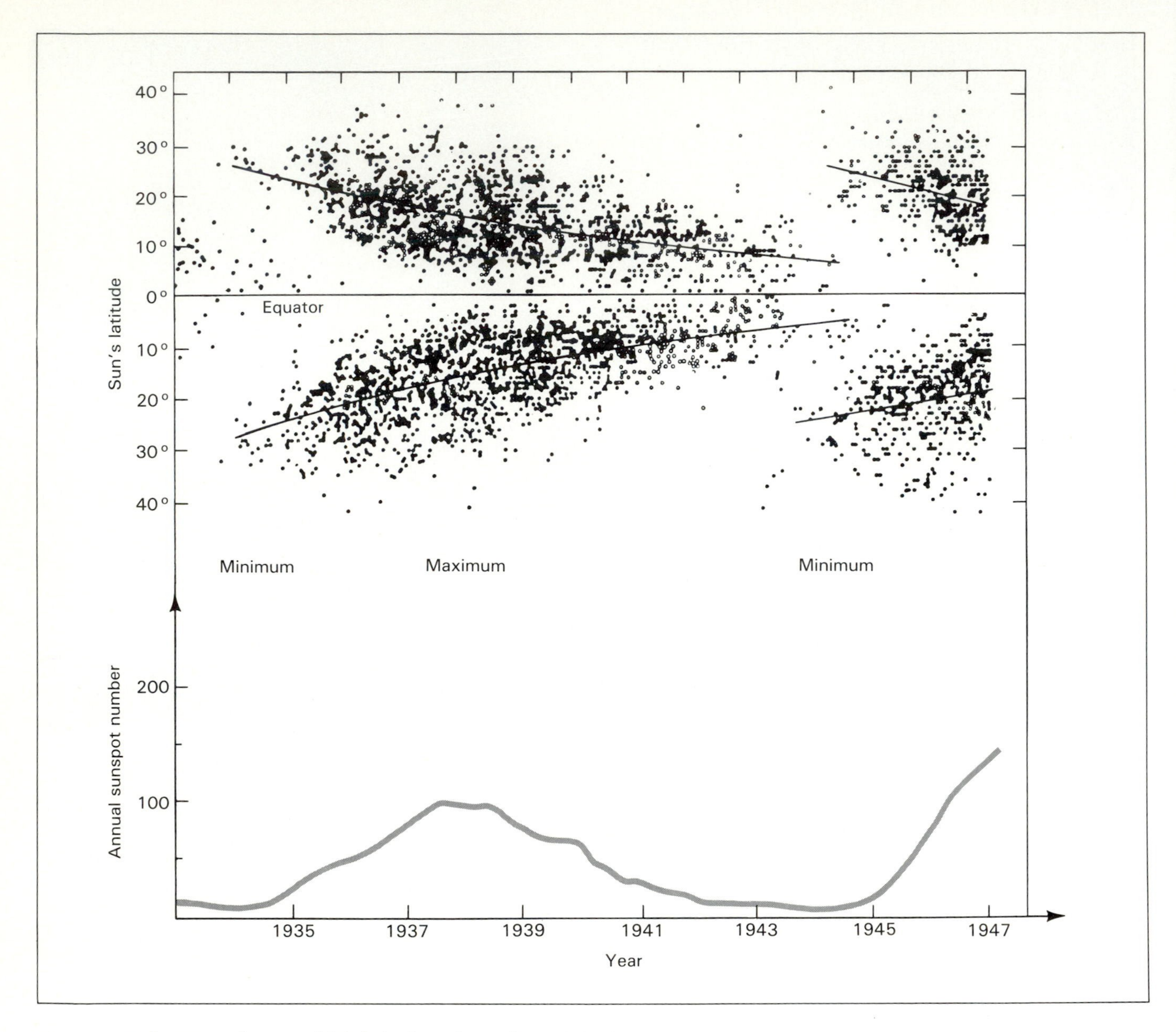

FIGURE 9.17 Sunspots cluster at high latitudes when solar activity is at a minimum. They then seem to reside at lower latitudes as the number of sunspots maximize. Finally, they are prominent near the Sun's equator when the solar cycle is again minimized. (The curve at the bottom is derived from the preceding figure.) The next solar maximum is expected in 1990.

mum), only a few spots are usually observed; they often seem to be confined to two narrow zones about 25 to 30 arc degrees to the north and south of the solar equator. Approximately four years into the cycle (time of **solar maximum**), the number of spots grows rapidly while moving to within about 15 to 20 arc degrees north and south of the equator. Finally, at the end of the cycle (solar minimum again), the total number decreases as the sunspot bands move to latitudes of about 10 arc degrees (see Interlude 9-1).

Individual sunspots do not move up or down in latitude; it's just that new sunspots appear closer to the equator as older ones at higher latitudes fade away over the course of the 11-year solar cycle. Complicating this picture further, the 11-year cycle seems to be part of a larger 22-year cycle (see Interlude 9-2), and furthermore the beginning of each new cycle seems to overlap the last phases of the preceding one. The whole process—number of spots and variation with latitude—repeats over and over, although, unfortunately, researchers are currently unable to understand either why the sunspot period is roughly 11 years in magnitude or why there is a period at all.

INTERLUDE 9-1 *Maunder Sunspot Minimum*

The 11-year periodicity of the solar sunspot cycle is far from regular. Not only does its period vary from 7 to 15 years, but the sunspot cycle is also known to have disappeared entirely during an earlier epoch of the Sun's history. Historical records show that the sunspots themselves were generally lacking during a long period of the late seventeenth century. In honor of the British astronomer who drew attention to these historical records, the 1645–1715 period of solar inactivity is called the Maunder Minimum.

The figure below shows the full extent of all the sunspot data recorded since the invention of the telescope. It's a simple extension of the data shown in Figure 9.16 and illustrates the average number of sunspots observed at any one time.

The lack of sunspots during the late seventeenth century cannot be attributed to a lack of interest in the Sun or to a shortage of equipment. Galileo and other astronomers had made beautifully detailed drawings of sunspots prior to the Maunder Minimum, and telescopes were in common use throughout seventeenth-century Europe. (Interestingly enough, and to the glee of French patriots, the Sun was indeed unblemished during the reign of Louis XIV of France, popularly known as *le Roi Soleil.*) Apparently, no spots at all were reported for many years at a time, and the total observed for the entire 70-year Maunder Minimum was less than the number normally occurring in a single year near solar maximum. Clearly, if the solar cycle had not shut down entirely, it was at least severely disrupted.

Additional historical records tend to verify this relative lack of solar activity. The corona was apparently less prominent during total solar eclipses, and Earth auroras were sparse throughout the late seventeenth century. Furthermore, the abundance of a radioactive isotope of carbon, assimilated (via photosynthesis) into trees in the form of carbon dioxide (CO_2), shows an increase for those tree rings corresponding to the period roughly 1640 to 1710. After all, if the Sun is inactive, its solar wind will not sweep away many of the galactic (cosmic-ray) particles that normally collide with nitrogen in our air and build up radioactive carbon in Earth's atmosphere. Thus the data show high levels of radioactive carbon to correspond to low levels of solar activity. What's more, other tree-ring anomalies also exist for the intervals A.D., 1440 to 1550 and 1100 to 1250, suggesting that long-term changes in solar activity may be common.

Lacking a detailed understanding of the sunspots, and especially for their cyclical behavior, we cannot easily explain how the solar cycle could shut down entirely. Most astronomers suspect the culprit to be related somehow to changes in the Sun's convection zone and/or its rotation pattern. But the specific cause of Ole Sol's century-long variations remains a mystery.

At any rate, all these recently discovered pieces of historical evidence clearly imply that the considerable solar activity of the past 250 years is not necessarily typical. The *active* Sun seems not as regularly behaved as scientists once thought.

On the other hand, be sure to recognize that this irregular activity contrasts greatly with the steadiness of the much more powerful *quiet* Sun. In fact, Earth's fossils imply that the bulk solar luminosity could not have changed by more than about 10 percent over the past half-billion years or so; if the Sun had changed by more than that, Earth's surface temperature would have made impossible life as we know it. Furthermore, geologists argue that Earth's water has been in the liquid state for several billion years, implying that the Sun has been reasonably steady throughout much of Earth's history. In modern times, the quiet Sun's luminosity has not changed by more than 1 percent (although in recent decades spacecraft seem to have detected an exceedingly small decrease in solar output). All in all, the quiet Sun seems to be the well-oiled, smoothly running machine scientists have assumed all along.

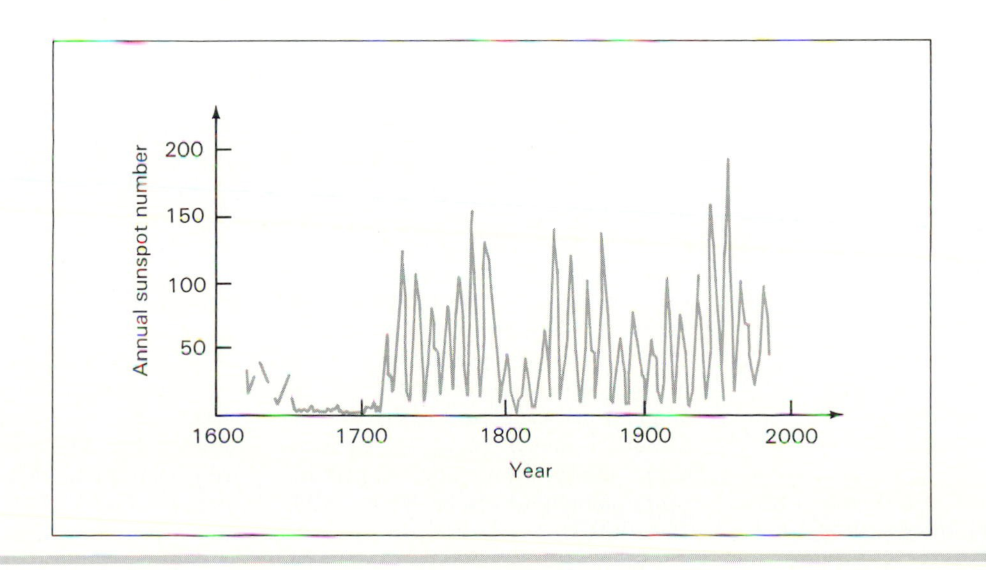

INTERLUDE 9-2 ***Solar–Terrestrial Relations***

Apparent connections between solar activity and terrestrial events have existed for so long that our Sun has often been worshipped as a god with power over human destinies. Claims for a connection between, for example, the variable Sun and Earth's weather have been made repeatedly throughout the past century, perhaps even before that. Only recently, however, has this subject become scientifically respectable—in other words, more natural than supernatural.

One such solar–terrestrial correlation suggests a cause for droughts on Earth. As mentioned elsewhere, the Sun proceeds through a complete rising-and-falling cycle of sunspot activity about every 11 years. For each cycle, however, the sunspots' magnetic polarity reverses in the two solar hemispheres; during one cycle, the north polarity of a spot leads the rotation pattern, during the next cycle, the south polarity leads. So there are really about 22 years between returns of the sunspots, including their magnetic properties. This is the so-called "double sunspot cycle."

Now, periods of climatic dryness on Earth do seem to be correlated with the double sunspot cycle. Near the start of the past eight 22-year cycles, there has been a drought—at least within the American middle and western plains from South Dakota to New Mexico. The droughts typically last three to six years, the last two coming in the early 1930s and the late-1950s—although the one expected in the 1980s did not occur as clearly as anticipated.

Other Sun–Earth connections include a chronological tie between solar activity and increased atmospheric circulation on our planet. As circulation increases, terrestrial storm systems deepen, extend over larger latitudes, and carry more moisture. The connection is complex, and the subject controversial, because no one has yet shown any physical mechanism—other than heat—that would allow solar activity to stir our terrestrial atmosphere. And without a better understanding of the physical mechanism involved, none of these effects can be incorporated into our weather forecasting models.

Solar activity might also influence long-term climate on Earth. For example, the Maunder Minimum (consult Interlude 9-1) seems to correspond fairly well to the coldest years of the so-called "little ice age" that chilled northern Europe during the middle of the current millennium. How the *active* Sun, and its abundance of sunspots, might affect Earth's climate is a frontier problem of terrestrial climatology.

A correlation that is definitely established and also better understood is one between solar flares and geomagnetic activity at Earth. The extra radiation and particles thrown off by the flares impinge on Earth's environment, overloading the Van Allen Belts, causing brilliant auroras in our atmosphere, and degrading our communication networks at the surface. These disturbances have been known for many years, but only recently have we associated them with the Sun's matter and energy blowing in the solar wind. We are furthermore only beginning to understand how the radiation and particles emitted by solar flares also interfere with terrestrial radars, power networks, and many other technological devices. Some power outages or blackouts on Earth are actually caused not by increased customer demand and not by malfunctioning equipment, but by solar flares!

We are currently unable to predict when and where the flares will occur. It's certainly to our advantage to know more about solar flares; the active Sun surely does affect our lives. Here, then, is another fertile area of astronomical research—one for which there are clear terrestrial applications.

SOLAR FLARES AND PROMINENCES: Regions near sunspots occasionally erupt violently, spewing forth large quantities of energetic particles into the surrounding corona. Figure 9.18 shows the result of one of these explosive events called a **prominence**. The largest prominences release a total of about 10^{32} ergs of energy, counting both particles and radiation; compare this to the total rate of solar energy emission—4×10^{33} ergs/second.

Many prominences display a reasonably well-defined "loop-like" structure, which are almost certainly governed by magnetic force fields that extend far beyond the surface before arching back. Some persist for days; others come and go erratically; still others (though rarely) surge up and then immediately fall back on themselves, all the while ejecting particles and radiation. Such prominences are composed of excited atoms, especially hydrogen, although ionized helium, calcium, iron, magnesium, and numerous other elements have been observed. The extension of a solar prominence typically measures 100,000 kilometers, which is, by comparison, nearly 10 times the size of planet Earth.

Flares are another type of solar activity observed low in the Sun's atmosphere near sunspots and prominences. Flares are even more impulsive and more violent than spots or prominences. They often flash across a region of the Sun in a matter of minutes, while suddenly releasing pent-up energy. Coordinated observations made by the *Solar Maximum Mission*—an Earth-

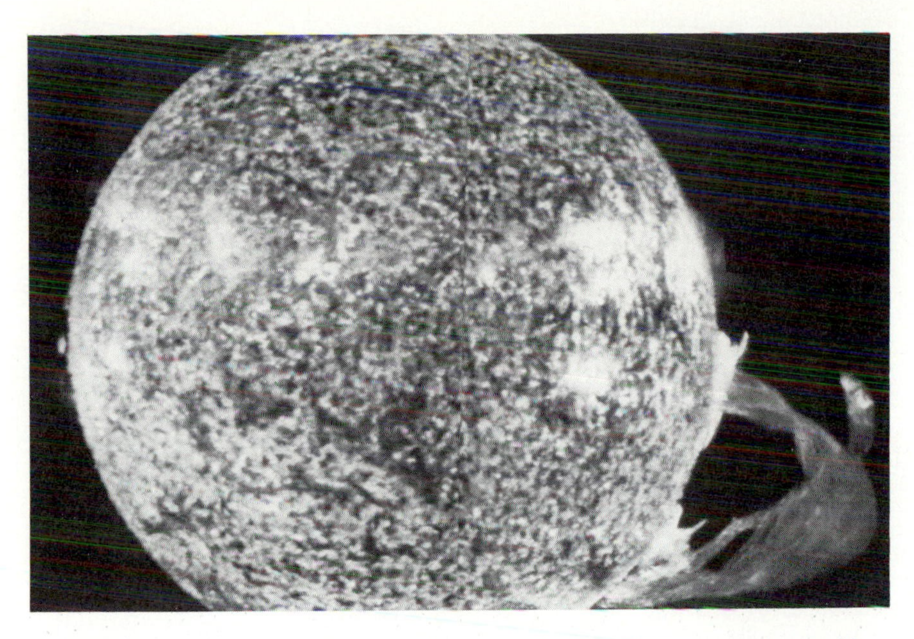

FIGURE 9.18 This image of an especially large solar prominence was observed by ultraviolet detectors aboard the *Skylab* space station in 1973.

orbiting satellite recently repaired by a NASA *Shuttle* astronaut—demonstrated that x-ray and ultraviolet radiations are especially intense when emanating from the extremely compact hearts of flares, where temperatures can reach 100 million kelvin.

Now recognized to be cataclysmic explosions, solar flares are the closest thing to lightning strokes on Earth. But astronomical lightning strokes to be sure! Instrumented spacecraft regularly measure flares to be among the most violent of all solar activity—so much so that some researchers prefer to view flares as bombs exploding in the lower regions of the Sun's atmosphere. Indeed, major flares release some 10^{27} ergs *per second*—the equivalent of some 1 million nuclear bombs unleashed simultaneously.

SOLAR MAGNETISM: Why are there spots, prominences, flares, and other disturbances on our Sun? What controls the plasma on the Sun's surface, occasionally though irregularly permitting that plasma to burst forth with considerable activity? We can address some aspects of these questions by noting a few facts that hint at their underlying nature.

The first fact is that the absorption spectral lines of a sunspot are notably wider than those observed anywhere else on the Sun. Careful analysis has shown that this additional line breadth is caused by a strong magnetic force field. Recall that magnetism was one of the line-broadening mechanisms studied in Chapter 2. Typically, the magnetic force field in a sunspot is about 1000 times the normal solar magnetism in the neighboring, undisturbed regions; sunspots also have several thousand times more magnetism than Earth. Sunspot magnetism can hardly be exaggerated; no other magnets of comparable power are known anywhere in the Solar System.

A second fact that hints at the true nature of sunspots is their grouping; they are almost always paired. Furthermore, one sense of magnetism is always associated with one member of a pair, the opposite sense with the other. That is, pairs of sunspots have opposite magnetic (north-south) polarity. Force fields emerge from one spot and reenter the other, magnetically breaking through the Sun's surface, and guiding the hot ionized gases along the arch connecting a sunspot pair.

Sunspots, then, are gigantic sites of concentrated magnetic force fields, created by the motions of the electrically charged particles comprising the gases of the hot solar atmosphere. The spots themselves are cooler because the abnormally strong magnetic force field tends to block (or redirect) the normal convective flow of hot gas to the surface of the Sun.

Apparently, the entire Sun has a magnetic force field much like Earth's, although much stronger. To imagine this invisible field, think of a bar magnet penetrating the solar interior perpendicular to the solar equator. Unlike the solid and molten Earth, the *gaseous* nature of our Sun greatly complicates the issue. For example, the nonuniform rotation of the Sun tends to distort its magnetic force field and even wrap it up into tangles.

Figure 9.19 shows an idealized schematic of such wrapping. As illustrated, the Sun's uneven rotation at the poles and equator eventually causes its original north-south force field to reorient itself in an east-west direction. Convection, whereby some magnetized gases upwell perpendicular to the surface, further complicates the pattern of solar magnetism. In some places, the force field becomes kinked like a knot in a garden hose; this causes an increase in magnetic strength. Occasionally, the strength of the magnetic force field becomes large enough to overwhelm the gravitational force field that normally holds the Sun intact. The result is a burst of hot matter out of the photosphere in the form of a flare.

Observations clearly show that streams of charged gases soar high into the solar atmosphere, all the while following the arching magnetic force fields connecting spots of opposite polarity. Indeed, most prominences have both ends of their arched loops associated with a single sunspot pair. Moreover, the magnetism of sunspots begins intensi-

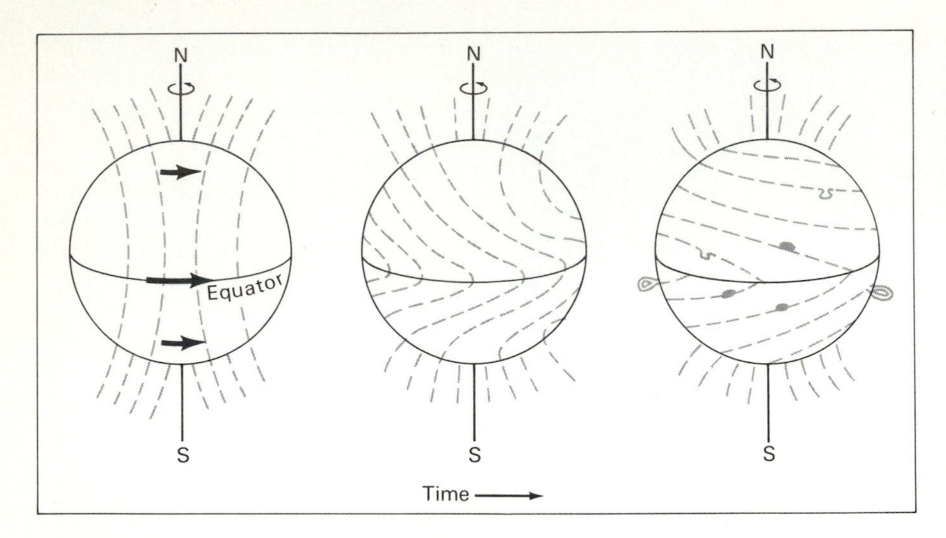

FIGURE 9.19 This schematic diagram illustrates the wrapping and distorting of the solar magnetic force field (dashed), thereby helping to create sunspots, flares, and prominences. Somehow, in these regions, magnetic energy is transferred into the energy of fast particles, flooding the Solar System with solar matter as well as solar radiation.

fying even before a spot becomes visible; the magnetism also persists long after a spot fades. For these reasons, we surmise that magnetism must be fundamental to solar activity. The spots, flares, and other outward manifestations of the active Sun are mere by-products.

There are problems, however. The periodic and transient kinks in solar magnetism may well account for the violent events of the active Sun. And qualitatively, the process outlined above seems plausible. Quantitatively, though, the process has not proved easy to decipher. Specifically, the origin of solar magnetism, the precise mechanism of flaring, and the 11-year (or 22-year) cycle of sunspot activity all remain currently unexplained.

Every new instrument we build and every new observation we make, not only answers a few questions, but also raises some more. That's how we learn. Studies of the active Sun are of particular importance partly because our parent star is of interest to us in its own right. Moreover, only in this neighboring object can we observe the relevant violence on a scale and in the detail needed to genuinely understand the mechanisms at work in vastly more energetic events far into the Galaxy and beyond.

SUMMARY

Astronomers regard our Sun as a spectacular and relatively nearby laboratory. This ball of gas that shines forth daily gives us an opportunity to study complex physical events that we cannot hope to simulate in experiments on Earth. Especially important, solar studies help us to learn more about the basic plasma processes occurring in a wide variety of astronomical objects throughout the Universe. Above all, our Sun enables us to study up close the fine-scale phenomena of a typical star—phenomena that pass unnoticed in stellar studies because most other remote stars in the nighttime sky, regardless of how powerful the telescope, are destined to remain rather undistinguished points of light.

Observations of the Sun have improved greatly in quality and scope during the past decade. Instrumented spacecraft, particularly those able to sample invisible radiation, are now yielding a much better view than ever could have been possible from ground-based observatories. With this greater factual information, however, comes increased complexity; a general consensus about our Sun is at hand, but specific theories capable of explaining detailed solar phenomena remain largely inadequate. Indeed, though our star is a beautiful and fascinating astronomical object, it is also an incredibly complex one.

KEY TERMS

active Sun
chromosphere
convection zone
corona
coronal hole
first-generation star
flare
granule
luminosity
photosphere
prominence
quiet Sun
radiation zone
second-generation star
solar constant
solar core
solar eclipse
solar interior
solar maximum
solar minimum
solar wind
spicule
sunspot
sunspot cycle

QUESTIONS

1. Explain in some detail how astronomers measure the temperature of the Sun's surface.
2. Cite the (macroscopic and submicroscopic) physical properties of the Sun's zone of convection, and describe the role of this region in the release of energy from our parent star.
3. Explain the physical process of solar granulation. Give two other examples of this process in nature.

4. Describe and briefly discuss the method used to find each of the following: (a) distance from Earth to the Sun, (b) size of the Sun, (c) luminosity of the Sun, (d) composition of the Sun, (e) mass of the Sun.
5. Pick one: The Doppler effect (a) explains the origin of so many lines in the Sun's spectrum; (b) helps prove that radiation sometimes acts as a particle; (c) is crucial in our understanding of stellar parallax; (d) measures P and S waves that travel through Earth's interior; (e) is used by cops to catch speeders on the highway.
6. Explain why, if stars are composed mainly of hydrogen, the dark absorption lines of this element are not especially prominent in our Sun. Be explicit. Also, why are the Sun's spectral features dark anyway?
7. What observational evidence establishes that the solar corona is very hot, on the order of 1 million kelvin? Explain.
8. How does the Sun appear as a sharply defined cosmic object despite the fact that it's nothing more than a ball of gas?
9. If the sunspots are thousands of kelvin hot, why do they appear dark?
10. Discuss how it is that solar magnetism is probably the prime source of much of the Sun's activity.

FOR FURTHER READING

Eddy, J., *The New Sun,* NASA SP-402. Washington, D.C.: U.S. Government Printing Office, 1979.

Frazier, K., *Our Turbulent Sun*. Englewood Cliffs, N.J.: Prentice-Hall, 1982.

Friedman, H., *Sun and Earth*. New York: Scientific American Books, 1986.

Leibacher, J., Noyes, R., Toomre, J., and Ulrich, R., "Helioseismology." *Scientific American,* September 1985.

Noyes, R., *The Sun, Our Star*. Cambridge, Mass.: Harvard University Press, 1983.

Paresce, F., and Bowyer, S., "The Sun and the Interstellar Medium." *Scientific American,* September 1986.

Wolfson, R., "The Active Solar Corona." *Scientific American,* February 1983.

10
SOLAR ENERGY: *A Recipe from Nuclear Physics*

Compared to the daily activity on Earth—storms, floods, geysers, even violent volcanoes and earthquakes—the spots and flares of the active Sun described in Chapter 9 are overwhelming. Even the steady emission of electromagnetic radiation from the inactive quiet Sun staggers the imagination.

We tend to take the Sun for granted. The next time you go outdoors on a sunny day, take note of the fact that solar energy constantly inundates our planet. Each day, this nearby cosmic object provides a steady stream of heat and light, those simple physical phenomena without which life could not exist.

This remarkable ball of gas—our Sun, the nearest star— is somehow able to produce large amounts of energy continuously. According to Earth's fossil record, the Sun must have done so for several billion years. How does it do it? What powers the Sun? What forces are at work in the Sun's core to produce such large quantities of energy? By what process does the Sun shine, day after day, year after year, eon after eon?

Answers to these questions are of basic importance to all of cosmic evolution. Without them, we can understand neither the astrophysical existence of stars and galaxies in the Universe, nor the biochemical existence of life on Earth.

The learning goals for this chapter are:

- to study how stars shine
- to understand the nuclear physics operating at the Sun's core
- to briefly appreciate how heavy elements can be produced in the hearts of stars
- to realize that matter and energy are interchangeable, and together obey the famous equation $E = mc^2$
- to appreciate one of the greatest problems of modern astronomy: our inability to test accurately the nuclear events recurring within the Sun

Emission Mechanisms

Consider for a moment the luminosity of the Sun. Recall that luminosity is the *rate* of energy emitted, that is, the amount of energy radiated per unit time. In Chapter 9 we estimated our Sun's luminosity to be 4×10^{33} ergs/second. This is a most important number when characterizing our Sun, but it's not enough if we are to appreciate the forces at work in its core. A knowledge of the solar mass, noted earlier to be about 2×10^{33} grams, is also useful when studying solar energy.

We can determine how well the Sun generates energy by dividing the solar luminosity by the solar mass, namely,

$$\frac{\text{solar luminosity}}{\text{solar mass}} = 2 \text{ ergs/gram/second.}$$

This value means simply that every gram of solar material yields about 2 ergs of energy in every second of time. A couple of ergs, remember, is not much energy. Humans expend a lot more than 2 ergs in the 1 second needed to get up out of a chair. Consequently, this rate of solar energy production per unit mass is not overpowering, or even large. In fact, a piece of burning wood generates about a million times more energy per unit mass per unit time than does our Sun. But there is another basic difference between the Sun and a piece of wood: the latter will not burn for billions of years. At best, a good log will burn for a few hours.

To appreciate fully the magnitude of the energy generated by our Sun, we must consider not only the luminosity/mass ratio but also, and more important, the total amount of energy generated by each chunk of solar matter *throughout the full duration of the Sun as a star.* In other words, we seek to evaluate the quantity,

$$\frac{\text{solar luminosity}}{\text{solar mass}} \times \text{solar lifetime.}$$

As previously noted, a large amount of circumstantial evidence suggests that the Sun is approximately 5 billion years old. Furthermore, the theory of stellar evolution, to be studied in Chapter 20, maintains that the Sun is currently a middle-aged object and should continue to emit steadily for about another 5 billion years.

To be conservative, though, let's suppose that we don't know the total solar lifetime of approximately 10 billion years. Let's instead calculate only what it would take to power the Sun from its origin to the present time. Our answer will then be independent of the theory and assumptions of stellar evolution; hence it will be a firm result based only on the many terrestrial facts (rocks, fossils, etc.) that imply the Sun to be at least 5 billion years old.

Converting 5 billion years to seconds, and evaluating the previous relationship, we find a value of 3×10^{17} ergs/gram. This number, then, is a measure of the average amount of energy radiated by every gram of solar material. It represents a *minimum* value, for more energy will be needed for each additional day the Sun shines. Should the Sun endure for another 5 billion years, as predicted by theory, this value would have to be doubled.

In contrast with the small luminosity/mass value of 2 ergs/gram/second, this energy/mass value is large. Nearly a billion billion ergs of energy must arise from *each* gram of solar matter in order to power the Sun throughout its lifetime.

These simple calculations demonstrate that the mechanism generating solar energy is not an explosive one that occurs rapidly. Whatever that mechanism is, it must be a rather slow, steady, time-released process that provides a *uniform* rate of lighting and heating of the many planets, including Earth.

FAMILIAR ENERGY SOURCES: Let's now investigate several of the energy emission mechanisms familiar to twentieth-century science. The objective here is to identify an energy process capable of producing at least 3×10^{17} ergs for every gram of matter consumed in that process.

We can begin by ruling out wood as a source of solar energy. Wood simply doesn't burn long enough. If the Sun were totally made of wood—even hard mahogany—it would last only a few thousand years since, as we've mentioned, wood explosively produces energy about a million times faster than the Sun. So we can confidently conclude that the Sun is not made of wood—big surprise, eh?

What about gasoline or other chemicals? Could the Sun be just a huge ball of flaming gasoline? A rather simple experiment that can be done with automobile fuel—although it's not recommended—readily demonstrates that 1 gram of highly refined gasoline yields only 10^{12} ergs. Furthermore, if the Sun were made entirely of gasoline, its burning time would be only about 10,000 years, far too short to account for the billions of years of constant heating of planet Earth.

Could the Sun be radiating simply because of its stored heat? After all, energy can be stored in any object by heating it, just as a piece of metal radiates when warmed by a fire. Perhaps the Sun gained lots of heat as its parent galactic cloud contracted long ago, and has been essentially cooling ever since by slowly releasing energy. Could the extremely hot solar core as well as its surrounding interior regions conceivably radiate enough stored energy to account for our calculated solar output of 3×10^{17} ergs/gram? The answer is no, for the subject of thermodynamics stipulates that the maximum amount of stored heat the Sun could radiate is about 10^{15} ergs/gram. Although close to our required value, this process still falls short by a factor of several hundred. Stored heat, then, contributes less than 1 percent of the total solar energy radiated.

Since the process of stored heat disagrees with the calculated solar output by a factor of only 300, you might wonder if we have the right energy source but the wrong time scale. In other words, if the Sun were 300 times younger than currently thought, the requirement of 3×10^{17} ergs/gram is reduced to about 10^{15} ergs/gram, thus matching the typical amount of energy produced by the stored-heat process. The problem with this argument is that the Sun would have to be as young as 5 billion years/300 or about 15 million years old. This is simply not the case, since radioactive dating proves the Earth, Moon, and numerous meteorites—and by strong implication the Sun—to be much older than tens of millions of years, in fact nearly 5 billion years old. Furthermore, the marine fossil record extends back nearly 4 billion years, implying that liquid water has existed on Earth for at least that long, which in turn implies that the Sun's luminosity must have been fairly constant for billions of years. Stored heat, then, although it comes close, cannot be responsible for the bulk of solar radiation.

A more exotic mechanism of energy production known to twentieth-century science is called nuclear **fission**. This is a process whereby nuclei of heavy elements break down into lighter ones. Recall from our previous discussion in Chapter 5 that the uranium nucleus and other radioactive nuclei like it are unstable. Accordingly, uranium decays, naturally and of its own accord, into other nuclei that are lighter and more stable. Of great importance, Figure 10.1 illustrates how the spontaneous decay of uranium also produces some energy.

Since many radioactive elements

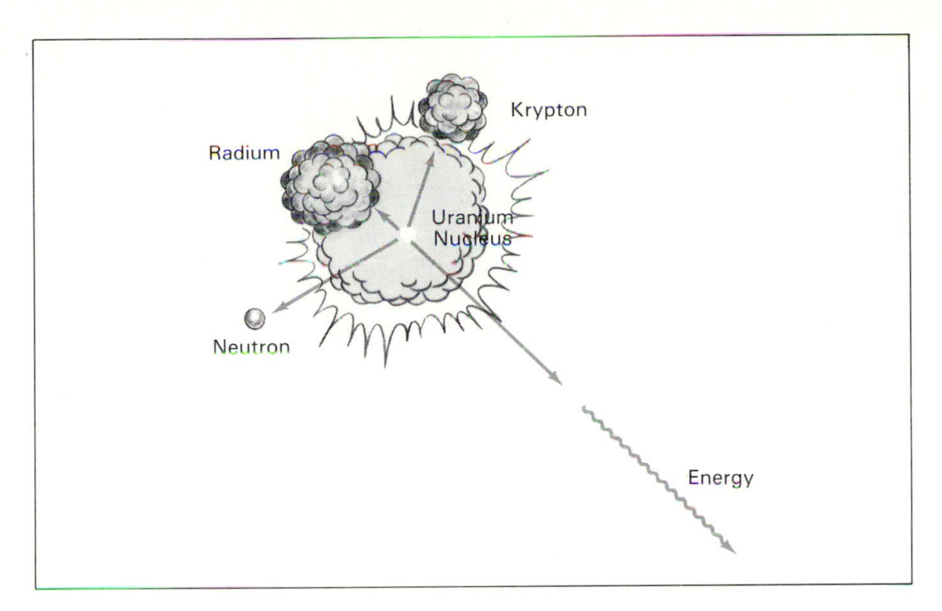

FIGURE 10.1 The natural decay of heavy radioactive nuclei yields energy in addition to light nuclei and several particle fragments.

have half-lives on the order of a billion years, they tend to release their energy rather slowly—much too sluggishly for natural fission to produce the required amount of solar energy. However, uranium can be *induced* to decay more quickly, thus speeding the release of energy. Neutrons usually provide the most efficient inducement, for when these subatomic particles collide with uranium nuclei, the radioactive decay process bolts into action, thereby releasing the pent-up energy. Furthermore, since each decay produces a neutron, as shown in Figure 10.1, these additional neutrons can interact with other uranium nuclei that in turn produce more neutrons, and so on. The result is a chain reaction and an ever-increasing yield of energy.

Induced fission of uranium and plutonium nuclei was responsible for the atomic-bomb explosions of World War II. The same mechanism is also used to power nuclear reactors around the world. Tests conducted with nuclear bombs and reactors demonstrate that the energy released per gram of uranium is about 10^{17} ergs/gram, impressively close to the 3×10^{17} ergs/gram value needed to account successfully for the observed production of solar energy.

Note, however, that this quantity really amounts to 10^{17} ergs/gram *of uranium*. But the Sun is not made of pure uranium or of any other fissionable material. Spectral-line observations clearly prove that hydrogen is by far the most abundant solar element, followed by several other *light* elements. The heavy, radioactive elements, such as uranium and plutonium, are very rare. In fact, the mass ratio of uranium to all other solar matter is only about 10^{-11}, or a billionth of 1 percent. Hence the extremely small amount of uranium contained in a gram of average solar material means that this induced fission process really produces only 10^6 ergs/gram of solar material.

Nuclear fission therefore falls far short of the needed 3×10^{17} value. Even the accumulated energy emitted by *all* the solar radioactive elements falls far short of our requirement. There's simply not enough radioactive matter in the Sun.

If we are to understand solar energy in terms of known twentieth-century science, only one other energy process could conceivably power the Sun. That process is nuclear **fusion**. It's the opposite of fission. Fission is the breakdown of a heavy nucleus into lighter fragments, whereas fusion is the combination of light nuclei into a heavier one. Both mechanisms change one kind of nucleus into another kind of nucleus, much like the dream of the Dark Age alchemists. And both mechanisms yield energy. The basic difference between them is that large quantities of *rare, heavy* elements are needed for fission, whereas large quantities of *abundant, light* elements are needed for fusion.

Let's examine fusion in some detail, for as we'll see, it indeed releases energy in the proper amount and at the proper rate to power the Sun over the required eons of time.

Fusion

Fusion mechanisms generally operate on the premise that two light nuclei can be fused into a heavier one, namely:

$$\text{nucleus 1} + \text{nucleus 2} \rightarrow \text{new nucleus} + \text{energy}.$$

In this type of reaction, the initial reactants are written to the left of the arrow and the final products to the right. The arrow itself should be read as "yields"; it symbolically represents nuclear events occurring on a subatomic scale. Note that fusion reactions involve *nuclei*, not atoms. And the equation above emphasizes that not only is a new, heavier nucleus produced during fusion, but energy is released as well.

Consider the simplest nucleus, that of the hydrogen atom. The nucleus of hydrogen contains only one proton and no neutrons. Bringing two such protons together, we can write symbolically,

$$\text{proton 1} + \text{proton 2}.$$

Since each proton has a charge, an electromagnetic force exists between them. Normally, this electromagnetic force is attractive, binding together positive and negative particles into atoms, positive and negative atoms into molecules, and positive and negative molecules into the matter of everything around us. But since the two protons considered here are of the *same* (positive) sign, the force is one of repulsion. The closer the two protons come together, the greater the magnitude of the repulsive force. This is a natural barrier against the combination of two like charges, much as you can feel between two north poles or two south poles of a magnet. This repulsiveness is not peculiar to hydrogen nuclei, for all atomic nuclei have a net positive charge and thus a natural desire to stay away from one another.

The basic problem in any fusion process, then, is the need to overcome this electromagnetic force of repulsion. How can it be done? Most easily by violently colliding protons together. In this way, one high-speed proton can momentarily plow into another proton, "squashing" it to within the exceedingly short range of the nuclear binding force. As noted in Interlude 2-2, this is the force that binds elementary particles within nuclei. With the exception of the so-called quark force holding quarks together inside the elementary particles, the nuclear force is the strongest of all forces known in nature. At distances less than about 10^{-13} centimeter, the nuclear force completely overwhelms the electromagnetic force. And since binding forces are always attractive, this nuclear force literally slams the protons together, fusing one with the other.

THE FIRST STEP: What sort of velocity do protons need in order to penetrate into the realm of the nuclear forces? On the planets, particle velocities are normally far less than that required for nuclear fusion; but in the somewhat abnormal environment of nuclear laboratories, we are learning to fashion a few exceptions, as noted in Interlude 10-1. In the Sun, particles have extremely high

INTERLUDE 10-1 ***Controlled Fusion on Earth***

Our civilization has known how to fuse nuclei—just like that occurring in the Sun—for some 50 years. Unfortunately, we have not yet mastered the fusion process, and the result of our efforts is a bomb—the awesome hydrogen bomb. Indeed, *uncontrolled* fusion in the form of thermonuclear weapons threatens the future of life on our planet. But if *controlled* nuclear fusion can be achieved, we would have an unlimited supply of energy capable of serving our civilization indefinitely.

Scientists understand nuclear physics well enough to have successfully tapped the nucleus to produce energy, but they have not yet learned how to harness that energy in a controlled reaction. The problem is that no furnace on Earth can maintain the incredibly high temperatures and densities needed for nuclear fusion. Thus, far, fusion can only be sustained in the heart of a star.

Generating useful energy by nuclear fusion means more than just harnessing the energy of the Sun. To do so, we need to effectively create a small Sun in the laboratory. As difficult as this may seem, researchers are moving steadily toward solving twin goals: *triggering* and *containing* hot plasma.

First, triggering a safe fusion reaction is a little like trying to ignite a wet log. The Sun uses enormous gravitational forces as its triggering device, while hydrogen (fusion) bombs use atomic (fission) bombs as triggers. And in recent years, sophisticated equipment has come close to reaching the physical conditions needed to trigger fusion. We might say that state-of-the-art researchers essentially have the log smoldering, although not yet burning.

No machine has yet produced sustained fusion power, but some have recently met one or another of the key requirements needed to bring it off. For example, experimenters currently induce microscopic hydrogen-bomb explosions by zapping miniature pellets simultaneously on all sides with large amounts of conventional energy. In this way, temperatures have reached 10 million kelvin for almost a microsecond. However, the energy produced in such fusion events has not yet equaled the amount of energy needed to run the equipment. In other words, the process uses more energy than it creates, which is not very cost-effective.

The second major obstacle to controlled fusion concerns the container that must confine and concentrate the plasma once the triggering mechanism is mastered. Walls for such ovens cannot be made from any known material, for none can withstand the 10 million kelvin or more needed for fusion. Modern techniques instead use magnetic force fields as wall-less containers of the charged plasma. In this way, magnetism in the laboratory plays the role of gravity in a star, exerting control over the ionized gas. Such magnetically controlled plasmas resemble laboratory versions of the Van Allen Belts. Successful confinement times, however, are far too short—on the order of milliseconds—to be called sustained.

Despite these recent breakthroughs, no one but a politician would claim that controlled nuclear fusion is imminent. The technological problems are formidable, and it's likely to be some time—perhaps the end of this century—before the current experimental machines can be converted into pilot plants and only eventually to commercial fusion reactors.

Yet the obstacles pale in view of fusion's vast potential. The fuel is plentiful—ocean water; and the products are nonradioactive—helium and energy. When the bright promise is realized, every gallon of ocean water will become the energy equivalent of 300 gallons of gasoline. In fact, all the U.S. energy needs for the next century could be supplied by tapping only 1 millimeter from the water in Boston Harbor. Controlled thermonuclear fusion can become the very foundation for an entire civilization.

velocities simply because the solar core is so hot. They are not heated by fusion, for that would be the cart before the horse. We seek here to determine the particle velocity or the amount of heat needed to *initiate* fusion. The heat generating the large particle velocities must have resulted from the gravitational infall of matter that originally formed the Sun.

Detailed calculations show that a gas temperature of 10 million kelvin is needed to slam the protons together fast enough to initiate fusion. At this temperature or higher, two protons interact, in the process changing into another proton, a neutron, and two additional elementary particles. Figure 10.2 is a schematic diagram of this explosive event, although nuclear physicists prefer to write the following equation to represent it:

$$\text{proton 1} + \text{proton 2} \rightarrow \text{proton 3} + \text{neutron} + \text{positron} + \text{neutrino 1}. \quad \text{(I)}$$

The newly produced neutron and proton immediately merge to form a deuteron particle. This deuteron particle normally comprises the nucleus of a special form of hydrogen, namely **deuterium**, which we also call "heavy hydrogen;" deuterium differs from ordinary hydrogen only by virtue of an extra neutron. (Atoms having more or fewer neutrons than usual are called **isotopes**.)

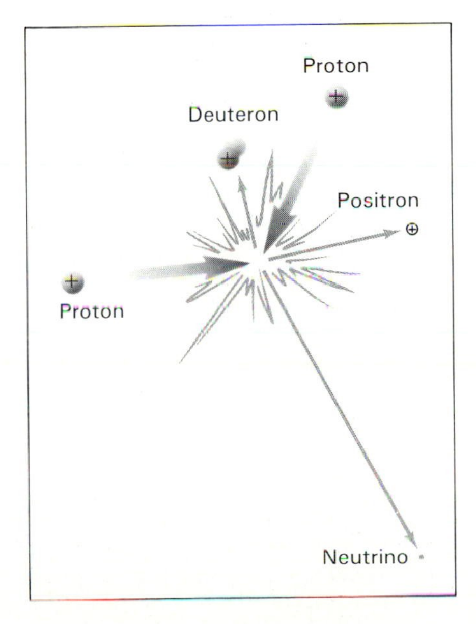

FIGURE 10.2 Schematic diagram of two protons colliding violently, thereby initiating nuclear fusion.

The **positron** particle is another name for a positively charged electron. Its properties are identical to those of a normal negatively charged electron with the single exception that a positron's charge is positive.

The final product of the reaction above, the **neutrino** particle, is one of the most mysterious constituents of nature. Its presence is required by theory in order to preserve some of the basic laws of nuclear physics. And it has been detected for fleeting moments in laboratory experiments. Derived from an Italian word meaning "little neutral one," neutrinos are virtually massless and chargeless, difficult concepts to appreciate sometimes. What's more, they move at (or nearly at) the velocity of light, and hardly interact with anything. Like ghosts, neutrinos are theoretically estimated to be able to penetrate freely, and without stopping, several light-years of lead!

The nuclear equation above is labeled (I) since the production of a deuteron by the fusion of protons is the first step in the fusion process inside most stars. Since two protons are needed to start the process, and since we'll see below that the process is cyclical, we call this type of nuclear fusion the **proton–proton cycle**. In stars, this reaction is not an isolated event occurring among a few protons here and there; gargantuan quantities of protons are fused within the core of the Sun each and every second.

Let's further examine the nature of the products. As mentioned, the neutrino interacts with virtually nothing. Created in proton fusion reactions deep in the solar core, neutrinos travel outward without being absorbed, gliding uninhibitedly through the convection zone, the chromosphere, the corona, and out into space. (Technically, only about 1 in every billion neutrinos is absorbed by the Sun.)

In contrast, the newly created positrons find themselves surrounded by a sea of electrons with which they interact almost immediately. Since the electron and positron are similar in every way except charge, scientists call them "antimatter pairs"; the positron is the antimatter counterpart of the electron, and conversely. **Antimatter** is simply a form of matter having an opposite charge than is normally the case.

Upon interaction, antimatter pairs mutually and violently annihilate each other, producing pure energy. There is no material debris from the wreckage—just energy. For an electron–positron annihilation, the resultant energy is of the very high frequency type, usually manifest in the form of gamma-ray photons. These are the photons that travel from the site of the nuclear reactions outward through the radiation zone of the Sun's interior, hesitating for nearly a million years mostly in the convection zone just below the photosphere, as explained in Chapter 9.

Fortunately for us, none of the solar radiation reaching Earth originates directly in the fierce nuclear reactions at the Sun's core. This absolutely lethal x-ray and gamma-ray radiation is "filtered out" by the solar convection zone, the top portion of which radiates the much lower-energy (visible and infrared) photons detectable at Earth.

Energy is produced in fusion reactions primarily because of the violent collisions among nuclei. It's important to note, however, that fusion itself is not really caused by the initial high velocity (or kinetic energy) of the protons. Sure, these high proton velocities, characteristic of a 10-million-kelvin gas, get the protons to the starting gate at the edge of the realm of the nuclear forces. But once there, it's the nuclear force that smashes the protons together so ferociously as to transform matter into energy—nuclear energy—which is eventually released as heat (from colliding products) and light (from electron–positron annihilation).

Gram for gram, proton fusion yields about as much energy as does uranium fission. But the beauty of fusion is that all stars have enormously larger quantities of hydrogen than any of the heavier radioactive elements.

THE SECOND STEP: The next step in the solar fusion process is the formation of an isotope of helium. It is synthesized in another nuclear process, symbolized by the equation

$$\text{deuteron} + \text{proton} \rightarrow {}^3\text{helium} + \text{energy}. \quad \text{(II)}$$

Here, a proton, of which there are copious amounts in the solar core where all atoms are stripped of their electrons, interacts with the deuteron particle produced in step (I). Actually, step (II) begins as soon as there are enough deuterons to sustain this new reaction. The main product is the rare isotope of helium—rare because it lacks one of the neutrons contained in a normal (two proton–two neutron) helium nucleus. Energy is also emitted in the form of gamma-ray photons.

The fusion event represented by equation (II) is not just a theory but a tried and true recipe. Scientists have verified this reaction numerous times, both during experiments in nuclear laboratories and during tests of thermonuclear warheads. Bombardment of heavy water (containing deuterium) by high-velocity protons really does form ^{3}helium nuclei.

THE THIRD STEP: The third and final step of the proton–proton cycle, also verified by direct experiments, involves the production of normal ^{4}helium nuclei. This step, however, can be much more complicated than the preceding step. The complexity arises because of the many different ways that heavy nuclei can interact. The most straightforward path toward the production of ^{4}helium is the fusion of two ^{3}helium nuclei created earlier in step (II):

$$^3\text{helium} + {}^3\text{helium} \rightarrow {}^4\text{helium} + \text{proton} + \text{proton} + \text{energy}. \quad \text{(IIIa)}$$

This equation represents the synthesis of a normal ^{4}helium nucleus plus two more protons. There are no neutrinos produced in this, the first of several possible third steps.

An optional path for step (III) is more complex, although its principal end product is also a normal ^{4}helium nucleus. This alternative process, which we call step (IIIb), begins by fusing ^{3}helium nuclei produced in step (II) with ^{4}helium nuclei either produced in step (IIIa) or already present from events that occurred in the earliest moments of the Universe:

$$^3\text{helium} + {}^4\text{helium} \rightarrow {}^7\text{beryllium} + \text{energy}.$$

The net products are the ^{7}beryllium nucleus plus an additional gamma-ray photon. The beryllium nucleus immediately interacts in turn with a free proton to produce the ^{8}boron nucleus:

$$^7\text{beryllium} + \text{proton} \rightarrow {}^8\text{boron} + \text{energy}.$$

The ^{8}boron nucleus is radioactively unstable, much like uranium. But unlike uranium, which takes about a billion years for half of it to decay, half of the boron nuclei spontaneously disintegrate every half second. The disintegration products are a pair of identical ^{4}helium nuclei, a positron, and a neutrino:

$$^8\text{boron} \rightarrow {}^4\text{helium} + {}^4\text{helium} + \text{positron} + \text{neutrino 2}. \quad \text{(IIIb)}$$

This neutrino (labeled number 2) has a lot more energy than the one (labeled number 1) emitted via step (I).

Which path will the fusion process follow—the more straightforward step (IIIa) or the more complex step (IIIb)? The answer depends on the temperature of the gas. If the temperature at the solar core is relatively "cool," say about 13 million kelvin, then path (IIIa) is preferred, whereas if the temperature is "warmer," say about 15 million kelvin, then path (IIIb) is preferred. You can understand the difference by noting that a higher temperature is needed to initiate the fusion of ^{3}helium with ^{4}helium than the fusion of two ^{3}helium nuclei.

Before examining the consequences of solar fusion, let's pause for a moment to take inventory. The *net* effect of the previous three steps, (I) through (III), is this: The combination of four hydrogen nuclei (protons) creates one ^{4}helium nucleus plus some radiation and either one or two types of neutrinos, depending on the path taken in step (III). Symbolically, we have

$$\text{four protons} \rightarrow {}^4\text{helium} + \text{energy} + \begin{cases} \text{two neutrinos 1} \\ \quad or \\ \text{two neutrinos, 1 and 2.} \end{cases}$$

The gamma-ray photons are slowly degraded while passing through the solar interior, while the neutrinos escape unhindered into space at the velocity of light. Helium, the heaviest end product, stays put in the core.

Energy Generated

How can we be certain that nuclear fusion yields enough energy to power the Sun, and by implication, other stars? Let's calculate the energy produced in the fusion process and compare it to our requirement of 3×10^{17} ergs/gram.

The first thing to note is that the total mass of the initial reactants is *always* larger than the total mass of the end products. That is, the mass sum of the particles on the left side of an equation for an energy-producing reaction always exceeds the mass sum of the right side. This statement is true for either fission or fusion reactions. What's more, it's true for any energy-producing reaction, whether it be the nuclear transformation of elements or the chemical burning of wood, coal, or oil. In any reaction, this mass difference is the key to the generation of energy, either controlled as in your automobile engine and the natural Sun, or uncontrolled as in a burning wood log and an exploding atomic bomb. Actually, the difference in mass is more than the key; it *is* the energy released.

Einstein proved at the beginning of the twentieth century that matter and energy can be interchanged. He generalized an ancient law of physics, known as the law of conservation of mass, wherein all matter was considered indestructible, into the law of **conservation of mass and**

energy. Here the *sum* of the mass and energy must always remain constant. In other words, the sum of the two can be neither increased nor decreased during any physical transaction. Physicists know of no exception to this law; as far as we can tell, it has never been violated.

Matter and energy, then, are interchangeable. They resemble different denominations of the same currency, much as do paper money and coins in our economy. The extent to which they can interchange is governed by the famous equation $E = mc^2$. This well-known symbolism is a shorthand way of writing the following equation:

$$\text{energy} = \text{mass} \times (\text{velocity of light})^2.$$

According to this formula, an object of mass m can literally disappear, provided that some energy (an amount equal to mc^2) appears in its place. For example, the origin of the energy in the above-mentioned matter–antimatter annihilation is the mass that vanished. Similarly, if magicians really made rabbits disappear, the magic profession would be on the list of endangered species; the result would be a puff of energy equaling the product of the rabbit's mass and the square of light speed—enough energy to destroy both the magician and everyone in the audience.

Similarly, material objects can, in principle, appear where none existed before. For example, many of the elementary particles that emerge as products in steps (I) through (III) of the solar fusion process are transformed energy. They are literally created from energy.

We can now calculate the energy created during step (I) of the proton–proton cycle. In the nuclear reaction,

$$\text{proton} + \text{proton} \rightarrow \text{deuteron} + \text{positron} + \text{neutrino 1},$$

the total mass of the two proton reactants is 3.34486×10^{-24} gram. The total mass of the deuteron and positron products is only 3.34412×10^{-24} gram. The neutrino, you'll remember, is virtually massless. Consequently, some mass has been lost—not much mass, but a measurable amount. The difference, 0.00074×10^{-24} gram, is transformed into energy. Multiplying this vanished mass by the velocity of light squared (to find mc^2) yields 7×10^{-7} erg. It takes more energy than that to bat an eyelid, but remember, this energy is released by the collision of only two very small protons.

This newly produced energy can become enormously larger in the case of stars that house astronomical numbers of protons. For example, suppose that all the protons contained within *one gram* of hydrogen are fused together, via steps (I), (II), and (III), in order to form ^{4}helium nuclei. Sensitive equipment would show that only 0.993 gram of helium results. The remaining 0.007 gram of matter is converted into energy, whose amount we can calculate by multiplying by the velocity of light squared. The answer, $(0.007 \text{ gram}) \times (3 \times 10^{10} \text{ centimeters/second})^2$, is 6×10^{18} ergs.

Now, since we initially considered only 1 gram of hydrogen, we note that the proton–proton cycle must generate 6×10^{18} ergs/gram. This is more than enough to meet the solar requirement of 3×10^{17} ergs/gram derived earlier.

Finally, be sure to recognize that the solar fusion process uses the most abundant of all elements, hydrogen. Neither uranium nor any other exotic atom or chemical is necessary. Only the *simplest* of all elements is required.

Helium Production

Knowing the rates at which the fusion reactions churn away, we can estimate the amount of hydrogen that is being transformed into helium. The proton–proton cycle does not operate everywhere in the Sun. As noted earlier, it depends critically on the temperature, and therefore occurs exclusively at the core where the plasma is at least 10 million kelvin.

Figure 10.3 illustrates the increase of helium and the corresponding decrease of hydrogen in the solar core. Three cases are drawn—the chemical composition of the original core of the Sun (a), the composition now (b), and the estimated composition some 5 billion years in the future (c). As shown, the helium content continuously increases, spreading out from the core as the Sun continues to shine. With time, the core's abundance of hydrogen will become depleted, and the location of principal burning will be forced to move to higher layers in the solar interior. In this way the fusion process operates throughout progressively larger solar volumes, converting hydrogen into helium more and more rapidly. The overall consumption of hydrogen accelerates partly because the nuclear fire spreads through more voluminous territory, and partly because the solar core gets hotter, as explained below.

Some 5 billion years in the future, hydrogen will have become depleted enough—at least at the core—to cause a serious problem for our Sun. After all, hydrogen is its primary fuel, and without it the nuclear fires will suffocate. Events thereafter are destined to occur at an increasingly rapid pace. The core will shrink under the relentless pull of gravity. Outer portions of the solar interior will transform hydrogen into helium at a faster and faster pace. With the foundation sinking and the roof burning, the Sun will become unstable while eroding its equilibrium between gravity pulling in and heat pressing out. After 10 billion years of steady burning, the Sun's emission will become erratic shortly before death.

Actually, the Sun will make one final attempt to repair its foundation. As the core shrinks, it inevitably compresses the helium ash to higher densities, thereby heating it to even greater temperatures by means of increased collisions. When the temperature reaches 100 million kelvin, 10 times that needed to fuse hydrogen into helium, the helium nuclei will begin fusing into even heavier nuclei. For example, three helium nuclei can effectively fuse under intense heat to yield a carbon nucleus plus more energy:

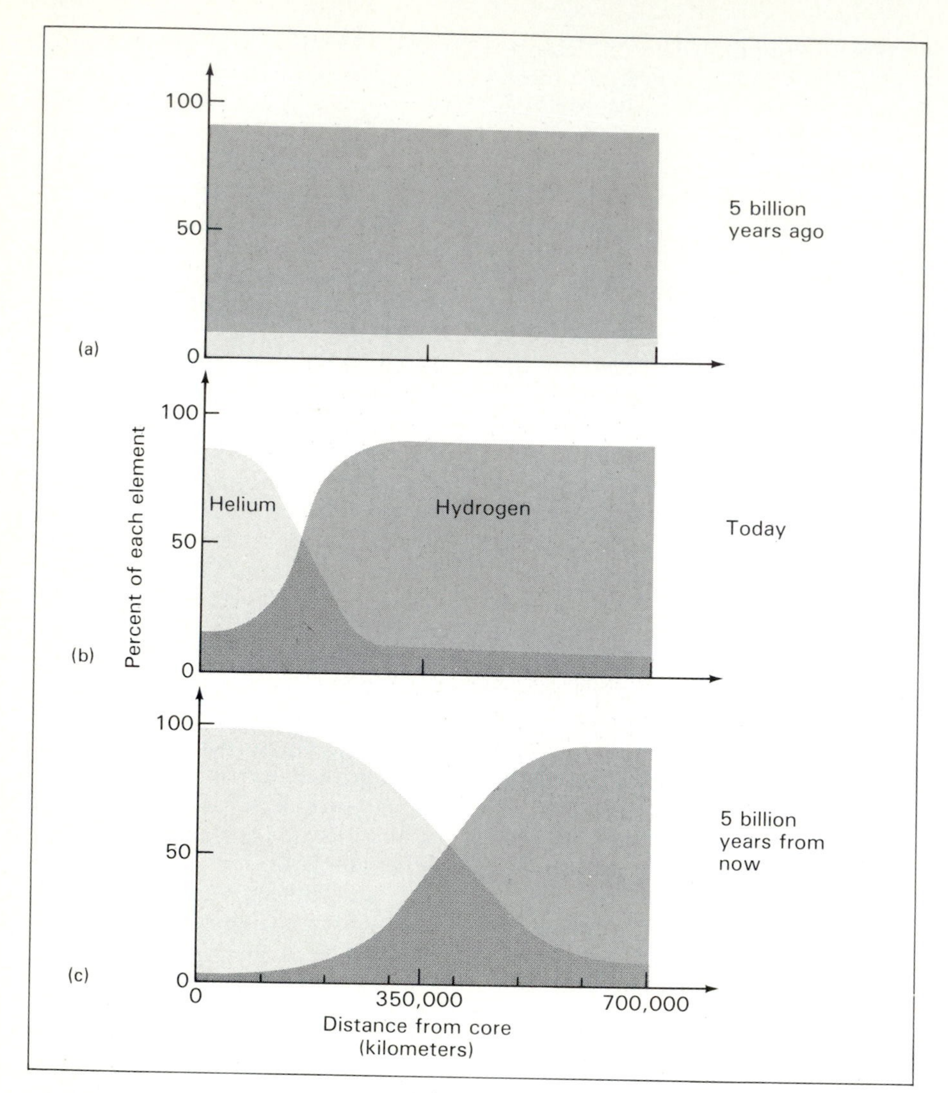

FIGURE 10.3 Theoretical estimates of the Sun's hydrogen (intermediate shading) and helium (light shading) composition (a) 5 billion years ago, (b) at present, and (c) 5 billion years in the future. Thus far, only about five percent of the Sun's total mass has been converted from hydrogen into helium. This change will speed up in the future as the nuclear burning core grows larger.

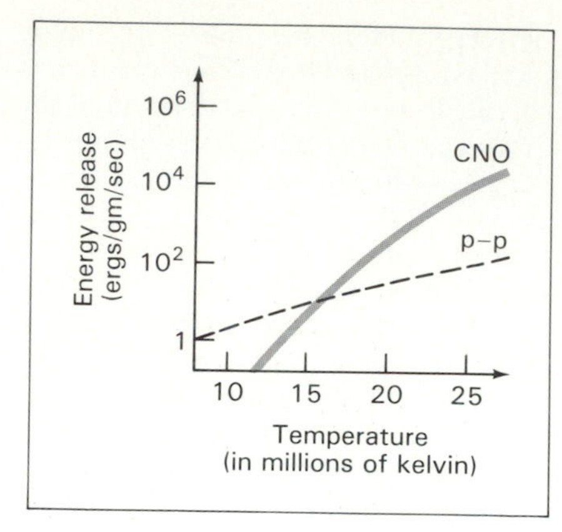

FIGURE 10.4 A comparison of the proton-proton and CNO nuclear cycles. Energy liberated for the former is greater at "low" temperatures, but the CNO process releases more energy at temperatures above some 16 million kelvin.

$$^{4}\text{helium} + {}^{4}\text{helium} + {}^{4}\text{helium} \rightarrow {}^{12}\text{carbon} + \text{energy}.$$

With the nuclear fires rekindled in the core, the Sun will once again stabilize its gravity-pressure equilibrium. But renewed burning will not last long, perhaps only a few hundred million years. It's not easy to stoke a 100-million-kelvin fire—even in the Sun.

Even heavier elements can be created in turn as some larger stars undergo successive cycles of gravitational shrinkage and core heating in their few remaining years before retiring to the graveyard of dead stars. This is essentially the way all the elements heavier than hydrogen and helium have originated in the Universe. It's a story that we shall relate in greater detail in Chapters 20 and 22.

Neutrino Experiment

No radiation can get cleanly out of the Sun's interior. Even the powerful gamma-ray photons fashioned in the solar core are absorbed and reemitted numerous times, bouncing their way mostly through the convection zone for about a million years before finally escaping the solar surface as "downgraded" visible or infrared photons. In the process of bouncing around, the photons heat the gas of the solar interior.

The neutrinos, which arise as byproducts of the proton–proton cycle, do travel cleanly out of the Sun because of their peculiar property of interacting with virtually nothing. They just zip out at the velocity of light, escaping into space a few seconds after being created at the core. The neutrinos do not heat the gas at all. On the contrary, they serve to cool it by swiftly carrying away energy.

Because the gamma rays are trapped within the Sun, and because the neutrinos are so hard to capture, astronomers have had trouble proving that the nuclear reactions above really do occur in the core of the Sun. Yet without experimental tests, any hypothesis will remain a theory, and only a theory.

Theoreticians are quite sure that the proton–proton cycle is now operating in our Sun. They have performed excruciatingly detailed "numerical experiments" on large computers and regard their results as reasonably accurate. The resulting models of the Sun's interior tem-

INTERLUDE 10-2 ***The CNO Cycle***

No known stars are made solely of hydrogen and helium. All (except the most primordial stars, not yet found) form with heavier elements already present within them. Our Sun is one of these later-generation stars, so-called because the heavier elements such as oxygen, nitrogen, carbon, and iron—spectroscopically observed in the Sun—could not have been made there. Instead, these "heavies" were cooked in the cores of massive stars that lived long ago; after fusing numerous heavy elements in their cores, these stars then died catastrophically by exploding their heavies into the surrounding interstellar space. We'll study these explosive mechanisms in some detail in Chapters 20 and 22. The important point here is that stars are born with small amounts of heavies already sprinkled throughout their interiors.

Under these circumstances, the proton–proton cycle may not be the only nuclear process operating in the Sun and other later-generation stars. Another fusion mechanism capable of converting ^{1}hydrogen into ^{4}helium uses the ^{12}carbon nucleus as a catalyst. It proceeds according to the following six steps:

I. ^{12}carbon + proton → ^{13}nitrogen + energy,
II. ^{13}nitrogen → ^{13}carbon + positron + neutrino,
III. ^{13}carbon + proton → ^{14}nitrogen + energy,
IV. ^{14}nitrogen + proton → ^{15}oxygen + energy,
V. ^{15}oxygen → ^{15}nitrogen + positron + neutrino,
VI. ^{15}nitrogen + proton → ^{12}carbon + ^{4}helium.

Aside from the radiation and neutrinos, note that the sum total of these six reactions is

^{12}carbon + 4 protons → ^{12}carbon + ^{4}helium.

Since the originally used ^{12}carbon reappears at the end of this cycle, the *net* result is the fusion of four protons into a single ^{4}helium nucleus. ^{12}Carbon acts merely as a catalyst, that is, a stimulant of change.

This alternative process yields the same result as the proton–proton cycle studied elsewhere in this chapter. To distinguish the two processes, the six steps above are termed the **CNO cycle**, since carbon (C), nitrogen (N), and oxygen (O) are the prime intermediate products. Both the proton–proton and CNO mechanisms are called cycles because they do just that: They each produce not only helium and radiation, but also deposit an equal amount of initial reactant (either hydrogen or carbon nuclei).

The repulsive electromagnetic forces operating in the CNO cycle are greater because the charges of the heavy-element nuclei are larger. Accordingly, higher temperatures are required to coerce the heavy nuclei into the realm of the nuclear force in order to ignite fusion.

Figure 10.4 gives a numerical estimate of the energy released by each cycle plotted against the gas temperature. The proton–proton cycle dominates at lower temperatures, up to about 16 million kelvin. Above this temperature, the CNO cycle becomes the more important fusion process.

Provided that the theoretical modeling of the Sun is correct and the core temperature is really 15 million kelvin, these curves predict that the proton–proton cycle is the dominant source of solar energy. The CNO cycle contributes no more than 10 percent of the observed solar radiation.

Stars more massive than our Sun often have core temperatures much higher than 20 million kelvin, making the CNO cycle the dominant energy emission mechanism.

perature, density, composition, and nuclear burning rates predict bulk properties that agree well with the observations. The observations, however, are almost exclusively confined to the solar exterior—the photosphere, chromosphere, and corona.

We say "almost" because one experiment does seem possibly suited to directly test the inner workings of the solar core. Unfortunately, the results of this experiment do not match the theoretical predictions. Let's examine this problematic experiment further.

As stated earlier, neutrinos interact with *virtually* nothing. There are a few exceptions, however, and since neutrinos are the only particles able to make a clean getaway from the solar core, scientists have placed a heavy importance on those few exceptions. (Actually, the fact that neutrinos avoid interaction with most matter works to our advantage; they are not likely to change their properties by interacting with anything between the Sun and the Earth, thereby giving us a good idea of the neutrinos' properties deep in the solar core.)

One of the elements with which some neutrinos do interact is chlorine (Cl). A few decades ago, researchers built a large tank near the bottom of a gold mine near Lead, South Dakota, and filled it with 400,000 liters (100,000 gallons) of a chlorine-containing chemical with the tongue-twisting name of tetrachloroethylene (C_2Cl_4). There's nothing special about this chemical—in fact, it's a common cleaning fluid used by dry cleaners—but it's cheap and easy to handle in large quantities. At 1.5 kilometers below ground level, the experimenters can be reasonably sure to avoid detecting neutrinos and other interfering elementary particles arising from nonsolar events. So here is a new type of astronomical instrument not discussed in Chapter 3—a **neutrino telescope**, the detector for which is a vast tank of liquid more than a kilometer (nearly a mile) down a mine shaft! Figure 10.5 shows a photograph of it.

Neutrino 1 that we spoke of earlier is predicted to have an energy too low to interact successfully with C_2Cl_4 or any other known material. But neutrino 2 is more energetic, should interact with C_2Cl_4, and therefore should be detectable. Given the size of the "telescope" and the expected physical conditions at the Sun's core, theory predicts that about one solar neutrino should be detected per day.

Neutrino astronomy is similar to high-energy astronomy—individual particles are counted one by one. The detection of a solar neutrino is cause for a minor celebration. However, for the first few years the experiment ran (in the late 1960s), no celebrations were held at all—no neutrinos were found. Some researchers took this null result to mean that the neutrino telescope was working improperly. But more sensitive apparatus installed in the 1970s has succeeded in detecting *some* neutrinos, although not as frequently as theoretically predicted. Neutrinos are currently detected about twice per week, rather than once per day. So the equipment seems to be working, but the total number of neutrinos is less than expected. Apparently, the only way we have of peering directly into the core of the Sun presents a major problem.

FIGURE 10.5 This swimming-pool-sized detector is a "neutrino telescope" buried underground in a South Dakota gold mine. (Note the man on the catwalk at left.)

CURRENT IMPASSE: Where are the neutrinos? Is the proton–proton cycle operating as theoretically advertised and as outlined earlier in this chapter? Do we *really* know what processes are at work deep in the hearts of stars? These are questions that solar astronomers are now wrestling with, as they attempt to resolve what looms as a serious threat to our knowledge of stellar fusion. What was once regarded as an almost fully understood phenomenon—the proton–proton production of solar energy—is once again under intensive study by theoretical and experimental astronomers alike.

Some theorists continue to question the accuracy of the experiment. The detection of solar neutrinos is an incredibly exacting task. Perhaps the correct number of neutrinos escapes the Sun, but somehow the equipment fails to detect them properly. Maybe the same penetrability of matter that allows us to use the neutrino as a diagnostic probe of the Sun's core is also frustrating our attempts to capture it. Even so, the recent detection of *some* neutrinos implies that the apparatus is capable of capturing these elusive particles. Furthermore, scores of technicians have examined every facet of the neutrino telescope, trying to find any instrumental flaws. They have not succeeded; the experimental gear seems well designed and well built.

On the other hand, many experimentalists (and some theorists as well) question the numerical models of the solar core. Recall that the predicted number of neutrinos relies on the assumption that the solar core temperature is 15 million kelvin. This temperature ensures that ^{4}helium is fused via step (IIIb) as well as step (IIIa). Step (IIIa) is the more probable reaction, but step (IIIb) should also occur to some extent, thus producing detectable number 2 neutrinos. However, if the Sun's core temperature is cooler by 10 percent, namely about 13.5 million kelvin, reaction (IIIa) completely dominates, permitting (IIIb) to proceed very seldom. As a result, ^{4}helium is still produced, but it's accompanied by fewer number 2 neutrinos.

Decreasing the temperature value used to model the solar core, then, is one way to justify the puzzling results of the neutrino experiment. However, the theoreticians refuse to concur; they assert that their detailed numerical models cannot possibly be in error by as much as 1.5 million kelvin.

Some scientists are trying to squirm out of the problem by proposing a compromise. They argue that while the numerical solar models are generally valid, the 15 million kelvin at the core represents an *average* gas temperature—that is, the core temperature averaged over the lifetime of the Sun. For example, perhaps the Sun pulsates slightly, changing its size and therefore its luminosity in a regular way. If so, the solar core might experience alternating periods of higher and lower temperatures. When the solar core temperature is, say, 13 million kelvin, ^{4}helium production via step (IIIa) completely dominates, whereas a temperature as high as 17 million kelvin would enhance the production of neutrinos via step (IIIb). Overall, the core temperature would average out to 15 million kelvin, in accord with the current solar models.

This idea presumes that our Sun is currently experiencing a low core temperature. No pulsations have yet been observed to match this idea. But, interestingly enough, since the periods of solar pulsation are expected to be on the order of tens to hundreds of millions of years, perhaps such cyclical heating and cooling of the Sun could have a concomitant effect on our terrestrial climate—and thereby partly contribute to some of Earth's ice ages (see Interlude 9-2 and Chapter 27).

The most currently promising idea suggests that the neutrino particles have more mass than recently thought. Although neutrinos are usually assumed to be massless, like photons, theorists have recently proposed that neutrinos perhaps have a minute amount of mass. How much? About 10,000 times less mass than electrons, which are themselves some 2000 times less massive than protons. So the amount of mass we're talking about is quite small—in fact, far smaller than for any other elementary particle described here. Nonetheless, even this small mass might enable neutrinos to change their properties, even to decay, during their 8-minute flight time from Sun to Earth. The problem is that experimentalists, using laboratory accelerators such as the one shown in Figure 2.2, have not yet been able to confirm the theoretical prediction that neutrinos have appreciable mass.

The stalemate between theorists and experimentalists continues today. Some suspect the experimental apparatus, others the numerical models. Still others argue that we don't yet know enough about the mysterious neutrino particle itself. Whichever the correct solution, the answer is unlikely to be an easy one. Chances are good that some completely new physical or astronomical insight is needed.

Even so, virtually all researchers concur—or at least hope—that the correct interpretation of the neutrino experiment will probably not tear apart the fabric of the proton–proton cycle. Almost everyone clings to the notion that the solar fusion process as described in this chapter is basically all right. Our understanding just needs to be fine-tuned.

In the extreme, should drastic measures be needed eventually to solve the solar neutrino puzzle, we may well have to return to the drawing board for one of the most fundamental scientific questions of all: How does a star basically tick?

SUMMARY

Although somewhat hard to imagine, our Sun is a huge nuclear reactor, churning forth daily vast quantities of radiation. Much like a controlled bomb, this star as well as all others in the nighttime sky convert matter into energy by means of an intricate series of nuclear reactions. The mechanism of solar energy generation is reasonably well understood, with one exception. A necessary by-product of the nuclear events in the Sun's core—the creation of neutrino subatomic particles—has not been clearly detected in the quantities suggested by theory. This discrepancy suggests a problem with our understanding either of the physics of elementary particles or of the astronomy of our Sun.

KEY TERMS

antimatter
CNO cycle
conservation of mass and energy
deuterium
fission
fusion
isotope
neutrino
neutrino telescope
positron
proton–proton cycle

QUESTIONS

1. Copious quantities of gamma rays would practically destroy Earth's atmosphere, not to mention life on Earth. Why, then, is this not a problem with the gamma rays produced at the Sun's core?

2. Briefly describe what goes in, what comes out, and how the process is able to supply the enormous amount of energy that the Sun has given off during its lifetime.
3. Describe the main steps by which energy, liberated at the core of the Sun, reaches the Earth.
4. Why do astronomers regard the Sun to be about 5 billion years old?
5. Contrast the processes of nuclear fission and nuclear fusion. What are the advantages of each? Why can't the Sun be powered by fission?
6. What do scientists mean by "numerical experiments"? Do you think such experiments are as reliable as an actual observational test? Explain.
7. In a paragraph or two, explain the essence of the solar neutrino problem.
8. Cite some of the physical properties of a neutrino particle.
9. Note, generally, and briefly explain, at least two solutions to the solar neutrino puzzle.
10. Aside from the possibility that Earth-based "neutrino telescopes" might be working incorrectly, how might the astrophysics of the Sun be changed in order to account for the unexpectedly low intensity of the detected solar neutrinos?

FOR FURTHER READING

CONN, R., "The Engineering of Magnetic Fusion Reactors." *Scientific American*, October 1983.

CROXTON, R., MCCRORY, R., AND SOURES, J., "Progress in Laser Fusion." *Scientific American*, August 1986.

WILLIAMS, G., "The Solar Cycle in Precambrian Times." *Scientific American*, August 1986.

11 STARS: *Red Giants and White Dwarfs*

Up to this point, we have studied the Earth, Sun, Moon, and planets. To continue our inventory of the material contents of the Universe, we must now move away from our local environment, toward the increasingly greater depths of space. In this chapter we take a large leap in distance and consider stars in general.

On a clear night, nearly 3000 points of light greet our eyes. Including the view from the other side of the world, nearly 6000 stars are then visible to the naked eye. The natural human tendency is to link the bright stars into geometrical patterns called **constellations**, which ancient astronomers named after gods, heroes, animals, and mythological beings. Figure 11.1 shows an example of a constellation especially prominent in the nighttime sky from October through March; this one depicts the outlines of a "hunter" in the constellation named Orion.

The origins of many constellations, and of their names, date back to the dawn of recorded history. In all, there are 88 constellations, most of which are viewable from the United States at some time during the year; many are identifiable without the aid of a telescope. In some sense, constellations are analogous to land continents or voting precincts that enable geologists or politicians to identify certain localities on planet Earth. Likewise, constellations help astronomers to specify bulk regions of the sky.

Be sure to note that many of the stars forming each constellation are not at the same distance from Earth. In other words, all the stars comprising any one geometrical pattern do not reside in the same plane of the sky, equidistant from Earth. Some stars of a particular constellation are relatively local, while others are quite distant. The apparent arrangements of stars belie the true distances of the stars comprising those constellations.

This text deemphasizes constellations, star names, star positions, and other peculiarities of practical astronomy. If this were a text on classical astronomy, celestial navigation, or even Greek mythology, constellations would form an integral part of our further study. But it's not. It's a text on the origin and evolution of things.

Here we strive for a deeper understanding of all material things. Our foremost interest concerns the *nature* of the stars comprising the constellations, as well as the myriads of more distant stars that we cannot perceive with our naked eyes. Rather than devoting time to a study of the peculiarities of individual stars, we'll concentrate on the physical and chemical properties of stars in general. If we are to succeed in describing an all-encompassing scenario of cosmic evolution, we must sometimes sacrifice specifics in favor of the bigger picture.

In attempting to understand this larger scheme of things, we'll find that distance, which we've already discussed to some extent in Chapter 4, is not the only useful property of stars. A wide variety of additional properties also characterize stars. In particular, we'll want to know something about the sizes of stars, their energetics, their temperatures, masses, and chemical compositions.

We ultimately want to know how

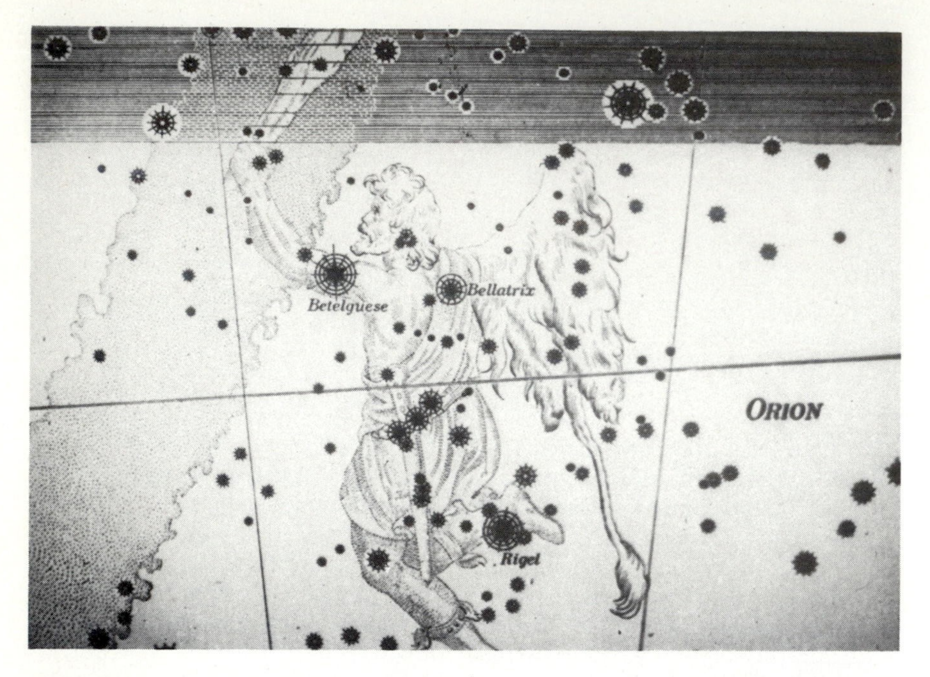

FIGURE 11.1 The geometrical appearance of stars gives rise to constellations like this one called Orion. Here, the brightest stars in this region of the sky are supposed to form the outlines of a "hunter," which is drawn fancifully for the benefit of the unimaginative. (An actual photograph of most of this same region is reproduced as Figure 19.13(a). You can orient yourself by identifying the line of 3 bright stars in the hunter's "belt.")

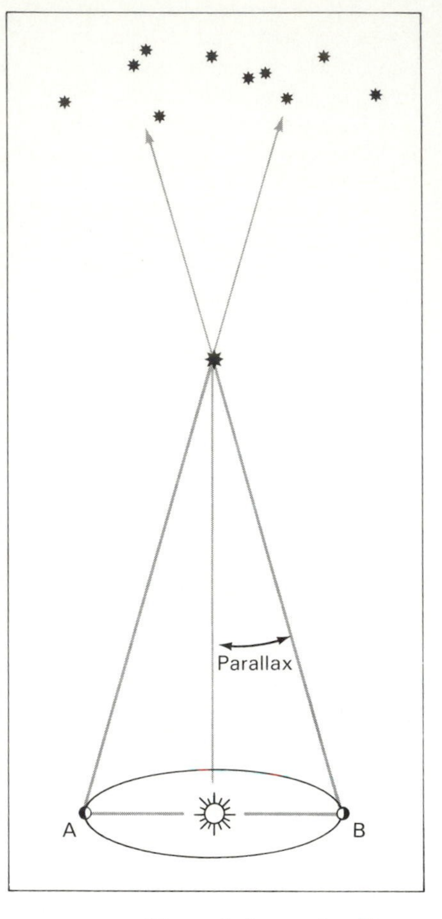

FIGURE 11.2 The technique of stellar parallax is used to measure the distances to some of the nearby stars.

other stars compare to our Sun. We especially need to know if our Sun is in any way special. Expressed more grandly, are we humans, the product of sunlight and basic molecules, in any way unique?

If we are to truly appreciate the cosmic evolutionary theme, we must seek to discover *any* property possessed by our Sun but not by other stars. Is there any stellar property that distinguishes our Sun from the billions upon billions of other stars scattered throughout our Galaxy and other galaxies beyond? If we are to evaluate the prospects for life beyond Earth by applying cosmic evolution to every nook and cranny of the Universe, we need to understand precisely how the basic properties of our Sun compare to those of all other stars.

The learning goals for this chapter are:

- to recognize that stars move through space
- to realize that the size of stars, with few exceptions, must be estimated indirectly from the laws of physics
- to understand that astronomers use two different types of brightness
- to know that stars are often categorized according to their surface temperatures
- to appreciate the well-defined correlation between luminosity and surface temperature for most stars
- to realize that our Sun is rather undistinguished when compared to all other known stars

Motion

APPARENT MOVEMENT: Figure 11.2 reminds us of the technique used to measure the distances to some local stars. The method of stellar parallax, described in Chapter 4, uses a baseline of twice the astronomical unit to observe the apparent shift of a star at 6-month intervals. The parallactic angle can then be measured for relatively nearby stars, provided that those stars show some displacement when projected onto a background of much more distant stars that show no such displacement. Recall that this method works only for stars having parallaxes greater than 0.03 arc second. The inverse relationship between parallax and distance then provides us with reliable estimates of distances to only about 30 parsecs, or equivalently about 100 light-years from Earth. However, most stars—even the vast majority of those seen with our naked eyes—reside far, far beyond this distance.

At various points throughout this chapter as well as in the next five chapters, we shall occasionally discuss many other techniques used to extend the distance scale beyond 100 light-years. To be sure, most of the Universe lies well beyond this rather "local" realm. But the next step in the distance scale requires a detailed understanding of stars. So let's first consider some of the basic properties of stars and then return to our discussion of distance.

If the stars were completely motionless in space, the parallactic shift of the nearby stars would be the only stellar movement seen in the sky. The stars would tend to wobble back and forth as Earth orbits the Sun each year. For example, the Alpha-Centauri star system closest to Earth would appear to wobble to and fro each year over an interval of about 1.5 arc seconds (i.e., double the 0.76-arc-second parallactic angle; see Figures 4.10 and 4.11). The wobble would be less pronounced for more distant stars: in fact, only 0.06 arc second for stars some 100 light-years from Earth. But the important point to recognize is that this apparent wobble is not the real motion of a star. Rather, the wobble is induced by the motion of Earth in its annual orbit about the Sun. It's only an apparent stellar motion.

GENUINE MOVEMENT: In addition to their apparent motions caused by parallax, stars have some real motion. In other words, stars really do travel through space. The annual movement of a star on the plane of the sky is called **proper motion**, and it's measured in terms of the angular displacement, much like parallax. As an example, Figure 11.3 compares two photographs. The photographs display identical parts of the sky, but each was taken at a different time. Specifically, each plate was made on the same date of the year, though separated by an interval of 22 years. As can be seen, Barnard's Star has genuinely moved during the 22-year interval, for, if the two photographs were superposed, this star's two images would not be in the same place. And since Earth was at the same place in its orbit when each photograph was taken, its displacement could not possibly be due to parallax caused by Earth's motion around the Sun. Thus we conclude that the observed displacement indicates real space motion.

Careful measurement of the two photographs shows that Barnard's Star has moved 227 arc seconds in the 22-year interval. Since proper motion is defined as the *annual* angular displacement, the proper motion of Barnard's Star must be 227/22 or 10.3 arc seconds. This is the largest known proper motion of any star. Only a few hundred stars have proper motions larger than 1 arc second.

FIGURE 11.3 Comparison of two photographic plates, taken 22 years apart, shows evidence of some real space motion for Barnard's star (denoted by an arrow).

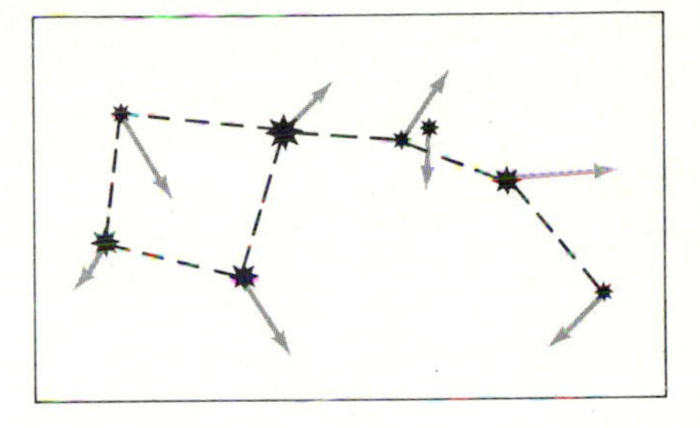

FIGURE 11.4 Schematic diagram showing the magnitudes and directions of the proper motions of the stars in the Big Dipper constellation.

Figure 11.4 illustrates the measured motions of some of the stars of of the familiar Big Dipper constellation. The orientation of each arrow shows the direction of each star's movement, while the length of each arrow is proportional to the magnitude of the proper motion of each star. As the arrows suggest, within another 10,000 years, this constellation will lose its obvious resemblance to a dipper.

It's important to realize that proper motion does not correspond to the total space motion of a star. Proper motion amounts to only one part of the real space motion—namely that component, called transverse motion, which is projected onto the plane of the sky, perpendicular to the line of sight. There is another component of motion, called radial motion, namely motion parallel to the line of sight. These motions can be translated into velocities as in the following example.

Figure 11.5 is an idealized sketch of the Alpha-Centauri star system in relation to our own Solar System. This alien star system is normally observed in a plane of very distant background stars, against which its proper motion has been measured to be about 3.5 arc seconds. Since Alpha-Centauri's distance is known, we can convert this angular displacement into a physical displacement. To do so, we use the geometrical analogies of Chapter 4 to show that at the 4.3-light-year distance of Alpha-Centauri, a proper motion of 3.5 arc seconds equals a physical movement

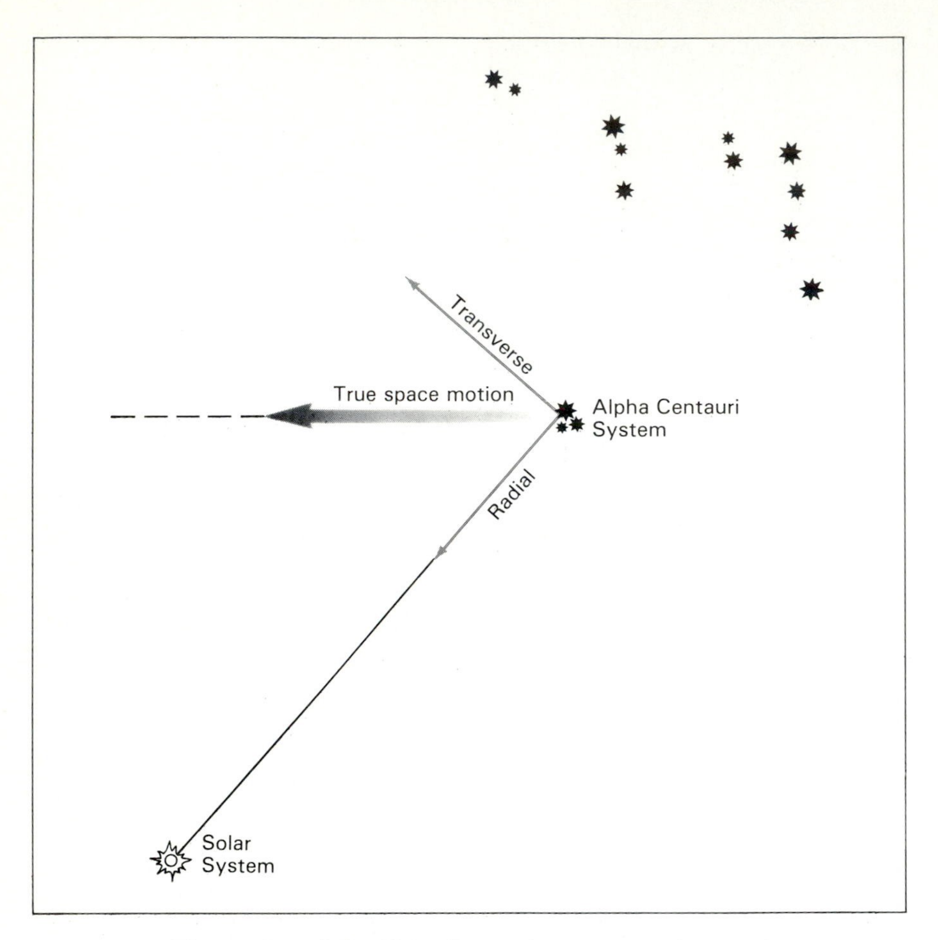

FIGURE 11.5 The motion of the Alpha-Centauri star system drawn relative to our Solar System.

of 0.00007 light-year. Or, since approximately 10 trillion kilometers equals 1 light-year, we can say that Alpha-Centauri has a projected motion on the plane of the sky of 700 million kilometers each year. Knowing, furthermore, that it takes a year to travel this distance, we can divide the distance by the time to find the star's velocity of 22 kilometers/second. Remember, though, this is only one component of Alpha-Centauri's true velocity through space. Velocity in the plane of the sky, that is, perpendicular to the line of sight from Earth as shown in Figure 11.5, is called transverse velocity.

The other component of motion, namely radial velocity, has already been encountered during the discussion of Doppler effect in Chapter 1. As depicted in Figure 11.5, this is the component of velocity along the line of sight from the Solar System to the Alpha-Centauri system. For example, by measuring the extent of the Doppler blue-shifted spectral lines emitted by Alpha-Centauri, astronomers have determined that this nearest star system has a radial velocity component of 20 kilometers/second. This component of motion is in the direction of the Solar System, but will the alien system collide with our own system some time in the future? No, the reason being that Alpha-Centauri also has some transverse velocity that will steer it clear of our Solar System.

We can find the true space motion of Alpha-Centauri by combining the transverse and radial velocities according to the age-old Greek theorem of Pythagoras—namely by taking $\sqrt{22^2 + 20^2}$, the answer to which is about 30 kilometers/second. Thus Alpha-Centauri's current motion will keep it clear of the Solar System for all time, as shown by the horizontal arrow in Figure 11.5. In fact, it won't even come close. Although currently 4.3 light-years away, the Alpha-Centauri system will get no closer to us than 3 light-years, and even that won't happen for another 280 centuries.

Now, we don't want to complicate this discussion more than necessary, but you should be aware of a practical problem. The actual displacement of a star is often more complex than that shown in Figure 11.5. As seen from Earth, proper motion—the real (transverse) space motion of a star through space—is often mixed with parallax, the apparent (to-and-fro) motion of a star caused by Earth's movement around the Sun. If, however, astronomers are careful to measure proper motion only at 1-year intervals (and that's why proper motion is defined as an *annual* angular displacement), Earth will be at the same location in its orbit, and the amount of stellar parallax will be zero.

Whenever studying the motions of stars through space, it's natural to wonder about the general movements of stars in the larger realm well outside our cosmic neighborhood. Do stars merely have chaotic motions with some going randomly this way and others that way, or are they all part of a large, organized traffic pattern? The answer is that when stars are studied individually within our local neighborhood, they indeed seem to move at random. There is no discernible pattern to their movement. But when stars are studied in groups as part of a much larger spatial domain, as will be done when we take a broader view in Chapter 13, the stars seem to partake of a grand design known as the Galaxy.

Size

Motions tell us nothing about the physics and chemistry of stars. To learn more about the legions of ordinary stars in the sky, we must study the intrinsic properties of the gas balls themselves. Let's begin with the size of stars. How big are they? Are they all as large as our Sun?

Dismayingly, we cannot easily determine even something as basic as the size of a star. After all, most of them are unresolvable points of light in the sky, even when viewed through the largest telescopes.

Even so, some stars are big enough and close enough to allow a *direct* measurement of their sizes. Figure 11.6 shows an example of one of them. Known by the peculiar name of Betelgeuse (which is also part of the Orion constellation of Figure 11.1), this star is hundreds of times larger than our Sun. It's a swollen star near death and can be seen in the nighttime sky to have a reddish tint.

Optical astronomers have now directly measured the size of several dozen stars. They usually use an optical interferometer to observe the angle subtended by a star's disk and then, knowing the star's distance, use simple geometry to compute its physical radius.

These exceptions aside, the sizes of the great majority of stars must be derived *indirectly* by applying two basic physical laws to interpret their observed radiation. Collectively, we shall refer to these fundamental relationships as the **radiation laws**. Let's consider them briefly, for they will be useful now and later.

Recall from Chapter 9 that an amount of radiation given in units of ergs/square centimeter/second can be converted into a simple quantity—luminosity—that expresses the Sun's emission in terms of ergs/second. Using these notions, we can approximate the first radiation law by writing the following proportionality:

$$\text{luminosity} \propto \text{energy/area/time} \times \text{radius}^2.$$

Here the term "radius" refers to the actual size of an object; in other words, its *real, physical* size, not any apparent size.

Another basic radiation law combines the amount of radiation emitted by an object and the surface temperature of that object. This law applies to a hot piece of metal, a glowing light bulb, or a star—any object whose emitted radiation obeys the Planck curves of Chapter 1. A careful analysis of any Planck curve yields the specific form of this second radiation law: The amount of energy/area/time is proportional to the fourth power of the object's surface temperature, namely:

$$\text{energy/area/time} \propto \text{temperature}^4.$$

We can now derive a most useful relationship by substituting this proportionality into the previous one. We then find that

$$\text{luminosity} \propto \text{radius}^2 \times \text{temperature}^4,$$

or, by manipulating,

$$\text{radius} \propto \frac{\sqrt{\text{luminosity}}}{\text{temperature}^2}$$

This is a useful relationship because it demonstrates that knowledge of a star's luminosity and temperature can yield an estimate of its radius. Recognize, though, that the radiation laws provide an *indirect* determination of stellar size. In doing so, we must use the relationship above to mathematically *calculate* a star's physical size.

Consider a few examples. A real star with the strange name, σ Ceti (also called Mira), has a surface temperature of about 3000 kelvin, and a luminosity of 1.6×10^{36} ergs/second. Note that this surface temperature is half, and the luminosity about 400 times, that of our Sun. The radius–luminosity–temperature relationship written immediately above would then yield a stellar radius of some 80 times that of our Sun. (Provided that all quantities are expressed in solar units, we can solve relationships like these rapidly: $\sqrt{400/0.5^2} = 80$.) That's a big star; in fact, we call such an object a "giant star." Technically, a **giant star** is any star much larger in size than the Sun. (And since the glowing color of any 3000-kelvin object is red, σ Ceti is actually a red giant star—a rather rare type of cosmic object more appropriately studied in Part III.) If our Sun were as large as this star, its inner atmosphere would literally engulf at least one of the terrestrial planets. In Chapter 20 we shall study how our Sun is now in the process of slowly evolving into just such a giant star, although it will not reach that stage for another 5 billion years.

Consider another example, this

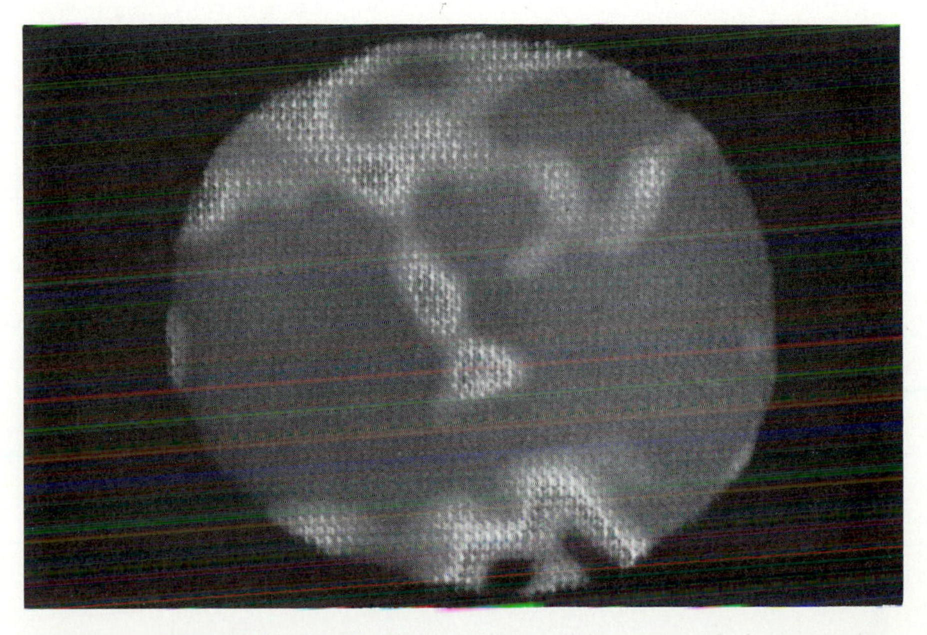

FIGURE 11.6 The swollen star Betelgeuse is close enough to directly resolve its size as well as some surface features thought to be storms similar to those occurring on the Sun. Betelgeuse is such a giant star (some 300 times the size of our Sun) that its photosphere roughly spans the size of Mars' orbit.

one the case of Sirius B, a companion to the brightest star (excepting the Sun) in the night sky. (Figure 20.11 shows a picture of this double-star system.) Sirius B's surface temperature has been measured to be nearly 12,000 kelvin, twice the value of our Sun; its luminosity is almost 10^{31} ergs/second, about two-thousandths of the solar value. Substitution of these quantities into the relationship above yields a radius of about 0.01 times the solar radius. (Again, we can derive this rapidly by using solar quantities: $\sqrt{0.002}/2^2 = 0.01$.) This star is then much hotter, although dimmer and smaller than our Sun. In fact, such a star is approximately the size of Earth. Such an object is known as a "dwarf star." Technically, a **dwarf star** is any star comparable to or smaller in size than the Sun. (And since any 12,000-kelvin object glows whitely, Sirius B is actually a white dwarf star; as for red giants, white dwarfs are more appropriately studied in the evolutionary context of Part III.)

As a rule of thumb, stellar sizes vary from 0.01 solar radius to 100 solar radii. There are a few known exceptions, however.

To evaluate the relationship above for stellar size, we need to have a good working knowledge of luminosity and temperature. Let's now consider in turn each of these basic quantities. By studying them, we'll learn a lot more about the general properties of stars.

Luminosity or Brightness

Earlier, we used the term "luminosity." We defined it as the total amount of energy radiated into space each second from a star's surface. In other words, luminosity is the *rate* of energy emission—not only the energy of visible light, but also the energy of any type of radiation from radio waves to gamma rays. Another term closely related to "luminosity," and one perhaps more familiar to nonscientists, is "brightness." "Intensity" is yet another. In this book we take the words "luminosity," "brightness," and "intensity" to be synonymous.

The concept of luminosity or brightness can sometimes be confusing, mainly because astronomers use two types of brightness. The luminosity used in the derivation of the radius–luminosity–temperature relationship above, and thus the luminosity that must be used to evaluate stellar size, is related to one of these brightnesses—something we call absolute brightness. To see what we mean by this, consider the following.

The stars in the sky have different brightnesses. Just look at them on a clear night; some *appear* bright, while others *appear* dim. Generally speaking, a star's dimness could be caused either by an intrinsic lack of emitted radiation, or by the large distance over which starlight must pass on its way to Earth. A star is dim, then, either because it's a weak emitter or a distant object (or both). Conversely, a bright star is either powerful or nearby (or both). Hence two factors bear heavily on stellar brightness—intrinsic emission and distance.

We can better appreciate this interrelationship between energy emission and distance by means of a familiar example. Figure 11.7 shows two photographs of the same scene taken at different times of the day. The top frame is a daylight photograph of a Harvard Square neighborhood, taken from Cambridge Common a few blocks west. The bottom frame shows the same area after dark. The first thing to note is that this particular scene was photographed while looking down a main boulevard and therefore has built into it a sense of distance. The other thing to note are the streetlights—not the blur of advertisements near the Square, but the individual lamps along the street—each of which emits an intrinsically equal amount of energy in the form of light radiation. Clearly, the more distant streetlights are dimmer; the most distant, dimmest of all. They *appear* less bright. The reason for this is not because the distant lightbulbs are any less powerful. Rather the dimness of the faraway lights, as compared to those nearby, results directly from their greater distance. Hence, in accord with our own experience, objects having similar intrinsic brightnesses *appear* fainter at greater distances. Conversely, relatively faint objects can sometimes appear fairly bright should they be

FIGURE 11.7 A neighborhood near Harvard Square photographed (a) during the day and (b) at night.

nearby. In fact, if a weakly emitting, ordinary flashlight were to be placed near the camera used to take the photograph in Figure 11.7(b), the flashlight could conceivably outshine some of the obviously more powerful, but more distant streetlights.

INVERSE-SQUARE LAW AGAIN: We need to inquire precisely *how* brightness diminishes with distance. Figure 11.8 schematically illustrates light leaving a star. Moving outward, the radiation must pass through imaginary spheres of ever-increasing radius surrounding the emitting source. As such, the total amount of radiation passing through any given area of a shell must decrease in proportion to the surface area of the ever-increasing sphere. Knowing also that the surface area of a sphere is proportional to the square of its radius, we reason that the apparent brightness of an object must be inversely proportional to the square of its distance. Thus, an illuminated object 2 times, 5 times, or 100 times farther away than normal will be dimmer than normal by amounts of 4 times, 25 times, or 10,000 times, respectively. In short, brightness weakens with distance, specifically with distance squared. Not just for stars, this inverse-square distance law applies equally well to flashlight bulbs, streetlight beacons, automobile headlights, and radio communications, as well as to *any* other type of radiation.

TWO DIFFERENT BRIGHTNESSES: To compare sources of light—any light, streetlight, or starlight—we need a common ground. This is especially important if we are ultimately to determine if our Sun is in any way abnormal. We need to compare *intrinsic* stellar properties, not *apparent* stellar properties.

Heedless of standards set by the light-bulb industry, astronomers use 10 parsecs as the common ground for stars. There's no good reason to use 10 parsecs, other than the fact that both it (10) and its square (100) are convenient numbers. This is the distance used to compare all stars.

Let's recapitulate for a moment. The **apparent brightness** is the intensity of a star as it naturally appears in the sky. It's the directly observed brightness—the energy that can be directly measured while gazing at a star. But, remember, brightness is affected by distance. Thus only the **absolute brightness** can provide a meaningful comparison among stars. By definition, *the absolute brightness equals the apparent brightness that a star would have if that star resided at a distance of 10 parsecs from Earth.*

Note that absolute brightness cannot be measured directly (unless a star fortuitously happens to be at a distance of 10 parsecs). Instead, absolute brightness is a mathematically calculated quantity, enabling us to judge stars from the common distance of 10 parsecs. In this way, the *intrinsic* brightnesses of all stars can be meaningfully compared.

Finding a star's absolute brightness or luminosity, then, is a twofold task. First, the astronomer measures the apparent brightness of a star. This is usually easy; any telescope can immediately yield the rate of energy detected. He or she then mentally—that is, by mathematically using the inverse-square law—places that star at a distance of 10 parsecs and calculates the value of the absolute brightness. It is this result that relates most closely to luminosity.

Note that relatively nearby stars can appear rather bright even if they are intrinsically weak emitters of radiation. Their apparent brightness is often quite large. But once such stars are mentally pushed out to the common ground of 10 parsecs, their brightness decreases. In short, the absolute brightness of nearby stars is less than their apparent brightness; they would appear dimmer at 10 parsecs than at their actual closer distance. A perfect example is the Sun, which is, of course, very nearby, astronomically speaking. Its apparent brightness is enormous, so bright in fact that it's hard to look at. But if the Sun were at a distance of 10 parsecs, it would be much less luminous; the Sun would remain visible to us here on Earth, but it would be much dimmer than many other stars in the evening sky. Because of its closeness, the Sun is the premier case of an astronomical object whose ap-

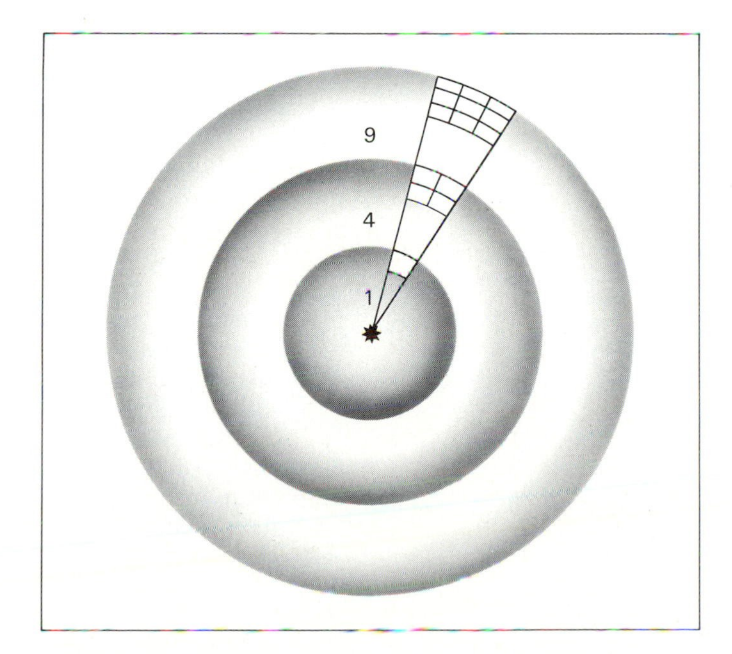

FIGURE 11.8 The amount of detectable radiation (apparent brightness) varies inversely as the square of the distance to an emitting object. As radiation spreads out from a source, its brightness steadily dilutes while encountering progressively larger surface areas (depicted here as imaginary shells).

parent brightness exceeds its absolute brightness.

On the other hand, stars that are very distant, say much farther away than 10 parsecs, usually have small apparent brightnesses. But when their distances are mathematically diminished to the standard of 10 parsecs, their absolute brightnesses can become quite large.

So, to our eyes, all the small stellar points of light in the sky are literally overwhelmed by the extremely bright Sun. That's why we can't see other stars during the daytime. But that's only because the Sun is so near. When all stars are mentally given the distance of 10 parsecs from Earth, many of those apparently faint stars become just as bright as the Sun. In fact, the absolute brightness of some visually faint but giant stars can far exceed that of our Sun.

Let's emphasize once again, then, that only the *absolute* brightness or luminosity can provide meaningful comparisons of the average physical properties of all stars.

SPECTROSCOPIC PARALLAX: A most important tool results from the interrelationships among apparent brightness, absolute brightness, and distance. The dependence of brightness on distance can be used in reverse, thus becoming the next rung of the universal distance scale shown in Figure 11.9. Specifically, if the absolute and apparent brightnesses of a star are known, the distance can be derived mathematically. Once we know two out of three related quantities, we can readily find the third. How is this done operationally?

The first step is easy. Measure the apparent brightness of a star as it appears on a photograph or through a telescope. The second step—obtaining the star's absolute brightness—is not as easy. Although not straightforward, techniques do exist for this, and we'll develop the main one before the end of this chapter. And the third step is to solve mathematically for the distance using the inverse-square law.

To clarify further the way in which absolute and apparent brightnesses can be used to determine distance, consider the analogy of a traffic light. All of us have a rough idea of the approximate brightness and size of a red traffic signal. But suppose that we're driving down an unfamiliar street and notice a red traffic light in the distance. Our knowledge of the intrinsic luminosity of the light often enables us to immediately begin making a mental estimate of its distance. A normal traffic light that's relatively dim must be quite distant (assuming it's not just dirty), but a bright one must be relatively nearby. The same sort of procedure could be used to estimate the relative distances to each of the streetlights in Figure 11.7(b).

The important point to understand here is that a measurement of the apparent brightness of a light source as well as some familiarity with its intrinsic properties can be used to estimate the distance to that light source.

For a star, the tricky thing is to find a measure of its absolute brightness without knowing its distance. Distance, after all, is what we seek here. Fortunately, absolute brightnesses or luminosities can be inferred from the spectral lines emitted by all stars. For that reason, this third rung of the distance scale is called **spectroscopic parallax**; the parallax term is a misnomer, but the adjective is valid since a spectroscope is needed to infer the luminosity.

To appreciate fully this new distance technique, we must first study the intricacies of stellar spectra. And

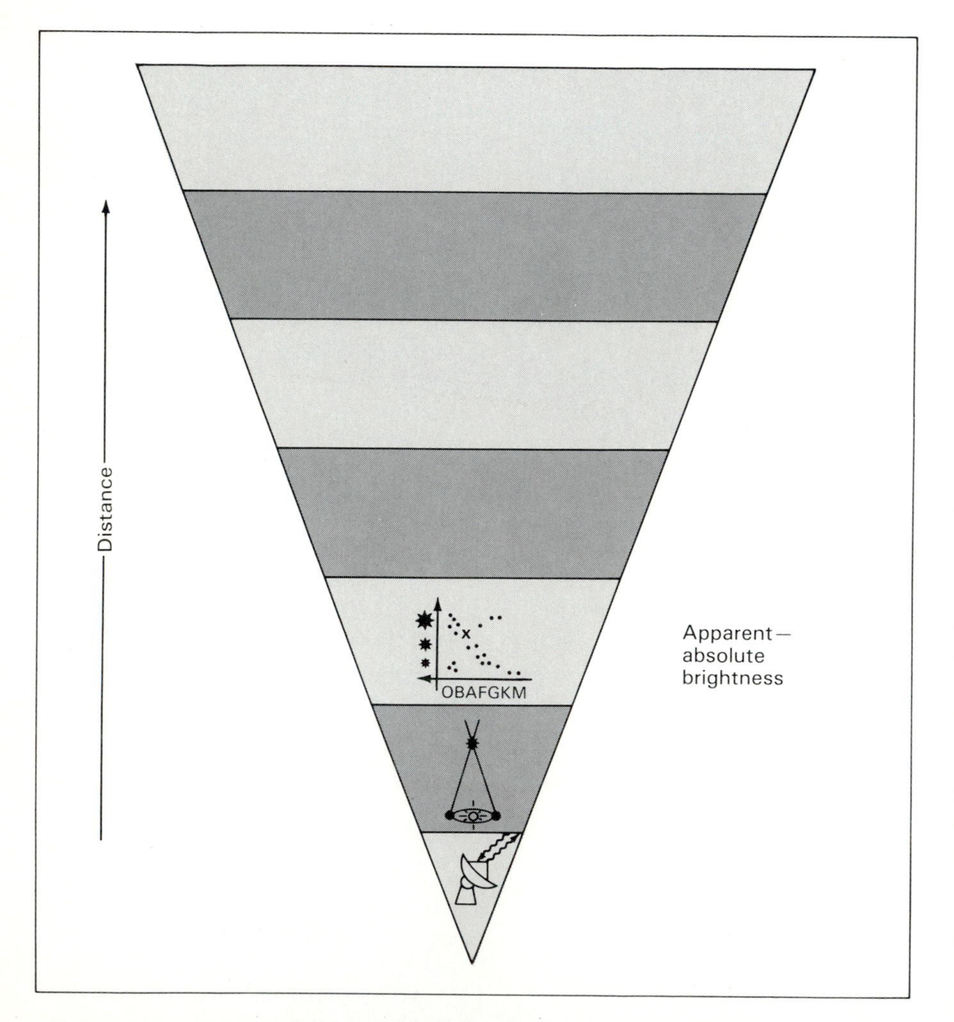

FIGURE 11.9 Knowledge of a star's absolute and apparent brightnesses can yield an estimate of its distance. This third rung, called spectroscopic parallax, can be used to find distances out to as far as individual stars can be clearly discerned, namely a few thousand light-years.

INTERLUDE 11-1 ***The Magnitude Scale***

A quick glance at the nighttime sky shows legions of stars, some bright, some dim. Without a telescope or some other piece of equipment, our human eyes can perceive only differences in brightness (and occasionally color). All other properties of stars must be measured with the use of telescopic apparatus of some sort. Still, brightness is a vital piece of information about any extraterrestrial object.

In this text we have quantified *brightness* in two ways. We have used an apparent brightness *observed* for some astronomical object when at its actual distance from Earth. Or it can be an absolute brightness *calculated* for that object when at the prescribed distance of 10 parsecs. The important point is that absolute brightness or luminosity is a useful quantity that measures the rate of energy emission. It's usually expressed in simple units, such as ergs/second, and therefore specifies the amount of energy emitted per unit time.

All astronomers use the brightness term, especially the more meaningful absolute brightness. But optical astronomers—those who practice the oldest type of visual astronomy—also use another term to describe brightness or luminosity. This term is called the *magnitude*. It was invented long ago (second century B.C.) by the Greek astronomer Hipparchus, who arranged the naked-eye stars into six groups. He let the brightest, and thus to him the most important star be of the first magnitude, and the faintest sixth magnitude. Of course, modern telescopes can collect and focus light from stars much dimmer than the faintest naked-eye objects, hence astronomers have extended this magnitude scale beyond the six groups originally envisioned. The large Palomar telescope, for instance, can detect objects as faint as twenty-fifth magnitude, which is equivalent to seeing a candle at about 15,000 kilometers (or about 10,000 miles). The orbiting *Hubble Space Telescope* can study objects as dim as thirtieth magnitude. The fainter the star, the larger the magnitude.

Unfortunately, there is no easy conversion between this ancient magnitude system and the modern luminosity system. The ratio in brightness between adjacent magnitudes is 2.5. This means, for example, that a difference of one magnitude equals a brightness difference of $(2.5)^1 = 2.5$, whereas a magnitude difference of five equals a brightness difference of $(2.5)^5 \simeq 100$, and so on. Another example is the brightness difference between the faintest stars seen with the naked eye and with the Palomar telescope; the 19 magnitudes of difference means that this large telescope can detect stars more than 10 million times fainter than can our eyes alone.

As with brightness, magnitudes come in two varieties. Apparent magnitude denotes a star's brightness when at its actual distance from Earth; absolute magnitude specifies that brightness calculated for a star at the prescribed distance of 10 parsecs, and thus refers to the *intrinsic* brightness of the star itself.

If you were not familiar with magnitudes before reading this interlude, or if you do not encounter them in your supplementary readings, ignore them. The magnitude system is a charming but confusing heirloom.

to do that we need to discuss the other quantity, namely temperature, that enters into the radiation laws discussed earlier. Then, toward the end of the chapter, we shall return to the issue of distance.

Temperature

The surface temperature of any star can be obtained rather easily. A few measurements of a star's brightness or intensity are sufficient. Recall from our study of electromagnetism in Chapter 1 that the intensity of radiation from a hot, glowing object depends both on the outside temperature of that object and on the frequency (or wavelength) at which the intensity is measured.

Figure 11.10 reminds us of the typical intensity–frequency plots describing the emission of radiation from all hot objects. For example, the 6000-kelvin surface temperature of the Sun is found by measuring its radiative intensity at different frequencies and then matching the observed curve with the theoretical Planck curve for a 6000-kelvin emitter.

The same technique can be used for any star, regardless of its distance. Actually, we need not even measure the entire Planck curve. Because the basic *shape* of this type of curve is well understood, only a few data points are needed to determine the proper curve and thereby to find the surface temperature. In practice, the astronomer simply observes a star's intensity at a couple of selected frequencies. This is done with telescope filters that block out all radiation except that within a narrow frequency range. For example, a B filter is often used to reject all radiation except for a certain range of blue light. Defined by international agreement to extend from 3800 to 4800 angstroms, this range corresponds to the frequency regime where photographic emulsions are most sensitive. Similarly, a V filter rejects all radiation except that within a narrow light range from 4900 to 5900 angstroms, an interval corresponding to that part of the spectrum to which human eyes are most sensitive.

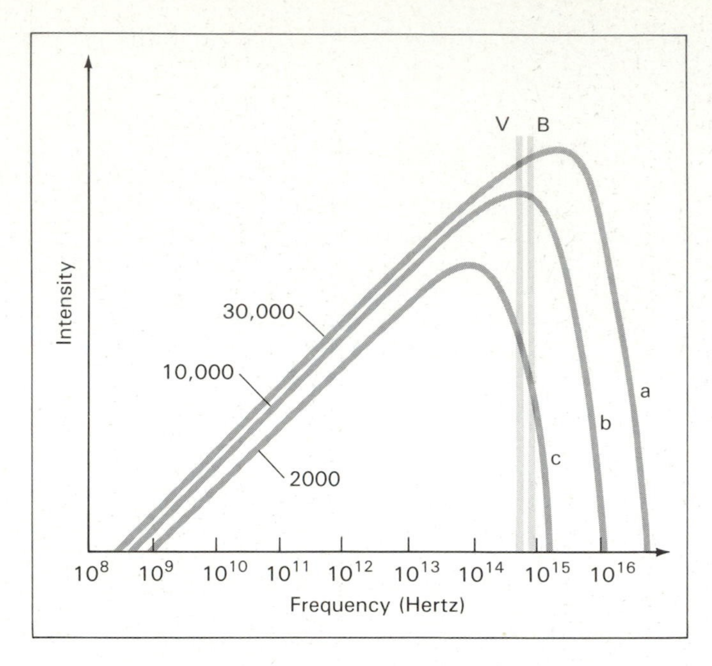

FIGURE 11.10 A set of Planck curves for different temperatures, along with the locations of the B and V filters.

Figure 11.10 shows how each of these filters admit to the telescope different light intensities for objects of different temperature. The intensities passing through those two filters, or any other two suitably separated filters, are all that's needed to specify the Planck curve uniquely and thus determine the surface temperature. This is a perfectly general technique. It works for all objects that emit radiation by virtue of their heat, whether they be glowing metals, flashlight bulbs, burning fires, or distant stars.

Consider a few examples. Suppose that we observe a particular star and find that light passing through the B filter is almost twice the intensity of that passing through the V filter. This case is shown by the curve labeled *a* in Figure 11.10. That Planck curve is constructed solely on the basis of those two intensity measurements. No other Planck curve could be drawn through *both* measured points. The curve corresponds to a surface temperature of 40,000 kelvin. Such a very hot glowing object has a bright bluish tint when observed without any restricting filters. This is a really brilliant, almost "electric" blue color, much like the extremely hot tip of a blowtorch flame.

Another example is one where the stellar intensity measured through the B filter equals that through the V filter. These two measurements define the Planck curve labeled *b* in Figure 11.10, a curve corresponding to a surface temperature of about 10,000 kelvin. When observed without filters, the characteristic color of such a glowing object is white because all the colors contribute nearly equally.

As a final case, consider a star whose intensity measured through the B filter is only one-fifth that measured through the V filter. The third curve, labeled *c* in Figure 11.10, shows how these two intensity values describe a Planck curve for a relatively cool 2000-kelvin emitter. No other temperature can fit these two data points. Because such stars emit very little high-frequency radiation, they often appear red when observed without filters.

Color, then, is generally synonymous with temperature, especially as regards a glowing object, as noted in Chapter 1. A red star is usually cool, often a few thousand kelvin. A blue star is usually hot, often several tens of thousands of kelvin. A yellow star, such as our Sun, has an intermediate temperature of several thousand kelvin. These intrinsic colors are not caused by Doppler red and blue shifts. Color is simply a convenient, though rough indicator of surface temperature. Looking up at the night sky, you should be able to tell your friends which are the hot and cool stars. To obtain numerical values for the temperatures of those stars, however, would require at least two intensity measurements at different wavelengths, as described above.

The first three columns of Table 11-1 list more detailed (yet still simplified) descriptions of the B and V relative intensities, the surface temperatures derived from them, and the dominant colors perceived in the absence of any filters. This type of non-spectral-line analysis of a star's intensity is known in the trade as **photometry**.

Classification

Color and temperature can be used alone to classify stars reasonably well. But in practice, a more detailed classification scheme is often used. This scheme incorporates additional knowledge of stellar physics and chemistry, which can only be obtained via spectroscopy, the study of spectral-line radiation.

DETAILED SPECTRA: Figure 11.11 and Plate 1(b) show typical examples of the visible spectrum of a single star. As usual, it extends from 4000 to 7000 angstroms, much like those of the Sun shown earlier in Figures 2.19 and 9.7. This spectrum, however, belongs to some other star, called Vega, which is one of the brightest stars in the sky.

Figure 11.12 compares the spectra of several different stars. For each, the overall appearance of dark absorption lines superposed on a background of continuous color resembles that of Figure 11.11. The precise pattern of lines has its differences, however. As can be seen, some stars

TABLE 11-1
Stellar Properties

B INTENSITY / V INTENSITY	SURFACE TEMPERATURE (KELVIN)	COLOR	PROMINENT ABSORPTION LINES	SPECTRAL CLASS
1.7	30,000	electric blue	Ionized helium intense Multiply ionized heavies Hydrogen dim	O
1.3	20,000	blue	Neutral helium moderate Singly ionized heavies Hydrogen moderate	B
1.0	10,000	white	Neutral helium very dim Singly ionized heavies Hydrogen intense	A
0.8	8,000	white-yellow	Singly ionized heavies Neutral metals Hydrogen moderate	F
0.6	6,000	yellow	Singly ionized heavies Neutral metals Hydrogen dim	G
0.4	4,000	orange	Singly ionized heavies Neutral atoms intense Hydrogen dim	K
0.2	2,000	red	Neutral atoms intense Molecules moderate Hydrogen very dim	M

FIGURE 11.11 A detailed visual spectrum of a star a little hotter than our Sun. The star is called Vega (or Alpha Lyrae), and, despite its distance of about 26 light-years, it is the fifth brightest star in the sky. Vega is an A-type star. [See also Plate 1(b).]

display strong absorption lines in the long-wavelength part of the spectrum. Others have their strongest lines at short wavelengths. Still others show absorption lines spread across the whole visible spectrum.

Most stars' spectra differ considerably from our Sun's. Chapter 9 discussed how different spectral patterns are caused more by differences in surface temperature than by differences in chemical composition. For example, the spectral-line pattern at the top of Figure 11.12 is just that expected from a star having a surface temperature of about 30,000 kelvin and a normal set of elemental abundances. The other spectra, from top to bottom of Figure 11.12, are arranged in order of progressively decreasing surface temperature. In other words, stars having very similar elemental abundances but very different surface temperatures emit spectral-line patterns like those shown in Figure 11.12.

The hotter (bluer) stars can excite and ionize many atoms and ions at their photospheric surfaces, either

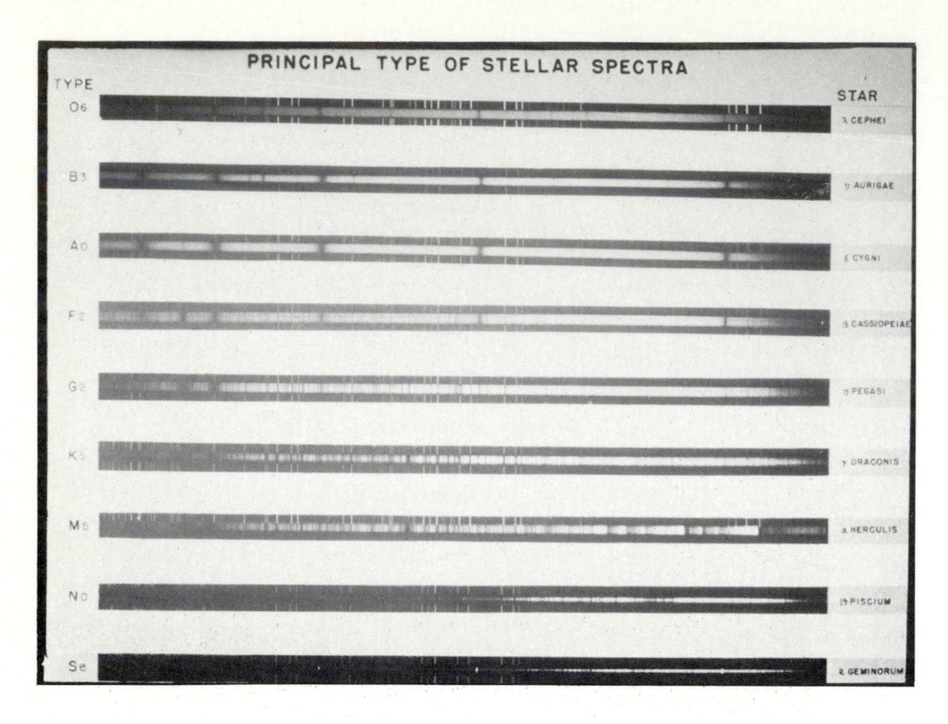

FIGURE 11.12 Comparison of nine spectra observed for different stars having a range of surface temperatures.

because each photon emitted from the surface of a hot star contains a lot of energy or because the higher temperature causes many collisions. Either radiation or collisions—and often both together—can excite and even ionize some strongly bound atoms. Spectra of stars having surface temperatures more than 25,000 kelvin usually show intense absorption lines of ionized helium; helium is a very tightly bound atom, and much energy is needed to strip an electron from it. Equally intense lines of multiply ionized heavier elements such as oxygen, nitrogen, and silicon are also observed. It takes a lot of energy (heat) to remove several electrons from atoms in order to multiply ionize them. The most abundant element, hydrogen, on the other hand, produces only weak absorption lines in the hottest stars. The reason is not that hydrogen is underabundant. On the contrary, there's plenty of hydrogen, but it's so excited that its effects aren't visible. Remember, from our discussion of spectroscopy in Chapter 2, that electron transitions between the second and third orbitals in hydrogen are responsible for the well-known spectral feature at 6563 angstroms. All other principal transitions in hydrogen, especially those among the higher-lying orbitals, produce spectral lines outside the visible domain. Hence electrons are surely transitioning within some of the stellar hydrogen atoms, but in the hottest stars, few of them do so between the second and third orbitals. (In fact, hydrogen is mostly ionized in the hottest stars.)

In cooler stars, the 6563-angstrom line of hydrogen is observed to be more intense. For example, for whitish stars having surface temperatures of about 10,000 kelvin, hydrogen is responsible for the most intense (darkest) absorption line. Conditions are just about optimized for electrons to change between the second and third orbitals in hydrogen. The strongly bound atoms—ones that need lots of energy for excitation or ionization—such as helium, oxygen, nitrogen, and so on, are hardly observed at all. This is due to an underabundance of energy, not of elements; the star's photons are just not energetic enough to boost tightly bound electrons up to the higher atomic orbitals. These same photons are, however, sufficiently energetic to ionize some other loosely bound atoms, such as calcium, titanium, and several other metals. To be sure, ions of these elements are observed in such moderately hot stars.

The spectrum of a yellowy star like our Sun, with a surface temperature of about 6000 kelvin, shows a wide variety of absorption lines from different elements. Few of the intense lines, however, arise from ionized elements. The Sun, relatively speaking, is just too cool. The hydrogen line at 6563 angstroms is no longer the most intense feature because, as with the very hottest stars, the gas in Sun-like stars does not have much electron traffic from the second to the third atomic orbital. But unlike the hottest stars where most of the electrons within hydrogen are in high-lying orbitals or even fully removed from the nucleus, hydrogen absorption lines are weak in cool stars because most of the electrons reside in the ground or first orbital. The majority of the electrons travel between the first and second orbitals, darkening the 1216-angstrom line, which, as noted in Chapter 2, lies in the invisible ultraviolet part of the spectrum. A small fraction of the electrons that have been excited to the second orbital will be additionally excited to the third orbital, thereby producing a visible 6563-angstrom line. But the line is not intense because only a small number of second-orbital electrons make it up to the third orbital before relaxing back to the first orbital.

Cooler red stars, having surface temperatures of a few thousand kelvin, still show extremely weak 6563-angstrom lines of hydrogen. The most intense lines in the spectra of these stars, however, are due to neutral heavy atoms—and weakly excited ones at that. Lines from ionized elements are not observed at all. In fact, the energy of the photons arising from the surface of some of the coolest stars is less than that needed to destroy fragile molecules. Consequently, many of the absorption lines observed in red stars are caused by molecules, still intact within the relatively cool confines of lower stellar atmospheres. We stress again a point made earlier: Although such molecular lines are often more intense than some atomic lines, this does not necessarily mean that the molecules are more abundant than the atoms. The intensity variations across stellar spectra result from the uniquely selective way that electrons shift within atoms and molecules. Very detailed analyses are required for scientists to derive precise values of both surface temperature and chemical composition.

Column 4 of Table 11-1 summarizes the most prominent kinds of spectral-line absorption features observed in a variety of different stars.

CATALOGING: Stellar spectra like those of Figure 11.12 were obtained for numerous stars even before the turn of the current century. Field stations around the world amassed spectra for several hundred thousand stars in both hemispheres. Often spectra were obtained for many stars on a single photograph, like that illustrated in Figure 11.13.

Early (1880 to 1920) researchers correctly identified some of the observed spectral lines, although this was at the time a strictly empirical comparison of stellar lines with laboratory lines. These workers had no firm understanding of how the lines were produced. Modern atomic theory had not yet been developed, and the correct interpretation of the line strengths as described above and in Chapter 2 was impossible at the time. These early workers went ahead anyway and classified stars according to their hydrogen-line intensities. They adopted an A, B, C, D, . . . scheme, thinking that A stars had more hydrogen than did B stars, and so on.

With the subsequent discovery, mostly during the 1920s, of the intricacies of the atom and of the causes of the spectral lines, astronomers realized that stars could be more meaningfully classified according to their surface temperatures. Accordingly, the original alphabetical classification scheme became jumbled. Confusing or not, this revised scheme is still used by modern astronomers.

In this new scheme, the hottest stars have the designation O (i.e., as in "oh," not "zero") mainly because they have weak absorption lines of hydrogen and thus were originally classified toward the middle of the alphabet. This stellar designation, or **spectral class** as it's officially called, is then followed by the letters B, A, F, G, K, and M. Surface temperatures corresponding to these classes are listed in column 5 of Table 11-1. Note, for example, that an A-type star has the most intense absorption lines of hydrogen, hence its high placement in the original alphabetical classification. (Indeed, the dark lines in the Figure 11.11 spectrum of Vega—an A-type star—are mainly due to hydrogen.)

These spectral classes can be remembered in the order of decreasing surface temperature by noting the time-honored mnemonic, "*O*h, *b*e *a* *f*ine *g*irl, *k*iss *m*e." No doubt, today's students should be able to do better than that! (Try this one: "*O*verseas *b*roadcast: *A* *f*lash! *G*odzilla *k*ills *M*othra!")

Actually, some stars have surface temperatures even less than 2000 kelvin. Like the M-type stars, these exceptionally cool stars show conspicuous evidence for molecular absorption lines. Such stars are rarely observed, although their low luminosity probably makes all but the nearest ones tough to see. In other words, cool stars could be quite numerous, but their dimness prohibits us from seeing many of them. These cool, dull, red stars were also known to the early catalogers, who, because of the apparent weakness of the hydrogen lines, designated them with the low alphabetical letters, R, N, and S. So you can wince harder while extending the above mnemonic to even greater heights of chauvinism: "Oh, be a fine girl, kiss me right now, smack."

FIGURE 11.13 Spectra of many stars can be observed simultaneously provided the starlight from all of them passes through a prism within the telescope before being recorded as a photograph such as this one.

HR Diagram

We've now studied the two most meaningful properties of any star: luminosity and surface temperature. Astronomers use these two quantities to classify the basic attributes of stars, much the same way that height and weight serve to classify the bulk properties of human beings. Furthermore, since we know that people's height and weight are tightly correlated—tall persons are usually heavier, short persons lighter, as illustrated in Figure

FIGURE 11.14 Two of the basic properties of most human beings—height and weight—are correlated. That is, the majority of the human population falls within a diagonal area (bounded by the shaded area). This is true since tall persons usually weigh more than short persons. Some exceptions—including midgets and basketball players—lie outside this well-defined range.

11.14—we might naturally wonder if these two stellar properties are also related.

In the second decade of this century, the Danish astronomer Hertzsprung and the U.S. astronomer Russell independently discovered just such a relationship between these two intrinsic stellar properties. They found a clear correlation when luminosity is plotted against temperature for any reasonably large sample of stars in any part of the sky.

Historical details aside, Figure 11.15 shows the essence of the way that Hertzsprung and Russell originally plotted these two quantities. In their honor, we refer to such a temperature-luminosity plot as the Hertzsprung–Russell diagram, or **HR diagram** for short. The vertical scale, expressed in units of solar luminosities, extends over a large range, from 10,000 to 0.0001 times our Sun's luminosity. The other basic property, surface temperature, is plotted horizontally, although in the unconventional sense of temperature increasing to the *left*. To change the horizontal scale so that temperature increases conventionally to the right would play havoc with historical precedent.

The two points plotted on the HR diagram of Figure 11.15 are just those values used for the earlier examples in this chapter's section on stellar size. One point represents the case for which the surface temperature is about half that of the Sun (namely, 3000 kelvin) and the luminosity about 400 times that of the Sun; this giant star's properties lie in the upper right-hand corner of the HR diagram. Since M-type stars with such low surface temperatures are reddish in color, the upper-right corner is termed the *domain of red giants*.

Similarly, we have also plotted the other case considered earlier, where the surface temperature was about double that of the Sun (12,000 kelvin) and the luminosity about 0.002 times that of the Sun. Such dwarf stars have properties in the bottom left-hand corner of the HR diagram. And recall that A-type stars have a white color since their surface temperatures are large enough to allow almost equal emission in all spectral colors from red to violet. Consequently, the bottom left-hand corner of the figure is called the *domain of white dwarfs*.

By way of a few more examples, let us stress that even stars having the same spectral class can differ widely in size and luminosity. The following list shows how four different stars having a constant 4000-kelvin surface temperature but a wide range in luminosity can have greatly different radii:

SURFACE TEMPERATURE (kelvin)	LUMINOSITY (units of solar luminosities)	SIZE (units of solar radii)	OBJECT
4000	0.001	0.07	Orange dwarf star
4000	0.1	0.7	Orange main-sequence star
4000	10	7	Orange giant star
4000	1000	70	Orange super-giant star

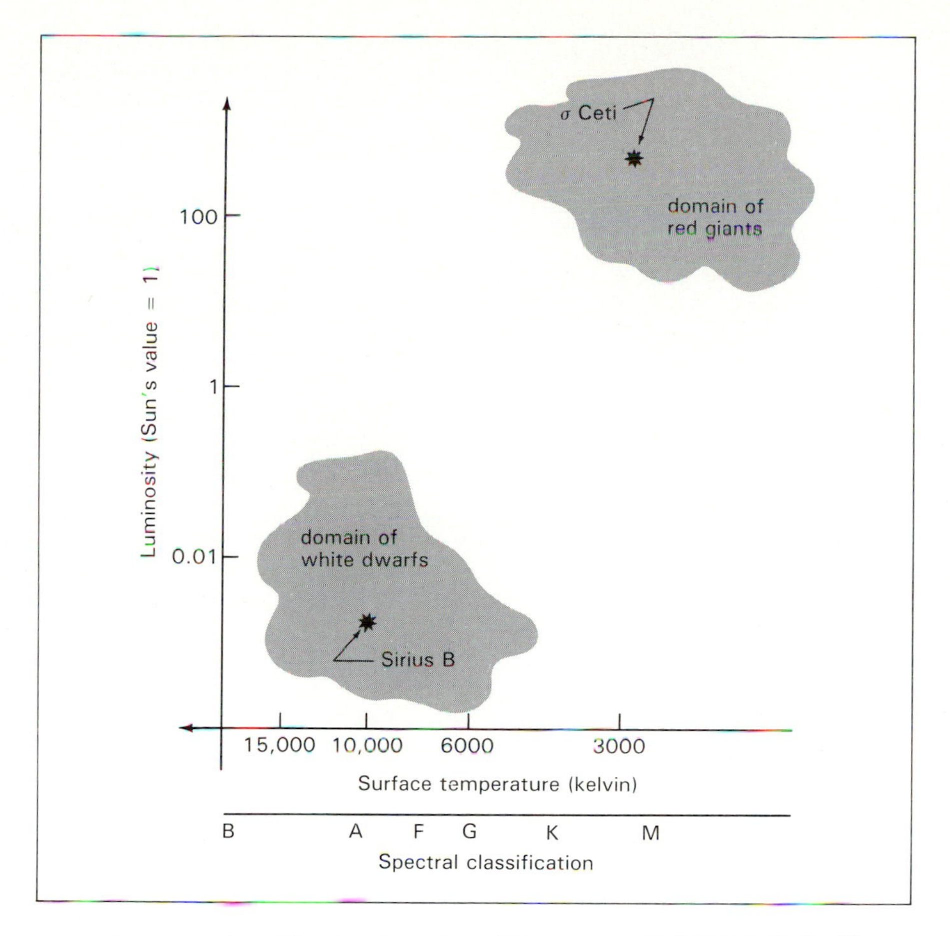

FIGURE 11.15 A plot of luminosity and surface temperature (or spectral classification), known as an HR diagram, is a useful way to compare stars. Plotted here are the values of the two stars studied earlier in this chapter.

This tabulation has been derived solely by means of the radius–luminosity–temperature relationship studied earlier in this chapter. More than any other, this relationship provides the key to an understanding of stars, and especially to the interpretation of HR diagrams.

For the most part, observations prove that stars have surface temperatures in the range 3000 to 30,000 kelvin. This small range, only a factor of 10, is dictated mainly by the rates at which nuclear reactions occur in stellar cores. On the other hand, the observed range of luminosity is large. Figure 11.15 accommodates this wide range in luminosity from 10^4 to 10^{-4} solar luminosities, an extension of eight orders of magnitude or a factor of a hundred million. Consequently, since the surface temperature varies only little from star to star, the radius–luminosity–temperature relationship implies that the luminosity depends mainly on a star's size. The larger the star, the greater its absolute brightness or luminosity. It must be that way since we defined luminosity as the total rate of energy emitted in all directions from all portions of a star's surface; stars with large radii have large surface areas and hence emit more energy per unit time. Thus all giant stars, whether they are red or not, have large luminosities. Conversely, all dwarf stars of any color have relatively small luminosities. So we can be confident that the observed differences in stellar luminosity are mainly due to differences in stellar size.

MAIN SEQUENCE: Despite our examples, which concerned primarily white dwarf and red giant stars, the great majority of stars are of neither category. The examples above were designed to give some feeling for the extreme properties of stars. Most stars, however, have more normal properties much like our Sun, and lie in a rather narrow regime on the HR diagram.

Figure 11.16 plots this narrow regime, called the **main sequence**. About 90 percent of all stars in our solar neighborhood, and probably an equal percentage elsewhere in the Universe, are main-sequence stars. About 9 percent are white dwarfs and 1 percent red giants.

Only the luminosity and surface temperature confine the stars to this narrow strip, which essentially defines the close correlation originally found by Hertzsprung and Russell. The main sequence can be used to specify the *average* properties of most stars, with the exception of a few giant and dwarf oddballs.

SOME REAL HR DIAGRAMS: An HR diagram can be constructed for any group of objects. Suppose, for example, that we consider all stars viewable within a given region of the sky. This need not be a special region; any arbitrary region will do, provided that there are enough stars to give a fair amount of data.

The first step is to determine the surface temperature (or spectral class) of each star. We could do this either by measuring the stellar intensity (or brightness) in the B and V bands and then fitting a Planck curve, or by observing a visual spectrum of a star and then finding its spectral class by identifying the relative intensities of certain prominent spectral lines. The photometric technique is the easier way if we want only an HR diagram, but the more laborious spectroscopic technique provides much additional information about each star's composition, motion, and physical state.

The second step is to determine the luminosity of each star. To do so is either easy or impossible. It's easy if a star's distance is known, enabling us to convert the measured apparent brightness into an absolute brightness or luminosity. But if a star's distance is unknown, that star can't be used in the construction of

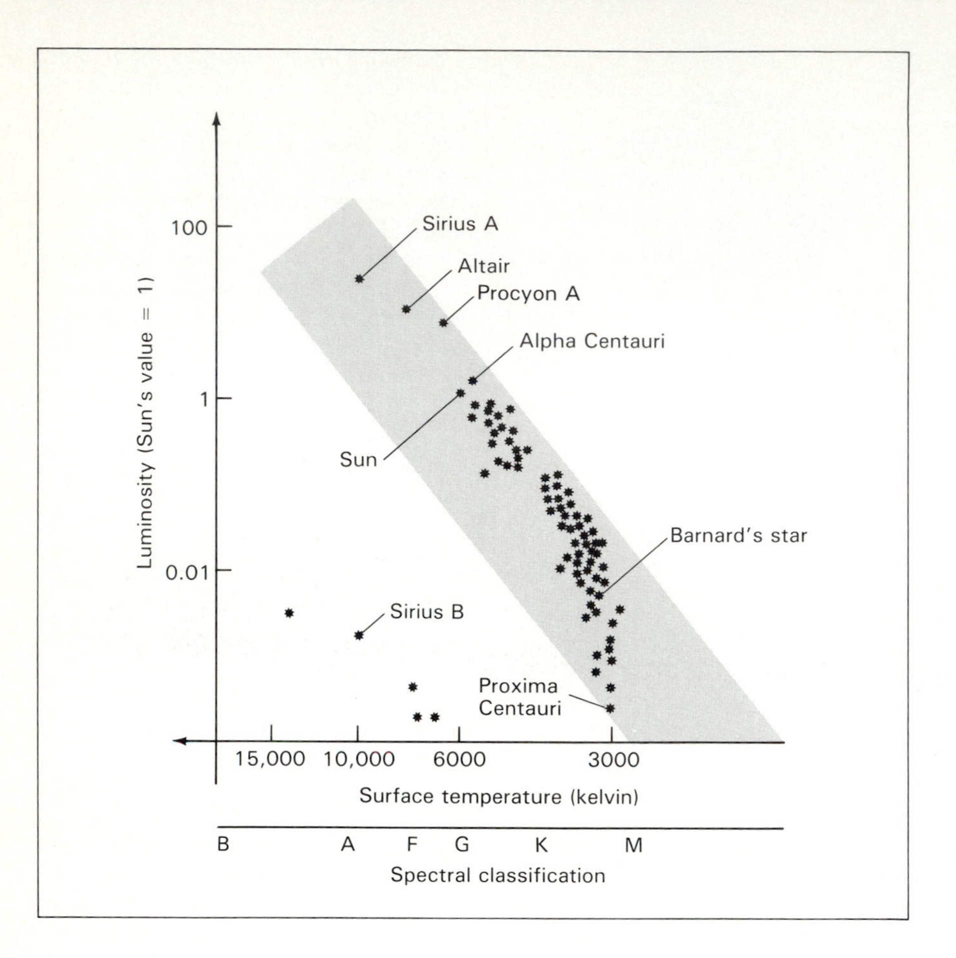

FIGURE 11.16 Most stars have "normal" properties within the shaded region known as the main sequence. The values plotted here are for stars within about 20 light-years of the Sun.

the HR diagram. That's why it's important initially to choose a region containing numerous stars whose distances are known.

Figure 11.17 shows an HR diagram for the sky's 100 brightest stars whose distances are known. Note the abnormally large number of very luminous stars toward the upper end of the main sequence. That's because very luminous stars can be seen at greater distances. In fact, of the 20 brightest stars in the sky, only six are within 10 parsecs of Earth; the rest are visible, despite their large distances, because of their large luminosities.

While the very luminous blue giants seem to be overrepresented in Figure 11.17, the low-luminosity red dwarfs are surely underrepresented. In fact, no dwarfs are present on the diagram. This is not surprising since such low-luminosity stars are difficult to observe from Earth. They're just too dim, since they radiate most of their energy in the in-

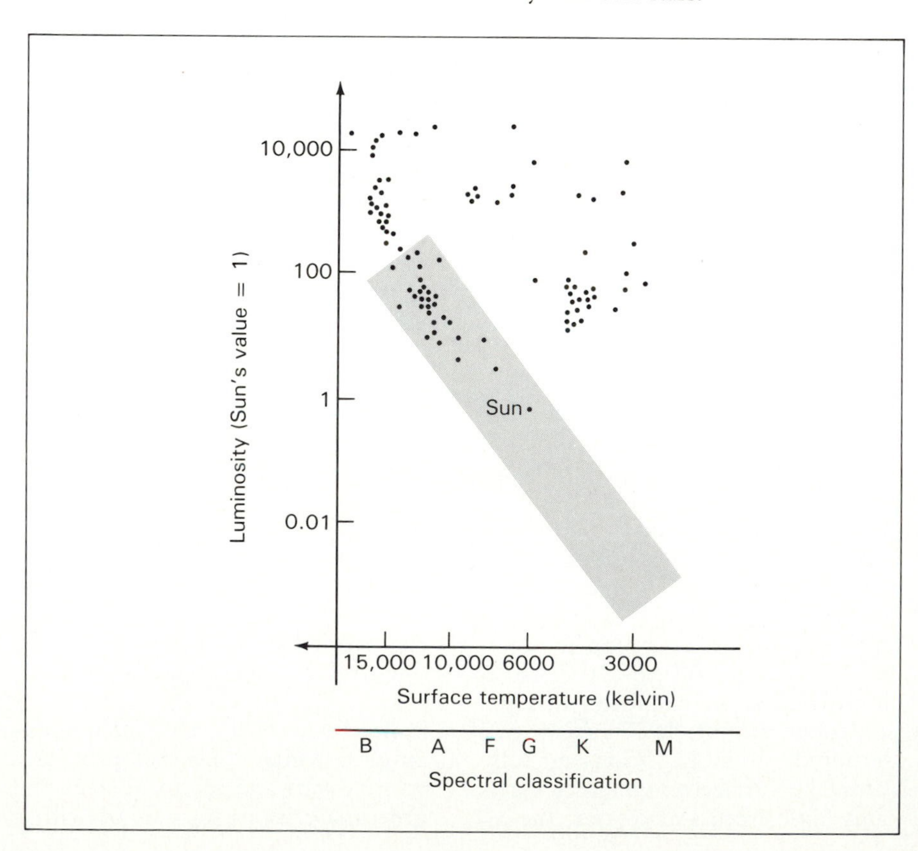

FIGURE 11.17 An HR diagram for the 100 brightest stars in the sky. Such a plot is biased in favor of more luminous stars since they can be seen more easily than dim stars.

INTERLUDE 11-2 ***Stacks and Stacks of Photographs***

Surely everyone has spent some time in a library where books are stacked row after row in some orderly fashion. In this way, various texts may be consulted for pleasure reading or reference work. Several of the world's observatories have not only stacks of books filed by author and title, but also stacks of celestial photographs filed by sky location and date. The largest collection of such stellar photographs is housed at the Harvard Observatory, where there are now nearly a million glass plates cataloged in a building suspended on springs to guard against earthquakes.

Prior to the invention of photography in the mid-nineteenth century, astronomical observations were made by visual impressions and hand sketches. Viewed by most astronomers at its inception as an idle diversion from the real study of the skies, photography has become a major tool, recording and quantifying observations. In effect, photography transformed observational astronomy from an art into a science. Foresight on the part of E. C. Pickering, who assigned funds and staff to this new pursuit, resulted in the large Harvard collection, which includes photographs regularly taken with several telescopes in both hemispheres for nearly the past century. He is shown in the accompanying photograph making his way to the summit of El Misti in Arequipa, Peru, the site of one of the most advanced observatories south of the equator around the turn of the century.

The early (nineteenth-century) photographic surveys concentrated mostly on the variability of individual stars; the cataloged plates are the best way to monitor long-term changes of stellar brightness. Toward the end of the nineteenth century, researchers realized that the spectroscopy of stars probably contains even more information than luminosity alone. Detailed spectral observations were then made for tens of thousands of stars by 1900, and for millions of stars to the present time.

The Harvard Photographic Collection is a rich source of astronomical lore, only part of which has yet been tapped. Numerous astronomers from around the world apply to analyze the plates, much as they routinely request observing time on modern telescopes. The cataloged plates allow astronomers to study the brightness of almost any visible cosmic object and the spectra of the brighter ones. Despite variations in the quality of the photographs because of observatory squabbles, funding difficulties, wars, shipwrecks, and the like, the detailed light histories have become greatly important in our attempts to unravel the intracacies of a wide variety of astronomical objects.

visible, infrared part of the electromagnetic spectrum.

Recently, astronomers have begun to realize that they have greatly underestimated the number of red dwarfs in the Universe. In fact, red dwarfs are now known to be the most common type of star in the sky; they probably comprise upward of 80 percent of all stars in the Universe. To see why, we must consider in the next section the mass of stars. But before doing so, we need to clean up one loose end.

DISTANCE AGAIN: As promised, we now return briefly to the issue of distance. Specifically, the technique of spectroscopic parallax, introduced earlier in this chapter, requires a means to find the absolute brightness (or luminosity) of a star. It is the HR diagram that can provide it. By taking a spectrum of a star, one can infer the star's surface temperature and thus one of its key properties on an HR diagram. If the star seems normal, we can *assume* that it lies on the main sequence. With these two properties determined—its surface temperature and thus its position on the main sequence—the star can then have only one possible absolute brightness. We can read the luminosity (or absolute brightness) directly from a plot like that of Figure 11.17.

If, by chance, the star in question is a giant or a dwarf, the determination of distance by this method will be incorrect. But since roughly 90 percent of all known stars are normal, our assumption will be correct and our distance valid at least 9 times out of 10. And all that is needed is a spectrum of a star—hence the adjective "spectroscopic" (the subsequent word, "parallax," being in this case a complete misnomer). This method of finding distance works well out to several thousand light-years, beyond which spectral measurements are difficult to secure for individual stars.

Mass

A reasonable question often asked concerns the spread of stars along the main sequence. Apart from the red giants and white dwarfs, what determines a star's position on the main sequence? The answer is mass.

The mass of a star is found by measuring its gravitational influence upon some other nearby object. Provided that we know the distance between the two objects, Newton's laws can be used to calculate the mass exerting the gravity force field. With this technique, we were previously able to derive the mass of Earth by watching the Moon or a man-made satellite orbit about it. Similarly, we discussed how the Sun's mass is found by studying the orbital motions of the planets. Unfortunately, though, it's currently impossible to measure the orbits of planets around any other star. Why? Because astronomers have not yet been able to detect any planets around any other star. Even so, there is a way to estimate the masses of many stars.

MULTIPLE STARS: Most stars are members of multiple-star systems. Many are **binary-star systems** containing two stars in orbit about their common center of mass and held together by their mutual gravitational attraction. Other stars are members of triple, quadruple, and even more complex systems. The most complex are the rich star clusters that often house thousands of individual stars spread over 10 to 100 light-years.

(The fact that our Sun is not part of a multiple-star system does not make it special. But it does mean that it is in the minority. If our Sun has anything at all uncommon about it, it may be its lack of stellar companions.)

Binary stars can be used to estimate stellar mass. "Visual binaries" have members bright enough and separated enough to be seen directly. Others can be indirectly perceived either by carefully monitoring the back-and-forth Doppler shifts of their spectral lines (called "spectroscopic binaries") or by observing a periodic decrease of starlight as one object passes in front of the other (called "eclipsing binaries"). Figure 11.18 illustrates how we can discover and study each of these types of binary systems.

Knowing the period of the binary stars as well as their orbital size, we have all that's needed to derive their combined mass. Specifically, recall that Kepler's third law, derivable from Newton's law of gravity, was expressed in Chapter 7 as

$$\text{orbital size}^3 \propto \text{orbital period}^2.$$

One of the proportionality factors in this relationship is the sum of the masses. As such, we can rewrite Kepler's third law as

$$\text{orbital size}^3 \propto (\text{mass of one object} + \text{mass of other object}) \times \text{orbital period}^2.$$

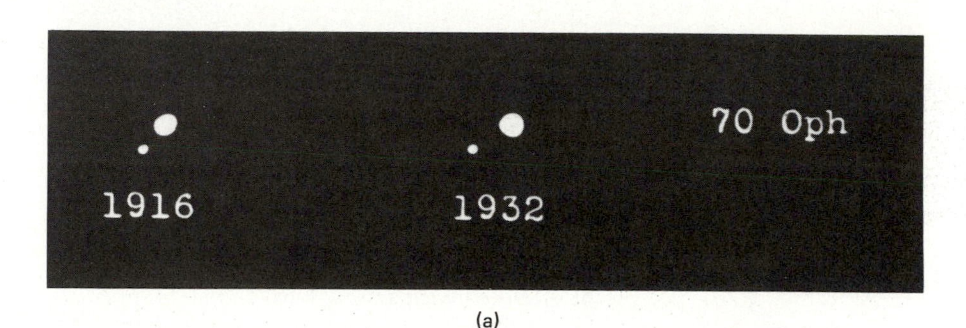

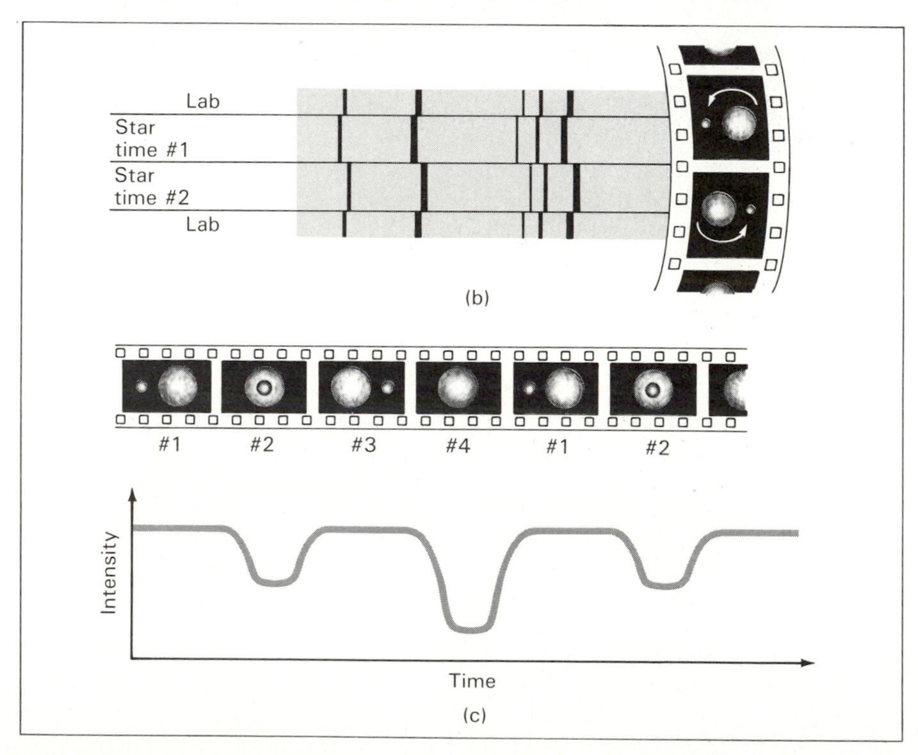

FIGURE 11.18 The periods and separations of binary stars can be observed directly (a) if each star is clearly seen. These properties can also be found indirectly (b) by measuring the periodic Doppler shift of one object relative to another, or (c) by observing the periodic decrease in starlight as one object orbits in front of another.

Knowing the period and separation of the members of a binary-star system, we can then mathematically calculate the sum of the stars' masses.

Additional observation leads to the individual masses of each star. For example, in any system of orbiting objects, each one orbits around a common imaginary point called the center of mass. The distance of each star from this center of mass yields the ratio of the star masses. Having both the sum of the masses and the ratio of the masses, we can estimate the mass of each star.

Consider the case of the nearby double-star system comprising the bright star Sirius A and the faint star Sirius B (pictured in Figure 20.11). Observations show that the sum of their masses is three solar masses; they further show Sirius A to have roughly twice the mass of its companion. From these facts, the only possible solution is that Sirius A contains 2 solar masses, and Sirius B only 1 solar mass.

Repeated observations have shown that stellar masses are roughly of the same order of magnitude as our Sun's. With few exceptions, their values range from 0.1 to 20 solar masses. Although this range is small, it's enough to distinguish the various types of stars along the main sequence.

Figure 11.19 is an HR diagram showing the masses of various stars along the main sequence. Note the clear progression from the low-mass red dwarfs to the high-mass blue giants. The hot O- and B-type stars are generally about 10 times more massive than our Sun, whereas the cooler K- and M-type stars contain a fraction of a solar mass.

AGE: Mass determines a star's position on the main sequence. That's because mass, more than any other property, basically governs a star's ability to radiate.

Recall from Chapter 10 that mass is a key factor in the nuclear reactions that power stars. In short, mass specifies the rate at which the nuclear fires burn in stellar cores: Very massive stars have so much matter pushing on their cores that the gas there becomes more compressed than in the core of our Sun. The increased pressure causes the core nuclei to collide more vigorously, hence increasing the gas temperatures while speeding the nuclear fusion rates. The opposite pertains to the least massive stars: Their core particles collide less frequently than in our Sun, and so their central temperatures and nuclear fusion rates are correspondingly smaller.

The rapid rate of nuclear burning deep inside a massive O- or B-type star releases vast amounts of energy per unit time, and thus guarantees high surface temperatures and luminosities. Accordingly, these massive stars endure for short times, since they use their resources quickly. Their nuclear reactions proceed so rapidly that their fuel is quickly depleted—despite their large mass. Paradoxically, the most massive stars run out of fuel quickest; they may be likened to large, inefficient gas-guzzling automobiles. Consequently, we can be sure that the O and B stars now observable in the sky are relatively young objects. Most of them are less than 20 million years old. Massive stars that are older have already exhausted all their fuel, and no longer emit large amounts of energy typical of the very hot and luminous O and B stars. They have, in effect, "died."

On the opposite end of the main sequence, the cooler K- and M-type stars have less mass than our Sun. With their core densities small, their proton–proton cycles churn away rather sluggishly, even more slowly than in the Sun's core. The small energy release per unit time then guarantees low luminosities and surface temperatures for these stars. Despite their small mass, however, they endure for long times; their low fuel consumption likens them to the more efficient compact automobiles.

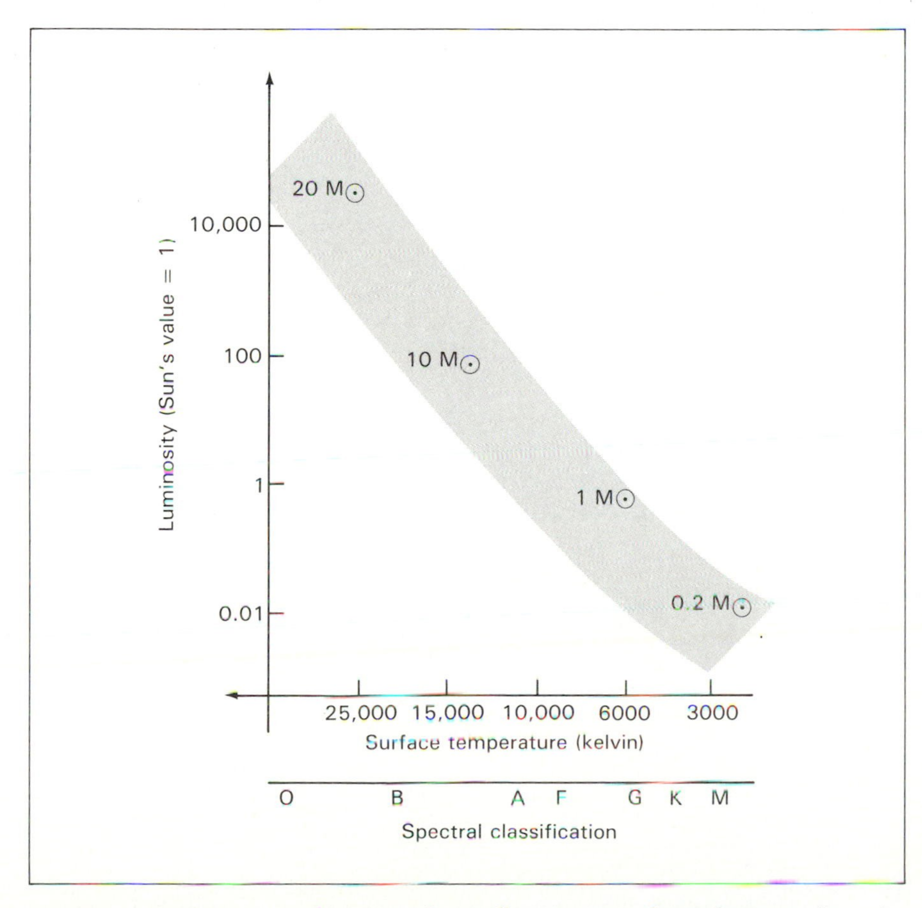

FIGURE 11.19 Mass, more than any other stellar property, determines a star's position along the main sequence. (The symbol, $M_\odot$, denotes solar masses.)

TABLE 11-2
Key Properties of Some Well-Known Stars

STAR	SPECTRAL TYPE	MASS (SOLAR MASSES)	CENTRAL TEMPERATURE (KELVIN)	ESTIMATED LIFETIME (YEARS)
Rigel[a]	B	10	30,000,000	2 million
Sirius A[b]	A	2.3	20,000,000	1 billion
Alpha-Centauri	G	1.1	17,000,000	7 billion
Sun	G	1.0	15,000,000	10 billion
Proxima-Centauri[c]	M	0.1	6,000,000	>1 trillion

[a] A bright star in the constellation Orion.

[b] The brightest star in the sky, except for our Sun.

[c] A companion star of Alpha-Centauri.

In fact, many of the K- and M-type stars now seen in the sky are expected to shine for at least a trillion years!

Table 11-2 compares some key properties for several well-known stars.

Star Clusters

We can construct a more representative HR diagram—one that lacks the bias mentioned earlier regarding Figure 11.17—by observing a group of stars known to have roughly the same distance. Called a star cluster, such a swarm can house anywhere from a few dozen to a million stars in a region many light-years across.

Figure 11.20(a) shows a rather loose cluster whose spectra can provide an estimate of the surface temperature of each star. The luminosity follows directly from measurement of the apparent brightness and knowledge of the cluster's distance. The result, shown in Figure 11.20(b), is an unbiased HR diagram for the whole cluster.

This HR diagram displays stars throughout the main sequence. Stars of all colors are represented. The blue stars must be relatively young, for they burn their fuel rapidly; many of the red stars are also probably young, although we cannot be sure for they are sluggish emitters of energy.

Other factors hint at the cluster's youth: an abundance of interstellar gas and dust not yet processed into stars, and a richness of heavy elements that could not have been created inside the cluster's stars but must have been cooked within ancient stars long since perished.

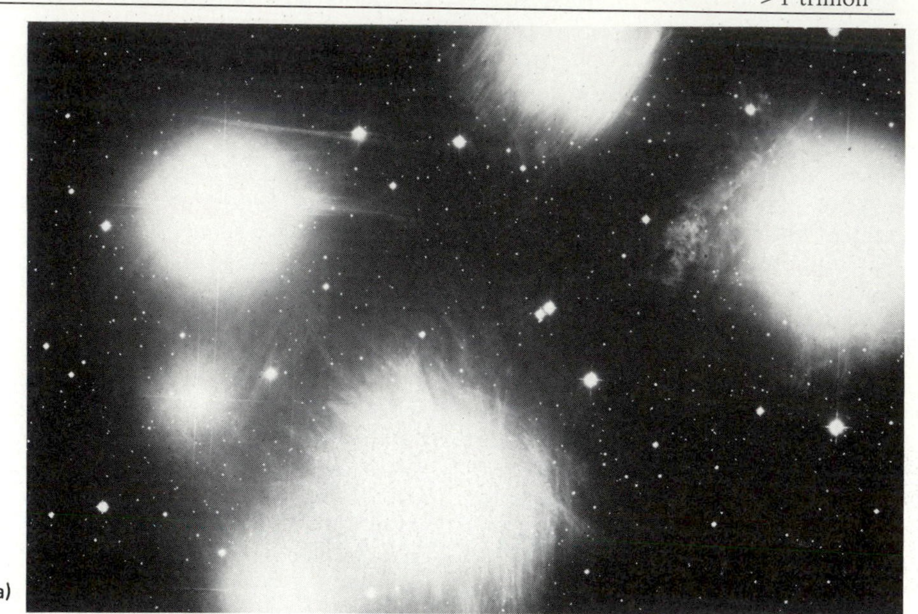

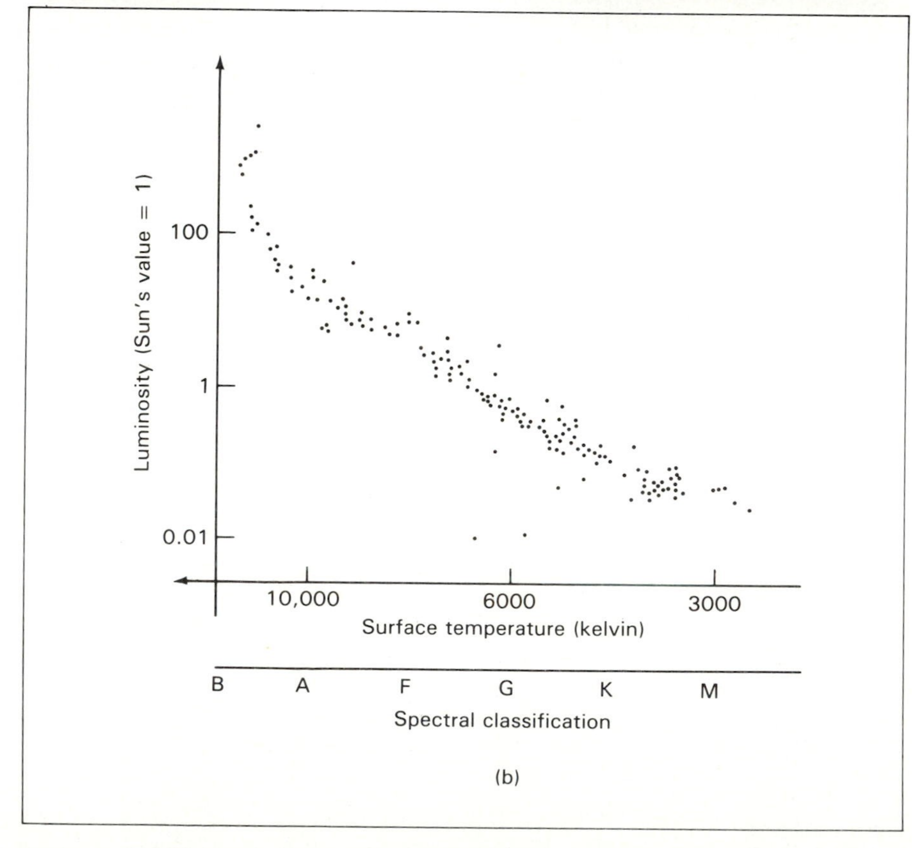

FIGURE 11.20 The stars of this well-defined galactic (or open) cluster (a) yield an unbiased HR diagram (b). This cluster has the name Pleiades (or M45); only naked eyes are needed to see its brighter stars, which are only about 400 light-years away.

(a)

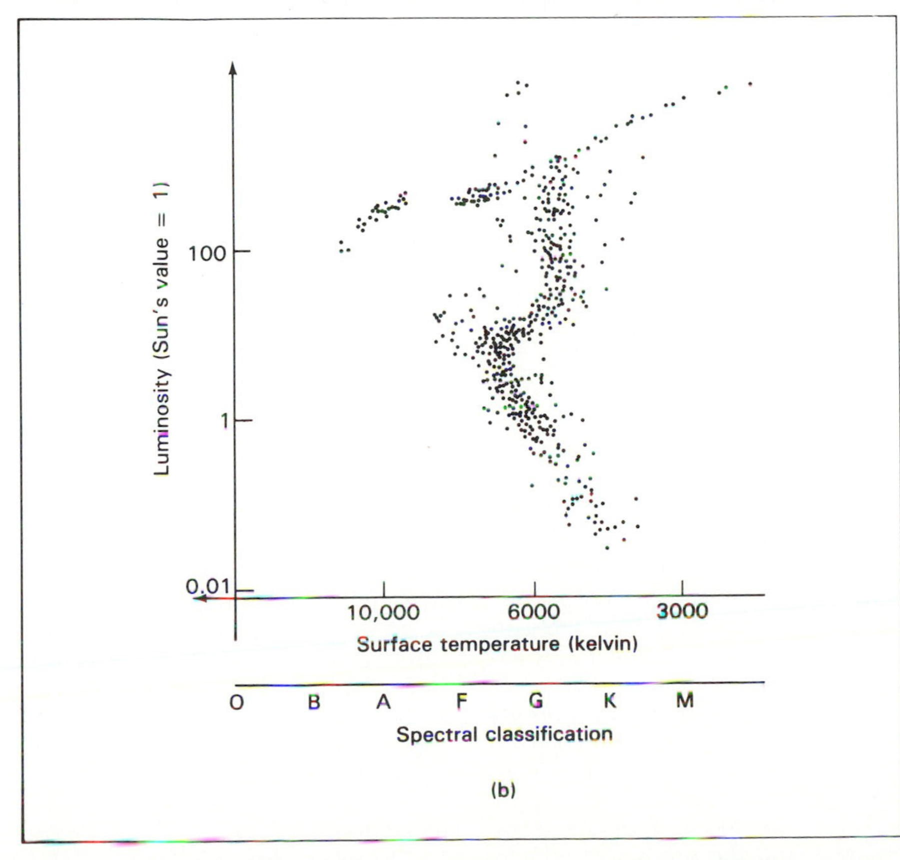

FIGURE 11.21 Photograph of a globular cluster (a) and an HR diagram for many (though not all) of its stars. This one is called M3 and resides approximately 30,000 light-years from Earth.

Such a star cluster, containing tens to hundreds of mostly young stars spread loosely over several light-years, is known as a **galactic cluster** or open cluster. Based on their stellar populations, most such clusters are less than a few billion years old; many are much younger.

Even our Sun may be within, or near the edge of, a loose galactic cluster centered some 70 light-years away. Several stars in the Big Dipper are also probably members of this same cluster.

In contrast to galactic clusters, astronomers know of another type of stellar swarm called a **globular cluster**. Figure 11.21(a) shows how a globular cluster is much more tightly knit than the loose groups of stars usually comprising galactic clusters. All globular clusters have this roundish shape, and often contain many thousands, and sometimes millions, of stars spread over about 100 light-years. Stellar motions within such a cluster mimic bees in a swarm. The entire assemblage is held together by gravity.

The most outstanding feature of all globular clusters is their relative lack of O- or B-type stars. This is clear from Figure 11.21(b), which is a plot of the HR diagram for the accompanying cluster. Low-mass red stars and especially intermediate-mass yellow stars abound, but high-mass blue stars are nearly absent. Apparently, globular clusters formed long ago, in the meantime the massive O- and B-type stars having exhausted their fuel and disappeared from view.

Other factors confirm that globular clusters are old: Spectra show few if any heavy elements, implying that these stars formed much earlier in the Universe when the heavies were underabundant. Also, a general deficiency of interstellar gas and dust among the stars implies that star formation has nearly gone to completion inside these clusters; this could have occurred only after an extremely long period of time.

On the basis of these and other facts, astronomers have estimated that all globular clusters average 12 billion years old. These are the oldest known stars in our Milky Way Gal-

INTERLUDE 11-3 *Stellar Activity*

Our Sun's activity is apparently not unique. Stars of all types have recently been found to be active and to have stellar winds. The importance and ubiquity of strong stellar winds became clear only through advances in ultraviolet and x-ray spaceborne astronomy as well as radio and infrared ground-based astronomy. For example, in the late 1970s and early 1980s, an orbiting observatory named *Einstein* (noted briefly in Chapter 3) quite unexpectedly detected x-rays from nearly all types of stars, thereby revealing that they too are surrounded by coronas having temperatures of 1 million kelvin or more.

Although it may be hard to appreciate while gazing at seemingly quiescent stars in the nighttime sky, apparently all stars sport active regions, including spots, flares, and prominences much like those of our Sun. Some stars exhibit "starspots" so large that an entire face of the star is relatively dark, while others display flare activity thousands of times more intense than that on our Sun. As in the case of our Sun, however, virtually all such activity is invisible, manifest only at wavelengths to which the human eye is insensitive.

In retrospect, we should hardly be surprised at this newly discovered stellar activity. By now, the doctrine that our Sun is an ordinary, average, highly typical star has become firmly entrenched in scientific thought. Still, it is exciting to see that philosophically reasonable viewpoint vindicated by direct observation.

What *is* surprising is the extent to which the most massive stars exhibit stellar winds. Ultraviolet observations made by both the *Copernicus* satellite and the *International Ultraviolet Explorer* satellite, both in Earth orbit, have shown key stellar absorption features to be shifted in wavelength by the Doppler Effect. In the case of some massive, hot, blue stars, their wind speeds often reach 3000 kilometers/second; such stars lose mass at rates up to a billion times that of our Sun's wind, thus shedding an entire solar mass in the—astronomically speaking—short span of 100,000 years. Such blasts are of gale force, even when we speak of "weather" in the Galaxy. Accordingly, the wind of the most luminous stars must cause profound effects, not only on the stars' immediate environment, including any attendant planets, but also on the subsequent evolution of the stars themselves.

All things considered, one of the most fascinating features of stars is their remarkable repertoire of activity. Magnetic fields, photospheric violence, flare-like effects, coronal holes, and stellar winds are part of the normal daily routine of essentially every type of star. Clearly, the atmospheres of most stars are far removed from the quiescent state implied by their steady visible image burning forth in the evening sky.

axy. As such, globular clusters are considered to be remnants of the earliest stages of our Galaxy's existence. For this reason, we can take the age of our Milky Way Galaxy to be roughly 12 billion years.

The Sun's Mediocrity

The HR diagram is a tremendously useful tool. It contains a wealth of information about all kinds of stars. You should study it carefully, along with the various arguments offered above to explain it. HR diagrams consolidate the many, seemingly diverse properties of stars, and make comparisons among them relatively easy.

Be sure to recognize that HR diagrams plot two stellar properties derived from radiation arriving at Earth *now*. Any HR diagram is essentially a graphical snapshot of a group of stars as we currently perceive them.

Studying such diagrams, we note a veritable zoo of different kinds of stars—main-sequence objects, red giants, white dwarfs, and so on. Each, at this moment, seems an entirely different type of object characterized by noticeably varied physical conditions. But these varied appearances belie their true nature. In reality, main-sequence and red-giant stars are not different kinds of stars, nor are red giants and white dwarfs. These names are simply convenient ways to identify stars at different stages of their evolutionary life cycles.

As noted earlier, stars are born, mature, and die much like people. Consequently, when we gaze at the sky, the various objects we see are each at different stages of their life cycles. Some are old, some young, some are just now reaching maturity. We'll never be able to watch a star move through its complete evolutionary phases. Lifetimes of humans—probably even of civilizations—are too short compared to the lifetimes of stars. We're stuck observing stars as they currently exist—at fleeting moments of their full lifetimes. Even astronomers, who might study in depth only a handful of stars during their entire professional careers, witness no more than a mere snapshot in the full life cycle of those stars.

Nonetheless, we can patch together an understanding of a star's "life" by studying stars of different ages, just as we could develop an understanding of human life by studying people of different ages, or as anthropologists gain an appreciation for ancient life by studying bones of different ages.

In the evolutionary part of our text, especially in Chapters 19 through 22, we'll be in a better position to appreciate the full life cycle of stars by studying the theory of stellar evolution. In doing so, we'll come to understand how the physical conditions of a star change dramatically with age. In short, the data points representing stars on an HR diagram change with time—main-sequence objects transform into red giants, and red giants in turn into white dwarfs.

For now, it suffices to note—in fact, it's important to realize—that the HR diagram demonstrates our Sun to be in no way special. As a G-type astronomical object with a 6000-kelvin surface temperature and an absolute emission rate of 1 solar luminosity, our Sun resides right in the middle of the main sequence (consult Figure 11.16). Thus in accord with our task originally set forth in the introduction to this chapter, we now conclude that our Sun is a very average star—a seemingly perfect example of magnificent mediocrity.

SUMMARY

Our main objectives in this chapter concerned an appreciation for the HR diagram, as well as for the two quantities that are correlated in it. In the process, we have learned much about stars—their typical sizes, luminosities, surface temperatures, colors, and masses. These are among the most important properties used to catalog stars, and, above all, to conclude that our Sun seems rather ordinary among the legions of stars in our nighttime sky.

We saw also that stellar brightnesses are apparent (just as they appear) and absolute (or genuine luminosities). Furthermore, it is distance that connects these two kinds of brightness. Thus we now have another observational technique to extend our subject of astronomy to greater realms in the Universe. Knowledge of both the apparent and absolute brightnesses will become an increasingly important way to study distant stars, clusters of stars, and galaxies of stars in the remaining chapters of Part II.

KEY TERMS

absolute brightness
apparent brightness
binary star system
color
constellation
dwarf star
galactic cluster
giant star
globular cluster
HR diagram
main sequence
photometry
proper motion
radiation laws
spectral class
spectroscopic parallax

QUESTIONS

1. Explain why, if stars are composed mainly of hydrogen, the absorption lines of this element are not especially prominent in our Sun. Be explicit.
2. Discuss the principles involved in measuring distance using the technique of spectroscopic parallax.
3. When you look at the night sky (with the naked eye), would you expect, on average, blue stars to be farther from or nearer to Earth than red stars? Why? (Neglect any absorption or scattering effects.)
4. Draw a labeled HR diagram for the stars of a typical globular cluster.
5. Compare and contrast the properties of globular clusters and galactic clusters, making certain to note their relative sizes, physical makeups, variabilities, positions in space, and ages.
*6. Consider a slowly orbiting, widely separated binary-star system some 46 light-years from Earth. The brighter star appears yellowish, whereas the dimmer one is distinctly bluish. (a) Which star has the hotter surface, and why? (b) Which star is intrinsically (absolutely) brighter? (c) If the yellow star lies on the main sequence, what kind of object is the blue star? Justify your answers.
*7. Two stars, one red and the other blue, have the same apparent brightness. (a) If they both lie on the main sequence, which one is farther away, and why? (b) If, instead, they have the same *absolute* brightness, and the blue star lies on the main sequence, what type of object must the red one be? Explain.
8. Distinguish the radial motion from the transverse motion of a bird in flight. Discuss the proper motion of such a bird. Compare the apparent size of the bird with its real, physical size. How can we determine the real size of such a distant bird?
9. When a nearby star is perceived to be reddish, what might that color indicate? Try quantifying your answer. Do the same for a yellow star.
10. Make a plot of the height and weight of the students in your class. Can you identify a "main sequence" or narrow interval in which these quantities are well correlated?

FOR FURTHER READING

HACK, M., "Epsilon Aurigae." *Scientific American*, October 1984.

KAFATOS, M., AND MICHALITSIANOS, A., "Symbiotic Stars." *Scientific American*, July 1984.

KING, I., "Globular Clusters." *Scientific American*, June 1985.

MARTIN, M., AND MENZEL, D., *The Friendly Stars*. New York: Dover, 1966.

NORTON, A., *Norton's Sky Atlas*. Cambridge, Mass.: Sky Publishing, 1985.

PAYNE-GAPOSCHKIN, C., *Stars and Star Clusters*. Cambridge, Mass.: Harvard University Press, 1979.

*SWIHART, T., *Astrophysics and Stellar Astronomy*. New York: Wiley, 1968.

12 INTER-STELLAR MATTER: Gas and Dust

We have now studied the basic properties of stars and planets. These objects are well-defined clusters of matter that are visible either because they emit light on their own accord (stars), or because they reflect light from their surfaces (planets). Stars and planets are dense collections of matter.

Space is populated not only with stars and planets. It also harbors a great deal of matter throughout the invisible regions separating the stars. Known as **interstellar space**, this is the dark realm "seen" among the stars in the evening sky.

A near-void, interstellar space contains extremely thin matter—approximately a trillion trillion times less dense than the average matter in either stars or planets. Only because interstellar space is so vast does it amount to anything at all.

Then why bother to study interstellar space? Aside from curiosity, we do so for two significant reasons. First, it's the region out of which new stars are born. And second, it's the region into which some old stars explode at death. In short, interstellar space is one of the most significant crossroads through which matter passes in the cosmic evolutionary scenario of our Universe.

The learning goals for this chapter are:

- to understand the physical and chemical properties of interstellar matter
- to appreciate the fact that, despite their near-void, the vast interstellar regions among the stars harbor gargantuan quantities of matter
- to understand the many experimental techniques used to probe the mostly invisible interstellar matter
- to recognize the signposts of stellar birth advertised by the hot, thin, glowing patches of matter called gaseous nebulae
- to know of the myriad clumpy "clouds" of slightly denser matter strewn throughout interstellar space
- to realize the recently discovered interstellar molecules and the potential role they play in the birth of stars, planets, and possibly life

Interstellar Matter

Figure 12.1 and Plate 4(a) show examples of a large region of space, a more expansive piece of universal real estate than anything we have studied thus far. The regions of brightness are simply congregations of innumerable stars. This *stellar* matter can be studied optically and we've discussed some of the relevant techniques in previous chapters. The regions of darkness are just that—regions that often obscure or extinguish background light. These are the regions among the stars that house **interstellar matter**, which cannot be studied as easily as stars. Their very darkness means that they cannot be studied at all by optical methods. There is, quite frankly, nothing to see!

A quick glance at Figure 12.1 shows the dark interstellar matter to be rather patchy, that is, irregularly spread through space. In some directions, the matter seems largely absent, permitting astronomers to study objects literally billions of light-years from Earth. In other di-

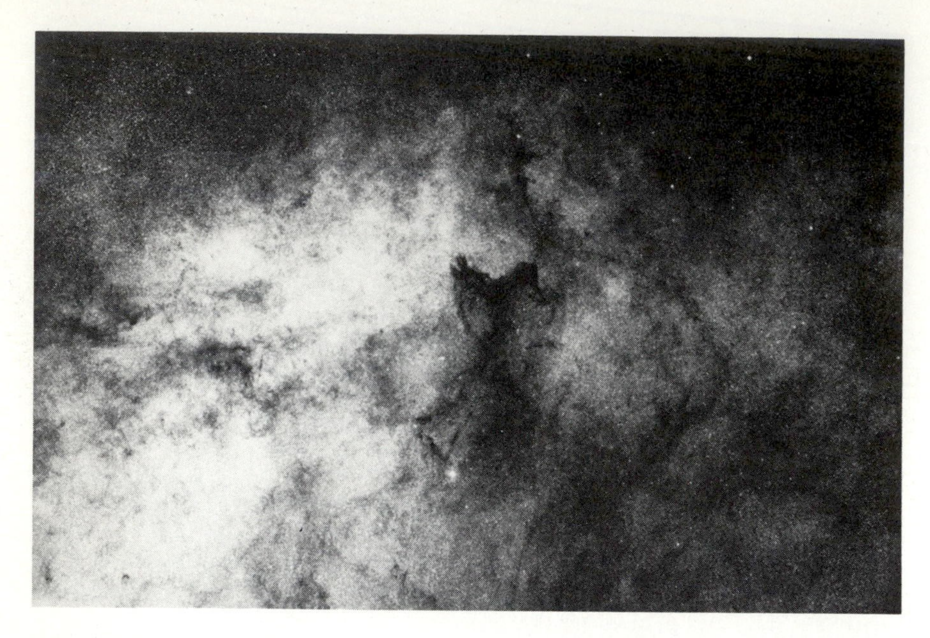

FIGURE 12.1 A wide-angle photograph of a great swath of space, showing regions of brightness (vast fields of stars), as well as regions of darkness (obscuring interstellar matter).

rections, the obscuration can be moderate, prohibiting visual observations of objects beyond more than a few thousand light-years. And still other regions are so heavily obscured that starlight from some relatively nearby stars can be completely extinguished before reaching Earth.

Clearly, matter exists free of the clutches of individual stars. Interstellar space is not empty. While at first seeming of nuisance value, interstellar matter is of vital importance in our understanding of the way matter changes. It's both cradle and grave—the stuff from which stars form, and the stuff to which many of them return at death. In short, interstellar space is one of the great proving grounds of universal change.

GAS AND DUST: We can broadly specify the nature of interstellar matter by classifying it into two simple components: gas and dust. Both are intermixed throughout all of space.

The gas component of interstellar matter is easy to understand. It's mainly composed of individual atoms of visual size about 10^{-8} centimeter. In lesser amounts, the gas also contains some relatively small molecules, each probably no larger than 10^{-7} centimeter in size. Regions with particles of such small size are completely transparent to nearly all types of radiation, including ultraviolet, light, infrared, and radio radiation. Gas alone does not block radiation.

The dust component is more complex. Dust is made of larger clusters of atoms—really mixtures of millions of molecules—not unlike the small specks of dust that often accumulate in household closets. Chalk dust is another good example, as are the microscopic particles of smoke, soot, or smog. The miniature particles of a terrestrial fog make an especially appropriate example—and they perhaps best illustrate the problem caused by dust: Light from distant stars just cannot penetrate the innumerable accumulations of interstellar dust any more than a car's headlights can illuminate roadside objects in a dense Earth-bound fog.

Comparisons of the obscuration or decrease of starlight in interstellar space with the scattering of man-made light in a terrestrial fog imply that the typical size of an interstellar dust particle must be about 10^{-5} centimeter—namely, comparable in size to the wavelength of light. As a rule of thumb, then, we can imagine interstellar dust particles to have about 1000 times larger size, or therefore about a billion times larger volume, than interstellar gas particles. Given these particle sizes, dusty interstellar regions should be partially transparent to infrared radiation, and completely transparent to long-wavelength radio radiation (consult Figure 3.19). However, dust is an especially effective blocker of high-frequency optical, ultraviolet, and x-ray radiation.

TEMPERATURE: The temperature of the "dark" interstellar gas and dust ranges from a few kelvin to a few hundred kelvin, depending on its proximity to a star or some other source of radiation. Generally, we can take 100 kelvin as an average temperature of a typical region of interstellar space. Compare this with the 273 kelvin at which water freezes, or 0 kelvin, at which all atomic and molecular motions cease. Thus much of interstellar space is cold—not cool—cold!

DENSITY: The density of interstellar matter is extremely low. It averages roughly 1 atom/cubic centimeter, although densities as great as 1000 atoms/cubic centimeter and as small as 0.01 atom/cubic centimeter have been found in special parts of interstellar space. Nonetheless, be sure to realize that even the densest interstellar region comprises a very thin gas, in fact still more tenuous than the best vacuums that we can make in laboratories here on Earth.

By making a small conversion in units, we can compare these interstellar densities with the more familiar stellar or planetary densities noted earlier in the text. Since 1 gram of matter contains about 10^{24} atoms, we can equivalently express an interstellar density of 1 atom/cubic centimeter as 10^{-24} gram/cubic centimeter. The typical interstellar gas density is then some trillion trillion times less than the average density of Earth.

So interstellar space is populated with gas so thin that, on average, no more than one atom exists in every cubic centimeter—only a single atom in each interstellar region the size of an acorn or a thimble. This is extraordinarily thin material, so scarce that harvesting all the matter in an interstellar region the size of Earth would barely yield enough matter to make a pair of dice.

How can matter so thinly distributed play a significant role compared to stars and planets that are so much denser? The key here is size. All together, interstellar space is vast. The typical distance among stars is far, far greater than the typical size of stars themselves. Stellar and planetary sizes pale in comparison to the size of interstellar space. Thus matter can accumulate, regardless of how tenuously spread, so as to play a role in the gargantuan realms of interstellar space.

Believe it or not, interstellar dust is even rarer than interstellar gas. On average, there are only about 10^{-12} dust particle/cubic centimeter, making it some trillion trillion trillion times less dense than Earth. How can such superthin matter diminish light radiation so effectively? The reason, again, is that interstellar dust is spread throughout enormous volumes of space—and that means great distances between Earth and astronomical sources of light. Even though we could more reasonably express the dust density in units of dust particles/cubic *kilometer,* typical interstellar regions often span vast numbers of kilometers. In fact, interstellar distances are routinely expressed in light-years, even thousands of light-years. For example, an imaginary cylinder 1 square centimeter on an end and extending from Earth to Alpha-Centauri would contain more than a million dust particles. Over huge distances, then, dust particles can accumulate slowly but surely to the point of blocking light and other types of short-wavelength radiation.

Note one other interesting feature about interstellar gas and dust. The ratio of the number of dust particles to the number of gas atoms is about 10^{-12}. This is called a *number*-density ratio, simply because it compares numbers of things. But since dust particles are accumulations of many atoms and molecules, the ratio of the mass of a dust particle to the mass of a gas atom is only about 10^{-2}. This is the *mass*–density ratio. In other words, although interstellar dust particles are few and far between, they collectively comprise about 1 percent of the mass of interstellar space. The remaining 99 percent of the interstellar mass is gas.

Despite their rarity, the dust particles make interstellar space—pound for pound—a *relatively* dirty place. Earth's atmosphere, by comparison, is about a million times cleaner than interstellar space. Our air is tainted by only one dust particle for about every billion billion (10^{18}) atoms of atmospheric gas. Another way of expressing this comparison is to note that if we could compress a typical parcel of interstellar space to equal the density of air on Earth, this parcel would contain enough dust to make a fog so thick that we would be lucky to see our hands held at arm's length in front of us.

COMPOSITION: Of what is interstellar gas and dust made? The composition of the gas is reasonably well understood. Spectroscopy of atoms has recently provided astronomers with a comprehensive knowledge of the elemental abundances of interstellar gas. Most of it, about 90 percent, is made of hydrogen atoms and molecules. Some 9 percent is helium, and the remaining 1 percent trace amounts of the heavier elements. Surprisingly, though, the abundances of several of the heavy chemical elements (at least their amounts in gaseous form) are much lower in the galactic clouds than in our Solar System and in stars. The most likely explanation for this unexpected finding is that substantial quantities of familiar elements such as carbon, oxygen, silicon, magnesium, and iron have been partly used to form the interstellar dust.

By contrast with the gas, the composition of the dust is currently not well understood. We have some infrared evidence for silicates (like those in terrestrial rocks), graphite (like that in lead pencils), and iron (like that in cooking skillets). And these materials are among those that are underabundant in the gas, lending support to the notion that interstellar dust forms directly out of the galactic gas. The dust probably also contains some "dirty ice," a frozen mixture of ordinary water-ice contaminated with trace amounts of ammonia, methane, and other assorted chemicals. At any rate, the dust is a microscopic solid—not a gas. Think of it this way: If you enlarged a dust particle about a million times, it might resemble a comet nucleus; a billion times, perhaps primitive Earth.

DUST SHAPE: Ironically, the shape of interstellar dust particles is better known than their composition. Although the minute atoms in the interstellar gas have a basically spherical shape, the dust particles do not. Individual dust particles—often called grains—are apparently elongated or rod-like, as shown in Figure 12.2. We have been able to infer this because the light emitted by some stars is dimmed and a bit peculiar. The observed light is peculiar because it's partially polarized.

Remember back in Chapter 1, we studied how polarized light is produced by passing unpolarized light through a Polaroid filter having specially aligned molecules. Regarding the present case of polarized interstellar light, we can imagine the source of the original unpolarized light to be some distant star, and the intervening Polaroid filter to be the interstellar space through which the light must pass on its way to Earth. If starlight passed through a region of *spherical* interstellar dust particles, the resulting light received at Earth couldn't possibly be polarized (see Figure 1.16). Thus the direct observation that the received light *is* polarized implies that the interstellar dust particles have an *elongated* shape to some extent, in analogy with the elongated slots of the Polaroid filter. Not only that, but the dust must also be aligned, as depicted in Figure 12.3.

The alignment of the interstellar dust is the subject of intense research among today's astronomers. The current view, accepted by most scientists, holds that the dust particles

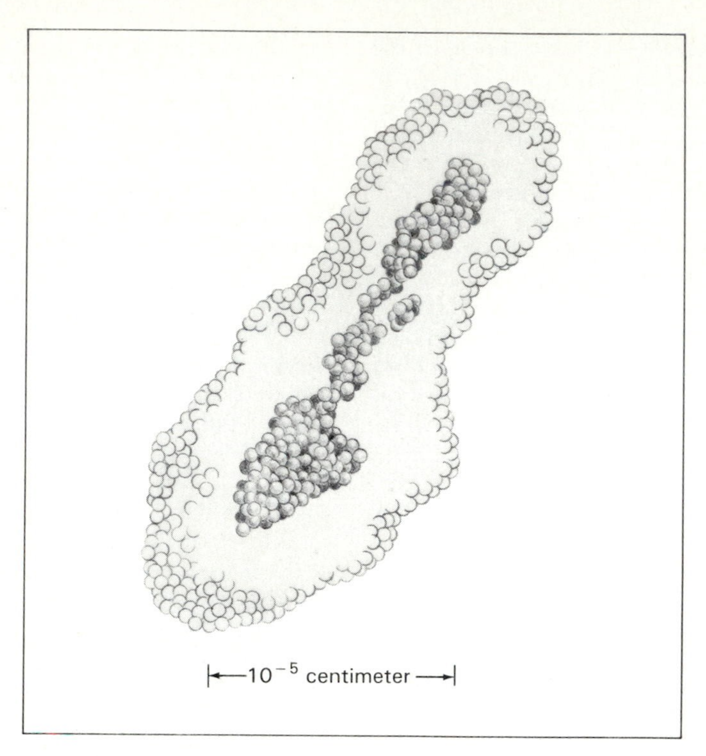

FIGURE 12.2 Schematic diagram of a typical interstellar dust particle. Some researchers suggest that such microscopic pieces of matter resemble miniaturized terrestrial planets—in composition, not in shape.

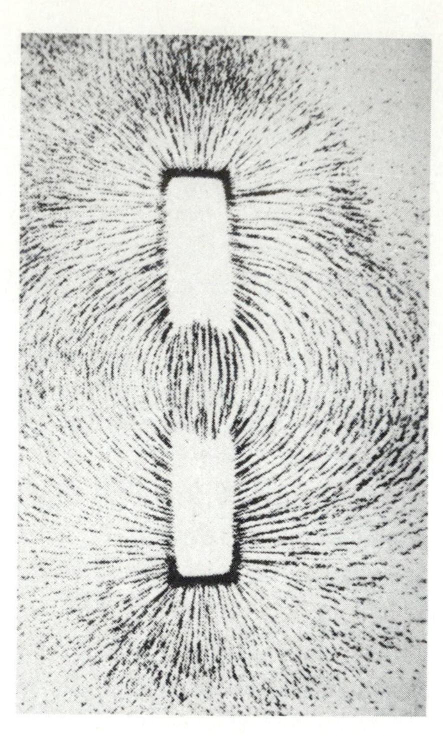

FIGURE 12.4 Interstellar dust particles, like the iron filings shown here, respond to a magnetic force field. The pattern was formed by sprinkling dust-sized iron scrap onto a paper underneath which was placed a horseshoe magnet.

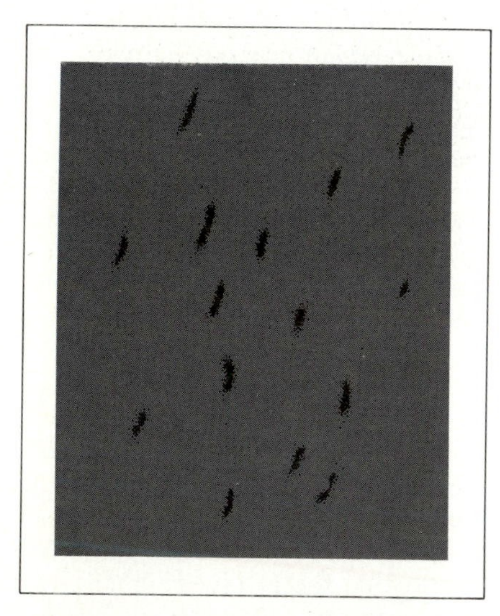

FIGURE 12.3 Schematic diagram of the probable alignment of dust particles in interstellar space.

are affected by a weak interstellar magnetic force field. How weak? Nearly a million times weaker than Earth's field. Containing a small amount of iron, each dust particle probably responds to the force field of magnetism in much the same way that the small iron filings of Figure 12.4 can be aligned by an ordinary magnet.

Of great usefulness, the analysis can be reversed: Measurements of the amount of blockage and polarization of light can indirectly yield information about the size, shape, and other geometrical properties of the interstellar dust particles. Much effort will doubtless be made during the next decade to better appreciate these rare, but ubiquitous interstellar contaminants.

Gaseous Nebulae

Now that we have a general idea of the basic contents and properties of interstellar space, let's examine some typical regions in more detail, noting especially how astronomers have observationally unraveled the nature of the matter contained within them.

Figure 12.5 is a mosaic of numerous photographs showing those parts of space where the stars, gas, and dust seem to congregate. The bright areas comprise myriads of unresolved stars. The dark areas represent vast pockets of dust, blocking from our view what would otherwise be a uniform distribution of bright starlight.

From our vantage point within the Solar System, this assembly of stellar and interstellar matter follows a bright band extending all across the sky. On a clear night, this band of patchy light is visible to the naked eye. And it's clearly seen in Figure 12.5. In the next chapter we'll come to recognize this band as the flattened disk—or galactic plane—of our much larger Milky Way Galaxy.

The dark dusty areas are neither regions lacking in stars nor holes through which we can see to infinity—views held by some researchers less than a century ago. They are simply foreground concentrations of interstellar matter, prohibiting us from seeing very far along the ga-

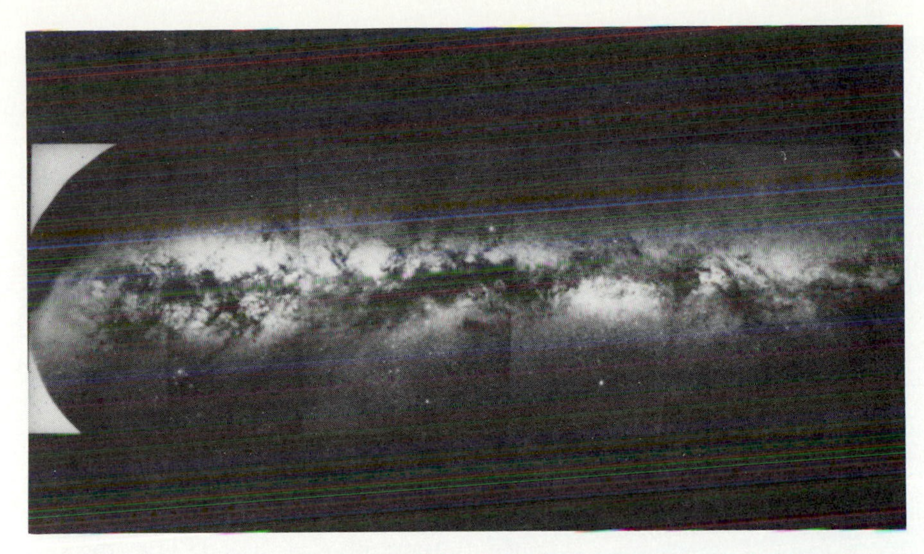

FIGURE 12.5 A mosaic of the band of the Milky Way Galaxy, known colloquially as the galactic plane. Photographed almost from horizon to horizon, and thus extending over nearly 180 arc degrees, this band displays heavy concentrations of stars, as well as interstellar gas and dust. (This view is roughly 20 times the expansiveness of that shown in Figure 12.1.) (See also Plate 4.)

lactic plane. Optically, with either naked eyes or large telescopes, we cannot observe galactic-plane objects more than a few thousand light-years distant.

Figure 12.6 shows a 12-degree swath of the galactic plane as photographed from Earth. (For comparison, the full Moon is only a $\frac{1}{2}$-degree across.) The view is rather mottled, with a patchy distribution of stars and interstellar debris. In addition, large fuzzy patches of light are clearly discernible. These fuzzy objects, labeled in Figure 12.7 as M8, M16, M17 and M20, correspond to the eighth, sixteenth, seventeenth, and twentieth objects in a catalog compiled by Messier, a seventeenth-century French astronomer. As can be noted, the stars, the fuzzy objects, and the dark obscuring matter are all concentrated around and along the galactic plane. Indeed, the galactic plane is the site of greatest concentration of almost all astronomical objects within our Milky Way.

The four fuzzy objects of Figure 12.7 are almost equidistant from Earth. The method of spectroscopic parallax discussed in Chapter 11 shows all of them to be 4000 to 5000 light-years away, near the limit of visibility for any object embedded in the dusty galactic plane. At that distance, M16 at the top left is approximately 1000 light-years from M8 at the bottom. So this figure still displays a rather large slice of interstellar space, although it's much smaller than the entire galactic plane shown in Figure 12.5.

We can gain a better appreciation for the various objects along the galactic plane by examining an even smaller region. Figure 12.8, for example, is an enlargement of only the region near the bottom of Figure 12.7. Now M20 is seen above, and M8 below, only a few arc degrees away. On the plane of the sky, about 100 light-years separate these two objects.

An even clearer indication of the interstellar matter along the galactic plane, especially near these fuzzy objects, can be seen in Figure 12.9, which is yet another enlargement of only the top of Figure 12.8. This is a close-up of the M20 object and of its immediate environment. The total area photographed measures some 50 light-years across, at the 4000-light-year distance of M20. (Consult also the magnificent Plate 5, for this M20 object is a real jewel of the night.)

Fuzzy objects such as M20 are termed **gaseous nebulae** (or "emission nebulae" by some astronomers). Absolutely beautiful sights when viewed in color under high resolution, they are among the most spectacular objects in the entire Uni-

FIGURE 12.6 A photograph of a small portion (about 12 arc degrees) of the galactic plane shown in Figure 12.5, displaying higher-resolution evidence for stars, gas, and dust, as well as several distinct fuzzy patches of light.

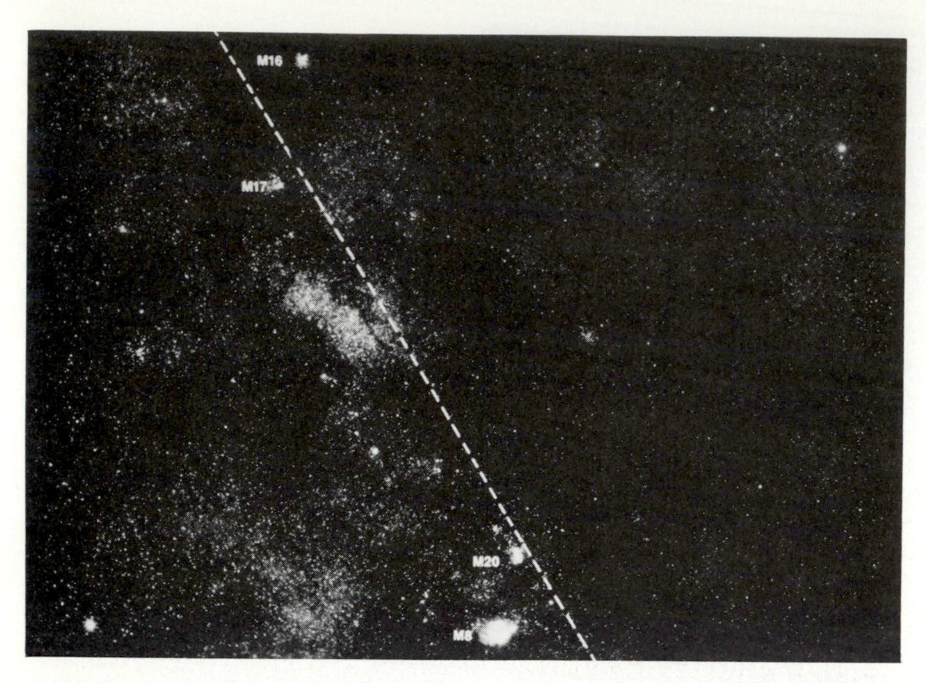

FIGURE 12.7 The same photograph as illustrated in the preceding figure, though here with the fuzzy patches marked relative to the middle of the galactic plane depicted by the dashed line.

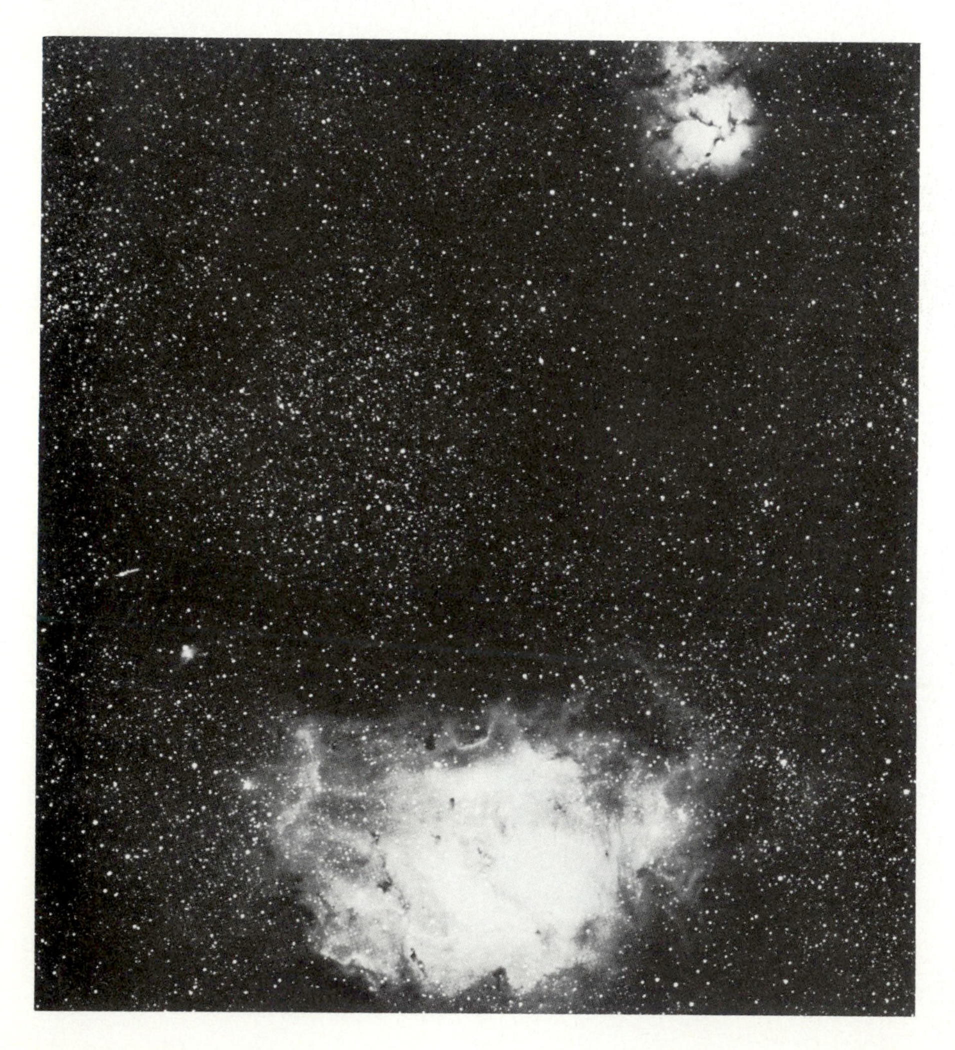

FIGURE 12.8 An enlargement of the bottom of Figure 12.7, showing the M20 object (top) and the M8 object (bottom) more clearly.

verse. We often find them adorning all manner of advertisements, wall posters, and the centerfolds of most astronomy textbooks—and this one is no different. But, remember, they appear only as small, almost indecipherable patches of light when viewed in the larger context of the entire galactic plane, like that depicted in Figure 12.5 or Plate 4(a). Perspective is crucial in astronomy.

Gaseous nebulae are regions of glowing, ionized gas. At or near the center of every one is at least one hot, O- or B-type star producing copious amounts of ultraviolet photons. Zipping away from the star, these photons collide with the surrounding gas, and ionize much of it. The gas then glows, emitting radiation as some of the electrons recombine with their parent atoms.

M20's reddish hue (see Plate 5) results from gargantuan quantities of hydrogen atoms constantly emitting light in the red part of the electromagnetic spectrum. Specifically, the red color in the M20 nebula—in fact in any nebula like it—results from the emission of radiation at 6563 angstroms. Other elements in the nebula also emit radiation as their electrons recombine, but since hydrogen is so plentiful, its emission overwhelms that from all other kinds of atoms. Tinged here and there with other colors, red usually dominates in gaseous nebulae.

Woven through the glowing nebular gas are several dust lanes of dark obscuring matter. Recent studies have demonstrated that this dust is mixed directly with the nebular gas and is not just unrelated foreground obscuration somewhere along the line of sight. M20 is a classic example of such an interstellar region having a mixture of gas and dust.

IONIZATION: Now let's investigate the nature of interstellar space that produces these stunning astronom-

FIGURE 12.9 A further enlargement of the top of Figure 12.8, showing only the M20 gaseous nebula, as well as its surrounding interstellar environment. The nebula itself is about 25 light-years in diameter and is colloquially called the Trifid Nebula because of the way that the clearly seen dust lanes trisect its midsection. (See also Plate 5.)

ical objects. As already noted, gaseous nebulae are regions of matter ionized by the very energetic ultraviolet radiation emitted by embedded O- or B-type stars. Only stars of these types have hot enough surface temperatures to radiate the high-energy photons needed to strip electrons from the atoms of the surrounding interstellar gas.

The top of Figure 12.10 is a schematic diagram of the physical process of ionization. This interaction of radiation and matter can be written in symbolic form as follows:

ionization process:
atom + photon → ion + electron.

Since hydrogen is overwhelmingly the most abundant element in the interstellar gas, in most cases this process amounts to,

hydrogen atom + photon →
proton + electron.

The result is a mixture of free electrons and free protons, as well as a lesser number of heavier ions. Such a totally ionized region, you will recall from Chapter 2, is termed a plasma—often considered the fourth state of matter.

Once formed, the central hot star or stars begin to emit radiation, causing the nebula to grow. The ionized region becomes progressively larger as an increasing number of ultraviolet photons interact with and are absorbed by the surrounding interstellar gas. We might prefer to view the ionization as "eating" its way into the normally neutral matter of interstellar space. Eventually, though, nebulae reach a limiting size, beyond which they cannot become any larger.

Figure 12.11 shows a schematic diagram of a gaseous nebula. Its final size can range from about 0.1 light-year for particularly compact nebulae, to some 50 light-years for rather extended ones. Their actual size depends on the number of hot stars as well as the interstellar gas density near those stars. Logically, the thinner the gas and/or the greater the number of central hot stars, the larger the resulting nebula. On average, the typical nebular size of about a few light-years is much larger—several orders of magnitude larger—than our Solar System. Typical nebulae are also much more massive, for they usually contain hundreds of times more ionized gas than in our Sun.

RECOMBINATION: You may be puzzled at the finite size of nebulae. If the embedded hot stars continuously emit radiation, then why don't nebulae grow indefinitely? The answer has to do with another atomic process operating in the nebular gas.

In addition to the ionization process outlined above, a recombination process is also at work in the plasma. This process is sketched at the bottom of Figure 12.10 and can be de-

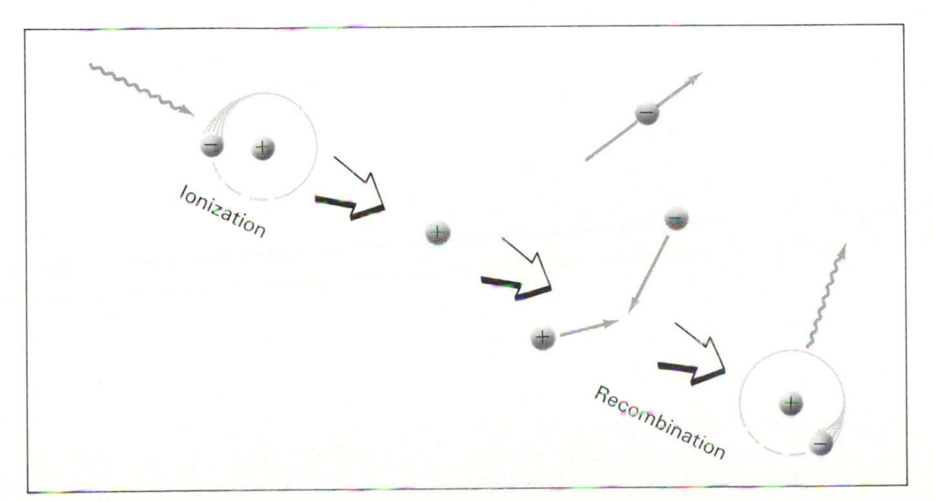

FIGURE 12.10 Schematic diagram of the ionization process, followed by the recombination process (drawn here for the hydrogen atom).

FIGURE 12.11 A gaseous nebula often resembles an island of hot, ionized, glowing interstellar matter embedded within an immense "sea" of cold, neutral, dark interstellar matter.

scribed symbolically by the following atomic reaction:

recombination process:
ion + electron → atom + photon,

or, as is most often the case,

proton + electron →
hydrogen atom + photon.

As can be seen, this recombination process is just the opposite of the ionization process. Here, free electrons are captured by protons (or by heavier ions), recombine with them, and produce once again intact atoms. In the process, photons are emitted as electrons cascade down the atomic orbitals within newly formed atoms. Recombination thereby competes with ionization—one process ionizing atoms, the other process neutralizing them—thus confining the plasma to a rather localized region of interstellar space.

NEBULAR SPECTRA: The photons emitted by means of this recombination process often escape the nebular gas. These photons fail to produce more ionized gas beyond nebulae, since they are much less energetic than the ultraviolet photons originally emitted by the embedded stars. Some of the nebular photons eventually reach Earth, where we can detect them with telescopes. In fact, only by means of these lower-energy photons can we learn anything about gaseous nebulae.

In a certain sense, we might find it useful to think of the radiation reaching us from gaseous nebulae as a downgraded form of stellar radiation. Much as in the convection zones inside stars, an embedded star's radiation is reprocessed or filtered during its passage through the surrounding nebular gas. Accordingly, the photons detected at Earth provide direct information about the nebular gas as well as indirect information about the embedded hot stars.

Just as we studied the spectra of ordinary stars and derived a wealth of information about their ionized stellar surfaces, we can also measure nebular spectra to infer a great deal of knowledge about ionized interstellar gas. Since at least one hot star resides near the center of the nebular gas, we might expect that the combined spectrum of the two—the star and the nebula—would be hopelessly confused. But they're not. The nebular spectrum can easily be distinguished from the stellar spectrum, for the physical conditions differ substantially in stars and nebulae. In particular, gaseous nebulae are made of hot thin gas which, in accord with our discussion in Chapter 2, yields detectable *emission* lines. Such nebular emission lines completely contrast with the absorption lines normally displayed by stars.

Figure 12.12 is a typical nebular emission spectrum spanning much of the visible wavelength interval from 4000 to 7000 angstroms. As noted in Chapter 2, we can most usefully extract information once the optical spectrum has been traced in detail. This figure also shows such a tracing, made by passing a narrow beam of light across the optical spectrum.

In theory, each line should exhibit a bell-shaped curve, as also discussed in Chapter 2, and noted again here in Figure 12.13. By way of reminder, such lines are characterized by an intensity at line center, a full width at half the maximum intensity, and a central frequency at which the line is peaked. In reality, though, observed spectral features are hardly ever this simple. Usually, as shown in Figure 12.12, real spectral lines have undesirable noise superposed on them; furthermore, many spectral lines often tend to overlap one another.

Figure 12.14 illustrates another nebular spectrum, this one observed at radio frequencies toward the M20 object. Messy as the lines are, we can still interpret them in terms of their basic bell-shaped curve, and thereby

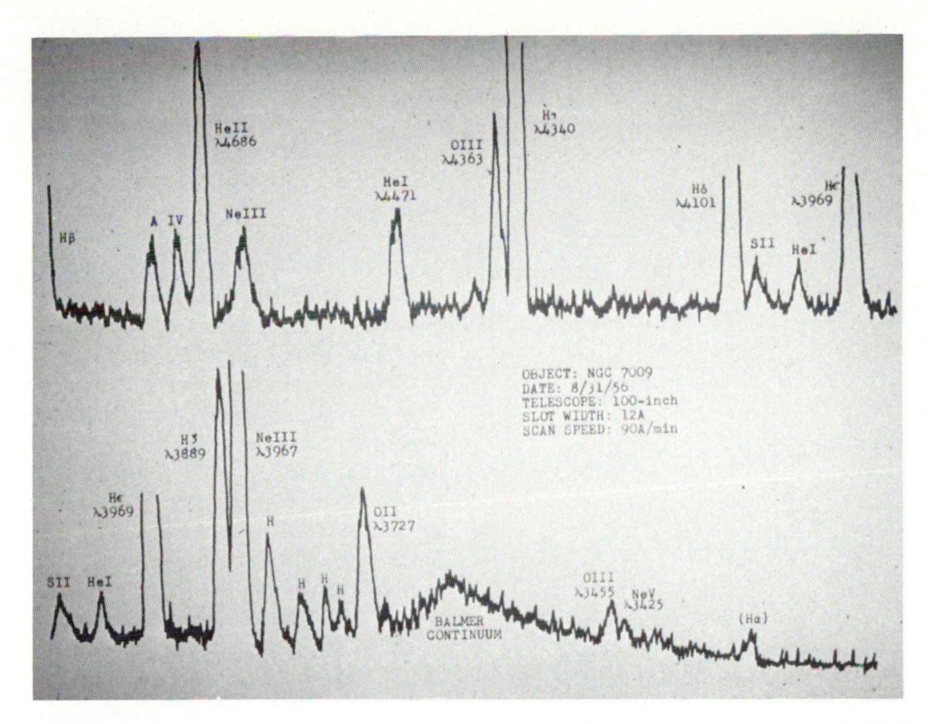

FIGURE 12.12 A nebular emission spectrum can be interpreted more accurately by analyzing its traced spectrum shown here.

specify each line's intensity, width, and centroid.

Line intensity is important when deriving elemental abundances. That's because intensity relates to the amount of gas contributing to the emission lines. In recent years, analysis of nebular spectra have given us reasonably good estimates of all the chemical abundances of the *ionized* interstellar gas. The results agree very closely with the cosmic abundances already noted in Figure 2.20: Hydrogen is 90 percent abundant, followed by helium at about 9 percent, while the heavier elements comprise the remaining 1 percent.

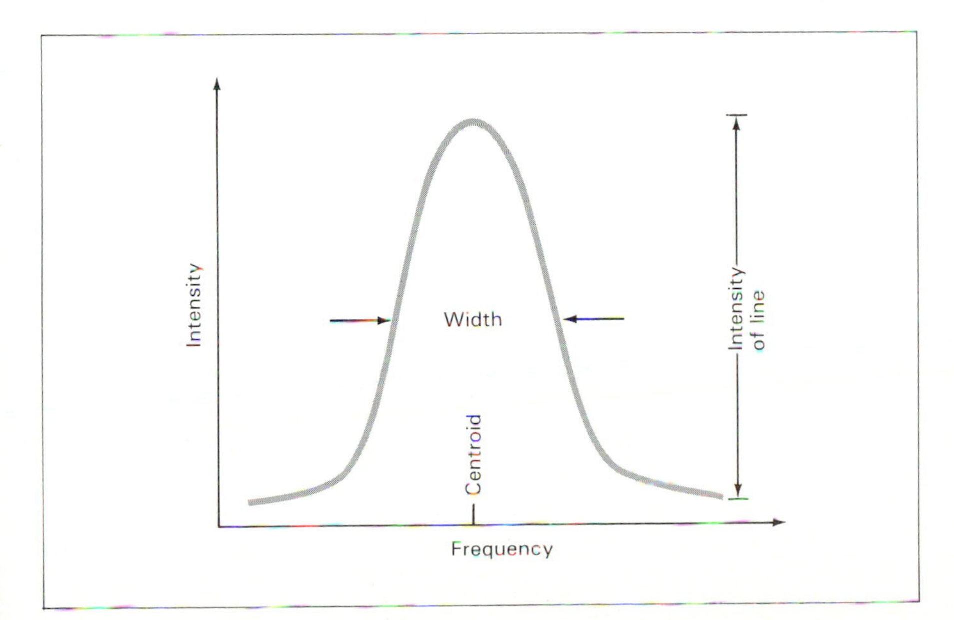

FIGURE 12.13 "Textbook" spectral lines resemble bell-shaped curves. They are characterized by an intensity, a width, and a centroid.

Knowing the total gas abundance, we can measure a nebula's size (and therefore volume) to get an estimate of its density. A typical result is about 100 particles/cubic centimeter, a very small value. Expressed another way, 100 particles/cubic centimeter equals about 10^{-22} gram/cubic centimeter, assuming that 1 gram of typical nebular plasma contains about 10^{24} atoms and ions. That's a density some 22 orders of magnitude thinner than planetary densities.

Spectral-line widths also provide much information. If all the nebular atoms and ions were at rest, each emitted photon of a given kind of atom or ion would have precisely the same energy, and the various spectral lines would seem exceedingly narrow, like spikes with no breadth. But nebular atoms and ions are not at rest. All gaseous nebulae contain enormous numbers of atoms and ions that have motions—random motions to and fro, back and forth. The hotter the gas, the greater the random motions. All real spectral lines, as we might suspect from Figures 12.12 and 12.14, have some breadth. And this breadth is mostly due to the motions of the emitting gas.

The random shufflings of the nebular particles cause some of them to be red shifted with respect to Earth, while others are blue shifted. Figure 12.15 reminds us of our earlier discussion of line-broadening mechanisms in Chapter 2. In a typical parcel of gas, only a small proportion of the atoms or ions are likely to have large red or blue shifts; greater numbers have small red or blue shifts; most have hardly any Doppler shift. Provided that sufficient numbers of emitting particles move every which way at random, the resulting spectral features will have the standard bell-shaped curve.

The width of a spectral line, then, depends on the particle motions, and this motion in turn depends mainly on the gas temperature. Hence we can turn the problem around by inferring the temperature of gaseous

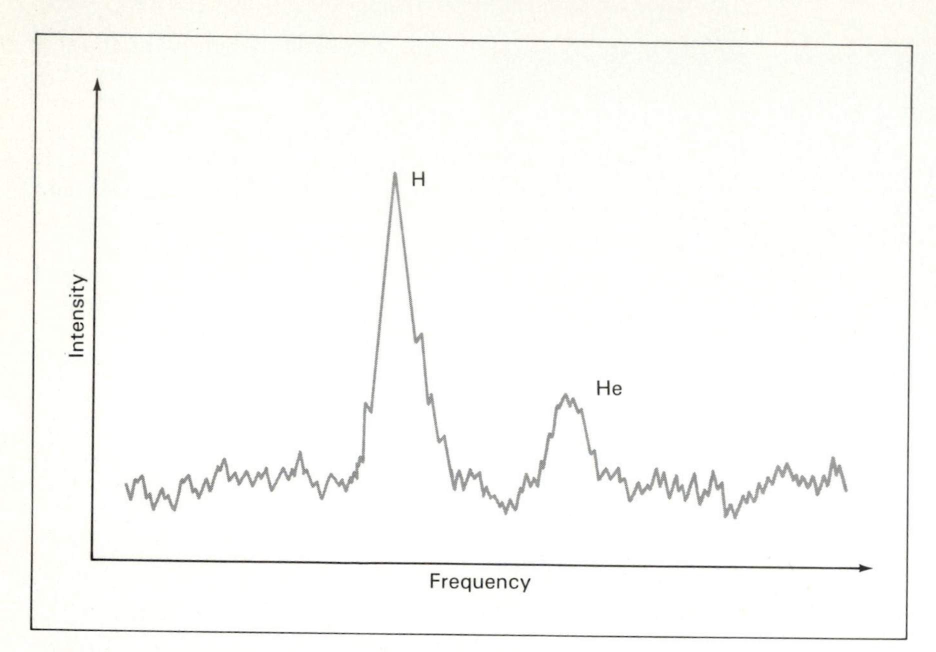

FIGURE 12.14 A real spectrum observed toward the M20 nebula at the central radio frequency of precisely 5009732060 Hertz (or 5.98834-centimeter wavelength). These signals result from large quantities of photons emitted as electrons change between orbital 111 and orbital 110 within the myriad hydrogen (H) atoms of the nebula. A weaker helium (He) spectral line can be seen at right.

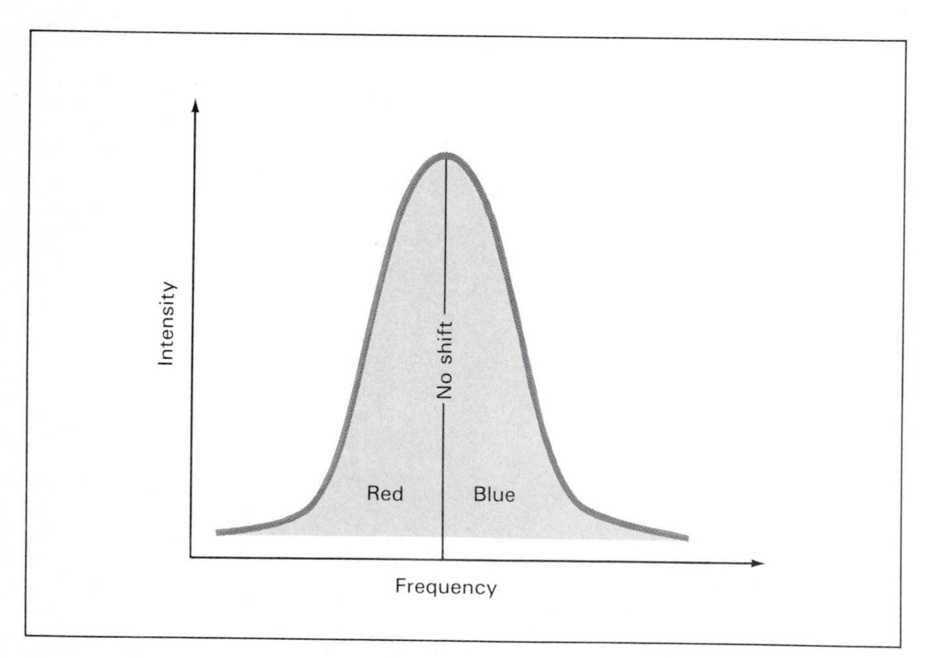

FIGURE 12.15 Red and blue shifts of individual nebular atoms or ions tend to broaden spectral lines.

nebulae from the observed widths of their emitted spectral lines. For all gaseous nebulae, the widths imply that the gas atoms and ions have temperatures of approximately 10,000 kelvin.

The third property of a spectral line, namely the line centroid, yields still more information about nebulae. The line center is especially useful when studying the bulk motions of a nebula. This is true because Doppler shift depends on radial velocity as discussed in Chapter 1. In fact, the extent of Doppler shift is a direct measure of velocity—in this case, radial velocity of the emitting gas atoms. Thus the frequency axis, normally labeled in units of hertz or angstroms, can also be expressed in units of velocity (e.g., kilometers/second). In this way we can measure a nebula's radial velocity relative to Earth. [Consult Plate 1 (d).]

Observations of spectral-line radiation from several positions across a nebula can also yield information about the internal motions of a nebula itself. Figure 12.16 is an example of the velocity changes across the M20 nebula. The drawn contours connect locations of similar velocity, much like pressure contours on a meteorologist's weather map connect regions of similar pressure. In the case of M20, the contours suggest some nebular rotation, although the actual motions are probably more complex than that.

Table 12-1 lists some of the vital statistics found for each of the gaseous nebulae of Figure 12.7.

Dark Dust Clouds

Gaseous nebulae represent only one small component of interstellar space. Most of space—more than 99 percent of it—is devoid of nebulae. Nor does it contain stars. Instead, the greatest volumes of interstellar space are just plain dark; take another look at Figure 12.5, or ponder the evening sky. The darkened realms are the most representative regions of interstellar space—regions where temperatures (about 100 kelvin) and densities (about 1 atom/cubic centimeter) are extremely small by terrestrial standards.

In this dark void outside nebulae and stars lurks another type of astronomical object. These are the **dark dust clouds**, or just interstellar clouds for short. We call such regions "clouds" because their amorphous morphology resembles terrestrial clouds in Earth's atmosphere. But don't be fooled. Inter-

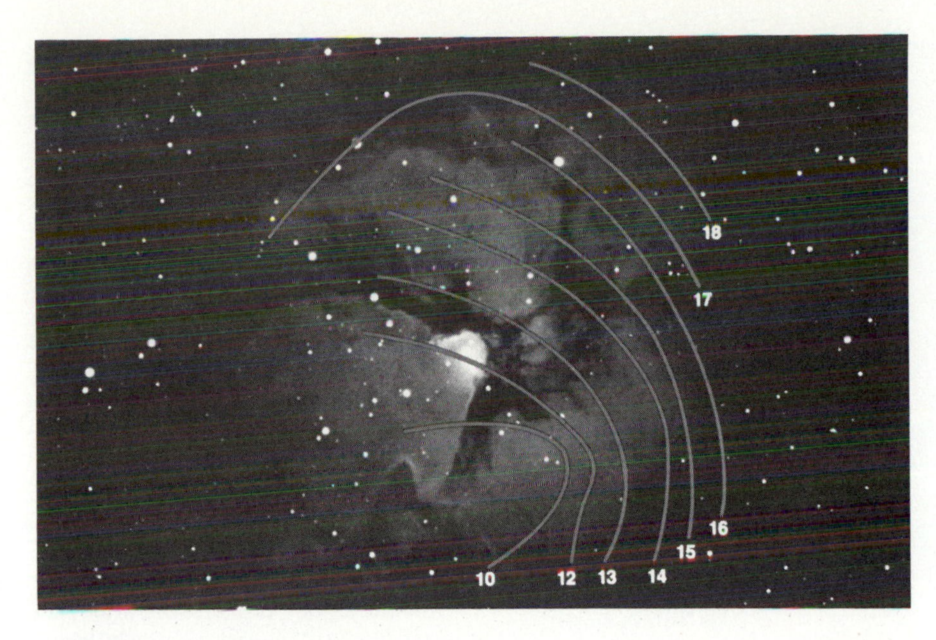

FIGURE 12.16 Velocity contours measured for the plasma in the M20 nebula. The labeled values have units of kilometers/second.

stellar clouds differ greatly from clouds on our own planet. And they are much, much larger than terrestrial clouds; most interstellar clouds are bigger than our Solar System, some even stretching several light-years across. Yet despite their huge size, all the interstellar clouds together comprise no more than a few percent of the entire volume of interstellar space.

So, recognize that dark dust clouds are largely misnamed. Yes, they are dark, but they harbor not only dust. These clouds also contain huge amounts of gas. In fact, the ratio of gas atoms to dust particles is still about 10^{12} in these clouds, just as for general interstellar space discussed earlier. And second, interstellar clouds resemble terrestrial clouds only in a most superficial sense. Interstellar clouds are colder, bigger, and less dense than anything familiar to us on Earth.

CLOUDS BLOCK LIGHT: Figure 12.17 is a photograph of a typical dust cloud of interstellar space. Pockets of intense blackness mark regions where the dust *and gas* are especially concentrated. This cloud takes its peculiar name from a nearby star, Rho Ophiuchi, and resides at the relatively nearby distance of about 1000 light-years. Measuring several light-years across, this cloud is only a minuscule part of the mosaic shown in Figure 12.5. Note especially the long "streamers" of (relatively) dense dust and gas; this cloud is clearly not spherical in any sense, seeming more like a randomly kinked blob with extended tentacles. Actually, that's probably a good description of it and others like it. Most interstellar clouds are very irregularly shaped.

We now realize that herds of these dark and dusty interstellar clouds are sprinkled throughout our Galaxy. Until recently, they could be studied only if they happened to obscure the light emitted by distant stars and nebulae. The dark dust lanes trisecting the M20 nebula of Figure 12.9 are a good example. There the dust is apparent only because it blocks the light behind it.

Figure 12.18 shows another example of dust that obscures background light. This illustration maps the extent of obscuration across the Orion Nebula, another nearby gaseous nebula. The location and extent of the dust was derived by comparing the undiminished intensity of radio radiation with the slightly diminished intensity of visible radiation. Remember, the radio map of a region such as this (see Figure 3.22) yields a *true* map of the nebular gas; long-wavelength radio radiation travels unhindered through any dust between the nebula and Earth. Light radiation, however, is scattered; none of it reaches Earth if the scattering is severe enough. The difference between these two impressions—the true radio map and the distorted optical image—provides an estimate of the amount of dustiness along our line of sight. Figure 12.18 is a map of that dustiness.

The earliest (1930s) awareness of interstellar dark dust clouds came from the details of the optical spectra of stars. The wide absorption lines normally formed in the lower atmospheres of stars are often accompanied by narrow absorption lines as well. The narrow lines immediately suggest cooler regions. Indeed, after penetrating its own hot atmosphere, light emitted by a star is often ab-

TABLE 12-1
Nebular Properties

OBJECT	DISTANCE (LIGHT-YEARS)	RADIUS (LIGHT-YEARS)	DENSITY (PARTICLES/CM^3)	MASS (SOLAR MASSES)	TEMPERATURE (KELVIN)
M8	3800	9	400	60	7500
M16	5000	14	200	600	8000
M17	4500	10	800	300	8700
M20	4300	12	100	150	8200

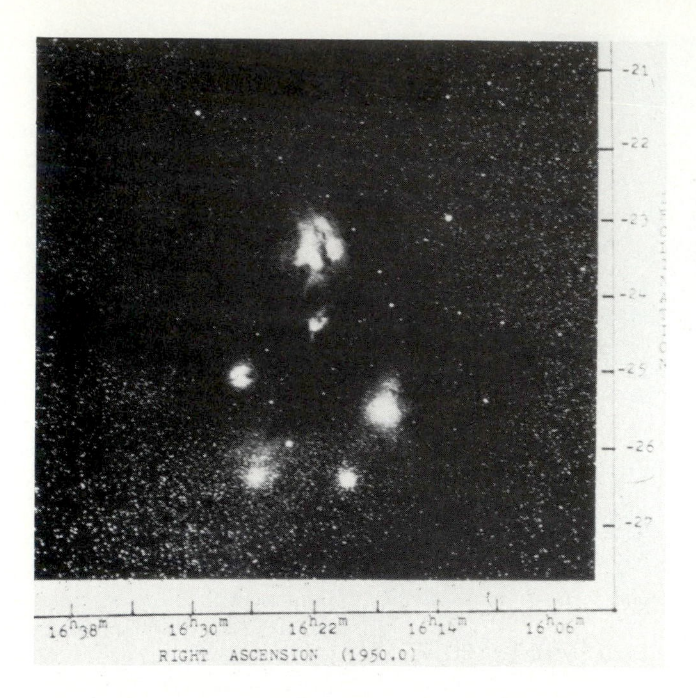

FIGURE 12.17 Photograph of a typical interstellar dark dust cloud. This cloud has the tongue-twisting name of Rho Ophiuchi and extends for several light-years.

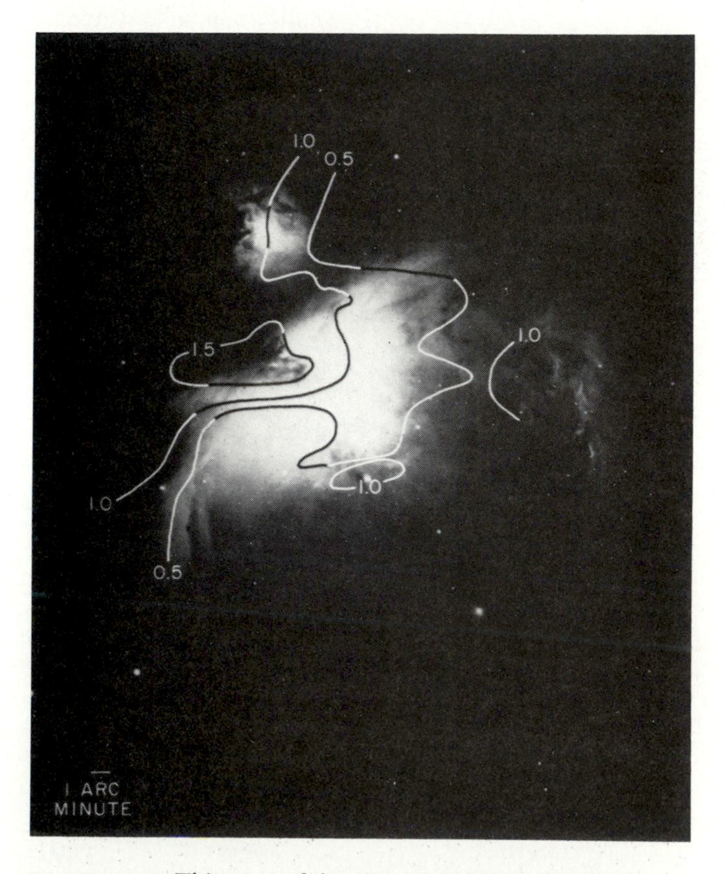

FIGURE 12.18 This map of the obscuring dust across the Orion Nebula was derived by comparing the optical [Figure 3.20(a)] and radio [Figure 3.20(b)] observations of the nebula. (The numbers on the contours merely denote the relative amounts of dustiness.)

sorbed by cold interstellar regions somewhere along the line of sight.

Figure 12.19(a) is a schematic diagram of a typical case. Light from a distant star passes through some interstellar clouds on its way to Earth. Such clouds need not even be close to the star, and usually they are not. Each cloud absorbs some of the stellar radiation in a manner that largely depends on the cloud's temperature, density, and elemental abundance.

Figure 12.19(b) depicts a tracing of a typical resulting spectrum. The narrow absorption lines allow us to derive many of the same physical properties obtained earlier from nebular emission lines or stellar absorption lines. For example, the intensity of the absorption—the amount of background starlight absorbed at a particular frequency—is a rough measure of the elemental abundance *in the absorbing clouds*. For most interstellar clouds, the relative abundances mimic once again the cosmic abundances typifying most astronomical objects (but consult Interlude 12-1). This is hardly surprising, for interstellar clouds are the regions from which stars and nebulae form. Since almost the beginning of time, most of the matter in the Universe has become fairly well mixed, repeatedly processed in and out of stars, nebulae, clouds, and so on. The matter content seems to be, on average, much the same for most astronomical objects, although we do know of some important exceptions such as the old globular clusters noted earlier.

The narrow widths of the absorption lines mean that the temperatures of the interstellar clouds are much less than the thousands of kelvin characterizing gaseous nebulae and stellar atmospheres. The lines are often so narrow as to imply gas temperatures of about 100 kelvin or less. None of this interstellar gas can possibly be ionized; it must be completely neutral matter. That's why the clouds are invisible; they are just too cold to emit any visible light.

We can find the density of an interstellar cloud by estimating the size and shape of the cloud (i.e., its volume) and then measuring the amount of the most abundant ele-

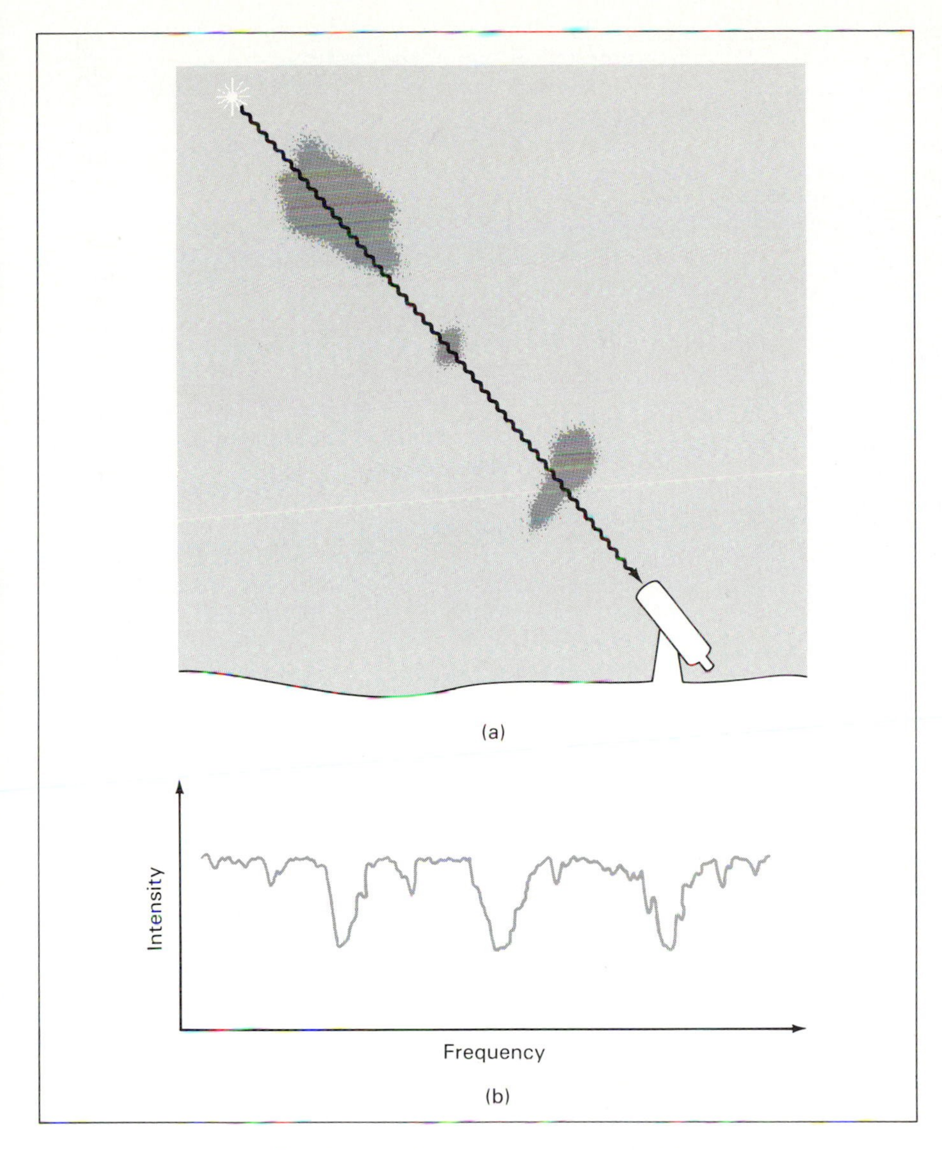

FIGURE 12.19 A simplified diagram (a) of some interstellar clouds between a hot star and Earth. Optical observations might show an absorption spectrum like that traced in (b). The wide, intense lines are formed in the star's hot atmosphere, whereas the narrow, weak lines arise from the cold interstellar clouds.

ment, hydrogen, contained within. Along any given line of sight, some clouds have densities of 10 atoms/cubic centimeter, while others reach 10^6 atoms/cubic centimeter. Clouds with the latter densities are called *dense* interstellar clouds by researchers, but you should recognize that even these densest interstellar regions are still as tenuous as the best laboratory vacuum. Actually, it is only because their density is larger than the usual 1 atom/cubic centimeter that we can distinguish clouds from the surrounding expanse of general interstellar space.

Finally, the absorption lines yield limited amounts of information about the motions of the interstellar clouds. For example, their various red and blue shifts enable us to use the Doppler effect to derive the clouds' velocities. However, details cannot be extracted. The basic problem is that this optical technique can examine interstellar clouds *only* along the line of sight to hot background stars. Astronomers need to see the stars in order to decipher the radiation passing through the clouds. Consequently, optical astronomers cannot easily map clouds by observing spectra at various places across the clouds; a foreground cloud can be studied only at those precise localities through which the veiled stars' light passes on its way to Earth.

The need to see stars through clouds also restricts this optical technique to relatively local regimes of a few thousand light-years from Earth. This is especially true in the plane of the Galaxy where the optical obscuration is heavy.

21-CENTIMETER RADIATION: To probe interstellar space more thoroughly, another method is clearly required. We need a more general, more versatile observational technique that does not rely on conveniently located stars or nebulae. In short, we need a method capable of detecting cold, neutral interstellar matter anywhere in space. Although this may sound impossible, such an observational technique does exist. The method relies on low-energy radio transmissions from the general interstellar gas.

This radio technique is crucial to our modern understanding of interstellar space, so let's consider it in some detail. Recall that the hydrogen atom has one electron orbiting about a single proton nucleus. Besides the electron's orbital motion around the central proton, electrons also have some rotational motion—that is, spin—about their own axis. The proton also spins as well. This model is roughly analogous to a planetary system where, in addition to the orbital motion of a planet about a central star, the planet (electron) and the star (proton) each rotate of their own accord.

When the spin of the electron parallels the spin of the proton—that is, when they rotate in the same sense—the hydrogen atom is energized a little more than when the electron and proton spin in opposite directions. Figure 12.20 illustrates these two energy states, each of which refers to the hydrogen atom's first orbital. The lower of these two energy states

INTERLUDE 12-1 *A Satellite Named IUE*

Much can be learned about the *local* regions of interstellar space by studying ultraviolet radiation. (This short-wavelength technique is restricted to local clouds because it requires hot background stars that are normally obscured except for those within a few hundred light-years.) And since Earth's atmosphere becomes increasingly opaque at wavelengths shorter than 4000 angstroms, such observations must be made from space.

One of the most successful satellites ever launched is called the *International Ultraviolet Explorer,* or *IUE* for short. Shown in an artist's drawing as Figure 3.34, the *IUE* satellite contains several on-board spectrometers capable of resolving narrow spectral features between about 1000 and 3000 Angstroms. This orbiting observatory is totally controlled from the ground; gas-powered jets on the spacecraft give the thrust needed to orient the telescope on command. On-board star trackers point the craft using a navigational map based on the known positions of the stars. Large solar panels provide the electrical power needed to operate the instruments and the telemetry systems that send the data back to Earth. The accompanying figure shows the *IUE* control center at NASA's Goddard Space Flight Center outside Washington.

The line-of-sight technique described in the text for optical observations can also be used for these ultraviolet observations. In other words, we must observe the ultraviolet spectrum of a hot star, carefully noting the spectrum for particularly narrow absorption lines caused by intervening interstellar clouds. By this means, many interstellar gases can be studied, including hydrogen (both atomic and molecular), carbon, nitrogen, oxygen, iron, and several others. Interestingly enough, some of these gases have slightly lower abundances in the galactic clouds than in our Solar System and in stars. The clouds' elemental composition, then, deviates, slightly but measurably, from those of the cosmic abundance scale plotted earlier in Figure 2.20. As suggested in one of the first sections of this chapter, substantial quantities of familiar elements such as carbon, oxygen, silicon, magnesium, and iron have probably been used to form interstellar dust particles. This idea is supported by infrared observations sensitive to the way that iron and magnesium silicates absorb radiation. If our interpretation of these ultraviolet-infrared measurements is correct, then the interstellar dust particles must account for at least a fraction of the elements missing from the interstellar gas.

Among its many other interesting findings, *IUE* confirmed evidence (discovered by an earlier ultraviolet satellite named *Copernicus*) for speeding clouds in space. Usually, interstellar clouds wander around with velocities of 20 kilometers/second or less. However, in some regions of space, the ultraviolet observations have clocked clouds and "sheets" of matter moving nearly 100 kilometers/second (or more than 200,000 miles/hour). The origin of these swift regions of matter is not well understood, but they may have been pushed around by explosions of old stars or by stellar winds like those escaping the Sun (consult Interlude 11-3).

Furthermore, ultraviolet observations of weak spectral lines from highly excited atoms imply that some parts of interstellar space are much thinner (0.005 atom/cubic centimeter) and hotter (500,000 kelvin) than expected. Although we currently lack a complete inventory, much of interstellar space not filled with clouds or nebulae might be made of extremely dilute, yet seething plasma that probably results from the concussions and expanding debris of ancient stars blown up long ago. Interstellar "bubbles" has been suggested as a name for such regions, as has "intercloud medium." Such a superheated "cosmic bubble bath" extends far into interstellar space beyond our local neighborhood, and even conceivably stretches into the vast intergalactic spaces among the galaxies.

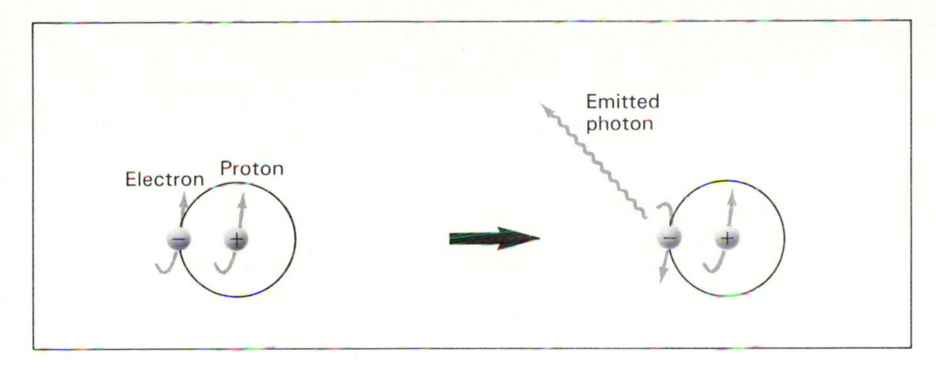

FIGURE 12.20 Schematic diagram of a ground-level hydrogen atom changing from a higher energy state (electron and proton spins parallel) to a lower energy state (antiparellel spins). The emitted photon carries away an energy equal to the energy difference of the two spin states.

is the least active one, so hydrogen atoms usually exist in this state. In fact, the great majority of the gargantuan quantities of hydrogen atoms throughout cold interstellar space are not excited in any way, and hence enjoy the lowest possible energy state.

On occasion, a hydrogen atom may become *slightly* excited—not excited enough to eject the electron thus forming an ion, not even excited enough to boost the electron to the second orbital. By "slightly" excited, we mean just enough to change the spins of the electron and proton from an antiparallel state to a parallel state. Hydrogen atoms can become excited in this way by absorbing very low-energy photons that just happen to be wandering through interstellar space—a **radiative process**. Or, hydrogen atoms can absorb a small part of another particle's kinetic energy by banging into it—a **collisional process**.

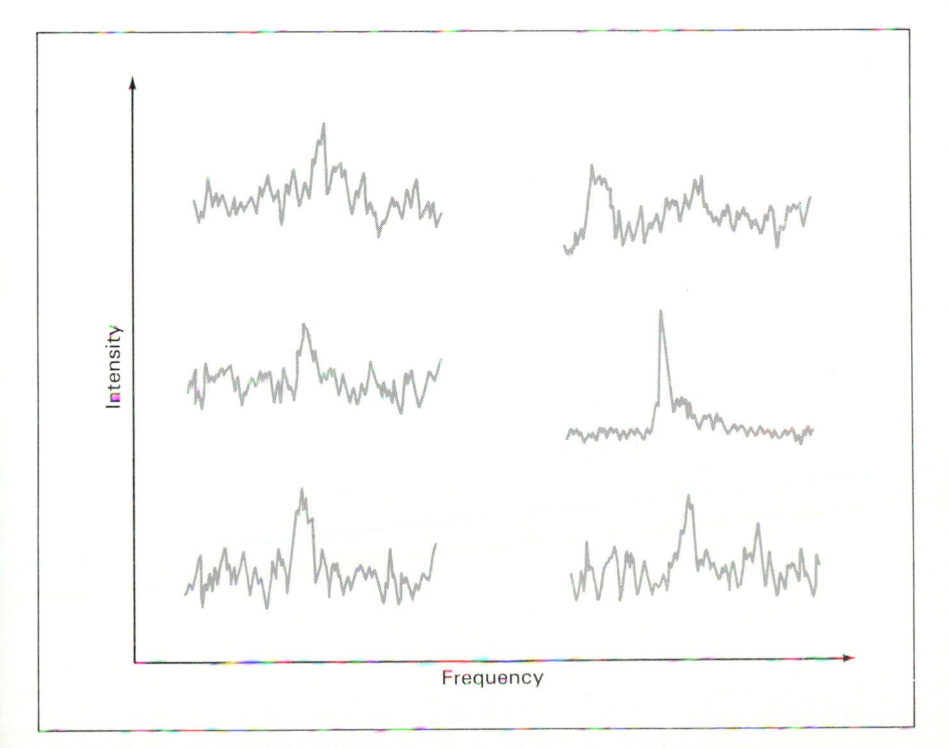

FIGURE 12.21 Typical 21-centimeter specral lines observed toward several different directions of interstellar space.

All matter in the Universe tries to achieve its lowest possible energy state, and interstellar gas is no exception. Even slightly excited hydrogen atoms strive for as little energy as possible. Consequently, after a certain amount of time, hydrogen atoms having the higher of these two ground states relax back to the lower state. That is, the electron suddenly and spontaneously flips its spin, becoming opposite to that of the spinning proton. As with any change of state, the transition from the higher-energy state to the lower-energy state releases a photon. And the laws of atomic physics demand that the energy of the emitted photon equal the difference between the two atomic energy states.

Very little energy is involved in the flip of an electron's spin. Because the energy of hydrogen atoms in the higher state scantly surpasses that of those in the lower state, the emitted photon carries extremely low energy. And since the energy of a photon is directly proportional to its frequency (consult Chapter 2), the frequency of the emitted radiation is also quite small. It amounts to 1.42 billion hertz; that's small compared to the million billion hertz for ordinary light radiation. Conversely, we could say that the wavelength of the radiation is rather long—about 21 centimeters, or roughly the width of this book. Waves of these dimensions belong to the radio portion of the electromagnetic spectrum and are thus detectable only with radio telescopes. Researchers refer to the spectral line resulting from this hydrogen-spin-flip process as the **21-centimeter line**.

Figure 12.21 shows some typical spectral profiles of 21-centimeter radio signals observed toward several different regions of space. These tracings are the characteristic signatures of cold, neutral hydrogen in various parts of our Galaxy. Free from the need to observe starlight, radio telescopes can "listen" to any interstellar region having enough gas to produce an observable signal. Notice the lack of symmetrical bell-shaped features. Such perfect 21-centimeter lines are hardly ever observed. Real spectral lines are jagged and multifaceted because the radio signals from numerous interstellar clouds are superposed, each with a different radial velocity, and each spread along the line of sight sampled by the telescope.

Analyses of 21-centimeter spectral profiles like those of Figure 12.21 verify that most of interstellar space has physical conditions much like those found by the previously discussed optical technique where local starlight passes through line-of-sight interstellar clouds. The 21-centimeter observations generally imply gas temperatures of about 100 kelvin, and densities somewhat less than the aforementioned clouds, namely about 1 atom/cubic centimeter.

Of great importance, the wavelength of this spectral-line technique—21 centimeters—is much longer than the typical size of dust particles. Accordingly, this radio radiation reaches Earth completely unscattered by interstellar debris. Indeed, the opportunity to observe interstellar space well beyond a few thousand light-years, as well as in directions devoid of background stars, makes 21-centimeter observations among the most important and useful in all of astronomy. We'll use this 21-centimeter-line technique in next chapter's study of our own Galaxy, as well as in many subsequent chapters while encountering truly distant astronomical objects.

Interstellar Molecules

In certain interstellar regions of cold (typically 20 kelvin) neutral gas, the densities reach as high as 10^6 particles/cubic centimeter. Until recently, such regions were simply regarded as abnormally dense interstellar clouds. But astronomers now recognize that these very dense clouds belong to an entirely new class of interstellar matter. Of special importance, the gas particles in these regions are not atoms at all; they are molecules. By far the most abundant particles are those of diatomic molecular hydrogen, H_2, but more complex molecules are also known to exist in smaller amounts. The predominance of molecules is the main reason why these relatively dense interstellar regions rate a special name—**molecular clouds**.

Only within recent years have astronomers begun to appreciate the vastness of these clouds. Most molecular clouds are huge, literally dwarfing gaseous nebulae, which were heretofore thought to be the most massive residents of interstellar space. Clearly, then, the molecular clouds must play an important role in our understanding of the darkened regions beyond the stars. To appreciate them fully, let's briefly review some molecular physics, and then examine how the molecules are used as diagnostic tools to probe these fascinating interstellar regions.

MOLECULAR PHYSICS: A molecule is a combination of atoms. They range from simple diatomic molecules made of two atoms to extremely complex molecules containing literally millions of atoms. Generally, on Earth, the simpler molecules are studied by chemists, while the larger ones fall into the domain of biologists. Lately, however, both small and large molecules have become of increasing interest to astronomers.

As already noted near the end of Chapter 2, molecules, like atoms, emit and absorb energy. They can absorb energy either by colliding with other molecules or atoms, or by capturing photons of radiation. In other words, molecules can be collisionally or radiatively excited much like atoms. Furthermore, like atoms, molecules prefer to relax back to their ground, or lowest-energy states whenever the opportunity arises. In the process, they emit radiation.

Energy states of molecules are more complex than those of atoms. Molecules can undergo internal electron transitions as do atoms, but molecules can also rotate and vibrate. They do so in very specific, quantized ways, obeying the complex laws of quantum physics. Consider a simple example.

Figure 12.22 illustrates a simple molecule rotating rapidly. After a certain amount of time that depends on its internal makeup, the molecule relaxes back to a slower rotational rate (or lower-energy state). As with atoms, this change causes a photon to be emitted. And this photon carries an energy equal to the energy difference of the two rotational states. It so happens that the energy differences between *rotational* states are generally small. Hence the emitted radiation is of the radio variety.

RADIO AND INFRARED PROBING: The emission of radio radiation from molecules is fortunate, for they are

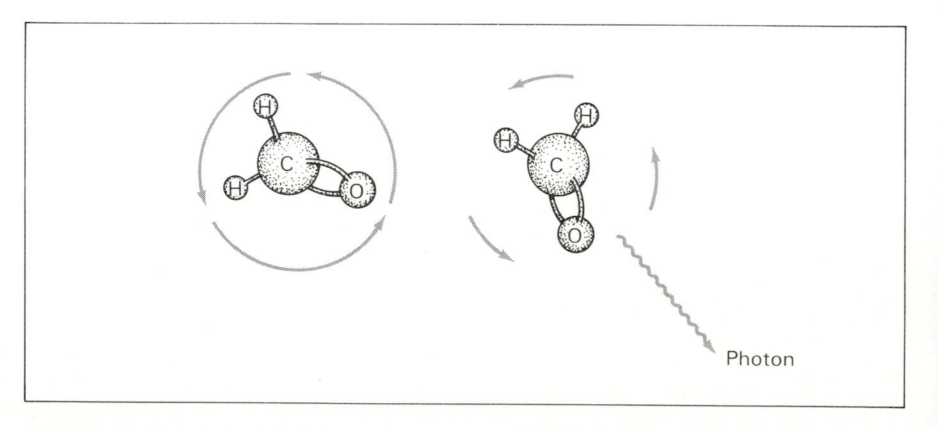

FIGURE 12.22 As a molecule changes from a rapid rotation (left) to a slower rotation (right), a photon is emitted that can be detected with a radio telescope. Depicted here is the formaldehyde molecule, H_2CO. (The length of the curved arrows is proportional to the spin of the molecule.)

invariably found in the densest parts of interstellar space. These are regions for which the scattering of radiation is great enough to prohibit ultraviolet, optical, and most infrared techniques that might ordinarily be used to detect changes in the energy states of the molecules. Only low-frequency radio radiation can make its way out of such clouds, eventually to be detected by large radio telescopes on Earth.

Why are molecules found only in the densest and darkest of the interstellar clouds? One possible reason is that the dust serves to protect the fragile molecules from destruction by the normally harsh interstellar environment. Another possibility is that the dust is a convenient catalyst that helps form the molecules. Most probably, the dust plays both these roles in molecular clouds.

Actually, astronomers have known of the existence of a few simple types of diatomic molecules in interstellar space for nearly a half-century. They include methylidyne (CH) and cyanogen (CN), whose spectral lines were originally detected superposed on the spectra of a few hot stars by the line-of-sight optical technique noted earlier in this chapter. Their narrow widths clearly demonstrate that the molecules reside in cold, neutral, interstellar regions.

For decades, theorists argued that the most important constituent of the densest interstellar clouds must be the hydrogen molecule (H_2). This particular molecule does not emit or absorb radio and optical radiation, only short-wavelength ultraviolet radiation. Hence theorists were not proved correct until Space Age rockets and satellites began observing lines of H_2 in the ultraviolet spectra of a few stars located near the edges of dense clouds. Instrumental problems still prohibit us from directly observing H_2 in the majority of dense molecular clouds, but this simplest-of-all molecule is thought to be the most abundant component of these regions.

Theorists also argued that interstellar space could not possibly house molecules more complex than these simple diatomic types. And on this point, they were proved wrong. During the past decade, spectral emissions caused by changes in the rotational states of many heavy molecules were detected by large radio telescopes. Molecules such as hydrogen cyanide (HCN), cyanoacetylene (HC_3N), ammonia (NH_3), water (H_2O), methyl alcohol (CH_3OH), and about 60 others are now known to populate interstellar space. Their abundances are small—generally 6 to 10 orders of magnitude less abundant than H_2—but the molecular clouds are large and dense enough so that photons emitted by many molecules accumulate to yield detectable signals. Hindsight suggests that the theorists misjudged this issue largely because they underestimated both the amount and the role of dust within the giant molecular clouds. "Astrochemistry" is a whole new subject, what with its unfamiliar conditions and unearthly temperatures and densities.

Apart from the potential biochemical implications sketched in Interlude 12-2, interstellar molecules have become very important tools in our current attempts to understand the evolutionary steps that lead to the birth of stars. We shall study these steps in some detail in Chapter 19, but for now consider an example of the ubiquity of the molecules.

INTERLUDE 12-2 ***Interstellar Biochemistry?***

In this chapter we have briefly discussed the pharmaceutical array of some rather complex molecules now known to populate the densest of the dark interstellar clouds. These molecular clouds contain surprisingly large quantities of organic (carbon-rich) matter, including formaldehyde (H_2CO; a popular cleaning fluid and preservative), ethyl alcohol (CH_3CH_2OH; interstellar booze), formic acid (H_2CO_2; typical of the moisture in ants and other terrestrial insects), and cyanodecapentayne ($HC_{11}N$; a 13-atom molecule not naturally found on Earth).

The greatest significance of the interstellar molecules is that they exist. Their very presence in the near void of interstellar space is profound. The molecular clouds provide sites where we can study whole new types of chemistry unfamiliar to us on Earth—namely, astrochemistry. After all, textbooks titled *General Chemistry* are not general at all, but are really books of Earth chemistry; the truly "general" or universal chemistry texts are only now being written based on the astronomers' findings in the wider extraterrestrial domain.

Most intriguingly, the densest molecular clouds *might* be biochemical "factories." We can say this because the mixture of energy and a few simple molecules—ammonia (NH_3), methane (CH_4), and water (H_2O)—produces an organic "soup" containing amino acids and other organic substances known to be precursors of life on Earth. It's a simple reaction, one routinely performed in chemical laboratories, and we'll study it in some detail in Chapter 25. The point is that it may well be occurring naturally in interstellar space.

We now know that ammonia, methane, and water vapor are widespread in our Galaxy, and clearly there's lots of energy, especially near stars. Indeed, all the ingredients for the reaction described above are present in numerous clouds throughout interstellar space. Furthermore, some of the exotic molecules identified in laboratory experiments of this type—for example, hydrogen cyanide (HCN) and cyanoacetylene (HC_3N)—are also observed in interstellar space. Two other of the observed interstellar molecules, methylamine (CH_3NH_2) and formic acid (H_2CO_2), can react to form glycine (NH_2CH_2COOH). Although not yet discovered in interstellar space, glycine is one of the key amino acids that help to form the huge protein molecules in all living cells.

Many of these molecules, as we'll see in Part III, play basic roles in terrestrial life. In fact, some of them combine into DNA, the extraordinarily complex molecule that transfers coded hereditary information from one generation of living systems to another. The prospect that interstellar matter contributes directly to the origins of life looms larger with the discovery of each and every new interstellar molecule.

Put in simple terms: There's something common to most of the complex molecules now known to populate interstellar space. That common denominator of galactic matter is the element carbon. But carbon is also the common denominator of life. Hence some researchers now feel that interstellar molecular clouds may be fertile areas where astrophysics and biochemistry merge.

Is it possible that life could exist in an interstellar cloud? Or, is a solid body, like a planet, really needed? In other words, can we think of anything that might prohibit the existence of life within an interstellar cloud—self-supporting systems that originate, eat, and evolve? Although the low interstellar temperatures and densities (even in the "dense" molecular clouds) would seem to make this idea unlikely (for lack of warmth and nourishment), this is nonetheless one aspect of extraterrestrial life we'll need to consider in greater detail in Part III when addressing the prospects for life beyond Earth.

Figure 12.23 shows some of the sites where formaldehyde molecules have been detected near M20. At practically every dark area sampled between M16 and M8, this molecule is present in surprisingly large abundance (although still far less common than H_2). Analyses of the spectral lines at myriad locations along the 12-arc-degree swath illustrated in Figure 12.6 prove that the physical conditions are pretty much the same everywhere (50 kelvin and 10^5 molecules/cubic centimeter, on average). And, interestingly enough, the nearly constant velocity of the spectral features suggests that a giant cloud of molecular matter bridges M16 to M17, and M17 to M20, and possibly M20 to M8.

These studies imply that several of the nebulae studied earlier in this chapter perhaps formed from what was, and for the most part still is, a truly vast interstellar cloud. At the average distance of these nebulae, some several thousand light-years from Earth, this cloud would span roughly 1000 light-years. It's surely not a homogeneous cloud, for molecules are detected only at those places that are especially dark and obscuring—in itself a remarkable demonstration of the strong correlation between the abundance of molecules and the presence of dust. The mass of this huge cloud, the full extent of which is still unknown, may be on the order of several hundred million times the mass of our Sun.

Figure 12.24 is a contour map showing the spread of formaldehyde molecules in the immediate vicinity of the M20 nebula. Like other contour maps, this one was made by observing radio spectral lines of the molecule at various places, and then drawing contours connecting regions of similar abundance. Too fragile to exist within hot gaseous nebulae, molecules instead thrive in the cooler surrounding regions of in-

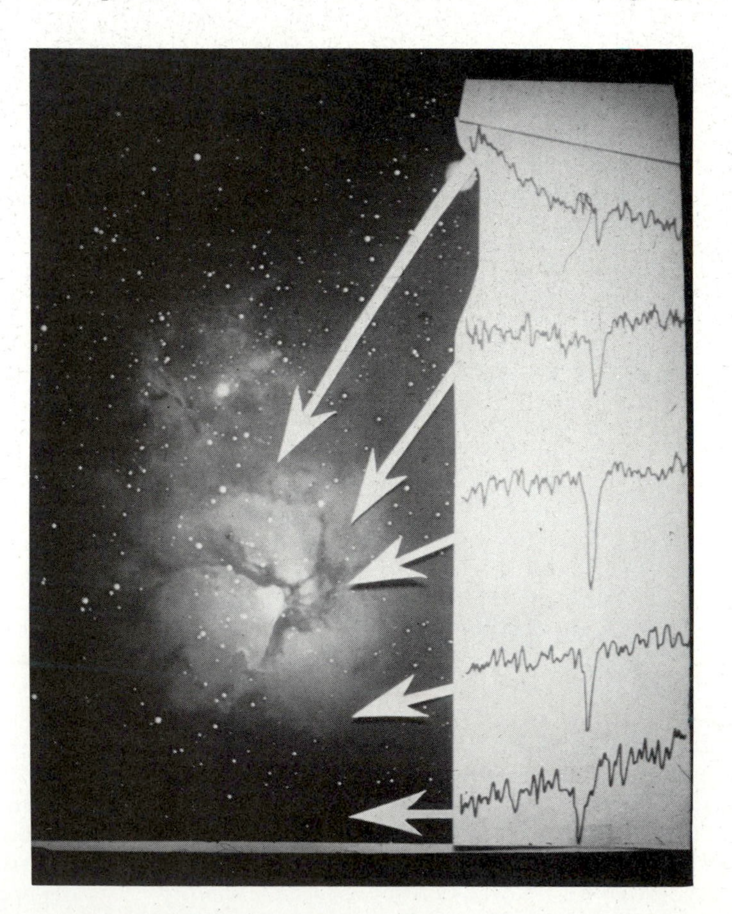

FIGURE 12.23 These spectral features show some of the sites where formaldehyde molecules are known to lurk near the M20 gaseous nebula originally illustrated in Figure 12.9. Note how the spectral lines are most intense both in the dark dust lanes trisecting the nebula and in the dark regions beyond the nebula.

FIGURE 12.24 Contour map of the amount of formaldehyde near the M20 nebula, demonstrating how this and other molecules are especially abundant in the darkest interstellar regions. As usual, the contour values increase from the outside to the inside.

terstellar space, particularly in the darkest regions shown in this figure.

Figure 12.25 is another map of the matter within a dark molecular cloud. Not a radio map, however, this figure displays the infrared radiation emerging from the Rho Ophiuchi cloud that was previously illustrated in Figure 12.17. (Consult also Plate 4.) Although the word "infrared" is synonymous with heat, and although such dark clouds are very cold, any object having even the most minute heat still emits some infrared radiation, albeit weakly. Sensitive sensors aboard the *Infrared Astronomy Satellite* captured this radiation, which is actually associated with the dust (and not the gas) within the cloud. Invariably, where the molecules are present in considerable abundance, so is the dust. Here, then, is another tracer of dark interstellar clouds and yet another way to study their physical conditions. As we shall see in the final section of this chapter, the molecules of gas and particles of dust probably go hand in hand. More to the point, dust is probably needed to help form the molecules, or to protect them, or both.

Over all, radio maps of interstellar gas (especially the widespread carbon monoxide molecule) and infrared maps of interstellar dust reveal that molecular clouds do not reside alone in space; rather, they comprise huge complexes, some spanning as much as 150 light-years across and harboring enough loose gas to make roughly a million stars like our Sun. The giant complexes, which number about a thousand in our Galaxy, are usually elongated parallel to the plane of the Milky Way; each complex contains numerous individual clouds some 30 to 60 light-years across. In turn, within these individual clouds lurk much more compressed and extremely dark "globules" often less than a light-year in diameter. It is in these compact cloud cores where star formation is now observed to be occurring, a subject to which we shall return in Chapter 19.

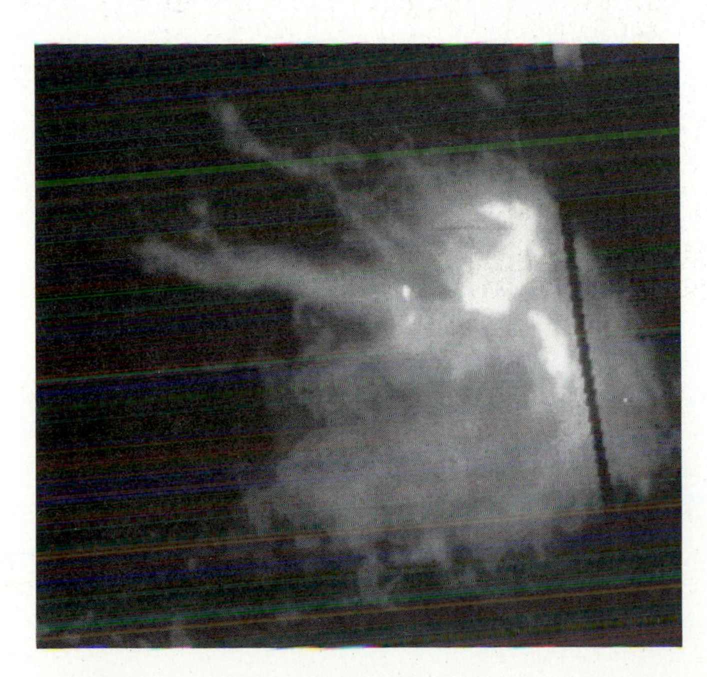

FIGURE 12.25 A filled contour map of infrared radiation detected from the dark interstellar cloud, Rho Ophiuchi, illustrated in Figure 12.17. Note that the infrared radiation, and therefore the dust that emits this radiation, displays a similar kind of tentacle structure as does the cloud visually. (The very bright source of infrared radiation near the top of the cloud comes from an embedded nebula which can also be seen optically and which is very hot.)

FORMATION MECHANISMS: The existence of such complex molecules in space has prompted a number of interesting questions. For example, how do the molecules form in the cold, near-vacuum conditions of interstellar space? If the larger molecules are built from clusters of smaller ones, what are the precise chemical steps in their formation?

Or, are the currently observed molecules mere fragments torn from much larger molecules yet to be discovered?

Current consensus suggests that the observed molecules are constructed from smaller atoms and molecules already present in interstellar space. Figure 12.26 sketches four plausible mechanisms for their formation.

The first mechanism stipulates that the molecules form in the outer layers of relatively cool stars. The conditions there are cool and tenuous by stellar standards, but hot and dense by interstellar standards. These conditions should be conducive to the formation of molecules. After all, as noted in Chapter 11, simple molecules are observed in the spectra of the coolest stars. The basic drawback to this mechanism is that it's hard to imagine a process that could gently expel these stellar atmospheric molecules into the wider realms of interstellar space; even if the expulsion process did not break apart the rather fragile molecules, the harsh radiation traveling through interstellar space would presumably do so before they had a chance to condense into dark clouds.

Two other formation mechanisms can conceivably occur directly in interstellar space far beyond stars. One involves the direct, though random combination of two atoms, A and B, in order to form molecule AB. This is a straightforward colliding, sticking, and clustering of two atoms of gas, according to the symbolic equation

$$\text{atom A} + \text{atom B} \rightarrow \text{molecule AB} + \text{heat}.$$

As noted, such gas-phase chemical ("exothermic") reactions produce not only a molecule but also some heat. And therein lies the problem. The heat has no place to go except back into the newly created molecule. Accordingly, most molecules break apart. They get cooked by their own heat of formation, and hence once again become atom A and atom B.

On the other hand, some gas-phase chemical ("endothermic") reactions *require* heat instead of producing it. These can be written symbolically as

$$\text{atom A} + \text{atom B} + \text{heat} \rightarrow \text{molecule AB}.$$

This type of reaction will not proceed unless heat is applied. But where are the two atoms going to find the needed heat in the cold void of interstellar space? Again, we have a problem, except perhaps in the immediate vicinity of stars (in which case the fragile molecules cannot be easily expelled without breaking apart, as noted above).

The final mechanism, probably operating directly in interstellar space, involves another body—a third microscopic object that can absorb the heat in the above heat-producing chemical reaction. In this way, the newly created molecule can remain intact. But what body could do this? The interstellar dust particles could, with the dust itself absorbing the heat produced by the chemical reaction. In other words, in addition to protecting the molecules from the harsh high-frequency radiation that ordinarily destroys them, the dust particles might also aid in forming the molecules. As with the other mechanisms, however, there's a problem here, too. How do the molecules escape from the dust particle's surface in order to mingle with the gas of interstellar space? The interstellar molecules are clearly detected in the gaseous state, not riding the backs of any solid dust particles. To date, researchers cannot understand how the fragile molecules detach from the dusty grains without breaking apart.

At any rate, from time to time, some H_2 molecules riding atop dust particles are hit by fast-moving charged particles (known as "cosmic rays" which we shall study in Chapter 13). This collision knocks off an electron, forming a hydrogen molecular ion, H_2^+. The H_2^+ then attaches to another H_2, making H_3^+ as well as a H atom. The H_3^+ then reacts in turn with carbon (C), nitrogen (N), and oxygen (O) to pro-

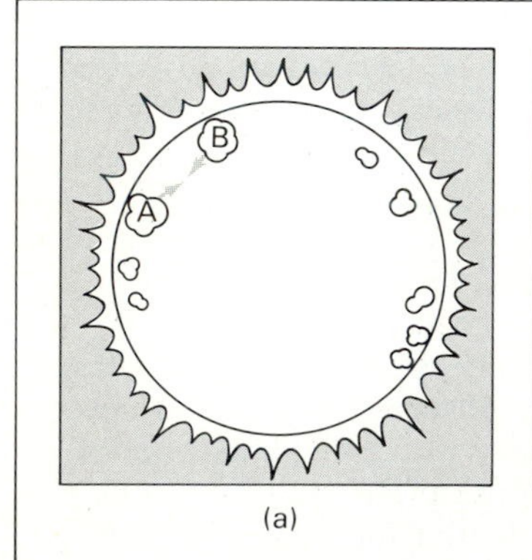

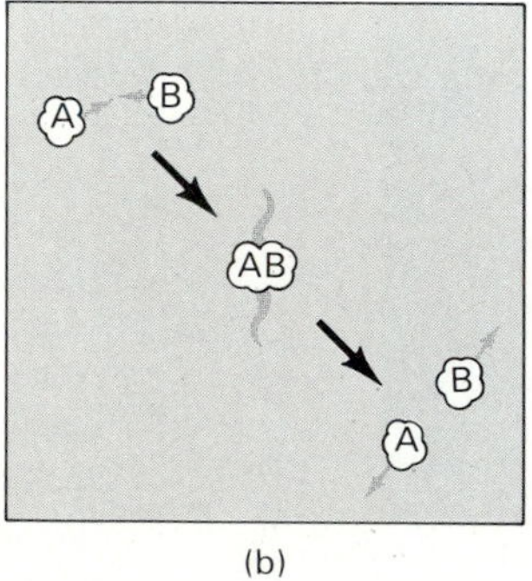

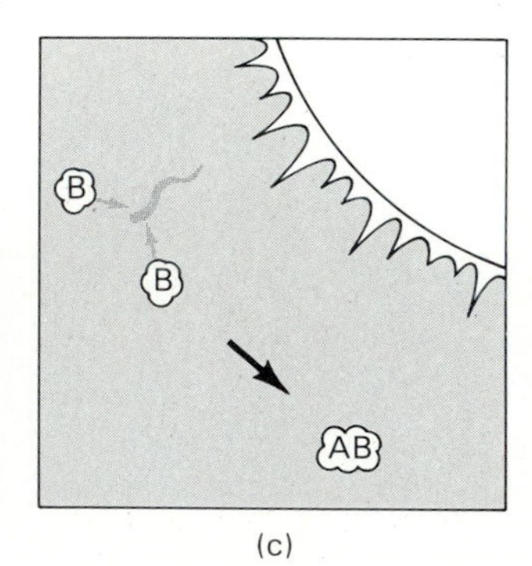

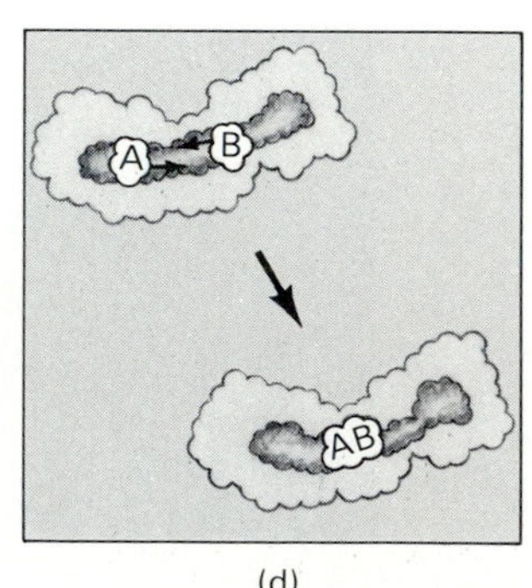

FIGURE 12.26 Schematic diagram of four plausible ways that molecules (AB) might form from atoms (A and B) in interstellar space: (a) chemical reactions in stellar atmospheres, (b) direct, though random atomic collisions of two atoms, (c) atomic collisions plus heat, and (d) chemical interactions on the surface of dust particles.

duce "species" such as the formyl ion (HCO^+), hydrogen cyanide (HCN), and formaldehyde (H_2CO). Further reactions result in CH_4, NH_3, H_2O, and so on. Complete chains of reactions are understood for the simplest molecules, but the formation of the heaviest molecules has proved difficult to decipher. Conceivably, some of the more complex interstellar molecules are actually fragments of even larger molecules not yet discovered, but how the huge molecules would form is a mystery.

Although the paragraph above implies that at least the basics of interstellar chemistry are known, we are left with the uneasy feeling that some of the key secrets of the interstellar molecules currently elude us. The very existence of the molecules now forces astronomers to rethink and reobserve interstellar space. In doing so, we are beginning to realize that this active and interesting domain is far from the void theoretically suspected not long ago. Regions recently thought to contain nothing more than galactic garbage—the cool, tenuous darkness among the stars—are now known to contain molecules. And some of these molecules seem complex enough to be considered biologically conducive to the seedbeds of life.

SUMMARY

The darkened realms of interstellar space harbor surprising amounts of matter. As most of it is quite cold by Earth standards, the interstellar gas and dust tend to radiate at long wavelengths to which our eyes are insensitive. Accordingly, the bulk of interstellar matter is invisible, and must be studied by techniques other than optical astronomy.

In addition to the cold, thin matter throughout the vast majority of space, the interstellar domain is populated with colder, moderately dense atomic clouds, with even colder and denser molecular clouds, and with hot, glowing patches of plasma surrounding young stars. Thus interstellar space is not a true void but is rich in huge quantities of thinly spread matter waiting to be processed into future generations of stars.

QUESTIONS

1. State the average density of each of the following regions: the Sun's core, the Sun in bulk, a dark galactic cloud, and interstellar space generally.
2. Define a gaseous nebula, being certain to note its physical and chemical properties as well as its average temperature and density.
3. Briefly explain the two main processes at work in gaseous nebulae.
4. Pick one: During the 1970s and early 1980s, observers discovered, quite unexpectedly, several dozen interstellar molecules mainly by the technique of (a) x-ray astronomy; (b) ultraviolet astronomy; (c) radio astronomy; (d) optical astronomy.
5. Explain, in some detail, why the spectra of typical stars show mainly absorption lines, whereas those of gaseous nebulae show mainly emission-line features.
6. What are the two different types of interstellar matter, and how do we know of their existence?
7. From which of the following regions might we expect to detect 21-centimeter radiation, and why? (a) a young, bright star; (b) a gaseous nebula surrounding a young, bright star; (c) a molecular cloud in which gaseous nebulae are normally embedded; (d) a galactic cloud where the contents are mostly in atomic form.
8. Name several reasons how we know that matter lurks beyond individual stars. Compare and contrast the physical and chemical properties of that interstellar matter.

*9. If two spectral emission lines have the same width but different intensities, which has been emitted from hotter gas?

10. Why do you think that the interstellar molecules are correlated so strongly with regions of greatest galactic dust? Explain.

KEY TERMS

collision process
dark dust cloud
gaseous nebula
interstellar matter
interstellar space
molecular cloud
radiative process
21-centimeter line

FOR FURTHER READING

BLITZ, L., "Giant Molecular-Cloud Complexes in the Galaxy." *Scientific American*, April 1982.

CHAISSON, E., "Gaseous Nebulas." *Scientific American*, December 1978.

*CHAISSON, E., "Gaseous Nebulae and Their Interstellar Environment," in *Frontiers of Astrophysics*, E. Avrett (ed.). Cambridge, Mass.: Harvard University Press, 1976.

*DALGARNO, A., "Chemistry of the Interstellar Medium," in *Frontiers of Astrophysics*, E. Avrett (ed.). Cambridge, Mass.: Harvard University Press, 1976.

GREENBERG, J., "The Structure and Evolution of Interstellar Grains." *Scientific American*, June 1984.

SCOVILLE, N., AND YOUNG, J., "Molecular Clouds, Star Formation and Galactic Structure." *Scientific American*, April 1984.

WYNN-WILLIAMS, G., "The Newest Stars in Orion." *Scientific American*, August 1981.

13
THE MILKY WAY GALAXY: A Grand Design

Many different states of matter are known to exist in space—stars, gas, dust, nebulae, and clouds. And we've studied each of them. Yet we wonder: Are stellar and interstellar matter spread in all directions indefinitely, extending outward forever? What is the grand design of matter in the vicinity of the Sun, and beyond? Do stars, gas, and dust reach infinity, or do they terminate some place in the distance? If they do end, at what distance from the Sun does matter become noticeably thinner? What lies beyond this point of termination?

Glancing from Earth in virtually any upward direction, we note that the nighttime sky seems more or less the same. Bunches of stars cluster here and there, but overall our immediate locale seems homogeneous. Equal volumes of space in the neighborhood of the Sun seem to contain pretty much the same number of stars and the same amount of interstellar gas and dust. In short, the evening sky appears fairly uniform.

But this is only a local impression. Ours is a provincial view. Once we consider very large dimensions of space—dimensions much, much larger than those spanning the distances among stars—the spread of stellar and interstellar matter changes. It becomes patchy and irregular, much like the wide-angle photographs illustrated in Chapter 12 (consult Figures 12.5 and 12.6). And eventually the matter thins to essentially nothing.

A brief answer to the foregoing questions is that there is indeed a finite material boundary or edge—a distance beyond which the spread of matter is so thin that stars, gas, and dust are virtually nonexistent. Inside this boundary lies the huge collection of stellar and interstellar matter generally termed a **galaxy**.

Figure 13.1 is a series of photographs of a typical galaxy. This is not the galaxy in which we live, but another collection of stellar and interstellar matter already met in the Overview—the Andromeda Galaxy. We've also seen it earlier in Chapter 1 and Plate 2, but take this opportunity to examine it more carefully.

The entire object pictured in Figure 13.1 resides about 2 million light-years away. Regardless of this great distance, Andromeda is the nearest major galaxy beyond our own. We display it here because our own Galaxy cannot be photographed. The ensemble of stellar and interstellar matter that includes the Sun—a system known as the **Milky Way Galaxy**, or just "the Galaxy" with a capital G—is too large for us to see in entirety. (The word *galaxy* is Greek for "milky way.") We live inside our Milky Way, and there's no way to move away some distance, look back, and take a snapshot.

The learning goals for this chapter are:

- to appreciate how the cepheid variable stars help to determine the size of our Galaxy
- to understand the techniques used to prove that the Earth–Sun system is not at the center of the Universe
- to appreciate how astronomers map the Galaxy by means of radio astronomy
- to know how the size and mass of the Galaxy compare to planets, stars, and star clusters

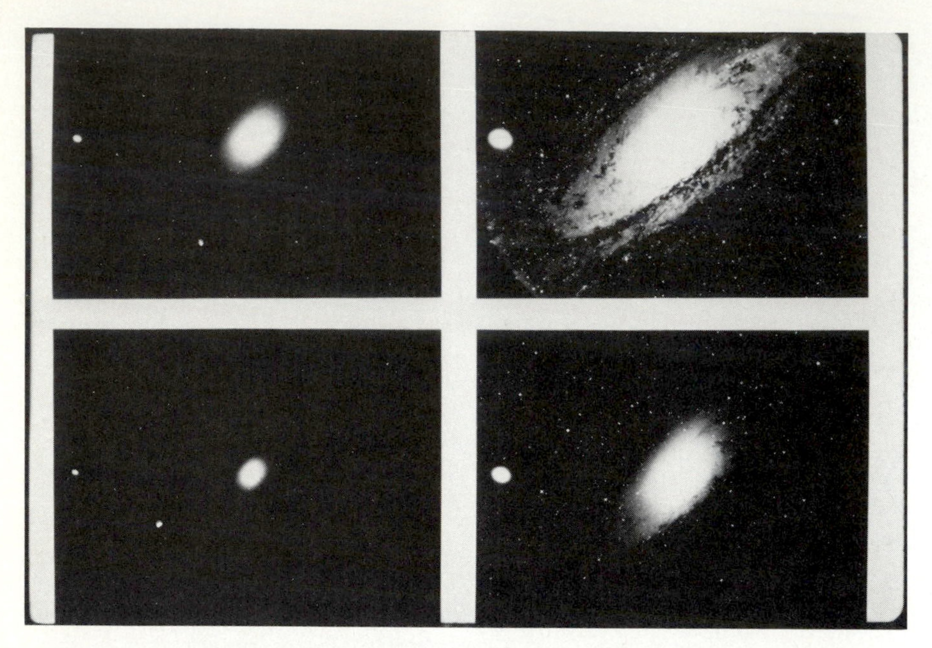

FIGURE 13.1 The Andromeda Galaxy probably resembles fairly closely the overall morphology of our own Milky Way Galaxy. Here, clockwise from top left, the exposure times were 5 minutes, 45 minutes, 30 minutes, and 1 minute, respectively. (See also Plate 2.)

- to recognize that cosmic-ray particles constantly bombard Earth, bringing us additional information about the nature of matter beyond our Solar System
- to realize the bizarre phenomena now being explored in the central regions of our Galaxy

Cepheid Variables

Studies of nearby stars, gas, and dust, as well as prudent comparisons with other galaxies such as Andromeda, give us a general idea of the overall spread of matter in our own Milky Way Galaxy. But the resulting picture is crude, for most stars are obscured from our view. And interpretations of the gas and dust are tricky, since interstellar space contains innumerable clouds with all sorts of irregular shapes and sizes. Better observational tools are needed if we are to fully appreciate the size and shape of our Galaxy. We now consider one of these tools.

PULSATING STARS: We've already noted the great efforts to catalog stars around the turn of the twentieth century. One by-product of that laborious research is the study of **variable stars**. These are stars whose luminosity changes with time, some quite erratically, others more regularly. In other words, a small fraction of stars do not emit radiation steadily as do our Sun and most other stars. Rather, variable stars seem to blink at us.

Figure 13.2 exemplifies the unsteady emission of a typical variable star whose luminosity regularly changes over a period of several days. The change in brightness is often a factor of 2 or 3. Such regular blinkers—ones that show a repeatable cycle of variations—are given a special name, **cepheid variables**, after the archetypal variable star known by the name Delta Cephei.

Cepheid variables have periods of optical variability that are completely reproducible, day after day. Recognize, however, that all cepheids do not have the same period of variability; periods of variation differ for different cepheid stars, although they remain highly regular for a given cepheid star.

Note also that each luminosity cycle of Figure 13.2 is not symmetrical. A steep rise in luminosity is followed by a slow decrease, which leads to another rapid rise, and so on. This is a common phenomenon. The shapes of all cepheid curves show this steeply rising and slowly falling pattern, suggesting that the increased brightness is caused by a mild but sudden swelling, followed by a slow decay prior to the next eruption.

Why swelling? Well, recall that luminosity is proportional to the square of a star's radius and to the fourth power of its surface temperature. Since spectroscopic studies of cepheid variables demonstrate that their surface temperatures change very little throughout an observed cycle, we can safely conclude that the brightness changes must result from changes in stellar radius. Indeed, stellar spectral lines exhibit periodic Doppler shifts, blue then red, directly confirming an out-and-in motion. In other words, the cepheids pulsate. They physically change their size by as much as 25 percent. Thus they regularly enhance their brightness simply because their size balloons rather suddenly.

Let's not get lost in the detailed nature of the cepheid variables. Rather, we want to study how they are used to infer something about the large-scale distribution of stars in our Galaxy. Our goal in this chapter is to extract only that information which might aid our understanding of our Milky Way Galaxy.

ANOTHER NEW YARDSTICK FOR DISTANCE: A most useful point to realize about the cepheid variable stars is this: their *absolute* brightnesses and their pulsation periods are highly correlated. Those stars that blink slowly, that is have long periods, are known to have large absolute brightnesses when averaged throughout their regular cycles. The converse is also true; rapid blinkers have small absolute brightnesses.

Figure 13.3 illustrates this basic correlation, derived for numerous cepheids not too far from Earth. Astronomers can prove this correlation for nearby cepheids whose distances are known and thus whose absolute

(a)

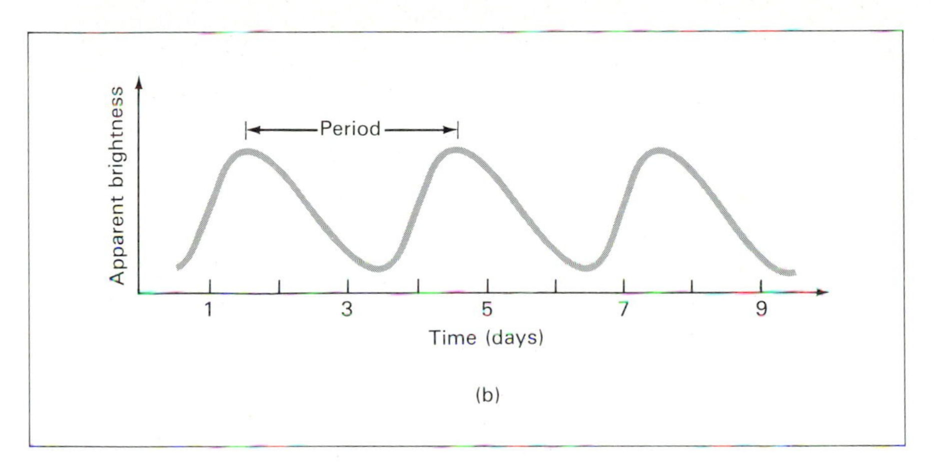

(b)

FIGURE 13.2 A typical Cepheid variable star (center) can be seen on these two nearly superposed photographs (a) taken on successive nights. Monitoring the star over the course of a week or so (b) yields a record of the star's changing brightness.

brightnesses (or luminosities) are calculable. And since we know of no exceptions to this general correlation, we give it the status of a law, often called the **period-luminosity law**. Furthermore, we assume it holds for all cepheids, near and far.

The beauty of this correlation is that a simple measurement of a star's period of pulsation is enough to estimate the star's distance. Here's how we do it. Just observe the star's period, and then read its absolute brightness from the plot in Figure 13.3. Since the apparent brightness is usually easy to measure, we can estimate the distance to that star by comparing its apparent and absolute brightnesses, as discussed in Chapter 11.

Cepheid variables, then, comprise an important rung in our universal distance ladder. This technique works well provided a cepheid can be clearly identified and its period of variability measured. In fact, this method enables astronomers to estimate distances as far as 10 million light-years, well beyond the range of stellar parallax (few hundred light-years) or spectroscopic parallax (few thousand light-years). Figure 13.4 extends our inverted pyramid to include this new method of determining distance.

SHAPLEY'S BOLD STATEMENT: Observations show that most cepheids reside in globular clusters, those tightly bound swarms of old red stars. By measuring their periods of pulsation, we can find distances to the cepheids, and hence to the clusters themselves.

Early in the twentieth century, the American astronomer Shapley used the cepheid technique to make two important discoveries. First, he proved that most globular clusters reside at great distances from the Sun, much farther than any of the nearby stars we've studied previously. Second, he proved that all the globular clusters are distributed nearly spherically in space, and he took this distribution to outline the vast collection of stars in our Milky Way Galaxy. Furthermore, because more globular clusters were found toward one direction than any other, Shapley concluded that our Sun does not reside in the middle of this huge collection of stars. It's not even close to the center.

Just as our planet is not central to the Solar System, our Solar System in turn is not central—or in any other way special—to that vast assembly of stars we now recognize as the bulk of the Milky Way Galaxy.

Figure 13.5 schematically illustrates how the globular clusters outline the **galactic halo** of our Milky Way Galaxy. This whole system of stars spans approximately 100,000 light-years, the hub of which we call the **galactic center**. Our Sun, about 30,000 light-years from the center, clearly resides in the suburbs of this truly huge ensemble of matter.

Shapley's bold interpretation that the globular clusters define the overall dimensions of our Galaxy was an enormous step forward in human understanding of our place in the Universe. Five hundred years ago, Earth was considered the center of all things. Copernicus argued otherwise, relegating our planet to an insignificant place far removed from the central Sun. Yet, almost as recently as a half-century ago, the pre-

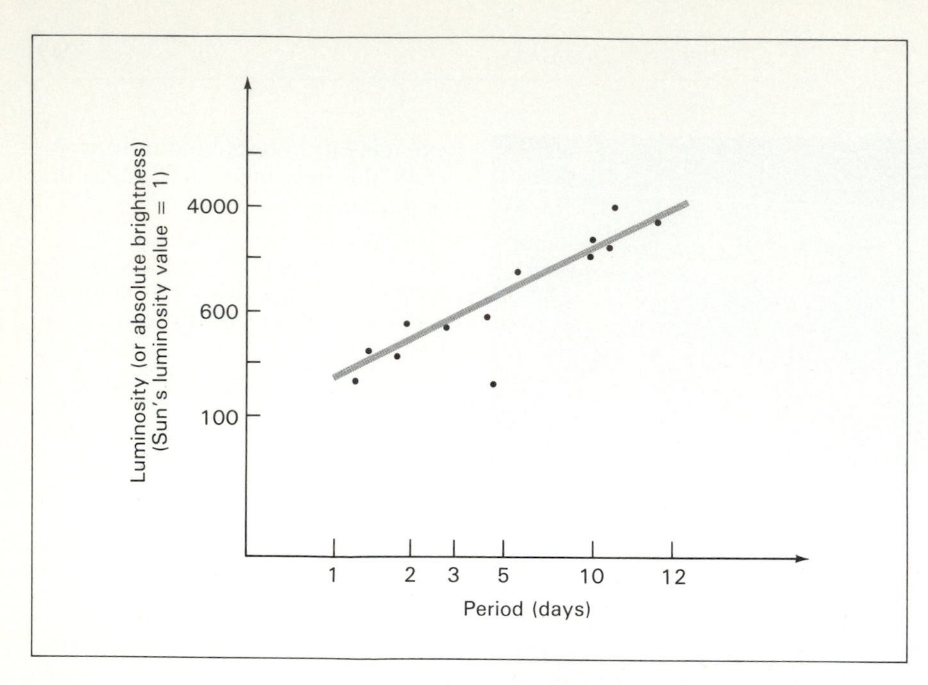

FIGURE 13.3 A plot of the pulsation period versus the average absolute brightness (or luminosity) for a group of Cepheid variable stars. The two properties are highly correlated.

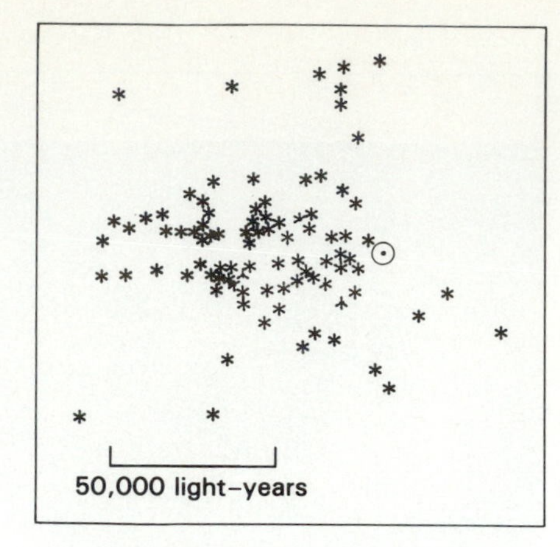

FIGURE 13.5 Our Sun (denoted by the symbol ☉) does not coincide with the center of the very large collection of globular clusters (denoted by asterisks). Instead, the Sun resides closer to the edge of the collection. The whole bunch of stars measures some 100,000 light-years across.

vailing view maintained that our Sun was the center of not only the Galaxy but also the Universe. Shapley showed otherwise; he demonstrated that our Sun has no special place in the Universe at all. He endowed our Sun with the same mundaneness some 50 years ago that Copernicus had done for Earth some 500 years ago.

So, astronomers of the first half of the twentieth century knew that our Sun is only one, rather undistinguished, member of a giant system of stars, gas, and dust comprising our Milky Way Galaxy. They furthermore recognized that our Sun is well outside the galactic center, which was otherwise invisible and mysterious. Yet not until the 1950s did scientists manage to develop another new tool that today enables us to accurately explore the full grandeur of our Galaxy. That new tool is spectroscopic radio astronomy.

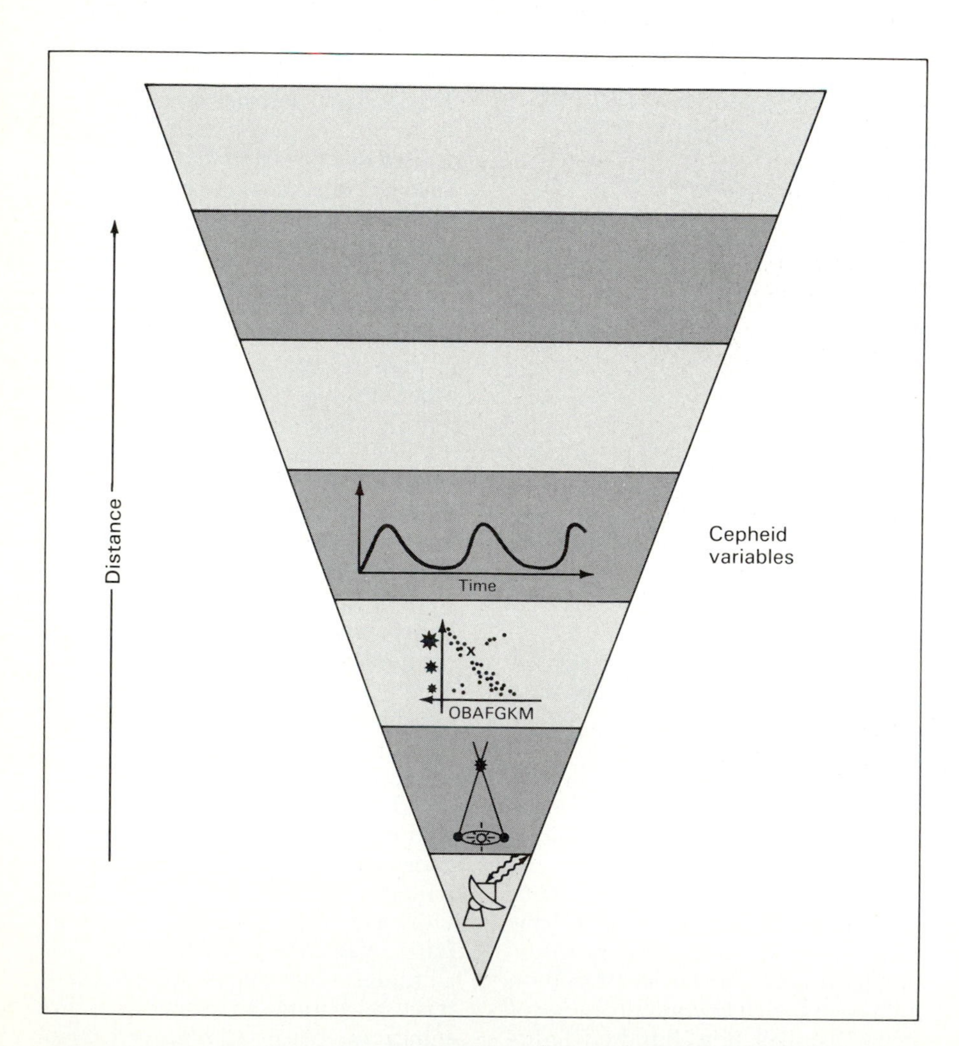

FIGURE 13.4 Application of the period-luminosity law for Cepheid variables enables us to determine distances out to about 10 million light-years.

Radio Studies of the Milky Way

The central regions of our Milky Way Galaxy cannot be studied optically. Looking toward the galactic center, we encounter just too much obscuring dust. Above and below the galactic center, some globular clusters can be identified and stud-

INTERLUDE 13-1 ***Early Computers***

Much of the early work in astronomy concerned monitoring stellar luminosity and analyzing stellar spectra. It's well known that much of this pioneering work was done by photographic methods. What's not well known is that most of the labor was accomplished by women. Around the turn of the century, a few dozen women created an enormous data base by observing, sorting, measuring, and cataloging photographic information that helped form the foundations of modern astronomy. Some of them went further, making several of the most basic astronomical discoveries often taken for granted today.

Shown in this 1910 photograph, several women are carefully examining star images, measuring variations in luminosity or wavelengths of spectral lines. Their data yielded information on hundreds of thousands of stars through direct eye inspection in the cramped quarters of the Harvard Observatory, a relatively mild environment compared with the often frigid area of the telescope dome. Note the plot of stellar luminosity changes pasted on the wall at the left; the cyclical pattern is so regular that it likely belongs to a cepheid variable. Known as "computers" (for there were no electronic devices then), these women were paid 25 cents/hour.

Since 1880, these women and others like them have performed much of the daily analysis required for what amounts to a century-long survey of the skies. Their first major accomplishment was a huge catalog of the brightnesses and spectra of some tens of thousands of stars, published in 1890 under the direction of Wilhaminia Fleming. On the basis of this monumental effort, several women made fundamental discoveries: In 1897, Antonia Maury undertook the most detailed study of stellar spectra to that time, enabling Hertzsprung and Russell independently to develop what's now called the HR diagram. In 1898, Annie Cannon proposed the Harvard Spectral Classification System noted in Chapter 11 and now adopted as the internationally standard method of categorizing stars. And in 1908, Henrietta Leavitt discovered the period-luminosity relation for cepheid variable stars, enabling Shapley later to recognize our Sun's rightful position near the edge of the Milky Way, as well as to measure the size and grandeur of our Galaxy.

ied, as depicted in Figure 13.5. But even through the world's largest optical telescopes, our view toward the interior of the Galaxy is forever restricted to objects less than a few thousand light-years distant.

With the advent of radio astronomy, matter anywhere in the Galaxy could be studied. The 21-centimeter radio emission from neutral hydrogen gas is especially useful in this regard, first because hydrogen is the most abundant element, and second because this long-wavelength radiation is unaffected by dust. We now extend Chapter 12's discussion of this most important observational tool, while simultaneously learning a great deal more about our Galaxy.

EXPLORING CLOUDS: Imagine observing with a radio telescope pointed away from Earth's surface. In any direction, there are likely to be several large interstellar clouds, as depicted in Figure 13.6. These clouds could be the moderately dense atomic variety or the very dense molecular type. For purpose of discussion, we assume here that these clouds contain substantial amounts of hydrogen atoms that experience occasional collisions and hence emit the 21-centimeter radio signal. Suppose also that Earth and the clouds are moving toward one another. The clouds need not be moving precisely in the direction of Earth, as long as their radial velocity component is approaching us.

At any one moment, some atoms are in the lowest-energy state (electron and proton spins antiparallel) awaiting a collisional or radiative boost to the higher-energy state (spins parallel); other atoms are already in the higher-energy state waiting to relax back to the lower-energy state; and still others are in the process of relaxing and emitting the 21-centimeter line. Thus because gargantuan numbers of hydrogen atoms populate such large interstellar clouds, we can be sure that large numbers of atoms emit this radio-frequency signal at any given moment. Indeed, our Galaxy is never quiet at 21-centimeter wavelength.

Figure 13.7 shows some typical examples of radio spectral lines observed toward a region of space containing several interstellar clouds. These spectral features are meant to correspond to the clouds depicted in the preceding figure. As for any spectral line, each radio feature has an intensity, width, and centroid, as discussed earlier in Chapters 2 and 12.

The interstellar clouds labeled A, B, and C in Figure 13.6 emit the blue-shifted spectral features of Figure 13.7. The lines are slightly blue shifted from the usual 21-centimeter wavelength because we've assumed in this example that the clouds have some component of their motion toward Earth. In other words, large-scale motion of a cloud can change our perception of the cloud's emission; each cloud genuinely emits at 21-centimeter wavelength, but we observe the spectral features to be Doppler shifted to a little less than 21 centimeters. Alternatively, we could say that the radio signals are not received at the standard frequency of 1.42 billion hertz but at slightly higher frequencies. Rather than specifying these new wavelengths or frequencies, we conveniently refer to these features as "Doppler-shifted 21-centimeter lines."

Note that although our discussion here concerns radio radiation, we still use the terms *red* and *blue* shift. Of course, there's no such thing as red radio radiation or blue radio radiation; these color terms are simply carried over from our earlier discussion on the Doppler effect of light radiation. Actually, astronomers use the terms "red shift" and "blue shift" to describe the Doppler change of spectral lines anywhere in the electromagnetic spectrum. The important point is that red-shifted spectral lines are emitted or absorbed by astronomical objects having some recessional motion. Conversely, blue-shifted spectral lines, such as those of our current example, arise in gaseous regions having approach velocities.

This 21-centimeter technique works well only for those galactic regions abundant in *atomic* gas. Until recently, such regions were regarded as the bulk of interstellar space. Now, however, as noted in Chapter 12, we have found molecules widely spread throughout the Galaxy. Consequently, 21-centimeter observa-

FIGURE 13.6 Schematic diagram showing several interstellar clouds whose radiation can be detected and studied with a radio telescope.

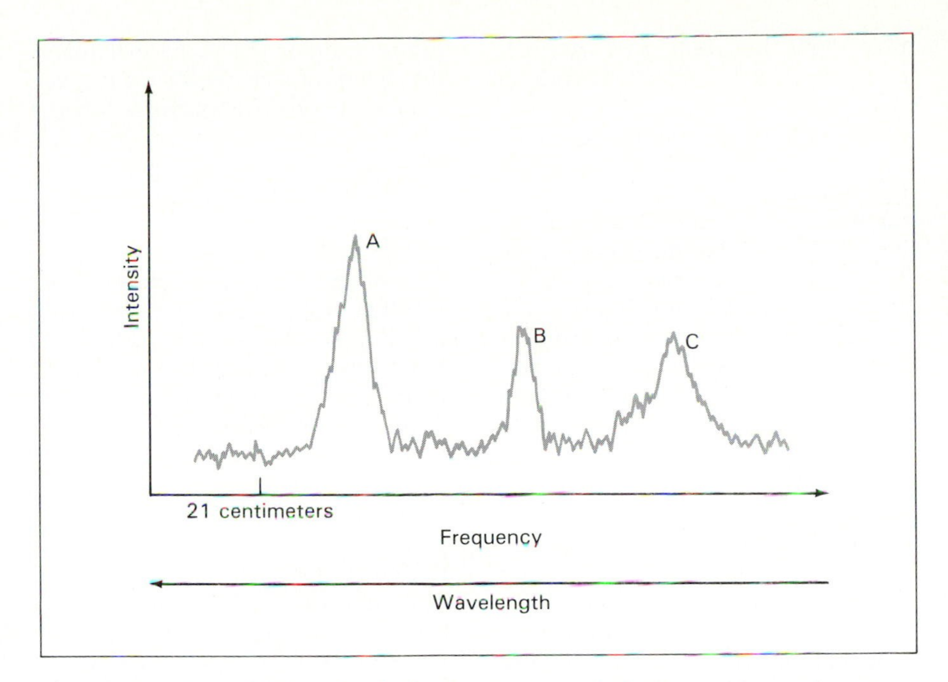

FIGURE 13.7 Typical radio signals from neutral atomic hydrogen observed at or near 21-centimeter wavelength. The spectral-line profiles labeled A, B, and C refer to the clouds of the preceding figure.

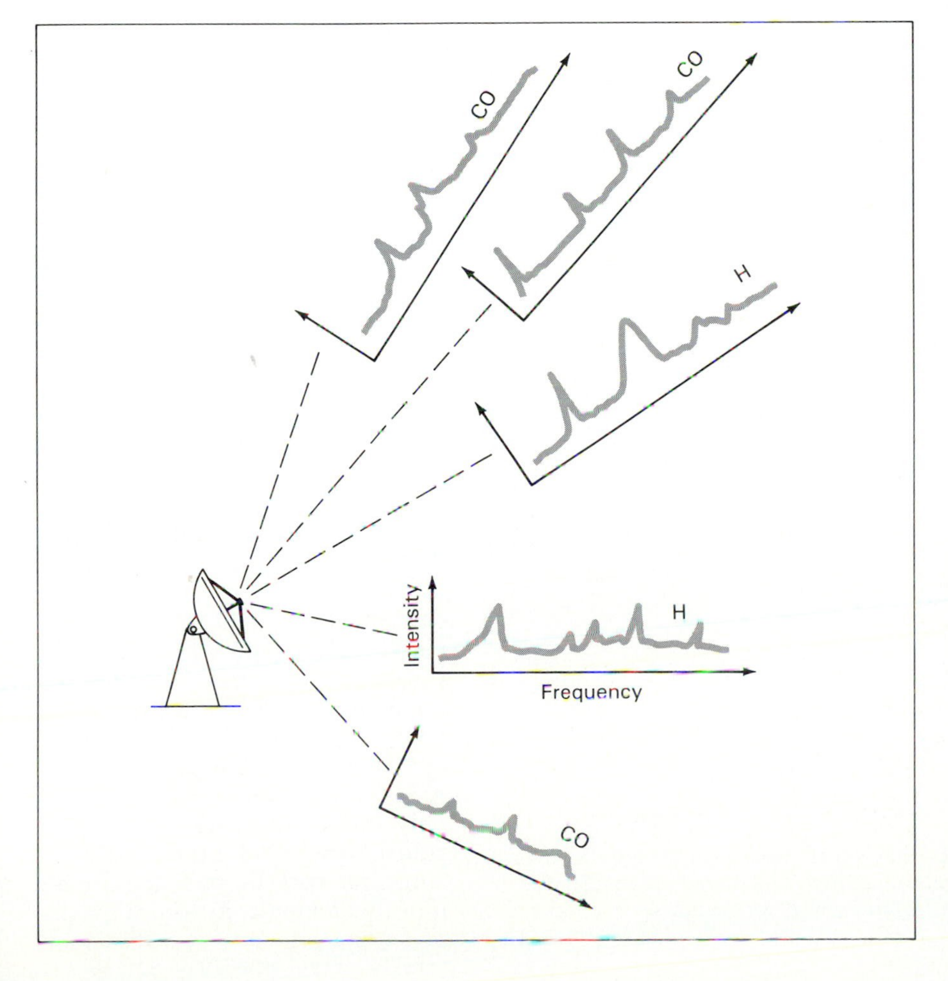

tions detect radiation mainly toward the moderately dense clouds having 10 atoms/cubic centimeter; the more dense molecular clouds—those with of order 10,000 molecules/cubic centimeter—are not easily detectable at 21-centimeter wavelength. Little or no atomic hydrogen inhabits molecular clouds; there this element is mainly bound up in the molecular form H_2, which, unfortunately, has no characteristic rotational, vibrational, or electronic transitions in the radio part of the electromagnetic spectrum.

Other molecules, fortunately, do emit radio radiation that we can use to study the spread of molecular gas throughout the Galaxy. The carbon monoxide (CO) molecule is most appropriate for this purpose, mainly because it's abundant in all molecular clouds (although still a million times less prevalent than H_2), and also because its characteristic spectral line occurs in a convenient part of the radio domain (about 0.3 centimeter wavelength). So carbon-monoxide research complements the neutral-hydrogen research.

Figure 13.8 exemplifies the wide variety of radio emissions detected from various clouds toward many different directions of our Galaxy. Together, the carbon-monoxide research and the neutral-hydrogen research provide an adequate understanding of the gas in much of the Milky Way.

MAPPING THE GALAXY: Years of radio observations and detailed analyses have given researchers an appreciation for the true spread of atomic and molecular gas in our Galaxy. The work is still under way, but the big picture has already emerged. The upshot is that the "global" spread of interstellar clouds displays an organized pattern on a grand scale.

The neutral hydrogen and carbon monoxide studies show that most of

FIGURE 13.8 Radio spectra of neutral hydrogen atoms (H) and carbon monoxide molecules (CO) observed in several different directions from our vantage point on Earth.

the galactic gas is centered far from our Sun. The overall spread of the gas coincides roughly with that outlined earlier by the globular clusters. Again, this special, central location is what astronomers call the galactic center, a region some 30,000 light-years from the Sun.

Figure 13.9 represents a sample of many of the known interstellar clouds of our Galaxy. Together, the clouds show clear evidence for pinwheel-type structures called **spiral arms**. Each arm seems to originate close to the galactic center. And one of them, as best we can tell, includes our Sun. More succinctly put, we can say that our Sun apparently resides in a vast spiral arm that emanates near the galactic center and wraps around a large part of the entire Galaxy.

Studies of our own Galaxy, and especially of many other galaxies, prove that the spiral arms are made of more than heavy concentrations of interstellar clouds. Young objects, such as O- and B-type stars, galactic clusters, and gaseous nebulae all reside in the spiral arms. Since neither these young objects nor interstellar gas and dust have been found in large quantities outside the arms, we can safely conclude that the spiral arms are the sites of recent star formation. These youthful objects could not have traveled too far from their places of origin, given their short lifetimes.

Unlike the globular clusters, the radio-emitting galactic gas is not spherically spread through space. Instead, the interstellar clouds are rather flatly distributed, looking a bit like a double Frisbee clamped together at the brims. Shown in Figure 13.10, the gas is confined to a thin plane or disk called the **galactic plane**. This is the plane that forms the band of the Milky Way seen overhead on moonless nights—a band that was illustrated in Figure 12.5. It is also delineated particularly well in the infrared image printed as Plate 4(b).

The full diameter of the visible part of our Galaxy measures some 100,000 light-years, meaning that our Sun resides more than halfway out from the center. Here, approximately 30,000 light-years from the center, the thickness of the plane is small by galactic standards—on average 1000 light-years across. This makes the average diameter of the plane some 100 times larger than its thickness. Don't be fooled, though; put these dimensions into perspective. Even if you could travel at the velocity of light, 1000 years would still be needed just to traverse the thickness of the galactic plane. The plane is thin by galactic standards, but it's absolutely huge by most standards familiar to humans.

The galaxy does bulge a little toward its center, as shown in Figure 13.10, reaching some 5000 light-years above and below the plane. But even here, the gas does not depart from the plane as much as the globular clusters. For contrast, these clusters are shown schematically in Figure 13.10 to have more spherical symmetry, while extending nearly 100,000 light-years fully around the galactic center.

As best we can tell, our Galaxy looks very much like the Andromeda Galaxy. It has an overall shape and size like Andromeda's and apparently contains similar proportions of stars, gas, and dust. Just how similar are the Milky Way and Andromeda Galaxies? We'll probably never know for certain. We live embedded within the Milky Way's galactic plane, and we can map the Galaxy from only one vantage point. It's much like trying to unravel the layout of walks, benches, and monuments on Boston Common while looking around from only one of the park benches. Of course, we could more easily examine the park by traveling around it or by looking at it from some distant vantage point, say a skyscraper

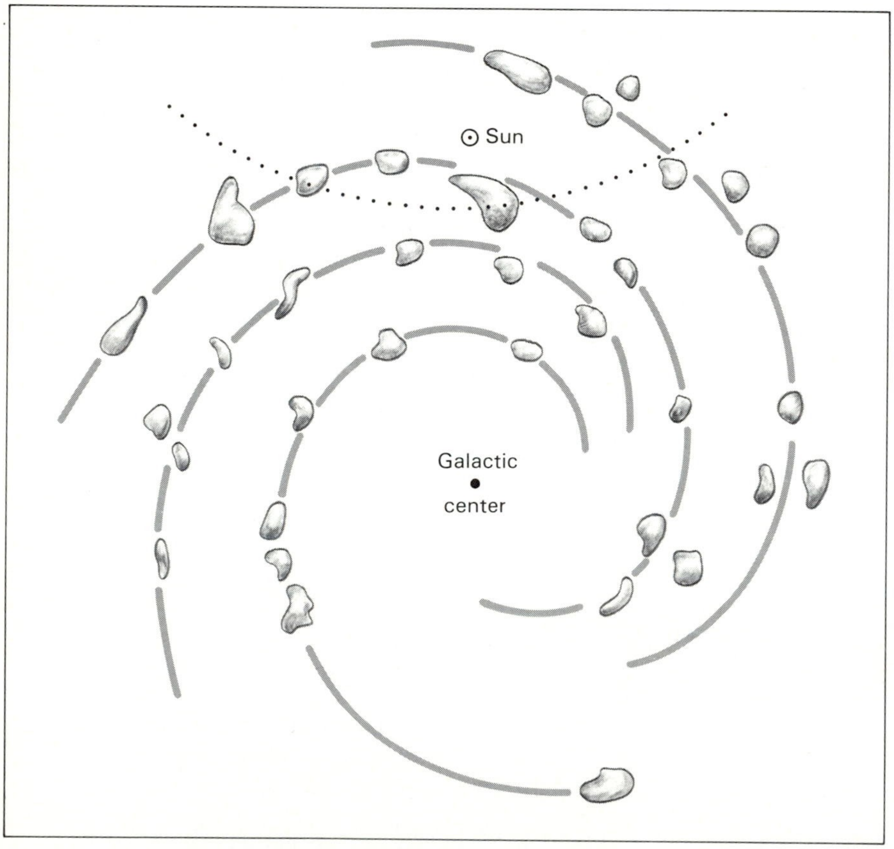

FIGURE 13.9 A map of many of the Milky Way's interstellar clouds suggests huge "spiral arms" emanating from near the galactic center. The dotted curve near the Sun outlines the limited range in distance beyond which we cannot see stars in the galactic plane (even with large optical telescopes).

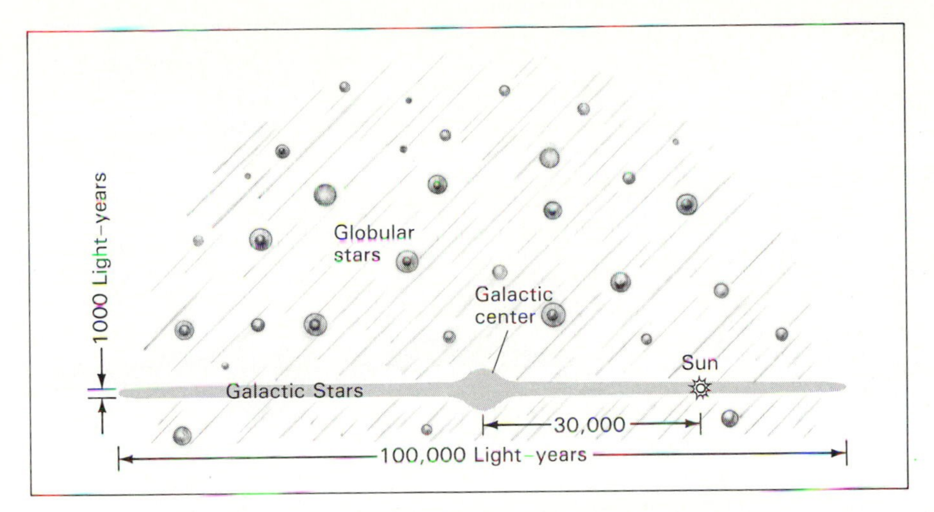

FIGURE 13.10 An edge-on schematic diagram of the interstellar gas and dust (shaded area), shown in relation to the globular and galactic star clusters. The numerical values give the primary dimensions of our Milky Way Galaxy in units of light-years.

FIGURE 13.11 An artist's conception of our Milky Way Galaxy, painted with an air-brush for realism. Indeed, this illustration is based on real data accumulated by legions of astronomers during the past few decades. Painted from the perspective of an observer 300,000 light-years above the galactic plane, the spiral-arms are at their best-determined positions, and all the features are drawn to scale (except for the oversized dot near the top, which represents our Sun for perspective). The two small blotches to the left are dwarf galaxies, called the Magellanic Clouds. They will be studied in the next chapter.

or helicopter. But we'll probably never be able to achieve such a direct, clear view of the Galaxy; the notion of traveling far enough away to look back at our Galaxy from some distant vantage point is destined to remain science fiction for a good long time, perhaps forever. We must remain content with the view from within, a particularly striking example of which is illustrated in Figure 13.11.

Red and Blue Stars

We can now better understand the structure of the Milky Way hovering above us in the evening sky. As noted in Chapter 12, our nightly view is dominated by a concentrated band of stars, nebulae, and interstellar clouds. This band or galactic plane comprises an inside view of the flat disk of our Galaxy.

The disk-shaped structure of a typical spiral galaxy is also clearly seen in the Figure 13.1 photograph of Andromeda. The blue stars reside almost entirely within the galactic plane, as can be strikingly seen in Plate 2(a). This is true whether looking at the band of stars surrounding us in our own Galaxy or examining the stellar population of faraway Andromeda. And since the blue stars are the O- and B-type objects that burn their fuel rapidly, we can surmise that the galactic plane must house the sites of current star formation. We can also now recognize the band of concentrated stellar and interstellar matter of Figure 12.5 to be the Milky Way's analog of the disk-shaped plane in Andromeda.

On the other hand, the reddish stars seem to reside throughout a spherical region in Andromeda. Careful study of Figure 13.1 shows the red stars to exist not only in Andromeda's flattened plane, but notably in its more spherical galactic halo extending well beyond the plane. The red stars, remember, are mainly old stars—especially those members of the aged globular clusters.

A similar pattern of red and blue stars can be noticed for our own Galaxy, even by viewing it from within. Figure 13.12 roughly sketches this twofold distribution. The local blue stars, observable to distances of a few thousand light-years before interstellar obscuration obliterates them from view, seem to reside in the band or plane of our Galaxy, as do the young galactic star clusters. However, the local red stars—especially the old globular clusters like those studied by Shapley—inhabit a more spherical region extending well above and below the galactic plane.

This dichotomy of halo red stars and plane blue stars can be understood if our Galaxy (or Andromeda) had the shape of a ball when the oldest stars formed in globular clusters about 12 billion years ago. The stars, gas, and dust must have originally extended at least 100,000 light-years all the way out through the galactic halo. During the past 12 billion years, rotation presumably flattened the bulk of the Galaxy, drawing loose gas and dust into a plane that now resembles a pancake. Currently, the Milky Way system concentrates virtually all its gas, dust, and younger stars in the narrow plane and central bulge, thus granting the plane a faint bluish tint. Similarly, the short-lived blue stars have long since burned out in the halo, leaving only the long-lived red stars, thus giving the halo a pinkish spherical glow. The halo is ancient; the plane is full of youthful activity.

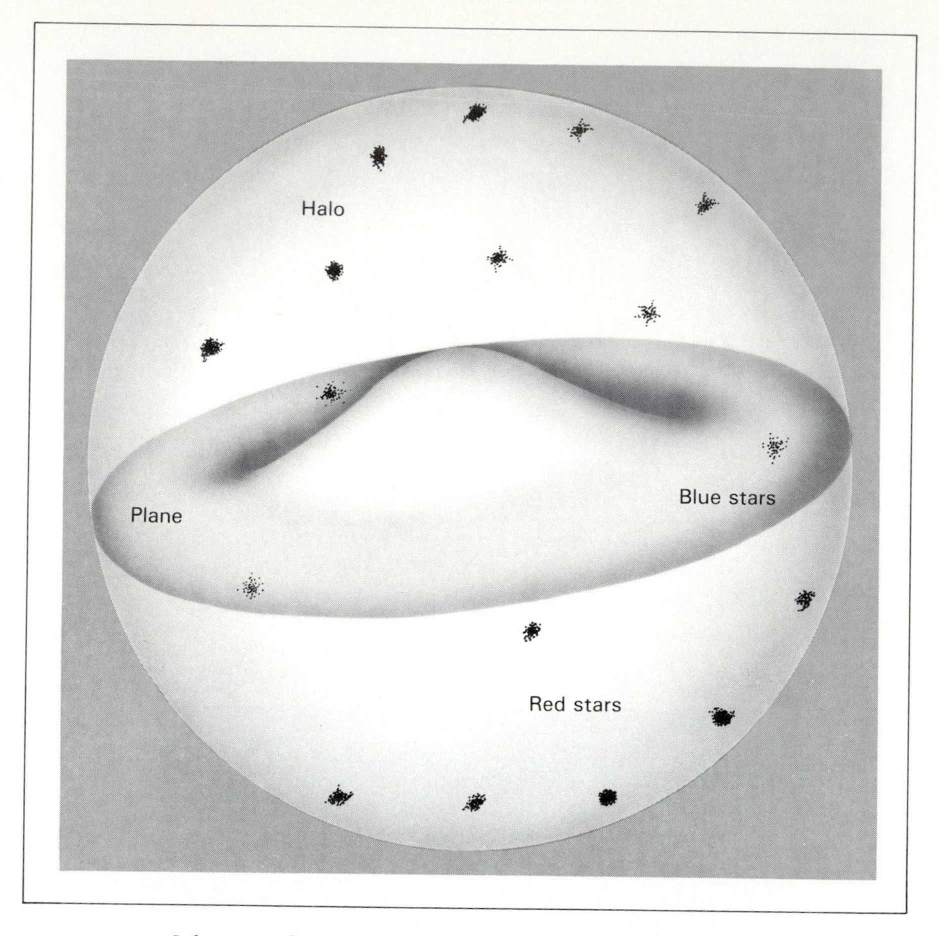

FIGURE 13.12 Schematic diagram of a galaxy, such as our own Milky Way or Andromeda, showing the different distributions of young blue stars and galactic clusters, as well as old red stars and globular clusters.

Mass of the Galaxy

THE GALACTIC TRAFFIC PATTERN: What about the dynamics of the matter in our Galaxy? Can we learn anything by studying the motions of the clouds and stars? In short, we return to our original question posed early in Chapter 11: Are the internal motions of the galactic constituents chaotic and random, or are they part of a gigantic traffic pattern? The answer depends on our perspective. While the motions of stars and clouds might seem random on a small scale (say, on the scale of a few light-years), those motions are indeed systematic on a large scale (say, on the scale of thousands of light-years). Expressed another way—and one that's generally applicable throughout astronomy—a large enough clump of space often displays some organization, whereas a clump too small shows mostly chaos.

To appreciate the organized movement of our galactic system, let's consider the spiral arms a little further. The gas and stars in the arms are detected, of course, only *from our vantage point on Earth* while in orbit about the Sun. Yet, while observing in various directions like those depicted in Figure 13.8, a clear pattern of Doppler motion emerges. In particular, all the spectral lines emitted from interstellar gas in the upper right quadrant of Figure 13.9 are blue shifted. On the other hand, all the interstellar regions sampled in the upper left quadrant are red shifted, as are those in the lower right quadrant. And as Figure 13.13 summarizes, those in the lower left quadrant are also blue shifted. In short, some galactic regions are approaching the Sun, while others are receding. The important point is that they are observed to do so in a reasonably systematic fashion.

Pondering this fourfold pattern of Doppler-shifted interstellar clouds, we are driven to an unmistakable conclusion: The entire system of galactic spiral arms is spinning. The whole Galaxy is rotating, although not uniformly. It's rotating *differentially*. By this, we mean that the entire Galaxy rotates not as a solid body; instead, some galactic parts rotate faster than others. Specifi-

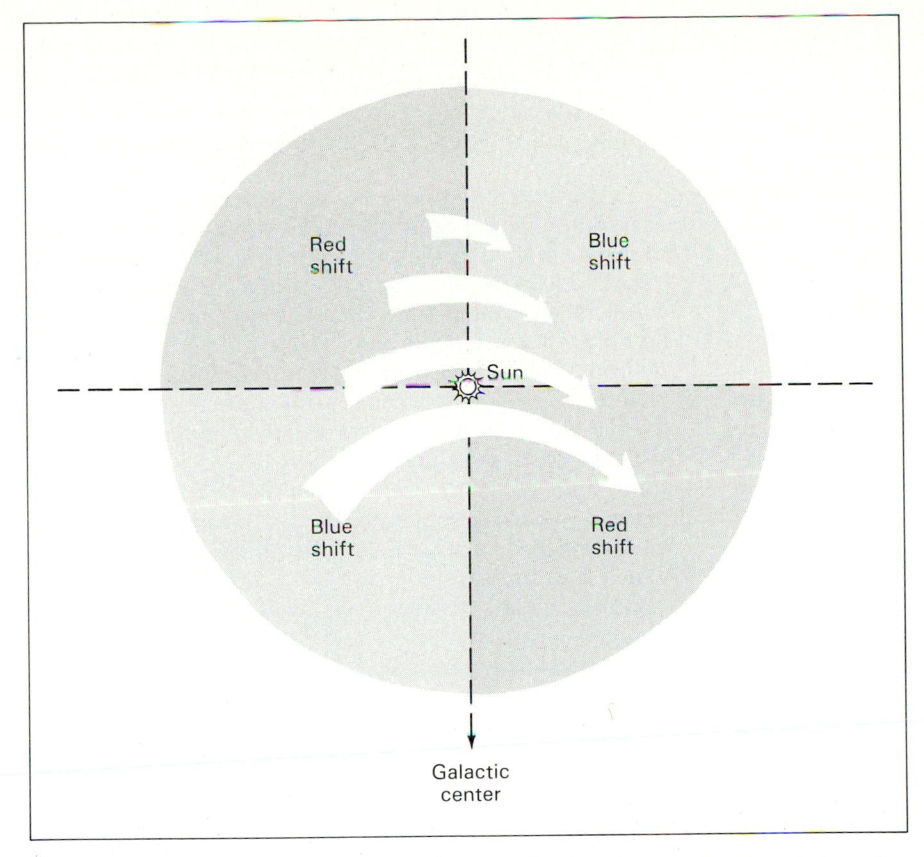

FIGURE 13.13 Schematic diagram of the four galactic quadrants in which interstellar clouds show systematic Doppler motions. Note that these quadrants are drawn (as dashed lines) to intersect at the Sun, not at the galactic center—for it is from the viewpoint of our own planetary system that our observations are made.

cally, the radio observations demonstrate that the inner parts of the Galaxy spin more rapidly than the outer parts.

The entire galactic plane, then, is circulating *around the galactic center*—not around the Sun. Overall, the pattern of differential galactic rotation is reminiscent of planetary orbital motion; the inner planets revolve more rapidly about the Sun than do the outer planets. Both systems of astronomical objects—the Solar System and the Galactic System—revolve at rates that depend on distance from the center of rotation. This is the case for the Andromeda Galaxy as well; spectroscopy proves the inner parts rotate quickly, the outer parts sluggishly. Accordingly, the stars in any galaxy, ours included, do not move smoothly together, but ceaselessly change their positions relative to each other while orbiting around the galactic center.

The similarity of the rotation patterns of our Galaxy and Solar System is more than analogical. It's substantive, for they both obey the same dynamical laws. As such, many of the principles of planetary motion discussed earlier can be used here to obtain a better understanding of galactic motions. As an example, we can apply Kepler's laws to the entire Galaxy.

Recall, from Chapters 7 and 11, that Kepler's third law combined the orbital period, orbital size, and masses of any two objects in orbit. We qualitatively expressed this proportionality as follows:

$$\text{(mass of one object + mass of other object)} \propto \frac{\text{orbital size}^3}{\text{orbital period}^2}.$$

In the case of our Solar System, the mass factor on the left includes the mass of the central Sun plus the mass of an orbiting planet. With the mass of any planet being small compared to that of the Sun, the planet's mass can be considered negligible, thus making it relatively easy to deduce the mass of only the Sun from a knowledge of orbital periods and sizes. The same is true for binary-star systems as discussed in Chapter 11; provided that one star of the binary system is much more massive than the other, we can take the mass of the smaller one to be negligible and thus use the proportionality above to derive the mass of the larger star.

Similarly, the mass factor above for our Galaxy is the sum of the mass of any galactic object whose period and distance are known *plus* the mass of the rest of the Galaxy. For example, if we can determine the orbit of the Sun around the galactic center, then we'll have both the orbital size and period needed to evaluate the relationship above. The mass thereby found would be the combined mass of the Sun and the Galaxy. But since our Sun is literally dwarfed by the Galaxy, we can regard the result to be the mass of the Milky Way Galaxy.

Neutral-hydrogen and carbon-monoxide observations provide both of the quantities needed to calculate the mass of the Galaxy in this way. Measurements show that the Sun orbits about the galactic center in approximately 250 million years; this is the "galactic year." (Note that since the Sun is about 5 billion years old, it must have orbited the galactic center about 20 times.) The second parameter needed is the separation between the Sun and the galactic center, a distance we've found earlier to be about 30,000 light-years. Substituting these quantities into the relationship above, we find that the mass of the Milky Way Galaxy is nearly 200 billion solar masses. That's 2×10^{11} times the mass of our Sun.

Although most stars of our Galaxy are obscured from our vantage point at Earth, we can estimate that about half of the Galaxy's mass is

INTERLUDE 13-2 ***Spiral Density Waves***

The bulk of our Galaxy is shaped like a thin disk with a bulge at the galactic center. Within the disk are "spiral arms" where the younger objects tend to reside.

This gargantuan collection of billions of stars and vast quantities of gas and dust is not static, for all the stellar and interstellar matter is rotating. At our distance of 30,000 light-years from the galactic center, the observed movement of the Sun means that about 250 million years are required for one complete "galactic orbit." That's a long time, of course, but not because the Sun is moving slowly; the Sun and the entire Solar System are gliding through the Galaxy at a 250-kilometer/second clip. Instead, the rotation period is long because the Galaxy is so huge.

Mentioned elsewhere, this rate of movement implies that our 5 billion-year-old Sun has cycled around the galactic center some 20 times. But even before our Sun's birth, the Galaxy existed—the globular clusters prove it, and simultaneously imply an age of about 12 billion years for our Galaxy. Hence the galactic matter at a distance of 30,000 light-years has orbited the galactic center at least 40 times since the birth of the Galaxy. The rapidly rotating inner parts of the Galaxy must have orbited many more times than that, the sluggishly rotating outer parts fewer times.

These arguments suggest a problem: If the Galaxy is a continuously rotating collection of stars and gases, how do the spiral arms retain their structural integrity? In other words, how do the arms avoid destruction in the face of repeated differential rotation? After a few revolutions, we might expect the pinwheel-like arms either to wrap up into a ball or to stretch out into thinner sheets of stars, gas, and dust. Yet the spiral arms persist—not only in our Galaxy, but also in many other galaxies to be studied in the next chapter.

This paradox can perhaps be resolved by carefully examining the types of objects in the spiral arms. The most obvious objects in the spiral arms are blue stars, galactic clusters, and gaseous nebulae. Each of these comprise O- and B-type stars having typically short lifetimes of about 10 million years or less, and hence will not endure long enough to make even one complete trek around the Galaxy. Thus we might expect the arms to disappear in one stellar generation. Yet the arms *do* exist, and they've presumably existed for quite some time. Since the grand spiral pattern seemingly persists, we must conclude that the arms are continuously replenished with new stars and matter from which they are made.

One popular, yet unproven, way to imagine replenishing the arms with new matter is to assume that the arms contain no permanent matter. As such, an arm should not be viewed as an assemblage of stars, gas, and dust moving intact. In this "density-wave theory" of spiral structure, the stars and gas clouds of a galaxy are envisioned to rotate slowly *through* a set of invisible gravitational waves. These waves delineate the spiral arms, allowing the matter of a galaxy to pass through them, much like a water wave on the open sea passes through water without actually pushing the water along. Instead, a water wave just builds up the water temporarily in some places (crests) and lets it down in others (troughs). A better analogy is a kitchen blender; as the metal arms whirl around, crests of juice or cake batter form at the top of the mixture.

Similarly, when the gravitational spiral wave encounters interstellar matter, the gas is compressed into

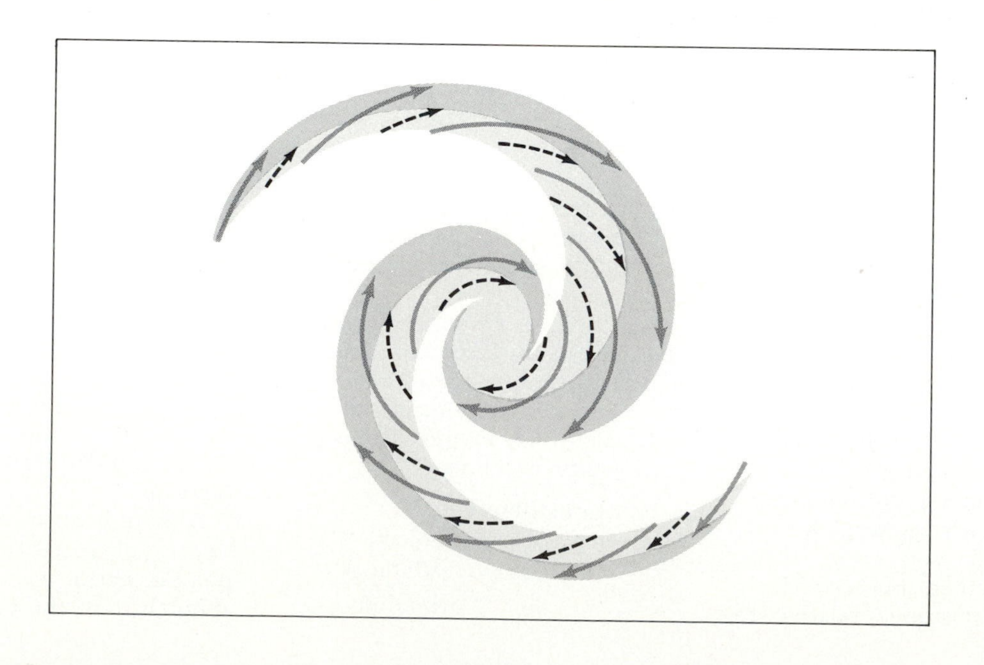

a crest of higher-than-normal density, which is likely to trigger the formation of new stars and nebulae. In this way, the spiral arms are formed and reformed repeatedly, without the arms themselves hardly moving at all (thus avoiding being wrapped up into a ball). As shown in the accompanying illustration, the (slowly) moving spiral-density wave (short dashed arrows) is outrun by the (faster) rotation of the matter (longer arrows), thereby creating the new (heavier shaded) arms in front of the old spiral arms.

An observer (such as ourselves) would then perceive the spiral arms to consist of compressed interstellar matter and mostly young objects—precisely what we do see. But since the brightest stars are the most massive ones, they have short lifetimes and hence perish rapidly. At a later time, new matter will have encountered the gravitational spiral wave, will have created new young stars and nebulae, and will give the (incorrect) impression that the spiral-arm pattern itself is spinning.

---

trapped in stars. Thus if every star has an average of 1 solar mass (big stars offset little ones), our Galaxy must house at least 100 billion stars. A hundred billion! That's a lot of stars—by anyone's standards. The Milky Way Galaxy is truly vast. It's vast in size, and it's vast in mass. Clearly, it's much bigger than anything we've encountered thus far in this text.

To further appreciate the Galaxy's enormity, recall the numerical example noted earlier in the Overview: To count from 1 to a billion at a rate of one number per second requires an entire human lifetime. Well, *just to count* the number of stars within our Milky Way Galaxy, we'd need at least 100 people counting for their whole lives! Another gee-whiz statistic is that if you had a penny for every star in our Galaxy, you would be a billionaire.

Unquestionably, then, our Galaxy is huge—nearly incomprehensibly so. Before proceeding, you should think carefully about the relative sizes of the Earth, the Solar System, and the Galaxy. We're progressing to distant realms at a rapid pace now. It's of utmost importance to place into proper perspective the sizes, scales, and distances studied thus far, lest you become confused when we make giant leaps forward in the next few chapters toward the very limits of the observable Universe.

THE GALAXY'S CORONA: Actually, the mass value computed above refers to the mass of stars, gas, dust, and all else residing inside the Sun's orbit around the galactic center—namely, most of everything within the spiral arms sketched in Figure 13.11. But astronomers now realize that the Galaxy outlined by the globular clusters and by the spiral arms is merely the "tip of the galactic iceberg."

In a recent and dramatic development, radio and more indirect visible-light observations are helping to discern a much more extensive halo of thin, cold, invisible matter that seems to dwarf the inner halo of globular clusters. This (three-dimensional) *extended* halo or **galactic corona** reaches well beyond the 100,000-light-year limit considered, until recently, to demarcate the typical dimension of our Galaxy. As we shall discuss in Chapter 14, such coronas have been inferred for several other galaxies as well, leading astronomers to suspect that most galaxies are much larger than outlined by their optical photographs.

These findings have caused an obvious revision of the large-scale features of our Milky Way. At face value, they demonstrate that astronomers have underestimated the size, scale, and mass of our Galaxy. Given the current status of the observations (which are still in progress), its size is at least double and probably triple (and possibly even larger yet) what astronomers thought a mere decade ago. We can be certain that our Galaxy is at least as large as the entire frame of the illustration in Figure 13.11; all the darkness in the figure, well beyond the outermost spiral arm, is apparently populated with invisible matter. The gas in this galactic corona is exceedingly thin, though, implying that our Galaxy's mass probably needs to be adjusted by hardly more than a factor of 2. Now totaling approximately 400 billion solar masses, our Milky Way must henceforth be considered a major spiral galaxy.

The corona's low density—at best that of the terribly tenuous medium among the dark clouds and between the spiral arms—probably explains why no one has ever observed a conventional star, cloud, or nebula in the corona; there's apparently not enough matter *per unit volume* to fabricate even the simplest structure, such as a flimsy cloud (see Interlude 14-1, however). An alternative that astronomers are considering is that the coronal material could be in the form of old, burned-out stars. This could also explain why nothing has been seen in the corona from our vantage point. In any case, because of the tremendously large total volume of such an extended halo (recall that volume scales as the cube of a linear dimension), astronomers now suspect that roughly as much dark matter lurks in the invisible galactic corona as exists in the visible spiral-arm disk itself.

SOME HIDDEN MASS? The previous calculation for the total mass of the Galaxy assumed that Newtonian gravitation and Keplerian motions are valid for the Galaxy as a whole. These are assumptions, yes, but assumptions that astronomers regard as reasonable. There is no reason why principles applicable to the Sun and planets should not also apply to a much larger (galactic) collection of stars.

On the other hand, this assumption implies that our Galaxy is plagued with a vexing problem that just will not go away. An inventory of matter within the Galaxy shows a disturbing discrepancy. Since we've already computed the total amount of mass contained within the Milky Way, and since we can estimate reasonably well the volume of such a lens-shaped Galaxy, we can divide the one by the other to find the density of galactic matter. The answer is about 0.15 solar mass/cubic parsec. (Note that the volume unit "per cubic centimeter" becomes obsolete when considering the entire Galaxy. The cubic light-year or cubic parsec is a more convenient unit.)

This value of mass density means that within every volume of space defined by a cube having 1 parsec on a side, there exists a little more than one-tenth the mass of our Sun. This is an *average* galactic density that takes into account a higher density near the galactic center and a lower density in the galactic suburbs (near the Sun). This density value includes all kinds of stellar and interstellar matter combined.

Now here's the interesting point—and the discrepancy. Direct observations of the average number of stars in any galactic volume yield a stellar density of about 0.05 solar mass/cubic parsec. Similarly, direct observations of the general galactic gas yield an interstellar density of also roughly 0.05 solar mass/cubic parsec. The observed dust is negligible, being about 100 times less massive than interstellar gas (or about 0.0005 solar mass/cubic parsec); its contribution is too small to worry about here.

Together, then, the *observed* stellar and interstellar matter amount to about 0.10 solar mass/cubic parsec. But this value is only two-thirds of the *computed* 0.15 solar mass/cubic parsec. The problem can be expressed more dramatically: Astronomers cannot account for fully one-third of the matter in our Galaxy. Apparently, there is some "hidden mass." (Note that this minor quandary is *not* the "missing-mass problem" that plagues much larger realms in the Universe; this major challenge to modern astronomy will be discussed in Chapter 14.)

As best we can tell, the Galaxy must house still more, as yet unobserved, matter. We can suggest several candidates, although none are well understood. The first candidates are the black dwarfs. As we'll see in Part III, low-mass stars die uneventfully simply by cooling down. Eventually, they become black dwarfs and cannot be seen. (These are the burned-out stars mentioned in the preceding section as candidates for the Galaxy's coronal matter.) Stars perish in this way, however, only after extremely long times. In fact, low-mass stars burn so slowly that some researchers suggest that *no* black dwarfs could currently reside in our Galaxy. After all, since the age of the Galaxy is "only" about 12 billion years, it would seem that not enough time has transpired for any black dwarfs to form.

Another possibility is that numerous black holes lurk within the galactic darkness, sucking up matter and thus making it invisible to us mere mortals outside the holes. This is a controversial issue, however. In Chapter 21 we shall study the nature of black holes and their peculiar powers that make matter disappear. For now, suffice it to state that the very existence of bizarre black holes is debatable, let alone that they provide "hiding places" for large amounts of galactic matter.

A third and perhaps most reasonable solution to the missing-mass problem concerns the interstellar molecular clouds. Discussed in Chapter 12, these invisible regions are only currently being mapped by radio techniques. It's entirely possible that we have underestimated their size and scale. They might be more ubiquitous, more extensive, and more massive than portrayed toward the end of Chapter 12. To solve the hidden-mass problem, however, these dense clouds will need to house a surprisingly large component of galactic matter, in fact an amount comparable to the total mass of gas within the stars themselves.

Finally, this problem might be solved only by discovering some completely new contributor to our Galaxy's mass. For example, should the ghostly neutrino particles have even a slight mass (say, thousands of times less mass than the already lightweight electron), they could weigh heavily on the galactic scale of things. Even with so little mass apiece, these elusive particles might well have been produced in copious amounts during events of the early Universe (in fact, more so than in stars now), thus populating our Galaxy with hoards of subatomic particles which astronomers have not yet inventoried. Together, such neutrinos could conceivably contribute as much as 0.05 solar mass/cubic parsec. However, the neutrinos' very elusiveness currently makes this idea tough to test, or even to determine if they have any mass at all.

Cosmic Rays

In addition to the collection of stars, gas, and dust, one other type of matter populates our Galaxy. These are the **cosmic-ray particles**, often called cosmic rays for short. Since they continuously collide with Earth the cosmic rays comprise our only sample of matter from outside the Solar System. There are not nearly enough of them in the Galaxy, however, to solve the hidden-mass problem noted in the preceding section. Let's first study their bulk properties and then speculate about their origin.

PROPERTIES: High-altitude balloons, rockets, and satellites have in recent decades enabled astronomers to discover the chemical composition and the energies of the cosmic-ray particles. Figure 13.14 schematically illustrates a technique used to record them, showing an actual track of some cosmic rays photographed while colliding with a balloon-borne detector.

Cosmic rays are subatomic in size, generally atomic nuclei and a small proportion of electrons. The nuclei of hydrogen atoms—protons—are the most abundant of the cosmic

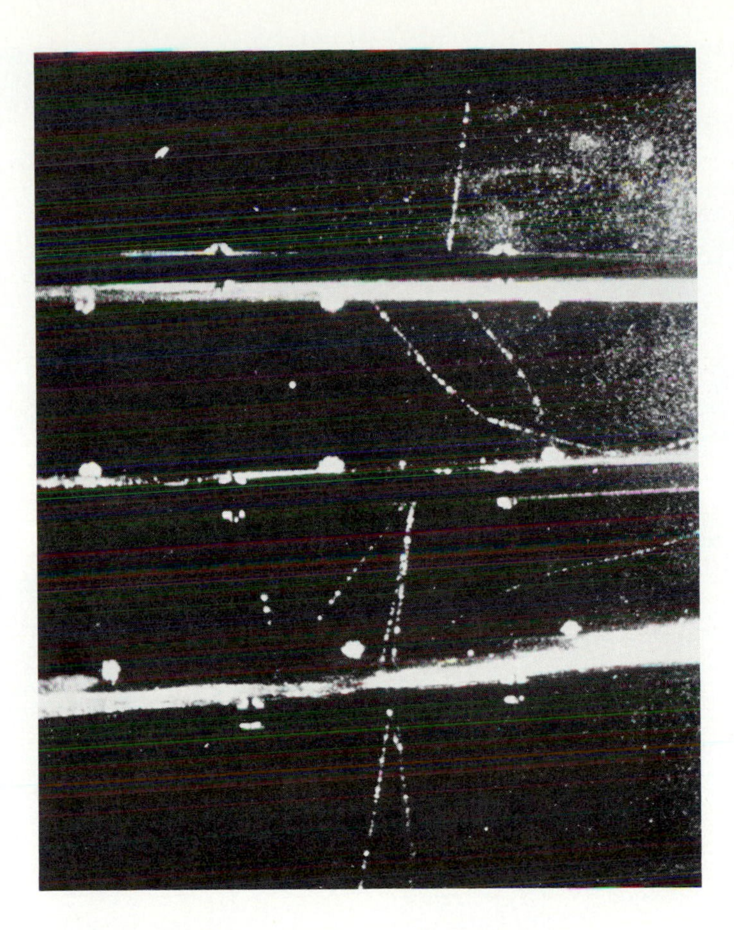

FIGURE 13.14 This track of a cosmic-ray particle, was recorded on photographic film as the high-energy particle crashed into an instrument aboard a high-altitude balloon. As implied here, when a galactic cosmic-ray particle hits our atmosphere, it creates a series of cascades or "showers" of many lower-energy particles. Large quantities of these lower-energy, unharmful "secondary" particles in turn collide with objects on Earth's surface, including humans, virtually all the time. Our bodies are peppered with them right now.

rays, followed by the nuclei of helium atoms as well as nuclei of an entire menagerie of heavier atoms. Not surprisingly, the protons amount to nearly 90 percent of all the cosmic-ray particles, while heavier nuclei account for about 9 percent. (Electrons contribute only 1 percent.) So, at least at the top of Earth's atmosphere where the cosmic rays are directly detected, their bulk cosmic abundances match those of all the known elements. There are some differences to be sure; the lighter cosmic rays are generally less abundant, and the heavier ones more abundant, than their atomic counterparts in most galactic objects. At any rate, an important point is that cosmic rays are subatomic *particles*; they are neither atoms nor "rays."

Cosmic-ray particles are very energetic. This energy is mostly in the form of kinetic energy, that is, energy of motion. By some unknown mechanism, cosmic rays are accelerated to extremely large velocities, much higher than our technology can accelerate subatomic particles in terrestrial laboratories. Virtually all of them, even the heaviest ones, travel close to the velocity of light. As a result, they must move rapidly from one region of the Galaxy to another.

The number density of cosmic rays is on the order of 10^{-9} particle/cubic centimeter in interstellar space. That makes them rare, although they are still more numerous than interstellar dust particles. The cosmic-ray particles, however, are much, much smaller than typical dust particles; furthermore, they zip rapidly through regions of otherwise sluggishly moving gas and dust.

Being charged, both the electrons and the nuclei tend to be deflected by the weak interstellar magnetic force field permeating the Galaxy. These charged particles are electromagnetically controlled by galactic magnetism in much the same way that Earth's magnetosphere controls the charged particles in the neighborhood of our planet. The majority of particles captured in the surrounding Van Allen Belts originate in the Sun and flow into the Solar System via the solar wind. But several experiments have shown indirectly that the genuinely galactic cosmic rays originate far beyond the Solar System and are much more energetic than the solar particles.

ORIGIN: What is the source of the galactic cosmic rays? The answer is currently uncertain. The circuitous and complex paths taken by the cosmic-ray particles while traversing the magnetized Galaxy prevent us from pinpointing their source by observing their direction of arrival at Earth. Indeed, careful searches have failed to show any preferential direction for the cosmic rays; they seem to be equally intense in all directions. Candidates for the origin of these cosmic rays include (1) violent events at the galactic center (discussed in the next section), (2) explosions of massive stars (see Chapter 20), and (3) escape from highly evolved white dwarf stars (see Chapters 20 and 21).

Most researchers currently favor the second of these explanations. In reality, each of these candidates probably contributes some particles to the observed hodgepodge of cosmic rays in our Galaxy. Some of the highest-energy cosmic rays might even originate far beyond our Milky Way Galaxy, a distant and ill-understood realm of the Universe that we shall consider in the next few chapters.

Center of the Galaxy

The entire Milky Way Galaxy is bound together by the mutual gravitational attraction of its stars, gas, dust and cosmic rays. Like a colossal pinwheel, the Galaxy rotates majestically around its totally enshrouded hub, some 30,000 light-years distant. The mysteriousness of this galactic center results largely from our inability to see it. The dark dust clouds within a few thousand light-years of our Sun effectively shroud what otherwise would be a stunning view of billions of stars tightly concentrated in and around the bulge of our Galaxy's midsection. Figure 13.15(a) shows the (optical) view we do have of the region of the Milky Way toward which the galactic center resides. This general direction is sometimes referred to as the Sagittarius region, so named for one of the constellations in that part of the sky.

With the help of infrared and radio techniques, we have recently begun to peer into the central regions of our Galaxy. Infrared observations indicate that the galactic-core environment harbors roughly 1000 stars/cubic light-year. That's a stellar density well over a million times greater than that in our solar neighborhood. Had any planets been associated with these galactic-center stars, they would probably be rapidly ripped from their orbits and obliterated as the stars must experience frequent close encounters and even collisions. Infrared radiation has also been detected from what seem to be huge clouds rich in dust.

Radio observations provide additional information. Figure 13.15(b) depicts contours of radio emission observed toward the core of the optical photograph in Figure 13.15(a). This radio radiation is the combined emission from dark clouds, gaseous

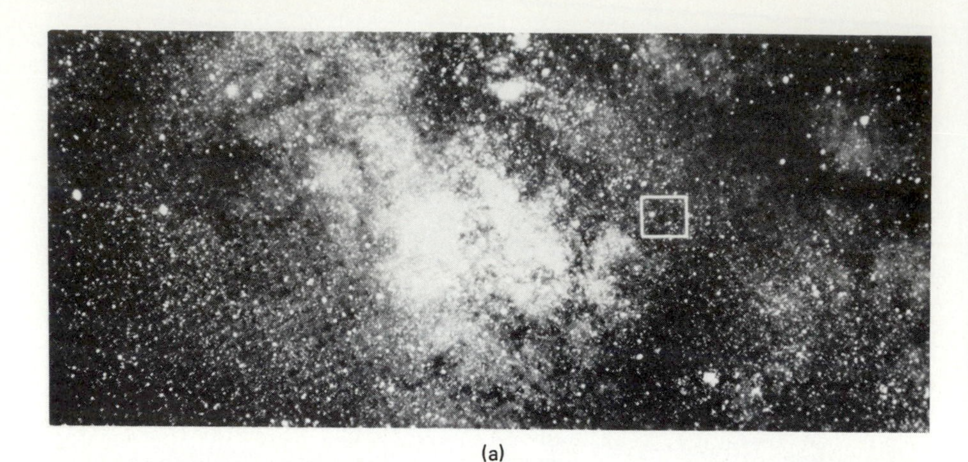

(a)

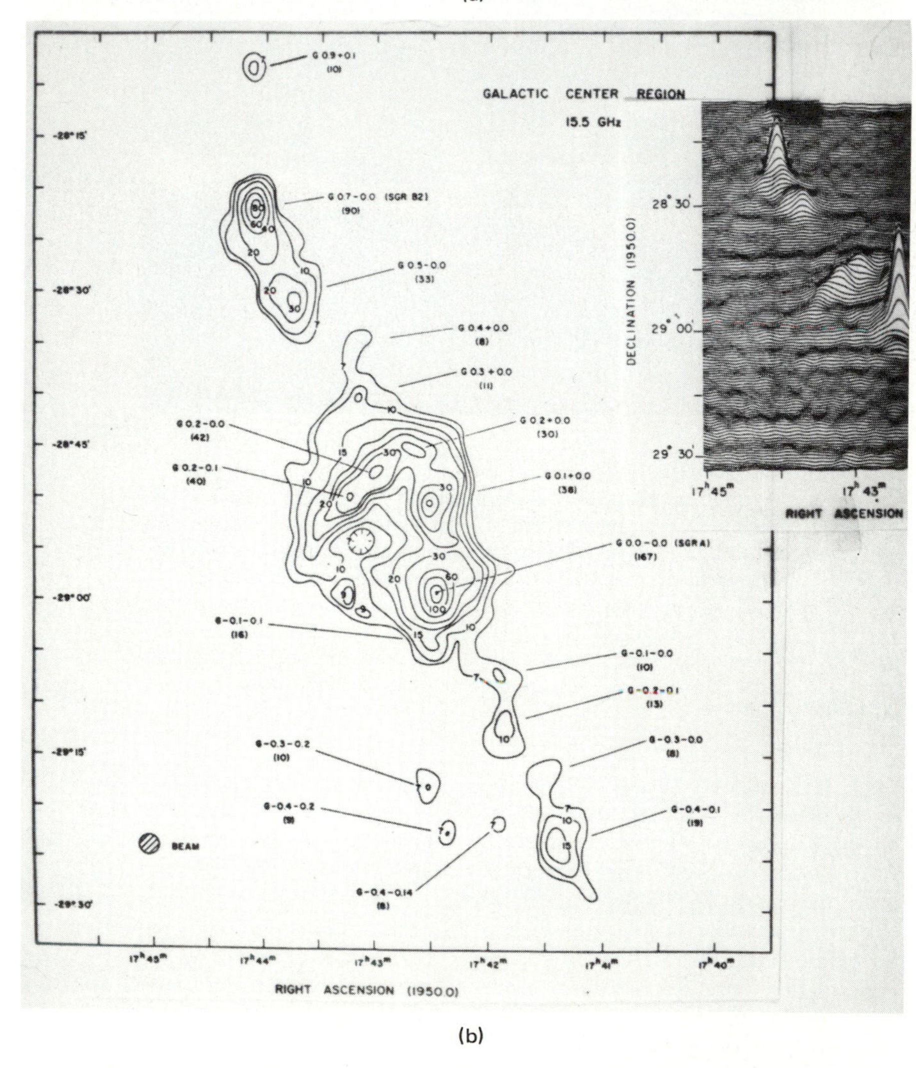

(b)

FIGURE 13.15 An ordinary photograph (a) of stellar and interstellar matter in the direction of the galactic center (outlined by the superposed box). Because of heavy obscuration, even the largest optical telescopes can see no farther than one-tenth the distance to the center. [This photograph is a southerly continuation of Figure 12.6. At top center of Figure 13.15(a), both the M20 and M8 nebulae can be seen.] A map of radio signals (b) emitted by matter within the superposed box surrounding the galactic center. The long-wavelength radio emission cuts through the galactic dust, providing an image of matter in the immediate vicinity of the galactic center, some 30,000 light-years away.

nebulae, and myriad sites of loose galactic gas. Similar maps, not shown, have been made of neutral hydrogen, carbon monoxide, and other molecular emission in this same region.

Is the galactic center's mystery unraveled by decoding radio and infrared radiation, or is this most intriguing galactic location still plagued by unexplained phenomena? The answer is yes on both counts. Astronomers now have a fairly good "road map" of the innermost regions, but we have yet to understand the precise mechanisms at work there.

Figure 13.16 places these findings into a simplified perspective. In a series of six airbrush conceptions, an artist has captured the salient results of recent long-wavelength radio and infrared studies of the Milky Way's heart. Each painting is centered on the Galaxy's core, and each increases in resolution by a power of 10.

Frame (a) renders the Galaxy's overall morphology, as painted in Figure 13.11. The scale of this frame measures about 300,000 light-years

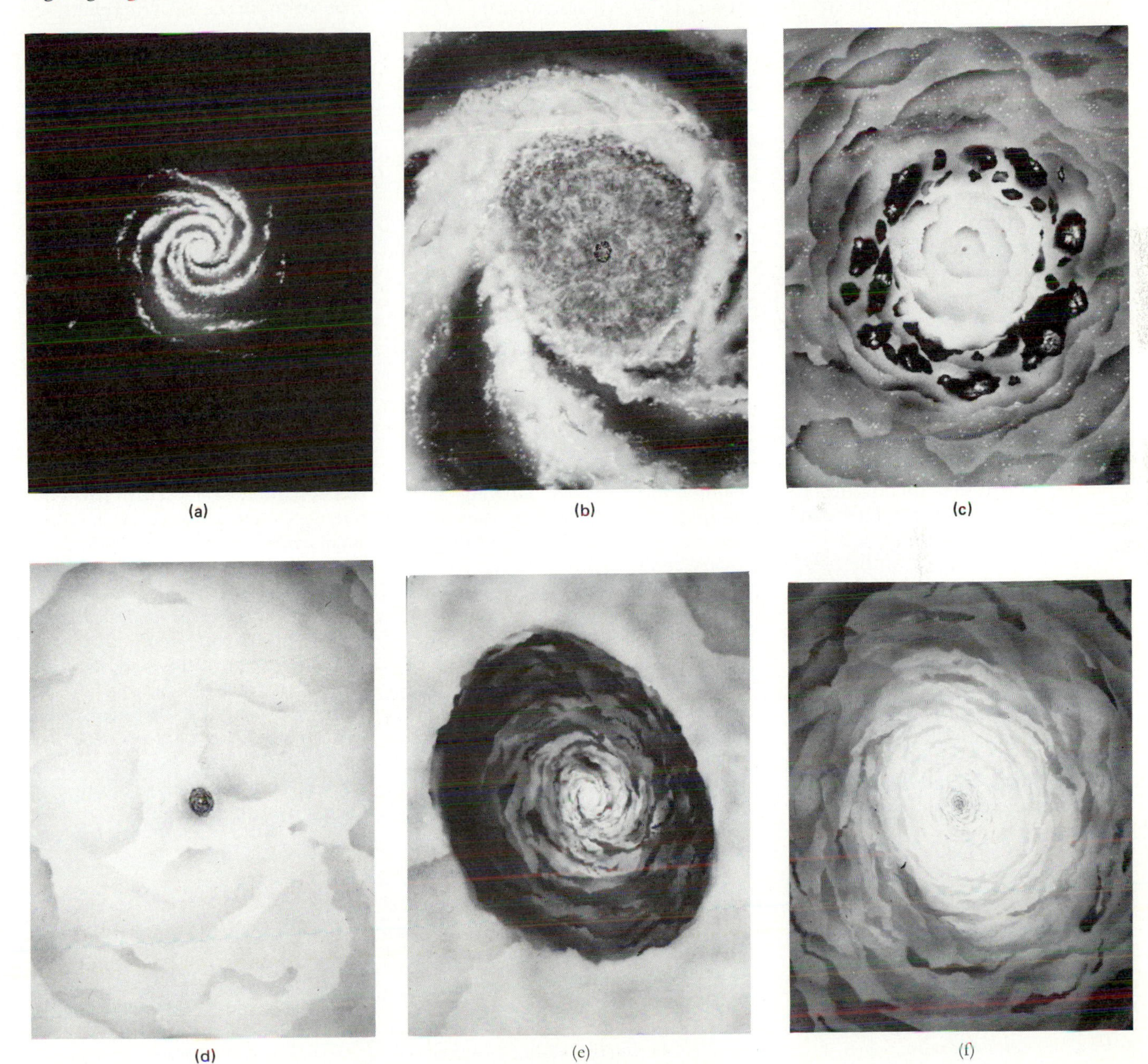

FIGURE 13.16 Six airbrush conceptions, each centered on the galactic core, and each increasing in resolution by a power of ten. Frame (a) is a reproduction of Figure 13.11, whereas frame (f) is an artist's rendition of a vast whirlpool within the innermost parsec of our Galaxy. [See also Plate 8(d).]

from top to bottom. Frame (b) spans a distance of 30,000 light-years from top to bottom and is nearly filled by the great circular sweep of the innermost spiral arm. Moving another 10 times closer, frame (c) depicts a ring of matter made mostly of giant molecular clouds and gaseous nebulae. This entire flattened, circular feature, about 1000 light-years in diameter, rotates rapidly (with speeds around 100 kilometers/second); the ring is also known to be expanding (at a comparable velocity) away from the center, much like a cosmic smoke ring. In frame (d), a pinkish ionized gas surrounds the reddish heart of the Galaxy. (All these colors, by the way, despite the region's invisibility, are inferred from the gases' temperature and density.) The source of energy producing this vast cloud of plasma is currently unknown, as is the case for the expanding ring in the previous frame, although the two may be related. Frame (e), now spanning 30 light-years, depicts a tilted, spinning whirlpool of hot (10,000-kelvin) gas that marks the core of the Galaxy. The innermost sanctum of this gigantic whirlpool is painted in frame (f), where a swiftly spinning, white-hot disk of superheated (million-kelvin) gas nearly engulfs an enormously massive object too small in size to be pictured (even as a minute dot) on this scale.

The consensus among astronomers today is that the galactic center is an explosive region—and one quite unlike anything discovered anywhere else in our Galaxy. The best evidence for the explosiveness is the ring of matter noted in frame (c). Based on the observed expansion velocity of that ring, we can surmise that a major explosion probably occurred at or near the Galaxy's center some 10 million years ago. The violence must have been considerable, since estimates for the amount of matter now in the ring average 100 million times the mass of our Sun. Other suggestive, though less well-documented observational evidence for additional rings of matter at differing distances implies that titantic explosions might be a regular phenomenon at the center of our Galaxy. Further observations have revealed a surprising group of parallel filaments about 150 light-years long and streaming out of the galactic center. The filaments are apparently formed by particles trapped in a magnetic field—the same effect that, on a smaller scale, is thought to cause sunspots. [Consult Plate 8(d).]

The problem—or the mystery—concerns the cause of the explosions. Whence might the explosive energy arise? What is the source of the violence capable of sending a *hundred million* solar masses hurtling outward toward the galactic suburbs? We can only speculate about the answers at this time. A leading contender for the cause of the violence includes, for example, a supermassive black hole, millions of times more massive than our Sun. Not that a hole itself need be emitting matter and energy, lest it violate some basic ideas about black holes, as discussed in Chapter 21. Instead, modern ideas focus on the vast disk or doughnut of matter being drawn toward the hole by its enormous gravity. The outer parts of such a region might become gravitationally unstable as infalling matter accumulates, possibly causing periodic expulsions every 10 million years or so—a galactic quake of sorts. Suffice it to note here that a compact, though massive concentration of matter is needed in some form or another to hold together the swirling maelstrom in the galactic center, lest the gaseous whirlpool itself disperse outward into the rest of the Galaxy. We shall return to this issue in the last section of Chapter 21 after we have taken up the study of black holes.

If our knowledge of the galactic center seems sketchy, that's because it *is* sketchy. Frankly, astronomers are still learning to grope in the dark, to decipher the clues hidden within invisible radiation. In particular, we are only beginning to appreciate the full magnitude of this entirely novel realm deep in the heart of the Milky Way. In some respects, our research should not yet even be called mature science. Rather, it's exploration—but absolutely fascinating exploration, enabling us to return from our telescopes with tales of the monuments at the core of our galactic system.

Our Galaxy's center is unquestionably problematic, but these same issues will become even more troublesome when we encounter the central regions of some other galaxies. Compared to the really violent galaxies, to be studied in the next few chapters, we shall come to recognize that the center of our own Galaxy is quite peaceful.

SUMMARY

The stars, gas, and dust of our Milky Way partake of a giant pinwheel pattern that spirals around a central hub some 30,000 light-years away. The entire Galaxy is disk-shaped, measuring at least 100,000 light-years across. Our Sun, in the suburbs, is just one of approximately a hundred billion other stars strewn throughout the galactic system.

Radio and infrared astronomy have played, and are continuing to play, crucial roles in our studies of the Milky Way. Mapping the spiral arms, probing the interstellar clouds, and exploring the galactic center, astronomers often find that long-wavelength techniques are needed to unravel the dark and dusty realms of our Galaxy. Even so, several important issues remain unsolved, among them the hidden-mass problem, the origin of the spiral-arm pattern, the source of the cosmic rays, and the nature of the galactic center.

KEY TERMS

cepheid variable
cosmic-ray particle
galactic halo
galactic center
galactic plane
galaxy
Milky Way Galaxy
galactic corona
period-luminosity law
spiral arm
variable star

QUESTIONS

1. Explain the period-luminosity law, citing its relevance as a distance-determining technique.
2. Pick one: The connected properties of Cepheid-variable stars that are used to derive distances are (a) period and color; (b) mass and color; (c) mass and luminosity; (d) period and luminosity.
3. Briefly explain two observational methods used to determine the size and shape of our Milky Way Galaxy.
4. What evidence do astronomers have—either directly or indirectly—that our Galaxy has a spiral structure?
5. Sketch a face-on and edge-on view of our Galaxy, the Milky Way. Include a distance scale, the position of the Sun, and indicate where you would expect to find young and old stars.
6. Why can't we take a photograph of our own Milky Way Galaxy? Speculate about whether Earthlings might ever be able to do so.
7. Why can't we see the galactic center? How do we know anything about it then? Explain.
8. How do astronomers estimate that there are about 100 billion stars in our Milky Way Galaxy?
9. What are cosmic-ray particles? Where do they come from? How do they help us learn something about the Universe?
10. Cite and briefly discuss some reasons that the galactic center is an explosive region.

FOR FURTHER READING

BERENDZEN, R., HART, R., AND SEELEY, D. *Man Discovers the Galaxies*. New York: Science History Publications, 1976.

BOK, B., "The Milky Way Galaxy." *Scientific American*, March 1981.

BOK, B., AND BOK, P., *The Milky Way*. Cambridge Mass.: Harvard University Press, 1981.

GEBALLE, T., "The Central Parsec of the Galaxy." *Scientific American*, July 1979.

MACKEOWN, P., AND WEEKES, T., "Cosmic Rays from Cygnus X-3." *Scientific American*, November 1985.

MATHEWSON, D., "The Clouds of Magellan." *Scientific American*, April 1985.

MEWALDT, R., STONE, E., AND WIEDENBECK, E., "Samples of the Milky Way." *Scientific American*, December 1982.

14 NORMAL GALAXIES: *Local Group and Beyond*

Descendants of our civilization may never become advanced enough to journey far away from our Milky Way Galaxy to see its full grandeur. The big picture of our swarm of starlight floating proud and silent in the near void of space may forever elude us. Yet from our vantage point at Earth, it *is* possible to study the full extent of other galaxies—colossal star systems far beyond our own Milky Way. To do so, we must observe the skies perpendicular to the plane of the Milky Way, thereby avoiding the obscuration associated with our own Galaxy.

Figure 14.1 shows a region of space containing myriad objects looking strangely unlike stars. Many have a fuzzy, lens-shaped appearance, often resembling a disk rather than the clear, bright, spherical image usually associated with stars. Kant, the eighteenth-century German philosopher, regarded each of them as individual "island universes" well beyond the confines of our Milky Way Galaxy. To label each of them a "universe" is an obvious semantics problem, but he was correct in arguing that these nonstellar patches of light reside outside our Galaxy.

Large telescopes have since revealed the blurry beacons of Figure 14.1 to be entire galaxies. They are all well beyond our Milky Way. Silently and majestically, the galaxies twirl ever so slowly in the far reaches of the Universe—huge pinwheels of radiation, matter, and perhaps life—giving us simultaneously a feeling for the immensity of the Universe and for the mediocrity of our position in it.

FIGURE 14.1 A cluster of many galaxies, each galaxy housing hundreds of billions of stars. This one, called the Coma Cluster, lies nearly 400 million light-years away.

(a)

(b)

FIGURE 14.2 These two galaxies are good examples of the architecture of our own Milky Way Galaxy, of the Andromeda Galaxy, and of many of the other objects shown in Figure 14.1. The M81 galaxy (a) is a "face-on" spiral system, whereas NGC4565 (b) is an "edge-on" spiral system.

Take another, detailed look at Figure 14.1. Think deeply about the message burned into this photograph. Recognize that each of these galaxies mimics those shown in Figure 14.2. Each is a vast collection of matter comparable to our Milky Way, measuring some hundreds of thousands of light-years across. Each is a gravitationally bound assemblage of stars, gas, dust, cosmic rays, and radiation much like our own Galaxy or Andromeda. In fact, *each* galaxy harbors more stars than people who have ever lived on Earth. Indeed, by examining Figure 14.1 and carefully pondering its implications, you will find it among the most mind-boggling photographs printed in this book.

In short, it is the galaxies that trace out the immense realms of the Universe. It is also the galaxies that best remind us that our position in the Universe is no more special than that of a boat adrift at sea.

The learning goals for this chapter are:

- to understand the basic properties of each of the different types of normal galaxies
- to understand several new distance techniques that enable astronomers to probe realms of the Universe far beyond our Milky Way
- to appreciate the way that galaxies are spread through space
- to recognize how matter clusters atop the hierarchy of material coagulations in the Universe
- to appreciate the different techniques used to derive the mass of faraway galaxies
- to know Hubble's law and how we can use it to derive distances to the most remote objects in the observable Universe

Galaxy Classification

Looking out from Earth, astronomers can see countless additional galaxies. Figure 14.1 is only one of many such photographs showing legion upon legion of remote galaxies. Naturally, we wonder what types of galaxies they are. Do they all have roughly the same size and shape as our Milky Way? Or do galaxies display an array of odd shapes and different sizes? If so, can we categorize galaxies much as we did earlier for stars?

Most of the objects that are easily identified as galaxies have spiral shapes much like our Milky Way or the neighboring Andromeda. Clearly, that's the case in Figure 14.1; the obvious galaxies have a central bulge from which emanate thin spiral regions, or "arms," chock-full of stars. But Figure 14.1 also displays numerous galaxies that do not have an obvious spiral shape. In reality, galaxies have many morphologies, and the spiral one is not the most abundant type of galaxy in the Universe.

The American astronomer Hubble was the first to categorize the galaxies comprehensively. Working with the then recently completed 2.5-meter optical telescope on Mount Wilson in California in the 1920s, he demonstrated basically four distinct categories of normal galaxies.

ELLIPTICALS: The first type is the **elliptical galaxy**. Most abundant of all the different types of galaxies in the Universe, they vary from highly elliptical shapes resembling footballs to nearly spherical shapes like beach balls. Figure 14.3 displays this variation in shape from the slightly elliptical E0 class to the almost-flat E7 class. Elliptical galaxies with intermediate elongations are classified E1 through E6.

Regardless of shape, most elliptical galaxies contain roughly equal numbers of stars as our Milky Way Galaxy. Here and there, exceptionally large or small elliptical galaxies are known to exist, though by-and-

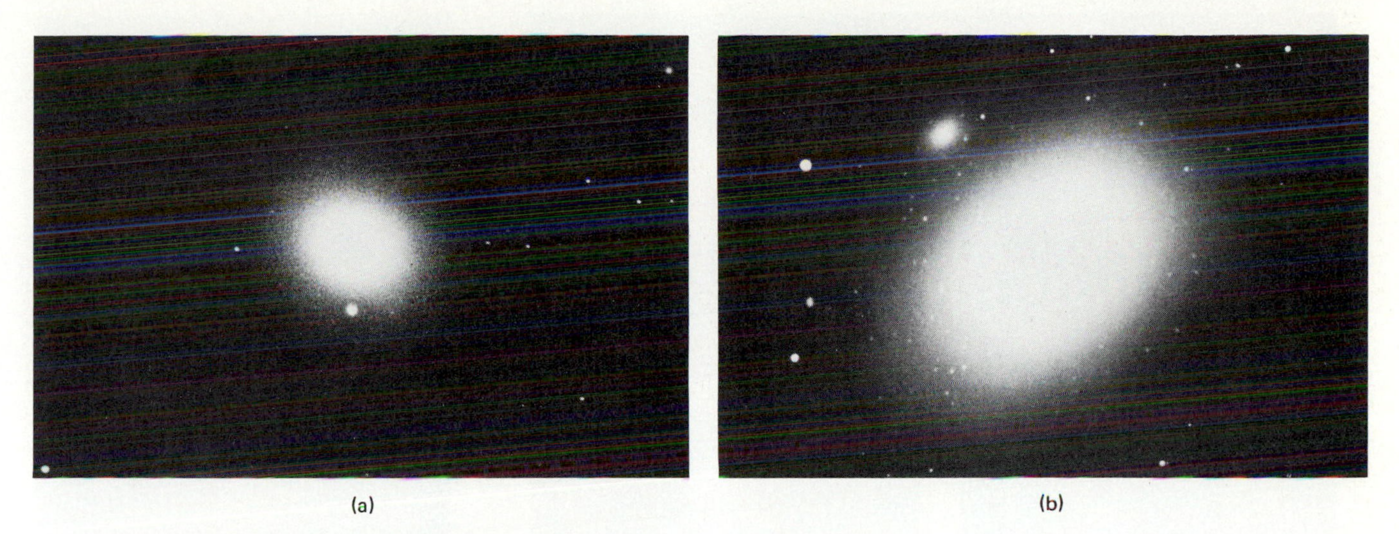

FIGURE 14.3 Variation in shape among different elliptical galaxies. M49 (a) is a nearly spherical elliptical galaxy, whereas M84 (b) is a slightly more elongated elliptical galaxy. Both of them lack spiral structure and both show no evidence for interstellar matter among their member stars.

large, most of them harbor the usual 100 billion stars scattered across some 100,000 light-years. The elliptical galaxies show no spiral arms; they simply appear to be undistinguished, though monumental globs of stars.

Spiral arms are not the only trait that elliptical galaxies lack. They seem to have little or no dust that would normally obscure parts of them from view. What's more, they seem to be severely deficient in interstellar gas. In many ways, the elliptical galaxies resemble the Milky Way's globular clusters; both types of object lack interstellar gas and dust, and both show no evidence of current star formation. The E0 and E1 elliptical galaxies even look like globular clusters. But be sure not to confuse these two types of objects. Globular clusters are clearly residents of our Milky Way Galaxy, and they are known to reside in the halos of other galaxies as well. Elliptical galaxies are truly vast aggregates of matter, much, much larger than typical globular clusters.

Comparison of elliptical galaxies with globular clusters is instructive. Their similar physical properties suggest that the elliptical galaxies are old. More precisely, they are made of old stars—red, low-mass stars. No blue stars are observed. Although they undoubtedly formed eons ago, all the massive blue stars in elliptical galaxies have presumably exhausted their fuel. This complete extinction of blue stars strongly implies that star formation no longer occurs in elliptical galaxies. The signposts of youth are missing. Furthermore, 21-centimeter radio emission from neutral hydrogen gas is, with few exceptions, completely silent for these galaxies. Evidently, all, or nearly all, the interstellar gas within elliptical galaxies was swept up into stars long ago, leaving no loose gas and dust for the continued formation of future generations of stars.

SPIRALS: The lack of gas, dust, and star formation in the elliptical galaxies contrasts sharply with the abundance and activity of interstellar matter within the second galaxy class—**spiral galaxies**. Here we also encounter a variety of shapes. Some spiral galaxies have a large central bulge around which emanate fairly tightly wrapped spiral arms. Others have a more open pattern of spiral arms protruding from an intermediate-sized central region. And still others have a rather small center and long, stringy arms, making it sometimes difficult to recognize these galaxies as spirals. Figure 14.4 samples these variations among spiral galaxies. Their nomenclature ranges from Sa for the tightly wound spirals, through Sb, and onto Sc for the loosely wound spirals.

Good photographs of typical spiral galaxies clearly show obscuration, indicating lots of dust. And 21-centimeter radio radiation is reasonably intense from spiral galaxies, betraying large amounts of interstellar gas. Furthermore, stars are still forming in spiral galaxies, especially within the arms where the interstellar density is enhanced. The galaxy M51 in Figure 14.5 and Plate 7(a) demonstrates how the young blue stars delineate the spiral arms of many spiral galaxies.

M51 is a perfect example of a spiral galaxy in which interstellar gas and dust are still present, thus granting such an object continued vitality. This doesn't mean that the spirals are necessarily young. Rather, they are still rich enough to provide for continued stellar birth.

Not all spirals can be seen as clearly as M51. Many are observable only at oblique angles, not face-on as is the case in Figure 14.5. Andromeda, for example (consult Figure 13.1), is almost certainly a spiral galaxy with plenty of interstellar gas, dust, and star formation in its spiral arms, but its angle of inclination prohibits a clear view of those arms. NGC4565 is an even better ex-

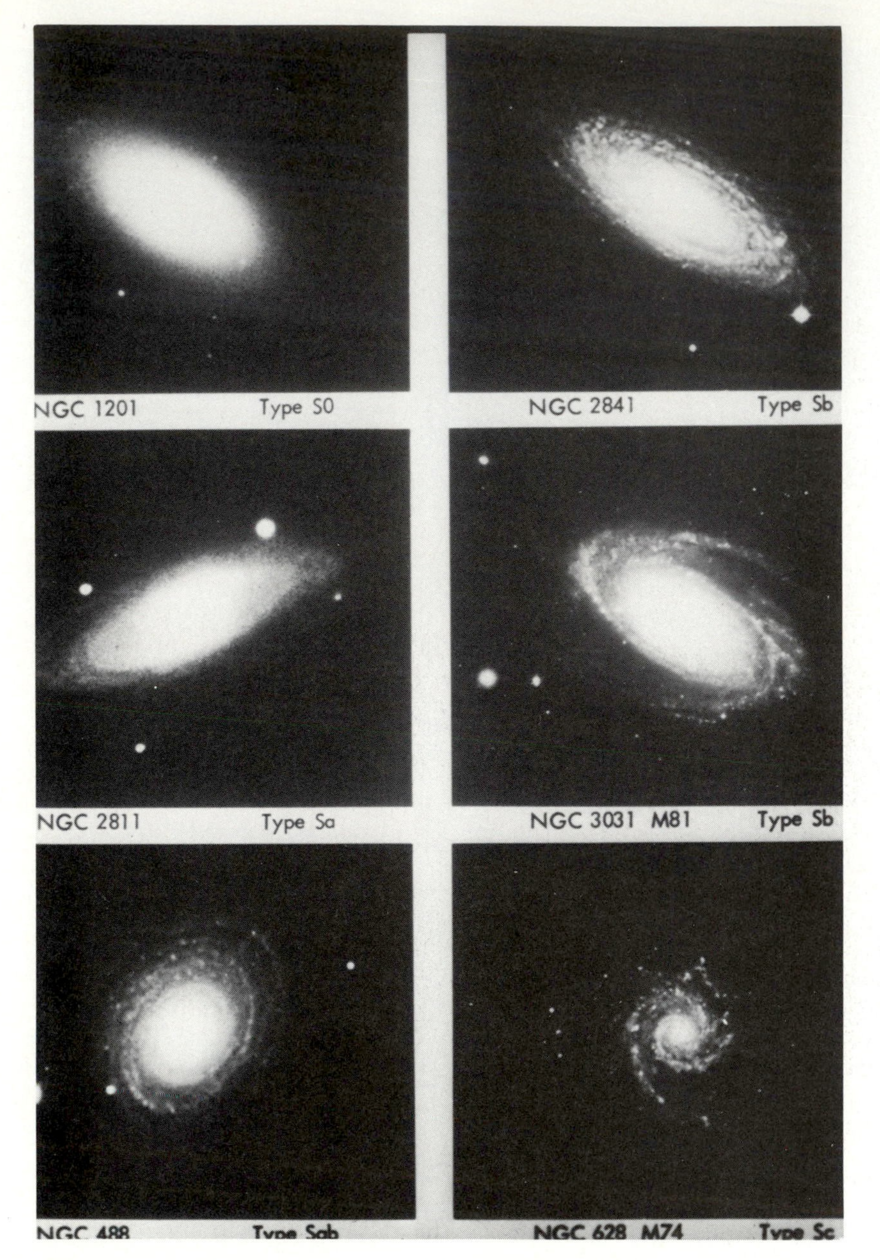

FIGURE 14.4 Variation in shape among different spiral galaxies.

FIGURE 14.5 The spiral galaxy, M51, clearly shows young (blue) stars spread along its spiral arms. This so-called "Whirlpool Galaxy" is actually a pair of interacting galaxies approximately 35 million light-years away. [See also Plates 1(c) and 7(a).]

ample of an edge-on spiral galaxy; see Figure 14.2(b).

So spiral galaxies contain a mixture of star types. They have large numbers of reddish M-type stars, many of which are giant members of old globular clusters observed in the spherical halos, as noted earlier for Andromeda and our own Galaxy. Spirals also have numerous bluish O- and B-type stars, often demarcating the spiral arms and other regions of relative youth in the plane. Most of the matter in spiral galaxies, however, is wrapped up in average A-, F-, G-, and K-type stars; together, these stars dominate the light of spiral galaxies observed from Earth, giving these galaxies a whitish-yellowish glow.

Our own Milky Way system seems to have all the properties of a typical spiral galaxy. Together with neutral-hydrogen and carbon-monoxide radio observations as described in Chapter 13, the spread of galactic and globular star clusters in our Galaxy is the best evidence that our own Milky Way system is an ordinary spiral galaxy—probably an average Sb type.

BARRED SPIRALS: A third category devised by Hubble is termed the **barred-spiral galaxy**. Figure 14.6 shows the variation of these types of galaxies from the tight-knit SBa to the more loose SBc designation. The barred spirals differ from ordinary spirals mainly by the presence of a linear extension or "bar" of stellar and interstellar matter passing through their centers. It is from near the ends of this bar that the usual spiral arms emanate. In the case of the SBc category, it's often hard to distinguish between the extended bar and the spiral arms.

Figure 14.7 shows an example of a galaxy that almost certainly is of the spiral variety. Known for obvious reasons (just look at it) as the Sombrero Galaxy, a very clear line of obscuring dust can be seen along the plane of this galaxy. Although in

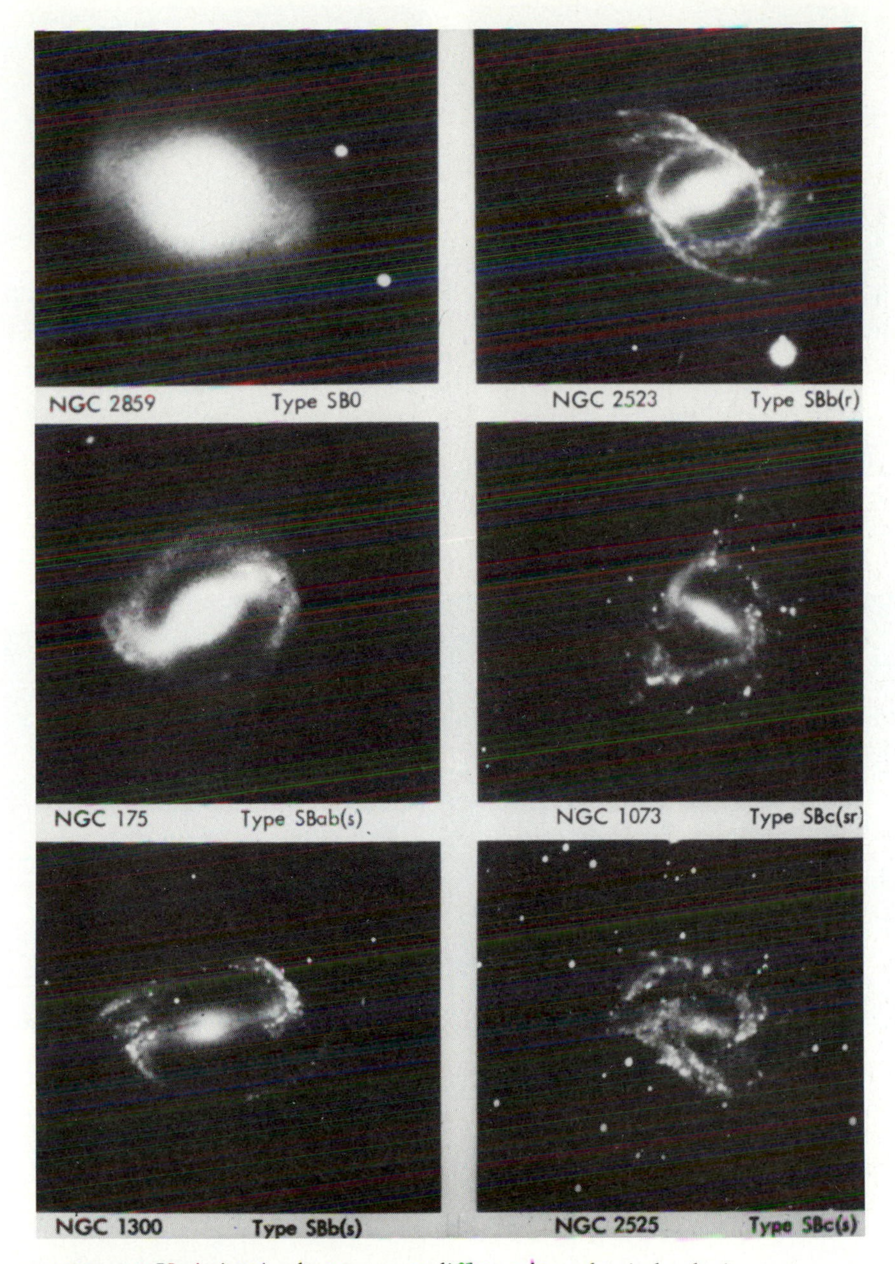

FIGURE 14.6 Variation in shape among different barred-spiral galaxies.

FIGURE 14.7 The "Sombrero Galaxy," a spiral system seen edge-on. Officially catalogued as M104, its dark band is composed of interstellar gas and dust.

this case we cannot be sure that the galaxy is a spiral or barred-spiral type, the unmistakable presence of dust rules out the elliptical type. Much of the near-spherical spread of light results from the older globular clusters in Sombrero's halo, whereas most of the interstellar matter lies in the plane of this striking galaxy.

Often, then, astronomers cannot distinguish between spirals and barred spirals, especially when a galaxy happens to be oriented with its flat galactic plane edge-on toward Earth. Because of the physical and chemical similarities of spiral and barred-spiral galaxies, some researchers don't even bother to distinguish between them. Others, however, regard their slightly different structures to be important, suggesting basic differences in the dynamical conditions that led to their formation eons ago.

IRREGULARS: The fourth and final galaxy class is a catchall category—**irregular galaxies**. Figure 14.8 shows some examples of these strangely shaped galaxies, known to be rich in interstellar matter. They are named irregular galaxies largely because of their peculiar visual images; they cannot easily be placed into any of the foregoing three categories.

Figure 14.9 and Plate 6(b) are photographs of the **Magellanic Clouds**, a pair of famous irregular galaxies. These "satellite" or dwarf galaxies orbit around our Milky Way and have been drawn to proper size and scale to the left of Figure 13.11. Similar "satellite" galaxies can also be seen in photographs of the Andromeda Galaxy in Figures 0.3 and 1.2, as well as Plate 2.

Irregular galaxies tend to be smaller than the other three galaxy types, hence their adjective "dwarf." They often contain about 10^8 to 10^{10} stars, considerably less

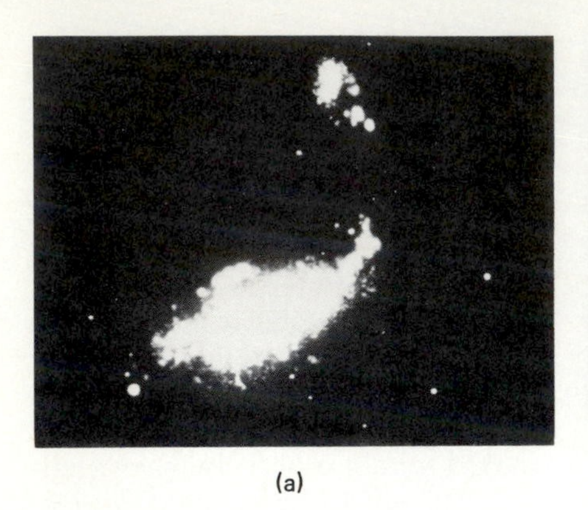

(a)

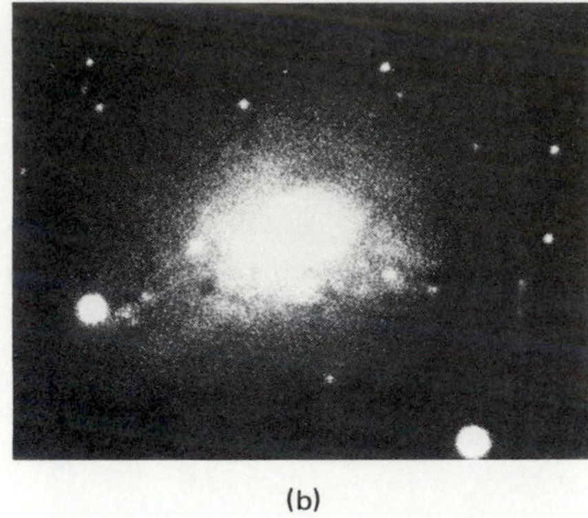

(b)

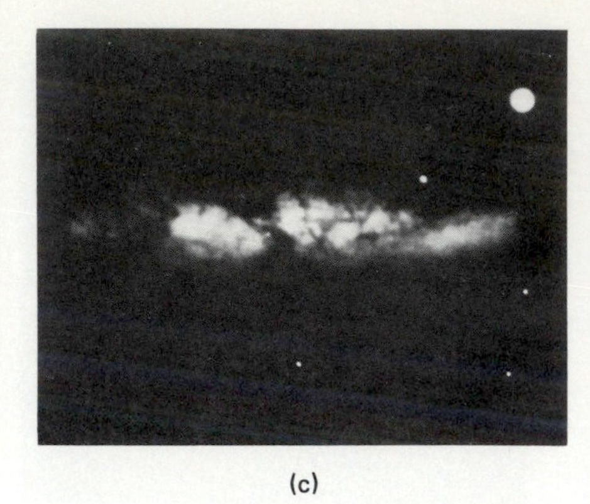

(c)

FIGURE 14.8 Photographs of some irregular and peculiar galaxies. The oddly shaped galaxies, NGC4485 and NGC4490 [both shown in (a)] might be related. The peculiar galaxy, NGC1275 (b), depicts what seems to be a system of long filaments exploding outward into space. And M82 (c) likewise seems to show evidence of explosiveness, although interpretations of peculiar galaxies remain debatable.

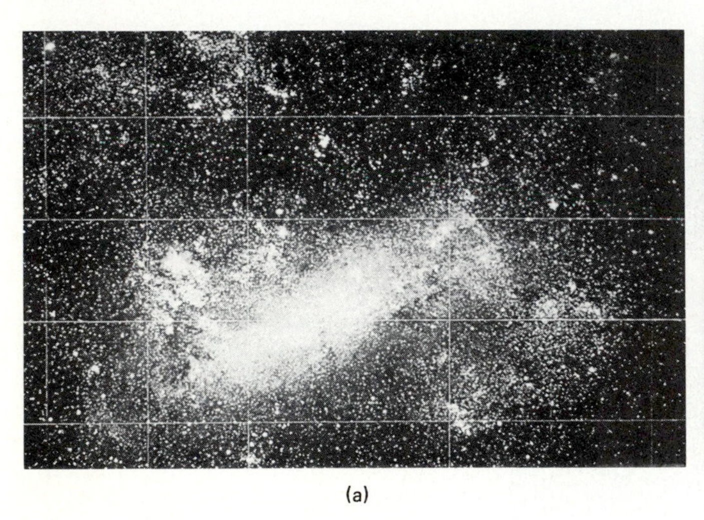

(a)

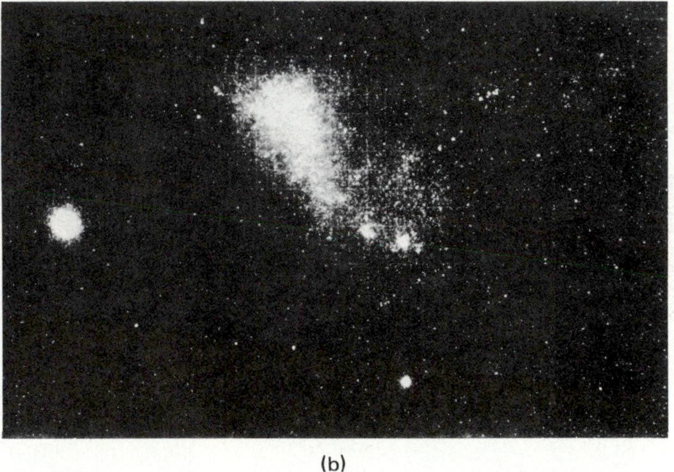

(b)

FIGURE 14.9 These irregular galaxies, called the Large (a) and Small (b) Magellanic Clouds are dwarf "satellite" galaxies that are gravitationally bound to our Milky Way system. [See also Plate 6(b).]

than the 10^{11} to 10^{12} stars normally populating both elliptical and spiral galaxies. Furthermore, the irregular galaxies are almost always allied with a "parent" galaxy from one of the other three categories.

NO HR DIAGRAM FOR GALAXIES: Having earlier found a convenient classification scheme for stars, namely the Hertzsprung–Russell diagram discussed in Chapter 11, we naturally wonder if there might be a similar overall pattern or evolutionary scheme relating the various types of galaxies. In other words, is there any convenient way to link the different galaxies much like the HR diagram relates red stars to blue stars and dwarfs to giants? The answer is no, as best we can tell. Astronomers have thus far failed to discover any underlying physical process among all the galaxies that would enable us to relate one type of galaxy to any other. We shall return to this issue in Part III when we discuss the possible evolution of galaxies.

Galaxy Distribution

Obviously, Earth is finite; beyond it stretches the tenuousness of interplanetary space. Furthermore, recall that our Solar System is also of finite size. That is, it has an end point beyond which we know of no additional planets. This is not to say that Pluto is definitely the most distant planet in our Solar System, but simply that we have no evidence to the contrary. That end point is the limit of the Solar System beyond which lies the near-vacuum of interstellar space. We then proceeded in Chapter 13 to discuss the point at which stars and interstellar matter seem to terminate well beyond our Solar System. That termination defines the limits of the Milky Way Galaxy beyond which lies the virtual void of **intergalactic space**.

INTERLUDE 14-1 ***The Clouds of Magellan***

Far to the south and out of viewing range of inhabitants of most of the northern hemisphere, there reside two nearby galaxies called the Magellanic Clouds. These dwarf irregular galaxies are gravitationally bound to our own Milky Way Galaxy; they not only slowly orbit our Galaxy, but also accompany it on its general trek through the cosmos.

The Large Magellanic Cloud and its companion, the Small Magellanic Cloud, can easily be seen with the naked eye from any location south of Earth's equator. Looking much like dimly luminous atmospheric clouds, they have undoubtedly served as sources of celestial wonder to residents of the southern hemisphere since the dawn of civilization. Even so, these clouds are named for the sixteenth-century Portuguese voyager Magellan, whose round-the-world expedition first brought word of these giant fuzzy patches of light to the European civilizations in the northern hemisphere.

Studies of cepheid variables within the Magellanic Clouds show them to be approximately 175,000 light-years from the center of our Galaxy. Figure 13.11 shows their position, size, and scale relative to the Milky Way. Close-ups of the Large Magellanic Cloud in Figure 14.9 and Plate 6(b) illustrate its distorted shape, although some researchers claim that they can discern a single spiral arm. Whatever its structure, direct observations prove this irregular galaxy to have lots of gas, dust, and blue stars, indicating youthful activity and presumably current star formation.

Over the years, various radio studies have claimed evidence of a possible bridge of hydrogen gas connecting our Milky Way to the Magellanic Clouds, but more observational research will be needed to establish this beyond doubt. More recently, some astronomers have argued that we must reassess the status of the Magellanic Clouds as galaxies. They maintain that if large quantities of invisible matter do in fact lurk in the darkness of the galactic corona well beyond the outermost spiral arms (consult Figure 13.11), then perhaps these Clouds are not galaxies at all—not even dwarf galaxies. Rather, should this view prevail, the Magellanic Clouds are merely rich clumps of galactic matter far from the Galaxy's center. Only further research, directed toward a more complete mapping of the full extent of the Milky Way's corona, will tell if the Clouds are genuine galaxies or merely forlorn regions of star formation in an otherwise structureless halo.

Having studied the different types of galaxies beyond our Milky Way, we naturally wonder about their distribution in space. How are the galaxies spread through the expansive tracts beyond our Milky Way? Do they reside everywhere, all the way out to the very limits of the observable Universe? Or is there, as noted for each of the regimes studied thus far, some terminal point beyond which the galaxies are no longer observed?

REVIEWING THE DISTANCE SCALE: To answer these questions properly, we must first find the distances to the galaxies. Let's review for a moment the various techniques capable of determining distances in astronomy. The bottom part of the inverted pyramid of Figure 14.10 summarizes the distance scale we've already discussed. Following the initial radar technique, we studied stellar parallax, a method developed extensively in Chapter 4. This second technique allows us to measure distances confidently out to about 100 light-years. At the time we discussed parallax, such a distance seemed great but in the last few chapters, we've come to recognize that 100 light-years is extremely local compared to the 100,000-light-year size of our Milky Way Galaxy.

The third technique, used to extend our knowledge of distances well beyond the local realm, requires us to compare the absolute and apparent brightnesses of individual stars. Recall from Chapter 11 that apparent brightness is the directly measured rate of energy emission, whereas absolute brightness is the intrinsic emission of any object when assumed to be at a distance of 10 parsecs from Earth. Knowing that the intensity of radiation diminishes as the square of the distance traveled by that radiation, we can estimate the true distance provided that we know both kinds of brightness. You will recall that the operational procedure for measuring the absolute brightness of a star is first to determine its spectral class (or surface temperature), and second, assuming it to be a main-sequence star, to read the absolute brightness from the HR diagram. This so-called spectroscopic parallax technique enables us to find distances as far as several thousand light-years from Earth—just about the greatest distance at which we can identify and study ordinary stars in the galactic plane. Such distances, though, are still well within our local niche of the Milky Way Galaxy.

A fourth technique of distance determination uses the cepheid variable stars, as discussed in Chapter 13. Since this method also requires a comparison of absolute and apparent brightnesses, it resembles the third one described in the paragraph above (and located one notch below on the inverted pyramid). The cepheids provide a welcome advantage, however, for they enable us to determine a star's absolute brightness without having either to know its spectral classification or to as-

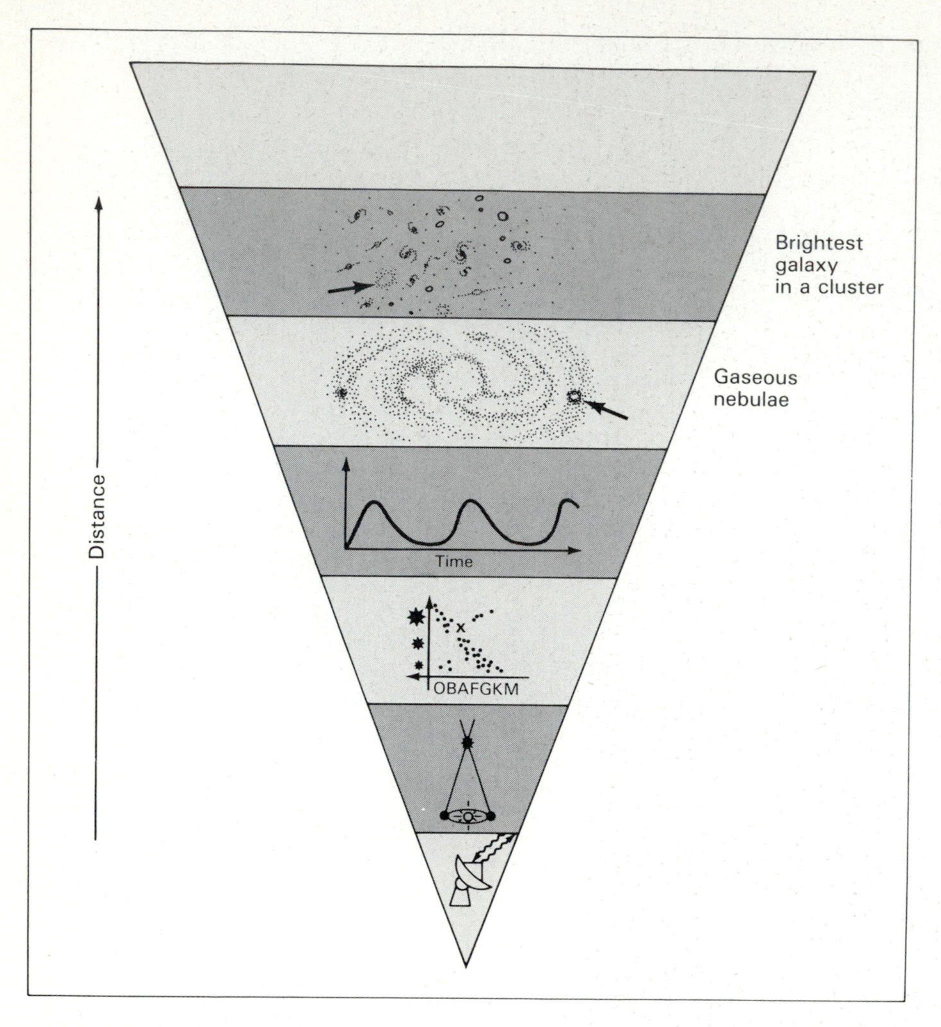

FIGURE 14.10 The inverted pyramid summarizes the various distance techniques used to study different realms of the Universe. One more rung on this distance ladder will be added in the final section of this chapter.

sume its location on the main sequence.

These general techniques utilizing absolute and apparent brightnesses have been discussed up until now for stars of our Milky Way. But they hold equally well for stars in other galaxies. By comparing these two types of brightness, we can estimate the distance to any object.

Thus we can extend our distance yardstick to the faraway galaxies provided that we can identify an individual cepheid variable star and measure its period. This technique assumes only that the period-luminosity law derived for the cepheids in our Galaxy holds true for cepheids in other galaxies. This is tantamount to saying that we assume the laws of physics to be the same everywhere. And since we have no reason to suppose otherwise, this is a most reasonable assumption.

The cepheid technique has become a powerful tool, enabling us to find distances as far as the best telescopes can resolve individual cepheids and hence measure their apparent brightness and pulsation period. With the best telescopes, the abnormally bright cepheids can be seen out to a distance of approximately 10 million light-years.

THE LOCAL GROUP: Now we're in a better position to answer the question posed earlier: Upon studying the cepheid variable stars within other galaxies, what does the spread of galaxies look like? In other words, how is matter distributed within 10 million light-years of our Milky Way?

Figure 14.11 sketches all the known major astronomical objects within a few million light-years of the Milky Way. Our Galaxy is shown, of course, with its two satellite galaxies, the Magellanic Clouds. The Andromeda Galaxy, a spiral system very much like our own, is also depicted in Figure 14.11. This system resides some 2 million light-years from the Milky Way, and, as noted earlier and shown in the figure, Andromeda also has a pair of minor satellite galaxies orbiting about it.

Observations have pinpointed several more galaxies within a few million light-years. In all, some 20 galaxies populate our Galaxy's neighborhood, including a few giant spiral galaxies among many dwarf irregular and dwarf elliptical galaxies. Some of these dwarf systems may be abnormally large star clusters (and thus not galaxies at all) that escaped from the giant galaxies. Their true nature is currently unclear.

A group of galaxies held together by their mutual gravitational attraction is called a **galaxy cluster** (or cluster of galaxies). The one including our Milky Way is given a special name, the **Local Group**. As de-

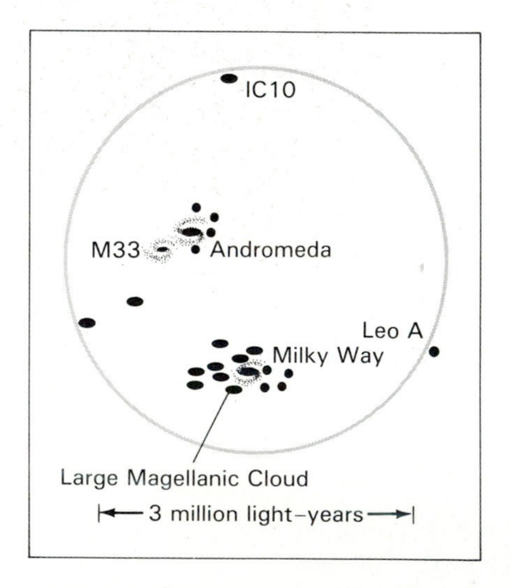

FIGURE 14.11 Schematic diagram of the Local Group of some twenty galaxies within approximately 3 million light-years of our Milky Way Galaxy. Only a few are well-established spirals; most of the rest are elliptical and irregular dwarf galaxies.

picted in Figure 14.11, the diameter of this well-defined galaxy cluster spans roughly 3 million light-years. The words "vast," "huge," and "gargantuan"—even the slang adjective "astronomical"—take on their real meanings when describing galaxy clusters.

Slow down for a moment to comprehend the contents of this section. In particular, be sure to note two things. First, recognize that we've suddenly made a great jump in spatial dimensions, from the 100,000-light-year size of our Milky Way Galaxy to the 3,000,000-light-year size of the Local Group of galaxies. Second, recognize that our Milky Way is not at the center of the Local Group. Not only is Earth not the center of our Solar System, and the Sun not the center of our Galaxy, but our Galaxy is also not the center of the much larger Local Group.

At every stage of our text, we fall prey to human nature and strive to discover a special place for ourselves in the Universe. But at every stage we fail. On all scales, our environment seems in no way special or unique.

Beyond the Local Group

Although our Local Group contains merely 20 or so galaxies, we can be sure that many more such galaxies exist in the Universe. The single photograph reproduced as Figure 14.1 reveals several hundred galaxies. And scores of similar photos have been taken for different regions of the sky. In all, astronomers estimate some hundred billion other galaxies roam the observable Universe.

Interestingly enough, all the other galaxies are *well* beyond our local galaxy cluster. For millions of light-years beyond the edge of the Local Group, there seems to be nothing—no galaxies, no stars, no gas and dust—just empty intergalactic space!

The cepheid distance technique has limited usefulness beyond the Local Group. As best we can tell, few galaxies reside between 3 and 10 million light-years away. Virtually all the other known galaxies are beyond 10 million light-years—too far for the cepheid technique to work. Cepheid stars embedded within very distant galaxies simply cannot be observed well enough, even through the world's largest telescopes, to permit measurement of their luminosity and period.

To extend our knowledge of the spatial distribution of galaxies, we need another technique—one that's capable of determining really great distances. In short, we need a "yardstick" valid beyond 10 million light-years. Fortunately, there is such a method.

DISTANCE ONCE AGAIN: The fifth technique used to determine distance is based on the same intercomparison of apparent and absolute brightnesses studied earlier. Since ordinary stars and cepheid stars appear too small and weak at great distances, we need to find a class of astronomical objects for which the angular size exceeds those of individual stars. In this way, we can observe, identify, and study such objects even within faraway galaxies. But what objects are larger than stars? That's easy—gaseous nebulae. Indeed, gaseous nebulae are much larger than individual stars, and their brightly glowing gases often help them stand out in distant galaxies.

Like the other distance techniques lower down on the inverted pyramid of Figure 14.10, we need to make an assumption for this technique to work. The assumption is that gaseous nebulae in other galaxies are pretty much the same as those in our own Milky Way Galaxy. In particular, we assume that the absolute brightness of *average* gaseous nebulae within our Milky Way equals that of *average* gaseous nebulae in any other galaxy. This assumption, along with a measurement of a nebula's apparent brightness, then yields an *estimate* of the distance to the remote galaxy in which the nebula is embedded.

The word "estimate" is italicized here to emphasize that distances are increasingly uncertain beyond the Local Group. Gaseous nebulae in our Milky Way range widely in luminosity, and to assume that the intrinsic luminosities of nebulae observed in distant galaxies equal those of average Milky Way nebulae is a tricky maneuver. It's not a poor assumption, though. Nor is this distance technique incorrect or unreliable. The gaseous nebula technique has been calibrated and tested using galaxies within our Local Group; and here the technique yields about the same distances as the cepheid technique. Nonetheless, you should be aware that, as we consider progressively greater realms of the Universe, distances become less certain—largely because the validity of the assumptions needed to determine these distances becomes less clear.

How far out into space in this fifth distance technique valid? Approximately 100 million light-years, for only to this distance are gaseous nebulae individually resolvable with the largest optical telescopes. The fifth rung of the inverted pyramid of Figure 14.10 notes this limitation.

OTHER GALAXY CLUSTERS: Do any galaxies exist between 10 and 100 million light-years? The answer is yes. For example, another cluster of galaxies resides some 50 million light-years from our Milky Way. Known as the Virgo Cluster, it resembles the Local Group in that both comprise well-defined volumes of space harboring numerous entire galaxies. However, the Virgo Cluster doesn't contain just 20 galaxies as for our Local Group; it houses approximately 2500 individual galaxies. Imagine, 2500 other galaxies all clustered together in a tight-knit group, each galaxy having 100 billion or so individual stars. You should now be rapidly gaining a feeling for the immensity of the Universe—and for the virtual insignificance of our place in it.

The Virgo Cluster is not the only group of galaxies within 100 million light-years of us. Astronomers have used the gaseous-nebula technique to identify several other such galaxy clusters. Figure 14.12 schematically illustrates many well-defined clusters sprinkled across our cosmic

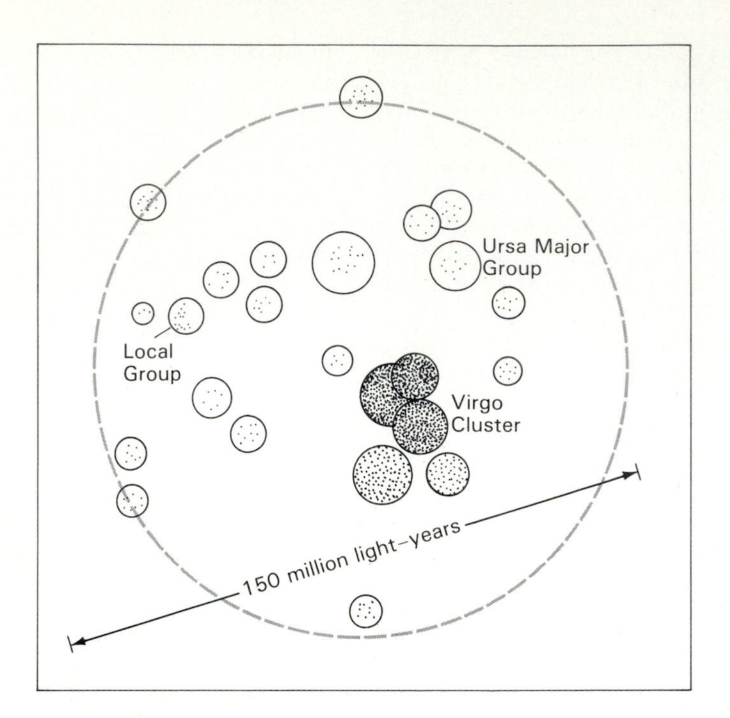

FIGURE 14.12 Schematic diagram of the relative locations of many galaxy clusters in our part of the Universe. Our Milky Way is only one of these dots and our Local Group only one of the clusters of dots.

neighborhood. Such a "map" proves that galaxies are not evenly spread throughout space—at least for distances on the order of 100 million light-years or so.

We might reasonably ask: Is there any matter of any kind outside clusters? The answer is apparently no. Astronomers have never found even a hint of intergalactic matter beyond any of the well-defined galaxy clusters. Evidently, when all the galaxy clusters formed eons ago, they did so very efficiently, literally sweeping up all the matter within any given location in the Universe. Matter definitely exists outside individual galaxies *within* a particular cluster, but apparently nothing roams the spaces between the galaxy clusters. (In fact, there is some evidence that the dark matter in any one galaxy of a galaxy cluster can stretch so far beyond its visible image that it might merge with the invisible extensions of neighboring galaxies.)

Be sure to view the galaxy clusters as astronomical "objects" in their own right. In much the same way that galaxies are collections of stars, galaxy clusters are collections of galaxies. Indeed, galaxy clusters comprise a most loftly level in the hierarchy of material coagulations within the Universe—elementary particles, atoms, molecules, dust, planets, stars, galaxies, and now galaxy culsters.

CLUSTERS OF CLUSTERS? The obvious next question is: Does the Universe have even greater groupings of matter, or do galaxy clusters top the cosmic hierarchy? Astronomers are currently uncertain. Many researchers argue that galaxy clusters themselves are clustered, thereby shaping titanic **galaxy superclusters**. But the evidence is partly conjecture, and hence subject to some debate. If future observations confirm and strengthen this idea, then our Local Group along with several other galaxy clusters comprise a supercluster centered near the large Virgo Cluster. The size of this possible supercluster, sketched in Figure 14.12, spans some 150 million light-years. More than 1000 times larger than the size of our Milky Way Galaxy, galaxy superclusters—if they really exist—are truly mammoth.

THE EMERGING BIG PICTURE: Let's recapitulate for a moment. We live on planet Earth. Our planet orbits the Sun. The Sun is in turn just one of hundreds of billions of stars in the suburbs of our Milky Way Galaxy. And our Galaxy is furthermore only one of many members of the Local Group, which is in turn just an undistinguished galaxy cluster near the edge of what might be an even larger galaxy supercluster. We can find no special place for our Earth, our Sun, our Galaxy, or our Local Group. Evidently, mediocrity reigns throughout.

Such is our niche in the Universe.

Deep Space

The tentatively proposed galaxy supercluster certainly contains a huge number of individual galaxies. The exact number is unknown, but it's probably on the order of 10,000 separate galaxies.

Still more galaxies populate the Universe—a lot more. To be sure, the great majority of known galaxies exist beyond the rough edge of our galaxy supercluster. Figure 14.13 is a long-exposure photograph of one such remote cluster. Called the Corona Borealis Cluster, this gravitationally bound aggregate is much farther away than 100 million light-years, for no gaseous nebulae can be identified within any of its individual galaxies. Yet this rich cluster is only one of myriad other large and distant groups of galaxies scattered throughout the observable Universe. On and on, the picture is much the same; the farther we probe deep space, the more we find galaxies, clusters of galaxies, and perhaps clusters of clusters of galaxies. . . .

DISTANCE YET AGAIN: Astronomers cannot easily study the most distant galaxies. Truly remote galaxies are often small, dim, and hard

FIGURE 14.13 The Corona Borealis Cluster contains huge numbers of galaxies and resides roughly a billion light-years away.

to resolve. Galaxy experts talk about only rough quantities, including approximate distances.

As noted earlier, gaseous nebulae are reasonably good indicators of distance to about 100 million light-years. To determine distances to objects beyond this limit, such as members of the Coma Cluster, yet another technique is needed. This sixth distance technique does not rely on the properties of any specific objects within galaxies. The galaxies are simply too far away, hardly resembling more than mere points of light. Astronomers find it tough enough to identify the galaxies as galaxies, let alone study cepheids or nebulae within them. Like the third, fourth, the fifth distance techniques (see Figure 14.10), this sixth technique uses an apparent-absolute brightness comparison to infer distance. As such, we must assume the absolute brightness (or luminosity) of an entire galaxy. You might think that this is tricky, and indeed it is. But it's not entirely guesswork; if it were, the technique would be useless.

Here's how the technique works. First, an observer identifies the *brightest* galaxy within a galaxy cluster. An absolute brightness is then assumed for that galaxy, based on the notion that the brightest galaxy in any galaxy cluster has approximately the same absolute brightness as the brightest galaxy in our own Local Group. Once the apparent brightness of the remote galaxy is measured, the distance is then derived in the usual fashion. The technique works quite well for rich clusters of galaxies because the brightest galaxies within each of the nearby galaxy clusters do seem to have equal absolute brightnesses. In other words, our assumption is perfectly valid for galaxy clusters within 100 million light-years. We can only surmise its validity beyond this distance.

Even if our central assumption is always correct, be sure to realize that this sixth technique yields only the *average* distance to any galaxy cluster. It cannot be used reliably to determine the different distances to each of the member galaxies within a given galaxy cluster. Nor can we use it to find, even roughly, the distance to *any* galaxy that is not part of a galaxy cluster. We simply have no way of knowing such a galaxy's absolute brightness, making its derived distance no better than a poor guess.

This method is then valid to whatever distance our largest telescopes can identify and measure the brightest galaxy within a galaxy cluster. This limiting distance turns out to be on the order of 4 billion light-years. Beyond that, individual galaxies within a galaxy cluster often fade into the soft glow of light emitted by all the members of the cluster. As such, we cannot derive distance by any aspect of the apparent-absolute brightness technique. Some researchers refer to this 4 billion light-year distance as the **galaxy horizon**.

Is there any way to know how the galaxies are distributed in space out to this galaxy horizon? Figure

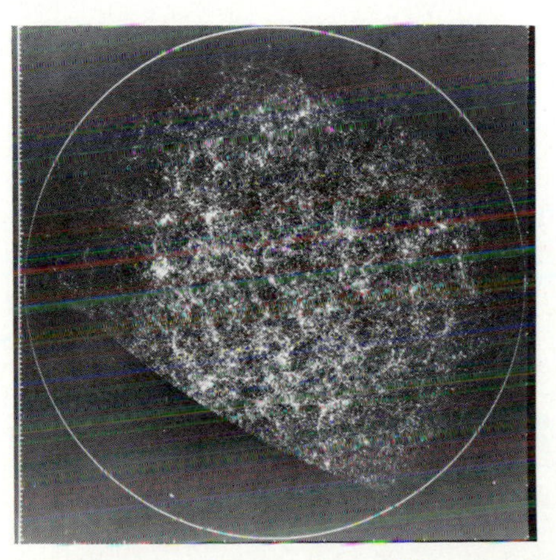

FIGURE 14.14 A computer-generated map of the million brightest galaxies observed from the Northern Hemisphere. Most of these galaxies are within a billion light-years of us. (The boundary at the lower left results merely from the telescopes' inability to see below the horizon; this boundary has no physical significance in the Universe.)

14.14 is a computer-generated map of what some call "the broadest view of the biggest picture." The small white dots depict the million brightest galaxies as observed from the northern hemisphere. (Galaxy counts in the blank area to the bottom right are seriously distorted by our own Milky Way and are therefore not included.) Gravitational clustering of the galaxies is strong, as most galaxies reside in tight clumps that in turn possibly form superclusters. In the absence of other strong forces, gravity will always bring matter into a ball; that's why we live on a ball, stars are balls, and many galaxies, too.

The galaxies plotted in Figure 14.14 also display a striking lacework of long, stringy groups. Many galaxies tend to lie in great sheets or filaments, leaving vast regions apparently devoid of all matter. Although their origin is uncertain, these linear patterns may not be real; they might be artifacts of the way the data were analyzed. The human eye, under great physiological stress, tends to interpolate linearly, and can sometimes manufacture long structures that are not real. A good analogy may well be the case of "canals" on Mars (see Interlude 7-2); at the turn of the century, observers strained to discern surface features on that planet, and as a result, mistook numerous unrelated craters as evidence of Martian-engineered canals.

If the galaxy filaments shown in Figure 14.14 do not turn out to be of a similar artificial nature, researchers will be faced with the somewhat annoying problem of galaxies having clustered into unexpectedly lengthy structures as well as the expected roundish blobs. One possibility is that many galaxies are spread along the surfaces of vast bubbles in space. The interior of these gigantic bubbles could then account for the apparent voids of matter; and the galaxies would only seem to be distributed like beads on strings because of the way we view them on the rims of the bubbles. What might be the origin of this "sudsy" or "frothy" fabric of the Universe? Perhaps the first-generation stars that formed early in the Universe were extremely large, in fact more massive than any stars now known. They would have raced through their nuclear cycles, exploded like bombs, and suddenly released tremendous amounts of energy. The subsequent shock waves might well have piled matter and debris along the spherical rims of such blasts, possibly even triggering the formation of galaxies at the edges of huge cavities. Perhaps.

Does the galaxy horizon mark the end of material objects in the Universe? Not at all. We have reached hardly one-fourth of the full size of the Universe. And, yes, there are myriad objects far beyond several billion light-years. But, we're now moving too fast in our intellectual rush toward the limits of the observable Universe. All the while, we're missing some interesting information about the nature of our cosmic "neighborhood" within a billion or so light-years. Let's pause for one section to consider once more the galaxies and galaxy clusters, prior to making our final leap in distance toward the end of the chapter.

Galaxy Masses

Total mass is the one physical property we've not yet mentioned about galaxies and galaxy clusters. How can we find the masses of such huge astronomical systems? Surely, we can neither count all their stars nor estimate their interstellar content very well. The galaxies are just too complex to take direct inventory of their material makeup. Instead, we must rely on indirect techniques.

INDIVIDUAL GALAXIES: Two such techniques employ Kepler's third law, the method used to estimate the mass of our own Milky Way Galaxy in Chapter 13. By observing each of two ends of a galaxy with a telescope and spectrometer, we could measure the Doppler shift of the spectral lines and hence infer the rotational period of the galaxy. If we know the size of the galaxy, the mass then follows directly from the proportionality noted several times previously:

$$\text{mass} \propto \frac{\text{orbital size}^3}{\text{orbital period}^2}.$$

To use this technique, a galaxy must appear large enough on the sky to permit us to focus on each end separately, as shown in Figure 14.15(a).

In many cases, galaxies are too distant and therefore appear too small for such measurements to be made. Still, we could use a modified version of Kepler's third law to estimate the masses of binary galaxies (two galaxies orbiting each other) in much the same way that binary-star masses were found in Chapter 11. Again, the orbital size and the rotational period of one galaxy around the other are needed to infer mass in this way, as sketched in Figure 14.15(b).

The results from both methods show that individual galaxies have masses much like our Milky Way. Most spirals and ellipticals generally contain between 10^{11} and 10^{12} solar masses, although the irregulars often contain less, about 10^8 to 10^{10} solar masses.

GALAXY CLUSTERS: Another technique can be used to derive the *combined* mass of all the galaxies within a galaxy cluster. Really a statistical technique, it's useful when the foregoing two methods fail (either because a galaxy appears too small or because it's not part of a binary system). This technique, called the **Virial method**, can be explained as follows.

Consider the gas particles composing Earth's atmosphere. Because the air is heated (usually about 300 kelvin at sea level), its atoms and molecules have some motion. On warm days, these particles move more rapidly than on cool days. In either case, the atmosphere remains intact; it doesn't disperse or fade away. What holds it to the Earth? Gravity, of course. Gravity keeps our atmosphere from escaping by constantly pulling back on each of the gas particles. Furthermore, gravity depends on mass. Thus, we conclude, as we did in Chapter 5, that our atmosphere is gravitationally

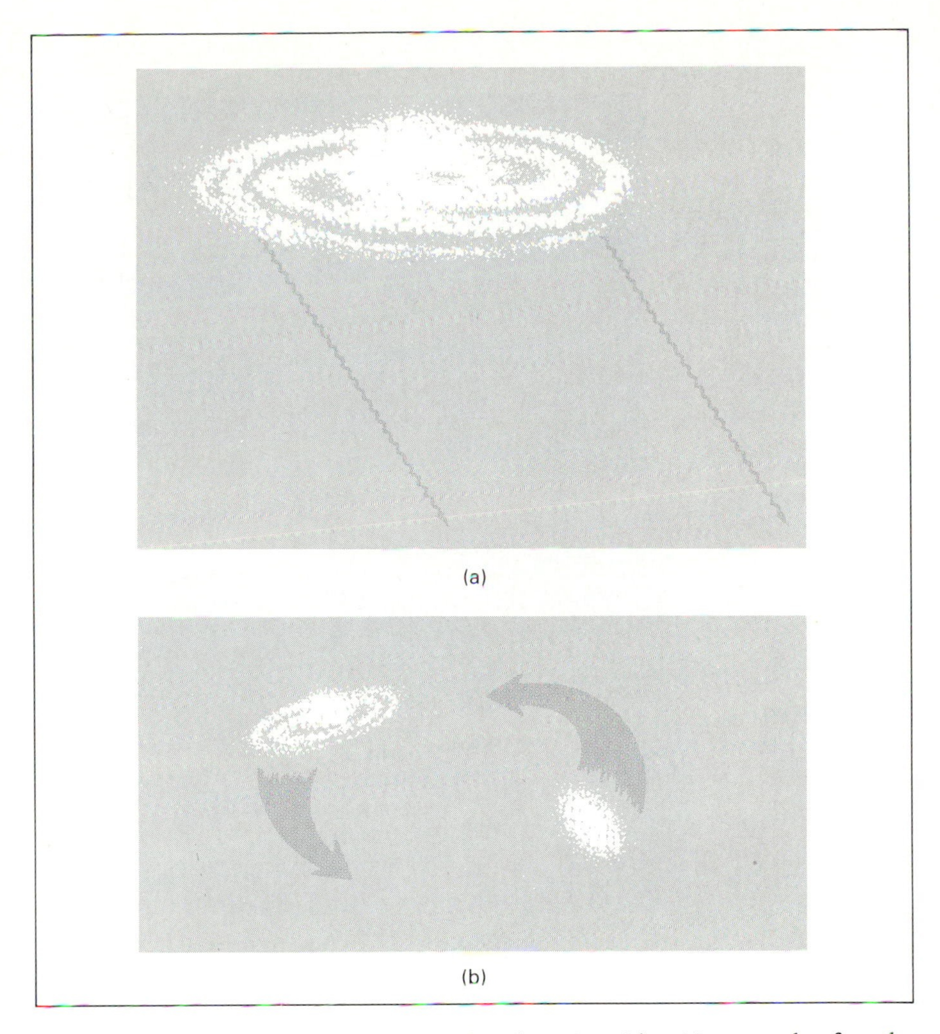

FIGURE 14.15 Galaxy masses are derived by observing either (a) two ends of a galaxy of known size, or (b) the orbital period of one galaxy about another.

bound to the massive Earth; the individual particle motions are inadequate to escape the pull of gravity. With these concepts, we could use the Virial method to estimate the mass of Earth needed to retain its atmosphere.

This example of atmospheric particles (or "cluster" of particles) is analogous to the case of galaxy clusters. Each—planetary atmospheres as well as galaxy clusters—involves particles of matter having random motions. In the astronomical analog, the "particles" are enormous—they are the galaxies themselves.

We stress that the Virial method of mass determination is only a statistical technique. That is, it yields the correct answer only when enough galaxies (say, a dozen or more) provide a sufficiently large statistical sample. The Virial method does not work for a handful of galaxies or less, but should work quite well for large galaxy clusters housing hundreds and sometimes thousands of individual galaxies.

Thus we can estimate the combined mass of an entire cluster by calculating the gravitational pull needed to bind all the galaxies of a given cluster. In other words, we ask: How much mass (or gravity) is needed to overcome the effect of individual galaxy motions (just like air-particle motions), thereby preventing the cluster (or atmosphere) from dispersing? After all, most galaxy clusters appear to be well-defined, tight-knit groups, suggesting that they are indeed gravitationally bound together; if the clusters were not bound, their member galaxies would long ago have dispersed into intergalactic space.

All this sounds reasonable—until we apply the Virial method. We then encounter a problem—in fact, one of the major mysteries of modern astronomy: The mass needed to bind the galaxy clusters is larger than expected. The calculated value is some 10 to 100 times larger than that found by adding the masses of all the invisible galaxies observed within any cluster. Stated another way, apparently a good deal more mass is needed to bind the galaxy clusters than we can find observationally. Some mass seems to be missing, or at least hiding.

THE MISSING-MASS PROBLEM: The "missing-mass" (or "hidden-mass") problem applies, then, not only to our own Milky Way Galaxy (as noted in Chapter 13) but especially and more grandly to galaxy clusters. The problem is not just a local issue; it's widespread, probably pervading the entire Universe. If so, then upward of 90 percent of the Universe could be composed of inherently dark matter.

What could be the source of the missing mass within the galaxy clusters? Where could it hide? The answer is currently unknown. Whatever the solution, the missing matter cannot be a simple accumulation of smaller amounts of dark matter within the individual galaxies, whatever the source of that problem may be—faint red dwarfs, invisible black dwarfs, black holes, molecular clouds, massive neutrinos, or halo and coronal material. For every piece of detected galactic matter, some 10 to 100 times more matter remains undetected. It's simply unrealistic to think that astronomers have underestimated galaxy masses by a factor of as much as 100.

A more reasonable solution to this grand missing-mass problem is diffuse, intergalactic matter among the galaxies of the clusters. This possibility does not imply the existence of intergalactic matter in the vast tracts of the Universe outside of galaxy clusters. In fact, no such "extracluster" matter has even been found. This potential solution sug-

INTERLUDE 14-2 ***Colliding Galaxies***

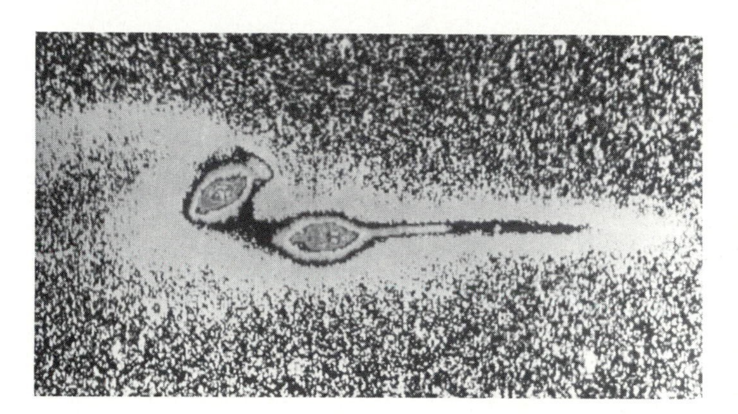

Contemplating the galaxy clusters, especially the rich ones like Virgo with its thousands of member galaxies, it's hard to avoid the impression that collisions must be common among the galaxies. Just as gas particles collide in Earth's atmosphere, or hockey players in an enclosed rink, the random motions of the galaxies within a galaxy cluster could conceivably induce phenomenal collisions among these giant coagulations of matter.

Indeed, galaxies do collide. Direct observational evidence proves that they do so quite often. For example, each of the accompanying images shows what seems to be sets of interacting galaxies. One image is an optical photograph of five galaxies in the constellation Serpens; connecting clouds or bridges seem to link some of these galaxies. The other image is a computer-enhancement of the pair of galaxies, NGC4676A&B (also known as "Playing Mice"), which seem to show streams of connecting gas apparently generated by the interaction between the two galaxies. Whether these galaxies are genuinely colliding or only experiencing a close encounter cannot easily be determined. No human being could witness an entire collision, for a typical collision takes perhaps millions of years for completion.

You might think that a collision among giant galaxies would create a mind-boggling crunching of matter, complete with spectacular explosions and superlative fireworks. Surprisingly, they don't. Such collisions, in fact, are rather quiescent. Aside from some slow tidal rearrangements of loose interstellar matter (consult Figure 18.13), the stars within each galaxy more or less just slide past one another. The reason is that stars hardly ever collide; they are, after all, small objects by cosmic standards. Although we have plenty of direct photographic evidence for galaxy collisions, no one has ever succeeded in witnessing or photographing a collision between two stars—not even in our own Milky Way.

There's good reason for this seeming paradox. To understand it, recall that the galaxies within a typical cluster are bunched fairly closely. The distance between adjacent galaxies in a given cluster averages about a million light-years, which is only about 10 times greater than the size of a typical galaxy. This doesn't really give them much room to roam around without bumping into one another. By contrast, stars within any galaxy are spread considerably thinner. The average distance between stars in a galaxy is about 5 light-years, which is millions of times greater than the size of a typical star. Hence stellar collisions are extremely rare within any one galaxy. Even when two giant galaxies collide, the star population merely doubles, leaving ample space for the stars to meander without sustaining much damage. The stellar and interstellar contents of each galaxy are undoubtedly rearranged by the gravitational tides induced by collisions, and the merged interstellar matter likely experiences some shock waves that probably trigger bursts of star formation, but there are no spectacular explosions. The two galaxies just sort of glide through one another without causing too much commotion.

gests only that intergalactic gas is spread as "intracluster" matter throughout the darkness inside galaxy clusters.

Until recently, astronomers had no observational evidence for intergalactic matter either inside or outside of galaxy clusters. Now, satellites orbiting above Earth's atmosphere have begun to detect substantial amounts of x-ray radiation in the direction of some galaxy clusters. There are few x-ray signals when pointing away from these clusters, so we can be fairly certain that this high-energy emission originates in the clusters themselves. These observations imply the existence of copious amounts of invisible though hot gas, heretofore unseen optically or unheard in the radio. The radiation seems to be thermal in nature, with the Planck curve peaked in the high-frequency, x-ray domain; this suggests gas temperatures of about 100 million kelvin.

How much matter have the x-ray satellites found? Generally, the observations suggest that nearly as much matter exists outside the galaxies as within them. Thus this increases by only a factor of 2 the estimated mass of the galaxy clusters. But to solve the missing-mass problem, we need to find factors of 10 to 100—that is, 10 to 100 times more dark matter outside the galaxies than bright matter within them. Although the x-ray observations are tantalizing, they fail to solve perhaps the most vexing problem in all of astronomy. And until it is solved, our inventory of matter will remain incomplete.

In noting several of the basic problems astronomers still face, don't get the impression that these problems are completely intractable. Throughout this book, we strive to impart a feeling for some of the many modern astrophysical issues that are not quite solved. Instead of making extreme statements claiming that galaxy clusters are well understood or that galaxy clusters are currently unfathomable, we attempt here and elsewhere to offer a balanced appraisal of the research frontiers. At any rate, you can be sure that legions of theoreticians and observers are fiercely trying to solve these basic dilemmas, thereby literally, and somewhat astonishingly, unlocking secrets of the Universe.

Hubble's Law

Our previous analogy of galaxies in a cluster to molecules in an atmosphere was chosen to stress the random motions of the individual galaxies within clusters. You might expect that on the largest possible scale, the clusters themselves also have some random, disordered motion—some moving this way, some that. This is not the case at all, however. On the largest possible scale, unclustered galaxies and galaxy clusters have some nonrandom organized movement.

UNIVERSAL RECESSION: Virtually all the galaxies observed in the Universe seem to be moving away from us. Individual galaxies that are not part of galaxy clusters, are steadily receding. Even the galaxy clusters, where the individual member galaxies do move randomly, have some recessional motion. An analogy might be a jar full of fireflies that has been heaved away. The fireflies within the jar have some random motions due to their individual whims, but the jar as a whole, like a galaxy cluster, has some directed motion as well.

How do we know the galaxies share this net, directed motion away from us? The answer is easy, since spectroscopy proves that galaxies' spectral lines are red shifted. Figure 14.16 shows some representative examples of optical spectra observed toward several galaxies. Interpreted as a Doppler effect, these shifts indicate that the galaxies are steadily receding. Furthermore, the extent of the red shift increases progressively from top to bottom in the figure. And since the distance also increases in the same manner—from top to bottom—we conclude that there must be a connection between Doppler shift and distance. This trend of greater red shifts for objects farther away holds valid for virtually all galaxies in the Universe. (Two galaxies within our Local Group, including Andromeda, and a few galaxies in the Virgo Cluster display blue shifts and hence have some motion toward us, but this results from their random motions within these nearby galaxy clusters.)

So not only are the galaxies receding, but they recede at velocities proportional to their distances. A linear relationship—a perfect correlation—connects velocity and distance. The greater the distance of an object from us, the faster it recedes.

Figure 14.17(a) shows a diagram of recessional velocity plotted against distance for the galaxies of Figure 14.16. Figure 14.16(b) is a similar plot for numerous galaxies within 4 billion light-years. Diagrams like these were first made by the American astronomer Hubble in the 1920s and hence bear his name; the resultant statistical fit (solid line) is sometimes called **Hubble's law**. We could make such a diagram for

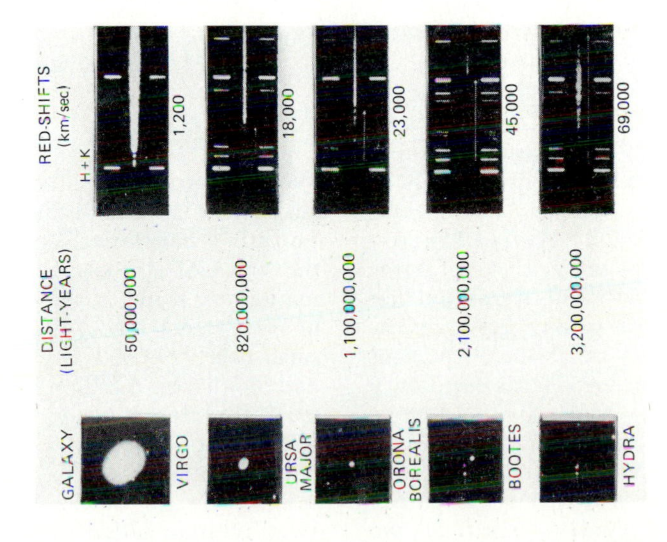

FIGURE 14.16 Optical spectra (at right) observed toward several different galaxies (at left). The extent of red shift (denoted by arrows at right) and the distance to each galaxy increase from top to bottom. [See also Plate 1(c) and (d).]

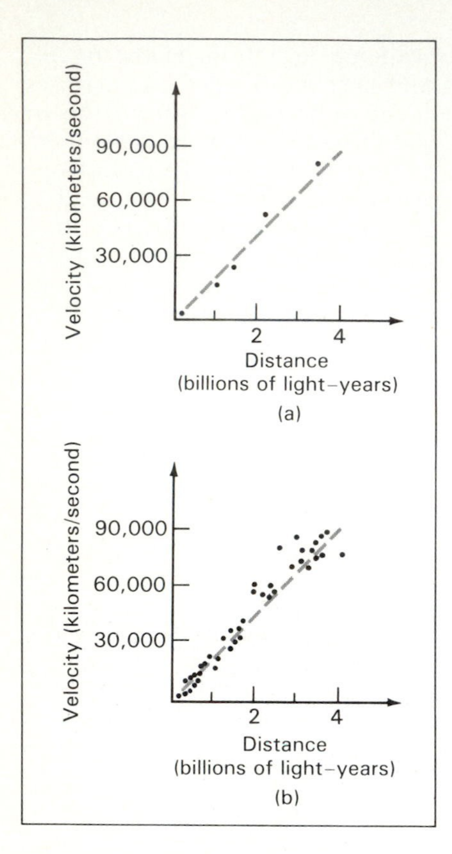

FIGURE 14.17 Plots of recessional velocity against distance (a) for the galaxies shown in Figure 14.16 and (b) for numerous other galaxies within 4 billion light-years.

any group of galaxies provided that their distances can be found by one of the previously discussed techniques and their radial velocities measured by spectroscopy.

Note the implications of Figure 14.17(b). As can be read directly from the plot, galaxies at the remote distance of 4 billion light-years speed away with velocities of nearly 90,000 kilometers/second. This is a fair fraction of the velocity of light. The ratio 90,000/300,000 means that such a distant galaxy has a recessional velocity of about 30 percent of light velocity. That's fast, very fast.

AN OBSERVATIONAL FINDING: Let us stress that Hubble's law is an *empirical* discovery. By "empirical" we mean one based strictly on observational results. Its central relationship—a statistical correlation between velocity and distance—is well documented to at least 4 billion light-years. But there is no basic physical reason for this relationship. No law of physics demands that all galaxies recede. And no physical law requires distance and velocity to be correlated. Consequently, astronomers are currently unsure if this relationship holds true for astronomical objects beyond 4 billion light-years. In this sense, then, it's not really a "law" at all.

Don't be confused here. There are good and basic reasons for the red shift as an indicator of recessional velocity. This is the Doppler effect studied in Chapter 1; the larger the spectral-line shift, the greater the net motion between the observer and the observee. But the Doppler effect in no way relates velocity and distance. In particular, the Doppler effect does not predict Hubble's law at all. Hubble's law is strictly a compact way of noting the observational fact that any galaxy's recessional velocity is directly related to its distance from us.

HUBBLE'S CONSTANT: We can quantify Hubble's law to make it more useful. This is relatively simple since velocity and distance are linearly related. The data of Figure 14.17 obey the following proportionality:

$$\text{recessional velocity} \propto \text{distance}.$$

Or, without the proportionality sign, we could write the equation

$$\text{recessional velocity} = \text{Hubble's constant} \times \text{distance}.$$

Here the proportionality factor between velocity and distance is called **Hubble's constant**. We can derive the value of this constant by estimating the slope of the dashed line in Figure 14.17(b). The slope measures nearly 90,000 kilometers/second divided by 4 billion light-years, or 22 kilometers/second/million light-years. Thus for every additional million light-years of distance from us, astronomical objects race away with an added speed of some 22 kilometers/second.

This is the best current value for Hubble's constant. We say "current" because Hubble's constant is a statistical solution to the data plotted in the Hubble diagram. Over the years, newer methods (and better calibration of older methods) used to determine distance have repeatedly forced astronomers to revise the value of Hubble's constant; 50 years ago it was thought to be some 10 times larger. And even today, astronomers suspect that the value of Hubble's constant might be skewed a bit by the drift of our Local Group toward the Virgo Cluster of galaxies; this net drift amounts to about 600 kilometers/second and might be nothing more than the random motion of our Local Group in the outskirts of the larger local supercluster. At any rate, modern researchers regard the current value to be accurate at least to within a factor of 2 and thus do not anticipate the need for further major revision.

Throughout the past few decades, astronomers have striven to refine the accuracy of Hubble's diagram and the resulting estimate of Hubble's constant. That's because Hubble's constant is one of the most fundamental quantities of nature. It specifies the rate of movement of the grandest contents of the Universe. As such, Hubble's constant will be the cornerstone of our study in Chapter 16 of the origin, structure, and destiny of the entire Universe.

For now, suffice it to note that Hubble's law implies that all the galaxies emanate from a point—perhaps the site of an explosion at some time in the remote past. (Admittedly, it would seem that we are at this point—at the center of the Universe—but in Chapter 16 we shall see how relativity theory maintains this to be incorrect.) The more distant an object is from us, the greater the force with which it must have been initially expelled; the faster-moving galaxies are by now farther away *because* of their high velocities. Bomb fragments form much the same pattern in the aftermath of an explosion.

The recessional motions of the galaxies prove that the entire Universe itself is in motion. It's not steady and inactive. Nor do its con-

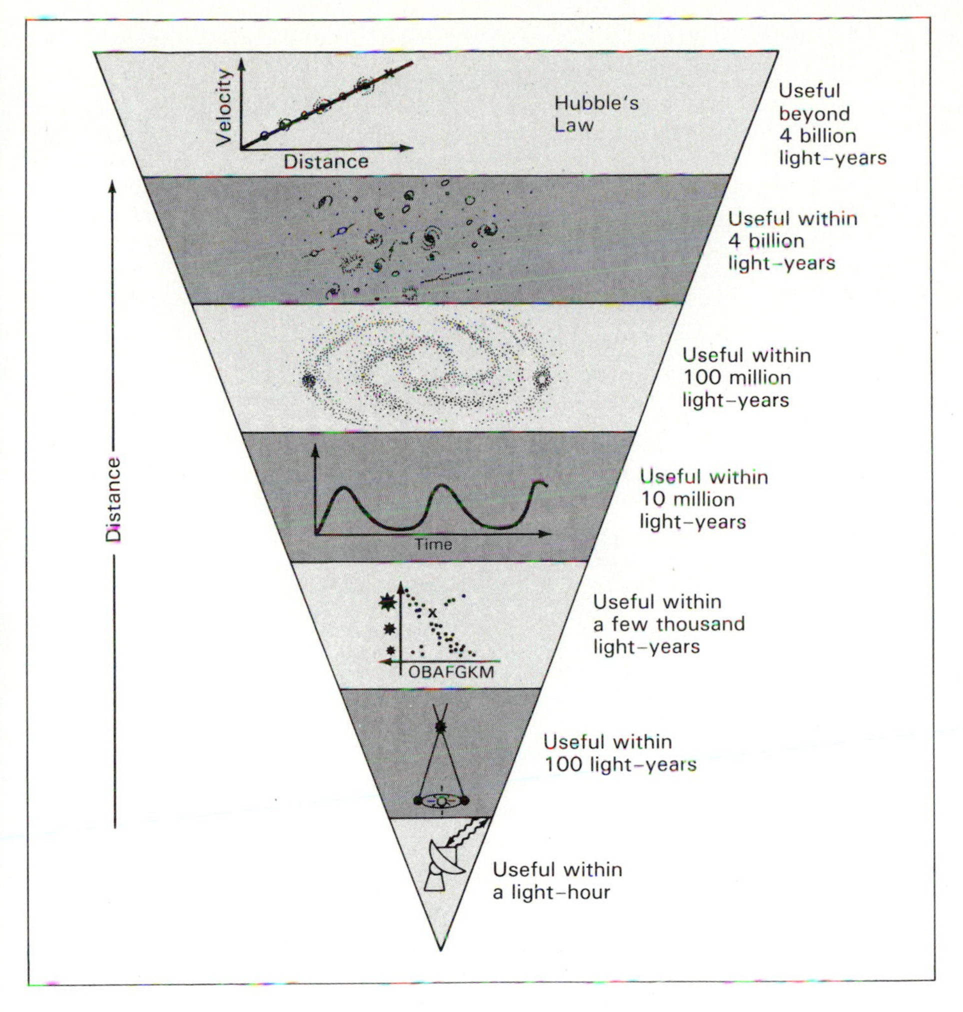

FIGURE 14.18 Extrapolation of Hubble's law beyond 4 billion light-years tops the inverted pyramid of distance techniques. This last method is used to find the distances of astronomical objects all the way out to the limits of the observable Universe.

tents move randomly on the largest scale. The Universe is expanding, and expanding in a directed fashion. In short, it's evolving. This does not mean that Earth, the Solar System, or the galaxies themselves are physically ballooning in size. These groups of rocks, planets, and stars are gravitationally bound and are not expanding. Only the largest framework of the Universe—the ever-increasing distances separating the galaxies and especially the galaxy clusters—manifests this expansion.

DISTANCE ONE FINAL TIME: One of the most useful aspects of Hubble's law is its potential for predicting distances of astronomical objects beyond 4 billion light-years. In other words, if Hubble's law holds valid for objects beyond the galaxy horizon, we can derive distances to the most remote objects by simply measuring their recessional velocities.

Operationally, the method works like this. An astronomer measures the red shift of an object's spectral lines. The extent of the shift is then converted to velocity by means of the Doppler relationship of Chapter 1. Having only the object's velocity, the astronomer then finds the object's distance by extending the plot of Figure 14.17(b).

Use of Hubble's law in this way tops the inverted pyramid of distance techniques. Sketched in Figure 14.18, this seventh method assumes that Hubble's law, certainly valid within 4 billion light-years, can be extrapolated to greater distances. If this assumption is correct, Hubble's law enables us to measure arbitrarily great distances in the Universe. However, some controversy surrounds the validity of this "law" beyond 4 billion light-years. We'll discuss this issue more in Chapter 15.

Do objects really exist beyond 4 billion light-years? Apparently so. Myriad red-shifted objects have recessional motions larger than 30 percent of the velocity of light. Some move as fast as 50 percent of light velocity. Others seem to travel even faster. Should extrapolation of Hubble's law be valid—that is, should the red shift indirectly measure distance, regardless of how great—some of these objects must reside at the farthest realms of the Universe.

The most distant object thus far observed in the Universe has the peculiar catalog name of QO051-279. Its extremely high red shift implies a recessional velocity 93 percent that of light. Hubble's law then predicts that QO051-279, solely on the basis of its observed red shift, is nearly 14 billion light-years away. Astonishingly, this object resides as close to the limits of the observable Universe as astronomers have yet been able to probe.

Before concluding, let's stress once again that, because the velocity of light is a finite number, time is needed for light or any kind of radiation to travel from one point in space to another. The radiation that we now see from distant objects originated long ago. Incredibly enough, the radiation that astronomers now detect from QO051-279 left that object some 14 billion years ago, well before our planet, our Sun, and even our Galaxy came into being.

So, be sure always to remember: Looking out into space is equivalent to looking back into time. *Looking out is looking back.* Indeed, studies of the most remote astronomical objects provide an important key to our understanding of the earliest epochs of the Universe.

SUMMARY

Galaxies come in only a few broad shapes and sizes. Many have spiral patterns such as our own Milky Way, but most galaxies seem to be elliptically shaped without any arms. On and on, the galaxies are observed out to distances of at least 4 billion light-years. Grouped into galaxy clusters, entire galaxies form titanic swarms much as stars assemble within the galaxies themselves. Such galaxy clusters, however, present one of the thorniest problems in astronomy; large amounts of undiscovered matter probably resides in the clusters, although no one yet knows what kind of matter that might be.

Several new distance techniques were introduced in this chapter. You should now be familiar with the entire distance scale in the Universe—from radar transmissions in the very local realm to Hubble's law for the farthest reaches of the observable Universe. This is a good time to review the size and scale of all objects thus far discussed. Be sure to place our home planet into its cosmic perspective.

KEY TERMS

barred-spiral galaxy
elliptical galaxy
galaxy cluster
galaxy horizon
galaxy supercluster
Hubble's constant
Hubble's law
intergalactic space
irregular galaxy
Local Group
Magellanic Clouds
spiral galaxy
Virial method

QUESTIONS

1. Cite and briefly discuss recent observational evidence suggesting that our Galaxy, and presumably other individual galaxies as well, are a good deal larger (in size and mass) than heretofore thought.
2. Briefly explain the method of determining astronomical distances to objects residing beyond the "galaxy horizon."
3. Pick one: Compared to elliptical galaxies, spiral galaxies are (a) dead and dust free; (b) dead and dusty; (c) full of activity and dust free; (d) full of activity and dusty.
4. Describe the bulk distribution of matter within a few hundred light-years from us.
5. Explain why galaxy collisions are thought to be more frequent than star collisions.
6. When must astronomers resort to the so-called Virial method in order to derive galaxies masses? Explain this method, as well as the contemporary problem that results from application of it.
7. Explain the steps an astronomer takes to measure the distance to an object at, say, 8 billion light-years from Earth.
8. Explain Hubble's law. Is it really a "law"?
9. How do astronomers determine the masses of galaxies?
10. List the differences between the three main types of galaxies in terms of their shapes, stellar populations, and amount of gas and dust.

FOR FURTHER READING

de Boer, K., and Savage, B., "The Coronas of Galaxies." *Scientific American,* August 1982.

Ferris, T., *Galaxies.* San Francisco: Sierra Club Books, 1982.

Hodge, P., "The Andromeda Galaxy." *Scientific American,* January 1981.

Hodge, P., *Galaxies.* Cambridge, Mass.: Harvard University Press, 1986.

Krauss, L., "Dark Matter in the Universe." *Scientific American,* December 1986.

Mitton, S., *Exploring the Galaxies.* New York: Scribner, 1976.

Rubin, V., "Dark Matter in Spiral Galaxies." *Scientific American,* June 1983.

Sandage, A., "The Red Shift." *Scientific American,* September 1956.

15
ACTIVE GALAXIES: *Limits of the Observable Universe*

In Chapter 14 we developed a notion for our location within the larger Universe. We found our Milky Way Galaxy to be just one normal galaxy among many others that together form the Local Group. Beyond, other galaxy clusters abound, each a gravitationally bound group of galaxies within a reasonably well-defined volume of space. Galaxy clusters comprise among the largest assemblages within the hierarchy of universal matter, although some suggestive evidence points toward a clustering of clusters of galaxies, commonly known as galaxy superclusters.

Astronomers discovered this hierarchy of matter by studying progressively larger segments of space beyond our Milky Way. Let's now consider a really large chunk of the Universe—much larger than our 3-million-light-year-sized local galaxy cluster, even larger than our 150-million-light-year-sized suspected supercluster. In this way we can really begin to appreciate the way matter spreads throughout the grandest realms of the Universe.

To do this, imagine a truly large cube, not a centimeter or even a parsec on a side, but a cube having dimensions of 300 million light-years—clearly, a huge piece of universal real estate. Now ask, How many galaxies probably populate such large cubes? Figure 15.1 is a schematic diagram of such a counting exercise. The answer is approximately 4000 galaxies, some of them clustered, some not. This is true regardless of where in the Universe the cube is placed. Of course, we might find a few more or less than 4000 galaxies depending on the precise lo-

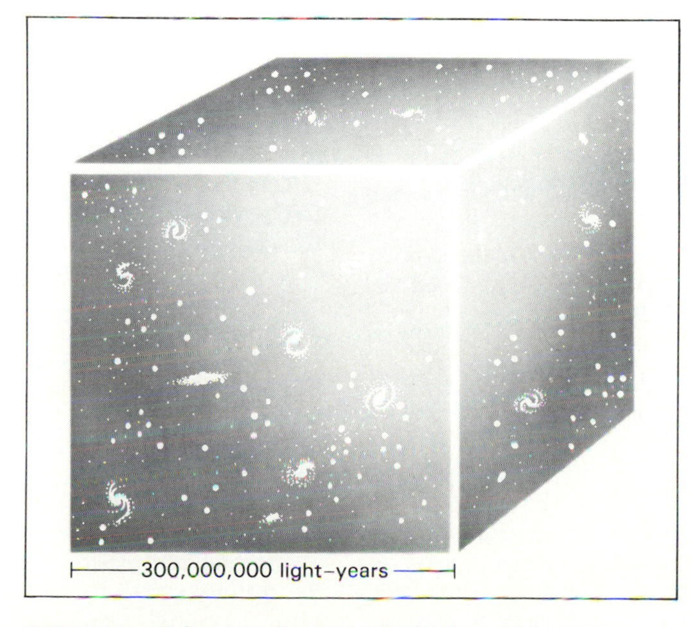

FIGURE 15.1 Schematic diagram of galaxies within an enormous cube, 300 million light-years on a side.

cation of our hypothetical cube. But, by and large, all volumes of space this big contain an average of 4000 galaxies.

Now, the number 4000 is not significant. It's the idea behind the number that's important. This counting exercise demonstrates that the density of galaxies eventually becomes fairly uniform; clustering does not occur indefinitely.

The hierarchical structure of matter in the Universe is an important concept. Let's stress it once more succinctly: For scales smaller than about 300 million light-years, matter possibly clusters into galaxy superclusters, smaller than 100 million light-years into galaxy clusters, smaller than 300 thousand light-years into galaxies, smaller than 1 light-year into stars, and so on all the way down to the elementary particles. But for scales larger than a few hundred million light-years, astronomers currently have no clear observational evidence for additional clustering. The Universe on the grandest scale seems to be even, homogeneous, and smooth.

These statements surely hold true for matter within the approximately 4-billion-light-year distance that roughly defines the galaxy horizon. But does this homogeneity persist beyond? What's it like beyond the galaxy horizon? Do the objects near the limits of the observable Universe resemble those nearby?

The learning goals for this chapter are:

- to understand the basic differences between active and normal galaxies
- to gain a feeling for the prodigious amounts of radiation emitted by the active galaxies
- to recognize the salient features of the powerhouse radio galaxies
- to appreciate the arguments favoring the nonthermal mechanism as the prime power source for some of the active galaxies
- to understand the plethora of problems currently facing astronomers in their quest to fathom the quasars

Beyond the Galaxy Horizon

The notion that objects exist beyond the galaxy horizon assumes that we can extend Hubble's law to distances greater than 4 billion light-years. Since we have no reason to suspect otherwise, objects having large red shifts are taken to reside beyond several billion light-years. But what type of objects exist in these outer limits? Are they spiral, elliptical, and irregular galaxies—close cousins of the usual material coagulations that populate the local Universe? Well, not exactly.

Occasional spirals or ellipticals have been found as far away as 3 or 4 billion light-years. And some of these types of galaxies probably exist beyond this artificial horizon. Yet if they do, they're tough to recognize.

Beyond the galaxy horizon, the basic nature of astronomical objects—whatever they are—seems different. By and large, truly distant objects are more active, to a certain extent more violent. By activity or violence, we mean that their absolute brightnesses or luminosities are much greater than most of the "nearby" spiral and elliptical galaxies discussed in Chapter 14.

This predominance of energetic objects at great distances might be an observational effect. It might partially result from our inability to detect relatively weak normal types of galaxies at great distances. Remember, the radiation emitted from all astronomical objects, regardless of their size, shape, or energetics, decreases as the square of the distance. Even with the very best telescopes, we should expect to observe preferentially the most energetic and powerful galaxies at the really great distances.

The intrinsic weakness of signals from very distant normal galaxies can only partially explain the predominance of energetic objects beyond the galaxy horizon. There's something basically different about the faraway objects themselves. It seems that the *radiative character* of the really distant objects (as well as some scattered local ones) differs fundamentally from the **normal galaxies**. We use the adjective "normal" here to describe those galaxies that emit the accumulated radiation of large numbers of stars—namely, those galaxies studied in Chapter 14.

A normal galaxy's radiation spreads from the radio to the gamma-ray part of the electromagnetic spectrum in much the same way as for ordinary stars. Most of the radiated energy is of the visible type since, after all, the Planck curve for most stars peaks in that part of the spectrum. As Figure 15.2 shows, normal galaxies emit a good deal less infrared and radio radiation. For example, our entire Milky Way has a luminosity of 10^{44} ergs/second at optical frequencies, but only 10^{38} ergs/second at radio frequencies.

The powerful and distant objects, often called **active galaxies**, contrast with the normal galaxies, not only because they emit radiation profusely, but also because they do so unlike stars. As shown in Figure 15.2, the radiation observed from active galaxies is not peaked at optical frequencies. To be blunt about it, the emission of radiation from some active galaxies is so inconsistent with the accumulated emission from myriad stars that astronomers are unsure if they really have any stars! Perhaps we shouldn't even call them galaxies.

The abnormal power and strange character of the distant astronomical objects suggest that the earlier epochs of the Universe were more violent. Remember, looking out in space equals looking back in time. Since physical conditions were undoubtedly different in earlier epochs of the Universe than they are now, we shouldn't be surprised that the remote astronomical objects that emitted their observed radiation long ago differ from the nearby ones that emitted their radiation much later. What *is* surprising—in fact, downright astounding—are the huge amounts of energy radiated from some of the most powerful objects. Their total release of energy often stretches scientific theory to its limits.

To place into proper perspective

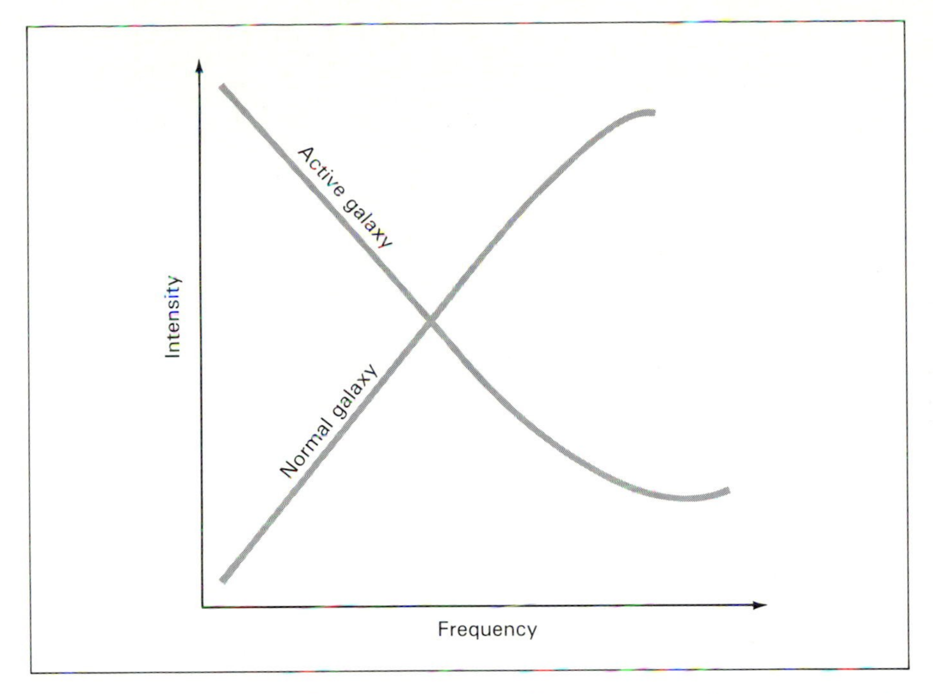

FIGURE 15.2 The nature of the energy emitted from a normal galaxy is completely opposite that of an active galaxy. (This plot illustrates the general spread of intensity for all galaxies, but is so oversimplified as to pertain to no individual galaxy.)

the prodigious radiative abilities of the active galaxies, let's compare the energetics of some familiar explosive events.

Cosmic Violence

We consider in this section the intrinsic energies of several well-known sources of radiation. The most powerful of these serve as benchmarks against which we can judge the emissive properties of active galaxies. Since luminosity measures the amount of energy radiated per unit time, the total energy emitted must be, as noted in Chapter 10, the product of an object's luminosity and its lifetime:

$$\text{energy} = \text{luminosity} \times \text{lifetime}.$$

In what follows, we use this equation to calculate the net emission of radiation for any object at all wavelengths. The resultant "total energy budget" is the best measure of an object's activity.

Our first example is a source of great terrestrial activity—a thermonuclear weapon. A one-megaton nuclear blast produces about 10^{23} ergs/second. Of course, the blast is just that—an instantaneous event that cannot be sustained for a long time. It lasts no more than a microsecond. Hence, despite the large luminosity of a nuclear event at the moment of fusion, the total amount of energy released is almost puny by cosmic standards. The total energy budget of a nuclear bomb is then 10^{23} ergs/second $\times$ 10^{-6} second = 10^{17} ergs.

Solar flares represent larger-scale energy production, although they too are nearly insignificant compared to the active galaxies studied in this chapter. The largest such flare typically produces about 10^{27} ergs/second. And they usually last no longer than minutes, making their total energy output over the lifetime of the flare about 10^{30} ergs.

The entire Sun's luminosity is larger, about 10^{33} ergs/second—the equivalent, at any one moment, of about a billion nuclear bombs. And since our Sun burns for billions of years, the total amount of energy released by stars such as the Sun is reasonably large. Multiplying this luminosity by the Sun's lifetime yields a total energy budget of approximately 10^{50} ergs. This is a good rule of thumb to remember: 10^{50} ergs generally equals the total amount of energy released by *any* star. Even the O- and B-type high-mass stars, which often have luminosities as high as 10^{38} ergs/second, live for short durations and hence produce only about 10^{50} ergs. On the other hand, the low-mass M-type stars have only a fraction of the Sun's luminosity, but they live longer, thereby yielding a similar energy budget. Hence all stars are just about equally prodigious emitters, especially when totaled over their lifetimes.

An even greater source of energy is the central region of our Milky Way Galaxy, as well as those of other normal galaxies. These are regions of great star density with many more stars per unit volume than in the outer spiral arms, as can be verified by noting any of the (especially spiral) galaxies shown in Chapter 14. Such central regions emit a total luminosity of about 10^{40} ergs/second. This amounts to a large total release of energy (about 10^{57} ergs) since, being made mostly of stars, galaxy centers sustain their luminosities for long durations.

The entire Milky Way Galaxy has an even larger luminosity. This is hardly surprising, since the Galaxy contains huge numbers of stars. Since about 10^{11} stars inhabit our whole Galaxy, and since each star emits an average of 10^{33} ergs/second, the accumulated luminosity then equals some 10^{44} ergs/second. Be assured, it's not easy to measure the Galaxy's total luminosity since we reside within it. But our Milky Way has been studied carefully in some parts of the electromagnetic spectrum (radio and optical), and at least surveyed in all other parts (infrared, x-ray, ultraviolet, and gamma ray). It's unlikely that additional large sources of luminosity lurk in the darkness.

TABLE 15-1
Energetics of Various Objects

OBJECT	LUMINOSITY (ERGS/SECOND)	TOTAL ENERGY (ERGS)
Nuclear bomb	10^{23}	10^{17}
Solar flare	10^{27}	10^{30}
Sun	10^{33}	10^{50}
Galactic central region	10^{40}	10^{57}
Milky Way Galaxy	10^{44}	10^{62}
Most energetic normal galaxy	10^{45}	10^{63}
Seyfert galaxy	10^{46}	
Radio galaxy	10^{46}–10^{52}	lifetimes unknown
Quasar	10^{46}–10^{53}	

The value of 10^{45} ergs/second is generally considered the maximum luminosity of all normal spiral and elliptical galaxies. It's typical of the luminosity of the giant elliptical galaxies that seem to be the most energetic of the normal galaxies. However, since we know of no dead galaxies, we have little idea of galaxy lifetimes. Hence we cannot easily calculate the total energy released by an entire galaxy throughout its lifetime. Still, keeping in mind the 12-billion-year age of our Milky Way, we can reason as follows: A normal galaxy emitting at an average rate of 10^{44} ergs/second for, say, several tens of billions of years, could release a total energy of at most 10^{62} ergs. Thus we extend our rule of thumb: 10^{50} ergs makes a normal star, whereas 10^{62} ergs makes a normal galaxy.

Table 15-1 summarizes the various levels of activity thus far studied. All the objects and their emission processes discussed up to this point are reasonably well understood. This is not the case for the more active galaxies, however. All the active galaxies are more luminous than 10^{45} ergs/second. And as we've said earlier, the great majority of them populate regions of space beyond the 4-billion-light-year galaxy horizon.

Many active galaxies are observed as mere points of dim light having large red shifts. Their apparent brightnesses, coupled with the inverse-square law, means that they must be hugely powerful emitters. If they weren't so powerful, they would be completely unobservable; their distances are simply too large. In fact, it is their great distances that preclude detailed study of these most intriguing objects.

Fortunately, a small percentage of active galaxies reside inside the galaxy horizon. And although these relatively nearby objects are still well beyond our Local Group, they are close enough to permit astronomers to make detailed observations.

In the next few sections we discuss some of the key properties of these "neighboring" active galaxies. The idea is that if we can understand the nearby active galaxies, then perhaps we can extrapolate our knowledge to understand the truly remote objects residing near the limits of the observable Universe.

Seyfert Galaxies

One type of active galaxy is termed the **Seyfert galaxy**, named after the American optical astronomer who first systematically studied them in the 1940s. They represent one class of large astronomical objects whose properties seem to lie midway between those of normal galaxies and those of the most violent of the active galaxies.

Figure 15.3 shows a series of optical photographs of a typical Seyfert galaxy. With a short exposure, the object looks very much like an elliptical galaxy—a fuzzy round patch of light. But this image is much smaller than an elliptical galaxy. Often measuring no more than 10 light-years in diameter, this faint blob is really the brilliant center of a much larger object. Moderate time exposures begin to show some evidence for a larger object—or at least faint wisps of visible radiation irregularly spread beyond the central blob of light.

The full extent of a typical Seyfert galaxy, including the tight spiral arms, totals about 100,000 light-years, approximately the same size as most other spiral galaxies. A cursory inspection of a long-exposure photograph of a Seyfert (cf., Figure

FIGURE 15.3 Photographs of a Seyfert galaxy (NGC4151) after a short exposure (left), a moderate exposure (center), and a long exposure (right).

15-3) reveals nothing strange about it. In this sense, they seem to resemble normal galaxies. However, more careful studies of Seyfert galaxies reveal some peculiar physical properties when compared to those of normal spiral galaxies.

First, a Seyfert's spectral lines are usually red shifted by moderate amounts, implying that these galaxies generally lie beyond most of the detectable normal galaxies. Most Seyferts seem to reside near the galaxy horizon at distances of 3 to 5 billion light-years, although at least one is known to be as close as 100 million light-years. The spectral lines themselves are abnormally wide, implying either that the gases are tremendously hot or that they are moving at nearly 1000 kilometers/second. The lines bear no resemblance to those produced by ordinary stars, but they do mimic those recently observed toward our Milky Way's center.

Second, maps of Seyfert energy emission show that most radiation emanates from small central regions. At first glance, you may find this hardly surprising since we now know our own Milky Way's center to be energetic. But the entries in Table 15-1 demonstrate that our galactic center emits less than 1 percent of the Milky Way's total energy output. Our Milky Way Galaxy or the Andromeda Galaxy would have to suddenly brighten their central regions nearly a million times to become comparable to a Seyfert galaxy. This predominance of radiation from a Seyfert's center, coupled with the fact that the typical size of the central region is so small, clearly makes this first type of active galaxy peculiar.

So the central regions of Seyfert galaxies are the seats of their activity. The absolute luminosity of a typical central region alone is approximately 10^{46} ergs/second. Comparison with the values in Table 15-1 then shows that the rather small Seyfert center *itself* is typically 100 times more energetic than our *entire* Milky Way Galaxy. Clearly, Seyfert galaxies are abnormal astronomical objects.

A third peculiarity of Seyferts concerns the type of radiation they emit. Much radiation arising from their central regions is of the low-frequency radio and infrared variety, although many Seyferts are also strong x-ray emitters. Remember, the word "luminosity" is not restricted to optical radiation only; luminosity is used to describe radio, x-ray, or any type of electromagnetic radiation. Copious emission of low-frequency radiation without tremendous optical output may seem contrary to intuition because, after all, the Planck curve clearly stipulates that glowing, hot matter like that contained in stars must emit most of its radiation in the higher-frequency optical and ultraviolet domains. Figure 15.2 displayed this fact earlier. Don't be confused on this point: In many cases, enough optical radiation is emitted to make active galaxies visible; even so, the bulk of their radiation must not be released by groups of stars as in normal galaxies. The excess emission of low-frequency radiation seems to be a common phenomenon among most active galaxies.

A fourth and final abnormality of Seyfert galaxies is that their radiation is time variable. That is, extensive monitoring of Seyfert radiation over long periods of time has proved that their luminosity varies. Figure 15.4 shows an example of luminosity variations for a typical Seyfert galaxy. These radiative changes are unlike anything in our Milky Way.

Be sure to recognize that these Seyfert variations involve not just small amounts of luminosity as in the case of the variable stars within normal galaxies. The luminosity changes of Seyfert galaxies are many orders of magnitude greater than the cepheid stars. On average, the 10^{46}-erg/second typical luminosity can double or halve within a fraction of a year. Furthermore, the data plotted in Figure 15.4 mean that this huge radiative output rather suddenly turns on and then off within a single year.

These rather rapid outbursts compel us to conclude that the source of energy emission must be quite compact. For astronomers to be able to

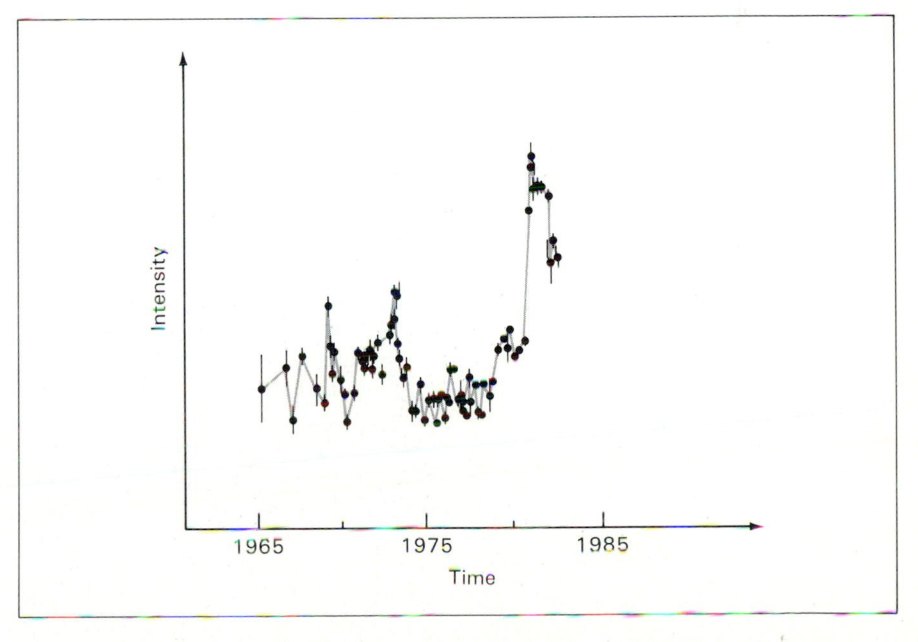

FIGURE 15.4 The irregular variations of a Seyfert galaxy's luminosity during the past two decades. Since this Seyfert, called 3C84, emits most profusely in the radio part of the electromagnetic spectrum, these observations were made with large radio telescopes. The optical and x-ray luminosities vary as well.

detect a regular, coherent pattern of variations (as we do), the source of radiation must be smaller in size than the variation it produces. Otherwise, the intensity variations would be blurred, in contrast to the sharp variations observed. And since the rise and fall of the observed radiation usually occurs within a year, we can confidently conclude that the emitting region is less than a light-year across. Expressed another way, an entire galaxy hundreds of thousands of light-years across could not possibly produce the rapid variations observed. A typical Seyfert galaxy spans some 100,000 light-years and would therefore require 100,000 years for some *cause* on one side of the galaxy to produce an *effect* on the other side. The entire galaxy could never synchronize its front-to-back emission to produce such rapid and coherent time variations. The variations would be smoothed out in time, and we wouldn't observe them. In short, an object cannot flicker in less time than radiation can cross it.

High-resolution maps made with radio interferometers have confirmed this cause-and-effect argument by proving that the variable, strongly emitting region resides within the very center of each Seyfert. These observations suggest that the peculiar region cannot be larger than a light-year across. That's an *extraordinarily* small region in view of the fact that it emits a hundred times *more* radiation than our *entire* Milky Way Galaxy. Therein lies the crux of the problem.

At the least, Seyfert galaxies are experiencing huge explosions within their hearts. The time variations and the large radio luminosities together strongly imply violent nonstellar activity. Actually, the *nature* of the phenomena probably resembles those of our own Milky Way's center as noted in Chapter 13. But the *magnitude* of the phenomena in the Seyfert central regions must be much greater—almost a million times greater—than the rather mild explosions within our own galactic center.

Thus Seyfert galaxies clearly differ from the normal spiral or elliptical galaxies within our local realm of the Universe. Much of a Seyfert's radiation is of the radio and infrared type; there's generally a hundred times more radiation than normal; and most of it arises from an extremely small region of space nearly a million times smaller than the full extent of the Seyfert galaxy itself.

Radio Galaxies

Seyfert galaxies are not the only type of active galaxy in the Universe. Observations during the past two decades have uncovered a wide variety of extremely energetic sources that seem to resemble galaxies of some sort. The rather sudden realization in the 1960s that the Universe is a lot more active than previously thought resulted largely from the Space Age surge of technology, especially radio astronomy. The theoreticians were almost totally unprepared for the discovery of extraordinarily weird and powerful objects.

In this section we consider some bulk properties of the active radio galaxies. We call them galaxies, but it's not clear if they really are in the true sense of the word; they might not contain individual stars, but just vast quantities of gas and dust. There are essentially two classes.

CORE-HALO RADIO GALAXIES: The first class is often called a **core-halo radio galaxy**. Like the Seyfert galaxies, these objects emit most of their intense radio radiation from what seems to be an extremely small (less than a light-year) region somewhere inside the core (or center). Their luminosity often reaches as high as 10^{50} ergs/second, or a million times greater than our entire Milky Way Galaxy. In addition, this first class of radio galaxies emits weaker radio radiation from an extended halo (or envelope) surrounding the core. The halo usually measures about 100,000 light-years across, much like a normal galaxy. However, these objects display no spiral arms—just a spherical blob of additional radiation.

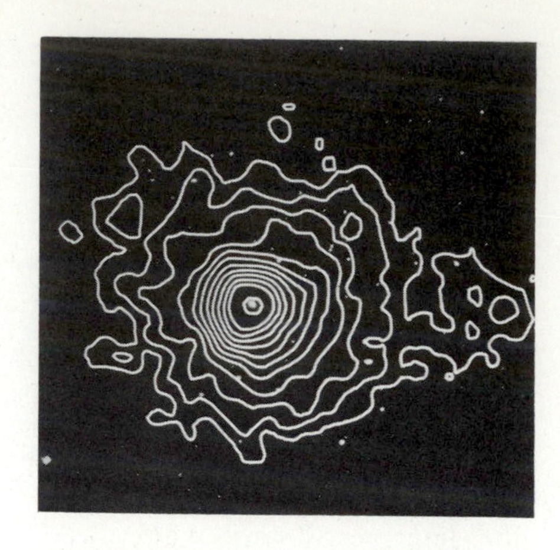

FIGURE 15.5 Radio map of a typical core-halo radio galaxy.

Figure 15.5 shows a map of the radio radiation emitted by a typical core-halo galaxy. Sometimes the halo appears to envelop the core spherically, while other objects have a more distorted halo. Many of these objects are also observed visually, although the optical luminosity is generally weak.

Figure 15.6 shows two optical photographs of a core-halo radio galaxy, along with a contour map of its radio and x-ray emission. This object is called M87; it's the eighty-seventh object in Messier's catalog (although we can be sure that this eighteenth-century Frenchman had no idea what he was really looking at! Nor perhaps do we!). M87 is roughly 50 million light-years distant, one of the closest active galaxies, and a prominent member of the Virgo galaxy cluster mentioned in Chapter 14. Its proximity and activity have made it one of the most intensely studied astronomical sources. A long time exposure yields a large ball of fuzzy light like that shown in Figure 15.6(a). The extent of this fuzzy emission (or halo) is about 100,000 light-years across. However, a short time exposure, like that shown in Figure 15.6(b), displays a compact central region only a few hundred light-years in diameter. The halo cannot be seen at all in such short time exposures.

What *can* be seen beyond the core is a long thin jet *of matter* ejected from M87's inner sanctum. Observations show the jet to be about 5000

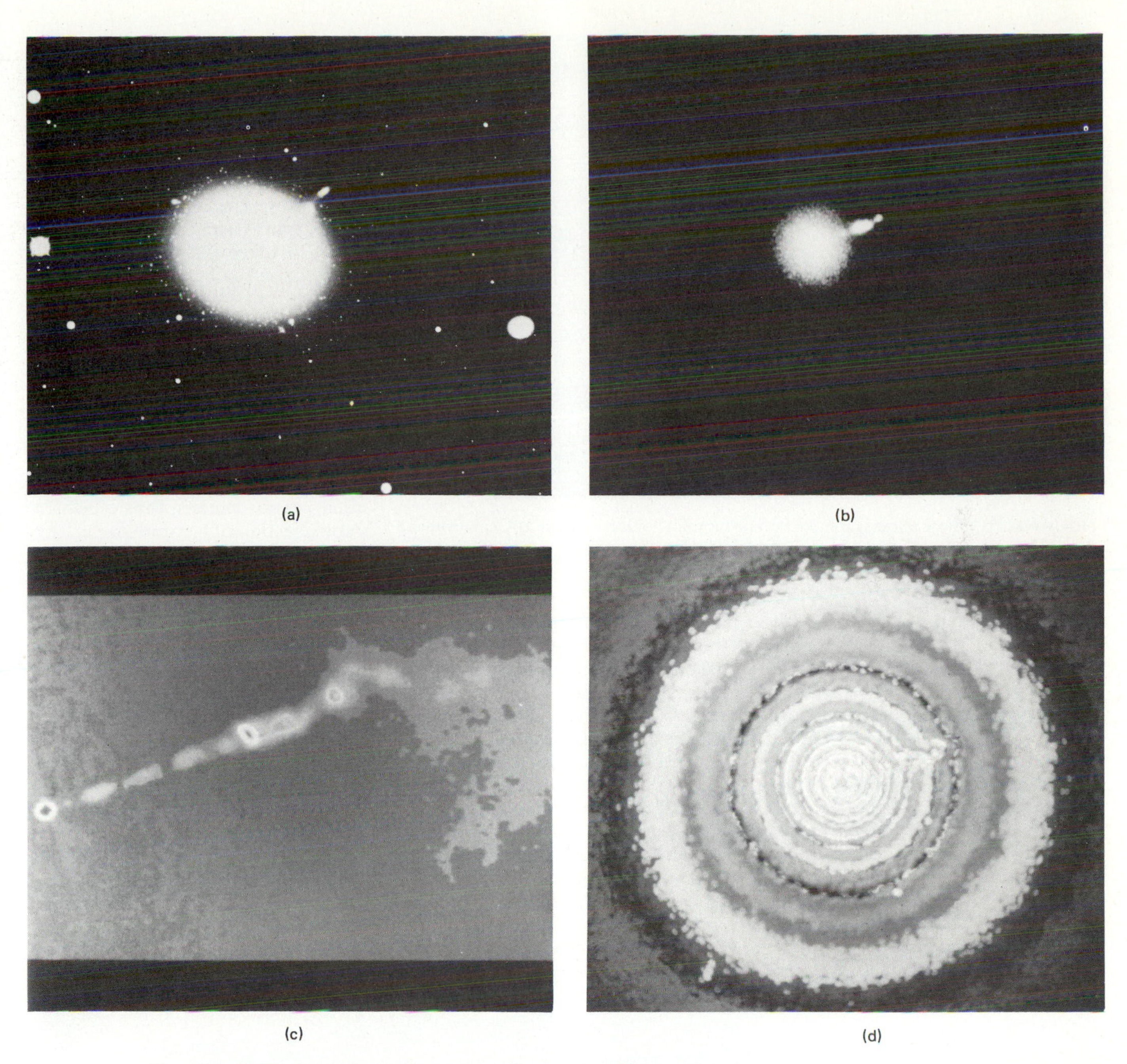

FIGURE 15.6 The giant M87 galaxy (also termed Virgo A) is displayed here (a) as a long optical exposure of its halo and embedded central region, (b) as a short optical exposure of its core and an intriguing jet of matter, (c) as a radio image of its jet [on a slightly expanded scale than in (b)], and (d) as an x-ray image of (mostly) its core and halo, although its jet can also be clearly discerned at upper right.

light-years in length and traveling outward at a velocity of nearly 25,000 kilometers/second (which is almost a tenth of the speed of light). This jet, which has been imaged in the radio and x-ray domains as well [consult Figure 15.6(c) and (d)], is most intriguing. In fact, it has become one of the keys to our understanding of these energetic radio galaxies. Its significance will be revealed in the next section.

LOBE RADIO GALAXIES: The other class of radio galaxies displays a somewhat different shape. Like the Seyfert and core-halo galaxies, this second class of radio galaxies emits most of its radiation in the long-wavelength part of the spectrum. But unlike the others, hardly any of this radiation arises from the central regions. Most of the radio radiation comes from giant lobes or extensions of gas well beyond the center of the radio galaxies. For this reason, these objects are known as **lobe radio galaxies**. Fortunately, several of these strangely active objects are relatively nearby, enabling astrono-

FIGURE 15.7 Optical photograph of the astronomical object known as Centaurus A, one of the most massive and peculiar galaxies known.

mers to study them at reasonably close range.

Often these objects display some optical indication of a galaxy midway between their lobes. These optical images, such as the typical one shown in Figure 15.7, measure the usual 100,000 light-years or so across. This lobe radio galaxy has the name Centaurus A and resides about 15 million light-years from us. A peculiar-looking object, it seems as though it's bisected by a band of dark matter, yet little in its optical image reveals its true strangeness.

In the 1960s when astronomers were first having trouble explaining the emissive properties of the active galaxies, it became fashionable to suggest that these objects are the sites of spectacular galaxy collisions. Because of its appearance, Centaurus A was considered prime evidence in support of this hypothesis, suggesting that the band of dark matter was a spiral galaxy, viewed edge-on, in the process of colliding with an ordinary elliptical galaxy. Another object, Cygnus A, shown in Figure 15.8 and known to be about a billion light-years away, also became good candidate for a collision between two galaxies. This source really does look like two galaxies in collision.

As we know from Interlude 14-2, such a collision hypothesis is not entirely unreasonable since the average distance among galaxies is rather small compared to their overall size. This interpretation, however, is by no means certain, for either Centaurus A or Cygnus A. In fact, the current consensus is that many of the galaxies, previously taken to be colliding objects, are probably not undergoing collisions at all. Even if they were, such galaxy collisions are incapable of producing energy in the required amount. Instead, objects such as Centaurus and Cygnus are more probably single galaxies that have suffered explosive events in their central regions, thus causing a rapid ejection of fast-moving matter out into space.

This view is supported by high-resolution radio observations of the extended lobes of gas. For example, the Centaurus A object displays only slightly perceptible radio emission from the location of the optical image, while most of its radio radiation arises from giant lobes well beyond this optical image. As shown in Figure 15.9, these completely invisible lobes are aligned almost perfectly with the central object positioned on or near an imaginary straight line joining the centers of the two radio sources. Cygnus A has a similar double-lobe pattern, which is displayed in Figure 15.8(b).

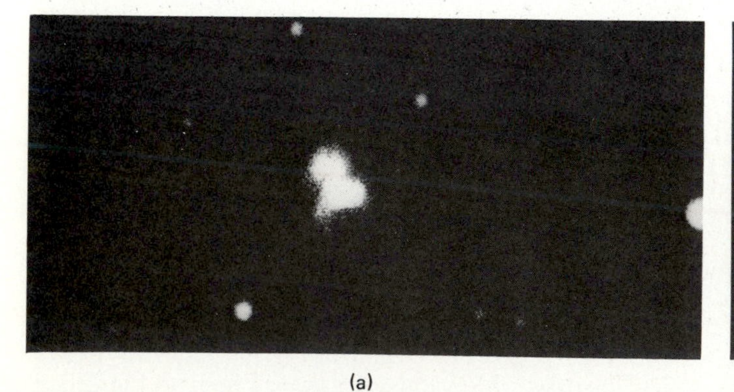

(a)

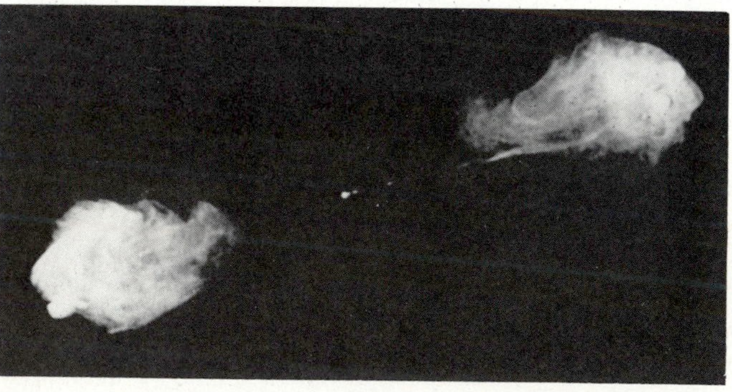

(b)

FIGURE 15.8 Cygnus A visually appears to be two galaxies in collision, although it's not at all clear that's what's happening (a). On a much larger scale (b), this cosmic object displays radio-emitting lobes to each side of the optical image. Be sure to put these regions into proper perspective: The optically seen galaxy in (a) is no larger than the small dot at the center of the "radiograph" (b).

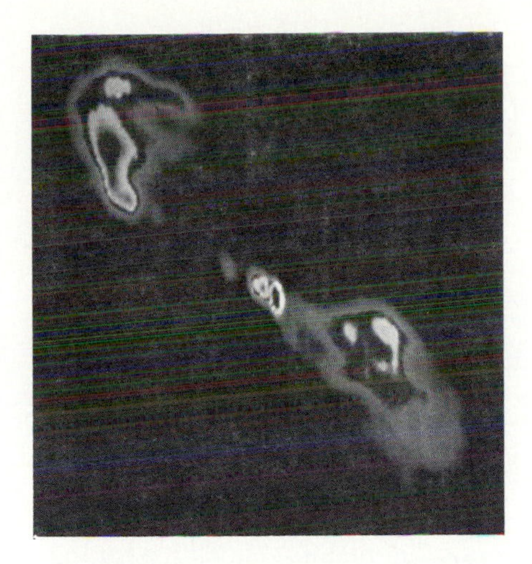

FIGURE 15.9 Lobe radio galaxies, such as Centaurus A shown here, have giant radio-emitting regions extending a million light-years or more beyond the central galaxy observable optically. The lobes are completely invisible, and must be observed with radio telescopes. The scale of this figure is much larger than that of Figure 15.7, as all of Figure 15.7 would fit into the dark gap between the two radio lobes of this figure. The orientation of these two figures is the same, however, as the radio lobes emerge perpendicular to the dark band of Figure 15.7.

For all such objects, the lobes are enormous, *each* one often at least three or four times the typical 100,000-light-year size of a normal galaxy. They vary in size and shape from source to source, but maintain their precise alignment in nearly all cases. An entire lobe radio galaxy usually spans a million light-years from end to end. (Centaurus A, in particular, measures 3 million light-years.) Typically 10 times the size of our Milky Way Galaxy, these objects are big—very big!

An explosive origin for the Centaurus lobes is further supported by recent radio observations that display an additional pair of lobes. These secondary lobes are smaller (about 50,000 light-years in length) than those described above and are closer to the visible galaxy. As shown in Figure 15.9, both pairs of lobes share the same high degree of linear alignment, and therefore generally agree with the hypothesis that periodic explosions eject matter from the central visible galaxy.

Some objects have even larger lobes than Centaurus A. Figure 15.10 shows a radio map of a gargantuan lobe radio galaxy. This one has the name 3C236 since it's the 236th object listed in the Third Cambridge Catalog of radio sources. 3C236 is currently the largest object known in the Universe, extending more than 10 million light-years across its lobes. Remember, our Local Group is "only" 3 million light-years in diameter. Hence this *one* active galaxy is about 100 times larger than our Milky Way Galaxy and several times larger than the size of the entire Local Group.

Admittedly, the very largest of these lobe radio galaxies are rare, but literally hundreds of "smaller" such objects abound with mere million-light-year extents. Enough of them have now been studied to realize that they do not represent an uncommon phenomenon. Many of them are within the galaxy horizon, although most are farther away, as betrayed by the red shifts of spectral lines emitted by the optical objects midway between the lobes. And some display intriguingly shaped lobes, such as the case of NGC1265 shown as Plate 7(b).

The geometry of each lobe radio galaxy clearly suggests explosive events that expelled vast quantities of matter in opposite directions from the central source. The lobes apparently contain matter too tenuous to emit optically, but like the gas of interstellar space, they do so at radio frequencies. Indeed, it is at the low frequencies that these objects are immensely powerful. Their luminosity, which mainly arises in the lobes, sometimes reaches as high as 10^{52} ergs/second. This means that some lobe radio galaxies release energy at radio frequencies at a rate some million times greater than does our own Galaxy at *all* frequencies. We might think that the sheer size of the lobes—10 to 100 times the size of our Galaxy—might well account for the abnormally great emissive properties of these strange beasts. To a certain extent, this is true. But the enormous magnitude of the *radio* energy, and especially the lack of a viable mechanism for the ejection of

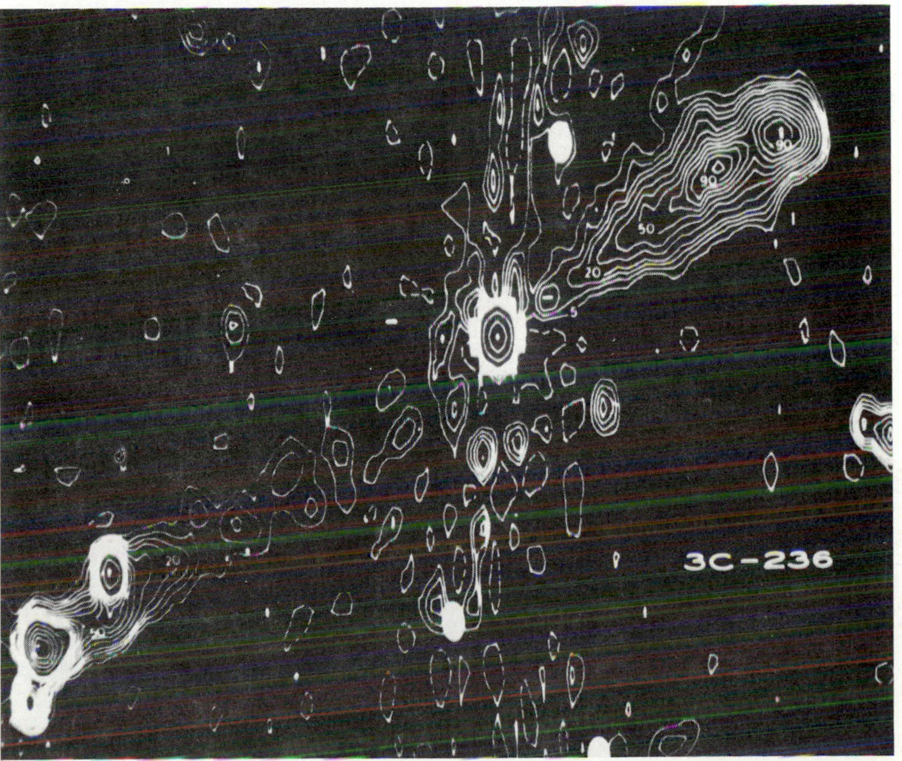

FIGURE 15.10 Radio map of 3C236, currently the largest known object in the Universe. This object is almost 2 billion light-years away and its lobes extend well over 10 million light-years. [See also Plate 7(b).]

the lobes, have thus far frustrated attempts to understand their true nature.

Emission Mechanisms

Understating the situation, we might say that active galaxies are not well understood. The behavior of active galaxies is clearly contrary to that expected from vast collections of stars. The lobe radio galaxies, in particular, are among the most enigmatic objects in the Universe. In short, the active galaxies represent a considerable dilemma for astronomy in particular and science in general.

The principal questions concerning the Seyfert and radio galaxies include: Why is much of the emitted radiation peaked at the low frequencies, especially in the radio and infrared domain? How can, in some cases, vast quantities of radiation arise from relatively small regions of space? And what is the origin of the extended radio-emitting lobes? To be truthful, astronomers have been able to advance an adequate explanation for only the first of these questions.

ENERGY REQUIREMENTS: To get a feeling for the prodigious emissive abilities of the active galaxies, consider for a moment an object having a typical luminosity of 10^{50} ergs/second. This is much larger than the emission of radiation from our Milky Way Galaxy but smaller than some of the most potent lobe radio galaxies. As noted earlier, the total energy output of any object—terrestrial or extraterrestrial, light-bulb or active galaxy—equals the product of its luminosity and lifetime. If, in the case of a 10^{50}-erg/second object, the total lifetime is, for example, on the order of the 10-billion-year duration of our Sun, some 10^{67} ergs of total energy is released during the history of this object. (When making such a calculation, don't forget to cancel the units properly; as a rule of thumb, roughly 10^7 seconds equals 1 year.) The requirement for the existence of a moderately emitting radio galaxy is then simple—10^{67} ergs. It's simple in the sense that the calculation is easy; it's not as simple trying to justify the emission of such a large amount of energy.

Just how large is 10^{67} ergs? Well, compare it to something we've already studied, for example, the thermonuclear fusion of two protons in the Sun's core. We know that stars such as the Sun cannot possibly be the source of the active galaxies' energy, for stars emit mostly visual radiation and hardly any radio radiation. Nonetheless, the proton–proton cycle is an instructive example, since it's an energy process with which we're familiar. Calculations made in Chapter 9 showed that energy unleashed during a hydrogen fusion process equals approximately 10^{33} ergs/second. And assuming that the Sun burns for 10 billion years, as expected, we found that our Sun's total energy budget *throughout its lifetime* is about 10^{50} ergs.

Since a typical active galaxy requires 10^{67} ergs of total energy, and individual stars contribute "only" 10^{50} ergs, we conclude that some 10^{17} stars are needed to *equivalently* power a typical active galaxy. That's nearly a million trillion stars—all in one object! And since the mass of most stars averages that of our Sun, we might expect such an object to contain on the order of 10^{17} solar masses. Quite frankly, astronomers have never found a single example of an object anywhere near that massive in the entire Universe. Not even the largest galaxy clusters are that massive.

Another way to appreciate the fantastic energy requirement of 10^{67} ergs is to compare it to a normal galaxy. For example, since our whole Galaxy emits about 10^{61} ergs (see Table 15-1), the luminosity of a typical active galaxy is equivalent to that of a million Milky Ways—all emitting simultaneously! And remember, we are considering here the typical or average active galaxy; the *most* active radio galaxies emit more than 100 times greater than this average, thereby requiring the equivalent emission of nearly a billion Milky Way Galaxies. Furthermore, recall that many active galaxies emit mostly *radio* radiation, and often from a *compact* region about a light-year in diameter.

The basic problem here is that no one has yet been able to imagine a million Milky Way Galaxies all squashed into a space no larger than a light-year. It's even hard to imagine *one* Milky Way Galaxy compacted into a light-year-sized region. Clearly, the energy requirements of the active galaxies are terrifically large, and the geometries of their emitting regions equally puzzling. Proton–proton reactions are simply not energetic enough. Even the accumulated output of trillions of stars densely packed into a small region of space falls far short of what's needed.

The situation is not hopeless, however. Surprisingly, stellar emission is not as efficient as many people think. Energy is converted from mass alright, but a lot of mass is left over; the fusion cycle is only a few percent efficient even when operating at its full capacity to yield the heaviest elements. Other exotic processes are known to be more efficient, and thus produce more energy for a given amount of matter. To illustrate the severity of our problem, let's consider the most efficient of all energy mechanisms.

The most efficient energy process known is **matter–antimatter annihilation**. The product of such an interaction is pure energy, with nothing left over—no mass, no ashes, nothing. The amount of energy derived equals mc^2, where m is the total mass of the matter plus antimatter and c is the velocity of light. For example, as an extreme scenario, we might imagine mixing a 1-solar-mass star made of matter with another equally massive star made of antimatter. Astonishingly, even this process falls short of our (10^{67} ergs) goal; twice the mass of the Sun times the velocity of light squared equals only 10^{54} ergs. Furthermore, this energy is released as a burst of gamma rays, that type of radiation on the opposite end of the spectrum from radio waves.

Even with this most efficient matter–antimatter annihilation mechanism, our requirement of 10^{67} ergs for a typical active galaxy means that the equivalent of 10^{13} stars or alternatively 100 Milky Way Galaxies must be annihilated. Not even our generation raised on Space Age gimmickry could imagine completely and totally annihilating 100 Milky Way Galaxies!

These examples are meant to emphasize the extent to which theoreticians have been pressed in order to understand the amount of energy emitted by the active galaxies—provided that those galaxies really do emit for such long periods of time. On the other hand, maybe we should reconsider the lifetime of these objects. Perhaps they do not emit at the luminosity level of 10^{50} ergs/second for such a long time. For example, if they have lifetimes much less than a billion years, their total energy budget would be correspondingly lessened.

Still, we cannot shorten their lifetime too much. Observations prove that the radio galaxies must have been emitting such high luminosities for at least a million years. The large blobs of matter presumably ejected from the lobe radio galaxies are typically separated by about a million light-years. Some objects have smaller lobes, some larger, but a million light-years is an average separation. Even if the lobes were ejected at the unlikely maximum velocity of light itself, they would still take about a half-million years to reach their current positions astride the central optical galaxy. In reality, most lobes cannot be moving at velocities anywhere near that of light lest astronomers easily detect their movements with sensitive radio telescopes here on Earth. Thus we conclude that most lobes have probably existed for much more than a million years.

Where do we stand? The average active galaxy requires the energy *equivalent* of complete annihilation of 100 Milky Way Galaxies. Note that we're not suggesting that matter–antimatter annihilation is a feasible emission mechanism. We know it isn't, in the first place because matter and antimatter could not remain separated for long within a galaxy; once they came together, they would instantaneously blow the whole galaxy to smithereens. In the second place, the annihilation process does not produce large amounts of radio radiation. Clearly, we need a physical mechanism capable of slowly releasing copious amounts of radio energy over at least as long as a million years. What are we to do then? Can these active galaxies be explained?

A NEW KIND OF RADIATION: Astronomers are not entirely ignorant of the physical processes occurring within the prodigious active galax-

INTERLUDE 15-1 *BL Lac Objects*

In the 1920s an object thought to be a variable star was discovered in the constellation Lacerta; as the ninetieth variable found in that part of the sky, astronomers gave it a two-letter code, BL Lacertae. Not until the 1970s, when it became clear that this object was a strong radio source, did anyone question its classification as a star. As more objects such as BL Lac were found, astronomers began to realize that they had stumbled across a whole new class of extragalactic sources. BL Lac-type objects, sometimes colloquially called "blazars," seem star-like in a telescope, vary greatly in brightness, and are also powerful and compact radio sources.

Astronomers place BL Lac objects into a special class simply because, unlike any other type of active galaxy, they originally displayed no spectral lines by which their red shifts could be determined. In recent years, extremely faint spectral lines strongly suggest that BL Lac objects reside at great distances, thus making their energy production nearly as formidable as for the quasars. Furthermore, observations have proved that these blazars reside at the centers of relatively normal (elliptical) galaxies.

The weak spectral lines might be the key needed to unravel the nature of the BL Lac objects, and by implication other active galaxies as well. Whereas quasars and other active galaxies show a mix of thermal and nonthermal emission properties, BL Lac objects seem to be powered almost exclusively by the nonthermal polarizing mechanism. Evidently, the thermally heated gas that normally generates spectral lines is swamped by the nonthermal process of electrons interacting with magnetism. Accordingly, BL Lac objects seem to offer the best available chance to study the "bare machine" responsible for energizing all the active galaxies.

In many ways, the BL Lac objects possibly represent a "link" or transition phase between radio galaxies and quasars. This is most evident from studies of the nonthermal properties of each kind of source. The luminosities of the compact radio sources inside BL Lac objects span the whole range from the relatively weak cores observed in radio galaxies to the stronger ones found in quasars. Perhaps, as we shall study in Chapter 16, there exists a time-sequence whereby the ancient quasars somehow use up much of their gas to power a "central engine," eventually evolving into the weaker yet more erratic BL Lac objects, afterwhich further decay yields the rather inactive cores amidst the giant lobes of the radio galaxies.

All things considered, the theory of active galaxies embarrassingly lacks far behind the observations. Even so, as for our normal Milky Way Galaxy, only further observations of their galactic centers will likely clarify the true nature and interrelationships of the varied, powerhouse active galaxies.

ies. Within the past two decades, observations have provided some inkling of the basic cause of the emitted radiation.

An important clue regarding the true nature of the radiation results from detailed studies of the jet in the M87 object shown in Figure 15.6. The key here is that the optical emission from that jet is polarized. The radio emission from the entire core-halo object is also polarized. What do we mean by polarization? Recall that in Chapter 1 we described the passage of light through a medium (a Polaroid filter) containing aligned molecules. Light passing through such a medium becomes polarized, a special form of radiation whereby the electromagnetic planes of oscillation are parallel. The same sort of thing happens in interstellar space. In Chapter 12 we studied the case of starlight passing through the presumably elongated and aligned interstellar dust particles. As described, interstellar polarization is most prominent in the galactic plane of the Milky Way. But when looking perpendicular to the plane, toward the distant galaxies, the polarization is negligible. That's because our line of sight to a galaxy encounters little interstellar dust within the thin galactic disk. Nor do we know of any dust or other polarizing matter beyond our Milky Way—an intergalactic medium—that could cause a distant galaxy's radiation to become polarized while passing through space.

Therefore, we conclude that the polarization of the radiation from M87's jet is caused not by any intervening matter along our line of sight, but by some intrinsic process causing the emission of the radiation itself. What kind of a physical process can possibly yield polarized radiation? Surely not the emission process producing radiation from a parcel of heated gas. Light bulbs, or atomic bombs, or ordinary stars do not intrinsically emit polarized radiation.

Part of Figure 15.11 reminds us of the **thermal radiation** released by any heated object. Whether from a light bulb or a star, this kind of ther-

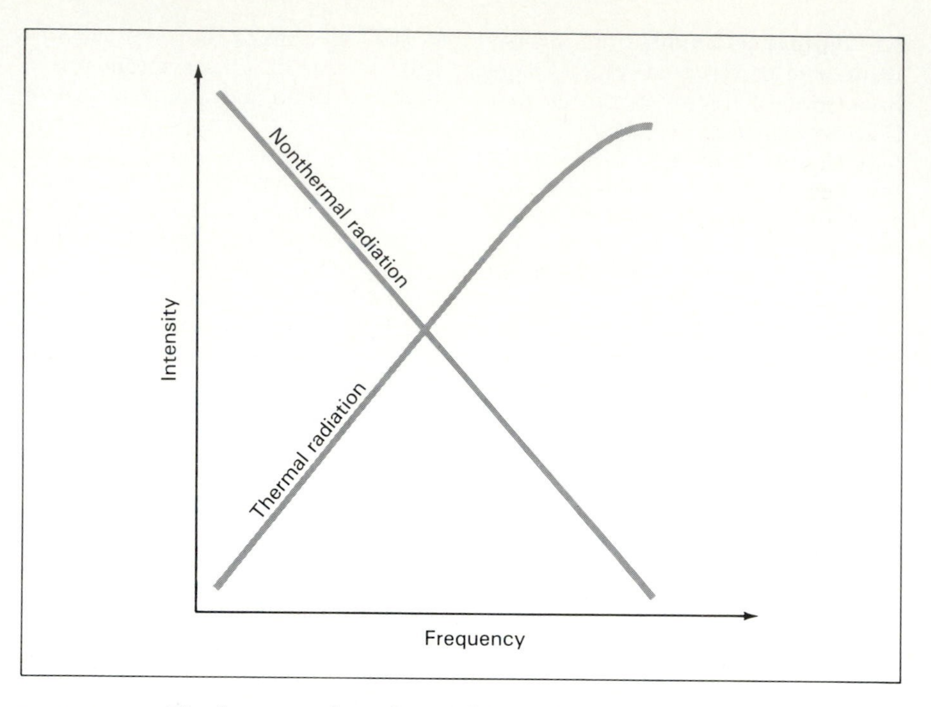

FIGURE 15.11 The frequency dependence of thermal and nonthermal radiation. Hot objects emit more intensely at high frequencies, whereas nonthermal emitters radiate most intensely at low frequencies. (Compare with Figure 15.2.)

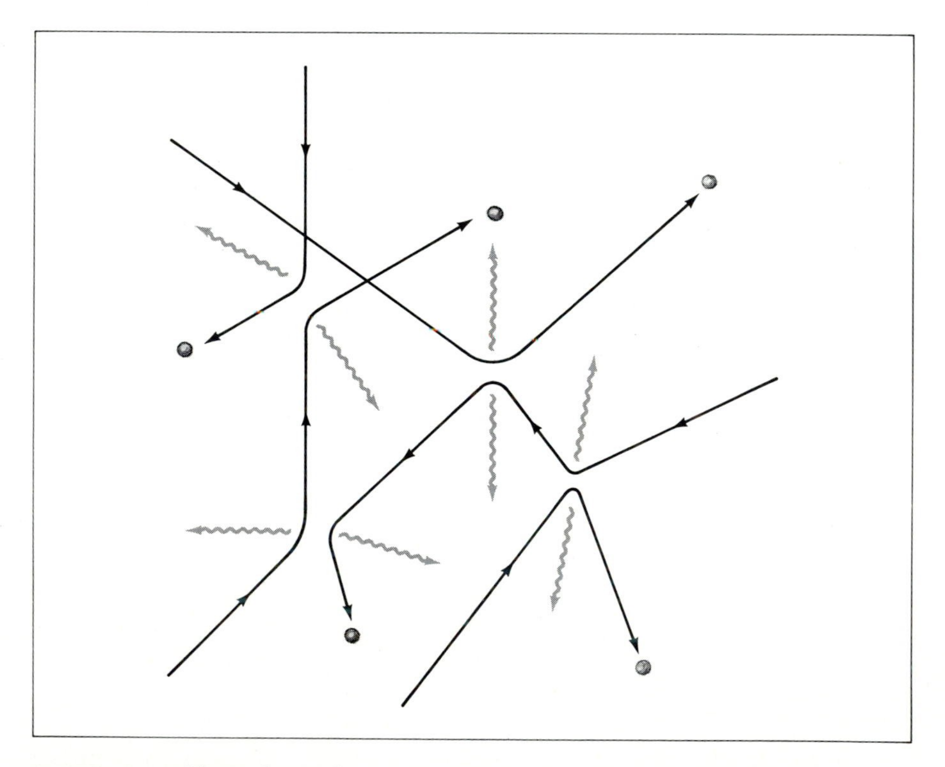

FIGURE 15.12 Charged particles emit unpolarized thermal radiation (wavy lines) upon interacting with one another. Whenever a charged particle changes the magnitude or direction of its velocity, radiation is released. Thermal radiation of this kind arises from many of the objects studied earlier: light bulbs, stars, nebulae, in short any heated substance.

mal process results simply and straightforwardly from the random interactions among electrically charged electrons, protons, and ions. Figure 15.12 sketches this process. Regardless of temperature, this emission obeys the Planck curve and thus usually peaks at the high-frequency end of the spectrum. Furthermore, the emission is always unpolarized.

To explain the luminosity and polarization of the active galaxies, we need to find a completely different kind of emission process—a distinctly nonstellar process. Is there any mechanism known to produce radiation (1) that is most intense at the low frequencies, and (2) that is polarized? Fortunately, there is. That mechanism, called **nonthermal radiation**, involves electrons interacting with a magnetic force field. No stars are involved, nor is any heat per se (hence the name "*non*thermal"). The radiation simply arises from elementary particles within magnetized regions of space. The frequency dependence of this nonthermal mechanism is sketched in part of Figure 15.11.

Presumably, magnetism pervades the Universe—not only the Earth, Sun, Solar System, and Galaxy, but also other galaxies as well. And although the magnetic force fields of large galaxy-like objects are thousands of times weaker than Earth's, magnetism can still play a significant role, especially when its effects accumulate across an entire galaxy.

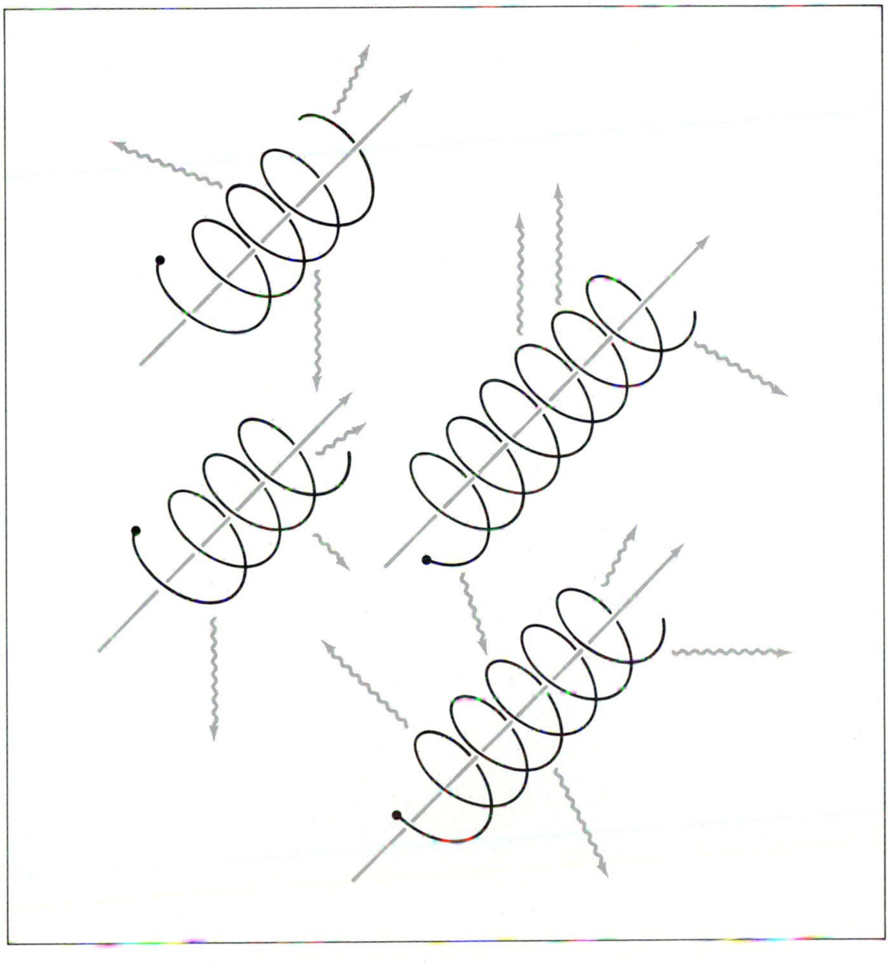

As depicted in Figure 15.13, whenever fast-moving electrons enter a region of magnetism, they tend to spiral around while emitting radiation. Laboratory experiments have shown that magnetism slows the particles, causing some of their kinetic energy to change into radiative energy. The higher the electron velocity and/or the stronger the magnetic force field, the greater the energy released. The physics of the process, which is exceedingly complex, predicts that the emitted radiation is intrinsically polarized and peaks in the radio part of the spectrum (consult Figure 15.11). These predictions have been confirmed in the laboratory. And since the radiation accumulates in proportion to the number of encounters with the magnetic force field, huge magnetized lobes having many, many electrons can become potent emitters of intense radio radiation.

All in all, the mechanism of nonthermal radiation gives us a glimpse of the types of events responsible for the emissions of most active galaxies, provided that they do not emit much longer than a million years. Apparently, a series of explosions repeatedly accelerates electrons to speeds close to that of light itself, after which large clumps of plasma move outward, forming the distorted envelopes of the core-halo radio galaxies or the extended blobs of the typical lobe radio galaxies. However, the most violent of the active galaxies cannot be understood, their rapid variability is almost impossible to fathom, the origin of the repeated explosions is currently unknown, and the cause of the acceleration of the high-velocity electrons is downright mystifying. Without doubt, the active galaxies comprise one of the ripest areas for research in all of modern astronomy.

FIGURE 15.13 Charged particles, especially fast electrons, emit polarized nonthermal radiation (wavy lines) while spiralling within a magnetic force field (straight lines). This process is not confined to active galaxies. It occurs, on smaller scales, when charged particles interact with magnetism in Earth's Van Allen Belts, when charged matter arches above spots on the Sun, and when cosmic rays traverse our normal Galaxy, as noted in chapters 5, 9, and 13, respectively.

Quasi-Stellar Objects

Formidable though they may be, the active galaxies are not the most energetic objects in the Universe. Astronomers know of an additional, extraordinarily powerful class of active astronomical objects—objects so hard to fathom that they threaten to render inadequate the laws of physics as currently known. These are the superluminous, highly redshifted *quasi-stellar sources*—**quasars** for short.

Not content just to rival the energy emission difficulties of the active Seyfert and radio galaxies, quasars actually extend those difficulties. The main problem is that the radio and optical radiation observed from quasars often varies on the order of weeks, sometimes even days. Thus the huge luminosities of the quasars—generally 10^{46} to 10^{53} ergs/second, assuming that the red shift is a valid indicator of distance—must often arise from regions hardly more than a light-day across. A light-day, remember, is 1/365 of a light-year, or not much larger than the size of our Solar System. The quasar emission mechanism—whatever it is—must then operate within a surprisingly small realm of space—not small compared to terrestrial standards, but almost unbelievably small considering that most quasars are more than 1000 times as energetic as our entire Milky Way. Can you imagine 1000 or more normal galaxies all packed together into a region hardly larger than our Solar System? That's an indication of the compact energetics needed to appreciate the herculean quasars, unquestionably the most mysterious objects in all the Universe.

Figure 15.14 is an optical photograph of a typical quasar, this one called 3C275.1. The photo was taken through a large telescope, as no quasar can be seen with the naked eye. This quasar is not terribly distinguished, nor does it have much observable structure. Nor should we expect to see much, given its 7-billion-light-year distance! Even very long time exposures of many hours show nothing more than a ball of fuzzy light—much like a star, hence the name "quasi-stellar."

Of the hundreds of quasars now known, few show more than a ball of luminous fuzz. Figure 15.15 is an optical photograph of one such quasar, labeled 3C273. Its red shift implies a Doppler velocity of about 50,000 kilometers/second. Should this velocity be a true indicator of distance according to Hubble's law, then, at 2.5 billion light-years away, this quasar is one of the nearest yet discovered. The jet of luminous matter, reminiscent of the one in M87, extends nearly a half-million light-years.

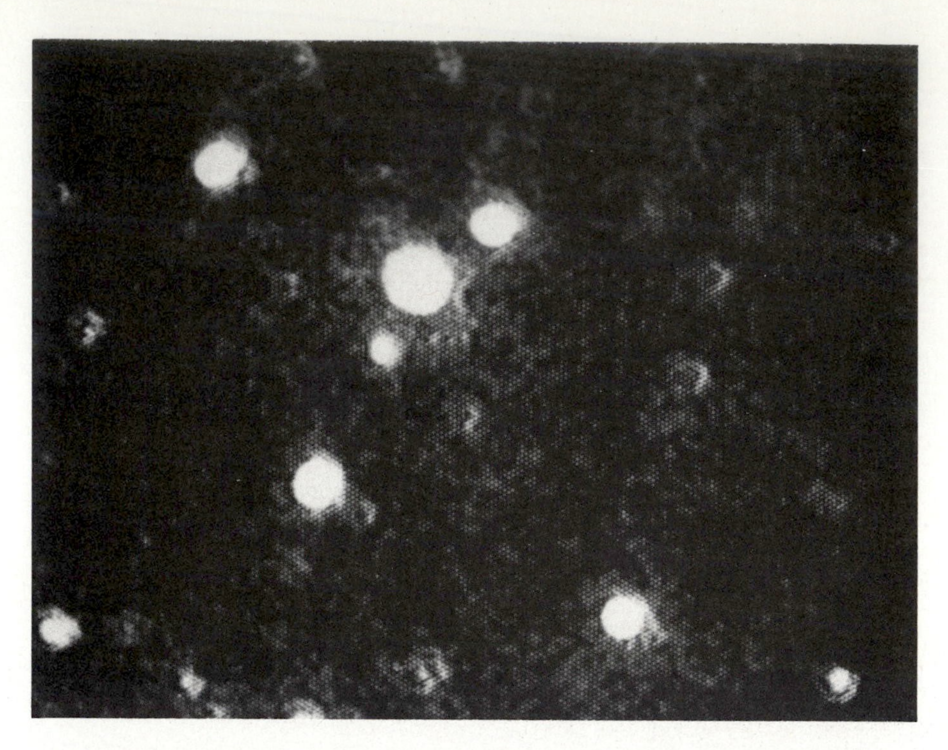

FIGURE 15.14 Optical photograph of a typical quasar, known as 3C275.1 (the brightest object near the center). Its red shift places the object some 7 billion light-years away.

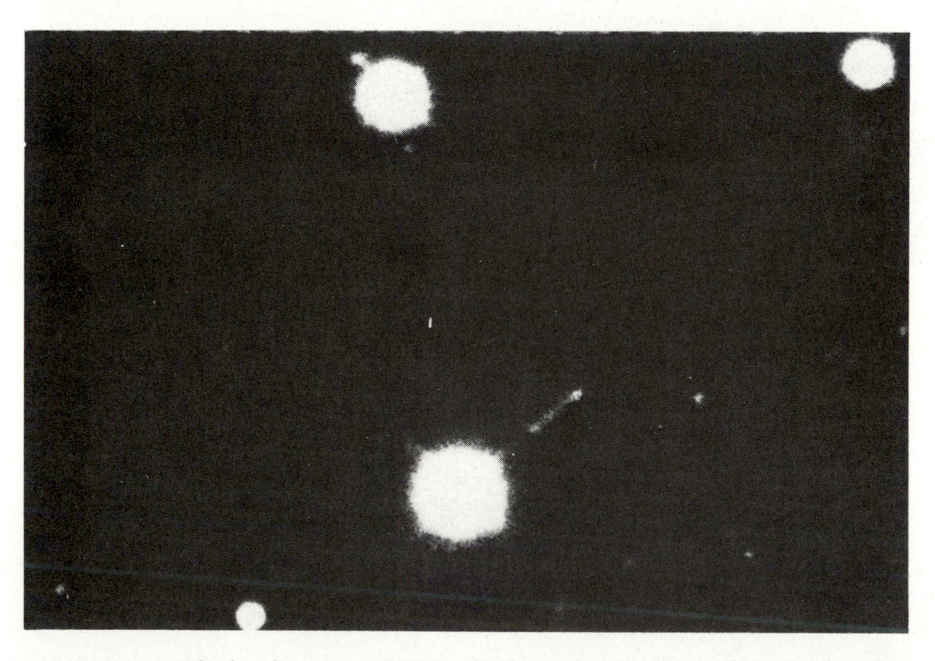

FIGURE 15.15 The bright quasar, 3C273, displays a luminous jet of matter, but the main body of the quasar remains starlike in appearance.

INTERLUDE 15-2 ***Faster-Than-Light Velocities?***

The compact sources of emission within radio galaxies and quasars vary not only in intensity but also in structure. Some quasars recently mapped with very-long-baseline interferometers have displayed dramatic changes in structure, often on time scales as short as months. Provided that the quasars are at cosmological distances, the movements of their various interior components seem to exceed the velocity of light (which is strictly forbidden in the currently known fabric of physics).

For example, the accompanying illustrations show three radio images of the core of quasar NGC315 made several years apart. The interior of this quasar is dominated by two large blobs of gas. Furthermore, these blobs seem to have moved over the course of time. Knowing the distance over which the blobs have apparently moved, and knowing the time they took to do so, we can infer their velocities. Astonishingly, the result is approximately 10 times the velocity of light.

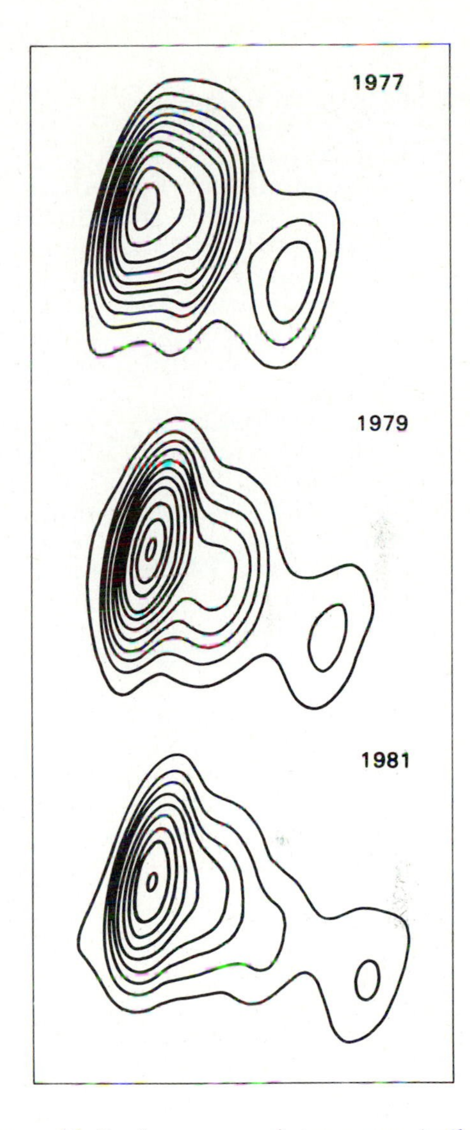

The notion that the speed of light is the highest attainable velocity is basic to modern physics; scores of predictions made assuming this to be true have been verified during the last several decades. Hence astronomers are almost universally agreed that some reinterpretation of the apparent quasar interior motions is needed. The simplest alternative assumes that quasars are not at cosmological distances; at relatively local distances from us, the interior quasar blobs would not physically move too much within the quasar. This reinterpretation thus avoids the notion that the velocities of the individual blobs move faster than light. But—and this is a very big "but"—if the quasars are at relatively local and not cosmological distances, the observed red shifts of their lines cannot be a distance indicator. We'd be forced to find some other reason for the large quasar red shifts and, as already noted elsewhere in this chapter, that's not easy. It may be impossible.

Scientists have proposed several alternative solutions to account for the apparent separations of the interior quasar blobs; these theoretical models require the quasars neither to be local objects nor to have faster-than-light speeds. For example, the most straightforward model suggests that the observed changes in quasar structure are not at all caused by actual motions of the interior blobs but rather, by variations of the radio intensity of *stationary* blobs. In other words, the interiors of quasars may, for some reason, behave as radio analogs of Christmas trees with different bulbs periodically blinking on and off, thereby giving the impression of bulbs in motion, whereas in actuality each of the bulbs is at a fixed position. Movie-theater marquees provide another familiar example of how flashing, stationary lights can simulate motion. To an unsuspecting viewer, fixed quasar blobs that wax and wane can similarly create an *illusion* of moving blobs of radiation. Another alternative model presumes the inferred motions to be another sort of illusion or projection effect produced by blobs moving almost precisely along our line of sight at slightly less than light velocity.

None of the alternative models is simple. They often require contorted geometries, and no one model is agreed upon by all researchers. In truth, we still lack a definite explanation for this puzzling phenomenon. At any rate, the current consensus is that quasars are at cosmological distances, their interiors are exceedingly complex and not at all well understood, but that there is no compelling need to resort to faster-than-light motions.

Many of the quasars can be observed in the radio, infrared, optical, ultraviolet, and x-ray parts of the electromagnetic spectrum, and some have even been observed to be releasing gamma rays. Often, quasar radio radiation arises from outlying regions, much like the core-halo and lobe radio galaxies studied earlier. In other cases, the radio emission is confined to the central optical image. And like the active galaxies, the radiation is usually polarized. As best can be determined, the average properties of quasars are similar to those of radio galaxies.

The nonthermal mechanism of high-velocity electrons spiraling around magnetic force fields can reasonably account for the gross features of the energy emitted, namely polarized radio radiation. (Some quasars do not emit detectable radio radiation, although their optical radiation is still of the nonthermal variety.) But as for the radio galaxies, the crux of the problem concerns the *magnitude* of the energy emission. This is especially difficult in the case of quasars, since their radiation originates in exceptionally small and presumably compact regions. The origin of the energy capable of accelerating the electrons to such high velocities within such a small volume is wholly mysterious.

Theoretical models of quasars postulate a layered structure surrounding a compact central energy source. Farthest out are the clouds of cool gas that were ejected by the quasar at earlier epochs and that are now studied by observing the absorption lines of heavy elements. Inside, emission features are formed; these spectral lines, somewhat like those arising in the central regions of normal galaxies, including our own Milky Way, extend over a broad wavelength range, suggesting that numerous clouds revolve in rather small orbits around a central region housing a very large mass. At the center is a compact energy source—the "central engine"—responsible for the acceleration of particles and for the generation of high-energy radiation by a complete combination of processes that we have yet to fathom in detail. Many researchers in the astronomical community are currently leaning toward a collection of black holes or a single supermassive black hole, a topic that will be studied further at the end of Chapter 21.

QUASAR RED SHIFTS: As noted above, quasars have spectral lines that are greatly red shifted. There are no known exceptions. In fact, some quasars have such large red shifts that their spectra were unexplained for several years after their initial discovery in the 1960s.

The most straightforward interpretation of the red shifts is based on the Doppler effect. If correct, most quasars are traveling away at more than 100,000 kilometers/second; several are receding at nearly the velocity of light itself. Should Hubble's law be valid beyond the galaxy horizon, these great velocities imply great distances. With this assumption in mind—an assumption thought valid by most researchers—we can conclude that virtually all the quasars are beyond the galaxy horizon, or in today's jargon, at "cosmological distances."

Quasars are almost invariably positioned on the Hubble diagram in the upper right corner. Plotted in Figure 15.16, they have in nearly all cases larger velocities than virtually every normal galaxy used to establish the Hubble law. Extrapolating that law means that most quasars must reside at much greater distances than any of the other astronomical objects studied thus far in our text.

Quasars seem to be especially abundant in the distance interval between 6 and 13 billion light-years. It's unlikely that any other type of object resides at distances much beyond the quasars, for many quasars already race away at velocities close to that of light. For example, one of the fastest quasars yet discovered recedes at about 90 percent the velocity of light; if Hubble's red shift–distance relationship holds true, this object, called OH471, is some 13 billion light-years away.

A hypothetical object receding with a velocity *equal* to that of light would, assuming it obeys Hubble's law, be expected to be some 15 billion light-years from us. This 15-billion-light-year value is often taken to be the limit of the observable Universe–a sort of **universal horizon**. Provided that the velocity of light

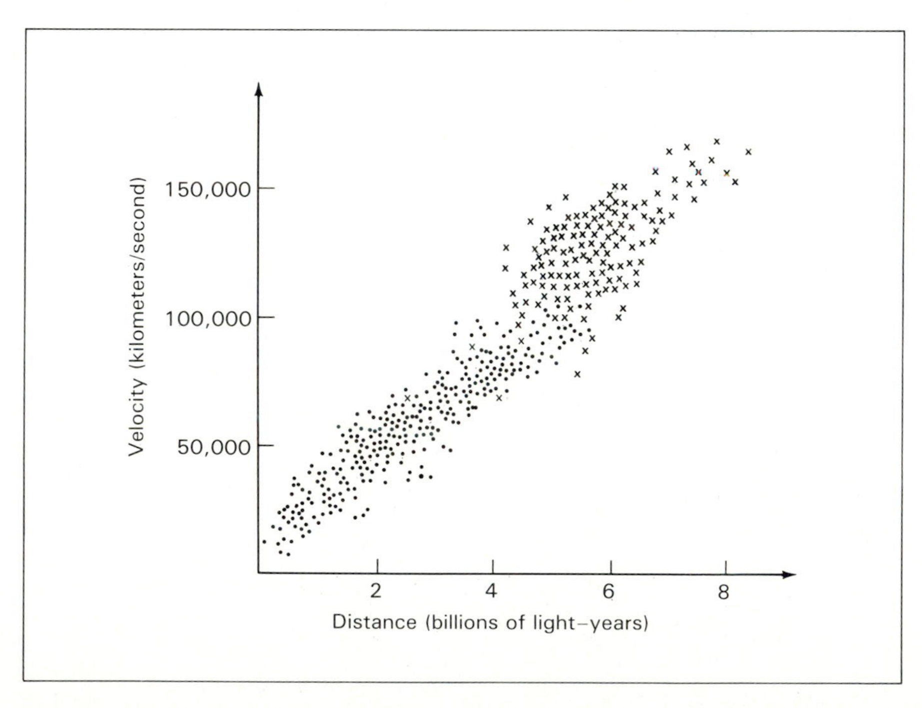

FIGURE 15.16 The Hubble diagram, including many quasars (x's) at the far distances.

really is the ultimate speed limit in the Universe, we cannot hope to perceive anything beyond.

THE ENERGY PROBLEM: Let's inquire more deeply about the mysterious quasars that seemingly extend to the limits of the observable Universe while radiating huge amounts of energy. We can ask several questions: Why are the quasars so special? What really makes them tick? Are the quasars just another type of exotic object within our varied Universe? Or are they basically different from anything we know?

The large quasar red shifts force us to one of two conclusions: Either the quasars obey Hubble's law, are truly very distant, and hence intrinsically most luminous; or they do not obey Hubble's law, can be relatively local, and hence comparatively dim. If the former is true, we need to unravel the origin of their emission. If the latter is true, we need to determine the origin of their red shifts. Let's examine the former case first, namely the energy problem.

Based on their position in the Hubble diagram, quasars might be regarded as normal galaxies having, for one reason or another, *some extra luminosity*. This extra luminosity is needed if the quasars are at cosmological distances. Otherwise, astronomers could never detect them. At face value, then, this first possible solution would seem to be a straightforward interpretation of the quasars that populate the upper end of the Hubble diagram. It's straightforward in the sense that the red shift is truly an indicator of distance as well as velocity. It's not a straightforward interpretation, though, when we try to justify the extra luminosity needed to perceive the quasars at such great distances.

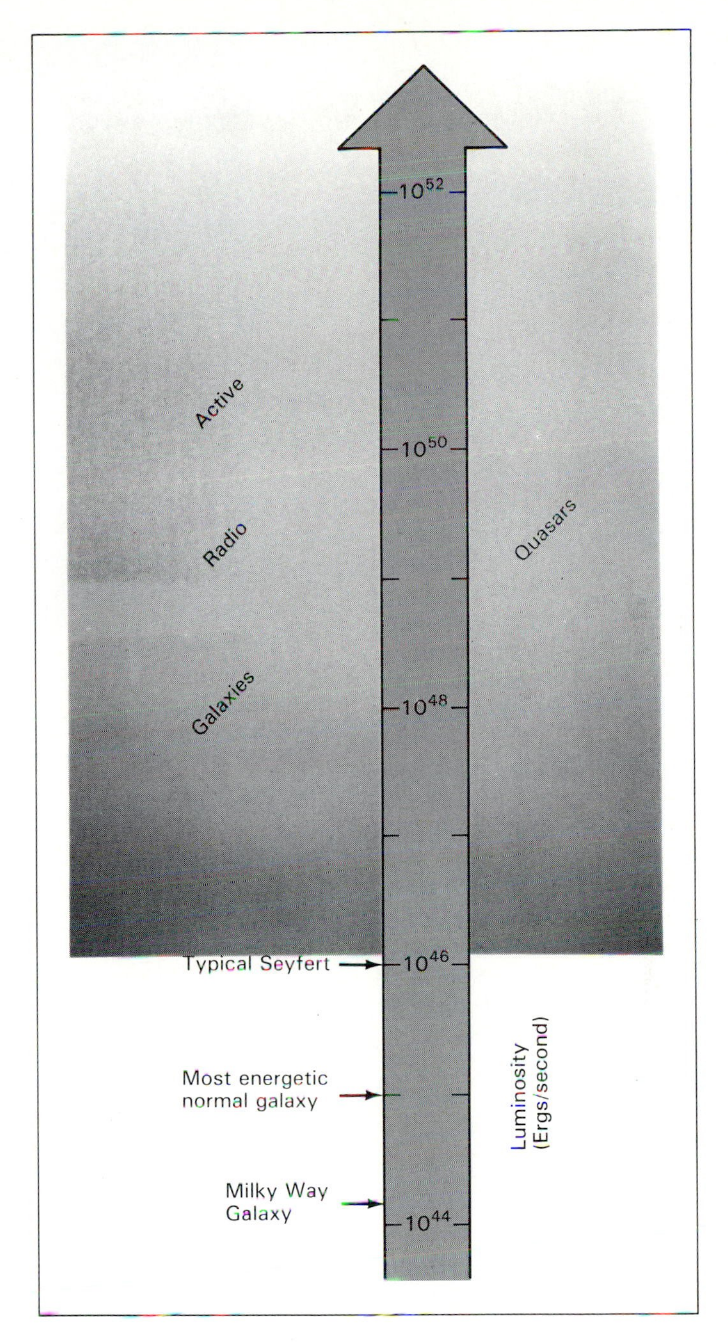

FIGURE 15.17 Typical luminosities of many types of large objects known in the Universe.

Figure 15.17 compares some typical luminosities for many cosmic objects thus far studied. Noted there are the Milky Way Galaxy, other normal galaxies, active radio galaxies, and Seyferts, as well as some quasars. The largest luminosity quasars, plotted toward the top of this figure, were discovered mostly by examining old photographs. As noted in Chapter 11 (especially Interlude 11-2), observers throughout the past century have routinely photographed the sky in order to monitor the properties of ordinary stars. Unbeknownst to the original observers, many of the specks of light captured on their photographic patrols are not stellar but quasi-stellar in origin.

Once a radio, x-ray, or some other observation indicates that something unusual is occurring within a certain region of the sky, a definite procedure is followed by modern researchers. First a spectrum is made of the object with one of the world's largest optical telescopes. Should the suspect object be confirmed as a quasar by measuring its large red shift, then the astronomers consult a comprehensive collection of photographic plates extending back to the birth of photography. The "light history" of the quasar—over the past 100 or so years—can then be derived by analyzing the many plates taken each year.

Figure 15.18 demonstrates how such light histories often reveal evidence for variations in a quasar's optical radiation. In frame (a), a 1937 photograph clearly shows an optical object marked by an arrow. Its large red shift and other emissive peculiarities prove its identification as quasar 3C279. This photograph contrasts with the one in frame (b), taken in 1976, when the quasar had nearly disappeared. Frame (c) shows the recent "light history" of quasar 3C279 derived from a large number of photographs taken since 1930. Its great distance and apparent brightness imply that in 1937 this faint speck of light was *intrinsically* the most luminous object ever observed in the Universe. Pouring out 10^{53} ergs/second at that time, 3C279 reigns champion to the top of Figure 15.17.

We can easily postulate that if quasar luminosities are roughly a million times those of normal galaxies, then perhaps they contain a million times more stars, gas, and dust. In other words, maybe quasars are just a million times more massive than our Galaxy and therefore would naturally emit a million times more radiation. Extra mass implies extra luminosity. This would indeed be a perfectly reasonable explanation of their position in the Hubble diagram—if that's all we knew about quasars. But we know more. In particular, we know that a good deal of their radiation is often emitted as radio waves, is usually polarized, is time variable, and comes from a small source. It is these additional observations that make quasar radiation so puzzling. So quasars as pumped-up normal galaxies having some extra luminosity just doesn't seem to work.

COULD QUASARS BE LOCAL? An alternative interpretation assumes that the quasars are not at cosmological distances. In other words, perhaps the quasars do not obey Hubble's

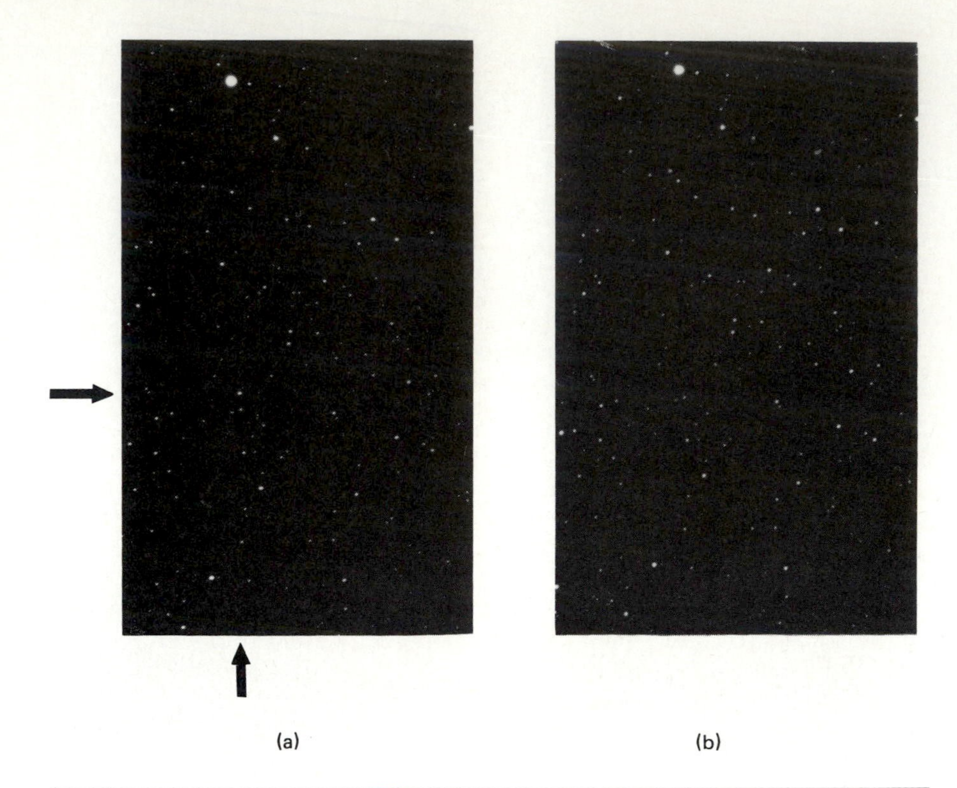

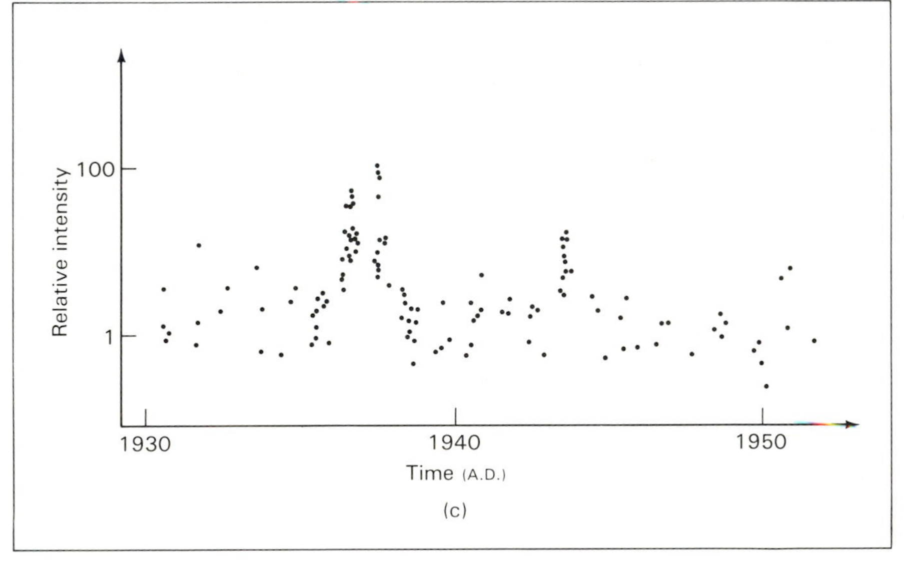

FIGURE 15.18 Two photographs showing (a) quasar 3C279 (at the intersection of the arrows) whose luminosity in 1937 made it the most intrinsically brilliant object known in the Universe, and (b) the same quasar in 1976 when its luminosity was much diminished. Frame (c) shows this quasar's variations observed optically since 1930. Its 1937 outburst gained 3C279 a place in the Guiness Book of Records.

FIGURE 15.19 An *apparent* cluster of five galaxies, four with similarly red-shifted spectral lines, and a fifth (bottom right) with a much different shift.

law. Perhaps they are relatively normal galaxies that have *some extra velocity*. This suggestion permits the quasars to be relatively local, at least within the galaxy horizon. This is a most intriguing prospect, for if the quasars are local, their calculated intrinsic energy outputs are much smaller and thus more easily understood. The mechanism of emission remains of the nonthermal variety, producing lots of polarized radio radiation from compact regions. But the inverse-square-distance law guarantees that the luminosity of a "local" object would be greatly diminished compared to the same object at cosmological distances.

Is there any evidence that the quasars might be local? Well, consider the following puzzle. Figure 15.19 shows an example of several astronomical objects seemingly clustered together. Called Stefan's Quintet, the five objects *appear* to be gravitationally clumped together, thereby forming a small galaxy cluster. Many such cases have been photographed on the sky. And as we might expect, each object in the cluster usually has a similar red shift. But the cluster shown here is unusual. Only four of the galaxies have comparable red shifts; the fifth one has a considerably different shift. Spectroscopic studies of four of the galaxies show them to have radial velocities of about 6000 kilometers/second. Similar studies of the fifth object suggest that it moves at a mere 800 kilometers/second.

Should all five objects in Stefan's Quintet be gravitationally bound in space—that is, genuinely clustered at roughly the same distance from us—the red shift cannot possibly be a correct indicator of distance. Hubble's law would seem to be incorrect—or at least invalid at large velocities. For here is a case, *if* all the objects in Figure 15.19 are really clustered together, where one red shift clearly disagrees with all the others.

Don't be confused here: Such observations do *not* invalidate the Doppler effect although they could invalidate Hubble's law. In this alterative explanation, measurements of red-shifted spectral-line features are still judged as valid velocity indicators. But that velocity may not necessarily indicate distance. In other words, the Doppler effect can still be correct, even though Hubble's law may be incorrect.

A small group of astronomers have argued that the object in Figure 15.19 with the discordant red shift could well have been ejected from one of the other normal galaxies. The discordant red shift of the oddball object would still result from motion, but that motion is theorized to arise from the ejection of one normal galaxy by another normal galaxy. None of the objects would need to be more than a billion light-years distant. If this be the correct solution, the discordant red shift is not at all caused by the grand expansion of the Universe. Hubble's law would break down, and the anomalous galaxy of Stefan's Quintet would *not* be at the distance suggested by its red shift.

Although none of the objects shown in Figure 15.19 is a quasar, the same argument could conceivably be made for quasars as well. If Hubble's law were found to be invalid for a galaxy, it could well be invalid for quasars. Instead of being at cosmological distances, the quasars would be local objects ejected with extra velocities. And if they are local (e.g., within the galaxy horizon), their apparent brightnesses and relatively small distances would imply that their luminosities are much smaller than if they're far away.

For example, if quasar OH471 is 130 million rather than 13 billion light-years away, its deduced luminosity would be reduced by the factor 100^2, from 10^{50} ergs/second to 10^{46} ergs/second. The local hypothesis for quasars would then soften—in fact, eliminate—most problems regarding their emissive qualities. For all practical purposes, the mystique of the quasars and the headaches of the theorists would disappear almost overnight. Should the foregoing reasoning be correct and be applicable to all quasars, we could conclude that quasars are no more exotic than most other galaxies studied previously; they would probably have properties intermediate between the normal and active galaxies.

Most astronomers, however, do not regard the quasars as local. They argue that the foregoing interpretation of the objects in Figure 15.19 is unconvincing. Despite the apparent clustering of five galaxies, the fifth object with the discordant red shift is theorized to reside at a completely different distance from the other four. The oddball object lies only coincidentally along the same line of sight toward the other four clustered galaxies. That is, the apparent clustering of five normal galaxies is an illusion; in reality only the four high-red-shift galaxies are clustered. The fifth is completely unrelated.

Which solution is correct? The answer lies with statistics. How many such groupings can be found on the sky? If we can find only a few such clusters, the apparent association of an abnormally red-shifted galaxy here and there may well be only a coincidence. After all, the Universe contains uncountable myriads of objects, and we should not be surprised if our perception shows some coincidences. However, should we discover a large number of such groupings—in particular, should a cluster of normal galaxies neighbor every quasar—the "local" quasar hypothesis might well be important.

At the present time, observations suggest that normal galaxy clusters do *not* have one or more members with discordant red shifts. In fact, there is a convincing argument *against* the quasars being high-velocity local objects ejected from normal galaxies: No quasar has ever been observed to have a blue shift. Clearly, if the local hypothesis is correct, astronomers should have discovered some blue-shifted quasars. This hypothesis cannot—except for one special case—explain why all the quasars are red shifted, thereby receding without exception from our viewpoint.

The one special case suggests that quasars could have been ejected from the central regions of our Milky Way Galaxy. This radical idea implies that quasars are very local—so local as to be in the halo of our own Galaxy! In some ways this model can justify the quasar red shifts, for we would expect all such quasars to be moving away from us. However, the model cannot in any way justify the amount of energy needed to eject so many objects with such fast velocities from the center of our Milky Way. Besides, if the quasars really are both very local and rapidly moving, we should expect to witness some of this movement from year to year. Yet none of the quasars has ever displayed any noticeable proper motion or change in position.

Another objection to any such local model for quasars is the recent observation that a few quasars seem to be members of galaxy clusters. Also, these quasars have the same red shift as the galaxies with which they are allied. The similarity in both position *and* red shift is not likely to be a coincidence. The implication is that these quasars and their galaxy clusters are gravitationally bound, and together they recede from us with very nearly the same velocity. If this be true, the red shift of these quasars must be a true indicator of Doppler motion. If this motion is in turn a good indicator of distance for the galaxies—and astronomers are confident it is—then it must also be a good indicator of distance for the quasars as well. Furthermore, what's true for a few quasars is probably true for all quasars. Which brings us back to our original dilemma: The high-red-shift quasars are most likely receding at large fractions of the velocity of light, residing at fantastically great distances from us, and emitting extremely large luminosities.

Table 15-2 summarizes the fundamental quasar dilemma: Are they local or are they distant?

GRAVITATIONAL RED SHIFT: For completeness, we should discuss one further possible interpretation of quasar red shifts; this one eliminates the need for both high velocities and great distances. Summarized on the third line of Table 15-2, this model assumes that both Hubble's law and the Doppler effect are incorrect for quasars; as such, the observed red shift says nothing about either velocity or distance. But what could conceivably cause the large red shifts other than velocity? The answer is that a dense and massive object could produce what's known as a **gravitational red shift**.

A sufficiently massive object, having a very large density, can produce a red shift regardless of that object's velocity and distance. Even a nearby object at rest can conceivably cause its spectral lines to be red shifted, provided that it's sufficiently massive and compact. The shift arises because of the work done by the photons of radiation while moving away from the object. Even though photons are massless, the interchangeability of mass and energy assures that they possess some "equivalent mass." This equivalent mass equals E/c^2, where E is the photon's energy and c is the velocity of light. Thus photons lose some of their energy while struggling out of the gravitational force field of any object. And since we argued in Chapter 1 that energy is also proportional to the frequency of radiation, a decrease in energy means a

TABLE 15-2
The Quasar Dilemma

WHAT ARE THEY?	WHERE ARE THEY?	REMARKS
Galaxies plus extra luminosity	Cosmological distances	Doppler effect valid Hubble's law valid Study evolution for a link: quasars → Seyferts → normal galaxies (to be addressed in Chapter 18)
Galaxies plus extra velocity	Local distances	Doppler effect valid Hubble's law invalid Seek physical connection between galaxies and quasars
Not galaxies at all	Unspecified	Doppler effect invalid Hubble's law invalid Speculate wildly

decrease in frequency. Thus photons change their frequency, that is, become red shifted, while escaping from a strong gravitational force field.

In principle, photons emitted by any object are gravitationally red shifted, for they all must work to get away from the object. Whether they originate in a star or a flashlight, photons are affected by gravity. In practice, though, a measurable red shift requires an extremely strong gravitational force field—so strong that only supermassive, superdense objects are expected to produce them.

Are quasars sufficiently massive and dense? Surely they are massive, but probably not overly compact. Their spectral lines strongly imply a diffuse, hot gas whose density is rather low (about 10^5 particles/cubic centimeter). Astronomers infer this from quasar spectral lines, which are too narrow to explain on the basis of gravitational red shift. Thus we are stuck with interpretations that rely on the validity of the Doppler effect. There being no really good alternative explanation for quasar red shifts, they must be attributed to large recessional motions.

A Consensus

So there we stand. If quasars are normal galaxies plus some extra luminosity—that is, at cosmological distances—then their energy output is truly huge and cannot currently be explained. On the other hand, if quasars are normal galaxies plus some extra velocity—that is, at local distances—then their luminosity is less but their great velocities are unexplainable. Either solution presents substantial difficulties.

The current consensus among astronomers is that the quasars really do reside at great cosmological distances. They share in the expansion of the Universe, and they populate the upper end of the Hubble diagram. They are powerful, and they are mysterious. Is there any way to understand them? Quite frankly, if there is, that way is currently unclear. In a nutshell, no one has yet figured out how the radiative equivalent of a thousand normal galaxies can originate from an object having less than one-trillionth the volume of a normal galaxy.

SUMMARY

Active galaxies are among the strangest and most intriguing of all astronomical objects. Their properties differ from those of normal galaxies in nearly every category. The radiation from active galaxies is powerful, polarized, often time variable, and often of the low-frequency type. Though astronomers now have a general idea of the explosive, nonthermal events giving rise to these properties, a detailed understanding of the active galaxies still eludes us.

The extraordinarily powerful quasi-stellar objects are even more troublesome, stretching scientific theory to the limits of human ingenuity. Of all the types of objects in the Universe, the quasars most threaten to render inadequate—even to topple—the laws of physics as we know them. On the other hand, should we eventually come to understand the quasars as truly distant and prodigious emitters, analysis of their radiation might grant us the chance to probe some of the earliest epochs of the Universe—a time now long gone, but a time for which the quasars house hints and clues to their origins, and perhaps to ours as well.

KEY TERMS

active galaxy
core-halo radio galaxy
gravitational red shift
lobe radio galaxy
matter–antimatter annihilation
nonthermal radiation
normal galaxy
quasar
Seyfert galaxy
thermal radiation
universal horizon

QUESTIONS

1. Normal galaxies are huge accumulations of stars. Cite the main observational reason that makes astronomers unsure that active galaxies really have any stars.
2. Draw a labeled sketch of a lobe galaxy, and explain briefly the emission process thought to be operative within this type of active galaxy.
3. Compare and contrast the main properties of Cepheid objects and Seyfert objects, making certain to note their relative sizes, physical makeups, variabilities, positions in space, and ages.
4. Pick one: Electromagnetic radiation from radio galaxies has a (a) thermal spectrum and comes from the galaxy nucleus; (b) thermal spectrum and comes from a tight-knit halo; (c) nonthermal spectrum and comes from the galaxy proper; (d) nonthermal spectrum and comes from extended lobes.
5. Give a basic observational reason why neither stellar fusion nor matter–antimatter annihilation can account successfully for the radiative power of active galaxies.
6. Speculate about the origin of the vast size of the galaxy-like object often found at the center of the rich galaxy clusters (such as the M87 beast at the core of the Virgo Cluster).

*7. In the last few years, improved technology has enabled astronomers to obtain detailed spectra of quasars. Often, in addition to the quasar's emission and absorption lines, another set of absorption lines are observed, typical of those caused by a cloud of hydrogen gas. This "other set" of lines always has a smaller red shift than the quasar's lines. (a) Give two possible reasons suggesting that the cloud of hydrogen gas is closer to us than the quasar. (b) An alternative possibility is that the hydrogen cloud is at virtually the same distance from us as the quasar. If so, how could the difference in red shifts of the cloud and the quasar be explained?

8. Even more than active galaxies, quasars present modern physics with a major energy problem. Describe the two key properties of quasars that lead to this statement, how astronomers know that quasars have these properties, and why it is that *both* properties are required to cause the problem.

9. Can the nonthermal (synchrotron) mechanism explain the emissive properties of most active radio galaxies? How about the most violent active radio galaxies? What problems remain to be explained about these active radio galaxies?

10. Contrast the "local" and "cosmological" theories for quasars. Which is currently favored by the majority of astronomers today? Why?

FOR FURTHER READING

*Brecher, K., "Active Galaxies," in *Frontiers of Astrophysics,* E. Avrett (ed.). Cambridge, Mass.: Harvard University Press, 1976.

Burns, J., and Price, M., "Centaurus A: The Nearest Active Galaxy." *Scientific American*, November 1983.

*Davis, M., "Galaxies and Cosmology," in *Frontiers of Astrophysics*, E. Avrett (ed.). Cambridge, Mass.: Harvard University Press, 1976.

Ferris, T., *The Red Limit*. New York: William Morrow, 1977.

Margon, B., "The Origin of the Cosmic X-Ray Background." *Scientific American*, January 1983.

Osmer, P., "Quasars as Probes of the Distant and Early Universe." *Scientific American,* February 1982.

Verschuur, G., *The Invisible Universe Revealed*. New York: Springer-Verlag, 1987.

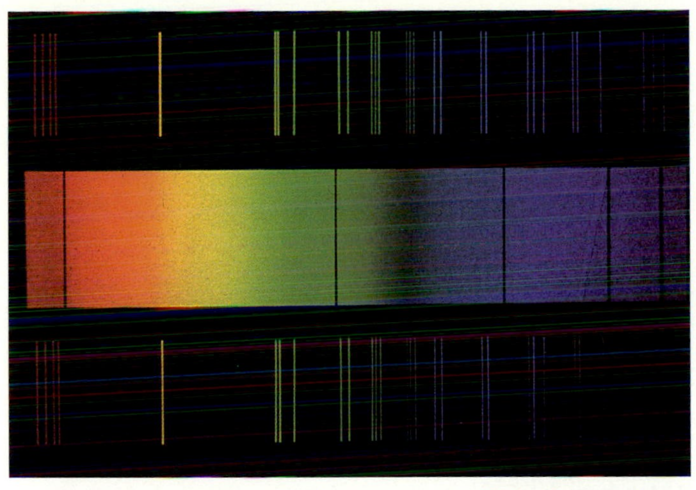

PLATE 1(a)—Continuous color spectrum with hydrogen absorption lines (center); comparison emission spectra, mostly from excited iron atoms (top and bottom).

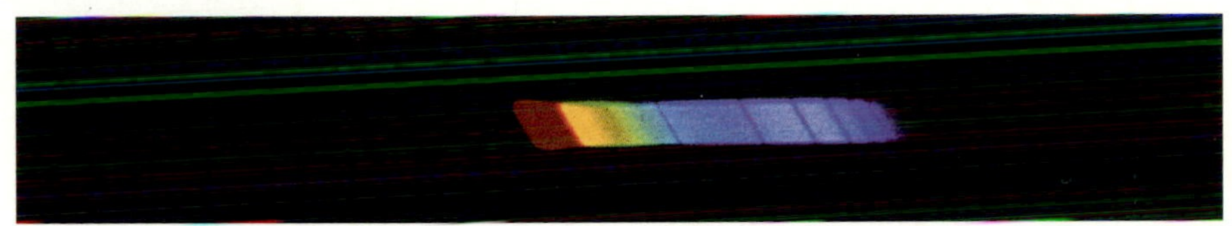

PLATE 1(b)—Color spectrum of the A-type star Vega (Alpha Lyrae) showing dark absorption lines superposed.

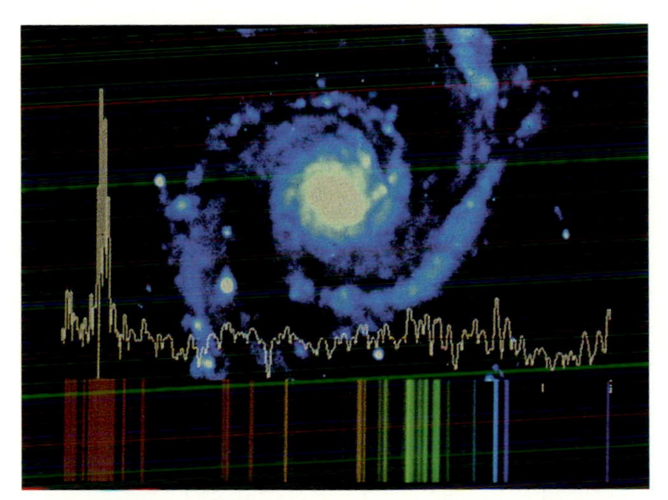

PLATE 1(c)—Emission spectrum from a typical galaxy, with the traced spectrum clearly showing the intensity of the 6563-Angstrom line of hydrogen.

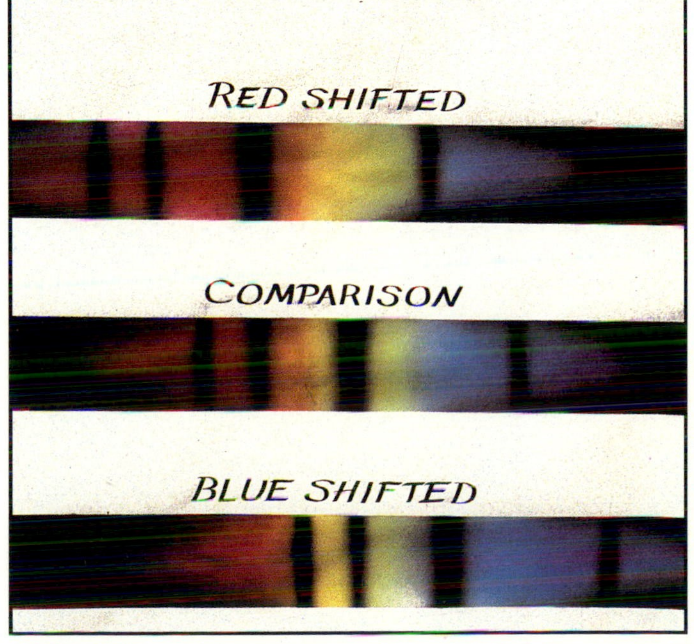

PLATE 1(d)—Artist's conception of a typical dark-line spectrum (center), observed red shifted (top), and blue shifted (bottom).

PLATE 2(a)—The Great Galaxy in Andromeda, M31 (real color).

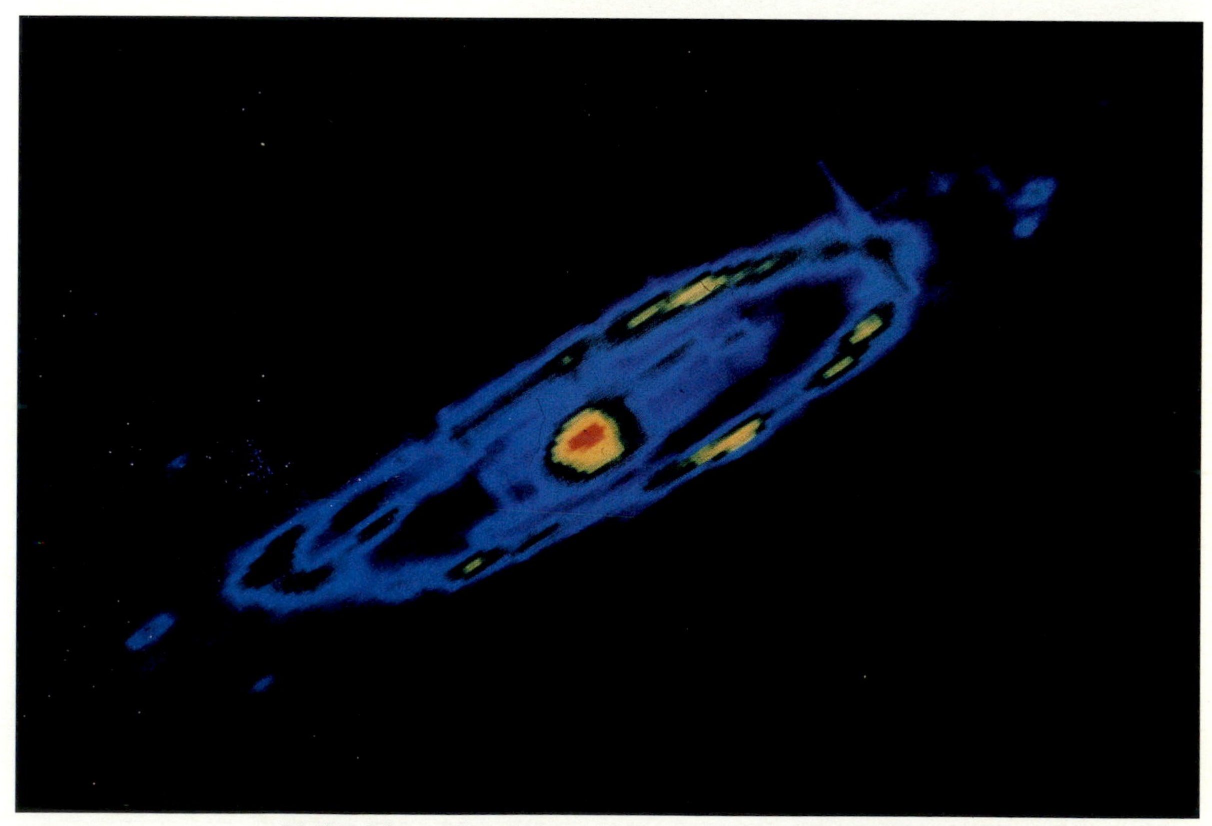

PLATE 2(b)—Infrared image of the Andromeda Galaxy (false color).

PLATE 3(a)—Montage of Jupiter (upper right) and its four Galilean moons: Io (upper left), Europa (center), Ganymede (lower left), and Callisto (lower right).

PLATE 3(b)—The surface of Mars (real color).

PLATE 4(a)—Star fields along some 60 arc degrees of the Milky Way's galactic plane, and centered on the Galaxy's center (real color).

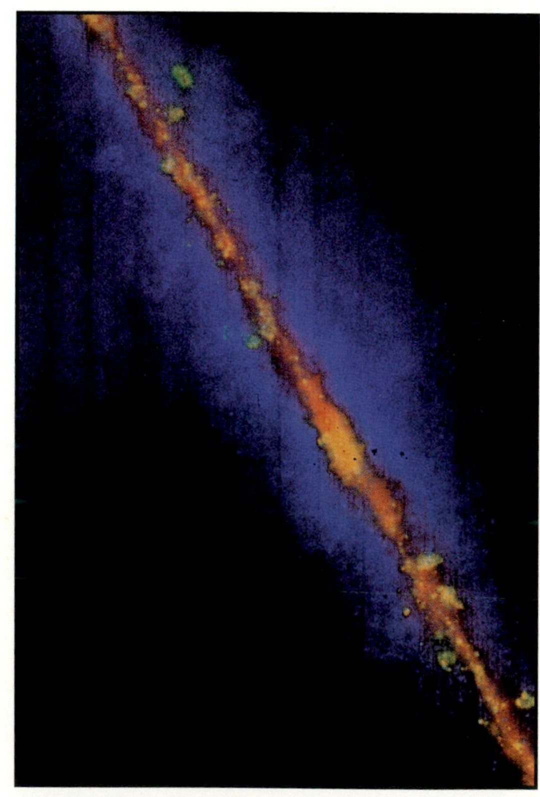

PLATE 4(b)—Infrared image of the same part of the Milky Way's galactic plane as in Plate 4(a) (false color).

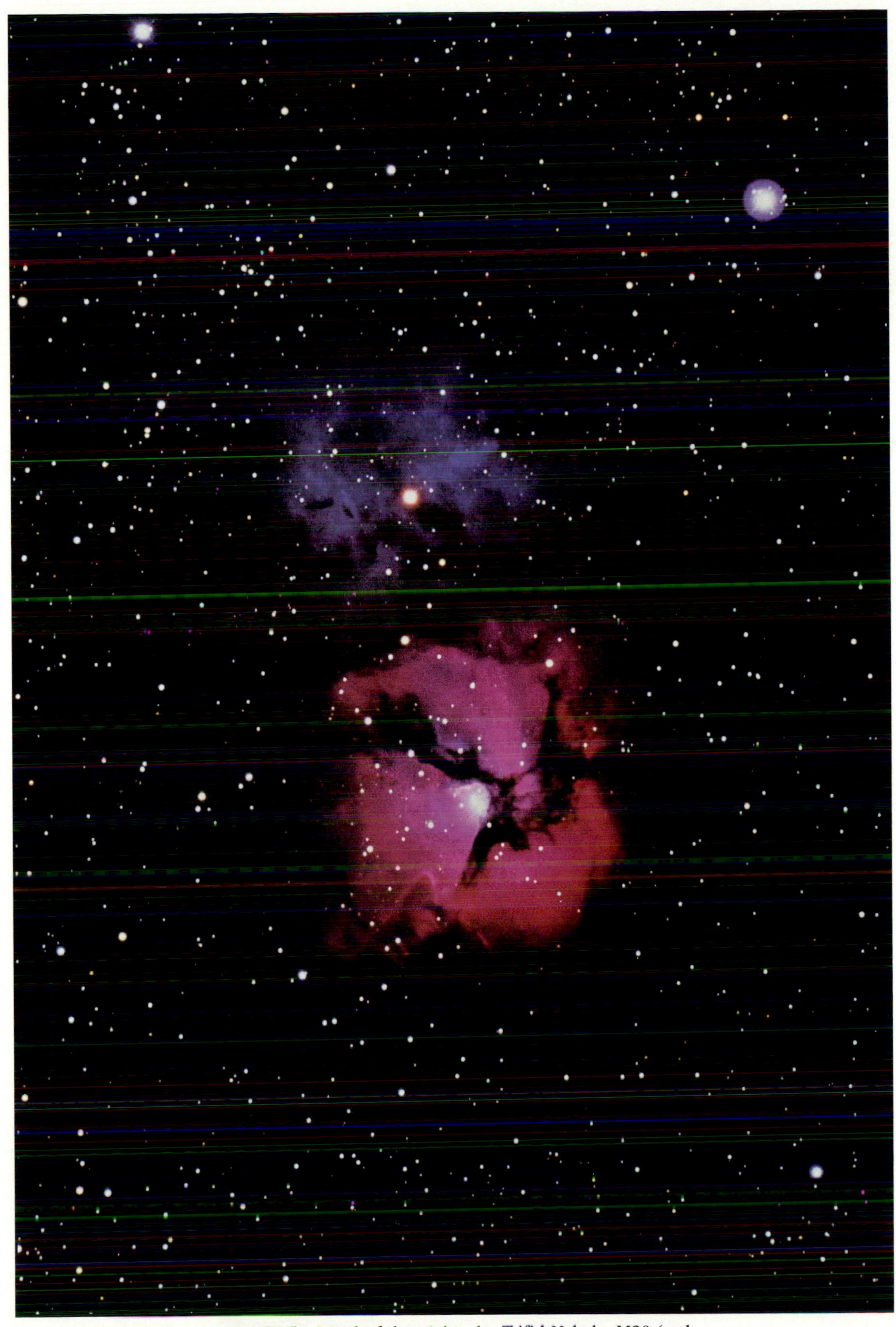

PLATE 5—Jewel of the night, the Trifid Nebula, M20 (real color).

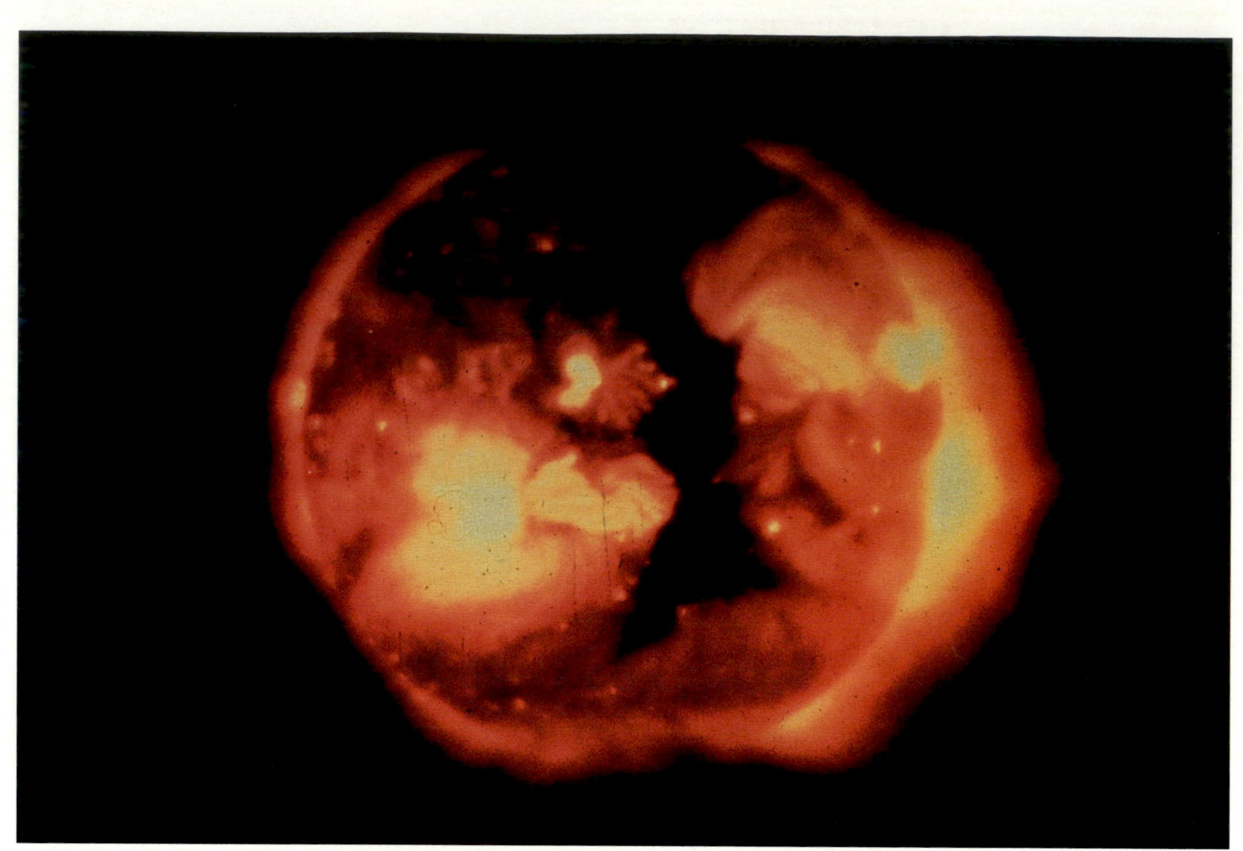

PLATE 6(a)—X-ray image of the Sun (false color).

PLATE 6(b)—Large Magellanic Cloud, showing supernova SN1987A at right center (real color).

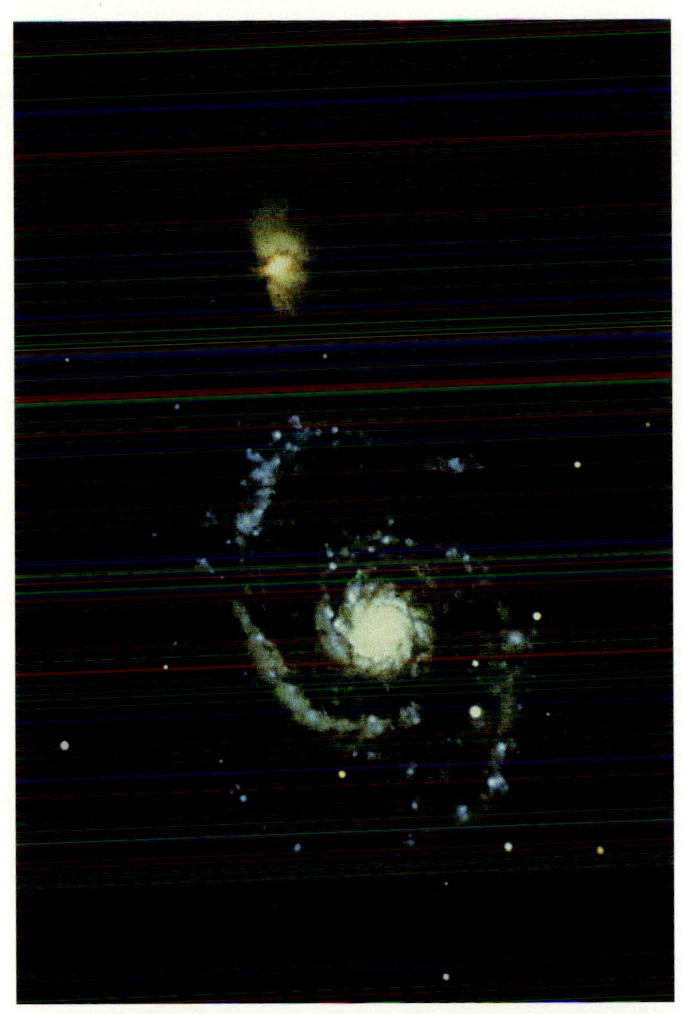

PLATE 7(a)—Whirlpool Galaxy, M51 (real color).

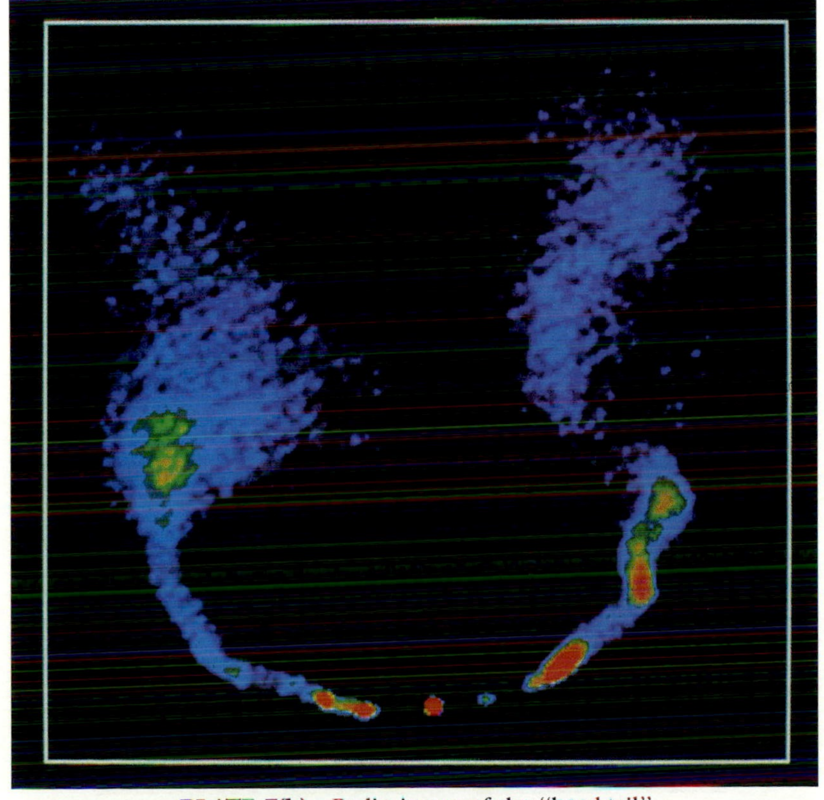

PLATE 7(b)—Radio image of the "head-tail" galaxy, NGC1265 (false color).

PLATE 8(a)—Crab Nebula, M1, a supernova remnant (real color).

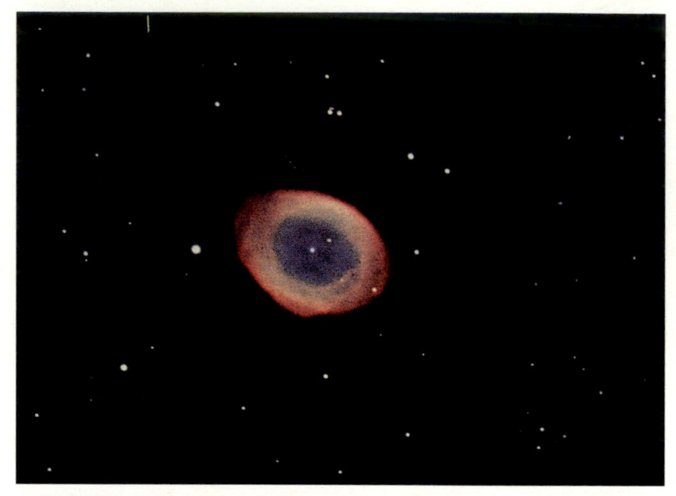

PLATE 8(b)—Ring Nebula, M57, a planetary nebula (real color).

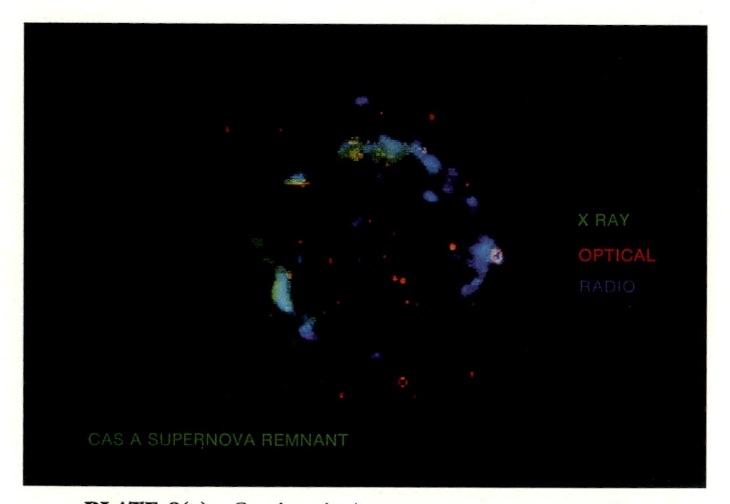

PLATE 8(c)—Cassiopeia A, a supernova remnant is illustrated here as three images—in visual light, radio waves, and x-rays (false color).

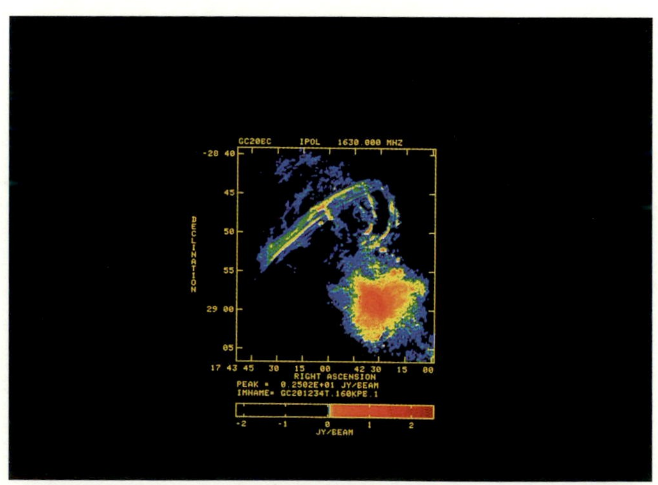

PLATE 8(d)—Radio image of the Milky Way's center, showing a system of filaments extending some 3 light-years across the mid-plane of the Galaxy (false color).

16 UNIVERSE: *Open or Closed?*

We've now reached the limits of the observable Universe, having encountered a wide variety of objects en route. We've studied the bulk properties of planets, stars, galaxies, all the way out to the distant quasars. Here we tackle the entire Universe, describing the central features of the biggest picture of all.

We strive for still more answers. What is the size of the Universe? What is its structure, its shape? What are the origin and fate of the Universe? Such questions comprise the subject of **cosmology**, the study of the Universe on the grandest scale, and especially its origin, evolution, and destiny.

Be sure to place things into proper perspective. When considering the grand properties of the Universe, the smaller contents such as planets and stars—even galaxies, to a certain extent—become irrelevant. To the cosmologist, planets are of negligible importance, stars only point sources of hydrogen consumption, and galaxies mere details. In the much broader context of all space, familiar objects shrink in comparison.

Time also pales in significance when compared to eternity. An interval of a million years becomes a wink of an eye in the cosmic scheme of things. Even a billion years encompasses a rather short duration in the context of all time.

Indeed, all of time plays an important role in this, the last chapter of Part II. The subject of cosmology necessarily involves issues of time, change, and evolution. Thus this chapter forms both a conclusion to the Space Format of the book and an introduction to the Time Format of Part III.

To appreciate cosmology, we must widen our view to include all of space and all of time. In short, now is the time to think big!

The learning goals for this chapter are:

- to know the central tenets of relativity theory
- to appreciate how the notion of curved spacetime can replace the idea of gravity
- to recognize that we do not reside at any special place in the Universe
- to know the age of the Universe and how it's derived
- to understand the various evolutionary models of the Universe
- to realize that astronomers are now actively observing to test different theories of the Universe

Perspective

Cosmic activity permeates our Universe. So does quiescence. Perspective determines which dominates. Examined casually and in bulk, astronomical objects usually display stability. Higher resolution, however, often reveals some violence. Generally, the larger the perspective, the more stable things seem. For example, that our planet is ruptured by quakes and volcanoes is obvious to those of us who live on it and witness its daily activity up close; but from afar (like in those "Earth in space" photos such as Figure 7.20), Earth is a picture of tranquility. Similarly, telescopic studies of our Sun show it to be peppered with bright flares, dark spots, and surface explosions; but to the naked eye, the Sun and most stars assume a rather peaceful, steady appearance (consult Figure 0.2).

We might then expect the whole Universe—the biggest picture of all—to be a pillar of stability. While there are surely pockets of violence tucked here and there throughout the fabric of the Universe, we might anticipate the largest possible perspective to be the epitome of perfect peacefulness. Not so, however.

The entire Universe is not at all calm and stable. The empirical "law" of Hubble strongly suggests that the Universe is expanding. On the largest scale of all, it has some bulk motion, some awesomely ordered activity. To be sure, the Universe is changing with time—in short, evolving.

We'd like to know whether the Universe will continue to expand forever or whether it will eventually stop. If it eternally expands, there will be unimaginable time available for the continued evolution of matter and life. Alternatively, should the Universe contain enough matter, the combined pull of gravity could bring the expansion to a halt, and even reverse it into contraction. Several questions then come to mind: At what time in the future will the Universe cease expanding? If it does start to contract, what will happen upon eventual collapse of the entire Universe? Will the Universe just end as a small, dense point much like that from which it began? Or will it bounce and begin expanding anew? Perhaps the Universe has bounced before. Maybe we inhabit a cyclically expanding and contracting Universe—one having a continuous array of cycles though never a true beginning or end.

These are the basic large-scale fates of the Universe in bulk: It can expand forever; it can expand and then contract to a virtual point and end; or it can cyclically expand and contract indefinitely. Each model represents a working hypothesis—a theory based on available data. But unless we can take the third step in the scientific method and experimentally test these various models, we cannot know which one is correct, if any.

Fortunately, we live at a time when astronomers are actually subjecting the possibilities to observational tests. Our experiments, together with the theories underlying these experiments, seek direct answers to many of the above questions. This is the task we describe in this, the final chapter of Part II.

Relativity Theory

BASIC TENETS: Cosmologists have two primary tools. One is the combined motions of all the galaxies. The fact that the galaxies are observed to recede clearly leads to a model of an expanding Universe. It's not a terribly difficult concept to understand.

The second cosmological tool is **relativity**, a theory of physics that describes the dynamical behavior of matter and energy under peculiar circumstances. Though its details are not easy to understand, the basic concepts of relativity theory are relatively simple. Its foundations are rather straightforward, provided that we are willing to forgo common sense and human intuition.

Relativity is simple in its symmetry, its beauty, its elegant ways of describing grandiose aspects of the Universe. Sure, it employs higher mathematics—advanced calculus and beyond—to quantify its application to the real Universe. Yet everyone should strive to gain at least a nonmathematical feeling for some of the underlying concepts of relativity theory. Such an understanding is the basis for an appreciation, albeit only a qualitative one, for some of the weird effects encountered while studying in Part III the origin of the Universe, the bizarre black holes, the prospects for interstellar spaceflight, and even (in this chapter) models of the entire Universe.

Relativity theory has two main tenets, both proposed in 1905 by a German-Swiss-American physicist named Einstein; together they lead to history's most famous equation, $E = mc^2$, studied earlier. The first tenet maintains that the laws of physics are the same everywhere and for all observers. Regardless of where a person is, or how fast he might be moving, the basic physical laws are identical.

The second tenet of relativity is that there is a fourth dimension—time—which is equivalent to the usual three spatial dimensions. In other words, we are familiar with three dimensions of space; an object's position can be described as either left or right, either up or down, and either in or out. Three dimensions are sufficient to describe *where* any object is in *space*. A fourth dimension of *time* is needed to describe *when*—either past or future—an object exists in that space. By coupling time with the three dimensions of space, Einstein was able to reconcile long-standing, though subtle problems with Newton's ideas about nature; he did so by postulating that the velocity of light is an absolute constant number at all times and to all observers, regardless of when, where, and how radiation is measured. In fact, space and time are so thoroughly intertwined within Einstein's view of the Universe that he urged us to regard these two quantitites not as space *and* time, but as one—**spacetime**.

(The concept of spacetime is not terribly tricky. We use this basic idea often throughout our lives, although we usually don't recognize it. After all, when arranging to meet someone, we must specify not only the place but also the time. Otherwise, we would never rendezvous successfully in order to create the event of meeting at the same place and at the same time.)

Replacement For Gravity

Several important consequences of relativity theory can be qualitatively explained only by analogy. For example, consider the pair of situations illustrated in Figure 16.1. As sketched there, suppose that we are in an elevator having no windows. When it rises, we feel the floor pushing, especially on our feet. It's easy

INTERLUDE 16-1 ***Weird Consequences of Relativity***

When the basic tenets of relativity—especially the speed of light being constant—are carried to their logical conclusions, we find some awfully bizarre predictions. In short, when objects travel at speeds close to light velocity, they do not behave as we might expect from our commonsense experience. That's because common sense corresponds to our everyday familiarity with people, automobiles, trains, aircraft, and the like, all of which move very much slower than light velocity; even our fastest-moving spacecraft travel at speeds of about 100 kilometers/second, which is less than a tenth of 1 percent of light's speed of 300,000 kilometers/second. Instead, a very fast-moving object's size, mass, and time are predicted to change greatly compared to those same properties when at rest.

Take size, for example. Imagine a spaceship cruising horizontally past us at low velocity—that is, with a speed much less than light velocity. In principle, we could devise an experiment to measure the length of the spaceship. Provided that the experiment is done carefully, our estimate of the spaceship's length, while passing us rather slowly, would equal that measured by the travelers on board the spaceship. Everyone agrees, and common sense prevails, because the ship's speed is said to be "nonrelativistic."

On the other hand, should the spaceship pass us at a much faster velocity, say with a motion comparable to light velocity, then peculiar ("relativistic") effects occur. For example, our estimate of the ship's length would disagree with that measured either by the on-board travelers or by us when the spaceship cruised by more slowly. Instead, we stationary observers would measure a smaller spaceship; its length would have shrunk. The greater the ship's velocity, the greater the amount of shrinkage. This is not an optical illusion. It is a genuine shortening of the spaceship in its direction of motion—not an overall shrinkage of the entire ship's size; just in the direction of motion.

The most intriguing aspect of this weird effect is that the effect itself is relative—that is, the peculiar consequences depend on point of view. For example, the travelers aboard the spaceship measure no change in the length of their spaceship. Regardless of what we measure as their ship's speed, the travelers themselves have zero velocity with respect to their own spaceship. After all, they are riding in it. To them, the spaceship as well as any objects within it obey the commonsense notions familiar to us all. But once the travelers observe objects outside their spaceship, their viewpoint changes: They realize their ship's great speed, and they attribute peculiar effects to the outside objects. In particular, as the travelers peer through their ship's windows at the rest of us standing stationary, they would measure all objects to be smaller in the direction of the ship's motion. In fact, *they* would perceive *us* as thinner individuals!

You may want to ask: Who is correct? Which objects are really shortened, ours or theirs? The answer, according to Einstein's relativity theory, is that both observers are correct. The weird consequences of relativity theory apply to any and all observers, provided that there is a very large relative velocity between them. The central issue is one of relative velocity—hence the name "relativity theory."

Note that these special relativistic effects occur only when relative motions are comparable to light velocity, that is, at least some 250,000 kilometers/second. Even velocities as large as half that of light, namely 150,000 kilometers/second, are insufficient to produce any noticeable shortening in the direction of motion.

Length is not the only physical property changed by very high velocities. From our stationary viewpoint, relativity theory predicts that the spaceship's mass increases. Even the travelers aboard become more massive, and the faster the ship goes, the more massive they and their ship become—relative to our perspective. As with length, however, the physical mass is unchanged to the on-board travelers. Since they have zero velocity relative to their own ship, they would measure their ship and everything on board to be completely normal, and not at all overweight. It is *they* who measure *us* to have become more massive.

Time is also altered by very high speeds. Provided we, the stationary observers, could see a clock aboard the fast-moving spaceship, we would measure that clock to be ticking more slowly than an identical clock beside us. In the hypothetical case of the spaceship traveling *at* the velocity of light, we would not be able to detect any passage of time on the ship's clock. Relative to us, time on the spaceship would have stopped—not only mechanical clocks but presumably biological ones as well. We could then claim, from our viewpoint, that the travelers had achieved immortality!

These effects, bizarre though they seem, are not idle predictions of an untested theory. Although we cannot observe spacecraft having speeds close to light velocity, our civilization has built devices capable of boosting elementary particles to fantastically high velocities. For example, laboratory accelerators such as the one shown in Figure 2.2 can boost electrons to velocities equal to approximately 0.999999999992 times the velocity of light. As a result, from our viewpoint outside the accelerator, the relative motion is very great and the electron should gain in mass. Indeed, such super-swift electrons have been measured to be nearly 100,000 times heavier than their normal mass when at rest. And this agrees precisely with that predicted by Einstein's theory.

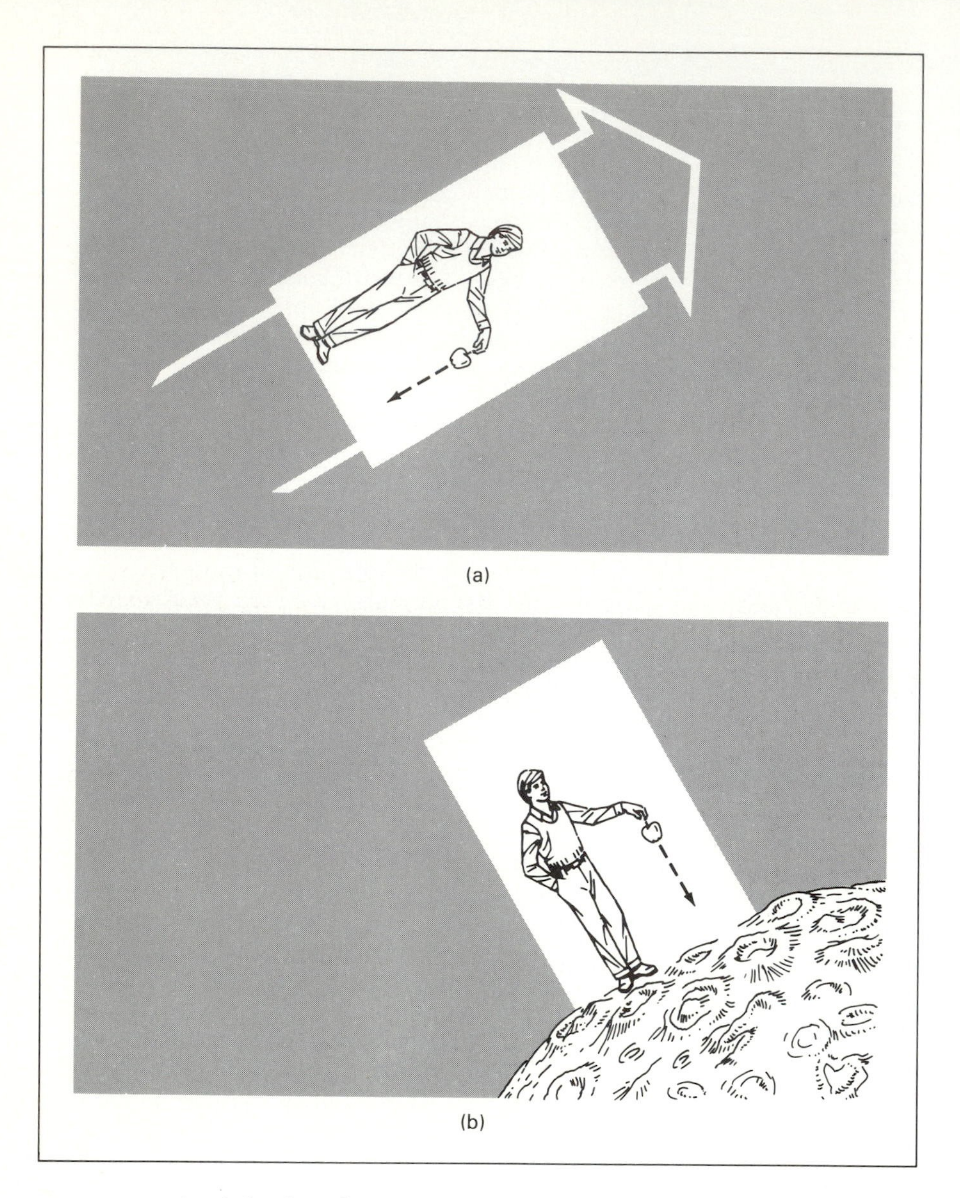

FIGURE 16.1 A windowless elevator accelerating (a) through outer space in the absence of gravity is indistinguishable from one at rest (b) in the presence of gravity.

to attribute this pushing sensation to the upward acceleration of the elevator.

Now imagine that such a windowless elevator exists in outer space far from Earth. Normally, we would experience the weightlessness made familiar by watching the astronauts floating around where there are no net gravitational force fields. But if we *did* experience a sensation of pushing on our bodies, we could draw one of two conclusions: We could argue that the elevator is accelerating upward in the absence of gravity, thus pinning us to the floor. Or we could maintain that the elevator is at rest in the presence of gravity, which is pulling us from below. In the first case, the rapid acceleration causes the sensation of pushing (or pulling); in the second case, it is gravity that actually does the pulling.

Without looking at objects outside the hypothetical elevator, there's no way to determine which of these explanations is correct. If we could look out, we would have no trouble establishing whether the elevator is really at rest or really accelerating. *Relative* to the Universe outside the elevator, we could easily assess the real status of that elevator.

The important point is this: The effect of gravity on an object and the effect of acceleration on an object are indistinguishable. In other words, uniform acceleration imitates completely the behavior of a uniform gravitational force field. Scientists call this the **principle of equivalence**: Gravity and the acceleration of objects through spacetime can be viewed as conceptually and (almost) mathematically equivalent. Consequently, Einstein postulated as unnecessary the Newtonian idea of gravity as a force that pulls. Not only is that idea unnecessary, but Newton's theory is also today known to be less accurate than Einstein's. Let's briefly see how the notion of acceleration can replace the commonsense idea of gravitation.

Relativity predicts that mass alters the nature of spacetime. Bypassing the details, we can say that mass curves spacetime. In other words, matter itself effectively shapes or "warps" the geometry of spacetime. To appreciate these ideas, recall our earlier statements about curved space in Interlude 4-1. Remember, in particular, that ordinary Euclidean geometry is valid when the extent of curvature is zero—that is, when spacetime is flat. Even when spacetime is only slightly curved, Euclidean geometry of flat space is approximately correct. Thus, in the absence of matter, the curvature of spacetime is zero, and an object in flat space moves uniformly in a straight line. Newtonian dynamics and Euclidean geometry are quite satisfactory, for all practical purposes, wherever spacetime is not appreciably curved. Flat space is not a hypothetical situation, for beyond the galaxy clusters little or no matter presumably exists, thus ensuring only slight spacetime curvature.

On the other hand, the geometry of spacetime is strongly warped near massive objects. It's not the object or the surface of the object that's warped—just the near-void of spacetime in which the object is embedded. The larger the amount of matter at any given location, the larger the extent of curvature or warp of spacetime at that location. Furthermore, the extent of warp lessens at greater distances from a massive object. As with gravity, both the amount of matter and the distance from that matter specify the magnitude of spacetime curvature. But since this view of warped spacetime is more accurate than the conventional view of gravity, to be precisely correct in all possible situations, the world view of Newton must be replaced by that of Einstein.

Yet, surely you ask, How can a curve replace a force? The answer is that the geometry of spacetime influences celestial travelers in their choice of routes much as Newton imagined gravity to hold an object in its path. Just as a pinball cannot traverse a straight path once shot along the side of a bowl, so the shape of space causes objects to follow curved paths. And any object whose motion changes direction, even though its speed might remain steady, is said to be accelerated.

For example, as is depicted in Figure 16.2(a), Earth accelerates while orbiting the Sun—not because of gravity, as Newton maintained, but because of the curvature of spacetime, as Einstein preferred. (As inhabitants of Earth, we don't feel our planet accelerating, but it is since it constantly changes its direction of motion.)

To appreciate this Einsteinian view of spacetime, consider an analogy. Not an example, an analogy. Imagine a pool table with a playing surface made of a thin rubber sheet rather than the usual hard felt. As suggested in Figure 16.2(b), such a rubber sheet would become distorted should a large weight be placed on it. A heavy rock, for instance, could cause the sheet to sag or warp. The otherwise flat rubber sheet would become curved, especially near the rock. The heavier the rock, the greater the curvature. Trying to play billiards, we would quickly find that balls passing near the rock are deflected by the curvature of the tabletop.

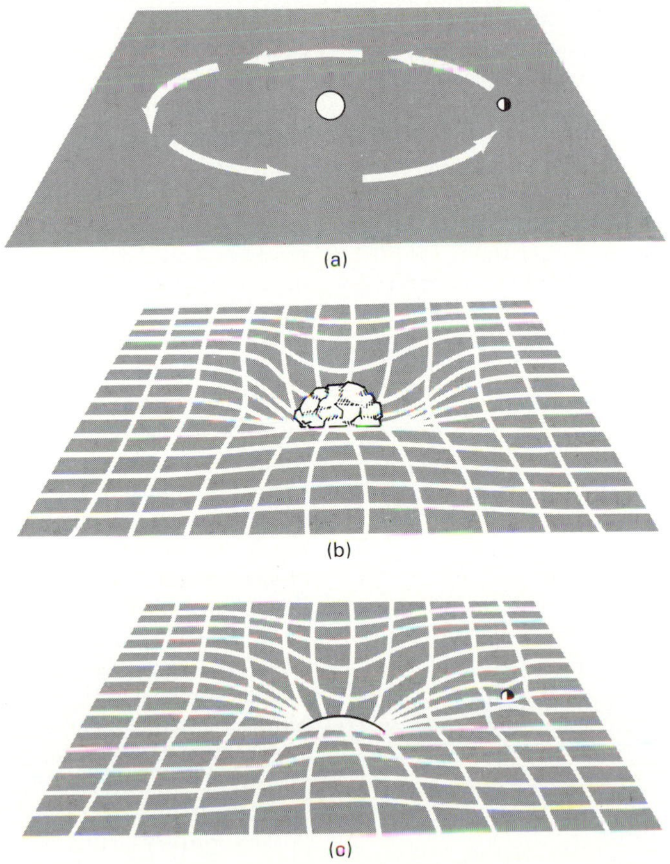

In much the same way, radiation and material objects are deflected by the curvature of spacetime near massive objects. As suggested in Figure 16.2(c), Earth is deflected from a straight-line path by the slight spacetime curvature created by our Sun. The extent of the deflection is large enough to cause our planet to circle repeatedly, or orbit, the Sun. Similarly, the Moon or a baseball responds to the spacetime curvature created by Earth, and they too move along curved paths. The deflection of the Moon's path is not too large, causing our neighbor to orbit Earth endlessly. The deflection of a small baseball is much larger, causing it to return to Earth's surface.

The commonsense notion of gravity, then, is just a convenient word for the natural behavior of objects responding to the curvature of spacetime. Accordingly, we can use a knowledge of spacetime to predict the motions of objects through space and time. More powerfully, we can turn the problem around: By studying the accelerated motion of any object, we can learn something about the geometry of spacetime near that object.

And so it is with the whole Universe. When we seek the size, shape, and structure of the entire Universe—the biggest picture of all—we need to take account of the net effect of spacetime curvature caused by each and every massive object in the cosmos. By studying the accelerated motions of various parcels of matter within the Universe, we can learn much about the net curvature

(a)

(b)

(c)

FIGURE 16.2 Spacetime can be visualized to be curved near the massive Sun (c) in much the same way that a rubber sheet curves when a heavy rock is placed on it (b). The rock and Sun themselves are represented by the unshaded area in each case. The response of a billard ball to the rock's dimple in the rubber sheet, or of the Earth to the Sun's warp in real spacetime, is much like the conventional view of our planet orbiting the Sun under the common-sense influence of gravity (a).

of the whole Universe. This we shall do later in this chapter.

SPACETIME CURVATURE: To illustrate further the curvature of spacetime, consider the following hypothetical case. Suppose that three planets are inhabited by equally advanced civilizations capable of launching rockets. Earth can be one, since we now have the required technology. Mars and Jupiter can be the other two planets. Of these, Mars is the least massive and Jupiter the most massive. Suppose furthermore that the inhabitants of each planet possess identical rockets. And for sake of discussion, let's assume that these rockets can achieve only a fixed amount of thrust (energy) at launch, after which they glide freely through space.

When the rockets are launched from each of the three planets, the shapes of their paths differ. In the Newtonian view of space, the rocket paths are determined by the gravitational interaction between the rocket and each planet. In the Einsteinian view of spacetime, these paths are determined by the response of the rocket to the spacetime warp created by each planet.

Figure 16.3(a) depicts a possible path of a rocket launched from the most massive planet, Jupiter. As shown, the initial thrust was chosen in this case to be large enough to place the rocket into an elliptical orbit. Like gravity, whose strength decreases with increasing distance from a massive object, the curvature of spacetime is also greater close to the massive planet. The rocket accordingly speeds up (or accelerates) when close by, and slows down (or decelerates) when far away. General relativity thus agrees with the Keplerian laws of planetary motion studied in Chapter 7; relativity maintains that rockets move faster (accelerate) when close to massive objects because the rockets actually respond to the greater degree of spacetime curvature there.

An ellipse, a "closed" geometric path, is only one possible type of motion. It's the trajectory of minimum energy, so labeled because a rocket in such an orbit does not have enough energy to escape the planet's influence.

Rockets can have other paths as well. Figure 16.3(b) shows a typical path taken by an identical rocket after launch from planet Earth. The same thrust used to launch the Jupiter rocket into an elliptical orbit is now great enough to propel the rocket entirely away from Earth. Less energy is used in the launch from Earth than from Jupiter, and thus more energy can be imparted to the motion (i.e., kinetic energy) of the Earth rocket. The rocket literally escapes the influence of Earth because, as a Newtonian classicist would say, Earth has less gravitational pull than Jupiter. Alternatively, Einsteinian relativists maintain that such a rocket escapes from Earth because our planet Earth warps spacetime much less than does Jupiter. The two views—Newtonian and Einsteinian—predict almost identical paths for the rocket as it recedes toward regions of spacetime progressively less curved by planet Earth.

The path shown in Figure 16.3(b) is called a parabolic trajectory. This is the type of flight path taken by human-made spacecraft that have been probing other planets of our Solar System in recent years; such a trajectory exemplifies a more "open" geometry. An object traveling along a parabolic path has more energy than one on an elliptical trek. This is true either because the initial thrust needed to achieve a parabolic trajectory is large or because the mass of the parent object from which the launch is made is small. In the particular example considered here, the rockets are identical, so the increased energy of the parabolic case results from the comparatively small mass of Earth.

Even while receding far from its parent planet, a rocket is still affected by the pull of gravity (or the warp of spacetime) created by the mass of that planet. Although large only in the immediate vicinity of the planet itself, Earth's influence over the rocket never diminishes to zero. Mathematical analyses predict that in the idealized absence of all other astronomical objects, such a parabolically launched rocket should approach infinity—and will have a velocity of zero when it gets there. (Theoretically, the rocket just barely makes it and then stops! But since nothing can ever *really* reach infinity, this is tantamount to saying that the rocket will continue to recede forever.)

The parabolic path contrasts

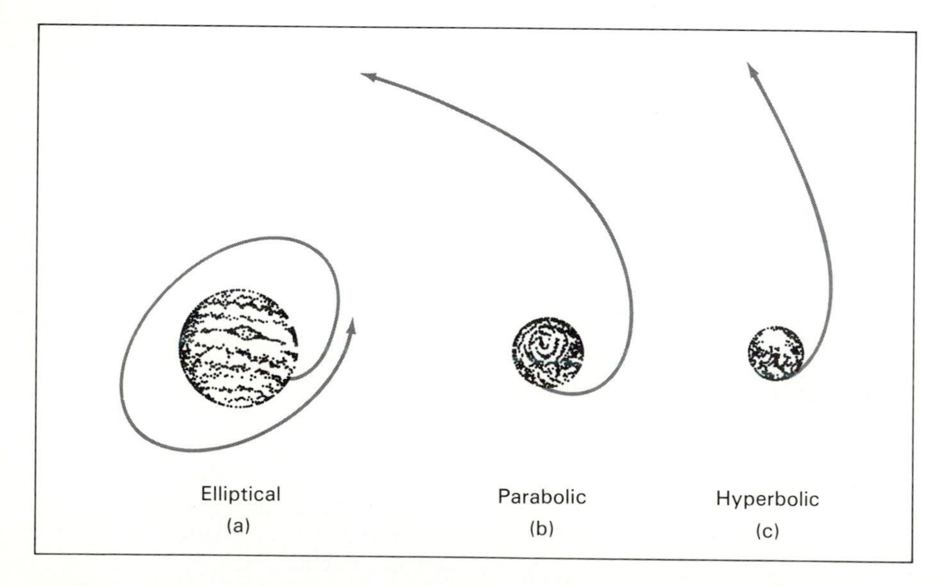

FIGURE 16.3 The paths of identical rockets launched from three different planets—(a) a very massive Jupiter, (b) a moderately massive Earth, and (c) a less massive Mars—can be explained either in terms of gravitational pulls of the planets on the rockets, or in terms of spacetime curvature that causes the rockets' paths to change.

slightly with another type of trajectory conceivably taken by a rocket. Figure 16.3(c) shows this third path, also "open" in form and called an hyperbola. In our example it's the path taken by a rocket launched from the least massive of the planets considered, Mars. The hyperbolic path closely mimics the parabolic path, but they differ a little in energy content. Since less energy is used to launch identical rockets from Mars than from Earth, the receding Martian rocket has more energy of motion. And with the least amount of gravity or spacetime warp near Mars, the hyperbolically launched Martian rocket approaches infinity with little hesitation. (Having more energy than the parabolically moving rocket, the hyperbolically moving rocket will theoretically reach infinity with some finite velocity larger than zero. The rocket will not stop there, so says the mathematics; it will continue moving beyond! In reality, no object can reach infinity, let alone go beyond. Apart from the academic language of the mathematician, then, you should simply recognize that parabolically and hyperbolically moving objects recede *toward* infinity forever.)

The three cases above are convenient ways to describe the motion of any object in terms of its energy content and of its response to the spacetime curvature. These will be useful analogies when we consider the essentials of cosmology, for then the "object" will be the entire Universe itself. We shall return to these ideas after the next section.

Cosmological Principle

By infusing the basic tenets of relativity throughout a full-blown (mathematical) development of the theory, researchers have learned to map the varied ways that matter warps spacetime. This is the area where relativity becomes notoriously complex; in this book what we glean from their ponderous calculations can only be subjective. The results in a nutshell are the so-called Einstein field equations—a group of a dozen or so equations that must be solved simultaneously to determine how the Universe is structured.

Although on the one hand these equations are extremely hard to solve, on the other hand they contain remarkable symmetry. Much like works of art, they often inspire a sense of wonder, a certain awe. The complexity arises largely because, in addition to the field equations specifying the geometry of the Universe, the relativist must also solve several other formulas—called the geodesic equations—to determine the response of individual objects to the curvature of spacetime at any location in the Universe. In the case of the planets, that object might be a rocket like the one in the preceding section; in the case of the entire Universe, our view necessarily broadens and the object becomes, for example, a galaxy.

A STATIC SPHERE: Since Einstein created relativity, he was better positioned than anyone else to apply his equations to deduce the nature and structure of the cosmos. In 1917, his field equations predicted that the whole Universe is indeed largely curved. He found that the flat geometry of Euclid just won't do when studying the bulk properties of the whole Universe. Unfortunately, Einstein's solution can be cast only in terms of unimaginable four-dimensional spacetime. It's fully imaginable mathematically, but quite unimaginable conceptually.

The essence of his solution can nonetheless be visualized by using another analogy. No one has ever built a viewable model of anything in four dimensions, so in this analogy we suppress one of those four dimensions. For sake of argument, suppose that we consolidate the three dimensions of space into only two dimensions of space. Then, with time as the remaining dimension, we can construct a three-dimensional analog of Einstein's four-dimensional Universe.

Figure 16.4 is a sketch of our analogy—a three-dimensional sphere. Here, all of space is spread *on the surface* of this sphere. In other words, all three dimensions of space have been consolidated into two dimensions, and these two dimensions exist on the surface of the sphere. The remaining dimension—time—is represented by the radius of the sphere.

Be sure to realize that in this analogy, the Universe and its contents should *not* be considered to be distributed inside the sphere. Rather, they are distributed *just on its surface*. All three dimensions of space are warped—in this special case, into a perfect sphere—because of the net influence of all the matter within every astronomical object. Thus all the galaxies, stars, planets, life forms, and even all the radiation exist only on the surface of the sphere of this modeled Universe.

Now, since the radius of this model sphere represents time, we are forced to conclude that this spherical analog grows with time. After all, as discussed in previous chapters, we observe the galaxies to be receding. As time marches on, the radius of the sphere increases and so does its surface area (much like an inflating balloon). In this way, our three-dimensional analog agrees with the observational fact that the Universe is expanding.

Actually, in 1917, Einstein didn't know that the Universe is expanding. Hubble and other astronomers did not establish the recession of the galaxies until the 1930s. Einstein's own field equations had allowed an expansion (or contraction) of the Universe, but he didn't believe it. He was probably fooled by the (then) still-popular Aristotelian philosophy that few things change. So he tinkered with the field equations, introduced a fudge factor that just offset the predicted expansion, and thus forced the model Universe to remain static. Einstein was wrong in doing this; indeed, he later declared it to have been the biggest mistake of his scientific career.

This error did not prevent Einstein and other relativists from uncovering many notable features of curved spacetime. One of the most important features is termed the **cosmological principle**—the notion that all observers perceive the Universe in roughly the same way re-

FIGURE 16.4 A sphere is one way to visualize a model of the entire Universe. Here, the three dimensions of space are consolidated onto the surface of the sphere, while the fourth dimension, time, is represented by the radius of the sphere.

gardless of their actual location. Put slightly more technically, the Universe is presumed homogeneous and isotropic.

To grasp the cosmological principle, consider a sphere again. It can be any sphere, so let it be Earth. Imagine ourselves at some desolate location on Earth's surface, for instance in the midst of the Pacific Ocean. To make the analogy valid, we must confine ourselves to two dimensions of space; we can look east or west, and north or south, but we cannot look up or down. This is the life of a fictional "flatlander"—a person who can visualize only two dimensions of space. Perceiving our surroundings, we note a very definite horizon everywhere. The surface *seems* flat and pretty much the same in all directions (even though we know it really to be curved). Accordingly, we might get the impression of being at the center of something. But we're not really at the center of Earth's surface at all. *There is no center on the surface of a sphere.* Such is the cosmological principle: There is no preferred, special, or central location on the surface of a sphere.

Similarly, regardless of our position in the real four-dimensional Universe, we observe roughly the same spread of galaxies as would be noted by any other observer from any other vantage point in the Universe. Despite our observation that galaxies surround us in the sky, this doesn't mean that we reside at the center of the Universe. In fact, if our spherical analogy is valid, there is no center of the Universe. Nor is there any edge or boundary. The case of a flatlander roaming forever on the surface of a three-dimensional sphere is completely analogous to a space traveler (or any radiation) voyaging through the real four-dimensional Universe. None of them ever reaches a boundary or an edge. Proceeding far enough in a given direction on the surface of the sphere, the traveler (or radiation) would eventually return to the starting point, just as Magellan's crew proved long ago by circumnavigating planet Earth.

In much the same way, if four-dimensional spacetime is structured according to this spherical analogy—and it might be—then a beam of light or an astronaut could conceivably travel one direction, only to return at some future date from the opposite direction. Figure 16.5 shows the bizarre case of an observer shining a light beam in one direction; if he waits long enough, that same beam might eventually hit him in the back of the head.

AN EXPANDING SPHERE: Today, we know that the Universe is not at all static. The recessional motions of the galaxies make indisputable the fact that the Universe is expanding. Led by the 1920s efforts of the Russian meteorologist Friedmann and the Belgian priest Lemaître, modern relativists seek more realistic models of the Universe by coupling the Einstein field and geodesic equations together with the observed rate of universal expansion. In this way, observations of galaxy recession become a boundary condition—a constraint helping us to refine our latest cosmological models.

Note that the cosmological principle is valid even though the Universe is expanding; no surface of any expanding sphere, like that of any static sphere, has a center, edge, or boundary. To see this, imagine a sphere again, although now one that can swell. For example, visualize the entire Earth to be expanding, causing the surface area of our planet to increase as time progresses. Standing on such a hypothetically expanding "Earth" we would see familiar objects moving away; all surface objects, whether they be trees, houses, or mountains, would appear to recede. Now, more than ever, we may want to conclude that our position is special—that we exist at the center of some explosion. But we do not. Our position is no more special than anyone else's on the

FIGURE 16.5 If the geometry of spacetime is warped as originally envisioned by Einstein, then a beam of light, launched in one direction, might return someday from the opposite direction.

sphere's surface. In fact, everyone everywhere on the expanding surface would observe their surroundings to be receding. Who is correct? Everyone is correct. Recessional motions are observed from *any and all* positions on the surface of an expanding sphere.

Figure 16.6 further illustrates this all-important cosmological principle. Each frame shows a group of spirals painted onto the surface of a balloon. The spirals are meant to represent galaxies, whereas the balloon represents the Universe. Together the frames mimic a movie—a hypothetical film strip of numerous galaxies at successive times in the history of the Universe. Concentrate on the 3 spirals in the frame at the left. Imagine yourself as a resident of one of these spirals and note your position relative to other nearby spirals. As the balloon inflates and the frames are progressively followed from left to right, the other galaxies appear to move away as the Universe expands.

Here's the important point: Regardless of which galaxy we inhabit, we would note that all the other galaxies are receding. To appreciate this, focus on any other spiral in Figure 16.6. Now again trace the positions of all the spirals while viewing the frames from left to right. The galaxies appear to recede for any observer in the Universe. Nothing is special or peculiar about the fact that all the galaxies are receding from us. This is the case for *all* observers everywhere. Such is the cosmological principle: No observer anywhere in the Universe has a privileged position.

And so it is in the real four-dimensional Universe. Despite the fact that we observe galaxies receding while gazing out into space, this is not a peculiarity of our vantage point. All observers anywhere in the Universe would see essentially the same sort of galaxy recession. Neither we nor anyone else reside at the center of the expanding Universe. Indeed, there is no real center in *space*. There is no position that we can ever hope to identify as the location from which the universal expansion began.

There is nonetheless a center in *time*. This is the origin of time, and it corresponds in our three-dimensional spherical analogy to the sphere's having zero radius. In other words, at the beginning of the Universe, the three-dimensional sphere was a point. It had a radius of zero. This was the origin of time. We can think of it as the edge of time. But there is no edge in space.

Finally, recognize that no one really knows if the cosmological principle is absolutely correct. Astronomers have adopted it as a working hypothesis, largely because it greatly simplifies the Einstein field equations. All that we can say is that the cosmological principle seems consistent with all observations made thus far.

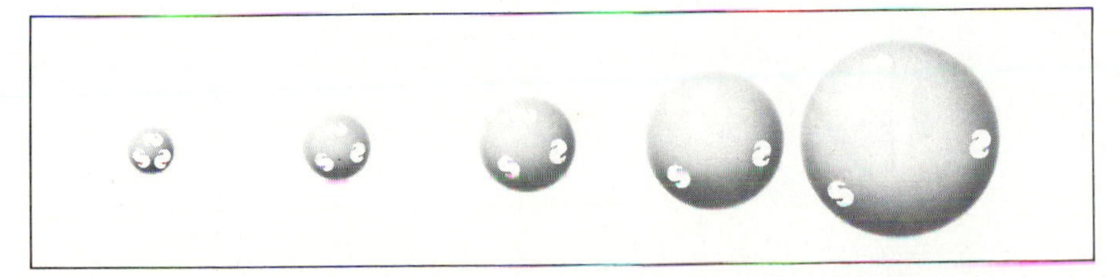

FIGURE 16.6 All the spirals painted on the surface of a spherical balloon recede from one another as the balloon inflates from left to right. Similarly, galaxies appear to recede from one another regardless of the galaxy inhabited. As they do recede, the distance increases between any two galaxies, and the rate of increase is proportional to the distance apart. Thus, our analogy agrees with Hubble's law in that velocity is proportional to distance.

Age of the Universe

Visualizing the past when the sphere was much smaller, we can ask an obvious question: When did the sphere have zero radius—a mere point? In other words, how long ago were all the contents of the Universe squashed into a single point? More fundamentally, when did time begin?

To appreciate answers to these questions, imagine that time can be reversed. Mentally reverse the expansion of the Universe by contracting it at the same rate as we currently observe it expanding. The galaxies would come together, eventually they would touch, and finally they would mix. If we can estimate how long it would take for the whole Universe to shrink to a single point, we'll then have a measure of the age of the Universe.

Actually, we can quantify this problem in very simple terms. Since the distance traveled by any object equals the product of its velocity and the time traveled, Hubble's law of Chapter 14 can be reexpressed as

$$\text{velocity} = \text{Hubble's constant} \times \text{velocity} \times \text{time}.$$

Canceling the velocity terms from each side of the equation, we find that

$$\text{time} = \frac{1}{\text{Hubble's constant}}.$$

The inverse of Hubble's constant is thus a measure of the total time interval during which the cosmic objects have been receding from one another. It's the time needed for the expelled debris of universal matter to reach the places at which they are now observed. This amount of time is then the answer to the fundamental question: How long ago in the distant past was all the matter in the Universe concentrated within a single, small region of space? It is a measure of the duration of the Universe's expansion—quite simply, the age of the Universe.

Numerically, the inverse of Hubble's constant yields a value of approximately 15 billion years. Thus the singular, compact region of space often associated with the origin of the Universe must have "exploded" about 15 billion years ago.

The range of possible error in this age is considerable since Hubble's constant is not known precisely. As noted in Chapter 14, its value is affected by the motion of our Local Group, which is currently uncertain and especially hard to measure accurately. In fact, the rate of galaxy recession depends critically on the rather uncertain distances of the farthest galaxies, which even the largest telescopes have trouble observing. (The velocities of galaxies, near and far, are by comparison relatively easy to measure.) Currently, some researchers argue that the Universe could be as young as 10 billion years, while others maintain it is nearly as old as 20 billion years. An error of several billion years may seem large, but the difference between these extremes is only a factor of 2, really quite good for an order-of-magnitude subject like cosmology. As a compromise, we adopt 15 billion years as the approximate age of the Universe—a remarkable finding in and of itself.

Evolutionary Models

At the origin of time, the Universe began to expand. Like air inflating a balloon, the galaxies recede, and the Universe expands. Figure 16.7 graphs this simple fact. By plotting the size of the Universe against cosmic time, we can graph the widest possible temporal perspective. By "size of the Universe" we mean either (in principle) the total four-dimensional region of spacetime in which the galaxies reside, or (in practice) the average distance separating the galaxy clusters. Either notion is valid, but only the latter can be observationally measured.

The curve drawn in Figure 16.7 suggests two basic facts: The Universe began with an explosion of some sort, popularly called the **big bang**, after which its size increased with time. As noted earlier, the Universe expands at a rate that depends on the density of matter contained within it. After all, each clump of matter in the Universe gravitationally pulls on all other clumps of matter. Since this gravitational force field is always attractive, it tends to counteract the expansion. So a tightly packed Universe, namely one that's reasonably dense, causes a strong net gravitational pull, and a potentially significant slowing of the universal expansion. Even a rather tenuous Universe houses huge quantities of matter and would therefore be expected to eventually slow the rate of expansion, hence curve the solid line as drawn in Figure 16.7.

(Note that we have returned to the notion of gravity. Although warped spacetime is a more valid concept, we'll use the more familiar term "gravity" whenever it makes the argument easier to understand.)

The phenomenon of universal expansion somewhat resembles the three rocket cases considered earlier in this chapter. Each rocket's escape from its parent planet was governed by the planet's mass. Earth, for example, pulled strongly on the launched rocket and was able to slow the rocket's escape; a less massive planet, Mars, slowed the rocket even less; the most massive planet, Jupiter, exerted the strongest pull on the rocket and was able to halt its escape. The parallel between the orbital motion of a small rocket and the cosmic dynamics of the entire Universe is quite a good one. As for rockets, there are three possible models of a

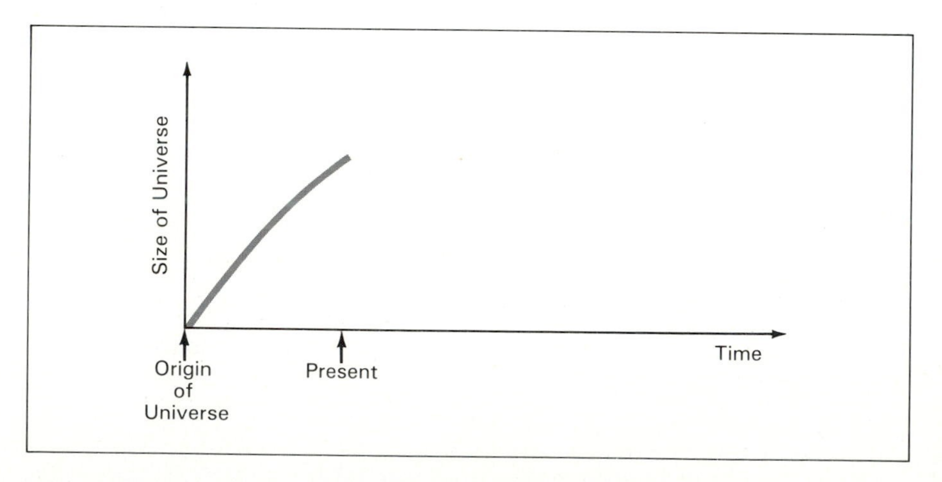

FIGURE 16.7 A graph of the size of the Universe plotted against cosmic time. The drawn curve represents the expansion of the Universe from its origin to the present.

dynamic, changing Universe. We now consider each of them in turn.

AN OPEN UNIVERSE: The first model Universe is one that evolves from a powerful initial "bang"—an explosion of some sort that occurred at the origin of time. The Universe then expanded from what must have been an extremely dense primeval clump of energy. As time progressed, space diluted the primeval stuff within the Universe, causing the average density of matter to decline. In this first model, there is insufficient matter to gravitationally counteract the expansion. Accordingly, the Universe simply expands forever, with the density of matter thinning out eventually to nearly zero. This type of possible Universe is analogous to the rocket of Figure 16.3 moving away from Mars; there is insufficient mass to halt the outward motion of either the rocket or the Universe. As this Universe will theoretically arrive at infinity with some finite (nonzero) velocity, some researchers refer to this case as the hyperbolic model of the Universe. Represented diagrammatically in Figure 16.8, its spacetime is curved like that of a saddle (see Interlude 4-1).

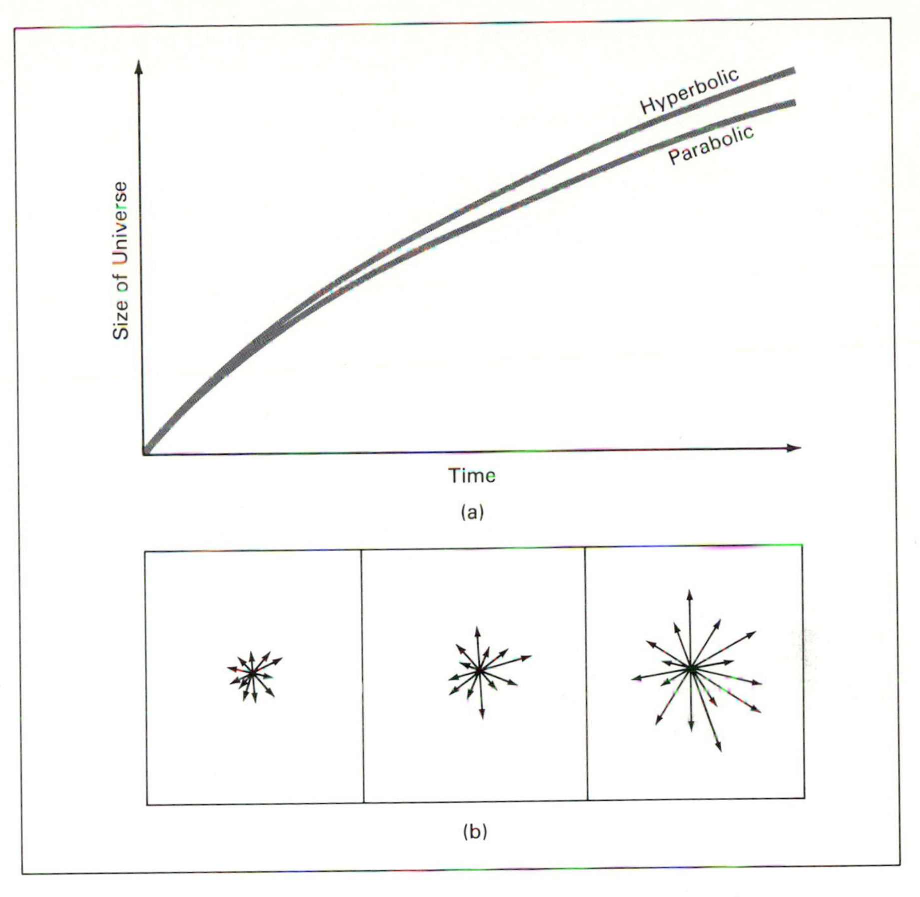

FIGURE 16.8 Both the hyperbolic and parabolic galaxy motions imply an open Universe. As illustrated by the curves in (a), an open Universe will expand forever. The frames below in (b) diagram the change of this type of Universe with time.

A hyperbolic model is said to imply an **open universe**. It's "open" in the sense that the initial bang was large enough, and the contained matter thin enough, to ensure that this type of Universe will never stop expanding. Despite the fact that matter everywhere pulls on all other parts of the Universe, this type of Universe will never collapse back on itself. There's simply not enough matter.

Of course, the Universe can never really become infinitely large. An infinite amount of time is required to reach infinity. This is just the mathematician's way of stating that the hyperbolic, or open, Universe will continue expanding forever. Properly stated, an open Universe *approaches* infinity.

Should this model be correct, the galaxies will recede forever. With time, for an observer on Earth, they will literally fade toward invisibility, their radiation becoming weaker with increasing distance. Eventually, even some of the closest galaxies will be so far away as to be hardly visible. Someday, all the galaxies might become unobservable; they will be too old and too far away, their radiation too faint. Our Milky Way Galaxy will then be the only matter in the observable Universe. All else, even through the most powerful telescopes, will be dark and quiet. And even beyond that in time, the Milky Way too will someday peter out as the hydrogen in all its stars changes into heavy elements. This type of Universe, including all of its contents, eventually experiences a "cold death." All the radiation, matter, and life in such a Universe are destined to freeze.

Another open universal model is possible, this one closely mimicking the parabolic trajectory of the Earth-launched rocket discussed in Figure 16.3. In this case, the accumulated matter is just sufficient eventually to halt the expansion—but only after reaching infinity. This type of Universe also expands forever, though it barely makes it to infinity before running out of energy. But since nothing can ever really reach infinity, this parabolic model, like the hyperbolic model, comprises an open Universe. It will expand forever, as shown in Figure 16.8. Its net spacetime is uncurved, that is flat, as described in Interlude 4-1.

A CLOSED UNIVERSE: There is yet another plausible model for the Universe. Like the open cases just considered, this model expands with time from a superdense original point. Unlike the open cases, however, this model contains enough matter to halt the universal expansion before reaching infinity. That is, the outwardly expelled matter has been losing momentum ever since the initial bang, so much so that the galaxies will eventually stop at some time in the future. Astronomers everywhere—on any planet within any galaxy—would then announce that the galaxies' radiation is no longer red shifted. The cosmological principle guarantees that this new view will prevail everywhere. The bulk motion of the Universe, and of the galaxies within, will be stilled—at least momentarily.

The expansion may well stop, but the inward pull of gravity does not. Gravitational attraction is relentless. Accordingly, this type of Universe will necessarily contract. It cannot stay motionless. Nothing fails to change. Astronomers everywhere will then announce that all galaxies have blue shifts.

In broadest terms, the contraction of this type of Universe is a mirror image of its expansion. Figure 16.9 shows its general dynamics. Not an instantaneous collapse, it's rather a steady movement toward an ultimate end, requiring just as much time to fall back as it took to rise.

This type of model in many ways resembles our earlier Jupiter analogy, for which the gravitational pull was great enough to cause the rocket path to become elliptical. Having a similar geometrical pattern, a universal model containing enough matter to reverse the expansion is often called an elliptical Universe. We also sometimes refer to it as a **closed universe**—closed because it represents a Universe finite in size and in time. It has a beginning and it has an end. Its net spacetime mimics that of a sphere, as discussed earlier in this chapter and in Interlude 4-1.

The expansion–contraction scenario of a closed Universe has many fascinating (and dire) implications. From what must have been a huge value initially, density thins to a rather small value by the time the Universe begins contracting. Thereafter, the density returns to its huge value when, at some future epoch, all the matter collapses onto itself. Life, which has evolved steadily from simplicity to complexity during the expansion, will begin breaking down into simplicity again while inevitably heading toward its demise during the contraction. Why? Because toward the end of the contraction phase, the galaxies will collide frequently as the total amount of space in which they exist diminishes. Just as compressing the air in a bicycle pump or rubbing our hands causes heating via friction, collisions among galaxies will generate heat as well. The entire Universe will grow progressively denser and hotter as the contraction approaches the end. Near total collapse, the temperature of the entire Universe will have become greater than that of a typical star. Everything everywhere will have become bright, obeying the Planck curve for any thermal emitter—so bright, in fact, that stars themselves will cease to shine for want of contrasting darkness. This type of Universe will then shrink toward the superdense, superhot state of matter called a **singularity**, much like the one from which it originated. In contrast to the open Universe that ends as a frozen cinder, this closed Universe will experience a "heat death." Its contents are destined to fry.

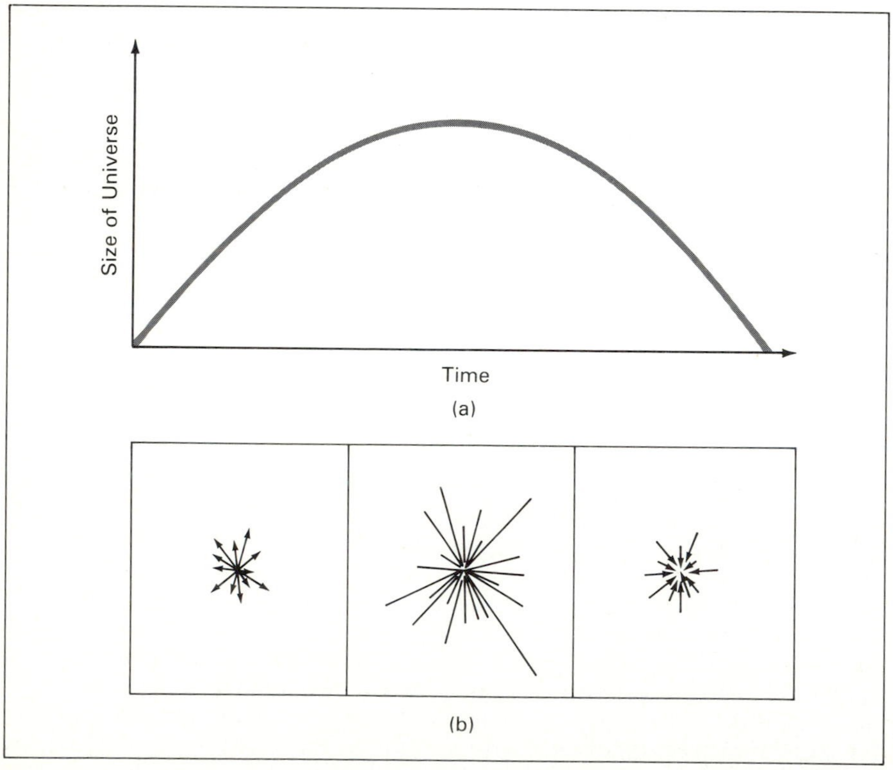

Size of Universe

Time

(a)

(b)

FIGURE 16.9 The elliptical or closed universal model (a) has a beginning, an end, and a finite lifetime. The frames in (b) further illustrate this type of Universe.

AN OSCILLATING UNIVERSE: Cosmologists are uncertain of the fate of a closed Universe on reaching the singularity. The Universe might simply end. Or it might bounce—into another cycle of expansion and contraction. Frankly, the mathematics of singularities are not yet understood. This ultimate state of matter poses one of the hardest problems in all of science. For the most part, scientists are experimentally and theoretically ignorant of the physics of singularities. Here's the problem.

With both density and temperature increasing as the contraction nears completion, pressure—the product of density and temperature—must increase phenomenally. The question as yet unanswered is: Will the Universe just end as a final minuscule speck, or will this pressure be sufficient to overwhelm the relentless pull of gravity, thereby

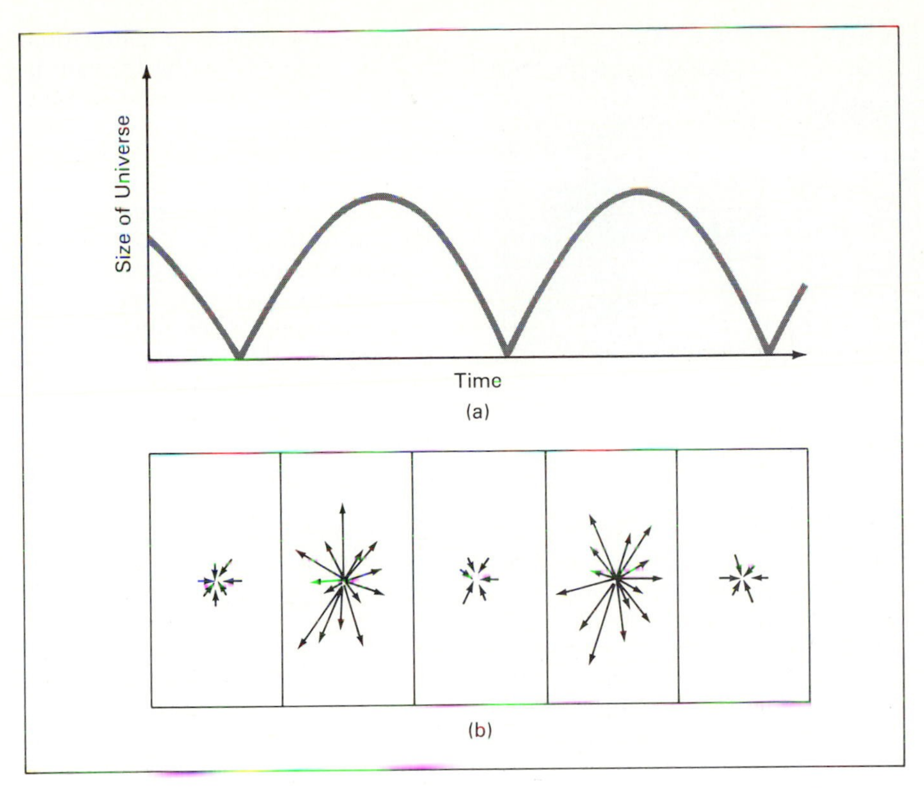

FIGURE 16.10 An oscillating Universe (a) compromises the basic features of the open (indefinitely long) and closed (beginning and end) models. The frames in (b) further illustrate this grand evolutionary scheme having neither beginning nor end.

pushing the Universe back out into another cycle of expansion and contraction? In other words, will a closed Universe bounce?

Such a cyclical Universe having many—perhaps an infinite number of—cycles of expansion and contraction is often termed an **oscillating universe**. Figure 16.10 diagrammatically illustrates this type of Universe, in some ways a hybrid of the open and closed models.

Many researchers see aesthetic beauty in this type of model Universe. Here, there is no need for a unique, once-and-for-all-time explosion—no need for a *big* bang. Nor does this model have a need for a definite beginning or a definite end. The oscillating model merely goes through phases, each initiated by a separate explosion or bang. Indeed, the oscillating model has many "bangs," each expansion a "day," each contraction a "night." But none of these bangs is unique, apparently none of the origins any more significant than any other. Furthermore, the idea of oscillation avoids the potential philosophical hangup of wondering about what preceded a unique big bang of a one-cycle closed Universe or of an open Universe.

Should the oscillating model be valid, we need not trouble ourselves with the concept of "existence" before the beginning of time. Such a Universe always was and always will be.

Other Models

All the foregoing models of the Universe stipulate evolution as their central concept—the Universe changes with time. They are derived from Einstein's theory of relativity, and they are favored in one form or another by the great majority of today's cosmologists. However, several other Universe models have been proposed during the past few decades. Most of them do not follow directly from relativity; some do not even call for change with time or embrace evolution as their guiding principle. Let's consider one of the more prominent ones, for until recently it was favored by some parts of the scientific community.

The "steady-state" model stipulates not only that the Universe appears roughly the same for all observers, but also roughly the same *throughout all time*. Its basic assumption is often called the *perfect* cosmological principle: To any observer at any time, the physical state of the Universe is the same. In other words, the average density of the Universe remains eternally constant. It holds *steady*.

The steady-state model was conceived as a powerful alternative to the various time-dependent or evolutionary models of the Universe. Its initial motivation was based as much on philosophy as on science. The oscillating Universe aside, many scientists and philosophers were (and still are) unwilling to concede that nothing could have existed prior to the unique big bang. To ask what preceded the origin of the Universe is a tricky question indeed. What existed prior to the big bang? Why was there a big bang? What or who caused it? These are questions unaddressable within the realm of modern science. When there are no data, the scientific method becomes a useless technique. Philosophies, religions, and cults of all sorts can offer theories to the *n*th degree, but science can say nothing.

The steady-state model avoids these thorny questions, as does the oscillating model. For these models, there is no beginning, and there is no end. The Universe just *is* for all time.

Steady-state cosmologists concede that the Universe is expanding, for they are unable to refute the observational fact that the galaxies are receding. But since the idea of an initial explosion is unacceptable to them, they are forced to propose that an unknown repulsive force pushes the galaxies apart. Even so, the steady-statists demand that the bulk view of the Universe—the average density of matter and the average

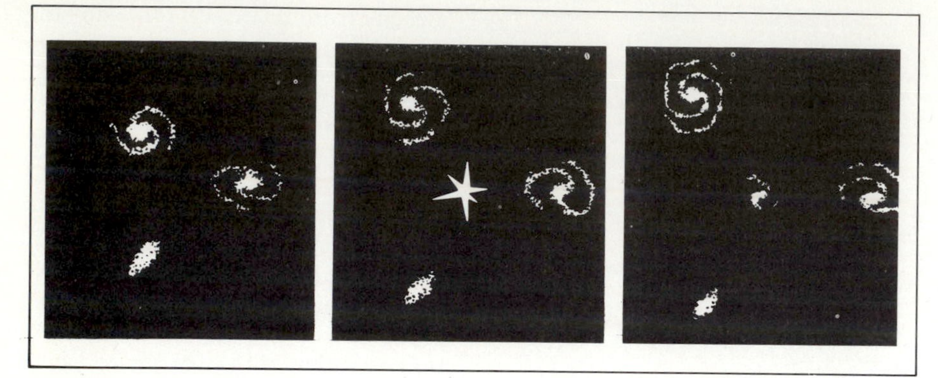

FIGURE 16.11 Given the recession of the galaxies, the steady-state model of the Universe requires the continual creation of new matter in order to keep the bulk appearance of the Universe steady for all time. Although the Universe expands—shown in these frames from left to right—the emergence of new matter (center frame) helps to keep constant the average distance between galaxies in the future (right) as in the past (left).

distance between galaxy clusters—remains constant forever. Accordingly, to offset the dilution of the density due to the galaxies' recession, the steady-state model requires the appearance of new matter in the Universe. As odd as it may seem, the steady-state theorists propose that this new matter is created from nothing.

Figure 16.11 suggests that despite the recession of the galaxies, the creation of additional galaxies in just the right amount can keep constant the number of galaxies per unit volume. In this way, the average density of matter in the Universe is preserved forever.

The major problem with the steady-state model is its failure to specify how the additional matter is created. Nor does it specify where. Some researchers theorize its infusion out beyond the galaxies in intergalactic space, whereas others prefer injection within the centers of galaxies. Not much new matter is needed to offset the natural thinning as the galaxies speed apart. Creation of a single hydrogen atom in a volume equivalent to that of the Houston Astrodome every few years would do it. Unfortunately, the sudden appearance of such a minute quantity of matter, either inside or outside of galaxies, is currently quite impossible to detect, and therefore to test.

Regardless of *where* matter is created, the real problem is *how* it's created. The sudden appearance of new matter from nothing violates one of the most cherished concepts of modern science—the conservation of matter and energy. A basic principle of modern physics maintains that the sum of all matter and all energy is constant throughout the Universe. Matter may be indeed created from energy (and energy from matter), but it's very hard to understand how that matter can be created spontaneously from nothing at all.

The grand puzzle of the steady-state model, then, is the process of material creation. Still, the lure of a Universe that always has existed and always will exist is strong, for it provides a way to skirt the need for a unique big bang and all the other sticky questions regarding the very start of an evolving Universe. All things considered, the big-bang theory is as troubling for a steady-state cosmologist to swallow as is this continual-creation idea for an evolutionary cosmologist. At any rate, we'll note in the next section that observations during the past decade have virtually eliminated the steady-state theory as a feasible model of the Universe.

Cosmological Tests

How can we distinguish among these various possible models of the Universe? The most straightforward way uses the process of elimination.

The steady-state model can be ruled out for at least two reasons. First, the spread of galaxies, especially the quasars, does not seem uniform throughout space. Provided that the red shift is a true distance indicator, the number of faraway quasars far outnumber those nearby. If any of us could have lived 10 billion years ago, when quasars were presumably the dominant cosmic objects, our view would have been filled with quasars—many more than now surround our vantage point at Earth. The perfect cosmological principle is clearly violated: The large-scale view of the Universe was not the same eons ago as it is now.

COSMIC BACKGROUND RADIATION: A second, rather fatal argument against the steady-state model accidentally resulted from an experiment. Observations made with radio telescopes always yield a signal regardless of the time of day or night. Unlike optical observations for which we sometimes encounter a complete void of light toward the dark, obscured regions of space, radio observations never fail to detect some radiation. Sometimes the radio signal is strong, especially when the telescope is aimed toward an obvious source of radio emission. At other times it's weak, particularly in regions devoid of all known radio sources. Yet, whenever the accumulated emission from all the known radio sources and from all the atmospheric and instrumental noise is accounted for, there remains a minute radio signal—sort of a weak hiss, much like static on a home AM radio or the "snow" seen on an inactive TV channel. Never diminishing or intensifying, this weak signal is detectable any time of the day, any day of the year, year after year; it apparently inundates all of space. What's more, it's equally intense in any direction of the sky—that is, isotropic. The whole Universe is seemingly awash in this feeble (low-frequency) radiation.

This omnipresent signal was discovered in the 1960s with the radio

FIGURE 16.12 This "sugarscoop" antenna, originally built to communicate with Earth orbiting satellites, was used to detect the 3-kelvin cosmic background radiation.

antenna shown in Figure 16.12, during an experiment designed to improve America's telephone system by using satellites. In their data, scientists unexpectedly noticed a bothersome radio hiss that just wouldn't go away. Unaware that they had detected a signal of cosmological significance, they sought many different origins for the excess emission, including atmospheric storms, ground interference, equipment short circuits, even pigeon droppings deposited inside the horn-shaped antenna. Later conversations with theorists enlightened the experimentalists as to the static's most probable origin. That origin is the fiery creation of the Universe itself.

This weak, isotropic radio radiation is widely interpreted as a veritable "fossil" of the primeval explosion that began the universal expansion long ago. The relic hiss is often termed the **cosmic background radiation**. Its existence is completely consistent with any of the evolutionary models of the Universe, but there is no clear role for it in the steady-state model.

The cosmic background radiation is presumed to be a remnant of an extremely hot phase of the early Universe—a Universe that has since cooled during the past 15 billion years or so. Regardless of whether the initial explosion was a unique big bang producing an open and infinite Universe, or even one of repeated bangs typifying a closed, finite, and oscillating Universe, the primeval superhot, superdense matter must have emitted (high-frequency) thermal radiation. In its fiery beginnings, the Universe almost certainly released extremely energetic gamma-ray radiation. But with time, the Universe expanded, thinned, and cooled. Consequently, the emitted radiation must have shifted from the gamma-ray and x-ray varieties normally associated with hot matter, down through the less energetic ultraviolet, visible, and infrared kinds, eventually becoming the radio waves usually released by relatively cool matter. Figure 16.13 illustrates the theoretically expected change of the Planck curve representing the average heat of the entire Universe.

The evolutionary models predict that some 15 billion years after the primeval explosion, the average temperature of the Universe—the relic of this big bang—should be quite cold, in fact no more than a few kelvin. To confirm the theory, researchers have carefully measured the intensity of the weak isotropic radio signal at many wavelengths. Figure 16.14 plots the results of these observations. The dashed line is the Planck curve that best fits all the data acquired during the past decade; confirming the theory, the curve shows a universal temperature of approximately 3 kelvin. Furthermore, this cosmic radiation really does seem to pervade the whole Universe, including Earth, the building and the room where you are now reading this. The amount of cosmic radiation absorbed by humans at any one time, however, is minuscule, totaling less than a millionth of the power emitted by a 100-watt light bulb.

So the existence of the cosmic background radiation, together with the spread of galaxies in space, discredit the steady-state theory as a feasible model of the Universe. Clearly, the Universe has changed with time; it hasn't been steady at all. The choice of the correct Universe model must then be made among the various evolutionary models. Other data are needed to sift through each of them.

CRITICAL DENSITY: The most straightforward way to distinguish between the open and closed models requires an estimate of the average density of matter in the Universe. More than anything else, density is what differentiates the closed model, wherein matter is packed tightly enough to gravitationally halt the universal expansion before reaching infinity, from the open model, wherein the density simply isn't large enough to bring the Universe back.

Researchers have computed the precise density of matter needed to halt the expansion just as the outer limits of the Universe reach infinity. For today's thinned-out Universe, the answer is roughly 10^{-30} gram/cubic centimeter. That's extraordinarily tenuous matter, equaling hardly more than 10^{-6} hydrogen atom/cubic centimeter (or equivalently 1 atom/cubic *meter*—or the density of a few atoms within a volume the size of a typical household

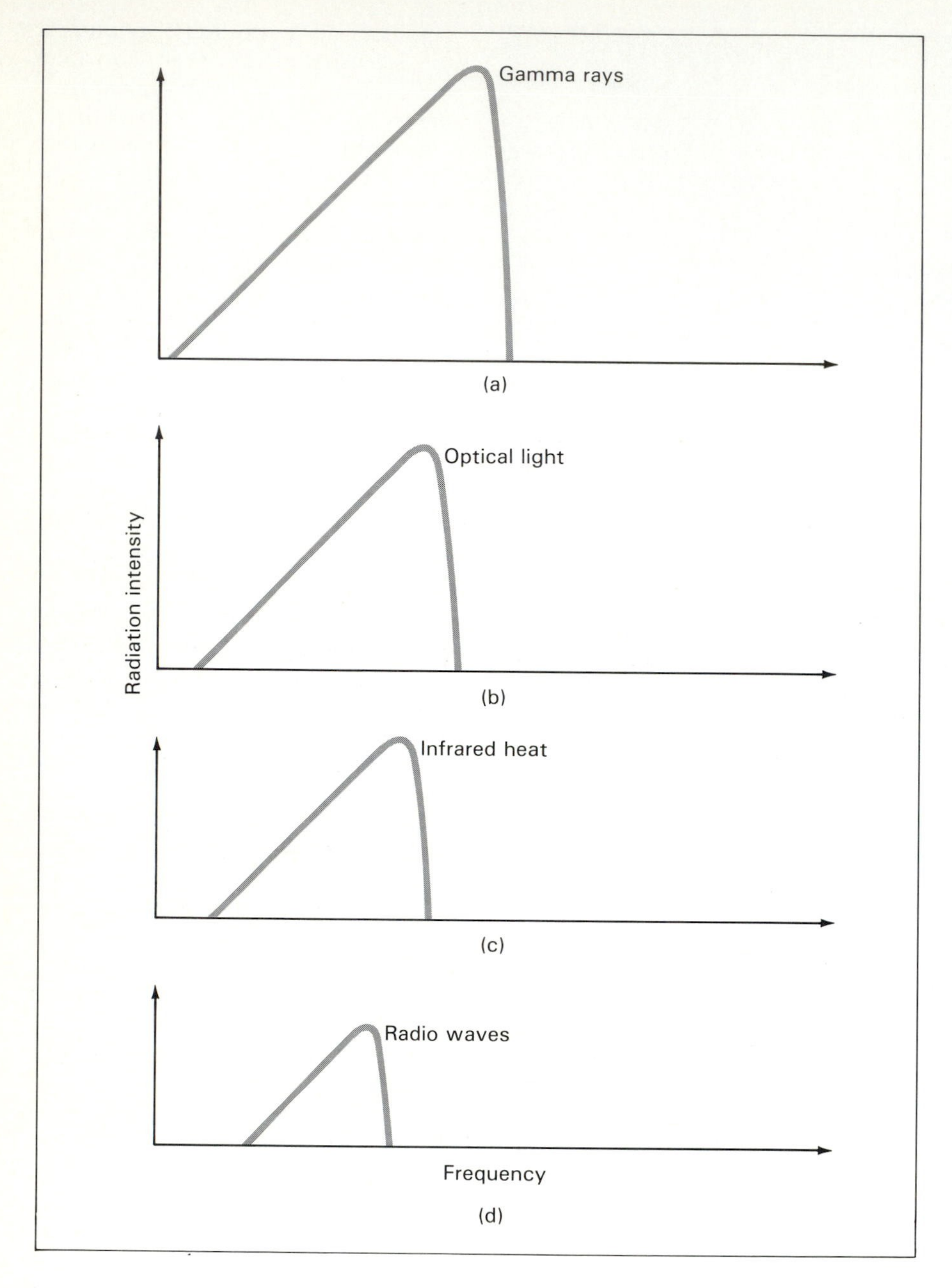

FIGURE 16.13 Theoretically derived Planck curves for the heat of the entire Universe (a) near the origin of the Universe, (b) some 5 billion and (c) 10 billion years after the origin, and (d) at present, approximately 15 billion years after the origin.

closet). But remember, this is an *average* density of the *entire* Universe—lumping the galaxy clusters where the matter is most concentrated with the intergalactic voids, where little if any matter resides. Officially called the **critical universal density**, this terribly small density corresponds to the parabolic model which is, so to speak, intermediate to the hyperbolic and elliptical models of the Universe.

We can then ask: What is the actual density of matter in the Universe? Is it more or less than this critical value? Cosmologists quantify these questions by letting the Greek symbol Ω denote the following ratio:

$$\Omega = \frac{\text{actual density}}{10^{-30}}.$$

If Ω equals 1 precisely, then the actual density equals the theoretically computed density above and the Universe obeys the parabolic model; it's open and will continue to expand forever. This model dictates the Universe to have no net curvature; it's a flat Universe governed largely by Euclidean geometry and Newtonian dynamics. Localized regions of spacetime, especially those near massive astronomical objects, are surely curved, but on the whole the accumulated curvature of spacetime is zero.

Should the value of Ω be less than 1, the Universe's matter is not dense enough to ever stop its expansion. This type of Universe is destined to expand forever, conforming to the open, infinite, hyperbolic model. On the other hand, if Ω exceeds 1, the closed, finite, elliptical Universe

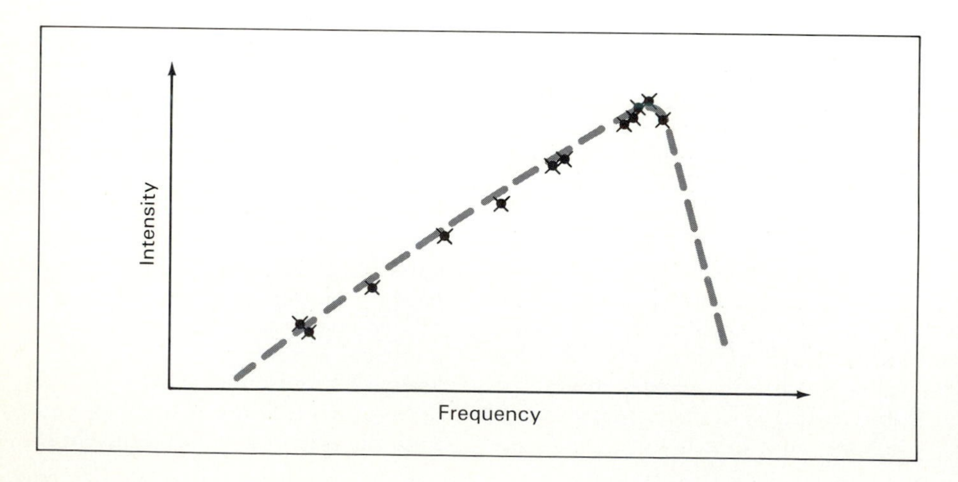

FIGURE 16.14 The observed intensity (X's) of the cosmic background radiation agrees well with that expected from theory [see frame (d) of Figure 16.13]. The dashed curve is the best fit to the data, consistent with a thermal temperature of 3 kelvin.

INTERLUDE 16-2 ***The Oldest Fossil***

"Looking out into space is equivalent to looking back into time." We've repeated that phrase many times now; we must always keep it in mind, especially when discussing cosmology. Just how far back in time can we probe? Is there any way to study the Universe beyond the most distant quasar? How close can we perceive the edge of time, the very origin of the Universe?

Somewhere beyond the realm of the farthest galaxy lies a primeval plasma from which those galaxies evolved. This is not a structured object of any kind—just a hot, dense gas uniformly filling the entire Universe. It's the "stuff" that existed prior to the formation of any galaxies, stars, planets, or any other organized cosmic object. Can we observe radiation from this most ancient gas? Yes, astronomers have already done so.

The cosmic background radiation detected by radio techniques is truly the "fossilized" remains of the primeval plasma. Although it now engulfs us, the radiation originated far, far away in time—about 14.9999 billion years ago or some 100,000 years after creation. Hence the radiation we detect is greatly red shifted. In fact, as shown in the accompanying diagram, radiation now reaching us from the extremely hot primeval plasma has been shifted clear across the electromagnetic spectrum. Intense gamma-ray radiation present soon after the birth of the Universe has been "downgraded" to rather feeble radio radiation during the past 15 billion years or so; the Planck curve for a gas temperature of more than a billion kelvin has been red shifted to the point where we now observe it to be about 3 kelvin. In short, the lethal high-energy photons that escaped the gas of the original "bang" have found their way to us as harmless radio waves.

So we can imagine the cosmic background radiation as a cooled (and thus radio-emitting) version of the primeval gas, as explained herein. But it's more correct to think of this isotropic radio hiss as the red-shifted glare of the intensely hot conditions prevalent shortly after the start of the Universe. We can then better appreciate that the cosmic background radiation comes from the most remote region of the Universe that astronomers have thus far been able to probe observationally—remote in space and remote in time.

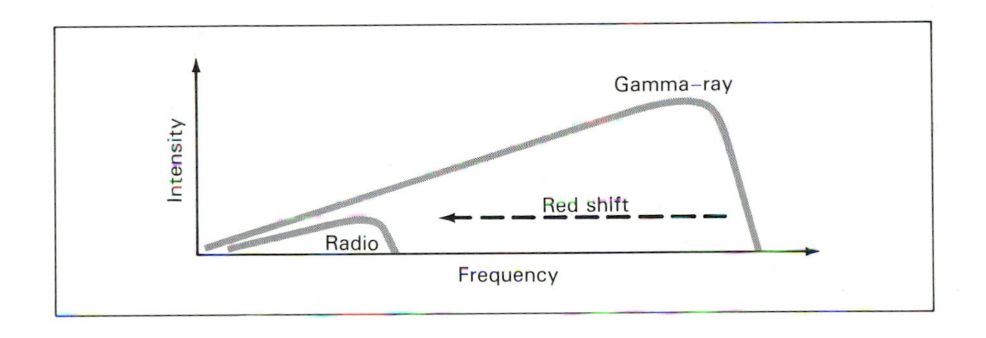

prevails and it will someday start contracting. In either case—Ω greater or less than 1—Einstein's relativity rules.

GALAXY COUNTING: Theory aside, how can we determine the value of Ω? At face value, it would seem simple. Just measure the average mass of each of the galaxies residing within any parcel of space, estimate the volume of that space, and calculate the total mass density. Having done this many times, astronomers usually find about 10^{-31} gram/cubic centimeter. As best can be determined, this calculation is independent of whether the chosen region contains only a few galaxies or a rich cluster of galaxies; the resulting density is about the same, within a factor of 2 or 3.

Galaxy-counting exercises thus favor $\Omega \cong 0.1$, implying that the Universe is open. If correct, the Universe must have originated from a unique big bang and will expand forever. Such a Universe has no end, but it definitely had a beginning.

However, an important caveat deserves mention here. As we noted a few chapters earlier, all the matter in the Universe might not be housed exclusively within brightly visible objects. Recent observations suggest that some invisible matter does exist beyond each of the galaxies. This is especially true of the dark regions within the galaxy clusters, as we noted in our Chapter 14 discussion of the "missing-mass" problem; there, x-ray observations have demonstrated the existence of heretofore unrecognized (hot but thin) chaotic gas. Furthermore, recent optical and radio observations are now hinting at the presence of huge invisible (cold but thin) halos of loose gas surrounding most galaxies.

Our estimate for Ω above did not account for any of this potential missing (or hidden) matter since the extent and amount of it is currently unclear. But if more than 10 times as much additional matter resides outside the galaxies as within the galaxies themselves, Ω would correspondingly increase by more than a factor of 10. That's why it's so very important to search for reservoirs of invisible matter skirting the galaxies. For if enough of it lurks in the darkness, Ω could exceed 1, forecasting the Universe to be closed. Whether such a Universe originates from a unique big bang prior to which nothing existed, and whether such a Universe ends without bouncing, cannot be addressed by taking this kind of an inventory.

The value of Ω as determined by this galaxy-counting method is thus quite uncertain at the present time. It cannot unambiguously distinguish between the open and closed models, although at face value it favors an open Universe destined to expand forever.

DECELERATION OF THE UNIVERSE: Astronomers have developed another observational test to determine the value of Ω. Like the method above, this second test attempts to estimate the average density of the Universe. It essentially relies on the fact that every piece of matter gravitationally pulls on all other pieces of matter. Specifically, this second test addresses the question: At what rate is matter everywhere causing the universal expansion to slow down? Put another way, how fast is the Universe decelerating?

As this cosmic slowdown is likely to be excruciatingly subtle, astronomers cannot realistically expect to measure it by watching the slackening motion of any one galaxy. But by thinking more broadly, there seems to be a statistical way to do it. If the Universe began in an explosive bang, it must have expanded rapidly at first, after which it gradually slowed down. The expansion of anything—an ordinary bomb, an atmospheric thunderclap, whatever—is always greater at the moment of explosion than at some later time. Hence, the recession of the galaxies should be larger for the faraway galaxies and somewhat smaller for those nearby. (Remember, looking out into space is equivalent to looking back into time.)

Figure 16.15 illustrates an idealized Hubble diagram like the ones we studied in Chapter 14. For the nearby objects, namely for galaxies whose radiation was emitted in relatively recent times, the usual Hubble relationship (solid line) yields a certain amount of recessional velocity per distance interval. As noted in Chapter 14, our best estimate of this value is currently 22 kilometers/second for every million light-years of distance. This is Hubble's constant, a measure of the rate at which the Universe now expands. Should the universal expansion rate never change, then Hubble's constant is a true constant for all time, and the solid line of Figure 16.15 can be linearly extended to arbitrarily large distances.

By contrast, if the Universe is slowing its expansion, this second test suggests that we might be able to look back sufficiently far into the past to measure a more rapid expansion rate at an earlier epoch. The dashed curve of Figure 16.15 illustrates how a greater expansion rate in earlier times would show up as a departure from the solid line of the usual Hubble diagram. A greater expansion rate in the past means that the most distant objects would have receded with greater velocities than those nearby. This is equivalent to saying that Hubble's constant might not be a constant at all. It must have been larger in earlier epochs. How much larger depends on the average density of matter in the Universe.

Figure 16.16 shows several different departures that could conceivably be measured by studying the distant galaxies. The solid line corresponds to the discredited steady-state model in which the rate of expansion is fixed (it's steady) and Hubble's constant is really constant. All the time-dependent or evolving models (dashed curves), however, depart from this solid line since all of them predict the universal expansion rate to change with time. Drawn as dashed curves in Figure 16.16, this departure is expected to be largest for the finite, closed model of the Universe, $\Omega > 1$; this is reasonable since the huge amounts of matter needed to close the Universe

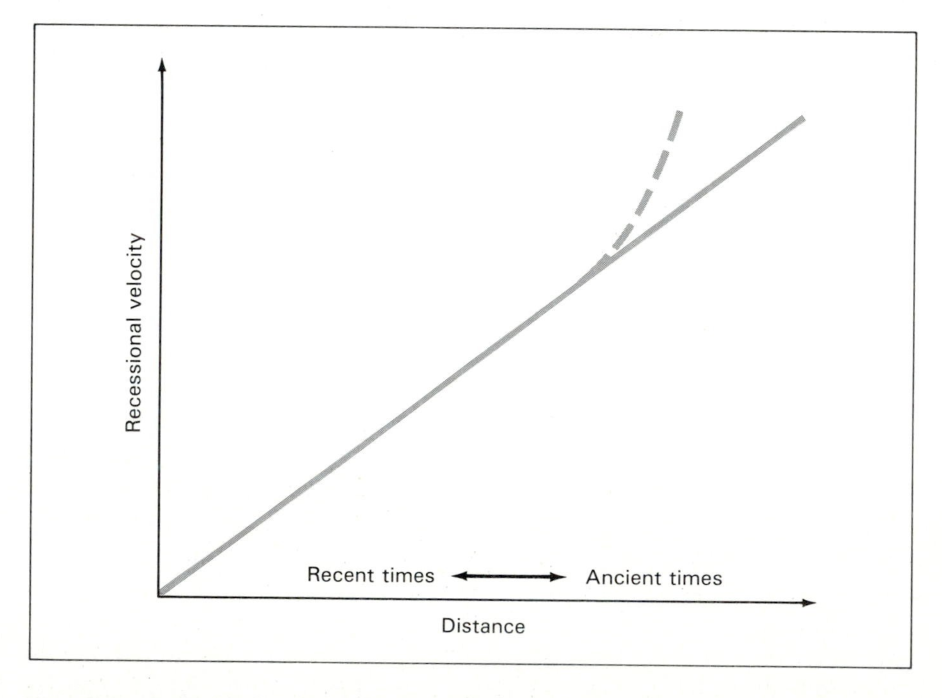

FIGURE 16.15 An idealized Hubble diagram showing how we might expect to detect some evidence for a deceleration of the Universe by observing a departure (dashed curve) from the usual Hubble relationship (solid line).

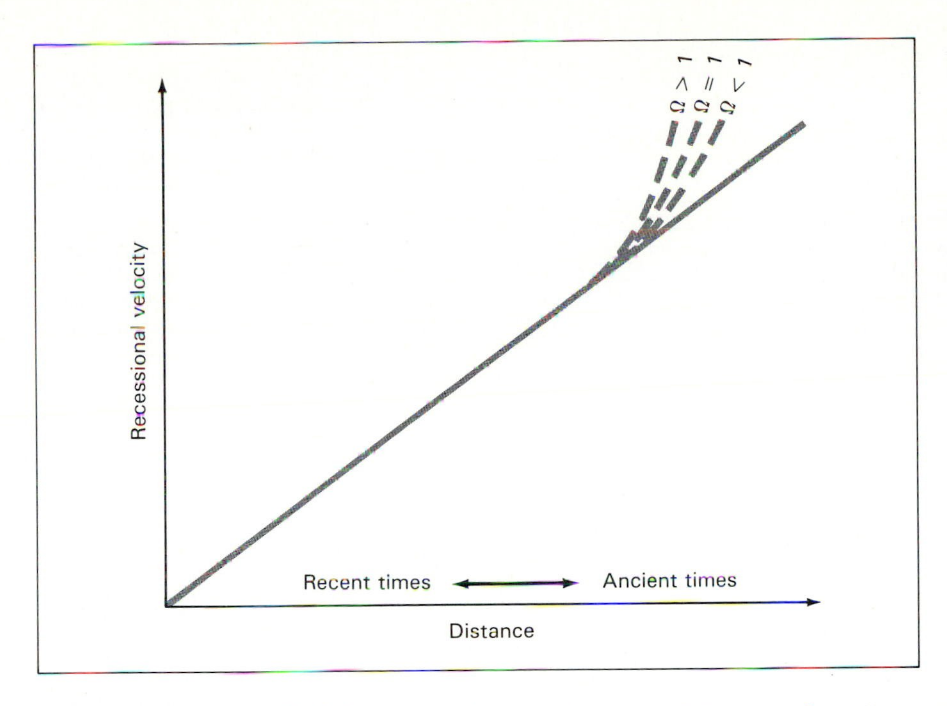

FIGURE 16.16 Another Hubble diagram showing the expected departure from the solid line for each of the evolving models of the Universe. (The departures of the dashed curves from the solid line are exaggerated for clarity.)

would have greatly decelerated the universal expansion over the course of 15 billion years. The infinite, open models, $\Omega \leqslant 1$, are expected to show smaller departures, for in these cases the deceleration of the Universe is less.

What do the data indicate? Is there any evidence for faster recessional velocities among the more distant galaxies? The Hubble diagrams that plot the observed velocities of numerous galaxies having reasonably well-determined distances (out to a few billion light-years) demonstrate no clear departure. Nonetheless, astronomers have extended those plots for a handful of more distant galaxies, assuming that red shift remains a valid indicator of distance. (Quasars which lie almost exclusively beyond the galaxy horizon cannot be used reliably in this test since their nature is still too controversial.) And, indeed, the most distant galaxies do have substantially greater recessional velocities than those nearby. The data seem to favor the curve $\Omega > 1$, although the accuracy of these observations is rather poor. The most distant galaxies are very faint, and observations of them are notoriously hard to make. Furthermore, the observations do not take account of any (as-yet undiscovered) evolutionary changes that might affect the brightness of galaxies over the course of time. Nonetheless, at face value, this second cosmological test suggests that the Universe is closed and finite. It contradicts the first test, unless substantial amounts of dark matter really do exist beyond the visible galaxies.

(A third cosmological test concerns a special form of hydrogen—called deuterium—produced in the early Universe; as discussed later in Chapter 22 and especially in Interlude 22-1, this test also implies that the Universe is open.)

SUMMARY

Whether we live in an open or closed Universe is currently unknown. The bottom line clearly suggests an evolutionary Universe; but its ultimate destiny remains concealed. Many cosmologists are inclined to say that we should expect a definite answer within a few years. This is perhaps overly optimistic, for the final solution requires the agreement of three often disparate groups of human beings. First, there are the theorists, those imaginative minds who invent the model Universes. They try to determine what the Universe is supposed to be like. Second, there are the experimentalists, constantly testing the theories, all the while extending their observations to more distant realms within our Universe. They try to determine what the Universe really is like. And third, there are the skeptics, who regard the models of the first group as mere speculation and the results of the second group as overinterpretation of the data without due regard for observational error. In the end, all three attitudes are helpful and necessary, for only by their cooperation and counteraction might we ever hope to approach the truth.

KEY TERMS

big bang
closed Universe
cosmic background radiation
cosmological principle
cosmology
critical universal density
open Universe
oscillating Universe
principle of equivalence
relativity
singularity
spacetime

QUESTIONS

1. State the two main tenets of relativity theory.
2. Explain the meaning of the phrase, "Looking out into space is equivalent to probing back into time." Does this mean that we could travel into the past?
3. Briefly but explicitly explain two different observational tests that aim to determine the ultimate fate of the Universe. Be sure to give some details concerning the current status (including uncertainties) of these tests.
4. Define the term "cosmology."
5. Describe the major missing-mass problem in astronomy and briefly discuss its relevance, if any, to the question of the ultimate destiny of the Universe.
6. Pick one: The overall structure of the Universe seems to be (a) clustering of galaxies, perhaps clustering of galaxy clusters, and thereafter uniformity on the largest scale; (b) clustering of matter on all distance scales thus far observed; (c) uniformity of matter on all distance scales; (d) clustering of matter in some realms, uniformity of matter elsewhere.
7. Name and discuss two pieces of observational evidence that argue against a steady-state model of the Universe and in favor of a big-bang evolutionary Universe. Be specific.
8. List, in order of progressively increasing distance, a series of techniques that allow us to measure the distance from the nearest star to the most distant quasar.
9. What is the relationship between Hubble's constant and the present age of the Universe? What are the uncertainties in this age?
10. How do astronomers know that Hubble's constant is not really constant?

FOR FURTHER READING

BARNETT, L. *The Universe and Dr. Einstein*. New York: New American Library, 1957

CHAISSON, E., *Relatively Speaking*. New York: W. W. Norton, 1988.

*FIELD, G., "The Mass of the Universe: Intergalactic Matter," in *Frontiers of Astrophysics*, E. Avrett (ed.). Cambridge, Mass.: Harvard University Press, 1976.

GOTT J., GUNN, J., SCHRAMM, D., AND TINSLEY, B., "Will the Universe Expand Forever?" *Scientific American*, March 1976. Freeman.

*HARRISON, E., *Cosmology*. Cambridge: Cambridge University Press, 1981.

MULLER, R., "The Cosmic Background Radiation and the New Aether Drift." *Scientific American*, May 1978.

SINGH, J., *Great Ideas and Theories of Modern Cosmology*. New York: Dover, 1970.

17 PARTICLE EVOLUTION: *The Early Universe*

What was it like at the start of the Universe? Precisely what happened at the origin of time? Can we say anything for certain about the very origin itself? What about the prevailing conditions during the first few moments of the Universe? How did those conditions change thereafter?

These are surely basic questions. And they are hard questions. Yet they are among the questions that nearly everyone thinks about at one time or another. Now, after more than 10,000 years of organized civilization, twentieth-century science seems ready to provide some insight regarding the ultimate origin of all things.

The solutions that astronomers have developed must be considered tentative and somewhat uncertain. After all, times long past are times long gone. We cannot be precise about the earliest epochs of the Universe, for such ancient history cannot be observed directly. Nonetheless, astronomers have constructed models—mathematical sketches based on a large body of relevant data dictating the shape and structure of our Universe. These models grant us some idea of the Universe's physical conditions roughly 15 billion years ago.

The learning goals for this chapter are:

- to appreciate the extremely hot and dense conditions present just after the origin of the Universe
- to understand how matter naturally emerged from the primeval fireball of radiation
- to realize how the simplest atoms originated in the early years of the Universe
- to appreciate some of the efforts now under way to unify all the known forces of the Universe
- to gain an inkling about current scientific models of creation itself

The First Few Minutes

To appreciate the earliest epochs of the Universe, we must think deeply about times long, long ago. We must strive to imagine what it was like long before Earth and Sun were created—even before any planet or star existed. Some people have trouble imagining such truly ancient times. Fortunately, there is a bit of a trick that can help us comprehend the earliest moment of the Universe.

A SYMMETRY ARGUMENT: The "trick" involves the natural symmetry of the closed model of the Universe. It is this: If you find it hard to reverse time mentally in order to appreciate the earliest epochs of the Universe that *have* occurred, you can, instead, visualize the physical events that *will* occur as a closed Universe nears the final phase of its collapse. The physical conditions near the end of a closed Universe are expected to be virtually identical to those near the start because the mathematics describing the contraction are a mirror image of the mathematics describing the expansion. In other words, the events that *will* occur just prior to the end of a closed Universe mimic those that *already* happened just after the Universe began. By using the laws of physics to predict the final events of a closed Universe, we can gain some inkling of the early aftermath of the univer-

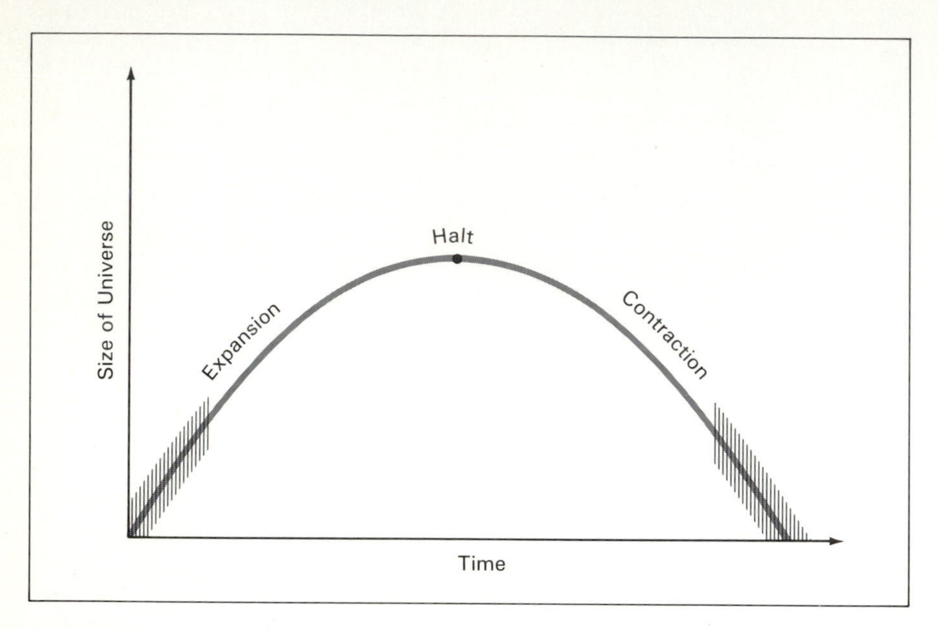

FIGURE 17.1 Schematic diagram of the expansion and contraction of a closed Universe. The shaded parts show how the physical conditions are roughly the same near the beginning and end of this model Universe.

sal "bang" approximately 15 billion years ago.

Figure 17.1 reminds us that if the Universe is closed, it will expand ever more slowly into the future, gradually coming to a halt. Once its contraction begins, the separations among the galaxies will decrease, causing more frequent galaxy collisions, larger amounts of friction, and higher temperatures for the whole Universe. The present cosmic background radiation will steadily rise as the Universe gradually compresses, just as any gas heats when the number of atomic collisions increases. (Witness the heating when compressing the air molecules inside a bicycle pump.) Eventually, the matter in such a contracting Universe becomes very hot and dense, culminating in a blazing, compact singularity. Such a bizarre region, having all the mass in the Universe concentrated within the volume of a pinhead, might be unable to exist in actuality. Even so, that's the state predicted by the currently known laws of physics.

The important point to note here is that there is complete symmetry between the origin and the end of time for a closed type of Universe. The superhot, superdense physical conditions expected to characterize the final singularity must have also typified the initial singularity from which the Universe originated. Even if the Universe is not closed and will never contract to this singular state, astronomers can use the closed model to understand theoretically the highlights of the earliest epochs of either a closed or open evolutionary Universe.

THE EARLY EPOCHS: Table 17-1 and Figure 17.2 summarize the physical conditions during six major epochs of the Universe. Specified in this table and figure are the general names of the epochs, the time domain of each epoch, the *average* density of everything in the Universe, the *average* temperature, and finally a brief description of the main physical events that probably dominated the Universe at the time.

These numerical values result from number-crunching exercises, using a mathematical knowledge of

TABLE 17-1
Major Epochs of the Universe

	EPOCH	TIME AFTER THE BANG	AVERAGE DENSITY (GRAMS/CUBIC CENTIMETER)	AVERAGE TEMPERATURE (KELVIN)	MAIN EVENTS
Radiation Era	Chaos	$<10^{-24}$ second	$>10^{50}$	$>10^{20}$	Unimaginable Big Bang
	Hadron	10^{-24}–10^{-3} second	10^{30}	10^{15}	Annihilation of heavy elementary particles produces fireball of radiation
	Lepton	10^{-3}–100 seconds	10^{10}	10^{10}	Annihilation of light elementary particles continues to produce fireball of radiation
Matter Era	Atom	100 seconds–10^{6} years	10^{-10}	10^{5}	Fireball diminishes; matter dominates and clusters into hydrogen and helium atoms
	Galaxy	10^{6}–10^{9} years	10^{-20}	300	Galaxies, quasars, and galaxy clusters form
	Stellar	$\gtrsim 10^{9}$ years	$\sim 10^{-30}$	~3	All galaxies have formed; many stars have formed; other stars are still forming, some accompanied by planets and life

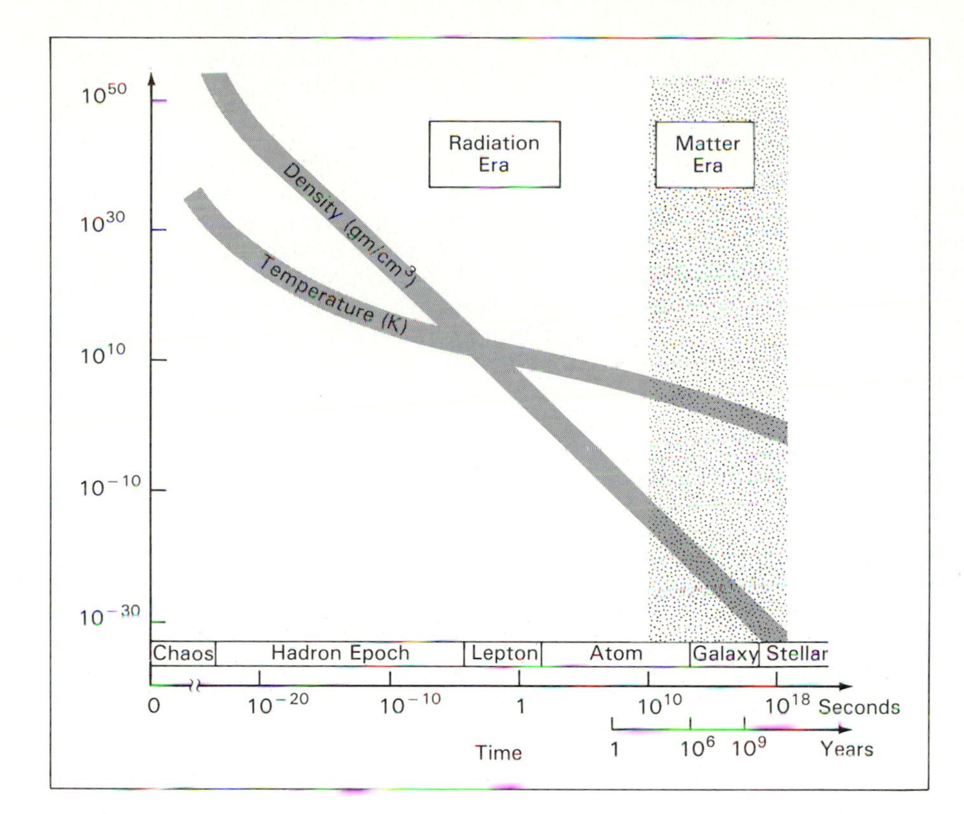

FIGURE 17.2 Plots of the average temperature and average density throughout the history of the Universe. (Numerical values for these plots are taken from Table 17-1.)

the laws of physics and a large computer. The calculations are complex, incorporating the essentials of much of what we know about the Universe. The objective of these "numerical experiments" is to derive the average density and temperature for the whole Universe at any time in its history.

Both the tabulated and plotted values of Table 17-1 and Figure 17.2 suggest that, in the beginning, there was chaos! Although we remain uncertain about creation itself (i.e., precisely zero time), some theorists argue that the physical conditions can be specified for some extraordinarily short time after the bang. For example, the laws of physics predict that a Universe as young as 10^{-24} second (i.e., a trillionth of a trillionth of the first second) would have an average density of roughly 10^{50} grams/cubic centimeter, and an average temperature of roughly 10^{20} kelvin.

It's hard for anyone to appreciate such youth, for 10^{-24} second is the amount of time light needs to cross a proton. Equally troubling are such large densities and temperatures; recall that the density of an atomic nucleus is "only" 10^{12} grams/cubic centimeter, the temperature of a star's core "only" 10^8 kelvin. Yet the tabulated values are the conditions predicted by the laws of physics as a closed Universe inexorably speeds toward its demise. Thus they are also, through the symmetry arguments above, the conditions prevailing at the earliest moments after the birth of the Universe.

The overall state of the Universe at this time was indescribably complex. Surely, much energy existed alongside exotic elementary particles of many types, but beyond that science can say little. We can merely suggest that the main event of this earliest epoch was simply unimaginable.

By most accounts, then, the Universe began with the explosion of an incredibly hot and dense state. Precisely what state, we cannot say. And why it exploded, we really don't know. *Why* the Universe began suddenly expanding some 15 billion years ago is a most difficult query—so formidable that scientists currently have no idea how to solve it. To understand why the Universe began expanding, or even more fundamentally, why the Universe exists at all, is currently beyond science. *Why* questions, as opposed to *what* and *how* questions (routine inquiries for scientists), require us to consider realms of inquiry outside the formalism of modern science—realms that cannot be addressed by means of the scientific method. Put simply, it is scientifically impossible to study events that might have occurred prior to the origin of the Universe, for there are no relevant data. Queries about pre-universal issues are tantamount to probing, not the origin of the Universe, but the origin of the origin.

To speculate about times before the origin of the Universe is simply not science. And it's hard to imagine that such a subject will ever become science. Accordingly, we necessarily confine our discussions in this book to the events that have occurred after the Universe did originate, for whatever reason. Indeed, cosmic evolution comprises a broad synthesis of the "whats" and "hows."

HADRONS AND LEPTONS: Within a microsecond (10^{-6} second) of existence, the Universe was filled with an entire array of subatomic particles of matter. This was a straight materialization—a creation—of matter from the energy of the primeval bang. No magic was involved, just a well-known and oft-studied nuclear process whereby elementary particles result from the decay of high-energy photons. (Consult Interlude 17-1.)

The second major epoch of the Universe is easier to comprehend, but it's still characterized by severely nonterrestrial conditions. Called the **hadron epoch**, the name derives from the fact that the heavy elementary particles such as protons and neutrons were the most abundant type of matter at the time. Such particles must have existed as free unbound entities; the hadrons surely collided and interacted with one an-

INTERLUDE 17-1 ***Antimatter in the Universe***

Atoms of ordinary matter are made of a positively charged heavyweight nucleus, surrounded by one or more negatively charged lightweight electrons. In Chapter 2 we noted that all the atoms found on Earth have this common structure; in Chapter 22 we will describe how the various kinds of atoms originated with their differing numbers of elementary particles.

Some theorists wonder why nature is not more symmetrical. Why, for example, should the heavy nuclei always have a positive charge, relegating the negative charge to only the lightweight electrons? They argue that the Universe would seem more philosophically pleasing if its basic building blocks had more symmetry in their charge and mass.

Around the mid-twentieth century, experimentalists in fact discovered several lightweight particles having a positive charge, as well as heavyweight particles having a negative charge. These antimatter particles are identical to ordinary matter particles in every way except charge. (Recall that we've already studied the positron, or positively charged electron, in Chapter 10.)

The appearance of matter and antimatter presumably occurred during the earliest moments of the Universe. Yet we don't observe much antimatter around us. Earth, the other planets, and the Sun all seem to be made of ordinary matter. Exceptions include the products of nuclear reactions churning away inside stars, a small fraction of the cosmic rays showering Earth each day, and microscopic fragments created during fleeting moments when elementary particles collide in nuclear laboratories here on Earth. But for the most part, virtually all the mass in the Solar System seems to be of the matter variety, with little trace of antimatter. If matter and antimatter were created in equal amounts from primordial energy at the start of the Universe, where is all the antimatter?

(Be sure to realize that antimatter does not imply antigravity. Particles of antimatter gravitationally attract one another just as do two or more particles of matter. The only property that distinguishes matter from antimatter is charge; the mass of every matter particle is identical to its antimatter opposite, and hence gravity invariably pulls while never pushing. Indeed, there is no such thing as "antigravity" known anywhere in the Universe.)

Nothing prohibits elementary particles of antimatter from combining into large clumps. Antihydrogen, antioxygen, anticarbon, and numerous other antiatoms could conceivably form antiplanets, antistars, and antigalaxies. The fact that we are unware of such large clumps of antimatter does not preclude their existence. Since atoms of antimatter emit and absorb precisely the same type of photons as do atoms of ordinary matter, there is no way to determine if, for example, a distant star is made primarily of matter or antimatter. Spectroscopic observations of a clump of antimatter would be identical to those made of a clump of matter.

So, despite living in a Solar System made primarily, if not totally, of matter, large pockets of antimatter could exist elsewhere in the Universe. It's even possible, though not likely, that the Alpha-Centauri star system or the Andromeda galaxy as well as numerous other stars and galaxies studied earlier could be made of antimatter. However, this is unlikely for two reasons, the first having to do with the subtleties of the decay of the X-boson elementary particle (discussed in the last section of this chapter). The predicted, though as-yet untested decay of this particle is not symmetric, suggesting that the Universe is populated with much more matter than antimatter.

A second reason implying the nonexistence of great clumps of antimatter in the Universe concerns its inability to coexist with ordinary matter. Recall from Chapter 15 that when a matter particle and its antimatter opposite collide, the explosive result creates pure energy of the gamma-ray type. Thus it would seem difficult if not impossible to imagine how matter and antimatter objects would remain sufficiently segregated, lest they naturally annihilate.

Before leaving this subject, one other point can usefully be made. In addition to the process of matter–antimatter "pair annihilation" just discussed, the opposite process—"pair creation"—can also occur. Provided that the temperature is at least billions of kelvin, collisions among gamma-ray photons can yield pairs of particles—for instance, a matter electron and an antimatter positron. This sort of "materialization" of matter from energy still obeys the basic law relating matter and energy; in this case, one (energy, E), simply changes into the other (mass m), the conversion once again obeying the equation $E = mc^2$, where c symbolizes the velocity of light.

other, for the density was extremely large in the Universe's first second of existence. But the temperature was also extremely high, thus prohibiting these particles from combining into atoms. The dominant action at this time was presumably the self-annihilation of hadrons into photons, thus creating a brilliant fireball of radiation. Lacking a good understanding of elementary particles, scientists know little about this mystifying period in the history of the Universe.

As the Universe expanded rapidly, it cooled and thinned. About a millisecond (10^{-3} second) after the bang, the superhot and superdense conditions suitable for hadron creation had nearly subsided, thus allowing the less abundant, lighter elementary particles such as electrons and neutrinos to dominate. The average density and temperature of this so-called **lepton epoch** had fallen to about 10^{10} grams/cubic centimeter and about 10^{10} kelvin—values ripe for the materialization of leptons from energy. These physical conditions are still excessive compared to those familiar to us on Earth, but they are a good deal less extreme than the chaotically dense and hot conditions present a fraction of a second earlier. Indeed, it's important to realize that once the Universe began expanding, it did so very rapidly, thus cooling and dispersing its contents quickly. By the time the first second had elapsed, the leptons were self-annihilating into photons, much as the hadrons had earlier. The radiative fireball of this cosmic bomb was still fueled with light and other types of high-frequency radiation.

The radiation density greatly exceeded the matter density in these first few minutes. Not only did the photons of radiation far outnumber the particles of matter, but also most of the energy in the Universe was in the form of radiation, not matter. As soon as elementary particles tried to cluster, fierce radiation destroyed them, thus precluding the existence of even simple atoms. For this reason, the first three epochs are often collectively called the **radiation era**. Whatever matter managed to exist did so as a thin precipitate suspended in a world of dense, brilliant radiation.

Later Epochs

CREATION OF ATOMS: As time elapsed, change continued. The fourth major epoch extends in time from about 100 seconds to about 1 million years after the bang. Midway through this **atom epoch**, the density had decreased to a value of about 10^{-10} gram/cubic centimeter, while the average temperature had decreased to about 10^5 kelvin—values much like those in the atmospheres of stars today. A key feature of the atom epoch was the steady waning of the original fireball; all the annihilations of hadrons and leptons had virtually ceased.

Even as the fireball diminished, a crucial change began—perhaps the most important change in the history of the Universe.

At the beginning of the atom epoch, radiation still overwhelmed matter. All space in the Universe was absolutely flooded with photons, especially light, x-rays, and gamma rays. As the Universe expanded in time, however, the photon density decreased faster than the matter density. The early fog of blinding light gradually dimmed, thus decreasing the early dominance of radiation. Sometime between the first few minutes and the first million years after the bang—the process of changing from the plasma state to the neutral state was gradual—the charged elementary particles of matter began clustering. Their own electromagnetic forces pulled them together; radiation could no longer break them apart as quickly as they combined. In effect, the dominance of radiation had been destroyed as matter gradually became neutralized, a physical state over which radiation has little control. Matter had, in a sense, overthrown the cosmic fireball while emerging as the main constituent of the Universe. Matter henceforth would dominate radiation. To denote this major turn of events, the last three epochs of Table 17-1 are collectively termed the **matter era**.

This newly gained dominance of matter over radiation is probably the most important event of all time. For the sake of simplicity, we shall take an average value of some 100,000 years as the time after creation when this grand event occurred.

So the onset of the Matter Era saw the creation of atoms. The influence of radiation had grown so weak that it could no longer prohibit the attachment of the lepton and hadron elementary particles that had survived annihilation. Hydrogen atoms were the first type of element to form when single electrons became electromagnetically bound to single protons. In this way, copious amounts of hydrogen were synthesized in the early Universe.

HEAVIER ELEMENTS: Hydrogen was not the only kind of atom formed early in the Matter Era. Throughout the first part of the atom epoch, the average temperature of the Universe exceeded the 10 million kelvin needed to fuse two hydrogen nuclei into helium via the proton-proton cycle. The Universe was cooling all right, but it took time for the average universal temperature to dip below this critical value. Consequently, helium atoms must have been created within the primordial fireball in the same way that they now form within stars.

Most researchers agree that only so much helium could have been produced in the early Universe. Despite some uncertainty about the temperature and density at the time, only about 1 helium atom formed for every 10 hydrogen atoms. Accordingly, all cosmic objects should contain at least 10 percent helium abundance. For two reasons, this is a minimum amount of helium that should contaminate virtually all objects in the Universe. First, physical processes such as the proton–proton cycle within stars have surely created additional amounts of helium well after the bang; helium is, in fact, forming at this moment inside stars. And second, since helium is chemically inert, there is no easy way to transform it into something else once it exists; helium atoms cannot even "hide" within other substances, like molecules, since helium does not easily combine with other elements.

Small rocky planets are exceptions. They usually have little helium because their gravity isn't strong enough to prevent helium atoms from escaping. Life is also an exception, for it originated from the matter of which our planet is made. At any rate, the fact that the oldest stars, especially the stars of globular clusters, contain just about 10 percent of their atoms in the form of helium lends support to the idea that most of the helium was indeed created in the early moments of an explosive Universe.

By contrast, elements heavier than helium could not have been produced in the early Universe. As we'll note in Chapter 22, the creation of heavy elements such as carbon, oxygen, and iron requires temperatures even higher than 10 million kelvin. Their synthesis also requires lots of helium atoms, since the heavier elements are built from lighter ones. The basic problem is that even though the helium atom production was in high gear during the start of the atom epoch, the average temperature was falling quickly. The Universe, after all, was rapidly expanding from its primeval explosion. Theoretical calculations suggest that by the time there were sufficient helium atoms to interact with one another to manufacture some of the heavier elements, the cosmic temperature had fallen below the threshold value (hundreds of millions of kelvin) needed for doubly charged helium nuclei to hit, stick, and fuse. The Universe just wasn't hot enough anymore to permit this.

MORE RECENT TIMES: During the fifth, **galaxy epoch**, gravity began pulling some of this matter together into enormous clumps. Apparently, the conditions were just right for the formation of galaxies. Indeed, all galaxies must have originated long ago, for observations suggest that no galaxies are forming at the present time. How do we know this? Because every galaxy seems to contain vast quantities of old stars. Not that we know of any dead galaxies. But astronomers are equally unaware of any young galaxies.

Cosmic objects that are most distant from us, such as quasars and remote galaxies, must have formed in the earliest part of this galaxy epoch. The radiation we now detect from these objects left them during the early stages of this fifth epoch. One of the prime objectives of modern astronomy is to interpret the radiation from the most distant objects in order to better understand the physical conditions that characterized the Universe long ago.

Note how the time spans of the final two epochs of Table 17-1 have greatly lengthened. As discussed earlier in this chapter, an important and rapid series of events occurred immediately after the bang, especially in the first few minutes of the Universe's existence. But once the Universe cooled and thinned enough to permit the creation of atoms, the major events slowed down. For example, galaxies did not just appear instantaneously. Nature needed time to gather enough atoms to create the galaxies. How much time? Probably a billion years or so.

Note also that by the middle of this fifth, galaxy epoch, the average density of the Universe had decreased by another factor of 10 billion, reaching some 10^{-20} gram/cubic centimeter. The average temperature of the entire Universe had also diminished to a relatively cool 300 kelvin. The whole Universe was steadily becoming thinner, colder, and darker.

PRESENT TIMES: Now, roughly 15 billion years since the bang, the average density of the Universe is approximately 10^{-30} gram/cubic centimeter. This is the critical value above which the Universe will eventually close and below which the Universe will remain forever open. And as discussed in Chapter 16, the average temperature of everything in the Universe is currently 3 kelvin. This is the cooled relic of the incredibly hot fireball that existed eons ago, the fossilized grandeur of an ancient era.

Present times are dominated by still more change, although at a much reduced pace. The main events in the sixth, **stellar epoch** concern stars—the formation of stars and the creation of heavy elements inside stars. Research during the past decade has provided direct observational evidence that stars are now forming within galaxies; the galaxies themselves no longer form, but the stars within them do. Furthermore, as islands of high temperatures, great densities, and brutal atomic collisions, stars are perfectly suited to generate the heavy elements. Indeed, both stars and heavy elements represent still new products of a vibrant, changing, 15-billion-year-old Universe.

RECAPITULATION: The major epochs in the history of the Universe are so important that we should review them once more. The artist's conceptions of Figure 17.3 might help.

In the earliest epochs, the Universe evolved dramatically. Flooded with blinding light, radiation dominated matter. The enormous number of photons, and especially the scattering of photons by electrons, produced a spectacularly bright fireball inside of which no atoms or molecules could have existed. Even if, by some miracle, you could have witnessed the Radiation Era, you would have seen only blinding fog in all directions—light radiation everywhere!

Rapidly at first, and then more slowly, the Universe expanded, cooled, and thinned. The physical conditions of temperature and density that guide all changes in the Universe had themselves undergone extraordinary change. Matter gradually began clustering into atoms, and atoms eventually into galaxies. From the start of the Matter Era, some 100,000 years after creation, matter has dominated radiation. And it has dominated radiation ever since, successively forming galaxies, stars, planets, and life.

The history of the Universe described here is the prevailing view among cosmologists. But not all share a consensus concerning events prior to the first few seconds of existence. The further we attempt to probe the past, the more uncertain our statements become. Accordingly, the temperature and density of the first instant of the Universe are quite uncertain, mainly because their values depend on incompletely understood interactions among the peculiar elementary particles. Depending on the intricacies of the theoretical model chosen, the descriptions of the events during the

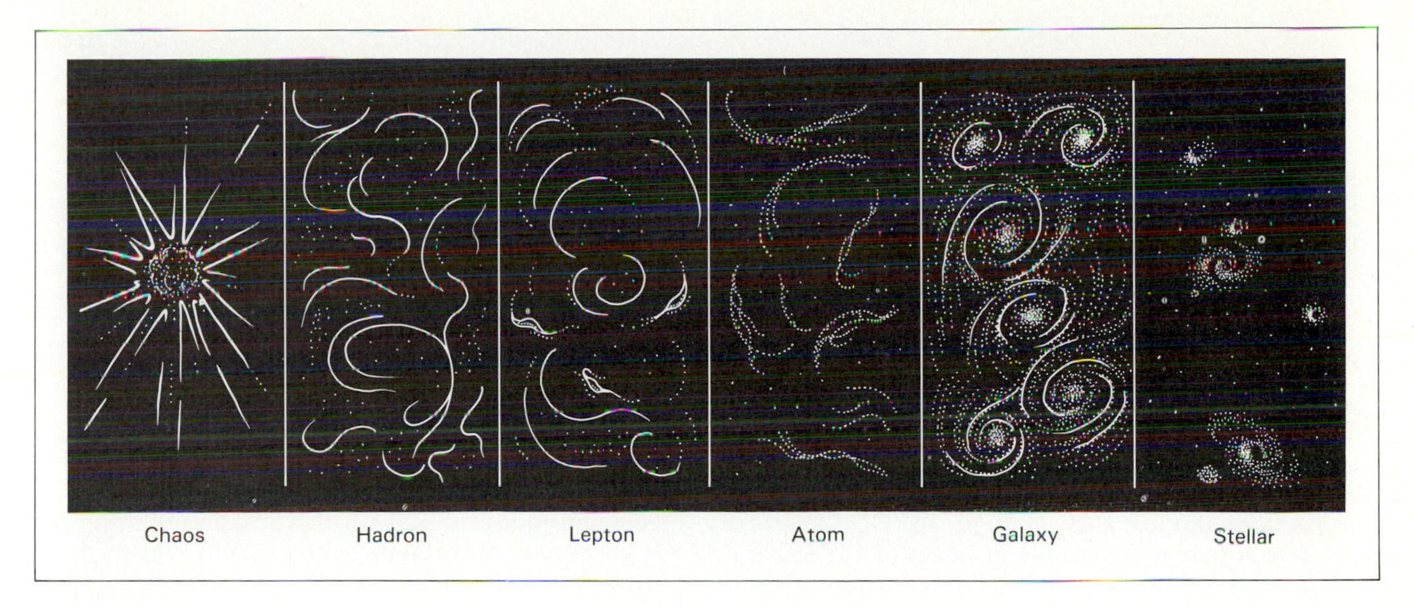

FIGURE 17.3 Artist's conception of the main events during the six major epochs of the Universe.

first few seconds can differ by several orders of magnitude. In virtually all models, however, the early Universe was very hot and dense, after which it cooled and thinned with the march of time.

Exploring Chaos

One of the currently great intellectual adventures in the subject of physics is the attempt to unify all the forces of nature (consult Interlude 2-2). During the past decade, research toward this objective has been partly successful, as two of the basic forces are now understood to be really one and the same force. This work has implications not only for the microdomain of physics but also for the macrodomain of cosmology. Most significantly, this aspect of cosmic evolution is beginning to provide great insight about the earliest epoch of the Universe—the time interval colloquially labeled "chaos" near the start of this chapter.

GRAND UNIFICATION: The electromagnetic force (that binds atoms and molecules) and the weak force (that governs the decay of radioactive matter and the emission of exotic particles such as neutrinos) have been merged by a new theory that asserts them to be different manifestations of one and the same force—an "electroweak" force. Since crucial parts of this theory have recently been confirmed by experiments conducted with particle accelerators, researchers are now striving to extend this unified theory to include the nuclear force (that binds elementary particles within nuclei). Furthermore, although physicists are unsure at this time how in turn to incorporate the fourth known force (gravity) into this comprehensive theory, there is reason to suspect that we are nearing the realization of Einstein's dream—to understand all the forces of nature as different aspects of a single fundamental force.

Here, briefly, is how the electroweak force operates. In submicroscopic (quantum) physics, forces between two elementary particles are represented by the exchange of another type of particle called a "boson"; in effect, the two particles can be imagined as playing a rapid game of catch using a boson as a ball. In ordinary electromagnetism familiar to us in our terrestrial surroundings, the boson is the usual photon that we have studied extensively to this point in the book—a bundle of electromagnetic energy that always travels at the speed of light. The new electroweak theory includes four such bosons: the usual photon, as well as three other particles, having the peculiar names of W^+, W^-, and Z^0. At temperatures below about 10^{15} kelvin—the range that includes almost everything we know about on Earth and in the stars—these bosons split into two families: the photon that expresses the usual electromagnetic force and the other three that carry the weak force. But at temperatures higher than this million billion kelvin, these bosons work together in such a way as to make indistinguishable the weak and electromagnetic forces.

Thus, by experimentally studying the behavior of this new force, we can gain some insight into not only the essence of nature's building blocks, but also the early epochs (especially the hadron epoch) of the Universe when the temperature was indeed great. In fact, the temperature was 10^{15} kelvin about 10^{-10} second after creation. Paralleling our earlier phrase, "observing out into space is equivalent to probing back into time," that we must always keep in mind macroscopically, we now have another, equally important phrase that pertains microscop-

INTERLUDE 17-2 ***Testing the GUTs***

A theory can never be proved correct; it can only be rendered incorrect. By experimenting, observing, and otherwise testing our various theories, we can gradually rule out the poor ideas, and thus by a process of elimination eventually gain a more refined approximation of the "truth."

How can we test the grand unified theory, including its implied inflationary change? The answer is that we must do so indirectly, for we shall never be able to directly recreate in the laboratory the energies of creation itself. After all, the world's most powerful accelerator, in Geneva, has only recently simulated, for the briefest of instants, a temperature of 10^{15} kelvin needed to test the electroweak theory. By contrast, the GUTs become operative at temperatures only in excess of 10^{28} kelvin. To accelerate elementary particles to the required huge energies (i.e., to achieve effectively such high temperatures in the laboratory), a machine would need to be built spanning the distance between Earth and the Alpha-Centauri star system several light years away. Furthermore, the operation of such a truly cosmic device for merely a few seconds would require an altogether unreasonable amount of power equal to several times the gross national product of the United States. Thus, while physicists have successfully simulated in the laboratory the physical conditions characterizing the hadron epoch, we currently have little hope to so reproduce chaos. Perhaps for the best, as this is one state of nature that seems literally too hot to handle.

So how do we test the grand unified theories? One apparent success of the GUTs is that they can generally account for the observed excess of matter over antimatter, as noted in Interlude 17-1. It so happens that the decay of the X bosons within the first 10^{-35} second of existence is expected to have created slightly greater numbers of protons than of antiprotons. This prediction can be tested in a straightforward way, for if protons can be created, they can also be destroyed. Using the GUTs, we can estimate the lifetime, or mean life expectancy, of the proton; it turns out to be more than a billion billion times the age of the Universe (i.e., roughly an incredible 10^{32} years)! This extremely long lifetime guarantees a very small probability of decay in any given time span. Nonetheless, it does suggest that protons are inherently unstable—not the immortal building blocks we once thought. Thus any one proton is theoretically in danger of decaying at any moment. In fact, since ordinary water is an abundant source of protons, most GUTs predict that on average one proton should decay per year in each ton of water.

Experiments are now under way to detect such events in huge quantities of water stored in tanks in deep underground mines much like that housing the neutrino telescope discussed in Chapter 10. There the water can be insulated from the spurious effects caused by other exotic particles reaching Earth from outer space. To date, no research group has yet confirmed the prediction that protons do indeed decay.

---

ically: "The higher the temperature, the better the probe of the early Universe."

To appreciate the nature of matter at even higher temperatures, and thus to study even earlier times closer to creation, physicists are now researching a more general theory that combines the electroweak and nuclear forces (but not yet gravity). Several versions of this so-called *grand unified theory*, dubbed GUT for short, have been proposed, although experiments have only recently begun capable of testing which (if any) of these theories is correct. Like the electroweak force just discovered, this grand force is expressed by a boson elementary particle, in this case called the X boson. It is, according to these grand theories, the very massive (and thus energetic) X bosons that play a vital role in the first instants of time.

COSMIC INFLATION: Imagine a time just 10^{-39} second after creation, when the temperature was some 10^{30} kelvin. At that moment, only one type of force other than gravity operated—the grand unified force noted above. According to the theory of such a force, the matter of the Universe must have exerted a very high pressure that pushed outward in all directions. (Classically, pressure is the product of density and temperature, meaning that in the early Universe when each of these quantities was large, the pressure must have been truly vast.) The Universe responded to this pressure by expanding in a regular way as described in the previous sections of this chapter; the temperature dropped as the Universe ballooned, in a manner inversely proportional to the size of the Universe. For example, as time advanced from 10^{-39} second to 10^{-35} second, the Universe grew another couple of orders of magnitude and the temperature fell to 10^{28} kelvin.

Now, according to most GUTs, this temperature—10^{28} kelvin—is special, for this value causes a dramatic change in the expansion of the Universe. In short, when matter is "cooler" than this temperature, the X bosons can no longer be produced; after 10^{-35} second, the energy needed to create such bosons was too dispersed, owing to the diminished temperature. Therefore, as the temperature fell below 10^{28} kelvin, the X bosons disappeared. And suddenly, instead of exerting an outward pressure, the matter of the Universe developed a huge inward tension.

Intuitively, we might reason that such an enormous inward force would halt the expansion of the Universe, or at least slow it down. However, the GUTs predict that the actual effect of the sudden demise of the X bosons was a surge of energy roughly like that released as latent heat when water freezes (an event that occasionally bursts a closed container in the process). After all, energy no longer concentrated enough to yield bosons was nonetheless available to enhance the general expansion of the Universe, in fact to cause it to expand violently or "burst" for a short duration. The youthful Universe, although incredibly hot, was quite definitely cooling and in this way experienced a series of such "freezings" while passing progressively toward cooler states of being. Perhaps the most impressive of all such transitions, the change from outward pressure to inward tension, caused a rapid acceleration in the rate of expansion. This period of rapid expansion has popularly been termed **inflation**. In a mere 10^{-35} second, the Universe swelled (or inflated) some 10^{20} times or more, from roughly 10^{-20} centimeter to about the size of a grapefruit.

As the inflationary phase ended some 10^{-35} second after creation, the X bosons had disappeared forever, and with them the grand unified force. In its place were the electroweak and nuclear forces that operate around us in our more familiar, lower-temperature Universe of today. With these new forces in control (along with gravity), the Universe once again experienced an outward pressure, and thus resumed its more leisurely expansion.

TOWARD CREATION: What about even earlier phases of the chaos epoch—earlier than 10^{-35} second? Can science probe, even theoretically, any closer to creation itself (which occurred by definition at time equal to zero seconds)? Efforts toward this end are currently hampered because physicists are unsure how to incorporate the force of gravity into the correct GUT. No one has yet found a way to invent a "super-grand" unified theory (or super-GUT); such an effort is tantamount to developing a quantum theory of gravity.

Our current knowledge of the behavior of gravity in microscopic domains implies that quantum effects will probably become important whenever the Universe is even more energetic than we have yet considered. The huge energies required could have prevailed only at times earlier than 10^{-35} second, when the Universe was even hotter and denser. Specifically, at a time of 10^{-43} second, when the average temperature was some 10^{32} kelvin, the four known basic forces are thought to have been one—a truly fundamental force operating at energies characterizing the earliest part of the chaos epoch. Only at smaller energies (i.e., at times after 10^{-43} second) would the more familiar four forces begin to manifest themselves distinctly, although in reality all four are merely different aspects of the single, fundamental, super-grand force that ruled at (or near) creation.

To penetrate even closer to creation is currently hardly more than conjecture. Nonetheless, many researchers have a "gut feeling" that once we have in hand the proper theory of quantum gravity, our understanding might automatically include a natural description of creation itself. In this respect it is not inconceivable that the primal energy emerged at zero time from literally nothing. This might be true because, even in a perfect vacuum—a region of space containing neither matter nor energy—particle–antiparticle pairs (such as an electron and its antiparticle opposite, the positron) are constantly created and annihilated in a time span too short to observe. Although it would seem impossible that a particle could materialize from nothing—not even from energy—it so happens that no laws of physics are violated because the particle is annihilated by its corresponding antiparticle before either one can be detected. Furthermore, for such events not to happen would violate the laws of quantum physics, which cite (via the so-called Heisenberg Uncertainty Principle) the impossibility of determining exactly the energy content of a system at every moment in time. Hence natural fluctuations in energy content must occur, *even when the average energy present is zero.*

Should any of the conjecture of the preceding paragraph be valid, the Universe might well have originated by means of an energy change that lasted for an unimaginably short duration—a "self-creating Universe" that erupted into existence spontaneously. This sort of "statistical" creation of the primal cosmic energy from absolutely nothing has been somewhat sacrilegiously dubbed "the ultimate free lunch."

SUMMARY

Our knowledge of the early Universe is uncertain. This should not surprise us, for the earliest moments of the Universe are gone with the expansion, perhaps forever lost to the march of time. Yet virtually all the cosmic data thus far accumulated imply an extremely hot and dense early Universe, growing cooler and thinner with time.

Indeed, it is the process of change among various physical conditions that has gradually promoted the successive formation of galaxies, stars, planets, and life forms. These changes—some well understood, others less so—form the essence of Part III.

KEY TERMS

atom epoch
galaxy epoch
grand unified theory
hadron epoch
inflation
lepton epoch
Matter Era
Radiation Era
stellar epoch

QUESTIONS

1. Describe the greatest change of cosmic evolution—the transition from the Radiation Era to the Matter Era.
2. Why do astronomers regard the currently observed 3-kelvin background radiation as a remnant of the superhot Big Bang?
3. If the various elementary particles of matter are created from packets of energy in the early Universe, explain why the more massive particles (hadrons) would have materialized before the less massive particles (leptons).
4. Imagine yourself able to exist during the lepton epoch of the Universe. Describe what you would observe around you.
5. Speculate about the extent to which science can explore (now and ultimately) the phenomenon of the very creation of the Universe.
6. In a paragraph or two, summarize the dilemma currently confronting scientists regarding the issue of antimatter in the Universe.
7. Explain what is meant by the "electroweak force," and recount why its recent experimental verification has scientists excited.
8. What is meant by the "grand unified theory"?
9. What are some of the difficulties encountered while trying to test grand unification and super-grand unification?

*10. Compare and contrast the extent to which we can probe the past (a) by observing distant galaxies, (b) by studying the cosmic background radiation, and (c) by working with tested theoretical models of the early Universe.

FOR FURTHER READING

Barrow, J., and Silk, J., "The Structure of the Early Universe." *Scientific American,* April 1980.

Crease, R., and Mann, C., *The Second Creation.* New York: Macmillan, 1986.

Davies, P., *Superforce.* New York: Simon & Schuster, 1984.

Green, M., "Superstrings." *Scientific American,* September 1986.

Guth, A., and Steinhardt, P., "The Inflationary Universe." *Scientific American,* May 1984.

Pagels, H., *Perfect Symmetry.* New York: Simon & Schuster, 1985.

Trefil, J., *The Moment of Creation.* New York: Scribner, 1983.

18 PRIMORDIAL EVOLUTION: *Galaxy Formation*

Individual galaxies probably contribute little to the architectural design of the Universe. Galaxies are essentially "along for the ride," much like scattered rocks which play a small role in the overall architecture of Earth as a planet. On the other hand, as we have seen in earlier chapters, galaxies can be used to probe the framework of the Universe, much as rocks can help us study the makeup of Earth. Galaxies resemble billiard balls that could be used to determine the shape of a tabletop, or golf balls that could be used to survey the surface of a putting green. It is in this sense that cosmologists, as noted earlier, use the radiation and motions of distant galaxies to unravel the state and structure of the Universe as a whole.

But how and from what did the galaxies originate? Did such vast coagulations arise from an early Universe containing a mixture of hot matter and intense radiation, or did they emerge later in time? Do galaxies form by gravitationally attracting already-made stars, or do stars form later within already-made galaxies? In other words, which came first, stars or galaxies? Furthermore, how do galaxies evolve, once formed?

Fortunately, we can address these and other questions pertaining to the Matter Era with more assurance than the rather uncertain events of the earlier Radiation Era. Even here, though, substantial puzzles remain about the details of the galaxy formation process. Scientists can address the problem and can identify its main difficulties, but they cannot yet solve it completely. Put more honestly: The origin of the galaxies comprises the greatest missing link in all of cosmic evolution.

The learning goals for this chapter are:

- to understand the basic problems plaguing our current knowledge of the origin of galaxies
- to gain a feeling for the roles of turbulence and gravitational instability in an evolving medium
- to appreciate the general outlines of the best current ideas regarding galaxy formation
- to realize that galaxies, like life forms, can evolve because of both intrinsic and environmental factors
- to recognize that galaxy evolution may hold a key to a better understanding of the largest objects in the Universe

Large Assemblies of Matter

The basic problem regarding galaxy formation centers about our present lack of a good observational understanding of the galaxies themselves. For example, galaxies show nothing analogous to the Hertzsprung–Russell diagram for stars. Galaxies can be classified according to their gross morphological appearances—spirals, ellipticals, barred spirals, or irregulars—as suggested in Chapter 14. Or by their total energy budgets—normal or active galaxies—as discussed in Chapter 15. But we have as yet no explanation for the observed properties of the galaxies in terms of, for example, the simple gas laws that describe our rather detailed understanding of stars.

Nonetheless, all galaxies share two common denominators. Together, these factors may help us understand the events that produced these most magnificent of all objects in the Universe.

First, as noted earlier in this book, no known galaxies are forming at the present time. Galaxies are surely now evolving as stars come and go within the galaxies, but no galaxies are currently originating from primeval matter. Furthermore, none seem to have formed within the past 10 billion years or so. Since all normal galaxies contain some old stars, and since most active galaxies are far away in space (and thus in time), we conclude that all the observable galaxies must have existed for a long time. Whatever the formation mechanism was, it was surely widespread in the early parts of the Matter Era. But if the galaxies formed as prolifically in the early Universe, why aren't they forming now?

A second common denominator derives from the observation that most galaxies house comparable amounts of matter. The masses of all individual galaxies, despite their shape and class, range between 10^9 and 10^{12} solar masses. There are no known genuine galaxies of much smaller mass, and none of much larger mass. They all seem to average to about 10^{11} solar masses, much like our own Milky Way Galaxy. Why should nature's grandest coagulations have such a narrow range of mass? What prohibits the construction of galaxies containing, for example, 10^{15} solar masses (although there are galaxy clusters that large)?

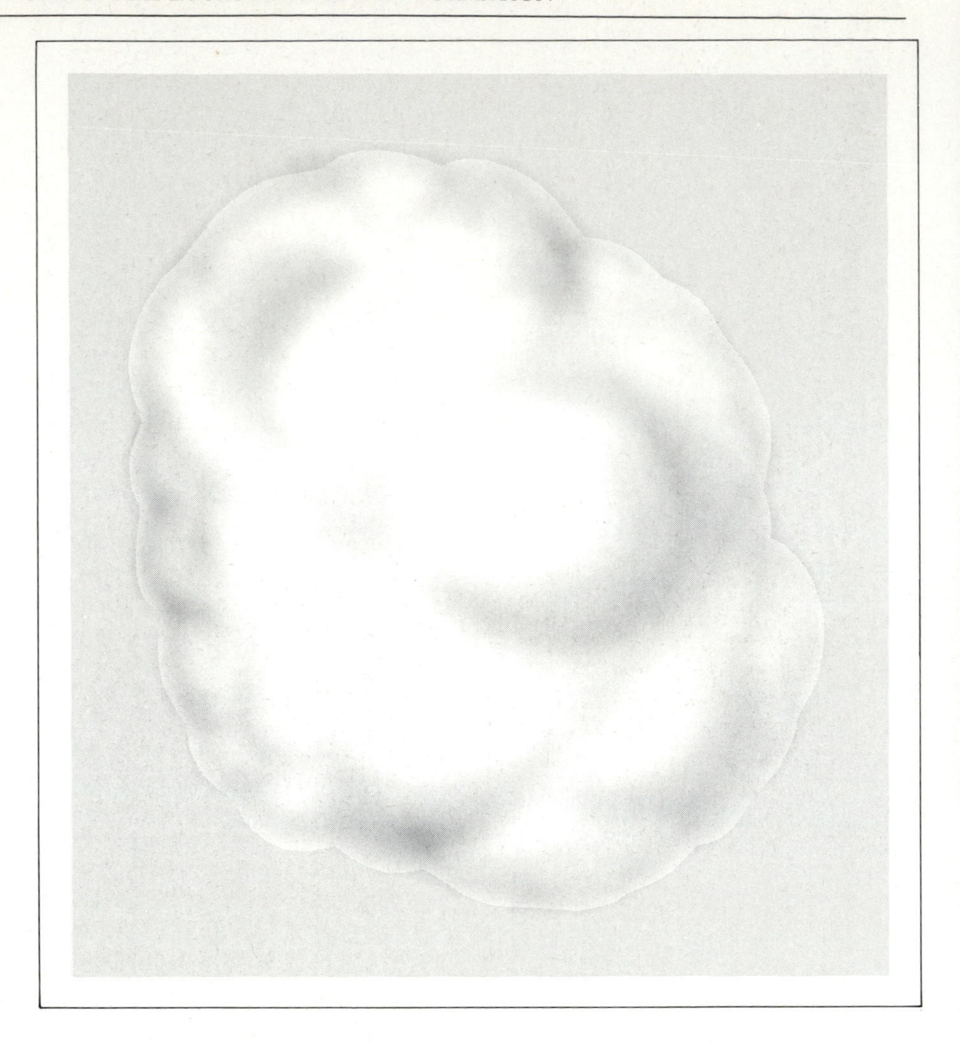

FIGURE 18.1 This artist's conception illustrates how a huge cloud of the early Universe might have fragmented to form several galaxies.

FRAGMENTATION: To address the question of galaxy formation, imagine a giant cloud of hydrogen and helium atoms embedded in a weakening sea of radiation, some tens of millions of years after the bang. This giant cloud should not be thought of as filling the entire Universe, only a small sector of it. The following description of events is keyed to Figure 18.1.

Physical conditions were changing rapidly at the time. The universal temperature and density had dropped considerably with the onset of the Matter Era. Radiation was no longer intense enough to break apart atomic matter. In fact, fully formed hydrogen and helium atoms were becoming sufficiently numerous to exert a collective influence of their own. Force fields, both nuclear and electromagnetic, bound elementary particles within atoms, while the gravitational force field in turn held the atoms within the giant cloud. All the known forces that currently direct the evolution of matter were even then operating well enough to grant the cloud some integrity of its own. Vast parcels of matter were becoming distinguishable from other segments, a state of affairs strongly contrasting with the stupendous and chaotic violence of the earlier Radiation Era.

Despite this stability, the initially homogeneous cloud would have surely experienced occasional gas fluctuations—slight local irregularities (i.e., inhomogeneities) in the gas density that came and went at random. No cloud, whether a fluffy cloud in Earth's atmosphere, an interstellar cloud in our Milky Way Galaxy, or the primordial cloud discussed here as part of the early expanding Universe, can remain completely homogeneous indefinitely. Such gas fluctuations initially are minute—as little as a tenth of 1 percent greater than the average density.

Here's how the gas inhomogeneities naturally arise. Each of the cloud's atoms has some motion, largely because of heat. Eventually, one atom somewhere in the cloud will accidentally move closer to another, making that part of the cloud just a little denser than the rest. The atoms may then separate and hence disperse this density fluctuation, or they may act together to gravitationally attract a third atom to enhance it. In this way, small pockets of gas can arise anywhere in a cloud simply by virtue of random atomic motions. This is how clouds break down into smaller clumps, a process often called **fragmentation**.

Provided that such density inhomogeneities further develop by gravitationally attracting more and more atoms, they could conceivably grow into groups of matter having the size of galaxies. Theoretical calculations support the idea that such random gas fluctuations could have created **protogalaxies**, that is, the forerunners of present-day galaxies. But—and this is an important "but"—these same calculations suggest that, at this rate, the galaxies would only now be forming. If these calculations are correct, we should be able to observe objects having a morphology somewhere between well-defined galaxies and sheer empty space. However, as noted earlier, astronomers have not found any galaxies forming in the present epoch; we have observed no peculiar regions caught in the act midway between full-fledged galaxies and intergalactic nothingness.

The basic problem here is that a great amount of time—at least 20 billion years—is needed for enough randomly moving atoms to coalesce into a large pocket of gas that can be rightfully called a protogalaxy. The long time required is not surprising in view of the absolutely gargantuan quantity of atoms in a typical galaxy—about 10^{11} stars per galaxy times roughly 10^{33} grams per star times about 10^{24} atoms per gram, or therefore about 10^{68} atoms in a normal galaxy. That's an awful lot of atoms to collect. Consequently, it takes a while for nature to do it at random. In fact, apparently nature could not have done it at random, as 20 billion years is longer than the current age of the Universe!

So, despite the fact that random enhancements in an otherwise homogeneous gas could have eventually produced galaxies, it's unlikely that the galaxies we now see originated strictly in this way. Still, the idea of naturally arising gas inhomogeneities remains a powerful concept, because it's a well-understood process not requiring any unknown forces or unique conditions. We shall return to it.

In all fairness, the problem of galaxy formation is currently a tough one for astronomers. Its solution has exasperated many brilliant minds. The origin of galaxies is an area of research that appears to theorists having fertile imaginations, especially those willing to make abnormal assumptions. It's one of the trickiest areas of astronomy to appreciate, for few hard facts are known about galaxies, and fewer still about the physical mechanisms that formed galaxies long ago.

One hard fact that is clearly known, however, is that galaxies do exist. And they exist in great numbers. Somehow they got there. Let's then consider in greater detail some of the specific galaxy formation mechanisms recently proposed by theoreticians.

Gravitational Clustering

Some researchers favor a radical viewpoint and have argued that perhaps the stars formed first, after which the stars clustered to make the galaxies. In their opinion, galaxies did not originate by means of the traditional idea of random inhomogeneities in the early cosmic fireball. Instead, they propose that the galaxies formed much later by gravitationally grouping stars, a model generally known as **gravitational clustering**. This unorthodox view, then, suggests that galaxies could indeed be forming at the present time, but partly from old stars that formed billions of years ago.

According to this theory, first the planets and then the stars originated from random gas fluctuations that developed in the early years of the Matter Era. These were slight enhancements in the gas density, much, much smaller than the ones noted in the preceding section. Once the stars had formed, gravity bundled them together, thus fashioning star clusters. These star clusters in turn migrated to form the galaxies, after which the galaxy clusters and galaxy superclusters formed in succession.

Gravitational clustering is attractive because of its hierarchical package: All big objects are successively constructed from small ones. This type of material buildup is just the reverse of the material breakdown expected if the Universe were to contract in the future. Apart from this pleasing theoretical symmetry, however, practical problems plague the concept.

One problem with gravitational clustering is that—we must repeat—there is no evidence for galaxies forming at present. The regions out beyond the galaxy clusters—the intergalactic medium—do not seem to contain much (if any) matter. Whenever and however the galaxies (and their clusters) did form, they apparently did so very efficiently, sweeping up almost all the matter available. In addition, strong theoretical arguments suggest that stars ought to be forming now within galaxies. As we'll see in the next chapter, these arguments have been verified within the past decade by observations of many sites in our own Milky Way where stars are known to be originating from the galactic hodgepodge of gas and dust. Yet another problem is that it's nearly impossible to imagine the formation of planets prior to the formation of stars. This is especially true since the heavy elements comprising many planets must first be synthesized in the cores of stars. The original stars, almost certainly, preceded the planets.

The crux of the issue here is that reasonably good observational evidence can virtually prove that stars are currently forming and that galaxies are not. Most modern arguments and all modern data point straight toward the notion of an

INTERLUDE 18-1 ***The Second Law of Thermodynamics***

Cosmic evolution can be defined in the simplest terms as the study of change. More specifically, it is the study of the changes in the assembly and composition of energy, matter, and life throughout the Universe. This broad subject attempts to trace a thread of understanding linking the evolution of primal energy into elementary particles, the evolution of those particles into atoms, in turn of those atoms into galaxies and stars, the evolution of stars into heavy elements, the evolution of those elements into the molecular building blocks of life, of those molecules into life itself, of advanced life forms into intelligence, and of intelligent life into the cultured and technological civilization that we now share.

As such, cosmic evolution stipulates that complexity arises from simplicity—we might say, order from chaos. Observations demonstrate that an entire hierarchy of structures has emerged, in turn, during the history of the Universe: energy, particles, atoms, galaxies, stars, planets, life, and intelligence. Yet this increase in complexity with the steady march of time bothers some people. Why? Because cosmic evolution, at face value, seems to violate the second law of thermodynamics.

The second law of thermodynamics is one of the most basic and cherished principles in all of physics. ("Thermodynamics" means "change of energy.") It dictates that randomness or disorder (technically called "entropy") increases everywhere. This is true because the law states that heat (i.e., thermal energy) always flows from a hot source to a cold source. And an inevitable result of this statement is that a house of cards, once built, will tend to collapse with time; by contrast, a random collection of playing cards is not likely to assemble itself into a structured "house." Similarly, water will of its own accord flow over a dam into a lake below, but has never been seen flowing back up to the top of the dam. These are classical examples of "closed systems" wherein events occur in one direction; nature is said to be irreversible or asymmetric.

However, we also have many examples of "open systems" whereby energy (and sometimes matter, too) can be introduced from outside the system. And this can make a great deal of difference. For example, a human being, by exerting some energy (and some patience, too, which also burns energy), can rebuild a house of cards. We could also use a water pump (which also requires energy) to transport water from a low-lying lake to above a high-lying dam.

The infusion of energy (or matter) into any system can potentially yield organized structures. Disorder (or entropy) can actually decrease within such open systems—which galaxies, stars, planets, and life forms are—even though that disorder is increasing everywhere else in the Universe. Such "islands of structure" do not violate the second law of thermodynamics, however, because the net disorder of the *system and its environment* always increases. The energy needed to run a water pump—or any such device of our industrial civilization—comes at the expense of the environment which is generally being ravaged to power our civilization; both digging for coal and burning it disorder the environment more than the order achieved by using the energy. The energy provided by humans to build a house of cards—or a table, chair, automobile, whatever—derives from the food we eat; we literally feed off our neighboring energy sources—locally plants and other animals, and more fundamentally the Sun.

For example, here's what happens in the food chain consisting of grass, grasshoppers, frogs, trout, and humans. According to the second law, greater disorder occurs at every stage of the food chain. At each step of the living process, when the grasshopper eats the grass, the frog eats the grasshopper, the trout eats the frog, and so on, a toll is taken on the environment. The numbers of each species required for the next higher species to continue decreasing entropy are staggering. In fact, the support of one human for a year requires some 300 trout. These trout, in turn, must consume 90,000 frogs, which yet in turn devour 27 million grasshoppers, which live off some 1000 tons of grass. Thus, for a single human being to remain "ordered" (namely, to live) over the course of a single year, we need the energy equivalent of tens of millions of grasshoppers or about 1000 tons of grass.

To be sure, humans are heat engines that temporarily—for roughly 70 years in the West—maintain the order of our bodies by creating greater disorder in the surrounding environment. Once we stop eating (or intaking energy), we die—and eventually decay into disordered chaos. The same is true of a galaxy or a star; during their formative stages, they increasingly order themselves by utilizing energy and dumping entropy into their surroundings. Eventually, after billions of years, they too will break down into disorder.

In this way, any form of matter or life can construct and order itself by exchanging energy (and entropy) with the ouside world. No laws of physics are violated. In particular, the evident order of the Universe around us does not at all contradict the basic principles of modern thermodynamics.

early formation of galaxies, followed by a later formation of stars and planets within those galaxies.

Gravitational Instability

Not willing to accept mildly radical theories like gravitational clustering or strongly radical theories suggesting that galaxies originate from the outgrowth of small dense objects, most astronomers today embrace some version of the random gas fluctuation concept discussed earlier. Remember, theory suggests that random gas fluctuations arose all right, but that their growth would have been too slow to form galaxies before now. If some way could be found to accelerate their growth, the problem might be solvable. One such way is to assume that the early Universe was very turbulent.

Turbulence was probably an important factor in the early Universe. By **turbulence** we mean the inevitable "confusion" or disordered motion of matter (the gas) within a rapidly moving medium (the Universe). All the atoms within the vast primordial clouds were set into motion, not only from the expulsion of the bang, but also from the heat of the fireball. As a result, the gas had some "directed" kinetic energy—outward from the ordered expansion of the Universe. It also had some "undirected" kinetic energy—irregular and random, from the disarrayed aftermath of the blazing inferno.

Turbulence and disorganization seem to go together; wherever there is turbulence, there are likely to be swirling eddies. That's because the turbulent flow of gaseous matter itself generates the eddies. In particular, turbulence may well have helped to drive the swirling eddies, thus accelerating the growth of density inhomogeneities beyond their natural development in the early Universe.

It's hard to imagine whirling eddies billions of years ago, but some familiar analogies might help. Figure 18.2 shows how turbulent eddies can be visualized by watching water swirl in the wake of a rock in a stream. Even better examples can be noted by moving your hand gently through water, or a teaspoon through coffee; swirling eddies naturally form in the wake of this turbulence.

FIGURE 18.2 Eddies naturally occur behind a rock in a stream or a moving canoe paddle.

Figure 18.3 shows another familiar example, in this case turbulent eddies within clouds of Earth's atmosphere. This is a photograph of

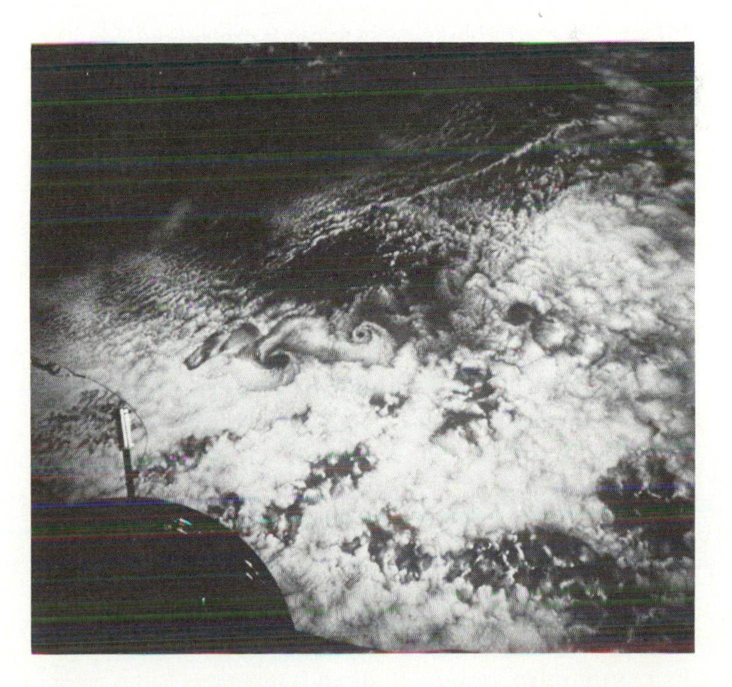

FIGURE 18.3 Satellite photographs of the top of the Earth's cloud layers clearly show atmospheric eddies. Such swirling vortexes are especially vivid when strong surface winds encounter a mountain, creating air turbulence. This photograph was taken by a *Gemini* spacecraft above Tenerife, where the eddies result from the Atlantic winds hitting the Canary Islands.

the *top* of the clouds, taken from an Earth-orbiting satellite. Weather observations show that such kilometer-sized swirls often come and go at random. These eddies become more pronounced whenever air currents are especially turbulent. Once in a while, such an eddy can accumulate large quantities of moisture and thus grow into a full-fledged hurricane hundreds of kilometers across.

In point of fact, the process we are contemplating for the origin of galaxies is not unlike that of the formation of rain clouds in our own atmosphere. Moisture-laden air, warmed by sunlight at the surface of the Earth, rises to higher levels of the atmosphere, which are normally cooler. There the air cools off and the moisture condenses into droplets of water. In a similar way, the primordial gas emerging from the Big Bang was cooled by the aging of the Universe to the point that gravitational forces could draw them together to form gas clouds that became galaxies.

Here is a case, then, where studies of a terrestrial phenomenon—Earth's weather—might help us understand one of the most troubling celestial problems. As suggested in Figure 18.4, planetary hurricanes mimic the overall morphology, the pancake shape, the differential rotation, and the concentration of energy within spiral galaxies. These several resemblances suggest that we may be able to learn something about galaxy formation via the study of hurricane formation. In particular, since most meteorologists agree that some sort of turbulent "priming" is needed to initiate a hurricane, the early stages of such storms could conceivably be used by astronomers to extract some clues about the elusive density fluctuations that gave rise to protogalaxies in the early Universe.

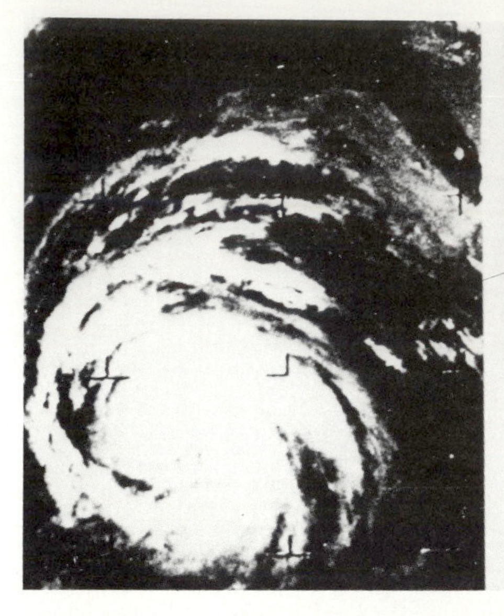

FIGURE 18.4 Terreestrial hurricanes seem to have several similarities when compared to spiral galaxies. At left is Hurricane Beulah in 1977, spanning about 1000 kilometers in the Gulf of Mexico. (Note Florida in the upper right corner.) At right is M51, the Whirlpool Galaxy, which spans nearly a billion billion kilometers (or some 100,000 light-years) across.

DEVELOPING EDDIES: Let's now return to the vast primordial cloud, imagined with the help of Figure 18.1. We can suppose it to be made exclusively of hydrogen and helium. And we can assume that it has some small-scale atomic motions here and there, as well as some large-scale gas turbulence.

Despite the constant cooling caused by the expansion of the Universe, the gas comprising individual eddies within the cloud will begin to heat up. It can't avoid doing so. That is, eddies are sites not only of turbulence, but also of increasing heat within a steadily cooling cloud. The heat results from friction caused by frequent collisions among the increasingly dense collections of atoms within each eddy. It's a little like the heat derived by rubbing our hands together on a cold winter day.

Where does the heat come from? The source is the cloud's gravitational potential energy. Recall the analogy about potential energy used earlier in Chapter 2. A pencil, held high above a tabletop, is said to have some gravitational potential energy relative to the tabletop; the pencil has even more potential energy relative to the more distant floor below. Potential energy is a measure of the energy that could *potentially* be gained should an object fall a certain distance; if the fall is caused by gravity, we call it gravitational potential energy. While falling, some of that potential energy is changed into kinetic energy—the energy of motion. Upon hitting the tabletop, the pencil of course stops falling. There the pencil has no potential energy relative to the tabletop, since the two are not separated by any distance; nor does the pencil have any kinetic energy since it has no motion on the table. During this process, the original potential energy of the pencil changed into kinetic energy while it fell. That kinetic energy was in turn changed into thermal energy (heat) when the pencil collided with the tabletop. The heat created is not large, but a sensitive thermometer could be used to measure it. No energy disappears; it just changes from one type into another.

Similar reasoning can be applied to a gas cloud. The atoms near the edge of the cloud have some gravitational potential energy relative to the center of the cloud. Provided that the cloud is massive enough, the combined gravitational pull of each atom on all the other atoms will force the cloud to contract. As the atoms at the cloud's edge fall toward the cloud's center, their potential energy changes to thermal energy. In short, the infalling cloud heats up.

This newly gained heat causes the individual atoms to increase their motions. In this way, thermal energy tends to buoy a cloud, that is, to retard its contraction. To some extent, heat counteracts gravity. That's why the Sun remains a large ball of gas; its great heat prevents gravity from collapsing it into a small ball.

The physical status of any gas cloud is mostly governed by the competition between its potential energy and its thermal energy. A cloud is unstable if its (inward-pulling) gravitational energy exceeds its (outward-pushing) thermal energy. Such a **gravitational instability** causes the cloud to contract. Some clouds can collapse rapidly if they are cool enough. If the conditions are reversed—should the thermal energy of the cloud exceed its gravitational energy—the cloud will expand, becoming larger and thinner. Some clouds can disperse entirely if they are hot enough. These opposing conditions can be expressed as two inequalities:

gravitational potential energy >
thermal energy (for contraction)

gravitational potential energy <
thermal energy (for expansion).

All gas clouds try to contract or expand until their gravitational energy equally balances their thermal energy. Such clouds have thus achieved an equilibrium. Apart from minor fluctuations in the gas density at small places here and there, clouds in equilibrium will remain that way unless they are otherwise influenced—that is, until they are heated, cooled, or pushed around in some way.

Since our main goal here is to understand the origin of galaxies, let's suppose that the conditions for contraction are satisfied. A gravitational instability sets in, since the inward pull of gravity exceeds the outward buoyancy of heat. As the cloud contracts a little, its density increases. This causes some heating. We might expect the newly generated heat to bring the cloud into equilibrium because, after all, the thermal energy would steadily increase until it equals the gravitational energy. If this were all that occurred, all clouds would easily stabilize. Neither galaxies nor stars could form—and you and I would not be here.

Eventually, individual eddies must lose some of their newly acquired thermal energy. Any heated object, gaseous or solid, for example, the Sun or the Earth, must unload some of its energy, lest it blow up. The eddies in the galaxy-forming clouds do it by radiating away some of their heat. In this way, a large cloud containing many eddies can cool even faster than can a normally homogeneous cloud. As it cools, the entire cloud contracts a little, thereby increasing the density and hence the heat within the eddies. The individual eddies and the whole cloud simultaneously radiate much of their newly gained energy into space, thereby allowing further contraction of both the parent cloud and the individual eddies. Provided that the thermal energy remains less than the gravitational energy, the cloud will continue to contract. On and on, this cycle of contracting, heating, radiating, cooling, and contracting proceeds. The cycle may operate at different speeds for each of the eddies, particularly since some eddies will be more successful than others at sweeping up additional gas from the parent cloud.

As sketched in Figure 18.5, it's easy to conceptualize a cluster of galaxies forming in this way, with each eddy becoming a member galaxy within that cluster. Alternatively, perhaps only one or a few galaxies formed within each of the vast primordial clouds in the early Universe, after which gravity gradually pulled the galaxies into the huge galaxy clusters like that shown in Figure 18.6.

A COMPLICATION: If the vast primordial clouds had some rotation—and surely parts of them would have, at least at the sites of the swirl-

FIGURE 18.5 Eddies of gas within a large primordial cloud would have experienced a cyclical process of contracting, heating, radiating, cooling, contracting, and so on. Those eddies that survived the process presumably became galaxies.

FIGURE 18.6 Nearly every point of light in this photographic time exposure represents a galaxy. Tight-knit galaxy clusters like this one, designated the Hydra cluster, generally confirm the galaxy-formation ideas discussed in the text. The Hydra cluster is more than 3 billion light-years distant.

ing eddies—the eddies must have changed their shapes as they contracted. Why? Because of the tendency of material in a rotating body to keep rotating. This is **angular momentum**, another one of those basic physical quantities, like matter and energy; specifically, it is the tendency of an object to keep spinning or moving in a circle.

To appreciate how angular momentum tends to change the shape of a contracting object, consider first its simpler counterpart, **linear momentum**—the tendency of an object to keep moving in a straight line. A truck and a bicycle rolling equally fast down a street each has some linear momentum. Trying to stop them, you would obviously find it easier to halt the less massive bicycle. Although each vehicle has the same speed, the truck has more momentum. Thus the linear momentum of an object depends on the mass of that object. It also depends on the velocity, for if there were two bicycles rolling down the street at different speeds, you would more likely be able to stop the slowly moving one. Consequently, linear momentum is defined as the product of both mass and velocity:

$$\text{linear momentum} = \text{mass} \times \text{velocity}.$$

Angular momentum is only a little more complex, and it applies to objects having some rotation or revolution. In addition to mass and velocity, angular momentum depends on the size of the object:

$$\text{angular momentum} = \text{mass} \times \text{velocity} \times \text{size}.$$

(The velocity term for linear momentum is velocity in a straight line, whereas that for angular momentum refers to the circular velocity of spinning or orbital motion.)

Each of these types of momentum must be conserved at all times. In other words, momentum must remain constant before, during, and after the physical change of any object. For example, if a spherical object having some spin begins contracting, the relationship above demands that the circular velocity grow larger. After all, the object's mass does not change during contraction, yet the size of the object clearly decreases. The circular velocity of the spinning object must therefore increase in order to keep unchanged the total angular momentum—the product of mass, circular velocity, and size—at all times.

Figure skaters know well the principle of angular-momentum conservation. As shown in Figure 18.7, when the arms are drawn in, the mass of the human body remains the same but the lateral size decreases. Hence the circular velocity must increase in order to conserve angular momentum, and the skater spins rapidly.

Figure 18.8 illustrates how continued contraction can force a cloud to rotate more and more rapidly. As the spin increases, the cloud gradually flattens. The greater the rotation, the flatter the cloud becomes. The flattening occurs since matter can contract more easily along the axis of rotation than perpendicular to it. In all, the whole cloud becomes smaller, particularly parallel to the spin axis.

In some cases, angular momentum can become large enough to counteract gravity. Any rapidly spinning object has a tendency to force matter away from its center, much like mud thrown from a rotating bicycle wheel. So heat is not the only agent capable of opposing gravity. Rotation can do so as well. In fact, it is rotation, not heat, that stabilizes galaxies once they are fully formed.

The broad outline of this contraction–flattening sequence for galaxy formation is supported by observations. As discussed in Chapter 13, young stars are spread throughout the plane of our Milky Way Galaxy, although none of them resides in the halo. Only old stars exist in the halo of our Galaxy, as well as other galaxies. Apparently, galaxies were more spherical in their youth when their first stars were forming, after which the galaxies flattened somewhat into the shapes now seen.

ONE MORE PROBLEM: The foregoing scenario of galaxy formation seems reasonable in words, but it runs into some problems when mathematics are applied to it. The trouble is this: Detailed calculations suggest that the time required for the clustering and contraction of the gas in a turbulent eddy is longer than the

FIGURE 18.7 Figure skaters, like Katerina Witt shown here winning an Olympic Gold Medal, know well that the spin of their "pirouette" can be increased by tucking in their arms. That is, as her lateral size decreases yet her mass remains the same, her circular velocity must increase. And just as her skirt flew outward, the primitive Solar System developed a disk perpendicular to its own spin axis.

typical time for the random dissipation of that eddy. In other words, the eddies of the early Universe are more likely to break up before they have a real chance to coagulate tightly. Turbulence produces eddies, thus enhancing the random gas fluctuations all right, but the eddies don't last long enough to form galaxies.

Eddies are ephemeral things, coming and going at random. They appear, disappear, and reappear at different parts of either a terrestrial atmospheric cloud of moist air or an extraterrestrial galactic cloud of primordial gas. Occasionally, a terrestrial eddy does indeed grow to form a hurricane. Similarly, on occasion an extraterrestrial eddy might presumably persist to form a genuine galaxy. But the expected rarity of their growth implies that turbulent eddies cannot be the sole solution to the problem of the formation of galaxies and of galaxy-like objects.

Primordial Fluctuations

As mentioned earlier, most researchers avoid radical theories of galaxy formation. They prefer to work with the basic notion of random gas fluctuations. But some way must be found to speed the growth of such fluctuations in the cooling and thinning primordial Universe. For if the Universe emerged from the Radiation Era as an initially homogeneous blob with matter spread evenly throughout all of space, it would seem that galaxies could not have formed in the time available. Turbulence apparently helps somewhat, but not enough to account for the multitude of galaxies now observed. Thus current research centers on other ways that might have further enhanced the growth of the gas fluctuations. Modern theorists have tried to do just this for years,

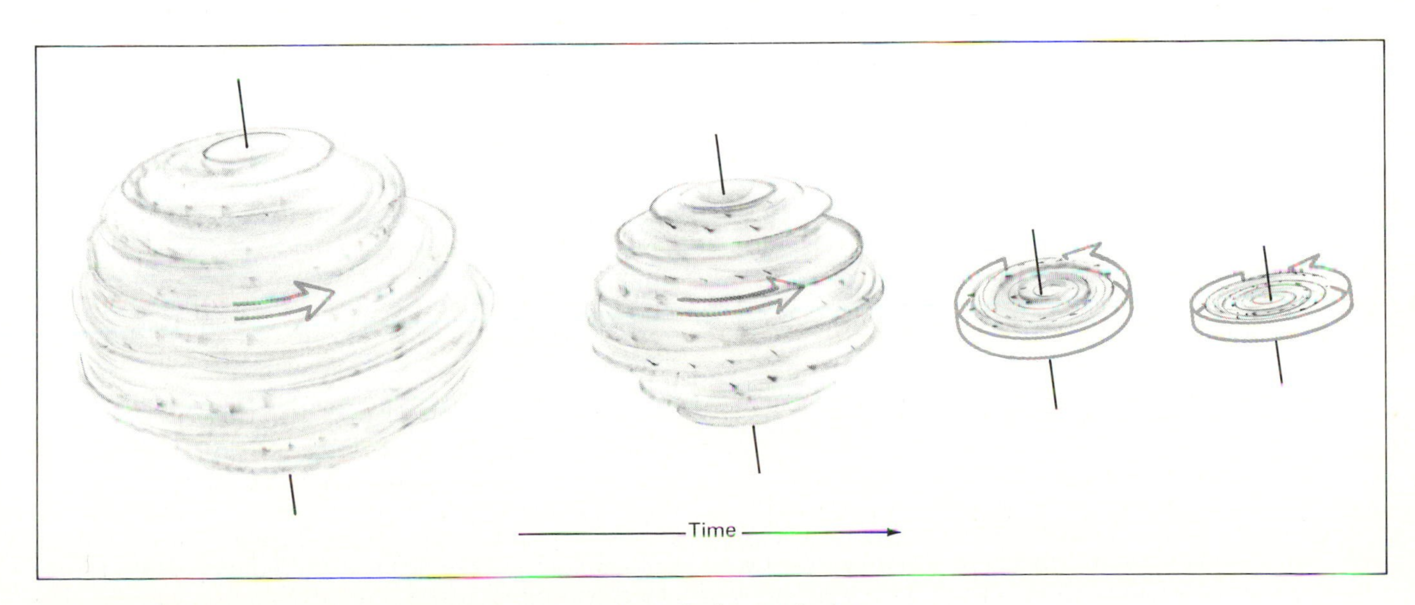

FIGURE 18.8 The conservation of angular momentum demands that a spinning and contracting gas cloud gradually flattens along its axis of rotation.

but they have not yet really succeeded.

The astronomical community nonetheless has two favorite scenarios for the formation of galaxies. Both these models, by no means accepted by everyone, postulate that the early Universe was *not* perfectly homogeneous, as suggested earlier in this chapter. Instead, the Universe is theorized to have been peppered, even at the start of the Matter Era, with many density fluctuations (consult Interlude 18-2). These already-formed pockets of gas could then have grown within the galaxy epoch to fabricate at least the basic outlines of today's galaxies. This mechanism of galaxy formation remains the familiar gravitationally induced cycle described in the preceding section, but the time needed to gather enough matter is shortened. To be sure, if gas density inhomogeneities were already present toward the end of the Radiation Era, the whole process would have been accelerated. Figure 18.9 illustrates the salient features of these galaxy formation scenarios, which we now discuss in slightly greater detail.

TWO LEADING MODELS: Analyzing the situation further, astronomers have found that before the change from the Radiation Era to the Matter Era—which occurred about 100,000 years after creation—two distinct types of fluctuations could have been present in the gas of the early Universe. The first type of fluctuation maintains a constant temperature throughout the eddy; these are technically termed "isothermal" fluctuations since there is no temperature change involved. The second, technically termed "adiabatic" fluctuations since no heat is transferred within the eddy, is a kind of disturbance that varies in step with the density of the gas. The behavior of the two types of disturbances differs greatly.

On the one hand, the variable-temperature (or adiabatic) disturbances within eddies containing less than a trillion solar masses would have died out with time; only those denser-than-average eddies housing more than 10^{12} solar masses could have survived the great change from the Radiation Era to the Matter Era. The reason for this lies in the behavior of the radiation trapped in the slightly denser eddies. Because it exerts pressure, the radiation causes an oscillation in the gas density, just as in a sound wave. But the trapping is not complete if the eddy is too small, and the radiation diffuses away after a few oscillations—and with it the eddy. Only if the eddy is large

INTERLUDE 18-2 ***Primeval Lumps***

One of the key predictions of the inflationary scenario described in Chapter 17—and thus another indirect test of the many different grand unified theories—concerns the origin of galaxies. Much as we have discussed throughout this chapter, small-scale fluctuations in cosmic matter are likely to be present at any given moment of the Universe. Even within the earliest moments of the Universe, quantum physics maintains that statistical fluctuations were ubiquitous. Accordingly, during the split-second inflationary period, any such surviving fluctuations could have been amplified to an extent now comprising entire galaxies and clusters of galaxies. Thus the growth of gravitational instabilities, greatly aided by the phenomenon of inflation, might well have gradually led to the formation of huge self-gravitating aggregations that top the hierarchy of matter in the Universe.

Should this idea be correct, then, surprisingly enough, the vast clumps of matter we see today as galaxies, galaxy clusters, and even potential galaxy superclusters are the remnants of quantum fluctuations prevalent when the Universe was a mere 10^{-35} second old.

Physicists have been able to calculate, tentatively, the size and scale of the density fluctuations at this extraordinarily early time. Their results to date seem to favor the onset of the variable-temperature (or adiabatic) disturbances that would have been large enough to survive—namely, greater than 10^{12} solar masses. This would lead to a scenario in which the first structures to form were far larger than individual galaxies. Such a formation scheme might help us understand why different types of galaxies would originate in different environments.

What is most needed here are observations capable of testing these many ideas of galaxy formation. The next generation of large telescopes capable of exploring the most distant and therefore the most ancient realms of the cosmos, especially NASA's orbiting *Hubble Space Telescope,* should soon be gathering data to elucidate the origin of galaxies by searching for gas fluctuations and even protogalaxies early in the Matter Era. And other instruments, especially the *Cosmic Background Explorer* satellite (to be orbited soon by NASA), should be able to observe directly any disturbances in the cosmic background radiation flooding the Universe; thus we stand a chance soon of perceiving the density fluctuations that survived the Radiation Era and therefore of better understanding how the suspected gravitational instabilities gave rise to galaxies long ago.

Somewhat ironically, with the physicists unable to build equipment on Earth sufficiently energetic to reproduce the earliest epoch of the Universe, it is the astronomers who, by studying the macro-realm, are beginning to provide tests, albeit indirect ones, of the grand unification of the micro-realm.

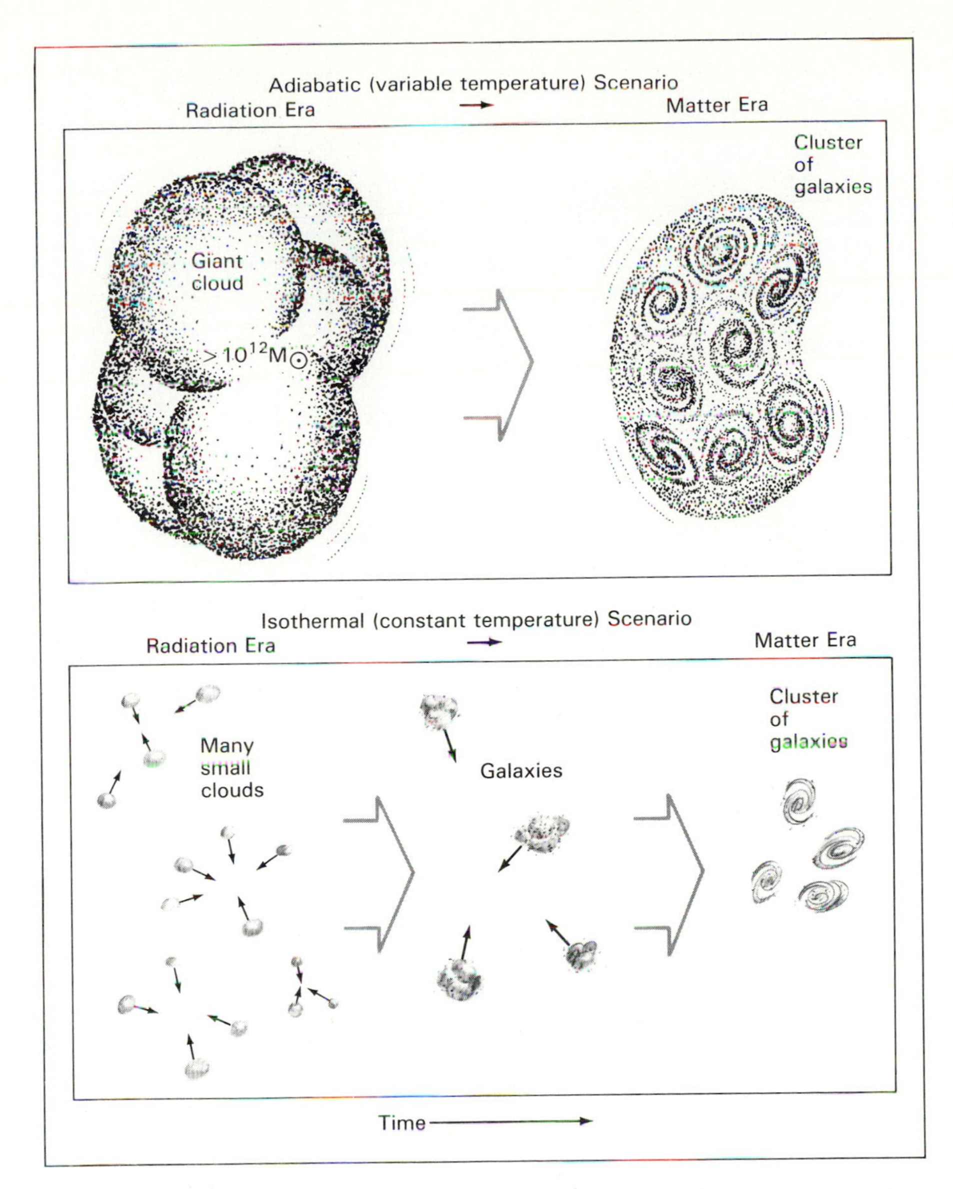

FIGURE 18.9 Schematic diagram showing how slight enhancements in the average density of matter might have led to the galaxies long ago, provided that such turbulent inhomogeneities had become established during the Radiation Era, before the Matter Era began.

enough can this type of oscillation survive until the Matter Era (and especially the galaxy epoch) begin.

On the other hand, constant-temperature (isothermal) disturbances embracing relatively small amounts of matter could have survived throughout the Radiation Era; the pressure of radiation in such fluctuations is constant throughout space anyway, and the matter is distributed quite independently.

In recent years, researchers have used powerful computers to explore these two very different types of primordial fluctuations. Thinking that constant-temperature fluctuations were more likely, they have spent much of their effort on them. As might be expected, moderate-sized clouds (although smaller than galaxies) form first, after which they agglomerate to form galaxies; as time passes, the galaxies themselves agglomerate to form larger units, such as galaxy clusters, sheets, and filaments. Because the results appear to mimic the real Universe, many astronomers have been encouraged by this work to think that galaxies really did form from constant-temperature fluctuations in the early Universe.

But closer inspection of the spread of galaxies in space suggests a problem. Observations of the vast clusters of galaxies clearly show a tendency for the elliptical galaxies to be found in regions where the numbers of galaxies are greatest, whereas spirals are usually found where the numbers of galaxies are low. If individual galaxies formed first and the galaxy clusters formed later, why would the ellipticals be segregated from the spirals? After all, the clustering ability of two or more galaxies depends only on gravity, and this long-range force does not care whether the galaxies are spiral or elliptical. These recent observations, along with some theory (see Interlude 18-2), imply that individual galaxies more likely formed only after the huge clouds of which they were a part were already quite compressed—the huge clouds being the parents of the immense galaxy clusters.

No one currently knows which if either scenario of galaxy formation is correct. Did constant-temperature fluctuations dominate in the early Universe, leading first to subgalaxy objects that later accumulated to form galaxies and finally clusters of galaxies? Or did variable-temperature fluctuations dominate, leading first to clouds far larger than galaxies, from which individual galaxies and eventually stars formed later? In the years ahead, astronomers will be searching for ways to discriminate between these possibilities.

Much of the fascination experienced by researchers studying galaxy formation derives from the fact that many of the theories cannot be proved incorrect. Many ideas remain possible, there being few experimental data to the contrary. Workers familiar with the sophisticated mathematics of notoriously tough subjects such as fluid dynamics, gas turbulence, and magnetohydrodynamics justify their interests by tinkering with the problem of galaxy formation. Yet, despite considerable efforts in the past decade or two, the specifics of a plausible galaxy formation process have thus far eluded discovery.

Galaxy Evolution

Regardless of how galaxies originated, either the formation process itself or some subsequent evolutionary process produced the sizes and shapes of the myriad galaxies now observed. There are loose, tight, and barred-spiral galaxies, each containing mixtures of old and new stars. There are small and large elliptical galaxies, each containing only old stars. And there are irregular and explosively active galaxies, not to mention the mysterious quasars, which may not house any stars at all.

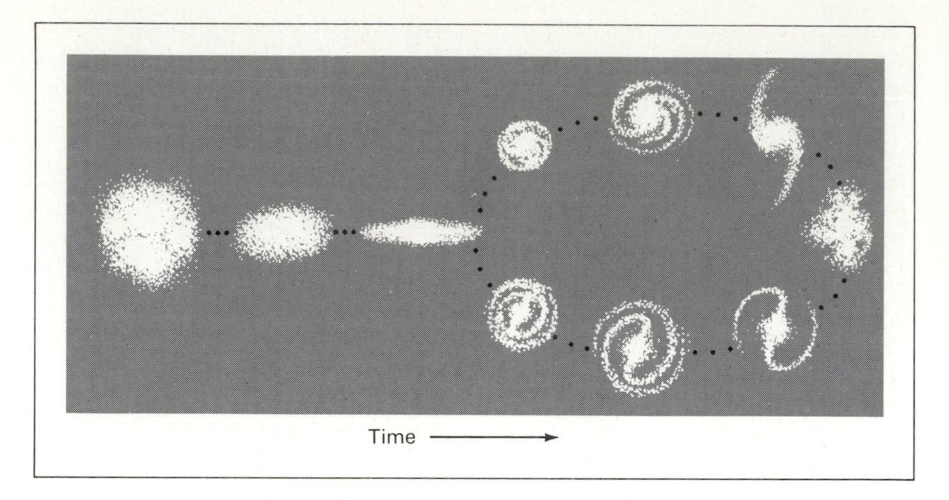

FIGURE 18.10 This idea for an evolutionary connection between elliptical, spiral, and irregular galaxies is now known to be incorrect.

NORMAL GALAXIES: Astronomers long ago suspected an evolutionary progression of galaxies starting with the near-spherical EO ellipticals, which became the flattened E7 ellipticals, eventually changing into closed spirals, followed by open spirals, and finally culminating in the irregular galaxies. Figure 18.10 is a schematic illustration of this evolutionary scheme. The idea here is that galaxies originate with a more or less spherical shape. As they grow older, their rotation tends to flatten them, first producing some ellipticity, and gradually some spiral arms, prior to their breaking up as aged irregular galaxies. This type of evolutionary hypothesis requires that all ellipticals be young and all irregulars old. But this is not the case at all. Observationally, elliptical galaxies are not young; they're old, nearly depleted of interstellar gas and dust, and displaying no evidence of current star formation.

On the other hand, we might argue that since ellipticals are clearly old galaxies, then perhaps the evolutionary scheme progresses in the opposite sense. Maybe irregulars are young and, having formed first, gradually evolve into ellipticals. It's easy to imagine loose spiral galaxies wrapping up into tighter spirals and eventually into elliptical galaxies. But there are problems here, too. First, we are hard pressed to understand how the beautiful spiral galaxies might have been fabricated from the distorted irregulars. Second, how can we reconcile this idea with the abundance of old (globular) stars in the irregular and loose spiral galaxies? Many of those stars lack heavy elements, thereby proving their old age. Simply put: If irregular galaxies and loose spirals are the starting point in any scheme of galaxy evolution, all of them should be young. But they are not. Virtually all irregulars and spirals house a mixture of old and new stars. The existence of old stars is just not consistent with the nature of a youthful galaxy.

The bottom line is that normal galaxies probably do not evolve from one type to another. Spirals do not seem to be ellipticals with arms, nor do ellipticals appear to be spirals without arms. In short, astronomers know of no parent–child relationships among the normal galaxies. Apparently, the various types of galaxies are more like cousins who trace their birth to the same ancestor—the varying turbulence of the primordial gases in the aftermath of the big bang.

INTRINSIC CHANGES: Many of the key issues concerning the evolution of astronomical objects mimic those of biological objects. Most students know (and if you don't, we shall study it in Chapter 26) that biological evolution among life forms on Earth is driven by internal (genetic) changes and by external (environmental) changes—and often by both working together. In the case of galaxy evolution, we can pose the same kinds of questions: Do galaxies evolve because of intrinsic changes—that is, by means of some process inherent to the galaxies themselves? Or do they evolve by responding to changes in their immediate environment?

The answers to these questions seem to be yes, on both counts, and again both kinds of change might well occur in a given galaxy. In this section we discuss briefly the first (internal) kind of change; in the next section we shall take up the issue of environmental change. Intrinsic evolution can perhaps best be illustrated by considering star formation and evolution, a topic we shall address at some length in the next chapters, but one about which we already know enough to illustrate some ideas that are central to galaxy evolution.

Star formation apparently proceeds at different rates in spiral and elliptical galaxies; after all, spirals currently contain large amounts of interstellar gas and dust, whereas ellipticals contain little. Discussed in Chapter 14, this fact is clear for two reasons: Dusty regions are associated with spiral and not elliptical galaxies, and 21-centimeter radio radiation from spiral galaxies is strong whereas that from ellipticals is often weak or absent (implying that atomic hydrogen is missing in the ellipticals). Astronomers suspect that early in the "life" of an elliptical galaxy, the star-formation rate was very high. The massive stars soon exploded (since they use their fuel more rapidly, as noted in Chapter 11), and the ensuing conflagration from many such explosions drove the remaining loose gas from the galaxy, thus eliminating the material needed for further star formation. We can envision such an outflow of gas as forming a "galactic wind," in analogy to the solar and stellar winds that are directly observed (and that were discussed in Chapters 9 and 11). Explosions of this sort (called "supernovae") occur frequently enough even today to keep ellipticals swept clean of interstellar matter.

By contrast, in spiral galaxies, stars might not have initially exploded frequently enough to cause a catastrophic purging of interstellar space, so a sufficient amount of interstellar matter remains today to support active star formation. Thus the differing amounts in interstellar matter in spirals and ellipticals conceivably results from the different initial rates of formation of massive stars, which later explode. Why the rates of star formation might have differed is an unsolved problem of galaxy evolution that can be addressed only by observing galaxies as they were long ago. Since "looking out is looking back," such observations are possible in principle by studying galaxies having great distances. In practice, however, these observations are difficult because of the faintness of the galaxies involved.

Nonetheless, x-ray astronomy seems to offer a way to test these ideas. The existence of galactic winds is quite consistent with current x-ray observations of rich clusters, especially those clusters in which most of the galaxies are ellipticals. In many such cases, astronomers have recently found hot, x-ray-emitting intracluster gas whose total mass and chemical composition agree with the expected accumulations of galactic winds from the various member galaxies of the cluster. Removal of loose gas from the galaxies is further aided by any intracluster gas; such intracluster gas can sweep matter from the galaxies as they move through it. Figure 18.11 is an x-ray image that dramatically illustrates how galaxies can be swept clean of any stray interstellar matter not bound in stars.

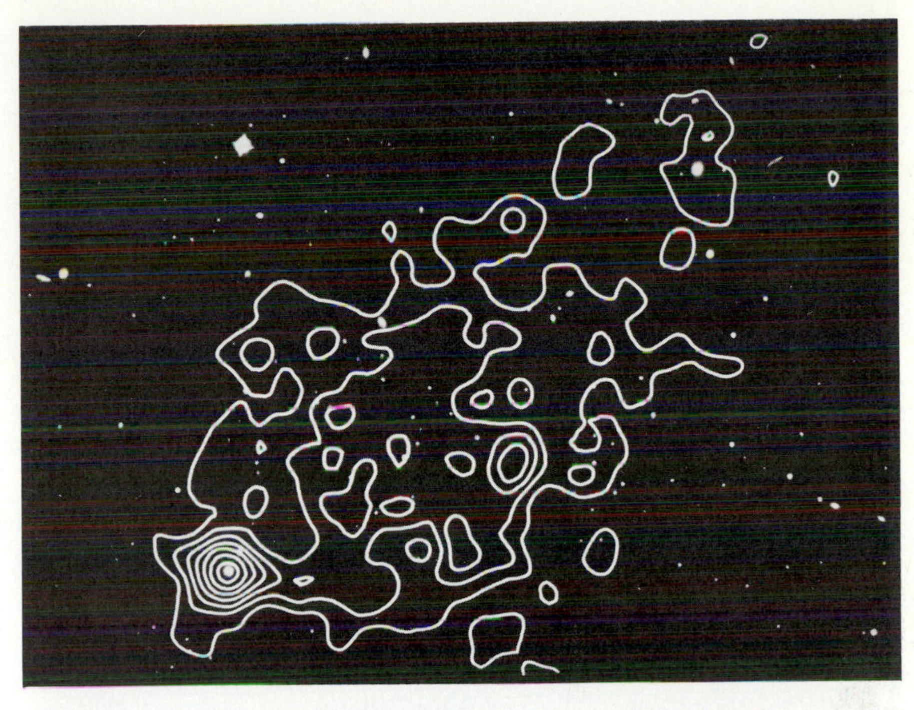

FIGURE 18.11 These x-ray contours superposed on an optical galaxy cluster, known as A1367, imply that its member elliptical galaxies have been swept clean of their loose interstellar matter. Instead of being associated with specific galaxies of the cluster, the x-ray radiation shows a broad distribution of intracluster gas throughout the cluster.

ENVIRONMENTAL CHANGES: The last point just mentioned—that galaxies are possibly being swept clean by some intracluster gas—is really an example of an environmental change. To be sure, astronomers have good reason to suspect that galaxies also evolve in response to environmental factors. As noted in Interlude 14-2, these kinds of changes are driven by gravitational interactions with other galaxies.

Take the case of spiral galaxies, which we now realize have huge, invisible halos and coronas (see Chapters 13 and 14) whose substantial mass is hidden in faint stars or in some other invisible form of matter. Such galaxies can orbit around one another, forming a "binary galaxy." As they do so, they tend to interact with each other's halos, one galaxy stripping the halo material from the other by tidal forces; the freed matter is then either redistributed within a common envelope or is entirely lost from the binary system. The orbits of the galaxies themselves are also likely to change their sizes and shapes. Moreover, if one galaxy of the pair has a low mass, it may end up spiraling into the other, finally merging with the more massive galaxy; since such events invariably make the more massive galaxy still more massive, the process is colloquially termed "galactic cannibalism" or even "galaxy gobbling." Such cannibalism might explain the fact that supermassive galaxies are often found at the cores of rich galaxy clusters. Having dined on their companions, they are now content to relax at their center. Figure 18.12 is an astonishing image that has apparently captured this process at work!

For example, consider two spherical galaxies, one a little smaller than

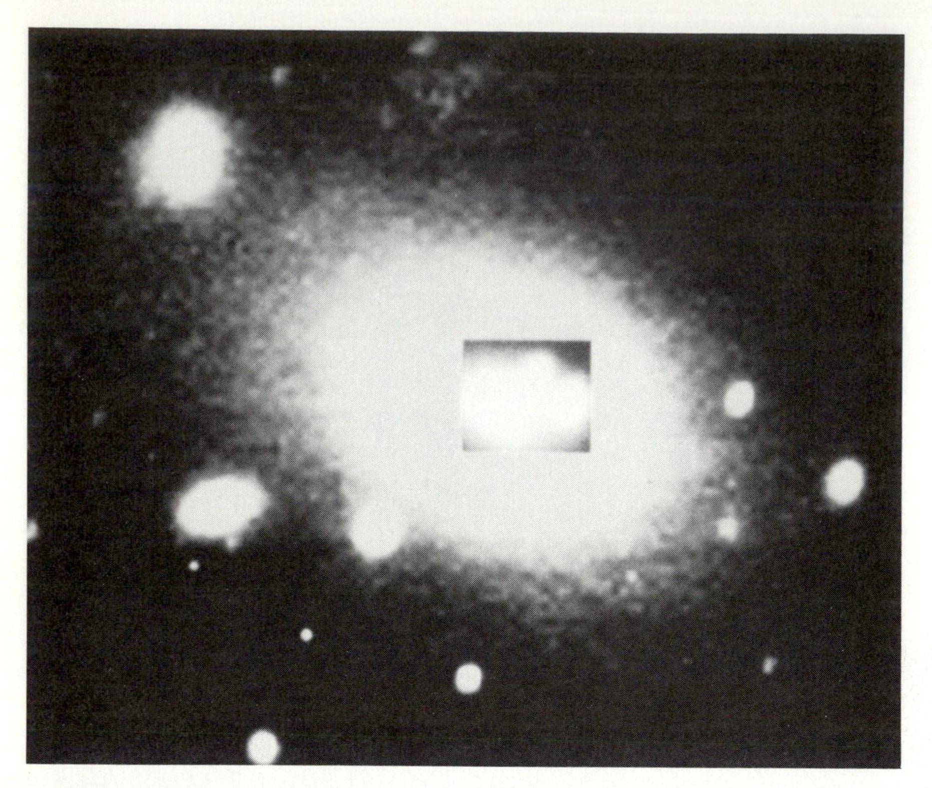

FIGURE 18.12 This composite (computer-enhanced) optical photograph of the galaxy cluster known as A2199 seemingly shows an example of galactic cannibalism. The central galaxy of the cluster is displayed with a superposed "window" revealing that within the core of the cluster are several smaller galaxies (3 bright images at center) which may be in the process of becoming trapped. Other, small galaxies swarm on the outskirts of the swelling galaxy, almost certainly about to be captured by the "central engine."

the other, though each having a mass comparable to our Milky Way Galaxy. Now let the two galaxies experience a close encounter. As depicted in the various frames of Figure 18.13(a), the smaller galaxy can substantially distort the larger galaxy. This figure is a computer-generated reenactment of the environmental changes produced exclusively by gravity. Note how the interaction causes the larger galaxy to sprout spiral arms where there were none before. The entire event required several hundred million years—the kind of accelerated evolution that modern computers can model in minutes.

Figure 18.13(b) is a photograph of a double galaxy having an uncanny resemblance to the final frame of Figure 18.13(a). Shown there are two galaxies having sizes, shapes, and velocities corresponding very closely with those of the objects in the computer simulation. The magnificent spiral galaxy is M51, popularly known as the Whirlpool Galaxy and shown in color as Plate 7(a). Its smaller companion is probably an irregular galaxy which, having drifted past M51 millions of years ago, managed to disturb M51.

This computer-generated encounter *might* be a valid model for the interaction of M51 and its companion. No one claims that the computer model accurately depicts the close encounter; nor does anyone suggest that M51 became a spiral galaxy because of such a gravitational rearrangement. Still, the computer rendition does demonstrate a plausible way that these two galaxies might have interacted millions of years ago, and how spiral arms might generally be created or enhanced by such interactions.

The M81–M82 system is another example of a galactic interaction that probably rearranged much matter in at least one of these galaxies. Shown in Figure 18.14, the two galaxies seem safely separated by several hundred thousand light-years of nothingness. But as also noted in the same figure, the overlaid radio contours of neutral hydrogen show clear evidence of invisible gas linking the two galaxies. The relative motions of the two galaxies, and of the intervening hydrogen gas, suggest that M81 swept past M82 about 200 million years ago, severely affecting the smaller M82 object. Unquestionably strong gravitational tides doubtless rearranged much matter in M82, triggering bursts of new stars, exploding other stars, and generally causing some activity in the central regions of M82.

These and other systems tend to support the newly emerging notion that some of the more spectacular changes occurring in galaxies perhaps result from their interactions with other galaxies. Still, such close encounters are random events and do not represent any genuine evolutionary sequence linking all spirals to all ellipticals and irregulars.

ACTIVE GALAXIES: An evolutionary link between normal and active galaxies seems more credible, although still controversial. A time sequence like that shown in Figure 18.15(a), proceeding from quasars to Seyferts (and BL Lac objects; see Interlude 15-1) to normal galaxies implies a continuous "winding down" of cosmic activity. Adjacent objects along this sequence are nearly indistinguishable from one another. For example, weak quasars have some things in common with some very active galaxies, while the feeblest active galaxies often resemble the most explosive normal galaxies. Perhaps all galaxy-sized objects started out more than 10 billion years ago as quasars, after which they gradually lost their emissive powers, becoming in turn Seyfert galaxies (or BL Lac objects) and eventually normal galaxies.

This idea stipulates that the observed quasars are actually ancestors of all the other galaxies. Far too distant for us to see any stars, the quasars are detectable only because of their tremendously energetic central

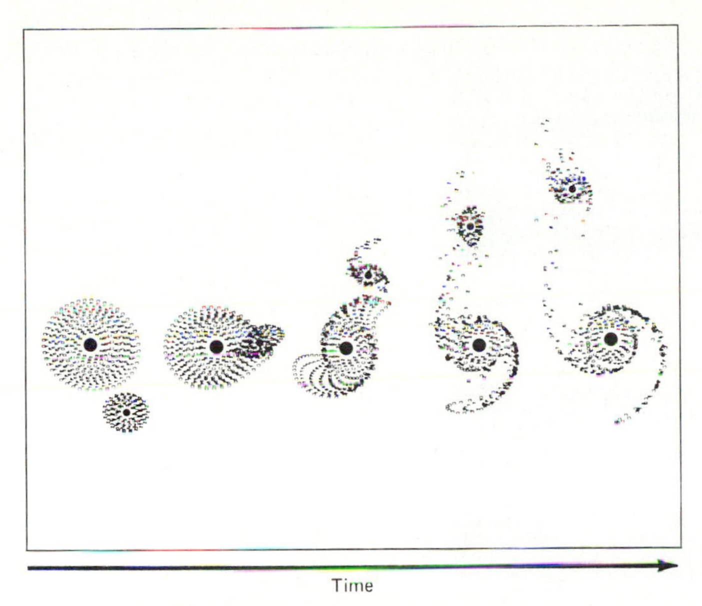

(b)

FIGURE 18.13 Galaxies might change their shapes long after their formation. In this computer-generated reenactment (a), two galaxies experience a close encounter over several hundred million years. As shown, the smaller galaxy has gravitationally disrupted the larger galaxy, thus changing it into a spiral galaxy. Compare the result of this computer simulation with a photograph (b) of M51 and its small companion, one of the most beautiful images of a spiral galaxy known. [See also Plate 7(a).]

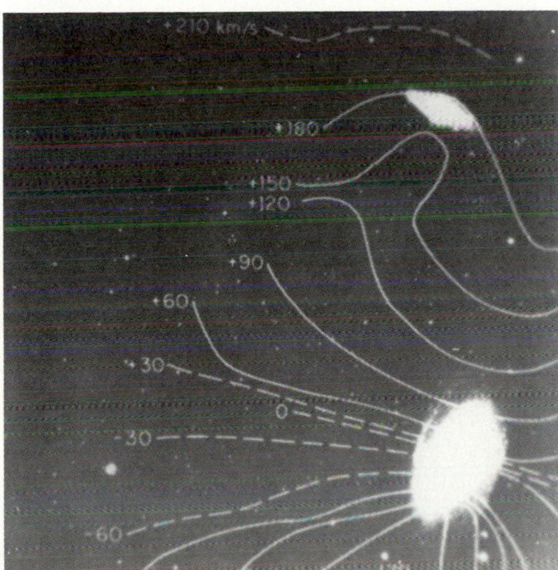

FIGURE 18.14 An optical photograph of the M81-M82 system, approximately 10 million light-years away, is overlaid here by a radio map of the system showing that M81 (bottom) and M82 (top) are wrapped within a common envelope of invisible hydrogen gas. (The numbers on the contours are the relative velocities of the connecting gas in kilometers/second.)

regions. Similarly, because of their great distances, we perceive them as they once were, in their blazing youth. As their central activity decays with time, quasars assume forms closer to those of more familiar and nearby galaxies. Should this idea be correct, even our Milky Way Galaxy was once a brilliant quasar.

Although attractive, this quasar → active galaxy → normal galaxy evolutionary idea is not at all proved. Some researchers argue that there is no evolutionary link at all. They suggest, as Figure 18.15(b) depicts, that the powerful quasars are merely extreme versions of the explosive phenomena observed in virtually all galaxies. After all, even the center of our own Milky Way is known to be expelling matter and radiation. The same can be said for active galaxies and quasars, though on a larger scale than for normal galaxies. Perhaps all these objects are part of the same galaxy family without there being any evolutionary sequence linking its members, just as there is no evolutionary link among different races within the human species. Each galaxy type or human race is distinctly different. One race of humans does not evolve into any other, and similarly one type of galaxy may not necessarily evolve into any other.

If no evolutionary sequence links these largest of all cosmic objects, perhaps all of them are essentially ordinary galaxies that formed long ago. Some of them were initially endowed with especially explosive central regions. Those able to display their explosiveness more than others for some as yet unknown reason are called quasars, while those hardly able to explode at all are called normal galaxies. Why the quasars explode most frequently and violently is not known. The answer

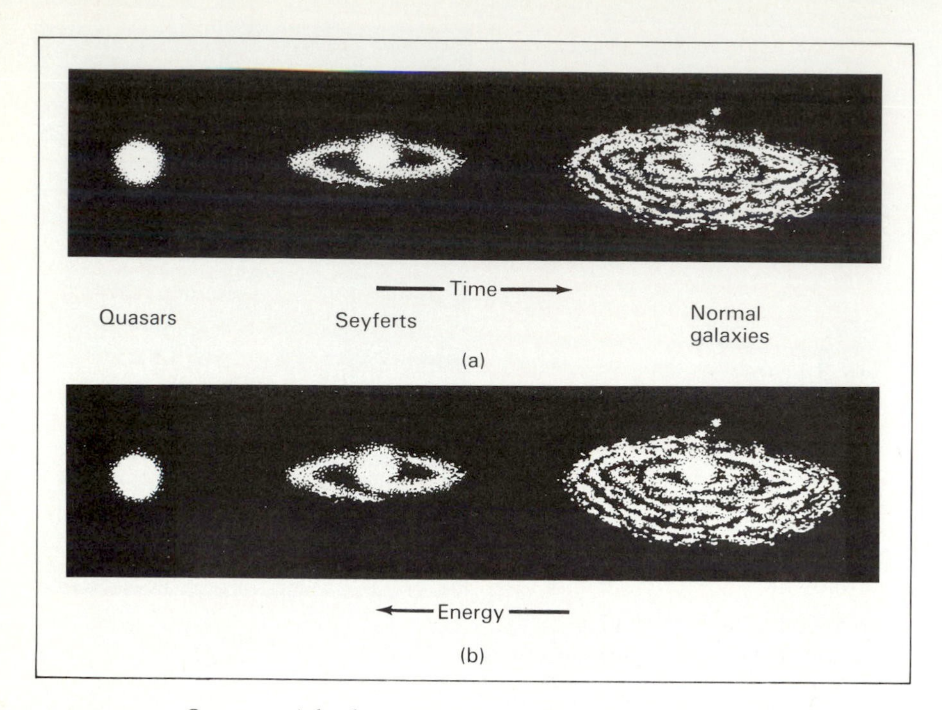

FIGURE 18.15 Quasars might decay over the course of time (a) into Seyfert galaxies or BL Lac objects, eventually becoming normal galaxies. Alternatively (b), all these huge objects may well be members of the same family with different ones having more explosive centers (quasars) than others (normal galaxies).

presumably lies buried within the relatively uncharted central regions of these objects.

It seems that future research on the centers of galaxies will provide the most help in unraveling the secrets of these gargantuan cosmic objects. Whether all galaxies actually change from one type to another, or some simply undergo repeated outbursts while remaining basically the same type of unevolved object, remains thus far unsolved.

SUMMARY

Aside from the creation of atoms, the formation of galaxies was the first great accomplishment of the Matter Era. However, our knowledge of the galaxies, especially their origin and evolution, is clearly inadequate. How each of them materialized, endowed with its peculiar shape and prodigious energy, remains problematic. To be sure, the subject of galaxy formation is the biggest missing link in the scenario of cosmic evolution.

The situation is not hopeless, however. The astronomical equipment scheduled to be built during the remaining years of the twentieth century will probably have a major impact on the origin and evolution of galaxies. Over the entire range of phenomena—from the earliest development of density fluctuations in the primordial Universe, through the slow conversion of interstellar gas into stars and the emergence of activity in the centers of galaxies—observations by novel instruments in the radio, infrared, optical, ultraviolet, x-ray, and gamma-ray regions are destined to provide a wealth of new information.

KEY TERMS

angular momentum
linear momentum
fragmentation
gravitational clustering
gravitational instability
protogalaxy
turbulence

QUESTIONS

1. How can we be sure that galaxies did not form via the chance accumulation of atoms?
2. Give some examples of how galaxies can change intrinsically as well as how they can change in response to environmental influences.
3. Compare and contrast the idea that stars form out of already-made galaxies, with that suggesting that galaxies form from already-made stars. Which scenario does the observational evidence favor?
4. In a few paragraphs, summarize the main reasons that galaxy formation has thus far eluded a complete solution.
5. Can the four major morphological types of normal galaxies be fit into an evolutionary sequence? Why or why not?
6. Explain how the evident order of the Universe around us is not a violation of the laws of thermodynamics.
7. Speculate about the similarities and differences between spiral galaxies and terrestrial hurricanes.
8. What is the relationship of gravitational energy and ther-

mal energy during the expansion or contraction of any object? Explain.

9. Define angular momentum and carefully explain how a shrinking object would tend to increase its spin.

*10. Compare and contrast galaxy formation by means of constant-temperature and variable-temperature fluctuations.

FOR FURTHER READING

ATKINS, P., *The Second Law.* New York: Scientific American Library, 1984.

BURNS, J., "Very Large Structures in the Universe." *Scientific American,* July 1986.

*GRIBBON, J., *Galaxy Formation.* New York: Wiley, 1976.

REES, M., AND SILK, J., "The Origin of the Galaxies." *Scientific American,* June 1970.

SILK, J., *The Big Bang.* San Francisco: W. H. Freeman, 1980.

SILK, J., SZALAY, A., AND ZEL'DOVICH, Y., "The Large-Scale Structure of the Universe." *Scientific American,* October 1983.

STROM, S., AND STROM, K., "The Evolution of Disk Galaxies." *Scientific American,* April 1979.

TOOMRE, A., AND TOOMRE, J., "Violent Tides between Galaxies." *Scientific American,* December 1973.

19 INTERSTELLAR EVOLUTION: *Star Formation*

We've studied galaxies before and we've studied stars before. Why return to discuss them again at the start of the evolutionary part of the book? The answer is that the early and late events in their histories bear heavily on two of the most important and exciting areas of modern scientific research.

First, understanding the formation and evolution of stars and galaxies should help elucidate the nature of that singular point of superdense, superhot matter from which the Universe arose. Our study of stellar evolution, especially of those bizarre regions known as black holes, will give us a better appreciation for the ultracompact singularity from which the Universe began.

Second, the subjects of star formation and stellar evolution will help us better understand the role stars play in the heating of planets, as well as in the origin and maintenance of life on those planets. In fact, star formation and evolution are crucially important in understanding a central feature of cosmic evolution—that the origin of life is a natural consequence of the evolution of matter.

This chapter, as well as the next 3 chapters, address the evolution of stars about which much is known. Indeed, although astronomers know little about galaxy evolution as discussed in the preceding chapter, we have a rather good understanding of **stellar evolution**, or the myriad changes experienced by stars as they originate, mature, and grow old.

The subject brings to mind a forest of trees. Some trees are young, some old, some deciduous, others evergreen; a few here and there are peculiar to say the least. Our study of the changes among stars are not dissimilar to the study of changing trees. By examining each of the various types of trees and stars, not only do we learn some things about those individual objects, but also, and perhaps more importantly, we explore intellectually the bigger picture of the forest of stars in the Universe beyond.

The learning goals for this chapter are:

- to recognize the many factors, such as heat, rotation, and magnetism, that compete against gravity in star-forming regions of interstellar space
- to appreciate the magnitude of the changes in density and temperature as interstellar clouds form stars or groups of stars
- to know in some detail the evolutionary sequence that likely led to our Sun
- to understand the wealth of observations supporting the theory of star formation
- to realize that interstellar space is so complex that all the details of prestellar evolution are currently hard to unravel

Gravitational Competition

Gravitational instabilities like those discussed in Chapter 18 can cause random fluctuations in density at various parts of any large cloud of gas. Although this process alone proved insufficient to group large chunks of matter to form galaxies, theory suggests that it should work quite well when bringing together smaller chunks of matter to form stars. Swirling eddies in interstellar space are cooler and denser than those of the primordial fireball, and hence are better suited to attract the mass required to form individual stars or clusters of stars.

How do gas fluctuations in interstellar space give rise to stars? To appreciate the answer to this question, consider a minute portion of a large interstellar cloud. Concentrate, as depicted in Figure 19.1, on just a few atoms. Each atom will have some random motion due to the cloud's heat, even if the cloud has a very low temperature. And each will be influenced somewhat by the gravitational pull exerted by all the other neighboring atoms. This force is not very large, owing to the small mass of each atom. Even when a few atoms accidentally cluster for a moment as shown in Figure 19.1, their combined gravitational pull is insufficient to bind them permanently into a distinct clump. Such an accidental cluster would disperse as quickly as it formed.

Now concentrate on a larger group of atoms. Imagine, for example, 50, 100, or even 1000 atoms, each gravitationally pulling on all the others. Would this many atoms exert a combined gravitational force field strong enough to prohibit them from dispersing as in the previous example? Just how many atoms are needed for gravity to bind the atoms into a tight-knit clump?

Answers to these questions cannot be found from a simple study of gravity alone. The correct solution depends not only on gravity, but also on several other physical conditions, such as heat, rotation, and magnetism. These additional agents tend to influence the evolution of an interstellar cloud; although they should not be regarded as antigravity, they do compete with gravity.

Let's discuss each of these competing agents in turn. First, there is *heat*. Analysis of radiation emitted by galactic regions proves that interstellar clouds have some heat, though not much. Most of their warmth derives from inevitable collisions among the atoms. More frequent collisions mean greater friction and thus more heat. It is heat that grants a cloud some buoyancy that tends to compete with gravity. In fact, heat is the main reason that the Sun and all the other stars don't collapse; the outward pressure of heated gases counteracts the inward pull of gravity.

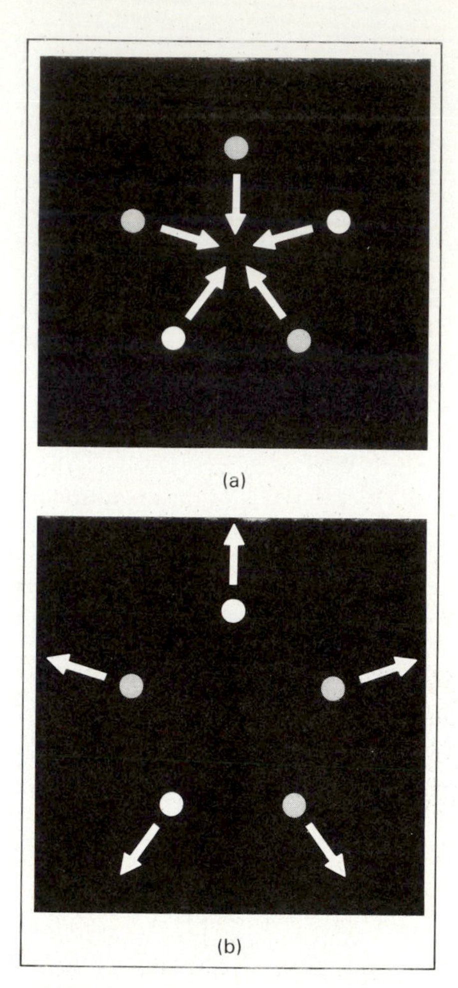

FIGURE 19.1 Motions of a few atoms within an interstellar cloud are influenced by gravity so slightly that their paths are hardly changed before (a) and after (b) an accidental, random encounter.

The heat contained within interstellar clouds is small by solar or even terrestrial standards. Observations of dark interstellar regions show that most interstellar clouds have temperatures below a few hundred kelvin. Some clouds have temperatures as low as 10 kelvin, and a few are known to be even colder. Consequently, thermal effects, which compete strongly with gravity once stars are formed, do not really play a large role until after interstellar clouds begin contracting and thus generating greater amounts of heat.

Rotation—that is, spin—can also compete with the inward pull of gravity. A contracting cloud having even a small spin (i.e., angular momentum) tends to develop a bulge around its midsection, as mentioned toward the end of Chapter 18. A flattening toward a disk-shaped object occurs because of the need to keep the total amount of angular momentum constant at all times during the contraction phase. As a cloud of constant mass contracts and thus decreases its size, its angular velocity necessarily increases, causing the cloud to spin faster and hence become more flattened perpendicular to its spin axis.

Figure 19.2 illustrates this important feature of rotation. As depicted, any rapidly rotating object exerts an outward force away from the center of the object; the faster the spin, the greater the force. We all feel this outward force while bearing the brunt of many circular rides at amusement parks. In the case of interstellar gas clouds, atoms near the periphery are particularly vulnerable to outward escape should the inward pull of gravity prove insufficient to retain them. It is in this sense that we can regard rotation as opposing the inward pull of gravity. Should the rotation of a contracting gas cloud increase enough to overpower gravity, the cloud would simply disperse,

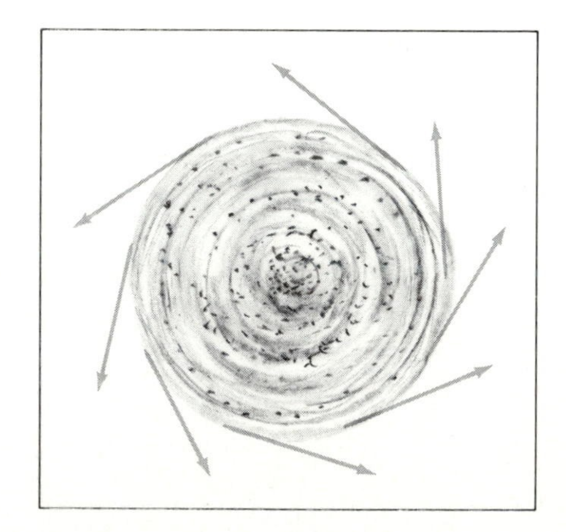

FIGURE 19.2 A rapidly rotating gas cloud tends to resist contraction. In this way, spin can compete with gravity.

sending its atoms back into the more tenuous surroundings. The familiar example of mud flung from a rapidly rotating bicycle wheel makes this point clear. The only way that a rotating interstellar cloud can preserve itself against outward dissipation is to continue to gather more and more atoms, thereby increasing the collective pull of gravity. The upshot is this: Rapidly rotating interstellar clouds need more mass for contraction toward star-like objects than do clouds having no rotation at all.

Magnetism can also hinder the contraction of a gas cloud. Just as Earth, the other planets, and the Sun have some magnetism, magnetic force fields surely permeate most interstellar clouds. As a cloud's contraction heats the atoms contained within, some atomic encounters become violent enough to ionize many of the atoms. The magnetic force fields, just as we noted in Chapter 5 for Earth's Van Allen Belts, can exert electromagnetic control over the interstellar charged particles and ions. If the magnetism is strong enough, the particles will be influenced more by the magnetic force field than by the gravitational force field. It is in this sense that magnetism can occasionally resist gravity.

This tug-of-war between gravity and magnetism often causes interstellar clouds to slowly contract in distorted ways. Since the charged particles and the magnetism are tied together, the magnetic force field itself follows the contraction of a cloud, as suggested by Figure 19.3. The charged particles and ions literally pull the magnetic force field toward the cloud's center, especially in the direction perpendicular to the magnetism. In this way, the strength of magnetism in a cloud can become much larger than that normally permeating general interstellar space outside gas clouds.

Observations made during the past decade show that real interstellar clouds are not very hot, spin only slowly, and are only slightly magnetized. But theory suggests that even minute quantities of any of these agents can compete effectively with gravity. Surprisingly small amounts of each can unite to alter considerably the evolution of a typical gas cloud.

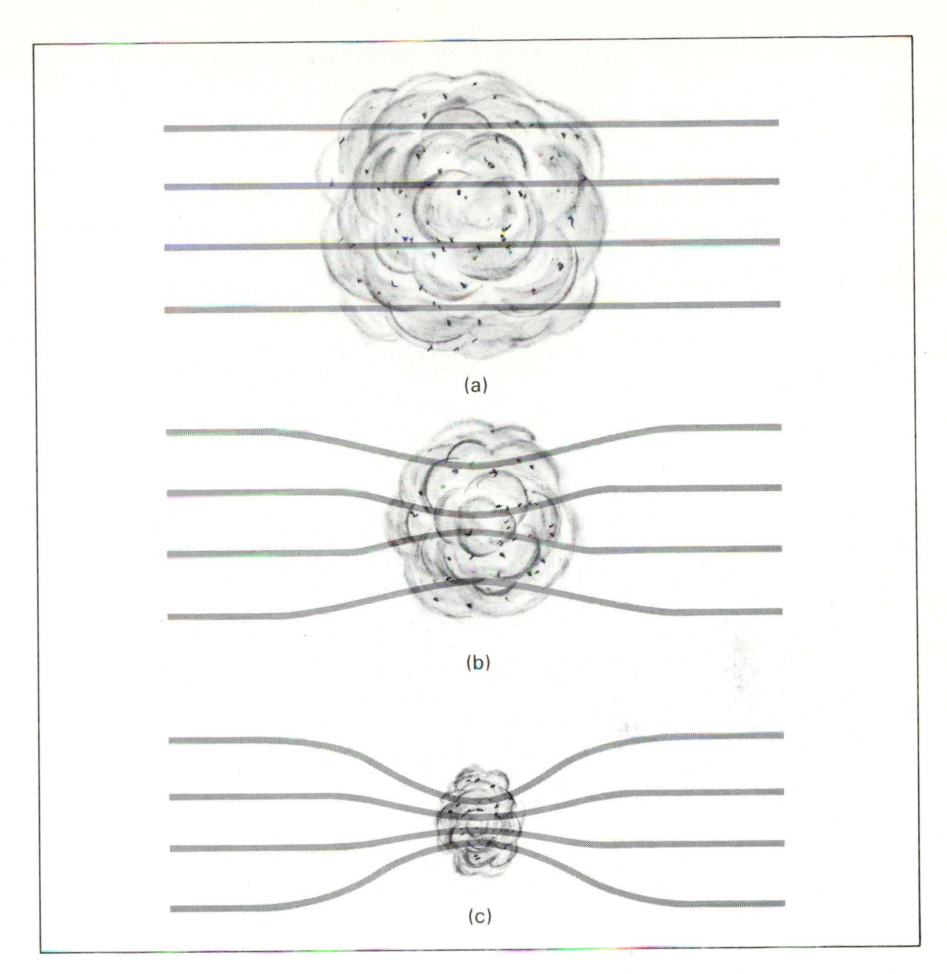

FIGURE 19.3 Magnetism can hinder the contraction of a gas cloud, especially in directions perpendicular to the magnetic force field (solid lines). Frames (a), (b), and (c) are schematic diagrams of the time evolution of a slowly contracting interstellar cloud having some magnetism.

We now return to our original question: How many hydrogen atoms need to be accumulated for the collective pull of gravity to prohibit a pocket of gas, once formed, from dispersing back into the surrounding interstellar space? (Real interstellar space is, of course, populated with lesser amounts of heavier atoms as well as some molecules. But the thrust of this problem can be made clear by treating the most abundant hydrogen atom as the sole inhabitant of the gas.) The answer, even for a cool cloud having no rotation or magnetism, is a very large number. Nearly 10^{57} atoms are required for gravity to bind a gaseous condensation. That's truly a huge number of atoms—much larger than the 10^{25} grains of sand on all the beaches of the world, even larger than the 10^{51} elementary particles comprising all the atomic nuclei in the entire Earth. It's large compared to anything with which we're familiar because there's simply nothing on Earth comparable to a star.

Converting 10^{57} hydrogen (H) atoms to a mass value,

$$10^{57} \text{ H atoms} \times \frac{2 \times 10^{-24} \text{ gram}}{1 \text{ H atom}} = 2 \times 10^{33} \text{ grams},$$

we see that this value is the approximate mass of our Sun. This is no coincidence. Our Sun is a very ordinary star. Most stars in our Galaxy have between 10^{56} and 10^{58} atoms, or equivalently masses between 0.1 and 10 times that of our Sun. The more massive stars probably formed in interstellar regions where heat, rotation, and/or magnetism competed strongly with gravity, requiring the clouds' eddies to attract more than the normal 10^{57} hydrogen atoms needed for successful gravitational contraction. Stars less massive than our Sun presumably formed in regions having little heat, rotation, and magnetism.

Formation of Sun-Like Objects

We can best study the specific steps of star formation by considering the Hertzsprung–Russell diagram studied earlier. Remember, this is a useful plot of stellar properties where surface temperature increases to the left, and luminosity increases upward. The luminosity scale in Figure 19.4 is expressed in terms of the solar luminosity (namely 4×10^{33} ergs/second), so that our G-type Sun is plotted at the intersection of the values, 1 solar luminosity and 6000 kelvin.

As noted in Chapter 11, most stars plotted on such an HR diagram fall along the main sequence. For roughly 90 percent of their lifetime, stars burn rather quiescently and hence do not change their physical conditions very much. Data points representing such stable, full-fledged stars remain stationary on the HR diagram.

Stars do change their properties, however, near the beginning and end of their existence. The HR diagram is a useful aid in describing these rather drastic changes. In the next chapter we shall describe the final death throes of stars; in this chapter we confine ourselves to the evolution of an interstellar cloud prior to a star's birth.

Table 19-1 specifies several evolutionary stages experienced by an interstellar cloud prior to the formation of an ordinary star such as our Sun. Characterized by varying central temperatures, surface temperatures, central densities, and sizes of the prestellar object, these seven stages trace the progress from a quiescent interstellar cloud to a genuine star. Specific numbers given in Table 19-1 and in the present discussion are valid only for the formation of stars having approximately the mass of our Sun. (In the next section we shall relax this restriction and consider the formation of any star.)

Stage 1 is just that of any ordinary interstellar cloud. Examples of these dark and dusty regions were displayed in Figures 12.5 through 12.9, 12.17, and 12.24. Many of these clouds are truly vast, often spanning tens of light-years across, or about 10^{14} kilometers. Temperatures are usually about 100 kelvin both within and at the edge of such large clouds, whereas densities are often not much more than 10 particles/cubic centimeter. Stage 1 clouds typically contain thousands of times the mass of the Sun in the form of cool atomic and molecular gas.

If a cloud is to become the birthplace of stars, it cannot remain as a stable, homogeneous blob. Interstellar clouds must eventually break up into subcondensations, often less than a light-year across. Theory suggests that fragmentation into smaller clumps of matter occurs naturally, because gravitational instabilities at various parts of an interstellar cloud force the development of inhomogeneities in the gas. A typical cloud can break up into tens, even hundreds, of fragments, each imitating the shrinking behavior of the cloud as a whole, albeit contracting even faster than the parent cloud.

In this way, interstellar clouds are thought to produce either numerous stars, each much larger than our Sun, or whole clusters of stars, each comparable to or smaller than our Sun. Indeed, there is no evidence for stars born in isolation, one star from one cloud. Most stars—perhaps even all of them—originate as members of star clusters. Those now appearing alone and isolated in space, such as our Sun, probably wandered

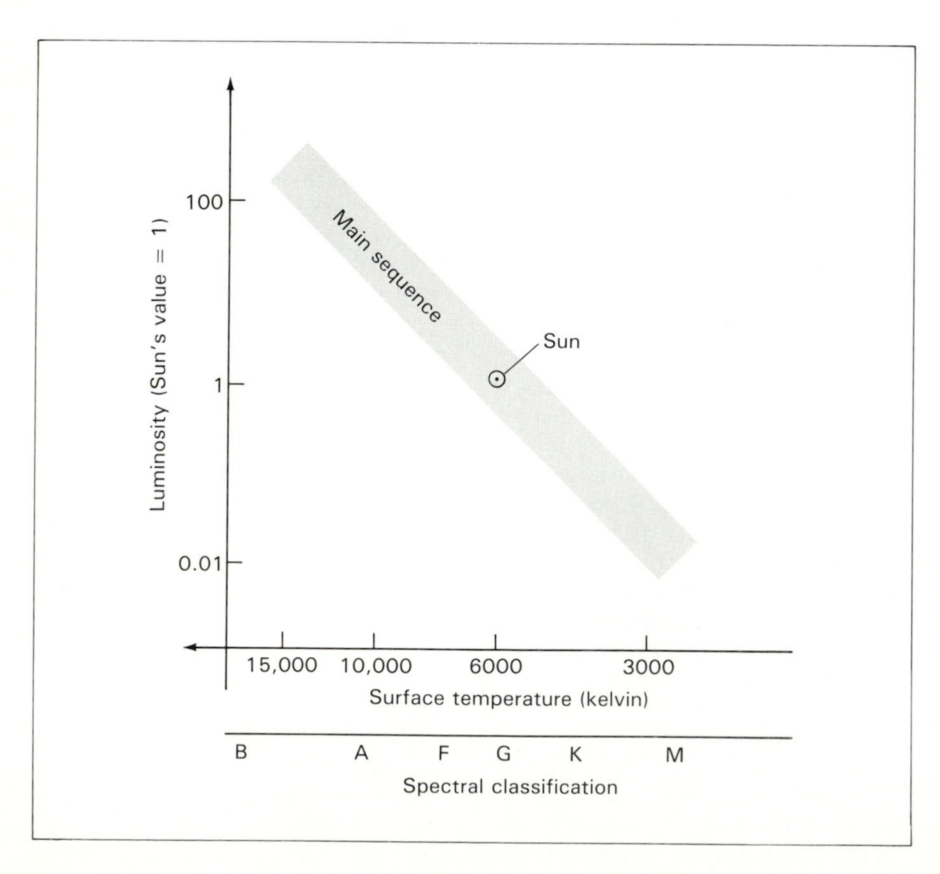

FIGURE 19.4 The HR diagram is a useful way to summarize the observed properties of stars.

TABLE 19-1
Protostellar Evolution of a Sun-Like Object

STAGE	CENTRAL TEMPERATURE (KELVIN)	SURFACE TEMPERATURE (KELVIN)	CENTRAL DENSITY (PARTICLES/CM^3)	SIZE (KILOMETERS)	OBJECT
1	100	100	10^1	10^{14}	Interstellar cloud
2	200	200	10^6	10^{12}	Cloud fragment
3	50,000	500	10^{12}	10^{10}	Cloud fragment
4	200,000	3000	10^{18}	10^8	Protostar
5	1,000,000	4000	10^{22}	10^7	Protostar
6	10,000,000	5000	10^{25}	2×10^6	Star
7	15,000,000	6000	10^{26}	1.5×10^6	Star

Note: 1 light-year = 10^{13} kilometers
Solar System diameter = 10^{10} kilometers
Earth–Sun orbit diameter = 3×10^8 kilometers
solar diameter = 1.5×10^6 kilometers
10^{24} atomic particles = 1 gram of matter.

away from a multiple-star system as the cluster dissolved.

Once a fragment takes on its own identity within an interstellar cloud, it then passes through a series of inevitable stages. It first begins to contract as gravity affects the ever-accumulating group of atoms. It literally shrinks under the onslaught of its own weight. As the protostar becomes more compact, the atoms collide more frequently, causing the gas fragment to warm.

Stage 2 in our evolutionary scenario represents the physical conditions of just one of the many small fragments that develop within a typical interstellar cloud. Estimated to span about a tenth of a light-year across, such a fuzzy, gaseous blob is still about 100 times the size of our Solar System. Temperatures both at the core and periphery of such a fragment have risen to roughly 200 kelvin, whereas the central density now measures some 10^6 particles/cubic centimeter. Such a region becomes hotter because gravitational potential energy of the gas particles converts into thermal energy as the fragment contracts. This newly gained heat causes the atoms to become agitated; hydrogen atoms in particular move around with velocities of about 1 kilometer/second in a 200-kelvin gas. These faster velocities ensure that the atoms collide frequently and violently. If such fragments are to continue to contract, they must constantly radiate away much of their newly generated heat, lest the cloud become stabilized against the relentless pull of gravity. After all, stars can't form in stabilized clouds.

Fragmentation might be expected to continue indefinitely, producing ever-smaller clumps that couldn't possibly form stars. Fortunately, the process halts before it's too late. Increasing gas density stops the process of fragmentation from reducing all parts of the cloud without limit into ever-smaller subunits. As stage 2 fragments compress their gas, they eventually become compact enough to prohibit radiation from escaping easily. With the cloud's natural vent partially blocked, the trapped radiation causes the temperature to rise, the pressure to increase, and the fragmentation to cease.

Several thousand years after it first began contracting, a typical fragment has shrunk in stage 3 to a gaseous object having a diameter roughly the size of our Solar System (still 10,000 times our Sun's size). Its central temperature has reached several tens of thousands of kelvin, temperatures greater than those within the hottest steel furnaces fashioned by our civilization on Earth. The temperature at the fragment's periphery has also increased, although not nearly as much as deep in the interior. In fact, the temperature at the surface of the fragment is expected to be just a little hotter than a comfortable living room, not much greater than several hundred kelvin. Because the density of matter inevitably increases at the core of the fragment faster than at its periphery, the outer surface of any contracting interstellar cloud is sure to be cooler and thinner than its interior. The central density is approximately 10^{12} particles/cubic centimeter, ensuring much more violent and frequent particle collisions at the core and thus producing a large temperature difference between these two zones of the fragment.

As the cloud fragment continues to evolve, its size diminishes, its density grows, and its temperature rises at both the core and the periphery. About 100,000 years after beginning its contraction, it reaches stage 4, where its center boils at more than 100,000 kelvin. Elementary particles, now mostly electrons and protons ripped from disintegrated atoms, are really whizzing around at very high velocities—on average, roughly 100 kilometers/second. Despite this veritable inferno, though, those particles are still far from the 10 million kelvin needed to ignite the proton–proton nuclear reactions to fuse hydrogen into helium.

Still larger than a Sun-like star, the gaseous heap at stage 4 has a diameter equal to about Mercury's orbit. Its surface temperature has risen to several thousand kelvin. For the first time, it's beginning to resemble a star. For these reasons, astronomers call such a stage 4 fragment a **protostar**—an embryonic object perched at the dawn of star birth.

Note that the time needed for the appearance of a protostar is only about 100,000 years. We say "only" because that's brief by cosmic standards. However, it's long by human standards, which partly explains why no one has even seen a protostar actually emerge from an interstellar cloud.

Once the cloud fragment reaches the protostellar stage, its surface temperature is high enough for the object's physical properties to be plotted on the HR diagram. Knowing from Chapter 11 that luminosity varies as the square of an object's size and as the fourth power of its surface temperature, we can calculate the luminosity. Surprisingly, it turns out to be several thousand times the luminosity of our Sun. As shown in Figure 19.5, the protostar is much more luminous than most other stars on the main sequence. You might find it paradoxical that even though the protostar has not yet begun its nuclear burning, it can have such a large luminosity. The reason is that despite a surface temperature only about half that of the Sun, a protostar is usually hundreds of times larger in size, thus making the total emitted luminosity very large indeed.

Figure 19.5 also depicts the approximate path followed by such an interstellar cloud fragment before becoming a protostar. Researchers often refer to this early evolutionary path as the **Kelvin–Helmholtz contraction phase**, so named after two European physicists who first studied the theory of contracting clouds about 100 years ago. Figure 19.6 is a series of artist's sketches of an interstellar gas cloud proceeding along this evolutionary path.

Protostars are still a little unstable; the outward pressure of heat does not quite balance the inward pull of gravity. Fortunately, the average temperature is still too low to make a protostar stable. We say "fortunately" because if the heated gas were able to counteract gravity before reaching the point of nuclear burning, there would be no stars. The nighttime sky would be abundant in dim protostars but completely lacking in genuine stars. And it's likely that neither we nor any other intelligent life forms could exist to appreciate them.

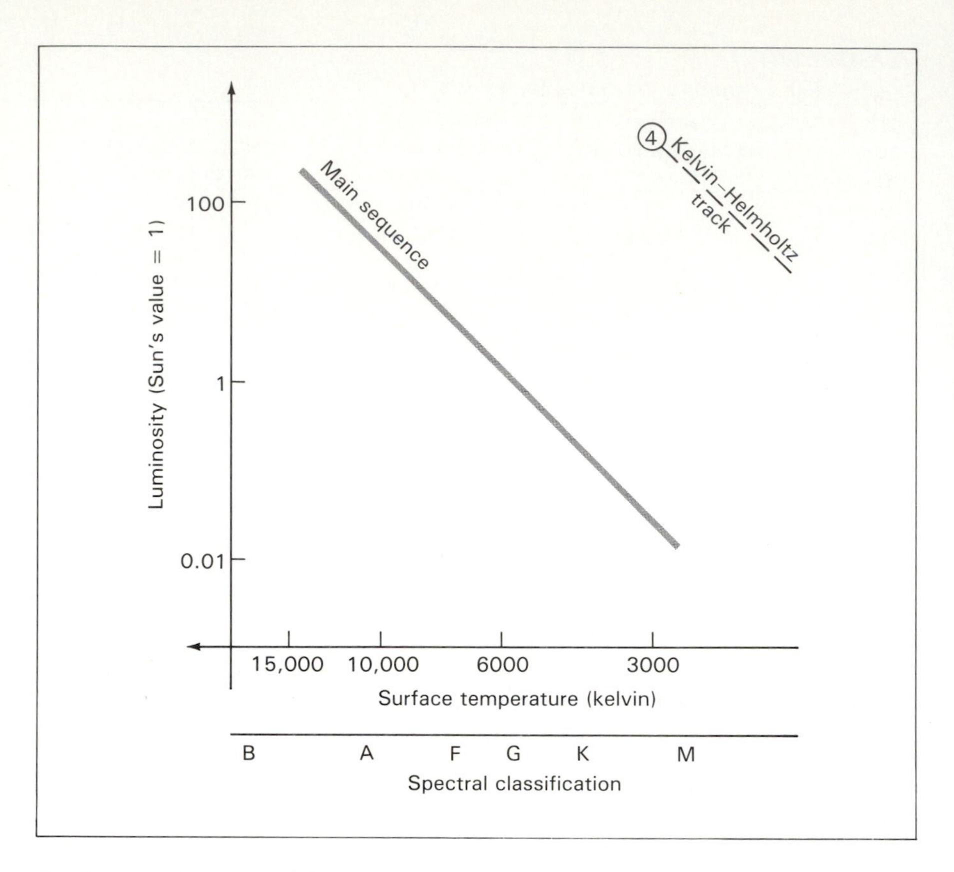

FIGURE 19.5 Diagram of the approximate evolutionary path followed by an interstellar cloud fragment prior to arriving, as a stage-4 protostar, at the end of the Kelvin-Helmholtz contraction phase.

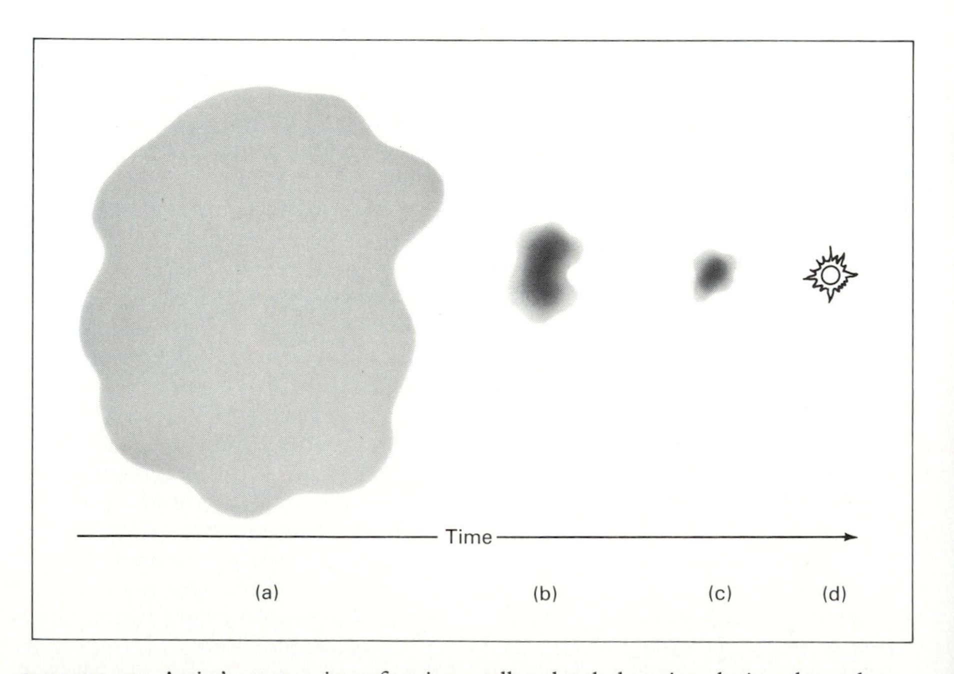

FIGURE 19.6 Artist's conception of an interstellar cloud changing during the early evolutionary stages outlined in Table 19-1. Shown are (a) a stage-1 interstellar cloud, (b) a stage-2 fragment, (c) a stage-3 smaller, hotter fragment, and (d) a stage-4 protostar. (Not drawn to scale.)

Thus protostars continue to contract, although now more slowly because heat is steadily becoming a powerful countervailing effect. By stage 5, a protostar's size has shrunk to nearly 10 times the size of our Sun. Its central temperature has reached about 1 million kelvin, although this is still not enough to initiate nuclear burning. To be sure, the core of the protostar is now mostly ionized because of the brutal collisions among the gas particles. But the protons still do not have enough thermal energy (i.e., high velocity) to overwhelm their mutual electromagnetic repulsions, and thus to slam into the realm of the nuclear binding force.

The protostar's surface continues to mimic the rise of the interior temperature, but only slightly; at stage 5, as Table 19-1 notes, the surface temperature has reached some 4000 kelvin. Despite this increase in surface heat, the protostar's luminosity does not continue to increase at stage 5. Although the temperature is increasing, the size is decreasing; both must be used to compute the luminosity. In this particular case, the square of the protostar's size decreases more rapidly than the fourth power of the temperature increases. The position of the protostar on the HR diagram has therefore moved down and to the left while changing from stage 4 to stage 5.

Events in a protostar's development happen more slowly as the protostar approaches the main sequence. The initial contraction and fragmentation of an interstellar cloud occur quite rapidly, but as the protostar nears the status of a full-fledged star, its time scale for change slows down. Heat is the cause of the slowdown, for even gravity must struggle to compress a hot object.

Some 10 million years after its first appearance, the protostar finally becomes a genuine star. It does this by continuing to contract under the relentless pull of gravity. At stage 6, when a 1-solar-mass object has shrunk to a size of about 2 million kilometers, its central temperature has reached 10 million kelvin. Now the heat is sufficient to ignite nuclear burning. Hydrogen nuclei (protons) begin fusing into helium nuclei.

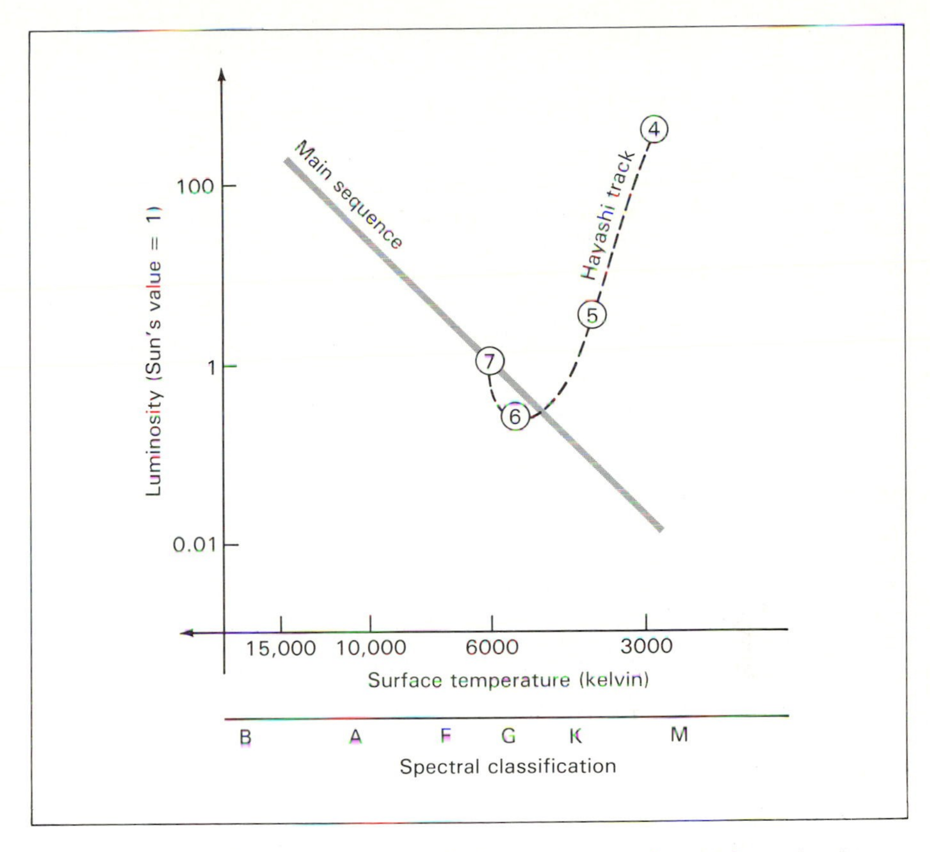

FIGURE 19.7 The changes in a protostar's observed properties are shown by the path of decreasing luminosity, from stage 4 to stage 6, often called the Hayashi track.

Stages 5 and 6 are plotted in Figure 19.7. Stage 6 is slightly below the main sequence mainly because the star's surface temperature at this point is a little less (5000 kelvin) than the Sun's. Here, again, the square of the newly formed star's size is decreasing quicker than the fourth power of its rising surface temperature. The object's changing size is the dominant influence in its evolution from stage 4 to stage 6, which accounts for the decreasing luminosity. This newly formed star at stage 6 is actually a little less luminous than our Sun at present.

To distinguish this later protostellar evolutionary path where luminosity decreases from the earlier Kelvin–Helmholtz contraction phase where the luminosity increases, astronomers often call it the **Hayashi track**. Named for a twentieth-century Japanese researcher who made major contributions to the theory of protostars, this evolutionary path is diagrammed in Figure 19.7.

Such a stage 6 star has yet to reach the main sequence, however. Although the outward push due to heat and the inward pull due to gravity are nearly balanced, this new born star is still a little unstable. One final, relatively slow adjustment must be made before the star completely settles onto the main sequence for 10 billion years of steady nuclear burning. As usual, gravity performs the task.

During the next 30 million years or so, the stage 6 star is squeezed just a little more. In making this slight adjustment, the central density becomes about 100 grams/cubic centimeter, the central temperature increases to 15 million kelvin, and the surface temperature mimics that increase by reaching some 6000 kelvin. It so happens that, in this case, the fourth power of the star's surface temperature increases more than the second power of its size decreases. Accordingly, at stage 7, a 1-solar-mass object finally reaches the main sequence just about where our Sun now resides. Pressure and gravity are finally balanced for this 1-solar-mass star. Its principal function thereafter is to consume hydrogen, thereby producing helium and releasing energy.

All the evolutionary events just described occur over the course of several tens of millions of years. Obviously a long time by human standards, this is still less than 1 percent of a 1-solar-mass star's lifetime on the main sequence. Once an object begins fusing hydrogen and establishes a gravity-in/pressure-out equilibrium, it pretty much burns steadily for a long, long time indeed.

Stars of Different Mass

The numerical values and the evolutionary paths suggested above for the birth of a 1-solar-mass object are not valid for the formation of stars having much more or much less mass. Temperatures, densities, and sizes of other prestellar objects exhibit similar trends of change, but the actual values differ, in some cases considerably. The time scales for these changes are also quite different.

For example, cloud fragments that eventually form more massive stars approach the main sequence along a loftier track on the HR diagram. Figure 19.8 shows how these more massive objects have both larger luminosities and larger surface temperatures at all stages. Furthermore, the pace at which these massive prestellar objects move through their evolutionary stages is a lot more rapid than for presolar objects. The most massive fragments contract into stars in a mere million years.

Mass is the cause of all these differences. Not surprisingly, massive fragments within interstellar clouds often produce massive protostars and eventually massive stars themselves. These big fragments initially have larger sizes, more gas particles,

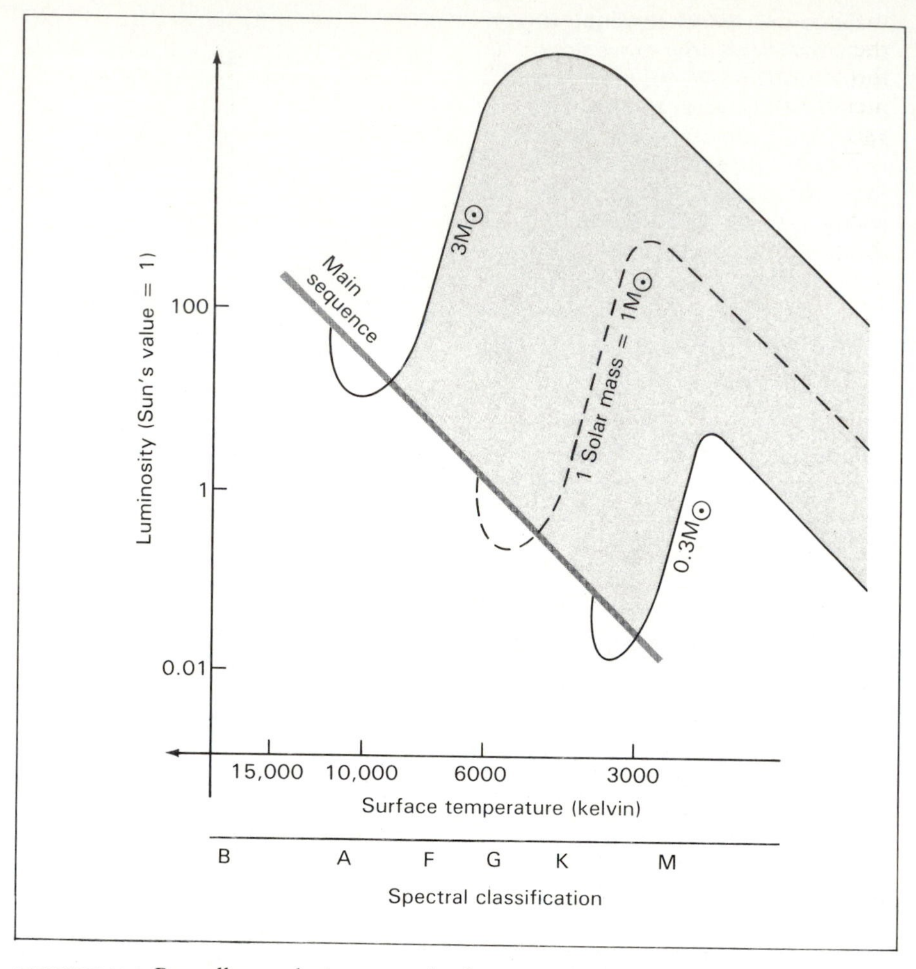

FIGURE 19.8 Prestellar evolutionary paths for stars either more or less massive than our Sun differ from those sketched in Figures 19.5 and 19.7.

and more frequent particle collisions than those of smaller cloud fragments. They thus heat up to the required 10 million kelvin more rapidly than do presolar objects.

This fast pace agrees with the paradoxical way in which, as noted in Chapter 11, the most massive main-sequence stars race through their lifetimes. This rapid evolutionary pace eventually causes problems for the biggest stars; just as hastily prepared foundations in our world often lead to collapse, these giant stars race through all parts of their life cycle, only to die catastrophically by exploding.

The opposite case prevails for stars and prestellar objects having masses less than our Sun. The cloud fragments that evolve into low-mass stars are generally smaller and thus have particle encounters less frequently. Not only do such fragments take a long time to become protostars, but these protostars also take their time changing into full-fledged stars. A typical M-type star, for example, requires nearly a billion years just to form.

DARK CLINKERS: Some cloud fragments are too small ever to become stars. Gaseous planets, such as Jupiter, are good examples. Jupiter's gases contracted all right under the influence of gravity, for the resultant heat is still detectable, as noted in Chapter 8. But Jupiter was unable to accumulate enough mass for gravity to crush its matter to the stage of a nuclear-burning star. The planet became stabilized by heat, rotation, and possibly magnetism, all opposing the pull of gravity. Thus Jupiter exists now as a chunk of gaseous matter that never evolved beyond the interstellar fragment stage. Although with a smaller size, Jupiter is literally stuck at stage 3 of our analysis.

The other Jovian Planets are similar failures. If they were still collecting gas, we could soon regard them as protostars. But virtually all the matter present during the formative stages of our Solar System is gone now, swept away by the solar wind of our Sun. Stranded in space, these giant planets will continue to cool, eventually becoming compact, dark clinkers.

Vast numbers of Jupiter-like objects may well be scattered throughout the Universe—fragments frozen in time somewhere along the Kelvin-Helmholtz contraction phase. Our technology is currently incapable of detecting them, whether they be planets associated with stars or interstellar cloud fragments far away from any star. (We can telescopically detect stars, and spectroscopically infer atoms and molecules, but astronomical objects of intermediate size are impossible to detect outside our Solar System.) In fact, galaxies *could* be chock-full of cold, dark objects ranging anywhere from pebble-size to Jovian-size without our knowing it. (Astronomers often joke about interstellar basketballs, but in fact objects of that size might well exist in great abundance.) Thus these dark clinkers might help solve the missing-mass problem plaguing our Milky Way Galaxy.

Observations of Cloud Fragments and Protostars

Details of the evolutionary stages described in the previous two sections were derived from "numerical experiments" performed with high-speed computers. The numbers listed in Table 19-1 and the evolutionary paths described in Figures 19.5, 19.7, and 19.8 are mathematical predictions of a multifaceted problem incorporating gravity, heat, rotation, magnetism, nuclear reaction rates, elemental abundances, and a few other physical conditions specifying the various states of contracting interstellar clouds. Only because of modern computer technology have theorists been able to construct such specific models. The accuracy of these models is only partly known, for it's currently difficult to test them observationally. Furthermore, each of the stages is not as clear-cut as suggested by the table; many of the stages probably overlap, just as with any gradually evolving system.

How can we test the theoretical predictions outlined earlier in this chapter? No human being has seen an interstellar cloud or a protostar move through all its evolutionary paces. In fact, the total lifetime of our civilization is (thus far) much, much shorter than the time needed to contract a cloud and form a star. We could never observe individual objects proceed through their full panorama of star birth. We can, however, observe different objects as they appear at different stages of their evolutionary cycle. Newly developed equipment now enables us to probe interstellar clouds, protostars, and very young stars approaching the stage of long-lived, main-sequence burning. By observing these different objects at various, often unrelated sites in our Galaxy, astronomers have observationally verified some of the prestellar stages described in the preceding two sections.

The method used is similar to that of archaeologists and anthropologists who dig up artifacts and bones at numerous unrelated places strewn across Earth's surface. Not having had the opportunity of living at the time of our ancestors, these scientists study the various remnants, trying to piece them together into an overall picture of human evolution. Similarly, astronomers probe various unrelated regions of our Galaxy, striving to understand how each region fits into an overall scheme of stellar evolution. It's much like a puzzle. The various terrestrial bones and extraterrestrial clouds are the pieces. The picture becomes clear only when each piece is found and oriented properly relative to all the other pieces.

Let's now consider some of the observational pieces of the stellar evolutionary puzzle.

EVIDENCE OF CLOUD FRAGMENTATION AND CONTRACTION: Figure 19.9 is an optical photograph of a rather typical region of interstellar space. The horizontal spread of stars, gas, and dust denotes the galactic plane, including numerous gaseous nebulae and interstellar clouds. (Consult Plate 4.)

Figure 19.10 is an enlargement of the small area outlined by the rectangle of Figure 19.9. Shown there is M20, the splendid gaseous nebula studied in Chapter 11 and displayed magnificently as Plate 5. Such brilliant nebular regions lit by stars formed long ago are of little interest to us here. But the presence of youthful O- and B-type stars producing glowing nebulae do alert us to the general environment where stars are likely to be forming. We might say that gaseous nebulae are signposts of star birth. Indeed, the dark interstellar regions near nebulae provide evidence for cloud fragmentation and protostars.

Recall from our earlier studies that the dense, dark, and dusty regions outside gaseous nebulae are not very hot; many have temperatures below 100 kelvin. Not surprisingly, the gas must be downright cold if these interstellar clouds are to be candidate regions for contraction and protostar development. Were the gas not cold, the necessary condition for contraction—gravitational potential energy exceeding the thermal energy—would not be met.

Prestellar objects at stages 1 and 2 are not yet hot enough to emit much infrared radiation. And surely no optical radiation arises from such dark, cool clouds; optical astronomers could observe star-forming regions forever and see little of interest, for there is, quite frankly, nothing to see in a dark cloud. Consequently, the best way to study the early stages of cloud contraction and fragmentation is to use radio telescopes to detect the radiation emit-

INTERLUDE 19-1 *Evolution Observed*

Astronomical objects generally evolve over enormously long intervals of time, making it almost impossible to study their changes during a human lifetime. Even the relatively short span of 30 million years needed to form a star such as our Sun equals roughly a million human generations—well more than have yet occurred.

Some stages of a star's evolution are nonetheless expected to occur extraordinarily rapidly by cosmic standards. One such stage is the sudden explosion of a massive star near death, an event we shall study in Chapter 20. Another rapid change is expected as a newly formed star reaches the main sequence. Given the huge number of stars in the sky, we should see a definite and sudden change in the properties of a few stars once in a while.

T Tauri stars comprise a class of very young stars on the verge of reaching the main sequence. Their peculiar name derives from the variable star labeled "T" in the constellation Taurus. The average properties of T Tauri stars lie somewhere between those of stages 5 and 7. During the past half-century, astronomers have noticed a few of them brighten greatly over the course of several years. They then remain at that increased brightness, although it's not certain for how long, for most of these young stars have been discovered only in the 1970s. The two accompanying photographs record the change of one such T Tauri star.

The image at left shows a region of interstellar space having a nebular streak (a little like a comet) and a faint star located at the tip of the nebula. The fan-shaped nebula, labeled NGC2261, is filled with dust (as well as gas) which reflects the light of the star, called R Monocerotis. The same field of view is shown at the right, photographed 3 years later. Here, the star has brightened and the nebula too. The star, considered a T Tauri variable, has retained this brightness ever since.

Currently, we have no adequate explanation for the sudden brightening of young stars like the one in this photograph. It could be caused by interstellar matter falling onto the newly formed star, causing flares on the star's surface, by external gas and dust being blown away thus making the new star more visible, or by an internal change in the star's nuclear burning rate. Whatever the cause might be, several T Tauri stars have undergone very definite changes in appearance on time scales shorter than a human lifetime. You can be sure that astronomers around the world are monitoring them closely, hoping that further observations will reveal the true reason for their extraordinarily rapid evolution.

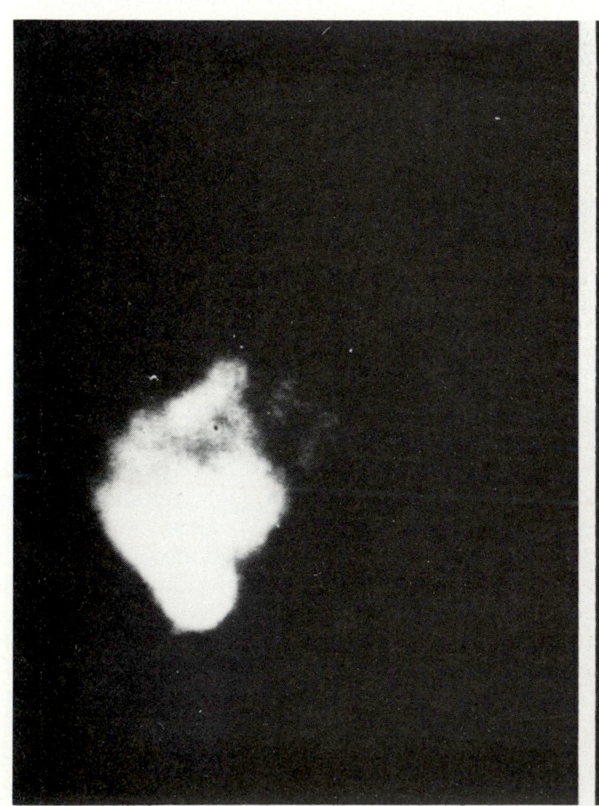

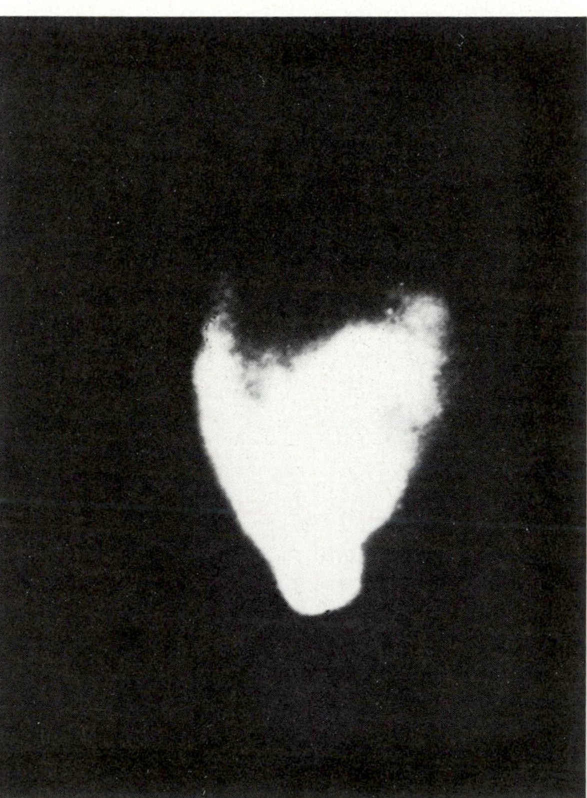

FIGURE 19.9 A wide-angle view along the plane of our Milky Way Galaxy (which extends diagonally, from upper left to bottom center). The small area outlined in the rectangle is enlarged in Figure 19.10. (See also Plate 4.)

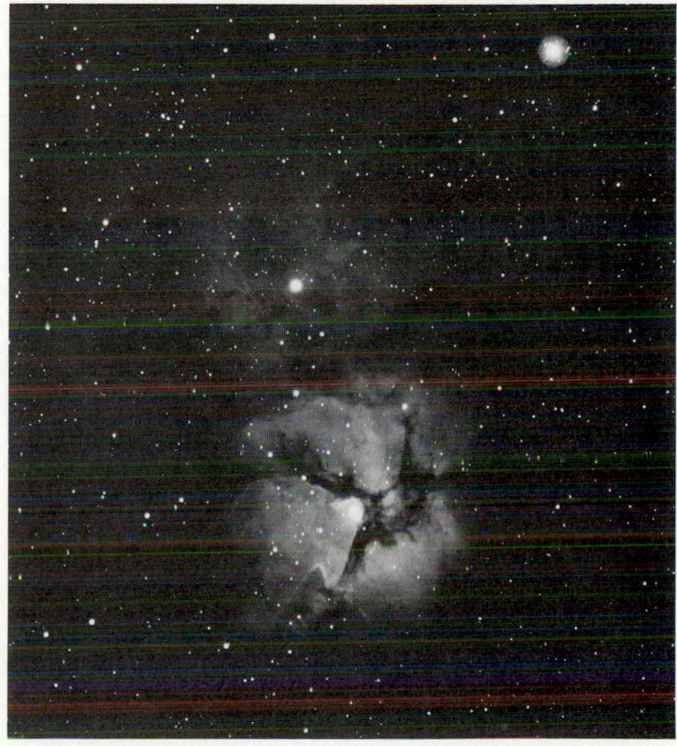

FIGURE 19.10 The interstellar region enclosed in the rectangle of the previous figure shows a beautiful gaseous nebula, M20, in addition to much dark surrounding matter. (See also Plate 5.)

ted or absorbed by one or more interstellar molecules. Only long-wavelength (radio and infrared) radiation can penetrate the dense, dark clouds, thus escaping the clouds and eventually (in minute quantities) being captured by telescopes on or near Earth.

The interstellar regions surrounding the M20 nebula in Figure 19.10 provide an example of galactic matter that seems to be contracting. The presence of much invisible gas there was illustrated in Figure 12.24. This earlier figure shows a contour map of the abundance of the formaldehyde (H_2CO) molecule. The presence of these molecules, as well as several other kinds of molecules, is widespread, especially throughout the dusty regions outside the right-bottom of the nebula. Analysis of the radio radiation shows that these molecules are especially abundant near a totally opaque dark region to the south of the nebula in Figure 11.24. Further analysis of the observations suggests that this region of greatest molecular abundance is contracting, fragmenting, and generally on its way toward forming a star or a cluster of stars.

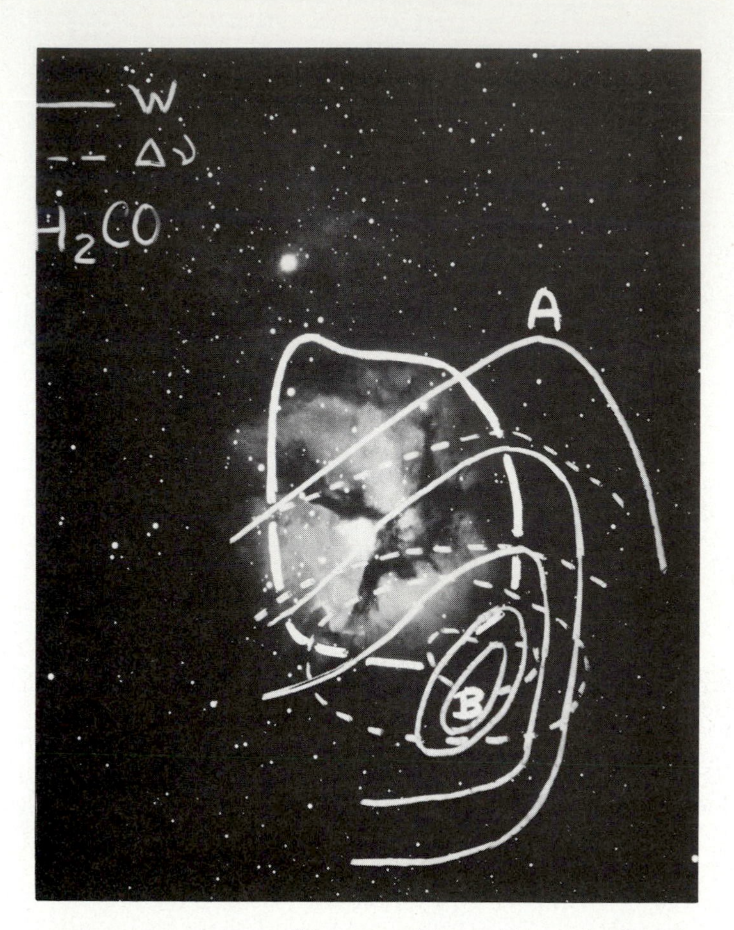

FIGURE 19.11 A map of the spectral-line width of the formaldehyde molecule's radiation toward the south of M20 reveals some evidence of a contracting interstellar cloud. The line-width contours increase in uniform steps from the outside to the inside of "blob B," as shown by the dashed curves here. They approximately coincide with the map of formaldehyde abundance (shown earlier in Figure 12.24, and also displayed here as solid contours).

How do we know that a portion of the dark cloud outside M20 is probably contracting? Well, recall from Chapter 2 that the widths of spectral lines are especially useful in revealing the motions of the regions sampled. Figure 19.11 shows a map of the spectral-line width of the formaldehyde molecule's radiation observed toward one of the darkest clouds to the south of M20. This map was made by measuring the widths of spectral lines at various places across the dark interstellar cloud. Contours were then drawn connecting places having spectral lines of equal width. Significantly, this line-width map peaks at roughly the same place as the molecular abundance map—at the totally opaque, dark region to the south of M20.

The similarity in the spreads of molecular motion (Figure 19.11) and molecular abundance (Figure 12.24) suggests that a small region of the large interstellar cloud engulfing M20 might be in a state of gravitational contraction. At the periphery of the region, we observe narrow spectral lines because the motion of infalling matter is mainly perpendicular to our line of sight; this causes little Doppler broadening of the spectral-line profiles. Toward the center of the dark region, infalling gas from the back and front of the cloud would be moving mainly parallel to our line of sight; this would produce Doppler-broadened line profiles of greater width, just as observed.

Other interpretations of the data are also possible. But the contracting interpretation just described is the simplest and most straightforward explanation of all the observations. Should it be correct, the M20 region provides a good piece of observational evidence for the beginnings of the evolutionary scenario outlined in Table 19-1. Only further observations will tell for sure.

The interstellar clouds in and around M20 provide tentative evidence for three broad phases of star formation. Figure 19.12 shows each of these general phases. In the northern part of this vast region, the nebular velocity of about 18 kilometers/second, deduced from the analysis of radio lines emitted by ions in the hot gas (see Figure 12.16), meshes nicely with the velocity of the adjacent molecular cloud. This is probably the average velocity of the gigantic, cool interstellar cloud engulfing much of the hot nebula. This

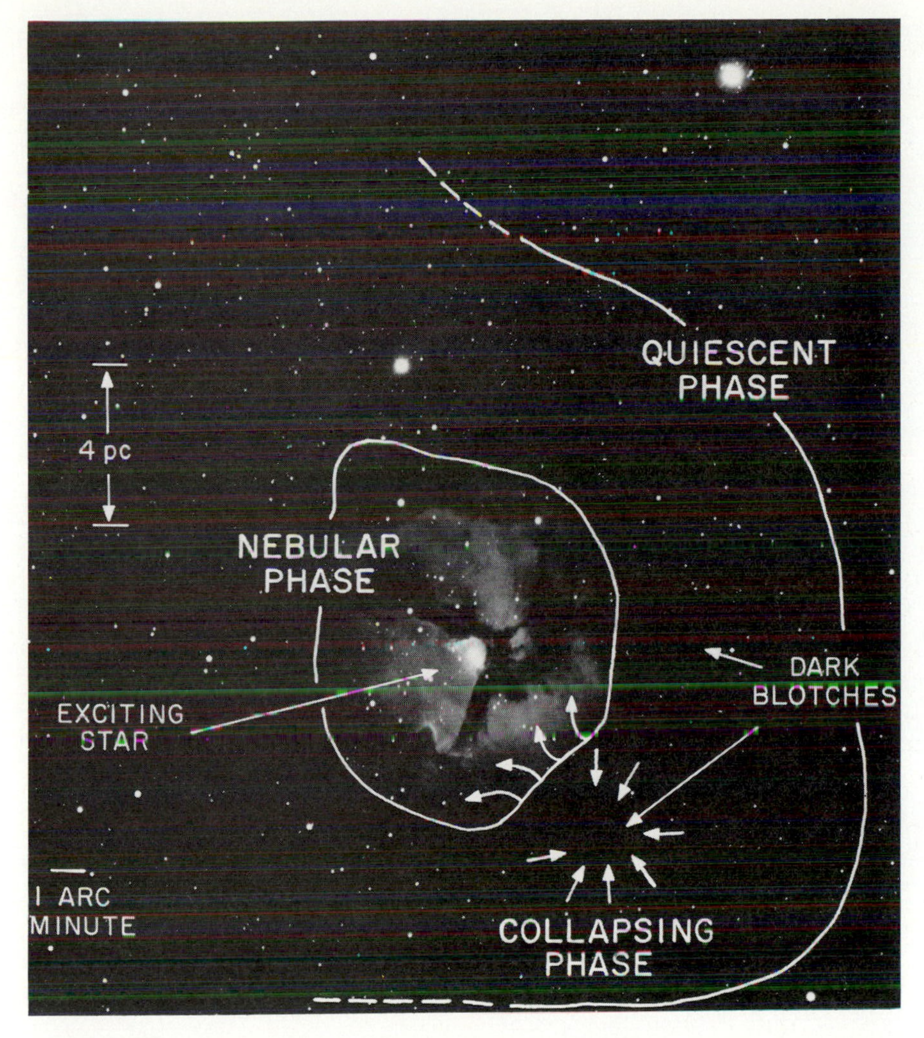

FIGURE 19.12 The M20 region shows observational evidence of three broad phases in the birth of a star: the inactive (or quiescent) phase, followed by the contracting (or collapsing phase), and finally the stellar (or nebular) phase.

inactive (or quiescent) phase is characterized by low densities and temperatures, in the range of about 100 particles/cubic centimeter and 20 kelvin.

Greater densities and temperatures typify smaller portions of this huge interstellar cloud. The totally obscured regions at which the molecular abundances peak (Figure 12.24) represent such denser and hotter fragments. Here the total gas density is observed to be at least 1000 particles/cubic centimeter. And the temperature measures about 100 kelvin. (Remember that these particle densities do not refer to formaldehyde molecules alone, but are total densities, mostly comprising hydrogen molecules, as noted in Chapter 12. The formaldehyde (and carbon monoxide) molecules are merely used by radio astronomers as convenient tracers of regions where molecules are abundant.) Less than a light-year across, the contracting (or collapsing) region noted in Figure 19.11 has a total mass of more than 1000 solar masses. This second broad phase of our star-formation sequence represents a cloud somewhere between stages 1 and 2 of Table 19-1.

The third phase, also shown in Figure 19.12, is exhibited by the star at the center of the M20 nebula itself. The glowing region of ionized gas results directly from a massive O-type star having formed there sometime within the past million years or so. Since the star is already fully formed, this final phase corresponds to stage 7 of our earlier evolutionary scenario.

Current observations of the M20 region do not display evidence for any of the intermediate evolutionary stages outlined earlier. Someday, when radio and infrared observations are made with better angular resolution, we might be able to study directly the details of the protostellar objects that presumably lurk in the darkest realms of these contracting fragments. In the meantime, we must examine other objects to find examples of stages 3 through 6.

EVIDENCE OF PROTOSTARS: Other regions of our Milky Way do show sketchy evidence for more advanced prestellar objects. The Orion complex, shown in the optical photograph of Figure 19.13, is one such region. Lit by several O-type stars, the bright Orion Nebula is partly engulfed by a vast molecular cloud. This dark cloud actually extends well beyond the 2 light-years bordered by the photograph and has been studied by means of the radio radiation attributed to carbon monoxide and formaldehyde molecules. These molecules are useful probes of interstellar gas having moderate density, about 1000 particles/cubic centimeter.

Many other molecules, such as hydrogen cyanide (HCN) and carbon monosulfide (CS), usually emit radiation from regions of even greater density. As shown in Figure 19.13, these molecules extend over only a small part of the molecular cloud, just behind the bright nebula. As we might expect, the molecules seem to delineate a fragment of the larger molecular cloud where the density is approximately 100,000 particles/cubic centimeter. (Again, as noted earlier, atomic and molecular hydrogen are by far the most abundant particles at all stages of early star formation.) The measured extent of the fragment is a little less than 1 light-year, has a temperature of nearly 200 kelvin, and hence can be identified as an object near stage 2 of Table 19-1.

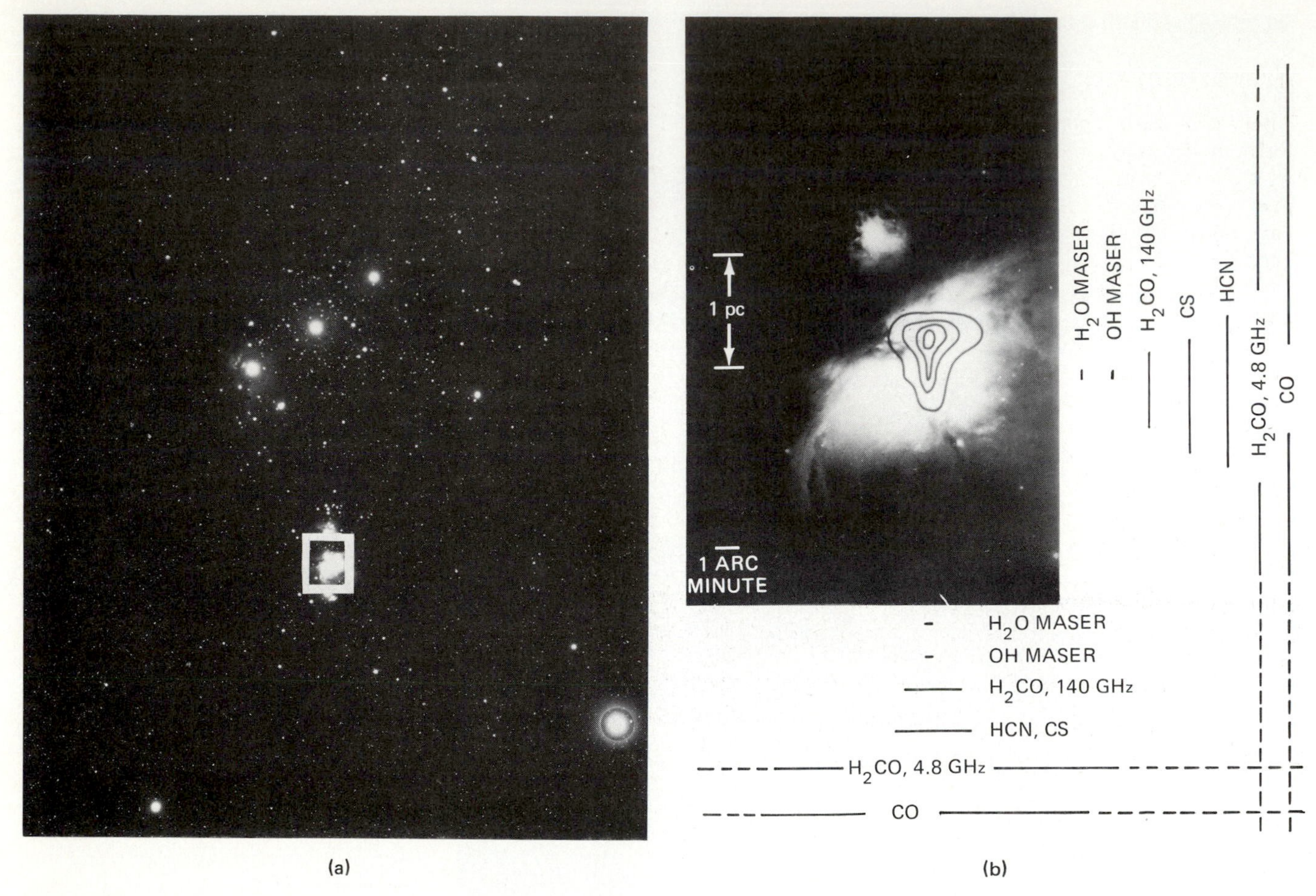

FIGURE 19.13 The Orion Nebula is a stunning gaseous nebula within the small rectangle in (a). This framed region is enlarged in (b), showing how Orion is partly engulfed by a vast molecular cloud, various parts of which are probably fragmenting and contracting, with even smaller sites resembling protostars. The various lines to the right and bottom of (b) depict the extent of molecular emissions from a dark cloud *behind* the bright nebula. (Review also Figures 3.22, 3.32, and 12.18.)

The Orion molecular cloud also harbors several smaller sites of intense radiation emitted by molecules under very special conditions (see Interlude 19-2). Molecules such as hydroxyl (OH) and water vapor (H_2O) have been found by radio techniques to be buried within the core of the cloud fragment. Their extent, shown in Figure 19.13, measures about 10^{10} kilometers, or 1000 times smaller than 1 light-year. This is just about equivalent to the full diameter of our Solar System. The gas density of these smaller regions is about 10^9 particles/cubic centimeter, and although the temperature cannot be estimated reliably, many researchers regard these regions as objects near stage 3. We cannot currently determine if these regions will eventually form stars more or less like the Sun, but it does seem certain that such intensely emitting regions are on the threshold of becoming protostars.

In a relatively recent development, strong winds have been found to be associated with potential protostars. Radio and infrared observations of hydrogen and carbon monoxide molecules in the same Orion cloud have revealed clouds of gas expanding outward at velocities approaching 100 kilometers/second. Furthermore, high-resolution, interferometric observations have disclosed expanding knots of water emission within the same star-forming region, thus linking the strong winds to the protostars themselves. The specific causes of these winds—like those of many genuine, fully formed stars, as discussed in Chapter 11—remain unknown. But astronomers will need to consider the implications for the early Solar System; after all, our Sun was presumably once a protostar, too.

When hunting for and studying objects at more advanced stages of star formation, radio techniques become less useful. The bulk of the exploration shifts to the infrared part of the spectrum because stages 4, 5, and 6 are expected to display increasingly high temperatures. As the Planck curve of thermal emission from warm protostars and young stars shifts toward shorter wavelengths, these objects should be observable largely in the infrared.

Sure enough, a most interesting

INTERLUDE 19-2 ***Interstellar Masers***

The word "laser" has become a common everyday term. It's actually an acronym for "light amplification by stimulated emission of radiation." Lasers are devices that emit a concentrated stream of light radiation in a vary narrow beam. Our civilization has only recently become smart enough to build such tools that rely on advanced technology and a good understanding of atomic and molecular physics. Lasers operate by using radiation to excite atoms and molecules in a gas or solid, after which that same kind of radiation stimulates the gas back to a lower-energy state. Provided that all the atoms or molecules return quickly and simultaneously to the lower state, a powerful packet of light is emitted from the gas. In this way, a tremendous burst of radiation results. Although laser emission is much more intense than normal (light bulb) emission, the wavelength or frequency of the emitted radiation is still uniquely characteristic of the particular atom or molecule excited.

Masers are similar to lasers, except that they produce microwave (radio) radiation rather than optical (light) radiation. We can build masers in our terrestrial laboratories, although they are very delicate machines, requiring special conditions and much patience to operate. When working properly, they are the best amplifiers known, much more effective than transistors.

Strange as it may seem, some parts of interstellar space are naturally suited to produce amplified microwave radiation. We know this because some interstellar molecules emit extremely intense radiation—much more intense than is possible from a normal collection of molecules energized only by random collisions with each other. In fact, some of the early observations of radio radiation from the hydroxyl (OH) molecule in the 1960s were so mysterious that puzzled researchers began calling the emitter "mysterium." Later identified as OH, the molecules emitting these powerful signals are amplified by the same maser process described above.

Apparently, some sites inside interstellar clouds have the special conditions required, first, to excite some molecules, and second, to stimulate them to emit intensely. In addition to OH, water vapor (H_2O) and silicon monoxide (SiO) emit microwave radiation in a maser-like fashion. The special physical conditions needed for maser radiation exist naturally in the vicinity of protostars—warm temperatures of about 1000 kelvin and high densities of about 10^{12} particles/cubic centimeter.

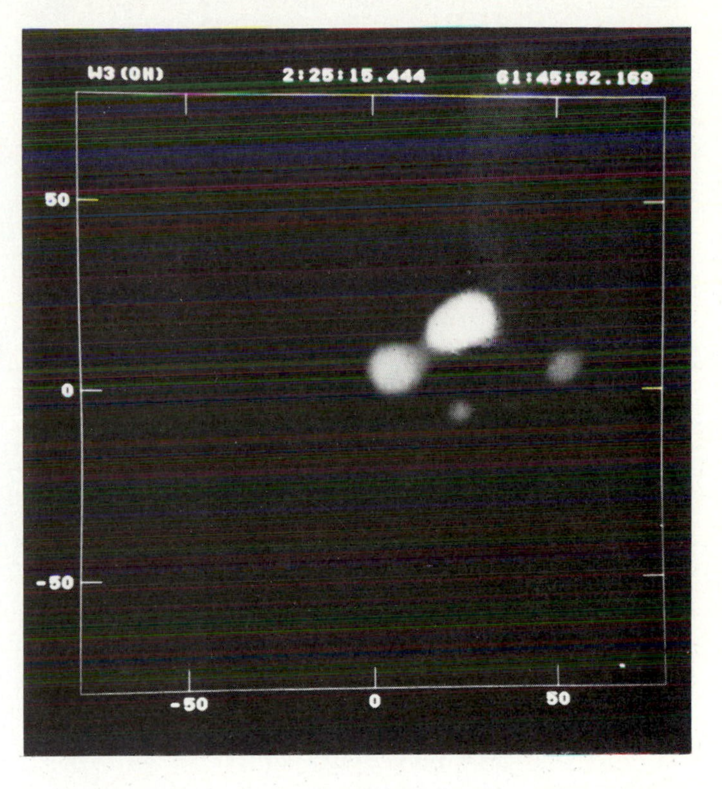

For example, the accompanying figure is a very high resolution map of several radio-emitting maser sources in a star-forming region known as W3. Each of these blobs has roughly Solar System dimensions and has been resolved here by the technique of very-long-baseline-interferometry discussed in Chapter 3. (The left and bottom scales refer to ± 50 milliarc seconds, or 0.05 arc second of angular measure.) In all probability, these molecular masers are protostars; at the least, they are about to become protostars. Such studies comprise one of the most exciting areas of modern astronomy.

object within the core of the Orion molecular cloud was detected by infrared astronomers during the 1970s. This compact infrared source is outlined only by contours in Figure 19.13, for only long-wavelength radiation can penetrate the cloud cores. Most astronomers agree that this warm, dense blob is a genuine protostar, poised on the verge of stardom.

The energy sources for the infrared objects seem to be optically luminous hot stars which, however, are hidden from view by nearby dark clouds. Apparently, some of the stars are already so hot that they emit large amounts of ultraviolet radiation, which is mostly absorbed by a surrounding "cocoon" of dust. The absorbed energy is then reemitted by the dust as infrared radiation. Some of the ultraviolet radiation heats and ionizes the accompanying hydrogen, whose emission can be observed in the radio domain. In particular, the intense OH and H_2O radiation (which escapes the region) enables the clouds to cool off despite the continual heating by the stars. Some of the cloud fragments are so massive that their own gravitation is trying to contract them further. As this tendency is resisted mainly by the random motions of the molecules, it is possible that the cooling provided by the escape of the intense radio radiation plays a significant role in the contraction process. That dust cocoons are invariably found in the dense cores of molecular clouds supports the idea that the hot stars responsible for their heating only recently emerged from the surrounding cloud.

Until the *Infrared Astronomy Satellite* was launched in the early 1980s, astronomers were only aware of giant stars forming in clouds far away. But *IRAS* showed that stars are forming much nearer than anyone knew, and some of these protostars have masses comparable to that of our Sun. Figure 19.14 shows a premier example of a solar-mass protostar—Barnard 5, which has an infrared heat signature just that expected of a warm blob somewhere along the Hayashi track.

Incidentally, *IRAS* also discovered, quite unexpectedly, a mysterious background emission throughout the Milky Way. This weak radiation is called "galactic cirrus" by some researchers since the infrared images of it resemble wispy cirrus clouds in Earth's atmosphere. Whether it is an important factor in the star-forming process or merely warm shells of matter left from previous stellar explosions—and whether the galactic cirrus might be a partial solution to the hidden-mass problem—is unresolved at this time.

Do not lose sight of the fact that all this protostar activity happens in the darkest regions of galactic gas and dust. Not unlike the mammals, the brightest stars are incubated in total darkness.

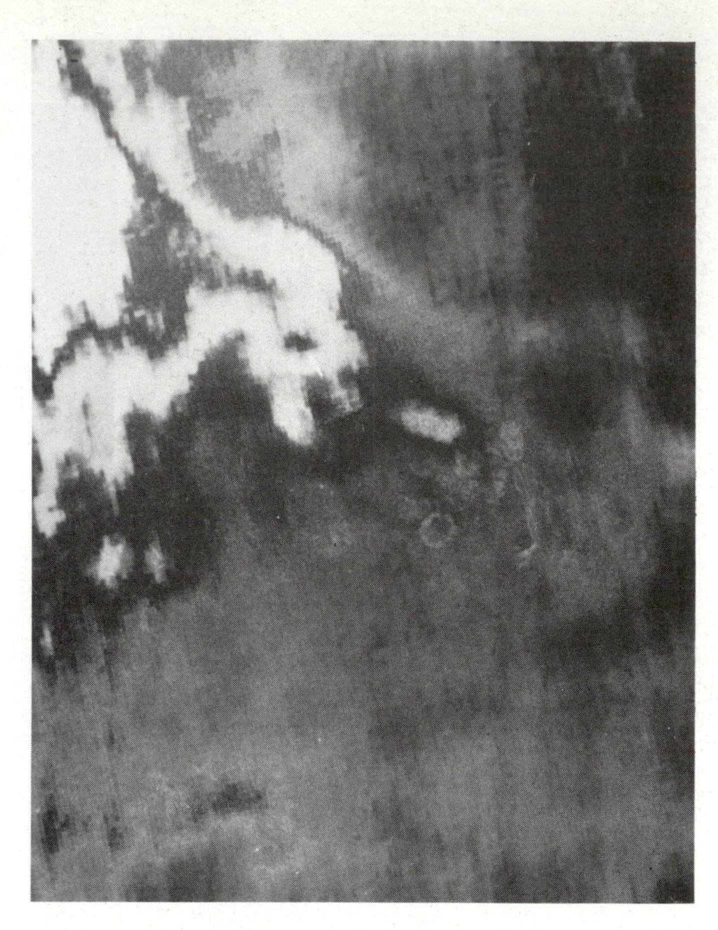

FIGURE 19.14 An infrared image of the nearby region containing the source Barnard 5 (arrow).

Complexities of Reality

The subject of star formation is much more complicated than the previous discussion suggests. Interstellar space at any one time is populated with all sorts of clouds, fragments, protostars, stars, and nebulae. As they all interact in some complex fashion, one type of matter doubtless affects the behavior of others.

For example, the presence of a gaseous nebula in or near a molecular cloud probably influences the evolution of the whole interstellar region. For any of the gaseous nebulae studied so far, we can easily visualize expanding waves of matter driven by the pressure of stellar ultraviolet radiation in the nebula. As the wave pushes outward into the surrounding molecular cloud, the interstellar gas would tend to become piled up or compressed, thus increasing the density of matter. Such a shell of gas, rushing rapidly through space, is known as a **shock wave**. It can push ordinarily thin matter into dense sheets, just as snow is swept up by the blade of a plow.

Many astronomers regard the passage of a shock wave through interstellar matter as the triggering mechanism needed to initiate star formation in our Galaxy. Calculations show that when a shock wave encounters an interstellar cloud, it races around the thinner exterior of the cloud more rapidly than it can penetrate its thicker interior. Shock waves do not blast a cloud from only one direction. They effectively squeeze it from many directions, as illustrated in Figure 19.15. Atomic bomb tests have experimentally demonstrated this squeezing: Shock waves created in the blast surround buildings, causing them to be blown

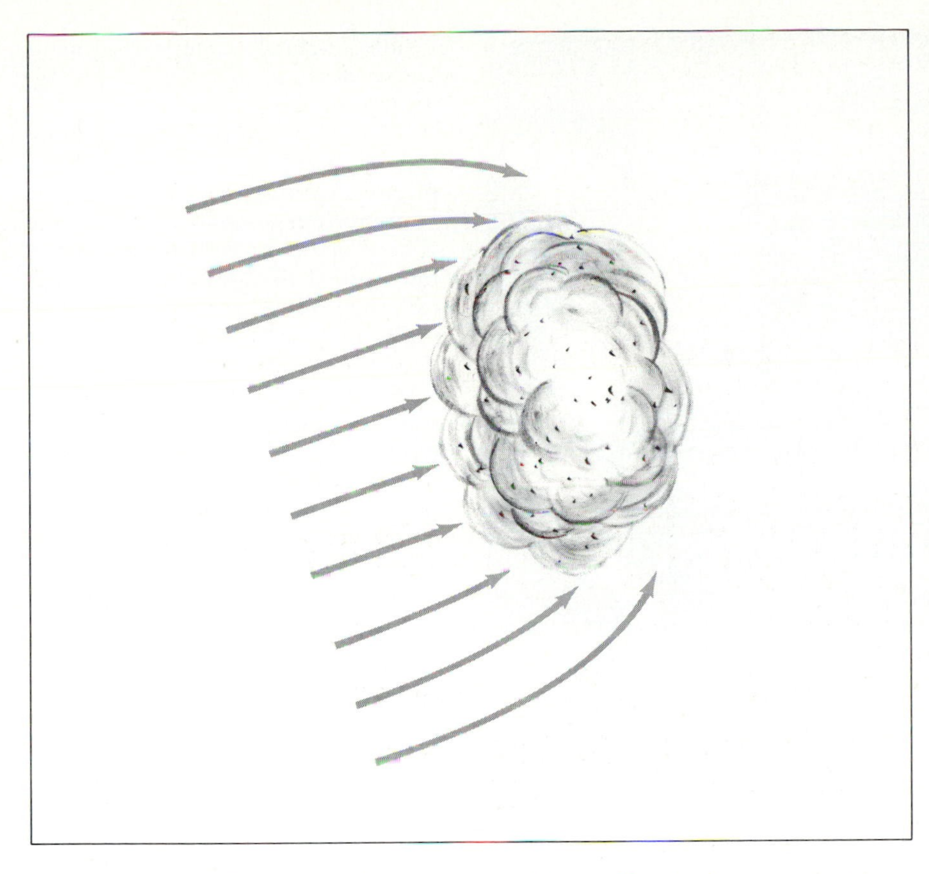

FIGURE 19.15 Shock waves tend to wrap around interstellar clouds, compressing them to greater densities, and thus possibly triggering star formation.

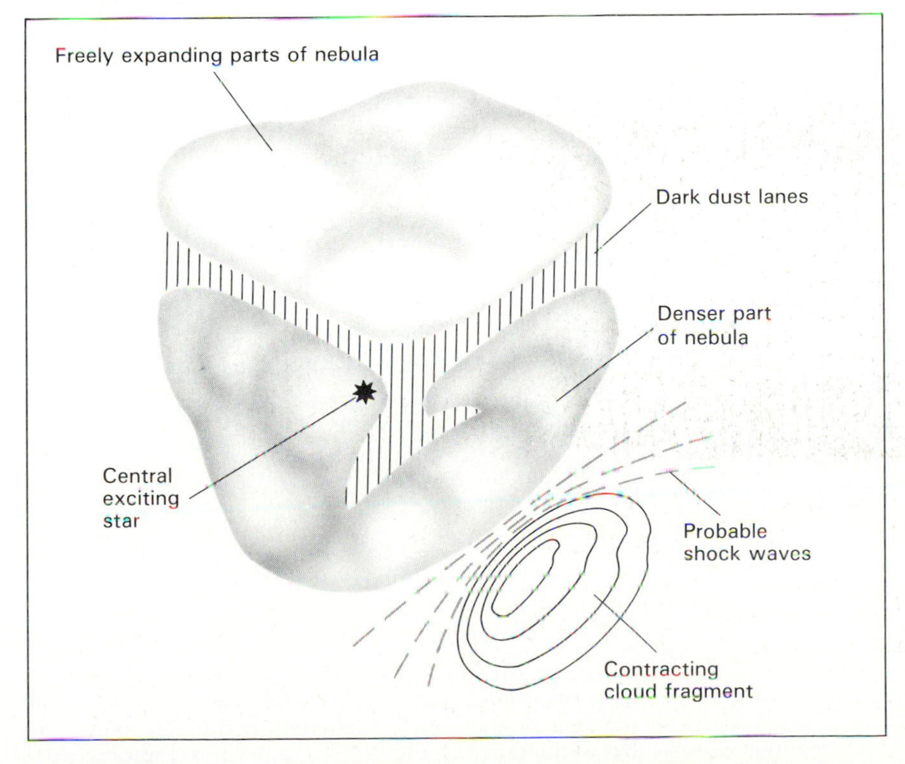

FIGURE 19.16 In this artist's conception, a cloud fragment is shown undergoing compression on the southerly edge of M20 as shock waves from the nebula penetrate the surrounding interstellar cloud. (Consult also Plate 5.)

together (imploded) rather than apart (exploded). Similarly, shock waves can cause the initial compression of an interstellar cloud, after which natural gravitational instabilities divide it into fragments that eventually form stars. Figure 19.16 suggests how this mechanism might be at work near M20.

Gaseous nebulae are not the sole generators of shock waves. At least two other sources are available—the spiral-arm waves that plow through the Milky Way as noted in Chapter 13, and the remnants of exploded stars (supernovae) to be discussed in Chapter 20. Perhaps each of these sources trigger star formation, but the last one is probably the most efficient way to pile up matter into dense clumps.

The photograph in Figure 19.17 shows a semicircular band of glowing gas in the region of the sky labeled Canis Major. The bright interstellar matter along the arc is almost certainly only part of a three-dimensional (spherical) shell, but we cannot be sure from such an optical view. Radio observations of the region reveal that other parts of the expanding shell are made mostly of invisible neutral hydrogen gas. The gas comprising the shell is thought to be the remnant of a stellar explosion that occurred long ago. How long ago? The size and expansion velocity of the shell's gas enable astronomers to infer that a star blew up about $\frac{1}{2}$ million years ago. Ever since, a wave of gas has been moving away from this point, piling up matter into high-density concentrations. Although the evidence is somewhat circumstantial, the presence of numerous O- and B-type young (and thus quick-forming) stars in the vicinity of this remnant do imply that the birth of new stars is often initiated by the violent, explosive deaths of old stars. Ironic, indeed, if the demise of old stars be the trigger needed to conceive new ones.

Wherever O- and B-type stars have formed, we can suppose that less massive stars are still in the process of formation. It takes longer for the less massive stars to form, and thus we should not expect to find many A-, F-, G-, K-, or M-type

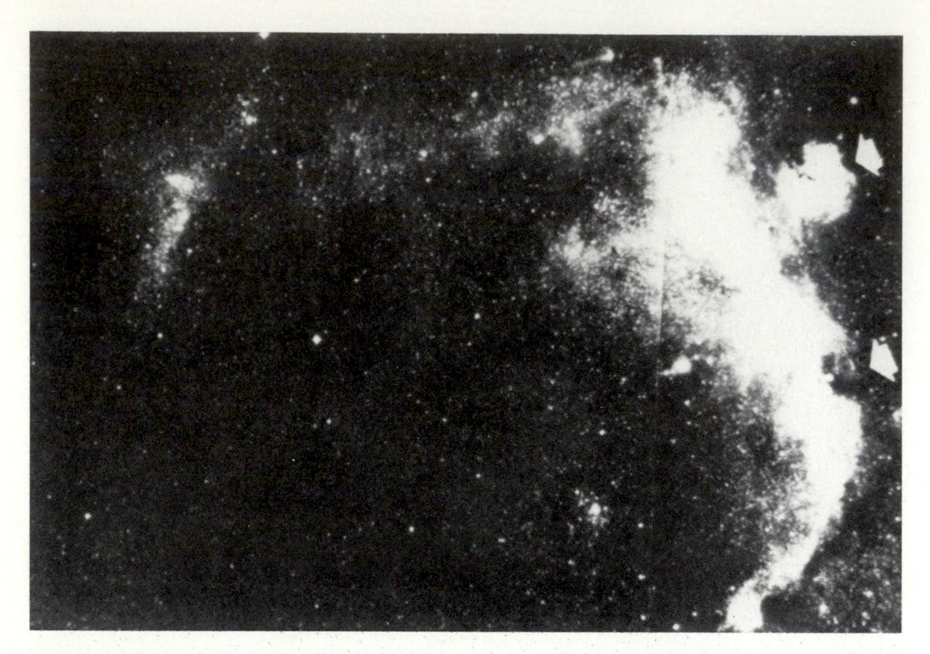

FIGURE 19.17 This arc of glowing gas is only part of a nearly complete shell of interstellar matter, which was probably ejected by a massive star that exploded nearly 600,000 years ago. Young stars are found on the inside edge of the shell, while additional stars are probably forming on the outer edge (marked by arrows) as the shell's shock wave piles up the matter.

stars, provided that the star formation mechanism really was triggered less than 1 million years ago. The whole Canis Major region, and many others like it, are probably vast breeding grounds—sites of invisible interstellar cloud fragments and protostars, as well as young, visible, massive stars.

This scenario of shock-induced star formation is even more complicated than described above. The complication arises because O- and B-type stars not only form rapidly, but as we shall note in the next chapter, they also die rapidly. Accordingly, massive stars that formed earlier by a passing shock wave will in turn produce, either by their expanding nebular gas or by their explosive deaths, additional shock waves in the region. In the meantime, other, less massive stars are still struggling to form. "Second-generation" shock waves, sketched schematically in Figure 19.18, can then cause yet more stars to form. Each O- or B-type star forms quickly, lives briefly, and dies explosively. Each generates a shock wave of gas that pushes outward into interstellar space, forming more stars that then explode anew as part of a continuous, sequential mechanism of star formation resembling a chain reaction. This idea is supported by the fact that alignments of stars are observed near the outside rims of various molecular clouds; groups of stars nearest such clouds appear to be the youngest, while those farther away appear to be older, pretty much like what is expected.

Take another look at the nighttime sky. Ponder all that cosmic activity while gazing upward some clear, lovely evening. After studying the material in this chapter, your view of the dark night sky is likely to be greatly changed. To be sure, even the darkness is dominated by change.

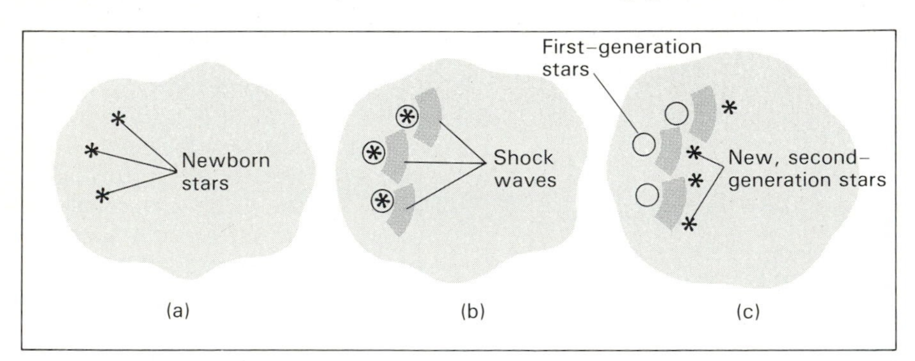

FIGURE 19.18 The process of star birth [asterisks in (a)] and shock waves [shaded areas in (b)], followed by more star birth and additional shock waves (c), is likely to produce a continuous cycle of star formation in many areas of our Galaxy. Like a chain reaction, old stars trigger the formation of new stars ever deeper into an interstellar cloud.

SUMMARY

The remarkable change from interstellar cloud to contracting fragment to protostellar blob to young star takes millions of years. Obviously a long time by human standards, this formation time is still less than a tenth of 1 percent of a star's complete lifetime. The entire process amounts to a steady metamorphosis, an evolution of sorts, a gradual transformation of a cold, tenuous, flimsy pocket of gas into a hot, dense, and round star. The prime factor in all this evolutionary change is gravity.

Despite their apparent diversity, astronomical studies continue to address a unifying theme: the varying interrelationships of the many components of interstellar matter. Comprising nothing less than a "galactic ecosystem," the evolutionary balance among these components might be as complex and delicate as that of life in a tidepool or a tropical forest. Only by being receptive to information from the Milky Way in all the electromagnetic wavelength bands in which it chooses to radiate can we hope to understand some of nature's cycles within our Galaxy.

KEY TERMS

Hayashi track

Kelvin–Helmholtz contraction phase

protostar

shock wave

T-Tauri star

stellar evolution

QUESTIONS

1. Name and discuss three external agents that compete against gravity in a star-forming interstellar cloud.
2. What physical property guarantees that a main-sequence star will not collapse under its own weight? Explain.
3. Why do interstellar clouds having thousands of times the mass of our Sun probably not form a single large star having that large amount of mass?
4. Why doesn't fragmentation continue indefinitely, constantly breaking down all matter into smaller and smaller objects?
5. Why do individual fragments within interstellar clouds contract quickly at first and then slow up as they reach stellar proportions?
6. Explain why the Hayashi track on an HR diagram runs diagonally opposite to the Kelvin–Helmholtz contraction phase.
7. Of all the physical properties of clouds, fragments, and stars, which one most controls the evolution of these objects? Justify your answer.
8. Discuss the experimental techniques used by astronomers to study protostars embedded in dusty regions of the interstellar medium.
9. Explain why Jupiter is almost, but not quite, a star.
10. Cite and discuss some observational evidence for stages 1, 4, and 6 of the pre-main-sequence evolutionary scenario.

FOR FURTHER READING

Boss, A., "Collapse and Formation of Stars." *Scientific American*, January 1985.

Dickinson, D., "Cosmic Masers." *Scientific American*, June 1978.

Habing, H., and Neugebauer, G., "The Infrared Sky." *Scientific American*, November 1984.

Lada, C., "Energetic Outflows from Young Stars." *Scientific American*, July 1982.

*Moran, J., "Radio Observations of Galactic Masers," in *Frontiers of Astrophysics*, E. Avrett (ed.). Cambridge, Mass.: Harvard University Press, 1976.

Zeilik, M., "The Birth of Massive Stars." *Scientific American*, April 1978.

20 STELLAR EVOLUTION: *From Middle-Age to Death*

The process of star birth produces large changes in the temperature and luminosity of cosmic objects. Colloquially, we say that such protostars "travel" around on the Hertzsprung–Russell diagram, as described in Chapter 19. Eventually, a newly born star reaches the main sequence, where it remains, nearly unchanged, for more than 90 percent of its complete lifetime.

At the other extreme, as stars begin to run out of fuel and die, their properties once again change greatly. Astronomers refer to these changes in highly evolved stars as "movement away from the main sequence." Dying stars "travel" along evolutionary paths that take them far from the main sequence.

M-type dwarfs and other low-mass stars burn their fuel so slowly that none of these could have yet left the main sequence of the HR diagram. Some of these stars burn steadily for a trillion years or more. On the other hand, the more massive stars, especially the O- and B-type stars, evolve away from the main sequence after only tens of millions of years of burning. Many of them are now observed in advanced stages of stellar evolution; their properties differ considerably from similar objects still on the main sequence. Many other massive stars must have perished long ago.

In this chapter we study the evolution of stars after they've passed through their main-sequence burning stage. Along the way, we'll meet some strange states of matter—weird objects that exist just prior to stellar death. These include planetary nebulae, red giants, white dwarfs, neutron stars, and black holes.

While studying these strange and often explosive states of matter, we must always keep in mind one of our primary objectives—to understand which of all the different kinds of objects in our Universe is best suited for the origin and evolution of life. In fact, if cosmic evolution is correct, life itself may be the strangest of all states of matter.

The learning goals for this chapter are:

- to appreciate the various types of strange objects resulting from the late-stage evolution of ordinary stars
- to recognize that the many different star-like objects seen in the Galaxy are basically the same type of object at different stages of evolution
- to realize that stars, like life forms, cannot last forever
- to understand the reasons for the gentle death of low-mass stars and for the explosive death of their high-mass counterparts
- to appreciate the many observations that help verify the theory of stellar evolution

Sun-Like Evolution

In this section we consider the detailed evolution of a Sun-like star—one having a mass similar to that of our Sun. The evolutionary paces described here will pertain to the Sun at some future time. But be sure to recognize that these are typical evolutionary changes experienced by all Sun-like stars nearing the twilight of their fusion cycles. Later in the chapter we shall broaden the discussion to include all stars, large and small.

MAIN-SEQUENCE EQUILIBRIUM: Gravity is always present wherever matter exists. Only some counteracting phenomena can prevent all astronomical objects from completely collapsing under the relentless pull of gravity. In the case of stars, that competing phenomenon is gas pressure caused by the heat of the raging inferno at the stellar core. Figure 20.1 diagrams the balanced equilibrium between the inward gravitational pull and the outward gas pressure. You should keep this simple diagram in mind while studying the various stages of stellar evolution described in this chapter.

Provided that a star remains in this equilibrium state, nothing spectacular happens to it. While a resident of the main sequence, its hydrogen fuel slowly changes into helium, its surface erupts in flares and spots, and its atmosphere ejects copious amounts of particles and photons capable of affecting any attendant planets. But by and large, stars do not experience sudden, large-scale changes while in equilibrium. Their average temperatures and luminosities remain fairly constant, and they would be expected to release energy indefinitely unless something drastic occurred. Eventually something drastic does occur.

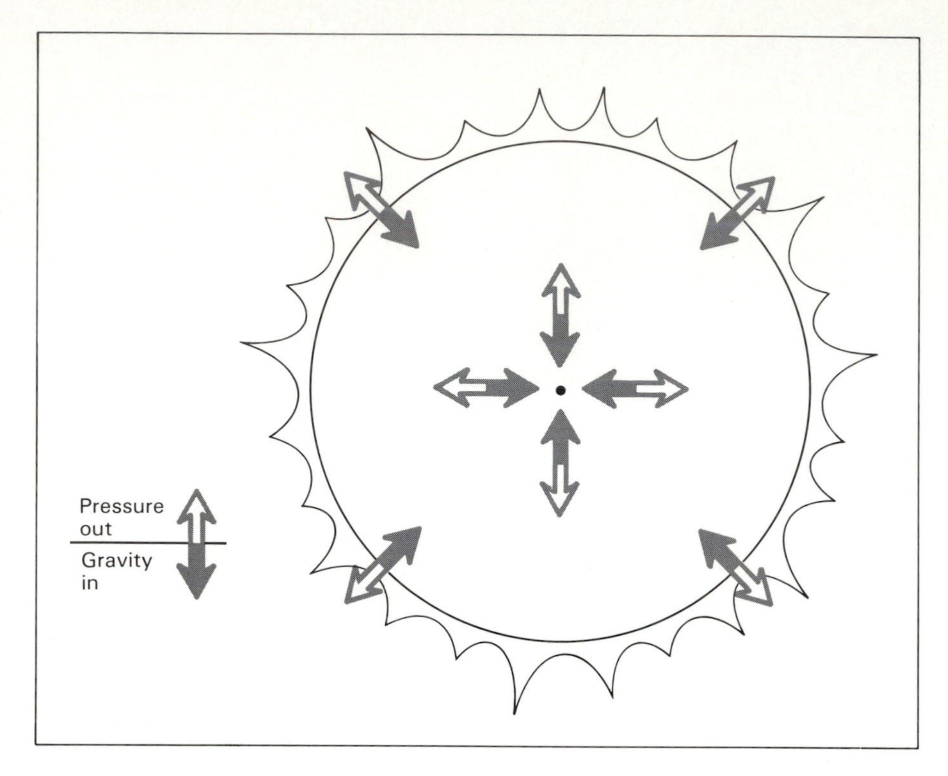

FIGURE 20.1 A steadily burning star on the main sequence has the inward pull of gravity counterbalanced by the outward pressure of hot gas. This is true at any point within the star, guaranteeing its stability.

DEPLETION OF HYDROGEN AT THE CORE: After approximately 10 billion years of steady burning, a Sun-like star begins running out of fuel. The hydrogen nuclei simply become depleted, at least in a small central core about one-hundredth of the star's full size. Not a sudden process, the depletion of hydrogen is slow and steady. But the consequences are drastic, for there comes a time when little hydrogen remains within the zone of 10 million kelvin. It's probably a bit like an automobile that can cruise along a highway at a constant velocity of 55 miles/hour for many hours without a care in the world, only to have the engine cough as the gas gauge reaches empty. Unlike automobiles, though, stars are not easy to refuel.

Widespread exhaustion of hydrogen in the stellar core causes the nuclear fires there to cease. Some hydrogen burning continues in the intermediate layers, above the core though well below the photosphere. But the core normally provides the bulk of the support in a star, establishing its foundation and guaranteeing its equilibrium. The lack of core burning thus creates an unstable situation because, although the outward gas pressure weakens in the cooling core, we can be sure that the inward pull of gravity does not. Gravity never lets up. Once the outward push against gravity is relaxed—even a little—changes in the star become inevitable.

If more heat could be generated, the star could possibly return to equilibrium. For example, were helium at the core to begin fusing into some heavier element such as carbon, all would be well once again, for energy would be created as a byproduct, thus reestablishing the outward gas pressure. But the helium there cannot burn–not yet anyway. Despite a phenomenal temperature of more than 10 million kelvin, the core is just too cold to fuse helium into any heavier elements.

Recall that a temperature of at least 10 million kelvin is needed to initiate the simpler hydrogen → helium fusion cycle. That's what it takes for two colliding hydrogen nuclei to get up enough speed to overwhelm the repulsive electromagnetic force field between two like charges. Otherwise, the nuclei cannot penetrate the realm of the nuclear binding force, and the fusion process simply doesn't work. Well, with helium, even 10 million kelvin is insufficient for fusion. Each helium nucleus, composed of two protons and two neutrons, has a net positive charge twice that of the hydrogen nucleus. Therefore, the repulsive electromagnetic force field is also larger, and more violent collisions are needed to fuse helium nuclei. To ensure the proper violence, tremendously high temperatures are required. How high? About 100 million kelvin.

A core full of helium at 10 million kelvin thus cannot generate energy via fusion. However, this region of helium ash does not remain idle for long. As soon as the hydrogen fuel becomes substantially depleted, the helium core begins contracting. It has to; there's not enough pressure to hold back gravity. This very shrinkage increases the gas density, thereby driving up the heat as particle collisions become ever more frequent.

The core temperature doesn't suddenly jump to the 100-million-kelvin value needed for helium fusion. Core contraction is a slow process. For a star the size and mass of our Sun, it heats slowly, taking several tens of millions of years to do so.

The increasingly hot core continues to heat the intermediate layers of the star's interior. It's a little like turning up a stove from warm to hot. Higher temperatures—now well over 10 million kelvin—cause hydrogen nuclei in the intermediate layers to fuse even more rapidly than in the core before. Figure 20.2 depicts this rather peculiar situation where hydrogen is burning at a fantastic rate around the nonburning helium ash.

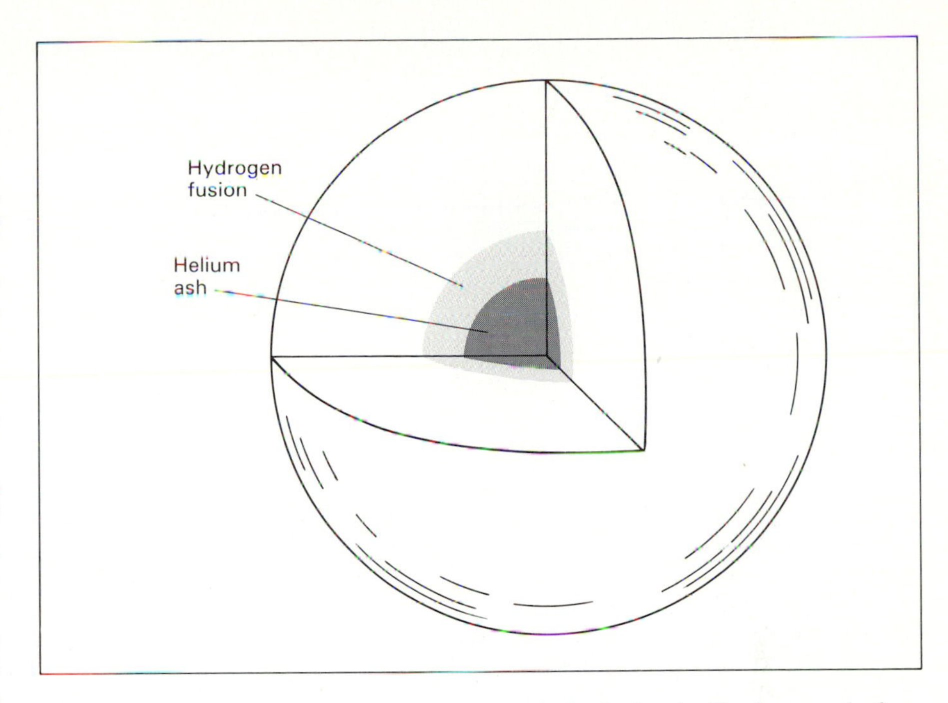

FIGURE 20.2 As a star's core becomes progressively depleted of hydrogen, the hydrogen fusion reactions continue to burn at intermediate layers, high above the nonburning helium ash.

The aged star is really in a predicament now. The overall conditions have clearly changed from the equilibrium that once characterized the main-sequence star. The core is unbalanced and shrinking, on its way toward generating enough heat for helium fusion. The intermediate and outermost layers are also unbalanced, fusing hydrogen into helium at a faster-than-normal rate. The gas pressure exerted by this enhanced hydrogen burning grows greater, forcing the intermediate layers and especially the outermost layers to expand. Not even gravity can stop them. Even while the core is shrinking, the overlying layers are expanding! The star, aged and unbalanced, is on its way to becoming a giant star.

Consider for a moment the observational consequences here. To a distant observer, the star would seem nearly 100 times larger than normal for a main-sequence star of its spectral type. Captured radiation would also show that the star's surface was 1000 kelvin cooler than normal. This is not to say that the act of either ballooning or cooling of an aged star could be observed directly. Theory suggests that the change from a normal main-sequence star to an elderly giant takes much longer (about 100 million years) than a human lifetime.

These large-scale changes in the disposition of an aged 1-solar-mass star can be traced on the HR diagram. Figure 20.3 shows the resulting path away from the main sequence. As illustrated, the luminosity of this giant star—a product of the square of the star's radius and the fourth power of its surface temperature—becomes about 100 times the current brightness of our Sun.

The second change—surface cooling—is a direct result of the first change—increased size. As the star grows larger, its sum total of heat is spread throughout a much larger stellar volume. Hence visible radiation emitted from such a cooling, yet still-hot surface shifts in color. Like a white-hot piece of metal that turns red while cooling, the whole star also displays a reddish tint. Over the course of time, a star of normal size and yellow color slowly changes into one of giant size and red color. The bright normal star has evolved into a brilliant, though cool **red giant star**.

MORE ABOUT RED GIANTS: Figure 20.4 compares the relative sizes of our Sun and a red giant star. The typical giant star is huge, having swollen to nearly 100 times its main-sequence size. By contrast, the helium core is surprisingly small, probably about 1000 times smaller than the entire star. This makes the core only a few times larger than Earth.

The density in the core is now huge. Continued shrinkage of the red giant's core has compacted its helium gas to approximately 10^5 grams/cubic centimeter. This value may be contrasted with the 10^{-6} gram/cubic centimeter in the outermost layers of the red giant star, with the 5 grams/cubic centimeter average Earth density, or with the 150 grams/cubic centimeter in the core of the present Sun. Owing to this greatly compressed helium state, about 25 percent of the mass of the entire star is packed into its small core.

So once the Sun exhausts the hydrogen fuel at its core, instability is

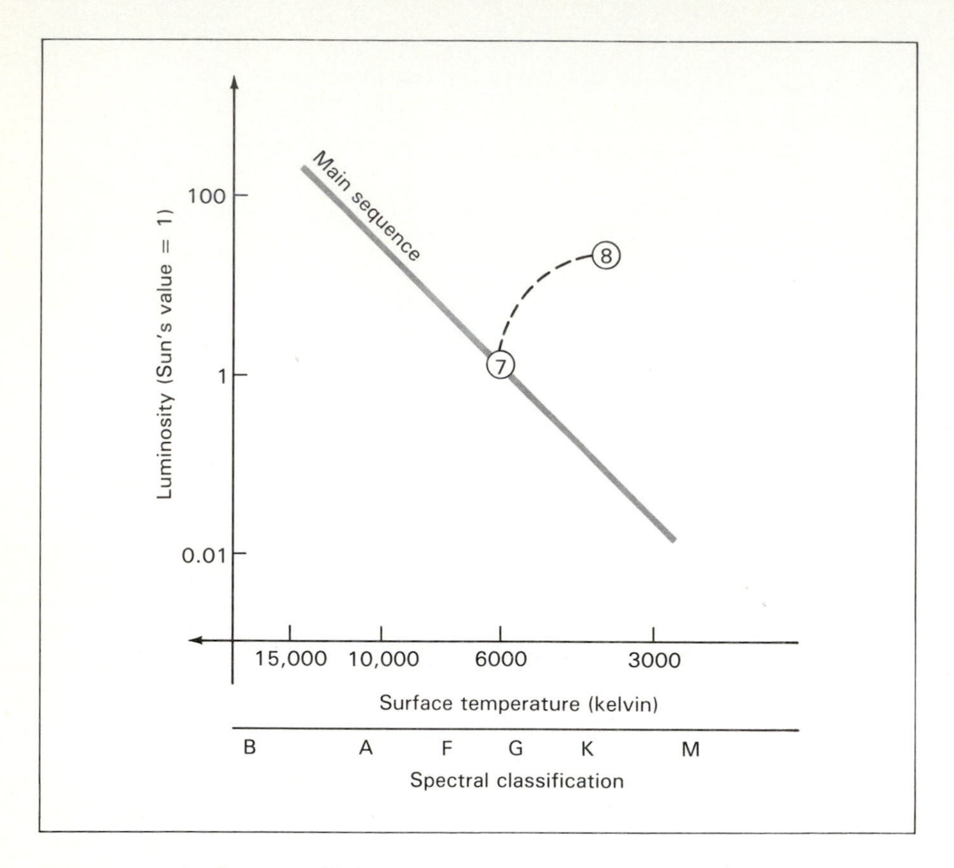

FIGURE 20.3 As the core of helium ash shrinks and the intermediate stellar layers expand, the star leaves the main sequence. Labeled stage 8, it's on its way to becoming a red giant star.

sure to set in. Its core will shrink. The overlying layers will expand. In short, the Sun is destined to become a hugely swollen sphere, in fact large enough to engulf many of the planets, including Mercury and Venus, and probably Earth and Mars as well.

Don't be alarmed. Provided that the theory of stellar evolution is reasonably correct as described in this chapter, we can be sure that our Sun will not balloon into this red-giant stage for several billion years in the future. Since stars having the mass of our Sun are expected to endure some 10 billion years, and since we take the Sun to have about the same 5-billion-year age as Earth, we conclude that our Sun should burn in steady equilibrium for about another 5 billion years. After that time, should civilization still reside on Earth, the consequences would be dire indeed!

Red giant stars are not figments of some theoretician's imagination. They really exist, scattered here and there about the sky. Even our naked eyes can perceive the most famous of all red giants—a star called Betelgeuse, a swollen, aged member of the constellation Orion.

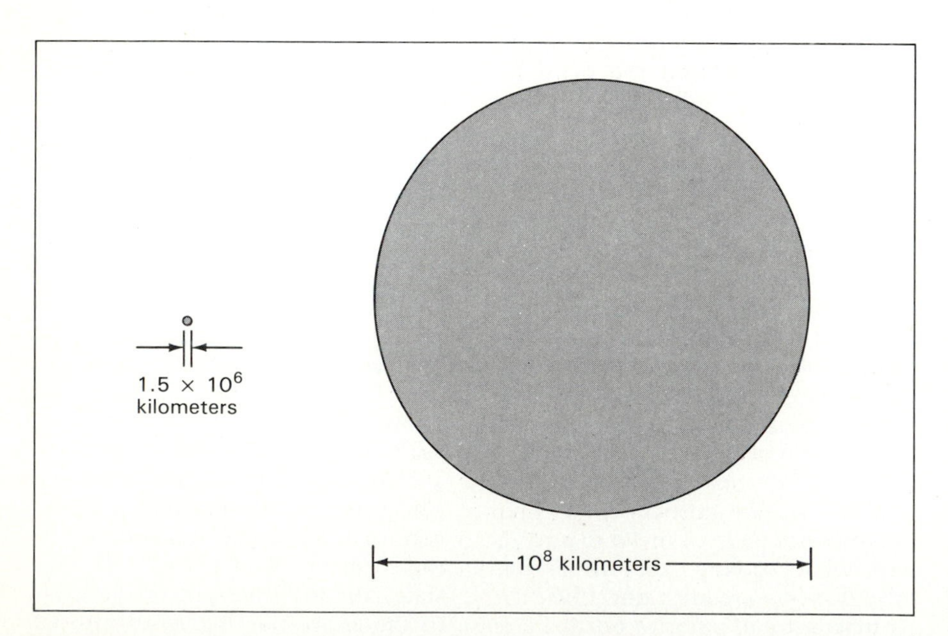

FIGURE 20.4 Schematic diagram of the relative sizes (to scale) of a normal G-type star when on the main sequence, and after it has ballooned to become a red giant star. The difference in size is approximately 70 times.

Helium Fusion

Should the unbalanced state of a red giant star continue unabated, the core would eventually implode, while the rest of the star receded in slow motion. Various forces and pressures at work inside such a decrepit star would literally pull it apart. However, this twofold shrinkage–expansion does not continue indefinitely. With a few hundred million years after the star leaves the main sequence, something else happens—helium begins to burn.

Once a density of 10^5 grams/cubic centimeter has been reached in the core, gas particle collisons will be violent and frequent enough to generate sufficient heat, via friction, to reach the 100 million kelvin needed for helium fusion. Helium nuclei then collide with one another, fusing into carbon nuclei, and igniting the central fires once again. Thereafter for a period of a few hours, the helium burns ferociously, like an uncontrolled bomb. A few hours is quite a short time—very short indeed—compared to the billions of years of a typical stellar lifetime. In fact, the onset of helium burning is such a sudden and rapid event in the history of a star that astronomers give it a special name: the **helium flash**.

Despite its brevity, uncontrolled fusion during the helium flash releases a flood of new energy. The energy is potent enough to push out the core matter somewhat, thus lowering the density and relieving some of the pent-up tenseness among the charged nuclei. This small expansive adjustment of the core halts the gravitational collapse of the star, returning it to equilibrium—an equilibrium once again between the inward pull of gravity and the outward push of gas pressure. After just a few hours of explosive helium burning, the flash subsides, and the slightly expanded star's core routinely changes helium into carbon at temperatures well above the required 100 million kelvin.

Once the helium → carbon fusion reactions have ignited, flashed, and stabilized the core, the hydrogen → helium fusion reactions burning in the layers above slow down a bit. We might imagine the star as having expanded its outer layers too rapidly, overshooting the distance at which the star achieves a comfortable gravity–heat equilibrium throughout. The entire star is then able to shrink a little, lessening its swollen appearance. This slight shrinkage of the outer layers causes the luminosity to decrease and the surface temperature to increase, reversing the star's evolutionary path once again, as shown in Figure 20.5. Like all evolutionary changes in the early and late stages of a star, this size adjustment is made quickly—in about 100,000 years.

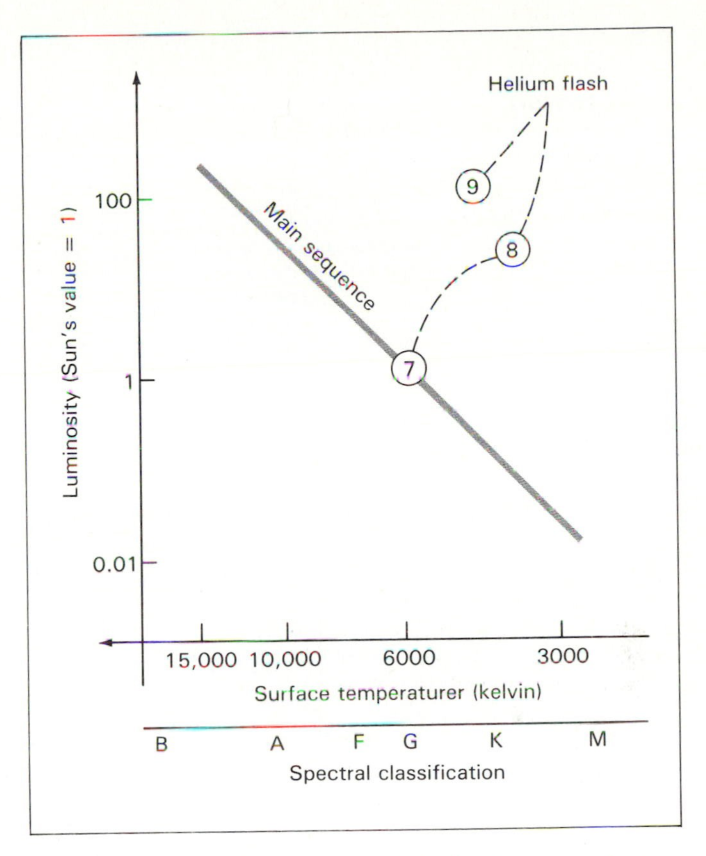

FIGURE 20.5 After a large increase in luninosity, a red giant star finally settles down into another equilibrium state at stage 9.

Although the time scales for stellar changes are relatively rapid throughout all stages of a star's birth from dust as well as its thrust toward death, we stress that these changes are still long compared to the duration of a human lifetime. Observers have little hope to watch an individual star move through all the evolutionary paces described here. Instead, we must rely on theoretical calculations or "numerical experiments" done with large computers to provide a good approximation of the post-main-sequence evolution of a star.

Table 20-1 summarizes a computer calculation done for a 1-solar-mass object. It's a continuation of the previous compilation listed in Table 19-1, except that the density units have been switched from particles/cubic centimeter to grams/cubic centimeter. The previous table ended with stage 7, a main-sequence object fusing hydrogen into helium over the course of some 10 billion years. The table here begins with stage 8, the evolutionary path away from the main sequence. Stage 9 describes an established red giant star fusing helium into carbon at its core.

As for the physical quantities listed in Table 19-1, those describing each of the stages of Table 20-1 cannot be specified with high accuracy. The temperature, density, size, and luminosity, as well as the precise evolutionary path, are not completely understood at this time. Each of these quantities depends on the initial conditions used for the mass and composition of a star, as well as on the rates of nuclear burning deep inside.

This reliance on computer modeling is exactly what makes the underabundance of solar neutrinos so disturbing. As noted in Chapter 10, the one experiment that bears directly on the physical events inside stars seems to disagree with the predictions for a 1-solar-mass star. Solar and stellar researchers, however, are reluctant to attribute the underabundance of solar neutrinos to any large conceptual errors in the theory of stellar evolution. Anyway,

TABLE 20-1
Late-Stage Evolution of a Sun-Like Object

STAGE	CENTRAL TEMPERATURE (KELVIN)	SURFACE TEMPERATURE (KELVIN)	CENTRAL DENSITY (GRAMS/CM^3)	SIZE (KILOMETERS)	OBJECT
8	50,000,000	4000	10^4	10^7	Post-main-sequence star
9	200,000,000	5000	10^5	10^8	Red giant star
10	300,000,000	50,000—inner core star	10^6	10^5	Planetary nebula
	—	3000—outer envelope	10^{-20}	10^8	
11	100,000,000	50,000	10^7	10^4	White dwarf star
12	Close to 0	Close to 0	10^7	10^4	Black dwarf star

it's useful to keep the neutrino experiment in mind, not because it's likely to lead to any major change in the stellar events outlined in this chapter, but because it illustrates some of the limitations that plague every aspect of cosmic evolution. Here in the stellar epoch, as elsewhere, astronomers and physicists know the broad outlines of many things, but the fine details are often yet to be understood.

The Carbon Core

The nuclear reactions in a star's helium core churn on, but not for long. Whatever helium exists in the core is rapidly consumed. The helium → carbon fusion cycle, like the hydrogen → helium cycle before it, proceeds at a rate proportional to the temperature; the greater the core temperature, the faster the reaction progresses. Under these excessively high temperatures, then, the helium fuel in the stellar core simply doesn't last long—no longer than a few million years after its initial "flash."

Buildup of carbon ash in the inner core causes physical phenomena similar to those in the earlier helium core. Helium first becomes depleted at the very center. Fusion then ceases there. In response, the carbon core shrinks and heats a little, causing the hydrogen and helium burning cycles to speed up in the intermediate and outermost layers of the star. These layers ultimately expand, such as they did earlier, making the star once again a swollen red giant. Figures 20.6 and 20.7 depict the star's interior and the evolutionary path followed during these latest events.

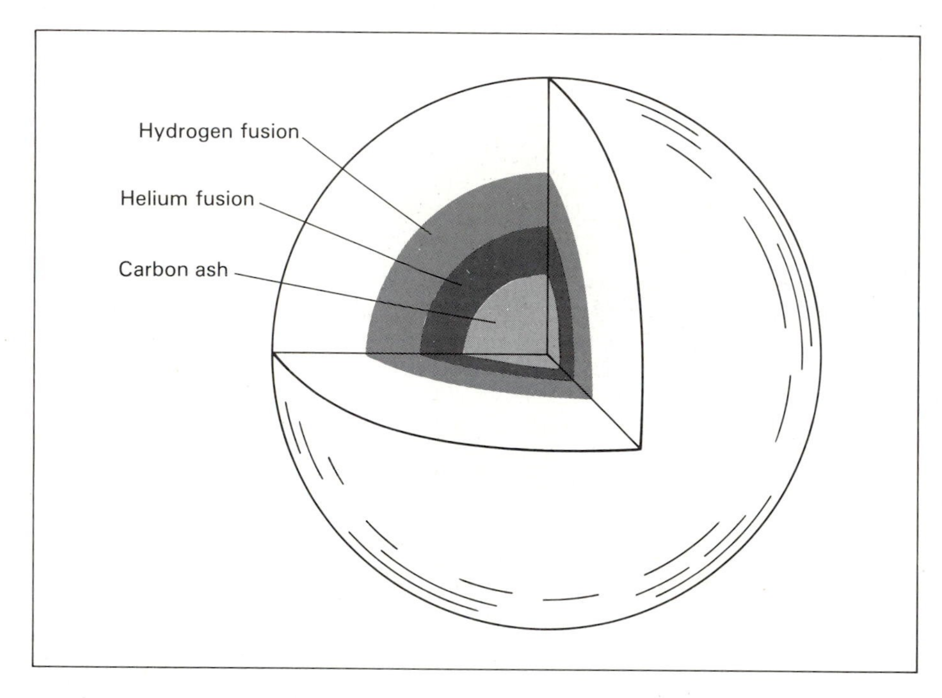

FIGURE 20.6 Within a few million years after the onset of helium burning, carbon ash accumulates in the inner core of a star, above which hydrogen and helium are still burning.

Provided that the core temperature becomes high enough for the fusion of carbon nuclei, or even a mixture of carbon and helium nuclei, still heavier products can be synthesized. Newly generated energy again supports the star, returning it to the usual equilibrium between gravity and heat.

This contracting–heating–fusing cycle is generally the way many of the heavy elements are cooked within the cores of stars. All elements heavier than helium are created within the final 1 percent of a star's lifetime. In Chapter 22 we'll study the detailed steps of elemental evolution. For now, though, we concentrate on the many interesting evolutionary changes experienced by stars prior to their death.

Mass Loss

As noted in Interlude 11-3, stars of all spectral types are now known to be active and to have stellar winds. Consider for a moment the highly luminous, hot, blue stars (of O- and B-type) that have by far the strongest winds. Observations of their ultraviolet spectra with telescopes on rockets and satellites have shown that their wind speeds (or gales!) often reach 3000 kilometers/second (or several million miles per hour).

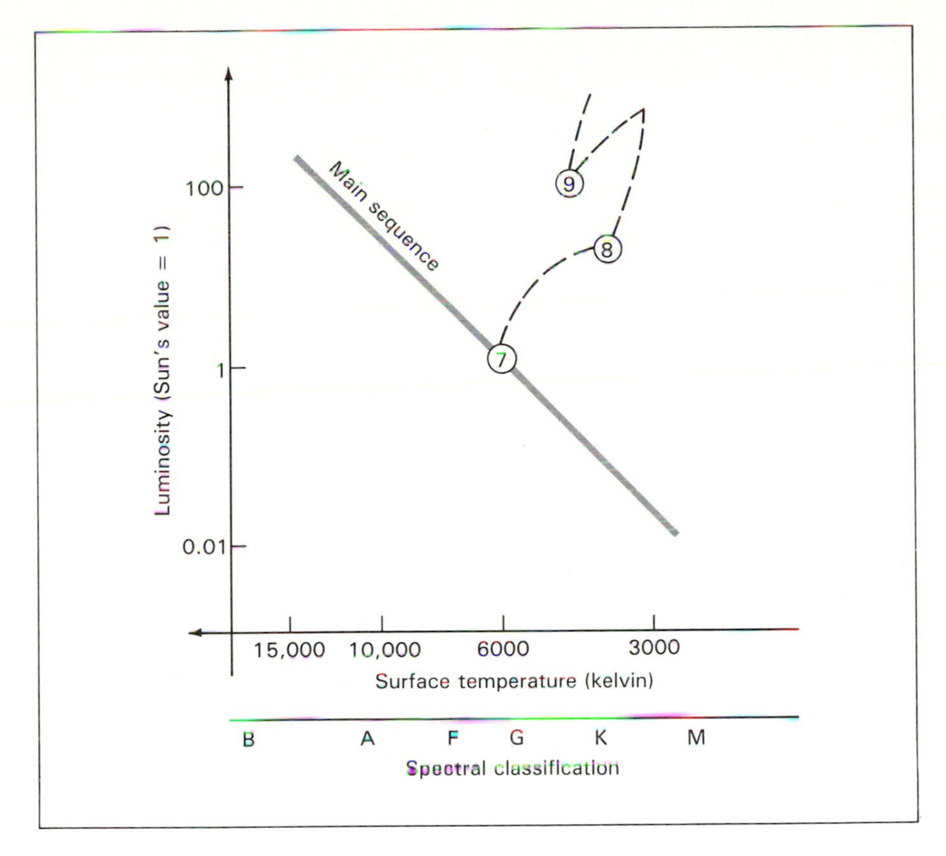

FIGURE 20.7 A carbon-core star eventually heads back toward higher luminosities for the same reason it evolved there in the first place: lack of nuclear burning at the core, causing contraction of the core and expansion of the overlying layers.

The corresponding mass-loss rates approach and sometimes exceed 10^{-5} solar mass per year; this is equivalent to an entire solar mass (perhaps a tenth of the total mass in the star) being carried off into space in the relatively short span of 100,000 years.

Observations made by the *International Ultraviolet Explorer* satellite currently orbiting Earth have shown that to produce such great winds, the pressure of hot gases in a corona (which drives the solar wind) does not suffice. Instead, the winds of the luminous hot stars must be driven directly by the pressure of the ultraviolet radiation emitted by these stars. (The same mechanism has been theorized to eject gas from the cores of some particularly active galaxies, a subject that we touched on briefly in Chapter 15.)

Such powerful stellar winds hollow out vast cavities in the interstellar gas, pushing outward expanding shells of galactic matter resembling those generated by exploded stars, as discussed both at the end of the Chapter 19 and toward the middle of this chapter. Aside from the well-known fact that copious quantities of ultraviolet radiation are available from luminous hot stars to drive the stellar winds, the details of the process are not well understood. Whatever is going on, it is surely complex, for the ultraviolet spectra of the stars tend to vary with time, implying that the wind is not steady. Apparently, instabilities of some kind or another are at the heart of the issue.

Observations made with radio and infrared as well as optical telescopes prove that luminous cool stars (e.g., K- and M-type giants) lose mass at rates comparable to those of the luminous hot stars; their wind velocities, however, are much lower, averaging 30 kilometers/second (or "merely" 70,000 miles per hour). Because luminous red stars are inherently cool objects (about 3000 kelvin surface temperature), they emit no detectable ultraviolet radiation, so the mechanism driving the winds probably differs from that in luminous hot stars; we can only surmise that gas turbulence and/or magnetic force fields in the atmospheres of these stars are somehow responsible. Unlike the hot stars, winds from these cool stars are rich in dust particles and molecules. Since nearly all stars more massive than the Sun eventually evolve into such red giants, these winds, pouring forth from vast numbers of stars, provide a major source of new gas and dust in interstellar space. Thus, the recently discovered stellar winds provide a vital link in the cycle of star formation and galactic evolution. As with the hot stars, astronomers are unsure what affect these winds and mass losses have on the subsequent evolution of the stars themselves.

Low-Mass Death

How do stars die? Again, we must rely partly on computer modeling and partly on what is observable in the sky. The problem, quite frankly, is that no one has witnessed a star die in our Galaxy since the invention of the telescope nearly four centuries ago. Guided by theoretical predictions of how stars behave near death, astronomers search the Universe, seeking evidence of objects resembling the predicted hulk.

All theoretical models suggest that the final stages of stellar evolution depend critically on the mass of the star. As a rule of thumb, we can say that low mass stars die gently, while high-mass stars die catastrophically. Depending on which astronomer you ask, the dividing line between a "low mass" and "high-mass" star can be anywhere between 2 and 4 times the mass of our Sun; we'll take 3 solar masses as an average value. Since we've already considered the evolutionary events of a 1-solar-mass star up to this point, let's continue our study of the relatively quiet death experienced by stars having less than 3 solar masses. Our Sun, of course, is one of these stars. We'll then return in a later section to study the explosive death of the more massive stars.

The preceding section contained an important qualification. This statement was made: *provided* that the core temperature reaches a high enough value, the carbon nuclei would begin burning. However, the required temperature for carbon burning is so high—600 million kelvin—that low-mass stars can never attain it. Here's why.

As the carbon core shrinks in response to the pull of gravity, the temperature and density must clearly increase. But before attaining the incredibly high carbon-burning temperature, the density reaches a value beyond which it cannot be compressed. At a density of 10^7 grams/cubic centimeter, the electrons' spheres of charge nearly interpenetrate one another. This is the maximum compression that a low-mass star can achieve, even in its core. There's simply not enough matter in the overlying layers of the smaller stars to bear down any harder.

Extraordinarily dense matter, a single cubic centimeter of this stellar core matter would weigh a ton on Earth. That's 1000 kilograms of matter compressed into a volume the size of a grape. Yet even at such high densities, collisions among nuclei are insufficiently frequent and violent to produce the phenomenal heat needed to fuse carbon into any of the heavier elements. Consequently, oxygen, nitrogen, iron, gold, uranium, and many other elements are not synthesized in low-mass stars.

PLANETARY NEBULAE: Such an aged star is in quite a predicament. Its carbon core is, for all intents and purposes, dead. Its intermediate layers continue burning hydrogen and helium. The zone of nuclear fire progresses steadily outward, causing the outermost layers to recede indefinitely. The star's size increases, its surface temperature drops, and its color gets even redder. Eventually, those outermost layers reach so far from the source of nuclear burning that they cool to a few thousand kelvin—cool enough for some electrons to recombine with nuclei, thus forming neutral atoms once again.

Computer calculations suggest that, in time, a weird-looking object results. We say "weird" because theory predicts it to have two distinct parts, both of which comprise stage 10 of Table 20-1. First, there is a small well-defined core of mostly carbon ash. Hot and dense, only the outermost part of the core still burns helium into carbon. Second, well beyond this core exists a spherical shell of cooler and thinner matter spread over a volume roughly the size of our Solar System. Weird or not, such objects are actually observed. Figure 20.8 and Plate 8(b) show an example of one of them, called a **planetary nebula**.

Both words in the name "planetary nebula" are misleading. The first word "planetary" implies that these objects are associated with planets in some way. They are not. The name originated in the eighteenth century when optical astronomers could barely distinguish among the myriad faint, fuzzy patches of light in the nighttime sky. Under poor resolution, some of them resembled planets. But later observations clearly demonstrated that the nebula's fuzziness results from a shell of warm, glowing gas. Modern telescopes fully resolve planetary nebulae, enabling astronomers to recognize their true nature.

The second word "nebula" might also be confusing, for it implies a kinship to gaseous nebulae. Although in some ways planetary nebulae resemble gaseous nebulae (as defined in Chapter 12), and both even undergo similar ionization–recombination processes, these two types of objects are distinctly different. Not only are planetary nebulae much smaller in size than gaseous nebulae, but they are also much older. Gaseous nebulae, remember, are sign posts of recent stellar birth. Planetary nebulae, on the other hand, are signposts of impending stellar death.

Nearly 1000 examples of planetary nebulae have been found in our

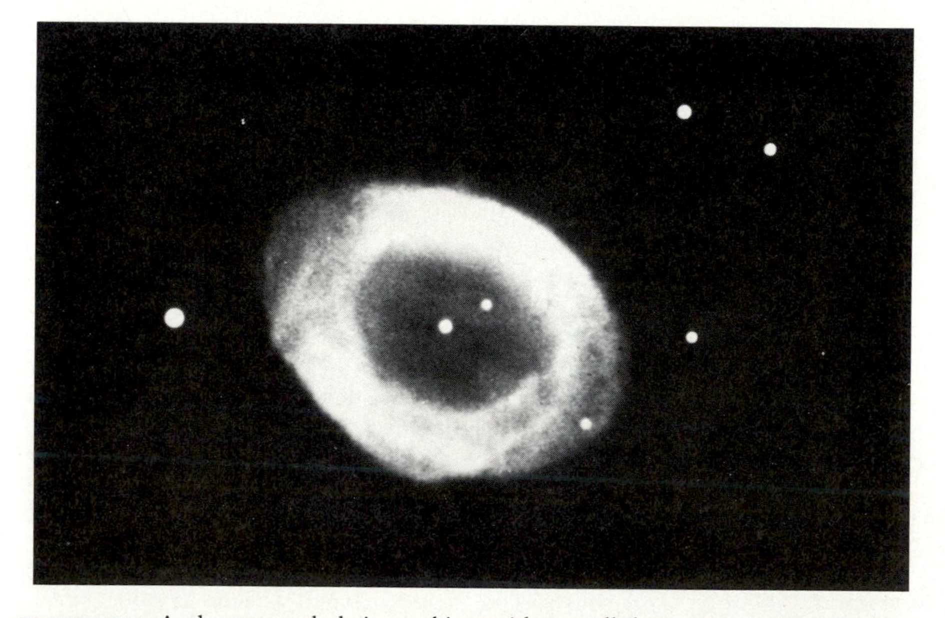

FIGURE 20.8 A planetary nebula is an object with a small dense core (center), and a large thin shell (or shells) of glowing matter. This one resides about 5000 light-years away in the constellation Lyra, and is called the Ring Nebula. Its apparent size equals about one-thousandth that of the Moon, though it's actually much larger than our Sun or Solar System. (In fact, it spans about 0.5 light-year.) Even so, the nebula is too dim to see well with the naked eye. (All the other stars shown are either foreground or background objects unrelated to the planetary nebula.) [See also Plate 8(b).]

Galaxy alone. During the past several decades, a few nearby planetary nebulae have been the focus of many interesting discoveries. One such finding proved that the shell is really that—a three-dimensional shell completely surrounding the core. Its halo-shaped appearance is only an illusion. The shell is actually a complete envelope that has been gently expelled from atop the core. It's so thin, however, that we can see it only at the edges where the emitting matter has accumulated along our line of sight. As shown in Figure 20.9, the shell is virtually invisible in the direction of the core.

Observational studies have also proved that the envelope is moving away from the core—literally being expelled, though gently. Spectroscopy has been used successfully to measure the radial velocities of the gas, and these motions agree with those predicted in the computer models of aged red giant stars.

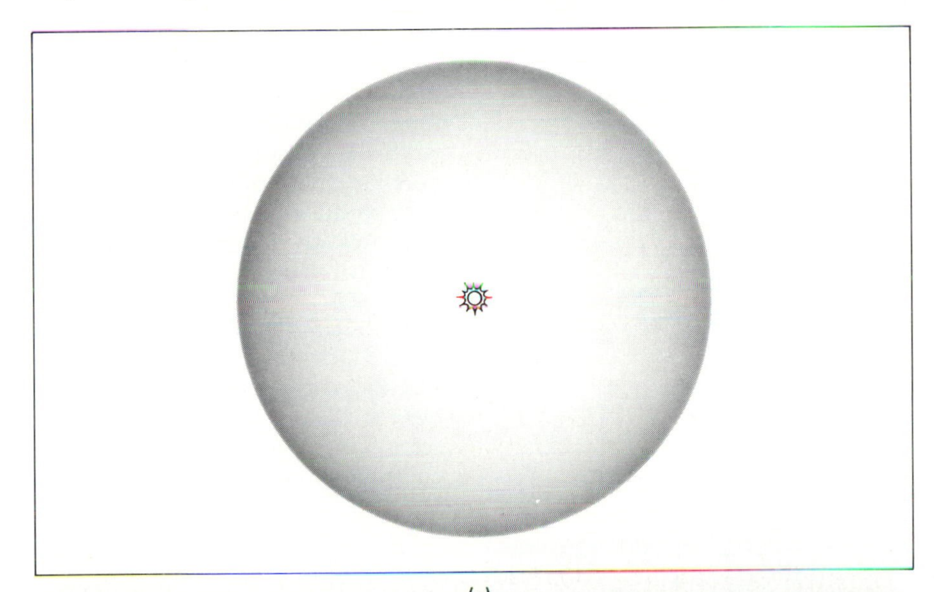

(a)

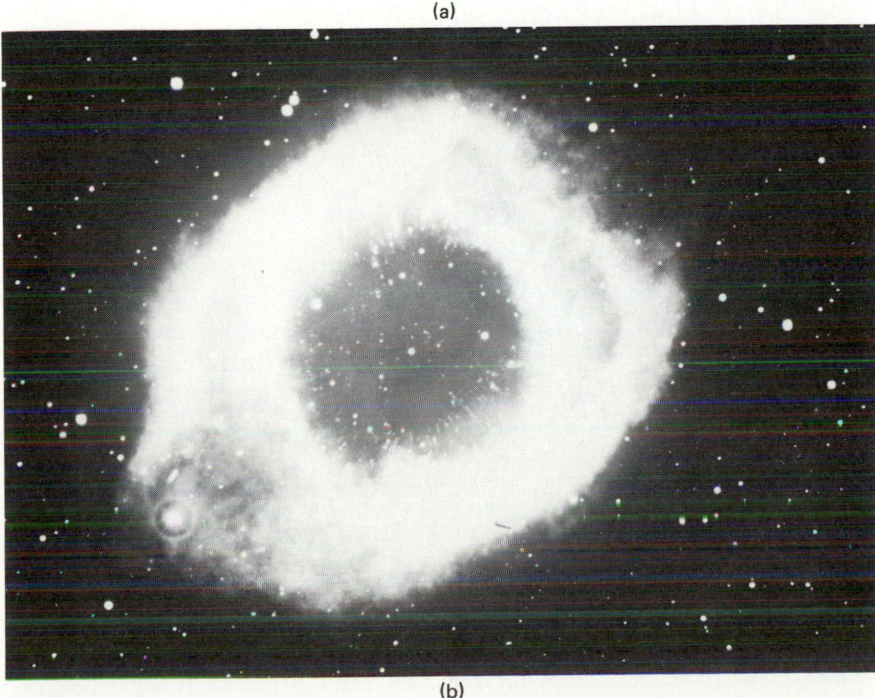

(b)

FIGURE 20.9 A planetary nebula (a), with a spherical outer shell or three-dimensional envelope, appears to the eye as a small star with a halo around it (b). The object in (b) is called the Helix Nebula and is only about 450 light-years away.

One final note on planetary nebulae: The recessional motions of a red giant star's outermost layers are initially caused by nuclear burning in the intermediate layers between the stellar core and its periphery. Later, the steady outward movement of the envelope results from the process of electrons recombining with newly formed atoms, thereby emitting photons. Repeating the process, these photons then ionize new atoms farther out, which eventually recombine to emit more photons, which in turn ionize new matter, and so on. This runaway process of atom ionization, electron recombination, and photon emission serves to steadily push portions of the gaseous envelope to greater and greater distances from the core.

MORE ABOUT WHITE DWARFS: Further discussion of the evolution of the expanding envelope is not very interesting. It simply continues to spread out with time, becoming more diffuse and cool, and gradually merging imperceptibly with interstellar space. This is one way, then, that interstellar space becomes enriched with additional helium atoms and possibly some carbon atoms as well.

Continued evolution of the core, or stellar remnant at the center of the planetary nebula, is more interesting. Formerly concealed by the atmosphere of the red giant star, the core first appears once the flimsy envelope has receded. Such a core is an extremely hot object, though rather dim because of its small size. Shining only by stored heat (and not by nuclear reactions), such dwarf stars have a white-hot appearance. In fact, the core remnant is generally not much larger than Earth; some are even smaller than our planet. Usually called a **white dwarf star**, such a core has properties listed at stage 11 of Table 20-1.

Part of the dashed line in Figure 20.10 depicts the evolutionary change from red giant to white dwarf. The large trek across the HR diagram results from the steady transformation of a large, cool (red giant) star into a small, hot (white dwarf) star. This large evolutionary

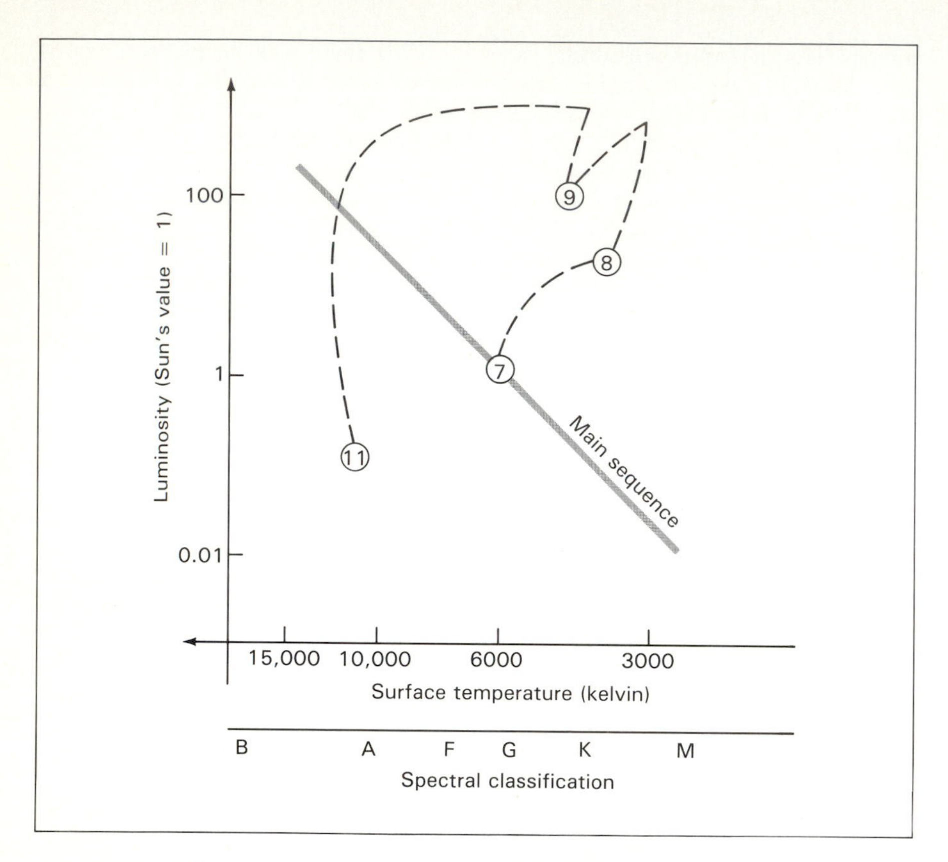

FIGURE 20.10 The change of a red giant star (stage 9) to a white dwarf star (stage 11) creates an evolutionary path clear across the HR diagram.

change between stages 9 and 11 is caused by the expansion and dispersal of a red giant's outermost layers to the point where the only thing remaining is its former core, namely the white dwarf star.

Not all white dwarf stars are found as cores of planetary nebulae. Several hundred have been discovered naked in our Galaxy, their envelopes expelled to invisibility long ago. Figure 20.11 shows an example of a white dwarf star that's part of a binary star system. The white dwarf, called Sirius B, is the faint companion of the much brighter star, Sirius A. Detailed observations show the white dwarf star to have the following properties, compared to the Sun:

mass = 1.05 solar masses,
radius = 0.008 solar radius,
luminosity = 0.002 solar luminosity,
surface temperature = 1.7 solar temperatures,
average density = 2,000,000 solar densities.

FIGURE 20.11 The Sirius B white dwarf star (speck of light at right) is seen here as a companion to the much larger and brighter Sirius A star. (The hexagonal shape of the image of Sirius A is not real; the "spikes" are artifacts caused by the support struts of the telescope.)

Note that Earth's size is 0.009 solar radius, making our planet larger than the star Sirius B! Thus this white dwarf star has more than the mass of our Sun packed into an object smaller than Earth; no wonder its density is about a million times those familiar to us in our Solar System.

DARK CLINKERS AGAIN: So astronomers are able to identify red giant stars, planetary nebulae, and white dwarf stars in the nearby cosmos. These objects seem to match fairly well the expectations of the theoretical calculations for aged low-mass stars. Once again, though, we should not expect to witness the act of envelope expulsion during the course of a human lifetime. Several tens of thousands of years are needed for a white dwarf to appear.

After this, nothing exciting happens to dwarf stars. For all practical purposes, they are dead. They continue to cool, becoming dimmer with time. Their temperatures and luminosities change according to the dashed line near the bottom of the HR diagram of Figure 20.10. Such elderly "stars" slowly transform from white dwarfs to yellow dwarfs and then red dwarfs, finally becoming black dwarfs—cold, dense, burned-out embers in space. This is stage 12 of Table 20-1, the graveyard of stars.

No one knows how many black dwarfs really exist in the Galaxy. That's not surprising since, after all, they are black and hence unlit. Even if these dark clinkers could somehow be detected, we would probably find that not many of them exist. The main-sequence lifetime of a low-mass star is long—comparable to or longer than the age of the Galaxy. Put another way, our Galaxy has not existed long enough for many low-mass stars to have evolved all the way from birth to death. Perhaps none has.

On the other hand, there is always a chance that astronomers are wrong. This is especially true in a subject such as stellar evolution, where some aspects of the theoretical calculations are hard to test observationally. Perhaps large numbers of undetected black dwarf stars do exist in our Galaxy, as well as in other galaxies. If so, they might help solve the minor hidden-mass prob-

INTERLUDE 20-1 ***Learning Astronomy from History***

Sirius A, the brighter of the two objects in Figure 20.11, appears twice as luminous as any other visible star, excluding the Sun. Its absolute brightness is not very large but, because its distance from us (9 light-years) is small, its apparent brightness is greater than any star beyond our Sun.

Sirius has been prominent in the nighttime sky since the beginning of recorded history. Ancient cuneiform texts of the Babylonians refer to the star as far back as 1000 B.C., while it's known to have had strong influences on the agriculture and religion of the Egyptians of 3000 B.C.

Since recorded observations of Sirius A go back several thousand years, we might have a chance to observe a slight evolutionary change, despite the long time scales usually thought necessary to produce such changes. The chance is bettered in this case because Sirius is so bright that even the naked-eye observations of the ancients should be reasonably accurate. Interestingly enough, the great books of recorded history suggest that Sirius A has indeed changed its stellar appearance. But the observations are confusing. Every piece of information about Sirius recorded between the years 100 B.C. and A.D. 200 claims that this star was red. (Earlier records of its *color* have not been found.) By contrast, modern observations now show it to be white, or bluish white, but definitely not red. So Sirius has seemingly changed from red to blue-white in the intervening years. The confusion is this: According to the currently accepted theory of stellar evolution, no star should be able to change its color in this way in such a short time. Any change of this sort should occur on a time scale of at least several tens of thousands of years, and perhaps a lot longer.

Astronomers have offered several explanations for the rather sudden change in Sirius A. These include the suggestions that (1) some ancient observers were wrong, and other scribes copied them; (2) a galactic dust cloud passed between Sirius and Earth some 2000 years ago, thus reddening the star much as Earth's dusty atmosphere often does for our Sun at dusk; (3) the companion to Sirius A, namely Sirius B, was a red giant and dominant star of this double-star system 2000 years ago and has since expelled its planetary nebular shell to reveal the white dwarf star which we now observe.

None of these explanations really seems plausible, for each is problematic: How could the color of the sky's brightest star be incorrectly recorded for hundreds of years? Where is the intervening galactic cloud now? Also, where is the shell of the former red giant? We are thus left with the uneasy feeling that the sky's brightest star doesn't seem to fit well into the currently accepted scenario of stellar evolution.

lem regarding our Milky Way, as noted in Chapter 13 (although they are not likely to offer a solution to the major missing-mass problem plaguing galaxy clusters, as discussed in Chapter 14).

High-Mass Death

Objects having a mass greater than 3 solar masses evolve along paths similar to their low-mass counterparts up through the red giant stage, with only one difference. All the evolutionary changes happen more rapidly for high-mass stars because their large mass enables them to generate more heat. And, more than anything else, heat speeds up all evolutionary events.

HEAVY ELEMENT CREATION: At the red giant stage, the core of a high-mass star is able to attain the 600 million kelvin needed to fuse carbon. Large mass is the key here. Massive stars have stronger gravitational force fields than solar-type stars, and the added gravity can crush matter in the core to a high enough density to ensure frequent and violent collisions among the particles. In this way, carbon can fuse with other nuclei, forming heavier, more complex nuclei within the cores of these massive stars. Each heavier type of nucleus requires a higher temperature in order for it to fuse with like nuclei and thus to produce ever heavier elements.

Figure 20.12 is a cutaway diagram of the interior of a highly evolved star of large mass. Depicted there are numerous layers where various nuclei burn. At the relatively cool periphery just below the photosphere, hydrogen fuses into helium. In the intermediate layers, helium, carbon, and oxygen fuse into heavier nuclei. Deeper down but outside the core, there reside neon, magnesium, silicon, and numerous other heavy nuclei. The core itself is shown to be full of iron nuclei, rather complex pieces of matter each containing 26 protons and 30 neutrons.

Each of the fusion cycles, during which nuclei for new elements are produced, is induced by periods of stellar instability; the core contracts, heats, fuses, becomes depleted of fuel, only to contract a little again, heat again, and so on. All the while, the star's central temperature increases, the nuclear reactions speed up, and the released energy supports the star for ever-shorter periods of time. For example, a huge star of 20 solar masses burns hydrogen for 10 million years, helium for 1 million years, carbon for 1000 years, oxygen for 1 year, silicon for a week, and iron for less than a day.

Once the inner core begins changing into iron, this sick and dying star

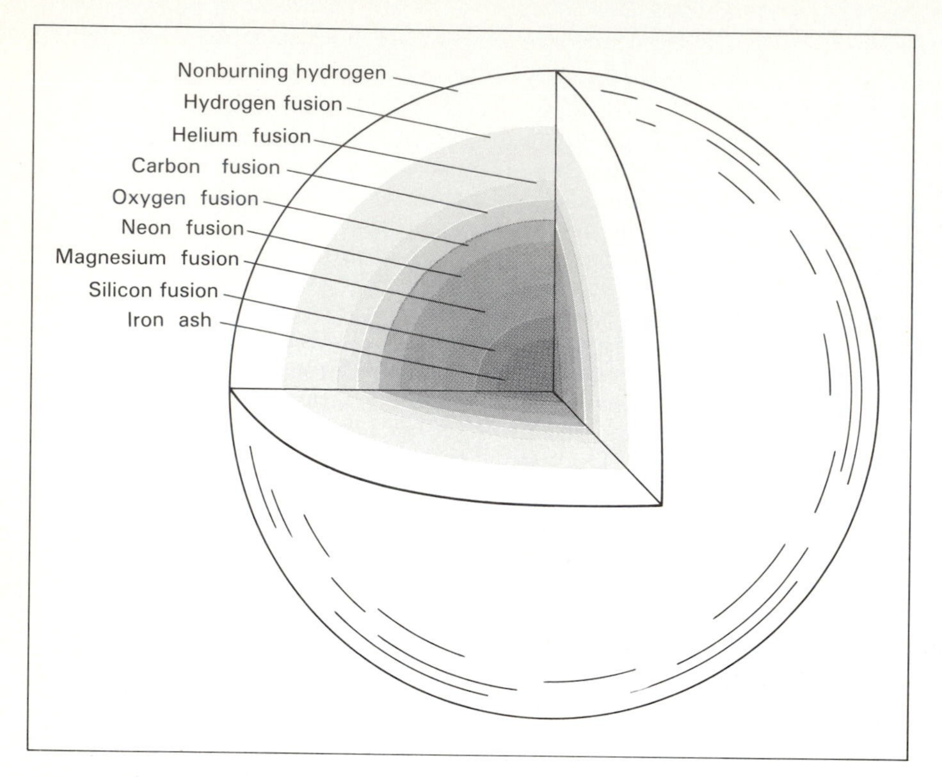

FIGURE 20.12 Cutaway diagram of the interior of a highly evolved star having a mass greater than 3 solar masses. Though this is a simplified sketch, in many ways the interior must resemble the layers of an onion.

runs into trouble. Nuclear fusion involving iron does not produce energy; iron nuclei are so compact that energy cannot be extracted. Even worse, iron nuclei consume energy, effectively robbing energy from the extremely hot core. To a certain extent, iron plays the role of fire extinguisher, suddenly damping the stellar inferno, at least at the core. With the appearance of substantial quantities of iron, the central fires terminate for the last time.

This situation is most unstable. A highly aged star of very large mass can no longer be supported by nuclei burning at its core. The star's foundation has been effectively destroyed. Equilibrium is completely gone, and potential for disaster clearly exists. Even though the temperature in the iron core has reached, by this point, several billion kelvin, the enormous inward gravitational pull of matter ensures catastrophe in the star's near future. Unless the nuclear fires continue unabated, trouble is certain for any star.

Once gravity overwhelms the pressure of the hot gas, the star implodes, falling in on itself. The implosion doesn't take long, perhaps only hours after the central fires are extinguished. Internal temperatures and densities then rise phenomenally, causing the star to rebound instantaneously. Part of the core is detonated on the rebound, jettisoning all the surrounding layers—including a variety of heavy elements—into neighboring regions of space. Much more violent than the expulsion of matter in planetary nebulae, in this event the star literally explodes. Such a spectacular death rattle is known as a **supernova**.

SUPERNOVAE: Nova is Latin for "new." Astronomers now recognize that novae are not really new stars at all. Their sudden brightening only made them appear so to the ancients. Figure 20.13 is a photograph of a **nova**—a star that rapidly brightens while expelling a small fraction of its matter. The origins of such stellar expulsions are not entirely understood, although they are probably caused by intense gravitational tides exerted on some stars within multiple-star systems. The result is a temporary instability causing a violent eruption of matter from the star's surface. Observed to brighten by about 1000 times the luminosity of our Sun, novae eventually dim back to normal after many months to a year. Some such stars have been observed to repeatedly brighten sev-

FIGURE 20.13 Small amounts of hot matter thrown from an old star can brighten the star. Novae, such as this one called Nova Persei (1901), experience small-scale expulsions of their surface gases.

eral times over the course of several decades.

A supernova is much more violent than a nova; the resulting explosion happens only once in a star's life, namely at death. The exploded stellar debris is hot and altogether can radiate a flash equaling nearly a billion solar luminosities. Just imagine this spectacular phenomenon: A single star can suddenly brighten to a billion times our Sun's luminosity within a few hours of the outburst. In the process, the galactic neighborhood is irradiated with plenty of potent energy and heavy elements.

As for any explosion, the initial flare up of luminosity from a supernova is expected to decrease steadily to a normal, preexplosion value. Luminosity and temperature of supernovae are not usually plotted on an HR diagram. Instead, "light curves" like that of Figure 20.14 are used to illustrate the change in luminosity. Such curves are expected to show a dramatic rise in luminosity over a few days, followed by a much slower decay of the observed light over the course of a year or more.

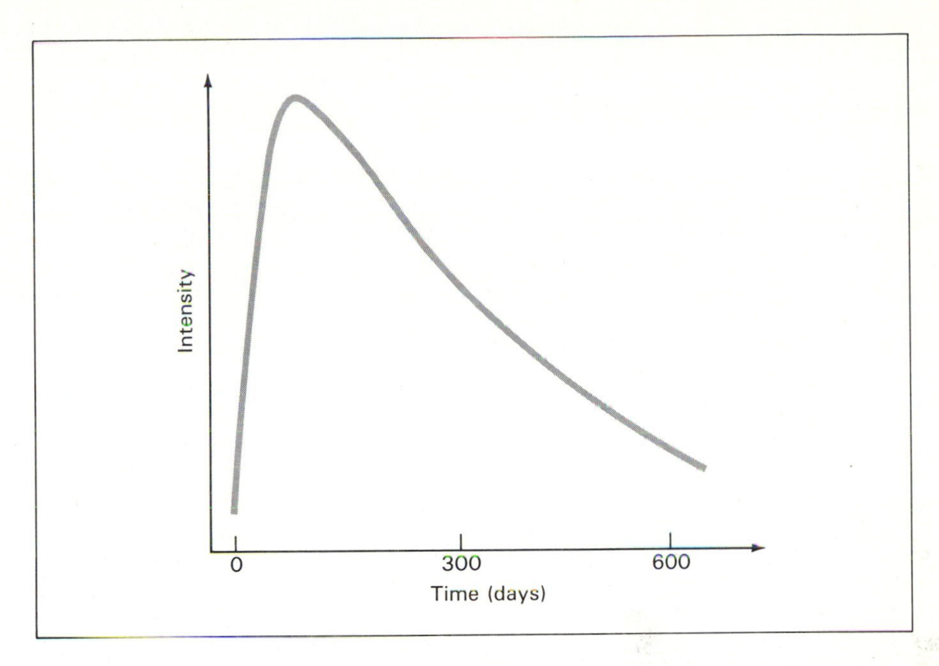

FIGURE 20.14 The "light curve" of a typical supernova. The maximum brightness or intensity can sometimes reach that of a billion suns.

Supernovae are not just idle predictions of theoreticians. We have plenty of evidence that cosmic explosions really have occurred in our Galaxy. We can detect their glowing remains, known as **supernova remnants**.

One of the most heavily studied supernova remnants, known as the Crab Nebula, is shown in Figure 20.15 and Plate 8(a). Its brightness has greatly dimmed now (and when these photographs were taken in the 1970s). But the original explosion was so brilliant that ancient manuscripts of Asian and Middle Eastern astronomers claim that its brightness rivaled that of the Moon in the year A.D. 1054. For nearly a month, this exploded star reportedly could be seen in broad daylight. Even the American Indians left engravings of the event in the rocks of what is now the southwestern United States.

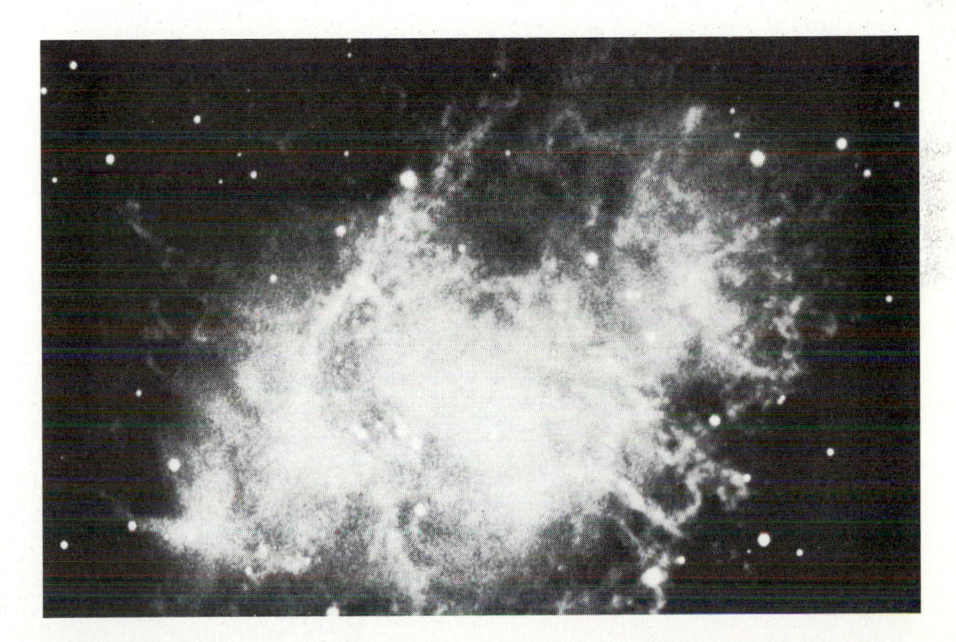

FIGURE 20.15 This remnant of an ancient supernova is called the Crab Nebula (or M1), for its appearance resembles that type of marine animal. It resides about 5000 light-years from Earth, and subtends an angle about one-fifth that of the full Moon. Since its debris is scattered over "only" a 10-light-year extent, the Crab is considered to be a young supernova remnant. [See also Plate 8(a).]

Certainly, the Crab Nebula has the image of exploded debris. What's more, the impression that matter was ejected has been proved within recent decades. Figure 20.16 is a superposition of a positive image of the Crab Nebula taken in 1960 and a negative image taken in 1974. If the filamentary structure of the gas were not in motion, the positive and negative images would overlap perfectly. But they do not. Clearly, some gas has moved outward in the intervening 14 years. By knowing the total distance traveled by the gas in this interval of time, a velocity of several thousand kilometers/second has been derived for the expelled de-

INTERLUDE 20-2 ***Nearby Supernovae***

Supernovae might be more than spectacular light shows. Should a massive star detonate in a neighboring part of our Galaxy, it could well produce high-energy radiation harmful to life on Earth. Supernovae literally rip off the topmost layers of massive stars, sending the debris flying into space as extremely fast-moving elementary particles; these are the cosmic rays studied in Chapter 13. Thus an understanding of the physical properties of the nearby stars is of more than just passing interest. An ability to predict the manner in which the nearby stars will die is downright critical. Of particular concern is the possibility of one of our neighboring stars exploding in a supernova, although we probably couldn't do much about it even if one did.

Only six galactic supernovae have been recorded within the past 1000 years. The figure shows the relative positions of these six known supernovae within our Milky Way. They are labeled by the year they first appeared. (The supernova Cassiopeia A apparently went unnoticed optically, although modern radio studies suggest that news of its explosion should have reached Earth midway through the seventeenth century. Plate 8(c) is dramatic proof that this supernova remnant, although invisible optically, can be studied at radio and x-ray wavelengths.)

Most astronomers assume that many more stars than these six have blown up in our Galaxy, but that their supernovae were missed because either they were too faint to be seen with the naked eye, or their enhanced brightness was hidden behind dark clouds in the galactic plane. All the Milky Way supernovae mapped above are in fact confined to our quadrant of the Galaxy, and each of them is at least 300 light-years outside the galactic plane.

Despite warnings elsewhere in this book, the danger to life on Earth is probably not very great, at least for the foreseeable future. Studies of the rate at which supernovae are thought to occur suggest that one can be expected within 30 light-years of our Sun only every 500 million years. Furthermore, none of the very nearby stars is massive enough to die catastrophically by exploding. Luckily for us, they all seem destined to die, as will our Sun, via the rather peaceful red giant—white dwarf route.

In a startling development, astronomers were recently treated to a spectacular supernova in the Large Magellanic Cloud, the Milky Way's companion galaxy some 175,000 light-years away (consult Interlude 14-1). Shortly after the explosion was seen by observers in Chile on 24 February 1987, practically all southern-hemisphere telescopes and every available orbiting spacecraft were focused on the object, called SN1987A. This was the most dramatic change observed in the Universe in nearly 400 years. Apparently, a 15-solar-mass, B-type supergiant star with the peculiar catalog name of SK-69°202 had detonated, thereby outshining all other stars in that galaxy, as shown in the accompanying figures before and after the explosion. [See also Plate 6(b).]

Among the important information provided by SN1987A are these:

- The light radiation that beamed forth displayed a rather baffling pattern. After fading gently, the supernova brightened rapidly; within hours of its initial detection, it had become about 100 times brighter than its progenitor, SK-69° 202. It then continued to increase, though more slowly, for the next few months, before fading in late May. This erratic behavior might be due to the decay of radioactive material supplied to the supernova's expelled cloud, as discussed in Chapter 22.
- As for invisible radiation, radio waves were picked up by Australian radio telescopes during the first few days after the explosion, but few have been detected since. This radio radiation was non-thermal, resulting from fast-moving electrons in a magnetic force field shortly after the outburst, after which the particles slowed and the field became dilute as the remnant spread out. After initially decreasing, the intensity of ultraviolet radiation steadily rose for a few months and was monitored by the *International Ultraviolet Explorer* satellite in Earth orbit. SN 1987A's ultraviolet radiation was probably blocked by dust and used to heat the debris during the first few days, after which the short-wavelength radiation shone through. Neither x-rays nor gamma rays were detected by an array of U.S., Soviet, and European spacecraft; this again might be due to the obscuration by dust and debris of the high-frequency radiation. Perhaps x-rays and gamma rays will yet arrive at Earth at a later date, as the dissipating shell of brilliantly glowing debris is known to be expanding at a fast clip of about 10,000 kilometers/second (or some 20 million miles/hour).
- A brief (about 13-second) burst of neutrinos was simultaneously recorded by underground detectors in Japan and the U.S., much like the instrument described in the last section of Chapter 10. The neutrinos are predicted to arise as the electrons and protons in the star's collapsing core merge to form neutrons and neutrinos, as discussed elsewhere in this chapter. Despite some unresolved details in SN1987A's behavior, detection of its neutrino pulse is considered to be brilliant confirmation of theory; in fact, this singular event might well herald a new age of astronomy, for this is the first time that astronomers have received information from beyond the Solar System by any means other than electromagnetic radiation.

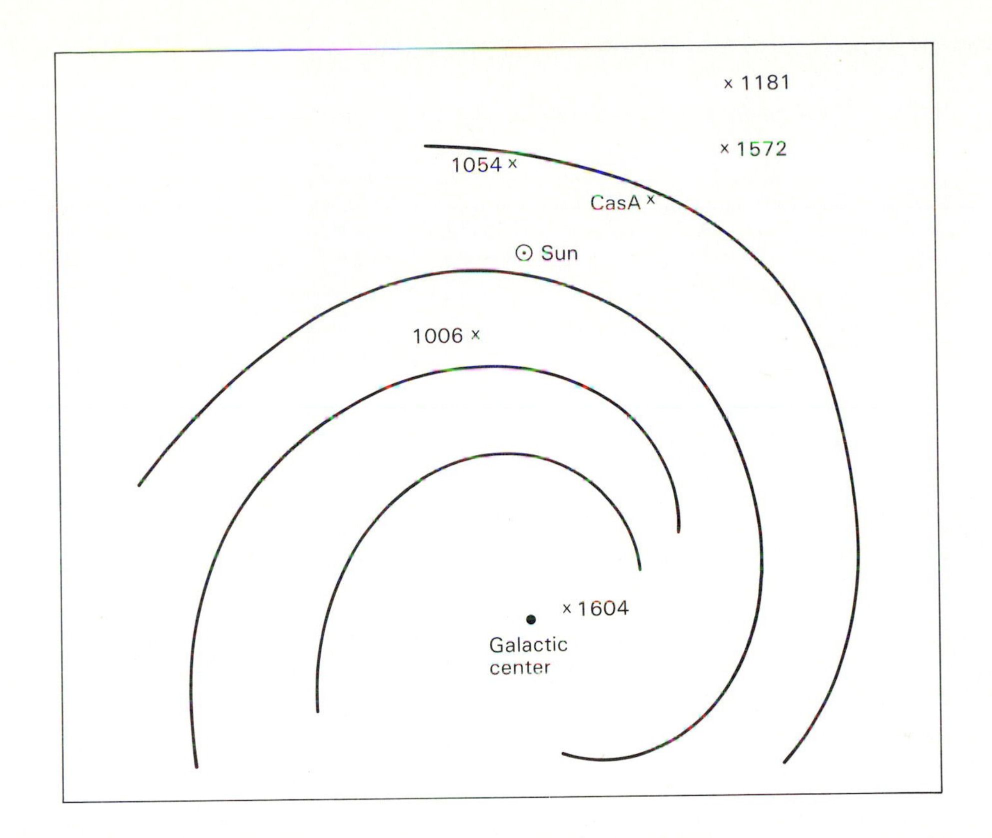
× 1181
× 1572
1054 ×
CasA ×
⊙ Sun
1006 ×
× 1604
Galactic
center

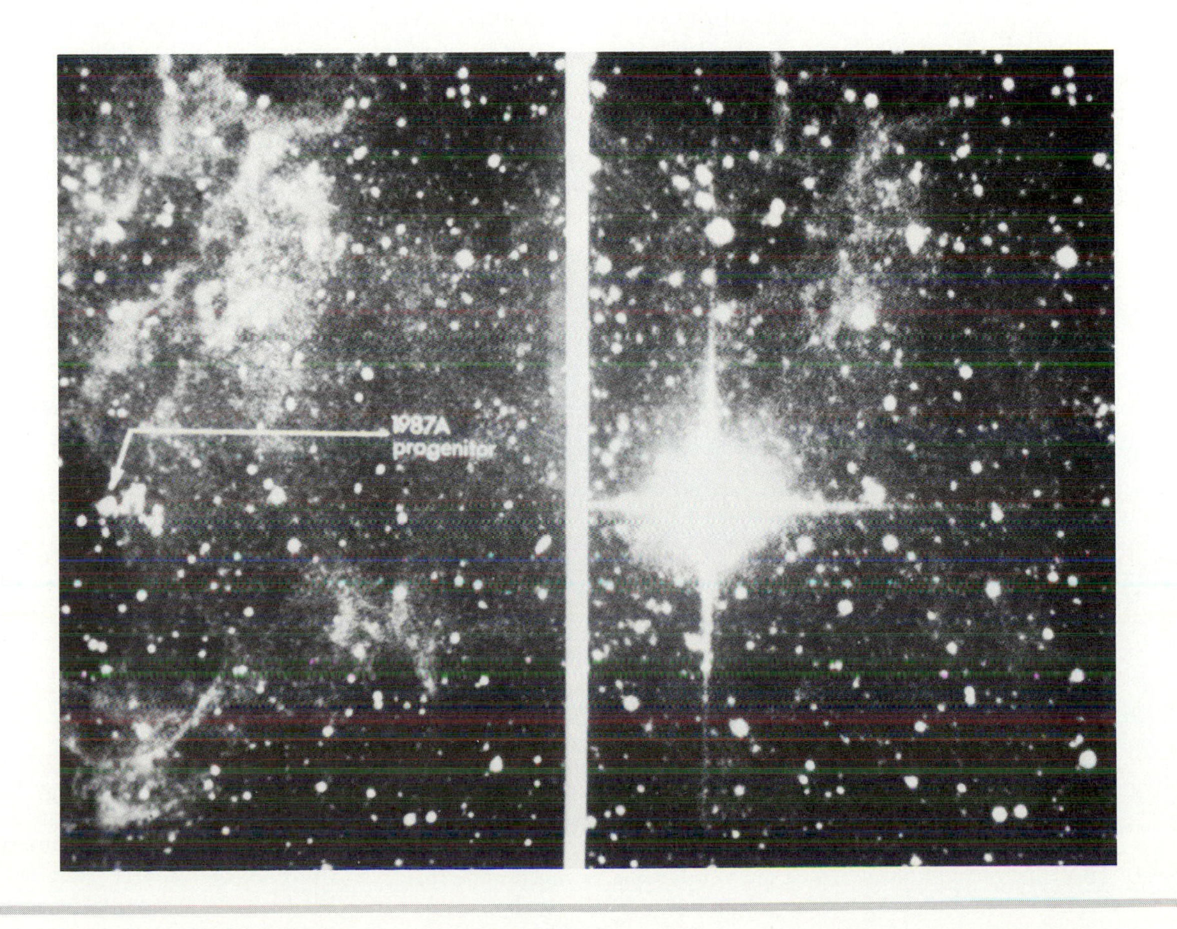
1987A
progenitor

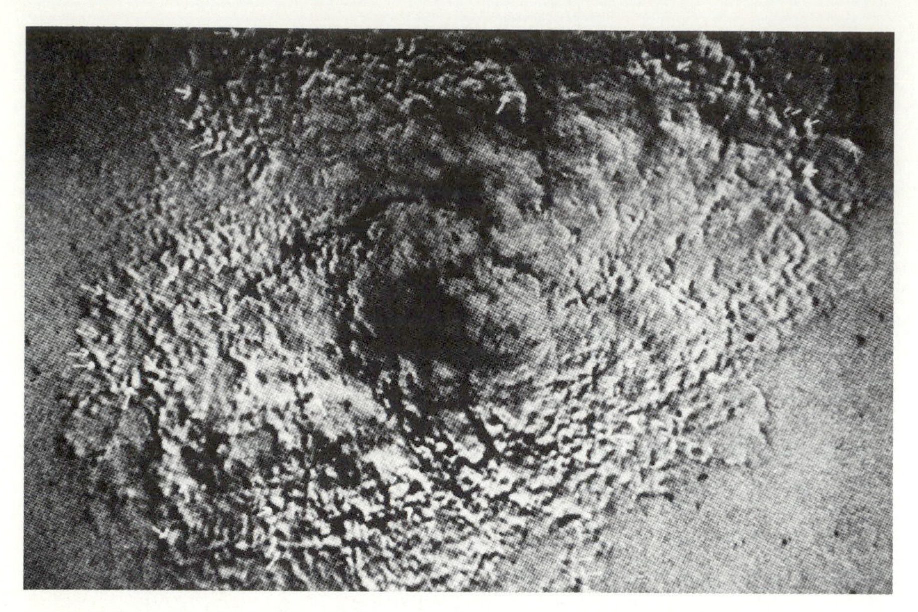

FIGURE 20.16 Positive and negative photographs of the Crab Nebula taken 14 years apart do not superpose with precision, proving that the gaseous filaments are still moving away from the site of the explosion.

bris. These outward motions confirm that an explosion must have occurred about 900 years ago.

OTHER SUPERNOVA REMNANTS: The nighttime sky harbors many remnants of stars that must have blown up well before the beginning of recorded history. Figure 20.17 is an example of one of these old supernova relics. Its expelled gases extend across more than 150 light-years. Measurements of the velocities of the expelled gas suggest that an explosion took place near the center of this object several hundred thousand years ago.

Figure 20.18 is another supernova remnant in our Milky Way. This one, called the Gum Nebula, has expansion velocities implying that a star blew up around 9000 B.C. Given the close proximity of the exploded star (only 1500 light-years), we can only speculate what impact such a bright supernova might have had on the myths, religions, and cultures of Stone Age humans.

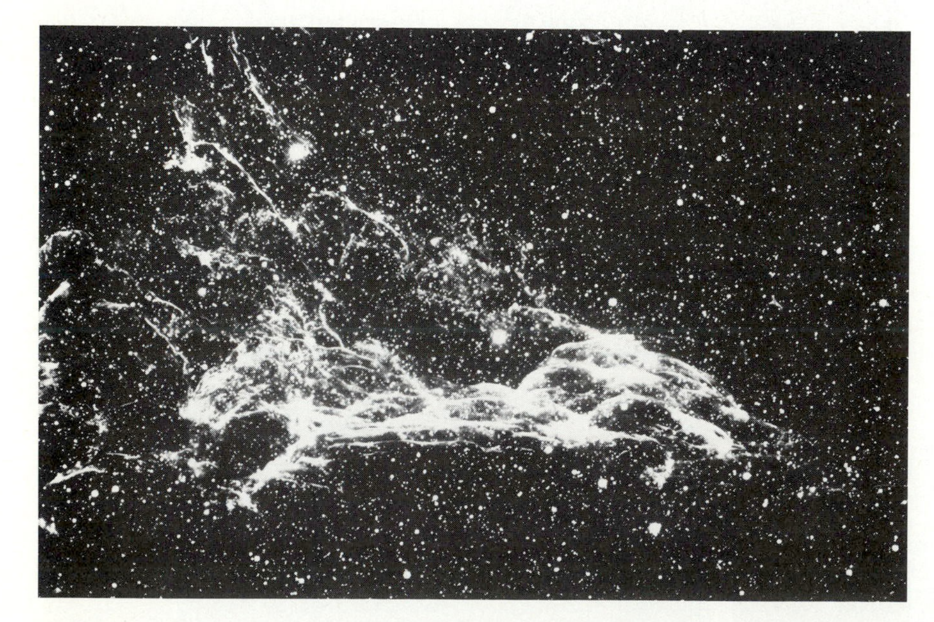

FIGURE 20.17 This supernova remnant, called the Veil Nebula, is much older than the Crab Nebula remnant. Note the complex gas motions, as the expelled matter pushes out into the surrounding interstellar space. [See also Plate 8(c).]

Furthermore, astronomers patrolling the skies with telescopes occasionally notice a sudden brightening of a small part of some distant galaxy. These observations enable them to derive a light curve like that shown in Figure 20.14, which in turn helps to refine the predictions of the theoretical models. For example, the photographs of Figure 20.19 were taken a few months apart and clearly show a sudden brightening large enough to rival the normal luminosity of an entire spiral arm of this distant galaxy. Nearly 100 such supernovae have been observed in other galaxies during the twentieth century.

Interesting enough, no one has ever observed with modern equipment a supernova in our own Galaxy. A viewable Milky Way star has not erupted in this way since Galileo first used a lens as a telescope in the year A.D. 1610. (However, the recent supernova sighted in the Milky Way's companion galaxy, the Large Magellanic Cloud, has been studied extensively; consult Interlude 20-2.)

The last supernova observed in our Galaxy caused a worldwide sensation in Renaissance times. The sudden appearance and subsequent fading of a very bright stellar object in the year 1604 helped shatter the Aristotelian idea of an unchanging Universe. This seventeenth-century supernova was one of the main reasons that people began to doubt the old-world philosophies of immutability; indeed, supernovae have helped to sow the seeds of the idea of cosmic evolution, a subject that is based completely on the concept of change.

The infrequence of supernovae is a bit disturbing. Knowing the rate at which evolutionary stages are theorized to occur and estimating the number of massive stars known in the Galaxy, we expect a galactic supernova to pop off in an observable

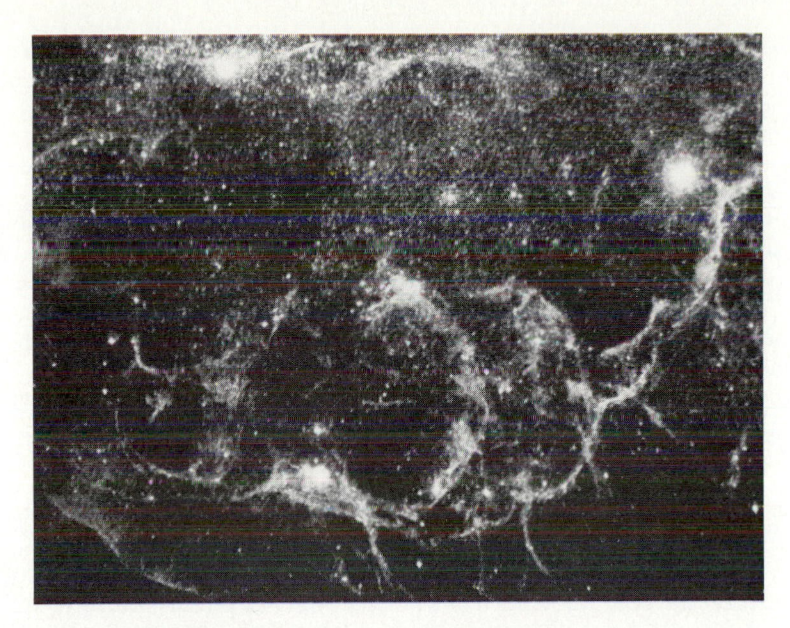

FIGURE 20.18 The glowing gases of the Gum Nebula supernova remnant are spread across an amazingly large 60 arc degrees. The closest edge of the expanding shell is only 300 light-years away.

FIGURE 20.19 A supernova can be seen exploding in this far-away galaxy at the moment the photograph on the right was taken. The photograph on the left is the normal appearance of the galaxy. [See also Plate 6(b).]

location (above the Milky Way's dusty plane) every 100 years or so. With a brilliance rivaling the Moon at night, it's unlikely that astronomers could have missed any since the last one nearly four centuries ago. Hence the Milky Way seems long overdue for a supernova. Unless massive stars explode much less frequently than suggested by the theory of stellar evolution, we should be treated to nature's most spectacular cosmic event any day now.

Strange States of Matter

What remains after a supernova explosion? Is the entire star just blown to bits and ejected into interstellar space? No, most theoretical models predict that some part of the star survives. As with planetary nebulae that expel matter less violently, supernovae also leave behind a remnant core. The matter within this severely compressed core comprises a strange state, unlike anything we've been able to find or make on Earth.

NEUTRON STARS: During the moment of implosion of a massive star—just prior to its explosion—all the electrons in the core violently smash into the protons. Free electrons were there all along, but the protons were freed when some of the heavy nuclei disintegrated. This results in the following elementary-particle reaction:

$$\text{electron} + \text{proton} \rightarrow \text{neutron} + \text{neutrino}.$$

This reaction proceeds throughout the collapsing core of a massive star, systematically converting within seconds all the electrons and protons into neutrons and neutrinos.

The neutrinos rapidly leave the scene at the velocity of light. They are suspected by many theorists to play a major role in the triggering of supernovae, for neutrinos transport much of the energy of the collapsed core to the outer layers of the star, deposit it there, and cause the rest of the star to get blown into space. (This is true despite our Chapter-10 statements about neutrinos interacting very little with matter; in neutron stars, the density is so large that even neutrinos are absorbed.)

The material debris of a supernova is much heavier than the neutrinos and thus departs at speeds much less than light velocity. Only the core remains intact as a ball of neutrons. Researchers colloquially call this core remnant a **neutron star**, although it's not really a "star" in the true sense of the word, for all nuclear reactions have ceased.

The theory of stellar evolution predicts that neutron stars are very small, although still massive. Composed purely and simply of neutron elementary particles packed together in a tight ball about 10 kilometers across, a neutron star is not much bigger than a typical asteroid or a terrestrial city (Figure 20.20). But with several times the mass of our Sun forced into a rather small volume, neutron stars must be incredibly dense. Their average density is estimated to reach 10^{12}, perhaps as

FIGURE 20.20 Neutron stars are not much larger than many of Earth's major cities. Here, in this fanciful comparison, a small part of a typical neutron star is juxtaposed against the Harvard Square area.

high as 10^{15}, grams/cubic centimeter. This is nearly a million to a billion times more dense than the already supercompact white dwarf stars. In fact, the density of a normal atomic nucleus is "only" 10^{14} grams/cubic centimeter.

Interestingly enough, such extraordinary density typifies that thought to have existed during the lepton epoch of the Universe (consult Table 17-1). Detailed study of these weird neutron stars might therefore enable scientists to understand better the physical conditions very close to the start of the Universe.

Neutron stars are expected to be solid objects, more like planets than stars. Provided that one is sufficiently cool, you might imagine standing on it. However, it wouldn't be easy; a neutron star's gravity is unbelievably strong. A 150-pound (70-kilogram) human being would weigh the Earth-equivalent of about 1 million tons (1 billion kilograms). Actually, standing wouldn't even be possible, for the severe pull of gravity would flatten you to the thickness of this piece of paper. In fact, gravity is so strong on a neutron star that the entire population of the world, if shipped there, would be crushed into a volume the size of a pea!

Furthermore, newly formed neutron stars must rotate extremely rapidly, with periods measured in fractions of seconds. And finally, any magnetic force field embedded in the original stellar core is amplified during the compression, reaching field strengths on the order of trillions of times those of Earth's magnetic force field (and even millions of times those in the hearts of solar flares). Strange objects, neutron stars represent states of matter unimaginably different from what we are used to.

PULSARS: Can we be sure that objects as strange as neutron stars really exist? The answer is "yes," for around 1970, radio astronomers made a remarkable discovery. They observed some astronomical objects emitting radiation as rapid pulses lasting for about 0.01 second apiece. Each pulse contains a burst of radiation, after which there is nothing. Then another pulse arrives. The time intervals between pulses are astonishingly uniform—so accurate that the repeated emissions can be used as a clock. Figure 20.21 is a typical recording of the periodic radio radiation from such a pulsating star, called a **pulsar** for short.

Several hundred of these objects are now known in our Milky Way Galaxy. Different pulsars have different time periods between successive radiative bursts, as well as different pulse durations. Yet each pulsar has its own characteristic pulse period and duration that repeat indefinitely.

In a few cases, pulsars seem to be directly associated with supernova remnants (although not all such remnants have a pulsar within them). The pulsar whose recording is shown in Figure 20.21 resides close to the center of the Crab Nebula remnant shown earlier. Observing the velocity and direction of travel of the Crab's ejected matter, we can work backward to pinpoint the location in space at which the explosion presumably occurred. There the supernova core remnant is expected to be located. And that's precisely the region in the Crab Nebula from which the pulsating signals arise. Apparently, pulsars are the remains of the once-massive stars.

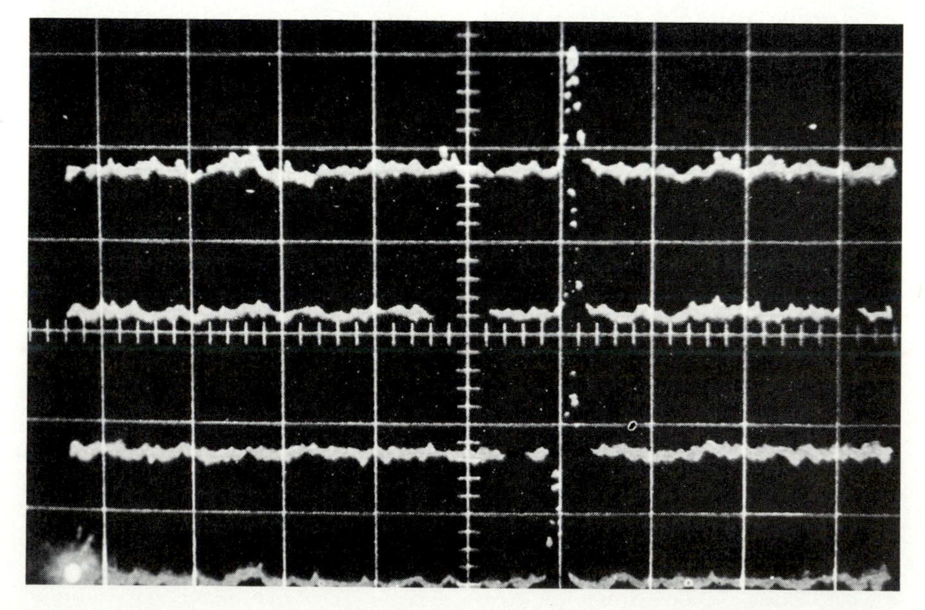

FIGURE 20.21 Pulsars emit periodic, very accurately timed, bursts of radition. This recording, taken from a video screen, shows the regular change in the intensity of the radio radiation emitted by one such object—the core remnant at the center of the Crab Nebula.

Most pulsars emit their pulses in the form of radio radiation. A few of these core remnants have been observed pulsing their visible, x-ray, and gamma-ray radiation as well. All these electromagnetic flashes are synchronized as we might expect if they arise from the same astronomical object. The period of most pulsars is usually short—about 0.03 to 0.3 second of time, depending on the pulsar. That's a flashing rate between 3 and 30 times per second. The human eye is insensitive to such quick flashes, making it difficult to observe the flickering of a pulsar either with the naked eye or even by using a large telescope. Fortunately, instruments can record pulsations of light that the human eye cannot. Figure 20.22 compares two optical photographs of the pulsar at the core of the Crab Nebula. In one frame, the pulsar is on; in the other, it's off.

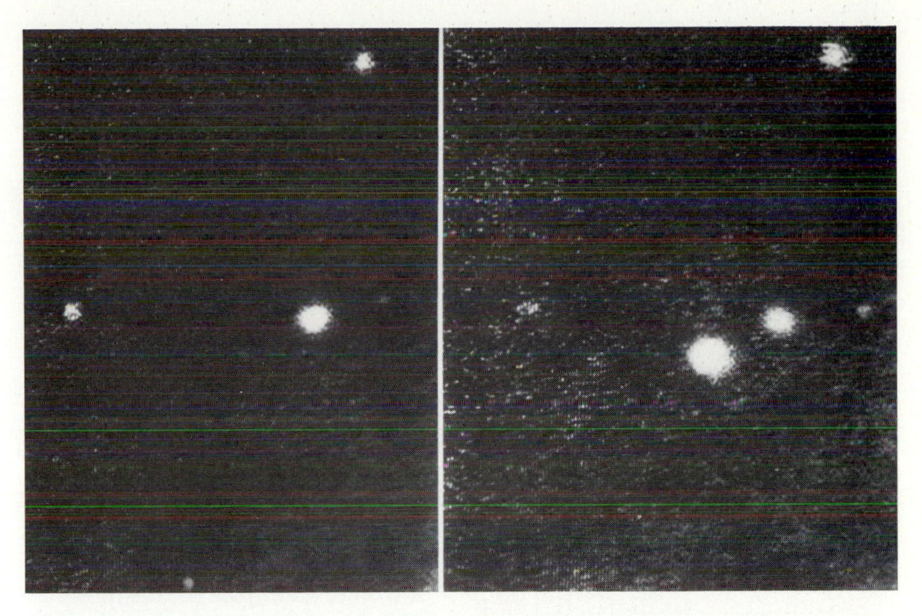

FIGURE 20.22 The pulsar at the core of the Crab Nebula blinks on and off about 30 times each second. At the left, it's off, while at the right, it's on.

Astronomers reason that the only physical mechanism consistent with such precisely timed pulsations is a small, rotating source of radiation. Only rotation can cause the high degree of regularity of the observed pulses. And only a small object can account for the sharpness of each pulse; radiation emitted by an object larger than about 10 kilometers would arrive at Earth at slightly different times, blurring the sharpness of the pulse. It's not surprising, then, that the best theoretical model of a pulsar envisions a small, compact, spinning neutron star that periodically flashes radiation toward Earth. Thus the terms "pulsar" and "neutron star" are virtually synonymous.

Figure 20.23 is an artist's conception of one such weird object, while Figure 20.24 outlines the important features of a leading theoretical model. According to this model, a "hot spot" on the surface of a neutron star, or in the atmosphere or magnetosphere above it, continuously emits radiation in a sort of narrow searchlight pattern. This "spot" could be a violent storm, enhanced by intense magnetism, much like the less energetic flares on our Sun or aurora on Earth. Or, it could be a localized region (especially near the magnetic poles) where charged particles are accelerated to extremely high energies, which in turn emit beams of nonthermal radiation like that sketched in Figure 15.13. Although the spot sprays radiation steadily into space, the star's spin rate of many times per second guarantees that the emitted radiation in any direction behaves like a hail of discrete bursts. The radiation sweeps through space like that from

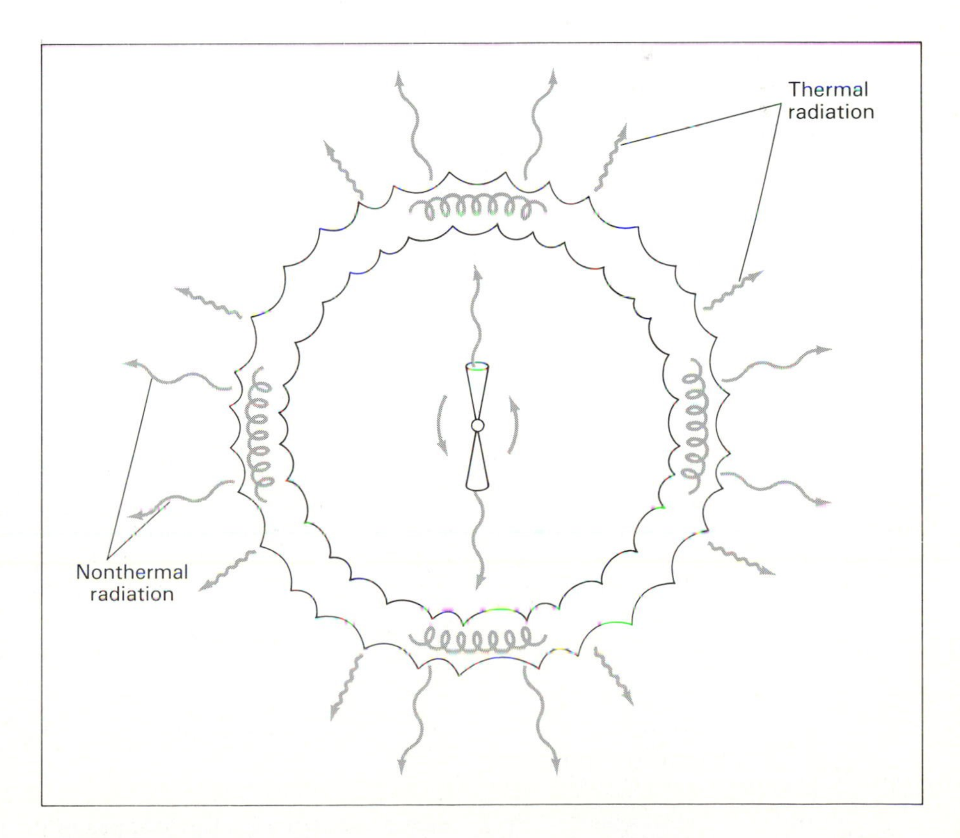

FIGURE 20.23 Artist's conception of a rotating neutron star (center) surrounded by a (three-dimensional) shell of glowing, radiating debris.

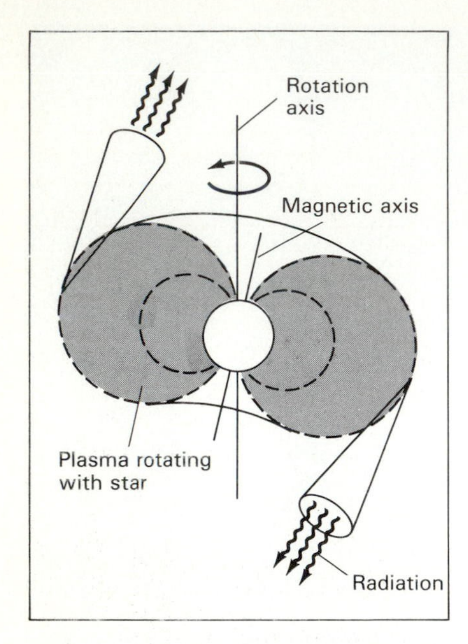

FIGURE 20.24 This schematic diagram of a popular theory of neutron-star emission can account for many of the observed properties of pulsars. As depicted, charged particles, accelerated by the magnetism of the neutron star, flow along the magnetic field lines, producing radio radiation that beams outward.

a revolving lighthouse beacon. Arriving at Earth, perhaps thousands of years later, it's observed as a series of rapid pulses. The duration of each pulse carries information about the source of activity on the neutron star, while the period of the pulses reveals the star's rotation. The details of the theoretical model are sketchy and controversial, for researchers have hardly any solid information about the behavior of matter having densities as large as a million tons/cubic centimeter.

More recently, a new class of pulsars has been found—very fast spinors called "millisecond pulsars." Although only a few are now known, these objects spin nearly 1000 times per second (i.e., with a period of about 0.001 second), which is almost the speed at which the outward (centrifugal) forces would break up the pulsar. Don't gloss over these numbers, for they suggest a phenomenon that borders on incredulity: a cosmic object of kilometer dimensions, yet having more mass than in our Sun, making nearly 1000 complete revolutions *every second.* One possible explanation for the high rotational velocity is that the neutron star has been "spun up," that is, it has drawn in huge amounts of matter from a companion star. The ongoing collision of this stream of matter, falling into the neutron star, is theorized to be the source of the extra "push" needed to make it spin faster than normal. If this is the case, however, it would seem that a supernova cannot be the source of the pulsar; not only does the fast spin seem too great for it to be simply the remains of a supernova, but a standard supernova explosion would be expected to blow apart a binary system. Astronomers are now searching the skies for other examples of this new class of pulsar. (Consult also Interlude 21-2.)

PUZZLING RELATED PHENOMENA: In a surprising development several years ago, intense bursts of x-rays were discovered from another major class of binary stars. Found near the central regions of our Galaxy and also near the centers of rich clusters of stars, these systems (called "x-ray bursters") emit thousands of times more radiation than our Sun, and in rapid bursts that last only a few seconds. The bursts seem to arise from weakly magnetized neutron stars that are members of stellar binary systems of low mass. In these systems, matter accreted by the neutron star accumulates on or near the star to a depth sufficient to suddenly commence nuclear reactions (due to the pressure of overlying matter). These reactions then release huge amounts of radiation in a burst of x-rays; after several hours of renewed accumulation, a fresh layer of matter produces the next burst.

In an equally startling discovery, optical astronomers found that the object with the catalog name "SS 433" ejects more than an Earth mass every year in two oppositely directly narrow jets of gas moving at approximately 25 percent of the speed of light. Periodic changes in the optical emission spectrum imply that the jets precess like a spinning top—like that depicted in Figure 20.25—which trace out a complete cone about twice a year. This interpretation has been confirmed by x-ray observations made from orbit and by images of SS 433 obtained by

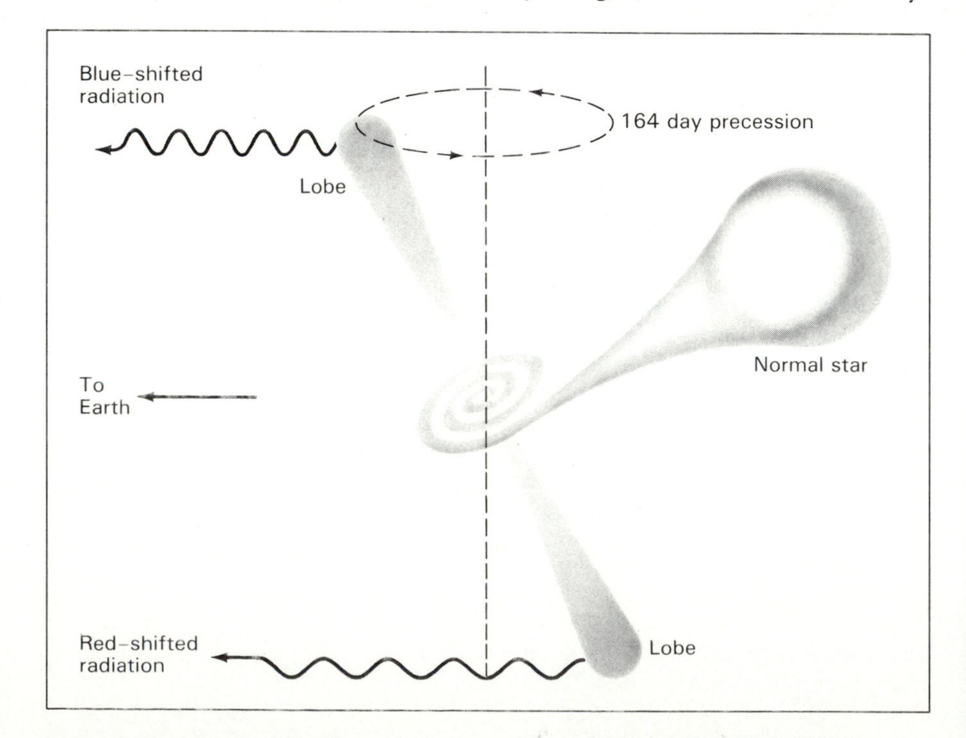

FIGURE 20.25 A schematic diagram of the peculiar object, SS 433. Its invisible tear-shaped lobes resemble those of the active radio galaxies (consult Chapter 15), but SS 433 is a much smaller phenomenon, being associated with a binary-star system in the Milky Way.

means of radio interferometry; the radio images demonstrate the invisible helical patterns of matter sprayed into interstellar space as the jets twirl around like a fireworks display. The generally accepted view of SS 433 is that it is a binary system with a compact object (probably a neutron star) orbiting around an ordinary star. The compact object emits a jet of high-energy matter from each pole, just as pulsars do. However, the origin of this peculiar system, the mechanisms that accelerate and align the matter into jets, and the cause of the precession are still unclear.

In an even more dramatic discovery, several military satellites detected gamma-ray bursts in space—bright, irregular flashes of gamma rays lasting only a few seconds. Although the poor spatial resolution of the gamma-ray equipment has precluded the optical identification of most of the gamma-ray sources, they probably resemble scaled-up versions of binary-star systems in which matter, falling onto a neutron star, experiences a sudden flash of thermonuclear burning. On the other hand, the process may not really be that simple, for, on 5 March 1979, the most intense gamma-ray burst ever recorded was observed from the direction of a known supernova remnant in the Large Magellanic Cloud; provided that it really did originate at the 175,000-light-year distance of this region beyond the outskirts of our Milky Way, the intrinsic luminosity of the burst would be so great as to defy explanation by any mechanism now known to science.

ULTRACOMPRESSION: Neutron stars are indeed peculiar objects. Theory, though, predicts that they are more or less in equilibrium, just like most other stars. In the case of neutron stars, however, equilibrium does not mean a balance between the inward pull of gravity and the outward pressure of hot gas. Neutron stars probably have no hot gas. Instead, the outward force is provided by the solid crystalline nature of the tightly packed neutrons. Existing side by side, the neutrons form a hard ball of matter that not even gravity can compress further—with one notable exception.

Suggestions have been made that there exist stellar core remnants with masses so large that the inward pull of gravity can in fact overwhelm even the seemingly incompressible sphere of pure neutrons. According to some theories, should enough matter be packed into an extremely small volume, the collective pull of gravity can eventually crush any countervailing phenomenon. In this case, gravity is envisioned to be so powerful that it can compress a massive star into an object the size of a planet, a city, a pinhead, a microbe, even smaller! The gravitational pull in the vicinity of these objects is thought to be so great that light itself would return to them much like baseballs return to Earth when thrown in the air. Such ultrastrange objects would be expected to emit no light, no radiation, no information whatsoever. Incommunicado, such a massive star catastrophically collapses into a hole—a hole perhaps no larger than a few centimeters across, but a hole into which all nearby matter falls, trapped by gravity perhaps forever. Astronomers call these most bizarre end points of stellar evolution black holes. We'll discuss their theoretically expected properties, and the evidence for their existence, in the next chapter.

SUMMARY

In recent years, astronomers have gained a reasonably good understanding of how stars proceed through the various evolutionary paces toward death—how they change their temperatures and densities while struggling to reestablish their equilibrium burning cycles. In contrast to our lack of knowledge regarding galaxy evolution, we know quite a lot about stellar evolution.

Furthermore, our evolutionary models are backed by large quantities of data—observations that have verified some of the odd states of matter expected theoretically. Huge red giants, dead white dwarfs, peculiar planetary nebulae, exploded supernovae, and ultradense neutron stars are all predicted by the theory of stellar evolution. And all are found to reside at myriad places in our Milky Way. On the whole, the subject of stellar evolution is one of the best developed and understood aspects of cosmic evolution.

KEY TERMS

helium flash
neutron star
nova
planetary nebula
pulsar
red giant star
supernova
supernova remnant
white dwarf star

QUESTIONS

1. Consider two stars that have just arrived on the main sequence. One has 20 times the mass of our Sun, whereas the other has 1 solar mass. (a) In 5 billion years, which, if either, star will probably still be on the main sequence? (b) If either star does leave the main sequence in 5 billion years, what type of object will it become? (c) Describe typical observations that could be made to study the ultimate fate of each object.

2. In a couple of paragraphs, describe chronologically the major evolutionary stages of a 1-solar-mass star from pre-birth to death. Your answer should include the stellar-energy mechanisms and time scales involved at each stage. Sketch the evolutionary path of this object on an HR diagram.
3. Compare and contrast the evolution of a 1-solar-mass star and a 10-solar-mass star. Draw a carefully labeled HR diagram as part of your answer.
4. A star is observed to explode as a supernova in the nighttime sky; its brilliance outshines the full Moon. As chief scientific advisor for the U.S. government, you have at your disposal the best equipment in our country's arsenal of observational apparatus. (a) Describe the observational tests you would undertake to ensure that the supernova's radiation is not directly harmful to life on Earth. (b) How would you explain such a spectacular event to a lingering Aristotelian who thinks that the Universe is immutable? (c) How might you monitor subsequent changes in the supernova's remnant?
5. Identify and briefly explain the stage of stellar evolution associated with each of the following observations: (a) optical emission from a bright, hot star and its surrounding three-dimensional shell of glowing gas; (b) x-ray emission from a region near an ordinary star not normally hot enough to emit x-rays on its own; (c) infrared emission from a dark, lukewarm blob embedded within an interstellar cloud of gas and dust; (d) bursts of radio and optical emission from a compact region at the center of an expanding diffuse cloud of gas; (e) ultraviolet spectral features showing evidence for strong "winds" emanating from a young star.
6. Pick one: Neutron stars have dimensions of (a) people; (b) cities; (c) states; (d) empires; (e) ideas.
7. Pick one: A neutron star is (a) a pre-black hole; (b) a rapidly rotating main-sequence object; (c) a post-white dwarf; (d) a ball of compressed elementary particles; (e) the latest Broadway craze.
8. Describe the physical characteristics of a neutron star: What is it made of? How big is it? How dense? What process might make neutron stars? Give some evidence for your answers.
9. Explain why, if M-type stars have so little mass, it is unlikely that any of them have ever "died" in the history of the Galaxy.
10. Draw to scale the typical sizes of a red giant star, a normal main-sequence star, a white dwarf star, and a neutron star. How would the size of the Earth compare?

FOR FURTHER READING

COOKE, D., *The Life and Death of Stars*. New York: Crown Publishers, 1985.

JASTROW, R., *Red Giants and White Dwarfs*. New York: W. W. Norton, 1979.

MITTON, S., *The Crab Nebula*. New York: Scribner, 1978.

SCHAEFER, B., "Gamma-Ray Bursters." *Scientific American*, February 1985.

SEWARD, F., GORENSTEIN, P., AND TUCKER, W., "Young Supernova Remnants." *Scientific American*, August 1985.

SHOKAM, J., "The Oldest Pulsars in the Universe." *Scientific American*, February 1987.

*STROM, S., "Star Formation and the Early Phases of Stellar Evolution," in *Frontiers of Astrophysics*, E. Avrett (ed.). Cambridge, Mass.: Harvard University Press, 1976.

21 BLACK HOLES: *The Ultimate Trap*

Our study of the evolution of stars has led to some unexpectedly strange objects, including white dwarf stars and neutron stars. These compressed states of matter are completely unfamiliar to us here on Earth. Yet perhaps the strangest of all states results from the catastrophic implosion–explosion of stars much more massive than our Sun. The aftermath of such a disaster is a most peculiar remnant having a great mass compressed into a microscopic volume. This is the so-called black hole.

A **black hole** is a region containing a huge amount of mass compacted into an extremely small volume. It's not an object per se so much as a hole, and one that's dark. These two factors—large mass and small size—guarantee an enormously strong gravitational force field. Why? Because one-half of the law of gravity stipulates that gravity is directly proportional to the mass. The other half dictates that gravity is inversely proportional to the square of the distance over which the matter is spread. Thus, because the distance term is squared, the gravitational force field grows spectacularly when a huge mass is compressed. Accordingly, spacetime is expected to be severely warped near black holes, requiring us to have some knowledge of relativity if we are to appreciate these totally foreign states of matter.

The learning goals for this chapter are:

- to understand how black holes fit into the overall theory of stellar evolution
- to appreciate how black holes can trap matter and radiation, prohibiting their escape
- to recognize that the bizarre phenomena near black holes result from the severe warping of their neighboring spacetime
- to understand the various tests used to probe the existence and behavior of black holes
- to know some of the regions where researchers suspect black holes are hiding

Reviewing Stellar Evolution

To place black holes into perspective, we need to review some of the physical properties of white dwarf and neutron stars. We'll then be better suited to introduce the central features of the bizarre black holes.

Figure 21.1 summarizes the evolutionary paths for all the stellar objects studied thus far. As depicted, an interstellar cloud begins contracting by gravitationally accumulating matter. The cloud gradually develops during the Kelvin–Helmholtz contraction phase, all the while progressively increasing its gas density, its atomic collisions, and its internal heat. A protostar eventually forms while evolving along the Hayashi track, after which a genuine star arrives on the main sequence. Once fully formed on the main sequence, stars exist as mostly unchanging balls of gas, the inward pull of gravity fairly well balanced against the outward pressure of hot gas.

The total stay on the main sequence is highly variable and depends mostly on the star's mass; contrary to intuition, the least massive stars endure the longest, some of them perhaps as long as a trillion years. Upon leaving the main se-

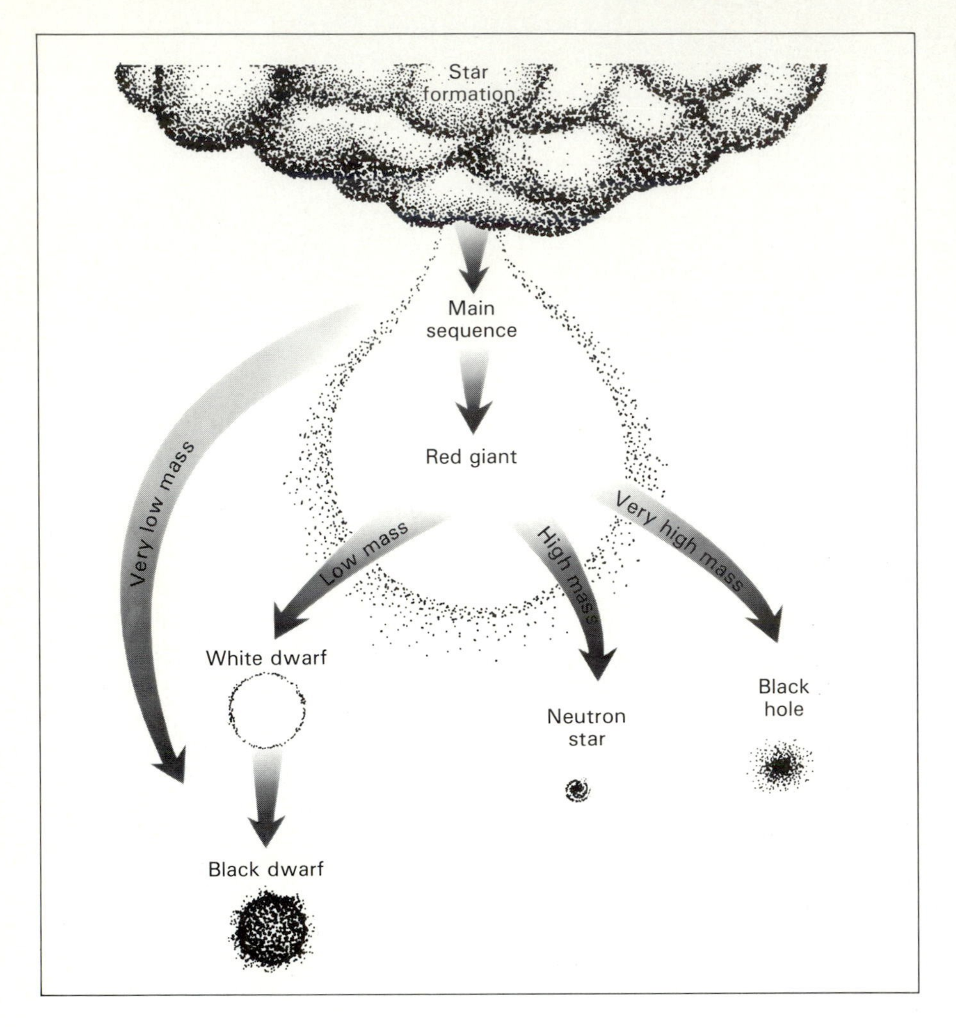

FIGURE 21.1 Summary of the evolutionary paths of all stellar objects. (Note how the lowest-mass M-type stars probably bypass the red giant stage, eventually cooling and shrinking to the black dwarf stage trillions of years later. The reason is that red dwarfs are too cool to burn even helium.)

quence, a star changes rapidly, becoming periodically unstable as it evolves toward death.

Late evolutionary stages depend mainly on the mass of the star. Low-mass objects, including our Sun, pass through the red giant stage, shedding their outermost layers as planetary nebular shells of gas, but leaving behind their cores, which in turn become white dwarf stars. The gravitational pull of the matter in a white dwarf star is counterbalanced by the basic impenetrability of the electrons in the star. White dwarfs can only cool off, and thus eventually become black dwarfs.

High-mass objects are not nearly as well behaved. Rather than gently dispersing layers of gas as in planetary nebulae, massive stars violently eject large amounts of their matter into interstellar space by means of supernova explosions. These stars, much more massive than our Sun, also leave at the supernova core an intact remnant known as a neutron star. Neutron stars counterbalance gravity by the sheer impenetrability of neutrons themselves.

DIVIDING LINES: As noted in Chapter 20, the dividing line between low-mass and high-mass stars is 3 solar masses. This value refers to the mass of the *original* star before any matter is expelled in a planetary nebula or supernova.

The theory of stellar evolution also makes specific predictions about the mass of the core remnant. Should the burned-out *core* be less than about 1.5 solar masses, the resultant object is expected to become a white dwarf. If the *core* mass is between 1.5 and 3 solar masses, the result is a neutron star; the added mass enables the star to crush its matter beyond atomic identity, driving the electrons into the protons and fashioning neutrons. And if the core remnant has a mass greater than about 3 solar masses, not even tightly packed neutrons can withstand the gravitational pull. Gravity wins out, and the object collapses perhaps forever.

To illustrate these differences, consider the nearby binary star system having the catalog name HD47129. Observed from Earth, its orbital size and period indicate that one of the two stars has a total mass of about 76 solar masses. This B-type star is one of the most massive stars known anywhere in our Galaxy. Provided that our arguments of the previous chapter are correct, this star will surely explode as a supernova someday. It's destined to leave a remnant core having a mass less than its original 76 solar masses. How much less is not known. If the future supernova explosion of this star ejects 74.5 or more of its original 76 solar masses, the core will become a white dwarf, as predicted for stellar remnants having less than 1.5 solar masses.

There might be a flaw in this argument. Supernovae probably do not eject such a high fraction of a star's original mass. A combination of theoretical arguments and recent observations of supernova remnants near our Sun imply that roughly half of the original mass is expelled to interstellar space. The exact amount is unknown, not only because the devastating effects of any explosion are hard to predict, but also because the total amount of scattered debris in observed supernova remnants is difficult to estimate. In any case, adopting as a rule of thumb that about half of the original star's matter is expelled during a supernova, we can expect that the future explosion of this 76-solar-mass star will produce a remnant core of approximately 30 to 40 solar masses. Too massive to be a white dwarf, it's also too massive to be a neutron star.

The dividing lines at 1.5 and 3 solar masses are uncertain because they ignore the effects of magnetism and rotation, two physical properties that are surely present in stars. Since these effects can compete with gravity, they must influence the evolution of stars. We cannot be certain to what extent the mass dividing lines will change, for no one really knows how the basic laws of physics might change in regions of very dense matter that is both spinning and magnetized. If anything, these dividing lines will shift upward when magnetism and rotation are included because even larger amounts of mass will be needed for gravity to compress stellar cores into neutron stars or black holes.

Disappearing Matter

What happens when a star's remnant core exceeds 3 solar masses? What makes this value so special? The answer is that this is the mass value for which gravity can (apparently) no longer be countered. Not even neutrons, touching one upon another, can halt the pull of gravity within such a massive, compact object. According to theory, such an object just continues collapsing, crushing matter to the dimensions of a point. The star catastrophically implodes without limit; apparently nothing can stop it.

How can we possibly appreciate such a seemingly ridiculous phenomenon? How can an entire star shrink to the size of an elementary particle, while presumably on its way to even smaller dimensions? Does this make sense? Well, this is what the detailed mathematics predict. Without some agent to compete against gravity, massive core remnants are expected to shrink to a singular point of infinitely small volume.

A complete analysis of the complex mathematics needed to understand the true nature of black holes is beyond the scope of this book. Nonetheless, we can usefully discuss a few qualitative aspects of these incredibly dense and bizarre regions of space.

IMAGINARY SQUEEZING: Consider first of all the concept of escape velocity—the velocity needed for any small object to escape from a large one. When discussing planetary atmospheres earlier in Chapter 5, we noted that the escape velocity is proportional to the square root of the planet's mass divided by the square root of its radius. For example, on Earth, with a radius of about 6500 kilometers, the escape velocity is nearly 11 kilometers/second. To launch any object—molecule, baseball, rocket, whatever—away from Earth, that object must move faster than 11 kilometers/second.

Imagine now a hypothetical experiment for which the apparatus is a three-dimensional vise. Let the vise be large enough to hold the entire Earth. Figure 21.2 should help you imagine this awful thought. Furthermore, imagine Earth being squeezed on all sides. As our planet shrinks under the onslaught, its density rises because the total amount of mass remains constant inside an ever-decreasing volume.

Suppose that our planet is compressed to one-fourth its present size. The proportionality noted above then predicts that the escape velocity is doubled. Any object attempting to escape from this hypothetically compressed Earth would need a velocity of about 22 kilometers/second.

Imagine compressing Earth still more. Squeeze it, for example, by an additional factor of 1000, making its radius hardly more than a kilometer. Accordingly, the escape velocity increases dramatically. In fact, a velocity of about 700 kilometers/second is needed to escape from an object having a radius of about 1 kilometer and a mass of about the entire Earth.

This then is the trend. As an object of any mass contracts, the gravitational force field grows stronger at its surface, mostly because of increasing density.

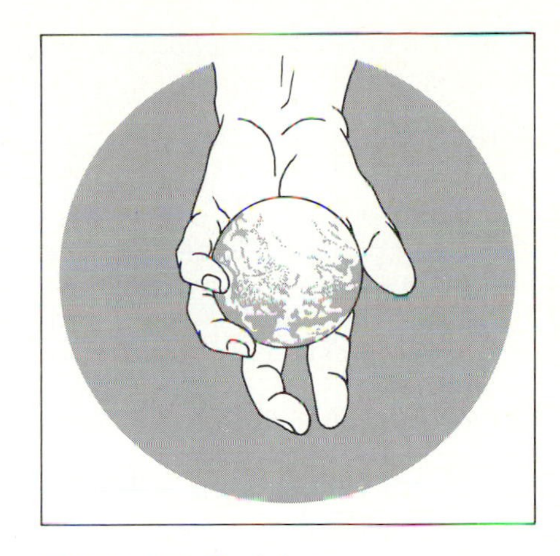

FIGURE 21.2 Earth being squeezed in a gigantic vise—just a thought!

If we could imagine Earth being further compacted, the escape velocity would rise accordingly. In fact, if our hypothetical vise were to squeeze Earth hard enough to crush its radius to about a centimeter, the velocity needed to escape its surface reaches 300,000 kilometers/second. Don't gloss over the number; this is no ordinary velocity. It's the velocity of light, the fastest velocity allowed by the laws of physics as we now know them.

So if, by some fantastic means, the entire planet Earth could be compressed to the size of a pea, the velocity for anything to escape must exceed the velocity of light. And since that's impossible, the compelling conclusion is that nothing—absolutely nothing—can escape from the surface of such a compressed "Earth."

Thus, if our planet were squeezed to less than centimeter dimensions, we could legitimately argue that knowledge of it would be lost to the rest of the Universe. After all, there would be no way for a launched rocket, a beam of light, or any type of radiation to get away. Such a compressed object would have become invisible and uncommunicative, for no information whatsoever could be exchanged with the Universe beyond. For all practical purposes, such a supercompact object can be said to have disappeared from the Universe!

REAL SQUEEZING: The example above is, of course, hypothetical. Nothing in the known Universe resembles a vise capable of squeezing the entire Earth to centimeter dimensions. But in massive stars, such a vise does in fact exist. It's known as gravity.

Gravity cannot crush Earth in this way because our planet simply houses too little mass; the collective gravitational pull of every part of Earth on all other parts of Earth is just not powerful enough. However, at the end of a star's life, when the nuclear fires have dwindled, gravity can literally crush a star on all sides, thereby packing a vast amount of matter into a very small sphere.

When stellar core remnants have more than 3 solar masses, the critical size at which the escape velocity equals the velocity of light is not, as for Earth, of centimeter dimensions. For typically massive core remnants, this critical size is on the order of kilometers. For example, a 10-solar-mass remnant would have a critical radius of nearly 30 kilometers. To be sure, it's no less a feat to compress an entire star to kilometer dimensions than to compress a planet-sized object to centimeter size. In the stellar case, however, we're not talking about a hypothetical situation using an imaginary vise. The relentless pull of gravity is truly strong enough to compress spent stars to extraordinarily small dimensions. The strong gravitational force field of massive stars is not at all hypothetical; it's real.

Astronomers have a special name for the critical radius below which any object is predicted to disappear in this manner. Called the **event horizon,** this size defines the region within which no event can ever be seen, heard, or known by anyone outside. Accordingly, the event horizons of Earth and of a 10-solar-mass star are 1 centimeter and 30 kilometers, respectively. (A quick rule of thumb states that the radius of the event horizon equals 3 kilometers multiplied by the object's mass, provided that mass is expressed in units of solar masses.)

So we might say that magicians could in fact make coins and other small objects disappear provided that they squeezed their hands hard enough. Even people could disappear if they could arrange to be compressed to a size smaller than 10^{-23} centimeter (which is much, much smaller than a proton). Gravity won't do it to us, though. Humans are just not massive enough. The collective gravitational pull of all the atoms in our bodies falls far short of the force needed to compress us to this minuscule size. Nor are there any technological means currently known that even come close to doing so—not even modern garbage compactors. Lucky for us!

On the other hand, perhaps someday, through some marvel of technology, our descendants might learn how to compact garbage to an almost incredibly small size. It would then disappear! Maybe some practical application will eventually result from black hole research after all.

The important point here is the following: Should no force be capable of withstanding the self-gravity of a dead star having more than 3 solar masses, such an object will naturally collapse of its own accord to smaller and smaller dimensions. Such a stellar core remnant will not even stop infalling at its event horizon. An event horizon is not a physical boundary of any type, just a communications barrier. The core remnant shrinks right on past it to ever-diminishing sizes, presumably on its way toward becoming an infinitely small point—a singularity. We say "presumably" because we cannot be sure that there isn't an as-yet undiscovered force capable of halting catastrophic collapse somewhere between the event horizon and the point of singularity. The subject needed to reveal any such force—quantum gravity—has not yet been invented.

WINK OUT: Here, then, is the observational sequence of events if these late stages of stellar evolution are correct. A very massive star ends its burning cycle by exploding as a supernova. Approximately half the star's original content is then ejected as fast-moving debris. Provided that at least 3 solar masses remain behind, the unexploded remnant core will collapse catastrophically, the whole core diving below the event horizon in less than a second. The core simply winks out—becoming not merely invisible, but literally disappearing—leaving a small dark region from which nothing can escape. This is the way a black hole is born as a blackened region of space. They are not really objects; they're just holes—black holes in space.

Properties of Black Holes

SPACETIME WARP: Modern notions about black holes rest solely on the theory of relativity. Whereas white dwarf and neutron stars are valid end points of stellar evolution even within the confines of the Newtonian theory of gravity, only the Einsteinian theory of spacetime specifies the physical properties of bizarre phenomena like black holes. As such, black holes should obey all the standard laws of relativity theory. In particular, the mass contained within a black hole is expected to warp space and time in its vicinity. Close to the hole, the gravitational force field becomes overwhelming, and the curvature of spacetime extreme. At the event horizon itself, the curvature is so great that spacetime folds over on itself, causing trapped objects to disappear.

Several props can help us visualize the curvature of spacetime near a black hole. Each way is, however, only an analogy. They are not really examples. The problem here, as earlier in Chapter 16, is our inability to work conceptually in four dimensions.

Our analogy is a pool table. Imagine the tabletop to be made of a thin rubber sheet, rather than the usual hard felt. As suggested by Figure 21.3, such a rubber sheet becomes distorted when a heavy weight, such as a rock, is placed on it. The otherwise flat rubber sheet sags or warps

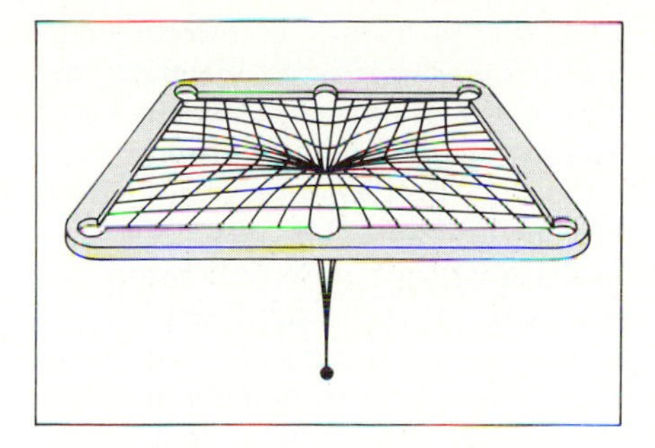

FIGURE 21.3 A pool table made of thin rubber membrane will sag with a weight on it. Likewise, the curvature of spacetime is said to be "warped" in the vicinity of a compact astronomical object.

(or curves), especially near the rock. The heavier the rock, the larger the curvature. Trying to play billiards, you would quickly find that balls passing near the rock are deflected by the curvature of the tabletop.

In such the same way, radiation and matter are deflected by the curvature of spacetime near a star. As depicted earlier in Figure 16.2 and discussed in Chapter 16, Earth's orbital path is governed by the slight spacetime curvature created by our Sun. In the case of very massive stars, spacetime is severely curved. And in the extreme, black holes curve spacetime more than any known object.

The formation of black holes and the severe warping of spacetime caused by them can be further appreciated by considering another analogy. This one requires us to imagine a large family of people living on a huge rubber sheet—a sort of gigantic trampoline. Deciding to hold a reunion, they converge on a given place at a given time. Like any reunion, this is an event in spacetime. However, as shown in Figure 21.4, one person remains behind, wishing not to attend the reunion; he can keep in touch by means of message balls rolled out to him along the rubber sheet. These balls are the analog of radiation traveling at the velocity of light, while the rubber sheet mimics the fabric of spacetime itself.

As the people converge, the rubber sheet begins to sag more and more. Their accumulating mass in a smaller place creates an increasing amount of spacetime curvature. The message balls can still reach the lone person far away in essentially flat spacetime, but they arrive less frequently as the sheet becomes progressively more warped, as in Figure 21.4(b) and (c).

Finally, when a large enough number of people have arrived at the appointed spot, the mass becomes too great for the rubber to support. As illustrated in Figure 21.4(d), the sheet closes off into a bubble which compresses the people into oblivion.

Until the end, message balls were able to reach the lone survivor, albeit at a slower and slower rate. But as the rubber sheet segregates the bubble of people (thereby forming an event horizon), the balls can no longer return to the person left behind. Regardless of the speed of the last message ball, it cannot quite outrun the downward stretching sheet.

This analogy (very) roughly depicts how a black hole is predicted to warp spacetime completely around on itself, thus isolating it from the rest of the Universe.

COSMIC CLEANERS, NO: Be sure to realize a basic feature about the nature of black holes: Black holes are not cosmic vacuum cleaners; they do not cruise around interstellar space, sucking up everything in sight. The movements of objects near black holes mimic those of any objects near a very massive star. The only difference is that, in the case of a black hole, objects skirt or orbit about a dark, invisible region. Neither emitted radiation nor reflected radiation of any sort emanates from the position of the black hole itself.

Black holes, then, do not go out of their way to drag in matter, but if some matter does happen to fall in via the normal pull of gravity, it will be unable to get out. Black holes are like turnstiles, permitting matter to flow in only one direction—inward—prompting some researchers to call them "ingestars," thus paralleling the terms pulsar and quasar. Swallowing matter, black holes con-

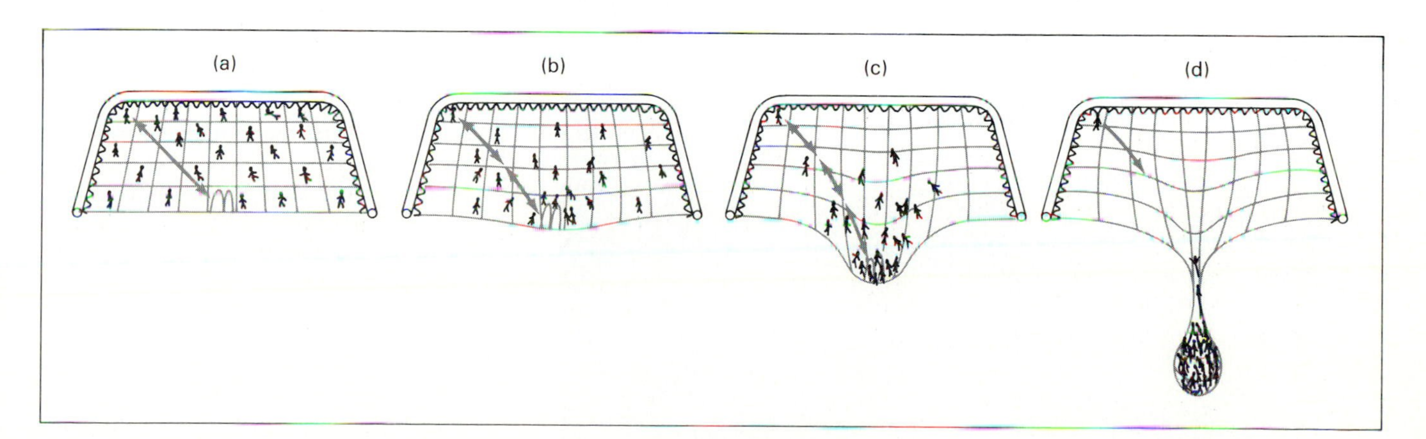

FIGURE 21.4 Any mass, even the small amount contained in people, causes spacetime to be curved. As all the people except one assemble at an appropriate spot, the curvature grows progressively larger in frames (a), (b), and (c). The people are finally dragged inside the sealed off bubble (d), forever trapping them as well as their message balls.

stantly increase their mass; their size also grows since the event horizon depends on the amount of mass trapped inside. If black holes really do exist in space, all of them are probably growing in mass and size as increasing quantities of matter accumulate below their event horizons.

COSMIC HEATERS, YES: One final point about the nature of black holes is this: Matter flowing into a black hole is subject to great tidal stress. An unfortunate person, falling feet first into a black hole, would find himself stretched enormously in height, all the while being squeezed laterally. He would be literally torn apart, for gravity would be stronger at his feet than at his head. He wouldn't stay in one piece for more than a fraction of a second after passing the event horizon.

As depicted in Figure 21.5, similar distortion and breakup apply to any kind of matter near a black hole. Whatever falls in—gas, people, space probes—is vertically elongated and horizontally compressed, in the process being accelerated to high speeds. The upshot is numerous and violent collisions among the torn-up debris, all of which yield a great deal of heating (via friction) of the infalling matter.

The rapid heating of matter by tides and collisions is so efficient that prior to submersion below the hole's event horizon, newly infalling matter emits radiation on its own accord. This is simple thermal radiation, emitted because the infalling matter has become hot—so hot, in fact, that the radiation is expected to be of the x-ray type. In effect, the gravitational potential energy of matter outside the black hole is converted to heat energy while falling toward the hole. Once the hot matter dips below the event horizon, the radiation ceases to be detectable.

Contrary to popular belief, then, black hole environments are expected to be *sources* of energy. Should the amount of infalling matter be large, regions just outside event horizons could be absolutely prodigious emitters of intense radiation. The ability technologically to tap this energy might well be a major milestone in the history of long-lived civilizations.

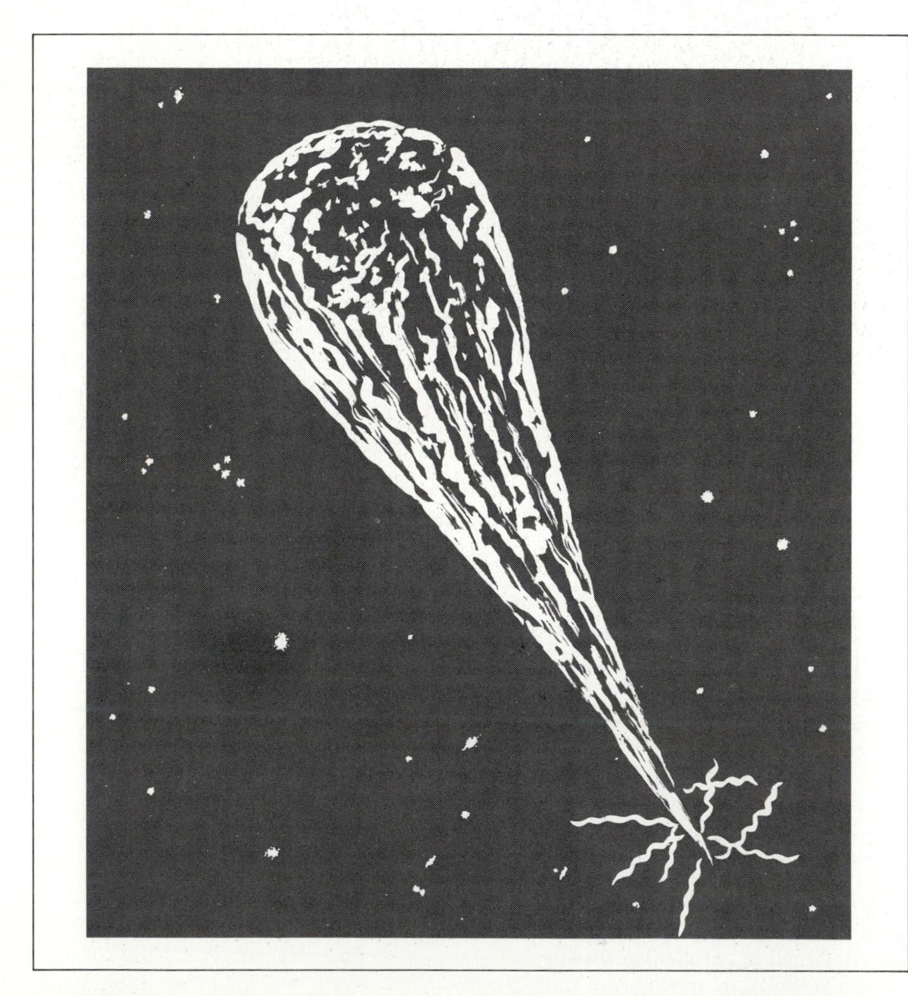

FIGURE 21.5 Any matter falling into the clutches of a black hole will become severely distorted and heated. In this sketch, an imaginary planet is being pulled apart by the gravitational tides of the black hole.

Space Travel Near Black Holes

Consider the prospects of exploring a black hole. It would be wise to conduct any studies from a safe distance. Otherwise, travelers getting too close may well find themselves being pulled past the event horizon by the hole's gravity—never to return.

Could we voyage anywhere near a black hole, then? One reasonably safe way to study a black hole up close is to orbit it carefully. A stable, circular orbit could get us quite close, even though we would be immersed in the hole's strong gravitational force field. After all, Earth and the other planets of our Solar System each orbit the Sun without falling into it; the Sun's gravitational pull offsets each planet's forward momentum, as described in Chapter 7. The gravity field around a black hole is basically no different, albeit stronger, and should permit stable orbits.

Even so, travel would be unsafe for humans should they venture (even in a stable orbit) too close to the hole. Extrapolation of human endurance tests conducted on American and Russian astronauts suggests that the human body cannot withstand stress more than 10 times the present gravitational pull we nor-

INTERLUDE 21-1 ***Mini Black Holes***

Descriptions of black holes in this chapter do not include the effects of magnetism and rotation. Once we learn how to account for these properties, the mass value dividing neutron stars from black holes will surely increase. Magnetism and rotation, remember, tend to compete with gravity. Thus these two effects will probably increase the minimum mass needed to form a black hole beyond the usual 3 solar masses. How much more is currently unknown, for we await some brilliant relativist to solve this most difficult problem.

The subatomic properties of matter comprise another feature so far neglected in theories of black holes. This could be a crucial issue because all the matter trapped within a black hole is predicted to collapse to the size of an elementary particle or less. To decipher the nature of matter deep down inside the event horizon, scientists will have to merge their knowledge of gravitational physics with that of subatomic physics—the yet-to-be-invented subject of "quantum gravity."

Some attempts to understand gravity on a microscopic scale suggest that black holes may not be entirely black after all. When subatomic physics is taken into account, it seems that matter and radiation might be able to escape from black holes. This might be true because when the phenomenon of pair creation (noted earlier in Chapter 16) occurs near a black hole, one particle could conceivably fall into the black hole before the pair of particles had a chance to annihilate. The other particle would then be free to leave the scene, making the black hole appear to the outside world as a source of radiation. According to these preliminary theories, black holes might not last forever. They slowly evaporate and finally explode, scattering their contents into interstellar space.

Just as for ordinary stars, the lifetime of a black hole depends on its mass. For conventional black holes having several times the mass of the Sun, a black hole is predicted to explode after many times the current age of the Universe. Thus the issue is moot for the usual type of black hole described elsewhere in this chapter; astronomers cannot expect to observe either their slow decay or ultimate explosion.

These new theories, however, also predict that the pressure in the early Universe may have been just right to compress pockets of matter into miniature black holes. For example, very small black holes having a mass of about 10^{15} grams are predicted to have a lifetime nearly equal to the current age of the Universe. Such a black hole, having the mass of a typical meteoroid or a terrestrial mountain, would have an event horizon about equal to that of a subatomic particle (about 10^{-13} centimeter). Provided that they do exist, such mini black holes should be exploding now, thus emitting intense bursts of gamma-ray radiation.

Although the theory remains to be proved, should astronomers detect this radiation— and efforts are now being made to do so—it will conceivably tell us something not only about the nature of black holes but also about the physical conditions near the start of the Universe.

mally feel on Earth's surface. This more stressful state would occur about 3000 kilometers from a 10-solar-mass black hole (which, recall, would have a 30-kilometer event horizon). Closer than that, gravity would tear the human body apart.

AN INDESTRUCTIBLE ROBOT: We need not send a human being to study a black hole. Automated space probes could be designed to withstand stressful conditions normally intolerable to humans. For purposes of discussion, let's imagine using only indestructible astronauts—mechanical robots of sorts. To make it interesting, let's send these robots toward the center of the hole. Watching from a distance in our safely orbiting spacecraft, we'll then be able to test the nature of space and time near the hole. In this way, the mechanical robot becomes a test particle whose behavior can be used to infer several things about black holes.

This case is hypothetical. We are imagining sending a robot toward a black hole, as shown in Figure 21.6. Dispatching robot spacecract may be in fact the only way we can ever study black holes directly. Such robots could be useful probes of theoretical ideas, at least down to the event horizon. After that, there is no way for a probe to return any information about its findings.

Suppose, for example, the robot in Figure 21.6 has an accurate clock and a light source mounted on its side. Distant humans, safely orbiting far from the event horizon, would be able to use telescopes to read the clock and measure the light. The information received should help us decipher the nature of spacetime in the vicinity of the hole as the robot astronaut travels ever closer to the hole. What might we discover?

First, recall our discussion in Chapter 15 concerning the behavior of light (or any type of radiation) in a strong gravitation force field. Photons use some of their own energy while moving away from the source of gravity. The photons lose energy because they must work to get away. They don't slow down at all; they just lose energy. And since a photon's energy is proportional to the frequency of its radiation, light that loses energy will have its fre-

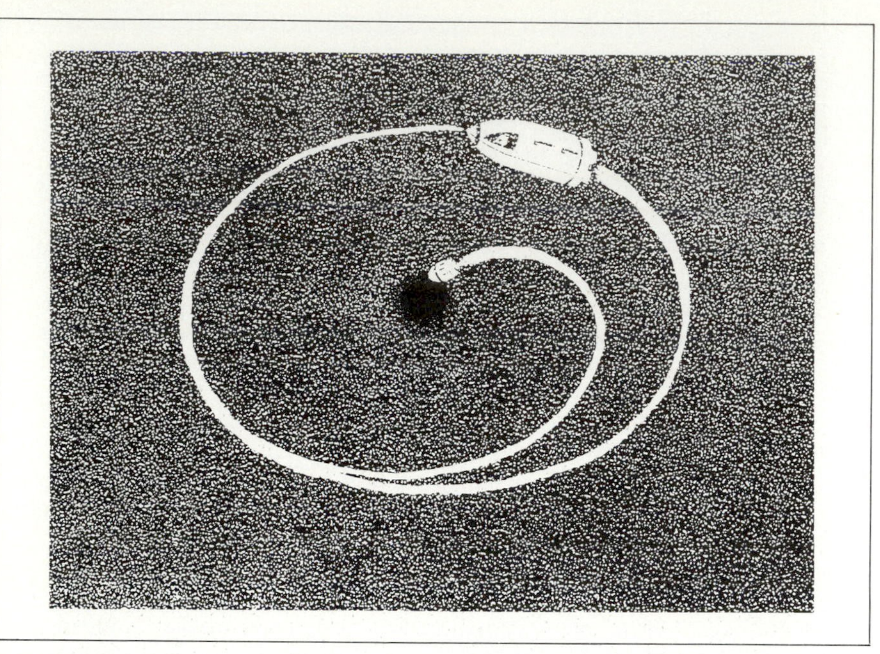

FIGURE 21.6 Robot astronauts can travel toward a black hole while performing experiments that humans, farther away, can monitor in order to learn something about the nature of spacetime near a black hole.

quency reduced (or conversely, its wavelength lengthened). Hence radiation emitted from the vicinity of a black hole will be red shifted by an amount depending on the strength of the gravitational force field. This is not a red shift caused by motion (Doppler effect); rather, it's a red shift induced by gravity.

Usually, an object's gravity is insufficient to shift radiation measurably toward the red. Sunlight, for example, is not red shifted by any detectable amount. A few white dwarf stars do show a slight reddening of their emitted light. And neutron stars should appreciably shift their radiation, but it is currently impossible to disentangle the effects of gravity and magnetism on the observed signals. Only near black holes is the gravitational pull so great that a red shift should be measurable, at least in principle.

So as photons travel from the robot's light source to the orbiting spacecraft, they become gravitationally red shifted. From the viewpoint of the safely orbiting humans, a green light, for example, would progressively become yellow and then red as the robot astronaut neared the black hole.

As the robot got close to the event horizon, radiation from its light source would become undetectable with optical telescopes. The radiation reaching the humans in the orbiting spacecraft would, by then, be lengthened so much that radio telescopes would be needed to detect it. Very close to the event horizon, the radiation emitted as light from the robot probe would have become shifted to wavelengths longer than conventional radio waves before reaching the humans.

Light trying to escape *from the event horizon itself* becomes gravitationally red shifted to infinitely long wavelengths. In other words, light uses all its original energy trying to escape from the edge of the hole. What was once light (on the robot) has no energy left upon arrival at the safely orbiting spacecraft. Theoretically, this radiation makes it to us—still moving at the velocity of light—but with zero energy. The originally emitted light radiation has become red shifted beyond our perception. Neither humans nor our equipment could detect it.

How about the robot's clock? Assuming that the distant observers in the safely orbiting spacecraft can read it, what time does it tell? Is there any observable change in the rate at which the clock ticks while moving deeper into the gravitational force field?

Einstein's theory of relativity (consult Interlude 16-1) suggests that, from the viewpoint of the safely orbiting spacecraft, any clock close to the hole would operate more slowly than an equivalent clock onboard the spacecraft. The clock closest to the hole would operate slowest of all. Upon reaching the event horizon, the clock would stop altogether. It would be as if the robot astronaut had found immortality! All action would become virtually frozen in time. Consequently, an external observer could never really witness an infalling astronaut sink below the event horizon. Such a process would take forever.

From the viewpoint of the infalling robot, however, relativity theory predicts no strange effects at all. To the computer onboard the indestructible robot, the light source hasn't reddened and the clock keeps perfect time. In the frame of reference of the infalling robot, everything is normal.

Nothing prohibits the robot traveler from passing right through the event horizon of a black hole. Neither the laws of physics nor those of biology physiologically constrain objects from passing through an event horizon. There's nothing resembling a brick wall at the event horizon; it's only an imaginary boundary in space. Travelers passing through might not even know it—which undoubtedly becomes a problem, for they are not going to get out! Once inside the event horizon, only velocities greater than that of light can ensure escape. And since that's impossible, any and all things—spaceships, people, light, information of any sort—become trapped.

DEEP DOWN INSIDE: No doubt, you are wondering what lies within the event horizon of a black hole. The answer is simple: No one knows.

Some researchers suggest that the

inner workings of black holes are irrelevant. Experiments could conceivably be done by robots sent "down under" to test the nature of space and time inside an event horizon, but that information could never reach the rest of us outside the black hole. Apparently, theories of the insides of black holes can never be put to the experimental test. Anyone's theory is as valid as anyone else's.

Perhaps the inner sanctums of black holes represent the ultimate unknowable. For that very reason, though, other researchers argue that it is of utmost importance to unravel the nature of black holes, lest we someday begin to worship them. Large segments of humankind have often revered the unknowable, worshipping that which cannot be tested experimentally. This is still the case for many of Earth's societies.

HOW MUCH DO WE KNOW? What sense are we to make of black holes? Do all these strange phenomena really occur? The basis for these weird predictions is the relativistic concept of mass warping spacetime—a phenomenon already tested to be a surprisingly good approximation of reality (consult Chapter 16). The larger the mass concentration, the greater the warp, and thus the stranger the observational consequences. Perhaps.

Some researchers argue that relativity is incorrect, or at least incomplete, when applied to black holes. Admittedly, it seems nonsensical to claim that very massive astronomical objects will collapse catastrophically to infinitely small points. Not even the wildest imaginations can visualize this phenomenon. Perhaps the current laws of physics are inadequate in the vicinity of a singularity. *At* the point of singularity, in fact, we are certain that relativity is incomplete and probably absurd. On the other hand, maybe matter trapped in black holes never does reach a singularity. Perhaps matter just approaches this most bizarre state, in which case relativity theory may still hold true. Scientists just don't know yet.

Observational Evidence

Theoretical ideas aside, let's examine how black holes might be observed. After all, if they cannot be found observationally, they might not even exist. Perhaps the computer modeling of the final stages of stellar evolution is incorrect in predicting that massive stars leave remnant cores larger than 3 solar masses. Maybe all of the original star is blasted to smithereens. Or perhaps yet another undiscovered force exists capable of competing with gravity despite the extreme conditions of ultracondensed matter expected in black holes.

Is there any observational evidence for black holes? More to the point, can we prove that invisible objects really do exist?

TESTS THAT DON'T WORK: Despite the fact that black holes are invisible, they are expected to remain strong sources of gravity. Accordingly, astronomers should be able to test for a hole's existence by studying its associated gravitational force field. For example, the motion of a spacecraft could conceivably be used to probe the outer environment of a black hole. Any craft outside a black hole should behave dynamically just as though there were a massive, visible object at the site of the hole. In other words, most conventional evidence of a black hole disappears, but its gravity persists.

Thus we could, at least in principle, maneuver a spacecraft into an orbit at a safe distance outside a region of otherwise total darkness. This is the case discussed in the preceding section and illustrated in Figure 21.6. Nothing would likely be perceived at the orbit's focus, but the hole's gravity would control the spacecraft's motion, much as the Sun guides Earth in its orbit. The spacecraft and the black hole then comprise a truly odd couple—a peculiar version of the binary-star systems studied in Chapter 11.

With the spacecraft safely in position, Kepler's third law could be used to infer the hole's mass. Knowing the orbital size and period of the spacecraft about the hole, we could compute the combined mass of the hole and the spacecraft, just as outlined in Chapters 11 and 13. Since the spacecraft's mass will surely be negligible compared to the black hole's, the orbital properties thus directly yield the mass of the black hole. In this way, the existence and mass of a black hole can be inferred without actually observing anything deep down inside it.

Another way that black holes might be detected, again at least in principle, also involves a binary system, as sketched in Figure 21.7. Here this test requires us to observe a small black hole while it passes in front of a much larger and visible companion. The observable effect we might expect would be a minute dot of blackness gliding across an otherwise bright star. But this would be extremely hard to see; the 12,000-kilometer planet Venus is barely noticeable when transiting the Sun, so a kilometer-sized object moving across the image of a faraway star would be virtually invisible.

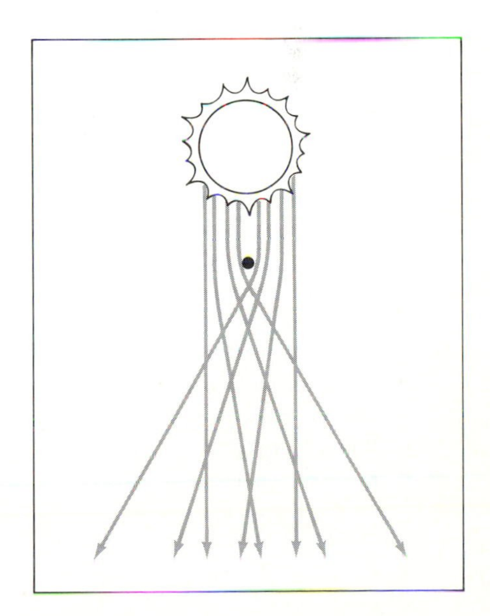

FIGURE 21.7 The gravitational bending of light around the edges of a small, massive black hole makes it impossible to observe the hole as a black dot superposed against the bright background of its stellar companion.

Actually, this test is not even as clear-cut as just described. The observable effect would not really be a black dot superposed on a bright background. The problem is more complex because the background starlight would be deflected while grazing the edges of the black hole on its way to Earth. The effect is much the same as the bending of distant starlight around the edge of the Sun, a phenomenon that has been repeatedly measured during solar eclipses throughout the last several decades; the bending occurs since photons have an equivalent mass (as described toward the end of Chapter 15) and mass is affected by the Sun's gravity. In the case of the Sun, the bending amounts to about 1.75 arc seconds, an extremely small angular deflection. But since this angle is proportional to the density of the deflecting object, the deflection of radiation should be much larger near small, massive black holes. So our perception of a black hole in front of a bright companion star would not show a neat, well-defined black dot. Instead, the bending of light around all sides of the hole would make its image seem fuzzy. Recent studies have shown that such a blur would be virtually, impossible to observe either with modern telescopes, or with any equipment we are likely to have in the foreseeable future.

The upshot is that these two methods of studying black holes are fine in principle but not in practice. Our civilization does not have the capability to maneuver spacecraft into the neighborhood of suspected black holes, even if we knew their exact locations. Nor do we have telescopes sophisticated enough to detect the fuzzy and diluted images of black holes passing in front of bright stars.

A BETTER TEST: The idea that some black holes might be members of natural binary-star systems leads to a promising indirect test of their existence. Although black holes cannot be seen directly, they are sure to block some light from their visible companions. Once again, as was the case around the turn of the century when they were the rage among astronomers, variable stars have become a fashionable and important subject in astronomy.

Figure 21.8 shows a light curve of a typical binary-star system. Data of this sort can yield the total mass of the system provided its orbital period and size are known. Recall from our study of double stars in Chapter 11 that each star of a binary system need not be seen directly. We need only observe the regular light variations of one star to infer both the existence of an unseen companion as well as some of the inherent properties of the invisible object. (Measurements of the Doppler velocity of only one star of a binary-star system are also enough to infer the existence and some properties of an invisible companion.)

Our Milky Way Galaxy harbors many such binary-star systems for which only one object can be seen. In the majority of cases, the invisible companion is probably small and dim, nothing more than an M-type star hidden in the glare of an O- or B-type partner. In other cases, dust or other interstellar debris probably shroud one object, making it seem invisible even when using the best equipment of our high-tech society. In either case, the invisible object is not a black hole.

However, a few binary systems have peculiarities implying that one of their members is indeed a black hole. Some of the most interesting discoveries, made only during the 1970s and 1980s by Earth-orbiting satellites, involve binary systems that emit large amounts of x-ray radiation. This high-frequency radiation cannot easily penetrate dust, making it unlikely that galactic debris has camouflaged one of the partners. Furthermore, in a few cases, the mass inferred for the x-ray emitting invisible object equals several solar masses, thus effectively ruling out small, dim stars. Here's a description of one such binary system.

CANDIDATE HOLES: Figure 21.9 shows the area of the sky where astronomers have reasonably good evidence for a black hole. This part of the sky was known to the ancients as the Cygnus region, a name still used today. The rectangle outlines the celestial system of interest, some 7000 light-years from Earth. The main features of many observations of this system are as follows:

1. Spectroscopic observations of visible radiation show that the bright object—a blue B-type giant star with the catalog name of HDE226868—is only one member of a binary-star system whose orbital period (5.6 days) and size (20 million kilometers in diameter) are well determined.
2. Other spectral-line observations suggest that hot gas is flowing from the bright star toward an

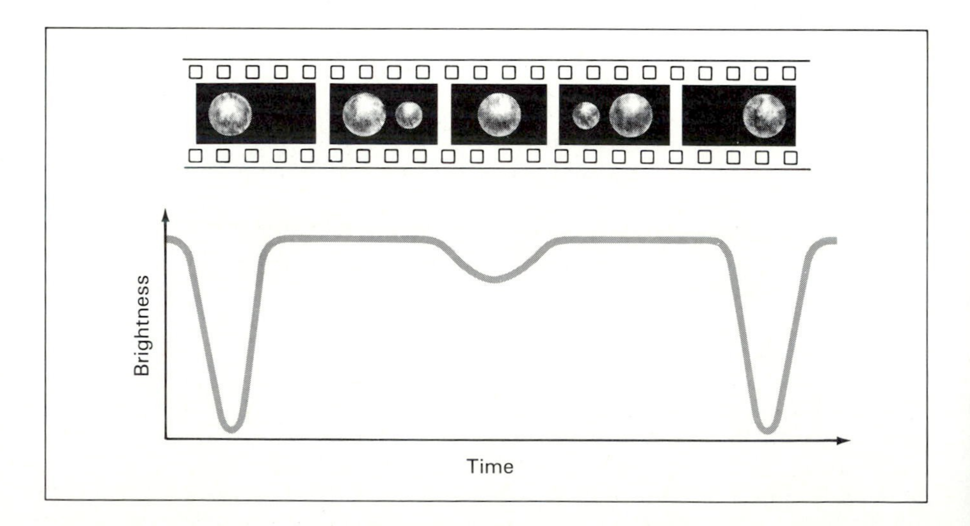

FIGURE 21.8 Light curve of a typical binary-star system. [See also Figure 11.18(c).]

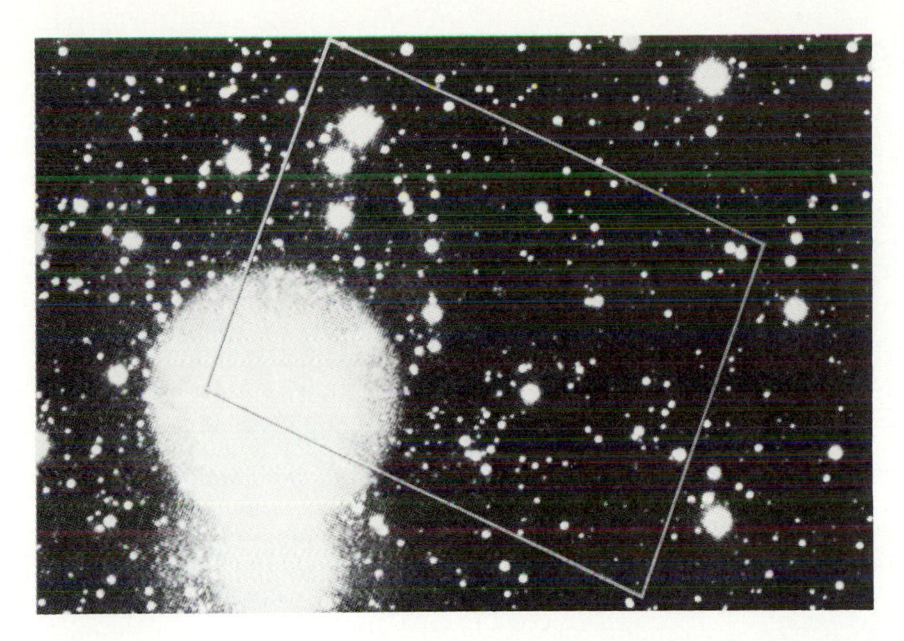

FIGURE 21.9 The brightest star in this photograph is a member of a binary system whose unseen companion, called Cygnus X-1, is thought to be a good candidate for a black hole. (The rectangle outlines the field of view illustrated in the next figure.)

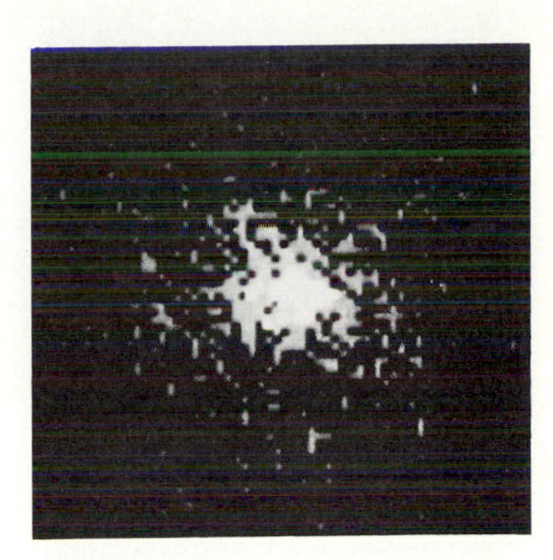

FIGURE 21.10 Invisible x-rays emitted near the Cygnus X-1 source can be analyzed by changing the detected x-rays into electronic signals, which can then be viewed on a video screen from which this picture is taken. (The field of view here is outlined by the rectangle in the previous figure.)

unseen companion, called Cygnus X-1.

3. This invisible object has a mass between 5 and 10 solar masses, as derived from knowledge of the binary system's orbital size and period.

4. X-ray radiation, emitted from the immediate neighborhood of Cygnus X-1, suggests the presence of scalding gas, perhaps as hot as a billion kelvin (see Figure 21.10).

5. Rapid time variations of this x-ray radiation imply that the size of Cygnus X-1 itself must be less than 100 kilometers across.

These general properties suggest that the invisible x-ray-emitting companion might be a black hole. But since invisible black holes cannot be seen or imaged in any way, they can't really be illustrated either; even so, Figure 21.11 is an artist's conception of this intriguing Cygnus X-1 region. As shown, most of the gas drawn from the visible star (which is probably evolving toward the red giant stage) ends up in a Life-Saver-shaped disk of matter. Some of this gas inevitably streams toward the black hole, becoming superheated and x-ray emitting just before being trapped below the event horizon.

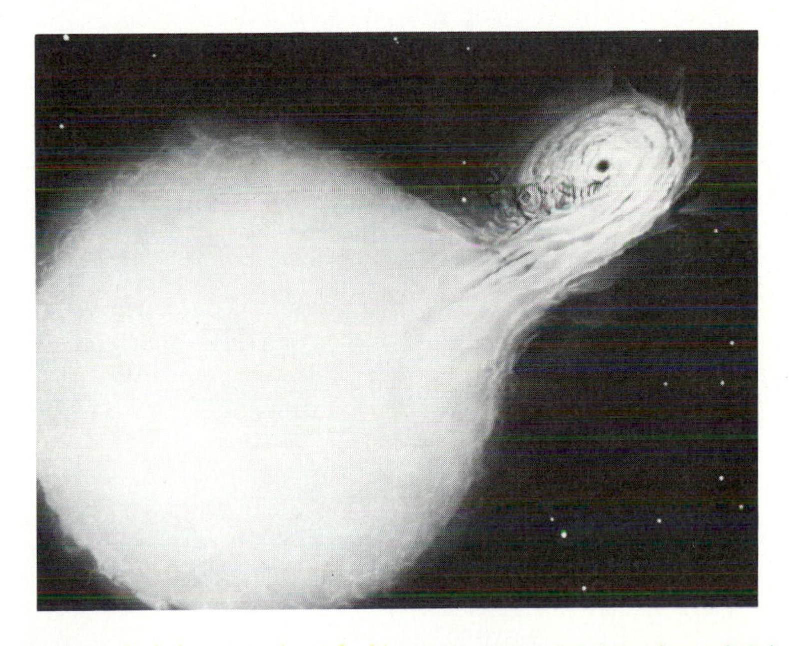

FIGURE 21.11 Artist's conception of a binary system containing a large, bright, visible star, and an invisible, x-ray emitting black hole. This particular drawing depicts the Cygnus X-1 region discussed in the text.

A few other similar candidates for black holes have been proposed in recent years. For example, the third x-ray source discovered in the Large Magellanic Cloud—called LMC X-3—is such an invisible object. And like Cygnus X-1, it is suspected to have nearly 10 solar masses. In this case, LMC X-3's visible companion star seems to be distorted into the shape of an egg, apparently the result of the intense gravitational attraction of the suspected black hole.

However, a potential problem plagues these interpretations. Cygnus X-1, as well as a few other sus-

INTERLUDE 21-2 *Gravity Waves and Lenses*

Electromagnetic waves are common everyday phenomena. Introduced in Chapter 1 and studied throughout the book, electromagnetic waves are well-known cyclical disturbances that move through space, transporting energy from one place to another. Whether they are radio, infrared, visible, ultraviolet, x-ray, or gamma-ray radiation, electromagnetic waves are essentially caused by changes in the strengths of electromagnetic force fields. Oscillations of charged electrons within a broadcasting antenna, for example, generate electromagnetic waves. Measurements of these waves agree perfectly with the theory of electromagnetism that predicts them.

The modern theory of gravity—Einstein's theory of relativity—also predicts cyclical disturbances that move through space. A **gravity wave** is the gravitational analog of an electromagnetic wave. Gravity waves, or gravitational radiation, result from changes in the strengths of gravitational force fields. In principle, every time an object of any mass accelerates, gravity waves should be emitted at the velocity of light. But in practice, no one has ever detected such gravity waves.

Although gravity waves are predicted by relativity theory, their existence is still not guaranteed. The formal mathematics of theoretical physics predict many things that disagree with physical reality. For decades, many scientists regarded the ideas of black holes and neutron stars as absurd. But these and other exotic phenomena are now gaining acceptance, since experimental evidence is hard to deny.

Gravity waves will not be easy to capture, even if we knew how to do it. Part of the trouble arises because theorists are still arguing about what kinds of astronomical objects should produce intense gravity waves. Leading candidates include (1) the merger of a binary-star system, (2) the collapse of a star into a black hole, and (3) the collision of two black holes. Each of these possibilities involves the accelerations of huge masses. In this way, the strength of the gravitational force field should change drastically. Other astronomical objects are also expected to emit gravity waves, but only changes in large amounts of mass will produce gravity waves of great intensity.

Of these three candidates, the first one probably presents the best chance to detect gravity waves, at least for the present. Binary-star systems sometime experience orbital decay, as the two stars slowly spiral toward one another. Their combined gravitational force

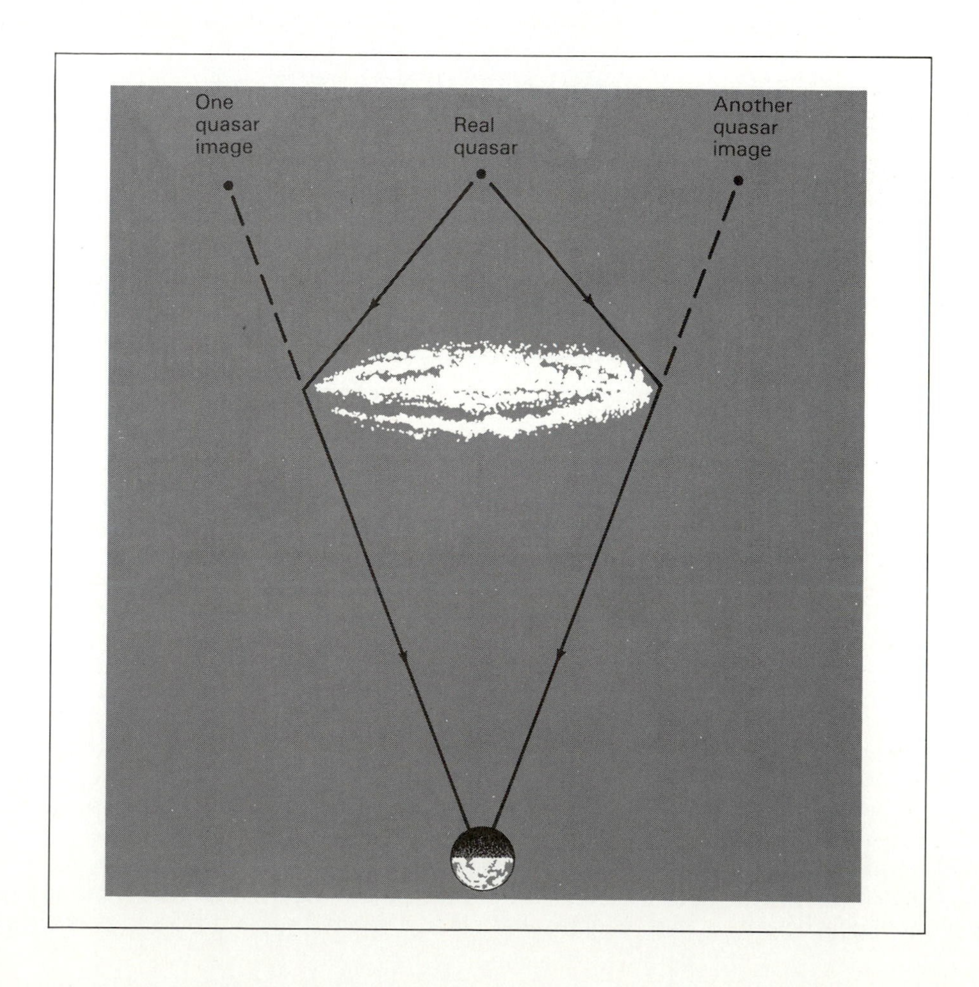

field thus changes, and the result should be the emission of gravitational radiation.

Interestingly enough, a slow but steady decay in the orbit of a recently discovered binary-star system is now being monitored. This galactic binary system is unique since it contains a neutron star that radiates bursts, just like any other pulsar. This system is also unique since the periodic Doppler shift of the pulsar's radiation proves that the orbit is slowly shrinking in size. As the two objects in the binary system gradually near one another, their orbital energy slowly changes into gravitational energy. This energy should theoretically be radiated away as gravity waves. However, no such waves have yet been detected.

We need to develop experimental tools to detect gravity waves from a variety of astronomical objects. Radiation is energy, and energy is information. As such, gravity waves should contain a great deal of information about the physical events in some of the most exotic regions of space. The discovery of gravity waves could herald a new age in astronomy, in much the same way that invisible electromagnetic waves, unknown a century ago, revolutionized classical astronomy and led to modern astrophysics.

Regarding a different subject, another bizarre prediction of relativity theory might well have been recently confirmed. Astronomers seem to have found several cases of what is called a "gravitational lens." Adjacent images of what at first sight appear to be identically twin quasars might be the result of a single quasar's radiation being gravitationally bent by some object (probably a galaxy) between us and the quasar. (Even the spectra of the two images are identical.) As sketched below, this effect would mimic the way that a glass lens can refract light and produce multiple images.

Gravitational lensing follows directly from Einstein's theory, which predicts that light, or any type of radiation, bends while passing through warped spacetime near a massive object. Recall from our discussion of the gravitational red shift near the end of Chapter 15 that photons possess an equivalent or effective mass equal to E/c^2, where E symbolizes the photon's energy and c is the velocity of light.

Thus radiation should be deflected ever so slightly while grazing a sufficiently massive object. In this way, as the sketch illustrates (in a greatly exaggerated manner), a distant quasar's radiation could be bent while passing through an intervening galaxy. It is this galaxy that acts as a gravitational "lens"; and if the intervening galaxy has a strong gravitational force field, multiple images of a single celestial object can result.

pected black holes in binary systems like it, have masses close to the neutron star–black hole dividing line. When the effects of rotation and magnetism are someday fully included in the theory, there's a chance that the dark objects in question could turn out to be more "ordinary." In short, they might be merely dim and dense neutron stars, not black holes at all.

OTHER BLACK HOLE CANDIDATES: Many researchers argue that other kinds of regions show better evidence for black hole candidates than some of the binary-star systems just described. Current exploration of the center of our Milky Way Galaxy is especially interesting in this regard. Although the central parts of our Galaxy are totally obscured by interstellar dust and thus cannot be studied with optical telescopes, radio and infrared observations of the innermost few hundred light-years have yielded spectacularly unexpected results. Data obtained around 1980 imply the presence of rapidly rotating hot gas, suggestive of a colossal whirlpool at the very center of our Galaxy. (A review of the last section of Chapter 13 would now be appropriate, especially Figure 13.16.)

The curves of Figure 21.12 depict two different theoretical models for the motion of the matter within 300 light-years of our galactic center. The solid line is computed assuming that the behavior of the obscured matter there is a mere extrapolation of the well-studied matter in the outer parts of the Galaxy near the Sun. As denoted by the solid line, the rotation velocity of the stars, gas, and dust is predicted to decrease toward the galactic center. The innermost matter is expected to be spinning, although sluggishly.

However, radio and infrared observations strongly imply that the gas in the core region is not spinning slowly. Instead, as illustrated by the data points in Figure 21.12, the observations show a dramatic rise in rotation velocity toward the center. Apparently, the very heart of the Milky Way is spinning furiously; the closer we probe toward the very center, the faster the matter there swirls, much like a whirlpool of water approaching a drain.

This discovery was quite a surprise to observers and theorists alike. It's surprising because of a simple, yet puzzling problem: How does such a vast galactic whirlpool maintain its structural integrity? After all, regions of rapidly rotating matter produce strong outward forces tending to push the gas away, much like those casting mud from the edge of a spinning bicycle wheel. Unless some other force pulls back on the galactic-center whirlpool, its gas should be flung into the outer parts of the Galaxy. In other words, the problem is this: How can such a huge vortex of matter stay intact without breaking apart? After eliminating a long shopping list of possibilities, very strong gravity seems to be the only viable way to hold together such a large quantity of matter.

Figure 21.13 depicts a simplified

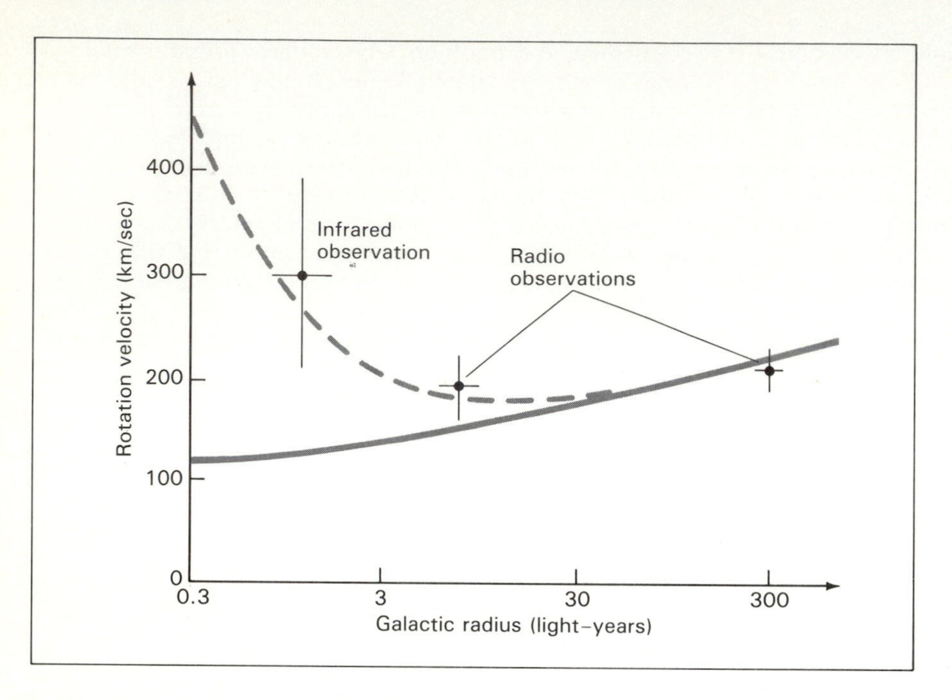

FIGURE 21.12 A plot of the rotation velocity of the stars, gas, and dust near the center of our Galaxy. The solid line is a theoretical prediction based on an extrapolation of the motions of matter near the Sun. The dashed curve is a model of the normally expected gas motions plus the presence of a whirlpool of loose gas held together by a compact, central object having a mass of about 5 million solar masses.

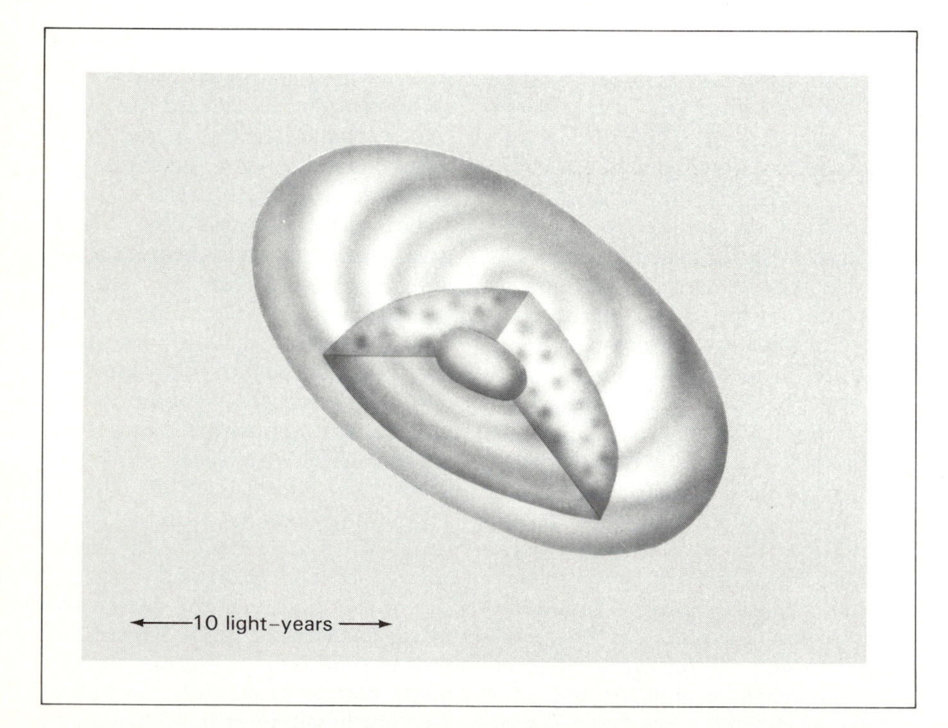

FIGURE 21.13 Artist's conception of a simplified model that can explain many of the radio and infrared observations of the central regions of our Galaxy. Even on this scale—the whirlpool spans about 300 square light-years—the suspected supermassive black hole embedded in the core would be no larger than the period at the end of this sentence.

model that can possibly account for most of the observations. As shown, a region of hot, thin ionized gas (much like that in ordinary gaseous nebulae) surrounds a much smaller core of hotter, denser gas. This inner core, around which the gas we can observe swirls, is thought to contain—and this is the punch line—a tremendously compact object having a mass of approximately 5 million suns all packed into a region no larger than our Solar System. Why so much mass? Because that's the amount needed for gravity to keep the whirlpool of gas from breaking up and dispersing. This model permits (in fact requires according to Kepler's Laws) an increase in gas velocity toward the galactic center, as shown by the dashed curve in Figure 21.12. As can be seen, the model agrees quite well with the observed data points.

Although the details are controversial at this time, astronomers have reached a consensus that a supermassive, ultracompact "something" resides at the very center of our Milky Way Galaxy. And since its mass is much too large to be mistaken for a neutron star, nor can it easily be an anomalously rich star cluster, the hub of our Galaxy would seem to be a huge black hole in space.

HOLES IN OTHER GALAXIES: Recent observations suggest that supermassive objects also lurk in or near the central regions of a few other nearby galaxies. The evidence here is much the same as for our own Galaxy, with gas and stars in the innermost regions of several active galaxies observed to be rapidly whirling. For example, observations of M87 (see Figure 15.6) imply an extremely compact object of even greater mass than in the case of our normal Galaxy. In fact, several *billion* solar masses are inferred within its central few light-years. Perhaps these core whirlpools are remnants of the turbulent eddies that helped form the galaxies in an earlier epoch, as discussed in Chapter 18.

We may find that the center of every galaxy is inhabited by a supermassive black hole. Normal galaxies such as our own probably have relatively small black holes of "only" millions of solar masses. More active galaxies such as M87 might have larger black holes, perhaps on the order of billions of solar masses. The great energetics and explosiveness of the active galaxies might naturally arise from matter falling into the clutches of such supermassive black holes. After all, in the process of consuming matter, black holes accelerate and heat the gravitationally infalling gas, causing it to radiate great amounts of energy before entering oblivion.

It's even possible that the most energetic objects in the Universe—the titanic quasars—could be powered by hypermassive black holes that regularly swallow whole stars. In fact, the energy released by a typical quasar is equivalent to the complete conversion into energy of an amount of matter equal to that in our Sun *every year*. Although we currently have little observational evidence to support it directly, this idea is a mere extension of the foregoing arguments. If true, then black holes in quasars would presumably be even more massive than the billion-solar-mass objects implied for the active galaxies.

Clearly, as noted several times earlier, an understanding of the powerhouse galaxies lies buried deep within their hearts. We can only await future explorers to discover, unravel, and share their secrets with us.

SUMMARY

Unless atronomers can find direct, or very convincing indirect, evidence for the existence of black holes, neither of which we currently have, then the whole concept of black holes may well turn out to be a figment of the human imagination. The nature of matter, energy, space, and time deep down inside event horizons might be no more significant than a challenging and amusing academic problem devoid of reality.

On the other hand, the Universe did emerge from what would seem to have been a bare singularity some 15 billion years ago. Black hole singularities might be the keys needed to unlock an understanding of the creation state from which the Universe arose. By theoretically studying the nature of black holes, especially by experimentally seeking their existence and properties, we may someday be in a better position to appreciate *the* most fundamental issue of all—the origin of the Universe itself.

KEY TERMS

black hole

event horizon

gravity wave

QUESTIONS

1. Describe what is meant by the term "black hole" using the concept of escape velocity.
2. How might black holes form?

*3. Identify and explain the stage of stellar evolution associated with each of the following observations: (a) optical emission from a bright, hot star and its surrounding three-dimensional ring of glowing gas; (b) radio emission from a dark, cool blob embedded within an interstellar cloud of gas and dust; (c) x-ray emission from a region near an ordinary star not normally hot enough to emit x-rays on its own accord; (d) bursts of radio and optical emission from a compact region at the center of an expanding diffuse cloud of gas.

4. Name and discuss the important physical effects, not yet studied in great detail by astronomers, which could alter the 3-solar-mass dividing line between neutron stars and black holes.
5. Discuss the prospects for an object "disappearing from the Universe."
6. Explain how, if black holes cannot emit anything, they can be considered prodigious sources of energy.
7. Explain the gravitational red shift. Why is it expected to alter radiation so much near a black hole?
8. Speculate about the physical conditions deep down inside a black hole.
9. Discuss some of the best candidates for black holes.
10. Assess the observational evidence to date for black holes.

FOR FURTHER READING

Calder, N., *Einstein's Universe.* New York: Viking, 1979.

Callahan, J., "The Curvature of Space in a Finite Universe." *Scientific American*, August 1976.

*Cameron, A., "Endpoints of Stellar Evolution," in *Frontiers of Astrophysics*, E. Avrett (ed.). Cambridge, Mass.: Harvard University Press, 1976.

DeWitt, B., "Quantum Gravity." *Scientific American*, December 1983.

Hawking, S., "The Quantum Mechanics of Black Holes." *Scientific American*, January 1977.

Kaufmann, W., *Black Holes and Warped Spacetime.* San Francisco: W. H. Freeman, 1979.

Shipman, H., *Black Holes, Quasars, and the Universe.* Boston: Houghton Mifflin, 1980.

22 ELEMENTAL EVOLUTION: *Origin of Heavy Matter*

In this chapter we study the specific mechanisms that create the elements during the evolution of stars. Nuclei of elements from lightweight helium to heavyweight iron are slowly cooked within the cores of massive stars. Only later, when near death, do stars produce elements more complex than iron. Many of the heaviest elements are ejected during the death throes of supernovae, thereby enriching interstellar space from which other stars, rocky planets, and intelligent life can later arise.

The subject of elemental evolution is complex because we need to apply nuclear physics to astronomy. Generally, the nuclear reactions play two major roles. They are responsible first for the production of energy in stars, and second for the synthesis of heavy elements within those same stars.

Fortunately, we've already had an introduction to this exotic subject while studying the problem of solar energy generation in Chapter 10. But the nuclear physics covered here involves more detail. We begin by reviewing some background material, after which we shall gradually increase the complexity of this most important astronomical process. Indeed, the origin of the elements is a crucial part of our study of cosmic evolution.

The learning goals for this chapter are:

- to review the cosmic abundances of all the elements
- to recognize that of all the known elements, only hydrogen and helium were created in the earliest epochs of the Universe
- to understand the specific nuclear steps needed to synthesize the heavy elements
- to appreciate some of the observations that tend to confirm the production of heavy elements in massive stars

Progression of Ideas

All things around us—including tables, chairs, books, pencils, air, people—are rather complex when analyzed chemically. The Greeks regarded these various types of matter as mixtures of simple, familiar structures. Combinations of their four basic "elements"—air, earth, fire, and water—were used to describe the nature of everything during much of recorded history. Later, the Greeks imagined microscopic entities called "atoms," which they regarded as invisible, indivisible, indestructible, and uncreatable substances. This was a step in the right direction.

A step in the wrong direction occurred with the return, during the Dark Ages, to the terrestrially familiar building blocks, including water, dirt, air, oil, sulfur, salt, and so on. Medieval charlatans known as alchemists bamboozled the public by claiming they could change abundant metals such as iron and tin into rare elements such as gold and silver.

Not until the nineteenth century was the concept of atoms resurrected to explain the basic building blocks of matter. Whether this matter was solid, liquid, or gas, these atoms were considered changeless hardly more than 60 years ago.

Review Figure 2.10, which illustrates the periodic table of all the known kinds of atoms or chemical elements. Recall that they currently number more than 100. Arranged in order of increasing complexity, the number at the upper left corner of each box equals the number of protons, while that to the upper right corner equals the sum of protons and neutrons within the nucleus of each element. Most of these elements were discovered toward the end of the eighteenth century and throughout the nineteenth century.

Well into the twentieth century, scientists recognized that these elements are not changeless. They discovered that nuclei can transform from one kind to another. Such changes or **nuclear transformations** require interactions among elements on a scale even smaller than that of atoms; they involve nuclear reactions among elementary particles within atoms, rather than chemical reactions among atoms themselves. To a certain extent, modern nuclear physics has verified the dreams of the medieval alchemists. Nuclear laboratories can now change uranium into gold by smashing uranium nuclei with very fast moving subatomic particles. The cost of operating the technology to do so, however, is much greater than the value of the minute amount of gold produced.

The first experimental evidence that elements (really nuclei) can change from one type to another occurred some half-century ago during the basic research that led ultimately to the development of the atomic bomb. Such bombs use uncontrolled fission reactions that break apart heavy nuclei into lighter nuclei, as explained in Chapter 10. Similar, though controlled thermonuclear fission reactions in nuclear reactors now power some modern factories, naval submarines, and electrical utilities. More important, the opposite process—nuclear fusion, that is, the production of heavier nuclei from lighter ones like that occurring in ordinary stars or hydrogen bombs—might eventually provide lots of cheap energy needed to power our technological civilization. Indeed, the fusion process studied in this chapter is likely to be increasingly important as Earth's fossil fuels become depleted in the years ahead; consult Interlude 10-1.

A Brief Inventory

Let's review at the outset the terminology used here so that no one gets lost in the ensuing complexity. In particular, we need to clearly distinguish atoms from ions from isotopes from nuclei from elementary particles.

TYPES OF MATTER: Recall that any atomic nucleus is extremely small compared to the size of the electron orbitals about it. Virtually all the mass of an atom is concentrated within its central nucleus. Electrons are negatively charged and protons positively charged so that ordinary neutral atoms have equal numbers of electrons and protons. While in hydrogen, one electron orbits about one proton, heavier elements quickly become more complex with neutrons coexisting in the nucleus alongside the usual protons. When an atom has a net charge because the number of electrons and protons is unequal, then most researchers refer to this charged atom as an ion.

Atoms of different kinds—that is, elements—are distinguished by their proton number or by their electron number. In all, we know of some 104 different elements. These vary from the simplest element, hydrogen (1 proton), to the most complex, kirchotovium (104 protons). (Unnamed elements having 105, 106, 107, and 108 protons have been proposed by one group of researchers, but they have not yet been confirmed by others.) Each element has a fixed number of protons and neutrons within its nucleus, as well as its usual complement of orbiting electrons. These are normal atoms.

Sometimes, however, the number of neutrons in a given element differs slightly from its usual value. These are abnormal atoms termed "isotopes" in Chapter 10. Having a few more or a few less neutrons than normal, isotopes play important roles in the creation of the heavy elements. Many isotopes are radioactively unstable, and hence decay into more stable isotopes or ordinary atoms after certain amounts of time.

Table 22-1 lists a brief inventory of the currently known atomic building blocks of nature. The 81 stable elements comprise the bulk of the list. All 81 elements are found in Earth's air, land, and sea.

In addition, 10 radioactive elements naturally reside on our planet. These elements are termed "natural" because they are created in nature (actually in stars). Like all radioactive elements, they have certain half-lives, after which half of the total amount of that element decays into something else. Even though their half-lives are very long (millions and billions of years typically), the steady decay or disappearance of these natural radioactive elements explains their scarcity on Earth, in meteorites, and in lunar samples. They are not observed in stars, for there are too few of them to produce detectable spectral lines.

Besides these 10 naturally occurring radioactive elements, some 11 more radioactive elements can be artificially produced under special conditions in nuclear laboratories on Earth. The debris collected after nuclear weapons tests also show evidence for these artificial radioactive elements. Unlike the naturally occurring radioactive elements, these artificial (human-made) ones decay

TABLE 22-1
Atomic Building Blocks of Nature

Elements	
Stable	
Terrestrially found	81
Promethium (made in lab only)	1
Radioactive	
Naturally occurring	10
Technetium (observed in stars only)	1
Artificially (lab) made	11
	104
Isotopes	
Stable	
Terrestrially found	280
Radioactive	
Naturally occurring	~70
Artificially (lab) made	~1500
	~1850[a]

[a] This number increases each year as nuclear physicists discover new short-lived radioactive isotopes.

into other elements quickly (much less than a million years). Consequently, we should not be surprised that they are extremely rare or virtually nonexistent in nature, except presumably in the hearts of stars where they are created.

Rounding out our table of elements, there is one other—promethium—formed only as a by-product of nuclear laboratory experiments. And still another unstable element—technetium—is created only in stars. Neither of these is found naturally on Earth.

There are many more isotopes than elements. The element carbon, for example, is known to have 10 isotopes, 8 of them radioactive. In all, we know of nearly 300 stable isotopes, approximately 70 naturally occurring radioactive isotopes, and more than 1500 artificially produced radioactive isotopes. The total number of isotopes is close to 2000 and growing larger each year as nuclear researchers discover more rare and exotic isotopes of the known elements.

Are these approximately 2000 elements and isotopes the absolute basic building blocks of matter? The answer is "no," of course. We know that atoms are made of even more basic entities—electrons, protons, and neutrons. And as noted in Interlude 2-1, these in turn are probably composed of smaller, even more basic particles. Physicists are currently probing the composition of protons and neutrons by smashing them together in the most powerful accelerators in the world. Although the resultant debris should tell us something about the internal makeup of these subatomic particles, the results of these experiments are not yet clear. At any rate, to understand the origin of the elements, we need not worry about the detailed structure of matter below the level of the proton, neutron, and electron.

ABUNDANCE OF MATTER: The central question addressed in this chapter is the following: How did the complexity now surrounding us arise from the simplicity that must have prevailed immediately after the start of the Universe?

A key ingredient in our understanding of heavy-element creation is the fact that experiments in nuclear laboratories have proven that larger elements can be built from smaller ones. This is usually accomplished by fusing together two or more relatively light nuclei in a violent collision. Among the collision's debris of energy, elementary particles and light nuclei, are some heavier nuclei as well. Hence we can naturally theorize that all the heavy elements have been built up from the lighter elements. In this scheme, the ultimate source of the heavy elements is the lightest and simplest one—hydrogen.

To test this idea, we need to consider not only the list of different kinds of elemental atoms and isotopes (Table 22-1), but also the observed abundances of these elements. The cosmic abundance scale, shown earlier in Figure 2.20, is reproduced and extended here as Figure 22.1. Recall that this curve is derived largely from spectroscopic studies of many stars including the Sun. Table 22-2 is another way of summarizing the essence of Figure 22.1. (The various isotopes of all the elements are included in both Figure 22.1 and Table 22-2.) Any theory proposed for the creation of the elements must match these observed abundances, the most obvious feature being that the heavy elements

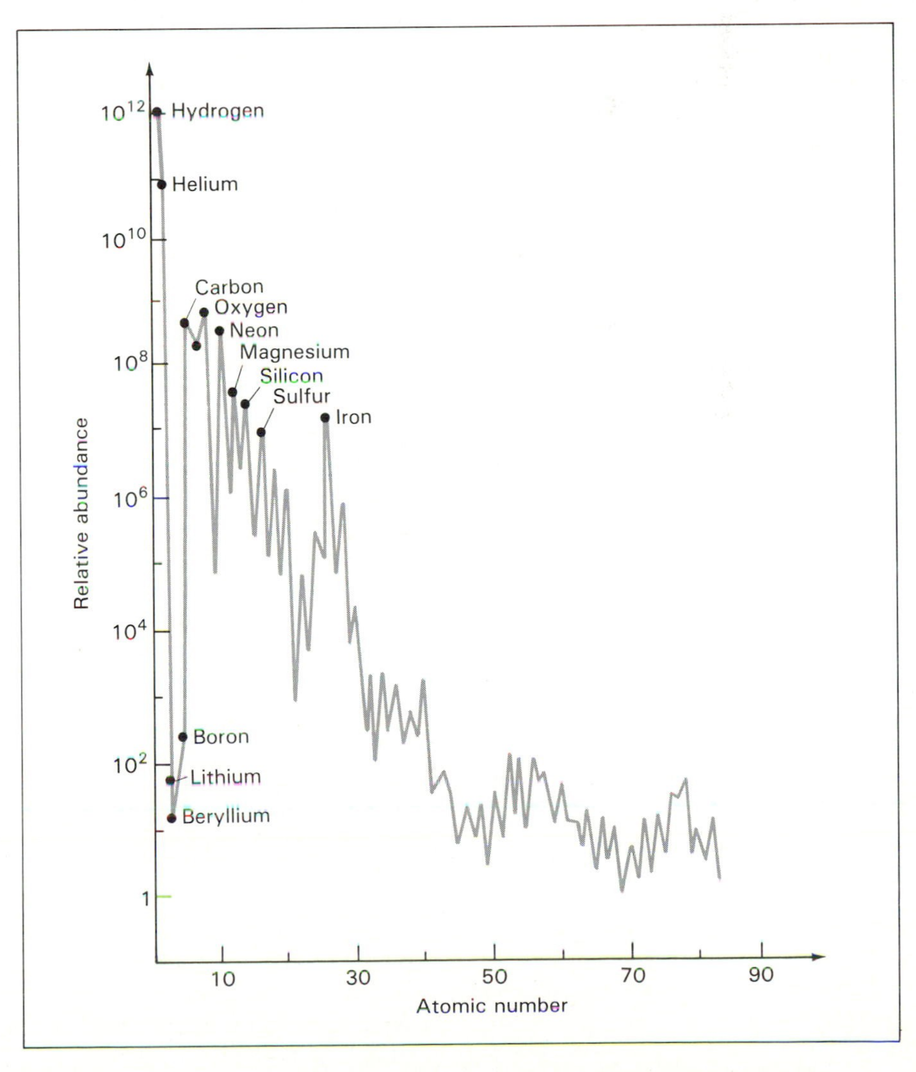

FIGURE 22.1 A summary of the cosmic abundances of the elements and their isotopes.

TABLE 22-2
Cosmic Abundances of the Elements

ELEMENTAL GROUP	PERCENT ABUNDANCE BY NUMBER[a]
Hydrogen (nuclear particle = 1)	90
Helium (nuclear particles = 4)	9
Lithium group (7 ≤ nuclear particles ≤ 11)	0.000001
Carbon group (12 ≤ nuclear particles ≤ 20)	0.2
Silicon group (23 ≤ nuclear particles ≤ 48)	0.01
Iron group (50 ≤ nuclear particles ≤ 62)	0.001
Middle-weight group (63 ≤ nuclear particles ≤ 100)	0.00000001
Heaviest-weight group (100 < nuclear particles)	0.000000001

[a] The total does not equal 100% because the helium abundance is uncertain.

are much less abundant than most light elements.

Approximately 10 hydrogen atoms exist for every helium atom. Together, these two lightest atoms comprise roughly 99 percent of all matter in the observable Universe. Next comes a group of slightly heavier elements, namely those having from 7 to 11 particles in their nuclei: lithium, beryllium, and boron. Even summed together, this so-called lithium group has an extraordinarily small abundance. Slightly heavier elements, such as carbon, nitrogen, oxygen, and neon, are more abundant than this lithium group, though still much less abundant than the lightest hydrogen and helium elements. The same can be said for the silicon and iron groups listed in Table 22-2. The middle-weight and heaviest-weight groups—those with 63 or more nuclear particles—have progressively lesser abundances.

These relative abundances show an interesting peculiarity. Despite the overwhelmingly large amount of hydrogen and helium throughout the Universe, not much of either of these elements exists on Earth. That's because the Terrestrial Planets do not typify the average matter in the Universe. For the most part, the various abundances of rocky planets do not mimic those of stars or galaxies. Earth's small gravity and large solar heating combine to make our planet rather inhospitable for hydrogen and helium. These light gases are unable to stick around. Instead, most of the elements surrounding us are of intermediate mass—lots of silicon in rocks, nitrogen in air, and carbon in people. Planets resemble abnormal-abundance islands in a giant sea of normal cosmic abundances. This is especially true for the rocky Terrestrial Planets, less so for the gassy Jovian Planets. The bigger and more distant planets of our Solar System have stronger gravity and less solar heating than the smaller, interior planets, and thus have elemental abundances more like those listed in Table 22-2.

Primordial Nucleosynthesis

We can attempt to reconcile the observed cosmic abundances in two ways. We can theorize that all the elements were created shortly after the explosion that started the Universe, in which case the entire process is called **primordial nucleosynthesis**. Or, we can argue that all the heavies are produced exclusively within stars, in which case the process is called **stellar nucleosynthesis**. Let's consider each of these possibilities in turn.

EARLY UNIVERSE: Imagine the physical conditions just after the start of the Universe. As described in Chapter 17, the temperature of matter and the density of radiation were very high. Neutrons were abundant, as were many other elementary particles, all racing out from the explosion. With this backdrop in mind, consider the following scenario.

A free neutron existing alone in space (and not part of a neutron star) cannot remain that way for long. In about 11 minutes, it decays into an electron and a proton:

neutron → electron + proton + neutrino.

Excepting the neutrino particle, note that the products comprise hydrogen; although the early Universe was too hot for the electron and proton to couple together and make an atom, it is interesting to realize that hydrogen nuclei first came into being in this way. Given the early-Universe conditions, this proton would have then combined with another free neutron (before it decayed) in order to produce a deuteron particle:

proton + neutron → deuteron.

This deuteron particle is nothing more than the nucleus of deuterium, the heavier, isotopic form of hydrogen. Deuterium is an isotope because it contains one extra neutron than normal hydrogen, but no additional protons. This deuteron particle can in turn combine with another free neutron to form a triton particle:

deuteron + neutron → triton.

This product is the nucleus of the heaviest form of hydrogen, another isotope called tritium. Like the free neutron, this triton particle is unstable and hence decays into a special form of helium:

triton → ^{3}helium.

This isotopic nucleus of helium can then interact with yet another free neutron, this becoming a more stable form of helium:

^{3}helium + neutron → ^{4}helium.

In addition to repeated neutron capture of this sort, protons can also be captured to yield both forms of helium. Such "captures" of neutrons and protons would have almost certainly occurred in the early epochs of an evolutionary Universe. These reactions are thus feasible ways of synthesizing hydrogen and helium nuclei. Eventual capture of electrons by these nuclei would have produced the lightest elements.

INTERLUDE 22-1 ***Helium and Deuterium as Speedometers***

Astronomers are unsure how much helium was produced in the early Universe. There remains the nagging difficulty mentioned earlier regarding the slow decay of triton, a necessary step in the formation of ^{4}helium. Doubtless some ^{4}helium was created in the hot and dense primordial fireball, for all the helium now observed could not have been produced in stars. There's just too much of it around.

Part of the problem lies in the uncertainty of the rate at which the Universe expanded during the earliest epochs. Put another way, we are unsure how fast the early Universe cooled and thinned. Time is of the essence here, for the slower the expansion, the greater the chance that elementary particles would have interacted with one another to synthesize helium. Most cosmologists suggest that there was probably enough time to create one ^{4}helium nucleus for every 10 to 20 ^{1}hydrogen nuclei. This compares favorably with the observed value of about 1 for every 12.

If there were no other way to produce helium, astronomers could conceivably use its observed abundance to turn the problem around and thereby infer the rate of universal expansion. In this way, the cosmic helium abundance might be able to tell us something about the early Universe. But the problem is not that simple. ^{4}Helium nuclei are also produced in the hot and dense cores of stars. And it's virtually impossible to unravel the relative contributions made to helium production by either primordial or stellar nucleosynthesis.

Deuterium, a heavy isotope of hydrogen, is also created in the early Universe, as we have seen elsewhere in this chapter. And, unlike helium, deuterium is not likely to be produced in stars. Therefore, our models of primordial nucleosynthesis should enable us to predict the amount of deuterium produced in the aftermath of the bang; on average, these models suggest one deuterium atom for each 100,000 hydrogen atoms. Slightly more deuterium could have been synthesized if the Universe expanded just a bit faster than a certain rate, and slightly less deuterium if the Universe expanded just a little slower. In effect, the formation of deuterium depends upon the density of matter.

Observations of deuterium should then provide a test of the universal expansion rate. In turn, this should tell us if the Universe is open or closed, for if the rate is fast, the galaxies should recede forever, whereas if the rate is slow, the Universe will eventually contract. In recent years, some limited observations of deuterium have been made, especially by orbiting satellites that can capture the deuterium's strongest spectral feature which is emitted in the ultraviolet part of the spectrum. Although the interpretation of the observations both for interstellar matter and for our Solar System is somewhat uncertain, the results at face value imply that the Universe is open.

PROBLEMS: Several difficulties arise when we try to use these kinds of nuclear events to understand the creation of the heavy elements. One problem is that the decay of the triton particle into the ^{3}helium nucleus takes roughly 12 years. That's a short time in the cosmic scheme of things, but in the earliest epochs of the Universe, the physical conditions must have changed rapidly. The Universe expanded greatly during the first decade after the bang, severely diluting the density of neutrons. Accordingly, the last reaction noted above, namely neutron capture by isotopic ^{3}helium to form ordinary ^{4}helium, is improbable. This nuclear event would have occurred all right, although only in a few places and rather infrequently at that.

Another problem with the scenario above is that, among all those elements found in nature, elements of mass 5 and 8 are missing. We know of no stable atoms or isotopes having those mass values. If the neutron capture scheme outlined here is viable, it would be difficult to create elements much more complex than helium if elements of mass 5 and 8 had to be skipped.

There is a third objection to primordial nucleosynthesis. If the heavies were created by repeated neutron capture in the early Universe, all stars should have identical elemental abundances. Reality shows otherwise, however, for we observe different elemental abundances in different objects. *On average,* all stars have pretty much the same cosmic abundances as those listed in Table 22-2, but there are small, yet unmistakable elemental differences among all the stars. This is especially true when comparing the young galactic cluster stars that are rich in heavy elements with the old globular cluster stars that are lacking in heavy elements.

A fourth problem concerns the temperature and density of the early Universe. As the heavy elements are produced, the electromagnetic charge of their nuclei would inevitably increase. Progressively greater force would then be needed to fuse the heavies into even heavier nuclei. Higher temperatures and greater densities would normally be able to do it, but the early Universe was rapidly cooling and thinning, as depicted in Table 17-1. This is precisely the opposite change as that needed to create heavy elements. Thus, the process of primordial nucleosynthesis, which requires increasing temperatures and densities, is inconsistent with the trend of an expanding Universe where these quantities are decreasing.

So, many problems plague the heavy-element scenario of primordial nucleosynthesis. This process would have had time to form some helium, but not enough time to form any of the heavier elements. Events at the start of the Universe happened too rapidly.

Stellar Nucleosynthesis

Astronomers now recognize that better places for the creation of heavy elements are stars themselves. The interiors of stars are regions where both temperature and density are high. In the midst of a Universe that is cooling and thinning, stellar temperatures and densities remain high, and even increase as stellar cores evolve. In short, stars are pockets of increasing temperature and density within a Universe of decreasing temperature and density.

The production of heavies in stars—stellar nucleosynthesis—seems more reasonable not only because primordial nucleosynthesis is riddled with problems, but also because of much favorable observational evidence. For example, the ages of stars and their evolutionary paths suggest that heavies are now being created deep inside stars. Spectroscopy confirms this idea. Also, differences in heavy-element abundances between the older globular cluster stars and the younger galactic cluster stars clearly imply that the heavies are slowly produced over time. Furthermore, stars are the only places known where temperatures and densities remain high over long durations, allowing the heavies to be processed steadily long after the start of the Universe.

From this point on in our chapter, we study the specific reactions that lead to heavy-element production by means of the usual stellar burning cycle of contracting, heating, expanding, cooling, contracting, and so on. As noted in Chapter 21, the heavy elements are then synthesized during each successive period of stellar contraction.

HYDROGEN BURNING: Stellar nucleosynthesis begins with the proton–proton cycle studied in Chapter 10. Provided that the temperature and density conditions are suitable—at least 10 million kelvin and 100 grams/cubic centimeter—the following nuclear reaction proceeds automatically:

$$^{1}\text{hydrogen} + {}^{1}\text{hydrogen} \rightarrow \text{deuteron} + \text{positron} + \text{neutrino} + \text{energy}.$$

(Remember, these are *nuclear* reactions where the symbols refer to nuclei, not atoms. ^{1}Hydrogen, for example, represents a proton. Once the nuclei are created, they can relatively easily capture the appropriate number of electrons in order to form different kinds of atoms.) Recall from Chapter 10 that the positron particle produced here immediately interacts with a nearby free electron, thereby producing high-energy radiation via matter–antimatter annihilation. The neutrino particle rapidly escapes, carrying with it some energy, but playing no direct role in stellar nucleosynthesis. Only the deuteron (a proton–neutron blend) sticks around to participate in further reactions.

In stars, ordinary helium is made by means of a two-step process. First, an isotope of helium is formed in the nuclear reaction

$$\text{deuteron} + {}^{1}\text{hydrogen} \rightarrow {}^{3}\text{helium} + \text{energy},$$

after which further interaction of these isotopic nuclei produce the normal form of helium:

$$^{3}\text{helium} + {}^{3}\text{helium} \rightarrow {}^{4}\text{helium} + {}^{1}\text{hydrogen} + {}^{1}\text{hydrogen} + \text{energy}.$$

The validity of these fusion reactions has been directly confirmed in nuclear experiments conducted in laboratories around the world during the past few decades.

The use of six hydrogen nuclei (protons) and the return of two of them along with a ^{4}helium nucleus means that four hydrogen nuclei are needed to create a single ^{4}helium nucleus. This is hardly surprising in view of the fact that hydrogen has a one-particle nucleus (a proton), while ^{4}helium has a four-particle nucleus (2 protons and 2 neutrons).

Remember, the combined mass of the reactants always exceeds that of the products, ensuring that energy is produced during all these nuclear fusion reactions. This energy is the heat that counteracts gravity to prevent stars from collapsing.

HELIUM BURNING: As a star steadily increases the amount of helium within its central regions, the core gradually begins to cool for lack of nuclear burning among the helium nuclei. The inward pull of gravity slowly overwhelms the diminished heat pushing outward. The core regions thus contract a little, increasing the density of nuclei to a value near 100,000 grams/cubic centimeter. More frequent collisions among the various nuclei build up the heat again, until the temperature reaches about 100 million kelvin. Such hot and dense physical conditions automatically ignite the fusion of nuclei according to the following nuclear reaction:

$$^{4}\text{helium} + {}^{4}\text{helium} \rightarrow {}^{8}\text{beryllium} + \text{energy}.$$

The elementary particles in a ^{8}beryllium nucleus do not attract each other very strongly. As an unstable isotope, it decays almost immediately (in a trillionth of a second) into a more stable nucleus. This extraordinarily short lifetime of ^{8}beryllium accounts for the fact that there is no element number 8 in the periodic table of the elements; ^{8}beryllium does not naturally persist in nature. Nonetheless, the ^{8}beryllium isotope has been detected for fleeting moments in laboratory accelerators, so we know a little bit about it.

Deep within the cores of evolved stars, the density is so great that some ^{8}beryllium nuclei can collide with other nuclei before changing into something else. After all, stars are so large that huge quantities of helium are fusing all the time. At any given moment, there are likely to be small amounts of ^{8}beryllium just produced, but not yet decayed. If a ^{4}helium nucleus happens to collide with an intact ^{8}beryllium nucleus before it has had a chance to decay, a

INTERLUDE 22-2 ***Lithium, Beryllium, and Boron***

As best we can tell, the light weight nuclei, lithium, beryllium, and boron, cannot be created by any of the processes mentioned previously. As shown in Figure 22.1 and Table 22-2, all three of these elements are surprisingly underabundant compared to neighboring elements of similar mass. Lithium has the additional problem of not being able to withstand the heat of normal stars; it breaks apart at temperatures greater than a few million kelvin.

Thus it seems unlikely that nuclei of these light elements were created in stars, but we can't be absolutely certain because the cores of stars cannot be directly observed. Yet it seems equally unlikely that these light elements could have been appreciably created in the earliest moments of the Universe; the high temperature and density decreased too rapidly following the bang.

Fortunately, several other processes are known which could conceivably form lithium, beryllium, and boron. Perhaps the most important of these processes is called "spallation," a word meaning "to break up or reduce by chipping." This mechanism occurs not inside hot stars but in the much cooler interstellar space. As fast-moving galactic cosmic rays bombard some of the carbon group of elements present in interstellar space, small amounts of the lithium group seem likely to be produced. Since the cosmic-ray particles have energies many times higher than that needed to break nuclear bonds, such collisions tend to split apart especially carbon, nitrogen, and oxygen nuclei, leaving a residue of lithium, beryllium, and boron. These nuclear fragments will then surround themselves with electrons to form new types of atoms.

The spallation process is very inefficient, in fact one of the slowest in nature; over the course of 100,000 years, a mere gram of beryllium is produced within each volume of interstellar space equal to a cubic astronomical unit. Along with their intolerance of high temperatures, this inefficient mechanism explains why these lightweight elements are rare, in fact comparably scarce as the heaviest-weight group of elements more massive than iron.

Rapidly streaming cosmic rays thus apparently play a significant role in evolution, completing the process of elemental evolution which the stars could not. Indeed, as we shall see in subsequent chapters, these ubiquitous particles also participate in chemical, biological, and neurological evolution, triggering mutations and genetic change while generally acting as the motor of evolution.

very stable nucleus of carbon is created:

$$^{8}\text{beryllium} + {}^{4}\text{helium} \rightarrow {}^{12}\text{carbon} + \text{energy.}$$

Note that the net result up to this point is that three ^{4}helium nuclei are required to synthesize one ^{12}carbon nucleus. This again makes sense, for three nuclei containing two protons and two neutrons apiece are then needed to form one nucleus containing six protons and six neutrons.

CARBON BURNING: This type of helium-capture process can continue in order to construct heavier elements. For example, provided that the temperature is at least 600 million kelvin, any ^{12}carbon nucleus colliding violently with another ^{4}helium nucleus will produce a stable oxygen nucleus:

$$^{12}\text{carbon} + {}^{4}\text{helium} \rightarrow {}^{16}\text{oxygen} + \text{energy.}$$

With each expansion–contraction cycle of a star, the temperature and density increase at the core, thus enabling ^{4}helium nuclei to be captured by progressively heavier nuclei. If the temperature did not grow larger, that is, if the velocities of the nuclei did not increase, the particle collisions would not be violent enough to overwhelm the larger electromagnetic repulsions associated with the heavier nuclei. Fortunately, gravity relentlessly contracts stellar cores, causing higher and higher temperatures. In this way, heavier elements such as ^{20}neon, ^{24}magnesium, and ^{28}silicon are eventually produced, along with energy, as stars move far from the main sequence. Nucleosynthesis becomes relatively straightforward, especially for those nuclei divisible by 4. Because such "helium capture" reactions are so frequent, elements numbered 4, 12, 16, 20, 24, and 28 stand out as abundant "peaks" in the Figure 22.1 chart of cosmic abundances.

Helium capture is not the only type of nuclear reaction occurring in stars. As more nuclei of different kinds accumulate in aged stars, a great variety of nuclear reactions become possible. Some nuclei capture neutrons, protons (hydrogen nuclei), and deuterons, in addition to helium nuclei. The result is a family of many nuclei having masses intermediate to those mentioned above. In fact, laboratory studies confirm that nuclei such as ^{19}fluorine, ^{23}sodium, ^{31}phosphorus, and a wide variety of others are created by the steady capture of small increments of mass and charge. Their abundances, however, are not as great as those nuclei produced directly by helium capture. That's why many of these elements (not divisible by 4) reside in the "troughs" of the Figure

22.1 chart of cosmic abundances.

Some Complications

With the appearance of ^{28}silicon in the core of a star, another nuclear process begins to complicate the rather simple ^{4}helium capture scheme. A competitive struggle ensues between the continued capture of helium nuclei to produce even heavier nuclei, and the tendency of the heavier nuclei to break down into simpler ones. The cause of this breakdown, called **photodisintegration**, is simply heat. By this point, a star's core temperature has reached an unimaginably large value of about 3 billion kelvin. The density is also enormous, having roughly millions of grams/cubic centimeter. Even nuclei have difficulty remaining intact within such a raging inferno.

Consider an example of typical constructive and destructive events thought to occur in highly evolved stars. Heat is so intense that some of the ^{28}silicon nuclei break down into seven ^{4}helium nuclei. Another nearby ^{28}silicon nucleus, not yet itself photodisintegrated, then suddenly captures one or several of these ^{4}helium nuclei. This type of process leads to even heavier nuclei including ^{32}sulfur, ^{36}chlorine, ^{40}argon, ^{40}calcium, ^{48}titanium, and ^{52}chromium. In the special case where all seven helium nuclei from a photodisintegrated silicon nucleus are suddenly captured by a nearby undestroyed ^{28}silicon nucleus, a much more massive nucleus of ^{56}nickel is instantaneously created:

$$^{28}\text{silicon} + {}^{4}\text{helium} + {}^{4}\text{helium} + {}^{4}\text{helium} + {}^{4}\text{helium} + {}^{4}\text{helium} + {}^{4}\text{helium} + {}^{4}\text{helium} \rightarrow {}^{56}\text{nickel} + \text{energy}.$$

This two step process—photodisintegration followed by direct capture of some or all the disintegrated ^{4}helium nuclei—is known as the **alpha process.** The name derives from the fact that ^{4}helium nuclei are often called alpha particles.

A further complication enters the picture here. The most stable nickel nucleus is ^{60}nickel, not ^{56}nickel as produced in this reaction. In a very small amount of time, the ^{56}nickel isotope decays first into a ^{56}cobalt isotope and then into a normal ^{56}iron nucleus. Any unstable nucleus will just continue to decay until stability is achieved. We know this much from experiments in nuclear physics laboratories. As far as nucleus number 56 is concerned, there are none more stable; ^{56}iron is the most stable of all the known nuclei.

IRONCLAD STABILITY: Iron's 26 protons and 30 neutrons are bound together more strongly than the particles in any other nucleus; as such, iron is said to have the greatest nuclear binding energy. Any nucleus with greater or fewer protons or neutrons has less nuclear binding energy and thus cannot be quite as stable as the ^{56}iron nucleus. This enhanced stability of iron, as well as of several other nuclei having similar complexity, explains how some of the heavier nuclei in the iron group have become more abundant than several lighter nuclei. That this is indeed the case can be seen by noting the substantial cosmic abundance of iron and its neighbors in Figure 22.1.

The unusual stability of the ^{56}iron nucleus also endows it with another important property. The ^{56}iron nucleus is unable to fuse with other nuclei to produce more energy. In fact, a fusion process involving iron *consumes* energy. This is a fundamental change in the overall scenario of stellar nucleosynthesis. Nuclear fusion in a highly evolved massive star inevitably produces an iron core that not only fails to produce energy, but also begins to rob the star of some of its previously gained energy. In a certain sense, the iron nuclei act as fire extinguishers, quenching nuclear fires in the cores of massive stars. This is the stage at which massive stars really start to panic; unable to stabilize, they begin to collapse.

Stars reaching the stage at which iron is synthesized have long left the main sequence. Nuclear burning is under way at all interior layers of such stars. Figure 22.2 is a cutaway

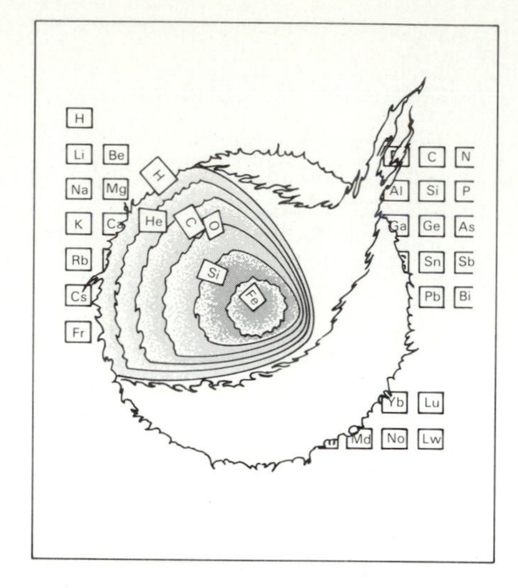

FIGURE 22.2 Artist's conception of the interior of a highly evolved massive star, showing the many internal layers of nuclear burning.

diagram of such a multiple burning star. The peripheral layers produce helium via the basic proton–proton cycle, the intermediate stellar layers synthesize a variety of heavier-than-helium nuclei via helium-capture mechanisms, and the deep core layers house the sites of the complex alpha-process that creates heavy nuclei up to and including iron.

Heaviest Elements

So the usual mechanism of helium capture does not create nuclei heavier than iron. Electromagnetic repulsion between helium and heavy nuclei, both positively charged, is too great for fusion to occur even in the unimaginably hot cores of aged stars. But if the very heavy nuclei at the bottom of the periodic table of the elements are to be synthesized—and clearly they must be somehow because we observe them to exist—then some other nuclear process must operate. Fortunately, we know of another such process, and it involves the capture of uncharged neutrons.

Deep in the interior of stars that are massive, highly evolved, and collapsing, the conditions are ripe for the elementary particles to play a role. Their numbers swell fantastically as some nuclei succumb to the onslaught of heat and pressure during collapse. Neutrons, in particular, can easily interact with other heavy nuclei that have not yet broken down; without any charge, neutrons have no electromagnetic repulsive barrier to overcome. For example, an iron nucleus can successfully capture a single neutron to form a relatively stable isotope, ^{57}iron:

$$^{56}\text{iron} + \text{neutron} \rightarrow {}^{57}\text{iron}.$$

This is then eventually followed by another neutron capture,

$$^{57}\text{iron} + \text{neutron} \rightarrow {}^{58}\text{iron},$$

producing a relatively stable ^{58}iron isotope, which can capture yet another neutron to produce an even heavier isotope:

$$^{58}\text{iron} + \text{neutron} \rightarrow {}^{59}\text{iron}.$$

^{59}Iron, however, is known from laboratory experiments to be radioactively unstable. It decays in about a month into ^{59}cobalt, after which this neutron-capture process resumes. In this way, further neutron captures can progressively form much heavier nuclei, including those of silver in any (non-U.S.) coins in our pockets, or gold in the rings on our fingers. These nuclear reactions can continue all the way up to ^{209}bismuth, for instance:

$$
\begin{aligned}
^{59}\text{cobalt} + \text{neutron} &\rightarrow {}^{60}\text{cobalt} \\
^{60}\text{cobalt} + (\text{decay}) &\rightarrow {}^{60}\text{nickel} \\
&\quad \vdots \\
&\quad {}^{99}\text{technetium} \\
&\quad \vdots \\
&\quad {}^{108}\text{silver} \\
&\quad \vdots \\
&\quad {}^{122}\text{tin} \\
&\quad \vdots \\
&\quad {}^{197}\text{gold} \\
&\quad \vdots \\
&\quad {}^{209}\text{bismuth.}
\end{aligned}
$$

^{209}Bismuth is the heaviest product possible in this scheme of successive neutron captures; it's the heaviest nonradioactive nucleus known.

You might be surprised to learn that this neutron-capture process is not very fast by nuclear physics standards. Nuclei do not rapidly cascade through the entire series of reactions that transform them from iron to bismuth, even though the tremendously hot and dense cores of massive stars would suggest a feverish pace. Instead, each capture of a neutron by a nucleus typically takes about a year. Consequently, researchers sometimes refer to this "slow" neutron-capture mechanism as the **s-process**.

OBSERVATIONAL EVIDENCE FOR HEAVY-ELEMENT CREATION: The scenario of stellar nucleosynthesis stipulates that the heavies are created exclusively within stars. How do we know that stars really do produce heavy elements in this way? How can we be sure that this scenario is correct? We're assured by a convincing piece of indirect evidence, and one item of direct evidence.

First, the rate at which various nuclei are captured and the rate at which they decay are known from laboratory experiments performed during the 1960s and 1970s. When all these rates are incorporated into a computer program, which also takes account of the temperatures, densities, and initial compositions at many positions within a typical star, the relative amounts of each kind of nucleus produced matches fairly closely those listed in the table of cosmic abundances. This is especially true for the elemental abundances up through iron. Thus despite the fact that no one has ever directly observed stellar nuclei in the act of creation, we can be reasonably sure that the theory of stellar nucleosynthesis makes sense in view of our knowledge of nuclear physics and stellar evolution.

Second, the presence of one kind of nucleus—^{99}technetium—provides direct evidence that heavy-element formation really does occur in the cores of stars. Laboratory measurements show the technetium nucleus to have a radioactive half-life of about 200,000 years. This is a very short time astronomically speaking, hence the reason no one has ever found even traces of naturally occurring technetium on Earth; it all decayed long ago. (It can be manufactured and studied, however, in nuclear laboratories.) On the other hand, the observed presence of technetium in numerous red giant stars implies that it must have been synthesized via neutron capture within the past several hundred thousand years. Otherwise, we wouldn't observe it in stars. Spectroscopic evidence for technetium is taken by many astronomers as proof that the s-process really does operate in stars.

THE VERY HEAVIES: The s-process allows us to understand the synthesis of nuclei up to and including ^{209}bismuth. But we've yet to reach some of the heaviest nuclei such as those of ^{232}thorium, ^{238}uranium, and ^{242}plutonium. Since the s-process terminates at bismuth, there must be yet another nuclear mechanism that produces the heaviest nuclei of all. Is there such a process known? Indeed there is; we call it the **r-process**, where "r" stands for "rapid." Another neutron-capture mechanism, the r-process operates quickly because it occurs literally during the moment of explosion as a massive star undergoes a supernova outburst.

Supernovae rebound like giant coiled springs after the nuclei themselves halt the stellar collapse, spewing forth many heavy elements created within. At the moment of explosion and for about 15 minutes thereafter, the density of neutrons dramatically increases as some heavy nuclei break apart during the explosion itself. Jammed into many of the light- and middle-weight nuclei, the neutrons help to create the heaviest of the known nuclei. Interestingly enough, then, the heaviest of the heavies are actually produced after stars have died.

Because the time available for synthesizing these heaviest nuclei is so brief, they never become as abundant as nuclei up to and including iron. The cosmic abundances (Table 22-2) of elements heavier than iron are a billion times less abundant than hydrogen and helium.

Again, astronomers have some evidence that this r-process really does occur in nature—at least indirect evidence. Figure 22.3 is a typical light curve of a supernova, displaying its dramatic rise in luminosity at

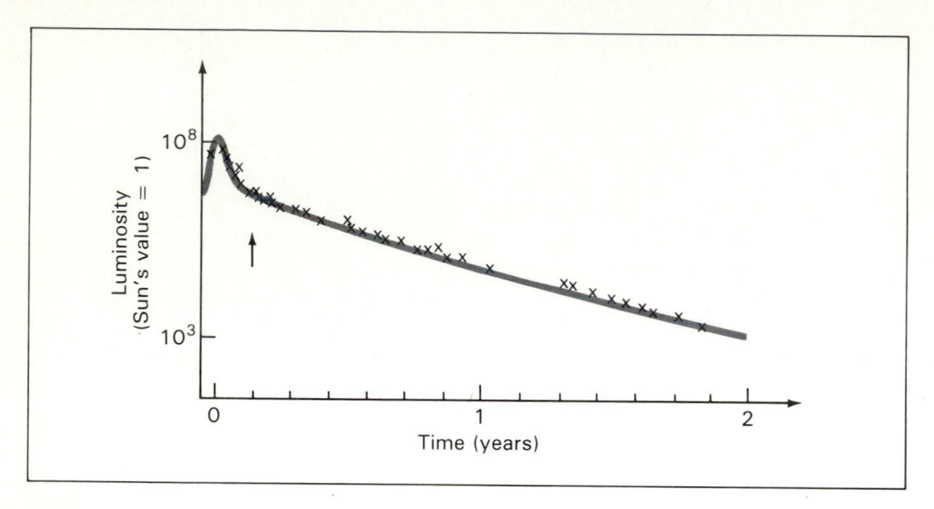

FIGURE 22.3 The light curve of an actual supernova, showing not only the dramatic increase and slow decrease in luminosity, but also the characteristic change in the rate of decay (arrow) about 2 months after the explosion. (The supernova, labeled IC4182, was observed in a faraway galaxy in 1938.)

the moment of explosion followed by a characteristically slower decay. Depending on the initial mass of the exploded star, the luminosity might take from several months to many years to decrease to its original value. But the *shapes* of the decay curves are pretty much the same for all exploded stars.

Detailed analyses show these decay curves to have two distinct features. As can be seen from Figure 22.3, the luminosity first decreases rapidly, after which it continues decreasing but at a slower rate. This change in the luminosity decay invariably occurs about two months after the explosion. Regardless of the supernova or the intensity of the outburst, the break in the decay curve always seems to happen some eight weeks after the explosion.

The peculiar twofold nature of this curve is not well understood, but many researchers suspect the answer lies in the rubble of hydrogen (fusion) bomb tests. Indeed, the rise, decay, and overall shape of the luminosity curve of a hydrogen bomb blast mimics quite closely that of most supernovae.

Figure 22.4 plots the decay of radioactivity measured for a group of elements created during human-made thermonuclear explosions here on Earth. Several curves are shown, each one representing a different heavy element known to have an unstable radioactive nucleus. Note, for example, the decay curve of the highly radioactive nuclei ^{56}nickel and ^{56}cobalt. These unstable isotopes, found in considerable abundance among the debris of nuclear-bomb tests, have half-lives of 6 days and 78 days, respectively; together, their decay (and thus their release of energy) mimic the light curves observed toward supernovae. In fact, a gamma-ray spectral feature attributed to radioactive cobalt was tentatively identified in a supernova that occurred in a distant galaxy during the 1970s. Other, even heavier, radioactive isotopes are also produced in nuclear explosions, although most have longer decay times bordering on years.

Thus some astronomers have suggested that the luminosity curves of supernovae can be explained by accumulating the radioactive decays of many unstable nuclei. Indeed, the similarity in the luminosity change of hydrogen bomb blasts and of supernovae is considered more than coincidence; it's taken as indirect evidence that the r-process really does work just after the death of massive stars.

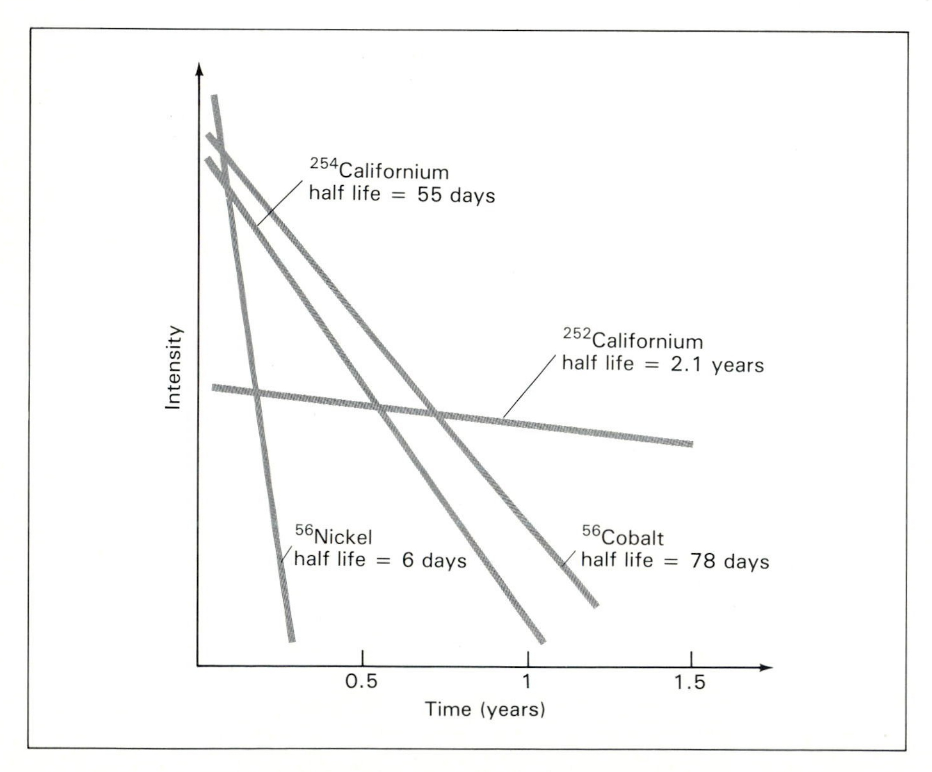

FIGURE 22.4 The characteristic radioactive decay curves for several very heavy nuclei found among the debris of hydrogen bomb tests on Earth.

SUMMARY

Although hydrogen and most of the helium in the Universe were created in the earliest epoch immediately after the Big Bang, the heavier elements were made later inside the cores of hot, dense stars. Indeed, that is where the heavies continue to be made today.

The theory of stellar nucleosynthesis predicts elemental abundances that agree remarkably well with the observed cosmic abundances. The story of the origin and evolution of the elements forms a complex though absolutely essential part of cosmic evolution. Without the heavies, rocky planets, intelligent life, and curious beings could not exist.

KEY TERMS

alpha process
nuclear transformation
photodisintegration
primordial nucleosynthesis
r-process
s-process
stellar nucleosynthesis

QUESTIONS

1. Cite, and explain briefly, three reasons why elements heavier than helium could not have been produced in appreciable amounts during the early Universe.
2. Noting that the element gold is quite rare, describe the process responsible for the production of this precious metal.
3. Carbon, nitrogen, and oxygen are among the three most important elements involved in life as we know it. What astronomical processes were responsible for the formation of each of these elements?
4. In a few paragraphs, cite both the highlights of the theory of stellar nucleosynthesis and the observations that tend to confirm the theory.
5. Describe the role played by iron nuclei in the stellar fusion cycle.
6. Briefly describe the chain of events leading to the creation of elements heavier than iron in massive stars.
7. Distinguish chemical reactions from nuclear reactions. What are the main differences?
8. What crucial role does the element ^{99}technetium play in our understanding of stellar nucleosynthesis?
9. Discuss a main piece of observational evidence that tends to support the production of heavy elements via the s-process.
10. Comment on the following statement: All the heavy elements within our human bodies have participated in at least one supernova explosion.

FOR FURTHER READING

BETHE, H., AND BROWN, G., "How a Supernova Explodes." *Scientific American,* May 1985.

FOWLER, W., *Nuclear Astrophysics.* Philadelphia: American Philosophical Society, 1967.

VIOLA, V., AND MATHEWS, G., "The Cosmic Synthesis of Lithium, Beryllium, and Boron." *Scientific American,* May 1987.

WOOD, J. (ed.), *The Cosmic History of the Biogenic Elements and Compounds,* Washington, D.C.: U. S. Government Printing Office, 1986.

23 NEBULAR EVOLUTION: *Origin of the Solar System*

Some of the heavy elements created by means of the nuclear events discussed in Chapter 22 are expelled into interstellar space, enriching the dark regions of all galaxies. There the heavies mingle with the primordial hydrogen-helium gas, eventually clustering into large planet-sized balls of matter. The ways and means that planets collect the cinders of burned-out stars is the subject of this chapter.

The origin of the planets and their moons is a complex and as-yet unsolved subject. Most of our knowledge of the Solar System's formative stages emerges from studies of interstellar clouds, fallen meteorites, and Earth's Moon, as well as from the various planets observed with ground-based telescopes and interplanetary space probes. Ironically, studies of Earth itself do not help much, since our planet's early stages eroded away long ago. Meteorites are providing perhaps the most useful information, for they have preserved within them traces of the solid and gaseous matter uneroded from the early Solar System. Radioactive dating of all meteorites uniformly suggests that the Solar System, including Sun and Earth, formed approximately 4.5 billion years ago. Lunar studies confirm this date.

Since astronomers have no good evidence for planets anywhere beyond our Solar System, we concentrate in this chapter on the origin of the planetary system in which we live. The learning goals are:

- to appreciate the ordered architecture of our Solar System
- to understand that planets are most likely formed as ordinary by-products of star formation
- to realize the role played by dust in the currently accepted model of Solar System formation
- to recognize that the origin of Earth's Moon is unknown
- to understand why the Terrestrial Planets differ greatly from the Jovian Planets
- to know that astronomers are currently unaware of any other Earth-like planets orbiting any other star

Model Requirements

Any model capable of explaining the origin and architecture of our planetary system must adhere to the known facts. We know of seven outstanding properties of our Solar System in bulk. They can be summarized as follows:

1. *Each planet is relatively isolated in space.* The planets exist as rather independent entities at progressively larger distances from the central Sun; they are not bunched together. In very rough terms, each planet tends to be twice as far from the Sun as its next inward neighbor.

2. *The orbits of the planets are nearly circular.* In fact, with the exception of Mercury and Pluto, each planetary orbit closely describes a perfect circle. The slight orbital eccentricity of Mercury, the innermost planet, probably results from great tidal stresses exerted on it by the nearby Sun. As for the outermost Pluto, some researchers regard this planet as a comet or an escaped moon of Neptune, thereby accounting for its large orbital eccentricity (see Figure 8.20), but this is only conjecture.

3. *The orbits of the planets nearly all lie in the same plane.* Each of the planes swept out by the planets' or-

bits are accurately aligned to within a few degrees. Put another way, the Solar System has the shape of a very thin disk. Again, Mercury and Pluto are slight exceptions, probably for the same reasons as noted above.

4. *The direction of the planets' revolution in their orbits about the Sun (counterclockwise as viewed from Earth's north) is the same as the Sun's rotation on its axis.* Virtually all the angular momentum in the Solar System—the planets in their orbits and the Sun's spin—seems to be systematically distributed. In other words, the central Sun and its family of planets all move with considerable unison.

5. *The direction in which most planets rotate on their axes also mimics that of the Sun's spin.* An exception to this statement is Venus, which spins opposite (retrograde) that of the Sun and the other planets. Uranus is another exception to this general property, for its poles lie in the plane of its orbit. Pluto might be tilted toward us, too.

6. *Most of the known moons revolve about their parent planets in the same direction that the planets rotate on their axes.* Some moons, like those of Jupiter, resemble miniature solar systems, revolving about their parent planet in roughly the same plane as the planet's equator. This again suggests uniformity throughout many aspects of our planetary system.

7. *Our planetary system is highly differentiated.* The inner Terrestrial Planets are characterized by high densities, moderate atmospheres, slow rotation rates, and few or no moons. By contrast, the outer Jovian Planets (save Pluto) have low densities, thick atmospheres, rapid rotation rates, and many moons.

All these observed facts, when taken together strongly suggest a large amount of order within our Solar System. The whole system is not a random assortment of objects spinning or orbiting this way or that. Consequently, it hardly seems possible that our Solar System could have formed by the slow accumulation of already-made interstellar "planets" casually captured by our Sun over the course of billions of years. The overall architecture of our Solar System is too neat, and the ages of its members too uniform, to be the result of random chaotic events. The overall organization points toward a single formation, the product of an ancient but one-time event.

Nebular Contraction

One of the earliest heliocentric models of Solar System formation is termed the **nebular theory**. Its origin is often attributed to the eighteenth-century German philosopher Kant, but he merely elaborated upon a proposal made a century earlier by the French philosopher Descartes. In this model, a giant swirling region of matter, called in this chapter "the primitive Solar System," is visualized to form planets and their moons as a natural by-product of the star formation process. But these philosophers did not work out the mathematical details of the model; their proposals were essentially ideas, and untested ones at that.

Later in the eighteenth century, a French mathematician-astronomer named Laplace tried to explain this type of model in a quantitative way. He was able to show mathematically that the conservation of angular momentum demands that the contracting matter of an interstellar fragment spin faster. A decrease in the size of a rotating mass must be balanced by an increase in its rotational speed, much like a high diver who somersaults quickly by tightly curling his body. As shown in Figure 23.1, the fragment eventually flattens into a pancake-shaped primitive Solar System, for the simple reason that gravity can pull matter toward the center of the region more easily along its spin axis than perpendicular to it. As such, we can begin to understand the origin of some of the ordered architecture observed in our planetary system today—the circularity of the planets' orbits, their coplanar distribution, and many of the other properties listed in the preceding section.

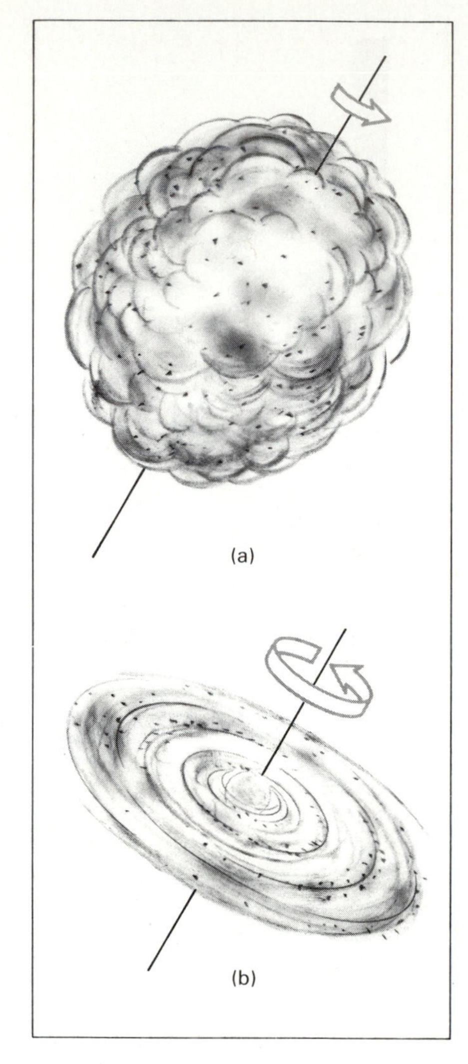

FIGURE 23.1 Conservation of angular momentum demands that a contracting, rotating cloud (a) gradually speed its rate of rotation. Eventually (b), the primitive Solar System resembles a giant pancake.

Continued contraction of the primitive Solar System forces the entire region to spin ever faster. Near the fringe, the outward centrifugal push eventually exceeds the inward gravitational pull, as shown in Figure 23.2. The push creates a flattened ring of gaseous matter that breaks away from the rest of the primitive Solar System, much like mud being thrown from the edge of a spinning bicycle wheel. After the rest of the system shrinks a little more, another ring of matter is deposited inward of the first. Progressing in this way, a whole series of rings can be formed around the central protosun.

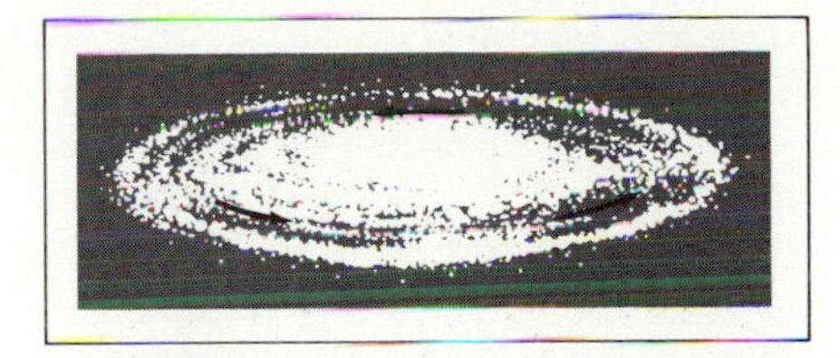

FIGURE 23.2 Artist's conception of a ring of matter breaking away from the continued contraction of the primitive Solar System.

Figure 23.3 is an artist's illustration of such a scenario. At various distances from the protosun, the effect of rotation overwhelms gravity, causing rings to be created. Each ring then clumps into a **protoplanet**—a forerunner of a geniune planet. By this scheme, several outer planets might develop while the interior of the primitive Solar System still contracted to shape the inner planets and the central Sun.

This description of the changes experienced by a shrinking interstellar fragment is rather straightforward. It follows directly from the obedience of a parcel of galactic gas to the known laws of physics. Even two centuries ago, Laplace got the description essentially correct. Yet, in modern times, as we learn how to program computers to account for subtle aspects of the problem, we have found some fatal flaws.

FATAL PROBLEMS: As nice as this nebular theory may seem, it has its share of difficulties. Detailed computations show that matter in a ring of this sort would not necessarily accumulate to form a planet. In fact, modern computer calculations predict just the opposite: The rings would tend to disperse. The protoplanetary matter is too warm, nor does any one ring have enough mass to bind its own matter into a ball. Gravitational clustering of interstellar matter is one thing; it works reasonably well in forming stars because, after all, there's a vast amount of mass contained within a typical galactic cloud. But coagulation of a warm, protoplanetary ring is another thing; there's not nearly enough matter for gravity to gather it into a planet-sized ball. Instead of coalescing to form a planet, the computations predict that the ring would break up and fade away.

A second problem with the nebular theory concerns angular momentum. Despite the fact that our Sun contains about 1000 times more mass than all the planets combined, the Sun possesses a mere 2 percent of the total angular momentum of the Solar System. Jupiter, for instance, has a lot more angular momentum than our Sun. Because of its large mass and great distance from the Sun, Jupiter possesses about 60 percent of the Solar System's angular momentum. All told, the four Jovian Planets account for nearly 98 percent of the total angular momentum of the Solar System. By comparison, the lighter Terrestrial Planets have negligible angular momentum.

The problem here is that the nebular theory predicts that the Sun should command most of the Solar System's angular momentum. After all, the Sun has most of the system's mass. Why, then, shouldn't it have most of the system's angular momentum? This is especially true since contracted objects are expected to increase their spin rate, in analogy with the figure skater who spins faster while bringing in her arms. Expressed another way, we could say that if all the planets, with their large amounts of orbital angular momentum, were hypothetically placed inside the Sun, it would spin on its axis about 100 times as fast as it does at present. Instead of rotating once in about 30 days, the Sun would spin once every several hours. And that would be fast enough to deform the whole Sun into a thin disk.

Thus the slow spin rate of our Sun is currently a bit of mystery. As we'll see, this problem plagues not only the nebular theory, but also many other models proposed for the origin of our Solar System.

FIGURE 23.3 The nebular theory envisions the formation of rings (a) of gaseous matter at various distances from the central protosun. Eventually, the rings clump into planets (b).

Close Encounters

These and other difficulties with the nebular theory forced researchers to consider alternative models. One such model is called the **collision theory** or, more popularly, the "close-encounter" theory. In this model the planets are visualized as the products of hot, streaming debris torn from the Sun during a near collision or close-encounter with another passing star. Figure 23.4 shows an artist's conception of the flaming streamers produced by such a near collision. Some streamers are surmised to remain gravitationally bound to our Sun, to be captured into orbits about it, and to condense eventually into planets. Despite the

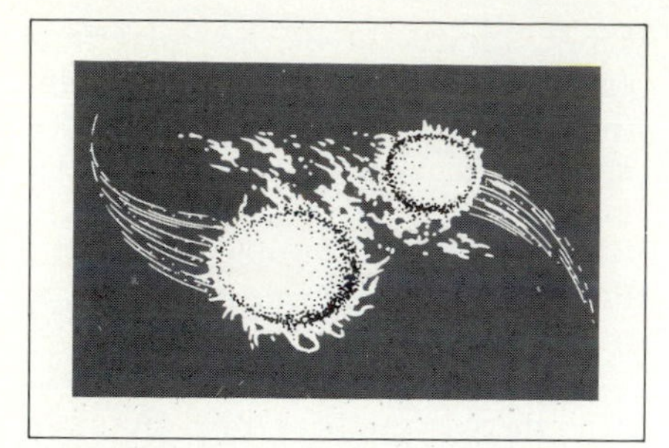

FIGURE 23.4 A close encounter between two stars would surely cause flaming matter to be torn from each star. According to the collision theory, some of these streamers eventually formed planets.

phenomenal tides undoubtedly accompanying the near collision of two stars, this scheme neatly accounts for the common orientation of the planets' orbits and the Sun's spin, as well as the close planar alignment of all the planets.

Although this model was also proposed during the eighteenth century, enthusiasm for it was rekindled about 100 years ago when it became clear that some minor exceptions plague the ordered architecture of our planetary system. Although we now know of several such exceptions, as noted in the first section of this chapter, astronomers in the mid-nineteenth century discovered that one of the moons of Neptune rotated retrograde—that is, opposite all the others. The overall simplicity of the planetary system thus broke down, giving a boost to catastrophic models like this collision theory—models that invoke accidental and unlikely celestial events.

NEAR-FATAL PROBLEMS: Although the collision theory has some qualitative points in its favor, it has some shortcomings as well. Foremost among them is that a near collision between two stars is highly improbable. Stars are large by terrestrial standards, but still minute compared to the typical distances that separate them. For example, the Sun is about 1 million kilometers in diameter, whereas the distance to the nearest (Alpha-Centauri) star system, is nearly 100 million million kilometers. Probability theory then suggests that given the number of stars, their sizes, and their typical separations, not more than a handful of such close encounters are likely throughout the entire expanse and history of the Milky Way Galaxy. Galaxy collisions are frequent (see Interlude 14-2), but stellar collisions are extremely rare.

The improbability of such collisions does not, of course, prove that the collision theory is wrong. After all, our Solar System could be the foremost—even the only—example of this extraordinarily uncommon phenomenon. Should this theory be correct, we can justifiably conclude that our planetary system is a rare type of astronomical region. Very few stars would be expected to have planets, and the chances for extraterrestrial life diminish accordingly.

Besides the small chance of collision, several other difficulties plague the collision theory. First, the angular momentum puzzle besetting the nebular theory is also a problem here.

Second and more formidable, it's hard to understand how hot solar gas torn from the Sun could contract; hot gases usually disperse. Consequently, although such a near collision between two stars might occasionally happen, the hot fragments are unlikely to have formed planets. Some of the hot streamers would surely fall back into the Sun. Others, because of their high temperatures, would tend to disperse even more quickly than the cooler gases in the rings of the nebular theory.

A third quandary concerns the nearly circular orbits traced by each of the planets. There is simply no way to justify how clumps of debris, tidally ripped from the Sun to form the planets, should end up orbiting the central Sun in near-perfect circles. The collision theory cannot explain this observed fact even qualitatively.

Gaseous Condensation

Yet another model of how the Solar System formed is known as the **condensation theory**. Currently favored by most astronomers, this idea is really a fancy version of the nebular theory discussed earlier. It mixes all the good features of the old nebular theory with our recently revised assessment of interstellar chemistry. Theorists can now concoct a condensation model that alleviates several of the problems noted above.

For example, recall that the first problem with the nebular theory concerned the inability of the loose, ringed matter to cluster into a tight-knit protoplanet. The gas in each of the rings possessed too little mass and too much heat to initiate gravitational contraction. But new insight has been added only within the past decade or so. Astronomers now recognize that the dust grains in interstellar space are not just a nuisance. Each of these microscopic dust particles probably formed by collecting thousands of atoms and molecules scattered throughout the Milky Way. Subsequent cooling changed them from gas to solid. The result is that our entire Galaxy is strewn with miniature chunks of icy and rocky matter having sizes of about 10^{-5} centimeter.

These dust grains play an important role in the evolution of any gas. Mixed with warm matter, dust helps to cool it by efficiently radiating heat away from protoplanetary blobs. Furthermore, the dust grains speed up the process of collecting enough atoms to form a planet; the grains act as microscopic platforms to which other atoms can attach in order to fashion larger and larger balls of matter. (This is thought to be the way that raindrops form in Earth's atmosphere; dust and soot in the air act as "condensation nuclei" around which water molecules cluster by surface tension.) Put another way, the presence of dust virtually guarantees that gaseous matter will cluster by cooling below the point at which outward-pushing pressure can effectively compete with inward-pulling gravity.

By postulating an especially (yet not unrealistically) dusty interstellar cloud 5 billion years ago, we can be sure that this dust-grain cooling occurred before the matter had a chance to float away. In this way, mixtures of gas and dust are expected to contract into planets rather than escape back into interstellar space.

THE CURRENT SCENARIO: To trace the formative stages of a planetary system such as ours, modern models stipulate the following broad scenario. Imagine a dusty interstellar cloud fragment measuring about a light-year across. Intermingled with the usual majority of hydrogen and helium atoms, the cloud harbors some heavy-element gas and dust, an accumulation of ejected matter of many past supernovae. Gravitational instabilities start the fragment contracting to a size of about 100 astronomical units, after which turbulent gas eddies form at various locations throughout the primitive Solar System. What caused the cloud instability in the first place? Perhaps the passage of a galactic spiral arm or even a concussion from a supernova explosion, as noted in Interlude 23-1; the specific reason is unknown.

Unlike the nebular theory sketched above, this condensation theory predicts no rings. As depicted in Figure 23.5, turbulent gas eddies would have naturally appeared, disappeared, and often reappeared at various places throughout the primitive, rotating Solar System—the bulk of which, by this time, would have flattened into a Frisbee-shaped disk. To a certain extent, these eddies resemble miniature versions of the gas-density fluctuations studied earlier in the case of galaxy and star formation.

Provided that an eddy was able to sweep up enough matter while orbiting the protosun, including a rich enough mixture of dust to cool it, gravity alone would ensure the formation of a planet. The entire process can be likened to a snowball thrown through a fierce snowstorm, growing bigger while encountering

INTERLUDE 23-1 *A Traumatic Birth*

Most astronomers accept some version of the condensation theory for the origin of our Solar System. According to this idea, once the contraction phase begins, planets are expected to form as natural by-products of star formation. This is especially true in cases where single stars form. Still, as noted in Chapter 19, a triggering mechanism of some sort is needed to initiate the contraction of an interstellar cloud fragment. In the case of our Solar System, some evidence suggests that the trigger was a shock wave from a nearby supernova.

Heavy elements in the primitive Solar System were accumulated from matter ejected by many supernovae. By the time the Solar System began taking shape, the various elements and isotopes should have been well mixed, homogeneously peppering the hydrogen–helium gas with supernova debris. Normal elements would be expected to differ from place to place in the Solar System, a fact that we know to be true when comparing, for example, the composition of the Sun, Earth, and some of the other planets. But the isotopes of a given element should have their same relative abundances whether part of a planet, a moon, a meteorite, or the Sun. In fact, however, this is not the case; the relative isotope abundances are not uniform throughout our Solar System. Compared to other matter in the Solar System, meteorites that have fallen to Earth often contain certain isotopes that are anomalously abundant. And remember, meteorites are among the oldest, unevolved pieces of matter known in our Solar System.

Analyses of small dust grains embedded in some meteorites show that isotopes of carbon, nitrogen, oxygen, magnesium, neon, and xenon differ considerably from those in Earth's rocks. Since scientists know of no *chemical* way to change one kind of isotope into another kind, we can surmise that some of the meteoritic grains must have had a different origin from most of the matter in meteorites as well as in the rest of the Solar System.

Conceivably, a nearby massive star exploded just before the Solar System began contracting some 5 billion years ago. The ejected debris of that supernova reached our parent galactic cloud, but did not have time to become thoroughly mixed before the planetary system formed. In this way, the abundance anomalies observed in meteorites can be explained as contaminants from this most recent supernova. Dating of the meteoritic grains support this idea, as they suggest the supernova blazed forth less than a few million years before the meteorites condensed into solid rock. Accordingly, some researchers argue that the appearance of the contaminant grains just prior to the Solar System's formation is more than a coincidence. They maintain that these grains may well be the direct debris of a supernova explosion, the concussion from which helped create our Solar System.

If this idea is valid, the demise of old stars may be the trigger needed to conceive new ones. To be sure, observers have long noted an abundance of young stars in the vicinity of supernova remnants of our Galaxy. The act of stellar death ensures a recurring fertilization of interstellar space, out of which emerge in turn later-generation stars as well as planets. In time, the stars' elemental ashes provide the seedlings for life itself.

On and on, the cycle churns. Stellar build up, break down, change. Dust to dust, and to dust some more. A kind of cosmic reincarnation.

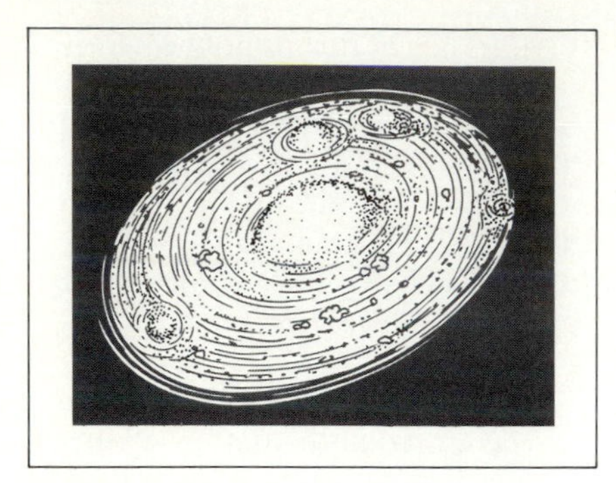

FIGURE 23.5 Gas eddies forming in a dusty cloud. According to the condensation theory, some eddies eventually became planets while the one big one in the middle became the Sun.

more snowflakes. Larger and larger particles accreted into asteroid-sized **planetesimals**, which in turn collided with one another, shattered, and further reassembled into even larger bodies. In this way, individual planets could be fabricated by accumulating much of the matter at varying distances beyond the protosun. Assuming that the "sweeping" (or accretion) process was reasonably efficient throughout the disk, we can begin to understand how our present Solar System has come to exist as a collection of rather small planets orbiting throughout an otherwise empty region of space.

Many of the natural satellites or moons of the planets presumably formed in similar fashion, as still smaller eddies and planetesimals condensed in the vicinity of their parent planets. Some of the smaller moons might well have been "chipped off" their parent planets during collisions with asteroids (see Interlude 23-2); others might be captured asteroids themselves.

Mathematical modeling suggests that nearly 100 million years would have been needed for the primitive Solar System to evolve nine protoplanetary eddies, dozens of protomoons, as well as the big protosolar eddy at the center. Roughly a billion years more would have been required to sweep the system reasonably clear of interplanetary trash.

A POINT OF CONTENTION: The weakest link in the condensation theory is, once again, the very small amount of angular momentum of our current Sun. All mathematical modeling requires the Sun to have been spinning very fast in the earliest epochs of the Solar System. Somehow, the Sun must have lost most of its spin momentum. Although the precise way it did so is unknown, we can surmise that the Sun probably transferred much of its spin angular momentum to the orbital angular momentum of the planets.

Some researchers speculate that the solar wind, moving away from the Sun into interplanetary space, could have robbed the Sun of some of its initial angular momentum. The early Sun probably had more of a solar gale than the gentle breezes our spacecraft now measure. High-velocity particles leaving the Sun, especially the charged particles escaping through the flares and spicules described in Chapter 9, surely followed the magnetic force field of the Sun. As the rotating magnetic field of the Sun tried to drag the ions of the inner solar nebula with it, the ionized gas might have acted as a brake on the Sun's spin. Although each particle boiled off the Sun carries only minute amounts of the Sun's momentum, over the course of 5 billion years, the vast numbers of escaping particles could have robbed the Sun of much of its initial spin momentum. If so, then even today, our Sun continues to slow its spin.

Other researchers prefer to solve the Sun's momentum problem by postulating that the primitive Solar System was much more massive than the present-day system. They argue that the accretion process was not entirely successful during the system's formative stages. Matter not captured by the Sun or the planets may well have transported much angular momentum while escaping back into interstellar space. This proposal is tough to test, since the escaped matter would be well beyond the range of our current robot space probes. (Perhaps the remote "Oort Cloud" of innumerable comets is the "escaped" matter.)

Despite some controversy as to how this momentum quandary can best be resolved, nearly all astronomers agree that some version of the condensation theory is correct. The details, however, have yet to be worked out. They form the essence of a most troubling problem now being addressed, both by exploration and calculation, at several observatories around the world.

The Formative Stage

Diversity of physical conditions in the earliest years of the Solar System is probably responsible for the large contrast between the Terrestrial and the Jovian Planets. Their basic differences in content and structure can be understood reasonably well by means of the condensation theory of the Solar System's origin. Indeed, it is here that the adjective "condensation" derives its true meaning.

THE ROLE OF HEAT: As the primitive Solar System contracted via gravity, it heated and flattened. Dust grains broke apart into molecules, and they in turn into excited atoms. Since the density, and hence the collision rate, was greater close to the protosun, matter there would have become hotter than in the outlying regions of the primitive Solar System. Calculations suggest that while the gas temperature was some several thousands of kelvin near the core of the contracting system, it would have been only several hundred kelvin some 10 astronomical units away, out where Saturn now resides.

Such a gas cannot continue to heat indefinitely, lest the region blow up. Like any hot gas, the primitive Solar System must have released some of its newly gained energy as infrared radiation. Even as the protosun continued heating upon contraction, the outer regions of the primordial system cooled. As a result, heavy elements several astronomical units from the center of the protosun must have condensed (crystallized) from their hotter gas phase to their cooler solid phase. (The same process occurs on Earth today, although on a smaller scale, as raindrops, snowflakes, and hailstones condense from moist, cooling air.)

INTERLUDE 23-2 ***Origin of Earth's Moon***

The origin of Earth's Moon is uncertain. Several theories have been advanced to account for it. One theory suggests that the Moon condensed as a separate object near the Earth, and in much the same way as did our planet. The two objects then essentially formed a binary-planet system, each revolving about their common center of mass. Although favored by many astronomers, this idea suffers from a major flaw: The Moon has a lower density and a different composition from Earth, making it hard to understand how both could have originated from the same protoplanetary blob.

A second theory maintains that the Moon condensed far from Earth and was then later captured by it. In this way the density and composition of the two objects need not be similar, for the Moon presumably materialized in an alien region of the primitive Solar System. However, the objection here is that the Moon's capture would be an extraordinarily difficult event; it might even be an impossible one. Why? Because (except perhaps for the oddball Pluto) the mass of our Moon relative to that of Earth is larger than for any other moon of any other planet. It's not that our Moon is the largest natural satellite in the Solar System; but it is unusually large compared to its parent planet, the Earth. Mathematical modeling suggests that it's unreasonable to expect Earth's gravity to have attracted our Moon in exactly the right way to capture it during a close encounter sometime in the past.

A third theory stipulates that the Moon originated out of the Earth itself. The Pacific Ocean basin has often been mentioned as the place from which protolunar matter may have been torn—the result of outward (centrifugal) forces and tidal effects (from the Sun) on a young, mostly molten, and rapidly rotating Earth. As absurd as this idea may seem, the early results of the *Apollo* missions seemed to favor it. Both the lunar composition and density were found to mimic those of Earth's mantle, just below the crust; the matter comprising both the Moon and the Pacific basin is basalt largely devoid of iron. However, there remains the fundamental mystery of how Earth could have possibly ejected an object as large as our Moon.

Today, many astronomers favor a variation of this third, "mother-Earth" theme. This idea—often called the "impact theory"—postulates a collision by a huge, Mars-sized object with a youthful and molten Earth. The collision caused not a direct impact as much as a glancing blow across Earth's edge, dislodging matter from our planet into Earth orbit where it assembled into a basalt-rich Moon. However, mathematical simulations of such a catastrophic event imply that most of the bits and pieces of splattered Earth would not have necessarily achieved a stable orbit; they would have more likely either escaped Earth completely or fallen back into our planet.

None of these theories is entirely satisfactory. Each suffers from a major flaw. Yet one of them or some version of them, it would seem, must be correct. Unfortunately, direct samples of lunar matter collected during the American and Soviet missions have failed to settle one of humankind's most ancient questions. Perhaps the formation of our Moon was the product of circumstances so rare that we will never be able to unravel the details of its birth. Indeed, the origin of Earth's Moon is such a frustrating subject that some researchers have been forced, in desperation, to suggest that the Moon cannot possibly exist!

Figure 23.6 plots the temperature gradient across the primitive Solar System, prior to the formation of any planets. Although high toward the center, the temperature is much lower farther from the protosun. With the passage of time, the temperature decreased at all locations, except at the very core where the Sun was forming. Everywhere beyond the protosun, atoms returned to their unexcited ground states, after which some of them collided and stuck to form molecules, which in turn coagulated to form dust grains once more.

You might think it amusing that, although there were plenty of interstellar dust grains early on, nature saw fit to destroy them only to rebuild them again later. However, an important change occurred in the process. Initially, the galactic gas was evenly peppered with an array of all sorts of dust grains. When the dust later reformed, the mixture was much different, for the condensation of solid dust from hot gas depends on the temperature. In other words, the act of contractive heating served to sterilize the entire region, thus setting the stage for a Solar System highly diversified in planetary composition.

JOVIAN PLANETS: In the middle and outer regions of the primitive planetary system, beyond several astronomical units from the center, the temperature would have been ripe for the chemical condensation of several abundant gases into solids. At temperatures of a few hundred kelvin or less, reasonably abundant elements such as carbon (C), nitrogen (N), and oxygen (O), combined with the most abundant element, hydrogen (H), to form some well-known simple chemicals, including ice crystals of water (H_2O), ammonia (NH_3), and methane (CH_4). (Helium is an inert element and does

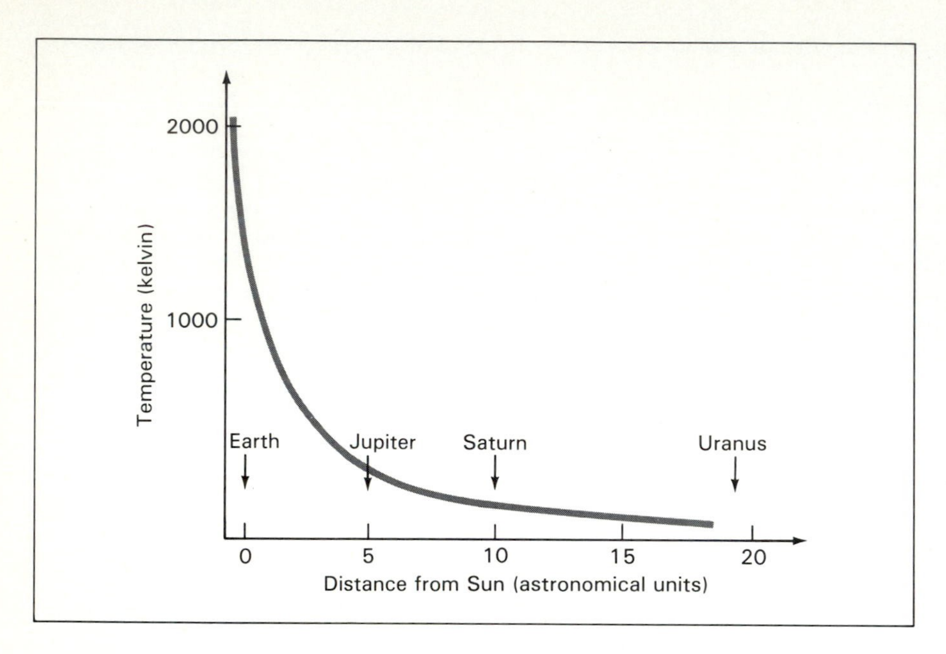

FIGURE 23.6 Theoretically computed variation of temperature across the primitive Solar System.

not combine chemically with other atoms.) Consequently, the ancestral fragments destined to become the Jovian Planets were formed under rather cold conditions by gravitational instabilities much like those discussed earlier for the formation of galaxies and stars.

Microscopic grains orbiting throughout the nebular disk gradually collided and stuck together, fabricating increasingly larger aggregates of ice in much the same way that fluffly snowflakes can be compressed into snowballs. Together with leftover hydrogen and helium atoms trapped by the strong gravitational pull of these protoplanets, gassy and icy compounds now comprise the bulk of the Jovian planets. There is little doubt that these massive planets formed in much the same way as our Sun. None of them is quite massive enough, though, to kindle nuclear burning, the hallmark of any star.

TERRESTRIAL PLANETS: In the inner regions of the primitive Solar System, the average temperature would have been about 1000 kelvin at the time when condensation from gas to solid began. The environment there was simply too hot for ices to survive. Instead, many of the abundant heavier elements such as silicon, iron, magnesium, aluminum—the heavy metals—would have combined with oxygen to produce iron oxides, silicates, and a variety of other rocky materials. These rocky grains gradually coalesced into objects of pebble size, baseball size, basketball size, and larger. The bigger they grew, the quicker gravity helped them to coalesce, sweeping more and more matter from the surrounding regions of the flattened nebular disk. Mostly through collisions, boulder-sized chunks evolved into (multikilometer-sized) planetesimals which eventually accumulated into genuine planet-sized objects. The bulk of the formation process probably took several million years. Such accretion of condensed, rocky solids thus explains the composition of the Terrestrial Planets.

Very abundant light elements such as hydrogen and helium, as well as many other gases that failed to condense into solids, would have surely escaped from these small protoplanetary objects. The inner planets' surface temperature was too high, and their gravity too low, to prevent gases from escaping. What little hydrogen and helium did manage to stick around was probably blown away later into interstellar space by the wind and radiation of the newly formed Sun (especially during its abnormally energetic T Tauri phase). What remained were a few rocky planets, each cool, hostile, and largely devoid of an atmosphere.

Why the myriad rocks of the asteroid belt between Mars and Jupiter failed to accumulate into a planet remains a mystery. Perhaps nearby Jupiter's huge gravitational force field caused the early rocks to collide destructively rather than accumulate constructively; huge Jovian tides on the planetesimals there would have also helped to prevent the development of a planet. Or, perhaps a planet did form, but then blew up for some unknown reason. The first explanation is currently favored by contemporary astronomers.

Planetary Systems Elsewhere

If our Solar System resulted from an encounter of two stars, and if this is the only way to form planetary objects around stars, then Earth and the other planets would be rare phenomena. The great majority of stars would not have any planets for the simple reason that stars hardly ever collide or experience a close encounter. As noted, though, a number of difficulties plague the collision theory, and most astronomers are inclined to reject it.

The leading theory for the origin of our Solar System holds that the events that produced it are not at all unique. Many stars—especially single stars that are not members of clusters—are expected to have a planetary system of some sort. Even if only 1 percent of all the stars in our Galaxy have planetary systems, that still leaves billions of stars with planets. And each star would probably have more than a single planet associated with it.

Theory is one thing, but observation is another. Is there any factual evidence for planets circling about other stars? Unfortunately, the answer is currently ambiguous. Certainly, there is no proof for Earth-like planets orbiting any other star.

The light reflected by any planet orbiting even one of the closest stars would be too faint to detect with the very best equipment now used by astronomers; an Earth-sized planet circling the nearby Sun-like star Tau Ceti—a mere 10 light-years distant—would have a (reflected) luminosity of only a billionth that of the star and would appear so close (less than an arc second) to its parent star as to be lost in the glare of the star. Possibly sometime in the 1990s, when large optical telescopes are operating in orbit above Earth's blurry atmosphere, astronomers might be able to detect some Jupiter-sized planets circling some relatively nearby stars. But even then, such stars would probably be at the threshold of visibility, and the observations might not yield *direct* evidence for other planetary systems.

Two types of *indirect* evidence suggest that planets orbit about a few of the nearest low-mass stars. Figure 23.7 summarizes the first piece of evidence, showing that high-mass stars generally spin much faster than low-mass stars. A sharp discontinuity in spin angular momentum occurs at about 2 solar masses, suggesting that stars with less than this mass rotate more slowly than expected. Like the Sun, most low-mass stars seem to have lost much of their original spin. The lost angular momentum could instead be stored in the orbital motions of associated planets. After all, planets of our Solar System account for most of the system's angular momentum, prompting speculation that most low-mass stars also have planets.

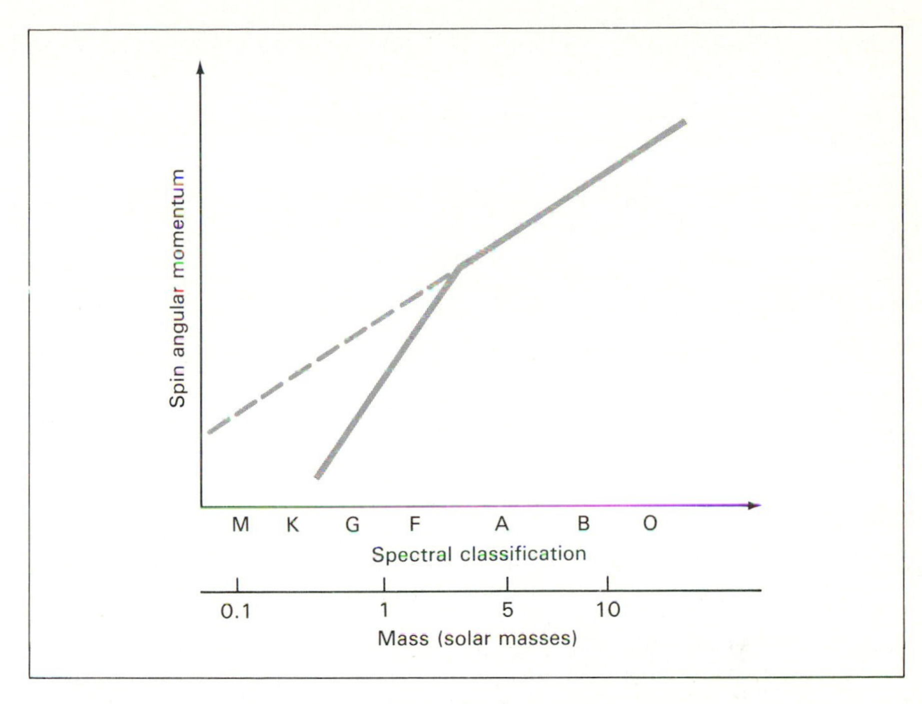

FIGURE 23.7 Low-mass stars are observed to spin more slowly than high-mass stars. What's more, the low-mass stars (solid line) show a good deal less rotation than expected (dashed line), prompting speculation that G-, K-, and M-type stars have planetary systems.

The second piece of indirect evidence for the existence of planets near other stars derives from the gravitational influence that planets are expected to have on stars. A large planet orbiting a low-mass star should produce a slight change in that star's motion, even though the planet itself is invisible to us. Figure 23.8 illustrates the expected effect on a nearby star; such an invisible object gravitationally pulls first one way and then the other during its annual orbit. The result is a slight back-and-forth shift or wobble in the path of the visible star. (To be understood properly, this wobble must be distinguished from any parallax induced by Earth's motion around the Sun, as explained in Chapter 4.)

Figure 23.9 shows the observed

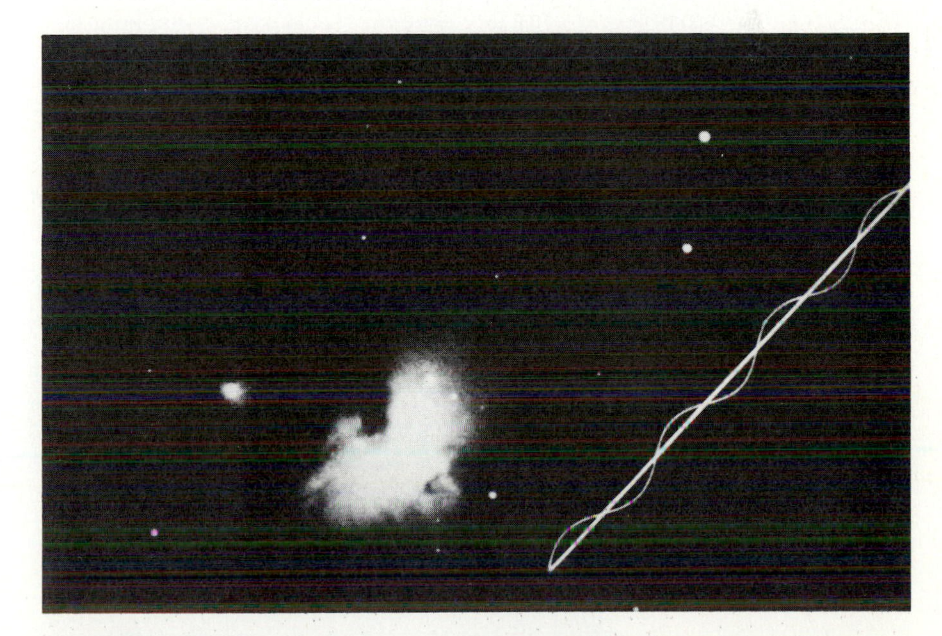

FIGURE 23.8 Theoretically computed wobble in the path of a nearby, low-mass star having a Jupiter-sized object orbiting about it. The straight line is the path the star would normally take in the absence of any planets, whereas the wavy curve is the resulting wobble in the star's motion caused by the to-and-fro gravitational pull of a small unseen planet. For a dwarf star some 10 light-years away, its expected deviation from a straight line would be a minute 0.005 arc second.

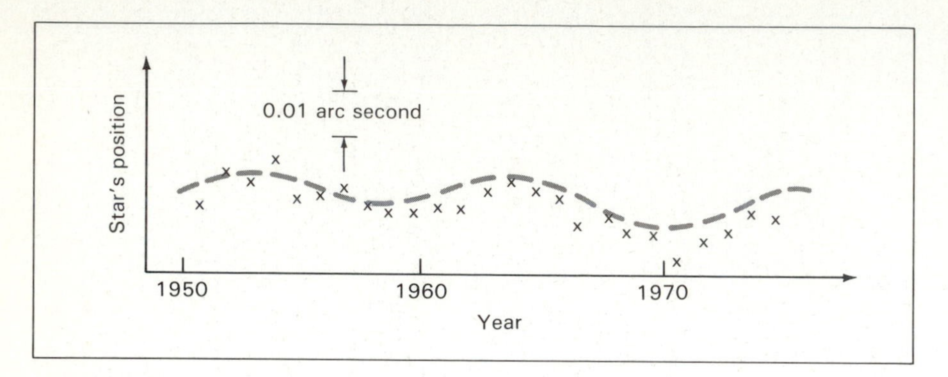

FIGURE 23.9 Barnard's star seems to show a wobble, suggesting that one or more large planets might be orbiting about it. But the interpretation is controversial, and even some of the data uncertain.

changes in the path of Barnard's Star, a heavily studied star having a mass of only about one-tenth that of our Sun. These positional changes of the star, merely 6 light-years away, were derived from observations made during the past several decades. The small wobbles in the star's path suggest that at least two objects accompany it, each having approximately the mass of Jupiter. A few other stars that are both nearby and of low mass show similar wobbles, but their interpretation is currently ambiguous and subject to much criticism. In each case, the wobble could alternatively be caused by a small and dim companion star. Some astronomers even contend that the observations are unreliable since much of the data base was accumulated with older telescopes decades ago.

Finally, in a more recent and exciting development, the internationally built *Infrared Astronomy Satellite* observed several nearby stars (including Vega and Fomalhaut) to be emitting about 100 times more infrared radiation than expected. The most reasonable explanation is that these stars are surrounded by warm dust or debris emitting invisibly in the infrared. In a related observation, Figure 23.10 is a visible image of the region around the well-known star Beta Pictoris; note how the warm matter seems spread out in a disk. This matter is composed of myriad dust particles mostly millimeter in size, and might well be partaking of the first stage of planetary formation. The image certainly resembles more or less what our own Solar System must have been like in its formative stages about 4.5 billion years ago. However, there is currently no way to ascertain whether the particles are primordial matter that could develop into planets or are merely unpromising bits of interplanetary litter. Even so, that these particles might materialize into the first known planetary system outside our own is an exciting prospect indeed.

It's fair to say that astronomers currently have no direct evidence, and little indirect evidence, for the existence of planets orbiting around any other star. For all we know, the Universe could be teeming with rocky and icy planets or even basketball-sized objects near or far from nearly every star. Our civilization has not yet invented the equipment needed to make an inventory of small, compact, dark objects residing in even nearby space.

We are left with the notion that if planets do form as natural by-products of star formation, then space must be infested with them, just as it bristles with stars in our Galaxy, and galaxies beyond. But the feeling is an uncertain one, for we don't know for sure.

FIGURE 23.10 A computer-enhanced photograph (taken from a mountaintop observatory in Chile) of an apparent disk of warm matter surrounding the star Beta Pictoris, some 50 light-years away. Most of the star itself is blocked by an instrument called a coronograph which is designed to detect faint halos around bright objects. The full extent of the disk, which is seen edge-on from our perspective, measures about 1000 astronomical units—roughly 10 times the diameter of Pluto's orbit.

SUMMARY

Current consensus has it that the genesis of a planetary system like our own is a natural, probably frequent, outgrowth of the birth of a star. Precisely how the small atoms of gas and grains of dust managed to coalesce into planets and moons remains, however, one of the great unsolved problems of science. What's most troubling is our inability to test the geological record of the first half-billion years of Earth's history. Matter from this critical time domain, which would ordinarily provide clues to the environment in which our planet was born, is missing, having been literally melted and eroded away long ago. Apparently, the only way to recover those clues to whatever did happen here roughly 5 billion years ago is to engage in a vigorous program of exploration throughout our Solar System.

KEY TERMS

collision theory
condensation theory
planetesimal
nebular theory
protoplanet

QUESTIONS

1. Explain the role played by dust in the modern (condensation) model for our Solar System's origin.
2. Choose two general observational properties of our Solar System and explain how the condensation theory can account for them.
3. Pick one: We observe a general decrease in density of the planets with increasing distance from the Sun because (a) the inner planets condensed at higher temperatures than the outer planets; (b) the inner planets condensed at lower temperatures than the outer planets; (c) the inner planets eddied as rocks were gravitationally pulled closer to the Sun; (d) the inner planets fashioned higher-density rocks through geological activity; (e) all things seem denser up close.
4. Discuss two ways in which O- and B-type stars might have influenced the development of life on Earth.
5. Trace the steps, starting with its origin, by which carbon became incorporated into the Earth.
6. Summarize the status of the search for planets surrounding stars other than our Sun.
7. Compare and contrast the general properties of the Terrestrial Planets and the Jovian Planets.
8. Summarize the main reasons why astronomers have discarded the nebular and collision theories for the origin of our Solar System.
9. State what is known as the "angular momentum problem" in our Solar System.
10. Explain the current impasse regarding the origin of Earth's Moon.

FOR FURTHER READING

*Ringwood, A. E., *Origin of the Earth and Moon*. New York: Springer, 1979.

Sagan, C., and Dryan, A., *Comet*. New York: Random House, 1986.

Schramm, D., and Clayton, R., "Did a Supernova Trigger the Formation of the Solar System?" *Scientific American*, October 1978.

*Ward, W., "The Formation of the Solar System," in *Frontiers of Astrophysics*, E. Avrett (ed.). Cambridge, Mass.: Harvard University Press, 1976.

Whipple, F., *Orbiting the Sun*, Cambridge, Mass.: Harvard University Press, 1981.

24 PLANETARY EVOLUTION: *Early Earth History*

Earth formed nearly 5 billion years ago as a relatively cold collection of burned-out stellar cinders. It heated and became thoroughly molten, partly at its surface because of fierce meteoritic bombardment but mostly from within owing to the radioactive decay of some of its heavier elements. During its first billion years of existence, our planet had geologically differentiated, its crust had solidified, and it had probably lost its original atmosphere. Change was rampant.

Subsequent change to Earth and its environment has produced mountain ranges, oceanic trenches, and atmospheric rejuvenation. Its surface has evolved, as has its atmosphere. The newly emerging subject of comparative planetology alluded to in Chapter 7 helps us put these changes into perspective. And we see how the stage was set for the emergence of life several billion years ago.

The learning goals for this chapter are:

- to understand how Earth's geological activity has shaped its mountains, oceans, and other surface features
- to know the arguments favoring the idea known popularly as "continental drift"
- to appreciate how Earth's atmosphere has evolved over the course of time
- to realize that Earth's magnetosphere is subject to regular reversals
- to recognize how some early evolutionary events on Earth differed from those on the other Terrestrial Planets

Surface Evolution

ACTIVE SITES: The lithosphere or rocky surface of Earth—that part of our planet most familiar to us—is interesting not only aesthetically because of its beauty but also scientifically because of its activity. Indeed, as noted in Chapter 5, Earth is geologically alive today. Its interior boils and its surface changes.

Figure 24.1 shows two clear indicators of lithospheric activity. These are volcanoes where molten rock and hot ash upwell through fissures or cracks in the surface. They are examples of present-day activity, as are earthquakes that occasionally occur when the crust suddenly dislodges under great pressure.

Major volcanoes and earthquakes are relatively rare events these days—rare enough to make them the lead item on news broadcasts around the world. Despite their current scarcity, geological studies of rocks, lava, and other surface features imply that surface activity must have been more frequent and perhaps more violent long ago.

For example, Figure 24.2(a) shows evidence for ancient cracks—crustal fractures that seem to divide enormous segments of Canadian rock. The rock to the left of the crack has been dated by radioactive methods to be about 3 billion years old, while the rock to the right is only 1 billion years old. Two vast slabs of rock of greatly different age, yet they lie side by side. Their juxtaposition suggests some large-scale jostling of surface rock.

Another example of past activity lies along the coast of Scotland, as shown in Figure 24.2(b). As illustrated, these cliffs are made of lay-

(a) (b)

FIGURE 24.1 Photographs of active volcanoes on Mount Kilauea in Hawaii and Mount St. Helens in the state of Washington. Kilauea (b) seems to be a virtually ongoing eruption, whereas St. Helens (a) was a rare catastrophic eruption (equivalent to about 1000 Hiroshima-type atom bombs), shown here blowing its top on 18 May 1980.

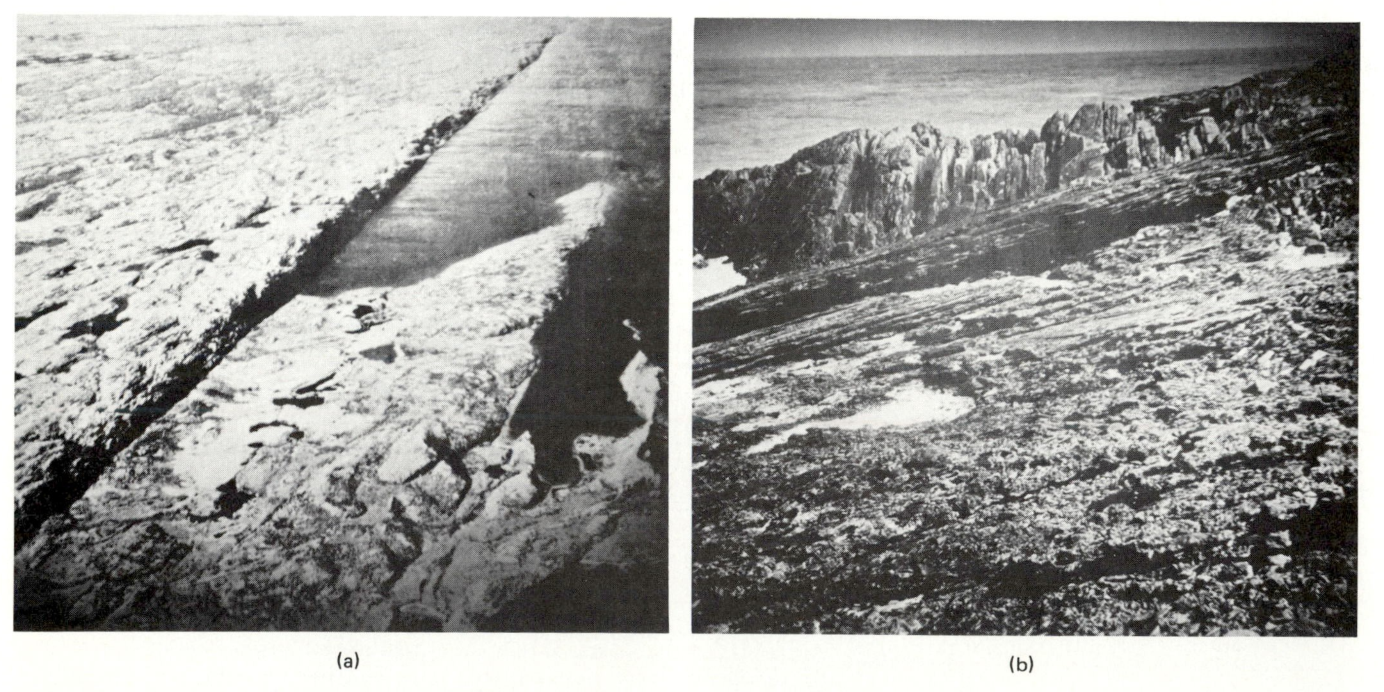

(a) (b)

FIGURE 24.2 Ancient surface activity is exemplified (a) along a fault near Hudson's Bay in Canada, and (b) along the coast of Scotland.

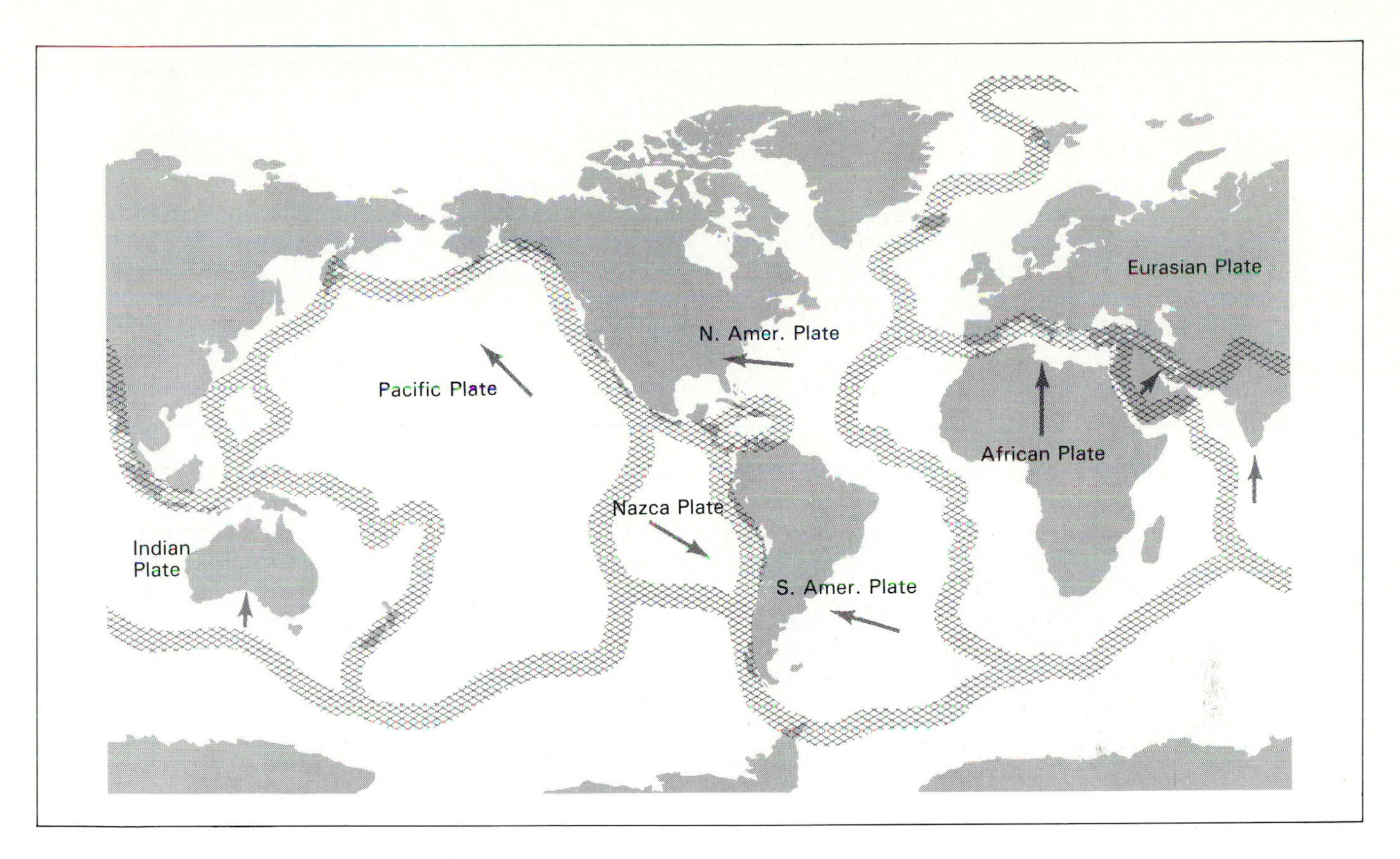

FIGURE 24.3 Hatched areas represent active sites where volcanoes or earthquakes have occurred in the twentieth century. Taken together, the sites outline the "plates" that drift around on the surface of our planet. Arrows show the general directions of the plate motions.

ered rock, just as we might expect from centuries of sedimentary growth on the ocean floor. However, atop this more-or-less horizontally layered rock lies vertically layered rock. Apparently, some type of surface activity thrust one part of this rocky cliff up onto the other.

These are just two examples of past activity at or near the surface of Earth. Numerous other cases are scattered across our globe. Erosion by wind and water has undoubtedly wiped away much of the evidence for ancient activity, but modern exploration has documented the sites of most of the recent activity.

Figure 24.3 is a map of the currently active areas of our planet. The hatched areas represent myriad sites of volcanism or earthquaking. Nearly all these sites have experienced surface activity within this century, some of them causing much damage and loss of lives. As can be seen from the map, these sites are especially abundant along the west coast of the United States, throughout the Aleutian Islands off the coast of Alaska, down along the Andes Mountains, across the Japanese Islands, up through India, as well as throughout much of Turkey, Greece, and the Aegean Sea.

SURFACE MOVEMENT: The most intriguing aspect of Figure 24.3 is that the active sites are not spread evenly across our planet. Instead, the sites trace well-defined lines of activity—called **faults** or zones of weakness where crustal rocks dislodge (as in earthquakes) or mantle rocks upwell (as in volcanoes). In the mid-1960s, it became clear that these faults are really the outlines of gigantic "plates" or slabs of Earth's surface.

Most startling of all, these plates seem to be slowly sliding—literally drifting around the surface of our planet. In doing so, these plate motions have created the surface mountains, the oceanic trenches, and many other large-scale features strewn across the face of planet Earth. In fact, plate motions have shaped the continents themselves.

Popularly known as "continental drift," the study of plate movement and the reasons for it is technically termed **plate tectonics**. Tectonics derives from the word "architecture," meaning to build or construct—in this case, mountains and oceans via the movement of plates.

A popular misnomer is that the plates are the continents themselves. In many cases, plates are made mostly of continental landmasses. But in other cases, they are made of a continent in addition to a large part of an ocean. Still other plates are mostly ocean, with the seafloor acting as the slowly drifting plate; the oceanic water merely fills in the

INTERLUDE 24-1 ***Changing Estimates of Earth's Age***

Estimates of the age of the Earth have increased dramatically since the middle of the seventeenth century. The most widely quoted of the early estimates is attributed to Anglican Archbishop James Ussher, who in the 1760s used the Bible to reason that Earth had been created in 4004 B.C. (October 26 to be exact). Other eighteenth-century researchers, however, preferring to examine the Earth itself, were convinced that our planet must be a good deal older than this 6000-year-old value suggested by the biblical account.

Although few scholars attempted a precise judgement, in 1778 a leading French naturalist, Georges Buffon, argued that Earth is of the order of 75,000 years. More extravagantly, even heretically for the time, he reportedly maintained in his unpublished studies that Earth was more like 3 million years old. For much of the nineteenth century, most geologists were prepared to accept Earth time as spanning millions of years. As for a specific value for the age of our planet, many were content to see, in the words of a Scottish geologist, James Hutton, "no vestige of a beginning, no prospect of an end."

Lord Kelvin was an exception. This British physicist was familiar with the newly emerging subject of thermodynamics (the science of heat) and he actually made an attempt to *calculate* Earth's age. Arguing in 1864 that any gravitationally contracting object cools at a certain rate, he reasoned that our planet would have been molten hot sometime between tens of millions and hundreds of millions of years in the past. While Kelvin did the calculation correctly, he was unaware of a most significant source of energy on Earth—radioactivity.

Not until the turn of the present century did the French researchers, Henri Becquerel and Marie and Pierre Curie, isolate uranium from pitchblende, and in turn study how that heavy element decays into several lightweight elements. By 1904 Lord Rutherford, the father of atomic physics, had pointed out the potential usefulness of radioactive elements and their "daughter" products in dating Earth materials (consult Interlude 5-2).

Soon thereafter, radioactive methods gave Earth ages variously in the range of 1, 2, and 3 billion years. And since the 1950s, as our understanding of nuclear physics advanced and laboratory techniques became more refined, so has progressively older rock been found on our planet. Today, the oldest rocks are those of Greenland and Labrador, dated to be approximately 3.8 billion years. Thus, our planet must be at least that old. Furthermore, since Earth is highly differentiated—heavies mainly at the core, lightweight elements at or near the crust—it must have been in the molten (nonrocky) state longer ago than that. A combination of thermodynamic arguments, the age of meteorites and the lunar highlands, and theoretical studies of the Sun's evolutionary rate all suggest that planet Earth is close to 4.5 billion years old.

This episode in the changing estimates of Earth's age is a good example of how the scientific method, though affected at any given time by the subjective whims and human values of various researchers, does lead to a definite amount of objectivity. Over the course of time, many groups of researchers checking, confirming, and refining experimental tests will neutralize the subjectivism of individual workers. Usually one generation of researchers can bring much objectivity to bear on a problem, although some particularly pivotal concepts—such as Earth's old age—can become swamped for long periods by cultural and institutional biases such as tradition, dogmatism, and even politics.

Today, with an open mind and a readiness to change our theories during the course of numerous experimental and observational tests, scientists maintain that nature yields a certain measure of objectivity through unraveled facts, thus granting us a progressively better "approximation of reality." It is in this sense that science claims to make progress, both in quantitative terms of a greater amount of knowledge and in qualitative terms of a fuller, more accurate knowledge.

depressions among the continents. For the most part, whole continents ride as mere passengers on much larger plates.

Figure 24.4 is a photograph of a typical plate. This one contains all of India as well as much of the Indian Ocean. It's called the Indian Plate, and it also includes all of Australia and its surrounding south seas.

We might wonder why these large drifting regions are called plates. The reason is simply that despite their large horizontal (surface) size, their vertical thickness is comparatively small. As shown in Figure 24.5, the thickness of a typical plate is approximately 50 kilometers, usually less than one-twentieth of their full horizontal size. Indeed, the plates resemble upside-down dishes.

Examining Figure 24.3 again, we can note most of the currently known plates of the world. For example, the major boundary separating the North American Plate from the Eurasian Plate is a thin strip of activity in the middle of the Atlantic Ocean. Discovered after World War II by oceanographic ships studying the geography of the seafloor, this giant fault is called the Mid-Atlantic Ridge. It extends, like a seam on a giant baseball, uninterrupted all the way from Scandinavia in the North Atlantic to the latitude of Cape Horn

FIGURE 24.4 Spacecraft photograph of the subcontinent of India and much of the Indian Ocean. All this (and more) comprises the Indian Plate.

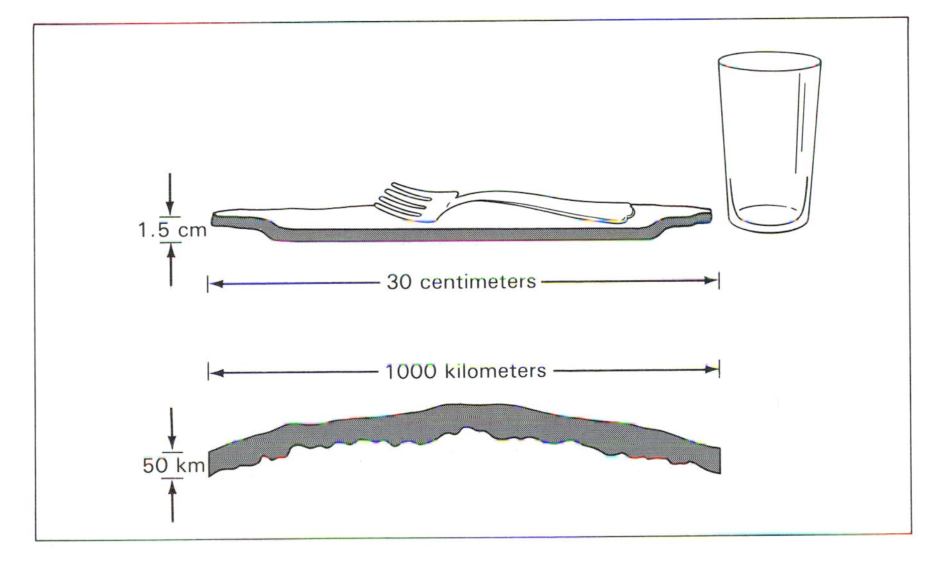

FIGURE 24.5 Physical dimensions of typical plates.

at the southern tip of South America. Consequently, this ridge also separates the South American Plate from the African Plate. The only part of the entire ridge that rises much above sea level is the subcontinent of Iceland.

As the plates drift around, we might expect collisions to be routine. Indeed, they do collide. But unlike two automobiles that collide and then stop, the surface plates have enormous momentum. They do not stop easily. Instead, they just keep crunching into one another.

Figure 24.6(a) shows a typical example of a collision currently occurring between two continental land masses. The resulting folds of rocky crust, clearly seen in the figure, create mountains—in this case, Mount Everest. Figure 24.6(b) is a photograph of the Himalayan mountain range, of which Everest is a part. This entire mountain system results from the Indian Plate thrusting northward into the Eurasian Plate. It is doing so right now.

The theory of plate tectonics suggests that the plates have not stopped after some initial and ancient movements. Rather, they still drift at the present day, albeit at an extremely slow rate. Typical velocities of the plates amount to less than a few centimeters/year. This is generally too small to measure directly, though a few of the major plates have had their drifts estimated indirectly.

Recognize also that during the course of Earth history, each plate has had plenty of time to move large distances, even at their sluggish pace. For example, a drift rate of only 2 centimeters/year means that two continents can separate by some 4000 kilometers over the course of 200 million years. That's surely a long time by human standards, but it actually represents only about 5 percent of the age of the Earth.

Figure 24.7 shows another interesting example of current plate movement—the northern part of Italy colliding with a small southern part of the gigantic Eurasian Plate. The resulting wreckage is the famous Alps Mountain Range. To be truthful, we are uncertain if the African Plate (containing much of Italy) is thrusting northward or the Eurasian Plate is moving southward (or both). There is no doubt, however, that these plates have some movement toward one another. Together, they are slowly reshaping the Alps, even at this present moment.

As noted above, some plates do not contain a large continent. For example, the southeastern part of the Pacific area, called the Nazca Plate, has no land mass. Accordingly, its seafloor crust rides much lower than the west coast of continental South America. As shown in Figure 24.8, the theory of plate tectonics suggests that the Nazca Plate is sliding underneath the South American Plate. The interface between these two plates is a well-known site of much

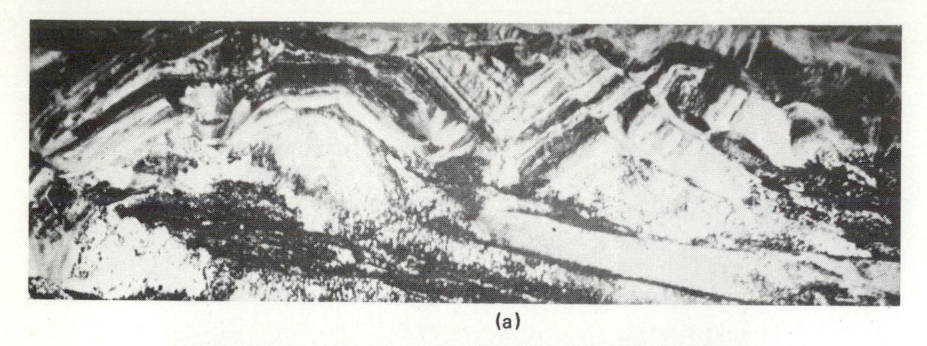
(a)

(b)

FIGURE 24.6 Mountain building is largely caused by plate collisions. Here, the folding of rock is clearly seen on one side of Mount Everest (a), which is only a small part of the entire Himalayan mountain range (b) midway between India and Central Asia.

geological activity. Even today, the region is plagued by earthquakes. Evidently, two gigantic plates cannot easily slip by one another, even if one lacks a continent. Although we have no observational data yet to prove it, circumstantial evidence strongly suggests that the crust of the Nazca Plate is indeed thrusting downward into the mantle, while the South American Plate is slightly uplifted; the South American Plate is apparently overriding the Nazca Plate. This, then, is another way that mountain chains are built—in this case, the Andes Mountains all along the west coast of South America.

Not all plates experience head-on collisions. As noted by the arrows of Figure 24.3, many plates slide or sheer past one another. A good example is the most famous active region in North America—the San Andreas Fault in California. Illustrated in Figure 24.9, this fault delineates much earthquake activity largely because the Pacific and North American Plates are rubbing past one another; they are not quite moving in the same direction and not quite at the same speed. The two plates are in contact and, like a poorly oiled machine, their motion is not steady and smooth. Rather, it's jerky, suddenly and sharply moving each time the pressure to drift overwhelms the friction to stay put.

EVIDENCE FOR CONTINENTAL DRIFT: Several pieces of data support the theory of plate tectonics. **Geography**—the study of positions, shapes, sizes, and numerous other qualities of Earth's continents—supplies the first bit of evidence. Figure 24.10 suggests how all the continents neatly fit together like

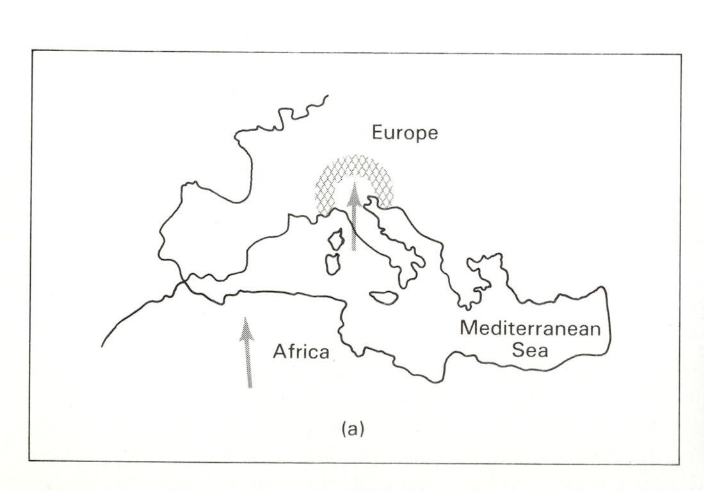

(b)

FIGURE 24.7 Italy, mostly a part of the African Plate, is thrusting northward into the Eurasian land mass (a), thus creating the Alps (b).

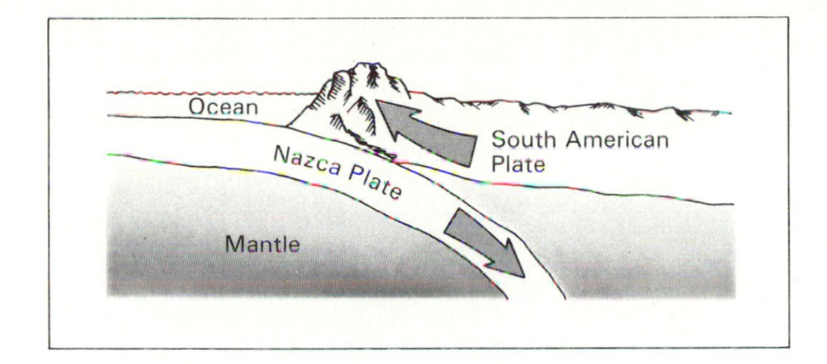

FIGURE 24.8 The southeastern part of the Pacific area—called the Nazca Plate—is apparently being overrun by the western part of the South American Plate. As a result, the Nazca Plate is diving into the mantle, while uplifting the Andes Mountains along the edge of the South American Plate.

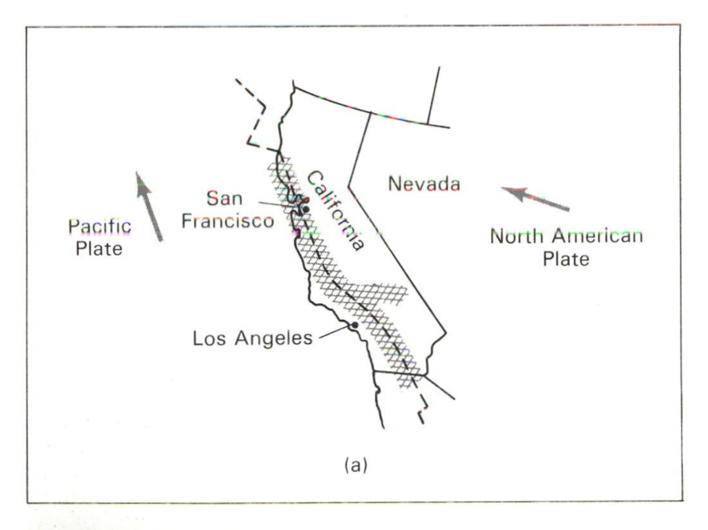

FIGURE 24.9 The San Andreas Fault (hatched) results from the North American and Pacific Plates sliding roughly past one another (a). The fault line separating the two plates [and shown dashed in (a)] can be seen clearly in a high-altitude photograph (b) of the San Francisco Bay area.

pieces of a puzzle. The fit is really a pretty good one, suggesting the existence of a single huge landmass sometime in the past. Note especially how the Brazilian coast meshes nicely with the Ivory Coast of Africa. In fact, most of the southern hemisphere landmasses fit together remarkably well. The fit is not as good in the northern hemisphere; we shouldn't be terribly surprised, given all the extra "debris" in the North Atlantic—Iceland, Greenland, and the British Isles.

Sometime in the past, then, a single gargantuan landmass probably dominated our planet. Geologists call this ancestral supercontinent Pangaea, meaning "all lands"; the southern part has the special name Gondwanaland, the northern part Laurasia. As suggested by Figure 24.10(b), these two major segments of Pangaea were probably partly separated by a V-shaped body of water called the Sea of Tethys. The rest of the planet was presumably covered with water.

The existence of such an ancient supercontinent, and its subsequent breakup, explain several heretofore peculiar discoveries. For example, when the first climbers reached the summit of Mount Everest in the 1950s, they found some fossils of fish. How could marine fossils get to the highest point on Earth? Plate tectonics provides the answer. Whatever caused Pangaea to break apart also sent its continental fragments drifting. At some point, India began its slow motion northward across the Sea of Tethys. In the process, fossils of marine life deposited at the bottom of Tethys were apparently pushed up alongside parts of the Eurasian landmass to form the Himalayas.

Plate tectonics has slowly reshaped the surface of our planet. In some cases, the seafloor has been literally thrust to the top of the world. In other cases, gigantic underwater mountain chains have slowly emerged. In still other cases, entire subcontinents have apparently submerged.

Modern geologists now regard the geographical outlines of the present continents as strong evi-

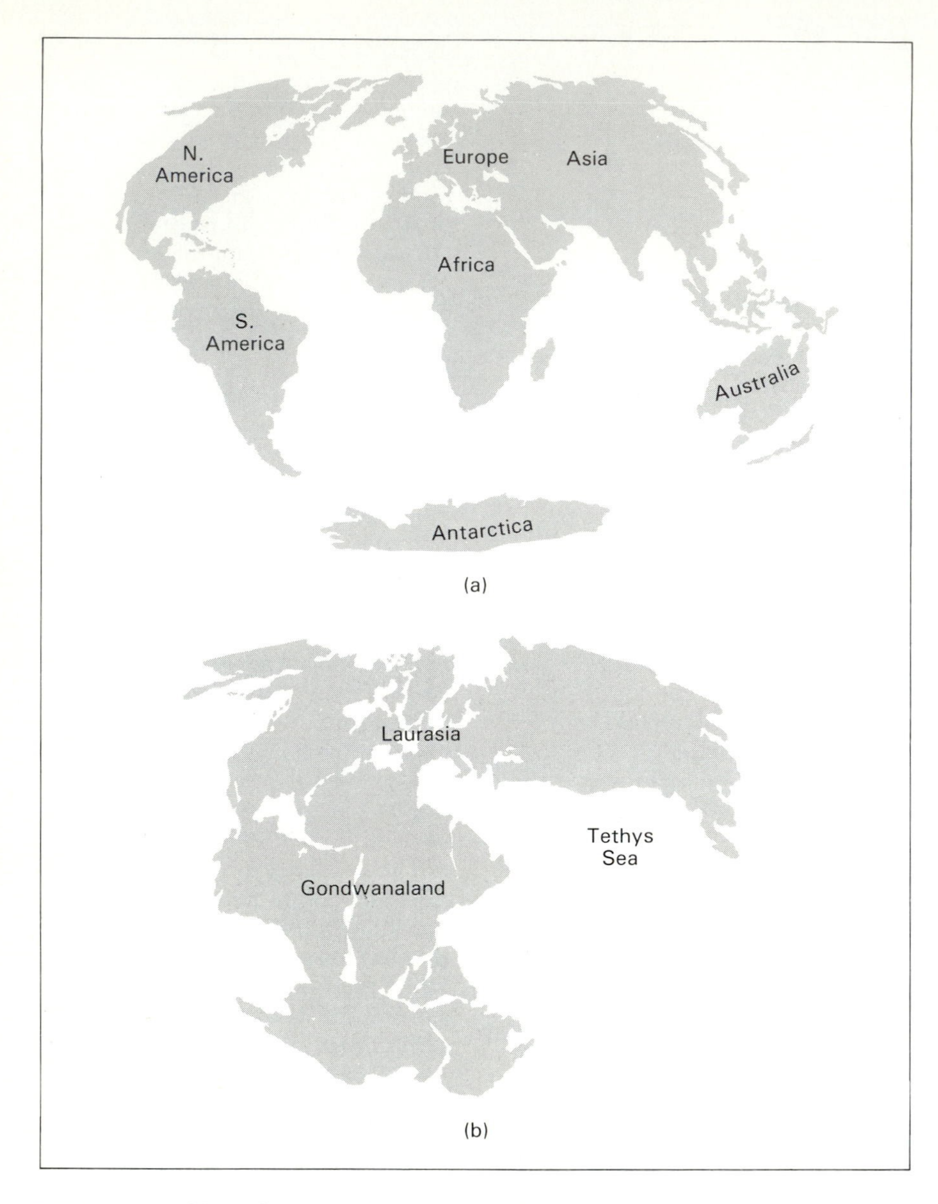

FIGURE 24.10 Given the currently estimated drift rates and directions of the plates, we can trace their movements back into the past. About 200 million years ago, they would have been at the approximate positions shown in (b). The continents' current positions are shown in (a).

dence for the ancestral continent of Pangaea. When did it exist? The current locations of the continents, along with some indirect estimates of their drift rates, suggest that Pangaea was the major land feature on Earth approximately 200 million years ago. Dinosaurs, which were the dominant form of life then, could have sauntered from Russia to Texas via Boston without getting their feet wet.

About 20 million years thereafter—that is, about 180 million years ago—Gondwanaland and Laurasia began dividing, probably near what we now call the Gulf of Mexico. About 150 million years ago, Gondwanaland itself broke into various pieces—South America, Africa, and Australia, as we now know them. Shortly thereafter, Laurasia split, perhaps more violently for some reason, thereby producing North America, Europe, and the "debris" in the North Atlantic Ocean.

Actually, this geographical puzzle was unraveled early in the twentieth century by a German meteorologist named Wegener. No one believed him, however. Until about the mid-1960s, nearly all geologists thought it preposterous that large segments of rocky crust could be drifting across the surface of our planet. Their views changed rapidly, however, when several additional pieces of data suddenly became available.

The second piece of evidence favoring the theory of plate tectonics comes from **paleontology**—the study of the fossilized remains of dead organisms. (The prefix "paleo" derives from the Greek, meaning ancient or old.) Fossils of a creature called the Mesosaurus reptile, extinct for nearly 200 million years, have been uncovered at only two locations on Earth. One place is a small part of the Brazilian coast, while the other is on the west coast of Africa. These two places are precisely where the continents apparently once meshed as part of the ancestral supercontinent of Pangaea.

Figure 24.11 shows a reconstruction of what the reptile probably looked like, as well as a sketch of the only two regions at which its fossilized remains have been found. If Africa and South America were always separated by the great expanse of the South Atlantic Ocean, these reptilian creatures could have hardly survived a swim between coasts. Even if they could have achieved that feat, the chances are slim that they would have departed and landed at exactly those parts of the continents that geographically mesh. A more reasonable conclusion is that Africa and South America were once joined, and that this reptile lived in a small region in the midst of Gondwanaland.

A third piece of evidence comes from **oceanography**—the study of the ocean's motion, history, and physical and chemical behavior. Many of the active sites submerged beneath the ocean form a giant system of undersea cracks. For example, the Mid-Atlantic Ridge mentioned earlier has been carefully mapped by underwater cameras lowered from oceanographic vessels. Furthermore, during the 1970s, robot submarines managed to retrieve samples of ocean floor at a va-

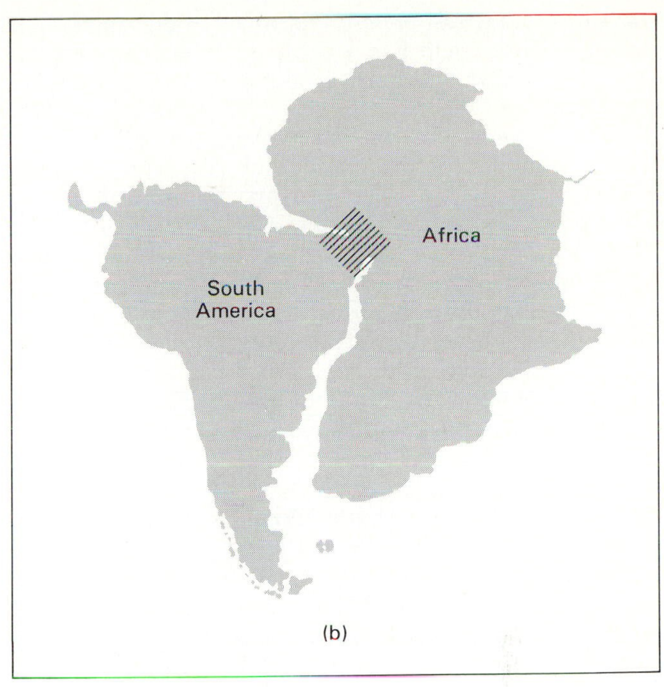

FIGURE 24.11 The fossilized remains of the Mesosaurus reptile (a) have been found at only two places on Earth: on the east coast of Brazil and on the west coast of Africa (b), just where the shading suggests the two continents overlapped about 200 million years ago.

riety of places on both sides of this mountain range. As depicted in Figure 24.12, matter on the ocean floor closest to the underwater ridge is relatively young, while material farther away is noticeably older. The ages vary consistently on both sides of the ridge; the greater the distance from the ridge, either to the east or west, the older the sea-floor matter.

These observations support the idea that hot matter upwells from a crack all along the Mid-Atlantic Ridge. In this way, the plates are pushed apart. The North and South American Plates are thus expected to be moving generally westward, while the Eurasian and African Plates drift eastward. This is exactly the trend implied by the geographical fit above; the plates on both sides of the Atlantic Ocean have presumably been drifting apart for the past 200 million years.

Dating of the ocean bottom confirms this idea. We have no evidence that any part of the Atlantic Ocean seafloor is older than 200 million years. Submerged rocks close to the east coast of America and the west coast of Europe or Africa are nearly 200 million years old; those closest to the Mid-Atlantic Ridge are only a few million years old. Thus the variable age of the bottom of the Atlantic Ocean offers important support for the basic tenets of plate tectonics.

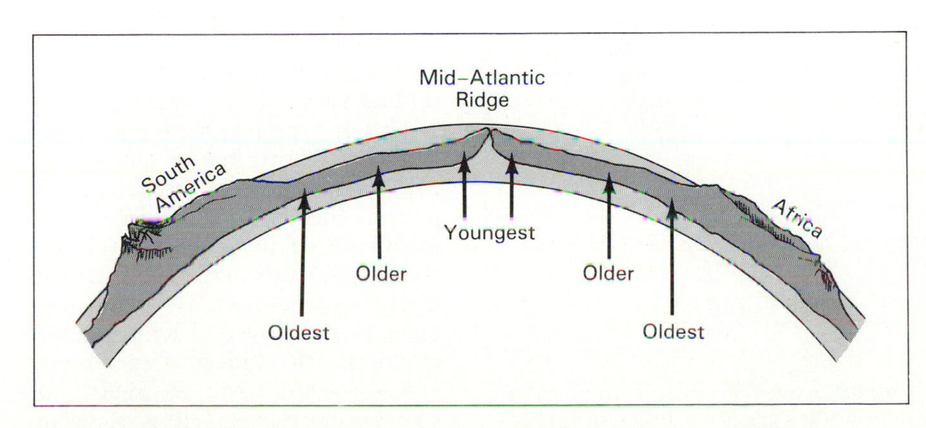

FIGURE 24.12 Samples of ocean floor retrieved by oceanographic vessels are youngest close to the Mid-Atlantic Ridge and progressively older farther away.

Finally, a fourth bit of evidence also favors the idea of plate tectonics, although here the data are not as clear as the previous three. This evidence derives from **paleomagnetism**—the study of ancient magnetism. Everyday experience tells us that iron is magnetic; in fact, any metal containing even a small amount of iron ore is usually magnetized. However, when iron is heated to a temperature of about 1000 kelvin, it loses its magnetic properties. Thus hot basalt (impregnated with traces of iron), upwelling from cracks in the oceanic ridges, is not magnetic. As basalt cools, however, magnetism becomes established in its iron atoms. In effect, each iron atom responds to Earth's magnetic force field like a compass needle; specifically, the iron aligns itself with the orientation of Earth's field *at the time of cooling*. When the basalt solidifies to form hard rock shortly thereafter, it fixes the orientation of the embedded iron. Accordingly, the ocean-floor matter has preserved within it a history of Earth's magnetism during past times. This is what we mean by fossilized magnetism.

Figure 24.13 is a schematic diagram of a small portion of the ocean floor. As shown, the current magnetism of Earth is oriented in the familiar north-south fashion. When samples of ocean floor are examined close to the Mid-Atlantic Ridge, the iron deposits are oriented just as expected for a north-south field. This is the "young" basalt that upwelled in recent times. Samples retrieved far from the ridge, however, often have their iron deposits oriented at odd angles relative to the usual north-south field. This is the older basalt that upwelled in earlier times.

Most researchers theorize that these different magnetic orientations resulted from the twisting, turning, and drifting of each plate. Working backward, we can use the embedded iron to infer the past positions of the plates, as well as the north and south poles. Much like working with giant, mobile pieces of a very large jigsaw puzzle, geologists realign parts of the ocean floor to a common north-south direction, and in so doing, infer the approximate drifts of the plates during the past 200 million years. When this is done, these paleomagnetic data also tend to support the idea of a single ancestral supercontinent on our planet.

WHAT DRIVES THE PLATES? Oceanographic expeditions of the past decade offer clear evidence that hot basalt is rising through a crack all along the Mid-Atlantic Ridge. Dating techniques imply that this has been a regular process for the past 200 million years. And submarine observations show that the ridge is still active today. The upshot is that the Atlantic seafloor is slowly growing, pushing apart the North American and Eurasian Plates, as well as the South American and African Plates.

Researchers assume that the other plates on Earth's surface are also pushed around by matter upwelling from similar, though only partially explored oceanic ridges. Oceanographers are now retrieving and studying rock samples from the bottom of the Pacific Ocean and Caribbean Sea. During the next decade, we should gain a better understanding of all the major underwater cracks through which hot matter oozes from Earth's interior.

Even so, we are faced with an important question. What is the origin of the seafloor spreading? In other words, what really drives the plates? The answer is probably convection—the same type of physical process we encountered earlier (see especially Figure 5.6).

Figure 24.14 is a cross-sectional diagram of the top few hundred kilometers of our planet's interior. It depicts roughly the region in and around the Mid-Atlantic Ridge. There the ocean floor is covered with a layer of sediment—dirt, sand, and dead sea organisms that have fallen through the seawater for millions of years. Below the sediment lies about 50 kilometers of granite, the low-density rock comprising the continents. Deeper still lies the upper mantle made of hot, but not molten basalt.

This is a perfect setting for convection: Warm matter underlying cool matter. The warm basalt wants to rise, just like hot air in our atmosphere. It does so through any cracks and faults in the granite crust. Every so often, such a crack may open in the midst of a continental landmass, producing a spectacular volcano such as Mt. St. Helens, or a geyser like that at Yellowstone National Park. However, most of the

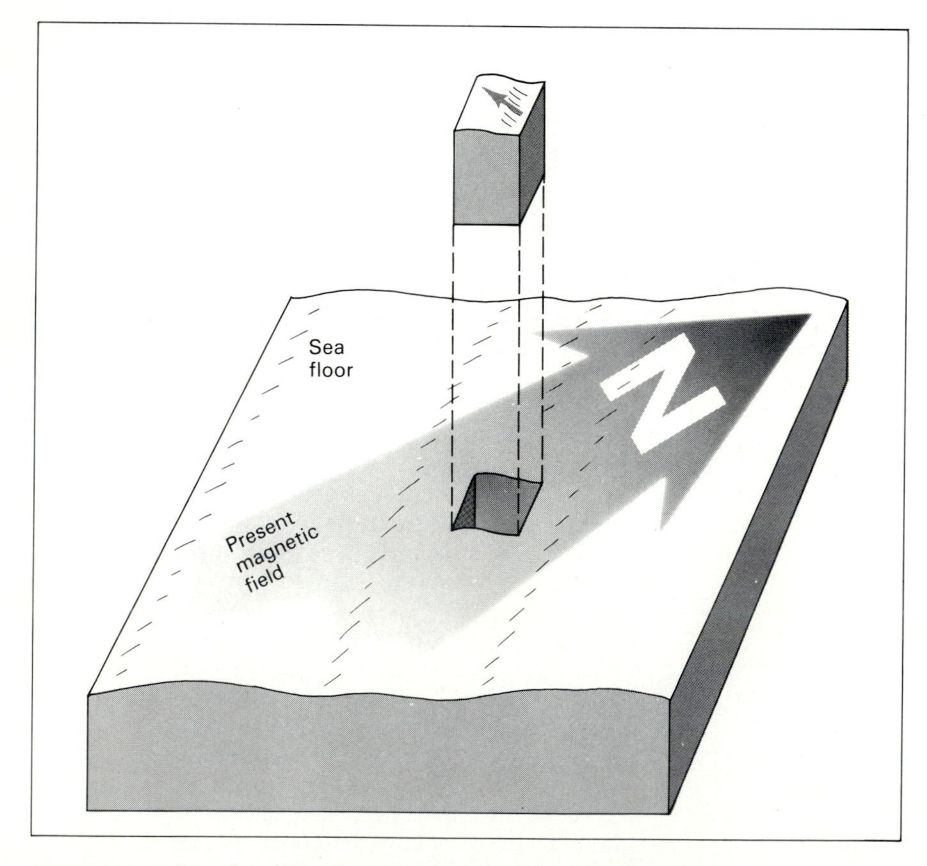

FIGURE 24.13 Samples of basalt retrieved from the ocean floor often show Earth's magnetism to have been oriented differently from the current north-south magnetic force field.

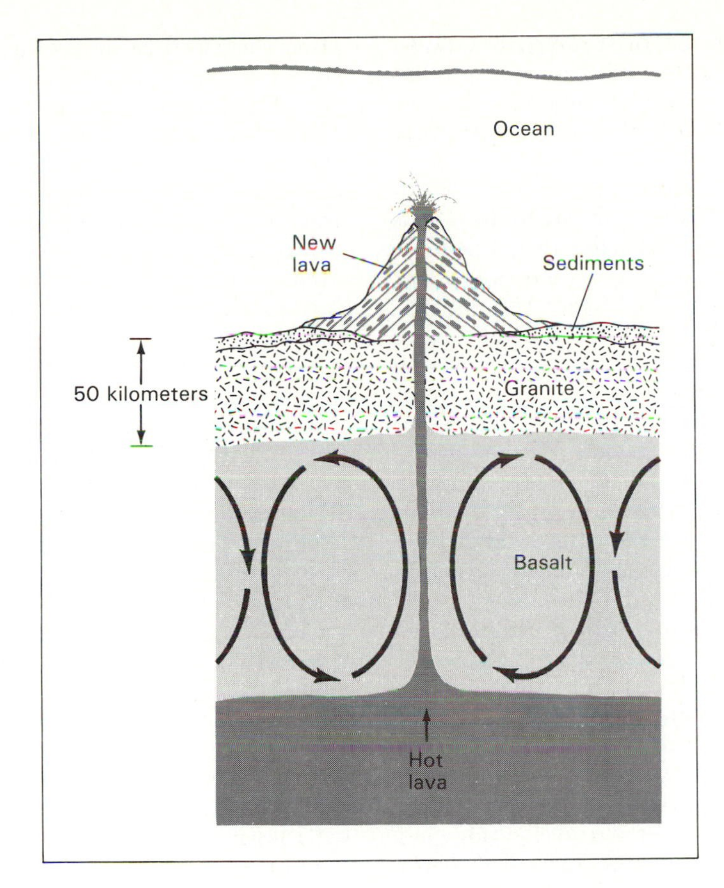

FIGURE 24.14 Plate drift is probably caused by convection—in this case giant circulation patterns in the upper mantle.

major crustal faults known today are submerged below water. The Mid-Atlantic Ridge is a prime example.

Not all the warm basalt in the upper mantle can squeeze through the cracks and fissures. Some gets pushed back down. In this way, ascending and descending matter cause large circulation patterns to become naturally established within the upper mantle as depicted in Figure 24.14. These circulation patterns are extraordinarily sluggish, since they involve the movement of molten rock. We can visualize the basalt as warm butter or asphalt, slowly circulating below a crack and taking probably millions of years to complete one convection cycle. Some researchers conjecture that this buttery basalt acts like a lubricant, enabling the thin plates to slide or drift across the surface of our planet.

FUTURE DRIFTS: Finally, we can inquire about the future positions of Earth's giant land masses. Since the plates are currently moving, we can use their estimated magnitude and direction of drift to predict where the continents will be in the years ahead.

Figure 24.15 charts the expected locations of the well-known continents 50 million years from now. This prediction assumes that warm basalt will continue to upwell through the Mid-Atlantic Ridge, thus widening the Atlantic Ocean. The Pacific Ocean will contract considerably; there being no large landmass on the Pacific Plate, this oceanic plate will presumably continue to be overridden by other continental plates such as the South American Plate and the Eurasian Plate.

Australia (which is probably part of the Indian Plate) will continue its northerly movement toward the Eurasian landmass, destined for massive collisions not long thereafter. India will continue to thrust northward, building the Himalayas to possibly greater heights.

The Mediterranean Sea is doomed because of the African Plate's northerly motion. This same drift guarantees great sking in the Alps for

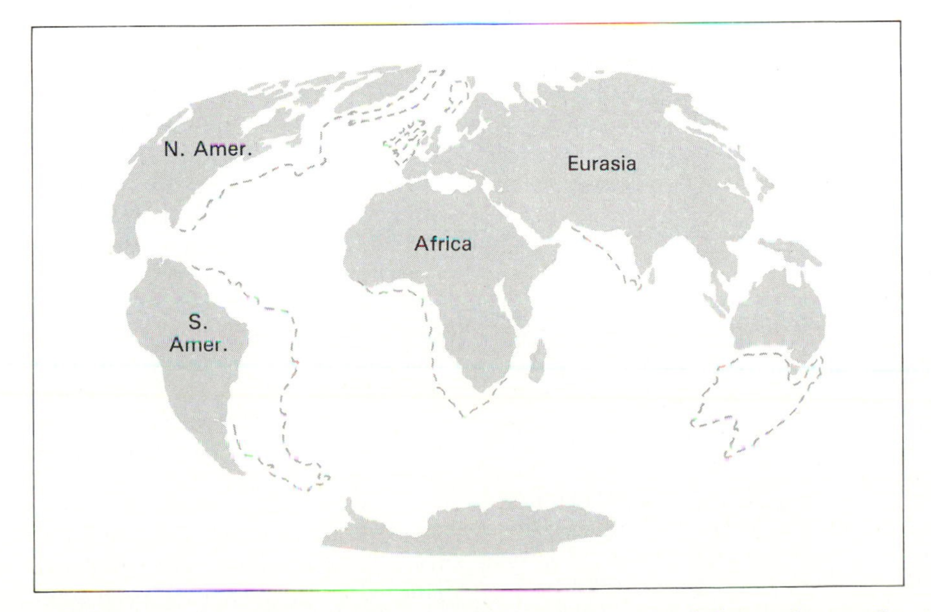

FIGURE 24.15 A prediction of where the plates will have carried the continents of our planet about 50 million years from now. The dashed contours outline the continents' current positions.

millions of years (provided that the climate doesn't change).

And southern California, as part of the Pacific Plate, will be torn away from the North American Plate, with Los Angeles becoming a suburb of San Francisco in 15 million years, before being dumped into the Aleutian Trench some 40 million years in the future.

Earth's Magnetic Reversals

The study of fossilized magnetism—paleomagnetism for short—has become a useful tool with which to study the magnetic properties of Earth. Within the past decade, researchers have used the paleomagnetic traits of numerous seafloor samples to support the theory of plate tectonics.

In addition, recent paleomagnetic studies have led to a remarkable discovery: The north and south poles of our planet have flip-flopped back and forth over the years. The north magnetic pole, now in the Arctic region, has been occasionally in the Antarctic region, while the south magnetic pole has occasionally resided near what we now call terrestrial north. These flip-flops have occurred many times in the past, although by human time scales it's not a frequent happening. Seafloor samples imply that the north pole has been in the Arctic region for the past 700,000 years or so.

Figure 24.16 shows that samples retrieved on both sides of the Mid-Atlantic Ridge display alternate strips of magnetic orientation. Seafloor matter in the shaded areas contains iron magnetized in the conventional north-south orientation; this suggests that Earth's magnetic force field was then as we know it now. In the unshaded areas, the magnetized seafloor matter generally has a south-north orientation, suggesting that the north and south poles were reversed at the time that this matter cooled and solidified. Since we know the seafloor is progressively older farther from the ridge, we can roughly date the times that Earth's magnetic force field changed its orientation.

As such, the seafloor can be imagined as a gigantic tape recorder. Hot matter first upwells through the crack in the Mid-Atlantic Ridge. The matter then cools and solidifies, preserving a record of Earth's magnetism at that time. The matter spreads out away from the ridge, all the while pushing apart the North American and Eurasian Plates, as well as the South American and African Plates. In this way, geologists have recently unraveled the history of the Atlantic Ocean's formation and growth.

How often has Earth's magnetic force field reversed its orientation? The oceanic data suggest at least a dozen times in the past 10 million years. What could have caused such reversals? While researchers don't yet know for sure, apparently something occasionally upsets the steady spin of Earth's liquid metal core, which is the probable source of our planet's magnetism. Some researchers have speculated that the culprit might well be the collision of Earth and a cosmic object such as a comet or asteroid. Such a catastrophic event could probably change the core's spin by sloshing the liquid trapped there.

Whatever the cause, a magnetic reversal is not likely to be an instantaneous event. Some time is probably needed for magnetism gradually to weaken and finally disappear. Also, some additional time—perhaps on the order of hundreds of years—might be needed for the magnetic force field to become reestablished. If so, then each time a flip occurs, the magnetosphere (including the Van Allen Belts) far above Earth's surface disappears for a rather long time. Without the protective "umbrella" of a magnetosphere to deflect or trap the charged particles incident on our planet, biological systems *might* be harmed.

Is there any evidence to suggest that life may have been so harmed in the past? Possibly. Fossil records of ancient life forms show that, every so often, once-abundant plants and animals suddenly became extinct. No one knows why they perished so rapidly. Even the dinosaurs, which reigned supreme on our planet about 100 million years ago, seem to have disappeared within a relatively short period of time.

Recent studies of seafloor matter suggest a connection between the extinction of certain categories of life ("species") and a reversal of Earth's magnetism. Apparently, a magnetic flip is the death sentence for *some* species. However, there is a positive note here as well: Other fossils found within the seafloor samples show that some new species emerged for the first time.

No one really knows what effect a magnetic force field reversal might have on the surface of our planet. But a general consensus seems to be at hand: While the magnetosphere is collapsed, the extra influx of high-energy particles would tend to increase the amount of radioactive

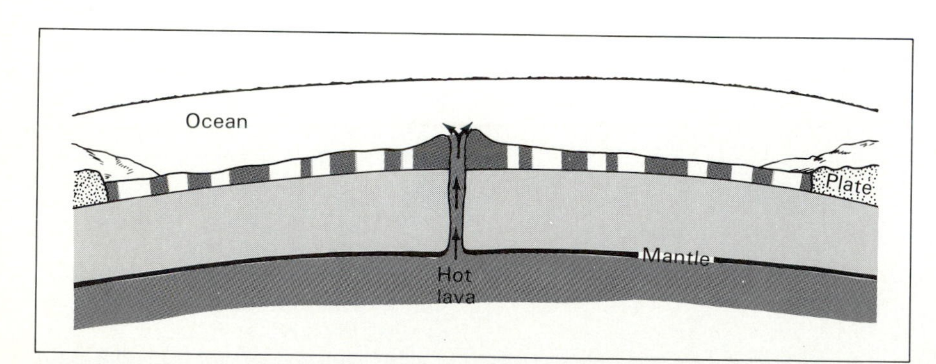

FIGURE 24.16 An entire series of alternately magnetised matter roughly parallels the Mid-Atlantic Ridge. In this simplified diagram, shading designates north-south orientation as we know Earth's magnetic force field today, unshading the opposite orientation.

atoms in our atmosphere. These atoms are then absorbed by plants, which in turn are eaten by animals, including humans. Although this higher level of radiation within plants and animals is unlikely to kill them directly, the normal course of biological evolution (to be studied in Chapter 26) would surely be disrupted. Reproductive errors from generation to generation ("mutations") would clearly increase, changing the basic biological molecules ("genes") in some living systems.

Yet, contrary to popular belief, not all mutations are bad. Some are beneficial, enabling living systems to adapt better to a changing environment—in short, to evolve. Mutations act as the motor of evolution. And when the magnetosphere has temporarily collapsed, that motor apparently runs faster than normal.

As we shall come to recognize in Chapter 26, whatever species dominates life at any one time does so largely because that species enjoys a nearly optimum equilibrium with its environment. It's best suited for its natural surroundings. An analogy might be a lantern slide that's in very good focus. A mutation (or a slight change of the slide's focus) is likely to harm the dominant species (or the projected image). By contrast, less-than-dominant species (or slides out of focus) might profit from a period of increased mutations by bettering themselves. But the dominant species at any one time is more likely to change for the worse.

Humans are now the dominant species on Earth.

Changing Atmosphere and Oceans

To consider how feasible it may have been to form life on Earth, we need to first examine the origin and evolution of our planet's atmosphere and oceans. To do so, we must go back several billion years since the oldest fossilized life forms are about 3.5 billion years old. In nearly every respect, the conditions on early Earth must have differed substantially from the world we now inhabit.

ATMOSPHERIC EVOLUTION: Earth's original atmosphere almost certainly contained all the most abundant elements—hydrogen, helium, nitrogen, oxygen, neon, carbon—as well as a long list of trace elements. If Earth formed as a by-product of the condensation theory discussed in Chapter 23, the gaseous composition of our primitive atmosphere probably mimicked those of the general cosmic abundances (see Figure 2.20). This primitive atmosphere did not stick around very long, however. Because the early surface of Earth became much hotter than today's, we can be reasonably sure that many of the original atmospheric gases—probably all of them—must have escaped into space. Gravity just couldn't hold back the hot gases of primordial Earth; the average molecular velocity exceeded the escape velocity of Earth, as discussed in Chapter 5. (Actually, the surprising scarcity of several inert gases—those unable to react with other chemicals—such as neon, argon, krypton, and xenon, despite their observed abundance in the Sun, is the best evidence that Earth failed to retain its original atmosphere. Apparently, none of the Terrestrial Planets did.)

Despite the escape of Earth's primordial atmosphere, a thin gaseous environment (air) surrounds our planet today. We wouldn't be here if it didn't. Hence Earth must have acquired more gases, called a **secondary atmosphere**, at a later date. (And these secondary gases in turn evolved, largely because of the presence of plants, as explained in Chapter 26, to become the air we now breathe; perhaps we should call our current atmosphere a "tertiary atmosphere.")

As primordial Earth gradually began to cool, its molten surface solidified into rock. Intense heat trapped below the crust had to get out somehow. The result was surely volcanoes, geysers, earthquakes, and a variety of other geological events that literally blew off steam and pent-up heat through cracks in the surface. This "outgassing" of heat and gas happens even today, although at only a few locations on Earth, and rather infrequently at that. But several billion years ago, this type of geological activity was surely more widespread and frequent. Calculations suggest that over the course of Earth's history, enough gas was exhaled from Earth's interior to create much of our current atmosphere. Observations of modern volcanoes show that lots of gaseous water vapor (H_2O), carbon dioxide (CO), and nitrogen (N_2) would have undoubtedly emerged, along with vast quantities of ash and dust. Smaller amounts of hydrogen (H_2), oxygen (O_2), carbon (C), and other chemicals also would have been released during these planetary eruptions.

The origin of our current atmosphere, then, is terrestrial outgassing. To be literally correct, the atmosphere is perhaps still forming as present-day volcanoes occasionally sputter gas from Earth's interior. Today's atmosphere was not, however, derived directly from a mixture of interstellar gases. Thus the composition of Earth's secondary atmosphere differs considerably from the cosmic abundances of the elements. (Jupiter and the other Jovian planets have atmospheres rich in hydrogen, helium, and other light gases; these planets are large enough to have retained their primitive atmospheres, which *were* formed directly from interstellar matter.)

PRIMORDIAL OCEANS: Since the atmosphere and oceans of planet Earth are so closely linked, they almost certainly originated from the same source—our planet's interior. As regards the oceans, geologists argue that as the surface cooled sufficiently, the first pools of water appeared. After all, steam (hot, gaseous H_2O) is a primary component of volcanically vented matter.

The debate among geologists concerns the rate of ocean formation. How long did it take? We are uncertain whether Earth's mantle outgassed the global oceans all at once early in our planet's history; leaning a little on popular astronomy ter-

INTERLUDE 24-2 ***Gaia***

Earth's atmosphere today—often called the biosphere, for this sphere's skin is full of life—has a potentially remarkable power. It might well be more than a mere by-product of volcanoes and life. That our air contains 21 percent oxygen and 0.03 percent carbon dioxide might not be accidental (nor supernatural). Despite the extreme differences in chemical composition of the air, land, and sea around the globe, Earth seems to harbor just the optimum amount of atmospheric gases, and furthermore seems to keep them within the narrow tolerances necessary to sustain life.

For example, with no oxygen, there could be no respiration (as we know it); even just a few percentage points less oxygen would cause most animals to die. By contrast, with just a little more oxygen—say 25 percent rather than 21 percent—the whole living world would burst spontaneously into flames. Similarly, without carbon dioxide, plants would die, and life (as we know it) would vanish from Earth. However, with just a bit more than today's carbon dioxide, so much heat would be trapped in the air by the greenhouse effect that our planet would probably resemble Dante's inferno.

The inhaling and exhaling of these and other atmospheric gases by the animal, plant, and bacterial kingdoms seem to help keep our planet livable. Similar life-sustaining "coincidences" also pertain to the watery component of our planet, especially, for example, the salt content of Earth's oceans; every year, rain washes roughly 500 million tons of salt from the land into the sea, yet were the salinity of the waters to increase by just a few percent, virtually all marine creatures would die. What keeps the salinity levels, the air composition, the surface temperature, and many other properties of the biosphere within such narrow ranges necessary for life? It is as though the many varied life forms and the planet itself are intimately bound and work as a whole. If true, then life shapes Earth as well as Earth's shaping life. To a certain extent, we might say that life manages its environment; it keeps our planet "healthy."

According to this theory—termed the Gaia hypothesis after the Greek goddess, Mother Earth—the hydrosphere, lithosphere, and biosphere work harmoniously to maintain an environment favorable for the continued existence of life. Much as our human bodies counteract a variety of rude shocks from outside, from sweating (thus cooling the skin of excess heat) to shivering (thus warming the body by releasing heat from muscles), this theory suggests that life itself compensates for environmental changes and threats—that life regulates the atmospheric gases in order to keep Earth's surface conditions steady. It's as though the whole biosphere were alive—a single, unified organism that unconsciously maintains the optimal conditions for its own life.

Opponents maintain that Earth did not outgas in order to produce the atmosphere. Gases simply erupted from vents in a hot Earth, and gravity kept most of them from drifting out into space. Nature is not a planner; nature is just what happens. Furthermore, the critics claim that the Gaia hypothesis is essentially anti-Darwinian: As we shall see in the next few chapters, classical Darwinism places a heavy emphasis on competition among life forms; by contrast, Gaia implies that living things cooperate for the common good.

In the conservative and contentious world of science, the Gaia idea has not won many converts. The institution of science is always on the lookout for anthropocentric sentiments, always wary of attributing human qualities to rocks, streams, winds, and trees. For Gaia's proponents, this reluctance presents hardly a worry, however, for most creative ideas are met with scathing criticism initially. Institutions are loathe to change—and not just religious institutions.

minology, geologists call this the "big-burp" theory. Alternatively, the oceans might have taken much time to form, having secreted from Earth's interior in a series of individual events. A minority of researchers argue that some (though not all) of the waters of Earth could have resulted from melted comets that collided with our planet in great numbers early in our planet's history.

Most probably, a large ocean originated early, and then, as the rate of outgassing declined, a global recycling system began to operate; waters locked in rocks were expelled back into the oceans whenever rocks were heated like those near volcanoes or suboceanic faults and ridges. To be sure, much of today's oceans have been recycled, but not quite all. In recent years, waters have been sited emanating from certain submarine vents. Thus small increments of "juvenile" waters still originate directly from Earth's mantle, constantly adding to the world's supply.

GASEOUS REACTIONS: Earth's oceans and secondary atmosphere would have gradually stabilized. As activity on the early planet subsided, the atmosphere cooled, enabling gravity to prohibit its further escape to outer space. Outgassed nitrogen partly reacted with other gases, and partly stayed put where it now comprises the largest fraction of our air. Gaseous water vapor changed into liquid water, which further rained down on Earth's oceans. And outgassed carbon dioxide reacted with silicate rocks in the presence of water to form limestone.

Shaded by Earth's secondary atmosphere, some of the outgassed chemicals would have further interacted with one another. For example, while the actual chemistry is more complex, we can summarize the basic changes with the following idealized equations:

three hydrogen molecules ($3H_2$) + one nitrogen molecule (N_2) → two ammonia molecules ($2\ NH_3$),

two hydrogen molecules ($2H_2$) + one carbon atom (C) → one methane molecule (CH_4),

two hydrogen molecules ($2H_2$) + one oxygen molecule (O_2) → two water molecules ($2\ H_2O$),

four hydrogen molecules ($4H_2$) + one carbon dioxide molecule (CO_2) → one methane molecule (CH_4) + two water molecules ($2\ H_2O$).

In these and other ways, simple gases would have reacted spontaneously to create the slightly more complex molecules of ammonia and methane among others. By spontaneously, we mean that the gases collided, stuck, and reacted on their own, unaided by any outside forces or energies. Electromagnetic force fields among the electrons in the simple chemicals on the left side of the equations above persuade those chemicals to combine readily, thus forming more stable molecules on the right side of the equations.

Of great importance, Earth's early atmosphere was apparently lacking in free oxygen gas. Whatever pure oxygen existed on primitive Earth would have quickly vanished either in the above reaction with hydrogen to make more water, or with surface minerals to form oxides such as rust and sand now found throughout the crust of our planet. Breathable oxygen is abundant in today's atmosphere, but it arose only much later when plants blossomed across the face of our planet. This lack of free oxygen gas in Earth's early atmosphere is crucial to the experimental work on the origin of life to be discussed in Chapter 25.

COMPARATIVE PLANETOLOGY AGAIN: We stand to learn a good deal about our own planet's early evolution by contrasting its properties with those of some of the other Terrestrial Planets. Recall from Chapter 7 that a synthesizing effort is now under way by planetologists—those scientists who compare a broad range of properties recently discovered among the varied worlds in our solar neighborhood. What makes Earth so different from the other planets? How is it that we alone have blue seas of air and water? Why do we alone have a gentle climate? And why is Earth the only world in the Solar System (as best we know) that is a home to life?

Consider, for example, the atmospheres of Venus, Earth, and Mars. Although almost certainly endowed at birth with comparable amounts of hydrogen, carbon, and oxygen, each of these three nearby Terrestrial Planets have evolved differently from one another. These differences derive largely from their varying sizes and distances from the Sun. Figure 24.17 depicts their major current atmospheric constituents.

Of these three neighbors, our inward sister Venus receives the most solar energy (roughly twice as much as Earth). Although water is nowhere to be found on this planetary hot house today, early in its history, when the Sun shone less brightly (about 70 percent of its present luminosity 4 billion years ago), Venus might have had extensive oceans, rivers, and lakes. As the Sun slowly increased its output (while evolving from stage 6 to stage 7 as noted in Chapter 19), the planet progressively heated and its water boiled off. In the meantime, Venus' volcanoes continued to vent much carbon dioxide into its atmosphere. And without the water to transform carbon into rocky carbonates such as chalk, limestone, or coral (as is the case on Earth), the CO_2 levels on Venus rose unchecked. This led to a "runaway" greenhouse effect, as described in Chapter 5, allowing solar energy to penetrate the thickening atmosphere, yet blocking some of its outgoing infrared radiation—all the while making the surface of Venus too hot to support even primitive life.

By contrast, Earth is far enough from the Sun to have retained its water. Our planet is furthermore able to recycle its carbon through plate tectonics (which Venus might

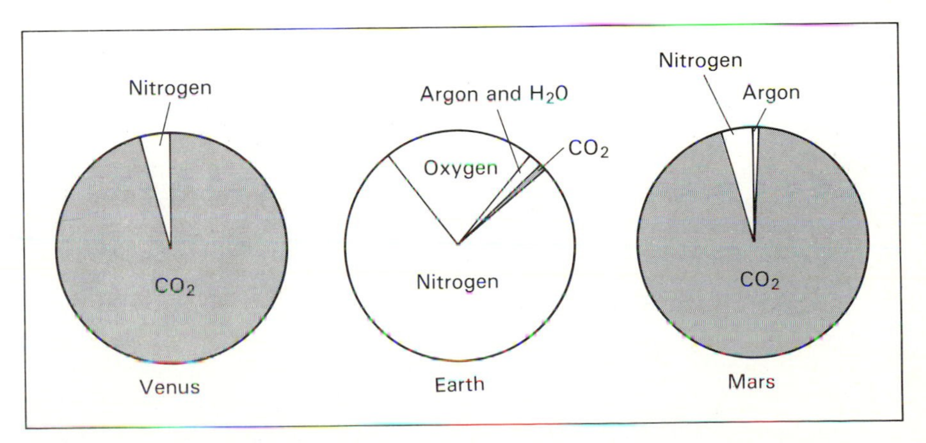

FIGURE 24.17 A comparison of the relative atmospheric gas compositions of Venus, Earth, and Mars, as we know these planets today. In absolute terms, Venus and Earth have nearly the same amount of CO_2, emphasizing the delicate balance between a poisonous and a life-sustaining environment.

not have); CO_2 outgases from terrestrial volcanoes (even today), but it merely dissolves in the sea and eventually reacts with oceanic rocks to help form the crust, thus avoiding the buildup of this noxious gas. The prolonged presence of water enabled marine organisms to evolve, and they now serve as an effective means of further removing CO_2 from the air (in order to make skeletons, which later fall to the seafloor and solidify into more rock). Long ago, an atmospheric steady state—a chemical balance of sorts—was apparently reached: Volcanism regularly vents CO_2 to the atmosphere, whereupon it is continuously trapped in rocks and plants. A small percentage of CO_2 gas in our air nonetheless drives a small greenhouse effect—thereby raising our average surface temperature above the freezing point of water, as noted in Chapter 5. Our climate is thus more moderate than Venus', although humans are beginning to tinker with the delicate balance. We are industrially polluting our air as well as deforesting the land, in the process causing both the CO_2 content and the global temperature to rise; these are measured facts.

Mars, too, probably once had a moderate climate and liquid water on its surface. But it is a good deal smaller than Earth, hence had trouble holding onto its original atmosphere. And given that its tectonics never really got going (nor is it likely they ever will), Mars is unable to generate much of another atmosphere. The result, despite the high percentage of CO_2 gas, is a slight greenhouse effect at best. Water there is completely frozen, as Mars is (perhaps permanently) in the grip of an ice age.

The contrast between Earth's early atmosphere and that of Titan today is even more instructive. Recall that Saturn's big moon is rich in methane and nitrogen gas as well as in several carbon-based compounds. Under the action of sunlight, these gases undergo a complex series of chemical reactions, producing sort of a hydrocarbon smog. Most significantly, these chemical reactions and the organic matter they produce are thought to resemble those that materialized in Earth's atmosphere billions of years ago, before the advent of living things and an oxygen-rich air. Titan seems to be a chemical "factory" awaiting our exploration, and one that might teach us a wealth of information about the vital prebiological steps that apparently led to life on our planet long ago.

SUMMARY

Each of the special attributes of any planet—tectonic dynamism, liquid water, free oxygen, life itself, among many others—depends on that planet's size and position in the Solar System. Planet Earth is large enough to have remained hot and thus continues to experience a variety of surface activity. Yet it has cooled enough to allow gravity to bind outgassed air and water to its surface. Viewed from a broad perspective, not much is stable on Earth; virtually everything evolves, although much of that developmental change these days is slow and often subtle.

Our distance from the Sun and our atmospheric blanket combine to keep Earth's surface temperature suitable for water to be liquefied—an appparently critically important factor in any environment hospitable to life as we know it. As has been stated many times—and not just from an anthropocentric viewpoint—Earth is the right body in the right place at the right time.

KEY TERMS

fault
geography
oceanography
secondary atmosphere
paleomagnetism
paleontology
plate tectonics

QUESTIONS

1. How does the composition of Earth differ from that of the Universe in general, and why?
2. In a couple of paragraphs, describe Earth's primordial surface and atmosphere.
3. Describe the role probably played by volcanoes in the early history of Earth.
4. Data from recent space probes seem to suggest that the atmosphere of Venus is primordial, that is, it is Venus' first atmosphere. How might this distinguish Venus from Earth?
5. Cite, and explain briefly, at least three pieces of evidence that favor the notion of plate tectonics on Earth.
6. Pick one: Studies of continental drift probably will not aid (a) earthquake predictions; (b) ICBM targeting; (c) our understanding of mountain building; (d) our study of Earth's earliest history, (e) our knowledge of environmental change.
7. Why are earthquakes more prevalent in Turkey than in Brazil, and in California than in Vermont?
8. What might Earth's surface be like without an atmosphere? Justify your answer.

9. Name at least six areas of the world where the surface is currently active. Discuss the probable reasons for this activity.

10. How do astronomers reason that if the atmospheres of Mars and Venus were interchanged, both planets might be more "Earth-like."

FOR FURTHER READING

BONATTI, E., "The Rifting of Continents." *Scientific American*, March 1987.

CLOUD, P., *Cosmos, Earth, and Man*. New Haven, Conn.: Yale University Press, 1978.

CLOUD, P., "The Biosphere." *Scientific American*, September 1983.

GROSS, M., *Oceanography*. Englewood Cliffs, N.J.: Prentice-Hall, 1987.

HURLEY, P., "The Confirmation of Continental Drift." *Scientific American*, April 1968.

SCHNEIDER, S., "Climate Modeling." *Scientific American*, May 1987.

25 CHEMICAL EVOLUTION: *The Origin of Life*

Throughout this book we emphasize that our place in the Universe seems in no way special. Neither our planet and star nor our place in the larger cosmos is unusual. The only possible exception is that Earth is the only place where we know life definitely exists. Even so, we have searched carefully for life hardly anywhere else in the Universe.

Having discussed many of the astronomical and physical (or "astrophysical") concepts needed to understand the origin and evolution *of matter*, we now consider some of the biological and chemical (or "biochemical") concepts needed to appreciate the origin and evolution *of life*. The synthesis of our understandings of matter and life comprises the essence of cosmic evolution.

This chapter emphasizes the origin and nature of early life here on Earth. We shall examine the efforts of scientists who study life's basic molecules and cells. And we shall evaluate the work of those who study the dead remains of primitive life often found embedded in ancient rocks. Together, these two subjects—biochemistry and paleontology—are providing a rich understanding of how life emerged and evolved eons ago.

Of course, some gaps in our understanding hinder our complete appreciation of life's history, just as some missing links hamper our current knowledge of galaxies, stars, and planets. But each day brings new discoveries, new tests, and further refinement of our modern ideas of biochemical evolution. And with these advances come greater objectivity, and progress too, in our search to know reality.

The learning goals for this chapter are:

- to appreciate the lack of clarity separating that which is living from that which is not
- to realize that life is more complex than any nonliving piece of matter
- to know the central ideas of modern chemical evolution
- to understand cells and the building blocks of life that comprise cells
- to appreciate some of the laboratory experiments that simulate the steps that probably led to life's origin on Earth

Definition of Life

We begin our study of life by asking the obvious question: What is life? However, even this seemingly simple question is difficult to answer. Definitions of life are not easy to develop.

Answers to the companion question, What is matter?, are a little easier. Physicists have unraveled the basic nature of matter reasonably well during the past several decades. As noted and studied in earlier chapters, elementary particles cluster into atoms, atoms into molecules, molecules into gases, rocks, mountains, and so on, through the hierarchy of matter up to and including galaxy superclusters. Matter is the basic stuff of the Universe, and we have a fairly good idea how it operates from quark to quasar.

Life is harder to appreciate. Whereas most physicists agree upon a definition of matter, most biologists have not been able to reach an agreement concerning a definition of life.

SOME PROPERTIES OF LIFE: We can gain some understanding of life by noting its operational properties. In other words, by studying a few of life's attributes, we can perhaps begin to appreciate its important features—especially those features that help to distinguish life from matter. For example, we might suspect that living systems differ from nonliving systems because *the whole is greater than the sum of the parts of which it is made*. An individual cell dies when removed from a living organism of which it was a part. The interactions of that cell with other parts of the whole living organism are vital to the cell's health. On the other hand, should a single cell be nourished in a comfortable laboratory environment having just the right temperature and density—a so-called cultured medium—the cell could once again flourish outside its original living organism.

At first glance, then, we might want to consider the italicized property above as a peculiarity of life. But on second thought, this property is not at all restricted to life, for it is also a property of matter. To see this, imagine removing a small part of a star normally fusing hydrogen into helium. The extracted chunk of matter would no longer release nuclear energy, for it would immediately disperse into space and grow cold. Yet if the extracted chunk were surrounded by additional matter having the appropriate temperature and density, it would once again shine as brightly as before.

These statements are not meant to suggest that stars are somehow "living." Quite the contrary. It is precisely because we can be sure that incredibly hot stars cannot possibly be living that this comparison demonstrates how tough it is to define life. Thus we cannot claim that the "whole being greater than the sum of its parts" is a property solely of living systems. This property applies equally well to many objects that are not living. A watch, for instance, is surely more than the sum of the gears, springs, and wheels (or silicon chips and integrated circuits) of which it's made.

We might then suggest that the *ability to heal itself* is a peculiar property of a living system. A shallow cut on a finger, for example, usually heals without delay and the system goes on living. On the other hand, the aforementioned star from which a small chunk of matter was extracted would eventually "heal" itself. The star would adjust a bit, eventually attaining a new equilibrium between the inward pull of gravity and the outward pressure of hot gas. Having resumed its original spherical shape, the star would then go about its business of shining as a perfectly normal, though slightly smaller star.

We might claim that living systems have a special property that allows them to *react to unforeseen circumstances*. But the aforementioned star did not expect to have a small part hypothetically extracted. Yet any star would react quite adequately to this unexpected occurrence.

Many people feel that the *ability to reproduce* is a special property of a living system. Yet we can imagine a contracting protostar which, because of faster and faster rotation, divides into two separate protostars. In this way, angular momentum can sometimes be considered an agent of replication, or at least subdivision. Admittedly, this example probably occurs rarely; yet it has undoubtedly happened numerous times in the billions upon billions of years since the start of the Universe. Some of the binary stars throughout our Milky Way Galaxy may well have been created in just this way.

Surely, there must be some property associated with life and only life. Another possibility is that living systems can *learn from experience*. Indeed, most living organisms remember what happened to them in the past. Yet some nonliving systems can also remember, and even learn from experience. Computers, for instance, are now able to play mental games such as checkers and chess. And when a well-programmed computer makes a mistake, it does not forget it. These machines can store mistakes in their memory, never to be made again under the same circumstances. Accordingly, few humans can beat our best computers at chess, and no one can beat them at checkers. So we conclude that computers—hunks of metal matter, or at least their "software"—can learn from experience, much like living systems.

Finally, we can suggest that life is characterized by an entire *hierarchy of functions*. Much of the activity of living systems is controlled by chemicals called hormones; hormones in turn are controlled by organs called glands; glands by brain cells, and so on. Surely such hierarchies characterize all living systems from the simple amoeba through the sophisticated human. In similar fashion, though, we can think of nonliving matter as being controlled by a hierarchy of functions. For example, the motion of the Moon is controlled by Earth; Earth's motion in turn is controlled by the Sun; the Sun by the Galaxy, and so on through the galaxy superclusters. Many material things, then, have entire hierarchies that somewhat resemble living systems.

The important point here is that we cannot easily specify any property applicable to life, and only life (see, however, Interlude 25-2). Apparently, under special circumstances, common properties of life can also apply to matter. In short, there seems to be no clear dividing line between nonliving systems and living systems—no obvious distinction between matter and life.

COMPLEXITY IS THE KEY: Living and nonliving systems, then, do not seem to differ *in kind*. We cannot easily distinguish among their basic properties. However, living and nonliving systems do differ *in degree*. All forms of life are more complex than any form of (nonliving) matter.

As a result, it seems reasonable to theorize that life is merely an extension of the complexities of matter. If correct, then everything surrounding us—galaxies, stars, planets, and life—comprises a grand evolutionary pattern encompassing all things in the Universe, including ourselves. Indeed, this is the essence of the subject of cosmic evolution.

With these preliminaries in mind, let us examine some of the theories proposed to account for life as an assembly of nonliving things.

Various Theories

The question of life's origin has engaged the minds of humans since they first contemplated the nature of their place on Earth and in the Universe. The subject often elicits emotion—first, because it involves ourselves, and second, because we do not yet have a comprehensive account of the specific steps that led to life on our planet.

A NONSCIENTIFIC IDEA: Many people have been raised to believe certain principles, one of which is that life originated by means of a God or gods. However, the theological or philosophical idea that life resulted from such a *super*natural process is a *belief.* Admittedly, it might be a perfectly good belief, but it remains just that—a belief—for there are no clear data supporting the creation of life by some supernatural being or beings. We have no data whatsoever suggesting that someone or something deposited already-formed life on planet Earth long ago. Furthermore, we have no known way to experimentally test the idea that a divine being instantaneously created life.

The belief that life suddenly arose by means of some vitalistic process is outside the realm of modern science. The logical approach of the scientific method, which is completely dependent on experimental and observational tests, cannot be used to study any supernatural idea for the origin of life. Accordingly, none of these ideas can be currently proven or disproven. They seem destined to remain beliefs forever.

THREE SCIENTIFIC PROPOSALS: Several other theories of the origin of life do not require the intervention of supernatural beings. Each of these theories relies on natural (not supernatural) principles, and each can be tested experimentally; these theories are thus based on science rather than on theology or philosophy. Only one of these has thus far survived the test of time, criticism, and debate.

One theory of the origin of life on Earth is termed **panspermia**, meaning "germs everywhere." This idea maintains that microscopic living organisms came to our planet from outer space. For example, meteorites or comets, perhaps containing cells or simple organisms, might have landed on Earth at some time in the past. Such primitive cells then evolved over billions of years into the more advanced forms of life now on Earth.

The basic tenet of panspermia is that primitive life, which originated someplace else, was deposited on Earth's surface by means of a collision with some other object that already harbored life. However, the current consensus among space scientists is that unprotected primitive life would probably be unable to survive the harsh environment of outer space. High-energy radiation and high-velocity particles in interplanetary and interstellar space would almost surely destroy any form of life riding on the backs of small astronomical objects.

(Extreme versions of the panspermia idea abound, among the strangest of which is that life on Earth arose from the garbage dumped here eons ago by extraterrestrial voyagers! Likewise, extraterrestrials might have deliberately seeded our planet, if only because of missionary zeal. These and other bizarre versions of the panspermia theory have fueled science-fiction writers for decades.)

A related aspect of the panspermia idea has, in recent years, become popular. With the rash of discoveries of molecules in interstellar space (consult Interlude 12-2), some researchers have proposed that not necessarily life itself, but the basic ingredients for life arrived on Earth embedded in comets or meteorites. These molecules could have then acted as the seeds that gradually evolved into life by natural chemical events.

At any rate, even if the notion of panspermia someday becomes a more promising idea for the origin of Earth life, it still does not qualify as a valid theory for the origin of life itself. Panspermia merely displaces the question of life's origin, transferring it to some other, unknown location in the Universe.

Another theory of the origin of life—one that directly addresses the ultimate origin of life itself—is called **spontaneous generation**. Here, life is thought to emerge suddenly and fully developed from peculiar arrangements of nonlife. This theory was popular as recently as a century ago, although only because people were misguided by their senses. For example, small worms often appear on decaying garbage, and mice sometimes seem to squirm spontaneously out of dirty bed sheets. Such phenomena were claimed as evidence for the spontaneous generation of new life from the decayed remains of old life. However, although these observations were correct, the interpretations of the observations were grossly invalid. Until about a century ago, people just did not realize that flies often lay eggs on garbage, after which the eggs hatch to become worms. Similarly, mice do not originate in dirty sheets; that may be where they like to hide, but sheets are not the source of their existence.

The theory of spontaneous generation was proved incorrect when scientists began carefully to monitor and control laboratory experiments. The nineteenth-century French chemist Pasteur, in particular, was one of the first researchers to conduct experiments under sterilized (clean) conditions. By using specially designed chemical equipment, he was able to show that any parcel of air contains microorganisms and other unseen contaminants. Without special precautions and close observations, living matter can easily come into contact with nonliving matter, thus giving the illusion that life originates spontaneously in places where no life had existed before. However, by heating the air and thus destroying the microorganisms, Pasteur thoroughly disproved the idea of the spontaneous generation of life. Once sterilized, air remains free of life, even microscopic life, indefinitely.

A third theory of life's origin is known as **chemical evolution**, namely those pre-biological changes that transformed simple atoms and molecules into the more complex chemicals needed to produce life. Favored by most scientists today, the central idea of chemical evolution stipulates that life arose from nonlife. In this sense, the theories of chemical evolution and spontaneous generation are similar. The main difference is that chemical evolution does not necessarily occur spontaneously; instead, it proceeds gradually, building complex structures from simple ones. Thus this modern theory suggests that life originated on Earth by means of a rather slow evolution of nonliving matter. How slowly we are unsure.

Estimates of the time scale over which chemical evolution operated can be obtained by studying **fossils**—the hardened remains of dead organisms whose outlines or bony features are preserved in ancient rocks. For example, Figure 25.1 illustrates two pieces of sedimentary rock discovered several years ago in the soil of South Africa. In these photographs, taken through a microscope, the rock has been magnified many times. Shown there is clear evidence for the fossilized remains of individual cells—the simplest known form of life. Radioactive testing proves that the age of the rock is 3.5 billion years. This is presumed to be the length of time that the fossils have been buried, making them some of the oldest fossils ever found. Apparently, the cells became trapped while the rock was solidifying.

Knowing that Earth originated 4.6 billion years ago and that the oldest rocks crystallized from their early molten state about 4 billion years ago, we can conclude that life must have originated roughly 1 billion years after Earth formed, and no more than $\frac{1}{2}$ billion years after Earth's crust became cool. Since even older, as yet undiscovered fossils probably lie buried somewhere in Earth's rocks, we surmise that the most primitive forms of life may have taken hardly more than a few hundred million years to evolve chemically from nonlife. They might have required a lot less time.

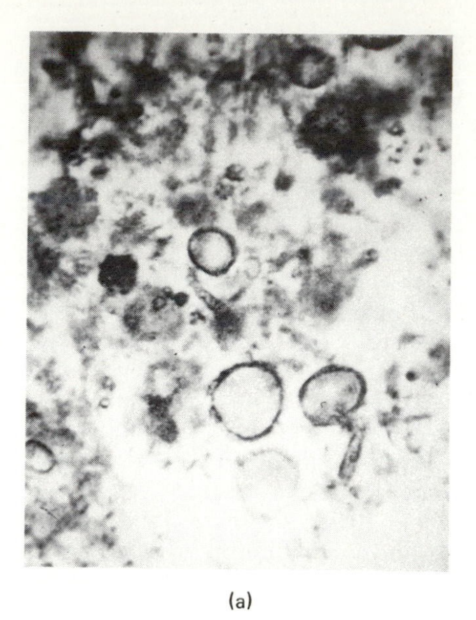

(a)

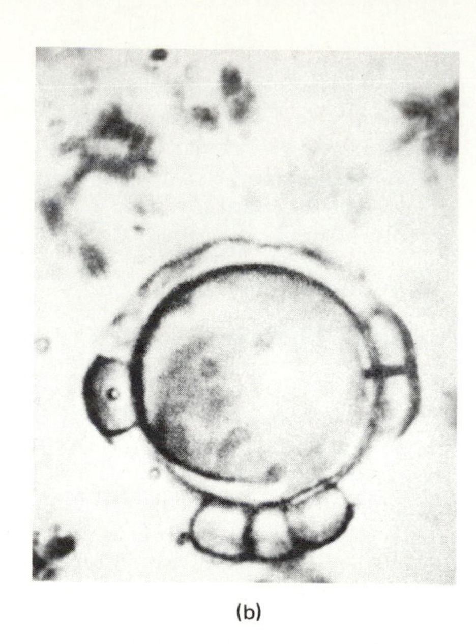

(b)

FIGURE 25.1 Two photographs, taken through a microscope, of now-dead cells that lived approximately 3.5 billion years ago. The scale of the entire frame in (a) is about 10^{-2} centimeter, whereas that in (b) is about 10^{-3} centimeter. Note the semi-permeable membrane surrounding the ancient cell in (b).

Biochemistry in a Nutshell

In this section we study the building blocks that contributed to the formation of life on Earth. We begin by examining the molecular nature of contemporary life. In this way we can better appreciate the complex ways in which matter must have clustered to form life in the first place.

CELLS: To understand the very basic properties of life, we need not study an organism as complex as the entire human body. From primitive creatures to intelligent humans, all living systems are made of **cells**, the simplest type of clustered matter having the common attributes of life—birth, metabolism, and death. We can therefore appreciate much of chemical evolution by understanding the construction of this simplest of all living systems, the cell.

Figure 25.2 is a sketch of a microscopic cell, normally about 10^{-3} centimeter in diameter, or some 10,000 times smaller than the size illustrated. About 1000 such cells would fit within the period at the end of this sentence. Most (but not all) cells have a nucleus close to their center. This biological nucleus is the most complex part of any cell, containing trillions of atoms and molecules; it should not be confused

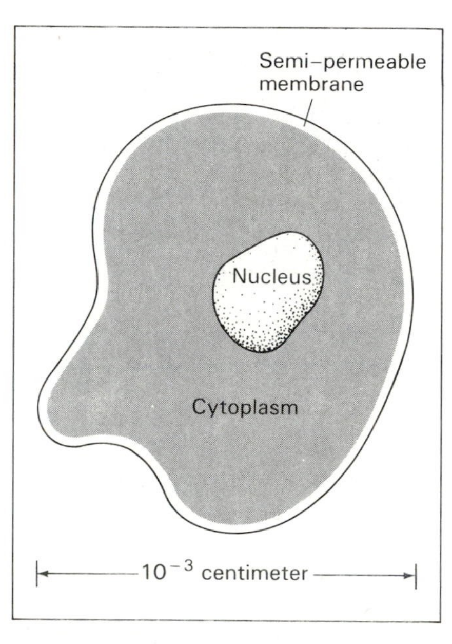

FIGURE 25.2 Schematic, though simplified, diagram of a single cell.

with the much smaller atomic nuclei produced in the cores of stars. Resembling the yolk of an egg, the biological nucleus is surrounded by a thick fluid of less complexity called "cytoplasm." The whole unicellular life form is encased within a semipermeable membrane through which atoms and molecules can pass in and out.

Cells, then, are the simplest form of life. However, they are vastly more complex than the simplest form of matter—elementary particles within atoms.

One of the very simplest creatures of all—the amoeba—consists of only one cell. More advanced organisms usually contain huge clusters of cells. For example, a grown human is built of nearly 100 trillion (or about 10^{14}) microscopic cells. These cells comprise the guts, skin, bones, hair, muscles, and all other parts of our bodies. Throughout the human body, their density averages 200 million cells/cubic centimeter.

Over the course of time—every second, in fact—large numbers of cells are destroyed via the normal process of aging and death. All living systems are nonetheless able to maintain a reasonably constant size and appearance throughout adulthood. This means that, while some cells are dying, others must be forming. Our bodies, and the bodies of all other living creatures, continually manufacture cell nuclei, cytoplasm, and membrane. To appreciate the mechanism whereby living systems sustain themselves during life, we must examine the two basic building blocks of life—the amino acids and the nucleotide bases.

AMINO ACIDS: The dominant ingredient of the cytoplasm is **protein**. The word *protein* derives from the Greek, meaning "of first importance." It's not the name of a particular substance, but rather a term for an entire class of molecules. There are tens of thousands of different proteins in the human body alone.

Proteins are known to contain large quantities of the element carbon. In fact, 50 percent of the dry weight of our bodies is carbon. Such carbon-rich "organic" substances strongly contrast with things obviously not living, such as a slab of concrete or a pinch of salt. Those things are said to be "inorganic," for they are made mostly of minerals. Inorganic substances contain no proteins, and their carbon content often amounts to less than about 0.1 percent of their total weight.

So carbon atoms play an important role in living systems. They play a vital role in the construction of proteins.

Of what are the proteins composed? Besides having lots of carbon, is there a common denominator among the myriads of different proteins found throughout the many varied cells of contemporary life? The answer is yes, for experiments have shown that proteins are made of a rather small group of molecules, called **amino acids**. In nature, there are only 20 of these structural units, although chemists have artificially made many others. When linked together in various ways, these acids comprise the millions of different proteins found throughout Earth's life—not just human life; all life. Amino acids are one of the two basic building blocks of life.

Figure 25.3 is a schematic diagram of a typical amino acid. All 20 kinds of amino acids have the same backbone or inner structure so marked in the figure. A carbon atom lies at the center of the backbone. On all four

INTERLUDE 25-1 ***The Virus***

The central idea of chemical evolution is that life evolved from nonlife. But aside from biochemical intuition, and laboratory simulations of some key events on primordial Earth, do we have any direct evidence that life developed from nonliving molecules? The answer is "yes."

The smallest and simplest entity that sometimes appears to be alive is called a **virus**. We say "sometimes" because viruses seem to have the attributes of both nonliving molecules and living cells.

Virus is the Latin word for "poison," an appropriate name since viruses are a cause of disease. Although they come in many sizes and shapes, all viruses are smaller than the typical size of a modern cell. Some are made of only a few hundred atoms. In terms of size, then, viruses seem to bridge the gap between cells that are living and molecules that are not.

Viruses contain both protein and DNA, although not much else—no unattached amino acids or nucleotide bases by which living organisms normally grow and reproduce. How, then, can a virus be considered alive? When alone, it cannot; a virus is absolutely lifeless when isolated from living organisms. But when in contact with living systems, a virus has all the properties of life. Viruses come alive by injecting their DNA into the cells of healthy living organisms. The genes of a virus seize control of a cell and establish themselves as the new master of chemical activity. Viruses grow and reproduce copies of themselves by using the free acids and bases of the invaded cell, thereby robbing the cell of its usual function. Rapidly and wildly, some viruses multiply, spreading the disease and, if unchecked, eventually killing the invaded organism.

Viruses, therefore, cannot be classified as either living or nonliving. Life seems to shade almost imperceptibly into nonlife. And the viruses apparently exist within this shaded, uncertain realm.

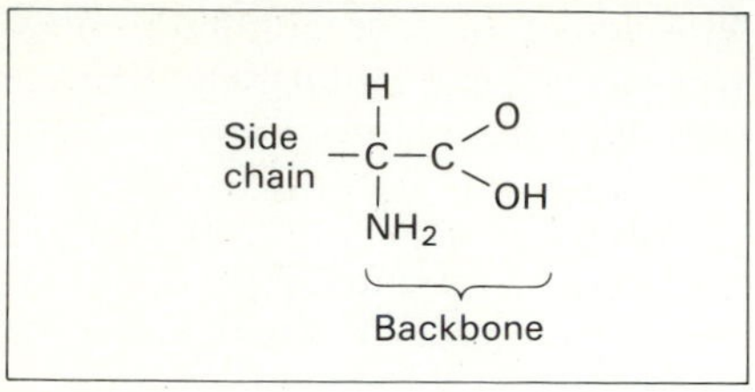

Side chain

Backbone

FIGURE 25.3 Schematic diagram of the molecular backbone for all 20 kinds of amino acids. The "side chain" can represent any one of 20 different types of molecules, each type granting an amino acid its chemical character.

FIGURE 25.4 A complete list of the 20 natural amino acids that participate in life. Each one has an identical structure, except for the side chain which differs for each kind of amino acid.

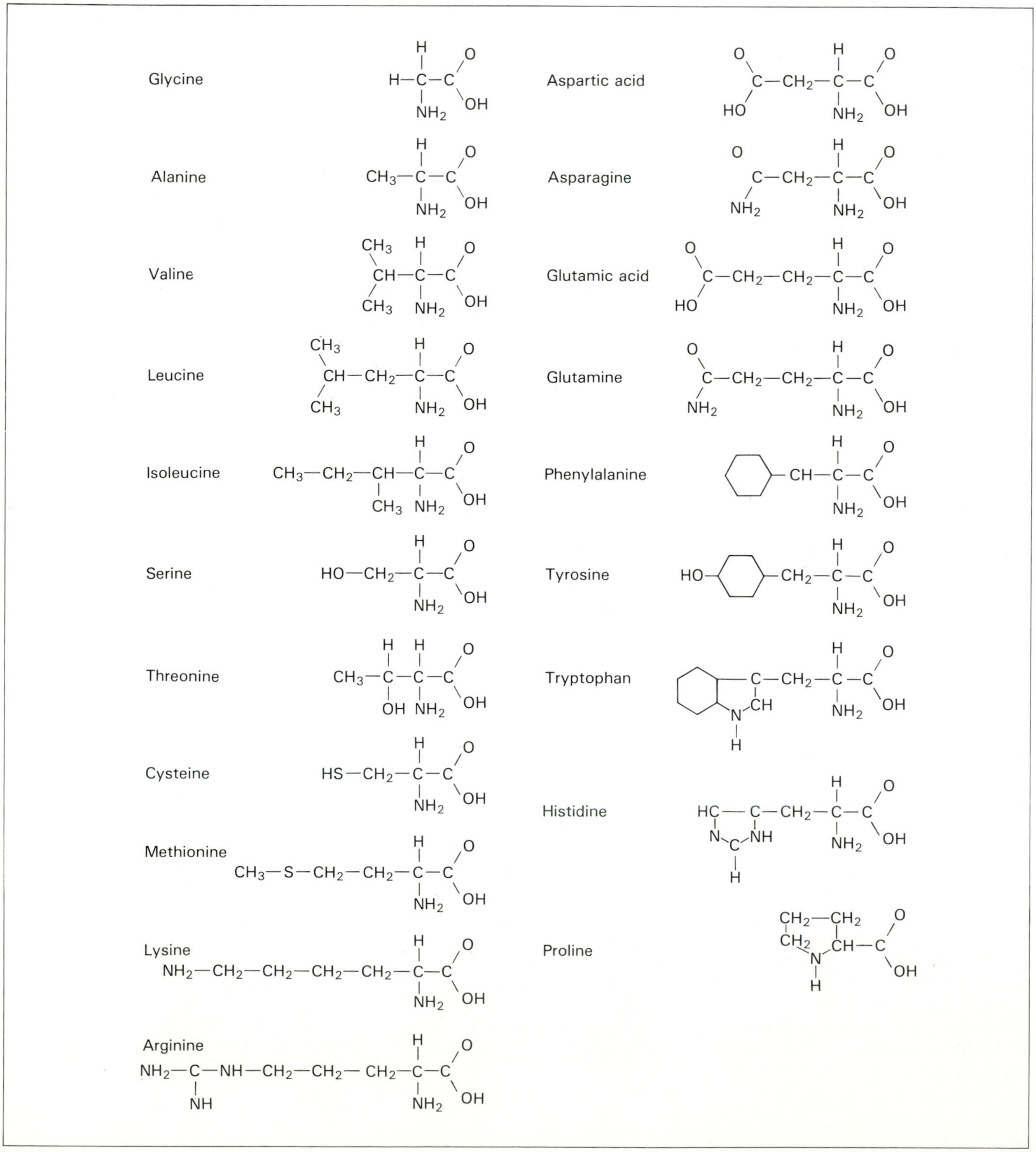

Glycine

Alanine

Valine

Leucine

Isoleucine

Serine

Threonine

Cysteine

Methionine

Lysine

Arginine

Aspartic acid

Asparagine

Glutamic acid

Glutamine

Phenylalanine

Tyrosine

Tryptophan

Histidine

Proline

sides of this central carbon, other atoms or small molecules are clustered. They are bonded at different angles to the central carbon atom by means of cohesive forces produced by the electric charges of each of the atoms and molecules.

As shown in Figure 25.3, a single hydrogen atom is attached to one side of the central carbon atom. To another side, a carboxyl group of atoms, designed COOH, represents a four-atom molecule which is itself bonded together by electric charges. An amino group of atoms, NH_2, is also attached to the central carbon atom; this molecule forms the basis of the more well-known ammonia molecule, NH_3.

A fourth molecule attaches to another side of the central carbon atom. Here we call it a "side chain," although there is no specific atom or molecule having that name. Instead, the side chain is shorthand notation for any one of 20 different molecules that can attach themselves to the backbone. For the simplest amino acid, known as glycine, the side chain is the simplest possible attachment—hydrogen, H. A slightly more complex amino acid, alanine, has a side chain made of the methyl molecule, CH_3. Figure 25.4 is a complete list of all 20 amino acid molecules that participate in life, showing the structure of the side chain in each case.

These 20 molecules are the only kinds of matter of which *all* proteins are constructed. Since the backbone is identical for all amino acids, we can surmise that the different side chains are responsible for the character of the amino acid molecules. In fact, the physical and chemical behavior of an amino acid depends largely on the structure of its side chain.

On a larger realm, the physical and chemical behavior of a protein—which is a long, stringy accumulation of amino acids—depends not only on the *number* of amino acids, but also on the *ordering* of the amino acids comprising that protein. Figure 25.5 illustrates how two glycine amino acids can be coupled to form the simplest possible protein. As shown, an electromagnetic link (also called a "chemical bond") requires the removal of hydrogen (H) from one of the glycine amino acids, and oxygenated hydrogen (OH) from the other. This amounts to the removal of some water (H_2O).

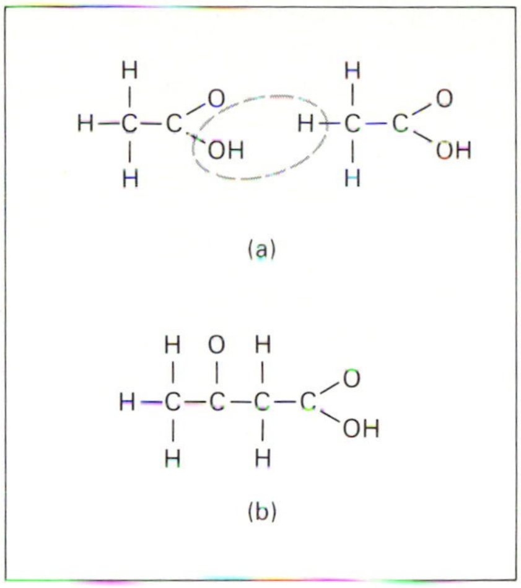

(a)

(b)

FIGURE 25.5 When water (H_2O) is removed (a), glycine amino acids can be linked together to form a simple "protein" (b).

The linking of two or more amino acids in this way is called **dehydration condensation**—dehydration because a water molecule has been removed, and condensation because the two-acid molecules are condensed or bonded together. The result is a strong molecular link between the two glycines.

Actually, chemists know of no real protein as simple as two amino acids, for such a small molecular cluster exhibits none of the function normally associated with proteins. Hence the combination shown in Figure 25.5 is only hypothetical and has no specific name (other than the chemists' classification of "polypeptide"). One of the smallest proteins in real life, called insulin, has 51 amino acids linked together like pearls on a necklace. In comparison with the atoms discussed in earlier chapters, this simplest amino acid has a mass equal to several thousand hydrogen atoms.

Figure 25.6 shows a model of another well-known protein. Called hemoglobin, this protein is a key component of human blood, and contains nearly 600 amino acids linked together by multiple bonds. In all, some 19 different kinds of amino acids are used in hemoglobin. The biological function of this protein (as well as all others) is highly specific, much as we know from experience with blood transfusions

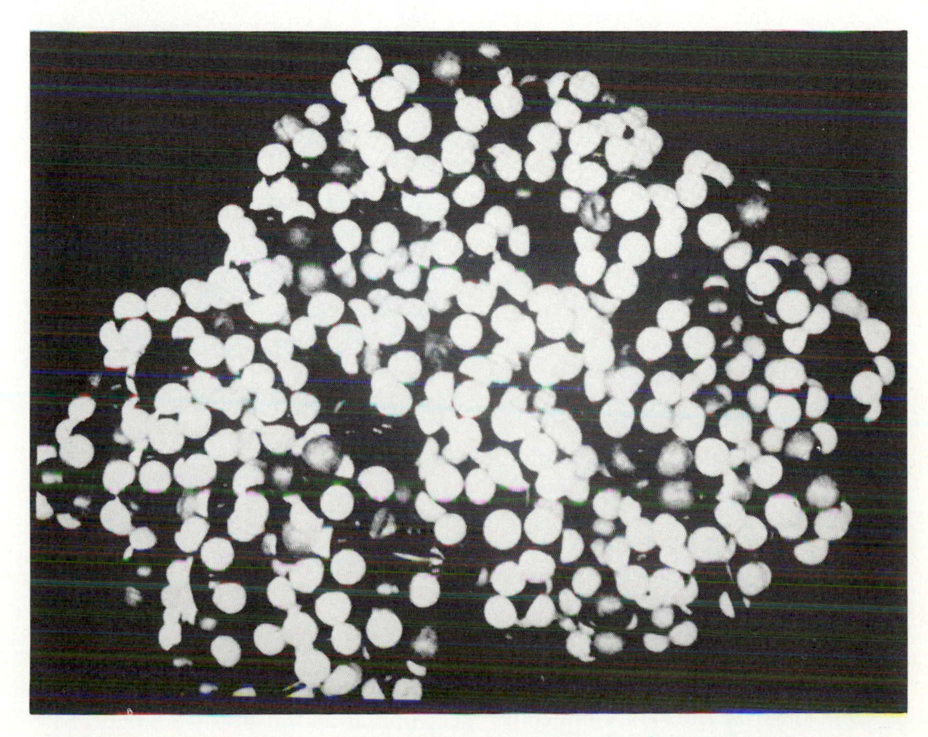

FIGURE 25.6 A model of the hemoglobin protein found in human blood.

that blood of one type cannot serve as a substitute for blood of another. The differences among various blood types result from the ordering of the amino acids along the protein.

On an even larger realm, proteins give some character to cells, and cells in turn to entire living organisms. Ultimately, then, the overall character of life depends on the number and ordering of the amino acid side chains. This numbering and ordering is what distinguishes man from mouse. Since the amino acids are few and relatively simple, the nature of life itself cannot be overly complex—at least at the microscopic level.

NUCLEOTIDE BASES: Knowing what proteins are, we return to our original question: How are proteins manufactured in living organisms in order to keep them alive? Specifically, what chemical process serves to link amino acids together to replenish dead cytoplasm in all living systems? Whatever the process, it must be important, since the production of protein is absolutely vital to an organism's well-being—not random proteins, but exactly the right kinds of proteins, with their amino acids strung along in precisely the right order. To understand how protein is constructed, we need to become familiar with another of life's basic ingredients—the **nucleic acids**.

Nucleic acids, like proteins, are long chain-like arrangements of molecules, most of them rich in carbon. Their name derives from the fact that they were first found in the biological nuclei of cells. Also like proteins, the nucleic acids are made of only a small number of key compounds. Known as **nucleotide bases**, these components are the second group of life's basic building blocks. The role of the bases can be illustrated by considering a nucleic acid familiar to most people: deoxyribonucleic acid, or **DNA** for short.

Figure 25.7(a) is a schematic diagram of the structure of the DNA molecule. Figure 25.7(b) shows a photograph of a model of it. As can be seen, most of the DNA molecule is made of a long string of 4 nucleotide bases. There is a fifth base used in the construction of other nucleic acids, although not in DNA. These 5 kinds of bases play the same role for nucleic acids as do the 20 kinds of amino acids for proteins. The names of the bases are adenine (A), cytosine (C), guanine (G), thymine (T), and uracil (U).

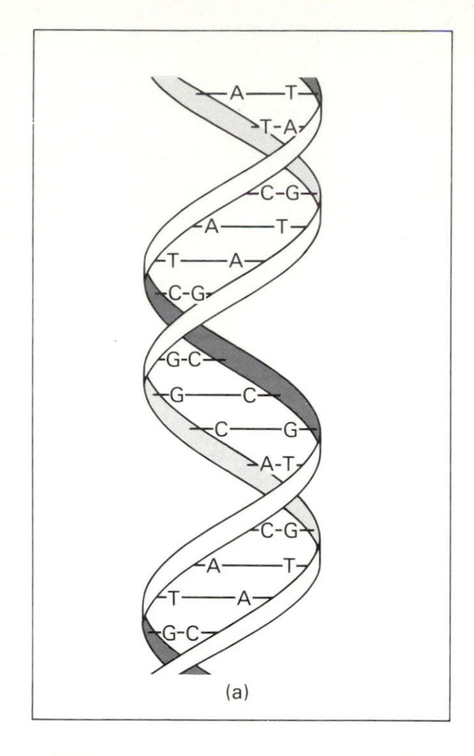

(b)

FIGURE 25.7 Schematic diagram (a) and photograph of a model (b) of the DNA molecule, the most famous of the nucleic acids. The symbols A, C, G, and T refer to various nucleic acids illustrated in the next figure.

Figure 25.8 depicts the chemical structure of each of the nucleotide bases. They are slightly more complex than the amino acids that make up proteins. As shown, the central part of each base forms a ring-like structure of atoms. Most of these rings are made of carbon (C), although there are some occasional nitrogen atoms (N). A ring is a complex molecule that has essentially bent around to attach to itself; it becomes a little more stable that way.

Although DNA at first sight seems to be a most complex molecule, it's really nothing more than a cluster of the 4 types of bases shown at the top of Figure 25.8. Examining Figure 25.7 carefully, we see that these bases are the "rungs" of a long, thin assembly resembling a twisted "ladder." Each rung of the DNA molecule is made of two interconnected (or paired) bases, giving this nucleic acid its famous double-helix structure. Experimental evidence shows, however, that all 4 nucleotide bases do not link together equally well. Cytosine always links with guanine, forming one of the two possible base pairs, while adenine links only with thymine, forming the other. Figure 25.9 is a chemical schematic of each of these DNA base pairs. The structure of the rings and especially their electromagnetic force fields make incompatible any other combinations.

The two sides or uprights—called strands—of the DNA "ladder" also help structure the DNA molecule. Shown in Figure 25.7, these strands are made of sugars (carbon–hydrogen as well as phosphorous–oxygen compounds) which serve to link the all important base pairs.

DNA is only one of many different kinds of nucleic acids, but it stands above all the rest because of one outstanding capability. DNA can make a copy of itself—in effect, self-replicate. To understand this remarkable property, refer to Figure 25.10, which is a schematic diagram of the DNA molecule in the nucleus

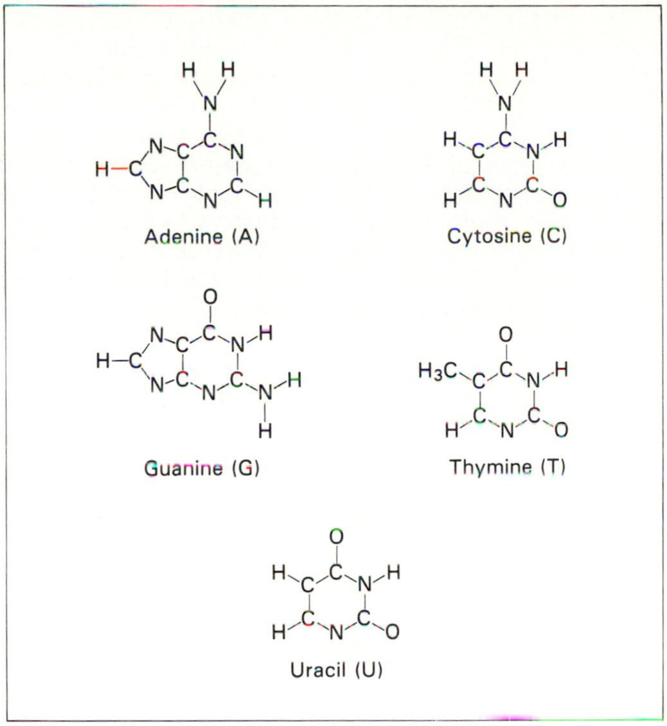

Adenine (A)

Cytosine (C)

Guanine (G)

Thymine (T)

Uracil (U)

FIGURE 25.8 Chemical diagrams of each of the five nucleotide bases.

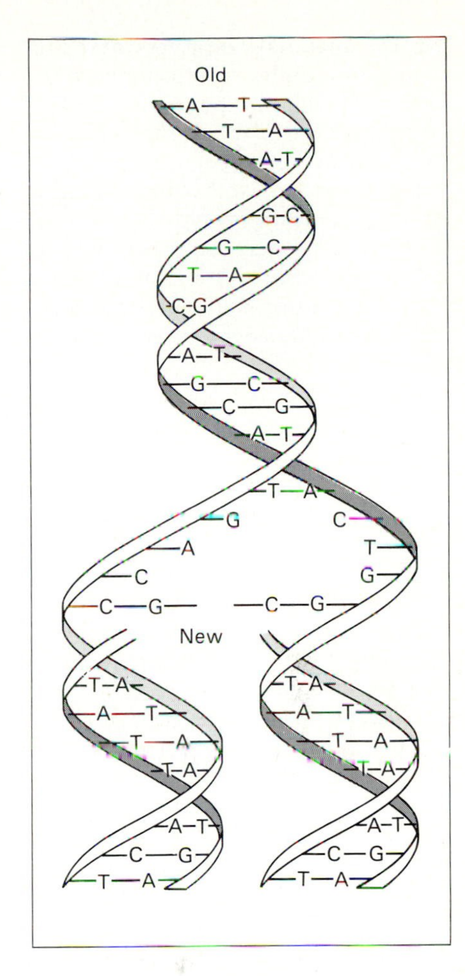

FIGURE 25.10 Schematic diagram of a DNA molecule undergoing replication in the nucleus of a cell.

of a cell. Just prior to the division of a cell, the DNA molecule splits apart by unzipping right up the middle of the ladder. Nucleotide bases floating freely in the cell nucleus then link (with the help of a catalyst called an "enzyme") with each of the broken strands. The result is two DNA molecules, where previously there was only one. The fact that cytosine can link only with guanine, and adenine only with thymine, ensures that the two "offspring" replicas are identical to the original "parent" DNA molecule. The newly assembled DNA molecules then retreat to opposite sides of the cell nucleus, after which the cell divides into two, with each new cell housing a complete set of DNA molecules.

Preservation of the exact structure of the original DNA molecules is the most important feature of replication. All the information about the specific duties of that type of cell—whether a blood cell, hair cell, muscle cell, or whatever—passes from an old cell to a newly created one. Accordingly, the biological function of the "daughter" cell remains identical to that of the "parent" cell. In this way, DNA molecules, whose functional units are called **genes**, are responsible for directing this inheritance from generation to generation.

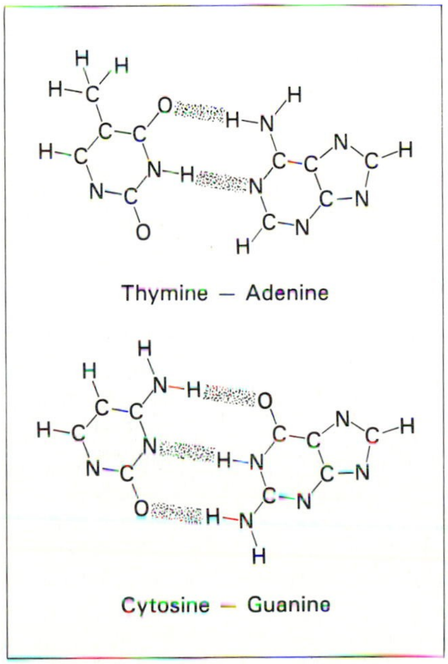

Thymine — Adenine

Cytosine — Guanine

FIGURE 25.9 Chemical diagrams of each of the two base pairs in the DNA molecule. The stippled area depicts where the link is located between the nucleotide bases.

As for the amino acid sequences in proteins, the *ordering* of nucleotide bases as well as their *number* are all-important in the construction of the nucleic acids. The sequence of the bases along a nucleic acid molecule specifies the physical and chemical behavior of that particular gene. Similarly, all the genes of a living system comprise a **genetic code**—an encyclopedic blueprint of the physical and chemical properties of all the system's cells and all its functions. This code we shall study more in the next chapter. For now, realize that the structure and behavior of a living organism depends chiefly upon the nucleic acid molecules in the many nuclei of its cells, for these are the material structures that are passed on or inherited from one generation of cells to the next.

In analogy with another type of information storage—this book, for example—the individual bases might be considered words, the base pairs a sentence, and the whole DNA molecule a book of instructions. The words and sentences must be in the right order to give meaning to the book. An entire library of such instructional books then forms the genetic code for all the varied functions performed by any living organism.

The nature of all living creatures is ultimately prescribed by the structure of their DNA molecules. The DNA molecules specify, not only how one type of organism differs from another in both structure and personality, but also how the physical and chemical events inside a cell properly coordinate so that the overall activity of the cell is as it ought to be.

At first glance, it would seem impossible that DNA could dictate the behavior of all the myriad life forms in the world today. After all, DNA has only 4 types of nucleotide bases. But DNA is the largest molecule

INTERLUDE 25-2 *Life's Left-Handedness*

The structure of many molecules can come in two forms that are mirror images of each other. The chemical formula of the molecule is the same in both cases, but the orientation of some of the molecule's atoms is reversed left for right and right for left. For example, as shown below, two forms of the alanine amino acid are possible; these are mirror images of one another, much like your left and right hands are mirror images, as are the left- and right-handed screws also shown below.

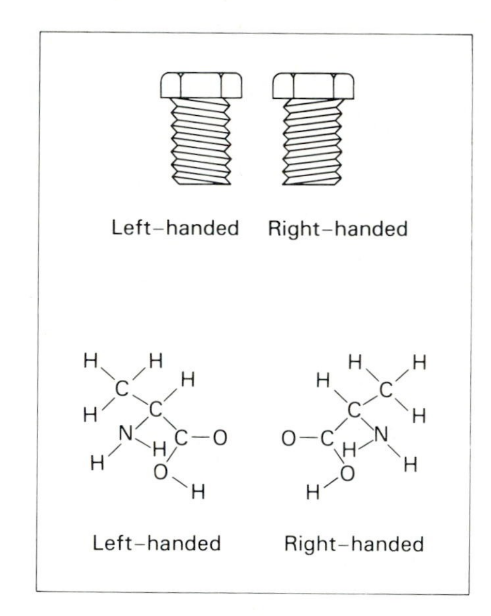

A molecule's orientation can be determined by passing through it polarized light (see Chapter 1), in which case the plane of polarization will rotate to the left or right. The mirror-image configurations are then said to be "left-handed" or "right-handed" (just as with the screws above). Most interesting, the basic molecules of life—especially the 20 amino acids comprising the structure of all proteins—are exclusively of the left-handed form in terrestrial organisms. This property is one of the most striking characteristics of life as we know it. Life's preference for left-handedness is puzzling because amino acid molecules synthesized in the laboratory invariably show a 50–50 mixture of left- and right-handed configurations. Furthermore, should a right-handed amino acid stray into a living organism, the catalysts that control protein production will quickly destroy it. Why terrestrial life selected only left-handed amino acids is one of the great unsolved mysteries of chemical evolution.

One possibility is that the first organism just happened, by chance, to be left-handed. If life arose only once on Earth, all its descendants would also be left-handed. An alternative possibility is that both left- and right-handed organisms originated, perhaps on different occasions billions of years ago, but that left-handed life developed a sufficient advantage to eliminate all competitors. The ability to synthesize an extra amino acid or vitamin, for instance, might have provided such an advantage.

Yet another intriguing idea interfaces physics with biology. In brief, life's left-handedness might result from one of nature's basic forces. Recall from Interlude 2-2 that the weak nuclear force operates on a size scale smaller than nuclear dimensions, and thus is often dismissed as unimportant to atomic physics, let alone molecular biology. However, as discussed in Chapter 17, the weak force has now been coupled to the electromagnetic force, which biologists refer to as the "life force." And since some weak-interaction events studied in nuclear laboratories show a preference for one handedness over the other (more elementary particles spin clockwise than counterclockwise during weak-force events), there might well be a small difference in total energy between left- and right-handed molecules. If this is true, and if the radiation that drove chemical evolution on Earth was polarized, the left-handed amino acids could have been preferred. This is only speculation, but a good example of interdisciplinary science at the boundaries of physics, chemistry, and biology.

known. In advanced organisms such as humans, the DNA molecule can have as many as 100,000,000 bases or 10 billion separate atoms, making the molecule nearly a meter long if extended end to end. In the analogy above, where a DNA base equals a word, a single DNA molecule would become the equivalent of a 100-page manuscript. Consequently, huge numbers of possible combinations of bases guarantee a vast arrray of diverse living creatures, each with a different appearance, style, and personality. Yet at the microscopic level, all creatures—without exception—are basically made of the same few amino acids and nucleotide bases. This is why we say that the 20 acids and 5 bases comprise the building blocks of all life.

The common molecular structure pervading all life on Earth is the best evidence that every living thing arose from a single-celled ancestor billions of years ago.

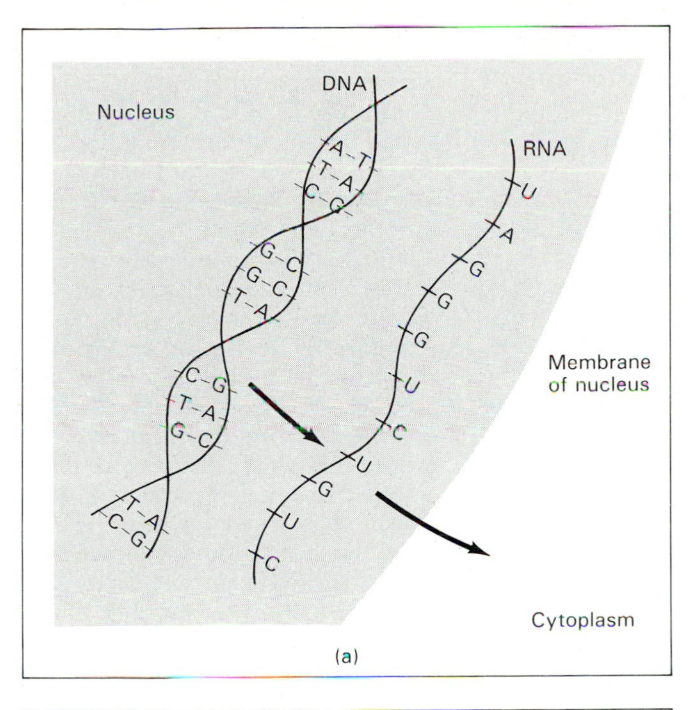

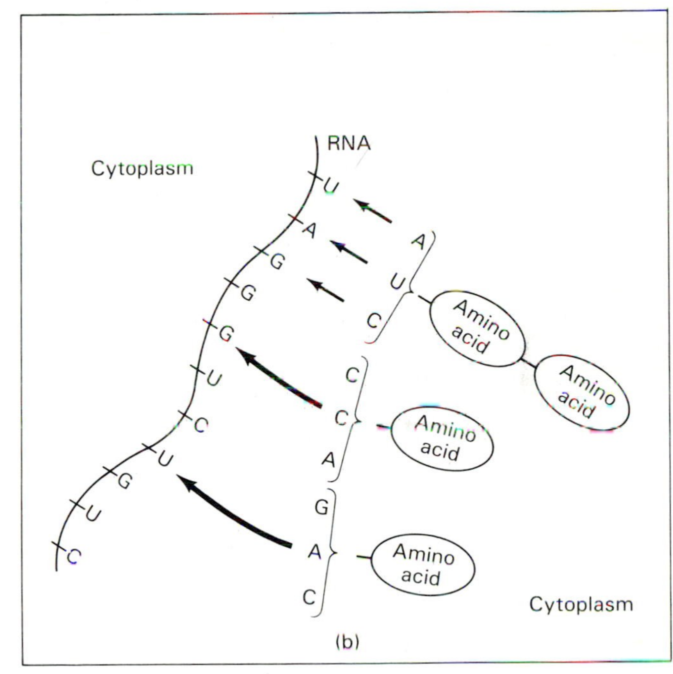

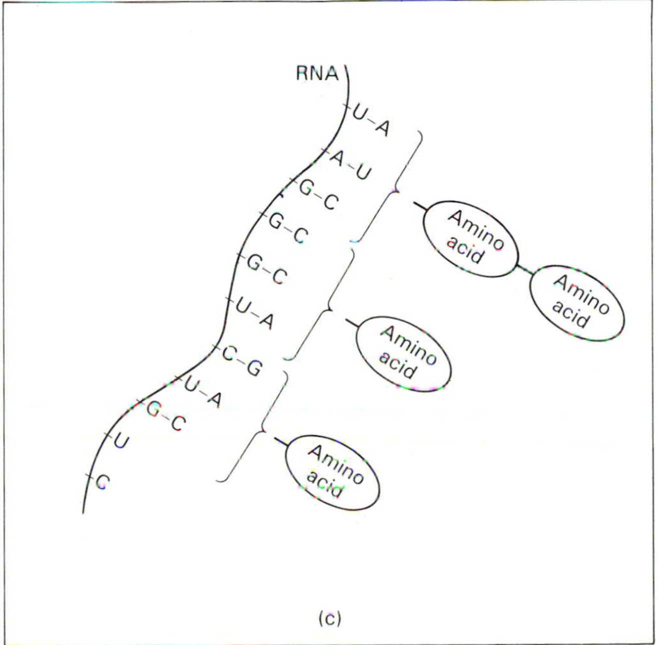

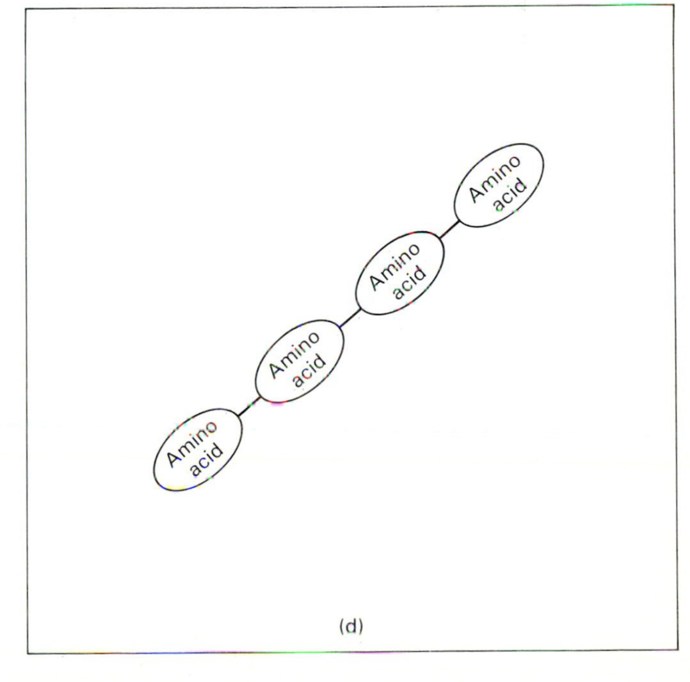

FIGURE 25.11 This simplified sequence of steps, (a) → (d), shows how proteins are made when the RNA molecule attracts, and then helps to link together, just the right amino acids.

PROTEIN SYNTHESIS: To return to our original concern, the continuous production of the cytoplasm's proteins makes heavy use of the cell's nucleic acids. The sequence of events, illustrated in Figure 25.11, is as follows. Just prior to cell division, the DNA molecule sends a relative molecule, called RNA, out of the biological nucleus and into the cytoplasm [Figure 25.11(a)]. "RNA" stands for ribonucleic acid—a smaller, single-stranded version of the normally double-stranded DNA molecule (wherein, for RNA, the thymine nucleotide base is replaced by the uracil nucleotide base). The RNA molecule acts like a messenger, carrying instructions from the DNA molecule. Once in the cytoplasm, single-stranded RNA attracts amino acids to its uncoupled bases [Figure 25.11(b)]. Only certain amino acids can successfully attach to the RNA bases, since the electromagnetic force fields of RNA's bases attract some amino acids, while repelling others. In other words, although different kinds of amino acids collide with the RNA molecule, only a minority successfully stick to the nucleotide bases of RNA. After some time—usually a few microseconds—the single-stranded RNA molecule fills its strand with an entire complement of the proper amino acids. These amino acids also attach to one another by means of their own electromagnetic force fields [Figure 25.11(c)]. Finally, when the long chain of amino acids is fully assembled along the entire length of the messenger RNA molecule, the chain detaches and drifts off into the cytoplasm of the cell [Figure 25.11(d)]. A protein has thus formed. Again, this is no ordinary protein created at random. Rather, it's a specific protein constructed according to the instructions provided by the RNA molecule. In this way, RNA acts as a prescription or template on which protein molecules are built—a template originating in the cell nucleus with the DNA molecule itself.

This, then, is a highly simplified account of the way that proteins are continually replenished in living organisms. All living systems grow and eventually become biologically stabilized in this same way. The entire process is occurring repeatedly in our bodies right now. Different organisms have different DNA structures, and therefore manufacture different kinds of protein. In fact, no two living organisms have the same DNA structure, except for identical twins.

Whether man, mouse, or daisy, the nucleic acids mastermind life, and the proteins maintain its well being.

Laboratory Simulations

The mixture of facts and (testable) ideas of the preceding section strongly suggests that life, although biologically, socially, and culturally complex, is rather simple physically and chemically. When broken down into its basic parts, the ingredients of life—any life: bacteria, ostriches, or humans—are never more complex than the 20 amino acids and the 5 nucleotide bases.

The question we now face is this: Could these basic ingredients have formed on Earth naturally—that is, as a natural consequence of the evolution of matter? If so, could these ingredients have then clustered to create life on our planet?

To consider how feasible it may have been to form the building blocks of life, we must again imagine the conditions on primordial Earth several billion years ago. Recall, in particular from Chapter 24, the gases in Earth's secondary atmosphere—principally ammonia, methane, carbon dioxide, water vapor, and molecular hydrogen. With time, these chemical products became the interactants of additional chemical reactions. These additional reactions, however, were not spontaneous; the molecules will not interact on their own. Laboratory experiments show that these simple molecules require some energy in order to combine further. This energy acts as a driver to produce bigger molecules. In fact, the application of energy yields molecules a lot more complex than those likely to form by chance in a collection of free atoms and simple molecules. These more complex molecules are the amino acids and nucleotide bases—the very building blocks of life.

SIMPLE EXPERIMENTS: Can we prove that the basic ingredients for life would have been produced under the primordial Earth conditions described in the preceding section? Furthermore, can we be sure that those nonliving building blocks could have fashioned a simple living cell? These questions can only be addressed in the laboratory, for today's atmosphere and surface differ greatly from those of the early Earth. Results of modern laboratory experiments imply that the answer to these questions is "yes."

Figure 25.12 shows a schematic diagram of the basic laboratory apparatus used to simulate Earth's early ocean and atmosphere. The whole contraption is no more complicated than a large test tube into which gases are placed. These gases—usually ammonia, methane, molecular hydrogen, and sometimes carbon dioxide—are meant to simulate the composition of our secondary atmosphere. The primordial oceans are simulated in this experiment by means of the flask of liquid water near the bottom of the apparatus. Upon heating this "ocean," water vapor rises to mix with the other gases in the "atmosphere," whereupon it eventually "rains" back down with any newly formed chemicals. A U.S. chemist by the name of Miller was the first person to undertake such an experiment in the 1950s, hence the reason that most such simulations are often today referred to as "Miller experiments."

The entire apparatus is shut tight, allowing all the gases to cycle around endlessly without escaping, much like the familiar evaporation–condensation sequence happening every day on Earth. In the absence of energy, these gases do just that—cycle through the machine unchanged. The gases refuse to react spontaneously with one another. For example, when molecules of methane and water vapor make contact, nothing happens. They need a little

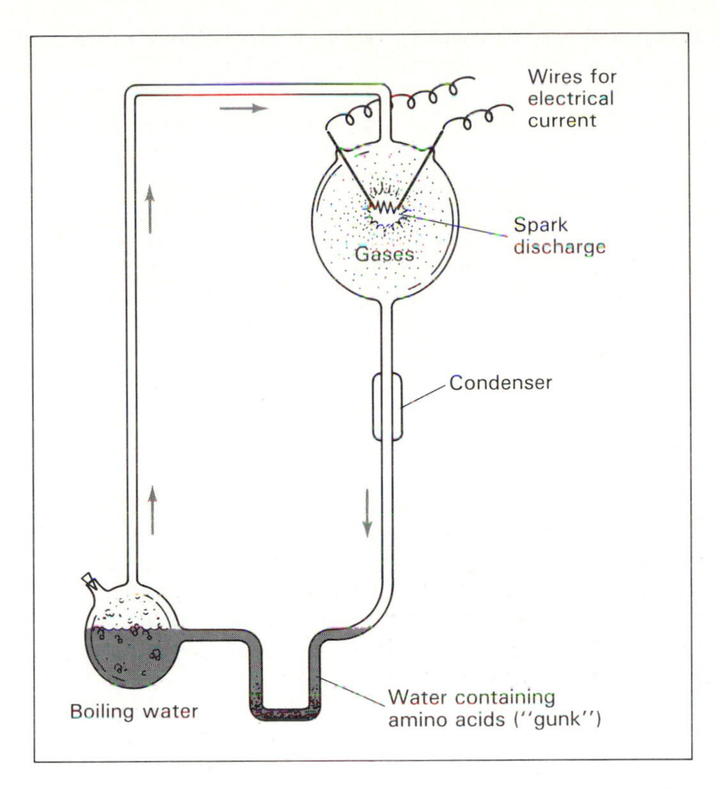

FIGURE 25.12 A laboratory apparatus designed to simulate the chemical activity in the ocean and atmosphere of primordial Earth.

help in order to react chemically. And that help is energy of some sort. Once energy is applied, some of the bonds within each of the gas molecules break apart, thereby allowing the liberated atoms and molecular fragments to reform as more complex molecules.

Sometimes, in order to speed up the reactions, researchers place into the apparatus greater gas abundances than are thought to have existed on Earth long ago. Or they increase the intensity of the energy source above the amount presumed present billions of years ago. In this way, the molecules' chance of colliding with one another improves enormously, allowing the experimental simulations to be completed in a few weeks. Quite frankly, researchers can't afford to wait several hundred million years to determine the outcome of their experiments.

After several days of energizing the gases, a thick, reddish-brown, soupy material appears in the trap at the bottom of the apparatus. Chemical analysis shows this slimy "gunk" to contain molecules much more complex than those put into the apparatus in the first place. Be assured, no worms or maggots crawl out of this primordial soup. But many of the molecular products are among the known precursors of life. They include many of the amino acids and nucleotide bases comprising the building blocks of all modern life. Chemicals such as formaldehyde (H_2CO) are also produced, as well as other molecules [e.g., hydrogen cyanide (HCN) and formic acid (H_2CO_2)] that are known to be among the basic ingredients for life as we know it. However, not all the acids and bases common to terrestrial life have yet been identified in the gunk. Even so, this "warm little pond," first theorized by the nineteenth-century British naturalist Darwin, is thought to be a pretty good simulation of Earth's early ocean into which heavy atmospheric molecules would have fallen, pulled down by relentless gravity.

The recipes for the successful creation of the pre-life acids and bases is not a very rigid one. Different researchers have widely varied the gas mixtures, energy sources, and "cooking" times. The result is always the formation of complex organic molecules, provided that there is no free oxygen gas present. With even small doses of oxygen in the test tube, the gases oxidize and the experiment fails to produce organic molecules. Ironically, although much of Earth's established life today requires oxygen, this gas was apparently a deadly poison during the formative stages of that very same life. This is why we have no evidence for acids and bases in the oceans of today's planet; there is too much oxygen around now.

A critical concern here is the amount and kind of energy needed to power these experimental simulations. Is it reasonable to suppose that enough of the right type of energy was present on the early Earth? As shown in Figure 25.12, the energy in the laboratory simulations is usually provided by electrodes that spark the gases in the test tube. In this case, the electrode sparks can be regarded as imitating atmospheric lightning. This spark discharge can also be thought of as an indirect simulation of several other types of energy undoubtedly present on early Earth. Besides atmospheric lightning, radioactive decay and volcanic activity surely must have provided energy in the form of heat. Cosmic rays were another source of energy on early Earth. Even thunder yields enough energy to have powered, in Earth's early atmosphere, some of the chemical reactions known to occur in the laboratory experiments. Meteoritic bombardment is a further source of energy; as huge rocks plow through the atmosphere, their friction often generates enough heat to ignite chemical reactions.

Most of these energy sources are localized, and hence were sufficiently intense to break or make molecular bonds only at isolated locations on early Earth. However, solar energy was widespread; it reaches every place on the surface of our planet. Although ordinary sunlight is not energetic enough to trigger chemical reactions, solar ultraviolet radiation is. And without oxygen gas on early Earth, the ozone layer could not have surrounded primordial Earth, thus easily permitting ultraviolet radiation to have reached the surface of our planet. Apparently, solar energy that sustains life now was also active in helping to create life billions of years ago.

Laboratory experiments such as these are significant because they demonstrate that the molecular building blocks of life could have been synthesized in any one of many different ways during the early history of planet Earth.

MORE STEPS TOWARD LIFE: Be sure to realize that the laboratory simulations produce only the building blocks of life, not life itself. The organic molecules found in the gunk are still much simpler than a single cell. Amino acids and nucleotide bases are in fact substantially less complex than even the proteins and nucleic acids found throughout contemporary life. How, then, were the acids, bases, sugars, and salts in this primordial soup initially assembled into proteins and nucleic acids? The answer is that these soupy organic substances must have been further clustered in order to permit stronger and drier interaction.

As illustrated in Figure 25.5, two amino acids can be linked to reach the next stage of complexity, provided that a water molecule is removed. Dehydration condensation of this sort can build up chain molecules into complex proteins by attaching any number of amino acids. Similarly, successive linkages of nucleotide bases, sugars, and salts can produce lengthy nucleic acids.

Heat, for example, could have evaporated some water from clusters of acids and bases, especially along the shoreline of a primordial ocean or a lagoon inlet. Repeated tides in and out of shallow waters could have led to a daily cycle of solar dehydration of molecules in a temporarily dried lagoon (during low tide), followed by further interaction of those moleules when washed into the open ocean (at high tide).

The opposite condition—cold—can also remove water molecules from an organic mixture. The freezing of water changes it from liquid to ice, thus allowing the acids or bases to become more concentrated and hence linked together. Repeated freezing and thawing could allow the buildup of progressively larger chain molecules.

A third mechanism can actually remove water while still in the presence of water. Although this sounds impossible, it happens all the time in living organisms; composed largely of liquid, the cells in our bodies routinely manufacture protein. They do it by using catalysts, that is, third-party molecules that speed up the process. Although the catalysts that now promote condensation reactions in modern life were probably absent in the primordial ocean, many researchers argue that other catalysts would have existed several billion years ago. Certain kinds of clay, for example, are thought by many researchers to have been the scaffolding needed to make larger organic molecules along the edges of oceans, lakes, and rivers.

Scientists are unsure if the first complex proteins and nucleic acids really did originate in any of these ways. It's unlikely that the fossil record will ever show the precise path whereby pre-life molecules

INTERLUDE 25-3 ***Oldest Life on Earth***

Scientists now recognize two general classes of life. One class is the lower form, mainly bacteria, while the other is the higher form, plants and animals. Each class has a completely different sequence of nucleotide bases along the RNA molecules of its members; yet most members of each class thrive in the current physical and chemical conditions of our planet.

Recently, researchers have proposed an entirely new class of life. Microscopic studies of small organisms suggest this third class of life has RNA sequencing and cell chemicals completely different from both bacteria and the higher life forms. This discovery, in and of itself, is quite important, because biologists want to understand all the life forms capable of surviving on our planet. Even more interesting, however, this third class of life may be the oldest form of life on Earth.

Known as "methanogens," these microorganisms thrive only in oxygen-free environments, such as on the seafloor, in sewage, or in the hot springs seeping through the crust of the Earth. Oxygen is poison to the methanogens. They stay alive by converting carbon dioxide and hydrogen into methane, which is the main chemical of natural gas.

Methanogens, then, can survive only in warm, oxygen-deficient conditions, much like those thought to have existed on our planet during its first billion years or so. This finding suggests that the methanogens originated and evolved on Earth prior to the formation of conventional bacteria. Accordingly, today's methanogens might be descendants of the most ancient class of life on Earth. They might be the best link to our "common ancestor"—that ultrasimple, original organism from which all forms of life have evolved.

By searching for any common properties among these three classes of life, scientists are striving to build a profile of Earth's original living entity, whatever it may have been. Then, and perhaps only then, can we understand precisely how life evolved from lifeless chemicals.

gradually clustered to form what might be genuinely called life. At any rate, heating, freezing, and catalyzing are all plausible agents for the self-construction of proteins and nucleic acids.

ADVANCED EXPERIMENTS: The simple cell remains extremely more complex than any of these pre-life molecules. To appreciate this very base of the evolutionary tree, chemists are currently trying to understand how the proteins and nucleic acids were in turn able to form more complex combinations of biological significance. Our understanding is limited in this area, however. Scientists have only been able to surmise that repeated interactions among the many molecules on primordial Earth could have eventually produced something resembling today's proteins, DNA, and simple cells.

More advanced laboratory simulations have been accomplished in recent years to support this view. Repeated energizing and dehydrating the simulated environment of primordial Earth produce organic molecules more complex than the amino acids and nucleotide bases. Of special interest are dense clusters of amino acids—protein-like substances called **proteinoid microspheres**. Only about a hundredth of a millimeter across, these are not well-known proteins such as insulin or hemoglobin, but simpler, protein-resembling compounds whose relevance to the origin of life is uncertain. Some researchers regard proteinoid microspheres as bona fide proteins; others are not so sure.

Figure 25.13 is a photograph of a few of these proteinoid microspheres, taken through a microscope. Direct chemical analyses confirm these microspheres to be dense clusters of organic matter floating in a mostly inorganic fluid. The main features of this figure can be visualized by shaking a mixture of oil and water and watching the globs of oil cluster on the surface of the water. Another good analogy is the grease that bonds together as droplets on the surface of cooled chicken broth.

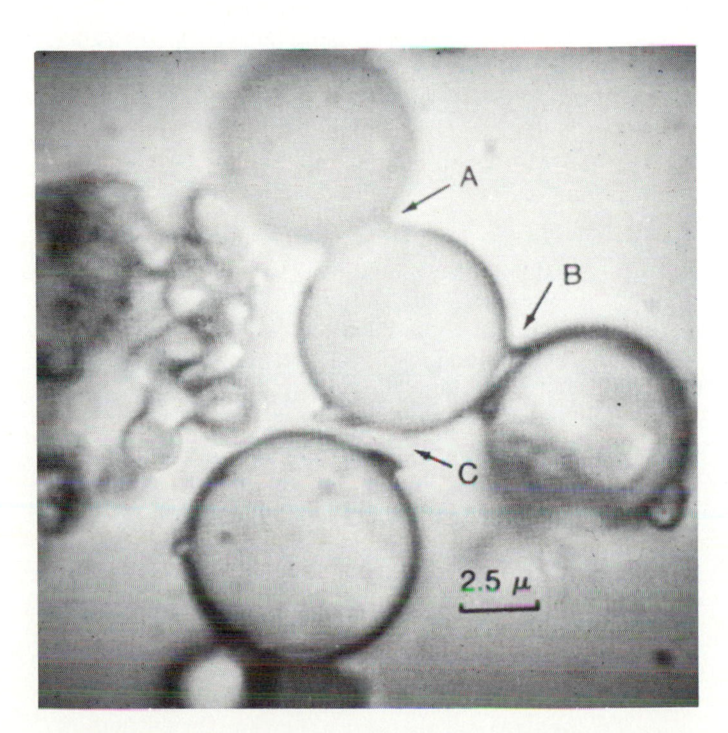

FIGURE 25.13 Photograph of proteinoid microspheres produced by repeated energizing and dehydrating the primordial soup. Seen here through a microscope, each microsphere contains a large concentration of amino acids. (The scale shown, 2.5 microns, equals 2.5×10^{-4} centimeter.)

In many remarkable ways, proteinoid microspheres produced in laboratory experiments resemble simple cells. The microspheres seem to possess a semipermeable membrane through which small molecules can enter from the outside in order to "feed" but through which larger molecules synthesized within cannot get out. Some discharge of wastes can be observed through the microscope but, by and large, there is a net intake of matter, in some ways mimicking modern living cells. Indeed, the microspheres are actually observed to become larger in the process.

Thus we can imagine the proteinoid microspheres as eating, growing, and excreting—much like a primitive metabolism, perhaps. Not only that but a slight jostling of the experimental apparatus to produce some turbulence in the fluid—the analog of early oceanic wave action—causes some of the larger microspheres to break apart into smaller units, demonstrating what some workers regard as a primitive form of replication. Some of the smaller microspheres disperse, much like the process of death. Others enlarge like their "parents," only to be broken up again in another act of "replication" (although these proteinoids surely lack enough information to direct their own replication from the primitive building blocks).

So the proteinoid microspheres have a primitive resemblance to simple cells. Some "eat," some "grow," some "reproduce," and some "die." Can the microspheres be called life? Probably not. Remember, we had problems defining life at the start of this chapter; we never did succeed. The dividing line between matter and life is simply not clear-cut.

The great majority of biologists argue that amoebas are definitely living but that the molecular contents of the organic soup are not. Proteinoid microspheres apparently lie somewhere in between. But if the microspheres are not at least progenitors of Earth's living systems—protolife—nature would seem to have played a very malicious joke on modern science.

SUMMARY

Theoretical considerations and experimental simulations suggest that life is a logical result of known chemical principles operating within the atomic and molecular realm. Furthermore, the origin of life itself seems to be a natural consequence of the evolution of those atoms and molecules.

However, these statements thus far lack absolute proof, since laboratory experiments have yet to fashion anything more sophisticated than life's precursors. To be sure, a large gap separates these preliminary stages of chemical evolution from the onset of biological evolution of simple living cells. This chapter has outlined a consensus concerning that blurred realm where chemical evolution ends and biological evolution begins.

KEY TERMS

amino acid
cell
chemical evolution
dehydration condensation
DNA
fossil
gene
genetic code
nucleic acid
nucleotide base
panspermia
protein
proteinoid microsphere
spontaneous generation
virus

QUESTIONS

1. Explain why chance, and chance alone, is probably not the only principle at work in the Miller experiment.
2. Describe the chain of events whereby (we reason) the newly formed Earth produced the first living cell. What criticisms can be leveled at our current understanding of the processes involved?
3. Describe two ways in which proteinoid droplets synthesized during laboratory experiments are similar to the oldest fossils of living cells, and two ways that they are different.
4. List, and explain in a phrase or sentence, at least three similarities between each of the following: laboratory-synthesized proteinoid microspheres, the oldest fossils, and modern blue-green algae.
5. How does the Miller experiment attempt to model the conditions on primordial Earth, and what criticisms might be made of these experiments?
6. Cite some other characteristics of life not mentioned in the text. Could these additional characteristics also be valid for material systems obviously not alive?
7. What is the main feature that distinguishes the spontaneous generation of life from the chemical evolution of life? Explain.
8. On the basis of what you know about the ages of the Earth, the oldest rocks, and the oldest fossilized life forms, would you conclude that life originated on our planet rapidly or sluggishly?
9. What are the two basic building blocks of life? How do they differ? Are they common to all life or just human life?
10. If the building blocks of life formed naturally via the action of ultraviolet energy and simple gases, why don't we see these greasy organic materials floating on the surfaces of today's lakes, rivers, and oceans?

FOR FURTHER READING

BARGHOORN, E., "The Oldest Fossils." *Scientific American,* May 1971.

BARRETT, J., ABRAMOFF, P., KUMARAN, A., AND MILLINGTON, W., *Biology*. Englewood Cliffs, N.J.: Prentice-Hall, 1986.

FOLSOME, C., *The Origin of Life*. San Francisco: W.H. Freeman, 1979.

*FOX, S., AND DOSE, K., *Molecular Evolution and the Origin of Life*. New York: Marcel Dekker, 1977.

MARGULIS, L., *Symbiosis in Cell Evolution*. San Francisco: W. H. Freeman, 1981.

MILLER, S., AND ORGEL, L., *The Origins of Life on the Earth*. Englewood Cliffs, N.J.: Prentice-Hall, 1974.

WEINBERG, R., "The Molecules of Life." *Scientific American,* October 1985.

26 BIOLOGICAL EVOLUTION: Fossils and Genetics

Our cosmic evolutionary scenario is really beginning to take shape now. From stellar atoms to planetary molecules, we have suggested a plausible method for the origin of life. Indeed, life's origin seems to be a natural consequence of the evolution of matter.

To fully grasp the spectacle of life, however, we must probe beyond chemical analyses of microscopic matter as studied in the previous chapter. Not even studies of living cells are enough to truly understand life. Most of modern life is much more complicated than a single cell. To decipher the complexity of life, we need to consider entire living organisms.

Much as we cannot hope to understand how an automobile works by grinding it up and chemically measuring its atoms and molecules, we cannot hope to decipher the complexity of life without studying whole living things. Indeed, the macroscopic features of entire organisms usefully complement the microscopic studies of their inner workings. And if microscopic studies of life forms comprise the realm of chemistry, then their macroscopic studies fall into the realm of biology.

In this chapter we study the host of plants and animals on our planet. Redwoods and rhinos, roses and reindeer, evergreens and elephants, uncounted more species. Where did they all come from? That they suddenly appeared intact from nothing is an interesting idea. But spontaneous (or miraculous) creation makes no sense scientifically, nor is there a shred of evidence to support it.

Together all the fossils chronicle an amazing story of life on Earth. Repeatedly, throughout the milennia, new life forms emerged while others perished. Some species survived for ages; others succumbed nearly as soon as they appeared. Incredibly, more than 99 percent of all life forms that once prospered are now extinct.

Only one thing seemingly remained constant throughout the eons of Earth's history: change. The phenomenon of change really does seem to have been the hallmark in the development of all structures—living or nonliving.

The learning goals for this chapter are:

- to appreciate further the important events that occurred during that blurred time domain when chemical evolution ended and biological evolution began
- to understand what the ancient fossils tell us about the oldest forms of life
- to understand how the recent fossils lend support to the modern theory of biological evolution
- to appreciate the role played by genetic mutations and environmental change in biological evolution
- to gain some feeling for the long evolutionary path that led from simple cells to advanced primates

Need for Food

One thing is required in any aspect of evolution, regardless of whether the evolution involves matter that is clearly living or matter that is clearly not. That requirement is energy.

Neither matter nor life can proceed from a simple to a complex state without absorbing some energy. Complex objects have some organization, and organization of any kind requires energy. Even when fully formed and highly evolved, no advanced form of matter, whether in stars or people, can sustain itself without a regular supply of energy. This energy is a fuel—a food of sorts.

Even for the laboratory simulation depicted in Figure 25.12, the need for energy is apparent. The energy derived from the spark discharge can be considered an "explosive food," since the energy was used to break bonds of the small molecules, thereby enabling them to re-form into more complex molecules. Part of the spark's energy is also absorbed. Much of it strengthens the chemical bonds needed to hold together—to organize—the new, more complex acids and bases. The organic soup floating on the surface of the primordial ocean thus became a tremendous storehouse of energy.

Repeated energizing—that is, repeated feeding—is needed to construct the proteinoid microspheres of Figure 25.13. Once formed, these organic droplets require even additional feeding to maintain their increasingly intricate molecular organization. They do so by absorbing nutritious amino acids and nucleotide bases admitted through their semipermeable membranes. The proteinoids then extract energy by breaking some of the chemical bonds among the atoms comprising those acids and bases. In this way, the proteinoids essentially "eat" by absorbing minute amounts of energy from their surroundings.

Why do the proteinoids obtain energy from their immediate environment? Why don't they continue to use one of the external forms of energy, such as solar radiation, atmospheric lightning, or volcanic activity? The answer is that the energy that helped form the proteinoids in the first place is often too harsh to sustain them later. As molecules become larger and more complex, they also become more fragile. They must eat and organize themselves by absorbing energy, but that energy must be slight and gentle. (It's a little like the difference between watering a plant and drowning it.) The small acids and bases able to pass through the miniature openings of a proteinoid's membrane contain just the right amount of energy. They enable proteinoids to survive without being subjected to the harsh external energy originally needed to produce them.

Although scientists have no direct evidence for the assembly of more advanced precursors of life, laboratory studies strongly support a two-step process like that outlined above: First, a heavy dose of energy was needed to build the precursors, after which milder energy was needed to maintain them.

Circumstantial evidence suggests that the proteinoids, once formed, were able to protect themselves from the uncontrolled energetic conditions that created them several billion years ago. This is reasonable since Earth was rapidly cooling at that time, becoming in the process much less geologically active. As time passed, volcanoes, earthquakes, and atmospheric storms would gradually have subsided. The amount of solar radiation reaching the ground would have also diminished as terrestrial outgassing thickened the atmosphere. Many of these prebiological substances probably found shelter under thin layers of water, which can absorb whatever harsh (ultraviolet) solar radiation did manage to penetrate the air.

From this point on, biologists can only presume that at least one proteinoid was eventually able to evolve into something everyone would agree is a genuine living cell. Be sure to recognize, though, that nothing yet discovered in the fossil record documents this pre-life evolutionary phase. Nor have laboratory simulations produced molecular clusters more complex than the proteinoids. Frankly, a large gap plagues our *direct* knowledge of the precise events that occurred between the synthesis of life's precursors and the appearance of the first bona fide cell.

Primitive Cells

What were the first living cells like? We do not know for sure. Most likely, they were tentative microscopic entities—fragile enough to be destroyed by strong bursts of energy, yet sturdy enough to reproduce, thereby giving rise to generations of descendants.

One thing is certain: The first cells—sometimes called **heterotrophs**—somehow had to find enough energy to continue living and organizing themselves. They presumably did so while floating on or near the ocean surface, absorbing the acid and base molecules of the rich broth of the organic ocean. This extraction of energy via the capture and chemical breakdown of small molecules is known as **fermentation**, a process still employed by some primitive bacteria on Earth today. (The change of grain into alcohol in beer casks is a familiar example.) But the primitive heterotrophs could not have indefinitely fed on the organic matter from which they originated. After all, the continual passage of time produced important changes in the environment.

As Earth cooled, several of the energy sources capable of producing the acids and bases began to diminish. Lesser amounts of solar ultraviolet radiation reached Earth's surface; geologic and atmospheric activity declined; and other gases thickened the atmosphere. Laboratory experiments show that these changing conditions are not conducive to the continued production of heterotrophs' food supply (which is why we don't see a thick film of organic acids and bases floating on today's oceans and rivers.)

Whereas originally there had been plenty of juicy organic molecules on which the heterotrophs could feed, later times saw fewer food sources for greater numbers of heterotrophs. Consumed more rapidly than it was replenished, the organic soup gradually thinned. Accordingly, the primitive cells had to compete with one another while scrounging for the decreasing supply of nourishing

INTERLUDE 26-1 *Is Life Inevitable?*

This book suggests that life is a natural consequence of the evolution of matter. Given the laws of physics and chemistry, as well as the proper ingredients and a lot of time, the subject of biology seems to arise naturally. The path from atoms to molecules to life seems straightforward enough, given our knowledge of modern science.

An important, though as-yet unanswered question concerns the direction and nature of the path from matter to life. Is there only one way that complex matter can eventually lead to life? Or, are there many ways that molecules can cluster to create life eventually? Either of these choices would be consistent with the basic ideas of cosmic evolution. But the chances for life elsewhere in the Universe depends critically on which of the two cases shown below is correct.

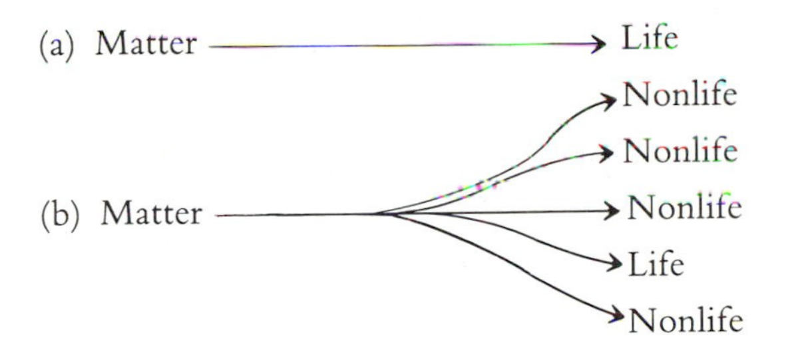

Case (a) depicts a single path from matter to life. Provided that the environmental conditions are not adverse, matter just steadily changes, becoming more and more complex, until eventually a single cell materializes which we call life. We have no way of knowing how long this process of chemical evolution usually takes. The time scale probably depends mostly on the surrounding physical and chemical conditions. Temperature, density, energy, and raw materials all play an important role in the development of life. Furthermore, there are probably many false starts where life almost forms—or even does so temporarily—only to be destroyed quickly thereafter.

If life definitely appears once a certain complexity of matter is reached, we can be reasonably sure that life is not only a *natural* consequence of the evolution of matter but an *inevitable* one as well. The direct path schematically sketched by case (a) greatly increases the chances that life resides elsewhere in the Universe.

On the other hand, if matter can become complex in many ways, only one of which leads eventually to life, we are not justified in suggesting that life is inevitably produced from matter. Life is a natural consequence of the evolution of matter, yes, but not an inevitable one. In other words, as sketched by case (b), vastly complex clusters of matter might form without ever reaching a system that can be judged "living." If so, the prospects for extraterrestrial life are poor.

The truth might also lie between these two extreme cases, with the chances for extraterrestrial life being neither good nor bad.

---

acids and bases. Eventually, the heterotrophs devoured every bit of organic matter floating in the ocean. The organic production of acids and bases via sunlight, lightning, or volcanoes simply could not satisfy the voracious appetite of the growing population of heterotrophs.

This scarcity of molecular food was a near-fatal flaw in life's early development. In short, it was a major ecological crisis. Had nothing changed, Earth's simplest life forms would have proceeded toward an evolutionary dead end—starvation. Earth would be a barren, lifeless rock. Fortunately, something did change.

Some primitive cells—the forerunners of plants—invented a new way to get energy. Accordingly, a unique opportunity for life emerged. This new biological technique used carbon dioxide (CO_2), the major waste product of the fermentation process. So, while the earliest cells were busy eating organic molecules in the sea and consequently polluting the atmosphere, more advanced cells were learning to use these pollutants to extract energy. In this case, the energy was not derived from the carbon dioxide gas, but from another well-known source—the Sun.

PHOTOSYNTHESIS: The key here is the molecule chlorophyll, a green pigment having its atoms arranged in such a way that light, striking the surface of a plant, is captured within the molecule. Advanced cells containing chlorophyll can extract energy from ordinary, gentle sunlight (not harsh ultraviolet radiation) by a process termed **photosynthesis** (*photo*, meaning "light"; *synthesis*, meaning "putting together"). This is a chemical reaction that uses sunlight to convert carbon dioxide and water into oxygen and carbohydrates. In simplified form, it's represented chemically by the formula

$$\text{carbon dioxide}(CO_2) + \text{water}(H_2O) + \text{sunlight} \rightarrow \text{oxygen}(O_2) + \text{carbohydrate}(CH_2O).$$

The oxygen gas escapes into the atmosphere, while the carbohydrate (sugar) is used for food. This, then, is another way a cell can "eat," or extract energy from its environment.

How did some cells develop photosynthesis? We again don't know for sure, other than to suggest that chance events changed the DNA structure in some early cells. Those whose DNA molecules allowed photosynthesis were then able to survive on solar energy. In short, those cells capable of photosynthesizing—called **autotrophs**—could adapt to the changing environment. They had an advantage, since they could persist on merely inorganic matter. They were clearly more fitted for survival during what was probably the first ecological crisis on our planet.

Photosynthesis freed the early life forms from total dependence on the diminishing supply of organic molecules in the oceanic soup. Fermentation within heterotrophs was no longer needed for survival. Early cells able to use sunlight overspread the watery Earth. In time, the autotrophs apparently evolved into all the varied types of plants strewn across our planet.

The photosynthetic process has survived to this day as plants routinely use sunlight to produce carbohydrates as food. All the while, the plants release oxygen gas that animals, including ourselves, breathe. Photosynthesis is, in fact, the most frequent chemical reaction on Earth. Each year, nearly 150 billion tons (almost a billion billion kilograms) of carbon dioxide mix with some 50 billion tons of water to produce about 100 billion tons of organic matter and another 100 billion tons of oxygen gas.

THE SIGNIFICANCE: By loss of their food source, the ancient and primitive heterotrophs were naturally selected to die. The better adapted autotrophs were naturally selected to live. Life on Earth was on its way toward using the primary and plentiful source of energy—solar radiation—in the most efficient and direct way possible. It all began some 3 billion years ago.

The operation of photosynthesis over eons of time is, by the way, partly responsible for the fossil fuels. Dead, rotted plants, buried and squeezed below layers of dirt and rock, have chemically changed over megacenturies into oil, coal, and natural gas. Such fossil fuels, with vast quantities of energy intact, have made industrial civilization possible. But they are virtually nonrenewable, at least over time scales shorter than tens of millions of years. Indeed, billions of years of energy deposits of dead plants will be depleted shortly. And, once again, things will have to change, just as they've changed in the past.

Ancient Fossils

The oldest fossils prove that photosynthesis occurred early in Earth's history. Shown in Figure 25.1 as well as in Figure 26.1, these remains seem to have a cellular structure resembling that of modern blue-green algae. Not very complex, these fossils lack well-developed biological nuclei—the technical term for which is **prokaryote**. Yet these life forms, of which the fossils are the remains, must have photosynthesized by some means or another, since chlorophyll is often formed in their immediate vicinity. Apparently, these oldest fossils are the remains of autotrophs that reproduced asexually by splitting in two.

Fossil remnants of the most primitive cells were discovered embedded in African and Australian sedimentary rock known to be about 3.5 billion years old. Hence the cells shown in the figure are presumed to have existed that long ago. Actually, there is no way of dating the fossils themselves; radioactive techniques are useless for carbon substances older than 40,000 years. But it seems inconceivable that recently living algae could have gotten so deeply encased in such old rock.

There are no known fossils of the most ancient heterotrophs. Small molecules can seep into all but the densest of rocks, so there is no good way of knowing if amino acids and nucleotide bases contaminating many rocks are as old as the rocks themselves. Even if there were techniques enabling us to search for prebiotic (before-life) organic matter, probably none would be found. The early heterotrophs probably devoured every bit of available organic matter, thus leaving absolutely no trace of the primordial soup anywhere on Earth. Consequently, we are unable to estimate either the amount of time needed for the autotrophs to have overwhelmed the

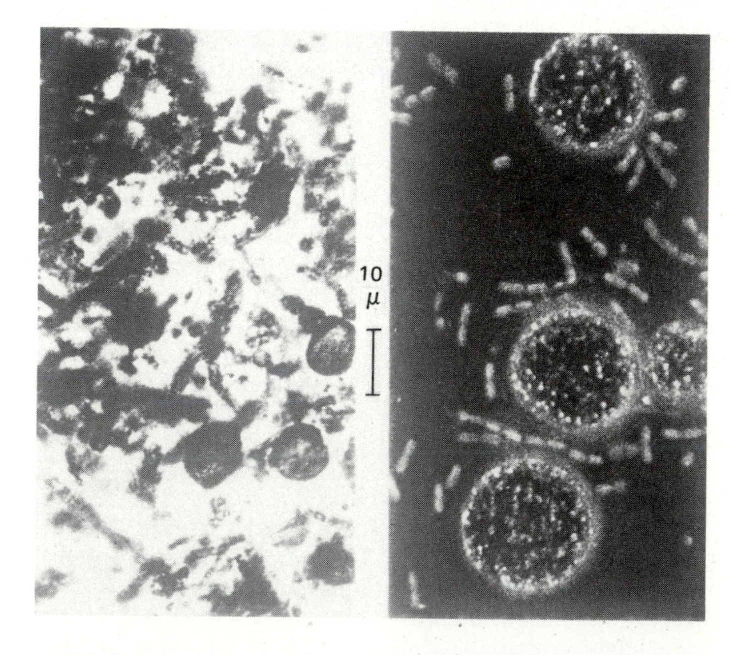

FIGURE 26.1 Fossils of some of the most primitive cells known, seen here through a microscope (left), are dated to be about 3.5 billion years old. In many ways, these fossils resemble the structure of modern algae (right)—that fuzzy blue-green moss often found beside lakes, streams, or even backyard swimming pools.

primitive heterotrophs, or for the primitive heterotrophs to have appeared in the first place. All that we can say with certainty is that life must have originated not more than 1 billion years after the formation of planet Earth. Life could have conceivably appeared earlier, but how much earlier can only be a guess.

Evidence of more recent, though still ancient, life has been unearthed in many places in our planet. The north shore of Lake Superior in Ontario is especially rich in ancient fossils, and the rocks there are radioactively dated to be nearly 2 billion years old. Embedded inside the limestone are layered clusters of algae—whole colonies of cells called stromatolites. Figure 26.2(a) shows the fossilized remains of one of these early life forms, created when primitive autotrophs clustered together and became trapped in sediment that later hardened into rock. Careful studies of this very old Canadian rock shows evidence for at least 12 distinctly different types of algae. All of them are extremely simple systems compared to the complexities of most of today's cells [consult, however, Figure 26.2(b)]. As best can be determined, these ancient cells still lacked well-defined biological nuclei, and thus were still prokaryotic. Yet despite their obvious clustering, each cell functioned on its own. In fact, all fossilized cells older than a billion years are called **unicells**, since each failed to collaborate with other cells nearby.

One-billion-year-old rocks in Australia contain extremely well-preserved fossils of autotrophs. Nearly 20 different types have been identified thus far, many similar in structure to modern blue-green algae. In addition the fossils of this period record the appearance of the first organized teams of cells—the ancestors of modern plants and animals. In short, sometime between 1 and 2 billion years ago, life had reached a whole new plateau—it had become complex and organized.

By about 1 billion years ago, life had already existed on Earth for well more than 2 billion years. Its basic cells had become 10 times larger, vastly more complex, and perhaps diverse in activity. Furthermore, fossilized cells of this age show clear evidence for **eukaryotes**—organisms having biological nuclei, including hereditary DNA molecules, and which reproduce sexually. Equally important, these billion-year-old organisms were beginning to enhance their survivability by working together as groups. Such groups are termed **multicells**, or clusters of cells that collaborated with one another. (For contrast, recall from Chapter 25 that each human being is a cluster of some 10^{14} such cells!)

But that was all there was a billion years ago. Primitive oceanic life flourished, but nothing else did. The fossils show no evidence that plants yet adorned Earth's landscape. No animals were crawling, swimming, or flying near the surface. Certainly there were as yet no men and women.

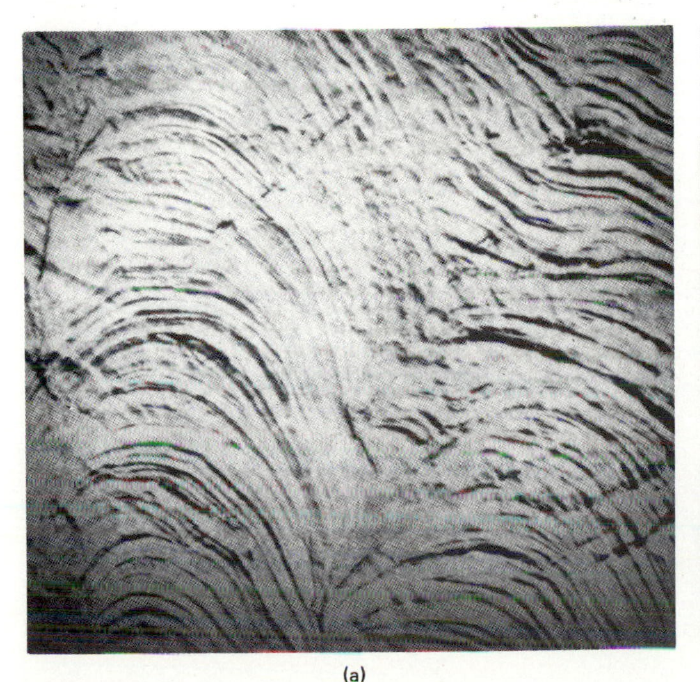

(a)

(b)

FIGURE 26.2 Blue-green algae cells sometimes cluster together to form underwater stromatolites. The fossilized ones shown in (a) were trapped within rocky sediments approximately 2 billion years ago. Not all stromatolites are fossilized: The club-shaped ones in (b) are alive and well under 3-foot-deep water near Shark Bay, Australia. Here, and very few other places, the water is too salty for the animals that normally graze on them. In size and form, modern stromatolites are much like their fossilized ancestors.

Oxygen-Breathing Organisms

The use of sunlight by cells was a double achievement of great importance for life on Earth. Not only did it provide an unlimited source of energy and assure a dependable supply of food, but it also drastically changed Earth's atmosphere by endowing it with lots of oxygen gas.

THE OXYGEN BUILDUP: Once photosynthesis began with the onset of the autotrophs more than 3 billion years ago, Earth's environment changed steadily. The release of oxygen gas into an atmosphere that previously had little or none of it had an enormous influence on the abundance and diversity of life on Earth. As noted in Chapters 1 and 5, the diatomic oxygen molecule (O_2) breaks down into two oxygen atoms when interacting with the Sun's ultraviolet radiation. High in the atmosphere, the oxygen atoms recombine to form large quantities of triatomic oxygen gas (O_3), also known as ozone. (Derived from the Greek, and meaning "to smell," this sharp-odored ozone gas can often be sensed near thermal copying machines that use ultraviolet radiation.) The ozone now almost completely surrounds our planet in a thin shell at an altitude of about 50 kilometers, effectively shielding the surface from further exposure to this high-energy radiation.

As the ozone layer matured, survival no longer depended on protection by a layer of water, or by some rock or other object acting as a barrier against the harsh realities of the real world. Life became possible on the surface of the water, and even on the surface of the land. Organisms were then free to spread out into all the available nooks and crannies on planet Earth. In short, life could invade areas where no life had existed before.

None of this happened overnight. The ozone layer took time to become thick enough to screen out most of the harmful ultraviolet radiation. The process was an accelerating one: The oxygen-producing autotrophs had an increased chance for survival and therefore replication. The more offspring they produced, the more oxygen they dumped into the atmosphere. And more oxygen meant more ozone, more protection from ultraviolet radiation, and a yet-enhanced chance for survival. But it still took time for the protective ozone to develop. How long? Perhaps 2 or 3 billion years after photosynthesis first began.

Some calculations suggest that the ozone layer did not become extensive enough to shield organisms from solar ultraviolet radiation effectively until about 600 million years ago. At about that time, the fossil remnants show that life rather suddenly became widespread and diverse. Prior to 600 million years ago, only primitive life forms existed. Time periods shortly thereafter saw a rapid surge in the total number and diversity of complex living organisms—a population explosion of the first magnitude.

Figure 26.3 diagrams the overall expansion of life in relatively recent times, as revealed by the fossil record. Since about 600 million years ago, the number and diversity of life forms have risen. The one major exception, represented by the dip around 200 million years ago, probably resulted from a widespread ecological crisis which we discuss in the next section.

What was responsible for this burst of biological activity? It was largely the presence of oxygen, which permitted a new, more efficient way for organisms to obtain energy for living. The first, most primitive life forms that ate via fermentation were superseded by more advanced life forms that developed photosynthesis as a means to manufacture food. Eventually, more advanced organisms—the forerunners of animals—began using oxygen to acquire the necessary nutrients. With oxygen, organisms could obtain more energy from the same amount of food. Combined with a fully extablished ozone shield, this global availability of oxygen meant that life was then able to survive and reproduce in all sorts of new habitats.

The previously harsh conditions under which early life had struggled were gone. Earth became a nice place in which to live. And the multicellular organisms of the time took advantage of their friendlier environment.

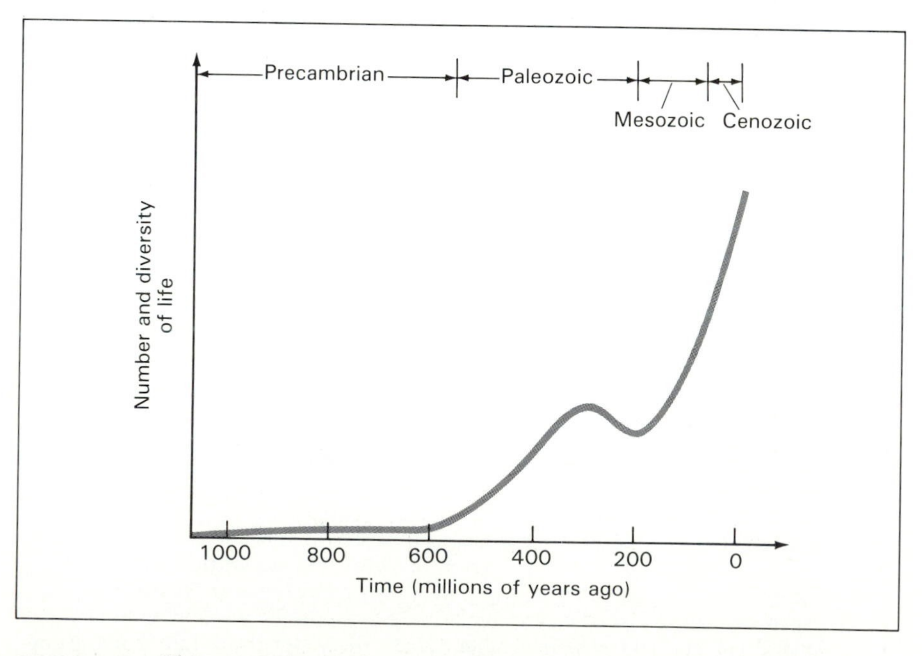

FIGURE 26.3 This curve shows the rise in the number and diversity of living things in the relatively recent history of planet Earth.

RESPIRATION: The chemical process whereby cells use oxygen to release energy is called **respiration**. Ingesting oxygen ("breathing") helps an organism to digest the carbohydrates in its body. The carbohydrates then decompose into carbon dioxide and water. The chemical process can be abbreviated by the formula;

$$CH_2O + O_2 \rightarrow CO_2 + H_2O + \text{energy}.$$

Comparison of this chemical reaction with the one written earlier in this chapter shows that respiration is just the reverse of photosynthesis. But there is an important difference. Whereas in photosynthesis energy must be absorbed to yield the foodstuff carbohydrates, in respiration much of that same energy is released as the oxygen breaks down the chemical bonds of those carbohydrates.

Today, these two major processes—plant photosynthesis and animal respiration—direct the flow of energy and raw materials in the living world. As shown in Figure 26.4, this energy for life flows only in one direction. The energy originates with the Sun, is absorbed in photosynthesis, is released by respiration and is consumed in the process of living. All the while, carbon dioxide, water, and oxygen are continually exchanged between photosynthesis and respiration. These materials are used over and over again, in a completely cyclical fashion; the plants use animal pollution, while the animals use plant pollution. Nature knows how to recycle.

As far as we know, this most efficient flow of energy—from the Sun to advanced organisms—contributed greatly to the veritable explosion of biological organisms several hundred million years ago.

Let's take a closer look at the fossilized remains of the oxygen-breathing life forms.

Recent Fossils

Life forms usually begin to decay as soon as they die. Once a means of gathering energy has ended, disorder sets in. Dead organisms—even their bones—decompose quickly, the former proteins becoming bad-smelling substances within a few days. This is the way in which all life forms—including humans—return to the planet the elements borrowed from it.

Some special environments can limit decay, including cold polar regions, high mountain tops, and deep ocean bottoms. Low temperatures and water burial tend to retard spoilage. For example, a living system having died along a stream or ocean shore might be buried under layers of sand and sediment while drifting down through the water. Volcanic lava is another material in which various life forms can be buried, in this case under layers of ash. In time, the sedimentary deposits of sand or lava become hardened into rock, entombing the remains of living systems. In this way, the bony structure of the ancient organism is occasionally preserved until later uncovered by natural causes (changes in Earth's crust) or human events (archeological expeditions). Figure 26.5 depicts examples of these two ways to discover fossils.

Studies of all the fossils thus far uncovered have enabled biologists to assemble a reasonably complete record. Using a variety of dating techniques, biologists can roughly determine when various organisms lived. More specifically, the fossil record shows that new life forms occasionally emerged, while others perished. Some types of life survived for long periods of time; others seem to have died as soon as they appeared.

As a rule of thumb, we can generally say that the oldest rocks have only simple life embedded in them, whereas the youngest rocks contain mostly complex life. Apparently, many forms of life were able to survive only by increasing their complexity.

Today, biologists know of at least 2 million different kinds of plant and animal life thriving on our planet. This number includes the entire range of current Earth life, from tiny invisible organisms to giant whales and redwood trees. And new kinds of life are constantly being discovered. Yet even with this enormous variety of existing life, the fossil record suggests that about 99 percent of all life forms that have ever resided on Earth are now extinct.

The fossil record relates the following history of relatively recent life on planet Earth. Once multicellular organisms learned to use oxygen nearly 1 billion years ago, they

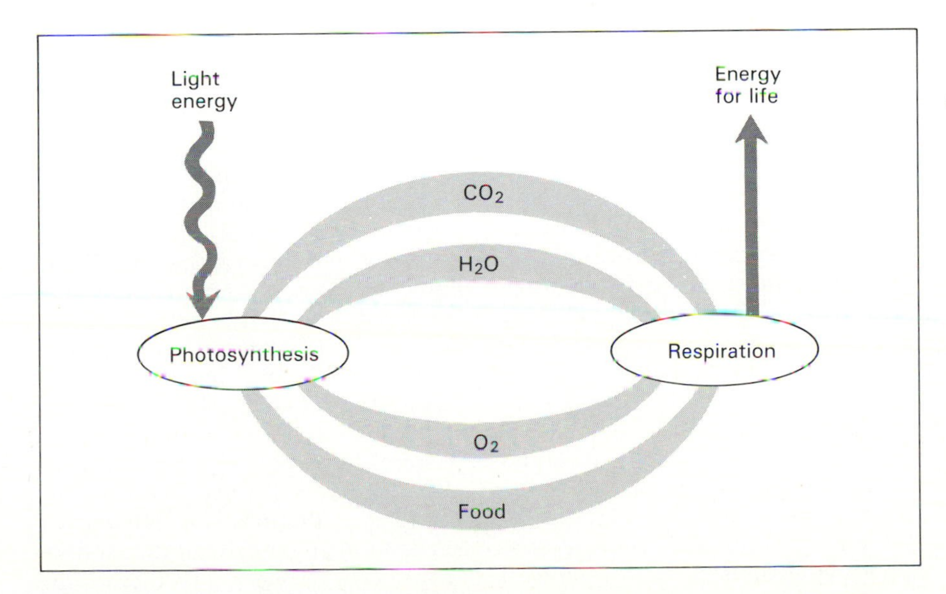

FIGURE 26.4 The cycle of photosynthesis (by the plants) and respiration (by the animals and some plants) converts sunlight into energy for many living things, including humans.

(a)

(b)

FIGURE 26.5 Fossils can become exposed when (a) erosion wears away layers of sediment or (b) humans excavate the land.

quickly evolved into highly specialized creatures. These oxygen-breathing animals—the earliest ancestors of men and women—swarmed in the sea, feeding on plants and one another. Some could only float on the water, others anchored themselves to undersea slopes, while still others had some mobility in the water. Almost all the creatures alive between 0.5 and 1 billion years ago had soft bodies. Hence the fossil record of the earliest respiratory organisms is very sketchy, for without bones or shells, little of them has remained intact to this day.

More than half of all fossils are called trilobites. Shown fossilized in Figure 26.6, these lobster-like creatures varied enormously. Some had heads, some apparently not; others had a dozen eyes, still others none at all. Most were quite small, measuring a few centimeters in length, although some stretched 50 centimeters from head to tail. All trilobites are now extinct, but biologists are reasonably sure that some version of them gave rise to all of today's animals.

As the years wore on, life-styles multiplied rapidly. Each type of organism responded to changes in the oceanic, continental, and atmospheric environment. Each attempted to adapt for better survivability. By some 500 million years ago, worms, clams, and snails ruled the world.

The biological "explosion" that must have occurred some 600 million years ago was partly the result of the establishment of the ozone layer and the cycle of photosynthesis and respiration. But it was also the result of creatures having developed primitive skeletons. For these reasons, fossil experts consider this to have been the start of a new age in the history of life on Earth. As Figure 26.7 documents, all Earth time preceding about 600 million years ago is called simply the Precambrian. Since then, there have been three additional broad ages: Paleozoic (Greek for "ancient life"), Mesozoic (or "middle life"), and Cenozoic (meaning "recent life"). The starting times of these ages are roughly 570 million, 225 million, and 65 million years ago, respectively.

PALEOZOIC: The Paleozoic fossil record shows that fish first appeared nearly 450 million years ago, the first amphibians and forests some 350 million years ago, and the first reptiles and insects approximately 300 million years ago. As depicted artistically in Figure 26.8, over the course of 100 million years or so, life had spread from its oceanic nursery; it had come ashore. Sea plants were probably the first life forms to migrate along the previously barren and rock-strewn coast. Some types of fish, which depended on this vegetation, apparently followed their source of food. Some may have invaded the land intentionally, whereas others may have been washed up onto shorelines by storms or left high and dry when shallow ponds evaporated. Those types of fish that successfully negotiated the land became four-legged, air-breathing amphibians; those that tried and failed became extinct. The details are sketchy because geologists have no clear record of where shorelines were so long ago. Remember, all this preceded the breakup of the ancestral continent of Pangaea that gave rise to the modern continents as we know them. The fossil record does provide overwhelming evidence for one set of changes: Descendants of these first

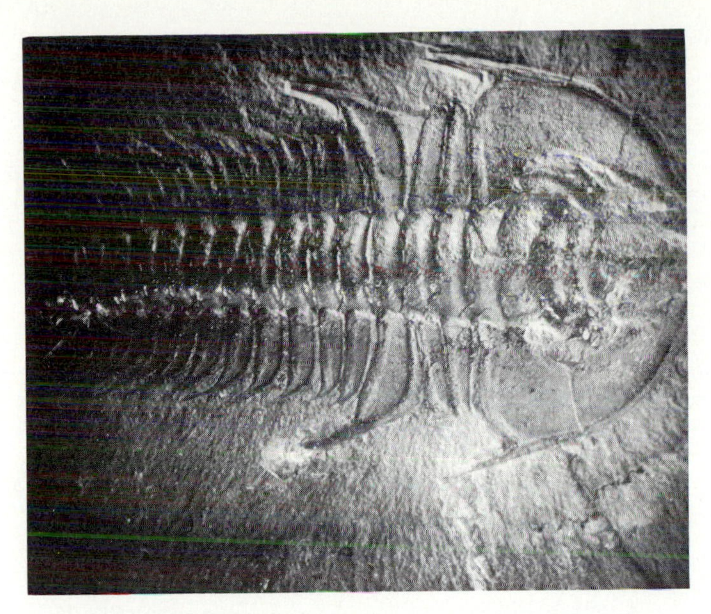

FIGURE 26.6 A fossil of a trilobite, an enormously widespread organism some 500 million years ago.

	Age	Period	Epoch	Time before present (millions of years)
Precambrian	Archean			4,600
	Proterozoic			2,500
Phanerozoic	Paleozoic	Cambrian		570
		Ordovician		500
		Silurian		425
		Devonian		395
		Carboniferous		350
		Permian		290
	Mesozoic	Triassic		225
		Jurassic		190
		Cretaceous		130
	Cenozoic	Tertiary	Paleocene	65
			Eocene	55
			Oligocene	38
			Miocene	26
			Pliocene	6
		Quaternary	Pleistocene	1.8
			Holocene	.01(11,000 yr.)

FIGURE 26.7 Major ages in the geological history of planet Earth. As shown, each age is sometimes divided into periods. For the most recent timespans, these periods are further subdivided into epochs.

shore plants became the world's first forests, and certain descendants of the amphibians eventually became the animals that lived in those forests.

By the end of the Paleozoic Age, life was firmly implanted in the sea, on the land, and in the air. By 200 million years ago, just as Pangaea was beginning to break apart, there existed a broad opportunity for living. The land in particular, with its green expanses and virgin forests, enabled animal life to proliferate with astonishing diversity. Types of life forms multiplied rapidly, so much so that the fossil record documents, for example, that there were at the time nearly 1000 different kinds of roaches; the household version—the common cockroach—is a direct, and very durable, survivor of the late Paleozoic Age.

All life on Earth had become dominated by the reptiles; this was a whole new life form that had, over millions of years, evolved from vertebrate amphibians. The conquest of the land was complete, as reptiles spread out to fill every conceivable niche on the planet. As ancestors of nearly every animal now on Earth, the reptiles of 200 million years ago had developed supple backbones, mobile legs, and keener brains than any other creature inhabiting Earth until that time.

MESOZOIC: The Mesozoic fossil record shows that many forms of life not only thrived but also evolved toward greater complexity. As depicted in Figure 26.9, plant life flourished, taking its most recent steps toward its modern forms. Flowers appeared in dazzling colors and rich scents, all for the purpose of attracting pollinating insects. And the first birds took flight, most as small as today's sparrows.

But the highlight of the Mesozoic Age was the first appearance of the mammals—warm-blooded animals able to derive body heat from digested food and thus stay comfortable in cold environments. Fossil evidence reveals that three types of mammals evolved throughout this 150-million-year age. The earliest mammals were probably ancestors of the present-day anteater—primitive creatures that had fur and nursed their young with milk, but, like reptiles, laid eggs instead of bearing live young. A second, more advanced group of mammals probably bore their young live like their descendants, the modern kangaroo and the koala bear; these young were so small and immature, how-

FIGURE 26.8 Artist's conception of the many forms of life that flourished on planet Earth during the Paleozoic Age.

ever, that they had to be incubated in a fur-lined pouch under their mother's belly. Toward the end of the Mesozoic Age, genuine mammals appeared, laying no eggs and needing no pouch for their young.

Details of the mammals' line of ascent during the Mesozoic Age are somewhat obscure, as they were completely overshadowed by the mightiest reptiles of all times—the dinosaurs. These monstrous creatures (see Interlude 26-2) gradually evolved from smaller reptiles until they devastatingly dominated the land approximately 200 million years ago. The dinosaurs included 25-meter-long and 25-ton hulks,

FIGURE 26.9 The Mesozoic Age saw a continued increase in the diversity of life forms, and not least the first appearance of the mammals—all of which, however, were completely dominated by the dinosaurs.

giant seagoing reptiles capable of devouring today's great white shark in one gulp, and fearsome bird-like creatures having wing spans comparable to those of modern fighter aircraft.

All these flying, swimming, and land-living prehistoric predators disappeared with bewildering abruptness near the end of the Mesozoic Age. The presence of the dinosaurs mysteriously vanishes from the fossil record. No one is certain why. Nor does anyone know how the mousy mammals were able to survive throughout the 100-million-year reign of terror brought on by these enormous beasts. Whatever the reason, it affected not only the dinosaurs but also many other forms of life. Fossil records demonstrate that about 65 million years ago, nearly half of all plants ceased to exist. Numerous kinds of mammals, reptiles, and birds also perished; it seems that all animals larger than 25 kilograms (i.e., about 50 pounds) disappeared at this time.

CENOZOIC: The beginning of the Cenozoic Age, roughly 65 million years ago, saw the appearance of an almost entirely new cast of characters. Pangaea had broken apart into the continents now familiar to us. The dinosaurs were gone completely. The earlier reptilian dominance over the mammals had been totally reversed. And the planet had returned to its pre-dinosaur tranquility. Clearly, mammals had taken over the world, although they had apparently done so by default. In a certain sense, the meek had indeed inherited the Earth.

Fossils as recent as 50 million years old show that most mammals had small brains, large jaws, and rather inefficient feet and teeth. Still, they were freely multiplying, swelling in numbers and diversity. As always, change was rampant. Ice ages came and went; continents split apart and drifted. In generation after generation, life forms constantly fine-tuned their daily routines for better survivability. Accordingly, many of these early mammals passed into extinction, to be replaced by better-adapted stocks.

In a relatively short time, the mammals had evolved into an amazing assortment of creatures. Some 40 million years ago, the ancestors of such modern mammals as the horse, the camel, the elephant, the whale, and the rhinoceros, among others, gradually ventured forth, although in shapes almost totally unrecognizable compared to their descendants of today. Most grew in size and improved their overall performance between 40 and 20 million years ago.

Darwinian Evolution

Many life forms have come and gone on planet Earth. Some were insignificant organisms, while others ruled the land, sea, and air. For hundreds of millions of years, our planet has housed a steady parade of new creatures, many of whom led in turn. Yet only the latest of these dominant life forms—men and women—are able to learn about those creatures that went before. Humans alone have been able to unearth and understand the amazing record of now-extinct and terribly bizarre life forms once prevalent on our planet.

What sense can be made of the vast array of past and present life on Earth? Is there any unifying logic linking it all? Well, classification is the first step in an attempt to discover the underlying causes of the abundance and diversity of life on our planet.

All current life as well as fossilized remains of ancient life can be broadly classified as plant or animal. These classes can, in turn, be divided into different **species**, a subclassification generally used to denote not only structural similarity but also the ability to mate and produce fertile offspring. For example, all humans are members of the same animal species, and all the many varieties of roses are of the same plant species.

Some life forms are hard to classify, however. Figure 26.10 shows a creature known as the Euglena, which is claimed to be an animal by zoologists (animal biologists) since it can move rapidly like an animal. But the Euglena is also claimed to be a plant by botanists (plant biologists)

INTERLUDE 26-2 ***Dinosaurs, The Great Failures***

Dinosaurs were no ordinary reptiles; they were nothing at all like the snakes, lizards, or crocodiles of modern times. Their name derives from the Greek words, *deinos* (terrible) and *sauros* (lizard). In their prime, roughly 100 million years ago, the dinosaurs roamed Earth with skill and power. Their fossils have been uncovered on all the world's continents, although not at the poles.

Until recently, the prevailing view claimed that dinosaurs were rather dumb—cold-blooded and small-brained. In chilly climates, or even at night, the metabolisms of these huge reptiles would have become sluggish, making it difficult for them to move around, secure food, and thus survive. However, a new and controversial view is now emerging. Recent studies of dinosaur fossils suggest that many of these monsters might have had large, four-chambered hearts, like those of mammals and birds. Such a heart could have pumped blood through organs, enabling dinosaurs to sustain a high level of physical activity. If these recent interpretations are correct, some dinosaurs were probably warm-blooded, and thus relatively fast-moving creatures. Also, although the dinosaurs clearly had small brains compared to those of today's mammals, they were still smart for their time. Indeed, no species able to rule Earth for more than 100 million years could have been too dumb. By comparison, humans have thus far dominated for little more than 2 million years.

Many explanations have been offered for the dinosaurs' complete and total demise. Devastating plagues, magnetic force-field reversals, increased tectonic activity, severe climate changes—perhaps triggered by supernova explosions or asteroid collisions—have all been proposed. Each of these ideas has some merit, although none is entirely convincing.

In recent years, the idea that a huge extraterrestrial object collided with Earth some 65 million years ago has become popular. According to this idea, an asteroid or comet struck the Earth, causing great quantities of dust (mostly its pulverized self) to become airborne, which in turn temporarily encircled the planet, darkened the atmosphere, shut down photosynthesis, and disrupted the base of the food chain by killing off many plants. A layer of clay enriched with the element iridium found in 65-million-year-old rocky sediments (at the boundary between the Cretaceous and Tertiary geologic periods) is the main piece of evidence supporting the asteroid-impact model. Although rare on Earth's surface (since most of it has sunk into the planet's core), the iridium in the clay layer is tens of times more abundant than in native rock, yet matches levels found in meteorites. This model is not without problems, however: If the iridium "rained down" out of the atmosphere, why is the layer of iridium so highly varied in abundance from place to place on Earth's surface? Where is the crater left by the impact? And if instead the asteroid landed in the ocean, thus leaving no crater, how did it uplift dust and debris high into the air? Perhaps, argue some opponents, the iridium layer was laid down by volcanoes and had nothing to do with an extraterrestrial impact.

For whatever reason the dinosaurs perished, dramatic environmental change of some sort was almost surely responsible. It would be useful to continue the search for the cause of their extinction, for there's no telling if that sudden change might strike again. As the dominant species on Earth, we are the ones who now stand to lose the most.

Despite our knowledge of dinosaurs, no human being ever saw one alive. Their remains lay hidden for about 65 million years before *Homo sapiens* in the early eighteenth century discovered their prior existence. Even so, only within the past few decades have we come to realize that we probably wouldn't be here had the dinosaurs not become extinct. Only when the dinosaurs disappeared did the spectacular rise of the mammals—including the ancestors of human beings—begin.

since it consumes energy much like a plant. Further research might resolve to which category it really belongs.

A few other very simple life forms also seem to be misfits, having little in common with either plants or animals. Bacteria, blue-green algae, and amoebas are especially good examples of small (10^{-5}-centimeter) unicellular life forms that generally lack structure within their cytoplasm; in fact, these so-called prokaryotes have no well-defined biological nucleus and hark back to the very earliest forms of life on Earth. As noted in Interlude 25-3, some researchers argue that these ultrasimple life forms deserve a separate classification, regarding them as a "lower" form of life. By contrast, multicellular plants and animals (all of whose cells have nuclei and are hence eukaryotes) are said to comprise a "higher" form of life.

Actually, understanding involves more than just categorizing life forms. Real creatures do not always match what is expected of a species. In other words, individual species usually show small, although noticeable variations from their "ideal" classification—slight changes from some standard specimen to which each organism may be compared. This is true of all species, whether they are now living or fossilized.

As with many aspects of matter in the Universe, the concept of change

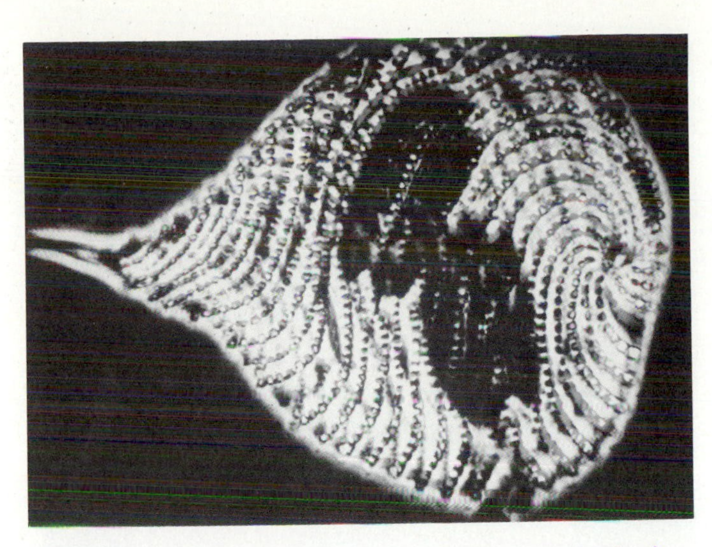

FIGURE 26.10 The Euglena is a microscopic organism having both plant and animal properties. The size of this creature is about 0.01 centimeter.

is central. In this case, subtle changes in life forms contain the key to our understanding of how life has developed over the course of time. We have entered the realm of **biological evolution**—the developmental changes, from generation to generation, experienced by life forms throughout the history of life on Earth.

THE BASIC IDEA: The theory of biological evolution, independently conceived by the nineteenth-century British naturalists Darwin and Wallace, can account for two outstanding features of the fossil record: (1) Living systems have generally become more complex with time, and (2) members of all species show some variation from their "ideal" category.

These two facts clash head-on with the age-old assumption that nature is unchanging. Like Copernicus and Galileo a few centuries before, Darwin and his colleagues faced the same kind of opposition made popular by Aristotle, who refused to concede that species change. But given the facts of nature, a static theory of life cannot possibly be correct. Everything changes with time, life included. The only feasible explanation is a dynamic, evolutionary one.

The central tenet of biological evolution is simple: Living things change, some for the better, others for the worse. Those that survive for a long time often change drastically, sometimes becoming whole new species. Some species become extinct; others arise anew. Organisms of similar structure have similar ancestry and are closely related. Those with very different structures have accumulated these differences over long periods of time and are therefore now only distantly related.

HOW IT WORKS: Biological evolution is not faith. It's fact. The fossil record no longer leaves room for any reasonable doubt that evolution does happen. Evolution is as securely founded as the revolution of Earth about the Sun.

On the other hand, the mechanism that causes evolution to occur remains a theory. But this theory is strictly the result of the scientific method: Observations were made of the fossilized remains of life; an idea was proposed to explain those facts; and subsequent experiments have served to strengthen and revise the intricacies of the theory during the past century.

What is the mechanism of biological evolution? How does it work? The prime mover is the environment—the physical conditions surrounding all living things. Temperature, density, air composition and quality, types of food, as well as the existence of natural barriers such as rivers, lakes, oceans, and mountains, are all influential environmental factors.

Now we know that Earth is not static. These environmental conditions on our planet are constantly, though slowly, changing. Biological evolution dictates that all life forms respond to their changing environment, inhibiting some traits while promoting others, but in any case yielding a vast diversity of species throughout the course of time. Accordingly, changes—in the environment, and in life—occur as a rule, not as an exception.

Observations show that although all species reproduce, few of them display gigantic increases in population; the total number of any one species remains fairly constant, there being no dramatic explosions of offspring. Furthermore, the process of reproduction is almost never perfect; offspring in each generation are hardly ever exact copies of their parents. The implication is that most offspring never survive to reproduce. All life must struggle and compete in order to endure.

Natural selection, an expression coined by Darwin himself, is the mechanism that guides life's evolution. It can be explained as follows: Since most members of the species exhibit some variation from their ideal standard, organisms having a variation particularly suited to their environment would be most likely to survive. They are quite naturally selected to live. By contrast, those organisms having unfavorable variations would be most likely to perish. They are naturally selected to die. In short, only those life forms able to adapt to a changing environment tend to survive to reproduce, thereby passing their favorable variations or traits on to their descendants.

Note that **adaptation** is a key feature of biological evolution—the positive response to a changing environment of an organism having some variation or trait that improves the organism's chance for survival and reproduction.

In successive generations, advantageous traits become more pronounced in each individual. Not only that, but the numbers of individuals having favorable traits also increase. Favored individuals generally produce larger families, as they and their offspring have greater opportunities for survival. Their favored descendants multiply more rapidly than those of their less advantaged neighbors, and over many generations, their progeny replace the heirs of individuals lacking the desirable trait.

Thus natural selection truly does imply the well-known phrase "survival of the fittest." It literally molds life forms. With the passage of sufficient time, the action of natural selection can greatly change the shape, disposition, and even the existence of individuals; old species disappear in response to changing conditions, while whole new ones arise.

FIGURE 26.11 A small variation among some of the creatures of a single species can sometimes be advantageous for survival. For example, moths survive best when they blend with their environment. Here, it is the dark moth that has an advantage. Can you spot it?

SOME EXAMPLES: Natural selection cannot be easily observed at work. Long passages of time are needed to note a distinct variation in any population of a species. Some experiments have been performed under laboratory conditions that mimic those of nature. Like the origin-of-life experiments, these are simulations that attempt to study the adaptation of life to a changing environment. The results support the theory of biological evolution via natural selection.

Here is the crux of one such laboratory experiment. Two sets of field mice, one set with dark fur and the other with light, were placed into a small barn along with an owl. The straw and ground cover were chosen to match closely the dark color of one set of mice. As such, the camouflage gave the darker-colored mice an environmental advantage to hide; the lighter-colored mice were clearly at a disadvantage. At the end of carefully controlled experiments, the owl had captured many more lighter-colored mice. When the ground cover was changed to lighter straw—corresponding to an environmental change granting the lighter-colored mice a greater opportunity to survive—the results were reverse; the owl captured the darker-colored mice more easily. This is an example of how small variations in one species can grant a competitive advantage. As might be expected in the real world outside the laboratory, the natural habitat of the light mice is cornfields, of the dark mice, forests. In each case, the mice have adapted to their environment, are naturally selected to live, and thus to reproduce their kind.

Examples of natural selection have also been noticed in nature's outdoor setting whenever an environmental factor changes exceptionally fast. For instance, early in the nineteenth century, the bark of all birch trees was nearly white, enabling light-colored ("peppered") moths to blend nicely with their environment. In their struggle to survive, they prospered around the trees against whose bark they were nearly invisible. By contrast, darker-colored moths stood out clearly against the white bark, and birds easily picked them off. By the turn of the current century, however, the bark of birch trees near manufacturing cities had become heavily soiled with the sooty pollutants of the Industrial Revolution. This environmental change had removed the competitive advantage previously enjoyed by the lighter-colored moths, as shown in Figure 26.11. The result is that few pale moths remain today, at least near industrialized areas; instead, grayish moths now possess the advantage of camouflage, enabling them to prosper, mate in peace, and freely reproduce. This is an example of how simple variations—in this case color—serve to guide natural selection within a changing environment. For some moths, in fact, a small change became an issue of life or death.

The common housefly is another example of how some members of a single species can adapt to a changing environment, giving them a better chance to be naturally selected for survival. Originally, the pesticide DDT was successful in killing houseflies. During the first few years of use, DDT killed almost all the flies; few flies could successfully adapt to this sudden environmental change caused by the chemical DDT in the air. A small minority, however, managed to survive because they possessed a chance variation or trait that made them resistant to this chemical. These survivors freely reproduced, thus passing the advantageous trait onto their descendants. Within a decade, the offspring of the oddball survivors outnumbered the original majority type of fly. Consequently, DDT has grown less and less effective over the years. Now most houseflies have inherited a resistance to DDT, and the pesticide is useless against these flies. The chemical DDT did not give this resistance to the flies; rather, it provided an environmental change enabling natural selection to go to work. To survive as a species, the housefly had to be able to adapt to the changing environment. Those that managed survived. Those that were unable to do so are long gone.

These last two cases are examples of evolutionary responses to environmental changes induced by humans—a whole new aspect of evolution whereby technologically equipped beings play the role of nature.

DIFFERENT TYPES OF SELECTION: Variations among members of a particular species can be illustrated in graphical form. Figure 26.12 shows a plot of the distribution of a particular trait among members of a given population of a species. This distribution is bell-shaped, implying that most members have some average properties, while fewer members have extreme properties to either side of the average. Known as the **gene pool**, this type of spread in variations can represent any trait of a species—hair color, eye color, size, shape, appearance, whatever. In the case of the species of mice or moths discussed above, the variation is their color. As shown by the graph, most moths are sort of tannish, although there are some white and brown ones.

Now imagine an environmental change. For the mice, consider a ground cover that becomes progressively darker. And for the moths, consider birch trees that gradually became dirtier. After natural selection has had a chance to take its course, the distribution of mice and moths will have shifted. Figure 26.13(a) schematically shows how the darker-colored mice or moths are favored for survival. The darker mice or moths will tend to increase in number; the others will tend to die off. Hence in this case, one of the extreme properties—dark color—is favored. Over a period of time, the gene pool representing the color of

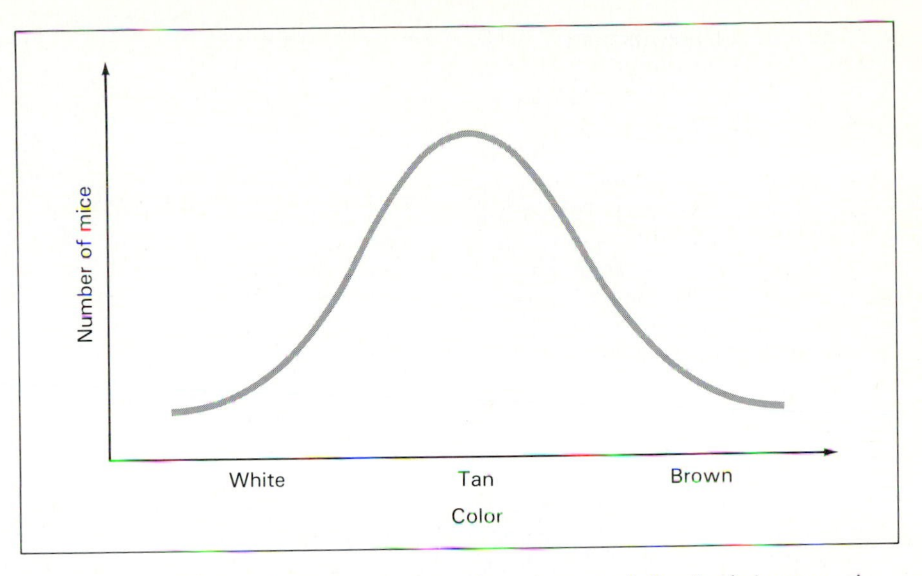

FIGURE 26.12 All members of any species show some variation in their properties (or traits). These properties can be height, weight, color, leg length, eyesight, or whatever. In the example chosen here, we have plotted the number of mice having different colors.

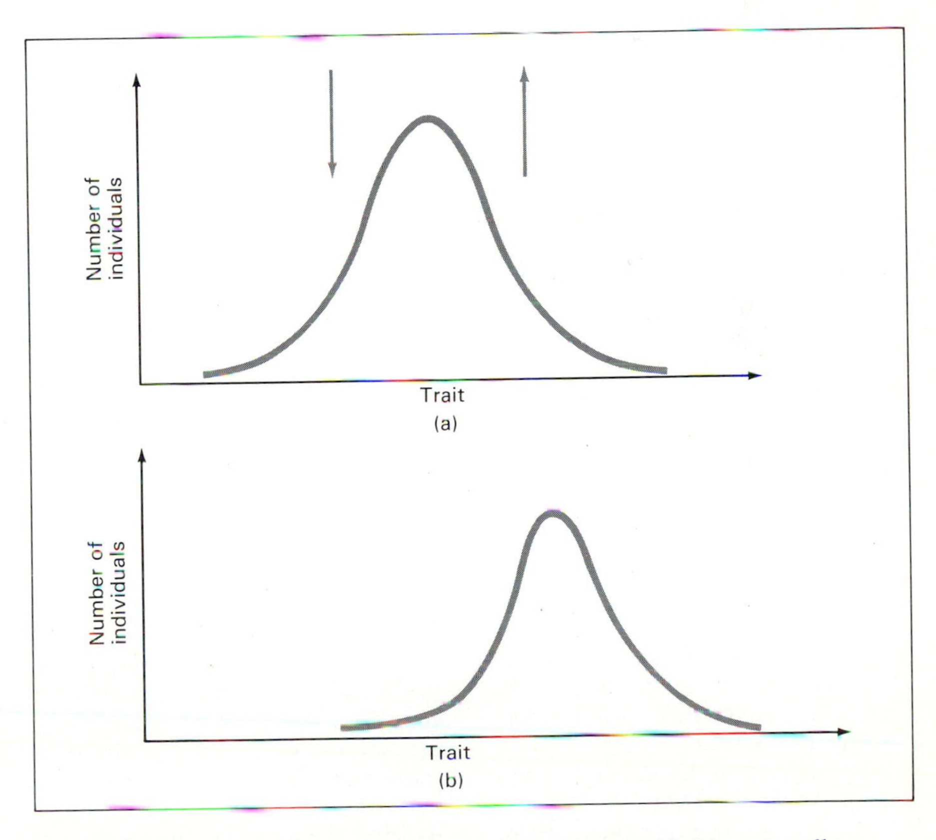

FIGURE 26.13 Directional selection enhances one extreme trait (a), eventually producing a shift in the distribution of that trait within the species (b).

all members of the mice and moth species will have shifted, as shown in Figure 26.13(b). This is known as *directional selection*.

A second kind of species change is called *stabilizing selection*. Figure 26.14(a) shows another spread of traits among members of one species. In the case illustrated, an environmental change causes both extremes to do poorly, while the average traits are enhanced. The result, after a period of natural selection, is the more peaked gene pool shown in Figure 26.14(b).

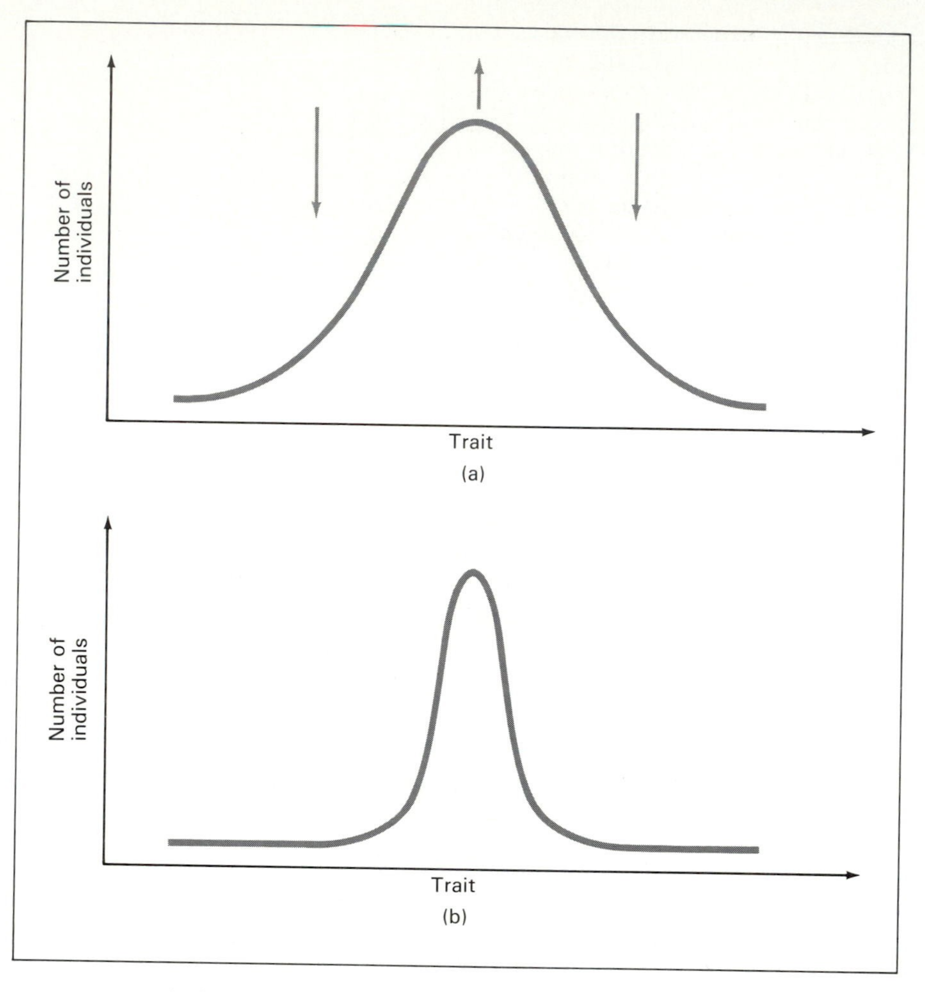

FIGURE 26.14 Stabilizing selection enhances the average traits of members of a single species (a), eventually producing a more peaked distribution (b).

Over long periods of time, chance variations in living things can accumulate. Hair color, eye color, size, shape, appearance, and a host of other traits all change as nature naturally selects for survival those life forms best adapted to the environment at any given time. Eventually, life forms come to differ considerably from members of their original species. In this way, the environment helps new species to evolve from old ones.

For example, members of a single species may be disrupted by some physical change in the environment. As in Figure 26.15, a new river might be gradually rerouted through an area inhabited by a species of butterflies. Should the river become wide enough to act as a physical barrier that cannot be crossed, butterflies on one side would then be prohibited from mating with those on the other side. In this way, the two populations of butterflies become completely isolated. Further changes in the environment on each side of the river can then produce, over long periods of time, some directional or stabilizing selection. Eventually, the two populations of butterflies will differ, in some cases considerably. Should the barrier be removed—if the river dries up, for example—the two populations of butterflies would be able to intermingle once again. Provided that they were separated long enough, however, they will be unable to interbreed. Accordingly, two new species of butterflies will then exist where previously there was only one. Furthermore, each new species will stake out its own claim or fill a separate niche, thereby coexisting peacefully (for the most part) within the new environment.

The change of a single species into two or more new species, usually driven by an environmental disruption of the above sort, is called **speciation**. It's the mechanism behind the diversification of all life.

Figure 26.16 shows how speciation or *disruptive selection* works. In this case, the physical barrier of a new river, or a recently upthrust mountain, can sometimes cause the average traits to do poorly while enhancing the extreme traits. One extreme trait may enable butterflies on one side of the river to increase their survivability; some other extreme trait may be useful for butterflies on the other side. The result, shown in Figure 26.16(b), is a complete disruption in the gene pool which effectively creates two new species.

An actual example of disruptive selection has occurred in the vicinity of the Grand Canyon. Two distinctly different populations of squirrels live on the north and south rims of the canyon. Shown in Figure 26.17, the Kaibib squirrels of the north rim have black bellies and white tails, whereas the Abert squirrels of the south rim have white underparts and gray tails. Both feed on pine-tree bark growing only on the 2-kilometer-high plateaus. The two populations are now separated, and presumably have been for thousands of years, by the intensely hot and dry environment of the canyon. But they have so many similarities that we can safely assume that their ancestors were once members of the same species.

There are many other examples of two or more slightly different spe-

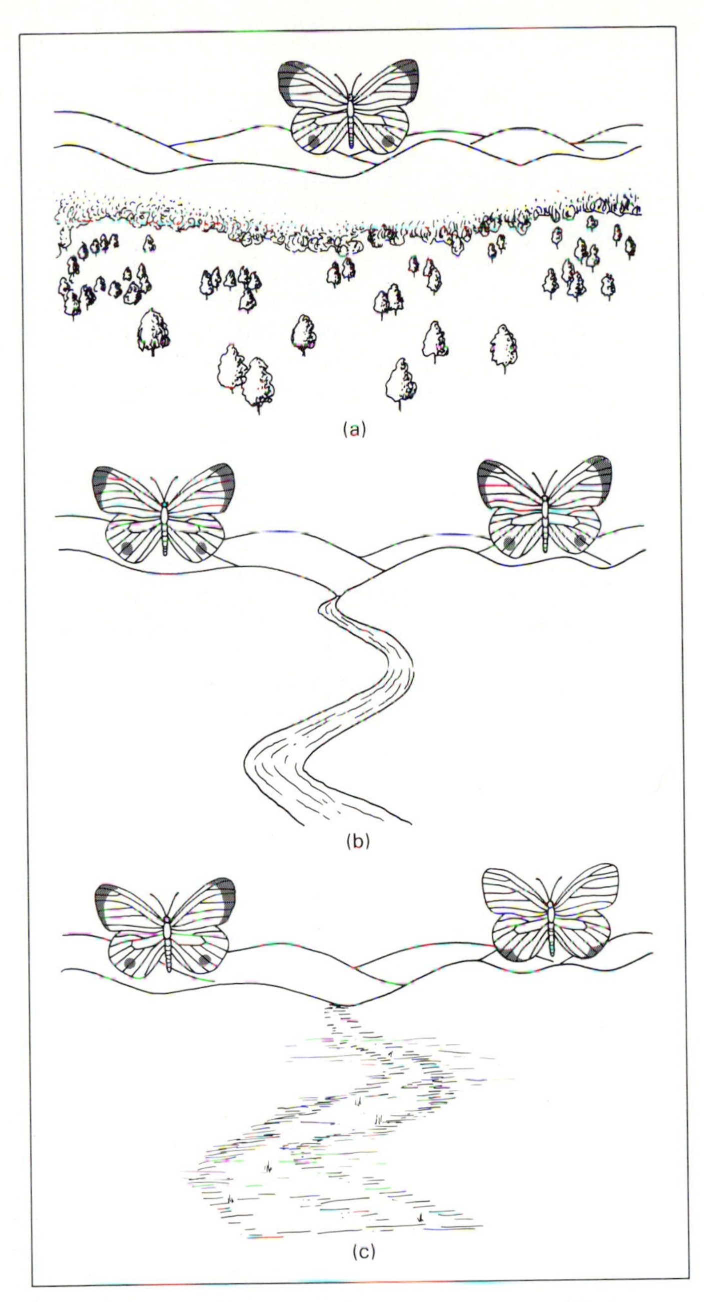

FIGURE 26.15 A species of butterflies which normally interbreed (a) can be disrupted by some physical barrier such as the appearance of a mountain or river (b). When the barrier disappears later (c), the populations of butterflies on the two sides might have changed so much that they can no longer interbreed.

cies, nearby but clearly isolated, and presumably sharing a common ancestral species. Biologists find more of them every day, including members of species that are not even separated by a physical barrier, but that for one reason or another do not freely interbreed.

One final point about these distributions of traits: When biologists refer to populations of species, they are really thinking in terms of gene pools. It's the genes or DNA structures of different populations that are isolated from one another and that gradually develop variations. Therefore, we need a better understanding of microscopic gene variations if we are to truly appreciate modern biological evolution. This is the subject of an upcoming section on Mendelian genetics. But before leaving this section on Darwinian evolution, let's pause for a few paragraphs to take up a currently popular though controversial alteration in classical Darwinism.

PUNCTUATED EQUILIBRIUM: The fossil record of the history of life on Earth clearly documents many periods of mass extinction—times when large fractions of all Earth life perished. In addition to the "great dying" some 65 million years ago that ended the reign of the dinosaurs (as well as three-fourths of all living things; consult Interlude 26-2), there have been other dark times in Earth's history. For example, about 440 million years ago, a vast number of animals then living in the sea vanished. Some 225 million years ago, as much as 96 percent of all animal species living in the sea or on the land became extinct. At this time, the last of the trilobites disappeared, as did all of Earth's ancient corals, most of its amphibian families, and nearly all of its reptiles; life itself was nearly extinguished on our planet.

More recently, some paleontologists have proposed that the great extinctions have been periodic, occuring roughly every 26 million years. One possible reason for this (hotly debated) periodicity in the fossil record is this: The remote Oort Cloud of comets (roughly 50,000 astronomical units away; consult Chapter 8) is disturbed each time our Solar System passes through interstellar clouds or spiral arms of the Milky Way, thereby causing numerous comets to be thrown out of their regular orbits and toward the Sun. Some of these comets rain down on Earth, disrupting the climate or otherwise disturbing our planet's environment, thus causing mass extinction of life on Earth. Another, astounding possibility is that our Sun has a companion star which, in a highly elliptical orbit, would bring it through the Oort Cloud when it is closest to the Sun, creating a disturbance that sends the comets sunward. Obser-

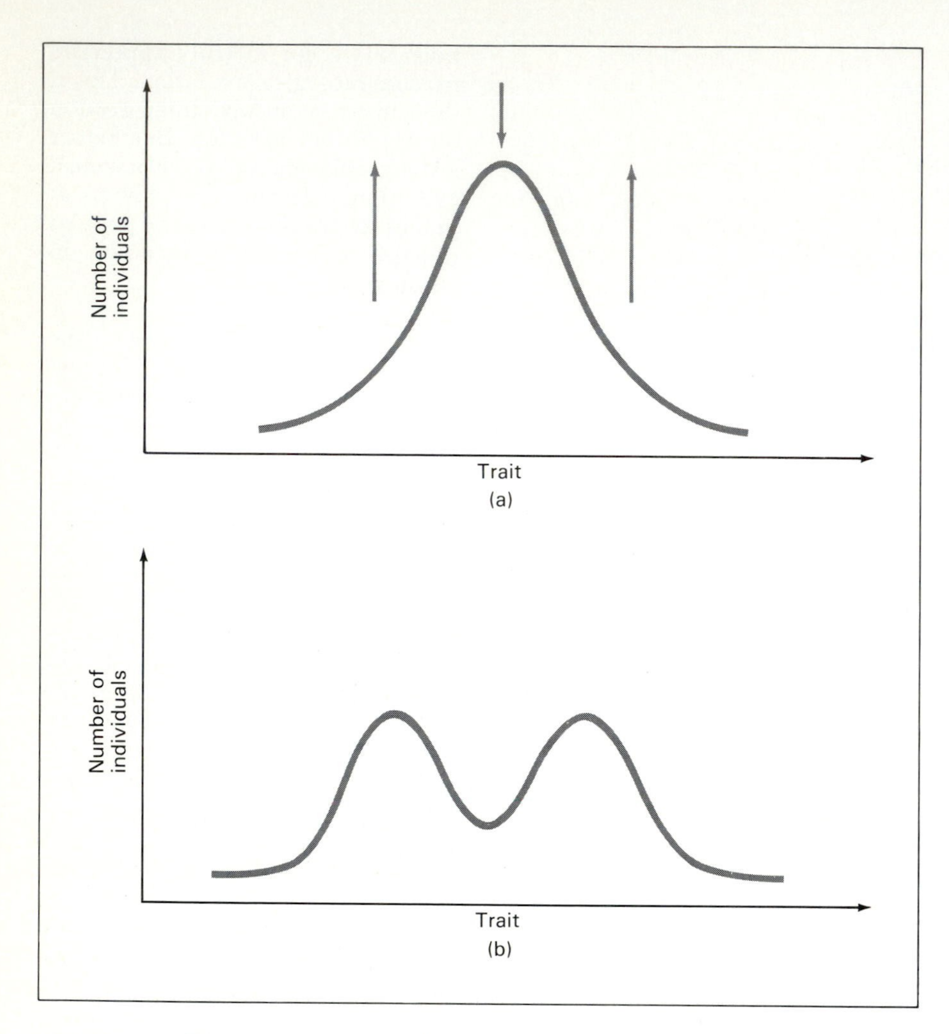

FIGURE 26.16 Disruptive selection enhances the extreme traits (a), eventually producing two completely new populations that can no longer interbreed (b). Two species now exist whereas previously there was only one.

FIGURE 26.17 Showing marked differences because of their isolation on opposite rims of the Grand Canyon, the Kaibib and Abert squirrels are probably descendants of a common species.

vational efforts using infrared equipment are now under way to search for this dim and hypothetical dwarf star—already proposed to have the name Nemesis, after the Greek goddess who relentlessly persecutes the excessively rich, proud, and powerful.

At any rate, whether comets and asteroids have regularly belted Earth sufficiently to cause mass extinctions, biologists are now aware that the rate or tempo of evolution has apparently not been steady throughout the history of life on Earth. Darwin himself stressed that natural selection operates *gradually*—with slow, steady, uniform changes occurring through time. He saw the development of species in much the same way that he saw environmental change on Earth—as a process of smooth, gradual change proceeding at a uniform rate. But paleontologists are now accumulating increasing evidence that biological evolution may well have operated more erratically—with occasionally rapid, even catastrophic changes occurring every so often.

The fossil record does not seem to confirm the gradualism of classical Darwinism. Very few fossils show smooth transitions from one form to another with myriad small, intermediate steps linking one species to another. Of course, one possible explanation is that the fossil record is incomplete—which it surely is. But perhaps nature does make jumps. Indeed, today's fossil record strongly implies that many species have remained more or less the same for long stretches of time, after which they rather suddenly underwent bursts of evolution in perhaps only a few hundred generations—too short to have left a continuous record in rock layers.

Thus a new idea has emerged, called **punctuated equilibrium**. According to this theory, life stays pretty much unchanged until something drastic happens, and then it changes fast. Life's "equilibrium" or "stasis" is said to be "punctuated" by rapid environmental upheaval.

The theory of punctuated equilibrium is a slight variation on classical Darwinism. It is not at all a violation of the basic mechanism of biological evolution; natural selection is fully the means whereby life is thought to change from species to species. Punctuated equilibrium merely emphasizes that the *rate* of evolutionary change is not gradual. Instead, the "motor of evolution" occasionally speeds up during periods of dramatic environmental change—such as cometary impacts, reversals of Earth's magnetism, and the like. We might say that evolution is imperceptibly gradual most of the time and shockingly sudden some of the time.

In point of fact, many scientists have not yet accepted the notion of punctuated equilibrium. Among other reasons, they suggest that the rate of evolution is all a matter of perspective: Those who examine the fossil record up close will see evidence for periods of rapid speciation, yet those who take a broader view over very long periods of time will see, on average, gradual change—just as Darwin maintained.

Mendelian Genetics

What is it that alters living systems to make members of a single species occasionally unable to interbreed? Basically, the microscopic gene is the culprit, for it's the genetic code (introduced in Chapter 25) that dictates if and how life forms reproduce. This is the subject of **genetics**, the study of gene structures and especially how they operate and change.

Pioneered more than a century ago by the Austrian monk, Mendel, genetics has become a good deal more complex than he could ever have imagined. Darwin himself would probably be surprised at the microscopic roots of biological evolution as we know it today—a modern synthesis of Darwinian and Mendelian ideas, often termed "neo-Darwinism."

Basically, hereditary error is a major factor promoting the evolution of living systems; it is, in fact, a prerequisite for evolutionary change. Note that we say hereditary *error*, not heredity itself, which is an agent of continuity, not change. **Heredity**, by definition, is the transmission of genetic properties from parents to offspring, thus ensuring the preservation of certain traits in future generations of a species. Otherwise, the basic life processes and body organs of each and every organism entering the world would have to be created from first principles. Normally, chemically coded instructions of the DNA molecules enable cells to duplicate themselves flawlessly millions of times. But occasionally, mistakes do occur at the microscopic level. Not even genes are immutable. Everything changes.

For reasons not completely understood, a DNA molecule can sometimes drop one of its bases during replication. Or it may pick up an extra one. Further, a single base can suddenly change into another type of base. Even such slight errors in the DNA molecule's copying process mean that the genetic message carried in the DNA molecule for that particular cell is changed. The change doesn't need to be large; even a change in a single nucleotide base among millions strung along the DNA molecule can produce a distinct difference in the genetic code. This, in turn, causes a slightly modified protein to be synthesized in the cell. Furthermore, the error is perpetuated, spreading to all subsequent generations of cells containing that DNA.

Microscopic changes in the genetic message, called **mutations**, can affect offspring in various ways. Sometimes the effect is small, and newly born organisms seem hardly any different. At other times, mutations can affect a more important part of a DNA molecule, inducing considerable change in the makeup of an organism. And at still other times, a single mutation can rupture a DNA molecule severely enough to cause the death of individual cells or even whole organisms.

Mutations are responsible for differences in hair color, eye color, body height, finger length, skin texture, internal structure, individual talents, and numerous other characteristics among a population of life forms of any given species. Virtually any aspect of the life of any organism can be modified by genetic mutations. Such mutations provide a never-ending variety of new kinds of DNA molecules.

Not all mutations are bad. Most of them do indeed create traits inferior to the previous generations—especially in many of today's highly evolved and exquisitely adapted organisms. But some mutations are favorable and serve to better the life of an individual. These can then be passed on to succeeding generations, making life more bearable for members of that species. Beneficial mutations act as the motor of evolution, steering life forms toward increasing opportunities to adapt further to the ever-changing environment.

What causes genes to mutate? Why do some DNA molecules occasionally replicate differently, although they may have copied themselves exactly for millions or even billions of previous cell divisions? Frankly, the precise causes are unknown, for they apparently arise from chance, indiscriminate events. Biologists have undertaken laboratory experiments with cells and have succeeded in increasing the numbers of mutations by artificial means, thereby helping to unravel the ultimate reasons for genetic change. The results so far show that the easiest way to enhance gene mutations is to treat reproductive cells with external agents.

Three of the most important mutation-inducing agents revealed in the last few decades are temperature, chemicals, and radiation. When cells are heated, or treated with industrially generated nerve gas or chemical drugs, mutations are clearly enhanced. In addition, ultraviolet and x-ray radiation seem to be particularly striking causes of genetic mutations. Such radiation has been present on Earth, in one form or another, since the origin of our planet. Radioactive elements embedded in rocks, cosmic rays bombarding Earth from outer space, and minute amounts of solar ultraviolet radiation reaching the ground all serve to prove that life evolved in a radiation-filled environment.

Generally, there is nothing wrong with immersion in radiation. We and other life forms probably wouldn't be here if the motor of evolution had not speeded change; without radiation, life itself might not have progressed beyond the primitive, unconscious, unicellular organisms drifting in the oceanic slime. Of some valid concern, however, is the fact that human inventions such as atomic bombs, nuclear reactors, and medical devices also release radiation. Intense doses of radiation can kill directly, although more subtle doses cause changes in the reproductive cycle that are then passed on to future generations. It's not clear that these human-induced mutations are in all cases harmful, but in the absence of evidence to the contrary, a healthy amount of skepticism is surely warranted.

We must not risk damaging the refined work of several billion years of organic evolution, for the long series of changes that evolution represents can never be repeated.

FIGURE 26.18 Chimpanzees (a) and gorillas (b), both members of the ape family, are known to be among the closest relatives of humans (c).

Paths Toward Humanity

A hundred years ago, the theory of biological evolution was intellectually and morally shocking. Few people accepted it; even many scientists of the late-nineteenth century failed to embrace it. The problem was not really the idea of evolution. Surely evolution occurs, and people a century ago knew it. Fossils were already abundant then, and farmers had for centuries bred crops and animals in a successful effort to develop healthy, disease-resistant foodstuffs.

The real problem was that people were disturbed to hear that humans had anything in common with apes. It seems that when ideas involve humans, vanity comes to the fore. Because of this same vanity, apparently, minor segments of our twentieth-century civilization still refuse to accept the basic ideas of biological evolution.

We now have a combination of fossil discoveries, chemical analyses, and behavioral studies virtually proving that of all modern (nonextinct) species of life now on Earth, the chimpanzee and gorilla (Figure 26.18) are our closest relatives. Note that we are not stating that humans have descended from apes; this is a common misunderstanding. Rather, modern science demonstrates that apes and humans have some common characteristics. In other words, a central idea of biological evolution stipulates that *both* modern apes and modern humans subsequently evolved from a common ancestor. We shouldn't be able to identify that ancestor from any of the currently living creatures on Earth, for genes and environments change over the course of millions of years. But such an ancestor should be part of the fossil record. Stated succinctly, our common ancestor more likely resides in a museum than in a zoo.

To discern our most recent ancestors and thereby trace the ways and means of relatively recent biological evolution, we must rely heavily on the fossil record. Generally, fossils of recent times are well preserved, enabling researchers to document evolution with reasonable accuracy. Not surprisingly, the older fossils are in poorer condition, often in pieces, and sometimes hardly recognizable. Placing the pieces back together is very much like a jigsaw puzzle. Trying to understand where and when the reconstructed fossils fit into the evolutionary line of descent is another puzzle.

TEETH AND SKULLS: Teeth and skull bones account for the majority of fossil discoveries ever since people began digging around for artifacts in Earth's rubble a few centuries ago. Teeth are the most enduring part of any life form because of their extremely hard enamel. Skulls are the most recognizable part, largely because they are more noticeable than arm or leg bones among sticks and other ground litter. Careful study of these and other bone fragments has now enabled researchers to distinguish between a wide variety of species. Consider a few examples.

Figure 26.19 illustrates three types of teeth. The molar at the top is that of a monkey (which is not a member of the ape family); it has two pairs of cusps or grinding bumps, for a total of four cusps. The teeth at the center and bottom of the figure are those of an ape and a human; most of their molars have five cusps, and none of the cusps are paired very well, forming instead a Y-shaped

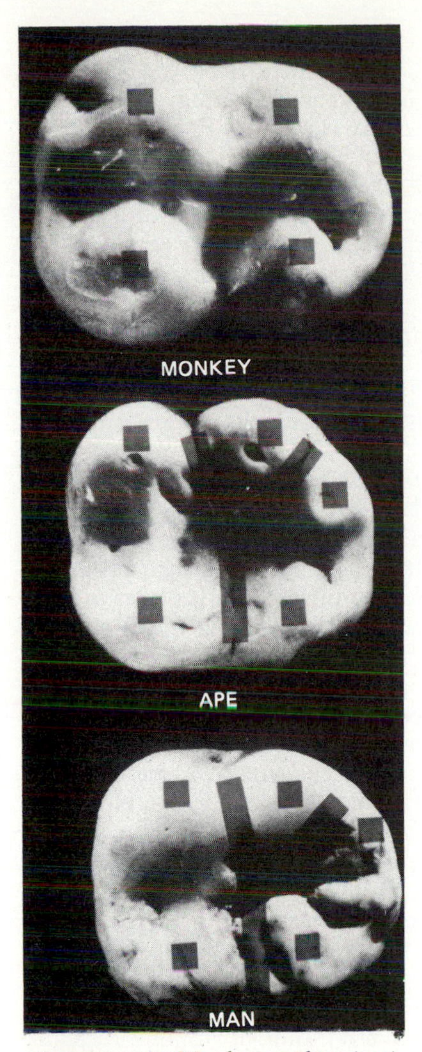

FIGURE 26.19 Monkey molars (top) have four cusps, usually paired, as marked by the small dark rectangles. Ape molars (middle frame) and human molars (bottom) have five cusps, each oriented around a Y-shaped pattern.

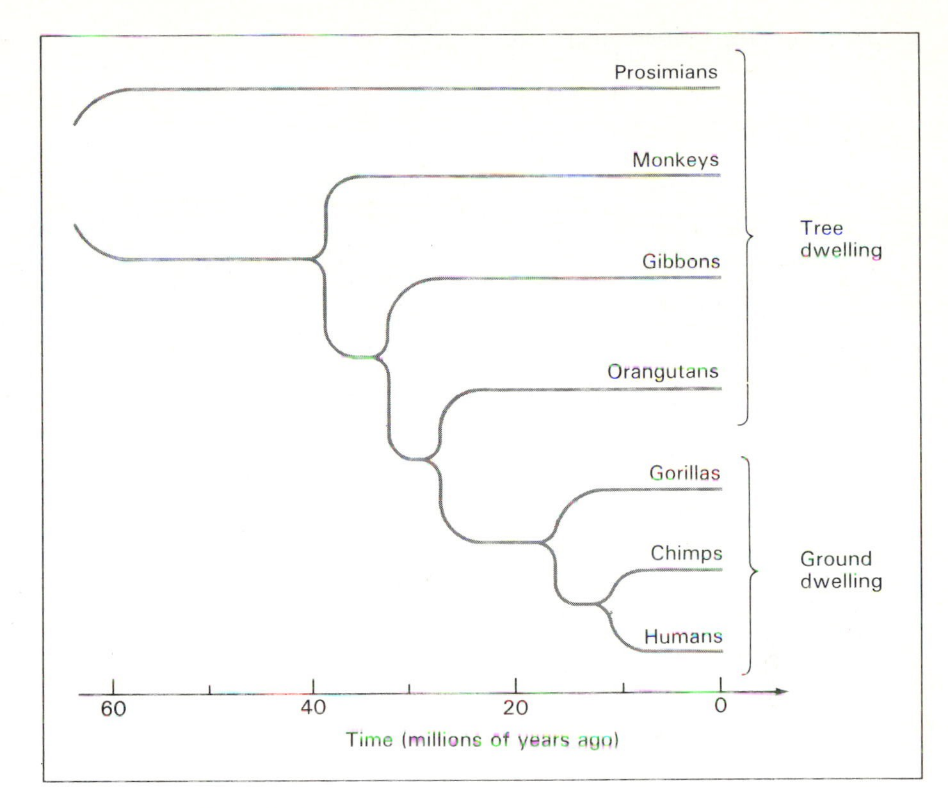

FIGURE 26.20 A rough sketch of recent evolutionary paths now agreed on by most researchers. Minor refinements may be needed as new fossils are unearthed.

pattern. This same Y-pattern characterizing our teeth is found in almost all mammals, now living or extinct. It is the more primitive of these two tooth patterns. But the line of ascent culminating in the modern monkey does not share this characteristic. At some point along the way, an ancestor of the monkey speciated, and moved along an evolutionary path different from that of ancestral apes and humans. Why this genetic change occurred no one knows. "Why" questions here, as elsewhere, are nearly impossible to answer. At any rate, detailed studies of numerous fossils prove that this change did occur. Furthermore, radioactive datings of the fossils suggest that it probably took place some time around 40 million years ago.

This is one way we can trace the various paths of evolution. As suggested by the sketch in Figure 26.20, it's a method of distinguishing the line of ascent that eventually led to modern monkeys from those that led to other species, such as modern apes and humans. Even so, analyses of individual teeth cannot distinguish between apes and humans. Another method is needed to document when or how this further speciation occurred.

Figure 26.21 illustrates one of the methods used by researchers to distinguish the line of ascent leading to modern apes from that leading to

FIGURE 26.21 An ape's jaw (left) differs distinctly from that of a human (right).

modern humans. Shown there are two jaws, that of a typical gorilla (an ape) at the top, and that of a typical woman (a human) at the bottom. Careful study demonstrates the following basic differences: Humans have smaller jaws than apes when compared against the total skull size of each; humans have an arched mouth roof (which can be felt with your tongue), whereas an ape's is flat; humans have a hyperbolic tooth pattern versus a rectangular one for apes, as can be clearly noted in the figure; and while both apes and humans have canine teeth, apes have greatly oversized ones that fit snugly into gaps within the opposite row of teeth when their jaws are shut. These are the kinds of subtle differences that help to determine where and when the human line of ascent deviated from the ape line. As suggested by Figure 26.20, this speciation seems to have occurred nearly 20 million years ago.

These are just two examples of the many techniques used to decipher relatively recent evolutionary paths. The entire picture of where and when all species originated and flourished on Earth is extremely complex, although the overall picture now seems to be in place after a century of research. In what follows, we outline a consensus for the relatively recent lines of ascent culminating in humanlike creatures.

RECENT EVOLUTIONARY HIGHLIGHTS: To appreciate our early ancestors, we must return to the start of the Cenozoic Age. At that time, approximately 65 million years ago, individual squirrel-like mammals like that of Figure 26.22 were increasing their chances of survival within a rather harsh environment. These were insect-eating creatures living primarily on the ground. The dinosaurs had vanished by that time, but life at ground level caused problems for these ancient mammals. The fossil record implies the existence of relatively large creatures who probably survived by preying on these mammals. Fortunately, though, sporadic mutations and a constantly changing environment granted some of these small mammals a chance to change their living patterns.

FIGURE 26.22 The insect-eating, squirrel-like mammal sketched here is widely regarded as the ancestor to humans, apes, monkeys, and a large number of creatures now inhabiting Earth. These land-dwelling prosimians, now extinct, lived approximately 65 million years ago.

At about this time, many of these mammalian species invaded the trees. They were undoubtedly searching for more food, while trying to escape the fierce competition prevalent on the ground. Some species found the trees an even rougher environment and thus became extinct. Other species found the trees to their liking, surviving famously. A few, like the Asian tree shrews of Figure 26.23, still thrive today. Their traits just happened to be adaptable to life in the trees. In fact, the trees became a whole new niche, helping to transform these creatures from ground-dwelling and insect-eating mammals to tree-dwelling and banana-eating prosimians. These pre-monkeys (or protomonkeys) are the least advanced members of the order of primates, a zoological category to which apes and humans belong.

FIGURE 26.23 The tree shrew, now living in many parts of Southeast Asia, resembles our ancestral tree-dwelling prosimians of some 60 million years ago.

The fossils document seemingly endless changes as many of the most successful early prosimians gained the best available niche. With time, over generation after generation of natural selection, paws gradually transformed into hands. Stubby claws eventually became flexible fingers. And the opposable thumb took shape as the environment promoted its prominence as a superior tool for maneuvering among the branches.

Remember, these were not anatomic changes that occurred as individuals grew during single lifetimes. Instead, these were genetic changes endured over the course of millions of years. Favorable mutations eventually gave those prosimians having good balance, keen eyesight, and dextrous hands and fingers a naturally increased opportunity for survival within their newly discovered tree-based environment. Those better adapted to this environment could reproduce more efficiently, thus passing these favorable traits on to their generations of offspring. Ability to jump, leap, swing, cling, as well as to gather food excelled differently among the many tree creatures. The result, again documented by the fossils, was widespread speciation causing the creation of an enormous variety of tree-dwelling species.

Figure 26.24 suggests how the evolution of accurate sight was a particularly important development. Trees are, after all, three-dimensional, unlike the flat two-dimensional ground. The advantageous trait of smelling on the ground gave way to that of seeing in the trees. Fossils show that over the course of millions of years and generation upon generation, mutations gradually brought the eyes of some of these tree-dwellers around toward the front of the head, thereby producing binocular, stereoscopic vi-

FIGURE 26.24 The head of some tree-dwellers (left) gradually transformed into that of monkeylike creatures (right). Genetic changes, followed by environmental adaptation, shortened the snout and brought the eyes around toward the front of the head, granting these creatures binocular vision and thus a distinct advantage some 50 million years ago. The shaded areas outline the field of vision.

FIGURE 26.25 Manipulative fingers and an opposable thumb gave some tree-dwellers another distinct advantage in the constant struggle for survival. Here, a monkey is shown grasping a branch while simultaneously reaching for food. This type of monkey is thought not to have changed much in the past 40 million years, and thus best resembles the advanced tree-dwellers of long ago.

sion. Eyes at the side of the head yield two independent fields of view, much like the flat perception gained when placing our nose and forehead against the edge of an open door. Better yet, try catching a baseball or hammering a nail with one eye closed; it's not so easy without depth perception.

The gradual shortening of the snout and the slow transformation of the eyes toward the front granted some early prosimians an overlapping field of view and thus more sophisticated vision. Clearly, these distant ancestors of about 50 million years ago had distinct advantages in the struggle for survival. They had become monkeylike creatures. As sketched in Figure 26.20, a major new evolutionary path had originated.

Fossil findings show furthermore that some species of monkeys gradually became larger. Again, a single generation of a given species did not suddenly balloon in size. Rather, sporadic mutations in their DNA molecules, spread over scores of generations, granted larger monkeys some advantages in the competition for survival. For example, larger, more aggressive males have clear superiority over smaller ones in the sexual competition for females. Also, bulky bodies usually provide additional protection from predators.

On the other hand, large size is not a total advantage. Sheer mass brings some problems, too. For example, bigger monkeys find it harder to hide, and they also need more food to survive. There are both advantages and disadvantages associated with all genetic changes. Only when the advantages outweigh the disadvantages is there an enhanced opportunity for living.

Figure 26.25 displays another combined advantage enjoyed by some early species of monkeys. The ability to grasp a branch securely while simultaneously extending an arm to secure food also provided an obvious advantage at the time. Those individuals unable to cling well enough to hold on, plunged, died, and became extinct. Those without long enough arms to reach the food likewise starved, died, and became extinct. The clear advantage was had by those creatures able to coordinate clinging and grasping *simultaneously*. Of course, being smart enough to repel attacks from an array of enemies also helped.

The lofty tree environment of 40 million years ago thus became a fairly comfortable niche for some species of monkeys. These well-adapted creatures could have probably remained in the trees indefinitely if a problem hadn't arisen. Fortunately for us, change stirred; otherwise, we wouldn't be here.

Leisurely life in the trees eventually became troublesome. The ancestral monkeys of 40 million years ago were so comfortably accustomed to their tree-dwelling environment that they multiplied faster than many other species residing in harsher environments. Time not used trying to survive can be most profitably used trying to reproduce. The sexual urge seems to be a basic biological tendency dating back literally hundreds of millions of years. The result was probably a population explosion, the type of crisis that's inevitably followed by a food shortage. Consequently, some of the prosimians survived only by using their limited ingenuity to discover

FIGURE 26.26 *Aegyptopithecus,* depicted here by an artist on the basis of 25-million-year-old fossils, is thought by many researchers to be the common ancestor of apes and humans.

new sources of food. For some, that meant leaving the trees and returning once again to the ground.

Some species of monkeys elected to stay in the trees. Most of these eventually became extinct, although some still survive in altered form today. Baboons, gibbons, orangutans, and many other modern tree-dwelling creatures are the descendants of the well-adapted monkeys that remained in the trees. Those prosimians that successfully came down out of the trees embarked on a whole new evolutionary path. This is the path toward the ape and human species.

Sketched in Figure 26.26, the fox-size *Aegyptopithecus* animal is currently taken to be the best candidate for the creature midway between the tree-dwelling prosimians and the streetwise humans—a common ancestor of sorts for modern apes and modern humans. The aegyptopithecines of 25 million years ago hung around forests, not venturing too far from the protection afforded by the trees. They foraged for food mainly on the ground, all the while gradually evolving larger brains, some rudimentary bipedalism, as well as the roots of the technical and linguistic skills needed for the development of culture more than 20 million years later.

At first notice, it seems foolish for the prosimians to have taken up residence in the trees about 65 million years ago if some of their descendants were just going to have to leave those trees several tens of millions of years later. But a critically important change occurred while they were aloft: They evolved and adapted to a very special environment. When the tree-dwelling prosimians returned to the ground, they were equipped with qualities that probably would not have been naturally selected had they stayed on the ground. With their manual dexterity and binocular vision, among other assets, they were far more advanced than any other type of life then on the ground.

Had our prosimian ancestors not taken refuge in the trees, they might have never experienced the need to foster these exquisite qualities, many of which we now use, for example, while writing or reading textbooks like this one. The 40-million-year detour into the trees was well worth the time and effort. By 30 million years ago, the ground-dwelling creatures had become dominant. In the process, they were also becoming more agile, more versatile, and smarter. They faced few roadblocks in the final rush along the evolutionary path that eventually led to a variety of peculiar animals now inhabiting our planet, including human beings.

SUMMARY

The most remarkable heavy-element assemblage on Earth is life. Plants and animals are widespread, both on the land and in the sea. Of particular interest, the heavy-element concoctions known as men and women have existed within only the last one-tenth of 1 percent of Earth's history.

Life everywhere now seems biologically adapted to the planet, but adaptation is a never-ending effort. Change is inevitable. Nothing remains immutable, nothing at all. What we can't see is often tough to believe. But we see so briefly in time; even the duration of our civilization is a mere flicker in the spectacle of cosmic change.

KEY TERMS

adaptation
autotroph
biological evolution
eukaryote
fermentation
punctuated equilibrium
gene pool
genetics
heredity
heterotroph
multicell
mutation
natural selection
prokaryote
photosynthesis
respiration
speciation
species
unicell

QUESTIONS

1. Briefly describe the major changes, including the reasons for those changes, that have occurred in Earth's atmosphere since the formation of our planet some 5 billion years ago.
2. Describe the concept of "speciation," also often termed "disruptive selection."
3. Pick one: According to the fossil record, the greatest of all extinctions of life on Earth occurred (a) toward the end of the Paleozoic around 225 million years ago; (b) toward the end of the Mesozoic around 75 million years ago; (c) at roughly the time that the dinosaurs disappeared; (d) during the last extensive period of glaciation; (e) during the American Civil War.
4. Describe the biochemical reaction that ultimately permitted marine creatures to leave the ocean and invade the land.
5. Explain why solar radiation cannot be used to feed the proteinoids.
6. Explain the origin of the fossil fuels now used to power our modern civilization.
7. What reasons can you suggest for the sudden increase in the number and diversity of fossils beginning about 600 million years ago? Justify your answer.
8. Name and compare the two main factors that lead to differences among organisms over the course of time.
9. Discuss some of the early ecological crises on our planet. How were they overcome?
10. Discuss the difference between the case where life is a natural consequence of the evolution of matter and the case where it is an inevitable consequence. Which adjective—"natural" or "inevitable"—do you think is correct here? Explain.

FOR FURTHER READING

ATTENBOROUGH, D., *Life on Earth*. Boston: Little, Brown, 1979.

*DOBZHANSKY, T., AYALA, F., STEBBINS, G., AND VALENTINE, J., *Evolution*. San Francisco: W. H. Freeman, 1977.

ELDREDGE, N., *Time Frames*. New York: Simon & Schuster, 1985.

KURTEN, B., "Continental Drift and Evolution." *Scientific American*, March 1969.

MAYR, E., "Evolution." *Scientific American*, September 1978.

RUSSELL, D., "The Mass Extinctions of the Late Mesozoic." *Scientific American*, January 1982.

STANLEY, S., *The New Evolutionary Timetable*. New York: Basic Books, 1981.

WILSON, A., "The Molecular Basis of Evolution." *Scientific American*, October 1985.

27 CULTURAL EVOLUTION: *The Rise of Humans*

By now, we can firmly grasp the working hypothesis that matter evolves into life. Scientists do not yet have absolute proof of this, but an abundance of circumstantial evidence makes it the most reasonable theory. Furthermore, everything known about earthly organisms suggests that complex life arises naturally from simple life.

Complex life includes ourselves. And we naturally wonder about the specific evolutionary path that led from our ancient ancestors to modern humans. Where did we come from? How did we get this way? What were the circumstances that produced our odd body shape? Specifically, what factors led to the development of our fabulous attributes of thinking with our brain, talking with our mouth, walking on our hind legs, constructing with our hands, and seeing with binocular vision?

Having studied an appreciation for the origins of matter and life, we thus arrive at another basic question: What are *we*? Not what is our Sun, our planet, or life itself. But who or what are these twentieth-century human beings? Everyone asks that question at one time or another. It's one of the most profound and interesting issues of all.

Human names and social security numbers do not concern us here, although admittedly they do tell others a little about ourselves. Instead, we seek a more general understanding of the origin of the human species. In a nutshell, each of us seems to be the product of many ancestral life forms—a cluster of genes inherited from all of them, and shaped partly by an environment that's partly ours, partly our parents', partly our parents' parents', and so on back in time.

Tracing back about 1000 years, each of us would have had more than a million ancestors, all alive at the same time. They were probably spread over much of the world, living in a diversity of environments.

Suppose that we go back another few thousand years. At that time, some of our ancestors could well have been members of the ruling class of ancient Egypt or Babylonia. But the bulk of our ancestors were probably slaves or peasants. Chances are good that they could neither read nor write. They were probably ignorant, superstitious, and cruel—primitive agriculturists at best. Few of them would have touched metal or operated a wheel. By modern standards, most of our ancestors of several thousand years ago were savages. They survived largely by hunting and gathering.

We find it hard to relate to our ancient forebears. But modern science suggests that we must. Evolution stipulates that we carry some of their genes in our bodies. Part of our features, shapes, desires, and attitudes, as well as our outlook on life and way of thinking, all derive to some extent from the genes of our ancestors, as well as from the environment in which they lived.

Answers to the fundamental questions, then, are evolutionary ones. They are answers that enable us to relate ourselves to all humankind, indeed to all living things. If we can find these answers, perhaps we can determine who we really are, how it is that we can walk upright, make tools, and speak to one another, as well as to think, speculate, and exhibit curiosity about ourselves and our Universe.

Long ago, our distant ancestors possessed none of these attributes. They were not human. They were small-brained creatures living in trees. Somehow they gave rise to humans. Somewhere in our ancestral past, there are links connecting creatures that clearly were human with creatures that clearly were not.

The learning goals for this chapter are:

- to understand some of the techniques used by researchers to sketch the relatively recent path of human evolution
- to recognize the central objectives of modern paleoanthropology
- to know our australopithecine forebears, perhaps the common ancestors shared by modern apes and humans
- to understand when and how technology and culture took root in human society
- to appreciate the many environmental and societal factors that helped make us human

Paleoanthropology

Anthropology is the scientific study of humans—their cultures, their development, their origins. The word derives from the Greek *anthropos,* meaning "man." Broadly conceived, modern anthropology mixes the results of fossil discoveries with the essentials of animal behavior in order to piece together a picture of twentieth-century men and women.

In this section we first concentrate on what the fossil record can tell us about human development. This subject is known by the tongue-twisting name **paleoanthropology**, the prefix *paleo* having been used in Chapter 24 to denote the study of old, fossilized organisms (paleontology) and of ancient magnetism (paleomagnetism); here, paleoanthropology means the study of prehistoric humanity. In a later section, the behavior of advanced life forms—studies covered by the allied subjects of social anthropology and sociobiology—will help us refine the evolutionary paths implied by the fossils. Together, studies of fossilized life and of contemporary life should enable us to appreciate the essence of **cultural evolution**—the changes in the ways, ideas, and behavior of society, especially among higher forms of life on Earth.

As noted earlier, the entire effort of unraveling human genealogy is often likened to the restoration of a gigantic mural painted over the course of millions of years. To the right, where the scene has been sketched in recent times by modern humans, the message is reasonably clear. Toward the left, the mural is soiled, peeling, and generally deteriorating. Painted long ago by our ancestors, the mural cannot easily be cleaned of dirt, nor can it be easily repaired to reveal the message once that dirt has been removed. Like any restoration, the process is a slow and painstaking one, done very deliberately to avoid destroying the mural and thus the message.

Anthropologists and archeologists studying old fossils need the virtue of patience in addition to an inquiring though unopinionated mind. It's very much like detective work, trying to decipher the entire story based on just a few hints. The solution here, however, means more than closing another crime case. It means knowing just that much more about the origins of human beings, a superbly satisfying objective.

We now concentrate on the fossil record of our relatively recent prehuman ancestors who lived during the past 20 million years. Be sure to place the findings of this section into time perspective. Although an extremely long time by human standards, the interval of 20 million years is less than a half of 1 percent of the total age of Earth.

OUR IMMEDIATE ANCESTORS: The late nineteenth century was an exciting time for geologists—a time full of discoveries and revelations about our planet. Field excursions and archeological excavations were only then becoming popular. Scientists began recognizing that the ground below their feet held clues to the nature of Earth, especially to former life on Earth. Among the most fascinating results were the first discoveries that there once existed such things as dinosaurs, the former rulers of our planet. More relevant to this chapter's task, ancient stone axeheads and crude implements of all sorts were being uncovered along rivers and within caves of western Europe. These were primitive tools, but tools nonetheless. Geologists found that many of them were nearly 100,000 years old. The question arose: Who had made these crude implements?

The first great fossil finds pertaining to prehumans were made a little more than 100 years ago. (By "prehumans" we mean ancestors of the hominid line, namely, humans and their extinct close relatives.) At about the time that Darwin published his study of biological evolution, a primitive-looking skull was found in a cave in the Neander Valley in Germany. This skull has a low, sloping forehead, a receding chin, and thick ridges over the eye sockets, but it still displays an overall "human-like" appearance. The German word for "valley" is *thal,* hence the origin of the name, Neanderthal Man, attached to the original owner of this skull. Although a bit odd compared to today's human skulls, there seems little doubt of its human origin. However, with only one such skull fossil known to exist at the time, it was easy to classify it as a deformed specimen of modern man, which was exactly the approach taken. Even as recently as 100 years ago, many scientists who embraced the concept of evolution regarding plants and nonhuman animals were still apparently unwilling to accept its relevance to humans as well.

Toward the end of the nineteenth century, more Neanderthal-type fossils were found in Germany. Based on these many discoveries, Figure 27.1(b) is an artist's reconstruction of what Neanderthal Man might have looked like. This figure also shows a similar reconstruction based on less primitive human-like skulls unearthed in numerous places scattered across France, especially

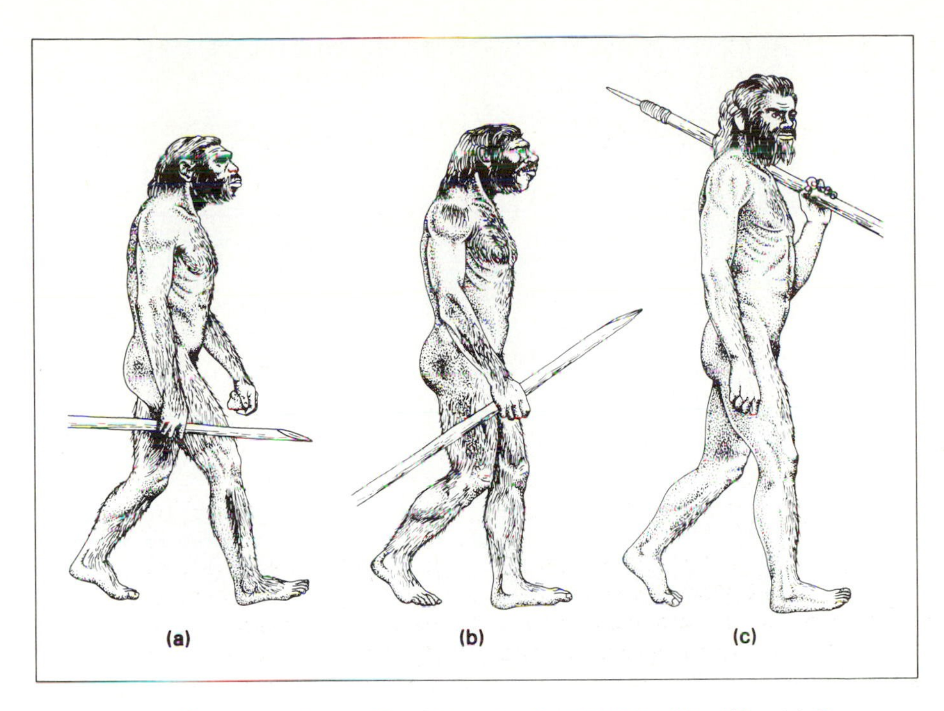

FIGURE 27.1 Reconstructions of Java Man (a), Neanderthal Man (b), and Cro-Magnon Man (c), each of them tool-using creatures known to have existed as long ago as nearly a million, 100,000, and 30,000 years ago, respectively.

near the village of Cro-Magnon. The latter skulls represent the group called Cro-Magnon Man, an entire subspecies of human ancestors [see Figure 27.1(c)]. Regardless of the scientific name, the important point is that many of these odd-looking, though clearly human-related, skulls were found alongside the ancient tools. The connection clearly implies that tool-making humans of some sort lived in Europe 100,000 years or so ago. Since the Cro-Magnon skulls are less primitive than those of the Neanderthal variety, some anthropologists regard Neanderthal Man as the ancestor of Cro-Magnon Man. Other researchers disagree, claiming that the Neanderthals represent a divergent branch of hominids that died out about 40,000 years ago. Whichever it was, and for whatever reason, the Cro-Magnons replaced the Neanderthals some 400 centuries ago. The critical question then becomes: Who were the ancestors of Neanderthal Man?

Rich fossil findings have traced our roots much farther back in time. Near the start of the present century, human-like skulls and teeth were discovered far from Europe, in an arid river bed in Java, a large island in Indonesia. These skulls, dated to be nearly 1 million years old, seem even more primitive than those of either Neanderthal or Cro-Magnon Man. Yet, as shown in the artist's reconstruction of Figure 27.1(a), the size, shape, and overall features of Java Man still resemble those of today's humans. Furthermore, the hole at the base of the Java-Man skull, through which the spinal cord passes, implies that these creatures must have stood erect. In other words, they were **bipedal**, namely they walked on two legs, unlike most apes which need to use their arms as crutches.

Astonishing findings early in the twentieth century, these ancient human-like fossils predictably drew a great deal of skepticism. Admittedly, it is hard to imagine that erect human-like creatures could have existed anywhere on Earth as long ago as a million years before the present. This is a terribly long time by human standards, equivalent to some 40,000 generations of human life. In fact, 1 million years is more than 100 times longer than all of the 7000 years of recorded history. Succinctly put, more than 99 percent of humankind's history is recorded only by the fossils.

Confirmation of these startling discoveries followed during the first decades of this century when many similar fossils were excavated at widespread locations throughout the temperate zones of planet Earth. Researchers have now uncovered numerous Java-Man skulls, as well as bones of Heidelberg Man in Germany, Peking Man in China, and a variety of other ancient, though clearly human-like fossils in Hungary, France, Spain, and Africa. Most of these fossils are on the order of $\frac{1}{2}$ million years old, although some are closer to a full million years old. Significantly, these are not skulls of apes. Nor are they skulls of apemen. They're skulls of humans—erect men and women who existed an awfully long time ago.

All the human-like fossils dated to be less than about 1 million years old closely resemble the skull and teeth of modern humans. For this reason, all of them are given the designation *Homo*, a Latin word meaning "man." To distinguish these older fossils from those of twentieth-century bones, a suffix is often added to this designation. For example, Neanderthal Man, Cro-Magnon Man, and fossils of other human creatures dated to have lived during the past 300,000 years are collectively given the name **Homo sapiens**, meaning "wise man." This is the same biological species as modern men and women, although some researchers prefer to endow the most recent humans of recorded history (including ourselves) with a special designation—*Homo sapiens sapiens*. (Undoubtedly another expression of human vanity, this designation of "very wise man" is a highly debatable label, given the multitude of global problems we have created for ourselves on twentieth-century planet Earth.)

By contrast, Java Man and other human fossils between roughly 0.3 and 1 million years old are collectively referred to by the species name **Homo erectus**, which is Latin for "erect man." They were definitely of human stock, walking upright and exhibiting considerable manual dexterity, but their brain volume was not as large and their tool use not as advanced as those of *Homo sapiens*.

MORE ANCIENT ANCESTORS: Finding human-like skulls as old as a million years does not really solve the central issue. They merely push back in time the key question: Who were the ancestors of *Homo erectus*?

The first clue came about 50 years ago, although it wasn't until about a decade ago that a reasonably clear line of ascent emerged. An anthropological expedition in the 1920s uncovered a fossilized skull having simultaneously some human and some ape characteristics. A large number of such hybrid skulls have since been found throughout the warm climate regions, especially on the African continent.

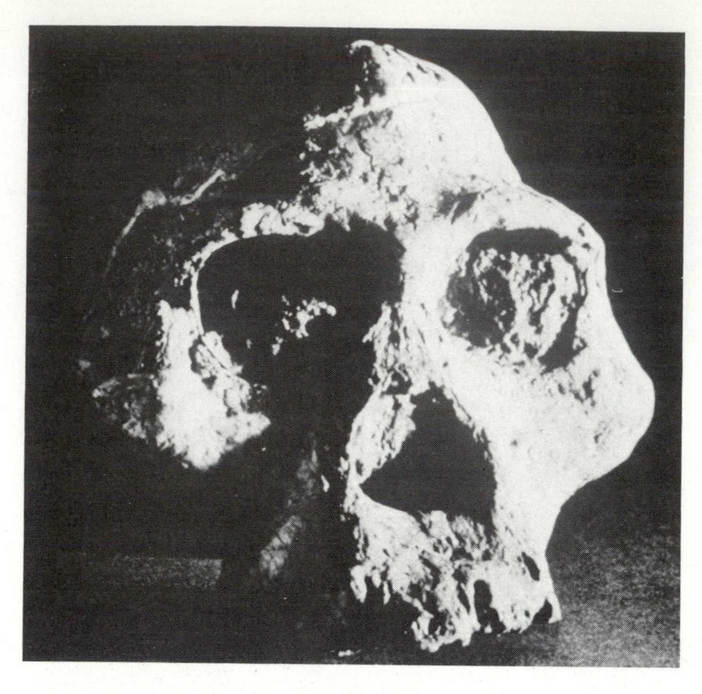

FIGURE 27.2 An *Australopithecus africanus* skull fossil.

Figure 27.2 pictures one of these skulls, after it had been carefully dug up, dusted off, and reassembled from pieces found in the sandy soil of southern Africa. Analysis shows it has the following curious blend of ape and human traits: an interior skull volume (brain capacity) larger than an ape's, although smaller than a human's; a jaw larger than a human's, although smaller than an ape's; a forehead resembling an ape's more than a human's; canine teeth more like those of a human than those of an ape; a hole at the base of the skull suggesting that this creature had walked upright, or nearly upright.

The mixture of bone qualities strongly suggests that this creature belongs someplace near the threshold of humanity. All fossils of this hybrid ape/human kind have subsequently been given the jaw-breaking Latin name of **Australopithecus**, meaning "southern ape." Unfortunately, the early findings in the sandy soil of southern Africa could not be dated; sand is not radioactive and tends to shift with time. But this discovery focused modern paleoanthropological research on the African continent, where for the most part it's been ever since.

Excavations during the past two decades have revealed many additional australopithecine skull and tooth fragments. Some of these findings have been made in the same southern African area, where soil cannot be accurately dated. In addition, though, numerous similar fossils have been uncovered all along the East African Rift Valley, a giant crack produced by plate tectonic activity of that large continent.

For example, about 20 years ago, an *Australopithecus* fossil was found protruding from a layer of volcanic ash along a dried-up river bed at Olduvai Gorge, Tanzania. Figure 27.3(a) shows this area, especially the ordered layers of volcanic rock that can be radioactively dated. Thus decades after its original discovery, the australopithecine fossils could finally be set in time. That date is approximately 2 million years ago, an age verified by other recent findings. Clearly, these protohumans (sometimes called apemen) inhabited our planet a very long time ago.

Be aware that the official designation of the 2-million-year-old skull remnants found at Olduvai Gorge is controversial. The discovering Leakey family of Britain and Kenya argue that these skulls belong to a species only mildly related to the australopithecines. In particular, the co-discovery of very elementary stone tools prompted them to suggest a new name, *Homo habilis*, or "handy man." However, the rock chips the Leakeys consider "tools" are most primitive, making it debatable just how skilled these creatures really were. Perhaps the *H. habilis* fossils are merely those of advanced australopithecines, and therefore not a new species deserving of the human status of *Homo*.

Sketched by an artist in Figure 27.4, the 2-million-year-old protohuman ancestors were no larger than 150 centimeters tall, weighing about 50 kilograms (roughly 5 feet and 100 pounds). They were surely smarter than any other life forms with which they shared the open plains away from the forests. Their brain was probably not large enough to have managed speech, although these creatures may well have communicated using a repertoire of grunts, groans, arm gestures, and other body movements. The more talented members surely possessed

(a) (b) (c)

FIGURE 27.3 Shown here are several sites where important fossil discoveries have been made in recent decades: Olduvai Gorge in Tanzania (a), Omo in Ethiopia (b), and Koobi Fora along the eastern shore of Lake Turkana in Kenya (c).

dexterous hands—not as good as ours, but good enough to build primitive stone tools—and probably keen eyesight. Whatever their full attributes, these creatures seem to have adapted well to their environment, for this is the key to survival.

More recent field work indicates that several variations of australopithecine creatures may have coexisted throughout Africa several million years ago. Hundreds of australopithecine fossils have now been categorized into at least two distinct species of protohumans. One of these species is characterized by a heavy jaw and large grinding teeth, suggesting that this species enjoyed a diet of mostly coarse vegetation, much like that eaten by modern gorillas. This more robust type is often called *Australopithecus boisei* or *A. boisei* or even *A. robustus* for short. The other species, called *Australopithecus africanus* or *A. africanus* or sometimes *A. gracile*, is of the originally discovered southern African variety. This species is characterized by a more slender jaw and smaller molars, suggestive of a more gentle anatomy, and thus of a class of protohumans that probably feasted on meat. These suggestions are just that—suggestions and not conclusions—but they do represent the prevailing view among anthropologists today.

Given these findings, we might naturally wonder if the observed differences in the australopithecine fossils could simply be variations of the same species. After all, today's humans display slight variations, as can be readily seen by contrasting thin track stars and husky football players. This interpretation doesn't make much sense, however, because most of the 2-million-year-old fos-

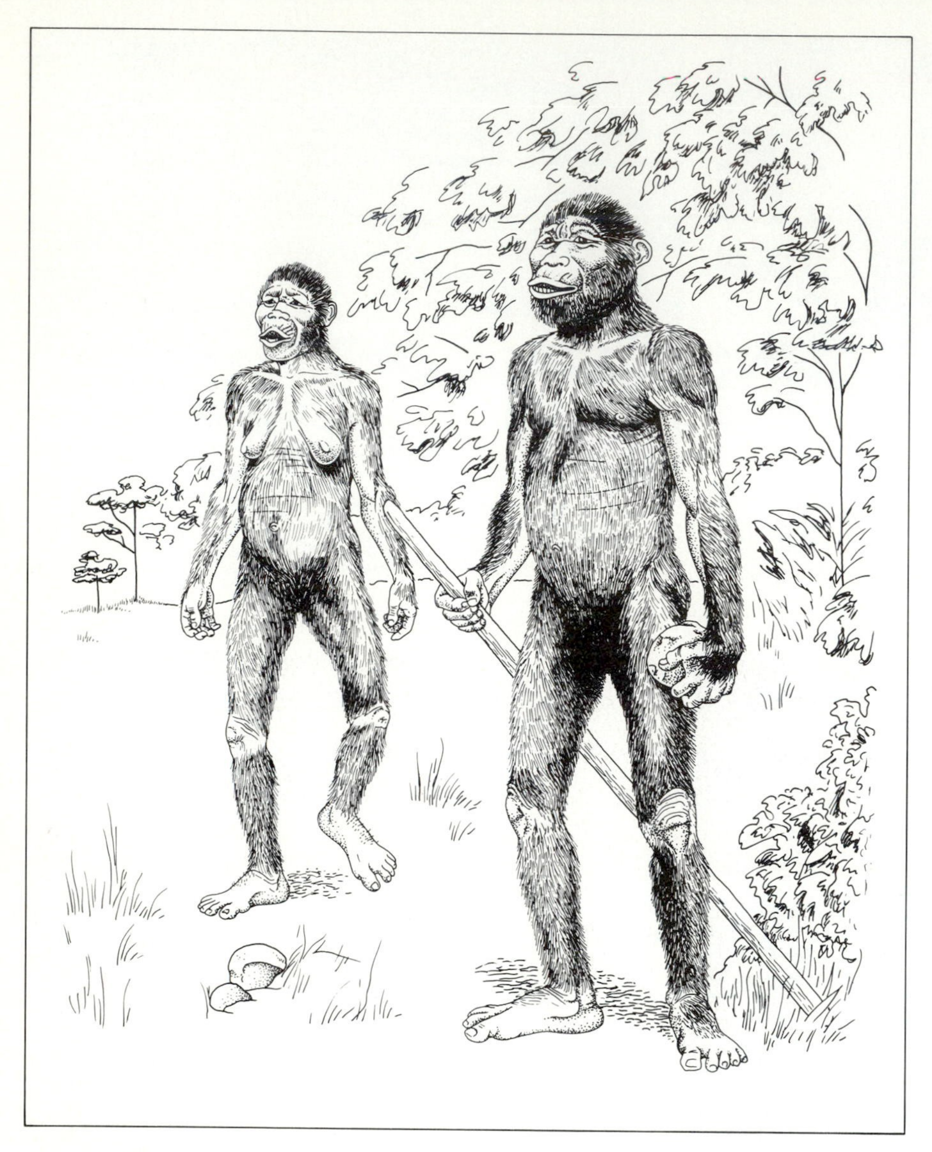

FIGURE 27.4 An artist's reconstruction of *A. africanus* emerging from the forest several million years ago.

sils of protohuman creatures fall into two distinct classes; either skulls and teeth are clearly big and oversized, or they are small and graceful.

Could these two types of australopithecine fossils simply correspond to male and female? Again, this interpretation is improbable because these two classes of fossils are hardly ever found at the same place within sedimentary rock. Unless there was something very strange about protohuman cultures that kept tribes of males separated from females, it would seem impossible that these classes correspond to sexual differences. Besides, males and females could not possibly reproduce their species if they remained segregated.

Thus, at least two species of protohumans, and quite possibly more, apparently coexisted on Earth a few million years ago. Presumably, only one of these species is our true ancestor. Further studies suggest which one.

Expeditions all along the East African Rift Valley have revealed much new information during the past 2 decades. Figure 27.5 shows the sites of several important discoveries, as well as some places where major groups of researchers are now operating. Besides the rich findings at Olduvai Gorge in Tanzania, several groups are continuing to trace the thread of our origins by examining fossils found along the shores of Lake Turkana (formerly Lake Rudolf) in Kenya. And before a (human) guerilla war in the mid-1970s prevented further digging, especially well-preserved fossils were found in easily datable volcanic rock at Omo, Ethiopia. See also Figure 27.3(b) and (c).

Among the recent discoveries at several of these sites, the most interesting is perhaps of something that's missing: No *A. boisei* fossils are less than 1 million years old. This more robust protohuman species rather suddenly disappears from the fossil record, implying a rapid extinction. The most popular explanation contends that competition between *A. boisei* and *A. africanus* was inevitable. Each biological niche can be filled by only one species, as noted in Chapter 26. Yet here were two protohuman species trying to make a go of it simultaneously. Somewhat surprisingly, the bigger, more robust species was the loser. Here's an explanation.

Despite *A. boisei's* larger size, vegetation was plentiful, producing a rather comfortable way of life. Such comfort, however, is not necessarily conducive to rapid evolution toward a technologically intelligent society. The smaller species was almost surely more versatile, quicker, and perhaps smarter. Only basic intelligence could help *A. africanus* capture the less abundant meat needed for survival. As a result, natural selection worked to help generations of *A. africanus* expand their brains, their capabilities, and their niche. Accordingly, *A. africanus* apparently crowded *A. boisei* right off the face of the Earth. This theory is supported, not only by the documented extinction of *A. boisei,* but also by the recent discovery that only *A. africanus* used primitive stone tools. We cannot be sure if these tools are simply a measure of *A. africanus'* proclivity for manual dexterity and gradual brain development, or if they were actually used as weapons to accelerate the demise of *A. boisei.*

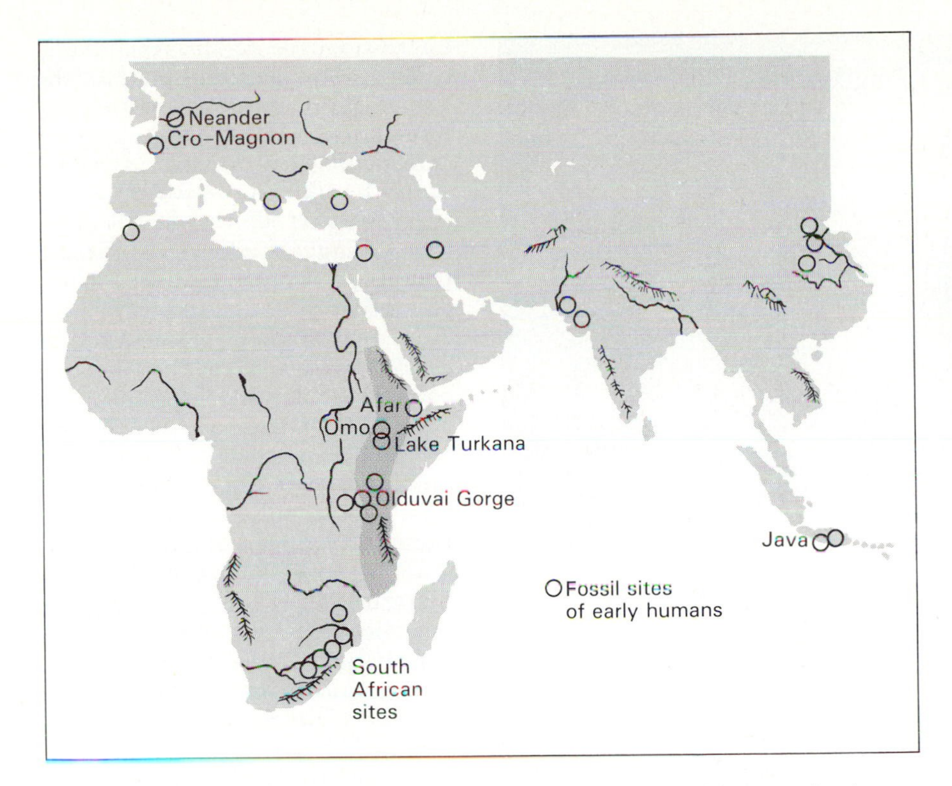

FIGURE 27.5 This map shows the locations (circles) of some of the sites that have produced a rich lode of protohuman fossils. The darkly shaped area depicts the East African Rift Valley, a region of rapid environmental change during the past several million years (and continuing today).

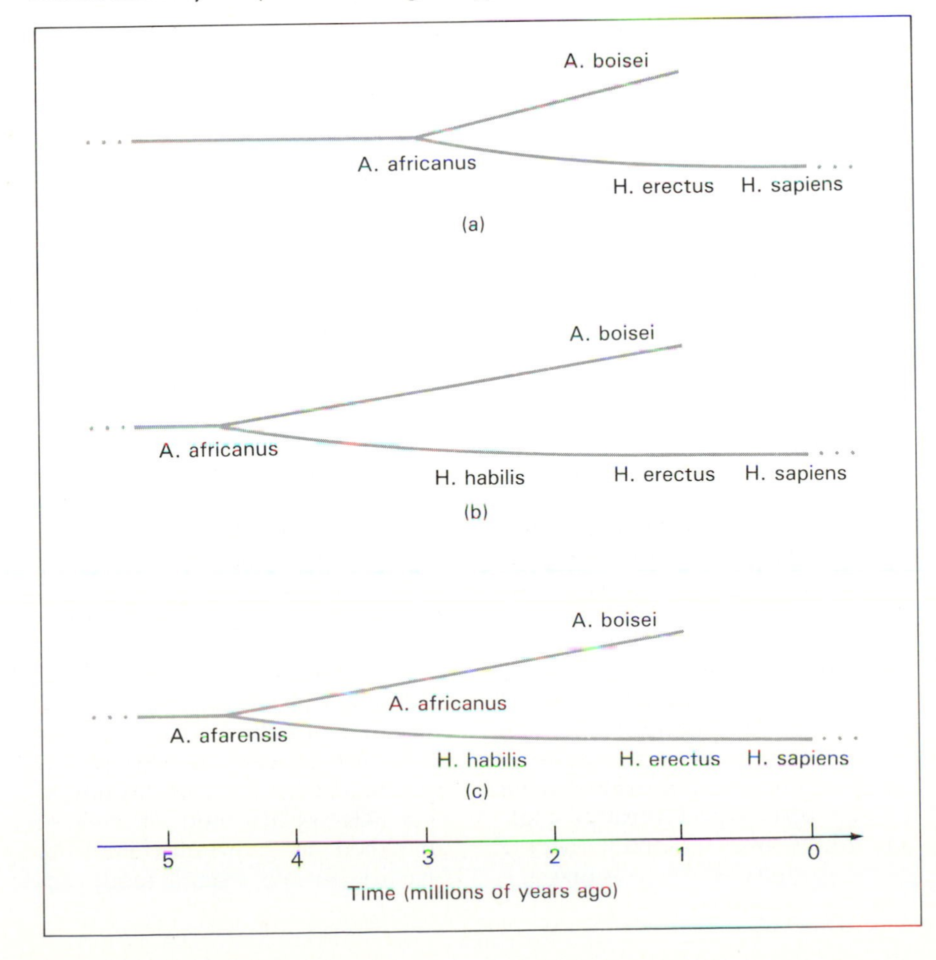

DIFFERING VIEWPOINTS: Figure 27.6(a) sketches the line of ascent just described. Recognize that as for many evolutionary scenarios at present, the details have yet to be worked out. Many have yet to be unearthed.

Accordingly, several researchers have proposed some slight variations in the evolutionary picture. Figure 27.6(b) shows the view preferred by scientists who feel that the tool-using creatures of 2 million years ago should be labeled *H. habilis,* as opposed to *A. africanus*. This mostly amounts to a question of semantics and is the reason why some workers contend that *Homo* dates back not much more than 1 million years, while others suggest that some species of *Homo* existed at least 2, and perhaps 3 or more, million years ago.

Figure 27.6(c) outlines another view that incorporates newer discoveries of skull, tooth, and bone fragments in the Afar lowlands of Ethiopia. These fossils, a representative example of which is shown in Figure 27.7, suggest that human-like creatures existed nearly 4 million years ago. Skulls having smaller brains and larger canine teeth than ours have been found near footprints preserved in hardened radioactive ash of this age. These data—the most famous of which is "Lucy," a partially complete skeleton of a 20-year-old female—furthermore suggest that these creatures must have stood erect. Based on these relatively recent findings, some researchers contend that these fossils are the best evidence for the creature that must have resided midway between apes and humans—a missing link of sorts. They suggest a whole new species, *A. afarensis,* as the common ancestor of *H. sapiens* as well as the extinct *A. boisei*.

Opponents who prefer Figure 27.6(a) argue that the Ethiopian fos-

FIGURE 27.6 Many evolutionary schemes are consistent with the sketchy data concerning human origins. The simplest one is shown in (a), alongside two other popular but slightly different versions (b and c).

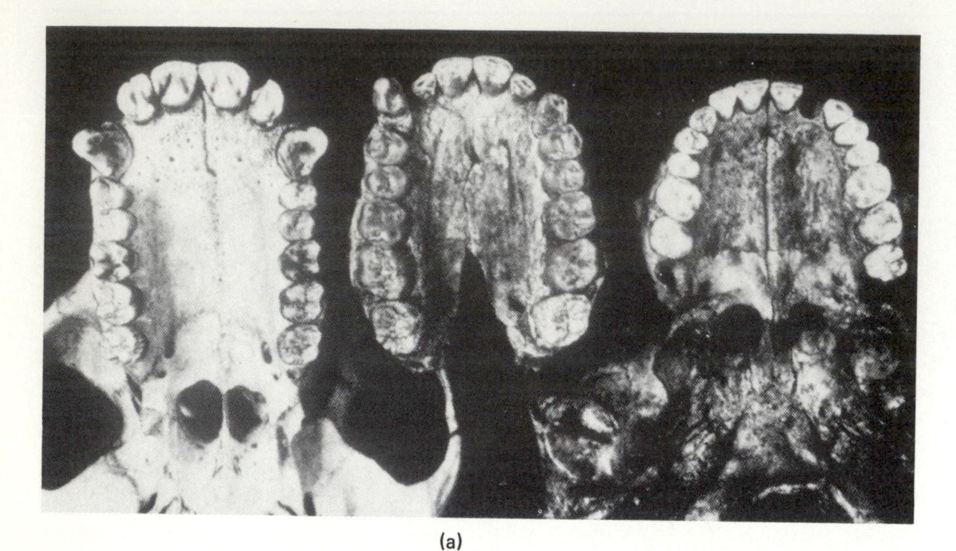

(a)

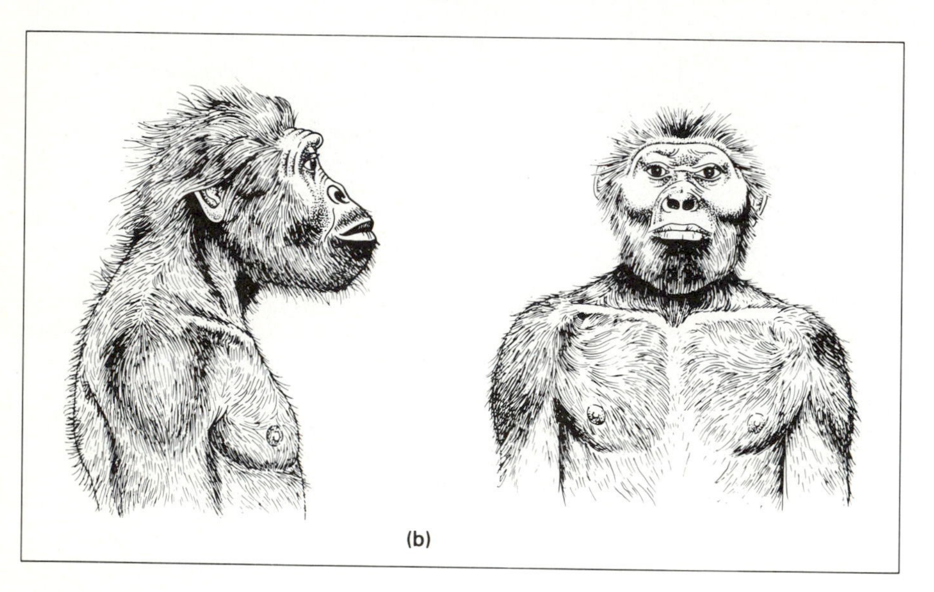

(b)

FIGURE 27.7 The jaw structure (a) of the *A. afarensis* species (center) seems to have qualities midway between apes (left) and humans (right). Frame (b) shows an artist's reconstruction of *A. afarensis,* based on a particularly rich trove of 4-million-year-old fossil remains.

sils simply belong to the *A. africanus* species, but acknowledge that that already distant ancestor must be pushed even farther back in time. Those preferring the evolutionary path sketched in Figure 27.6(b) argue that these new fossils comprise evidence for a more primitive version of *H. habilis*. Whichever evolutionary view is correct, these oldest human-like fossils virtually prove that our ancestors walked erect before their brain size enlarged appreciably. This will be an important factor when studying the evolution of the brain in Chapter 28.

Still other evolutionary paths generally agree with all the fossil data. A cynic might remark that there seem nearly as many possible paths as there are paleoanthropologists. The real problem here is that the current picture of human evolution is based on just a roomful of partially crushed skulls and broken teeth, most of them found scattered throughout East Africa, Asia, and central Europe. In fact, the whole lot thus far unearthed doesn't contain enough parts to reconstruct a single skeleton of an australopithecine; the oldest complete skeleton is that of a 60,000-year-old Neanderthal Man.

We're not trying to confuse the issue here. There is an overall evolutionary trend agreed upon by most anthropologists: *Australopithecus* → *Homo,* or near man → true man. The controversies, which often become heated and emotional, essentially concern details—specific dates, emergence of new species in Africa, Asia, or elsewhere, coexistence of various species, invention of tools, cooperative hunting, language, and other human-like qualities. These are important details to be sure. And the emotion is not surprising, for these issues involve our own origins. But until more fossils are unearthed, differing viewpoints will continue to flourish. This is perfectly fine, since each viewpoint is a slightly different theory, and each remains to be tested experimentally by collecting more fossils. This is the way science progresses.

OLDEST HUMAN-LIKE FOSSILS: The prevailing view that the *A. africanus* species is the best candidate for the link between modern humans and whatever ancestry we share with the apes is certainly instructive. But even if valid in every respect, it simply pushes the basic question still farther back in time: Who were the ancestors of the australopithecines? Here the answer becomes a lot more vague because the fossils are older, more scarce, and less well preserved.

A few discoveries have been made of fossils even older than the 4-million-year-old specimens of the Afar lowlands. Despite their age, they still seem to have some human-like properties. For example, Figure 27.8 shows bone fossils of tooth and jaw fragments found near Lake Turkana. The surrounding rock in which the fossils were embedded dates back approximately 5 million years. Most researchers agree that these remains belong to the *A. africanus* species, or whatever preceded *A. africanus,* although no one can be sure on the basis of a partially crushed jaw (and one arm bone not shown in the figure). In addition, a single 9-million-year-old molar tooth has been unearthed at a nearby location. Needless to say, a single tooth can-

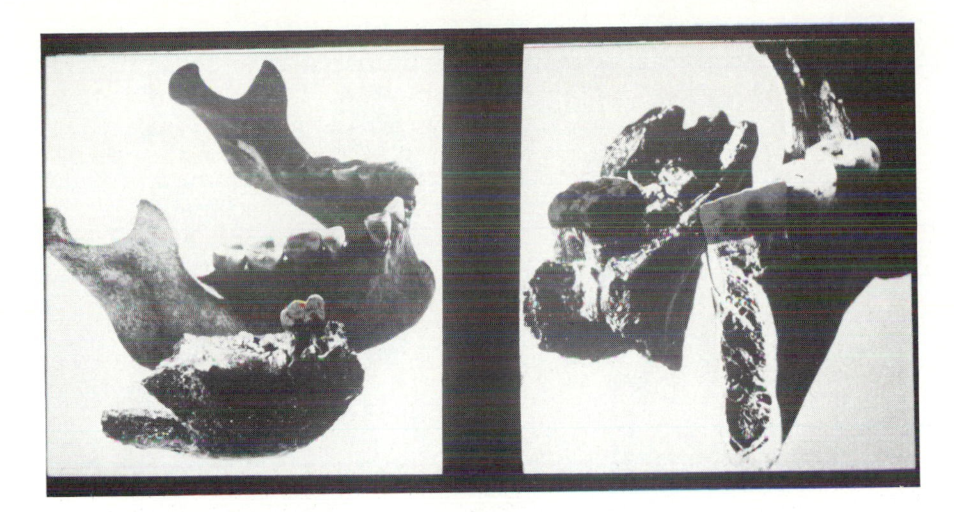

FIGURE 27.8 These 5-million-year-old tooth and jaw fragments are among the very few protohuman fossils ever found within the time interval between 4 and 10 million years ago. The fossilized jaw (foreground in left frame) closely resembles a human jaw placed behind it for comparison. However, the same fossil (at left in right frame) is clearly more robust than a human jaw at the right which has been cut to show its thinner cross section.

not be used to trace the ancestry of the australopithecines with much accuracy.

Apparently, creatures having some human qualities resided on Earth more than 5 million years ago. Even so, it's frustrating that so few protohuman fossils have yet been found for the period between 4 and 10 million years ago; some researchers call it "the gap." Be assured, Earth's soil harbors plenty of old fossils from this time domain, but these are usually the remains of animals not ancestral to humans.

Tooth and jaw fragments of the oldest known creatures having any resemblance to humans or protohumans were discovered in India, and later at several places in Africa and Europe. Figure 27.9 shows a representative sample of these fossils. Radioactive dating of the rocky dirt in which they were buried implies that these fossils are about 10 million years old (and possibly as young as 8 million years old). Despite their age, the jaws still seem to have a mix of ape-like and human-like qualities. The creature's brain capacity and posture are unknown, however, since a complete skull has never been found. Only a few such fossils exist, and none are very well preserved.

Even so, many anthropologists contend that this monkey-sized 10-million-year-old creature, called **Ramapithecus** in honor of the Indian god Rama, is probably the ancestor of the australopithecines—a sort of proto-australopithecine. Again, this is only conjecture based upon the meager data currently available; other researchers suggest that *Ramapithecus* is more likely the direct ancestor of our cousin, the orangutan. Much more fieldwork and interpretation are needed to pin down the details of our distant ancestors who roamed Earth some 5 to 10 million years ago.

Figure 27.10 summarizes the simplest version of current paleoanthropological findings regarding the relatively recent line of ascent toward humans. This figure is an enhancement of the evolutionary path labeled "humans" at the very bottom of Figure 26.20.

Precisely where and when one species changed into another cannot really by pinned down much better than that described above. After all, one life form gradually dissolves into another over the course of history. Put another way, the fossil record will never document an ape-like mother giving birth to a distinctly human infant, or an *A. africanus* mother giving birth to a *Homo erectus* infant. Evolution just doesn't work that way. Changes of this sort are usually slow and gradual, occurring over long, long periods of time.

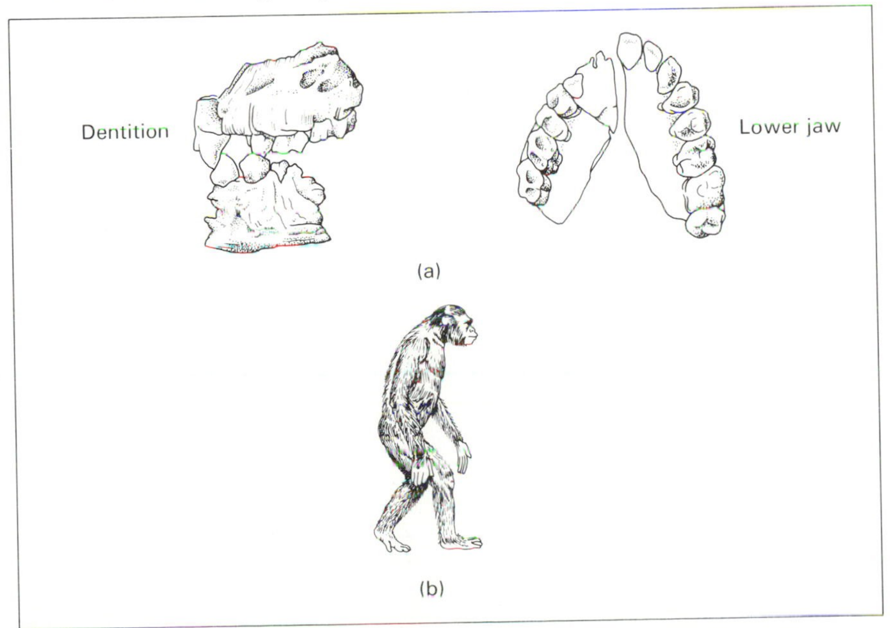

FIGURE 27.9 Scarce fossils (a) enable only a rough portrait of the 10-million-year-old *Ramapithecus* creature (b).

SOCIAL BEHAVIOR: Some day researchers will probably have enough fossils to prove the specific paths of evolution. Bones will ultimately show *what* evolved. But they are not as useful regarding *how* evolution occurred. To understand the reasons behind evolution in recent times,

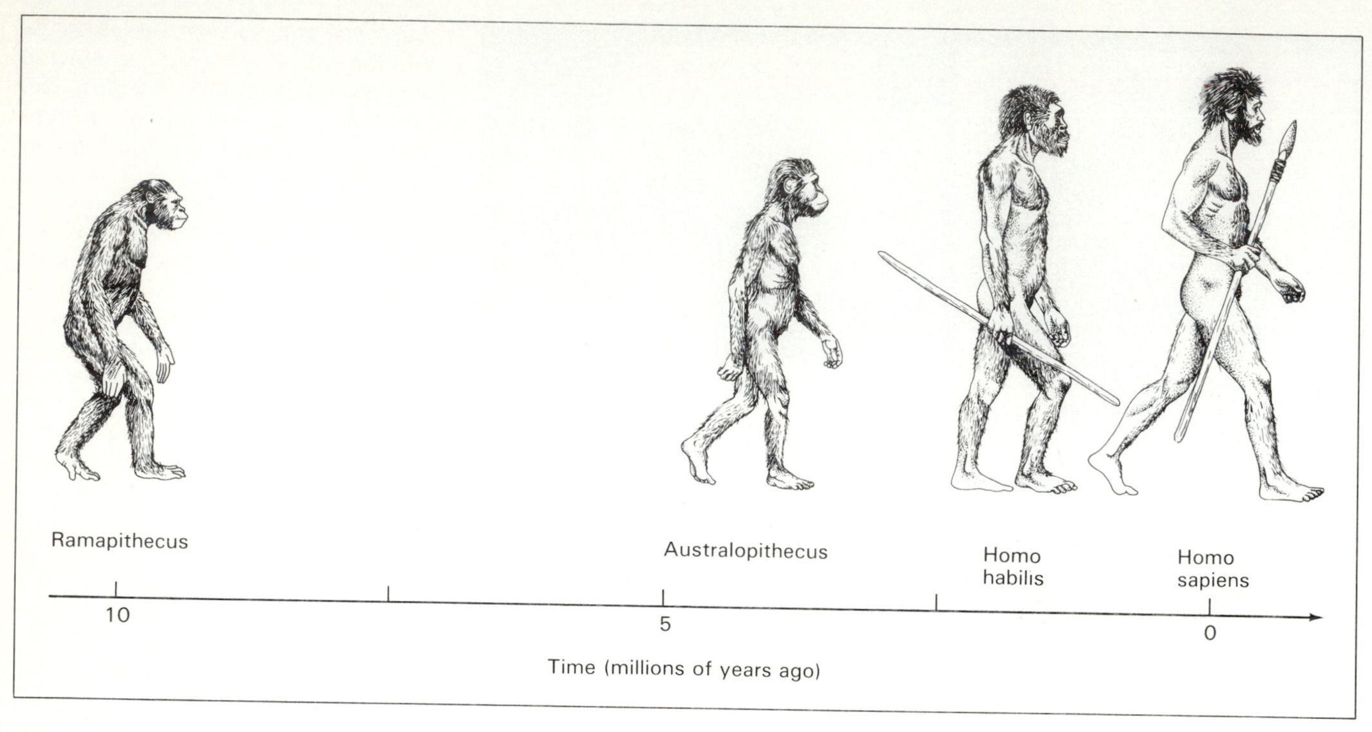

FIGURE 27.10 The relatively recent line of ascent toward humans.

scientists are now studying the behavioral patterns of our closest *living* relative—the ape.

How do we know that apes are so closely related to humans? One reason is the fossil record discussed in the preceding section, while a second reason is the genetic makeup of apes and humans. Laboratory studies of protein molecules can now be used to show similarities or differences among all animals. Using a variety of newly invented techniques, researchers can determine the exact sequence of amino acids along a chain-like protein. Comparisons of human proteins with those of, for example, a horse, a rat, or a frog show large differences in number and order of amino acids. But comparisons of human and ape proteins show very few differences. Especially significant is the fact that the average human protein is about 99 percent identical to that in chimpanzees. Some proteins, such as hemoglobin (blood), have exactly the same numbering and ordering of amino acids in humans and chimps. Thus of all the members of the ape family, chimpanzees have a genetic makeup closest to that of humans.

Chimps also have a life-style closest to our own. (Gorillas outwardly resemble humans more than do chimps, but their genetic structure is a little different, and their daily habits much different.) More than any other animal, chimps apparently resemble the ancestor from which other apes as well as humans ascended. This is a third way—behavior—that we regard apes and humans to be closely related. By studying the behavioral patterns of modern chimps, we might be able to discern a little bit of what life was like for our ancestors a few million years ago. Indeed, present attributes, adopted environment, and social behavior of the chimps might tell us something about the evolutionary events that led to the emergence of humans.

Earlier in this century, several attempts were made to study the lifestyles of caged chimps in zoos. But it soon became clear that the intricacies of ape society could be unraveled only in the wild. And "the wild" means just that; chimps and their ape relatives often live in remote places. Most are shy and totally unaccustomed to being watched by intruding humans. Many chimps inhabit inaccessible mountains, while others stay in the treetops of thick jungles. Reaching the appropriate places often proves as difficult for scientists as do the problems of which chimp properties should be studied and how to interpret the data once collected.

Organized fieldwork during the past few decades has enabled researchers to understand the behavior and social attitudes of chimps fairly well. Several revealing attributes have been noted. For example, chimps and other semi-bipedal (two-legged) apes are clearly more intelligent than monkeys and other quadripedal (four-legged) animals. Bidpedalism permits erect posture, thereby freeing the hands, and the resulting manual dexterity in turn provides a wealth of new opportunities for living. As shown in Figure 27.11, for instance, modern chimps have a habit of making an implement by stripping leaves from a tree branch, after which they insert it into a hole in a termite mound, remove it carefully, and systematically lick off the termites clinging to the branch.

FIGURE 27.11 Chimpanzees use sticks as tools to prod insects out of termite mounds, carefully extracting the stick and systematically eating the insects. Note also the chimp's reasonably erect posture.

Which was the cause and which the effect—ability to stand on hind legs or capacity to manipulate with hands—is currently an issue of dispute. Instead of the possibility noted above, that erect posture caused toolmaking, the situation might have been reversed: The need to manipulate food may have helped our human ancestors to become permanently upright over the course of millions of years.

It may turn out that these two critically important evolutionary traits—tool use and bipedalism—are so hopelessly intertwined as to preclude a determination of which was the original instigator. Actually, each trait might have contributed to the other in a complex way: Primitive bipedalism may have led to a small amount of manual dexterity, after which increased hand use accelerated the change toward more erect posture, which in turn fostered the development of even more sophisticated tools, and so on. This is an example of "positive feedback," whereby the development of one attribute stimulates another, causing further and faster mutual development. Such feedback reinforcement was probably a key mechanism during many phases of prehuman evolution.

The habits of modern chimps imply many things about our australopithecine forebears. Chimps are small enough to get around in trees, yet large enough to ward off most predators while on the ground. They can be especially formidable when traveling as part of a large group, as they often do.

Chimps eat meat and bird eggs, as well as small snakes, lizards, and insects. But their favorite diet is fruit, especially ripe figs. They also sometimes experiment with different foods, thus revealing some innate curiosity.

Basic chimp intelligence manifests itself in many ways. They have an open, free behavior, enabling them to try new things. In addition to using twigs as tools, chimps have been observed to use rocks to smash objects, to wave large branches overhead to scare away enemies, and to employ grass as a sponge to hold water.

Perhaps even more interesting than their expressions of curiosity, chimps seem to have some self-awareness. For example, when exposed to mirrors, chimps at first treat the image as if it were another chimp, but they then soon recognize it as themselves. Once thought to be inherent only to humans, self-recognition seems to be part of the intellectual repertoire of these remarkable apes.

Chimps are also known to be observant copycats. The young learn readily from their elders, as well as from human trainers. Although their lack of vocal equipment prohibits chimps from speaking, a few of them are now able to communicate symbolically with humans by means of sign language routinely used by deaf people. Most important, chimps have used such symbolic gestures to communicate with other chimps. Chimp-to-chimp communication of this type suggests that chimps' intelligence incorporates much communicative and learning ability, implying that they are really a lot smarter than any scientist would have argued a decade ago. Perhaps it's unfair to describe chimps only as copycats. Parrots and seals can also imitate, but there's a difference here: Other animals can be trained to imitate, whereas chimps seem to have the childlike ability to *learn* by imitation.

Chimps are furthermore sociable, although in a highly stratified way. All groups of chimps show a clear social hierarchy, comparable in many ways to twentieth-century human groups, whether in the military, business, academic, or political sector. One or a few males usually dominate a host of subservient chimps, thus ensuring some stability rather than the constant infighting that might otherwise arise in a completely free society. This social hierarchy does not, however, seem to stifle their curiosity or friendliness. As implied by Figure 27.12, some chimps display altruism while sharing food with other members of their group. They are thus not entirely self-centered, showing some affection for others in their group, even including chimps who are not their immediate offspring.

Chimpanzee society is so complex that about 15 years are needed before a newborn chimp reaches maturity. Like human adolescents, young chimps apparently require many years to learn everything necessary to become a full member of their social organization. In a certain sense, young chimps are schooled by their parents.

That chimps really do learn in their formative years demonstrates that environment plays a large role—at least among modern chimps. Other completely unrelated species, such as bees and ants, also have organized societies, but they don't really learn. Environment seems to have little bearing on insect knowledge. Laboratory studies show them lacking freedom of individual expression while going about their daily business; insects demonstrate little, if any, of the curiosity needed to try new things. Consequently, while insect society is definitely organized, insect behavior is not very complex; it's rather rigidly controlled. We can conclude that the social organizations of in-

FIGURE 27.12 Chimps often display altruism and affection toward other chimps. Here they are grooming each other.

sects are almost entirely programmed by genetics.

Because chimps learn so well, we cannot easily tell how much innate intelligence they really have. How much of their knowledge derives from their genes and how much from their cultural environment is simply unknown. We are now in the midst of a great debate concerning the relative importance of the gene and the environment. This debate regards not only the development of intelligence in chimps. As noted in Interlude 27-1, the gene–environment controversy affects all aspects of evolutionary theory, especially the cultural evolution of human beings.

Thus family life, altruism, love, and curiosity are attributes associated, not only with humans, but also with chimps and probably other animals as well. Many researchers contend that these qualities of goodwill are less apparent in chimps than in humans, but a glance at the (human) newspaper can introduce clear doubts.

On the other hand, we should not be fooled into thinking that chimpanzees are nice and gentle all the time. They resemble humans in yet another way, namely their occasional desire to exert unnecessary aggression. Some hostility within and among species is a normal, perhaps even essential ingredient of biological and cultural evolution. Without some aggression in the form of competition, few if any species could survive or adapt to a changing environment. But unnecessary aggression is another thing entirely.

Recent field studies in Tanzania show that some chimps occasionally murder other chimps for no apparent survival-related reason. Premeditated, gangland-style attacks were directed by a large group of male chimps on a smaller group of males and females that had previously broken away from the larger group. Over the course of five years, each member of the splinter group was systematically and brutally beaten. All died. Only young males initiated the attacks, which occurred only when the victims were isolated from the others. Hands, feet, and teeth were often used by the attackers, although occasionally fieldworkers noticed stones being deliberately thrown. The hope, of course, is that field studies like these will uncover the reasons behind not only chimp misdemeanors but human aggressions as well. Some researchers suggest that comparative studies of this sort may help guide the future survival of the human species, which, it would seem, can no longer tolerate further intraspecies aggression.

At any rate, despite current controversies about many details, behavioral studies of modern chimps have contributed greatly toward a consensus concerning the ascent of humans: As ape-like animals resembling chimps began leaving the forest for the open plains some tens of millions of years ago, they were probably forced to become more sociable in order to survive. The origins of our social organization may well have been shaped by this new, harsher environment in the open plains, where there would have been less food, reduced protection, and thus a greater need for group cooperation.

These hardships nonetheless gave our ancestors a chance to learn. Their sights and experiences grew over the course of millions of years. The limited knowledge of forest-living animals was replaced by the wider perspective of our plains-dwelling ancestors. This suddenly larger world created pressures to evolve larger brains capable of storing a dramatic increase of raw information.

In stating that this widening world view occurred "suddenly," note once again that we do not mean during the course of a single lifetime. Instead, migration from the trees to the plains was a renaissance of sorts that likely took a million years or so. Yet once it began, the race was on—a race to inhabit entirely new niches, to develop whole new ways of life, and eventually to become technologically intelligent.

INTERLUDE 27-1 *Human Nature*

The extent to which intelligence is genetically preprogrammed, as opposed to being enviromentally endowed, is most controversial. The gene versus the environment debate has forged a whole new interdisciplinary field of research. Called "sociobiology," it is the study of the biological basis of social behavior. This new subject attempts to determine the social instincts within any community of life forms by combining the basic principles of psychology, genetics, ecology, and several other seemingly diverse disciplines. The goal of sociobiology is to identify the inheritable traits that mold societies.

Sociobiology is an expansion of the study of classic biological evolution to include society. Another word for it might be social evolution. As such, the fitness of an individual creature is measured, not just by its own success and survival, but by the contributions he or she makes to the success of his or her relatives, namely those who share some of his or her genes. These contributions are often self-sacrificing ones, and thus can be classified under the general heading of altruism—unselfish devotion to the welfare of others—a fancy word for love. Whereas the catch phrase for classic biological evolution is "survival of the fittest *individual*," that for sociobiological evolution would be something like "preservation of an entire *society*."

Insect and other nonhuman societies seem to be founded on altruistic behavior. Extensive studies have virtually proved this to be the case for ants and bees, as well as some other animals. For example, wild dogs regularly regurgitate meals in order to feed their young; some species of birds postpone mating to help rear their siblings; "soldier" termites explode themselves, spraying poison over armies of ants, whenever a termite colony is attacked. This behavior is always the same, regardless of where and when the dogs, birds, or termites happen to act. They perform like programmed machines. Behavior so rigid and uniform has prompted many researchers to suggest that it might be determined genetically. If so, then each trait, act, or duty has its own gene, which is inherited in much the same way as body size, shape, and structure. The main role of these behavioral genes is to preserve the species.

Sociobiology is currently controversial largely because its proponents argue that its central tenets can be extended from insect societies to the societies of higher life forms, including humans. Problems always arise when scientists talk about our own species. The trouble stems from the fact that some aspects of human nature are often not what we think proper. The central issue is this: To what extent does human behavior mimic insect behavior and its largely genetic basis? In other words, which has the dominant influence over the actions of humans, nature (the gene) or nurture (the environment)? This is the root of the controversy—human understanding of human affairs.

Researchers generally fall into two groups, each conceding that environmental influences play the major role in human behavior. One group maintains that environment is the only important influence; behavioral differences among different humans are due solely to social, cultural, and political factors.

The other group contends that the gene has considerable importance. It weighs perhaps only 10 percent compared to the environment, but enough that many traits—for example, aggression, envy, guilt, sympathy, love, fear, intelligence, among others—are partly predestined in humans. If so, then changes in human behavior are limited because much of behavior is biologically dictated by the genes. This group presumes that, for example, the behavior of humans who go hungry to feed their children, or the behavior of persons who risk their lives to save a drowning swimmer, is not governed completely by free will. Instead, paralleling an insect's desire to preserve its own species, such behavior is an unconscious reaction built into our genes to ensure the survival of our own kind.

Should the basic ideas of sociobiology be correct, then the genes control much of human behavior—even when imprisoned within the bodies of life forms. In the extreme, life forms may exist for the sole purpose of perpetuating the genes— the selfish and unaltruistic genes.

Roots of Technology

Almost all researchers agree that our ancestors of the past few million years must have been **hunter-gatherers**, namely they lived by hunting and gathering food. The acquired traits of pursuing and eating meat were probably exported from the forest to the plains, whereupon they were enhanced because of the relative lack of fruit in the open plains. Although most inhabitants of modern civilization no longer regard themselves as hunter-gatherers, this was indeed the job description of all our ancestors from several million years ago until the rise of agriculture some 10,000 years ago.

How do we know that early humans, even the advanced australopithecines, hunted? The evidence is twofold, and both parts are found in the fossil record. First, scattered bones of a variety of large animals are often found near those of our australopithecine ancestors at numerous dwelling sites all along the East African Rift Valley. These animal bones do not comprise intact skeletons, but rather strewn debris suggestive more of a picnic than a natural death. Second, and more convincing, tools made from stones are often found alongside the remains of 2-million-year-old australopithecines, as well as of all the more recent human species. These stone artifacts have endured for millions of years, and it seems safe to conclude that even earlier ancestors might have used wooden tools that rotted away long ago.

Figure 27.13 displays several different kinds of stone tools discovered at Olduvai Gorge. Judging from their shapes, we can surmise that many of these egg-sized implements were used to chop, cut, and prepare food for easy consumption. Many others, though, suggest use as weapons. This is especially true of the rounded stones probably used to maim or kill when thrown at some animal, just as modern chimps occasionally do now. Other stones clearly resemble club heads, and they were probably used for exactly that—hunting by killing with a club of some sort. As noted earlier, these were perhaps the "tools" used by the advanced *A. africanus* (or *H. habilis*) to exterminate its relative, *A. boisei*, about 1 million years ago. Whatever their use, these primitive tools were in fact the beginnings of the technological society that all of us now share.

Stones were used not only for tools and weapons. They also provided the basis of early homes. A 2-million-year-old locality in Olduvai Gorge, for example, contains a circular stone structure theorized to have been the foundation of a hut of some sort.

This kind of primal stone work predated what is now popularly termed the Stone Age, an interval spanning from roughly 1 million years ago to approximately 10,000 years ago, depending on the place excavated. This time domain is distinguished by increasingly intricate stone tools, including various types of handaxes, cleavers, spatulas, and scrapers. A steady change from rather crude tools to more advanced ones can clearly be noted alongside the fossil record of biological species. Hence the beginnings of the Stone Age are customarily associated with the onset of *Homo*. This early toolmaking almost surely accelerated the evolution of the first true human beings.

FIGURE 27.13 A sample of some of the primitive stone tools unearthed at the australopithecine 2-million-year-old dwelling site at Olduvai Gorge.

To satisfy our need to categorize, anthropologists give labels to the times when different technologies were introduced into human life. Table 27-1 lists some well-known ages that document the steady rise of technology. The intervals are approximate, as many of the ages overlap. For example, natural metals were undoubtedly discovered in the course of working with some stone ores; crude metals were thus probably used well before the start of the Metal Age noted in the table. Similarly, humans knew much about the atom even prior to the development of thermonuclear weapons and the Atomic Age nearly a half-century ago. Furthermore, some researchers prefer to subdivide the ages listed, for instance by splitting the Metal Age into the Bronze, Iron, and Copper Ages. Thus one age was not simply and suddenly replaced by another; rather, the transitions were gradual much like any evolutionary process.

TABLE 27-1
Some Notable Technological Intervals

AGE	TIME INTERVAL (YEARS AGO)
Stone Age	1 million–10,000
Metal Age	10,000–500
Discovery Age	500–150
Industrial Age	150–50
Atomic Age	50–25
Space Age	25–present

The threshold of technology, then, is hard to pinpoint exactly, but it seems to have occurred more than a million years ago. The beginnings of cultural, as opposed to utilitarian, activities may be nearly as old, for brightly colored mineral pigments have been found alongside skeletons of the earliest of the true humans. Even the advanced australopithecines may have used ritual, as geometrically arranged pebbles are often found alongside their remains.

Only toward the end of the Stone

Age do more industrious and sophisticated undertakings become evident. Technological advances such as the construction of the wheel some 50,000 years ago, and the invention of the bow and arrow about 10,000 years ago, were matched by cultural advances such as the oldest deliberate burials in certain European and Asian caves nearly 70,000 years ago, and the beginnings of prehistoric art on the cave walls of western Europe about 30,000 years ago. For the most part, these were uniquely human inventions—cultural products of *Homo sapiens,* including Neanderthal Man and especially Cro-Magnon Man.

Climate

Throughout recent evolution, the environment continued to play an important role. To be sure, environmental changes act as the motor of evolution, allowing some life forms to adapt while forcing others to become extinct. And, on Earth, some of the most important environmental changes are caused by our climate.

Apart from seasonal changes taking place from month to month, and continental drifts occurring over millions of years, planet Earth apparently experiences global climatic changes that endure for thousands of years. For example, Figure 27.14 is a record of the average surface temperature during the past 500,000 years. These global climate data were obtained by a variety of methods, including analysis of core samples taken from the icy polar regions, of sandy sediments extracted from below the seafloor, and of land-based geological evidence for freeze–thaw cycles. (Less reliable data extend back hundreds of millions of years, but we'll concentrate here on relatively recent times.)

ICE AGES: As shown by the plot, our planet has cycled through numerous episodes of cool, dry climate—intermittent periods popularly known as **ice ages**. Although the data are sketchy, each cold ice age, as well as its opposite warm interglacial period, lasted roughly 50,000 years. We now reside in an interglacial period—a temporary thaw of sorts before heading back into the deep freeze.

What causes these cycles of heating and cooling on our planet? Some geologists contend that glaciation increases during the periods of global volcanic activity when ejected dust reduces the amount of sunlight penetrating Earth's atmosphere. Others suggest that the periodic reversals in Earth's magnetic force field cause the protective Van Allen Belts to disappear, thereby sporadically allowing unusually high doses of solar radiation to heat the ground and thus decrease glaciation. Still other researchers have noted that ice ages could be triggered on our planet by variations in the output of the Sun itself, by passage of Earth through an interstellar dust cloud, by deep circulation of Earth's oceans, or by any one of a long list of other proposals.

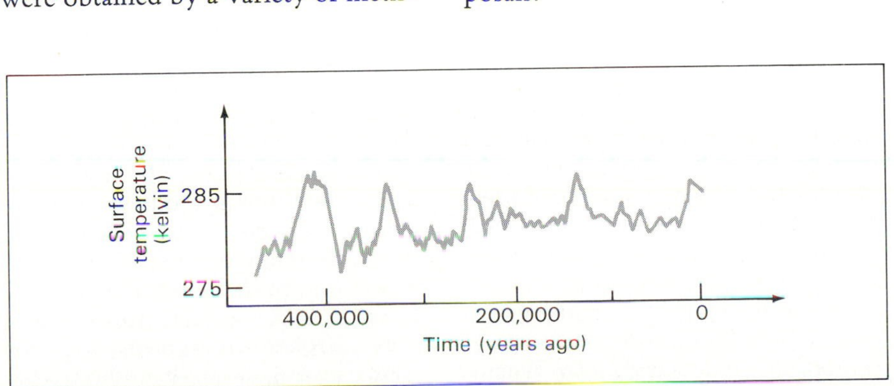

FIGURE 27.14 The average surface temperature of Earth has changed during the past half-million years. (Recall that 273 kelvin equals the freezing point of water.)

Recently, however, oceanographers have found convincing evidence to support yet another theory, dubbed the "Milankovitch effect" after the Yugoslavian geophysicist who proposed it earlier in this century. According to this theory, subtle though regular changes in Earth's orbit around the Sun initiate the ice ages. These changes are the combined result of three astronomical effects, each caused by the normal gravitational influence of other planets on the Earth: (1) changes in the *shape* (called "eccentricity") of Earth's elliptical orbit around the Sun, (2) changes in the *tilt* or wobble (called "precession") of Earth axis, and (3) changes in the *orientation* (called "obliquity") of Earth's tilt relative to its orbit, an effect caused by the ever-changing positions of the other planets.

Specifically, the shape of Earth's orbit changes approximately every 100,000 years, becoming more circular, then more elliptical. When the orbit is most elliptical, the Earth receives 20 to 30 percent more radiation when it is closest to the Sun than when it is most distant. Second, Earth wobbles on its axis every 26,000 years, thus changing the time in Earth's orbit when winter and summer ocur. Finally, over a period of about 40,000 years, the tilt of the Earth (relative to its orbital plane) varies from 22.1 to 24.5 arc degrees, the smaller tilt resulting in smaller temperature differences between winter and summer.

The first phenomenon would produce warmer summers and colder winters. The second affects the heat mostly during the summer months, and conceivably the amount of ice melting. And the third probably results in cooler summers with less ice melting. The combination of these three effects—for all three operate simultaneously—would sometimes cause abnormal solar heating (such as we are now experiencing); at other times, solar heating is distinctly reduced, producing widespread glaciation and a decline in global temperature.

This theory of an astronomically induced ice age is currently favored among the majority of scientists, mainly because samples of seafloor sediments show that during the past 500,000 years, tiny sea creatures called plankton have thrived at certain times, while barely surviving at others. Studies of the abundance of fossilized plankton known to prefer warm or cold water provide estimates of the prevailing water temperature during their lives. This inferred sea temperature correlates well with the expected heating and cooling of Earth by means of the combined astronomical effects just noted.

It seems, then, that slight peculiarities in Earth's orbital geometry are mainly responsible for triggering ice ages; whether they are the only trigger remains to be proved by further research.

Regardless of their cause, ice ages must have had profound effects on the evolution of life on Earth. The most recent major glaciation began about 75,000 years ago; the climate had returned to its present, "normal" state some 10,000 years ago. As depicted by Figure 27.15, at the peak of this ice age roughly 30,000 years ago, a kilometer-thick ice sheet extended from the North Pole far enough south to cover much of the northern United States, as well as much of northern and central Eurasia. Earth's overall surface temperature averaged 5 kelvin lower than today's, while the sea level, with much water locked up in ice bergs, was nearly 100 meters below current values.

The last 10,000 years have seen the glaciers retreat, the coastal plains flood, the vegetation climb toward northerly latitudes, and the ocean and atmosphere warm up. The change in climate from the peak of the most recent ice age to its present state occurred rapidly, by geological standards. These widespread changes in Earth's environment surely helped foster the many advances made by our human ancestors during this period of rich innovation. Humans were forced to adapt —to adapt biologically, culturally, and rapidly—to changes throughout the air, sea, and land. The motor of evolution had quickened.

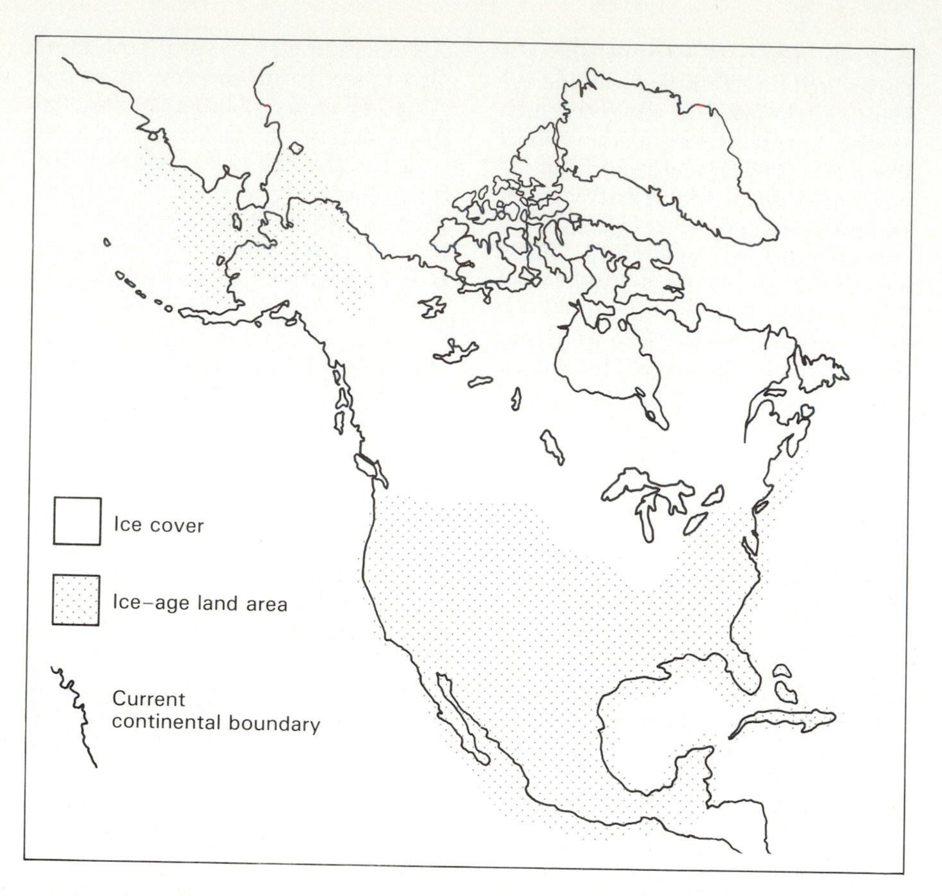

FIGURE 27.15 Glacial ice covered much of the Earth at the peak of the last major ice age, some tens of thousands of years ago.

EARLY AMERICANS: The ice ages themselves may have accelerated the migration and colonization of new lands. For example, we know that humans did not evolve in the Americas; there is no fossil evidence for ape-like creatures from which they could have ascended. Current anthropological consensus contends that humans arrived in the new world of the Americas as recently as a few tens of thousands of years ago—and possibly as recently as 12,000 years ago. Precisely how they did it is unknown, but during one of the more recent glaciations, enough water would have been wrapped up in ice to have permitted humans to walk dry-shod across the Bering Strait between what are now called Alaska and Siberia.

This is in fact the prevailing view; native North, Central, and South Americans are descendants of Asians who wandered here only a few hundred centuries ago. Once settled in the Americas, these migrants developed arts, languages, products, and numerous other cultural innovations. Whether the civilizations of the Americas experienced cultural evolution independent of those in Eurasia, or whether they had some as-yet undiscovered contact with them, is a central point of controversy in modern archeology.

Factors That Made Us Cultured

Changes that affected humanity were evolutionary, not revolutionary. They occurred by means of the usual adaptations to a gradually changing environment. But they happened in an increasingly rapid way. Broadened opportunities for living during the past million years or so has quickened the pace of evolution.

In the final section of this chapter, we examine briefly four important factors in human evolution: the discovery of useful fire, the development of meaningful communications, the invention of profitable agriculture, and the use of controlled energy. Undoubtedly other factors also helped to make us cultured humans. Fire, language, agriculture, and energy are meant to be representative of the more important ones.

FIRE: *Homo* has been using fire for light and heat for nearly a million years. Archeological expeditions of caves in France have revealed fossil hearths at least that old. The essence of fire, therefore, was basically appreciated a very long time ago, and was undoubtedly used to provide warmth in colder climates. But it seems that its broad usefulness went unrecognized until more recently. The practice of cooking food, for example, was probably developed only 100,000 years ago. Techniques needed to fire-harden spears and anneal cutting stones are likely to be still newer inventions. Significantly, the widespread use of fire was one of the last great steps in the domestication of humans.

Beginning about 10,000 years ago, humans learned to cast copper at 1000 kelvin and harden pottery at 1200 kelvin, as well as to extract iron from ore and change silica into glass at 1800 kelvin. Many other industrial uses were realized for fire, including the fabrication of new tools and the construction of clay homes.

Unfortunately, there is no agreement about the ordering of many of these basic discoveries. Just how one invention paved the way for another is so far unclear. For example, some archeologists argue that after heat and light, baking clay to make pots is the oldest organized use of fire—even predating the regular cooking of food. As depicted in Figure 27.16, pottery or ceramics is a technique whereby clay is changed back into stone by dehydrating and heating it to a high enough temperature to cluster the fine mineral particles. Others, however, argue that the need for pottery arose because cooking was already established; indeed, the earliest uses for pottery must have been for cooking and storing food.

FIGURE 27.16 Pottery is one of the oldest cultural uses of fire. This modern mud-brick village in Afghanistan features a pottery kiln in the foregound that is similar to those dating back some 10,000 years.

When inventions are intimately linked in this manner, which was the cause and which the effect is never quite clear. Both the motivation for and the exact time of an invention are difficult to establish. Some of the more crucial sequences may never be unraveled. Of one thing we can be sure: Many new mechanical and chemical uses of fire were mastered during the past 10,000 years, and a few may have been discovered even well before that.

The ancient clay, metal, and glass products can still be found in the bazaars and workshops of Afghanistan, China, Iran, Thailand, Turkey, and other generally Asian countries. More modern results of these early technologies are evident in the cities of steel, concrete, and plastic surrounding many of us in the twentieth century. Changes of this sort are not without problems, however; the basic need for fire has often been accompanied by grim aftereffects, not the least of which include environmental pollution and energy shortages only now becoming evident.

LANGUAGE: Development of language was surely another crucial factor in making us cultured. Early evolution of communications probably related to toolmaking and hunting; a complex feedback mechanism possibly led to the further development of each. After all, symbolic communication such as sign language and arm signals would have been an obvious advantage in big-game hunting. And it seems likely that communication of some sort must have been needed to convey such simple skills like tool use and toolmaking. Equally important, language ensured that experience, stored in the brain as memory, could be passed down from one generation to another.

Although there is no direct evidence, some type of primitive language could have been used a million or more years ago. More sophisticated language came much later, although again, it's hard to pinpoint its origins and evolution with any accuracy. Artifacts made by early humans provide some clues to the cognitive abilities and manual skills needed to develop such advanced techniques of communication as speaking and writing.

Excavations of caves, mostly in Europe, have uncovered a wealth of small statues and bones having distinctive markings or etchings. The oldest of these dates back about 50,000 years. Figure 27.17 shows a few of these stone artifacts, some of whose markings suggest that the statuettes had symbolic functions of some kind.

Since Neanderthal Man's larynx could not have possibly uttered all the sounds of modern speech, the etchings on these objects are thought by many researchers to represent a primitive sort of communication. This idea has been reinforced by recent microscope studies that show the same repertoire of distinctive

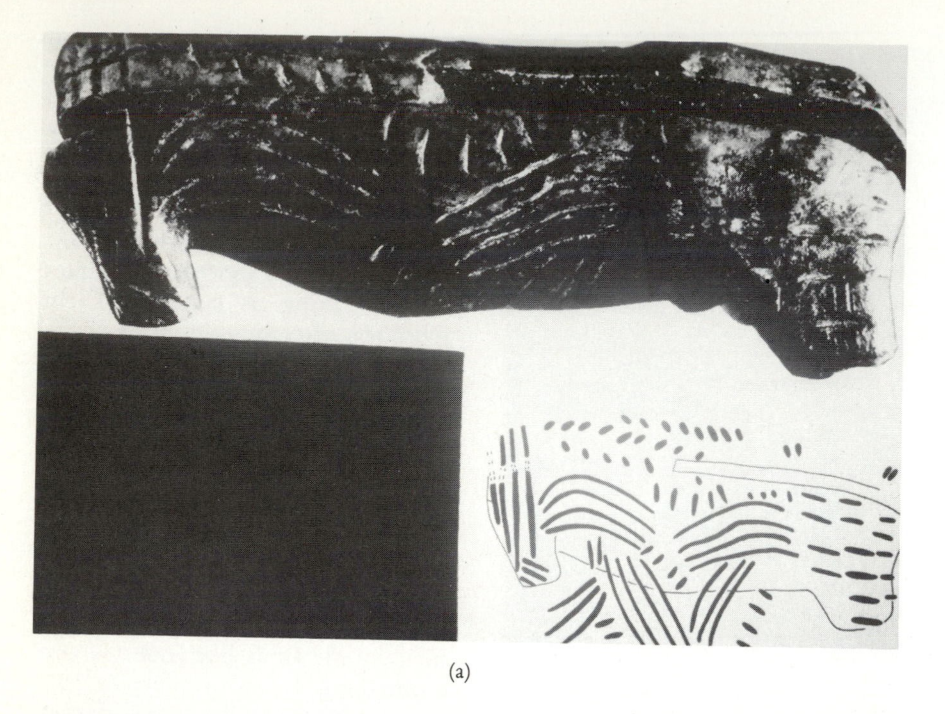

(a)

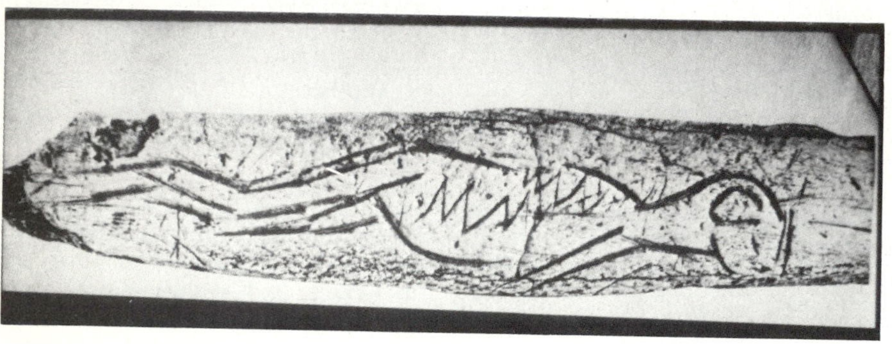

(b)

FIGURE 27.17 This stone artifact (a), dated to be about 30,000 years old, is one of several found by archeological expeditions to have precise engravings. The 3-inch statuette resembles a horse or reindeer, as interpreted by the sketch at lower right. Numerous bones like that in (b), associated with cultures that prospered in the interval 30,000 to 10,000 years ago, show deliberate markings or engravings. This 3-inch image on a reindeer rib resembles a reclining humanlike figure.

markings repeated on many of the statues and bones. In other words, the markings are neither accidental nor simply decorative; they go beyond just art and are apparently symbolic. These engraved artifacts are among the earliest known attempts to better the grunts and groans used not only now by modern chimps but also by our ancestral australopithecines.

Residing on a flat plain bordering the Persian Gulf in what was then called Mesopotamia (and now the Middle East; see Figure 27.18), the Sumerians were probably the first people to write texts. Nearly 6000 years ago, this ancient civilization had developed an intricate system of numerals, pictures, and abstract symbols. Thousands of Sumerian clay tablets have now been unearthed, and each shows that a stylus of wood or bone was used to inscribe a variety of characters. Estimates suggest that the Sumerians' basic vocabulary was large, having no fewer than 1500 separate signs. Although wild animals such as the fox and wolf, as well as technological aids such as the chariot and the sledge, are clearly depicted, these Sumerian texts remain largely undeciphered. Because they contain more than just pictures, the messages on these tablets represent an advanced stage in the evolution of the art of writing.

The Sumerian writings have survived mainly because they were inscribed on baked clay. If earlier civilizations were responsible for inventing some of the symbols, they probably wrote exclusively on papyrus or wood that would have decayed long ago.

The prevailing view among anthropologists is that writing evolved from the concrete to the abstract. Sometime between several tens of thousands of years ago and several thousands of years ago, the pictures and etchings on the bones and cave walls must have become increasingly schematic, utilizing a single symbol to represent an entire idea. This evolution could have been deliberate in order to speed the process of record keeping. Or it could have been the result of carelessness on the part of ancient scribes. Whichever, it suggests that pictures preceded symbolic writing, which in turn led to the character writing of modern times, much as in this book before you.

AGRICULTURE: The most recent one percent of human history—the past 10,000 years—has seen major and rapid cultural developments. The glaciers had retreated, creating warmer and wetter environments, thus allowing the land to flourish. Our hunter-gatherer forebears had spread to occupy, though sparsely, every part of the habitable globe except the Arctic and Antarctic. And they were fashioning implements and ideas to enhance their survival. But of all the factors that led to the rise of modern humans, the onset of agriculture was surely among the most important. Tilling the land made available a reliable source of food to feed the swelling numbers of people on Earth. In short, beginning with the invention of farming as well as the domestication of plants and animals some 10,000 years ago, hunter-gatherers rather rapidly changed into **agriculturists**.

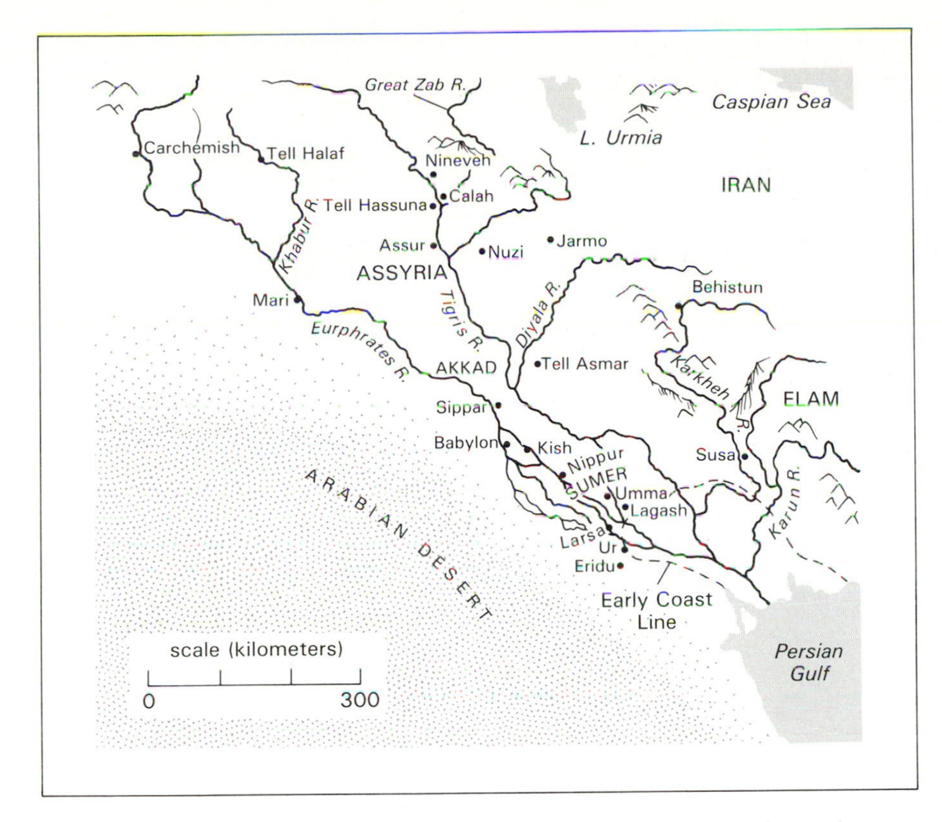

FIGURE 27.18 The cradle of modern civilization was probably located in and around the Tigris and Euphrates Rivers.

Archeological data show clear evidence for whole new methods of subsistence by 8000 years ago. Systematic crop planting and livestock raising near stable village settlements permitted large increases in the population. Not only did more people survive, but many more migrated to colonize almost every nook and cranny on the planet. The basic technique of agriculture spread rapidly from Asian and Aegean localities, and especially from the area now called Greece. Hunters gave way to farmers and herders. Change was rampant. Urbanization and ultimately industrialization were not far behind.

The environment continued to change, although not as slowly as nature would have had it. After all, humans themselves had become a factor. We had become the agents of change.

Civilization had moved into high gear. It had taken an awfully long time, after life originated, but highly organized and manipulating life forms had finally arrived. One thing led to another; life-styles multiplied. Aided by irrigation systems built alongside river valleys, the art of farming developed dramatically. The human population rose rapidly, especially in urban areas along waterways such as the Nile River in what is now Egypt, and the Tigris and Euphrates Rivers running through what are now Turkey, Syria, and Iraq.

Specialized crafts were refined to serve the people of these growing communities: Metalworks, fine ceramics, shipbuilding, and woodcuts all show up clearly in the archeological record beginning some 6000 years ago. This record is best documented for southwestern Asia (the Middle East), although surely practical and artistic progress occurred at other geographical places as well. Figure 27.19 schematically outlines some of the key cultural and technical innovations advanced by civilizations in this area, particularly Mesopotamia.

Although much of this preceded recorded history, the consensus is that urban societies and mature economies, as well as complex social and political systems, were the rule, not the exception. Agriculture, industry, and commerce were fully established several thousand years ago. And they have persisted to this, the twentieth century.

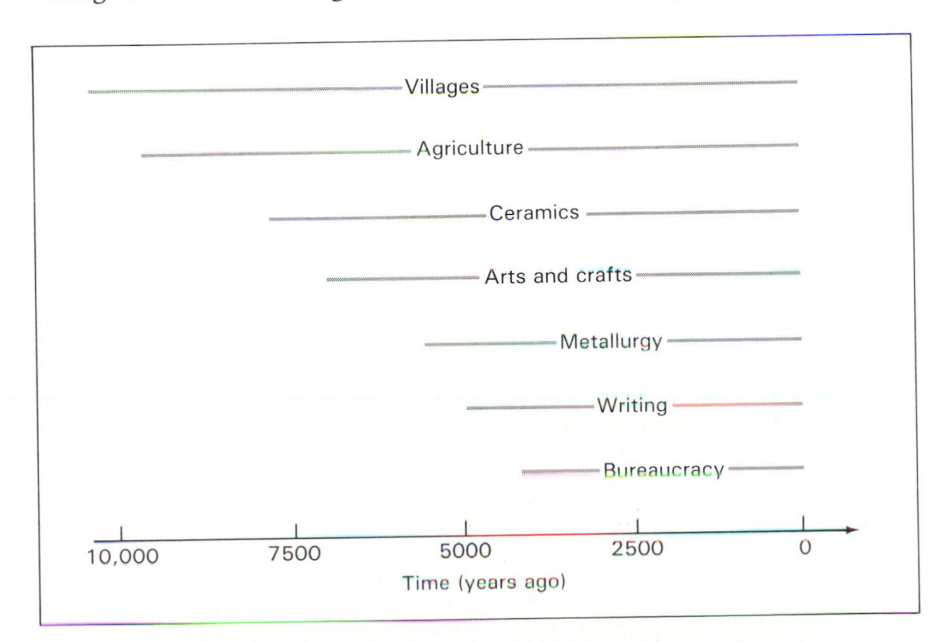

FIGURE 27.19 Some important intervals during which the ancient Mesopotamian civilization experienced numerous cultural and technological innovations.

ENERGY: Throughout the past million years or so, biological and cultural evolution have been inex-

INTERLUDE 27-2 *Ancient Science*

What are the origins of scientific research? Like most "origin" questions, this one is hard to pinpoint with any degree of accuracy. Even so, the statue and bone markings mentioned in this chapter may have had something to do with the onset of ancient science. Some of these etchings of several tens of thousands of years ago (see Figure 27.17) seem to correlate with the periodic lunar cycle, namely the appearance of the new Moon, quarter Moon, and full Moon. Some researchers reason that these were perhaps the first attempts to keep track of the seasons. Engravings of this sort, as well as large murals on the walls of caves, suggest that cave men and women of the late Stone Age were conscious of the periodic seasonal variations in plants and animals. While some archeologists prefer to interpret these deliberate bone markings as a simple arithmetical game, the above theory contends that these artifacts were among the very first calendars—in effect, scientific timekeeping instruments.

The earliest *written* record describing the use of scientific instruments does not appear until many thousands of years later. The 3000-year-old hieroglyphics partially document the Egyptians' knowledge of the sundial. Such an instrument is really a clock, telling the time of the day by noting the angle of the Sun's shadow. The oldest sundial excavated to date is a Greco-Roman piece of stone built about 2300 years ago. The Greek and Roman sundials reveal important refinements over the earlier Egyptian models, telling not only the hour of the day, but also the day of the year. Neither the length of an "hour" nor the construction of the calendar were the same as they are now, but these early scientific instruments were able to predict the orderliness of daily, seasonal, and yearly changes on Earth, thus identifying the basic rhythm of the farming cycle. Sundials, by the way, were among the instruments used by the Greek Eratosthenes to measure the size and shape of Earth in the last centuries of the pre-Christian era (consult Chapter 4).

Other examples of early humans' knowledge and use of the sky include megalithic monuments, the most famous of which are the pyramids of Egypt and Stonehenge in Britain. Stonehenge (at left in the accompanying figure) can be used to predict the first day of summer, and possibly solar eclipses as well, by means of the alignment of big stones with the rising and setting of certain celestial objects. Many other structures like it, although perhaps not as grand, are scattered across Europe, Asia, and the Americas. In Central America, for example, ancient "temple" pyramids (like that at Uxmal, Yucatan, in the center frame in the figure) have miniature portholes through which celestial objects, particularly Venus, can be viewed at propitious times of the year. Astronomers of the Middle Age Mayan civilization were essentially priests, with the destinies of individuals, cities, and even whole nations apparently determined by the position and movement of celestial objects across the sky.

By a thousand years ago, these Central American cultures had probably influenced many tribes of North American Indians roaming the plains of what we now call the western United States and Canada. The possibility of such cross-cultural ties has become strengthened by recent studies of numerous "medicine-wheel" structures (like that at right in the accompanying photograph taken in Wyoming) made of boulders arranged in various patterns of rings and spokes. Although the precise use of these stone configurations is unclear (since most Indians had no written language), some researchers suggest that these structures may have been used as calendars to mark the rising of the Sun at certain times of the year.

Thus, although modern scientific research, including instruments like the immensely useful microscope and telescope, dates back no more than about 500 years, we can be sure that pre-Renaissance people were adept in elementary astronomy, mathematics, mechanical engineering, and numerous other technical endeavors. Indeed, the roots of technology go way back. It seems that our ancestors were a lot smarter than many researchers, until recently, have been willing to acknowledge.

tricably interwoven. Their interrelationship is natural, for the development of culture bears heavily on one of those two factors affecting biological evolution—the environment. Cultural innovations enabled our immediate ancestors to circumvent some environmental limitations: Cooking allowed them to adopt a diet quite different from that of the australopithecines, while clothes and housing permitted them to colonize both drier and colder regions of planet Earth.

Similarly, present cultural innovations enable us—twentieth-century *Homo sapiens*—to challenge the environment. High technology allows us to fly in the atmosphere, to explore the oceans, even to journey far from our home planet. Change now quickens, and with it the pace of life. Culture apparently is a catalyst, speeding the course of evolution toward an uncertain future.

If there is any one trend that has characterized the evolution of culture, it is probably an increasing ability to extract energy from nature. Over the course of the past 10,000 years, humans have steadily mastered the wheel, agriculture, metallurgy, machines, electricity, and nuclear power. Soon, solar power will emerge in its turn. Each of these innovations has channeled greater amounts of energy into culture.

Indeed, the ability to harness larger energy sources is the hallmark of modern society. But it's also a cause of many of the sociopolitical problems in which we, twentieth-century humans, now find ourselves embedded.

SUMMARY

The precise path of human evolution during the past few million years is tricky to follow in detail. Whatever truly made us human involved the creative products of emotion and imagination—qualities difficult to define scientifically.

The causes of recent evolution include not only biological factors but cultural and technological ones as well. A complex feedback system has come into play among the increase in brain volume, the innovation of culture, the invention of improved technical skills, and the development of verbal communication and social organization.

Changes were very slow at first, but they have markedly accelerated within the past 100,000 years or so. Whatever the reasons, these many innovations have enabled *Homo* to enjoy unprecedented success as a life form on planet Earth. Indeed, we alone can technologically address the fundamental questions.

KEY TERMS

agriculturists
Australopithecus
bipedal
cultural evolution
Homo erectus
Homo sapiens
hunter-gatherer
ice age
paleoanthropology
Ramapithecus

QUESTIONS

1. Pick one: Neanderthal Man is classified as (a) an australopithecine; (b) *H. sapiens*; (c) *H. habilis*; (d) *H. erectus*; (e) an NFL lineman.
2. Could a technological civilization develop on a planet entirely covered with water? How, or why not?
3. Why does it seem likely that the gracile australopithecines developed intelligence faster than the robust australopithecines?
4. Is the process of natural selection, in the Darwinian sense, still at work in modern humans? Why or why not?
5. Why do researchers think that the primates developed binocular vision and manual dexterity when most animals did not?
6. Summarize the evidence that some australopithecines used primitive tools as long ago as a million years.
7. Explain why it's wrong to state that humans have evolved from apes.
8. Briefly discuss the evidence for the existence of several different types of australopithecines a few million years ago.
9. Pick one: Which of the following factors was most responsible for the increase in the rate of evolution during the past 10,000 years? (a) continental drift; (b) volcanic activity; (c) climatic change; (d) invention of tools; (e) cultural prowess of the Babylonians.
10. Describe how laboratory studies of genes complement information about evolutionary paths normally gained by analyzing fossils.

FOR FURTHER READING

Covey, C., "The Earth's Orbit and the Ice Ages." *Scientific American*, February 1984.

Ember, C., and Ember, M., *Anthropology*. Englewood Cliffs, N.J.: Prentice-Hall, 1985.

Johanson, D., and Edey, M., *Lucy*. New York: Simon & Schuster, 1981.

Pilbeam, D., "The Descent of Hominoids and Hominids." *Scientific American*, March 1984.

WASHBURN, S., "The Evolution of Man." *Scientific American,* September 1978.

WENKE, R., *Patterns in Prehistory*. Oxford: Oxford University Press, 1984.

WILSON, E., *On Human Nature*. Cambridge, Mass.: Harvard University Press, 1978.

28 NEURO-LOGICAL EVOLUTION: *The Onset of Intelligence*

One of the most remarkable aspects of life is an awareness of its surroundings. Unlike nonliving matter, life can monitor impressions from, and respond to, the outside world. Through life's various senses—hearing, seeing, smelling, touching, tasting—all organisms acquire and file enormous amounts of information. The extent to which beings are successful in doing so depends largely on their "complexity." Organisms manifest this complexity in terms of one exquisite piece of matter—the brain. The brain is the central clearinghouse of all behavior.

As you read these words, matter within your skull is full of activity. Silently, though efficiently, millions of nerve cells pass messages back and forth within your brain. They guide your eyes along this printed line. They quickly scan the shapes of the letters. And by matching them against memory, they allow you to recognize words.

Nerve cells constantly transmit signals within our brains, ordering our hearts to beat, our lungs to pump, and your hands to get ready to turn this page. The body's nervous system, of which the brain is the most important part, controls all mental and physical activity. In fact, every thought, feeling, or action begins in the brain. All human behavior is controlled by it.

Most interesting of all, these silent, unfelt activities inside your head make you aware that you are now reading about them. Astoundingly, the brain can contemplate the brain.

The human brain is the most complex clump of matter known. It's nature's most tantalizing, talented, and versatile creation. As best we can tell, our brains represent the ultimate example of the extent to which matter has evolved in the Universe. For this reason alone, we must study the origin and evolution of the brain.

The learning goals for this chapter are:

- to appreciate the basic geometry of the brain and of some of its control functions
- to understand the electrical and chemical activities inside the brain
- to recognize the ways and means that humans have come to possess such a magnificent clump of matter within our skulls
- to understand how intelligence can be judged
- to appreciate the idea that the human brain can contemplate itself

The Split Brain

Figure 28.1 shows two views of the human brain. Inside the insulating layers of skull bones, protective membrane, and shock-absorbing fluid, lies the wrinkled, pinkish-gray matter that keeps us alive. Looking like an oversized walnut, the brain actually has the dimensions of a large grapefruit.

The mass of a human brain averages about 1300 grams (or roughly 3 pounds). And since the density of the brain, like that of all body tissue, approximately equals that of water (1 gram/cubic centimeter), the average volume of the human brain is then about 1300 cubic centimeters. These values differ a little among modern humans, but for the most part, all human brains seem to have

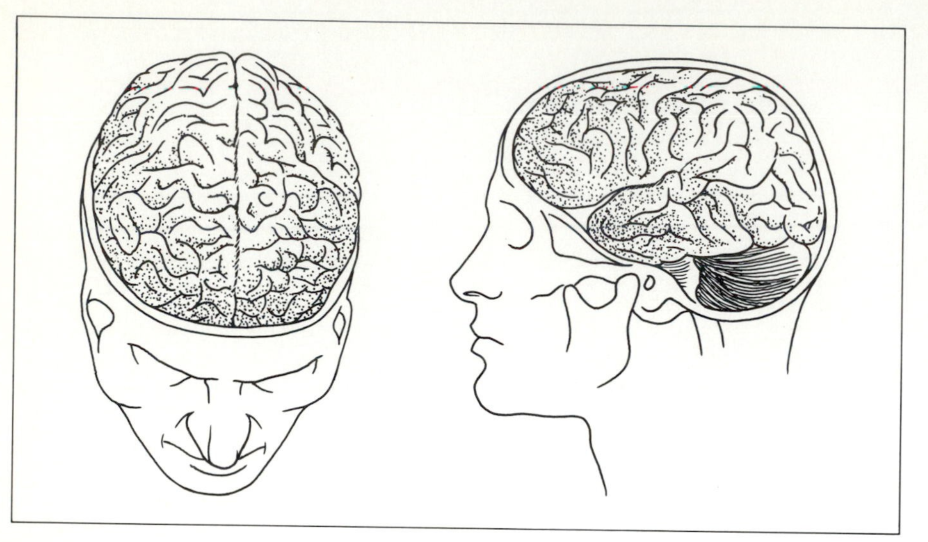

FIGURE 28.1 Schematic diagrams of the human brain.

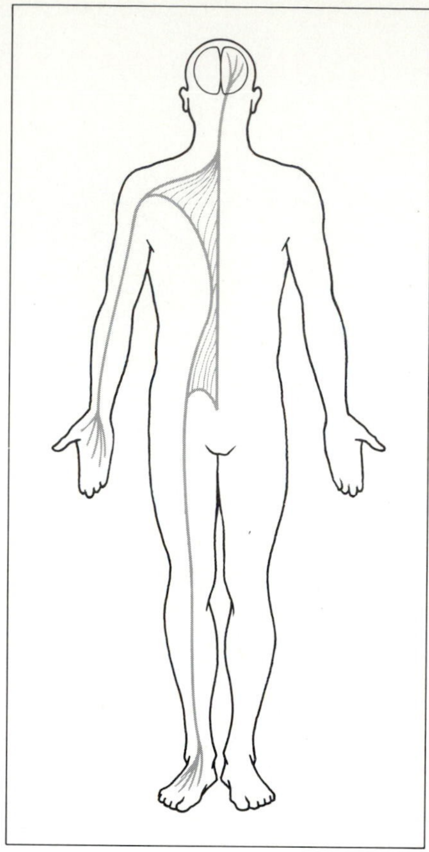

FIGURE 28.2 The nerves cross over in the spinal cord on their way from the brain to various bodily organs.

much the same size, shape, and chemical composition. Even geniuses like Einstein had brains that appeared no different from those of any other normal human being. The physical similarities of everyone's brain exemplifies the precision with which evolution designed our brain–body communication systems.

We can describe the brain in several ways. Generally, it's made of three components. First, the grapefruit-sized **cerebrum** is the domed, most prominent feature comprising the bulk of the brain; it contains the controls for the senses and muscles. Second, the plum-sized **cerebellum** resides at the bottom rear of the brain; it's largely responsible for coordinating and fine-tuning our body movements. And third, at the upper part of the spinal cord, there is the **brain stem** through which sensory and response information pass back and forth to the brain; this stem houses the mechanisms to control alertness as well as a variety of other voluntary and involuntary functions.

Interestingly, those parts of the brain responsible for controlling human behavior can be considered to be split vertically right down the middle. In other words, researchers often regard the cerebrum to be divided into two parts, each called a **cerebral hemisphere.** These hemispheres are symmetrical, paired much like many of our other body components—lungs, limbs, eyes, ears.

By conducting autopsies on the dead and by stimulating live patients with electrical currents, researchers have demonstrated that half the brain controls half the body. However, for some unknown reason, the nerves connecting the brain and body cross over in the spinal cord as depicted in Figure 28.2. This crossing ensures that each side of the brain presides over the opposite side of the body. For example, the act of writing by a right-handed person is controlled by nerves in the left hemisphere of the brain.

Despite their apparent symmetry, the two cerebral hemispheres often serve different functions. The capacity for speech and language, for instance, seems to reside in only one hemisphere—in 97 percent of all people, the left hemisphere. This concentration of verbal ability on one side of the brain makes it hard to talk while writing with a pencil in the right hand. Simultaneous writing and speaking put a double burden on the left hemisphere. Left-handed writers generally have an easier time of it.

Probes of the brains of accident victims have shown that the division of labor extends beyond language and hand use. For example, the left hemisphere is also the control site of many logical and analytical functions, such as understanding science and mathematics; the left side is furthermore associated with organization, social systems, and the ordering of everyday experiences. The right hemisphere controls less logical, although more creative functions, such as artistic and musical talents; ritualistic, mystical, transcendental, and supernatural qualities also originate in the right hemisphere. This specialization of brain functions is apparently valid for almost all humans; the same part of nearly everyone's brain senses and controls the same part of everyone's body.

The basis of modern education—the 3R's; reading, writing, and arithmetic—then amount to a training of only one of our hemispheres. For this reason, the left hemisphere is often regarded as the "dominant" brain, and the right hemisphere the "quiet" brain.

Knowing from experiments

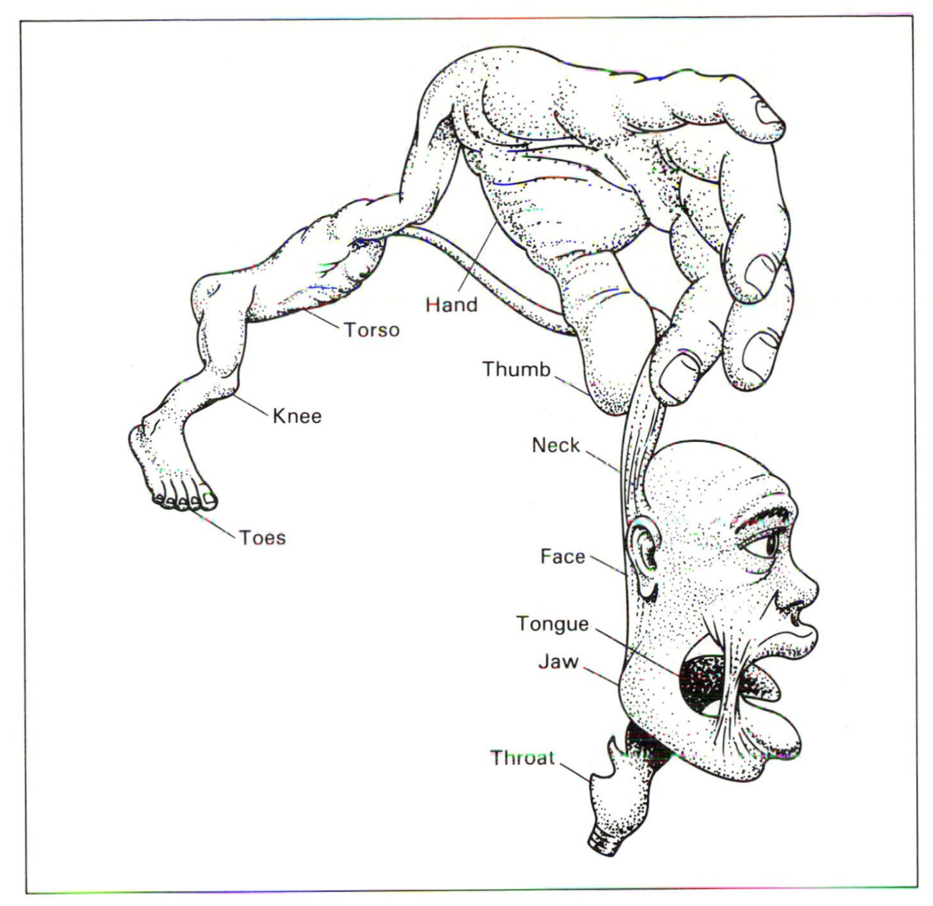

FIGURE 28.3 The distorted size of each human organ is meant to show the relative importance of brain-body signals. As illustrated in this diagram, the hand and mouth dominate.

which areas of the brain control which body functions, we can redraw a human in proportion to the amount of brain space devoted to each body function. As depicted in Figure 28.3, a large amount of brain area is used to control the fingers and mouth. Indeed, the ability to manipulate and to speak are precisely what sets us apart from most other animals. Our culture would never have developed without speech, our technology never without hands.

Neural Architecture

The matter within each hemisphere—the bulk of the brain itself—is called the **cortex.** Much like any other part of a living organism, cortical material is made of microscopic cells, in this case called nerves. The cortex is surprisingly large; if removed from the skull and stretched out flat, it would form a thin sheet about a meter on each side and one-third of a centimeter thick. In order to fit within the confines of the human skull, the cortex crumples up, producing deep folds and furrows. In this way, the surface area of the brain is tripled. If the cortical matter were not so wrinkled, our heads would have to be three times larger. And a head that large would require a larger body to support it.

PARIETAL LOBE: The deepest furrows in the crumpled cortex serve to further outline the surface of each brain hemisphere into four separate regions, or lobes. The parietal lobe, at the top of the head as shown in Figure 28.4, is the part of the brain that monitors the response of some of the body senses, especially touch and hearing. For example, the nerves in the parietal lobe keep track of the location of all the movable parts of the body. Hands, arms, feet, legs, tongue—all moving body parts—are monitored by the nerve cells in this lobe.

Note that we have not said that bodily functions are *controlled* here. They are just monitored here. The parietal lobe simply takes inventory of all bodily functions.

Everything that the human body touches or hears is collected and sent to the cells in the parietal lobe. On the way there, this information passes through the core of the brain—a region termed the thalamus. Positioned at the top of the brain stem, the thalamus acts as a central switchboard between the spinal cord and the brain's upper levels, collecting, sorting, and redirecting signals from the outside world to the appropriate part of the parietal lobe.

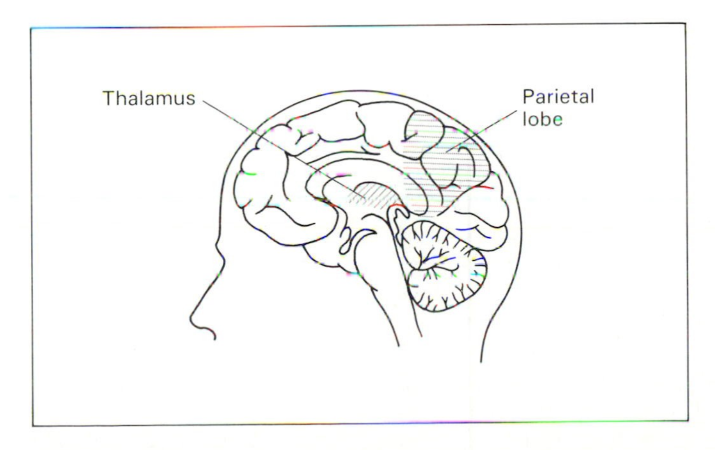

FIGURE 28.4 Matter in the parietal lobe monitors the response of the body senses.

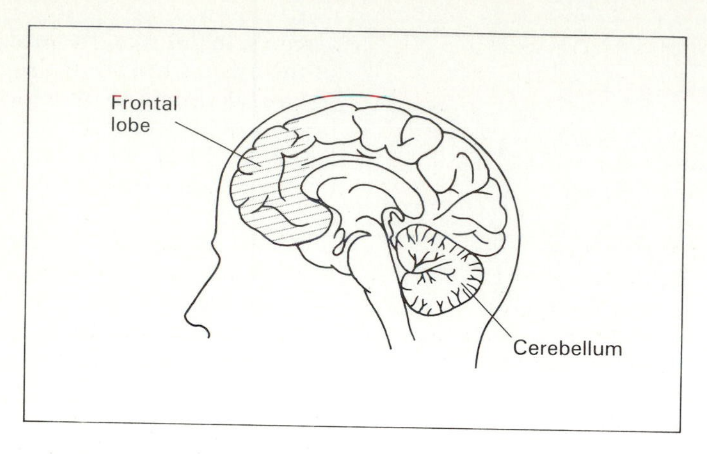

FIGURE 28.5 Matter in the frontal lobe controls the response of the body senses, and often works in conjunction with the cerebellum to fine tune those responses.

FRONTAL LOBE: Body functions are controlled by nerves within the frontal lobe, which lies toward the front of the skull, just behind the forehead. As depicted in Figure 28.5, all voluntary motion begins within a band of nerve cells that arch up over the top of the brain. Movements initiated here include not only those of the arms, legs, neck, and so on, but also the jaw, tongue, lips, and other motions needed for speech. In fact, all 600 muscles of the human body are controlled here.

The cortical matter in the frontal lobe does not work alone. Before sending messages to begin some motion, the brain must be aware of the body's present position. It gets its information through the sensory monitorings of the parietal lobe. The various senses gather data about the body, send it to the parietal lobe, and categorize it. Once this is done, the nerves of the frontal lobe can then send the proper commands for whatever body movement is desired.

In addition to its collaboration with the parietal lobe, the frontal lobe must also work with the cerebellum. Perched underneath both cerebral hemispheres, the cerebellum fine-tunes the various body movements. For example, it coordinates the many muscle activities of a teacher, who must often simultaneously think, speak, walk, and write on the blackboard.

So the frontal lobe works in conjunction with both the parietal lobe and the cerebellum. These coordinated brain activities give humans—and only humans—the ability to use tools, speak clearly, flip a coin, thread a needle, and lower themselves into chairs without falling flat on their face. All these actions require very delicate coordination among many muscles.

Don't get the wrong idea, though. Humans are not the only well-engineered life form. Certain parts of the brain are even more advanced in some lower organisms than in humans. In rats and mice, for instance, the parietal lobe seems to be developed to a remarkable extent, enabling these rodents to navigate through interiors of walls by means of sensitive whiskers.

Finally, the frontal lobe also receives and analyzes the sense of smell. Perhaps the oldest and most primitive sense (at least in the context of brain development), smells pass directly through the nose to the frontal lobe via a small tube jutting forward from the base of each cerebral hemisphere. Smell does not go through the "traffic-director" thalamus as do the other senses.

OCCIPITAL LOBE: Sight does pass through the thalamus. Visual sensations are collected by the eyes, organized by the thalamus, and then projected onto the rear of the brain. Since sight is probably the most delicate and reliable of all the senses, its cortical matter fills all of this third part of the brain, called the occipital lobe. At the rear of the cerebral hemispheres as drawn in Figure 28.6, each nerve cell in this part of the brain responds only to one tiny portion of whatever image is being visually perceived. Together, all the cells in this lobe build up a sort of mental mosaic which forms a picture. Sight, then, is essentially projected through the eyes and is perceived in the back of the brain. The eyes themselves do not see; the brain does.

Human eyesight is nature's best all-around vision. It is superior even to a hawk's in every respect except distance viewing. Human sight is better at that, too, when aided by such technological marvels as a telescope.

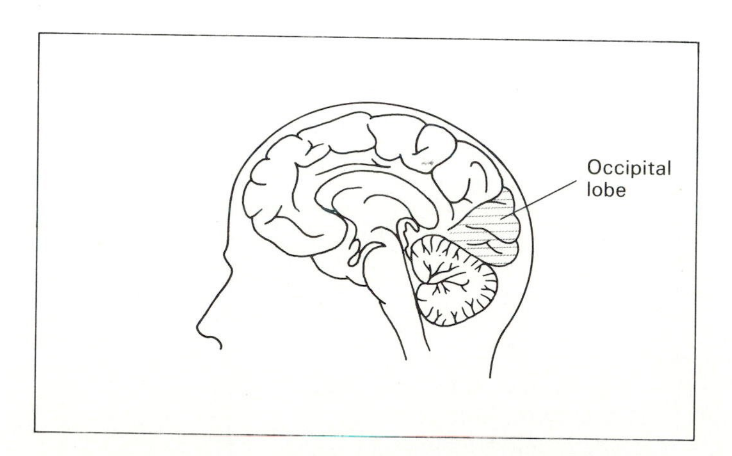

FIGURE 28.6 Matter in the occipital lobe helps form an image as information is projected *through* the eyes onto the back of the brain.

TEMPORAL LOBE: The fourth main part of each cerebral hemisphere is the temporal lobe. Shown in Figure 28.7, it's located at the side of the head, near the temple. Modern research has shown that the advanced function of hearing is controlled here in the temporal lobe. Some memory functions are also thought to be stored here.

Language understanding and general linguistic skills also originate in the temporal lobe. If words are to be read, they are scanned visually and compared with memory. If words are to be written, they are selected from memory in the temporal lobe, after which the frontal lobe commands the hand muscles to go into action. And if spoken language is required, the words are again selected from memory, but this time the frontal lobe activates muscles in the jaw, lips, tongue, and larynx.

AWARENESS: The brain stem, or upper part of the spinal cord, is a further important part of our central nervous system. The most primitive and essential behavior—breathing—is controlled here. Diagrammed in Figure 28.8, the cells in this part of the brain are connected by nerves in the spinal cord to the chest muscles responsible for expanding and contracting the lungs. This is why a broken neck is often fatal and a hangman's noose so efficient; when the nerves are severed, respiration stops.

Fortunately, breathing is prewired and does not require learning. It happens naturally while sleeping. And it occurs automatically after about 4 minutes of breath holding. So it's simply not possible to commit suicide by holding one's breath; the brain needs the vital flow of oxygen, and it apparently can overwhelm human desires to get it.

Alertness is also controlled by nerves in the brain stem. Three types of alertness are known, each located within different parts of the brain. As depicted in Figure 28.8, the lower part of the brain stem controls wakefulness, while the middle and upper parts control sleeping and dreaming.

The nerves in this part of the brain regulate the degree of awareness. The nerves here help to arouse the brain like an alarm clock, alerting it to the importance of incoming information. Like an emergency broadcast, the brain stem can preempt less essential activities of the upper brain in order to concentrate on things that are obviously more important. For example, being aware of fire in a house is clearly more important than what's on television or who's on the other end of the telephone.

The upper brain stem also seems to be the seat of human consciousness. The nerve cells here are affected more than any others by such things as drugs, rituals, and meditation. More and more, researchers are beginning to realize that the degree of alertness must be periodically minimized by means of sleep, meditation, or some other means. In this way, we gain mental refreshment more than physical rest.

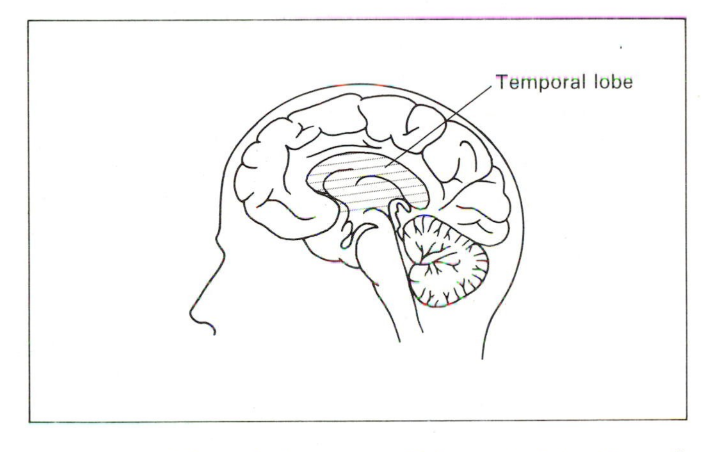

FIGURE 28.7 Matter in the temporal lobe controls hearing and language skills.

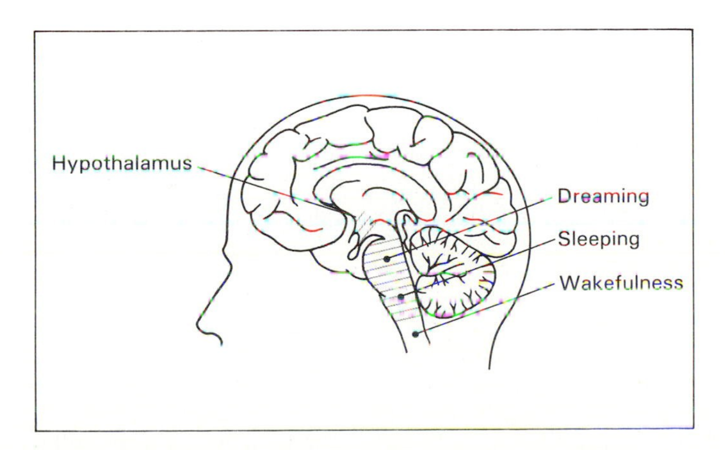

FIGURE 28.8 The upper part of the spinal cord, or brain stem, is the site of several important body functions, including alertness, survival, and possibly consciousness.

SURVIVAL: One final piece of important neural architecture lies just above the brain stem, as shown in Figure 28.8. Here resides a bean-sized region called the hypothalamus. Really another regulator of body function, the hypothalamus has responsibility for survival. In a sense, the hypothalamus is a thermostat, monitoring the temperature of the blood flowing to the brain. Normally kept at 310 kelvin (98.6 degrees Fahrenheit), the hypothalamus can command the body to shiver or sweat in order to maintain this comfortable body temperature.

The hypothalamus generates instinctive behavior aimed at self-preservation. It warns of hunger and thirst. It commands the body's response to stress, including quickened heartbeat, fluttery stomach, cold yet sweaty hands—all those features, for example, that students experience during exams. And, it also generates feelings of pleasure, anger, and fear. More than any other part of the brain, the hypothalamus is the complex center of human emotion.

One final note: Despite these details about our brain, most of the cerebral cortex is yet uncharted. Biologists have thus far been able to specify the inner workings of only about a quarter of the human brain. Somewhere in the unexplored labyrinths reside the superb yet puzzling attributes of thought, creativity, love, and foresight.

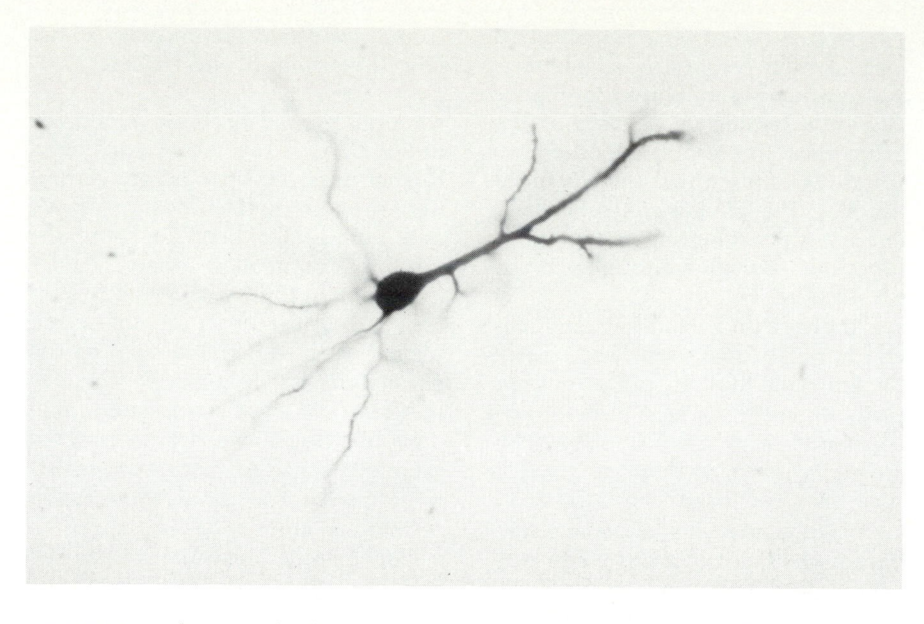

FIGURE 28.9 Photograph of a neuron taken through a microscope. (Scale of the image is about 10^{-2} centimeter.) The axon transmitter (upper right) and the dendrite receivers (lower left, mostly) are clearly evident.

Electrical Activity

As noted earlier, the brain is made of cells just like any other part of the human body. Because the brain is special, these nerve cells are given a special name, **neurons.** Each adult human has at least 10 billion neurons/cubic centimeter.

Like any cell, neurons come in assorted sizes and shapes, although most are microscopic. The bulk properties of each are nonetheless similar; in addition to the main cell body that contains the biological nucleus and manufactures protein in the usual way, neurons also have numerous long and wiry extrusions resembling roots of trees.

Figure 28.9 is a photograph of a neuron taken through a microscope. The cell body is clear, as are the extrusions. The main extrusion on one side of a neuron is called an **axon**; it acts as a transmitter of information, carrying signals away from the cell body. The network of extrusions nearer to the cell body on the neuron's other side are collectively called **dendrites** (from the Greek for "tree"); they act as microscopic antennae, picking up signals sent by other neurons and carrying those signals to the cell body.

Axons and dendrites enable neurons to communicate with one another in order to monitor and control the many diverse functions of an intelligent being. Together, neurons form an intricate network of trillions of intercommunicators, each performing a function either assigned by heredity or learned through experience.

SMART CHEMICALS: How does this communication system work? Much like a series of electrical circuits, each neuron passes along a signal from one place to another. When a neuron is stimulated by some external effect—a touch, sight, sound, smell, or taste—the arrangement of charges of some of the atoms and ions in that neuron changes. This redistribution of charges can suddenly change the voltage of a neuron, hence creating an electrical impulse. Neurons, in effect, act like chemical batteries, discharging rapidly in a burst of electricity. They can then recharge themselves in a fraction of a second. All this electrical activity requires energy—energy that is derived by absorbing oxygen during respiration.

Modern research has shown that the electrical signals travel through neurons with a velocity close to 0.1 kilometer/second. This velocity is much slower than the speed of light. But when 0.1 kilometer/second is equivalently expressed in more familiar units—namely 225 miles/hour—we begin to realize that these electrical signals zip around rather quickly compared to everyday velocities. We can imagine information traveling across a neuron, and from one neuron to another, like a very fast lighted fuse.

Why so fast? Well, speed is essential to get information to the brain and then back to the appropriate muscles in order to respond to the signal. The speed of information really depends on the diameter of the neuron. Some life forms that require extremely quick responses in order to escape their predators, and hence to survive, have developed much thicker neurons. Sea squids, for example, have neurons with diameters about 100 times those in humans, enabling them to coordinate movement away from a site of danger or toward a source of food with a system resembling jet propulsion.

Don't get the impression that all the neurons in our skulls are physically wired together. In fact, they are not; none of them are. Instead, a small gap, called a **synapse**, sepa-

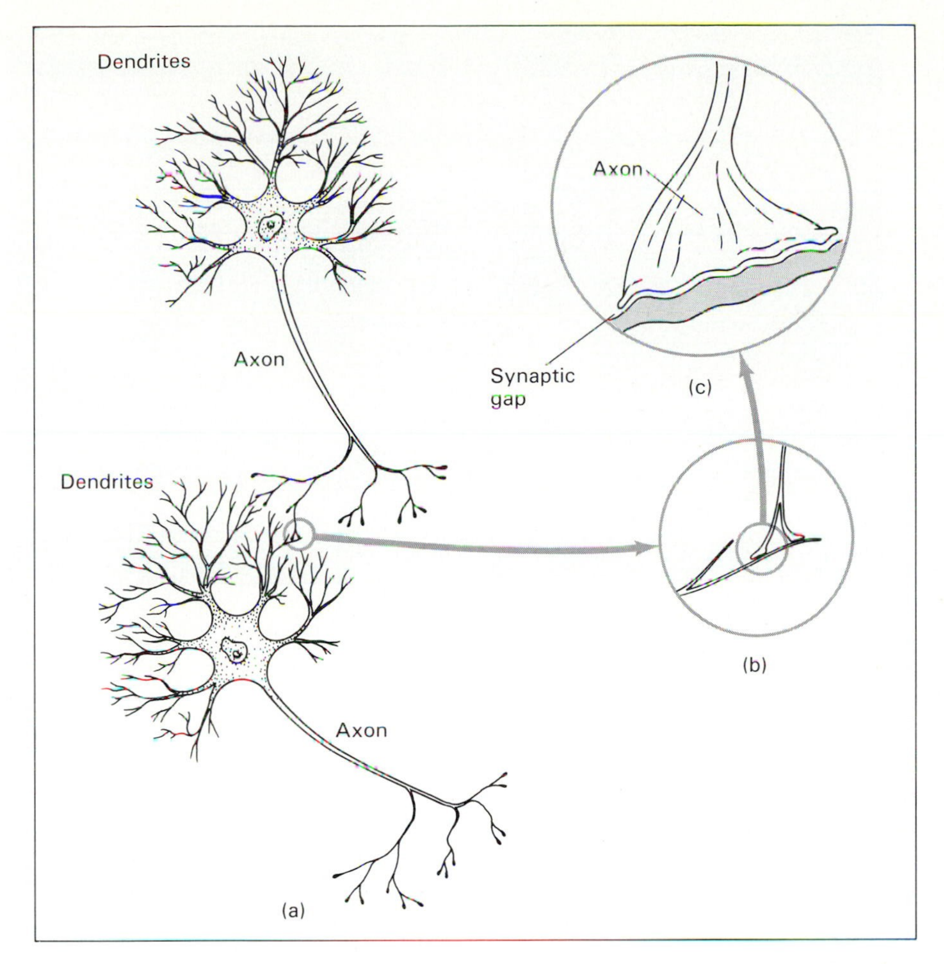

FIGURE 28.10 The axon of one neuron is not physically connected to the dendrites of another. Instead the neurons are separated by a microscopic synaptic gap across which chemicals are secreted. Frame (c) is a substantial magnification of the neuron interaction depicted in frame (a).

rates an axon of one neuron from a dendrite of another. Magnified many times in Figure 28.10, these synaptic gaps measure no more than about 0.00002 centimeter, or 500 times thinner than a human hair.

For two neurons to communicate, information must jump the gap between the axon transmitter of one neuron and the dendrite receiver of another neuron. However, this information is not passed along by transmitting electrical impulses across the synapse. Instead, the electrical impulse traveling the length of an axon induces that axon to excrete chemicals, known as neurotransmitters. The chemical then spreads across the synapse and causes a new nerve impulse to begin in the next neuron.

About a dozen neurotransmitting chemicals are now known. Each can, under certain circumstances, inhibit or enhance the voltage on a nearby dendrite. Consequently, this type of nonconnected wiring scheme provides enormous complexity—much, much more than would be possible if the neurons were physically connected to one another. Each neuron can have as many as 200,000 synapses, each of which might or might not cause an electrical impulse in a given circumstance. And since there are estimated to be approximately 500 trillion synapses in a typical human brain, the number of possible routes any electrical impulse can take is mindboggling—no pun intended!

Experiments have proved that the electrical impulses actually travel along a thin covering outside each neuron. Made of a fatty white substance called the myelin sheath, this covering apparently serves as insulation, much like the rubber wrapping around ordinary electric wires, preventing the network of neurons from short circuiting. Unfortunately, the myelin sheath is destroyed for some individuals; the disease called multiple sclerosis attacks myelin, allowing some short circuits to occur and causing jerky movements due to a lack of coordinated timing. In fact, it is the complete lack of myelin in newborn infants that prohibits their neurons from firing in a coordinated way; the result is a gradual ability to crawl and then walk, while the myelin grows during the first year or so after birth.

The neurotransmitting chemicals, then, determine how a person responds to any given circumstance. In a certain sense, the excretion of these chemicals dictate human behavior.

A final note of caution: The natural chemicals of our brains can be affected by what we breathe and eat. Poisons and drugs in particular—strychnine, tranquillizers, LSD, amphetamines, marijuana, and many others—change the brain's firing mechanisms, thus changing human behavior. Even caffeine, in coffee-cup doses, lowers our synaptic thresholds so that a tired nervous system, although mostly depleted of transmitters, can keep us alert just a little longer. Thus the chemistry of the synapse seems to be the key to many human ills, although biologists are really only beginning to explore these mysteries.

Development of Intelligence

Having gained an appreciation for our human brains, we now return to discuss how simpler life forms might have evolved the complexity inside our heads. In this section we speculate about the paths along which intelligence originated and developed. As for other aspects of biological and cultural evolution, our story relies heavily on the fossil record, as well as on studies of animal behavior.

The one-celled amoeba is among the most primitive forms of life known in today's world—perhaps the most primitive excepting the virus, which sometimes acts alive, as noted in Interlude 25-1. Roughly halfway in size between an atom and a human, the amoeba has poor "awareness" and coordination. It generally responds only at the point stimulated, communicating the information sluggishly through the rest of its body. Although amoebas have developed a crude nervous system, living systems that aspire to be more agile surely need quicker internal communication.

Other single-celled creatures have succeeded in evolving primitive intercom systems. For example, as shown in Figure 28.11, the microscopic paramecium has an array of oar-like hairs that enable it to move rapidly through water. The "oars" must act in a coordinated way, for if they functioned independently, the paramecium would never make any progress. In fact, the hairs are regulated by microscopic nerves that respond to a chemical emitted within the cell. In this way, messages can be transmitted swiftly and precisely from one part of a cell to another.

Paramecia clearly have more "intelligence" than amoebas. A paramecium has better coordination, and it also has a memory of sorts. An amoeba searches for food by drifting into water-plant algae; if it finds none, it often repeatedly gropes toward the same alga, even though that alga has no satisfactory food for the amoeba. The amoeba has no memory. A paramecium, on the other hand, having found no food near one alga, will back off and seek resources in a different direction—it retains momentary traces of experience.

Compared to the amoeba, then, the paramecium is a genius. But it's a genius operating in a watery world less than a few millimeters in extent. Paramecia are unaware of anything beyond. No single-celled creature can be much smarter, for it can develop no further.

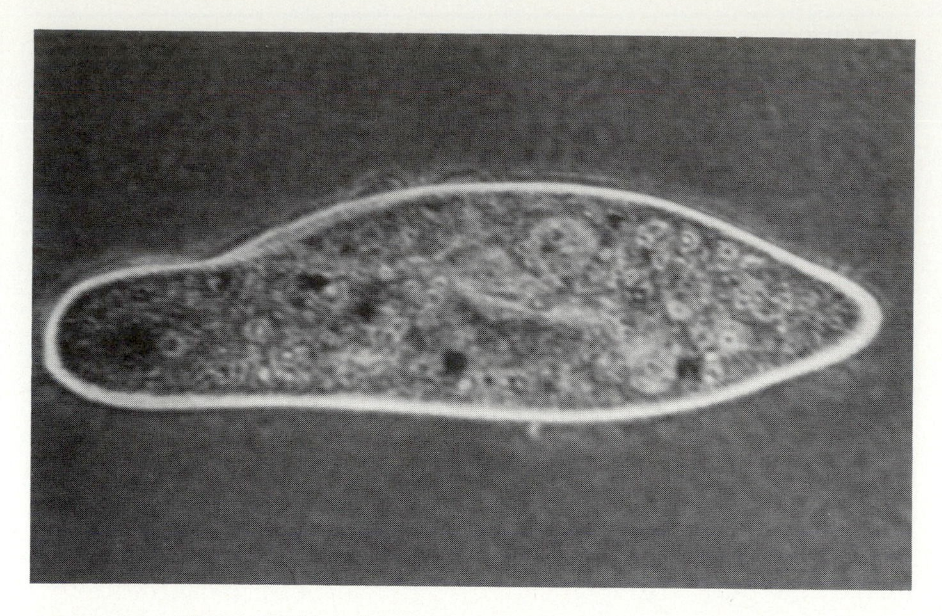

FIGURE 28.11 The paramecium has a primitive intercom system that helps it coordinate movement.

Despite the biochemical complexities of a single cell, it can develop only the simplest intelligence. To become smarter—to evolve an intricate nervous system—a single cell would require elaborate sense organs to inform it, as well as developed muscles to carry out its instructions.

Why can't there exist larger cells having these additional luxuries—perhaps equipped with miniaturized hands, eyes, and brain? The answer is that single cells cannot become much larger than the 0.0001-centimeter creatures described above. Should they try to do so, their surface areas would increase with the square of their size (1, 4, 9, 16, . . .). But the masses of the cells, which must be fed through the cells' membranes, would increase as the cube of their size (1, 8, 27, 64, . . .). Thus cells cannot become too large, lest they starve themselves.

Single-celled life forms, then, cannot become very smart. Their physical size prevents them from developing the many and complex organs needed for higher intelligence. Mutations have undoubtedly helped them try every conceivable means to do so for the past 3 billion years, but they have failed.

MULTICELLS: The road to greater intelligence required many cells. But not just a haphazard accumulation of many independent cells; clusters of millions of independent cells are no more intelligent than one. Consider a sponge, for example, much like those harvested for use in our bathtubs. Although a sponge is clearly multicellular, most of its cells act independently. A sponge has no central nervous system, and thus is really not much more "intelligent" than an amoeba. For some reason, sponges failed to profit by their multicellularity. As a result, they have produced no higher forms of life. The sponge is an example of a life form that long ago become an evolutionary dead end. A favorable mutation was needed to allow an accumulation of many cells to work together as a community.

Interactive, multicellular organisms have some clear advantages, not the least of which is that they avoid the surface–volume problem mentioned above. More important, groups of cells within a multicellular organism can develop specialized functions. This division of labor was one of nature's greatest inventions. One group of cells can become highly sensitive to foods; others become efficient in carrying oxygen; still others become tough muscular entities or protective skin casings. The net result is that each group of cells within a multicellular organism becomes more skilled in one specialty and less so in the rest. Accordingly, the total intelligence of an organism is vastly increased as cells, working as a team, become better able to protect themselves from predators and to obtain the food needed for survival.

The organism called hydra is a good example of a multicellular system that did evolve considerable intelligence. Sketched in Figure 28.12 and no larger than a toothpick, the modern hydra resembles a stalk of celery, being closed at the lower end and raveled into writhing tentacles at the upper end. In contrast to any sponge, a hydra can move its entire body in coordinated fashion to, among other things, avoid danger and seek food. In short, the cells within a hydra can communicate. Indeed, communication is the essence of organized intelligence.

Cells able to communicate—nerve cells—probably formed originally near the surface of multicellular life forms such as hydra. (More realistically, those life forms were hydra-like ancestors.) Being exposed, these cells had the greatest opportunities to sample their environment. But being near the surface also made them more vulnerable.

Thus natural selection apparently favored those hydra-like ancestors having deeply rooted nerve cells. Over the course of generations, these cells gradually retreated beneath the surface, but also kept their link to the environment by sprouting expendable tentacles. As suggested by Figure 28.13, these tentacles act as remote sensing devices to reach the body surface and beyond. These miniature octopus-like tentacles are the dentrites of modern neurons. (There are basically no differences in the neurons of modern hydra and modern humans, although the number and arrangement differ considerably.)

As evolution proceeded, the bulk of the neurons themselves retreated ever deeper within the multicellular organisms. Eventually, the buried neurons merged, thus forming a coagulated bunch of nerve cells. This was probably the first and most important step in the development of a central nervous system. Neuron clustering was one of the greatest of all evolutionary accomplishments. Once that barrier was crossed, sometime around a billion years ago, our hydra-like ancestors, as well as other sophisticated organisms like them, were on their way to generating all of Earth's brainy animal life forms, including humans.

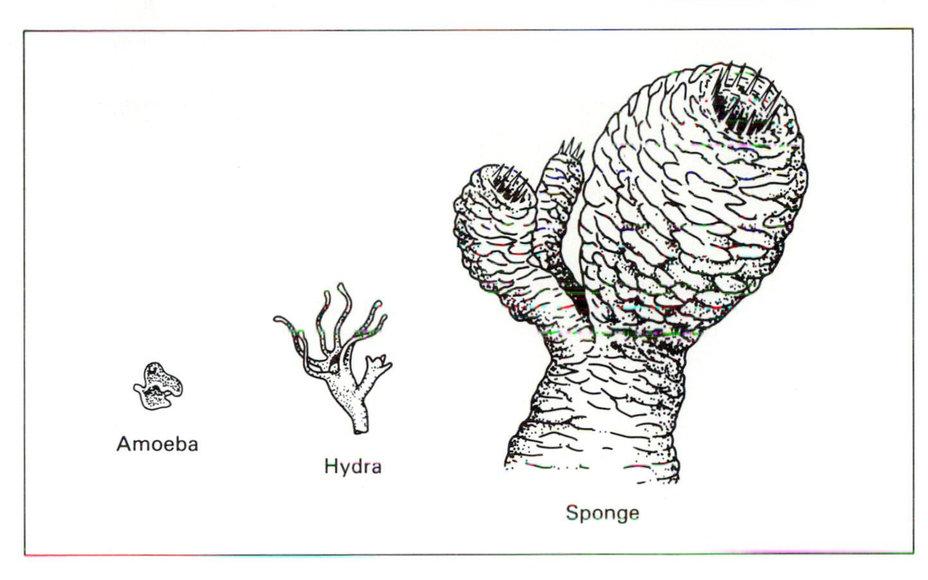

FIGURE 28.12 Hydra organisms possess considerable coordination. A hydra is shown here for comparison with an amoeba and a sponge.

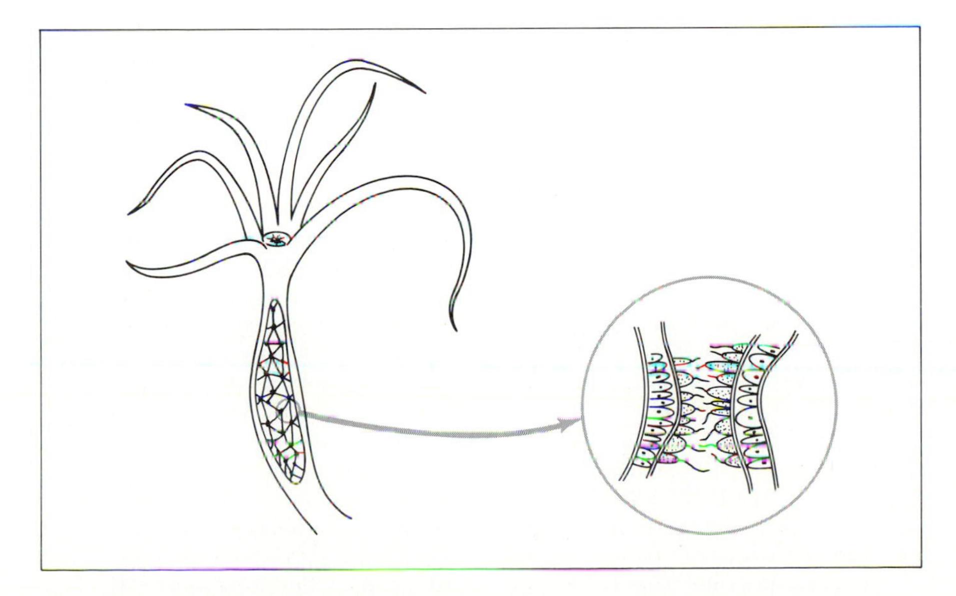

FIGURE 28.13 Mutations and natural selection caused the nerve cells to gradually retreat below the surface of early organisms. In turn, the clustering of such nerve cells led to a primitive central nervous system.

THE FIRST STEP, INVERTEBRATES: Let's take a brief look at what the fossil record says about the evolution of the brain. Figure 28.14 is an evolutionary diagram for the central nervous system. As shown, evolving organisms branch out in nearly every available direction, moving steadily toward greater complexity.

Most of the branches of Figure

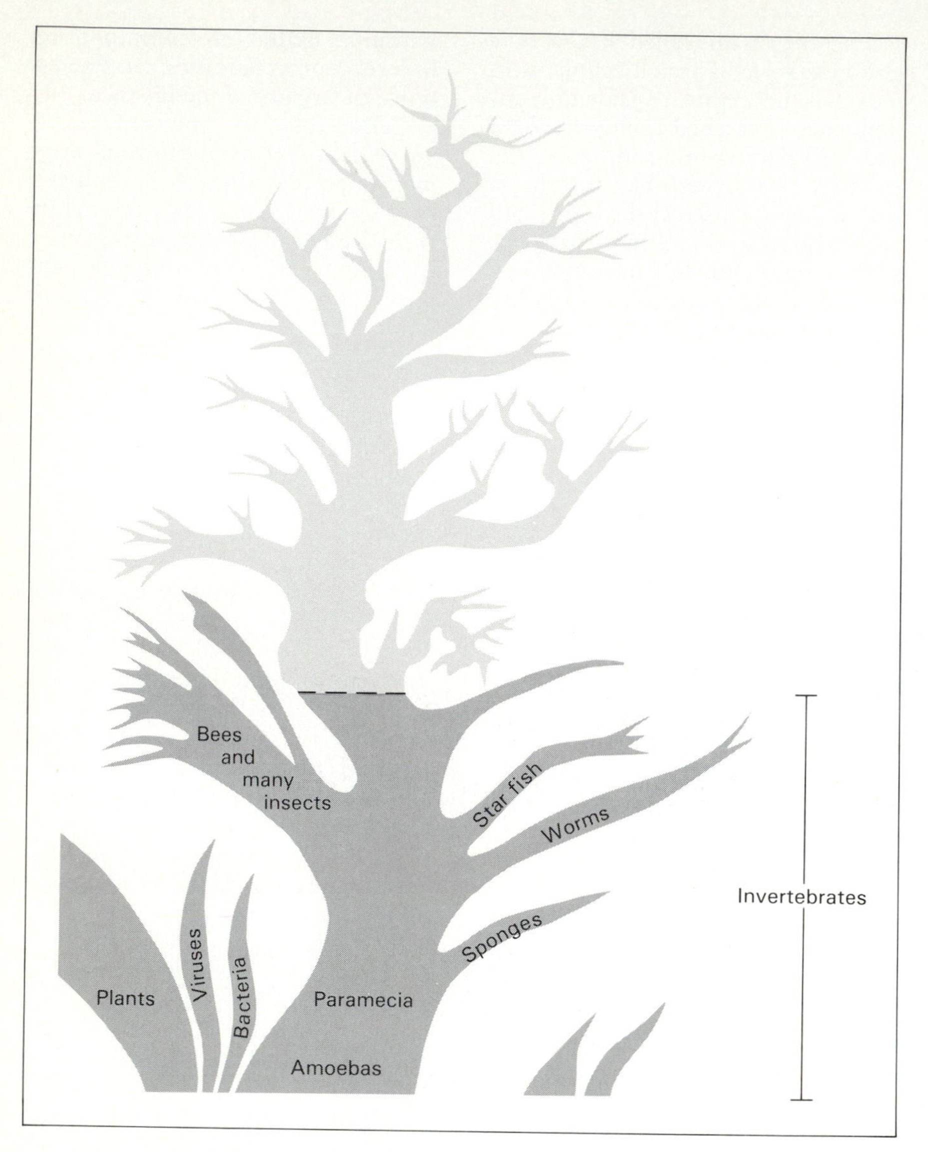

FIGURE 28.14 Schematic diagram of some early pathways that led toward greater neural complexity.

28.14, however, represent organisms that either became extinct long ago or survived only as dead ends. Apparently, they met some insuperable biological obstacle to progress at some point in their development. Some that persist to this day without progress include amoebas, paramecia, sponges, and worms of all sorts. These are the **invertebrates** or backbone-less organisms.

Many invertebrates are skilled and crafty in their domain. Spiders, for instance, are marvelously accomplished performers within their own environment; their nervous systems are complex and effective, and their sense organs are even more varied and subtle. Bees, wasps, ants, and moths also have superb bodies for dealing with their world. Some—especially bees and ants—even have sophisticated social organizations that rely on symbolic communication.

Virtually all these invertebrate animals have reached evolutionary dead ends. They are trapped in endless cycles of perfected day-to-day routines. Fossilized spiders of 100 million years ago show little variation from their modern descendants. In a sense, bees buzzing around the bushes are living fossils.

Invertebrates are simultaneously successes and failures. They are fabulously successful within their own small environment. The deerfly, for example, is reportedly the fastest animal; the flea can jump 100 times its own height. Successes for sure, for the invertebrates dominated Earth for nearly $\frac{1}{2}$ billion years. But failures too, because they neglected to develop the vertebral column of bones so conspicuous in fish as well as humans—bones that form the spinal column and skull.

As humans, we tend to take brains for granted. But the vast majority of animals are invertebrates, and thus can have no true brain, no *centralized* nervous system. As such, they cannot be creative, adventurous, or visionary—at least not as we have come to know these qualities.

THE SECOND STEP, VERTEBRATES: Humans and our fellow **vertebrates** (backboned fish, reptiles, and mammal relatives) are an exception to the great invertebrate failure. In a sense, vertebrates are a minor offshoot from the vast, teeming world of invertebrates. It seems that brains are the exception, not the rule.

Figure 28.15 shows some results of fossil studies of the vertebrates. Like the invertebrates, many vertebrates were apparently unable to utilize their sensory and motor organs to full capacity. As noted in the figure, a vast array of fish, amphibians, and reptiles, including modern versions of many birds, lizards, snakes, crocodiles, turtles, and many other vertebrates, dead-ended during the past $\frac{1}{2}$ billion years. A good number became extinct, and even the survivors seem to have been unable to decide on a division of authority between the "sight" and "smell" neurons.

Figure 28.16 is a sketch of the brain of a primitive fish, reconstructed from details left in the fossil record. Such fish lived several hundred million years ago and are among the simplest known true vertebrates. Although clearly crude, their brains nonetheless contained all the essentials present in modern fish as well as humans. Note the concentration of smell organs, causing a bulge toward the snout; these are the precursors of the much larger

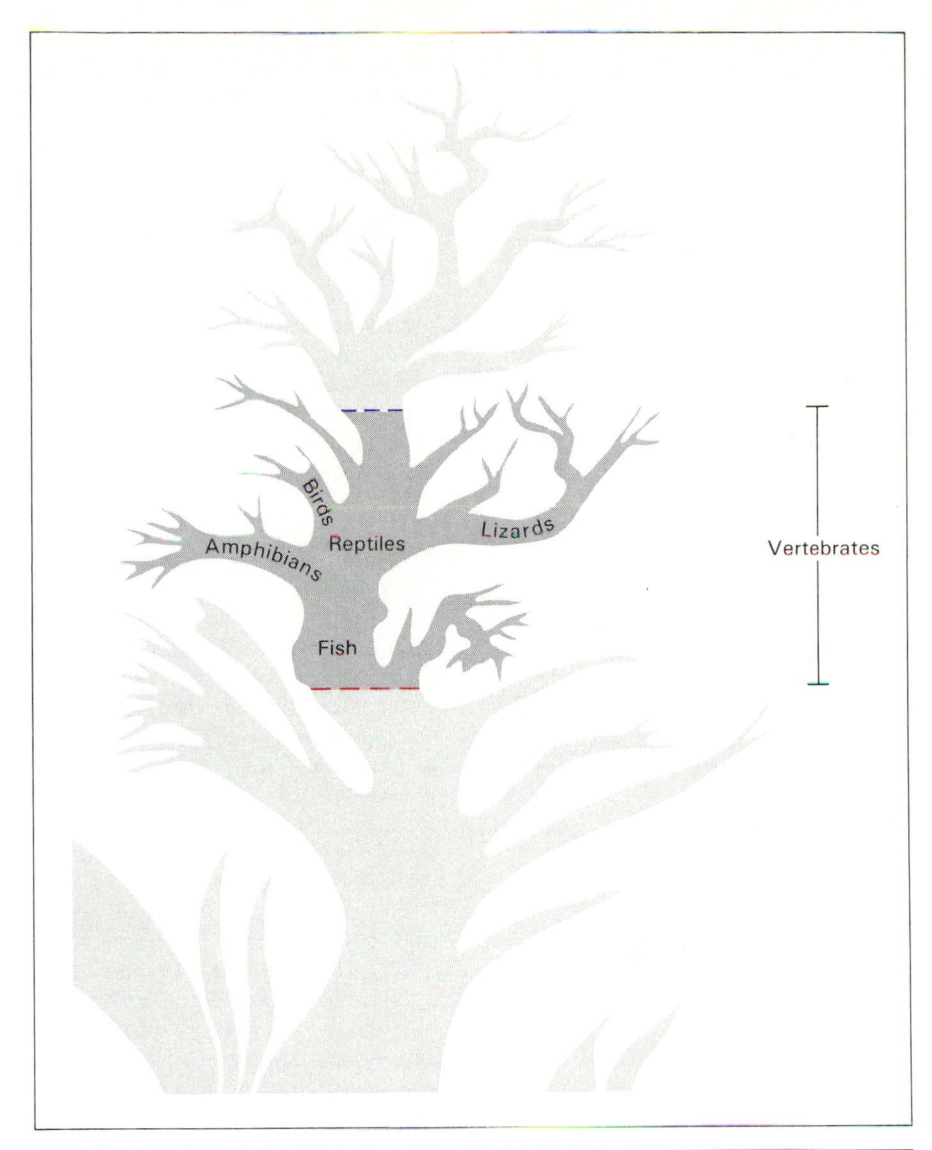

FIGURE 28.15 Schematic diagram of the neural development among vertebrate animals.

cerebral hemispheres in humans. Similarly, their eyes created another bulge farther back, precursor of our occipital lobe. Also shown are the lateral-line organs branching out to the side, precursors of our cerebellum. These ancient sense organs, although not in themselves as good as those of some modern invertebrates, were much more effective because of their connection to a unified central nervous system.

The development of specialized sense organs, and especially of their integration with a centralized brain, contributed to the intellectual dominance of the vertebrates over the invertebrates.

Sight certainly played a major role in the advancement of these early vertebrates. As depicted in Figure 28.17, a relatively large sight brain developed with time. Mutations simply gave an advantage to certain species of fish, enabling them to utilize improved eyesight to move, survive, and reproduce better in the water. The sense of sight did not rule unchallenged, however. The sense of smell remained a keen rival in the ever-refining development of Earth life several hundred million years ago.

This competition between sight and smell continued with time. When the amphibians transferred from the sea to the land, the flood of sight data probably overwhelmed even the sophisticated brain of these relatively advanced vertebrates. Smell input, on the other hand, was still within the grasp of such a brain. Accordingly, the first amphibians found smell to be of more practical use than sight. The fossil record suggests in fact that the occipital lobe shrank, while the cerebral hemispheres expanded with the march of

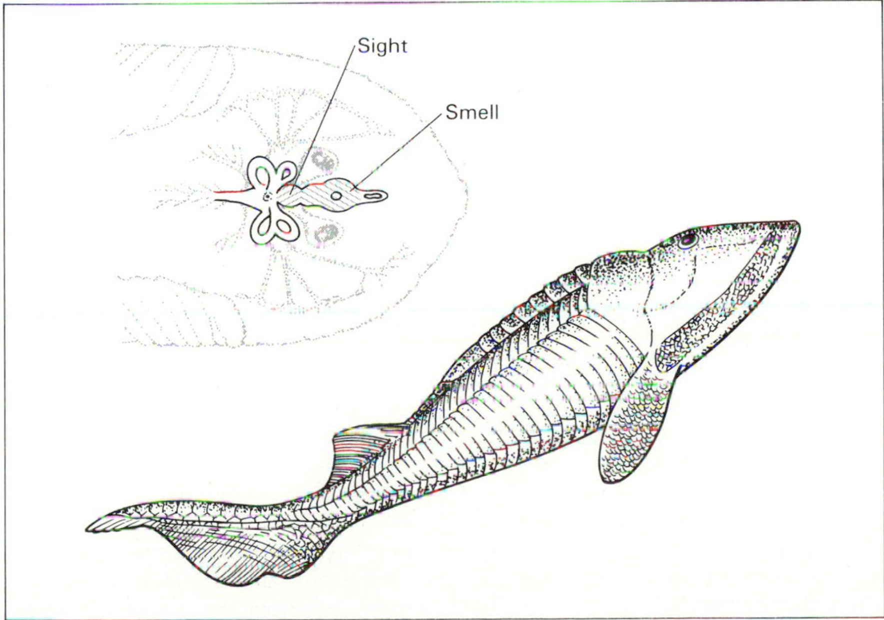

FIGURE 28.16 Schematic diagram of the brain of a primitive fish, the simplest known form of a true vertebrate. Such brains are reconstructed from fossils, and this one here highlights the brain matter devoted to the smell and sight senses.

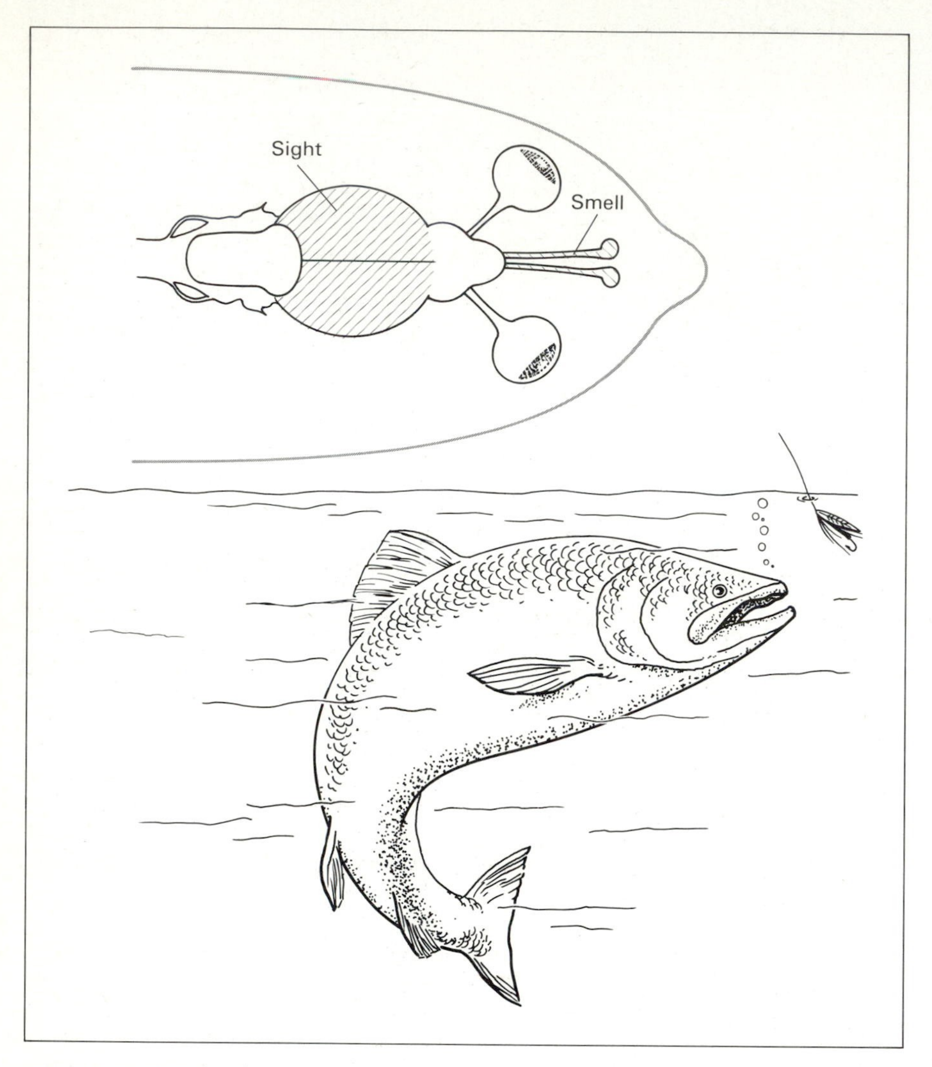

FIGURE 28.17 Sight played a major role for more advanced fish, thus enlarging the amount of brain matter devoted to that sense. Even so, the tubes providing a sense of small to the brain were still present in these fish of a few hundred million years ago.

time. Gradually, the sense of sight regained greater usefulness as the brain of the mammals grew larger through continued mutations and natural selection. The larger brains of the mammals were then able to cope with the full world of sight as well as sound. Those creatures having more advanced brains were thus better suited to survive in a changing terrestrial environment.

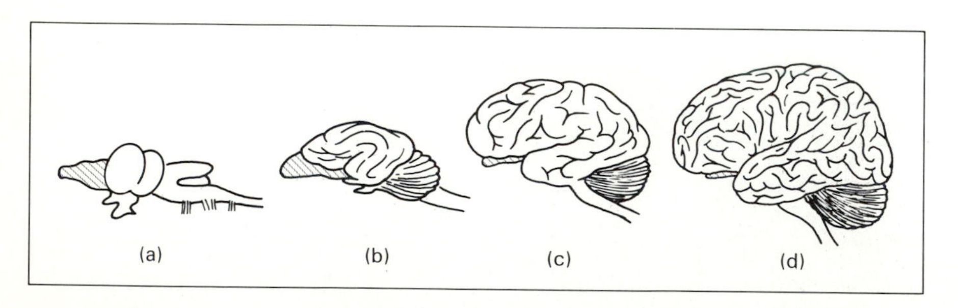

FIGURE 28.18 The sense of smell (shaded brain matter) declined in importance as life evolved toward more advanced forms. Shown here are schematic diagrams of brains for (a) a reptile, (b) a dog, (c) an ape, and (d) a human.

Figure 28.18 documents the decline of the importance of smell. As shown, the sense of smell was of greatest value to the lower vertebrates. But as the brain increased, other senses, such as seeing and hearing, became equally advantageous, and eventually more advantageous. Our much larger human cerebral hemispheres are indeed derived from the ancient smell brain, but the importance of this sense was long ago surpassed by sight, sound, and other general sensations.

THE THIRD STEP, MAMMALS: The final step in the evolution of the brain occurred in the mammals. Depicted schematically in Figure 28.19, there were once again many evolutionary dead ends—mutations that simply did not give much advantage to some species. However, other mutations did alter for the better the traits of some organisms, thus granting them a distinct advantage in the overall struggle for survival.

Figure 28.20 shows some examples of how favorable mutations, over millions of years and generations upon generations, caused some mammals to develop longer arms and gripping paws for leaping, swinging, and reaching food. Other mutations also gradually shifted the eyes of the early prosimians from the side of the head to the front of the head, thereby producing binocular, stereoscopic vision. The shortening of the snout and the movement of the eyes toward the front granted the early primates an overlapping field of view and thus three-dimensional vision. Likewise, the eventual development of longer arms and more dexterous hands combined with the more accurate eyesight to give these pre-monkey ancestors distinct advantages in the struggle for survival.

All the while, the cerebral cortex expanded, enabling mammals to develop increasing sense sophistication. The combined eye–hand–brain system was a powerful tool, not only for enhancing survival, but also evolving intelligence. The (current) culmination is the seemingly

FIGURE 28.19 Schematic diagram of the evolutionary pathways taken by various mammals. This increasing neural sophistication has granted an ever-better understanding of the Universe.

limitless manual skills of *Homo sapiens*.

Brain and Body Mass

ABSOLUTE BRAIN SIZE: As noted earlier, all modern humans—*Homo sapiens*—have brain capacities of about 1300 cubic centimeters. Some variations arise from person to person, but people known to have brains as small as 1100 and as large as 2000 cubic centimeters do not seem to show any clear behavioral differences.

On the other hand, most mental patients who have reduced intellectual abilities do have much smaller brains. Often measuring 500 cubic centimeters, the brain capacity of these mentally retarded adults roughly equals that of a normal 1-year-old child. Apparently, the brain mass can be so tiny that its function is much impaired, suggesting that a minimum brain mass is needed for "adequate" human intelligence as we know it. Once this threshold—probably around 1000 cubic centimeters—is surpassed, normal human behavior is possible.

What about our ancestors? Can we use the fossil record to estimate the brain size of some of the prehumans that paved the way for our existence? The answer is yes. Anthropologists have been able to sketch a rough outline of the recent evolution of the brain. They do so by estimating cranial volume of the fossilized skulls of our immediate ancestors, assuming that, as is now true for humans, apes, monkeys,

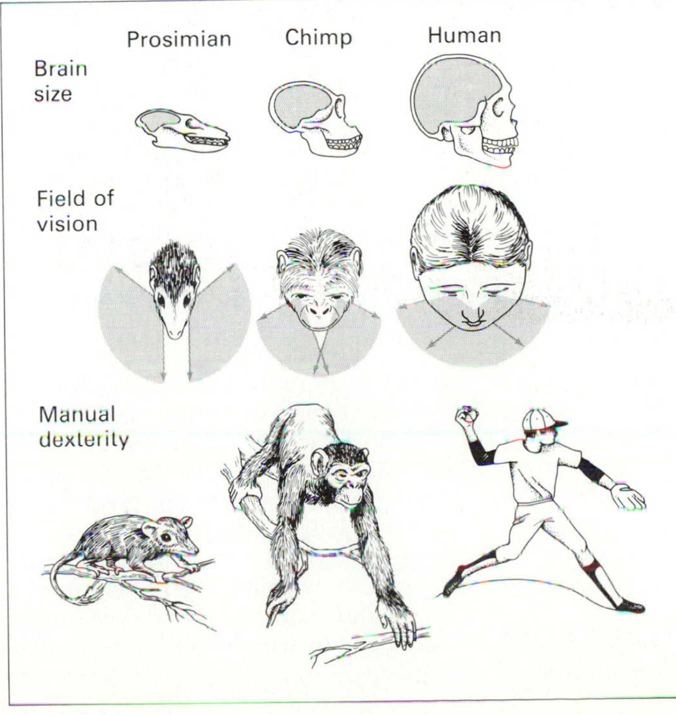

FIGURE 28.20 Favorable mutations, over long periods of time, gradually helped move the eyes from the side of the head to the front, change paws into hands, and lengthen the arms. The fossil record proves that the brain also got bigger.

TABLE 28-1
Comparison of Brain Volumes of Recent Ancestors

APPROXIMATE TIME AGO (MILLIONS OF YEARS)	SPECIES	AVERAGE BRAIN VOLUME (CUBIC CENTIMETERS)	BRAIN MASS / BODY MASS
3	*Australopithecus boisei*	500	0.011
3	*Australopithecus africanus*	500	0.020
2	*Homo habilis*	700	0.016
1	*Homo erectus*	1000	0.015
0	*Homo sapiens*	1300	0.022

and other mammals, the brain matter nearly fills the skull. Table 28-1 lists the results of these studies.

The partly bipedal and prehuman australopithecines of 3 million years ago had brain volumes averaging 500 cubic centimeters. This is comparable in size to the brain of a modern chimpanzee and is about a third the size of today's average human brain. Thus fossil evidence supports the idea that our ancestors could walk on two feet before they evolved large brains.

The first true men, (perhaps) *Homo habilis* of 2 million years ago, had definitely larger brain volumes. Fossil studies show that this ancestor was fully bipedal, and possessed an average brain volume of some 700 cubic centimeters. Not only that, but their fossilized skulls have a distinctly different shape from that of their forebears. Developed substantially were the frontal lobe behind the forehead and the temporal lobe above each ear, those brain regions regarded as sites of speech, foresight, curiosity, and surely many other behavioral qualities. Coupled with the fact that these ancestors were fully bipedal, the possibility that they might have made primitive tools implies at least two significant

INTERLUDE 28-1 ***Consciousness***

The epitome of culture may well be the ability to seek the truth about ourselves and our Universe. Who are we? Where did we come from? How do we fit into the cosmic scheme of things? These are some of the most fundamental questions humans can ask themselves. We are inquisitive, wondering creatures. And in this respect—the desire to seek the truth—we are unique among all living things.

The ability to wonder, to introspect, to abstract, to explain—the ability to step back, perceive the big picture, and study how we relate to that picture—is our most amazing quality. Just what is it that allows us, even drives us, to inquire as we have done in this text, to ask the basic questions, and to attempt to find answers to them? The answer seems to be embodied within that hard-to-define refinement of the mind called **consciousness.** Consciousness is that part of human nature that permits us to wonder.

Although not well understood, consciousness is often attributed not to matter, not even to living systems, but only to a special form of life—humans.

How did consciousness originate? When did humans become aware of themselves? Is consciousness a natural consequence of neurological evolution? Some researchers think so, but they can't yet prove it. Others object, suggesting that some specific, perhaps unlikely mechanism is required for its development; they argue that the capacity for imagery and imagination represents something more than just a continued accumulation of neurons.

Records of ancient history are sketchy, making it difficult to document the onset of self-awareness. Some researchers contend that consciousness, as we know it, is not manifest in ancestral records until about a few thousand years ago. This is about the time when some of the writings in ancient texts became abstract or reflective. Thus people were apparently wondering about themselves several thousand years ago, but it's unclear if this was the first time they began to do so.

If consciousness did originate that late in time, we must be prepared to assume that cultures can become highly refined without developing personal consciousness. Our historic ancestors would have had to invent just about every cultural amenity except consciousness, and to have lived until quite recently in a dreamlike, essentially unconscious state.

Other researchers argue that human consciousness developed long ago. If modern chimps which display rudimentary self-awareness, do, in fact, mimic our australopithecine forebears, then consciousness could have been a factor millions of years ago. That it evolved several tens of thousands of years ago, at about the time of the invention of the bow and arrow, is another popular assertion. Indeed, the development of the long-distance weapon could possibly be viewed as that giant step finally granting humans the freedom to innovate, to begin to evolve culturally, to wonder.

A compromise idea can possibly reconcile these two seemingly divergent views. For example, prehistoric humans may have developed a crude sense of consciousness as early as a million years ago, but only recently did they become sophisticated enough to reveal that sense of wonder and self-awareness in their writings.

changes in behavior—toolmaking and bipedalism—were accompanied by equally significant changes in brain volume. Whether this increased behavioral sophistication caused a larger brain or merely resulted from it is another one of those chicken-or-the-egg problems. At any rate, the fact that bipedalism freed the hands for tasks other than walking does suggest a causal link among upright posture, toolmaking, and brain size.

The fossil record also suggests that our closest relative, *Homo erectus,* had a brain volume just a bit less (roughly 1000 cubic centimeters) than those of many humans alive today. Large and small circular arrangments of stones found alongside the fossilized remains of this species suggest, furthermore, that our ancestors of a million years ago had domesticated fires and constructed homes outside of caves.

Comparisons of various cranial capacities listed in Table 28-1 clearly suggest that the human advances made in the last few million years are at least partly related to an increased total brain mass. New behavioral functions, increased neural specialization, and improved cultural adaptations apparently characterized the steady evolution from *Ramapithecus* through *Australopithecus* onward to *Homo habilis* and *Homo erectus,* currently culminating in *Homo sapiens.*

RELATIVE BRAIN SIZE: The absolute size of the brain is important, but it cannot be the sole indicator of intelligence. Small-bodied creatures such as spiders obviously have very small brains, whereas larger-bodied creatures such as elephants have much larger ones. Yet in many respects spiders act "smarter" than elephants. This is probably because spiders have a lot less body to monitor and control than elephants. In fact, much of the elephant's large brain consists of motor cortex—enormous numbers of neurons devoted to the process of enabling these huge hulks to put one leg in front of the other without tripping. Hence a better measure of intelligence is probably the *ratio* of brain mass to body mass.

Figure 28.21 compares the brain masses and body masses of various animals having similar overall size. (A given brain-to-body-mass ratio would display a diagonal line from lower left to upper right.) As shown, a clear gap separates reptiles from mammals. For any given body mass, mammals have a consistently higher brain mass. Brains of mammals are 10 to 100 times more massive than those of modern reptiles of comparable size. Similarly, the brain sizes of our prehuman ancestors (the early primates) also seem to have been more massive, relative to body mass, than those of all the other mammals.

Of all the organisms plotted in the figure, *Homo sapiens* is the creature having the largest brain mass for its body mass. The dolphins come next, followed by the apes, especially the chimpanzee.

We conclude that the brain-to-body-mass ratio provides a useful index of the intellectual capacity among different animals. The systematically different ratios of Figure 28.21 suggest that the evolution of mammals from reptiles about 200 million years ago was accompanied by a major increase in relative brain size and intelligence. These ratios furthermore suggest that the evolution of human-like creatures from the rest of the mammals a few million years ago was accompanied by an additional development of the brain.

Artificial Intelligence

Humans are currently the greatest intelligence on Earth. We are the only species able to communicate culturally and to construct technologically. We are the only ones capable of knowing our past and worrying about our future. Just how wise we are, however, is an issue of much debate.

Numerous suggestions have bee ı made that our days of intellectual dominance are coming to an end. "Smart" machinery—exemplified by the computer—already surrounds us. Is it only a matter of time before silicon-based computers surpass or even replace carbon-based humans? Might **artificial intelligence**—problem solving and deci-

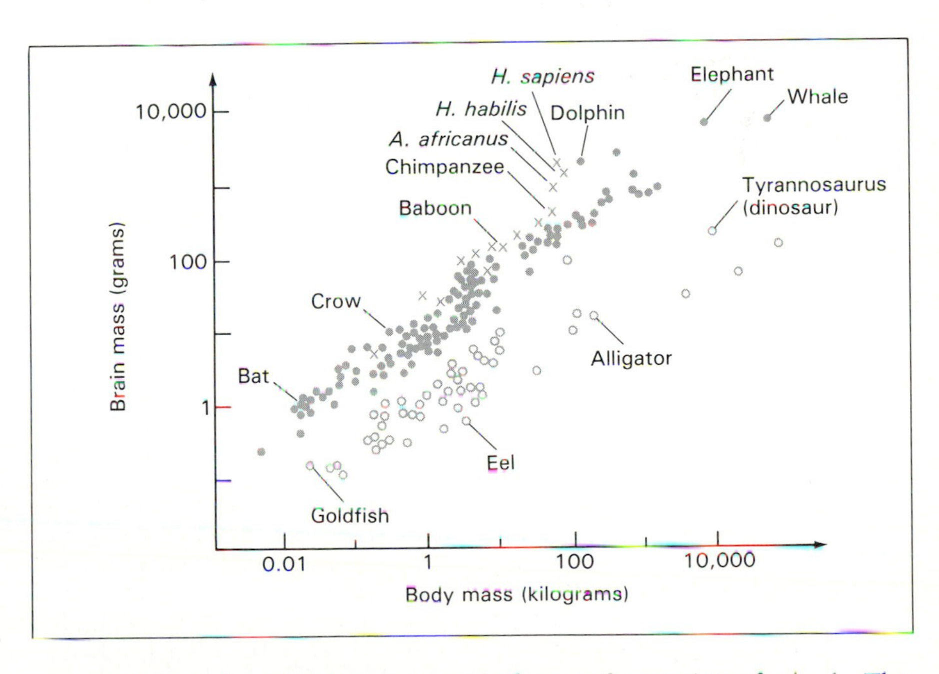

FIGURE 28.21 A plot of brain mass versus body mass for a variety of animals. The open circles represent reptiles (including some fish and dinosaurs), the filled circles represent mammals (including many birds), and the x's represent primates (including humans and their immediate ancestors).

INTERLUDE 28-2 ***How Smart Are the Dolphins?***

More than any other property, the brain most clearly distinguishes humans from other life forms on Earth. The development of speech, the creation of civilization, and the invention of technology are all products of the human brain's sophistication. But what about other forms of life? Are there other creatures now on Earth with comparable sophistication—animals having enough brain power to enable them to communicate, make tools, or fashion a society?

Many researchers argue that, next to humans, dolphins are the smartest animals now on Earth. Dolphins, along with whales and porpoises, are members of a family of mammals whose ancestors were once land-dwelling. Perhaps because of the keen competition among many amphibians 100 million years ago or so, the dolphins' ancestors returned to the sea, probably in search of food. Surely, they would have experienced some disadvantages in making such a seemingly backward move, but that ancestral decision probably saved them from extinction.

Dolphins, as we know them today, are well adpated to the sea. Their exceptionally strong bodies are streamlined for deep diving and speedy locomotion. They have extraordinary hearing, as well as an uncanny sonar system providing them with a kind of underwater vision; this advanced system of echo location, now being studied by human naval officials for military purposes, may employ an acoustical type of Doppler effect to map out the position and movement of objects in their environment.

Dolphins travel in schools or families, have a well-organized social structure, and assist each other when in trouble; some females help other dolphins give birth. They are not at all hostile, being abnormally friendly to other dolphins as well as to humans. In fact, dolphins seem to be the exception to the unwritten rule that all friendly species are inherently aggressive as well.

It's not easy to gauge dolphin intelligence. The criteria for human intelligence, controversial in their own right, can hardly be expected to apply to other species. Although the average dolphin brain is 1700 cubic centimeters, larger than a typical human brain, laboratory tests imply that dolphin intelligence lies somewhere between that of humans and that of chimpanzees. Indeed, as shown in Figure 28.21, dolphin brain-to-body ratios average 0.018, the closest value to humans for any non-extinct mammal of comparable size.

In addition to their unparalleled ability to navigate underwater, dolphins are known to communicate with one another by means of a series of whistles, quacks, squeaks, clicks, and other noises that often resemble Bronx cheers. Can we hope to communicate with them someday? Perhaps, but it will not be easy, since the human range of generating and hearing noise is relatively limited (50 to 500 hertz). Dolphins, on the other hand, have a much wider auditory range (2000 to 80,000 hertz). They produce and hear sounds in our communicative range, but to do so requires them to grunt and groan at frequencies lower (bass) than normal. Most of the sounds normally made by dolphins are inaudible to humans, making it improbable that their way of expressing meaning overlaps ours at all. Still, some researchers argue that dolphins in captivity have been trying to communicate with us for years. If so, the dolphins must be quite discouraged by our lack of response.

Interspecies communication will not be easy, whether with dolphins, whales, porpoises, or with chimpanzees or other members of the ape family. Empirical findings to date nonetheless suggest that there is some common ground for future cultivation of dolphin–human links. At the least, it seems that both parties are interested in such an enterprise.

sion making by machines in a way often done by humans—be the next stage in the ascent of cosmic evolution?

Engineers are now attempting to program computers to think, learn, and even create. The extent to which artificial intelligence might be advanced is not yet known. Although some researchers maintain that today's computers are nothing more than devices to serve our human wishes, everyone agrees that such machines can now perform mathematical calculations much faster than humans. Speed is truly their forte. In addition, computer memories are also becoming larger, enabling them to store vast quantities of raw information.

Programs written by humans, fed into today's most advanced computers, and further refined by the machine's memory now enable those computers to beat any human at checkers, and almost any human at chess. When the allowed time limit on moves is made short enough, computers can easily trounce even a chess grandmaster. Furthermore, robots capable of making limited choices on their own, without constant guidance from Earth, are now being built to explore the surfaces of other planets of our Solar System.

The key question now debated is this: Can a machine ever be built to think or understand completely on its own? Will computers someday not only answer key questions, but also ask them? Some researchers feel that such devices are inevitable, given the rate of technological growth. They contend that humans are essentially machines who will eventually build better machines to replace themselves. Others suggest that truly thinking computers can never be constructed for, they claim, the basic differences separating humans and machines cannot be overcome by any amount of technology. These critics argue that it's wrong even to suggest that computers work like human brains. The human qualities of will, intuition, emotion, and consciousness, and especially the possibility of a human "mind" in addition to our brain, all tend to suggest that today's computers are, by comparison, nothing more than incredibly fast morons.

At any rate, it will still be important to monitor the effect on human values, human ethics, and self-respect in a world where people *think* machines could someday dominate us.

SUMMARY

The matter comprising the human brain is more complex than anything in the known Universe. Many breakthroughs were needed along the road to intelligence, most notably the ability of cells to cluster and communicate sometime around a billion years ago. Thereafter, neural cells have evolved steadily to the point where we now use them to create an inquiring civilization, to unlock secrets of the Universe, and to reflect upon the material contents from which we arose.

Without a brainy seat of consciousness, galaxies would twirl and stars would shine, but no one or thing could comprehend the majesty of the reality that is nature. With a brain, we probe the past, striving to decipher our celestial roots, all the while searching for a better understanding of the cosmos and especially of our true selves. What emerges is nothing less than a cosmic heritage, a plenary view of who we are, where we came from, and how we fit into the universal scheme of all things material.

KEY TERMS

artificial intelligence
axon
brain stem
cerebellum
cerebral hemisphere
cerebrum
consciousness
cortex
dendrite
invertebrate
neuron
synapse
vertebrate

QUESTIONS

1. At least two modern human characteristics we now greatly value preceded (and were probably chiefly responsible for) the rapid rise in cranial capacity of our ancestors during the past few million years. Name and describe these characteristics.
2. Some neurobiologists argue that while the origin of life is rather probable given the right conditions, the origin of intelligence might not be. What is the basis of their reasoning?
3. Try balancing a pencil on one of the fingers of your left hand and then your right hand while simultaneously speaking. Can you notice a difference? Explain.
4. Take a sheet of paper about 1 square meter in size (about 3 feet on a side) and crumple it into a ball the size of your skull. Comment on the significance of this type of cortical geometry within the human brain.
5. Explain how the neurons within the parietal and frontal lobes work together.
6. Why does the fact that the neurons are not "hard-wired" (i.e., physically connected) make the brain even more versatile than if they were?
7. Explain why brain-to-body ratios are probably better indicators of intelligence than total brain mass alone.
8. What prohibits a single cell from growing to enormous size and eventually developing intelligence?
9. Explain how drugs can alter human behavior.
10. Name and discuss a few animals that apparently have one sense that is more advanced than in humans.

FOR FURTHER READING

ELLIOTT, H., *The Shape of Intelligence*. New York: Scribner, 1969.

HUBEL, D., "The Brain." *Scientific American*, September 1979.

JASTROW, R., *Until the Sun Dies*. New York: W. W. Norton, 1977.

MILNE, D., *et al., The Evolution of Complex and Higher Organisms*. Washington, D.C.: U.S. Government Printing Office, 1986.

SAGAN, C., *Dragons of Eden*. New York: Random House, 1977.

STENT, G., AND WEISBLAT, D., "The Development of a Simple Nervous System." *Scientific American*, January 1982.

VALENTINE, J., "The Evolution of Multicellular Plants and Animals." *Scientific American*, September 1978.

29 EXTRA-TERRESTRIAL LIFE: *Are We Alone?*

We have now studied the nature of matter and life. We've taken inventory of the material Universe and pondered the origins of intelligent life. In the process, we've come to realize that we inhabit no special place in the Universe. Every experimental test made to date suggests that we reside on an ordinary rock, orbiting about an average star, near the edge of a run-of-the-mill galaxy.

Thus the matter around us is not unique. But are *we* unique? Is life on Earth the only example of life in the Universe? These are tough questions, for the subject of extraterrestrial life is one for which there are no data. We can say only one thing for sure: Earth is the only place in the Universe where we know that life definitely exists.

We might like to think that cosmic evolution operates everywhere in the Universe. And it should if our text is correct. But to be truthful, we know of no other place in the whole Universe where life has arisen. This doesn't mean that there is no life beyond Earth. It means that if extraterrestrial life does exist, we haven't yet become smart enough to be aware of it.

The scenario of cosmic evolution would be greatly strengthened if we could prove that life—even elementary, nonintelligent life—originated somewhere beyond Earth. In this way, as noted in Interlude 26-1, life would be not only a natural consequence of the evolution of matter, but virtually an *inevitable* result of evolution.

The general case favoring extraterrestrial life can be summarized by noting the so-called **assumptions of mediocrity**. They are as follows: *Since* (1) life on Earth depends on just a few basic molecules, and *since* (2) the atoms composing these molecules are common to all stars, and *if* (3) the laws of science as we know them apply to every nook and cranny within the Universe, *then*, given sufficient time, life may well originate at numerous places in our Galaxy and galaxies beyond. In other words, given the enormity of space and the vastness of time, life must arise in many locations throughout the Universe.

The opposing view maintains that intelligent life on Earth is the product of extremely fortunate accidents—astronomical, geological, and biological events that are unlikely to occur anywhere else in the Universe. The idea of cosmic evolution is not challenged. In fact, cosmic evolution is judged to be correct in this alternative view. Life is still a natural consequence of the evolution of matter; it's just not inevitable. Simply stated: The steps leading to life—especially intelligent life—are assumed to be so rare as to make it unlikely that any living things reside beyond Earth.

The learning goals for this chapter are:

- to evaluate the chances of finding life in the Solar System
- to understand the main results of the search for life on Mars
- to realize that our type of life—one based on carbon, operating in water—is not the only possible biochemistry
- to appreciate the various probabilities used to estimate the number of advanced civilizations now residing in our Galaxy
- to understand some of the tech-

niques that humans might use to search for extraterrestrials and to communicate with them

Prospects for Life in the Solar System

PLANETS AND MOONS: Hardly anyone expected the American astronauts to discover any evidence for life forms, living or fossilized, while collecting lunar matter such as that shown in Figure 29.1. Still, some researchers argued that simple organic chemicals might have formed on the Moon's surface. If so, then we might be able to retrace the early steps of chemical evolution that led to life as we know it. Some carbon monoxide, carbon dioxide, methane, and a few other simple carbon-rich molecules were indeed found trapped in the lunar soil. But after careful analysis of the lunar samples, the amount of heavy organic matter was so small that no definite conclusions could be drawn.

The total amount of carbon-rich matter contained in lunar soil is so minute that some researchers suggest that none of it is native to the Moon itself. Instead, all of it could have been deposited on the lunar surface by impacting meteorites, or even carbon particles of the solar wind. Remember, the Moon lacks a protective atmosphere and magnetic force field and is thus subjected to fierce bombardment by solar ultraviolet radiation, the solar wind, and cosmic rays. Simple molecules, even those linked by strong carbon bonds, could not possibly survive in such a hostile environment.

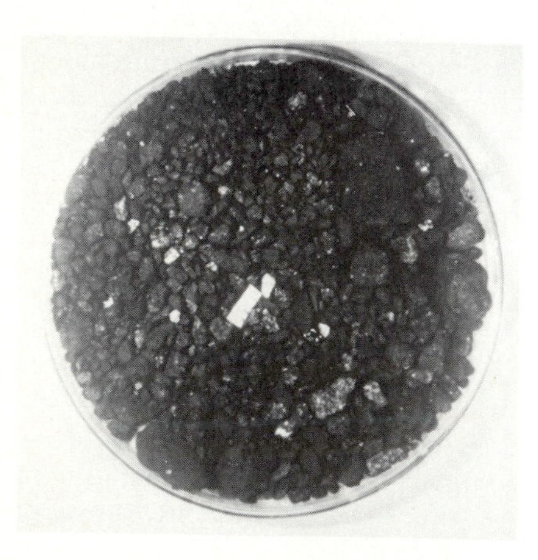

FIGURE 29.1 U.S. astronauts collected many samples of lunar soil during the several *Apollo* Moon landings. (The marks on the scale at center denote millimeter intervals.)

The planet Mercury is not a much better candidate to harbor life, or even the basic organic ingredients for life. Without much of an atmosphere, heat does not transfer well from one side of the planet to the other. Hence the surface is expected to be alternately very hot and then very cold every 176 days. To an even greater extent than on the Moon, the Sun's radiation would destroy any carbon bonding that might happen to develop on Mercury.

The outermost planet, Pluto, has the opposite problem. It's too cold. Being so far from the Sun, water could not possibly remain in the liquid state. Hence the building blocks of life, even if they exist on this frigid planet, would have no easy way to interact to form the larger organic chemicals needed to construct life.

The Jovian Planets, Jupiter through Neptune, are all quite far from the Sun and have cloud tops well below the freezing point of water. However, we must not be too quick to dismiss any possibility for life somewhere within these large gas balls. Surely, greater warmth exists deep down in their clouds. Both the greenhouse effect (which traps solar radiation) and internal sources of energy (at least inside Jupiter and Saturn) provide some heat. Consequently, some researchers argue that intermediate levels of the Jovian atmospheres might be warm enough for the development and maintenance of life.

Other researchers suggest that despite the heat, life is unlikely. To survive and prosper, life would have to remain at a stable atmospheric level. If it floated too high in the atmosphere, it would freeze; if it sunk too low, it would fry. Freezing would not necessarily destroy large molecules or organisms, but it would surely prevent their further development. (Refrigerators are technological devices that do not kill bacteria, but rather slow their metabolism.) Heat, on the other hand, is more dangerous than cold. Too much heat would physically destroy any carbon bonding.

Trace amounts of life's building blocks are known to exist in the atmospheres of Jupiter and Saturn. Ammonia, methane, hydrogen, and energy of many types are present. Still, a basic problem plagues the idea that these chemicals might produce life: As these molecules combine to form amino acids and nucleotide bases, and in turn larger polymers, the products become progressively heavier. These products are thus subjected to a larger pull of gravity, making them likely to sink. The heavier they become, the deeper they sink, and the greater the chance of destruction by heat.

It would seem unlikely, then, that life could originate and sustain itself at a comfortable level within the atmosphere of either of these giant planets. Nonetheless, future studies of their atmospheres might provide important clues to the early stages of chemical evolution.

Some of the moons of the Jovian planets are probably better candidates to house life as we know it. For example, Saturn's moon Titan, with its methane–ammonia–nitrogen atmosphere and its possibly solid surface, is a potential site for life, though the results of the 1980 *Voyager* fly-by suggest that its surface conditions are rather inhospitable for any kind of life familiar to us.

The chances for life on Venus approximate those on Jupiter and Saturn. The Venusian surface temperature and pressure are much too high to support life as we know it. However, the tops of the clouds are cooler and thinner. As for the Jovian planets, we can again reason that an intermediate level exists for which the physical conditions are more hospitable. The chemical conditions, on the other hand, are most undesirable, as Venus' atmosphere has large amounts of highly corrosive sulfuric acid. At any rate, organisms or even protoorganisms would probably be unable to stay at a constant altitude needed to avoid either freezing or cooking.

TESTS FOR LIFE ON MARS: For centuries, the planet Mars has been the center of speculation about extraterrestrial life. Of all the planets, its physical conditions most resemble those of Earth. Mars has a hard surface, and some mild pressure. The planet rotates in about 24.5 hours. And it has an atmosphere, although one made mostly of carbon dioxide, thereby ensuring some greenhouse heating; its surface temperatures average about 50 kelvin less than those at corresponding latitudes on Earth.

All things considered, Mars seems harsh by Earth standards. The scarcity of liquid water implies dehydration, the thin atmosphere ensures a continuous freeze–thaw cycle as the planet spins, and the lack of magnetism and an ozone layer allows the solar high-energy particles and ultraviolet radiation to reach the surface unabated.

To be sure, Earth life would have a hard time surviving on Mars. But in this section, we're not talking about Earth life; we're considering Martian life. Our views become somewhat tempered once we realize that any Martian life would have originated, evolved, and adapted in that environment. Continued adaptation could have presumably made native Martian life possible, especially in the past when the atmosphere was presumably thicker, the surface warmer and wetter. So the Martian environment is indeed harsh by our standards, but not entirely incapable of supporting life.

An early American *Mariner* spacecraft that traveled past Mars in the 1960s sent back disappointing pictures. The chances for life seemed bleak. Packed with numerous craters, Mars seemed to resemble the desolation of our Moon more than a comfortable abode for life.

However, the last in the series of *Mariner* spacecraft carried improved equipment and took sharper pictures while orbiting Mars in the early 1970s. As noted and illustrated in Chapter 7, the results were totally unexpected. *Mariner 9* showed Mars to have not only volcanoes and canyons, but also what resembled dried river beds or channels through which liquid of some sort must have flowed. The chances for life increased greatly, especially if that liquid was the water now trapped in the frozen polar caps. The possibility of finding evidence for life on Mars suddenly brightened—if not current life, then possibly extinct life that flourished at some previous epoch when the liquid flowed.

The search for life on Mars began when two American *Viking* spacecraft arrived at Mars in the mid-1970s. After dispatching a robot to the surface, each main craft continued orbiting the planet. Figure 29.2 shows (prior to launch) one of these robots that soft-landed on the Martian surface. They were programmed to sense life in several ways.

First, each robot carried a television camera to seek fossilized remnants of large plants and animals. It also scanned the horizon for any large-scale motion—trees blowing in the wind, giraffes scampering across the landscape, whatever. Neither fossils nor movements of any kind were seen.

Second, as can be noted in Figure 29.3 [see also Plate 3(b)], each robot was able to scoop up and then ingest Martian soil. Instruments onboard the robot could test for life by conducting a variety of chemical experiments. The results of these experiments were encouraging, although they remain ambiguous to this day.

Figure 29.4 outlines the main features of the first chemical test for life conducted by the *Viking* robots. This test was designed to detect even the most primitive plant-like organisms on Mars. Plants normally require gas, water, and light for sustenance, but they also grow exceptionally well if that gas is mainly carbon dioxide. The resulting chemical reaction is photosynthesis which yields oxygen, as discussed in Chapter 26. Plants were thought to be the best candidate for life because 95 percent of the Martian atmosphere is carbon dioxide gas. To check for plants, each robot inserted a small amount of Martian soil into one of its experimental chambers. This chamber contained carbon dioxide and some water, while a lamp bathed the interior with radiation that imitated Martian sunlight. This rich mixture of gas, water, and light should have enhanced the growth of any Martian plants. After an incubation period of a few days, this first test gave a positive result; oxygen gas was released. The soil *seemed* to

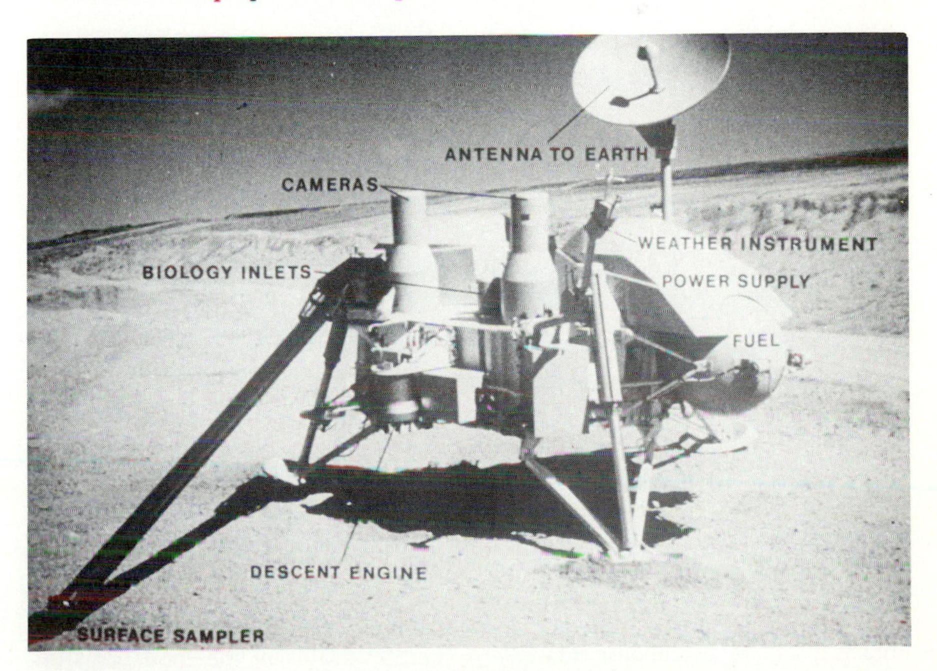

FIGURE 29.2 Many important features of the *Viking* robot are labeled in this photograph, taken prior to launch from Earth.

FIGURE 29.3 A trench dug by the "arm" of one of the *Viking* robots can be seen at the right. Soil samples were scooped up and taken inside the robot, where instruments tested them for chemical composition and any signs of life. [See also Plate 3(b).]

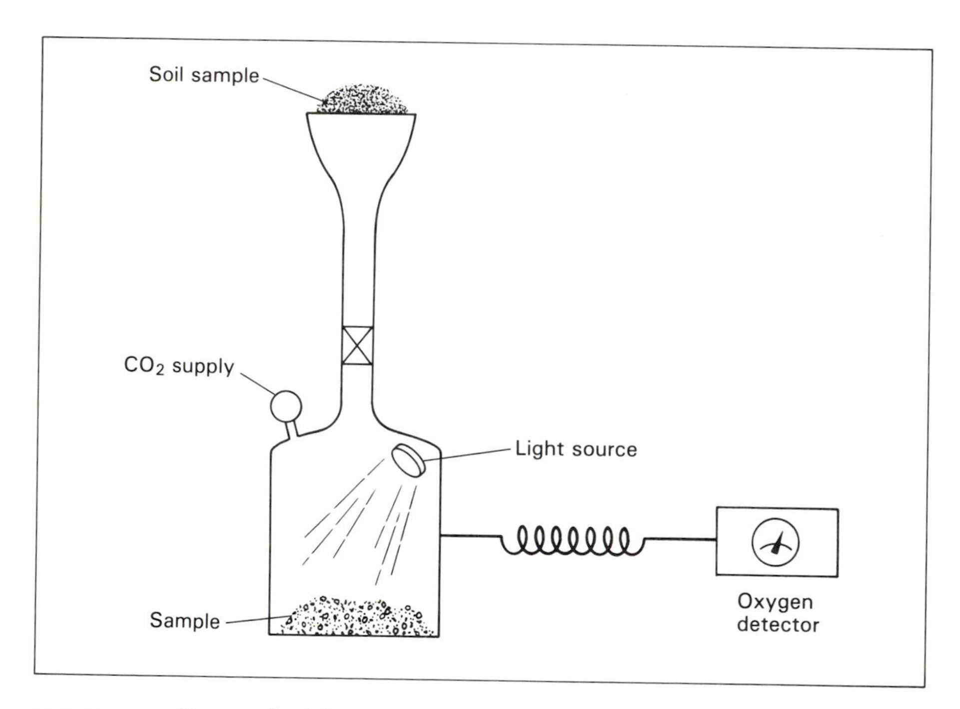

FIGURE 29.4 One test for life on Mars was designed to detect photosynthesis occurring in any Martian plants that were surrounded by a rich mixture of carbon dioxide, water, and sunlight.

contain plants or plant-like organisms of some sort.

The second chemical test was designed to detect the by-products of any microbial animals. Sketched in Figure 29.5, the essence of this test required moistening the Martian soil with a nutrient broth rich in amino acids and other food substances. This nutrient broth, transported to Mars by the *Viking* spacecraft, was not just ordinary foodstuffs. Specially prepared for the flight, some carbon atoms in the food were radioactive. If microbes exist in the Martian soil, they would eat the food. After digesting it, some of the basic ingredients of these consumed foods would be released as wastes. The wastes could then be measured by radioactive detectors on-board each Viking robot. As for the first chemical experiment, the results of this second test were also positive; radioactive carbon dioxide was detected. The implication was that Martian microbes *might* be present in the soil.

Figure 29.6 shows the basic outline of the third chemical test for life on Mars. This test was designed to detect any gases emitted as waste products by primitive forms of life now thriving on Mars. In the event that the Martian soil contained small plants or animals, the *Viking* robots were prepared to detect either carbon dioxide or oxygen gas. This was a spartan experiment; the third chamber contained no enriched carbon dioxide gas or any delicious food for the organisms. Only moisture was injected into the Martian soil.

Surprisingly, both gases were detected. Carbon dioxide was released in a slow and steady way, much as do animals on Earth. But the oxygen came off in a burst. This was puzzling since such a burst differs from the steady emission of oxygen by Earth plants. Furthermore, the sudden release of oxygen occurred shortly after the soil had been moistened. The burst suggested a chemical reaction among soil compounds—not a biological one involving life. But what kind of chemical reaction?

During the past few years, researchers have discovered that chemical reactions involving superoxide (oxygen-rich) compounds can mimic some of the expected gas emissions of life forms. Scientists didn't know this before the *Viking* mission. These reactions require abnormally large amounts of superoxides, suggesting that the Martian soil differs markedly from that thought possible when the *Viking* experiments were designed. The test results also imply an unusual series of chemical reactions, so strange as never to occur on Earth. At any rate, this third test implied that *each Vi-*

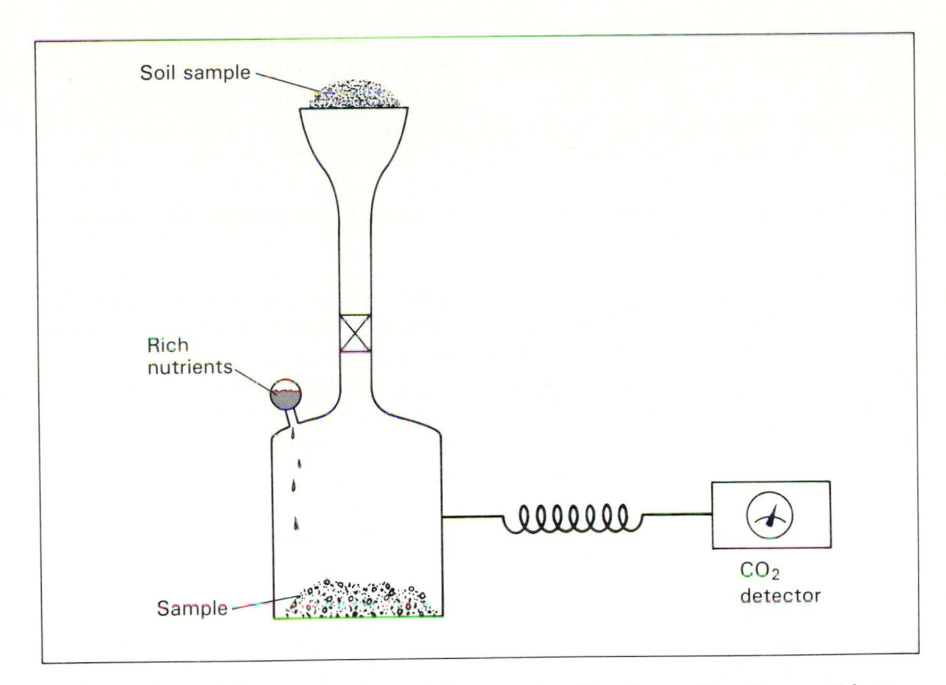

FIGURE 29.5 Another test for life on Mars used radioactive carbon from a rich nutrient broth. Assuming that any microbes in the Martian soil would consume and digest it, as they do food on Earth, radioactive gases in the form of released carbon dioxide would be detected.

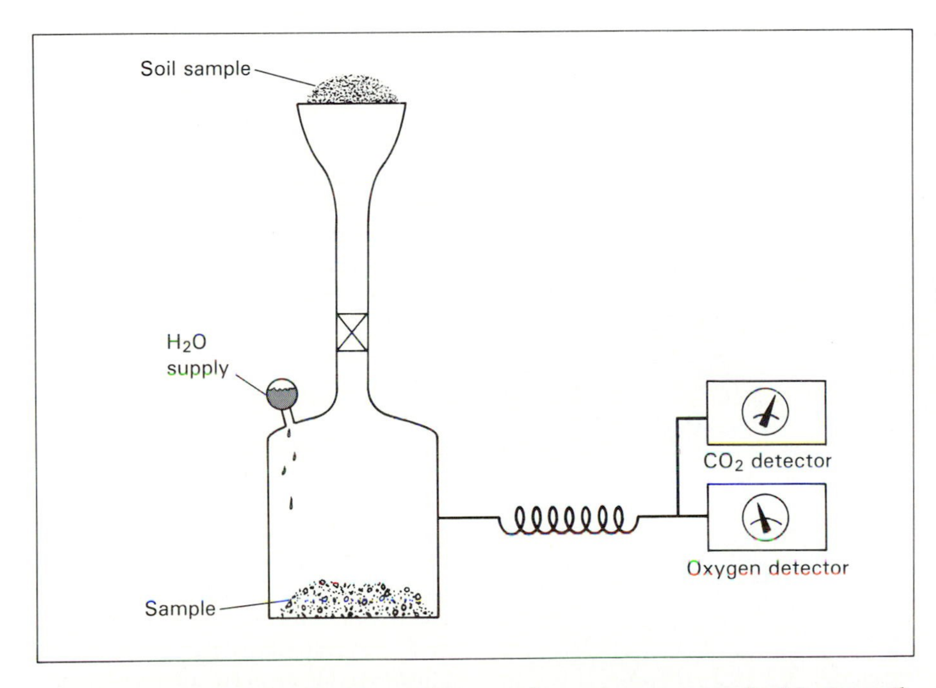

FIGURE 29.6 A third test for life on Mars sought waste gases once the Martian soil was moistened a little.

king test could possibly be explained by chemical reactions and not necessarily biological reactions. The results were unclear.

The chemical explanation is now favored by many researchers, not only because of the oxygen burst in the third test, but also because of a fourth test done by the *Viking* robots. This final experiment was designed to directly detect the organic molecules that form the basis for life as we know it. If life really exists on Mars, and in any way resembles that on Earth, the Martian soil should be abundant in organic molecules.

This fourth test failed to find any organic molecules, at least down to a level of one organic molecule for every million inorganic molecules; this is about the density of organic matter in a piece of Antarctic soil on Earth. Test four, then, does not absolutely disprove the existence of organic matter on Mars. In fact, this fourth experiment is a lot less sensitive than any of the first three. What it does suggest is this: If life does exist on Mars, it must be very thinly spread throughout the soil.

The consensus among biologists and chemists today is that Mars does not seem to house any life similar to that on Earth. Chemical, as opposed to biological, explanations for all the *Viking* tests are generally preferred by most scientists, even though peculiar chemical reactions are required to understand the gases measured in each test.

However, not all scientists agree. Some suspect that a different type of biology might be operating on the Martian surface. They suggest that Martian bugs capable of eating and digesting superoxides (rather than conventional organic matter) could easily explain all the results of the *Viking* tests. In addition, microbial life as we know it might reside in more habitable regions on Mars, such as near the moist polar caps. After all (partly for political reasons), the two spacecraft landed on the safest Martian terrain, not in the most interesting regions. Also, no one has yet searched for microscopic evidence of microbial fossils from an ancient Martian era.

The *Viking* robots were not prepared to probe any of these alternatives. A solid verdict regarding life on Mars may not be reached until we have thoroughly explored our intriguing neighbor. Apparently, secrets still lie on the reddish plains of this dusty planet.

LIFE AMONG THE DEBRIS: Besides planets and their moons, several other objects in our Solar System might house some form of life.

Comets, for example, are known to contain many of the basic ingredients for life, for instance ammonia, methane, and water vapor. And although comets are frozen, their icy matter warms while nearing the Sun. Such conditions would seem to favor the construction of heavy molecules, but few heavier than these have yet been observed. Thus comets are probably unlikely candidates for life itself, although they might be convenient "laboratories" where we can study the early phases of chemical evolution at close range.

Meteoroids also cruise about our Solar System. A small fraction of those that survive the plunge to Earth's surface—meteorites—contain organic compounds. These carbon-bearing meteorites are called **carbonaceous chondrites** and were noted briefly in Chapter 8. Shown in Figure 29.7, these meteorites contain many chondrules, which are pebble-sized granules having a few percent of their mass in the form of organic compounds. Often it's hard to determine if these large carbon-based molecules are really extraterrestrial remnants. In some cases, they may be mere terrestrial contaminants picked up while plowing through Earth's atmosphere, or accumulated while sitting for years on Earth's surface. Even perched on museum shelves, meteorites can become contaminated with small deposits of terrestrial carbon.

Fortunately, we know of a few meteorites containing organic matter that could not possibly be due to terrestrial contamination. The so-called Murchison meteorite, which fell near Murchison, Australia, in 1969, is an example of a carbonaceous chondrite that has been studied extensively. Located soon after crashing to the ground, this meteorite contains many of the well-known amino acids normally found in living cells. What's more, the meteorite's amino acids were found to be equally left- and right-handed, implying that they originated outside of Earth (consult Interlude 25-2). Other meteorites show evidence for parts of the nucleotide bases as well. These acids and bases are similar in kind and in relative abundance to those produced in the laboratory simulations described at the end of Chapter 25. And the density of amino acids in this meteorite is impressive; it exceeds that of many desert sands on Earth. However, nothing resembling the complexity of life itself has ever been found in a meteorite. Nor do meteorites contain any known fossils of ancient life forms.

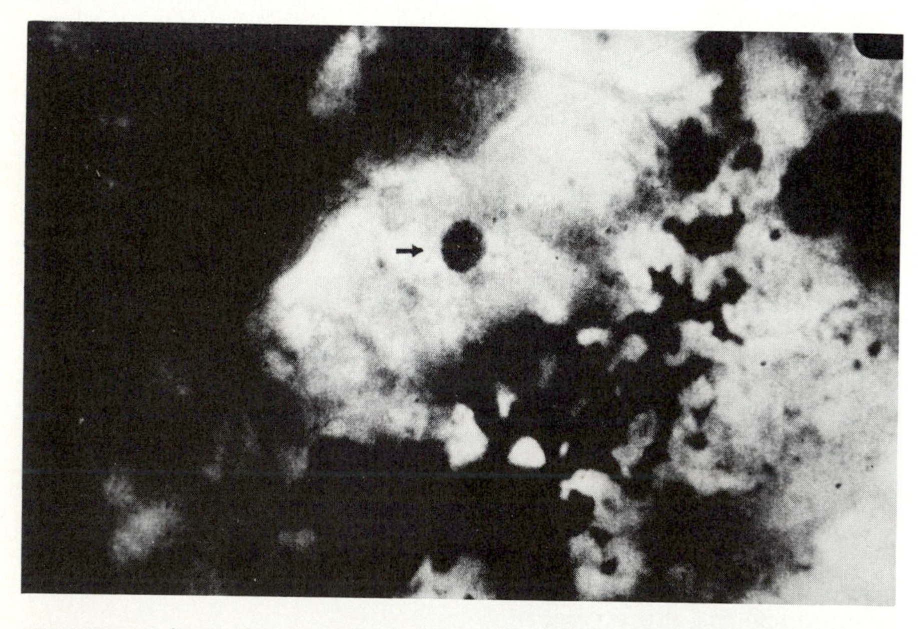

FIGURE 29.7 A portion of a wafer-thin slice of a carbonaceous chondrite. By passing different types of radiation through it, even light like that used here, researchers can study the meteorite's material composition. The arrow denotes a carbon-rich chondrule.

The moderately large molecules found in meteorites and in interstellar clouds comprise our only evidence that chemical evolution is occurring or has occurred elsewhere in the Universe. In each case, most researchers regard the organic matter to be prebiotic—that is, matter which could eventually lead to life, but which has not yet done so. Very few researchers think that the molecules in meteorites and interstellar clouds could be biotic—that is, the decayed remains of already established life that exists someplace else in the Galaxy.

Alternative Biochemistries

Several times in the preceding section, we couched our statements about extraterrestrial life with the qualifier "as we know it." Life as we know it is carbon-based life, operating in a water medium, with higher forms metabolizing oxygen. All forms of life on Earth—from bacteria to humans—share this same basic biochemistry. And on the basis of what we now know about the various chemical elements, carbon would seem to be the atom best suited to form the long-chain molecules needed for life. But are we being chauvinistic? How do we know that other biochemistries are not equally possible?

SILICON-BASED LIFE: Recall from Chapter 2 that carbon, with its four unpaired outer electrons, can form tight chemical bonds by sharing its electrons with other elements. In this way, as shown in Figure 2.7, ^{12}carbon atoms can achieve maximum stability by attracting other atoms to each of its four sides.

^{28}Silicon is a possible alternative to carbon. Listed in the same column as carbon in the periodic table of the elements, silicon also has four unpaired electrons in its outermost orbital. This similarity is especially important because, on Earth, silicon

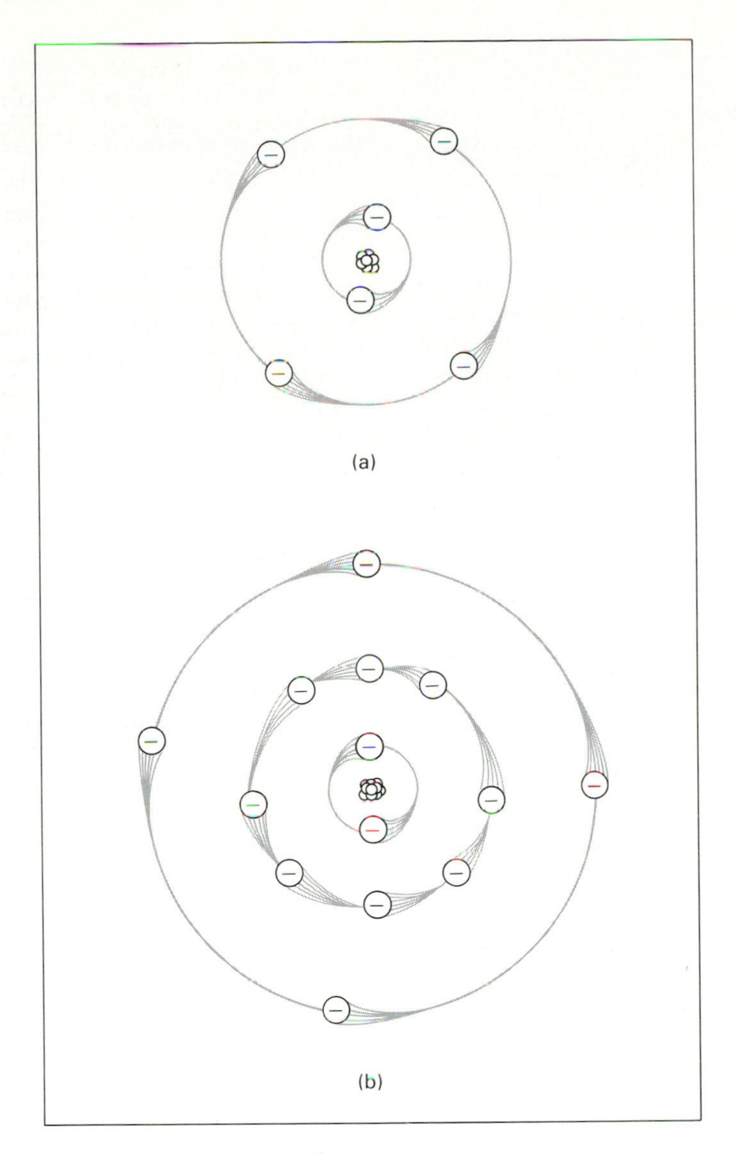

FIGURE 29.8 The orbital arrangements of electrons for (a) ^{12}carbon, and (b) ^{28}silicon.

is more than a hundred times more abundant than carbon.

Despite its great abundance and similarity to carbon's structure, silicon cannot bond to other atoms as well as carbon can. As shown in Figure 29.8(a), a carbon atom's four unpaired electrons normally reside in its second orbital. Since eight is the maximum number of electrons allowed in the second orbital of any atom, this orbital becomes full when carbon binds with other atoms on four sides. For this reason, a carbon chemical bond is among the strongest of all.

Silicon's four unpaired electrons, on the other hand, normally reside in its third orbital, as shown in Figure 29.8(b). The maximum number of electrons permitted in the third orbital of any atom is 18. Even though silicon might normally have atoms bonded to four of its sides, just like carbon, the silicon bond is not as strong as the carbon bond. Why? Because the outer orbital of silicon is unlikely to have a full complement of electrons, even when it's bonded to other atoms. Generally, carbon bonds are twice as strong as silicon bonds.

Of even greater importance, carbon links most strongly to other carbon atoms. This is especially true for diamond, which is made of carbon atoms bonded to one another, as sketched in Figure 29.9. In fact, diamond is the hardest substance known; hardness results from great

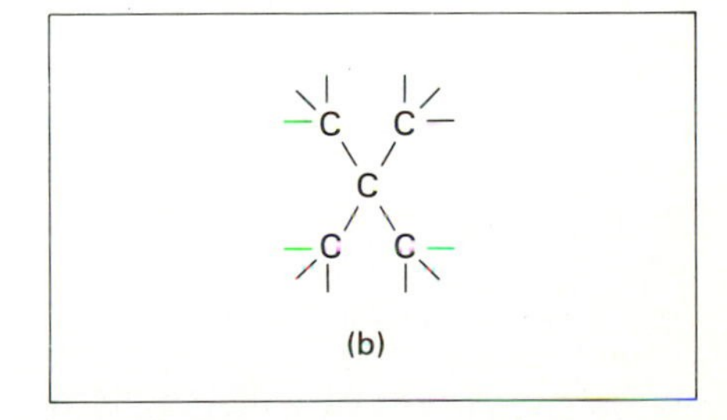

FIGURE 29.9 Diamond (a) is the hardest substance known. It's made of pure carbon atoms, each atom strongly bonded to four others (b).

bond strength. The carbon bond is also unaffected by water, giving it another advantage. Silicon, on the other hand, does not bond as well to other silicon atoms, and not well at all in the presence of many liquids. Chains of silicon are especially unstable in water; they break apart.

The fact that the carbon–carbon bond is stronger than the silicon–silicon bond, especially when immersed in liquid, is one reason that favors carbon-based life. Another reason is reluctance on the part of silicon to form double and triple bonds, which normally add great strength to a group of two or more atoms.

A third reason favoring carbon-based life is the high cosmic abundance of oxygen. When carbon (C) chemically reacts with oxygen (O), the result is carbon dioxide (CO_2). This is a gas and thus can easily combine with other compounds; in our case, humans exhale carbon dioxide after inhaled oxygen has reacted with the carbon in our bodies. When silicon (Si) reacts with oxygen, however, the result is quartz (SiO_2). This is a solid and cannot interact easily with other compounds. Besides, can you imagine a living creature exhaling solid quartz? Thus we should not be surprised that silicon plays no biochemical role on Earth, despite its large abundance on our planet.

Given the proper conditions on any planet, both carbon-based and silicon-based life might initially form. Other types of life might also emerge—perhaps based on the rare element germanium, which also has four electrons in its outer (fourth) orbital. However, carbon-based life would doubtless eventually eradicate all other types of life. Carbon clearly has greater bonding flexibility and strength and can adapt better to changing dry–wet conditions. On *chemical* grounds, then, carbon is best suited to act as the backbone of the long-chain molecules required for life.

Despite these strong statements, we shouldn't close our minds entirely to weird biochemistries. Some planets may have odd *physical* conditions that actually favor strange types of life. For example, high temperatures favor silicon bonding over carbon bonding. Silicon–oxygen bonds can withstand heat to 600 kelvin, and silicon–aluminum bonds to nearly 900 kelvin. Carbon bonding of any type breaks apart at such high temperatures, making carbon-based life impossible. This heat-resistant property of silicon is the prime reason that silicone compounds are often used as industrial lubricants. Even hot machinery runs smoothly with silicon-based grease.

Should silicon-based life arise on a hot planet somewhere in the Galaxy, its flexibility and adaptability would still be severely limited. Simple, primitive types of silicon-based life might well reside on such alien worlds. But based on everything known about chemistry, we would find hard to imagine anything as complex as intelligent life based on silicon.

NONWATER LIQUIDS: What about the medium in which life operates? Must it be liquid? Answers require us to speculate about the best way for complex organic molecules to move around and interact with one another.

A solid is a poor interaction medium—unless perhaps the solid be pulverized as powder; atoms and molecules within hard solids have little mobility unless they are on the verge of liquifying. Gases are also poor substitutes for liquids; a gas doesn't easily stay put unless restrained by gravity or in a container of some sort. This type of loose reasoning leaves the liquid phase as the most reasonable interaction medium. But do liquids seem best only because we ourselves are partly made of liquids? We again naturally wonder if our conclusion is chauvinistic.

What is the best type of liquid to enhance clusters of complex molecules? Must it be water? On the basis of what we know about life on Earth, we might conclude that this particular liquid is critically important. But is this chauvinism again? Could some other liquid substitute for water on another planet?

Several arguments favor water as the most reasonable liquid medium for life. The best reason is that the water molecule is made of two of the most abundant atoms—hydrogen and oxygen. According to our knowledge of cosmic abundances, these atoms are expected to be plentiful everywhere.

A second reason favoring water as a preferred medium for life is its widely separated freezing and boiling points. Recall that water remains in the liquid state for a large temperature interval (273 to 373 kelvin), allowing vital biochemical reactions to proceed anywhere in this range of temperature. (This range, however, depends on the pressure and is true as stated only for the conditions on Earth's surface.)

Another unique property of water is its reversal in density while cooling from 277 kelvin to 273 kelvin. Water ice is less dense than liquid water, a statement that is untrue for any other substance. Hence, as we all know, ice floats. If ice were denser than liquid water, ice would sink and water would freeze from the bottom up, just like other substances. Collecting at the bottom of a lake or ocean, the ice would hardly have a chance to melt. It wouldn't be long before entire bodies of water, including whole oceans, became solid blocks of ice. Fortunately, water's peculiar density property prohibits this, ensuring lots of terrestrial liquid in which molecules can freely interact.

Ammonia (NH_3) has often been suggested as a substitute for water. It's also made of abundant atoms. In many respects, ammonia resembles water. However, it's not an entirely appropriate replacement for water.

To remain liquid, pure ammonia must be colder than water. Its liquid range (on Earth) spans about 200 to 240 kelvin, although, as for water, this range is pressure dependent; room temperature, remember, is nearly 300 kelvin. The presence of ammonia might then enable life to prevail on a cold planet where water is normally frozen. But such low temperatures would inevitably cause biological reactions to slow down. Although admittedly billions of years are available, the rate at which molecules interact in ammonia to produce more complex life forms would be a lot lower than on watery Earth. Furthermore, ammonia does not have the peculiar density reversal near its freezing point, so large bodies of ammonia would likely freeze solid.

For these reasons, ammonia seems unlikely to act as a medium for advanced forms of life. Still, acknowledging nothing more than potential Earth-based chauvinism, ammonia and a few other chemicals cannot be absolutely ruled out as the basis of an alternative biochemistry.

Prospects for Life in the Galaxy

We can make some very rough estimates of the number of sites where life possibly exists beyond our Solar System. These estimates take account of many factors that affect the origin and evolution of life. To appreciate the method most often used, consider first a mundane example that has no bearing on extraterrestrial life.

AN EXERCISE IN PROBABILITY: Suppose that we wish to estimate the number of Swedes who are now living in Boston and who have both brown eyes and red hair. We could begin by consulting a recent population census, which states that about 10,000 Swedes live in Boston. This includes all Swedes, most of them probably having blue eyes and blonde hair. If we know the fraction of all Swedes having brown eyes to be 1 in 50, and the fraction having red hair to be 1 in 100, we can make an estimate of the number of Swedish Bostonians having both of these traits.

The problem can be formulated in terms of the following three simple multiplications—assuming that there is no natural linkage between hair and eye color (including, for example, no tendency for brown-eyed Swedes to dye their hair red):

number of Swedish Bostonians having both brown eyes and red hair = total number of Swedish Bostonians × fraction of Swedes having brown eyes × fraction of Swedes having red hair.

Evaluating the equation, we find $10{,}000 \times \frac{1}{50} \times \frac{1}{100}$, or 2 Swedes who have both brown eyes and red hair. We can't be sure that exactly two such Swedes live in Boston, for we didn't take an actual head count. Maybe there are three or four, or perhaps only one. Here we used statistical information and probability theory to help us make an *estimate.*

In a similar way, we can attempt to estimate the number of sites where life currently resides in the Universe. In formulating this problem, let's make two restrictions. First, let's confine our analysis to our own Galaxy. Second, let's stipulate these life forms to be not only intelligent but also technologically competent.

These restrictions are reasonable considering the time scales involved. Given our study of the oldest globular-cluster stars in Chapter 11, we concluded that our Milky Way Galaxy is about 12 billion years old. Even if a few billion years are needed for the most massive stars to populate interstellar space with enough heavy elements to construct planets, planets as old as 9 billion years could exist. If evolution proceeds at roughly the same rate throughout the Universe (a totally unverifiable assumption), such planets could have housed some form of life even before our Solar System originated. Provided that such life has survived, their civilizations could be vastly more advanced than ours.

Estimating the number of galactic civilizations is not a trivial exercise. It requires us to summarize a great deal of knowledge covered in Part III. A large amount of this knowledge (or ignorance!) can be compressed into a small space by formulating this problem much as we did in the case of the Swedish Bostonians. We represent the most important aspects of this extraterrestrial-life problem according to the following relationship. Since an early version of it was first devised at a radio observatory in Green Bank, West Virginia, we'll refer to it as the **Green Bank equation** (although it is also known as the Drake equation after the U.S. astronomer who first devised an early version of it):

number of technologically intelligent civilizations now present in the Milky Way Galaxy = rate of star formation averaged over the lifetime of the Galaxy × fraction of those stars having planetary systems × average number of planets within those planetary systems that are suitable for life × fraction of those habitable planets on which life actually arises × fraction of those life-bearing planets on which intelligence evolves × fraction of those intelligent-life planets that develop a technological society × average lifetime of a technologically competent civilization.

Figure 29.10 casts this equation into pictorial form. The figure illustrates how only a small fraction of star systems in the Milky Way are likely to generate the advanced qualities specified by the right-hand terms of the equation.

To evaluate each of the terms in this Green Bank equation requires familiarity with many fields of knowledge. For example, we need a good deal of astronomy to estimate the rate of star formation and the fraction of stars having planets. More astronomy, along with some biology, is needed to specify the main properties of an ecologically suitable planet on which life might originate. We need insight into biology, chemistry, anthropology, and neurology to estimate the chances of life, and then intelligent life, developing on any given planet. A great deal more anthropology, as well as archeology and history, are required to know what fraction of the intelligent-life-bearing planets actually create a cultured and technological society. And finally, the lifetime of such an advanced civilization depends on a large number of additional factors, including history, politics, sociology, psychology, and many other daily influences.

Be sure to recognize that the lifetime considered here is really the *technological* lifetime. By this we mean the longevity of an advanced civilization, starting from the time it gains the capability to explore its planetary system or to communicate across interstellar distances. For example, Earth's technological lifetime, thus far, is only a few decades.

The Green Bank equation amounts to an analysis in probability. The several terms in the equation are multiplied rather than added because each term is assumed independent of any other. As such, many of the terms could be less than 1. When multiplied together, several numbers equal to a little less than 1 can quickly yield a product that is very much less than 1. For example, if each of three fractions on the right side of the equation has a value of $\frac{1}{10}$, their product is $\frac{1}{1000}$. In other words, even though the individual chances are 1 out of 10 for the origin of life, of intelligence, and of technology, the combined chance of finding a planet with all three of these traits is only 1 in 1000.

Take note, then, that probability analyses of this sort can quickly lead to very small chances. Many researchers use this kind of reasoning to argue that advanced civilizations are not likely to exist anywhere else in our Galaxy. On the other hand, this type of pessimistic argument could also be used to claim that none of us should exist. After all, the chance that your parents, grandparents, great-grandparents, great-great-grandparents and so on, would have produced precisely you is extremely small. But here you are! Equally important, if you and your generation of humans did not exist, another statistically indistinguishable generation surely would.

Apparently, many different routes connect all states of matter and life. And it is the multiplicity of these different paths that makes possible highly unlikely events.

To estimate the number of civilizations *now* present in our Galaxy, we must provide a number for each term on the right side of the Green Bank equation. Unfortunately, many of the factors affecting each term are themselves quite uncertain. For example, many biologists argue that the probability for life or intelligence on any given planet is nearly impossible to assess with any degree of certainty. It's not that they necessarily regard the chances to be small for the develoment of life or intelligence. Rather, they argue that the science of chemical and biological evolution has not yet progressed to the point where meaningful estimates can be made.

Some researchers contend that we have no information on which to make the required estimates. Others disagree, arguing that science and technology provide *some* informa-

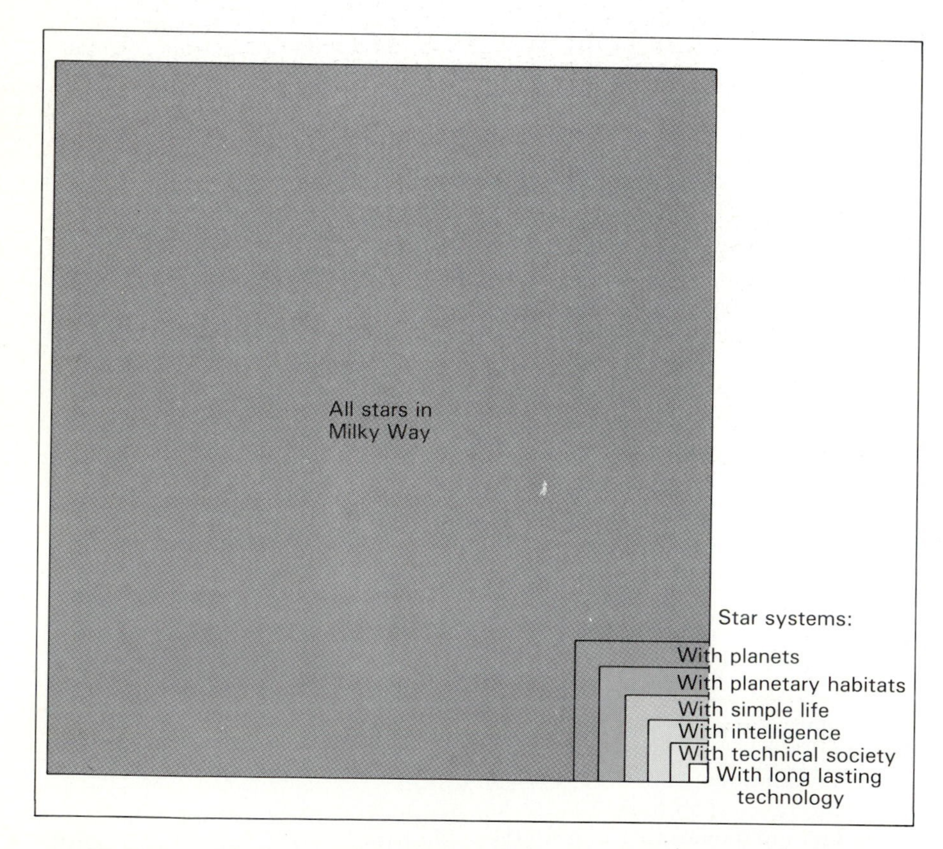

FIGURE 29.10 Of all the star systems in our Milky Way (represented by the largest box), progressively fewer and fewer have each of the qualities typical of a long-lasting technological society (represented by the smallest box at the lower right corner).

INTERLUDE 29-1 ***More Terms in the Equation?***

Some researchers feel that the Green Bank equation used in this chapter is an inadequate guide to the number of galactic civilizations. They suggest that many additional terms are needed to fully describe the evolution of a technological civilization from inanimate matter. For example, we could imagine additional terms with labels like "fraction of those planets bearing plant life that also develop animal life" or "fraction of those hunter-gathering people who eventually evolve language" or even "fraction of those intelligent civilizations that invent religion to help stabilize their societies." Numerous other terms could be invented—a new term for every conceivable evolutionary development. Accordingly, our equation would balloon into one having perhaps hundreds, even thousands of terms.

This argument is often used by researchers, usually biologists and chemists, to suggest that the prospects for extraterrestrial life are extremely small. An equation having hundreds of terms, with each term equal to less than 1, inevitably yields a very small result. For example, even if every one of 100 terms has a 9-out-of-10 chance, all 100 terms multiplied together yield a value of $(0.9)^{100}$, or approximately 0.00003. This somewhat pessimistic view suggests that because so many steps are required for the evolution of a technological society, the chance of *all* these steps randomly occurring anywhere else is so small as to make extraterrestrial life virtually impossible.

Optimists argue, on the other hand, that evolution is not entirely random. They furthermore argue that there are likely to be many, many paths from simple organisms to technological life. In other words, the exact sequence of steps in life's advancement on Earth may well be unlikely ever to occur more than once, but many alternative evolutionary paths permit the rise of more-or-less equivalent life forms. This optimistic argument suggests that technological civilizations are inevitable, given the proper conditions and enough time.

The optimistic researchers, usually astronomers and physicists, also argue that additional terms in the equation are not really needed. Adding more terms is the same as dividing each of the terms already used in this chapter into smaller steps describing the overall road to technological intelligence. Yet many of these additional factors leading to, for example, the onset of life or intelligence would seem to be theoretical certainties. Expressed another way, many if not most of the subterms representing any additional factors would have numerical values very close or even equal to 1.

Consider a specific example. In the Green Bank equation, we used a term labeled "fraction of life-bearing planets on which intelligence develops." This term could be subdivided to include many factors, including fractions that estimate the chances for the development of a primitive nervous system, eyes, learning ability, sociability, communications, dexterity, tools, fire, cities, science, and so on. But many researchers, who have studied each of these additional factors of intellectual development, maintain that almost all of them are virtually sure bets, assuming enough time for life to so develop. In fact, many of these factors are interconnected, the development of one almost surely leading to the development of others. Hence subterms representing each of these additional factors are practically equal to 1. The same can be said about the subdivision of other terms in our Green Bank equation. Regardless of the number of terms in the equation, if most of them are theoretical certainties, their product need not become very small.

The upshot is that an analysis of the Green Bank equation presented elsewhere in this chapter is probably independent of the number of terms used. Regardless of how we cast the equation, it amounts to a shorthand version of an extremely complex problem. Making the equation more complex by adding more terms does not really alter the final result.

The Green Bank equation is valuable in that it provides a framework for a discussion of the prospects for life elsewhere in the Universe. If, while trying to evaluate it numerically, we come across some rigorous argument to exclude extraterrestrial intelligence, we'll then be able to conclude that a search for galactic aliens is not a useful allocation of our time, effort, money, and resources. If, on the other hand, we cannot find any definitive arguments to exclude extraterrestrial life, our civilization will almost surely try to discover it sooner or later. Human nature seemingly compels us to make a complete inventory of the Universe, and that includes all radiation, all matter, and all life.

tion. At any rate, virtually everyone agrees that rigorous values are currently unknown for any of the terms in the equation. Values usually chosen are based on a good deal of personal insight (or prejudice) among a cross section of scientists.

Generally, the reliability of the estimate of each term in the Green Bank equation declines markedly from left to right. For example, our knowledge of astronomy enables us to make a reasonably good stab at the first term, namely the rate of star formation in our Galaxy. But it's much harder to evaluate some of the interior terms, such as the fraction of life-bearing planets that eventually develop intelligence. As for the term on the far right side of the equation, the longevity of technological civilizations is totally unknown. There is only one known example of such a civilization—that's us on planet Earth—and how long we'll be around before a natural or man-made catastrophe ends it all is impossible to tell.

Even though we cannot be accurate in judging most of the terms, it's still remarkable that we can formulate such a single equation to address this very general problem. Estimating the chances for advanced extraterrestrial life requires a knowledge of subjects ranging from astronomy and biology through anthropology and politics. There's probably no other problem like it. Trying to solve the equation exemplifies the true virtue of thinking about extraterrestrial life. It forces us to stretch our imaginations, to expand our minds. And it reminds us of all the beautifully woven interrelationships among virtually all the many disciplines of human knowledge.

Let's now consider the terms in the Green Bank equation. After we've discussed each of them, we'll return to the equation in order to evaluate it.

RATE OF STAR FORMATION: We can roughly estimate the average number of stars forming each year in the Galaxy by noting that at least 100 billion stars have shone during the course of the Milky Way's lifetime. Accordingly, this implies a star-formation rate of 100 billion stars/10 billion years, or approximately 10 stars/year.

Some researchers regard this value as an overestimate, especially since much galactic matter is already contained in stars. They argue that fewer stars are forming now than must have formed at earlier epochs of the Galaxy when much interstellar gas was still available.

On the other hand, we cannot very easily reduce this estimate too much, since radio astronomers have recently discovered numerous sites where galactic clouds now seem to be contracting to form stars. Furthermore, our estimate includes only stars now shining; those that formed long ago, and have since blown up, would tend to increase the star-formation rate.

Hence a value of 10 stars/year seems a reasonable number when averaged over the lifetime of the Galaxy. It's probably accurate to within a factor of 10.

FRACTION OF THOSE STARS HAVING PLANETARY SYSTEMS: To evaluate this term, we must estimate how rare or common planetary systems are. We discussed factors pertinent to this term at the end of Chapter 23. There we noted that despite the reported wobbles of some nearby low-mass stars, we have no clear evidence of any planet-sized objects beyond our Solar System.

Recall that the condensation theory is the currently best idea for the origin of the Solar System. As such, planets are imagined to form as byproducts of the star-formation process; recent optical and infrared observations of disks full of dust around some young stars tend to support this idea, as noted toward the end of Chapter 23. Recall also, however, that this theory is plagued by the Sun's unaccountably slow spin. Ironically, the fact that many other stars experience the same problem may well be the best evidence for the existence of planetary systems elsewhere.

Figure 23.7 showed that the low-mass F-, G-, K-, and M-type stars usually have less angular momentum than expected. As for our own Solar System, much of their angular momentum might have been transferred to surrounding planets. In other words, all low-mass stars might spin sluggishly because they have planets orbiting about them.

If this reasoning is correct, we can judge the O-, B-, and A-type star systems as poor sites for planets. Eliminating these three types of stars does not rule out large numbers of objects, however, because high-mass stars are scarce compared to low-mass stars in our Galaxy.

Provided the condensation theory is valid, we can then suggest that most stars have some planets. Accordingly, a value of 1 is probably a reasonable numerical result for this term in the equation. The critical assumption here is the validity of the condensation theory; if it's wrong, this term could be much less than 1.

NUMBER OF HABITABLE PLANETS PER PLANETARY SYSTEM: This term mainly concerns the range of temperature throughout a planetary system. Temperature, probably more than any other single quantity, determines the feasibility of life on a given planet. The temperature at the surface of a planet depends mostly on two things: the planet's distance from its parent star and the thickness of the planet's atmosphere. Planets having a nearby parent star (though not too close) or some atmosphere (though not too thick) are expected to be reasonably warm, much like Earth or Mars. Planets having neither, such as Pluto, will surely be cold by our standards. And planets having both, such as Venus, will be hot.

As shown in Figure 29.11, a three-dimensional zone of "comfortable" temperatures surrounds every star. Such a region is often termed a **habitable zone**. The extent of this zone depends on the spectral type of the star, the atmosphere of the planet, and the type of biology considered. Here we restrict ourselves to life as we know it, namely that which functions provided water remains in liquid form (273 to 373 kelvin). An F-type star, for instance, with a hot surface temperature, has a rather large zone in which temperatures are likely to be comfortable on a planet with a moderate atmosphere. Cooler stars, namely those of G, K, and M spectral types, have progressively smaller habitable zones.

(O-, B-, and A-type stars are not considered here because analysis of the previous term makes it unlikely that they have planets. Even if they do, these massive stars are not expected to last long enough for life to develop.)

We can appreciate habitable zones by imagining regions surrounding various sources of heat on Earth. For example, skaters on a frozen lake know well the range of distance surrounding a bonfire (analogous to a

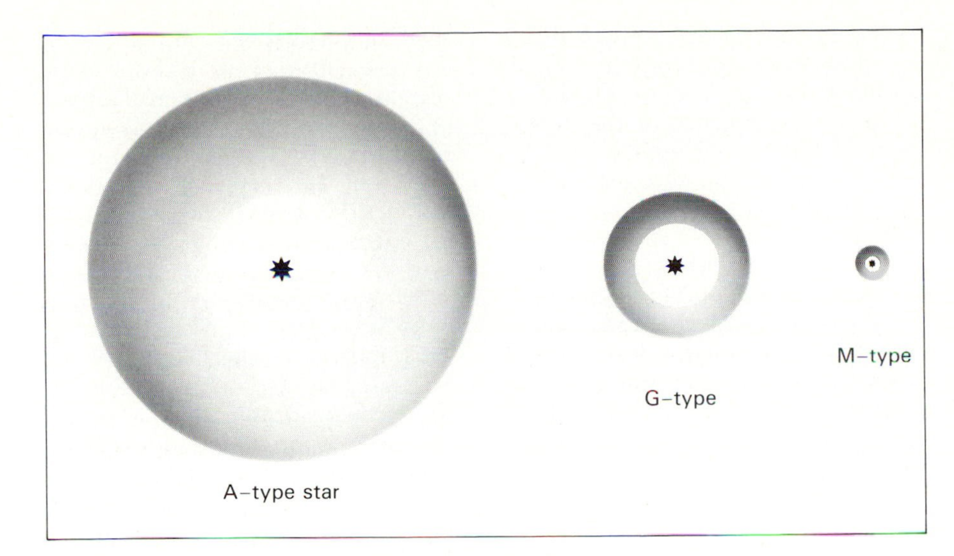

FIGURE 29.11 The extent of a habitable zone (shading) is much larger around a hot star than around a cool star. Be sure to recognize that the zones are three-dimensional, surrounding the central stars in all directions.

star) where the warmth is comfortable. This range depends somewhat on the amount of clothing worn (analogous to a planet's atmosphere). Not much heat is felt far from a bonfire, while just the opposite problem occurs when too close. Also, the larger the bonfire, the larger the zone of comfort, and the more people who can benefit from the heat of the fire.

To evaluate this term, we need to estimate the average number of planets expected within an ecologically comfortable zone. At the extremes, hardly any planets would be expected to reside in the small habitable zones about M-type stars, whereas several planets might coexist in the zone about an F-type star.

Another guideline is that we know for sure that one planet orbits within the habitable zone of a certain G-type star—namely, Earth orbiting about the Sun. Actually, three planets—Venus, Earth, and Mars—reside within (or close to) the habitable zone surrounding our Sun. Venus is hotter than we like because of its thick atmosphere and proximity to the Sun. Mars is a little colder than we like for just the opposite reasons. It's interesting to note that if Venus had Mars' thin atmosphere, and Mars had Venus' thick atmosphere, both of these nearby planets would have had surface conditions resembling those on Earth.

This term of the equation probably equals more than 1 for F-type stars but less than 1 for M-type stars. We might then ordinarily use 1 as a reasonable estimate for the average number of habitable planets orbiting any F-, G-, K-, or M-type star. But two other factors must also be considered in our assessment of this term.

The first additional factor concerns the M-type stars. Besides their having very small habitable zones, the majority of M-type stars undoubtedly formed in the Galaxy long ago. At that time, few, if any, heavy elements were available to fabricate rocky planets. We can thus eliminate M-type stars from further consideration: Those that are young probably have no habitable planets, while those that are old probably have no planets at all. Since M-type stars amount to about 80 percent of all known stars, our estimate of this term drops to 20 percent, that is, $\frac{1}{5}$.

A second factor rules out still more star systems as potential sites of extraterrestrial life. This argument depends on the observed fact that many stars in our Galaxy are grouped into close-knit clusters. Life probably could not develop in such multiple-star systems, even if the stars themselves have planets. Figure 29.12(a) and (b) shows how a planet within the simplest type of star cluster—a binary-star system—could maintain only certain stable orbits, during which time the planet would likely be either too hot or too cold. Even in the case of improbable, though stable "figure-8" orbits like that in Figure 29.12(c), uniform planetary heating over long periods of time—an apparent necessity for life—would be unlikely. These thermal problems worsen for star clusters having many stars.

Since approximately one-half of all known stars seem to be members of multiple systems, we can further reduce this term of the Green Bank equation. Multiplying $\frac{1}{5}$ by $\frac{1}{2}$, we obtain $\frac{1}{10}$. This number is a rough estimate for the average number of habitable planets per star. It doesn't

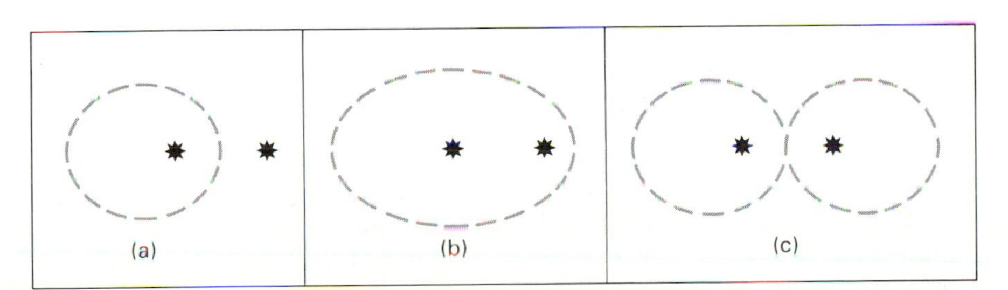

FIGURE 29.12 In binary star systems, planets are restricted to only a few kinds of orbits that are gravitationally stable. For example, a planet is shown here (dashed curves) (a) closely orbiting one of the two stars, which overheat it in the middle, (b) circulating about both stars in an elliptical orbit which takes it in and out of a habitable zone, and (c), interweaving between the two stars in a "figure-8" pattern, which again overheats it in the middle.

mean that one-tenth of a planet orbits all stars. Instead, our estimate suggests that 1 out of every 10 stars has a habitable planet. Clearly, single F-, G-, and K-type stars are the best candidates for stellar systems with habitable planets having life as we know it.

Before moving on to the next term in the equation, take note of a few cautionary remarks. Perhaps the analysis above is much too restrictive. Maybe cool M-type stars are more suitable candidates than the argument suggests. Or perhaps some planets could survive the rigors of a double-star system. Furthermore, other objects (like some of Saturn's and Jupiter's moons) might have thick enough atmospheres to remain comfortable outside the theoretically computed habitable zones. Should any of these statements be true, our estimate of $\frac{1}{10}$ is surely a conservative one; it could be increased.

Yet another reason suggests that the analysis above may substantially underestimate the number of extraterrestrial sites suitable for life. For example, it's not totally impossible that life could arise and survive in places much different from conventional planets. In view of the discussion in Interlude 12-2 about the prevalence of interstellar molecules, we can speculate about life inside interstellar clouds in the same way that we earlier noted the slim possibility for life in the atmospheres of Venus or Jupiter. Only the densest interstellar regions—the so-called molecular clouds—have physical conditions resembling those of planetary atmospheres. In all fairness, however, the lack of energy to provide warmth and the scarcity of matter to provide nourishment tend to make interstellar clouds unlikely abodes for life. If we're wrong, though, the prospects for extraterrestrial life would increase greatly.

FRACTION OF THOSE HABITABLE PLANETS ON WHICH LIFE ACTUALLY ARISES: This is the province of chemical evolution. As noted in Interlude 26-1, the critical question here is the following: Is life the *only* type of material complexity that can be expected in other habitable zones, or is life only *one* example of many possible types of complexity? In other words, is or is not life an inevitable consequence of the evolution of matter? Given the proper conditions, is life a sure bet, or is it quite rare?

Current research seeks to understand how complexity arises from simplicity. Much progress has been made in the past few decades, but a good appreciation for some of the most important chemical reactions still eludes us. The reason is that life itself is extraordinarily complex, much more so than galaxies, stars, or planets.

Consider, for a moment, the simplest known protein on Earth. This is insulin, which has 51 amino acids linked in a specific order along a chain. We can use probability theory to estimate the chances of assembling the correct number and order of amino acids for such a protein molecule. Since there are 20 different types of amino acids, the answer is $1/20^{51}$, which equals about $1/10^{66}$. This means that the 20 amino acids must be randomly assembled 10^{66}, or a million trillion trillion trillion trillion trillion, times before getting insulin. This is a great many combinations, so many in fact that we could randomly assemble the 20 amino acids trillions of times per second for the entire history of the Universe and still not achieve the correct ordering of this protein.

Larger proteins and nucleic acids would be much less probable *if chemical evolution operates at random.* And to assemble a human being would be vastly less probable, if we had to do it randomly starting only with atoms or simple molecules.

This is the type of reasoning used by some researchers to argue that we must be alone, or nearly so, in the Universe. They suggest that biology of any kind is a highly unlikely phenomenon. They argue that meaningful molecular complexity can be expected at only a very, very few locations in the Universe, and that Earth is one of these special places. And since, in their view, the fraction of habitable planets on which life arises is extremely small, the number of advanced civilizations now in the Galaxy must be even smaller. Of all the myriad galaxies, stars, planets, and other wonderful aspects of the Universe, this viewpoint maintains that we are among the very few creatures to appreciate the grandeur of it all. If their arguments are correct, we could be alone in the Universe.

But does chemical evolution operate at random, that is, by chance? Alas, there is another point of view. Several reasons suggest that the change from simplicity to complexity may not proceed randomly. The first reason is this: Of the billions upon billions of basic organic groupings that could possibly occur on Earth from the random combinations of all sorts of simple atoms and molecules, only about 1500 actually do occur. Furthermore, these 1500 organic groups of terrestrial biology are made from only about 50 simple organic molecules, including the amino acids and the nucleotide bases. This suggests that molecules critical to life are probably not assembled randomly by chance. Apparently, additional forces are at work at the microscopic level. Whatever they are, they remove some of the randomness.

Direct laboratory experiments also support this alternative view. Simulations that resemble conditions on primordial Earth are now routinely performed with a variety of energies and initial reactants (provided that there is no free oxygen). These experiments demonstrate that unique conditions are unnecessary to produce the precursors of life. Complex acids, bases, and proteinoid compounds are formed under a rather wide variety of physical conditions. And it doesn't take long for reasonably complex molecules to form—not nearly as long as probability theory predicts by randomly assembling atoms.

Furthermore, every time this type of experiment is done, the results are the same. The soupy organic matter trapped in the test tube always yields the same proportion of proteinoid compounds. If chemical evolution were entirely random, we might expect one type of compound one time, and another the next. Apparently, electromagnetic force fields govern the complex interactions of the many atoms and molecules, substituting organization for randomness.

Of course, precursors of proteins and nucleic acids are a long way from life itself. But the beginnings of life as we know it seem to be the product of less-than-random interactions among atoms and molecules. That's important to know. Just how unrandom—that is, how common—life itself might be, is unknown.

An important caveat deserves mention here. Even if life everywhere in the Universe is based on carbon chemistry and obeys the basic laws of biology familiar to us, we should not be foolish enough to think that organisms elsewhere would evolve to look like us. Life forms on other planets—even carbon-based organisms operating in a water medium—would experience a completely different set of environmental and genetic changes. The mechanism of biological evolution, with its mutations, natural selection, and adaptation operating over long durations of time, would guarantee little outward resemblance to life on Earth.

So what do we choose as a numerical estimate for the fraction of habitable planets on which life actually arises? Either we choose a number much smaller than 1, if we prefer to think that randomness or chance plays a big role. Or we choose a number close to or equal to 1, if randomness plays no appreciable role at all. The former view suggests that life arises naturally, though rarely, whereas the latter view maintains that life is inevitable given the proper ingredients, suitable environment, and long enough periods of time. No easy experiment can distinguish between these alternatives.

What we really need is a laboratory where the organic chemistry has been left alone for a few billion years. What transpires there could help us decide the degree of randomness inherent in the molecular reactions. Fortunately, some of the nearby planets or their moons provide us with just such a laboratory, and a most interesting episode in history will unfold as our spacecraft probe them for signs of life. In the minds of some researchers, the discovery of life on Mars, Jupiter, Titan, or some other object in our Solar System would convert the origin of life from an unlikely miracle to an ordinary statistic—to a value equal to or near 1 for this term of the equation.

In addition to the fact that "randomness" is not fully operational, we can offer other reasons to increase the prospects for extraterrestrial life. One of these is that extraterrestrial life could be based on something other than the carbon atom. We've noted in the preceding section that life "as we know it" is carbon-based life, operating in a water medium, with higher forms metabolizing oxygen. But once again, are we being chauvinistic by thinking that other types of biology are impossible? Perhaps we are, although we've also noted several reasons why carbon-based life has more strength, diversity, and adaptability than any other.

We have to wonder: Are we able to make an objective judgment independent of our own prejudices? After all, chemists study Earth chemistry, not general chemistry. And biologists study the only kind of biology they know. Perhaps the alternatives have not yet been sufficiently investigated. At any rate, should there be biochemistries other than the carbon-in-water type studied throughout this book, then the prospects for extraterrestrial life increase greatly.

FRACTION OF THOSE LIFE-BEARING PLANETS ON WHICH INTELLIGENCE ARISES: This is the province of biological evolution. Here we are concerned with the difficulty or ease with which multicellular organisms can arise. In particular, we wish to estimate the probability that a truly advanced organism will eventually develop a central nervous system or brain.

As discussed in Chapter 26, the fossil record demonstrates that evolution indeed occurs. Sporadic mutations in the genes of an organism allow it to adapt to gradual changes in the environment, enabling it to achieve the best available niche. Some types of organisms are naturally selected to die; others are naturally selected to thrive. Biological evolution via natural selection is a mechanism that generates highly improbable results.

Like other features of life, intelligence results from a series of adaptations. Organisms that profitably use those adaptations can develop more complex behavior, which in turn enables them to completely dominate a given niche. Only complex behavior can provide organisms with the *variety* of choices needed for their advanced development. Those that change in ways that allow them to function better in their environment are said to be "smarter."

What, then, are some of the factors that contributed to our complex behavior? Very basic traits, such as locomotion and food gathering, were probably not so important. These traits do not produce superior intelligence. Instead, they are needed by all animals just to ensure survival. Nature will not tamper with life's most essential needs.

Other factors helped generate the level of intelligence that we now share. The invention of primitive tools, for example, allowed our ancestors to inhabit a whole new ecological niche. As noted in Chapter 28, the fossil record of several million years ago strongly suggests a direct correlation between tool use and brain size.

Hunting, especially coordinated efforts by several hunters, was another important factor. It required much social cooperation and planned movement. The invention of hunting may have been the main factor that allowed our ancestors to leave the dense forests and venture into the open savannah.

Most important, social ties and basic intelligence were greatly enhanced by the development of language. By communicating, individuals could signal each other while hunting for food or seeking protection. In the opinion of many anthropologists, language was a key factor in the development of our brain—so important, in fact, that they suggest that when reduced to its essentials, human intelligence is synonomous with human language.

So again we are faced with a decision. What numerical estimate do we choose for the fraction of life-bearing planets on which intelligence of some sort eventually develops? The answer is not an easy one.

One school of thought maintains that given enough time, intelligence is inevitable. They argue that on every life-bearing planet, there is a niche labeled "intelligent life." Assuming that natural selection is a universal phenomenon, it would seem to be advantageous for an organism to try to fill it. In other words, provided that an organism could afford the extra weight and metabolism, intelligence is an obvious advantage. If this view is correct, the fifth term in our Green Bank equation equals or nearly equals 1.

Other researchers object. They argue that there is only one known case of intelligence, and that's life on Earth. For 2.5 billion years—from the start of life about 3.5 billion years ago to the first appearance of multicellular organisms about 1 billion years ago—life did not advance beyond the unicellular stage. Life remained simple, and dumb, but it survived. And thus far, it has survived a lot longer than any form of intelligent life. This alternative view suggests that while microscopic life may be widespread throughout the Universe, it hasn't necessarily advanced to the point of becoming intelligent. Should this extreme view be correct, this fifth term could be very small. If so, we might be the smartest life forms anywhere in the Galaxy.

FRACTION OF THOSE INTELLIGENT-LIFE PLANETS THAT DEVELOP A TECHNOLOGICAL SOCIETY: Intelligence is a most useful factor in the development of any species. In our case, we inherited several advantages from our reasonably smart ancestors of a few million years ago: A pollution-free environment, a sophisticated society, a good family life, a robust physique, and a taste for steak. Intelligence led to a whole new way of life—a reasonably comfortable way of life.

But, now, modern men and women are threatened with several global crises. The number of humans on Earth is increasing rapidly, and neither enough food nor enough energy can be distributed well enough to keep everyone content on a daily basis. As if these problems weren't enough, we also face the possibility of man-made disaster, especially nuclear warfare. Other planet-wide problems loom on the horizon. Our society now seems troubled with several potential problems, the likes and scopes of which Earth societies have never before experienced.

We can then ask: How did intelligent life change from the rather pleasant state of affairs left to us by our ancestors to the current predicament we now face? In other words, How did we mess it up so badly? The answer, apparently, dates back about 10,000 years. At that time, our recent ancestors invented agriculture, cities, states, empires. Above all, they created a technological civilization.

On the positive side, it is the rise of precisely this technological civilization that has given us the tools to unlock secrets of the Universe, as well as to search for extraterrestrial life. But technology also has its drawbacks, the chief one being that technology is a major source of each of our current global problems.

To evaluate the sixth term of our equation, we need to estimate the probability that intelligent life eventually develops technological competence. Should the rise of technology be inevitable, given long enough durations of time, this term is close to 1. If so, then at least one species on all life-bearing planets eventually develops a technological society.

If it's not inevitable—if intelligent life can somehow avoid developing technology—then this term could be much less than 1. The latter view envisions a Universe teeming with intelligent life, but very few among them ever becoming technologically competent. Perhaps only one managed it—us.

It's nearly impossible to distinguish between these two extreme views. We don't even know how many prehistoric Earth cultures failed to develop technology. We do know that the roots of our present civilization arose independently in several different places on Earth. These include Mesopotamia, India, China, Egypt, Mexico, and Peru. Since so many of these ancient cultures originated at about the same time, we might think the chances are good that some sort of culture will inevitably develop, given some basic intelligence and enough time.

But literary culture is one thing and technological culture another. Archeologists suggest that some of these ancient people never developed a technology. The Mayan civilization of Interamerica, for example, had sophisticated social and political organizations. They built primitive observatories, enabling them to study the motions of stars and planets with the naked eye. In fact, the Mayan calendar was more accurate than that of the Spaniards who conquered them a few centuries ago. Despite these accomplishments, archeological records show that the Mayans used neither the wheel nor metal. They built small toys with wheels, but not large carts or wheelbarrows useful in farming or herding. Either they never thought to use these technological aids, or they realized them and rejected them.

Regardless of how many ancient earthlings accepted or rejected technology, only humans developed it and now use it. This is a sticky point for some researchers. If technology is an inevitable development, they ask, why haven't other forms of Earth life also found it useful? The probable reason is that a given niche is usually filled by only one species. And the niche labeled "technological intelligence" is currently filled by *Homo sapiens*.

In an evolving society, we should expect only one species per niche. As an example, recall that the recent fossil record suggests the coexistence of several hominids angling for the same niche several million years ago. The inevitable result was competition, and the demise of all but one of those ancestral hominids. Competition between the various australopithecenes probably provided that great impetus in the survivor's drive toward superior intelligence.

So the fact that only one technological society exists on Earth does not imply that the sixth term in our Green Bank equation must be very much less than 1. On the contrary, it is precisely because *some* species will probably always fill the niche of technological intelligence that this term is probably close to 1.

One further point is worth noting. This sixth term could be decreased somewhat if most planets are completely covered by water. Why? Because technological intelligence is likely to develop only on the solid portion of a planet. Aquatic life may be intelligent, but it's hard to imagine how it could ever become technical. To discover the laws of applied physics, something resembling hands must be able to manipulate gears, pulleys, inclined planes, and the other rudiments of elementary technology.

This is not a criticism of the dolphins. There's no doubt about their intelligence. They probably admire the stars while poking their heads above water. Perhaps they even wonder if dolphins reside on other worlds. But unless they leave the water, they can never become technologically competent. Will they come out of the water to try to develop that technology? Probably not now, because we fill the niche of land-based technological intelligence. If they tried to evolve onto the land, they would soon enter into direct competition with us, and we would surely dominate them.

LIFETIME OF TECHNOLOGICAL CIVILIZATIONS: As if each of the terms discussed above were not uncertain enough, the last term in our equation is the most uncertain of all. How can we reliably determine the lifetime of a technological civilization? After all, we are the only known example of such a civilization. And the duration of Earth civilization is an issue of great debate.

One thing is certain: If the correct value for any one term in the equation is very small, then not many technological civilizations can now exist in the Galaxy. Even if the average lifetime of a civilization is a million (10^6) years, only one term with a value as small as 1 in a million ($1/10^6$) implies that we are alone in the Galaxy. For example, suppose that all the terms have optimistic values, with the exception of the evolution of intelligence. Assume that, for some reason, there is only 1 chance in a million that life on a suitable extraterrestrial planet will attain intelligence. Substituting this pessimistic value into the equation, along with the other optimistic values, we have

number of galactic civilizations =

$$\frac{10\text{ stars}}{\text{year}} \times \frac{1\text{ planetary system}}{\text{star}} \times \frac{0.1\text{ habitable planet}}{\text{planetary system}} \times \frac{1\text{ planet with life}}{\text{habitable planet}} \times \frac{0.000001\text{ planet with intelligence}}{\text{planets with life}} \times \frac{1\text{ planet with technology}}{\text{planets with intelligence}} \times \frac{\text{lifetime (in years)}}{\text{planets with technology}}$$

This long equation boils down to a simple answer, after we cancel all the units. That answer is 0.000001 × lifetime (in years). Accordingly, if the lifetime of a typically advanced civilization is a million years, the number of galactic civilizations equals 1. That's us! For there to be more than one civilization now in the Galaxy, the average lifetime of all civilizations must exceed 1 million years.

Be sure to recognize that even with this pessimistic estimate for the eventual onset of intelligence, thousands of other civilizations might be spread across the Galaxy, provided that their average lifetime is billions of years. Thus the average longevity of technological civilizations is a critically important factor in any analysis of extraterrestrial intelligence.

If any *two* terms in the equation are very small, then chances are slim that the Galaxy is teeming with intelligent life. For example, if there are many fewer planets than the condensation theory suggests, and if the development of technology is improbable, the multiplication of these terms implies that many civilizations cannot possibly now reside in the Galaxy.

The most optimistic estimate uses values close to or equal to 1 for each term. In this case, a most curious thing happens. If habitable planets are plentiful, and the development of life, intelligence, and technology inevitable, then our equation takes on the form

number of galactic civilizations =

$$10 \times 1 \times \tfrac{1}{10} \times 1 \times 1 \times 1 \times \text{average lifetime (in years)}.$$

Notice that all the interior terms cancel; $10 \times 1 \times \frac{1}{10} \times 1 \times 1 \times 1$ equals 1. Thus the number of technological civilizations now in the Galaxy equals the average lifetime of those civilizations. Thus this optimistic solution places great importance on the many factors affecting a civilization's lifetime. What are those factors? In particular, what are some of the factors that could terminate a civilization?

Answers to these questions are not easy. They might be impossible. To estimate the longevity of any civilization requires us to know how intelligent beings are likely to use technology. We can perhaps gain some insight by examining what the future holds for us on planet Earth.

No one can be sure that our civilization will last for a long time. Many of us might like to think that it can. And perhaps it will. But unlike any other time in history, humans now have the means to determine who on our planet survives and who doesn't. With technology, we probably even have the ability to destroy intelligent life on Earth. In fact, it's possible that our entire planet could be rendered completely lifeless in the near future.

One of the key factors affecting the lifetime of our civilization is the rate at which technology now forces us to change. Just in the past century, we've increased our speed of travel by more than 1000 times, and our speed of communication by roughly a million times. Within the past few decades alone, we've enhanced by several million times the speed of data processing, and the speed with which we fabricate new weapons.

Examined from afar, the problem seems paradoxical. Evolutionary change of matter created our Galaxy, Sun, Earth and life. And evolutionary change of life in turn created intelligence, culture, and technology. But now this very same phenomenon—change—seems to be threatening us. The reason is that we're in the driver's seat; humans have become the agents of change. And we are now forcing change more rapidly than nature could have ever produced.

Some people, unable to cope with rapid change, argue that technology is the main cause of many of today's pressing problems. Population growth, environmental pollution, depletion of natural resources, shortage of food and energy, possibility of nuclear war, and several other predicaments are now, or soon will be, threatening the existence of Earth civilization and perhaps all of Earth life itself. Some of these problems are in fact by-products of technology. Yet the saddest problem of all is that our social and political organizations seem unprepared to develop innovative responses needed to ensure our survival.

Our civilization is fast approaching definite natural limits. This does not necessarily mean an end to civilization itself. But the rapid changes brought upon ourselves do seem to be nearing an end. For example, we cannot transfer information faster than the light velocity of our current radio and telecommunications. We cannot travel around the Earth faster than orbital velocity, a speed already attained by our space probes and astronauts. We cannot increase energy production and atmospheric pollution beyond the point of melting the polar ice caps; a condition that we are rapidly nearing. And regarding weapons manufacture, we cannot be deader than dead!

Our civilization is in a transition period unlike anything Earth society has ever encountered before. What are we to do? As sketched in Figure 29.13, there are only a few possible routes into the future.

First, our civilization can fail to solve the worldwide problems now facing us. If so, then civilization ends for all practical purposes. And human life perhaps becomes extinct. This is the view of the doomy pessimists. Another civilization could perhaps arise from the remnants of such a dead civilization. But this solution admits that civilizations do not survive very long after inventing technology.

Second, our civilization can successfully solve the worldwide problems as each in turn becomes critical. This solution resembles someone successfully navigating a mine field. Taking one step at a time, we can use technology to solve our problems, to make progress, and thus to survive. This is the view of the technological optimists. This solution seems reasonable at face value. But deeper thought suggests a dilemma.

To solve any of the approaching global problems, our civilization must be willing to sacrifice. To alleviate any one problem requires our civilization to exercise some restraint, some constraint. And yet, as we solve them all—and to survive we must solve *every* global problem—the required restraint accumulates.

Should our civilization elect this second route into the future, all of us must therefore become less free to do what we want and more restrained to do specifically whatever enhances our survival. This second route seems to lessen democracy and to favor authoritarianism, all the while aiming us directly toward a state of "stagnation"—a social state that is hard to define but one where personal freedom, dignity, curiosity, and many other cherished human rights are diminished, perhaps even eliminated.

We might wonder: If curiosity dies, does intelligence die also?

We might also wonder: Do either of these two routes—extinction or stagnation—represent the normal course of cosmic evolution? Of course, we don't know.

What we do know is that any further development of our civilization—and thus any further attempt to extend our technological lifetime—will require a delicate balance between opposing dangers and temptations. On the one hand, we face the danger of extinction if even one global problem goes unsolved. On the other, we face the temptation

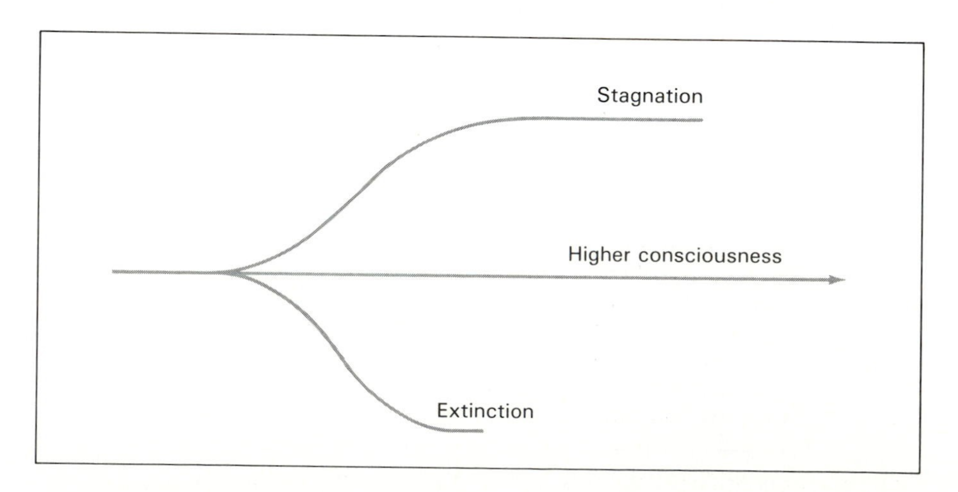

FIGURE 29.13 Our civilization will likely take one of three possible routes into the future.

of becoming a stagnated society because of the increased stabilization and restraint needed to survive.

Is there a narrow path between these two alternatives? Can we find some way to survive without becoming stagnated? Perhaps we can.

Some researchers argue that a search for extraterrestrial intelligent life may be just what we need to guide us along that narrow path. Why? Because an international program of searching for galactic civilizations will provide additional stimulation and competition needed to survive. It would boost our curiosity to enormous heights. And it might enable us to postpone, perhaps indefinitely, the stagnated society for which we seem destined sometime in the future.

Earth now stands on the threshold of gaining membership in the community of galactic civilizations—assuming that galactic civilizations really do exist in the depths of space. With the aid of modern equipment such as radio telescopes, robot space probes, and digital computers, our civilization is now capable of taking what may be the next great evolutionary leap forward—making contact with extraterrestrial intelligent life.

Provided that a galactic empire exists, we might profit greatly by communicating with its members. We have, after all, just crossed the threshold of technology. If advanced life really does reside beyond Earth, we are likely to be among the dumbest of intelligent civilizations in the Galaxy.

At the least, discovery of advanced galactic life will assure us that it is possible for technological civilizations to avoid doomsday—to survive.

At the most, by establishing an interstellar dialogue, our civilization may well be able to strive toward a

INTERLUDE 29-2 ***Where Are They?***

One of the greatest scientists of this century, the Italian physicist Enrico Fermi, once posed a straightforward query—namely, if intelligent extraterrestrials densely populate our Galaxy, then "Where are they?" Why don't we see some evidence of their advanced status—for example, interstellar radio communications, settlements on nearby planets or moons, unworldly stellar engineering projects, and the like?

An obvious possible answer to Fermi's question is that we are the only technologically intelligent creatures in the Universe. Advanced life forms might consciously decide to limit or even to terminate their technical growth, either out of economic necessity or ethical scruples. Or they might simply lose interest in pursuing grander objectives, in effect reach a stage of mental stagnation wherein curiosity is minimized. Yet another possible answer suggests that the Milky Way is ruled by a superintelligent civilization, in fact governed so masterfully that its inhabitants have placed aside our part of the Galaxy, as we do on Earth with wildlife preserves, so that they can learn more about cosmic evolution by studying us in our natural habitat—a proposal implying that we effectively live in a galactic zoo!

An even more sobering possibility is that technological civilizations fail to survive—a proposition not entirely unreasonable, for no civilization may be able to check its tendency to self-destruct over sufficiently long time scales. At virtually any given moment in cosmic history, the Universe might well be populated by one or only a few cosmic intelligences. An analogy is useful here: Imagine an ornate chandelier having a huge number of light bulbs. The chandelier is meant to represent the Galaxy, while the bulbs denote planets ecologically suited for the emergence of technological intelligence. Each bulb illuminates only when technology on a given planet surpasses some crucial threshold—such as radio communicative ability. Two factors generally determine whether the chandelier is blazing or dim: One concerns the vitality of the evolutionary process leading to technologically competent life—the extent to which evolution's hand twists far enough to screw in and light a bulb; while the other concerns the length of time each bulb stays lit—the longevity of a technical civilization. At the two extremes, the chandelier could be brightly lit with many glowing bulbs, indicating much intelligent activity on many planets, or it could be completely unlit, designating a technocultural void. Considering the many varied and speculative time scales affecting the evolution of intelligence and of culture—especially those for technological emergence and longevity—all the bulbs might eventually glow without any two bulbs' ever being lit simultaneously.

Perhaps the most reasonable of all answers to Fermi's question is that we might well be among the earliest of technologically oriented species to originate in the Galaxy—although this would result from evolutionary fortune rather than any anthropocentrism. The development of organized structures basically depends on the expansion of the Universe, for it is this grandest of all changes that establishes the thermodynamic conditions needed for those structures (consult Interlude 17-1). And since the onset of life and intelligence requires certain minimal (although broadly specified) times for their origin, many technological intelligences might arise and evolve more or less in parallel throughout the Universe. If true, then our Galaxy is potentially populated with myriad civilizations surprisingly close to us in their evolutionary development. As such, technological intelligence would only now be "coming on line" in the cosmos.

higher level of consciousness heretofore unimagined.

Recognize that contact itself will not necessarily provide us with instantaneous intelligence, although it might. Nor will extraterrestrials necessarily provide us with solutions to our worldwide problems, although they might. The suggestion now being made by some researchers is that *the very program of searching* will stretch our imaginations, widen our horizons, enhance our curiosity. The search itself becomes a method of survival.

Should this reasoning be valid, the establishment of a galactic culture may well be the normal route of cosmic evolution.

Ways to Make Contact

INTERSTELLAR SPACE FLIGHT: How can our civilization search for extraterrestrial life? By what means can we attempt to make this evolutionary leap forward?

One obvious way is to develop the capability to travel far outside our Solar System. By involving many nations in the space programs begun by the United States and the Soviet Union, our civilization might be expected eventually to develop the means to travel through interstellar space. However, such technology will not be achieved easily; it may not even be a practical possibility.

The basic problem with interstellar space travel is the quantity of fuel needed for long-duration flights. The damaging effect of galactic radiation is an additional problem, as are the loneliness and boredom of generations of humans having to spend their entire lives aboard a spacecraft.

Consider a few examples. A trip to the *nearest* objects beyond our Solar System, namely the Alpha-Centauri star system some 4.3 light-years away, would take about 25,000 years. This calculation assumes a constant flight velocity of 50 kilometers/second, which approximates the speed of our fastest unmanned space probes. That's a fast enough velocity to escape the Solar System, but 250 centuries would be needed to reach this closest star system. Contrast this duration with, for instance, the 15 centuries since the fall of the Roman Empire, or the 50 centuries since the construction of the Pyramids. Even if fuel for the spacecraft and food for the inhabitants were available, such a flight would take an incredibly long time by human standards.

In the example above, we assumed a spacecraft traveling at a velocity much less than the velocity of light. In fact, 50 kilometers/second equals only 0.02 percent of light velocity. We might imagine that our civilization could someday become smart enough to achieve flight velocities close to 100 percent of light velocity. If so, we might also expect that travel times could be reduced dramatically by means of special relativistic effects that occur at such high velocities.

Relativistic space flight seems nice in theory. Flight times could seemingly be reduced, as has been suggested by legions of science fiction writers. But in practice, there is a problem. Spacecraft traveling at speeds close to light velocity cannot be refueled. A reasonable-sized craft of, for example, 10^4 kilograms (10 tons) would need more than a billion kilograms (a million tons) of hydrogen fuel just to accelerate it to 10 percent of light velocity. These numbers assume that such a spacecraft derives its energy from a hydrogen–helium fusion reaction; the helium squirts out the rear, and the spacecraft lurches forward. To reach relativistic speeds, namely greater than 90 percent of light velocity, would require so much on-board fuel as to make this proposition absolutely ridiculous.

Attempts to use interstellar hydrogen gas as a spacecraft moves through space seem equally futile. To accelerate to 10 percent of light velocity would require a gigantic scoop kilometers across. And to reach relativistic velocities would require a scoop so large as to defeat completely this basic fueling problem.

Given the laws of physics as we currently know them, relativistic space flight seems destined to remain science fiction. Probably no intelligent civilization would or could do it.

These arguments do not necessarily prohibit traveling through interstellar space at much slower speeds. If our descendants can overcome the problems of radiation and boredom during long, long journeys, interstellar travel might someday become feasible with modest fuel supplies and speeds of a few percent of light velocity. To do so, however, would mean not just generations of space travel, but flights lasting on the order of thousands and even millions of years.

Equally possible, these practical problems might never be solved (or perhaps our civilization will not survive long enough to attempt interstellar flight). Future generations of Earthlings might well conclude that it's simply impractical to travel over large galactic distances. If so, we will be forever confined to our Solar System and a handful of nearby stars.

At any rate, we can hardly hope to use space flight to make contact with extraterrestrials. Even with the most optimistic estimates of the terms in the Green Bank equation, galactic civilizations are likely to be spread out like small islands within the vast sea of galactic space. For example, with all the interior terms of our equation maximized, and the average technological lifetime equal to 1000 years, we would conclude that 1000 advanced civilizations currently reside in our Galaxy. Given the size and shape of the Milky Way, several thousand light-years would separate any two adjacent civilizations. Such large distances force the word "neighboring" to take on a new meaning.

Even if the average lifetime of galactic civilizations is 1 million years, our most optimistic estimates suggest that each is separated by some several hundred light-years. To have a reasonable hope of successful contact, hundreds, perhaps thousands, of sorties would have to be launched toward candidate star systems. All in all, interstellar space flight is both impractical and uneconomical either at the present time or in the foreseeable future. It may never become feasible.

INTERSTELLAR PROBES: Other methods could be used to search for extraterrestrial intelligence in the Galaxy. One such method imagines the launch of many unmanned space probes. Each robot probe would be given a velocity sufficient to escape our Solar System. And each would eventually reach a star system thought to be a good candidate for intelligent life. There it would orbit the star, looking and listening for evidence of life on one of the planets.

Interstellar probes could be programmed, for example, to detect the leakage of electromagnetic radiation arising from the daily activities of a technological civilization. Such a probe might succeed immediately upon arriving, should an advanced civilization already be thriving. Or a probe might need to bug an alien star system for thousands of years before seeing or hearing any type of planetary activity resembling our television, FM radio, military radar, or whatever. Once the robots did detect any sign of intelligence, they would send a radio signal back to Earth, letting us know that a technological civilization has emerged.

Robot probes offer a couple of advantages in the search for extraterrestrials. Robots are neither bored by the long duration of the flight, nor harmed by the harsh radiation of interstellar space.

A disadvantage with this method of contact is that, once again, it would seem economically unfeasible for the foreseeable future. To bug all single F-, G-, and K-type stars within 1000 light-years of Earth would require about a million probes. Given the number of days in a year, we would need to launch one probe every day for nearly 3000 years. Aside from these formidable logistics problems, the cost of such a program would be staggering.

In a sense, our civilization has already launched several probes, although they lack the sophistication of the robots discussed above. Figure 29.14 is a reproduction of a plaque mounted on board the American *Pioneer 10* spacecraft launched in

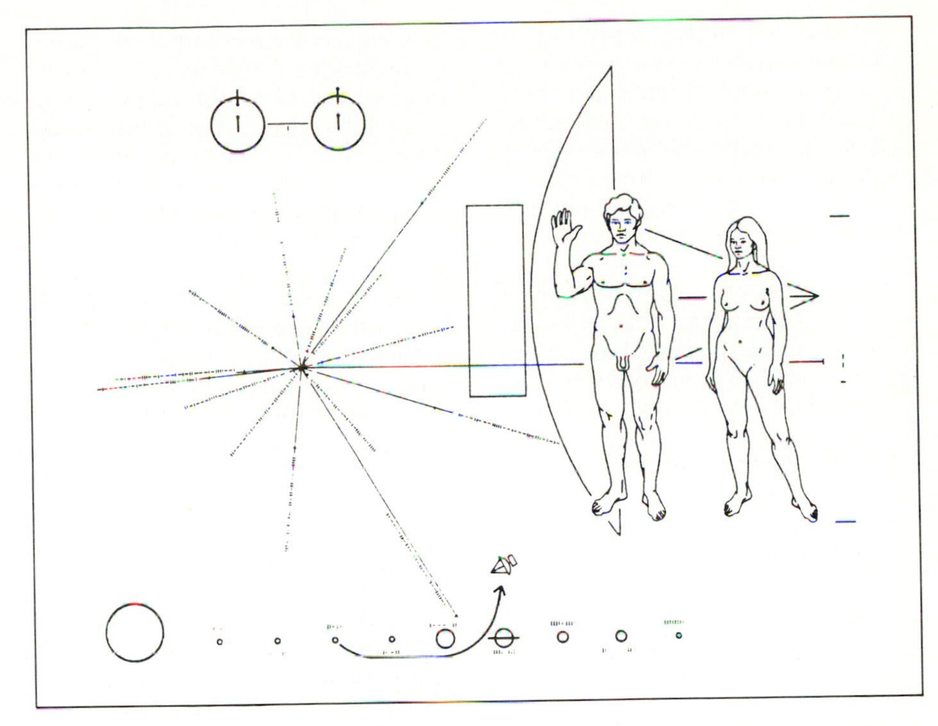

FIGURE 29.14 This is a replica of a plaque mounted onboard the *Pioneer 10* spacecraft. The important features of the plaque include: a scale drawing of the spacecraft, a man, and a women (right); a schematic diagram of the hydrogen atom undergoing a change in energy (top left); a starburst pattern representing various pulsars and the frequencies of their radio waves that can be used to estimate when the craft was launched (middle left); a depiction of the Solar System, showing that the spacecraft left the third planet from the Sun and passed the fifth planet on its way into outer space (bottom); and all the drawings have computer (binary) coded markings from which actual sizes, distances, and times can be derived.

the mid-1970s. Similar information was also included aboard the American *Voyager* spacecraft launched in 1978. After visiting the outer planets, these probes are on their way out of the Solar System, but they have no specific destination thereafter.

Even if these American spacecraft do accidentally encounter an alien star system housing an advanced civilization, these machines are incapable of reporting that news back to Earth. Should the civilization on the other end intercept the probe, they should be able to unravel most of its contents using the universal language of "mathematiceese." The caption to Figure 29.14 notes how the aliens might discover from where and when the *Pioneer* and *Voyager* probes were originally launched. They would then know that we are here (or were when the probes were sent), although we would be unaware of their existence.

So methods that rely on space travel, either manned or unmanned, do not seem to hold much promise in contacting extraterrestrial intelligent life. One can always argue that future technological breakthroughs may someday make such projects more favorable. Indeed, the microelectronic revolution now under way might eventually make unmanned probes more feasible. But on the basis of what we know now, long-distance space travel seems logistically and economically unlikely.

Aside from these practical problems, some scientists argue that it's not a good idea to signal extraterrestrials actively. As noted earlier, our recent emergence as a technologically competent civilization implies that we are now among the dumbest technological intelligences in the Galaxy. Any other civilization either that we discover, or that discovers us, will almost surely be more advanced than us. Consequently, a healthy degree of skepticism is warranted. After all, if you were lost in a jungle populated by unknown natives, you would be wise not to yell, scream, or generally bang on the trees to let them know of your presence. It's a lot safer to explore your environment quietly for a while, to listen to the sound of the drums, and to get a feeling for their intentions.

Some anthropologists have even speculated about the behavior of advanced galactic civilizations. If extraterrestrials even remotely behave like human civilizations on Earth, then the most advanced aliens might naturally try to dominate all others. Indeed, the "smarter" species have often taken advantage of others throughout the history of life on Earth.

On the other hand, the aggressiveness of Earthlings may not at all apply to extraterrestrials. Furthermore, the vast distances separating galactic civilizations would probably prohibit a civilization on one planet from physically dominating or enslaving a less advanced civilization on some other planet. In fact, physical contact among different cosmic civilizations may not be biologically desirable; what is healthy for one life form might be a disease to another.

RADIO COMMUNICATION: A third technique is much cheaper than either of the two methods discussed above and has many advantages. This method strives to make contact with extraterrestrials using only electromagnetic radiation. It does not involve hardware traveling through space. The electromagnetic technique is economically feasible, can be undertaken with existing equipment, and does not reveal our presence should some extraterrestrials be hostile.

The third method uses radiation as the fastest known means of transferring information from one place to another. Since light and other types of high-frequency radiation are hopelessly scattered while moving through dusty interstellar space, long-wavelength radio radiation seems the best way to transfer information in the plane of the Galaxy where civilizations are likely to be located.

To make this method work, radio telescopes on Earth would be used to passively listen for radio signals emitted by someone else. No radiation would be transmitted by our civilization toward distant star systems. Best of all, we have already invented the equipment needed to detect such radio signals. Some preliminary searches are now under way, thus far without success.

Radio searches of this type are not without problems, however. The foremost problem is that this method assumes that extraterrestrials are in fact broadcasting radio signals for one reason or another. If they are not, this search technique will fail. Another problem concerns the need to distinguish radio signals artificially generated by advanced civilizations from signals naturally emitted by interstellar gas clouds. Other problems include the direction in which to aim our radio telescopes, the frequency at which to tune our receivers, and the time of day at which to listen.

We could undertake such a search by following either of two strategies. One strategy attempts to eavesdrop on the radio radiation normally leaking from some planet while its civilization goes about its daily activities. Some idea of what we might expect can be gained by reversing the problem and examining the appearance of Earth from afar.

Figure 29.15 shows the pattern of radio signals unintentionally emitted into space by our civilization. From the viewpoint of some distant observer, the spinning Earth emits a bright flash of radio radiation every few hours. The flashes result from the periodic rising and setting of hundreds of FM radio stations and television transmitters (AM broadcasting is trapped below our ionosphere). Because the great majority of these transmitters are clustered in eastern United States and Western Europe, and because they emit their radiation parallel to the ground where people live, a distant observer would detect blasts of radiation leaking from Earth as our planet rotates each day. This radiation races out into space as a growing sphere of radio, television, and other electromagnetic signals (such as radar). It has been doing so since the invention of these technologies a few decades ago. In fact, Earth is now a more intense radio emitter than the Sun.

If any advanced civilizations reside within 30 light-years of Earth, we have already broadcast our presence to them. Whether or not the extraterrestrials have received the message, we do not yet know. Perhaps they have, but have concluded that we are not so intelligent, given the message content of our radio and television programs.

This eavesdropping strategy resembles the space probe technique discussed earlier, although here the "bugging" of other planetary systems can be accomplished from Earth using large radio telescopes. However, radio telescopes now in operation are probably sensitive enough to eavesdrop on only a few of the nearest star systems. Remember, the strength of radiation decreases as the square of the distance.

We currently have the engineering ability to build a radio system that could intercept the hodgepodge of radio, radar, and television signals wastefully leaked into space by rather distant civilizations—namely, ones like our own. Figure 29.16 shows the major features of a gigantic array of radio telescopes capable of eavesdropping on Earth-like civilizations within 1000 light-years of us. Called Project Cyclops, this huge machine hasn't been built yet. It may never be built, since the price tag for such a device is $10 billion (1971 American dollars). Its construction is currently improbable, given all the social, bureaucratic, and militaristic demands for taxpayers' money. Keep in mind, though, that this cost is roughly equivalent to the price tag of one American aircraft carrier or a fleet of Soviet warheads. It's simply a matter of where our society wants to place its priorities.

Another strategy might be used to search for extraterrestrial intelligent life. This method also uses radio radiation. However, here we must rely on a single, key assumption, the validity of which is totally unknown. The assumption is that at least one advanced civilization is

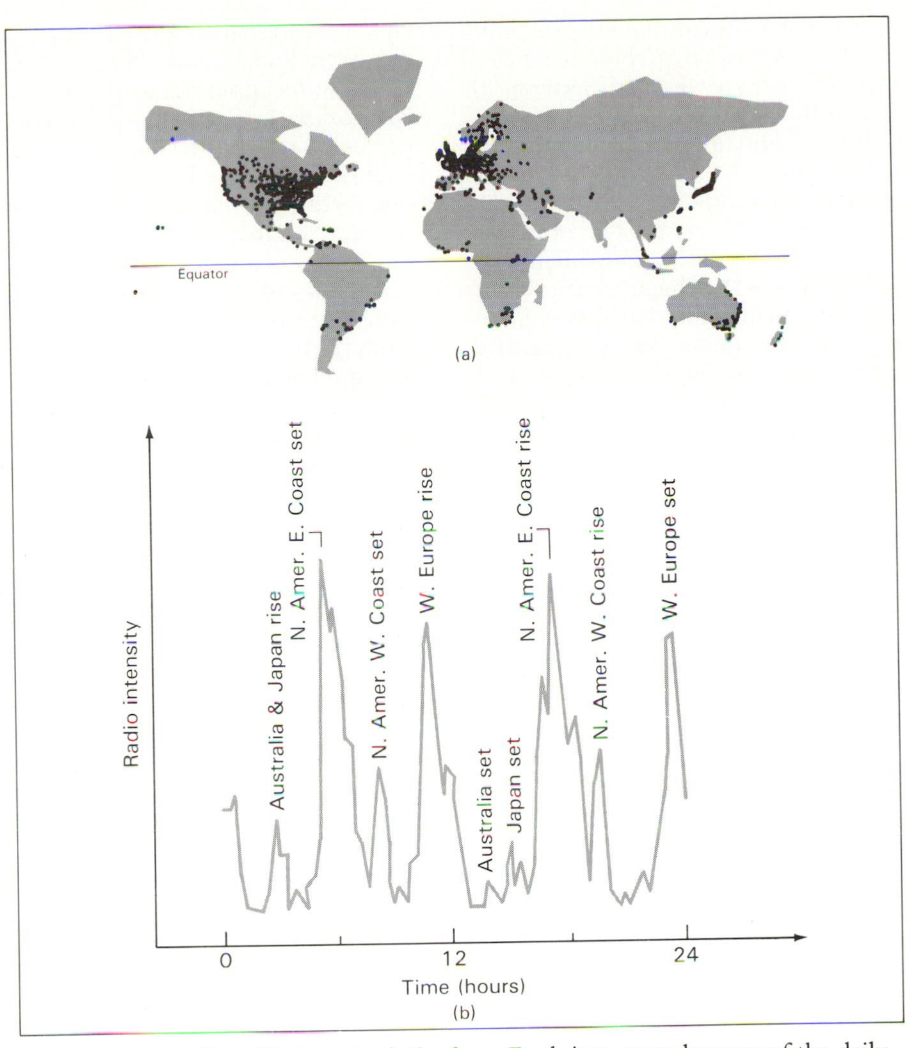

FIGURE 29.15 Radio radiation now leaks from Earth into space because of the daily activities of our technological civilization. Most of the radiation arises from television transmitters (dots) that strongly broadcast into space when on Earth's horizon (a). The particular pattern in (b) results from the sum of all Earth's television stations, as viewed from Barnard's star several light-years away.

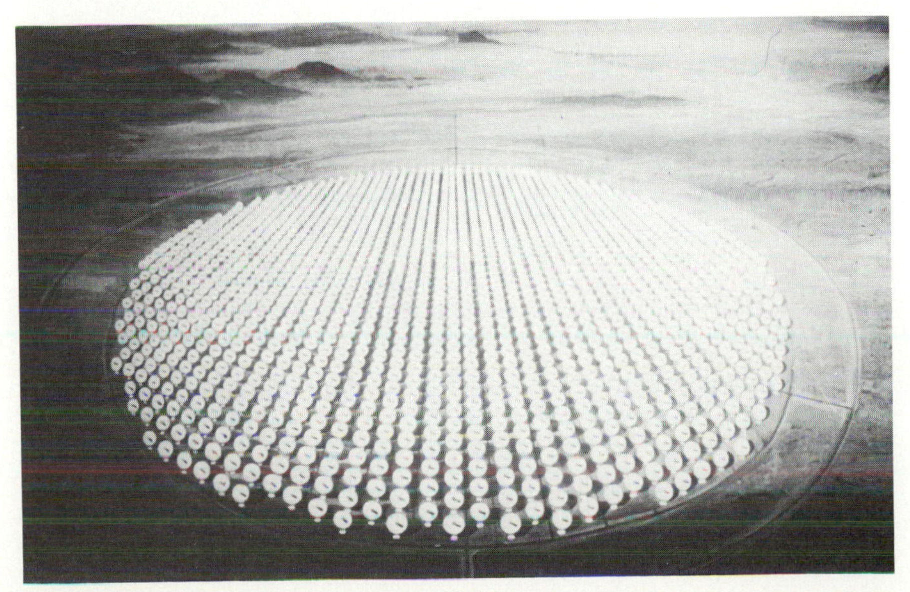

transmitting strong radio signals toward many star systems in the Galaxy—ours among them—actively hoping to contact us. Such a radio beacon, set up by an alien broadcaster specifically to attract our attention, might be detectable with radio telescopes already built on Earth.

Aside from this optimistic assumption, other concerns hinder this passive search strategy. For instance, in what direction should we listen? Even if we limit our targets to Sun-like stars, about 500 of them exist within only a few hundred light-years from Earth. Any one of them could have a planetary civilization that is transmitting toward us. To be thorough, we would need to sample every such star system.

The basic problem in estimating the number of extraterrestrial civilizations we could contact is not the extent to which a society develops technology. Rather, the real problem is the degree to which a society generates curiosity and maintains it for long periods of time. A two-way interstellar dialogue requires, not only technological competence, but also sufficient motivation on both ends—ours and theirs. How motivated is a society to communicate with other societies beyond its own home planet? How universal is curiosity? How persistent is it? Would the residents of every inhabited planet have a deep desire to know if others exist in the Universe?

Humans on Earth seem to have a genuine curiosity about the Universe. Since consciousness dawned, many humans have had a real need to know if extraterrestrial life lies beyond Earth. But how long does this curiosity last? Does technologically competent intelligent life remain curious indefinitely?

Curiosity may wax and wane over the ages. For example, on our planet, the Greeks and Romans were

FIGURE 29.16 This is a design for a gigantic array of a thousand interconnected radio telescopes, called Project Cyclops. Such a device could be used to eavesdrop on civilizations much like our own, provided they are within a thousand light-years of Earth.

curious; the barbarians of the Middle Ages apparently less so; now in post-Renaissance times, we are again highly curious.

Perhaps other advanced civilizations reside in the depths of space, but their societies have reached a stage of complete stagnation. Maybe their curiosity is gone forever. Again, we repeat our earlier query: If curiosity dies, does intelligence die also?

Distressingly, if the average technological lifetime is only 1000 years, then, as noted in the preceding section, the average distance between galactic civilizations would be on the order of a few thousand light-years. Hence, among our local realm of several thousand Sun-like stars, we would expect only one technological civilization. Since we are such a civilization, we would then be that one—the only one within this part of the Galaxy.

To have any real hope of success, the search strategy must be directed toward many additional candidate stars far beyond 1000 light-years. Over such vast distances, the two-way dialogue will not exactly be a snappy conversation. A reply to an initial "Hello" may take thousands of years.

As if the problem of where to search were not enough, there is another problem. At what frequency should we listen for their beacon? The electromagnetic spectrum is enormous; the radio domain alone is vast. To hope to detect a signal at some unknown radio frequency is like searching for a needle in a haystack. The technique would seem doomed to failure, unless we have some prior information concerning the likely frequencies on which aliens might transmit.

Fortunately, some basic arguments suggest that civilizations will probably communicate at a wavelength of nearly 20 centimeters. The basic building blocks of the Universe, namely hydrogen (H) atoms, naturally radiate near 21-centimeter wavelength. Furthermore, one of the simplest molecules, hydroxyl (OH), radiates near 18-centimeter wavelength. Together, these two substances form water (H_2O). Arguing that water is likely to be the interaction medium for life anywhere, some researchers have proposed that the interval between 18 and 21 centimeters is the best wavelength domain for civilizations to transmit or listen. Called the **water hole**, this radio interval might serve as an oasis where all advanced galactic civilizations conduct their electromagnetic business.

This water-hole frequency interval is only a guess, but it's supported by other arguments as well. In particular, the wavelength domain from 18 to 21 centimeters is precisely the part of the entire electromagnetic spectrum for which the galactic static from stars and interstellar clouds is minimized. Furthermore, the atmospheres of typical planets are expected to interfere least at these wavelengths.

Figure 29.17 shows the water hole's position in the electromagnetic spectrum. This figure also plots the amount of natural emission from our Galaxy and from Earth's atmosphere. As seen, the water hole is within the quietest part of the spectrum, namely where the intensity of these natural emissions is minimized. Thus the water hole seems like a good choice for the frequency of an interstellar beacon, although we cannot be sure of this reasoning until contact is actually achieved. Perhaps some other transmitting frequency is better for reasons unknown to us at this time.

A few radio searches are now in progress at frequencies in and around the water hole. Thus far, nothing resembling extraterrestrial signals has been detected. But that's not surprising, given the small efforts made to date. Attempts to detect extraterrestrials have been so brief in relation to the task at hand that they may be likened to a "Columbus" who, searching for a new route to the Indies, gave up about a mile off the coast of Spain. To maximize the chance for success, a dedicated effort is needed—one perhaps requiring hundreds of years of searching on a continuous basis. To have much hope of making contact, we will probably need to monitor millions of stars.

Many uncertainties plague this

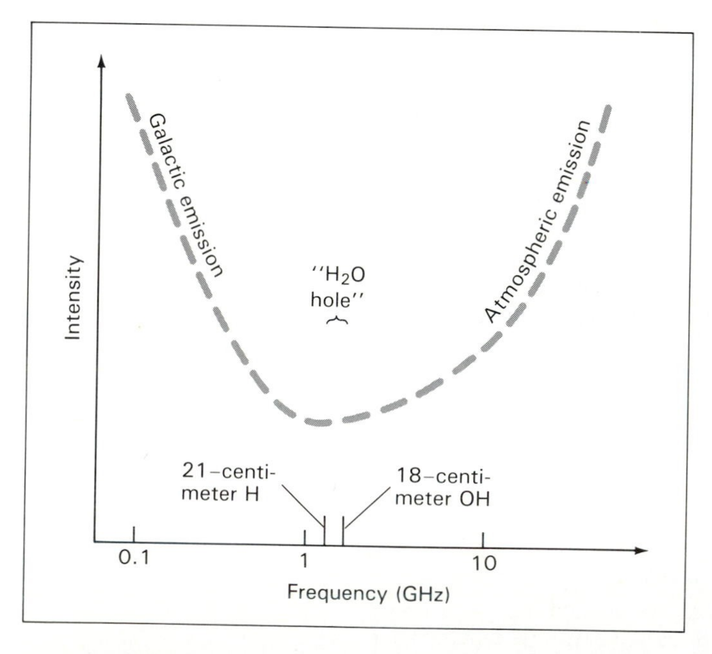

FIGURE 29.17 The "water hole" is bounded by the natural emission frequencies of the hydrogen (H) atom (21-centimeter wavelength) and the hydroxyl (OH) molecule (18-centimeter wavelength). The dashed curve sums the natural emissions of our Galaxy and of a planet's atmosphere (in this case Earth's). This sum is minimized near the water hole frequencies. Perhaps all smart civilizations conduct their interstellar communications within this quiet "electromagnetic oasis".

sort of search strategy. The greatest uncertainty concerns the *desire* of any civilization to transmit. From what we know, transmitting is boring, expensive, and potentially dangerous. Accordingly, perhaps everyone is just listening. If so, then the prospects for contact are dim; to create a dialogue, somebody's got to start talking.

Other uncertainties concern "language." Will galactic civilizations be able to find a common language? This puzzle includes not only the difficulty of matching the frequencies of transmission and reception, but also the potential problem of ever being able to understand or appreciate the content of the message. This may be especially troublesome if the transmitting civilization is much more advanced than ourselves. Even a few hundred years more progress would presumably give them technological prowess hardly imagined here on Earth. Extraterrestrials may use means of contact entirely foreign to us. We may be unable to detect their signals or decipher their code. They may be as uncommunicative with us as we are with, for example, ants or even dolphins. If so, then all civilizations are probably doomed to loneliness for as long as each survives.

On the other hand, if aliens are eager to contact emerging civilizations in our Galaxy, they will realize that less sophisticated means must be used. They would want to make the task as easy as possible for us, transmitting signals that inexperienced civilizations like ourselves could detect and decipher. Their language would probably be built around mathematics, since counting should be universal: 2 and 2 ought to make 4 everywhere.

Figure 29.18 exemplifies how mathematics can be used to send messages through space. In this case, no words are needed, just a picture.

All the many combined uncertainties cause some researchers to be pessimistic. They argue that a search for extraterrestrials is unreasonable and unwarranted. They claim that the assumptions needed to estimate the prospects for extraterrestrial intelligence contain too many unknowns.

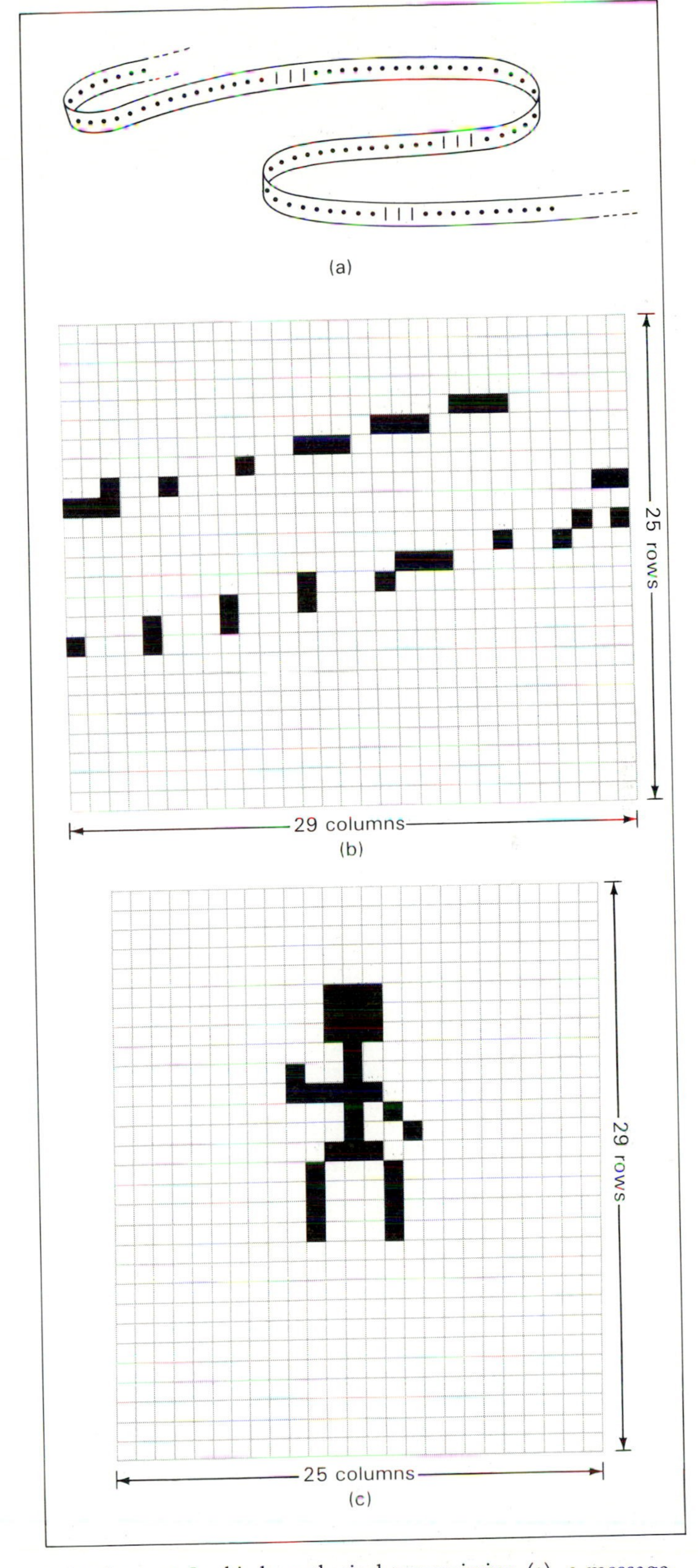

FIGURE 29.18 In this hypothetical transmission (a), a message is buried within a long series of digital information (dots and dashes, or ones and zeroes). The total number of digits is 725 which is a product of two prime numbers 25 and 29. When arranged in a geometrical pattern having 25 rows and 29 columns (b), the message is unrecognizable. But, when arranged in 29 rows and 25 columns (c), the message is clear.

The search strategy itself contains additional unknowns. They conclude that any expenditure of time, effort, or money for a search is unsupported by the meager evidence at hand.

Proponents argue that we have good reason to suspect that extraterrestrials exist somewhere, given the relative ease of our development on Earth. They admit that we have only a very small chance of making contact in the near future. But they argue that now is the time to test the theory that advanced civilizations inhabit the Galaxy. To fail to try, is to commit the cardinal sin of pre-Renaissance workers—thinking without experimentally testing. Failure to try might also cut prematurely short humankind's natural exploratory drive. Our longevity as a civilization may be shortened for the very reason that we didn't take up the search.

One thing is worth remembering: The space surrounding all of us could be, right now, inundated with radio signals from extraterrestrial civilizations. If we only knew the proper direction and frequency, we might be able to make one of the most startling discoveries of all time. The result would likely provide whole new opportunities to study the cosmic evolution of energy, matter, and life throughout the Universe.

SUMMARY

In this final chapter, we have generalized our theme of cosmic evolution. Our objective has been to assess realistically the chances that life exists elsewhere in the Universe. We concluded that Earth is probably the only place in the Solar System harboring life; the search for life on Mars has thus far yielded unclear results.

Intelligent life in the Galaxy depends on so many uncertain factors as to make it nearly impossible to know if we can ever contact extraterrestrial aliens. The central ideas of cosmic evolution would seem to favor such alien life, but we cannot currently prove its existence. We might be surrounded by star systems teeming with life much like that on Earth. Or, we might be alone in the Universe.

Thinking alone will never permit us to determine if we have galactic neighbors, or if we are destined to universal loneliness. Only thinking aided by experiments and observations—the hallmark of the scientific method—will help us to approach the truth.

KEY TERMS

assumptions of mediocrity
carbonaceous chondrite
Green Bank equation
habitable zone
water hole

QUESTIONS

1. State the "assumptions of mediocrity" often used by researchers to argue, on the basis of logic alone, for the prevalence of galactic aliens.
2. Summarize the advantages for a form of life based on the carbon atom rather than on any other chemical element.
3. Suppose that you had the means (and the inclination) to travel to any nearby region in search of life as we know it. Based on what you have learned in this text, describe and give reasons for the type of region you would choose as ideal for your visit.
4. List, and explain in a phrase or sentence, in addition to planets, at least two types of objects where complex organic molecules have been observed.
5. Describe the main findings to date regarding the search for life on Mars, including a brief explanation of at least three of the major tests conducted in situ by the U.S. *Viking* robot spacecraft. To what extent might there be microfossils buried in the Martian soil from an era long past?
6. Estimate, and briefly explain your reasoning for, the prospects for the origin and evolution of life near a main-sequence O-type star, a star cluster in the Milky Way's halo, and within a dark galactic cloud.
7. What is the significance of the Murchison meteorite?
8. Why are both high-mass and low-mass stars needed for the development of life as we know it on Earth?
9. What observational technique is currently envisioned to be the most advantageous method of discovering extraterrestrial civilizations, and why?
10. What is meant by the statement that we are now probably the "dumbest technological intelligence in the Galaxy"?

FOR FURTHER READING

Billingham, J. (ed.), *Life in the Universe.* Boston: MIT Press, 1981.

Goldsmith, D. (ed.), *The Quest for Extraterrestrial Life.* Mill Valley, CA: University Science Books, 1980.

Johnson, R. and Holbrow, C., *Space Settlements*, Washington, NASA SP-413, U.S. Government Printing Office, 1977.

McDonough, T., *The Search for Extraterrestrial Intelligence.* New York: Wiley, 1987.

Sagan, C. and Drake, F., "The Search for Extraterrestrial Intelligence." *Scientific American.* May 1975.

Shklovskii, I. and Sagan, C., *Intelligent Life in the Universe.* New York: Holden-Day, 1966.

GLOSSARY OF KEY TERMS

absolute brightness The brightness of an object assuming that it resides at a standard distance of 10 parsecs from Earth.

active galaxy A galaxy that radiates large amounts of energy quite differently in character than that of a normal galaxy.

active Sun The unpredictable, sporadic condition of the Sun in which radiation is emitted from occasionally violent eruptions.

adaptation The positive response to a changing environment of an organism having some variation or trait that improves the organism's chance for survival and reproduction.

agriculturist The main job description of our ancestors beginning some 10,000 years ago, namely those who mastered farming.

alpha process The capture of helium nuclei, thus producing a heavier nucleus. (Physicists call helium nuclei "alpha particles.")

amino acid An organic compound, of which 20 different types form the essence of all proteins comprising life on Earth.

angular momentum The tendency of an object to keep spinning or moving in a circle.

angular resolution The ability of a telescope to distinguish two adjacent objects on the sky, or to study the fine details on the surface of some object.

antimatter A form of matter having an opposite charge than is normally the case; for example, a positively charged positron is the antimatter opposite of a negatively charged electron.

apparent brightness The brightness of an object as it naturally appears in the sky.

arc degree A unit of angular measure of which there are 360 in a full circle.

arc minute A unit of angular measure of which there are 60 in 1 arc degree.

arc second A unit of angular measure of which there are 60 in 1 arc minute (or therefore 3600 in 1 arc degree).

artificial intelligence The study of problem solving and decision making by machines in a way often done by humans.

assumptions of mediocrity The reasoned idea that we are not alone in the Universe; rather, given the enormity of space and the vastness of time, life arises in many cosmic locations.

asteroid A small, rocky object revolving around the Sun, sometimes called a minor planet or planetoid.

asteroid belt A region of space between Mars and Jupiter where the great majority of asteroids are found.

astronomical unit The average distance between the Earth and the Sun.

atmosphere Those gases surrounding the surface of a planet, moon, or star.

atom A submicroscopic building block of much of matter, consisting of a positively charged nucleus and one or more negatively charged electrons.

atom epoch A period in the early Universe when some elementary particles began to cluster, thus fashioning the first atoms.

aurora Rapid and irregular displays of colorful light in the night sky of a planet, owing to the leakage of charged particles from the magnetosphere into the atmosphere.

Australopithecus The designation given to those prehuman creatures having a mixture of ape-like and human-like qualities, and who lived several million years ago.

autotroph Cells that survive on inorganic matter and an external energy source, such as photosynthesizing plants the world over.

axon The main extrusion of a neuron which acts as a transmitter of information.

barred-spiral galaxy A type of spiral galaxy having a linear extension or "bar" made of stars and interstellar matter passing through its center.

baseline The distance between two (or more) places from which observations are made.

big bang A popular term describing the explosive start of the Universe.

binary star system A system of two stars in orbit about their common center of mass, and held together by their mutual gravitational attraction.

biological evolution The developmental changes, from generation to generation, experienced by life forms throughout the history of life on Earth.

bipedal An adjective meaning "having two feet" and/or "an ability to walk on two legs."

black hole A region containing a huge amount of mass compacted into an extremely small volume, thus making its pull of gravity so strong that not even light can escape—hence its name.

blue shift The Doppler shortening of the wavelength of radiation (or the shifting of spectral lines toward smaller wavelengths) caused by some net motion of approach.

brain stem The upper part of the spinal cord through which sensory and response information passes back and forth to the brain.

carbonaceous chondrite A meteorite having embedded pebble-sized granules which contain significant quantities of organic (carbon-rich) matter.

cell The simplest type of clustered matter having the common attributes of life—birth, metabolism, and death.

cepheid variable A particular class of variable (pulsating) stars whose period of variation correlates well with its average luminosity.

cerebellum The plum-sized part of a brain to the bottom rear of the cerebrum, and which coordinates and fine tunes body movements.

cerebral hemisphere Either half of a vertically split brain.

cerebrum The domed, grapefruit-sized clump of matter comprising the bulk of a brain, and which contains the controls for the senses and muscles.

chemical compound A tightly knit cluster of elements, also sometimes called a molecule.

chemical evolution The chemical (i.e., prebiological) changes that transformed simple atoms and molecules into the more complex chemicals needed for the origin of life.

chromosphere The lower atmosphere of the Sun, between the solar photosphere and corona.

closed Universe A model Universe that stops expanding at some time in the future, after which it contracts to a point much like that from which it began.

CNO cycle An alternative to the proton-proton cycle whereby hydrogen is coverted into helium, in this case using carbon, nitrogen, and oxygen as intermediaries.

collecting area The total area of a telescope capable of capturing cosmic radiation.

collisional process An event involving a collision of objects; for example, the excitation of a hydrogen atom when hit by another hydrogen atom.

collision theory The idea, now out of favor, that the planets formed from hot streaming debris torn from the Sun during a near-collision or close encounter with another passing star.

color The visual perception of an object, which for a radiating object can often be considered an indicator of temperature.

comet A small ball of rock and ice from which emanates a long wispy tail of gas and dust while nearing the Sun.

condensation theory The idea, broadly accepted today, that the planets originated from gravitationally contracting and chemically condensing eddies as a natural by-product of the formation of the Sun.

consciousness That property of human nature generally, or of the mind specifically, that grants us self-awareness and a sense of wonder.

conservation of mass and energy A basic law of physics stipulating that the sum of the mass and the energy of a system must remain constant during any event.

constellation A geometric pattern of bright stars that appear grouped in the sky, and which are named after gods, heroes, animals, and mythological beings by ancient astronomers.

convection The physical upwelling of hot matter, thus transporting energy from a lower, hotter region to a higher, cooler region.

convection zone That region below a star's surface where matter physically upwells, thus transporting energy mechanically.

core The central region of a planet, star, or galaxy.

core-halo radio galaxy A class of active galaxies whose radiation arises primarily from a compact core but also from an extended halo surrounding the core.

corona The outermost, hot, thin atmosphere of a star.

coronal hole A region of relatively low gas density in the solar atmosphere through which much of the solar wind escapes.

cortex The bulk of the matter itself within a brain.

cosmic abundances A standard listing of the relative numbers of the various elements, determined by studies of the spectral lines in astronomical objects and averaged for many stars in our cosmic neighborhood.

cosmic background radiation A weak, electromagnetic (mostly radio) signal that permeates all of space, thought to be relic radiation of the big bang.

cosmic evolution The changes in the assembly and composition of energy, matter, and life in the Universe.

cosmic-ray particle A charged, subatomic particle of matter (not radiation) that races throughout interstellar space, and which regularly strikes Earth's atmosphere.

cosmological principle The idea that all observers, everywhere in space, perceive the Universe in roughly the same way regardless of their actual location.

cosmology The study of the Universe on the grandest scale, especially its origin, evolution, and destiny.

crater A bowl-shaped depression on the surface of a planet or moon, generally caused by meteorite impact, although sometimes by volcanic upwelling.

crater chain An almost straight alignment of several craters on the Moon, presumably caused by volcanic activity below a fault line.

critical universal density The density of matter above which the Universe is closed and below (or equal to) which the Universe is open.

cultural evolution The changes in the ways, ideas, and behavior of society, especially among higher forms of life on Earth.

dark dust cloud A region of interstellar space containing a rich concentration of gas and dust in an irregular but well-defined cloud that obscures the light from stars beyond it.

dehydration condensation The linking of two or more amino acids by means of removing water.

dendrite One of a network of extrusions of a neuron which acts as a receiver of information.

density The quanitity of something in a unit of volume; a measure of compactness.

deuterium A special form of hydrogen (an isotope called "heavy hydrogen") having a neutron as well as a proton in its nucleus.

differentiation The separation of heavy matter from light matter, thus causing a variation in density and composition.

DNA A long, helical molecule that resides in all biological nuclei and which has the remarkable property of being able to self-replicate; an acronym for deoxyribonucleic acid.

Doppler effect The apparent change in the wavelength (or frequency) of a wave, caused by line-of-sight motion of the source or of the observer (or both).

dwarf star Any star comparable to or smaller in size than the Sun.

earthquake A sudden dislocation of rocky matter near Earth's surface.

electric force field An attractive or repulsive influence that an electric charge exerts on some other charge (or on a magnet).

electromagnetic force The force that binds charges of opposite electrical charge and repels charges of identical electrical charge.

electromagnetic spectrum The entire range of all the various kinds of radiation; light (or the visible spectrum) comprises just one small segment of this much broader spectrum.

electromagnetism The phenomenon of electricity and magnetism studied together.

electron A negatively charged elementary particle that resides outside (but is bound to) the nucleus of an atom.

element A generic term denoting all the various kinds of atoms; all atoms of the same kind have identical chemical properties.

elementary particle A basic building block of atoms.

ellipse A distorted or elongated circle.

elliptical galaxy A galaxy having a spherical or elliptical shape, some more than others, and composed mostly of old stars and little interstellar matter.

energy The ability to do work.

erosion The wearing away of surface matter, usually by wind and water.

escape velocity The minimum speed needed for an object to escape the gravitational pull of a massive object.

Euclidean geometry The terrestrially familiar geometry of "flat space" that all of us learn in high school.

eukaryote A life form having a well-developed biological nucleus.

event horizon A region within which no event can ever be seen, heard, or known by anyone outside; also termed the "surface" of a black hole.

excited state A greater-than-minimum energy state of any atom; achieved when at least one electron of an atom resides at a greater-than-normal distance from its parent nucleus.

extraterrestrial An adjective meaning "beyond the Earth."

fault A zone of weakness where crustal rocks dislodge (as in earthquakes) or mantle rocks upwell (as in volcanos).

fermentation The extraction of energy via the capture and chemical breakdown of small molecules.

first-generation star A star made of primordial matter and thus having no heavy elements.

fission A nuclear process that releases energy when heavyweight nuclei break down into lightweight nuclei.

flare Violent solar activity often observed in the vicinity of a sunspot or prominence, and which releases pent-up energy in the form of electromagnetic radiation and huge quantities of charged particles.

fossil The hardened remains of a dead organism whose outlines or bony features are preserved in ancient rocks.

fragmentation Developing inhomogeneities in the gas density of a cloud which eventually breaks down into smaller clumps of matter within the cloud.

frequency The number of crests or cycles of a wave in a given unit of time.

fusion A nuclear process that releases energy when lightweight nuclei combine to form heavyweight nuclei.

galactic center The hub or core of the Milky Way, some 30,000 light-years from the Sun.

galactic cluster A loose collection of tens to hundreds of relatively young stars spread over several light-years, sometimes termed an open cluster.

galactic corona A three-dimensional extended region of very thin gas that reaches well beyond the limits of the galactic halo.

galactic halo A nearly spherical region that surrounds the Milky Way Galaxy and extends some 50,000 light-years from the galactic center.

galactic plane A relatively thin disk or plane in which most of the Milky Way's stars and interstellar matter now reside.

galaxy A colossal collection of billions of stars and interstellar matter, held together loosely by gravity. The Milky Way is one such galaxy.

galaxy cluster A group of galaxies held together by their mutual gravitational attraction.

galaxy epoch An early period in the history of the Universe when the galaxies formed.

galaxy horizon The roughly 4-billion light-year limit beyond which normal galaxies are difficult to discern.

galaxy supercluster A truly huge cluster of galaxy clusters, often stretching over a hundred million light-years or more.

gaseous nebula A region of ionized gas (plasma) surrounding one or more young, hot stars (sometimes termed an emission nebula).

gene A functional unit of any DNA molecule responsible for directing inheritance from generation to generation.

gene pool The spread or distribution of the variations or traits among a given population of a species; all the genes in a given population.

genetic code An encyclopedic blueprint of the physical and chemical properties of all of an organism's cells and all of its functions.

genetics The study of genes, especially how they operate and change.

geocentric An adjective meaning "centered on the Earth."

geography The study of positions, shapes, sizes, and numerous other qualities of Earth's continents.

geometry The study of the size, shape, and scale of things.

giant star Any star much larger in size than the Sun.

globular cluster A tight-knit collection of many thousands, sometimes even millions, of old stars spread over about a hundred light-years.

grand-unified theory An idea that three forces—the electromagnetic force, the strong nuclear force, and the weak force—are different manifestations of one and the same force.

granule A variable region of rising or sinking mass on the solar photosphere, thus giving the Sun a mottled appearance.

gravitational clustering The idea that large objects were built from small objects, including, for example, galaxies having originated by collecting already made stars.

gravitational force The force that holds matter together on a large scale, such as stars within galaxies, atoms within stars, and people on Earth.

gravitational instability A condition whereby an object's (inward-pulling) gravitational potential energy exceeds its (outward-pushing) thermal energy, thus causing the object to infall.

gravitational red shift The lengthening of radiation's wavelength owing to the mass from which the radiation has escaped.

gravity An abbreviated term for gravitational force or gravitational force field, which is basically the universal ability of all material objects to attract each other.

gravity wave The gravitational analog of an electromagnetic wave whereby gravitational radiation is emitted at the speed of light from any mass that undergoes rapid acceleration.

Greenbank equation A mathematical formula that attempts to evaluate the prospects for technologically intelligent life in the Milky Way Galaxy.

greenhouse effect The trapping of radiation by an atmosphere (or greenhouse), thus causing greater heating than would normally be the case.

ground state The minimun-energy state of any atom; achieved when all the electrons of an atom are as close to its nucleus as possible.

habitable zone A three-dimensional region of "comfortable" temperatures that surrounds every star.

hadron epoch A very early time in the history of the Universe when the heavy elementary particles such as protons and neutrons were the most abundant type of matter.

Hayashi track An evolutionary stage of a protostar about to become a main-sequence star, and named after the Japanese astronomer who first studied such changes in detail.

heliocentric An adjective meaning "centered on the Sun."

helium flash The rapid onset of helium fusion in a red giant star.

heredity The transmission of genetic properties from parents to offspring, thus ensuring the preservation of certain traits in future generations of a species.

heterotroph Cells that require organic matter for food, especially those primitive cells that survived by absorbing acids and bases floating on the primordial sea.

high-energy telescope A device capable of detecting x-ray and gamma-ray radiation from the cosmos.

highland Mountainous regions of the Moon elevated above the maria.

Homo erectus The species designation given to all human creatures who lived from roughly 0.3 to 1 million years ago; literally, the Latin means "erect man."

Homo sapiens The species designation given to all human creatures who lived during about the past 300,000 years, including ourselves; literally, the Latin means "wise man."

HR diagram A plot displaying the luminosities and surface temperatures (or spectral classes) of many stars, and named after astronomers Hertzsprung and Russell.

Hubble's constant The proportionality factor between the distance of a galaxy and the velocity with which it recedes; currently, its best estimate is 22 kilometers/second/million light-years.

Hubble's law An empirical finding linking the distance of a galaxy and the velocity with which it recedes.

hunter-gatherer The main job description of our ancestors during most of the past few million years, namely those who survived by hunting and gathering food.

hydrosphere The liquid part of a planet's surface, including any lakes, streams, oceans, and rivers.

ice age A period of cool, dry climate that intermittently plagues planets, causing, in the case of Earth, a long-term buildup of ice far from the poles.

inflation A period of extremely rapid expansion of the Universe shortly after the Universe originated.

infrared telescope A device capable of detecting cosmic electromagnetic radiation whose wavelength is a little longer than light.

intensity The magnitude or strength of a quantity (in this text, usually the strength of radiation).

interferometer A device that uses two or more telescopes to observe the same object at the same wavelength and at the same time, thereby achieving high angular resolution by studying the resulting interference.

intergalactic space Regions outside galaxies and especially galaxy clusters where matter has never been conclusively found.

interplanetary matter Debris in the great spaces among the planets of the Solar System.

interplanetary space Regions among the planets, moons, and related objects of the Solar System.

interstellar matter Sparse gas and dust in the vast domains among the stars.

interstellar space Dark regions among the stars of any galaxy.

inverse-square law A principle dictating the decrease in the strength of some quantity with distance: namely, rapidly as the square of the distance.

invertebrate A backbone-less organism.

invisible radiation Those kinds of radiation to which the human eye is not sensitive: for example, radio, infrared, and ultraviolet waves, as well as x-rays and gamma rays.

ion An atom with one or more electrons removed (or added), giving it a positive (or negative) charge.

ionosphere A radio-reflecting region high in Earth's atmosphere in which solar radiation has removed an electron from some of the atoms.

irregular galaxy A strangely shaped galaxy, often rich in interstellar matter, but apparently not a member of any of the major classes of spiral or elliptical galaxies.

isotope An atom having more or fewer neutrons than usual.

Jovian planets The four, big, gassy planets in the outer parts of the Solar System: Jupiter, Saturn, Uranus, and Neptune.

Kelvin-Helmholtz contraction phase An early evolutionary stage of an interstellar cloud fragment about to become a protostar.

Kelvin scale An internationally agreed upon temperature scale, equal to the Celsius (or centigrade) scale plus 273 degrees.

Kepler's laws Three principles, discovered empirically by a seventeenth-century German astronomer, that describe the motions of the planets in their orbits about the Sun.

kinetic energy The energy of an object due to its motion.

lepton epoch A very early time in the history of the Universe when the lightweight elementary particles such as electrons and neutrinos were the most abundant type of matter.

light The kind of radiation to which the human eye is sensitive.

light-year The distance traveled by light in a full year; equal to some 10 trillion kilometers.

linear momentum The tendency of an object to keep moving in a straight line.

lithosphere The solid part of a planet's surface, including any continents and seafloor.

lobe radio galaxy A class of active galaxies whose radiation arises from huge jets or lobes of invisible matter that extend well beyond a central, often visible galaxy.

Local Group The specific name given to the galaxy cluster that includes the Milky Way Galaxy as a member.

luminosity The rate of electromagnetic energy released from any object, sometimes called the absolute brightness.

lunar eclipse A darkening of the Moon by Earth's shadow when our planet blocks all or some of the Sun's rays normally falling on the Moon.

Magellanic Clouds Two close, irregular or "dwarf" galaxies that orbit our Milky Way Galaxy.

magnetic force field An attractive or repulsive influence that a magnet exerts on another magnet (or on a charged particle).

magnetosphere A region of space, usually high above a planet's atmosphere, where charged particles are magnetically deflected and/or trapped.

main sequence A narrow region on an HR diagram in which most stars fall.

mantle The interior of a planet, namely that matter below the crust yet above the core.

maria A Latin word (plural of mare) meaning "seas," often used to describe the dark (water-free) depressions on the Moon.

mascon A "mass concentration" (or gravity anomaly) known to exist at several locations just below the surface of the Moon.

mass A measure of the total amount of matter (or "stuff") contained within an object.

matter–antimatter annihilation A highly efficient process in which equal amounts of matter and antimatter collide and destroy each other, thus producing a burst of energy.

Matter Era A period in the history of the Universe (including now) when the density of energy contained within matter exceeds the density of energy contained within radiation.

meteor A heated, glowing object streaking through Earth's atmosphere, but not yet having hit the surface.

meteorite A meteoroid that manages to survive passage through an atmosphere to collide ultimately with the surface of a planet or moon.

meteoroid On average, a meter-sized boulder that has probably escaped from the asteroid belt and thus roams the Solar System.

micrometeorite A micrometeoroid that collides with the surface of a planet or moon having no appreciable atmosphere.

micrometeoroid On average, a millimeter-sized pebble that forms the bulk of the interplanetary matter scattered throughout the Solar System.

Milky Way Galaxy The specific galaxy to which the Sun belongs, so named because the stars in it can be seen overhead on a clear, dark night as a milky band running across the sky.

molecular cloud A relatively dense, cold region of interstellar matter where molecules are found in considerable abundance.

molecular velocity The average speed of the molecules in a gas of a given temperature.

molecule A tightly knit cluster of two or more atoms.

multicell A cluster of cells that collaborate with other cells.

mutation A microscopic change in the genetic message of an organism.

natural selection The process whereby only those life forms able to adapt to a changing environment tend to survive and reproduce, thereby passing their favorable variations or traits on to their descendants.

nebular theory The idea that the Solar System originated in a contracting, swirling cloud of gas that left behind a concentric series of rings from which the planets formed.

neuron A biological, or nerve, cell in a brain.

neutrino A neutral, weakly interacting elementary particle having little or no mass.

neutrino telescope A huge tank of an appropriate chemical capable of detecting neutrinos.

neutron A neutral elementary particle havng slightly more mass than a proton, and which resides in the nucleus of most atoms.

neutron star An extremely compact ball of neutrons having the mass of a star but a size smaller than a planet.

non–thermal radiation Radiation released by virtue of a fast-moving charged particle (such as an electron) interacting with a magnetic force field; this process has nothing to do with heat.

normal galaxy A galaxy that radiates energy much as expected from a large accumulation of stars.

nova A star that rapidly brightens while expelling a small fraction of its matter, after which it slowly fades back to normal.

nuclear force The force that binds atomic nuclei.

nuclear transformation Changes in atomic nuclei owing to the reaction of one nucleus with another nucleus.

nucleic acid A complex, long-chain molecule, mostly rich in carbon, that inhabits the biological nucleus of a cell.

nucleotide base One of five such molecules that comprise the essence of all nucleic acids that partake of life on Earth.

nucleus The positively charged core of an atom, comprising protons and (except for hydrogen) neutrons, around which electrons orbit.

oceanography The study of the ocean's motion, history, and physical and chemical behavior.

opacity A measure of the ability of a gas to absorb or block radiation; the opposite of transparency.

open Universe A model Universe that expands forever.

optical telescope A device capable of gathering visible radiation from the cosmos.

orbital A specific distance beyond an atom's nucleus at which an electron can orbit.

oscillating Universe A model Universe that continuously cycles between expansion and contraction.

ozone layer A layer in Earth's atmosphere rich in triatomic oxygen (O_3) molecules, which blocks high-frequency radiation.

paleoanthropology The study of prehistoric humanity.

paleomagnetism The study of ancient magnetism.

paleontology The study of the fossilized remains of dead organisms.

panspermia A theory that stipulates germs to be everywhere in the Universe, and that pimitive life on Earth originated when some of these germs came to our planet from outer space.

parallactic angle One-half of the angular shift (or parallax) of an object as seen from two different points of observation.

parallax The apparent shift in the position of an object as seen from two different points of observation.

parsec The distance of an object whose parallactic angle equals 1 arc second; also equals 3.3 light-years.

periodic table of the elements A systematic listing of all the known kinds of atoms.

period-luminosity law A correlation, true for Cepheid variable stars, connecting a star's period and its luminosity.

photodisintegration The breakdown of heavy nuclei by means of heat.

photoelectric effect A basic physical phenomenon that demonstrates radiation to be composed of particles.

photometry The study of a star's intensity or brightness.

photon A submicroscopic particle of radiation.

photosphere The visible portion of the Sun or any star, although this "surface" is made exclusively of gas.

photosynthesis The extraction of energy from ordinary sunlight when that light stimulates carbon dioxide and water to change into oxygen and carbohydrates.

Planck curve A standard plot of the frequency (or wavelength) and intensity of radiation emitted from a heated (thermal) object.

planet A rocky and/or gaseous body, generally much smaller and cooler than a star. The Earth is one such planet.

planetary nebula A twofold object comprising an old, yet hot white dwarf star surrounded by a thin, ionized, spherical shell of expanding gas.

planetesimal An asteroid-sized blob of matter that gradually collided with others in the formative stages of the Solar System, thus fabricating the planets.

plasma A state of matter wherein all atoms are ionized; a mixture of free electrons and free atomic nuclei.

plate tectonics The study of plate movements which are thought to be responsible for many of the major geological features on planet Earth; popularly termed "continental drift."

polarization A phenomenon whereby radiation displays a given orientation of its plane of wave oscillation.

positron A positively charged antiparticle of the electron.

potential energy The energy of an object due to its position.

prime focus The place to which a telescope initially directs its collected radiation.

primordial nucleosynthesis Element building that occurred in the early Universe when the nuclei of primordial matter collided and fused with one another.

principle of equivalence The idea that the pull of gravity on an object and the acceleration of that object (i.e., its gravitational and inertial forces) can be viewed as conceptually equivalent.

prokaryote A life form lacking a well-developed biological nucleus.

prominence An explosive eruption in which gases are suspended in a loop, apparently by magnetic forces, above the surface of the Sun.

proper motion Annual apparent motion of a star across the sky.

protein A term describing an entire class of carbon-rich molecules that inhabit the cytoplasm of cells.

proteinoid microspheres A microscopic protein-like cluster rich in amino acids.

protogalaxy A forerunner of a present-day galaxy, also sometimes termed a "baby galaxy."

proton A positively charged elementary particle that resides in the nucleus of every atom.

proton-proton cycle A series of nuclear events whereby hydrogen nuclei (protons) are converted into helium nuclei, releasing energy in the process.

protoplanet A forerunner or progenitor of a genuine planet.

protostar An embryonic condensation of interstellar matter perched at the dawn of star birth.

pulsar A compact, star-like object that emits rapid and periodic pulses of radiation, and which is thought to be a neutron star.

punctuated equilibrium The idea that life's species remain essentially unchanged for long periods of time, after which they change rapidly in response to sudden, drastic changes in the environment.

quark A fractionally charged, basic building block of protons, neutrons, and many other elementary particles.

quasar An acronym for *quas*i-stel*lar* source; a high red shift object whose image resembles a star but whose energy budget seems comparable to or larger than that of a normal galaxy.

quiet Sun The predictable condition of the Sun in which radiation is steadily emitted on a daily basis; especially evident when solar activity is minimized.

radar An acronym for *ra*dio *d*etection *a*nd *r*anging; a method of determining the distance of an object by transmitting a radio signal and measuring its echo (or reflection) from the object.

radial motion Motion along any line of sight.

radiation A form of energy that travels at the speed of light, of which light is a special kind.

Radiation Era A time in the early Universe when the density of energy contained within radiation exceeded the density of energy contained within matter.

radiation laws A set of basic principles stipulating how any hot object radiates energy.

radiation zone A region, such as the deep interior of the Sun, where radiation travels virtually unhindered by absorption.

radiative process An event involving the emission or absorption of radiation; for example, the excitation of a hydrogen atom when hit by a photon.

radioactivity The spontaneous decay of certain rare, unstable, heavy-weight nuclei into more stable light-weight nuclei, a natural by-product of which is the release of energy.

radio telescope A device capable of detecting radio waves from the cosmos.

Ramapithecus The designation given to a possible ancestor of both the australopithecines and humans, dating back roughly 10 million years.

red giant star An old, bright star, much larger in size and cooler than the Sun.

red shift The Doppler lengthening of the wavelength of radiation (or the shifting of spectral lines toward longer wavelengths) caused by some net motion of recession.

reflector A telescope that uses a polished, curved mirror to gather light and reflect it to a focus.

refractor A telescope that uses a transparent lens to gather light and bend it to a focus.

regolith A fine rocky layer of fragmentary debris (or dust) produced mainly by meteoroid collisions with the surface of the Moon.

relativity A theory of physics that describes the dynamical behavior of matter and energy under peculiar circumstances, especially at very high velocities and very high densities.

respiration A chemical process whereby cells use oxygen to release energy; a technical term for "breathing."

revolution The orbital motion of one object about another.

rille A ditch where molten lava once flowed.

rotation The spin of an object about its own axis.

r process Element building that occurs in highly evolved stars when a neutron is "rapidly" captured by a nucleus.

scientific method The investigative technique used by all natural scientists throughout the world. In general, some data or ideas are first gathered, then a theory is proposed to explain them, and finally an experiment is devised to test the theory.

secondary atmosphere Gases that a planet exhales from its interior after having lost its primary or primordial atmosphere.

second-generation star A star having some heavy elements and which is thus made of matter that has been previously processed through other stars.

seismic wave A systematic vibration emanating from the site of an earthquake and traveling through Earth's interior.

Seyfert galaxy A class of active galaxies that seems to have properties midway between normal galaxies and the most violent of the active galaxies.

shock wave A rapidly rushing shell of gas that tends to push aside and sometimes implode matter in its wake.

singularity A superhot, superdense state of matter, where the known laws of physics are likely to break down.

solar constant The amount of energy received from the Sun per unit area per unit time.

solar core The region immediately surrounding the center of the Sun where nuclear reactions release vast quantities of energy.

solar eclipse A blockage of light from the Sun when the Moon is positioned precisely between the Sun and the Earth.

solar interior The region of the Sun between its core and convection zone.

solar maximum The midst of a sunspot cycle when the numbers of sunspots are substantial.

solar minimum The beginning and end of a sunspot cycle when only a few sunspots are usually observed.

Solar System The collection of planets, moons, and related objects orbiting about the Sun. The Earth is one such member of the Solar System.

solar wind A stream of energetic particles of matter that constantly escapes the Sun.

spacetime A synthesis of the three dimensions of space and of a fourth dimension, time; a hallmark of relativity theory.

speciation The change of a single species into two or more new species; also termed "disruptive selection."

species A classification of life used to denote not only structural similarity but also the ability to mate and produce fertile offspring.

spectral class A classification scheme that groups stars according to their spectral lines or surface temperatures.

spectral line A radiative feature observed in emission (bright) or absorption (dark) at a specific frequency or wavelength.

spectroscopic parallax The third rung of the distance ladder, whereby a star's spectrum is used to estimate its absolute brightness and thus its distance.

spectroscopy An observational technique designed to disperse radiation into its component wavelengths in order to study in fine detail the way that matter emits or absorbs radiation.

spicule A short-lived jet or spike of gas that regularly expels particles of matter from near the surface of the Sun.

spiral arm Part of a pinwheel structure of young stars and interstellar clouds usually winding out from a galaxy's center.

spiral galaxy A galaxy having a spiral or pinwheel shape, some more than others, and composed of a mixture of old and young stars as well as loose interstellar matter.

spontaneous generation The theory, now out of favor, that life forms have suddenly emerged fully developed from peculiar arrangements of nonliving matter.

s process Element building that occurs in highly evolved stars when a neutron is "slowly" captured by a nucleus.

star A gaseous object so hot that its core is fusing lighter nuclei into heavier nuclei. The Sun is one such star.

stellar epoch A period in the history of the Universe (including now) when the stars form.

stellar evolution The changes experienced by stars as they originate, mature, and grow old.

stellar nucleosynthesis Element building that occurs in stars when nuclei collide and fuse with one another.

stellar parallax The measured parallax of a star when the baseline equals the diameter of Earth's orbit about the Sun.

sunspot Relatively cool, dark, disturbed regions on the surface of the Sun.

sunspot cycle The regular increase and decrease in the number of sunspots over the course of roughly 11 years.

supernova An explosive death of a massive star whose glowing debris produce for a short time a great brightening.

supernova remnant The remains of a supernova, namely glowing debris scattered over a light-year or more.

synapse A microscopic gap separating an axon of one neuron from a dendrite of another neuron.

temperature A measure of the heat of an object, namely of the average kinetic energy of the particles in the object.

terrestrial An adjective meaning "of the Earth."

Terrestrial Planets The four, small, rocky planets in the inner parts of the Solar System: Mercury, Venus, Earth, and Mars.

thermal radiation Radiation released by virtue of an object's heat; namely, by charged particles interacting with other charged particles.

triangulation An observational technique used to determine distance indirectly on the basis of a few simple measurements and some geometrical reasoning.

T-Tauri star A class of very young, often flaring stars on the verge of reaching the main sequence.

turbulence The disordered, irregular motion of matter, so complex as to defy description except in a statistical manner.

21-centimeter line A characteristic spectral feature of atomic hydrogen at 21 centimeters wavelength caused by a change in the orientation of the spin of its electron.

ultraviolet telescope A device capable of detecting cosmic electromagnetic radiation whose wavelength is a little shorter than light waves.

unicell A single cell that does not collaborate with other cells.

universal horizon The limit of the observable Universe.

Universe The totality of all space, matter, and energy.

Van Allen belts Zones of intense radiation surrounding Earth's midsection, caused by charged particles trapped in Earth's magnetic force field.

variable star A star whose luminosity changes with time, some quite erratically, others more regularly.

velocity of light The fastest speed that any thing can move, equal to some 300,000 kilometers/second.

vertebrate An organism having a backbone, and including fishes, amphibians, reptiles, and mammals.

Virial method A statistical technique used to estimate the mass of a galaxy cluster by observing the motions of its individual member galaxies.

virus The smallest and simplest entity that sometimes appears to be alive.

visible spectrum The narrow range of wavelengths in the electromagnetic spectrum to which the human eye is sensitive; namely, light.

volcano The site of hot lava upwelling from below the crust of a planet or moon.

water hole The radio domain between the 21-centimeter wavelength spectral line of hydrogen (H) and the 18-centimeter wavelength spectral lines of the hydroxyl molecule (OH).

wave A disturbance that moves from one place to another.

wavelength The distance between successive crests of a wave.

weak force The force that governs the change of one kind of elementary particle into another.

white dwarf star An old, dim star, much smaller in size and hotter than the Sun.

PHOTO CREDITS

The author is pleased to acknowledge the permissions kindly granted by the following sources for many of the illustrations.

National Aeronautics and Space Administration: Figures 0.1, 3.30, 3.32, 3.34, 3.35, 4.1, 6.1, 6.3, 6.5, 6.6, 6.7, 6.10, 6.11, 6.12, 6.13, 6.14, Int. 6-2, 7.11, 7.12, 7.13, 7.14, 7.15, 7.16, 7.17, 7.18, 7.20, 7.21, 7.22, 7.23, 7.24, 7.25, 7.26, 7.27, 8.1(b), 8.2, 8.3, 8.5, 8.6, 8.7, 8.8(b), 8.9, 8.10, 8.11, 8.12, 8.14, 8.15, 8.16, 8.18, 8.28, 8.30, 9.5, 9.12, 9.15, 9.18, Int. 12-1, 12.25, 18.3, 18.4, 19.14, 23.8, 23.10, 24.4, 24.6(b), 24.9(b), 29.1, 29.2, 29.3, 29.7, 29.16; Plates 2(b), 3(a), 3(b), 4(b), 6(a).

National Optical Astronomy Observatories: Figures 0.2, 0.3, 0.6, 0.7, 1.2, 1.3, 2.19, 3.3, 3.4, 3.9, 3.10, 3.20(a), 3.22, 3.28(b), 3.37(b), 8.1(a), 8.8(a), 8.13, 8.17, 8.19, 8.23, 9.6, 9.9, 9.11, 9.13, 11.6, 11.11, 11.20(a), 11.21(a), 12.9, 12.16, 12.18, 13.1, 14.1, 14.2, 14.3, 14.5, 14.7, 14.8, Int. 14-2, 15.3, 15.6(a),(b), 15.7, 15.8(a), 15.14, 15.15, 15.19, 18.4, 18.6, 18.11, 18.13(b), 18.14, 20.8, 20.9(b), 20.15, 20.17, 20.22; Plates 2(a), 5, 7(a), 8(a), 8(b).

National Radio Astronomy Observatory: Figures 3.13, 3.27, 3.28(a), 12.23, 12.24, 15.5, 15.6(c), 15.8(b), 15.9, 19.11; Plates 7(b), 8(d).

Harvard-Smithsonian Center for Astrophysics: Figures 2.17, 2.22, 3.31, 3.36, 3.37(a), 4.11, Int. 7-2, 8.26, 9.8, 11.1, 11.3, 11.12, 11.13, 11.18(a), Int. 11-2, 12.6, 12.7, 12.8, 12.12, 12.17, 12.23, 12.24, 13.2, 13.14, 13.15(a), Int. 13-1, 14.9, 15.6(d), 15.18, 18.11, 18.12, 19.9, 19.10, 19.11, 19.12, 19.13, Int. 19-2, 20.16, 21.9, 21.10; Plates 1(a), 1(b), 1(c), 8(c).

Massachusetts Institute of Technology (Haystack Observatory): Figures 3.17, 3.18, 3.20(b), 3.21, 3.22, 3.25, 12.14, 12.16, 12.18, 13.15(b).

Hale Observatories: Figures 9.7, 9.14, 12.1, 12.5, 14.4, 14.6, 14.13, 14.16, 19.17, 20.13, 20.19.

Miscellaneous: CERN, Figure 2.2; Sidney Fox, 0.8, 25.13; Elso Barghoorn, 0.9, 25.1, 26.1; A.E. Lilley, 2.24; Yerkes Observatory, 3.6, Int. 19-1; Sovfoto (via Jay Pasachoff) 3.7, 3.8; USSR Academy of Sciences, 7.19, 8.24(a); Max Planck Institute für Radioastronomie, 3.14; Max Planck Institute für Aeronomie, 8.24(b); National Atmospheric and Ionospheric Center, 3.15, 3.16; University of Cambridge, 3.26; U.S. Department of Energy, 6.4; D.E. Wilhelms and D.E. Davis, 6.15; Brookhaven National Laboratory, 10.5; Lola Judith Chaisson, 13.11, 13.16(a)–(f), 21.11, Plate 1(d); James Peebles, 14.14; Arno Penzias, 16.12; Alar Toomre, 18.13(a); Lick Observatory, 20.11; E.O. Wilson, 26.10, 26.11, 26.15; D.C. Johanson, 27.7; A. Marshack, 27.17; Woodruff Sullivan, 29.15 (adapted); European Southern Observatory, 20.2b, Plate 6(b); D. Malin, Anglo-Australian Telescope Board, Plate 4(a); Wyoming Travel Commission, Int. 27-2(c); AP/Wide World Photos, 18.7; U.S. Department of the Interior, 24.1; D.E. Brownlee, 8.27; Westerbork Observatory, 15.10; Leon Van Speybroeck, 3.12(d)–(g); Smithsonian Institution/University of Arizona, 3.11; Royal Observatory, Edinburgh, 20.18; Australian Information Services, 26.2; D.H. Hubel and J. Robbins, 28.9; Geza Teleki, 27.12; J. Moore, Anthro-Photo, 27.11; National Museums of Kenya, 27.13; Museum of Comparative Zoology, Harvard, 27.8; R.C. Wood, 27.3(a),(b); F. Woehr, Photo Trends, 27.3(c); Transvaal Museum, Pretoria, 27.2; Peabody Museum of Natural History, Yale, 26.19, 26.21.

INDEX

THE ELECTROMAGNETIC SPECTRUM

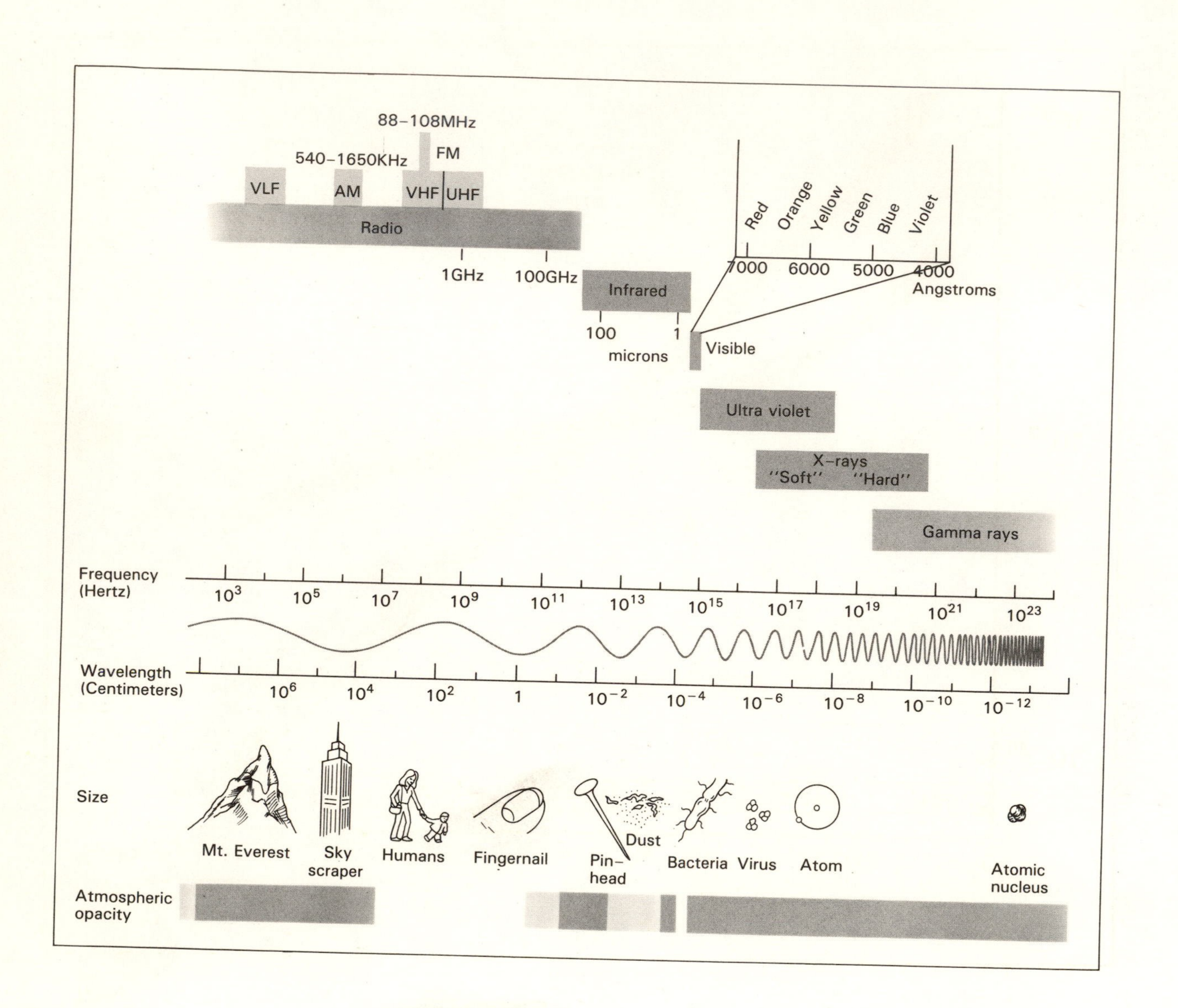